Rapid Guide to

Chemical Incompatibilities

D1189837

Rapid Guide to

Chemical
Incompatibilities

RICHARD P. POHANISH

STANLEY A. GREENE

VAN NOSTRAND REINHOLD

I(T)P® A Division of International Thomson Publishing Inc.

New York • Albany • Bonn • Boston • Detroit • London • Madrid • Melbourne
Mexico City • Paris • San Francisco • Singapore • Tokyo • Toronto

I⊤P®   Van Nostrand Reinhold is an International Thomson Publishing Company
The ITP logo is a registered trademark used herein under license

Printed in the United States of America

For more information, contact:

Van Nostrand Reinhold
115 Fifth Avenue
New York, NY 10003

Chapman & Hall GmbH
Pappellaee 3
69469 Weinheim
Germany

Chapman & Hall
2-6 Boundary Row
London
SE1 8HN
United Kingdom

International Thomson Publishing Asia
221 Henderson Road #05-10
Henderson Building
Singapore 0315

Thomas Nelson Australia
102 Dodds Street
South Melbourne, 3205
Victoria, Australia

International Thomson Publishing Japan
Hirakawacho Kyowa Building, 3F
2-2-1 Hirakawacho
Chiyoda-ku, 102 Tokyo
Japan

Nelson Canada
1120 Birchmount Road
Scarborough, Ontario
Canada M1K 5G4

International Thomson Editores
Seneca 53
Col. Polanco
11560 Mexico D.F. Mexico

1 2 3 4 5 6 7 8 9 10   COUWF   01   00   99   98   97   96

Library of Congress Cataloging-in-Publication Data

Pohanish, Richard R.
   Rapid guide to chemical incompatibilities / Richard R. Pohanish,
Stanley A. Greene.
     p.   cm.
   ISBN 0-442-02394-4
    1. Hazardous substances——Safety measures——Handbooks, manuals, etc.
  2. Chemicals——Safety measures——Handbooks, manuals, etc.   I. Greene,
Stanley A.   II. Title.

T55. 3. H3P644   1996
660 ' .2804——dc21

96—37063
CIP

http://www.vnr.com
product discounts • free email newsletters
software demos • online resources

email: info@vnr.com

A service of   I⊤P®

To Dina, Barb, and Jenny

CONTENTS

This book is intended to provide data about chemical incompatibilities and other hazards. It cannot be assumed that all necessary warnings and precautionary measures are contained in this work, and that other, or additional, information or assessments may not be required. Extreme care has been taken in the preparation of this work and, to the best knowledge of the publisher and the editors, the information presented is accurate. No warranty, express or implied, is made. Information may not be available for some chemicals; consequently, an absence of data does not necessarily mean that a substance is not hazardous. Neither the publisher nor the editors assume any liability or responsibility for completeness or accuracy of the information presented or any damages of any kind alleged to result in connection with, or arising from, the use of this book. The publisher and the editors strongly encourage all readers, and users of chemicals, to follow the manufacturers' or suppliers' current instructions, technical bulletins, and material safety data sheets (MSDSs) for specific use, handling, and storage of all chemical materials.

PREFACE

This manual was developed as a companion to *Rapid Guide to Hazardous Chemicals in the Workplace*, edited by Richard J. Lewis, Sr. The purpose of *Rapid Guide to Chemical Incompatibilities* is to fulfill the need for an easy-to-use, portable reference containing concise summaries or profiles of the known effects of many dangerously reactive substances commonly found in commerce.

Intended for safety professionals, environmental scientists, first-response personnel, transportation specialists, plant managers, and others, this book brings together from diverse sources one systematized, alphabetical compilation of binary chemical combinations reported to be dangerously reactive when accidentally mixed or subjected to physical changes.

Although intended to be helpful in preventing, or at least minimizing, the harmful effects of chemical accidents, this rapid guide will not address all possible contingencies that may be associated with storage or chemical mixtures, and should not be considered a substitute for the user's own knowledge or judgment. In compiling this manual the editors used various sources of information; and, occasionally, contradictory data were found in the literature. Consequently, the editors and the publisher strongly urge users to consult chemical manufacturers' and suppliers' technical bulletins, material safety data sheets, labels, and shipping and other documents related to protection from, and the safe handling and storage of, all chemical substances. Furthermore, users are cautioned that the absence of specific reaction information in no way implies that different materials, or combinations of materials, under any set of conditions, may be safely mixed. In like manner, users must weigh comments about the "violence" of particular reactions, which may be affected by factors such as the amount of material, physical properties, temperature, use of closed or restricted systems, and so on. In some cases very small quantities of contamination, or the presence of other materials in the working environment, may act as a catalyst and produce violent reactions such as polymerization, disassociation, and condensation.

As additional data become available, the editors will make appropriate additions and corrections to update and continually improve this manual. Although every effort has been made to compile and produce an accurate and useful manual, the editors know there is always room for improvement. Any comments or advice from readers regarding this book, as well as suggestions for future improvements, are welcomed. All correspondence should be submitted in writing.

HOW TO USE THIS BOOK

The term *incompatibility* is used to describe a wide range of chemical reactions that might include the generation of heat resulting from contact of a chemical with moisture; decomposition; the generation of toxic gases; the heating, overflow, and rupture of containers; polymerization; the formation of new and possibly more dangerous compounds; fire, detonation, and explosion; or any combination of these or other actions.

Using chemical names and Chemical Abstracts Service (CAS) numbers, readers can locate concise incompatibility profiles for thousands of commercial chemicals that may be found in the industrial workplace, frequently stored, and transported in bulk.

The simple keys to this manual are chemical names and identifying numbers. Chemical names, regulatory names, alternate names, commonly used "trivial" names, and in many cases, product and trade names, as well as numbers such as RCRA waste codes, can be looked up directly without the reader's having first to consult interim indices or charts. To avoid confusion with like-sounding names or multiple names with exact spellings, CAS registry numbers (in the format xxx-xx-x) have been provided. CAS numbers are unique identifiers assigned to all chemicals registered with the Chemical Abstracts Service of the American Chemical Society, and should always be used in conjunction with a substance name for positive identification. Occasionally, the expression "N/A" is used in the incompatibility profiles to designate a CAS number that could not be found.

A substance's incompatibility profile is based on the following information (as available):

- Incompatibility or reactions with common materials, including air and water.

- Incompatibility or reactions with other chemical substances.

- Incompatibility or reactions with structural materials such as metals.

- Incompatibility or reactions with protective materials such as plastics, rubber, and coatings.

- Information, as known, related to the ability of chemical substances to accumulate dangerous static electrical charges.

- The ability, when known, of a chemical (e.g., ethers) to form unstable and potentially explosive peroxides, or to cause polymerization.

Where a chemical mixes with air to form an explosive mixture, the flash point has been provided both in degrees Fahrenheit and Celsius. Flash point is defined as "The minimum temperature at which the liquid gives off sufficient vapor to form an ignitable mixture with air near the surface of the liquid or within the test vessel used. Data are apparatus- and procedure-dependent." However, in practice (e.g., outside the laboratory) ignition may occur at lower temperatures than those provided; and, it should also be noted that flash points found in the literature may differ for various reasons. When a conflict existed, the editors compared values from multiple sources and/or used the most conservative value. When a precise flash point is required, it should be established or verified by testing a technical-grade sample of the chemical substance.

ACKNOWLEDGMENT

Thanks are due to those individuals who reviewed and constructively commented on this work during its development and to the many scientists and contract employees who developed the various documents and public databases that provided much of the information appearing in the book. The author also acknowledges and thanks Nancy Olsen, Sharon Gibbons, Peter Rocheleau, and Constance Marsh MacDonald of Van Nostrand Reinhold for their help and encouragement.

- A -

A-10846 (67-68-5) Incompatible with boron compounds, halides, metal alkoxides, oxidizers.

A 361 (1912-24-9) Incompatible with strong acids.

AA-9 (25155-30-0) Reacts with acids, including fumes.

AA CAPTAN (133-06-2) Incompatible with tetraethyl pyrophosphate, parathion.

AAT (56-38-2) A combustible liquid. Mixtures with endrin may be explosive. Strong oxidizers may cause fire and explosions. Attacks some plastics, rubber, or coatings.

AATACK (137-26-8) Combustible solid (flash point 192°F/89°C). Strong oxidizers may cause fire and explosions; contact with strong acid or oxidizable materials produces toxic gases.

AATP (56-38-2) A combustible liquid. Mixtures with endrin may be explosive. Strong oxidizers may cause fire and explosions. Attacks some plastics, rubber, and coatings.

AATREX or AATREX 80W or AATREX HERBICIDE or AATREX NINE-O (1912-24-9) Incompatible with strong acids.

2-AB (13952-84-6) Forms explosive mixture with air. Incompatible with nonoxidizing mineral acids, acids, strong acids, organic acids, organic anhydrides, isocyanates, vinyl acetate, acrylates, substituted allyls, alkylene oxides, epichlorohydrin, ketones, aldehydes, alcohols, glycols, phenols, cresols, caprolactam solution, strong oxidizers.

ABESON NAM (25155-30-0) Reacts with acids, including fumes.

ABICEL (9004-34-6) Reacts with water, bromine pentafluoride, hydrogen peroxide, sodium hypochlorite, sodium nitrate, fluorine, or strong oxidizers.

ABSOLUTE ETHANOL (64-17-5) Forms explosive mixture with air (flash point 65°F/18°C). May accumulate static electrical charges, and may cause ignition of its vapors. Reactions may be violent with oleum, sulfuric acid, nitric acid, bases, aliphatic amines, isocyanates, oxidizers.

AC 3422 (563-12-2) Incompatible with alkaline formulations. Mixtures with magnesium may be explosive.

ACARIN (115-32-2) Incompatible with strong acids, caustics, aliphatic amines, isocyanates.

ACC 3422 (56-38-2) A combustible liquid. Mixtures with endrin may be explosive. Strong oxidizers may cause fire and explosions. Attacks some plastics, rubber, and coatings.

ACCELERATOR HX (5459-93-8) Forms explosive mixture with air (flash point 86°F/30°C). Incompatible with nonoxidizing mineral acids, acids, strong acids, organic acids, organic anhydrides, isocyanates, vinyl acetate, acrylates, substituted allyls, alkylene oxides, epichlorohydrin, ketones, aldehydes, alcohols, glycols, mercury, phenols, cresols, caprolactam solution, strong oxidizers. Attacks aluminum, copper, lead, tin, zinc and alloys.

ACEDE CRESYLIQUE (French) (1319-77-3) Incompatible with strong acids, oxidizers, alkalies, aliphatic amines, amides, chlorosulfonic acid, oleum. Liquid attacks some plastics, coatings, and rubber.

ACENTERIN (50-78-2) Fires and explosions may result from contact with strong oxidizers. Alkali hydroxides or carbonates cause decomposition.

ACETAAL (Dutch) (105-57-7) Forms explosive mixture with air (flash point −5°F/−21°C). Reacts violently with oxidizers. Forms unstable and explosive peroxides with heat and light. May accumulate static electrical charges, and may cause ignition of its vapors.

ACETAL (105-57-7) Forms explosive mixture with air (flash point −5°F/−21°C). Reacts violently with oxidizers. Forms unstable and explosive peroxides with heat and light. May accumulate static electrical charges, and may cause ignition of its vapors.

ACETALDEHYDE (75-07-0) Slowly polymerizes to paraldehyde. Explodes when mixed with iodine. Contact with strong bases can cause explosive polymerization. A strong reducing agent; reacts violently with combustibles, strong acids, caustics, ammonia, aliphatic amines, alkanolamines, aromatic amines, organic substances, halogens, oxidizers. Forms explosive peroxides with air. May dissolve rubber. May accumulate static electrical charges, and may cause ignition of its vapors.

p-**ACETALDEHYDE or** *para*-**ACETALDEHYDE** (123-63-7) Reacts with strong acids, caustics, ammonia, amines, oxidizers. Contact with acids produces acetaldehyde.

ACETALDEHYDE, CHLORO- (107-20-0) Incompatible with oxidizers, acids, water. May form a water-soluble polymer in storage.

ACETALDEHYDE DIETHYLACETAL (105-57-7) Forms explosive mixture with air (flash point −5°F/−21°C). Reacts violently with oxidizers. Forms unstable and explosive peroxides with heat and light. May accumulate static electrical charges, and may cause ignition of its vapors.

b-**ACETALDEHYDE OXIME or beta-ACETALDEHYDE OXIME** (107-29-9) Forms explosive mixture with air. Reacts violently with oxidizers. Forms explosive peroxides with air, acids. Attacks various alkali metals (i.e., lithium, sodium, potassium, rubidium, cesium, francium).

ACETALDEHYDE TRIMER (123-63-7) Reacts with strong acids, caustics, ammonia, amines, oxidizers. Contact with acids produces acetaldehyde.

ACETAL DIETHYLIQUE (French) (105-57-7) Forms explosive mixture with air (flash point −5°F/−21°C). Reacts violently with oxidizers. Forms unstable and explosive peroxides with heat and light. May accumulate static electrical charges, and may cause ignition of its vapors.

ACETALE (Italian) (105-57-7) Forms explosive mixture with air (flash point −5°F/−21°C). Reacts violently with oxidizers. Forms unstable and explosive peroxides with heat and light. May accumulate static electrical charges, and may cause ignition of its vapors.

ACETAMIDE, N,N-DIMETHYL (127-19-5) Forms explosive mixture with air (flash point 158°F/70°C). Incompatible with nonoxidizing mineral acids, strong acids, ammonia, isocyanates, phenols, cresols, halogenated compounds (above 185°F/85°C). Attacks some plastics, rubber, and coatings.

ACETATE C-7 (112-06-1) Incompatible with strong acids, nitrates.

ACETATE C-9 (143-13-5) Incompatible with strong acids, nitrates.

ACETATE d'AMYLE (French) (628-63-7) Incompatible with strong alkalies, strong acids, nitrates, strong oxidizers. Attacks some plastics, coatings, and rubber.

ACETATE de BUTYLE (French) (123-86-4) Forms explosive mixture with air (flash point 72°F/22°C). Reacts with water on standing to form acetic acid and *n*-butyl alcohol. Reacts violently with strong oxidizers and potassium *tert*-butoxide. Incompatible with caustics, strong acids, nitrates. Dissolves rubber, many plastics, resins, and some coatings. May accumulate static electrical charges, and may cause ignition of its vapors.

ACETATE de BUTYLE SECONDAIRE (French) (105-46-4) Forms explosive mixture with air (flash point 70°F/19°C). Reacts violently with oxidizers. Incompatible with strong acids, nitrates, potassium *tert*-butoxide. Attacks some plastics, rubber, and coatings. May accumulate static electrical charges, and may cause ignition of its vapors.

ACETATE de CUIVRE (French) (142-71-2) Incompatible with strong acids, nitrates.

ACETATE de METHYLE (French) (79-20-9) Forms explosive mixture with air (flash point 14°F/−10°C). Incompatible with strong acids, nitrates, oxidizers, bases. Attacks some plastics. May accumulate static electrical charges, and may cause ignition of its vapors.

ACETATE de PROPYLE NORMAL (French) (109-60-4) Forms explosive mixture with air (flash point 58°F/14°C). Incompatible with strong acids, nitrates, strong acids, strong oxidizers. Attacks some plastics, rubber, or coatings. May accumulate static electrical charges, and may cause ignition of its vapors.

ACETATE SALICYCLIC ACID (50-78-2) Fires and explosions may result from contact with strong oxidizers. Alkali hydroxides or carbonates cause it to decompose.

ACETATE de VINYLE (French) (108-05-4) Forms explosive mixture with air (flash point 18°F/−7°C). Polymerizes readily if not inhibited; heat can initiate reaction. Reacts violently with oxidizers. Incompatible with nonoxidizing mineral acids, strong acids, ammonia, aliphatic amines, alkanolamines. Also reacts with, 2-aminoethanol, chlorosulfonic acid, ethylene diamine, ethyleneimine, ozone, oleum, peroxides. May accumulate static electrical charges, and may cause ignition of its vapors.

ACETDIMETHYLAMIDE (127-19-5) Forms explosive mixture with air (flash point 158°F/70°C). Incompatible with nonoxidizing mineral acids, strong acids, ammonia, isocyanates, phenols, cresols, halogenated compounds (above 185°F/85°C). Attacks some plastics, rubber, and coatings.

***N,N*-ACETDIMETHYLAMIDE (127-19-5)** Forms explosive mixture with air (flash point 158°F/70°C). Incompatible with nonoxidizing mineral acids, strong acids, ammonia, isocyanates, phenols, cresols, halogenated compounds (above 185°F/85°C). Attacks some plastics, rubber, and coatings.

ACETEHYD (German) (75-07-0) Slowly polymerizes to paraldehyde. Explodes when mixed with iodine. Contact with strong bases can cause explosive polymerization. A strong reducing agent; reacts violently with combustibles, strong acids, caustics, ammonia, aliphatic amines, alkanolamines, aromatic amines, organic substances, halo-

3

gens, oxidizers. Forms explosive peroxides with air. May dissolve rubber. May accumulate static electrical charges, and may cause ignition of its vapors.

ACETEHYDE (75-07-0) Slowly polymerizes to paraldehyde. Explodes when mixed with iodine. Contact with strong bases can cause explosive polymerization. A strong reducing agent; reacts violently with combustibles, strong acids, caustics, ammonia, aliphatic amines, alkanolamines, aromatic amines, organic substances, halogens, oxidizers. Forms explosive peroxides with air. May dissolve rubber. May accumulate static electrical charges, and may cause ignition of its vapors.

ACETENE (74-85-1) Incompatible with strong acids, oxidizers, halogen acids, aluminum chloride, bromotrichloromethane, carbon tetrachloride, chlorine, ozone.

ACETIC ACID (64-19-7) Vapor forms explosive mixture with air (above 105°F/40°C). Reacts violently with oxidizers and bases. Incompatible with strong acids, aliphatic amines, alkanolamines, isocyanates, alkylene oxides, epichlorohydrin, acetaldehyde, 2-aminoethanol, ammonia, ammonium nitrate, chlorosulfonic acid, chromic acid, ethylene diamine, ethyleneimine, perchloric acid, permanganates, phosphorus isocyanate, phosphorus trichloride, potassium *tert*-butoxide, xylene. Attacks cast iron and other metals, forming flammable hydrogen gas.

ACETIC ACID, AMMONIUM SALT (631-61-8) Incompatible with sodium hypochlorite.

ACETIC ACID, *n*-**AMYL ESTER** (628-63-7) Incompatible with strong alkalies, strong acids, nitrates, strong oxidizers. Attacks some plastics, coatings, and rubber.

ACETIC ACID, ANHYDRIDE (9CI) (108-24-7) Forms explosive mixture with air (flash point 121°F/50°C). Water contact produces acetic acid. Reacts violently with potassium hydroxide or sodium hydroxide. Incompatible with nonoxidizing mineral acids, strong acids, ammonia, aliphatic amines, alkanolamines, aromatic amines, alcohols, 2-aminoethanol, aniline, bases, boric acid, chlorosulfonic acid, chromic acid, chromic anhydride, ethylene diamine, ethylenediamine, glycerol, hydrochloric acid, hydrofluoric acid, hydrogen peroxide, nitric acid, nitrogen tetroxide, oleum, oxidizers, perchloric acid, permanganates, sodium peroxide, sulfuric acid, water. Attacks some coatings, some plastics, and rubber. Corrodes iron and steel. May accumulate static electrical charges, and may cause ignition of its vapors.

ACETIC ACID, ANHYDRIDE with NITRIC ACID (591-09-3) Self reactive. Reacts explosively with active oxides such as mercuric oxide.

ACETIC ACID, BENZYL ESTER (140-11-4) Forms explosive mixture with air (flash point 216°F/102°C). Reacts with strong acids, nitrates, oxidizers.

ACETIC ACID, 2-BUTOXY ESTER (105-46-4) Forms explosive mixture with air (flash point 70°F/19°C). Reacts violently with oxidizers. Incompatible with strong acids, nitrates, potassium *tert*-butoxide. Attacks some plastics, rubber, and coatings. May accumulate static electrical charges, and may cause ignition of its vapors.

ACETIC ACID, 2-BUTOXYETHYL ESTER (112-07-2) Incompatible with strong acids, perchloric acid, oxidizers, nitrates.

ACETIC ACID, BUTYL ESTER (123-86-4) Forms explosive mixture with air (flash point 72°F/22°C). Reacts with water on standing to form acetic acid and *n*-butyl alcohol. Reacts violently with strong oxidizers and potassium *tert*-butoxide. Incompatible with caustics, strong acids, nitrates. Dissolves rubber, many plastics, resins, and some coatings. May accumulate static electrical charges, and may cause ignition of its vapors.

ACETIC ACID, *sec*-**BUTYL ESTER** (105-46-4) Forms explosive mixture with air (flash point 70°F/19°C). Reacts violently with oxidizers. Incompatible with strong acids, nitrates, potassium *tert*-butoxide. Attacks some plastics, rubber, and coatings. May accumulate static electrical charges, and may cause ignition of its vapors.

ACETIC ACID, *tert*-**BUTYL ESTER** (540-88-5) Incompatible with strong acids, oxidizers, and alkalies, and nitrates. Softens plastics.

ACETIC ACID, CADMIUM SALT (543-90-8) Incompatible with strong oxidizers, elemental sulfur, selenium, tellurium, strong acids, nitrates.

ACETIC ACID CHLORIDE (75-36-5) Forms explosive mixture with air (flash point 40°F/4°C). May accumulate static electrical charges, and may cause ignition of its vapors. Reacts violently with water, oxidizers, strong bases, alcohols (especially ethanol), dimethyl sulfoxide, phosphorus trichloride; forms corrosive hydrochloric acid with air.

ACETIC ACID, CHROMIUM SALT (1066-30-4) Contact with strong oxidizers may cause fire and explosions.

ACETIC ACID, CHROMIUM(3+) SALT (1066-30-4) Contact with strong oxidizers may cause fire and explosions.

ACETIC ACID, COBALT(2+) SALT (71-48-7) Contact with strong oxidizers may cause fire and explosions. Cobalt compounds may react with acetylene.

ACETIC ACID, COPPER(2+) (142-71-2) Incompatible with strong acids, nitrates.

ACETIC ACID, CUPRIC SALT (142-71-2) Incompatible with strong acids, nitrates.

ACETIC ACID, CYCLOHEXYL ESTER (622-45-7) Incompatible with strong acids, oxidizers, nitrates.

ACETIC ACID, DIMETHYLAMIDE (127-19-5) Forms explosive mixture with air (flash point 158°F/70°C). Incompatible with nonoxidizing mineral acids, strong acids, ammonia, isocyanates, phenols, cresols, halogenated compounds (above 185°F/85°C). Attacks some plastics, rubber, and coatings.

ACETIC ACID, 1,3-DIMETHYLBUTYL ESTER (108-84-9) Forms explosive mixture with air (flash point 113°F/45°C). Incompatible with strong acids, strong alkalies, nitrates, strong oxidizers.

ACETIC ACID, 1,1-DIMETHYLETHYL ESTER (9CI) (540-88-5) Incompatible with strong acids, oxidizers, and alkalies, and nitrates. Softens and dissolves plastics.

ACETIC ACID, ETHENYL ESTER (108-05-4) Forms explosive mixture with air (flash point 18°F/−8°C). Polymerizes readily if not inhibited; heat can initiate reaction. Reacts violently with oxidizers. Incompatible with nonoxidizing mineral acids, strong acids, ammonia, aliphatic amines, alkanolamines. Also reacts with 2-aminoethanol, chlorosulfonic acid, ethylene diamine, ethyleneimine, ozone, oleum, peroxides. May accumulate static electrical charges, and may cause ignition of its vapors.

ACETIC ACID, 2-ETHOXYETHYL ESTER (111-15-9) Forms explosive mixture with air (flash point 117°F/47°C). Incompatible with strong acids, nitrates, oxidizers. Attacks rubber.

ACETIC ACID, ETHYL ESTER (141-78-6) Forms explosive mixture with air (flash point 24°F/−4.4°C). Incompatible with strong acids, nitrates, oxidizers, chlorosulfonic acid, lithium aluminum hydride, oleum.

ACETIC ACID (ETHYLENEDINITRILO)TETRA- (60-00-4) Incompatible with sulfuric acid, bases, ammonia, aliphatic amines, alkanolamines, isocyanates, alkylene oxides, epichlorohydrin.

ACETIC ACID, FLUORO-, SODIUM SALT (62-74-8) Incompatible with strong acids.

ACETIC ACID, GLACIAL (64-19-7) Vapor forms explosive mixture with air (above 105°F/40°C). Reacts violently with oxidizers and bases. Incompatible with strong acids, aliphatic amines, alkanolamines, isocyanates, alkylene oxides, epichlorohydrin, acetaldehyde, 2-aminoethanol, ammonia, ammonium nitrate, chlorosulfonic acid, chromic acid, ethylene diamine, ethyleneimine, perchloric acid, permanganates, phosphorus isocyanate, phosphorus trichloride, potassium *tert*-butoxide, xylene. Attacks cast iron and other metals forming flammable hydrogen gas.

ACETIC ACID, HEPTYL ESTER (112-06-1) Incompatible with strong acids, nitrates.

ACETIC ACID, HEXYL ESTER (142-92-7) Incompatible with strong acids, nitrates, oxidizers.

ACETIC ACID, ISOBUTYL ESTER (110-19-0) Forms explosive mixture with air (flash point 64°F/18°C). Reacts with water on standing to form acetic acid and *n*-butyl alcohol. Reacts violently with strong oxidizers. Reacts with caustics, strong acids, nitrates. Dissolves rubber, many plastics, resins and some coatings. May accumulate static electrical charges, and may cause ignition of its vapors.

ACETIC ACID, ISOPENTYL ESTER (123-92-2) Forms explosive mixture with air (flash point 77°F/25°C). Incompatible with strong alkalies, strong acids, nitrates, oxidizers. Reacts violently with reducing agents. Attacks asbestos; softens and dissolves many plastics, rubber, and coatings.

ACETIC ACID, ISOPROPYL ESTER (108-21-4) Incompatible with strong acids, nitrates, oxidizers. Dissolves many plastic materials. Contact with steel causes slow decomposition.

ACETIC ACID, MERCURY(2+) SALT (1600-27-7) Heat and light cause decomposition.

ACETIC ACID, METHYL ESTER (79-20-9) Forms explosive mixture with air (flash point 14°F/−10°C). Incompatible with strong acids, nitrates, oxidizers, bases. Attacks some plastics. May accumulate static electrical charges, and may cause ignition of its vapors.

ACETIC ACID, 1-METHYLETHYL ESTER (108-21-4) Forms explosive mixture with air (flash point 40°F/4.4°C). Incompatible with strong alkalies, strong acids, nitrates, strong oxidizers. Dissolves rubber and many plastic materials. Contact with steel causes slow decomposition.

6

ACETIC ACID, 3-METHOXYBUTYL ESTER (4435-53-4) Incompatible with strong acids, nitrates, oxidizers.

ACETIC ACID, 1-METHYLPROPYL ESTER (9CI) (105-46-4) Forms explosive mixture with air (flash point 70°F/19°C). Reacts violently with oxidizers. Incompatible with strong acids, nitrates, potassium *tert*-butoxide. Attacks some plastics, rubber, and coatings. May accumulate static electrical charges, and may cause ignition of its vapors.

ACETIC ACID, 2-METHYLPROPYL ESTER (110-19-0) Forms explosive mixture with air (flash point 64°F/18°C). Reacts with water on standing to form acetic acid and *n*-butyl alcohol. Reacts violently with strong oxidizers. Reacts with caustics, strong acids, nitrates. Dissolves rubber, many plastics, resins, and some coatings. May accumulate static electrical charges, and may cause ignition of its vapors.

ACETIC ACID, NICKEL(2+) SALT (373-02-4) Incompatible with strong acids, strong oxidizers, sulfur, selenium.

ACETIC ACID, NICKEL(II) SALT (373-02-4) Incompatible with strong acids, strong oxidizers, sulfur, selenium.

ACETIC ACID, N-NONYL ESTER (143-13-5) Incompatible with strong acids, nitrates.

ACETIC ACID, 2-PENTYL ESTER (626-38-0) Incompatible with strong acids, nitrates, strong alkalies, strong oxidizers, heat. May soften or dissolve plastics.

ACETIC ACID, PHENYLMETHYL ESTER (140-11-4) Contact with strong oxidizers may cause fire and explosions.

ACETIC ACID, *n*-PROPYL ESTER (109-60-4) Forms explosive mixture with air (flash point 58°F/14°C). Incompatible with strong acids, nitrates, strong acids, strong oxidizers. Attacks some plastics, rubber, or coatings. May accumulate static electrical charges, and may cause ignition of its vapors.

ACETIC ACID, PROPYL ESTER (109-60-4) Forms explosive mixture with air (flash point 58°F/14°C). Incompatible with strong acids, nitrates, strong acids, strong oxidizers. Attacks some plastics, rubber, or coatings. May accumulate static electrical charges, and may cause ignition of its vapors.

ACETIC ACID, THALLIUM(I) SALT (563-68-8) Incompatible with strong acids, nitrates, thallium acetate.

ACETIC ACID, THALLOUS SALT (563-68-8) Incompatible with strong acids, nitrates.

ACETIC ACID, (2,4,8-TRICHLOROPHYNOXY)- (93-76-5) Sealed metal containers may burst in heat (above 316°F/158°C). Incompatible with sulfuric acid, bases, ammonia, aliphatic amines, alkanolamines, isocyanates, alkylene oxides, epichlorohydrin.

ACETIC ACID, VINYL ESTER (108-05-4) Forms explosive mixture with air (flash point 18°F/−8°C). Polymerizes readily if not inhibited; heat can initiate reaction. Reacts violently with oxidizers. Incompatible with nonoxidizing mineral acids, strong acids, ammonia, aliphatic amines, alkanolamines. Also reacts with 2-aminoethanol, chlorosulfonic acid, ethylene diamine, ethyleneimine, ozone, oleum, peroxides. May accumulate static electrical charges, and may cause ignition of its vapors.

ACETIC ACID, ZINC SALT (557-34-6) Incompatible with strong acids, strong bases, nitrates.

ACETIC ALDEHYDE (75-07-0) Slowly polymerizes to paraldehyde. Explodes when mixed with iodine. Contact with strong bases can cause explosive polymerization. A strong reducing agent; reacts violently with combustibles, strong acids, caustics, ammonia, aliphatic amines, alkanolamines, aromatic amines, organic substances, halogens, oxidizers. Forms explosive peroxides with air. May dissolve rubber. May accumulate static electrical charges, and may cause ignition of its vapors.

ACETIC ANHYDRIDE (108-24-7) Forms explosive mixture with air (flash point 121°F/50°C). Water contact forms acetic acid and considerable heat. Reacts violently with potassium hydroxide, sodium hydroxide, alcohols, amines. Incompatible with acids, ammonia, aniline, chromic anhydride, ethylene diamine, ethylenediamine, glycerol, hydrogen peroxide, nitrogen tetroxide, oxidizers, permanganates, sodium peroxide. Attacks some coatings, some plastics, and rubber. Corrodes iron and steel. May accumulate static electrical charges, and may cause ignition of its vapors.

ACETIC BROMIDE (506-96-7) Reacts violently with water or alcohols, forming hydrogen bromide.

ACETIC *sec*-**BUTYL ESTER** (105-46-4) Forms explosive mixture with air (flash point 70°F/19°C). Reacts violently with oxidizers. Incompatible with strong acids, nitrates, potassium *tert*-butoxide. Attacks some plastics, rubber, and coatings. May accumulate static electrical charges, and may cause ignition of its vapors.

ACETIC CHLORIDE (75-36-5) Forms explosive mixture with air (flash point 40°F/4°C). May accumulate static electrical charges, and may cause ignition of its vapors. Reacts violently with water, oxidizers, strong bases, alcohols (especially ethanol), dimethyl sulfoxide, phosphorus trichloride; forms corrosive hydrochloric acid with air.

ACETIC EHYDE (75-07-0) Slowly polymerizes to paraldehyde. Explodes when mixed with iodine. Contact with strong bases can cause explosive polymerization. A strong reducing agent; reacts violently with combustibles, strong acids, caustics, ammonia, aliphatic amines, alkanolamines, aromatic amines, organic substances, halogens, oxidizers. Forms explosive peroxides with air. May dissolve rubber. May accumulate static electrical charges, and may cause ignition of its vapors.

ACETIC ESTER (141-78-6) Forms explosive mixture with air (flash point 24°F/−4.4°C). Incompatible with strong acids, nitrates, oxidizers, chlorosulfonic acid, lithium aluminum hydride, oleum.

ACETIC ETHER (141-78-6) Forms explosive mixture with air (flash point 24°F/−4.4°C). Incompatible with strong acids, nitrates, oxidizers, chlorosulfonic acid, lithium aluminum hydride, oleum.

ACETIC METHYL ETHER (105-45-3) Incompatible with oxidizers, strong acids, nitrates.

ACETIC OXIDE (108-24-7) Forms explosive mixture with air (flash point 121°F/50°C). Water contact forms acetic acid and considerable heat. Reacts violently with potassium hydroxide, sodium hydroxide, alcohols, amines. Incompatible with acids, ammonia, aniline, chromic anhydride, ethylene diamine, glycerol, hydrogen peroxide, nitrogen tetroxide, oxidizers, permanganates, sodium peroxide. Attacks some coatings, some plastics, and rubber. Corrodes iron and steel. May accumulate static electrical charges, and may cause ignition of its vapors.

ACETIC PEROXIDE (79-21-0) A powerful oxidizer and extremely sensitive Explosive. Forms explosive mixture with air (flash point 100°F/38°C). Reacts violently with many substances, including acetic anhydride, combustibles, strong bases, ether solvents, olefins, organic matter, magnesium, metal oxides, heavy metals, metal chloride solutions, reducing agents, nickel, phosphorus, sodium nitride, zinc.

ACETICYL (50-78-2) Fires and explosions may result from contact with strong oxidizers. Alkali hydroxides or carbonates cause decomposition.

ACETIDIN (141-78-6) Forms explosive mixture with air (flash point 24°F/−4.4°C). Incompatible with strong acids, nitrates, oxidizers, chlorosulfonic acid, lithium aluminum hydride, oleum.

ACETILUM ACIDILATUM (50-78-2) Fires and explosions may result from contact with strong oxidizers. Alkali hydroxides or carbonates cause decomposition.

ACETOACETIC ACID, ETHYL ESTER (141-97-9) Incompatible with strong acids, nitrates, oxidizers.

ACETOACETIC ACID, METHYL ESTER (105-45-3) Incompatible with oxidizers, strong acids, nitrates.

ACETOACETIC ESTER (141-97-9) Incompatible with strong acids, nitrates, oxidizers.

ACETOACETONE (123-54-6) Forms explosive mixture with air (flash point 95°F/35°C). Reacts violently with strong oxidizers. Incompatible with aliphatic amines, alkanolamines, organic acids, isocyanates.

ACETOL (50-78-2) Fires and explosions may result from contact with strong oxidizers. Alkali hydroxides or carbonates cause decomposition.

ACETOMETHYLBENZENE (140-11-4) Forms explosive mixture with air (flash point 216°F/102°C). Reacts with strong acids, nitrates, oxidizers.

ACETON (Dutch) (67-64-1) Forms explosive mixture with air (flash point −4°F/−20°C). Reacts violently with chloroform. Incompatible with strong acids, aliphatic amines, chloroform, chromic anhydride, chromyl chloride, hexachloromelamine, hydrogen peroxide, nitrosyl chloride, nitrosyl perchlorate, nitryl perchlorate, permonosulfuric acid, potassium *tert*-butoxide. Unstable and explosive peroxides are formed with strong oxidizers. May accumulate static electrical charges, and may cause ignition of its vapors. Dissolves most rubber, resins, and plastics.

ACETON (German) (67-64-1) Forms explosive mixture with air (flash point −4°F/−20°C). Reacts violently with chloroform. Incompatible with strong acids, aliphatic amines, chloroform, chromic anhydride, chromyl chloride, hexachloromelamine, hydrogen peroxide, nitrosyl chloride, nitrosyl perchlorate, nitryl perchlorate, permonosulfuric acid, potassium *tert*-butoxide. Unstable and explosive peroxides are formed with strong oxidizers. May accumulate static electrical charges, and may cause ignition of its vapors. Dissolves most rubber, resins, and plastics.

ACETON (Polish) (67-64-1) Forms explosive mixture with air (flash point −4°F/−20°C). Reacts violently with chloroform. Incompatible with strong acids, aliphatic amines, chloroform, chromic anhydride, chromyl chloride, hexachloromelamine, hydrogen peroxide, nitrosyl chloride, nitrosyl perchlorate, nitryl perchlorate, permonosulfuric acid, potassium *tert*-butoxide. Unstable and explosive peroxides are formed

9

with strong oxidizers. May accumulate static electrical charges, and may cause ignition of its vapors. Dissolves most rubber, resins, and plastics.

ACETONCIANHIDRINEI (Rumanian) (75-86-5) Forms explosive mixture with air (flash point 165°F/74°C). Reacts violently with strong oxidizers. Heat may cause decomposition. Incompatible with nonoxidizing mineral acids, sulfuric acid, nitric acid, organic acids, caustics, aliphatic amines, alkanolamines, aromatic amines, organic anhydrides, allylene oxides, epichlorohydrin, caprolactam solution, ammonia, isocyanates, phenols, cresol.

ACETONCIANIDRINA (Italian) (75-86-5) Forms explosive mixture with air (flash point 165°F/74°C). Reacts violently with strong oxidizers. Heat may cause decomposition. Incompatible with nonoxidizing mineral acids, sulfuric acid, nitric acid, organic acids, caustics, aliphatic amines, alkanolamines, aromatic amines, organic anhydrides, allylene oxides, epichlorohydrin, caprolactam solution, ammonia, isocyanates, phenols, cresol.

ACETONCYAANHYDRINE (Dutch) (75-86-5) Not compatible with nonoxidizing mineral acids, strong acids, ammonia, isocyanates, phenols, cresol.

ACETONCYANHYDRIN (German) (75-86-5) Forms explosive mixture with air (flash point 165°F/74°C). Reacts violently with strong oxidizers. Heat may cause decomposition. Incompatible with nonoxidizing mineral acids, sulfuric acid, nitric acid, organic acids, caustics, aliphatic amines, alkanolamines, aromatic amines, organic anhydrides, allylene oxides, epichlorohydrin, caprolactam solution, ammonia, isocyanates, phenols, cresol.

ACETONE (67-64-1) Forms explosive mixture with air (flash point −4°F/−20°C). Reacts violently with chloroform. Incompatible with strong acids, aliphatic amines, chloroform, chromic anhydride, chromyl chloride, hexachloromelamine, hydrogen peroxide, nitrosyl chloride, nitrosyl perchlorate, nitryl perchlorate, permonosulfuric acid, potassium *tert*-butoxide. Unstable and explosive peroxides formed with strong oxidizers. May accumulate static electrical charges, and may cause ignition of its vapors. Dissolves most rubber, resins, and plastics.

ACETONECYANHYDRINE (French) (75-86-5) Forms explosive mixture with air (flash point 165°F/74°C). Reacts violently with strong oxidizers. Heat may cause decomposition. Incompatible with nonoxidizing mineral acids, sulfuric acid, nitric acid, organic acids, caustics, aliphatic amines, alkanolamines, aromatic amines, organic anhydrides, allylene oxides, epichlorohydrin, caprolactam solution, ammonia, isocyanates, phenols, cresol.

ACETONE CYANOHYDRIN (75-86-5) Forms explosive mixture with air (flash point 165°F/74°C). Reacts violently with strong oxidizers. Heat may cause decomposition. Incompatible with nonoxidizing mineral acids, sulfuric acid, nitric acid, organic acids, caustics, aliphatic amines, alkanolamines, aromatic amines, organic anhydrides, allylene oxides, epichlorohydrin, caprolactam solution, ammonia, isocyanates, phenols, cresol.

ACETONE, METHYL- (78-93-3) Forms explosive mixture with air (flash point 16°F/−9°C). Extremely flammable. Incompatible with sulfuric acid, nitric acid, aliphatic amines, strong oxidizers, potassium *tert*-butoxide, 2-propanol, chlorosulfonic acid, oleum.

10

ACETONIC ACID (598-8-3) Incompatible with strong acids.

ACETONITRIL (Dutch) (75-05-8) Forms explosive mixture with air (flash point 42°F/6°C). Incompatible with water (especially if acid or alkaline), acids, caustics, nitrating agents, indium, nitrogen tetroxide, *n*-fluoro compounds, sulfur trioxide, iron(III) salts of perchlorate, indium, perfluorourea, nitrogen-fluorine compounds. Reacts violently with strong oxidizers. May accumulate static electrical charges, and may cause ignition of its vapors.

ACETONITRIL (German) (75-05-8) Forms explosive mixture with air (flash point 42°F/6°C). Incompatible with water (especially if acid or alkaline), acids, caustics, nitrating agents, indium, nitrogen tetroxide, *n*-fluoro compounds, sulfur trioxide, iron(III) salts of perchlorate, indium, perfluorourea, nitrogen-fluorine compounds. Reacts violently with strong oxidizers. May accumulate static electrical charges, and may cause ignition of its vapors.

ACETONITRILE (75-05-8) Forms explosive mixture with air (flash point 42°F/6°C). Incompatible with water (especially if acid or alkaline), acids, caustics, nitrating agents, indium, nitrogen tetroxide, *n*-fluoro compounds, sulfur trioxide, iron(III) salts of perchlorate, indium, perfluorourea, nitrogen-fluorine compounds. Reacts violently with strong oxidizers. May accumulate static electrical charges, and may cause ignition of its vapors.

ACETONKYANHYDRIN (Czech) (75-86-5) Forms explosive mixture with air (flash point 165°F/74°C). Reacts violently with strong oxidizers. Heat may cause decomposition. Incompatible with nonoxidizing mineral acids, sulfuric acid, nitric acid, organic acids, caustics, aliphatic amines, alkanolamines, aromatic amines, organic anhydrides, allylene oxides, epichlorohydrin, caprolactam solution, ammonia, isocyanates, phenols, cresol.

ACETONYLACETONE (110-13-4) Forms explosive mixture with air (flash point 174°F/79°C). Strong oxidizers may cause fire and explosions.

3-(alpha-ACETONYLBENZYL)-4-HYDROXYCOUMARIN (81-81-2) Strong oxidizers may cause fire and explosions.

ACETONYL BROMIDE (598-31-2) Forms explosive mixture with air (flash point 113°F/45°C). Contact with oxidizers may cause fire and explosions.

ACETONYL CHLORIDE (78-95-5) Forms exposive mixture with air (flash point 40°F/7°C). Strong oxidizers may cause fire and explosions. May accumulate static electrical charges, and may cause ignition of its vapors.

ACETONYLDIMETHYLCARBINOL (123-42-2) Incompatible with strong acids (sulfuric, nitric), caustics, aliphatic amines, isocyanates, strong bases, oxidizers, alkali metals. Forms explosive mixture with air above 136°F/58°C. Attacks some forms of plastics, resins, and rubber.

ACETOPHEN (50-78-2) Fires and explosions may result from contact with strong oxidizers. Alkali hydroxides or carbonates cause decomposition.

ACETOPHENONE (98-86-2) Incompatible with strong acids, aliphatic amines, oxidizers. Forms explosive mixture with air (flash point 170°F/77°C).

ACETOSAL (50-78-2) Fires and explosions may result from contact with strong oxidizers. Alkali hydroxides or carbonates cause decomposition.

ACETOSALIC ACID (50-78-2) Fires and explosions may result from contact with strong oxidizers. Alkali hydroxides or carbonates cause decomposition.

ACETOSALIN (50-78-2) Fires and explosions may result from contact with strong oxidizers. Alkali hydroxides or carbonates cause decomposition.

ACETO TETD (137-26-8) Combustible solid (flash point 192°F/89°C). Strong oxidizers may cause fire and explosions; contact with strong acid or oxidizable materials produces toxic gases.

o-**ACETOXYBENZOIC ACID** (50-78-2) Fires and explosions may result from contact with strong oxidizers. Alkali hydroxides or carbonates cause decomposition.

2-ACETOXYBENZOIC ACID (50-78-2) Fires and explosions may result from contact with strong oxidizers. Alkali hydroxides or carbonates cause decomposition.

ACETOXYETHANE (141-78-6) Forms explosive mixture with air (flash point 24°F/−4.4°C). Incompatible with strong acids, nitrates, oxidizers, chlorosulfonic acid, lithium aluminum hydride, oleum.

1-ACETOXYETHYLENE (108-05-4) Forms explosive mixture with air (flash point 18°F/−8°C). Polymerizes readily if not inhibited; heat can initiate reaction. Reacts violently with oxidizers. Incompatible with nonoxidizing mineral acids, strong acids, ammonia, aliphatic amines, alkanolamines. Also reacts with 2-aminoethanol, chlorosulfonic acid, ethylene diamine, ethyleneimine, ozone, oleum, peroxides. May accumulate static electrical charges, and may cause ignition of its vapors.

ACETOXYL (94-36-0) Confined storage of dry chemical may lead to decomposition and explosion. A strong oxidant; extremely reactive. Fires and explosion may result from heat or contamination, and from contact with strong acids, combustible materials, oxidizers, acids, bases, alcohols, reducing agents, metals, metal oxides, amines, accelerators, methyl methacrylate, organic matter, lithium aluminum carbide, dimethyl aniline, amines, metallic naphthenates. May attack some plastics, rubber, and coatings. Protect containers from shock and friction.

2-ACETOXYPENTANE (626-38-0) Incompatible with strong acids, nitrates, strong alkalies, strong oxidizers, heat. May soften or dissolve plastics.

1-ACETOXYPROPANE (109-60-4) Forms explosive mixture with air (flash point 58°F/14°C). Incompatible with strong acids, nitrates, strong acids, strong oxidizers. Attacks some plastics, rubber or coatings. May accumulate static electrical charges, and may cause ignition of its vapors.

2-ACETOXYPROPANE (108-21-4) Forms explosive mixture with air (flash point 40°F/4.4°C). Incompatible with strong alkalies, strong acids, nitrates, strong oxidizers. Dissolves rubber and many plastic materials. Contact with steel causes slow decomposition.

alpha-ACETOXYTOLUENE (140-11-4) Forms explosive mixture with air (flash point 216°F/102°C). Reacts with strong acids, nitrates, oxidizers.

ACETYL ACETONE (123-54-6) Forms explosive mixture with air (flash point 95°F/35°C). Reacts violently with strong oxidizers. Incompatible with aliphatic amines, alkanolamines, organic acids, isocyanates.

ACETYL ANHYDRIDE (108-24-7) Forms explosive mixture with air (flash point 121°F/50°C). Water contact forms acetic acid and considerable heat. Reacts violently with potassium hydroxide, sodium hydroxide, alcohols, amines. Incompatible with acids, ammonia, aniline, chromic anhydride, ethylene diamine, glycerol, hydrogen peroxide, nitrogen tetroxide, oxidizers, permanganates, sodium peroxide. Attacks some coatings, some plastics, and rubber. Corrodes iron and steel. May accumulate static electrical charges, and may cause ignition of its vapors.

ACETYL BENZENE (98-86-2) Incompatible with strong acids, aliphatic amines, oxidizers. Forms explosive mixture with air (flash point 170°F/77°C).

ACETYL BROMIDE (506-96-7) Incompatible with water, steam, alcohols.

ACETYL CAPROLACTAM (1888-91-1) May form unstable peroxides. Reacts violently with oxidizers.

ACETYL CHLORIDE (75-36-5) Forms explosive mixture with air (flash point 40°F/4°C). May accumulate static electrical charges, and may cause ignition of its vapors. Reacts violently with water, oxidizers, strong bases, alcohols (especially ethanol), dimethyl sulfoxide, phosphorus trichloride; forms corrosive hydrochloric acid with air.

ACETYL CHLORIDE, CHLORO- (79-04-9) Incompatible with sulfuric acid, caustics, alkalis, alcohols, aliphatic amines, alkanolamines, ammonia, isocyanates, alkylene oxides, epichlorohydrin.

ACETYL DIMETHYLAMINE (127-19-5) Forms explosive mixture with air (flash point 158°F/70°C). Incompatible with nonoxidizing mineral acids, strong acids, ammonia, isocyanates, phenols, cresols, halogenated compounds (above 185°F/85°C). Attacks some plastics, rubber, and coatings.

ACETYLEN (74-86-2) A strong reducing agent that reacts violently with oxidizers. Forms explosive mixture with air (flash point 0°F/18°C). Forms shock-sensitive mixture with copper and copper salts, mercury and mercury salts, silver and silver salts. Reacts with brass, bromine, cesium hydride, chlorine, cobalt, cuprous acetylise, fluorine, iodine, mercuric nitrate, nitric acid, potassium, rubidium hydride, trifluoromethyl hypofluorite, sodium hydride.

ACETYLENE (74-86-2) A strong reducing agent that reacts violently with oxidizers. Forms explosive mixture with air (flash point 0°F/18°C). Forms shock sensitive mixture with copper and copper salts, mercury and mercury salts, silver and silver salts. Reacts with brass, bromine, cesium hydride, chlorine, cobalt, cuprous acetylise, fluorine, iodine, mercuric nitrate, nitric acid, potassium, rubidium hydride, trifluoromethyl hypofluorite, sodium hydride.

ACETYLENE BLACK (1333-86-4) Dust can form an explosive mixture in air. Oxidizers may cause fire and explosions.

ACETYLENE DICHLORIDE (540-59-0) Forms explosive mixture with air (flash point 36°F/2.2°C). Incompatible with strong bases, oxidizers, difluoromethylene, dihypofluoride, nitrogen tetroxide (explosive). Attacks some plastics, rubber, and coatings.

cis-ACETYLENE DICHLORIDE (540-59-0) Forms explosive mixture with air (flash point 36°F/2.2°C). Incompatible with strong bases, oxidizers, difluoromethylene, dihypofluoride, nitrogen tetroxide (explosive). Attacks some plastics, rubber, and coatings.

trans-ACETYLENE DICHLORIDE (540-59-0) Forms explosive mixture with air (flash point 36°F/2.2°C). Incompatible with strong bases, oxidizers, difluoromethylene, dihypofluoride, nitrogen tetroxide (explosive). Attacks some plastics, rubber, and coatings.

ACETYLENE, DICHLORO- (7572-29-4) Heat or air contact may cause explosion. Reacts violently with oxidizers and acids.

ACETYLENE TETRABROMIDE (79-27-6) Reacts with chemically active metals, caustics. Hot iron, aluminum, or zinc in the presence of steam may produce toxic vapors. Softens or destroys most plastics and rubbers.

ACETYLENE TRICHLORIDE (79-01-6) Caustics form a toxic and flammable gas. Reacts violently with chemically active metals. Contact with aluminum may produce a violent, self-accelerating polymerization reaction. Incompatible with nonoxidizing mineral acids, strong acids, organic acids, organic anhydrides, isocyanates, alkylene oxides, aldehydes, alcohols, glycols, phenols, cresols, caprolactam solution, epichlorohydrin, nitrogen tetroxide, metal powders, oxygen. May accumulate static electrical charges, and may cause ignition of its vapors.

ACETYLENOGEN (75-20-7) Water contact or moist air forms explosive acetylene gas. Incompatible with acids, oxidizers, hydrogen chloride, methanol, copper salt solutions, lead fluoride, magnesium, selenium, silver nitrate, iron trichloride, tin dichloride, sodium peroxide, stannous chloride, sulfur.

ACETYL ETHER (108-24-7) Forms explosive mixture with air (flash point 121°F/50°C). Water contact forms acetic acid and considerable heat. Reacts violently with potassium hydroxide, sodium hydroxide, alcohols, amines. Incompatible with acids, ammonia, aniline, chromic anhydride, ethylene diamine, glycerol, hydrogen peroxide, nitrogen tetroxide, oxidizers, permanganates, sodium peroxide. Attacks some coatings, some plastics, and rubber. Corrodes iron and steel. May accumulate static electrical charges, and may cause ignition of its vapors.

ACETYL ETHYLENE (78-94-4) Heat can cause polymerization. Strong oxidizers may cause fire and explosions.

ACETYL HEXANOLACTAM ACETYL CAPROLACTAM (1888-91-1) May form unstable and explosive peroxides. Incompatible with oxidizers.

ACETYL HYDROPEROXIDE (79-21-0) A powerful oxidizer and extremely sensitive explosive. Forms explosive mixture with air (flash point 100°F/38°C). Reacts violently with many substances, including acetic anhydride, combustibles, strong bases, ether solvents, olefins, organic matter, magnesium, metal oxides, heavy metals, metal chloride solutions, reducing agents, nickel, phosphorus, sodium nitride, zinc.

ACETYLIN (50-78-2) Fires and explosions may result from contact with strong oxidizers. Alkali hydroxides or carbonates cause decomposition.

ACETYLMETHYL BROMIDE (598-31-2) Forms explosive mixture with air (flash point 113°F/45°C). Incompatible with oxidizers.

ACETYL NITRATE (591-09-3) Self-reactive. Explosive reaction with active oxides such as mercuric oxide.

ACETYL OXIDE (108-24-7) Forms explosive mixture with air (flash point 121°F/50°C). Water contact produces acetic acid and considerable heat. Reacts violently with potassium hydroxide, sodium hydroxide, alcohols amines. Incompatible with acids, ammonia, aniline, chromic anhydride, ethylene diamine, glycerol, hydrogen peroxide, nitrogen tetroxide, oxidizers, permanganates, sodium peroxide. Attacks some coatings and some plastics and rubber. Corrodes iron and steel. May accumulate static electrical charges, and may cause ignition of its vapors.

ACETYL PEROXIDE (110-22-5) A powerful oxidizing agent that reacts violently with reducing agents. Unpredictable: shock- and heat-sensitive, self-reactive, and explosive; incompatible with ethers, solvents, organic materials, combustibles, diethyl ether.

2-ACETYL PROPANE (563-80-4) Forms explosive mixture with air (flash point 43°F/6°C). Incompatible with oxidizers.

ACETYSAL (50-78-2) Fires and explosions may result from contact with strong oxidizers. Alkali hydroxides or carbonates cause decomposition.

ACETYLSALICYLIC ACID (50-78-2) Fires and explosions may result from contact with strong oxidizers. Alkali hydroxides or carbonates cause decomposition.

ACETYL TRIBUTYL CITRATE Incompatible with strong acids, nitrates, oxidizers.

ACID AMMONIUM CARBONATE (1066-33-7) Reacts violently with strong acids. Reacts with bases, fluorine, magnesium, sodium hypochlorite.

ACID AMMONIUM CARBONATE, MONOAMMONIUM SALT (1066-33-7) Reacts violently with strong acids. Reacts with bases, fluorine, magnesium, sodium hypochlorite.

ACID AMMONIUM FLUORIDE (1341-49-7) Incompatible with alkalies, water (forms strong acid). Attacks glass, cement, and most metals.

ACIDE ACETIQUE (French) (64-19-7) Vapor forms explosive mixture with air above 105°F/40°C. Reacts violently with oxidizers and bases. Incompatible with strong acids, aliphatic amines, alkanolamines, isocyanates, alkylene oxides, epichlorohydrin, acetaldehyde, 2-aminoethanol, ammonia, ammonium nitrate, chlorosulfonic acid, chromic acid, ethylene diamine, ethyleneimine, perchloric acid, permanganates, phosphorus isocyanate, phosphorous trichloride, potassium *tert*-butoxide, xylene. Attacks cast iron and other metals forming flammable hydrogen gas.

ACIDE ARSENIEUX (French) (1327-53-3) Incompatible with fluorine, fluorides, sodium chlorate.

ACIDE BROMHYDRIQUE (French) (10035-10-6) Incompatible with aliphatic amines, alkanolamines, alkylene oxides, aromatic amines, amides, ammonia, ammonium hydroxide, bases, calcium oxide, epichlorohydrin, fluorine, isocyanates, oleum, organic anhydrides, sulfuric acid, sodium tetrahydroborate, strong oxidizers, vinyl acetate, water. Attacks most metals with the formation of flammable hydrogen gas.

ACIDE CARBOLIQUE (French) (108-95-2) Forms explosive mixture with air (flash point 174°F/79°C). Incompatible with strong oxidizers, strong acids, caustics, aliphatic amines, amides, oxidizers, formaldehyde, butadiene, calcium hypochlorite. Liquid attacks some plastics, rubber, and coatings; hot liquid attacks aluminum, magnesium, lead, and zinc metals.

ACIDE CHLORHYDRIQUE (French) (7647-01-0) Incompatible with acetic anhydride, aliphatic amines, alkanolamines, alkylene oxides, aromatic amines, amides, 2-aminoethanol, ammonia, ammonium hydroxide, calcium phosphide, chlorosulfonic acid, ethylene diamine, ethylene- imine, epichlorohydrin, isocyanates, metal acetylides, oleum, organic anhydrides, perchloric acid, 3-propiolactone, uranium phosphide, sulfuric acid, sodium hydroxide and other bases, strong oxidizers, vinyl acetate, vinylidene fluoride. Attacks most metals and some plastics, rubber, and coatings.

ACIDE CHROMIQUE (French) (7738-94-5); solution (1308-14-1) A powerful oxidizer. Reducing agents, combustibles, organic, or other oxidizable materials (e.g., wood, sulfur, aluminum, plastics) may cause fire and explosions. Attacks most forms of metals, cloth, leather, plastics, rubber, and coatings, and may cause spontaneous ignition.

ACIDE CYANHYDRIQUE (French) (74-90-8) Unless stabilized and maintained, samples stored more than 90 days are hazardous. Samples containing more than 2–5% water are less stable than dry material. Can be self-reactive, forming an explosive mixture with air (flash point 0°F/−18°C). Heat or contact with amines or strong bases can cause polymerization. Incompatible with acetaldehyde. Oxidizers may cause fire and explosions. Attacks some plastics, rubber, and coatings.

ACIDE 2,4-DICHLORO PHENOXYACETIQUE (French) (94-75-7) Decomposes in sunlight. Incompatible with strong oxidizers.

ACIDE ETHYLENEDIAMINETE TRACETIQUE (French) (60-00-4) Incompatible with sulfuric acid, bases, ammonia, aliphatic amines, alkanolamines, isocyanates, alkylene oxides, epichlorohydrin.

ACIDE FORMIQUE (French) (64-18-6) Forms explosive mixture with air (flash point 156°F/69°C). Incompatible with sulfuric acid, bases, ammonia, aliphatic amines, alkanolamines, furfuryl alcohol, hydrogen peroxide, isocyanates, alkylene oxides, epichlorohydrin. Attacks aluminum, cast iron and steel, some plastics, rubber, and coatings.

ACIDE NITRIQUE (French) (7697-37-2) A strong oxidizer that can react violently with reducing agents, combustible materials. Incompatible with many substances, including acrylates, aliphatic amines, alcohols, aldehydes, alkanolamines, alkylene oxides, anion exchange resins, aromatic amines, amides, bases, cresols, cyanides, cyclic ketones, epichlorohydrin, glycols, isocyanates, ketones, oleum, organic anhydrides, phenols, substituted allyls, sulfuric acid, strong oxidizers, terpenes. Attacks most metals and some plastics, rubber, and coatings.

ACIDE OXALIQUE (French) (144-62-7) Silver compounds may form explosive salts. Incompatible with caustics, furfuryl alcohol, mercury, silver, sodium chlorite, sodium hypochlorite.

ACIDE PERACETIQUE (French) (79-21-0) A powerful oxidizer and extremely sensitive explosive. Forms explosive mixture with air (flash point 100°F/38°C). Reacts violently with many substances, including acetic anhydride, combustibles, strong bases, ether solvents, olefins, organic matter, magnesium, metal oxides, heavy metals, metal chloride solutions, reducing agents, nickel, phosphorus, sodium nitride, zinc.

ACIDE PHOSPHORIQUE (French) (7664-38-2) Incompatible with aliphatic amines, alkanolamines, alkylene oxides, aromatic amines, amides, ammonia, ammonium hydroxide, bases, calcium oxide, epichlorohydrin, isocyanates; explosive with nitromethane, oleum, organic anhydrides, sulfuric acid, sodium tetrahydroborate, strong oxidizers, vinyl acetate, water. Contact with most metals produces hydrogen gas. Attacks some plastics, rubber, or coatings; incompatible with glass and ceramics.

ACIDE PICRIQUE (French) (88-89-1) Dry material is explosive. Impact-sensitive; must be protected from shock. Copper, lead, zinc and other metals, or their salts can form salts that are initiators and much more sensitive to shock than this chemical. Shock-sensitive salts include ammonium salts and calcium salts; the calcium salt may form when picric acid comes into contact with plaster and concrete. May accumulate static electrical charges, and may cause explosion. Aqueous solution is a strong oxidizer and a strong acid; reacts violently with reducing agents, combustibles, organics and easily oxidized materials, caustics, aluminum and other metal powders. Attacks metals.

ACIDE PROPIONIQUE (French) (79-09-4) Forms explosive mixture with air (flash point 126°F/52°C). Incompatible with sulfuric acid, strong bases, ammonia, aliphatic amines, alkanolamines, isocyanates, alkylene oxides, epichlorohydrin, oxidizers.

ACIDE SULFHYDRIQUE (French) (7783-06-4) Forms explosive mixture with air. Reacts violently with oxidizers. Attacks metals.

ACIDE 2,4,5-TRICHLOROPHENOXYACETIQUE (French) (93-76-5) Sealed metal containers may burst in heat above 316°F/158°C. Incompatible with sulfuric acid, bases, ammonia, aliphatic amines, alkanolamines, isocyanates, alkylene oxides, epichlorohydrin.

ACID GAS (124-38-9) Incompatible with acrylaldehyde, amines, anhydrous ammonia, cesium monoxide, lithium, metal dusts, potassium, sodium, sodium carbide, sodium–potassium alloy, sodium peroxide, titanium.

ACID LEATHER BROWN 2G (1300-73-8) Forms explosive mixture with air (flash point 206°F/96.7°C). Contact with strong acids may cause fire and explosions. Contact with hypochlorite bleaches produces explosive chloroamines. Incompatible with strong acids, organic acids and anhydrides, isocyanates, aldehydes. Attacks some plastics, rubber, and coatings.

ACIDO ACETICO (Italian) (64-19-7) Vapor forms explosive mixture with air above 105°F/40°C. Reacts violently with oxidizers and bases. Incompatible with strong acids, aliphatic amines, alkanolamines, isocyanates, alkylene oxides, epichlorohydrin, acetaldehyde, 2-aminoethanol, ammonia, ammonium nitrate, chlorosulfonic acid, chromic acid, ethylene diamine, ethyleneimine, perchloric acid, permanganates, phosphorus isocyanate, phosphorous trichloride, potassium *tert*-butoxide, xylene. Attacks cast iron and other metals forming flammable hydrogen gas.

ACIDO BROMIDRICO (Italian) (10035-10-6) Incompatible with aliphatic amines, alkanolamines, alkylene oxides, aromatic amines, amides, ammonia, ammonium hydroxide, bases, calcium oxide, epichlorohydrin, fluorine, isocyanates, oleum, organic anhydrides, sulfuric acid, sodium tetrahydroborate, strong oxidizers, vinyl acetate, water. Attacks most metals with the formation of flammable hydrogen gas.

ACIDO CIANIDRICO (Italian) (74-90-8) Unless stabilized and maintained, samples stored more than 90 days are hazardous. Samples containing more than 2–5% water are less stable than dry material. Can be self-reactive, forming an explosive mixture with air (flash point 0°F/−18°C). Heat or contact with amines or strong bases can cause polymerization. Incompatible with acetaldehyde. Oxidizers may cause fire and explosions. Attacks some plastics, rubber, and coatings.

ACIDO CLORIDRICO (Italian) (7647-01-0) Incompatible with acetic anhydride, aliphatic amines, alkanolamines, alkylene oxides, aromatic amines, amides, 2-aminoethanol, ammonia, ammonium hydroxide, calcium phosphide, chlorosulfonic acid, ethylene diamine, ethylene-imine, epichlorohydrin, isocyanates, metal acetylides, oleum, organic anhydrides, perchloric acid, 3-propiolactone, uranium phosphide, sulfuric acid, sodium hydroxide and other bases, strong oxidizers, vinyl acetate, vinylidene fluoride. Attacks most metals and some plastics, rubber, and coatings.

ACIDO(2,4-DICLORO-FENOSSI)-ACETICO (Italian) (94-75-7) Decomposes in sunlight. Incompatible with strong oxidizers; may cause fire and explosions.

ACIDO(3,6-DICLORO-2-METOSSI)-BENZOICO (1918-00-9) Incompatible with sulfuric acid, bases, ammonia, aliphatic amines, alkanolamines, isocyanates, alkylene oxides, epichlorohydrin.

ACIDO FORMICO (Italian) (64-18-6) Forms explosive mixture with air (flash point 156°F/69°C). Incompatible with sulfuric acid, bases, ammonia, aliphatic amines, alkanolamines, furfuryl alcohol, hydrogen peroxide, isocyanates, alkylene oxides, epichlorohydrin. Attacks aluminum, cast iron and steel, some plastics, rubber, and coatings.

ACIDO FOSFORICO (Italian) (7664-38-2) Incompatible with aliphatic amines, alkanolamines, alkylene oxides, aromatic amines, amides, ammonia, ammonium hydroxide, bases, calcium oxide, epichlorohydrin, isocyanates; explosive with nitromethane (75-52-5), oleum, organic anhydrides, sulfuric acid, sodium tetrahydroborate, strong oxidizers, vinyl acetate, water. Contact with most metals produces hydrogen gas. Attacks some plastics, rubber, or coatings; incompatible with glass and ceramics.

ACIDO NITRICO (Italian) (7697-37-2) A strong oxidizer that can react violently with reducing agents, combustible materials. Incompatible with many substances, including acrylates, aliphatic amines, alcohols, aldehydes, alkanolamines, alkylene oxides, anion exchange resins, aromatic amines, amides, bases, cresols, cyanides, cyclic ketones, epichlorohydrin, glycols, isocyanates, ketones, oleum, organic anhydrides, phenols, substituted allyls, sulfuric acid, strong oxidizers, terpenes. Attacks most metals, and some plastics, rubber, and coatings.

ACID ORANGE 24 (1300-73-8) Forms explosive mixture with air (flash point 206°F/96.7°C). Contact with strong acids may cause fire and explosions. Contact with hypochlorite bleaches produces explo-

18

sive chloroamines. Incompatible with strong acids, organic acids and anhydrides, isocyanates, aldehydes. Attacks some plastics, rubber, and coatings.

ACIDO OSSALICO (Italian) (144-62-7) Silver compounds may form explosive salt. Incompatible with caustics, furfuryl alcohol, oxidizers, mercury, silver, sodium chlorite, sodium hypochlorite.

ACIDO PICRICO (Italian) (88-89-1) Dry material is explosive. Impact-sensitive; must be protected from shock. Copper, lead, zinc, and other metals or their salts can form salts that are initiators and much more sensitive to shock than this chemical. Shock-sensitive salts include ammonium salts and calcium salts; the calcium salt may form when picric acid comes into contact with plaster and concrete. May accumulate static electrical charges, and may cause explosion. Aqueous solution is a strong oxidizer and a strong acid; reacts violently with reducing agents, combustibles, organics and easily oxidized materials, caustics, aluminum, and other metal powders. Attacks metals.

ACIDO (2,4,5-TRICLORO-FENOSSI)-ACETICO (Italian) (93-76-5) Sealed metal containers may burst in heat above 316°F/158°C. Incompatible with sulfuric acid, bases, ammonia, aliphatic amines, alkanolamines, isocyanates, alkylene oxides, epichlorohydrin.

ACIDUM ACETYLSALICYLICUM (50-78-2) Fires and explosions may result from contact with strong oxidizers. Alkali hydroxides or carbonates cause decomposition.

ACID PHOSPHORIQUE (French) (7664-38-2) Incompatible with aliphatic amines, alkanolamines, alkylene oxides, aromatic amines, amides, ammonia, ammonium hydroxide, bases, calcium oxide, epichlorohydrin, isocyanates, explosive with nitromethane, oleum, organic anhydrides, sulfuric acid, sodium tetrahydroborate, strong oxidizers, vinyl acetate, water. Contact with most metals produces hydrogen gas. Attacks some plastics, rubber, or coatings; incompatible with glass and ceramics.

ACIFLOCTIN (124-04-9) May accumulate static electrical charges, and may cause ignition of its vapors. Contact with strong oxidizers may cause fire and explosions.

ACIMETTER (50-78-2) Fires and explosions may result from contact with strong oxidizers. Alkali hydroxides or carbonates cause decomposition.

ACINETTEN (124-04-9) May accumulate static electrical charges, and may cause ignition of its vapors. Contact with strong oxidizers may cause fire and explosions.

ACINTENE A (80-56-8) Incompatible with sulfuric acid, nitric acid.

ACINTENE DP (138-86-3) Forms explosive mixture with air (flash point 115°F/46°C). Strong oxidizers may cause fire and explosions.

ACINTENE DP DIPENTENE (138-86-3) Forms explosive mixture with air (flash point 115°F/46°C). Strong oxidizers may cause fire and explosions.

ACL 85 (87-90-1) A powerful oxidizer. Forms explosive material with nitrogen compounds. Contact with organic materials or reducing agents may cause fire.

ACQUINITE (76-06-2) A strong oxidizer. Self-reactive. Incompatible with reducing agents, aniline in presence of heat, alcoholic sodium hydroxide, combustible substances, sodium methoxide, propargyl bromide. Liquid attacks some plastics, rubber, and coatings.

ACREHYDE (107-02-8) Forms explosive mixture with air (flash point −15°F/−26°C). Unstable and very reactive. In storage can form heat- and shock-sensitive compounds. Unless inhibited (usually by hydroquinone), readily forms explosive peroxides. Able to polymerize. A strong reducing agent. Reacts violently with oxidizers, strong acids, caustics, amines, 2-aminoethanol, ammonia, ammonium hydroxide, ethylene diamine, ethyleneimine, hydroxides, metal salts, oxidizers, sulfur dioxide, thiourea. Attacks metals: cadmium and zinc. May accumulate static electrical charges, and may cause ignition of its vapors.

ACRIDINE (260-94-6) Contact with strong oxidizers may cause fire and explosions.

ACROLEIC ACID (79-10-7) Forms explosive mixture with air (flash point 122°F/50°C). Forms explosive peroxides; light, heat and peroxides can cause polymerization. Incompatible with sulfuric acid, caustics, ammonia, amines, isocyanates, alkylene oxides, epichlorohydrin, oxidizers, toluenediamine, pyridine, methyl pyridine, *n*-methyl pyrrolidone, 2-methyl-6-ethyl aniline, aniline, ethylene diamine, ethyleneimine, 2-aminoethanol. Severely corrodes carbon steel and iron; attacks other metals. May accumulate static electrical charges, and may cause ignition of its vapors.

ACROLEIN (107-02-8) Forms explosive mixture with air (flash point −15°F/−26°C). Unstable and very reactive. In storage can form heat- and shock-sensitive compounds. Unless inhibited (usually by hydroquinone), readily forms explosive peroxides. Able to polymerize. A strong reducing agent. Reacts violently with oxidizers, strong acids, caustics, amines, 2-aminoethanol, ammonia, ammonium hydroxide, ethylene diamine, ethyleneimine, hydroxides, metal salts, oxidizers, sulfur dioxide, thiourea. Attacks metals: cadmium and zinc. May accumulate static electrical charges, and may cause ignition of its vapors.

trans-**ACROLEIN** (107-02-8) Forms explosive mixture with air (flash point −15°F/−26°C). Unstable and very reactive. In storage can form heat- and shock-sensitive compounds. Unless inhibited (usually by hydroquinone), readily forms explosive peroxides. Able to polymerize. A strong reducing agent. Reacts violently with oxidizers, strong acids, caustics, amines, 2-aminoethanol, ammonia, ammonium hydroxide, ethylene diamine, ethyleneimine, hydroxides, metal salts, oxidizers, sulfur dioxide, thiourea. Attacks metals: cadmium and zinc. May accumulate static electrical charges, and may cause ignition of its vapors.

ACROLEINA (Italian) (107-02-8) Forms explosive mixture with air (flash point −15°F/−26°C). Unstable and very reactive. In storage can form heat- and shock-sensitive compounds. Unless inhibited (usually by hydroquinone), readily forms explosive peroxides. Able to polymerize. A strong reducing agent. Reacts violently with oxidizers, strong acids, caustics, amines, 2-aminoethanol, ammonia, ammonium hydroxide, ethylene diamine, ethyleneimine, hydroxides, metal salts, oxidizers, sulfur dioxide, thiourea. Attacks metals: cadmium and zinc. May accumulate static electrical charges, and may cause ignition of its vapors.

ACROLEINE (Dutch or French) (107-02-8) Forms explosive mixture with air (flash point −15°F/−26°C). Unstable and very reactive. In storage can form heat- and shock-sensitive compounds. Unless inhibited (usually by hydroquinone), readily forms explosive peroxides. Able to polymerize. A strong reducing agent. Reacts violently with oxidizers, strong acids, caustics, amines, 2-aminoethanol, ammonia, ammonium hydroxide, ethylene diamine, ethyleneimine, hydroxides, metal salts, oxidizers, sulfur dioxide, thiourea. Attacks metals: cadmium and zinc. May accumulate static electrical charges, and may cause ignition of its vapors.

ACRYLALDEHYDE (107-02-8) Forms explosive mixture with air (flash point −15°F/−26°C). Unstable and very reactive. In storage can form heat- and shock-sensitive compounds. Unless inhibited (usually by hydroquinone), readily forms explosive peroxides. Able to polymerize. A strong reducing agent. Reacts violently with oxidizers, strong acids, caustics, amines, 2-aminoethanol, ammonia, ammonium hydroxide, ethylene diamine, ethyleneimine, hydroxides, metal salts, oxidizers, sulfur dioxide, thiourea. Attacks metals: cadmium and zinc. May accumulate static electrical charges, and may cause ignition of its vapors.

ACRYLAMIDE (79-06-1) Unless inhibited, ultraviolet light and heat (above 184°F/85°C) can cause polymerization. Incompatible with nonoxidizing mineral acids, strong acids, ammonia, oleum, oxidizers, isocyanates.

ACRYLAMIDE MONOMER (79-06-1) Unless inhibited, ultraviolet light and heat (above 184°F/85°C) can cause polymerization. Incompatible with nonoxidizing mineral acids, strong acids, ammonia, oleum, oxidizers, isocyanates.

ACRYLATE de METHYLE (French) (96-33-3) Heat, light and/or lack of appropriate inhibitor concentration can cause polymerization. Forms explosive mixture with air (flash point 27°F/−3°C). Incompatible with strong acids, oxidizers, aliphatic amines, alkanolamines.

ACRYLATE d'ETHYLE (French) (140-88-5) Forms explosive mixture with air (flash point 48°F/9°C). Atmospheric moisture and strong alkalies may cause fire and explosions. Unless inhibited (*note:* inert gas blanket not recommended), heat, light or peroxides can cause polymerization. Incompatible with strong acids, amines, oxidizers.

ACRYLEHYD (German) (107-02-8) Forms explosive mixture with air (flash point −15°F/−26°C). Unstable and very reactive. In storage can form heat- and shock-sensitive compounds. Unless inhibited (usually by hydroquinone), readily forms explosive peroxides. Able to polymerize. A strong reducing agent. Reacts violently with oxidizers, strong acids, caustics, amines, 2-aminoethanol, ammonia, ammonium hydroxide, ethylene diamine, ethyleneimine, hydroxides, metal salts, oxidizers, sulfur dioxide, thiourea. Attacks metals: cadmium and zinc. May accumulate static electrical charges, and may cause ignition of its vapors.

ACRYLEHYDE (107-02-8) Forms explosive mixture with air (flash point −15°F/−26°C). Unstable and very reactive. In storage can form heat- and shock-sensitive compounds. Unless inhibited (usually by hydroquinone), readily forms explosive peroxides. Able to polymerize. A strong reducing agent. Reacts violently with oxidizers, strong acids, caustics, amines, 2-aminoethanol, ammonia, ammonium hydroxide, ethylene diamine, ethyleneimine, hydroxides, metal salts, oxidizers, sulfur dioxide, thiourea. Attacks metals: cadmium and zinc. May accumulate static electrical charges, and may cause ignition of its vapors.

ACRYLIC ACID (79-10-7) Forms explosive mixture with air (flash point 122°F/50°C). Forms explosive peroxides; light, heat and peroxides can cause polymerization. Incompatible with sulfuric acid, caustics, ammonia, amines, isocyanates, alkylene oxides, epichlorohydrin, oxidizers, toluenediamine, pyridine, methyl pyridine, *n*-methyl pyrrolidone, 2-methyl-6-ethyl aniline, aniline, ethylene diamine, ethyleneimine, 2-aminoethanol. Severely corrodes carbon steel and iron; attacks other metals. May accumulate static electrical charges, and may cause ignition of its vapors.

ACRYLIC ACID AMIDE (50%) (79-06-1) Unless inhibited, ultraviolet light and heat (above 184°F/85°C) can cause polymerization. Incompatible with nonoxidizing mineral acids, strong acids, ammonia, oleum, oxidizers, isocyanates.

ACRYLIC ACID, BUTYL ESTER (141-32-2) Forms explosive mixture with air (flash point 120°F/49°C). Heat, sparks, open flame, light, or peroxides may cause explosive polymerization. Incompatible with strong acids, aliphatic amines, alkanolamines, halogens, hydrogen compounds, oxidizers (butyl acrylate + peroxide), sunlight, or other catalysts.

ACRYLIC ACID, ISOBUTYL ESTER (106-63-8) Forms explosive mixture with air (flash point 94°F/34°C). Incompatible with strong acids, aliphatic amines, alkanolamines.

ACRYLIC ACID, *n*-BUTYL ESTER (141-32-2) Forms explosive mixture with air (flash point 120°F/49°C). Heat, sparks, open flame, light, or peroxides may cause explosive polymerization. Incompatible with strong acids, aliphatic amines, alkanolamines, halogens, hydrogen compounds, oxidizers (butyl acrylate + peroxide), sunlight, or other catalysts.

ACRYLIC ACID, DECYL ESTER (2156-96-6) Incompatible with strong acids, aliphatic amines, alkanolamines, strong oxidizers, polymerization initiators. Attacks copper and its alloys, zinc, galvanized steel, alloys having more than 10% zinc by weight. Swells some rubbers, and softens some paints and coatings.

ACRYLIC ACID, *n*-DECYL ESTER (2156-96-6) Incompatible with strong acids, aliphatic amines, alkanolamines, strong oxidizers, polymerization initiators. Attacks copper and its alloys, zinc, galvanized steel, alloys having more than 10% zinc by weight. Swells some rubbers, and softens some paints and coatings.

ACRYLIC ACID, ETHYL ESTER (140-88-5) Forms explosive mixture with air (flash point 48°F/9°C). Atmospheric moisture and strong alkalies may cause fire and explosions. Unless properly inhibited (*note:* inert gas blanket not recommended), heat, light or peroxides can cause polymerization. Incompatible with strong acids, amines, oxidizers.

ACRYLIC ACID, 2-ETHYLHEXYLESTER (103-11-7) Forms explosive mixture with air (flash point 180°F/82°C). Incompatible with strong acids, aliphatic amines, alkanolamines.

ACRYLIC ACID, GLACIAL (79-10-7) Forms explosive mixture with air (flash point 122°F/50°C). Forms explosive peroxides; light, heat and peroxides can cause polymerization. Incompatible with sulfuric acid, caustics, ammonia, amines, isocyanates, alkylene oxides, epichlorohydrin, oxidizers, toluenediamine, pyridine, methyl pyridine, n-methyl pyrrolidone, 2-methyl-6-ethyl aniline, aniline, ethylene diamine, ethyleneimine, 2-aminoethanol. Severely corrodes carbon steel and iron; attacks other metals. May accumulate static electrical charges, and may cause ignition of its vapors.

ACRYLIC ACID, 2-HYDROXYETHYL ESTER (818-61-6) Incompatible with strong acids, aliphatic amines, alkanolamines. Attacks mild steel and tin.

ACRYLIC ACID, 2-HYDROXYPROPYL ESTER (999-61-1) Incompatible with strong acids, nitrates.

ACRYLIC ACID, INHIBITED (79-10-7) Forms explosive mixture with air (flash point 122°F/50°C). Forms explosive peroxides; light, heat, and peroxides can cause polymerization. Incompatible with sulfuric acid, caustics, ammonia, amines, isocyanates, alkylene oxides, epichlorohydrin, oxidizers, toluenediamine, pyridine, methyl pyridine, n-methyl pyrrolidone, 2-methyl-6-ethyl aniline, aniline, ethylene diamine, ethyleneimine, 2-aminoethanol. Severely corrodes carbon steel and iron; attacks other metals. May accumulate static electrical charges, and may cause ignition of its vapors.

ACRYLIC ACID, ISODECYL ESTER (1330-61-6) Incompatible with strong acids, aliphatic amines, alkanolamines, oxidizers.

ACRYLIC ACID, 2-METHYL- (79-41-4) Forms explosive mixture with air (flash point 152°F/67°C). A reducing agent; reacts with oxidizers. Forms unstable peroxides; can polymerize violently, especially if stored above 120°F. Incompatible with strong acids, caustics, ammonia, amines, isocyanates, alkylene oxides, epichlorohydrin.

ACRYLIC ACID, METHYL ESTER (96-33-3) Heat, light, and/or lack of appropriate inhibitor concentration can cause polymerization. Forms explosive mixture with air (flash point 27°F/−3°C). Incompatible with strong acids, oxidizers, aliphatic amines, alkanolamines.

ACRYLIC ALDEHYDE (107-02-8) Forms explosive mixture with air (flash point −15°F/−26°C). Unstable and very reactive. In storage can form heat- and shock-sensitive compounds. Unless inhibited (usually by hydroquinone), readily forms explosive peroxides. Able to polymerize. A strong reducing agent. Reacts violently with oxidizers, strong acids, caustics, amines, 2-aminoethanol, ammonia, ammonium hydroxide, ethylene diamine, ethyleneimine, hydroxides, metal salts, oxidizers, sulfur dioxide, thiourea. Attacks metals: cadmium and zinc. May accumulate static electrical charges, and may cause ignition of its vapors.

ACRYLIC AMIDE (79-06-1) Unless inhibited, ultraviolet light and heat (above 184°F/85°C) can cause polymerization. Incompatible with nonoxidizing mineral acids, strong acids, ammonia, oleum, oxidizers, isocyanates.

ACRYLIC EHYDE (107-02-8) Forms explosive mixture with air (flash point −15°F/−26°C). Unstable and very reactive. In storage can form heat- and shock-sensitive compounds. Unless inhibited (usually by hydroquinone), readily forms explosive peroxides. Able to polymerize. A strong reducing agent. Reacts violently with oxidizers, strong acids, caustics, amines, 2-aminoethanol, ammonia, ammonium hydroxide, ethylene diamine, ethyleneimine, hydroxides, metal salts, oxidizers, sulfur dioxide, thiourea. Attacks metals: cadmium and zinc. May accumulate static electrical charges, and may cause ignition of its vapors.

ACRYLNITRIL (Dutch or German) (107-13-1) Forms explosive mixture with air (flash point 32°F/0°C o.c.). Forms explosive peroxides; heat, light, caustics, silver nitrate, and peroxides can cause polymerization. Incompatible with strong acids, strong oxidizers, amines, 2-aminoethanol, bromine, chlorosulfonic acid, ethylene diamine, nitric acid, oleum, potassium hydroxide, sodium hydroxide, sulfuric acid. Attacks copper and copper alloys; attacks aluminum in high concentrations. May accumulate static electrical charges, and may cause ignition of its vapors.

ACRYLON (107-13-1) Forms explosive mixture with air (flash point 32°F/0°C o.c.). Forms explosive peroxides; heat, light, caustics, silver nitrate, and peroxides can cause polymerization. Incompatible with strong acids, strong oxidizers, amines, 2-aminoethanol, bromine, chlorosulfonic acid, ethylene diamine, nitric acid, oleum, potassium hydroxide, sodium hydroxide, sulfuric acid. Attacks copper and copper alloys; attacks aluminum in high concentrations. May accumulate static electrical charges, and may cause ignition of its vapors.

ACRYLONITRILE or ACRYLONITRILE MONOMER (107-13-1) Forms explosive mixture with air (flash point 32°F/0°C o.c.). Forms explosive peroxides; heat, light, caustics, silver nitrate, and peroxides can cause polymerization. Incompatible with strong acids, strong oxidizers, amines, 2-aminoethanol, bromine, chlorosulfonic acid, ethylene diamine, nitric acid, oleum, potassium hydroxide, sodium hydroxide, sulfuric acid. Attacks copper and copper alloys; attacks aluminum in high concentrations. May accumulate static electrical charges, and may cause ignition of its vapors.

ACRYLSAEUREAETHYLESTER (German) (140-88-5) Forms explosive mixture with air (flash point 48°F/9°C). Atmospheric moisture and strong alkalies may cause fire and explosions. Unless properly inhibited (*note:* inert gas blanket not recommended), heat, light, or peroxides can cause polymerization. Incompatible with strong acids, amines, oxidizers.

ACRYLSAEUREMETHYLESTER (German) (96-33-3) Heat, light, and/or lack of appropriate inhibitor concentration can cause polymerization. Forms explosive mixture with air (flash point 27°F/−3°C). Incompatible with strong acids, oxidizers, aliphatic amines, alkanolamines.

ACTIVATED CARBON or ACTIVATED CHARCOAL (64365-11-3) Incompatible with strong oxidizers, strong acids, oxides, unsaturated oils.

ACTIVE ACETYL ACETATE (141-97-9) Incompatible with strong acids, nitrates, oxidizers.

ACTYLBARYTE (7727-43-7) Explosions may result from contact with aluminum in the presence of heat. Incompatible with potassium, phosphorus.

ACTYLENOGEN, CALCIUM ACETYLIDE (75-20-7) Water contact or moist air forms explosive acetylene gas. Incompatible with acids, oxidizers, hydrogen chloride, methanol, copper salt solutions, lead fluoride, magnesium, selenium, silver nitrate, iron trichloride, tin dichloride, sodium peroxide, stannous chloride, sulfur.

ACYLPRIN (50-78-2) Fires and explosions may result from contact with strong oxidizers. Alkali hydroxides or carbonates cause decomposition.

24

ACYTOL (97-64-3) Presents fire or explosion hazard (flash point 115°F/46°C). Strong oxidizers may cause fire and explosions.

ADACENE-12 (6842-15-5) Forms explosive mixture with air (flash point 144°F/62°C). Incompatible with strong acids, oxidizers.

ADILAC-TETTEN (124-04-9) May accumulate static electrical charges, and may cause ignition of its vapors. Contact with strong oxidizers may cause fire and explosions.

ADIPIC ACID (124-04-9) May accumulate static electrical charges, and may cause ignition of its vapors. Contact with strong oxidizers may cause fire and explosions.

ADIPIC ACID, *BIS*(**2-ETHYLHEXYL) ESTER** (103-23-1) Incompatible with strong acids, nitrates, oxidizers.

ADIPIC ACID, DIBUTYL ESTER (103-23-1) Incompatible with strong acids, strong oxidizers, nitrates.

ADIPIC ACID, DIMETHYL ESTER (627-93-0) Incompatible with strong acids, nitrates, oxidizers.

ADIPIC ACID DINITRILE (111-69-3) May accumulate static electrical charges, and may cause ignition of its vapors. Decomposes at its flash point, forming poisonous cyanide gas. Forms explosive mixture with air (flash point 199°F/93°C). Incompatible with sulfuric acid, strong oxidizers.

ADIPIC ACID NITRILE (111-69-3) May accumulate static electrical charges, and may cause ignition of its vapors. Decomposes at its flash point, forming poisonous cyanide gas. Forms explosive mixture with air (flash point 199°F/93°C). Incompatible with sulfuric acid, oxidizers.

ADIPINIC ACID (124-04-9) May accumulate static electrical charges, and may cause ignition of its vapors. Contact with strong oxidizers may cause fire and explosions.

ADIPODINITRILE (111-69-3) May accumulate static electrical charges, and may cause ignition of its vapors. Decomposes at its flash point, forming poisonous cyanide gas. Forms explosive mixture with air (flash point 199°F/93°C). Incompatible with sulfuric acid, oxidizers.

ADIPOL 2EH (103-23-1) Incompatible with strong acids, nitrates, oxidizers.

ADIPONITRILE (111-69-3) May accumulate static electrical charges, and may cause ignition of its vapors. Decomposes at its flash point, forming poisonous cyanide gas. Forms explosive mixture with air (flash point 199°F/93°C). Incompatible with sulfuric acid, oxidizers.

4-ADP (92-67-1) Strong oxidizers may cause fire and explosions.

ADRONAL (108-93-0) Forms explosive mixture with air (flash point 154°F/68°C). Incompatible with strong acids, caustics, aliphatic amines, isocyanates. Attacks some plastics, rubber, or coatings.

AERO-CYANAMID, granular or special grade (156-62-7) Contact with any form of moisture causes decomposition, liberating acetylene and ammonia. Contact with all solvents tested also causes decomposition.

AERO LIQUID HCN (74-90-8) Unless stabilized and maintained, samples stored more than 90 days are hazardous. Samples containing more than 2–5% water are less stable than dry material. Can be self-reactive, forming an explosive mixture with air (flash point 0°F/−18°C). Heat or contact with amines or strong bases can cause polymerization. Incompatible with acetaldehyde. Oxidizers may cause fire and explosions. Attacks some plastics, rubber, and coatings.

AEROL 1 PESTICIDE (52-68-6) Contact with strong oxidizers may cause fire and explosions.

AEROSOL GPG (119-36-8) Forms explosive mixture with air (flash point 205°F/96°C). Incompatible with strong acids, nitrates, oxidizers.

AEROSOL SURFACTANT (119-36-8) Forms explosive mixture with air (flash point 205°F/96°C). Incompatible with strong acids, nitrates, oxidizers.

AEROTEX GLYOXAL 40 (107-22-2) Water contact causes polymerization. Incompatible with strong acids, caustics, ammonia, amines, chlorosulfonic acid, ethylene amine. Corrosive to metals.

AEROTHENE (71-55-6) Incompatible with strong caustics, strong oxidizers, chemically active metals, sodium and potassium. Attacks some plastics, rubber, and coatings.

AEROTHENE TT (71-55-6) Incompatible with strong caustics, strong oxidizers, chemically active metals, sodium and potassium. Attacks some plastics, rubber, and coatings.

AETHALDIAMIN (German) (107-15-3) Forms explosive mixture with air (flash point 104°F/40°C). Incompatible with chlorinated organic compounds, silver perchlorate, 3-propiolactone, mesityl oxide, ethylene dichloride, nonoxidizing mineral acids, strong acids, organic acids, organic anhydrides, isocyanates, vinyl acetate, acrylates, substituted allyls, alkylene oxides, epichlorohydrin, ketones, aldehydes, alcohols, glycols, phenols, cresols, caprolactam solution, strong oxidizers. Attacks aluminum, copper, lead, tin, zinc and alloys, some plastics, rubber, and coatings.

AETHANOL (German) (64-17-5) Forms explosive mixture with air (flash point 65°F/18°C). May accumulate static electrical charges, and may cause ignition of its vapors. Reactions may be violent with oleum, sulfuric acid, nitric acid, bases, aliphatic amines, isocyanates, oxidizers.

AETHER (60-29-7) Forms explosive mixture with air (flash point −49°F/−45°C). Incompatible with strong acids, strong oxidizers. Can form peroxides from air or light; may explode when container is unstoppered or otherwise opened. Being a nonconductor, the chemical may accumulate static electric charges that may result in ignition of vapor.

AETHYLACETAT (German) (141-78-6) Forms explosive mixture with air (flash point 24°F/−4.4°C). Incompatible with strong acids, nitrates, oxidizers, chlorosulfonic acid, lithium aluminum hydride, oleum.

AETHYLACRYLAT (German) (140-88-5) Forms explosive mixture with air (flash point 48°F/9°C). Atmospheric moisture and strong alkalies may cause fire and explosions. Unless properly inhibited (*note:* inert gas blanket not recommended), heat, light, or peroxides can cause polymerization. Incompatible with strong acids, amines, oxidizers.

AETHYLALKOHOL (German) (64-17-5) Forms explosive mixture with air (flash point 65°F/18°C). May accumulate static electrical charges, and may cause ignition of its vapors. Reactions may be violent with oleum, sulfuric acid, nitric acid, bases, aliphatic amines, isocyanates, oxidizers.

AETHYLAMINE (German) (75-04-7) Forms explosive mixture with air (flash point less than −0°F/−18°C). Incompatible with nonoxidizing mineral acids, strong acids, organic acids, organic anhydrides, isocya-

26

nates, vinyl acetate, acrylates, substituted allyls, alkylene oxides, epichlorohydrin, ketones, aldehydes, alcohols, glycols, phenols, cresols, caprolactam solution, strong oxidizers. Attacks aluminum, copper, lead, tin, zinc and alloys, some plastics, rubber, and coatings.

2-AETHYLAMINO-4-CHLOR-6-ISOPROPYLAM INO-1,3,5-TRI-AZIN (German) (1912-24-9) Incompatible with strong acids.

AETHYLBENZOL (German) (100-41-4) Forms explosive mixture with air (flash point 59°F/15°C). Incompatible with strong oxidizers, nitric acid.

AETHYLCHLORID (German) (75-00-3) Flammable gas. Forms explosive mixture with air. Contact with moisture produces hydrochloric acid. May accumulate static electrical charges, and may cause ignition of its vapors. Forms explosive mixture with air (flash point −58°F/−50°C). Contact with chemically active metals may cause fire and explosions. Attacks some plastics and rubber.

AETHYLENBROMID (German) (106-93-4) Reacts with chemically active metals, liquid ammonia, strong oxidizers. Heat and light cause slow decomposition. Attacks some plastics and rubber.

AETHYLENEDIAMIN (German) (107-15-3) Forms explosive mixture with air (flash point 104°F/40°C). Incompatible with chlorinated organic compounds, silver perchlorate, 3-propiolactone, mesityl oxide, ethylene dichloride, nonoxidizing mineral acids, strong acids, organic acids, organic anhydrides, isocyanates, vinyl acetate, acrylates, substituted allyls, alkylene oxides, epichlorohydrin, ketones, aldehydes, alcohols, glycols, phenols, cresols, caprolactam solution, strong oxidizers. Attacks aluminum, copper, lead, tin, zinc and alloys, some plastics, rubber, and coatings.

AETHYLENGYKOL-MONOMETHYLAETHER (German) (109-86-4) Forms explosive mixture with air (flash point 103°F/39°C). Heat or oxidizers may form unstable peroxides. Attacks many metals. Strong oxidizers cause fire and explosions. Strong bases cause decomposition. Attacks some plastics, rubber, and coatings. May accumulate static electrical charges, and may cause ignition of its vapors.

AETHYLENOXID (German) (75-21-8) Forms explosive mixture with air (flash point 20°F/−6°C). Incompatible with alkali metal hydroxides, highly active catalysts (e.g., anhydrous chlorides of iron, tin, or aluminum and oxides of iron or aluminum). Avoid contact with copper. Protect container from physical damage, sun, and heat. Attacks some plastics, rubber, or coatings.

AETHYLIDENCHLORID (German) (75-34-3) Incompatible with strong oxidizers, strong caustics. Attacks plastics and rubber.

AETHYLIS (75-00-3) Flammable gas. Forms explosive mixture with air. Contact with moisture forms hydrochloric acid. May accumulate static electrical charges, and may cause ignition of its vapors. Forms explosive mixture with air (flash point −58°F/−50°C). Contact with chemically active metals may cause fire and explosions. Attacks some plastics and rubber.

AETHYLIS CHLORIDUM (75-00-3) Flammable gas. Forms explosive mixture with air. Contact with moisture produces hydrochloric acid. May accumulate static electrical charges, and may cause ignition of its vapors. Forms explosive mixture with air (flash point −58°F/−50°C). Contact with chemically active metals may cause fire and explosions. Attacks some plastics and rubber.

AETHYLMETHYLKETON (German) (78-93-3) Forms explosive mixture with air (flash point 16°F/−9°C). Extremely flammable. Incompatible with sulfuric acid, nitric acid, aliphatic amines, strong oxidizers, potassium *tert*-butoxide, 2-propanol, chlorosulfonic acid, oleum (hydrogen peroxide + nitric acid).

AF 72 (9016-00-6) Incompatible with sulfuric acid, isocyanates.

AF 75 (9016-00-6) Incompatible with sulfuric acid, isocyanates.

AF 101 (330-54-1) Hydrolyzes in fairly strong acids.

AFICIDE (58-89-9) Not combustible, but may be dissolved in a combustible solvent. If solvent comes in contact with oxidizers, fire and explosions may result.

AGE (106-92-3) Forms explosive mixture with air (flash point 135°F/57°C). Explosive peroxides may be formed in light or air. Unless inhibited, may polymerize on contact with acids or bases. Strong acids, amines, oxidizers may cause fire and explosions. Attacks some forms of plastics, coatings, and rubber.

AGENAP (1338-4-5) Generally corrosive to metals.

AGENT 504 (112-30-1) Forms explosive mixture with air (flash point 180°F/82°C). Incompatible with strong acids, caustics, aliphatic amines, isocyanates, strong oxidizers.

AGRICIDE MAGGOT KILLER (F) (8001-35-2) Reacts with oxidizers, with a risk of fire or explosions. Attacks metals in the presence of moisture.

AGRICULTURAL LIMESTONE (1317-65-3) Incompatible with acids, alum, ammonium salts, fluorine.

AGRIDIP (56-72-4) Contact with strong oxidizers may cause fire and explosions.

AGRISOL G-20 (58-89-9) Not combustible, but may be dissolved in a combustible solvent. If solvent comes in contact with oxidizers, fire and explosions may result.

AGRITAN (50-29-3) Incompatible with salts of iron or aluminum and with bases. Do not store in iron containers.

A GRO (298-00-0) Mixtures with magnesium may be explosive.

AGROCERES (76-44-8) Forms hydrogen chloride gas with iron and causes rust above 165°F/74°C.

AGROCIDE (58-89-9) Not combustible, but may be dissolved in a combustible solvent. If solvent comes in contact with oxidizers, fire and explosions may result.

AGROCIDE 2 (58-89-9) Not combustible, but may be dissolved in a combustible solvent. If solvent comes in contact with oxidizers, fire and explosions may result.

AGROCIDE 6G (58-89-9) Not combustible, but may be dissolved in a combustible solvent. If solvent comes in contact with oxidizers, fire and explosions may result.

AGROCIDE 7 (58-89-9) Not combustible, but may be dissolved in a combustible solvent. If solvent comes in contact with oxidizers, fire and explosions may result.

AGROCIDE III (58-89-9) Not combustible, but may be dissolved in a combustible solvent. If solvent comes in contact with oxidizers, fire and explosions may result.

AGROCIDE WP (58-89-9) Not combustible, but may be dissolved in a combustible solvent. If solvent comes in contact with oxidizers, fire and explosions may result.

AGROCIT (17804-35-2) Heat, water, strong acids, and strong alkalies can cause decomposition and formation of toxic gas (oxides of nitrogen).

AGRONEXIT (58-89-9) Not combustible, but may be dissolved in a combustible solvent. If solvent comes in contact with oxidizers, fire and explosions may result.

AGROSOL S (133-06-2) Incompatible with tetraethyl pyrophosphate and parathion (emulsified concentrate formulations).

AGROTECT (94-75-7) Decomposes in sunlight. Incompatible with strong oxidizers; may cause fire and explosions.

AGROX 2-WAY or 3-WAY (133-06-2) Incompatible with tetraethyl pyrophosphate and parathion (emulsified concentrate formulations).

AGSTONE (1317-65-3) Incompatible with acids, alum, ammonium salts, fluorine.

AIP (20859-73-8) Contact with moisture (including atmospheric moisture) produces spontaneously combustible phosphine gas.

AK-33X (12108-13-3) Incompatible with strong acids, strong oxidizers.

AKROLEIN (Czech) (107-02-8) Forms explosive mixture with air (flash point $-15°F/-26°C$). Unstable and very reactive. In storage can form heat- and shock-sensitive compounds. Unless inhibited (usually by hydroquinone), readily forms explosive peroxides. Able to polymerize. A strong reducing agent. Reacts violently with oxidizers, strong acids, caustics, amines, 2-aminoethanol, ammonia, ammonium hydroxide, ethylene diamine, ethyleneimine, hydroxides, metal salts, oxidizers, sulfur dioxide, thiourea. Attacks metals: cadmium and zinc. May accumulate static electrical charges, and may cause ignition of its vapors.

AKROLEINA (Polish) (107-02-8) Forms explosive mixture with air (flash point $-15°F/-26°C$). Unstable and very reactive. In storage can form heat- and shock-sensitive compounds. Unless inhibited (usually by hydroquinone), readily forms explosive peroxides. Able to polymerize. A strong reducing agent. Reacts violently with oxidizers, strong acids, caustics, amines, 2-aminoethanol, ammonia, ammonium hydroxide, ethylene diamine, ethyleneimine, hydroxides, metal salts, oxidizers, sulfur dioxide, thiourea. Attacks metals: cadmium and zinc. May accumulate static electrical charges, and may cause ignition of its vapors.

AKRYLAMID (Czech) (79-06-1) Unless inhibited, ultraviolet light and heat (above $184°F/85°C$) can cause polymerization. Incompatible with nonoxidizing mineral acids, strong acids, ammonia, oleum, oxidizers, isocyanates.

AKRYLONITRYL (Polish) (107-13-1) Forms explosive mixture with air (flash point $32°F/0°C$ o.c.). Forms explosive peroxides; heat, light, caustics, silver nitrate, and peroxides can cause polymerization. Incompatible with strong acids, strong oxidizers, amines, 2-aminoethanol, bromine, chlorosulfonic acid, ethylene diamine, nitric acid, oleum, potassium hydroxide, sodium hydroxide, sulfuric acid. Attacks copper and copper alloys; attacks aluminum in high concentrations. May accumulate static electrical charges, and may cause ignition of its vapors.

AKTIKON (1912-24-9) Incompatible with strong acids.

AKTIKON PK (1912-24-9) Incompatible with strong acids.

AKTINIT A (1912-24-9) Incompatible with strong acids.

AKTINIT PK (1912-24-9) Incompatible with strong acids.

AKULON (105-60-2) Contact with strong oxidizers may cause fire and explosions.

ALANINOL (78-91-1) Forms explosive mixture with air (flash point 145°F/63°C). Incompatible with acids, organic anhydrides, isocyanates, vinyl acetate, acrylates, substituted allyls, alkylene oxides, epichlorohydrin, aldehydes.

beta-ALANINOL (156-87-6) Forms explosive mixture with air (flash point 175°F/79°C). Incompatible with acids, organic anhydrides, isocyanates, vinyl acetate, acrylates, substituted allyls, alkylene oxides, epichlorohydrin, aldehydes. Corrodes copper and its alloys.

ALATEX (75-99-0) Corrosive to iron, aluminum, and copper.

ALBOLINE (8012-95-1) Incompatible with nitric acid; oxidizers may cause fire and explosions.

ALBONE (7722-84-1) A powerful oxidizer; attacks many substances. Contact with most organic, readily oxidizable materials and reducing agents and combustibles causes fire and explosions. Contact with iron, copper, brass, bronze, chromium, zinc, lead, manganese, silver and other catalytic metals (and their salts), especially in a basic (pH 7 or above) environment, causes rapid decomposition with evolution of oxygen gas. Attacks, and may ignite, some plastics, rubber, and coatings. Decomposes slowly at ordinary temperatures and builds up pressure in a closed container. The rate of decomposition doubles for each 50°F(10°C) rise (1.5 times 50°F rise) in temperature, and decomposition becomes self-sustaining at 285°F (141°C).

ALBUS (10124-48-8) Reacts violently with halogens and metal salts of amines.

ALCIDE (10049-04-4) Water contact produces perchloric and hydrochloric acids with toxic and corrosive fumes. Explodes on contact with many substances, including carbon monoxide, hydrogen, mercury, potassium, phosphorus, strong caustics, sulfur, granulated sugar. A strong oxidizer. May decompose explosively from shock, friction, or heat: will detonate if heated rapidly to 212°F (100°C). Contact with reducing agents, dust and other combustible materials, organic matter, sulfur may cause fires and explosions. Attacks some forms of plastics, rubber, and coatings.

ALCOHOL (64-17-5) Forms explosive mixture with air (flash point 65°F/18°C). May accumulate static electrical charges, and may cause ignition of its vapors. Reactions may be violent with oleum, sulfuric acid, nitric acid, bases, aliphatic amines, isocyanates, oxidizers.

ALCOHOL, ANHYDROUS (64-17-5) Forms explosive mixture with air (flash point 65°F/18°C). May accumulate static electrical charges, and may cause ignition of its vapors. Reactions may be violent with oleum, sulfuric acid, nitric acid, bases, aliphatic amines, isocyanates, oxidizers.

ALCOHOL, DEHYDRATED (64-17-5) Forms explosive mixture with air (flash point 65°F/18°C). May accumulate static electrical charges, and may cause ignition of its vapors. Reactions may be violent with oleum, sulfuric acid, nitric acid, bases, aliphatic amines, isocyanates, oxidizers.

ALCOHOLS, MIXED Incompatible with strong acids, caustics, aliphatic amines, isocyanates, strong oxidizers.

ALCOOL ALLILCO (Italian) (107-18-6) May form unstable and explosive peroxides. Able to polymerize. May accumulate static electrical charges, and may cause ignition of its vapors. A strong reducing agent; reacts violently with oxidizers. Forms explosive mixture with air (flash point 70°F/21°C). Incompatible with strong acids, amines, isocyanates, carbon tetrachloride, chlorosulfonic acid, diallyl phosphide, oleum, sodium hydroxide, tri-*n*-bromomelamine, metal halides, caustic soda, sodium, magnesium, aluminum, and their alloys. Attacks some coatings, some plastics, and rubber.

ALCOOL ALLYLIQUE (French) (107-18-6) May form unstable and explosive peroxides. Able to polymerize. May accumulate static electrical charges, and may cause ignition of its vapors. A strong reducing agent; reacts violently with oxidizers. Forms explosive mixture with air (flash point 70°F/21°C). Incompatible with strong acids, amines, isocyanates, carbon tetrachloride, chlorosulfonic acid, diallyl phosphide, oleum, sodium hydroxide, tri-*n*-bromomelamine, metal halides, caustic soda, sodium, magnesium, aluminum, and their alloys. Attacks some coatings, some plastics, and rubber.

ALCOOL AMYLIQUE (French) (71-41-0) Forms explosive mixture with air (flash point 91°F/33°C). Incompatible with strong acids, caustics, aliphatic amines, isocyanates. Incompatible with alkali metals and alkaline earth forming hydrogen gas.

ALCOOL BUTYLIQUE (French) (71-36-3) Incompatible with strong acids, halogens, caustics, alkali metals, aliphatic amines, isocyanates.

ALCOOL BUTYLIQUE SECONDAIRE (French) (78-92-2) Forms an explosive peroxide in air. Ignites with chromium trioxide. Incompatible with strong oxidizers, strong acids, aliphatic amines, isocyanates, organic peroxides.

ALCOOL BUTYLIQUE TERTIAIRE (French) (75-65-0) Incompatible with strong acids, including mineral acids, strong oxidizers or caustics, aliphatic amines, isocyanates, alkali metals (i.e., lithium, sodium, potassium, rubidium, cesium, francium). Attacks many plastics and some coatings.

ALCOOL ETHYLIQUE (French) (64-17-5) Forms explosive mixture with air (flash point 65°F/18°C). May accumulate static electrical charges, and may cause ignition of its vapors. Reactions may be violent with oleum, sulfuric acid, nitric acid, bases, aliphatic amines, isocyanates, oxidizers.

ALCOOL ETILICO (Italian) (64-17-5) Forms explosive mixture with air (flash point 65°F/18°C). May accumulate static electrical charges, and may cause ignition of its vapors. Reactions may be violent with oleum, sulfuric acid, nitric acid, bases, aliphatic amines, isocyanates, oxidizers

ALCOOL METHYL AMYLIQUE (French) (108-11-2) Forms explosive mixture with air (flash point 106°F/41°C). Contact with alkali metals produces hydrogen gas. Incompatible with strong acids, caustics, aliphatic amines, isocyanates. Attacks some plastics, rubber, and coatings. May accumulate static electrical charges, and may cause ignition of its vapors.

ALCOOL METHYLIQUE (French) (67-56-1) Forms explosive mixture with air (flash point 52°F/11°C). Incompatible with strong acids, strong oxidizers, caustics, aliphatic amines, isocyanates, chromic anhydride, lead perchlorate, perchloric acid, phosphorus trioxide. May react with metallic aluminum at high temperature. Attacks some plastics, rubber, and coatings.

ALCOOL METILICO (Italian) (67-56-1) Forms explosive mixture with air (flash point 52°F/11°C). Incompatible with strong acids, strong oxidizers, caustics, aliphatic amines, isocyanates, chromic anhydride, lead perchlorate, perchloric acid, phosphorus trioxide. May react with metallic aluminum at high temperature. Attacks some plastics, rubber, and coatings.

ALCOOL PROPILICO (Italian) (71-23-8) Forms explosive mixture with air (flash point 74°F/23°C). Attacks some plastics, rubber, and coatings. Incompatible with strong acids, caustics, aliphatic amines, isocyanates.

ALCOOL PROPYLIQUE (French) (71-23-8) Forms explosive mixture with air (flash point 74°F/23°C). Attacks some plastics, rubber, and coatings. Incompatible with strong acids, caustics, aliphatic amines, isocyanates.

ALCOPOL O (119-36-8) Forms explosive mixture with air (flash point 205°F/96°C). Incompatible with strong acids, nitrates, oxidizers.

ALDEHYDE BUTYRIQUE (French) (123-72-8) Forms explosive mixture with air (flash point −8°F/−22°C). Incompatible with strong oxidizers, strong acids, caustics, ammonia, aliphatic amines, alkanolamines, aromatic amines.

ALDEHYDE C-6 (66-25-1) Forms explosive mixture with air (flash point 90°F/32°C). Incompatible with strong acids, caustics, ammonia, amines. Attacks some plastics, rubber, or coatings.

ALDEHYDE C-10 (112-31-2) Incompatible with strong acids, ammonia, caustics, amines. Attacks galvanized steel.

ALDEHYDE-COLLIDINE (104-90-5) Forms explosive mixture with air (flash point 155°F/68°C). Incompatible with acids, isocyanates, phenols, cresols.

ALDEHYDE CROTONIQUE (French) (123-73-9) A strong reducing agent. Forms explosive mixture with air (flash point 55°F/12.8°C). Readily converted by oxygen to peroxides and acids; heat or contact with many other substances may cause polymerization. Incompatible with strong oxidizers, strong acids including nonoxidizing mineral acids, ammonia, aliphatic amines, aromatic amines, 1,3-butadiene, strong bases. Liquid attacks some plastics, rubber, and coatings.

ALDEHYDE FORMIQUE (French) (50-00-0) Forms explosive mixture with air (flash point 122°F/50°C). Incompatible with strong acids, amines, strong oxidizers, alkaline materials, nitrogen dioxide, performic acid. Reaction with hydrochloric acid produces bis-chloromethyl ether, a carcinogen.

ALDEHYDE PROPIONIQUE (French) (123-38-6) Forms explosive mixture with air (flash point −2°F/−30°C). Incompatible with strong acids, caustics, amines. Reacts violently with strong oxidizers. Can self-ignite if finely dispersed on porous or combustible material. Heat or ultraviolet can cause decomposition. May accumulate static electrical charges, and may cause ignition of its vapors.

ALDEHYDINE (104-90-5) Forms explosive mixture with air (flash point 155°F/68°C). Incompatible with acids, isocyanates, phenols, cresols.

ALDEIDE BUTIRRICA (Italian) (123-72-8) Forms explosive mixture with air (flash point −8°F/−22°C). Incompatible with strong oxidizers, strong acids, caustics, ammonia, aliphatic amines, alkanolamines, aromatic amines.

ALDEIDE FORMICA (Italian) (50-00-0) Forms explosive mixture with air (flash point 122°F/50°C). Incompatible with strong acids, amines, strong oxidizers, alkaline materials, nitrogen dioxide, performic acid. Reaction with hydrochloric acid produces bis-chloromethyl ether, a carcinogen.

ALDIFEN (51-28-5) Explosion is caused by heat, friction, or shock. Contact with reducing agents or combustibles may cause fire and explosions. Forms explosive salts with ammonia or strong bases. May accumulate static electrical charges, and may cause ignition of its vapors.

ALDREX (309-00-2) Incompatible with concentrated mineral acids, acid catalysts, acid oxidizing agents, phenols, reactive metals.

ALDREX-30 (309-00-2) Incompatible with concentrated mineral acids, acid catalysts, acid oxidizing agents, phenols, reactive metals.

ALDRIN (309-00-2) Incompatible with concentrated mineral acids, acid catalysts, acid oxidizing agents, phenols, reactive metals.

ALDRINE (French) (309-00-2) Incompatible with concentrated mineral acids, acid catalysts, acid oxidizing agents, phenols, reactive metals.

ALDRITE (309-00-2) Incompatible with concentrated mineral acids, acid catalysts, acid oxidizing agents, phenols, reactive metals.

ALDROSOL (309-00-2) Incompatible with concentrated mineral acids, acid catalysts, acid oxidizing agents, phenols, reactive metals.

ALFANAFTILAMINA (Italian) (134-32-7) Oxidizes in air. Incompatible with nitrous acid, oxidizers, nitrates, organic anhydrides, isocyanates, aldehydes.

ALFA-TOX (333-41-5) Incompatible with water, copper-containing compounds, oxidizers, acids, or bases.

ALFLOC 7020 (2425-06-1) Incompatible with acids or acid vapor. Strong alkaline conditions contribute to instability.

ALFOL 8 (111-87-5) Forms explosive mixture with air (flash point 178°F/81°C). Incompatible with strong acids, caustics, aliphatic amines, isocyanates, strong oxidizers.

ALFOL-12 (112-53-8) Incompatible with strong acids, caustics, aliphatic amines, isocyanates, oxidizers.

ALGOFRENE TYPE 1 (75-69-4) Incompatible with chemically active metals. Attacks some plastics, rubber, and coatings.

ALGOFRENE TYPE 2 (75-71-8) Reacts violently with liquid aluminum. Incompatible with chemically active metals. Attacks some plastics, rubber, and coatings.

ALGOFRENE TYPE 5 (75-43-4) Incompatible with chemically active metals. Attacks some plastics, rubber, and coatings.

ALGOFRENE TYPE 6 (75-45-5) Moisture causes slow decomposition. Thermal decomposition results at high temperatures with alkalies or alkaline earth metals. Attacks some plastics, rubber, or coatings.

ALGOFRENE TYPE 67 (75-37-6) Reacts violently with strong oxidizers. Attacks metals in presence of moisture.

ALGRAIN (64-17-5) Forms explosive mixture with air (flash point 65°F/18°C). May accumulate static electrical charges, and may cause ignition of its vapors. Reactions may be violent with oleum, sulfuric acid, nitric acid, bases, aliphatic amines, isocyanates, oxidizers.

ALGYLEN (79-01-6) Caustics form a toxic and flammable gas. Reacts violently with chemically active metals. Contact with aluminum may produce a violent, self-accelerating polymerization reaction. Incompatible with nonoxidizing mineral acids, strong acids, organic acids, organic anhydrides, isocyanates, alkylene oxides, aldehydes, alcohols, glycols, phenols, cresols, caprolactam solution, epichlorohydrin, nitrogen tetroxide, metal powders, oxygen. May accumulate static electrical charges, and may cause ignition of its vapors.

ALIPHATIC PETROLEUM NAPHTHA (8002-05-9) Flash point −40°C to −86°F/−40°F to −66°C. Incompatible with nitric acid; oxidizers may cause fire and explosions.

ALKAMID (105-60-2) Contact with strong oxidizers may cause fire and explosions.

ALKARSODL (124-65-2) Corrodes common metals.

ALKOHOL (German) (64-17-5) Forms explosive mixture with air (flash point 65°F/18°C). May accumulate static electrical charges, and may cause ignition of its vapors. Reactions may be violent with oleum, sulfuric acid, nitric acid, bases, aliphatic amines, isocyanates, oxidizers.

ALKOHOLU ETYLOWEGO (Polish) (64-17-5) Forms explosive mixture with air (flash point 65°F/18°C). May accumulate static electrical charges, and may cause ignition of its vapors. Reactions may be violent with oleum, sulfuric acid, nitric acid, bases, aliphatic amines, isocyanates, oxidizers.

ALKRON (56-38-2) A combustible liquid. Mixtures with endrin may be explosive. Strong oxidizers may cause fire and explosions. Attacks some plastics, rubber, or coatings.

ALLENE (74-99-7) Forms explosive gas mixture with air. Can form explosive peroxide. Strong oxidizers may cause fire and explosions. Forms shock-sensitive compounds with copper, magnesium, silver, and their alloys. Copper or copper alloys containing more than 67% copper should not be used in handling equipment. Attacks some plastics, rubber, and coatings. May accumulate static electrical charges, and may cause ignition of its vapors.

ALLENE-METHYLACETYLENE MIXTURE (59355-75-8) Forms explosive gas mixture with air. Incompatible with strong oxidizers; copper alloys containing more than 6% copper may form explosive compounds. Attacks some plastics, rubber, and coatings. May accumulate static electrical charges, and may cause ignition of its vapors.

ALLERON (56-38-2) A combustible liquid. Mixtures with endrin may be explosive. Strong oxidizers may cause fire and explosions. Attacks some plastics, rubber, or coatings.

ALLILE (CLORURO di) (Italian) (107-05-1) Forms explosive mixture with air (flash point −25°F/−32°C). Violent polymerization and explosion may occur from heat, light, or contact with acid catalysts, ferric chloride, aluminum chloride, Lewis acids, or Ziegler catalyst (e.g., titanium tetrachloride + triethylaluminum monochloride). Incompatible with strong acids, amines, aluminum chloride, boron trifluoride, chlorosulfonic acid, ethylene diamine, ethyleneimine, ferric chloride, oleum, oxidizers, sodium hydroxide. Slow decomposition with moisture; attacks some coatings, plastics, and rubber. Corrosive to steel. May accumulate static electrical charges, and may cause ignition of its vapors.

ALLILOWY ALKOHOL (Polish) (107-18-6) May form unstable and explosive peroxides. Able to polymerize. May accumulate static electrical charges, and may cause ignition of its vapors. A strong reducing agent; reacts violently with oxidizers. Forms explosive mixture with air (flash point 70°F/21°C). Incompatible with strong acids, amines, isocyanates, carbon tetrachloride, chlorosulfonic acid, diallyl phosphide, oleum, sodium hydroxide, tri-*n*-bromomelamine, metal halides, caustic soda, sodium, magnesium, aluminum, and their alloys. Attacks some coatings, some plastics, and rubber.

ALLOMALEIC ACID (110-17-8) Incompatible with sulfuric acid, caustics, ammonia, amines, isocyanates, alkylene oxides, epichlorohydrin.

ALLTEX (8001-35-2) Reacts with oxidizers, with a risk of fire or explosions. Attacks metals in the presence of moisture.

ALLTOX (8001-35-2) Reacts with oxidizers, with a risk of fire or explosions. Attacks metals in the presence of moisture.

ALLUMINIO, CLORURO di (Italian) (7446-70-0) Contact with air or water produces hydrochloric acid. Water, alcohol, alkenes cause polymerization. Incompatible with nitrobenzene, organic material, and bases. Attacks metal in presence of moisture.

ALLYL AL (107-18-6) May form unstable and explosive peroxides. Able to polymerize. May accumulate static electrical charges, and may cause ignition of its vapors. A strong reducing agent; reacts violently with oxidizers. Forms explosive mixture with air (flash point 70°F/21°C). Incompatible with strong acids, amines, isocyanates, carbon tetrachloride, chlorosulfonic acid, diallyl phosphide, oleum, sodium hydroxide, tri-*n*-bromomelamine, metal halides, caustic soda, sodium, magnesium, aluminum, and their alloys. Attacks some coatings, some plastics and rubber.

ALLYL ALCOHOL (107-18-6) May form unstable and explosive peroxides. Able to polymerize. May accumulate static electrical charges, and may cause ignition of its vapors. A strong reducing agent; reacts violently with oxidizers. Forms explosive mixture with air (flash point 70°F/21°C). Incompatible with strong acids, amines, isocyanates, carbon tetrachloride, chlorosulfonic acid, diallyl phosphide, oleum, sodium hydroxide, tri-*n*-bromomelamine, metal halides, caustic soda, sodium, and magnesium, aluminum, and their alloys. Attacks some coatings, some plastics, and rubber.

ALLYLALDEHYDE (107-02-8) Forms explosive mixture with air (flash point −15°F/−26°C). Unstable and very reactive. In storage can form heat- and shock-sensitive compounds. Unless inhibited (usually by hydroquinone), readily forms explosive peroxides. Able to polymerize. A strong reducing agent. Reacts violently with oxidizers, strong acids, caustics, amines, 2-aminoethanol, ammonia, ammonium hydroxide, ethylene diamine, ethyleneimine, hydroxides, metal salts, oxidizers, sulfur dioxide, thiourea. Attacks metals: cadmium and zinc. May accumulate static electrical charges, and may cause ignition of its vapors.

ALLYLALKOHOL (German) (107-18-6) May form unstable and explosive peroxides. Able to polymerize. May accumulate static electrical charges, and may cause ignition of its vapors. A strong reducing agent; reacts violently with oxidizers. Forms explosive mixture with air (flash point 70°F/21°C). Incompatible with strong acids, amines, isocyanates, carbon tetrachloride, chlorosulfonic acid, diallyl phosphide, oleum, sodium hydroxide, tri-n-bromomelamine, metal halides, caustic soda, sodium, and magnesium, aluminum, and their alloys. Attacks some coatings, some plastics, and rubber.

ALLYLAMINE (107-11-9) Forms explosive mixture with air (flash point −20°F/−29°C). A strong base. Incompatible with acids, strong oxidizers, hypochlorites, nitrosyl perchlorate. Corrodes metals.

ALLYL BROMIDE (106-95-6) Forms explosive mixture with air (flash point 30°F/−1°C). Incompatible with oxidizers, heat.

ALLYLCHLORID (German) (107-05-1) Forms explosive mixture with air (flash point −25°F/−32). Violent polymerization and explosion may occur from heat, light, or contact with acid catalysts, ferric chloride, aluminum chloride, Lewis acids, or Ziegler catalyst (e.g., titanium tetrachloride + triethylaluminum monochloride). Incompatible with strong acids, amines, aluminum chloride, boron trifluoride, chlorosulfonic acid, ethylene diamine, ethyleneimine, ferric chloride, oleum, oxidizers, sodium hydroxide. Slow decomposition with moisture; attacks some coatings, plastics and rubber. Corrosive to steel. May accumulate static electrical charges, and may cause ignition of its vapors.

ALLYL CHLORIDE (107-05-1) Forms explosive mixture with air (flash point −25°F/−32°C). Violent polymerization and explosion may occur from heat, light, or contact with acid catalysts, ferric chloride, aluminum chloride, Lewis acids, or Ziegler catalyst (e.g., titanium tetrachloride + triethylaluminum monochloride). Incompatible with strong acids, amines, aluminum chloride, boron trifluoride, chlorosulfonic acid, ethylene diamine, ethyleneimine, ferric chloride, oleum, oxidizers, sodium hydroxide. Slow decomposition with moisture; attacks some coatings, plastics, and rubber. Corrosive to steel. May accumulate static electrical charges, and may cause ignition of its vapors.

ALLYL CHLOROCARBONATE (937-50-0) Forms explosive mixture with air (flash point 88°F/31°C). Incompatible with water (forms chloroformic acid and allyl alcohol), acids, oxidizers, caustics, amines, peroxides, allyl alcohol. Corrodes metals.

ALLYL CHLOROFORMATE (937-50-0) Forms explosive mixture with air (flash point 88°F/31°C). Incompatible with water (forms chloroformic acid and allyl alcohol), acids, oxidizers, caustics, amines, peroxides, allyl alcohol, Corrodes metals.

ALLYLE, CHLORURE d' (French) (107-05-1) Forms explosive mixture with air (flash point −25°F/−32°C). Violent polymerization and explosion may occur from heat, light, or contact with acid catalysts, ferric chloride, aluminum chloride, Lewis acids, or Ziegler catalyst (e.g., titanium tetrachloride + triethylaluminum monochloride). Incompatible with strong acids, amines, aluminum chloride, boron trifluoride, chlorosulfonic acid, ethylene diamine, ethyleneimine, ferric chloride, oleum, oxidizers, sodium hydroxide. Slow decomposition occurs with moisture; attacks some coatings, plastics, and rubber. Corrosive to steel. May accumulate static electrical charges, and may cause ignition of its vapors.

ALLYL EHYDE (107-02-8) Forms explosive mixture with air (flash point −15°F/−26°C). Unstable and very reactive. In storage can form heat- and shock-sensitive compounds. Unless inhibited (usually by hydroquinone), readily forms explosive peroxides. Able to polymerize. A strong reducing agent. Reacts violently with oxidizers, strong acids, caustics, amines, 2-aminoethanol, ammonia, ammonium hydroxide, ethylene diamine, ethyleneimine, hydroxides, metal salts, oxidizers, sulfur dioxide, thiourea. Attacks metals: cadmium and zinc. May accumulate static electrical charges, and may cause ignition of its vapors.

ALLYLEHYDE (107-02-8) Forms explosive mixture with air (flash point −15°F/−26°C). Unstable and very reactive. In storage can form heat- and shock-sensitive compounds. Unless inhibited (usually by hydroquinone), readily forms explosive peroxides. Able to polymerize. A strong reducing agent. Reacts violently with oxidizers, strong acids, caustics, amines, 2-aminoethanol, ammonia, ammonium hydroxide, ethylene diamine, ethyleneimine, hydroxides, metal salts, oxidizers, sulfur dioxide, thiourea. Attacks metals: cadmium and zinc. May accumulate static electrical charges, and may cause ignition of its vapors.

ALLYLENE (74-99-7) Forms explosive gas mixture with air. Can form explosive peroxide. Strong oxidizers may cause fire and explosions. Forms shock-sensitive compounds with copper, magnesium, silver and their alloys. Copper or copper alloys containing more than 67% copper should not be used in handling equipment. Attacks some plastics, rubber, and coatings. May accumulate static electrical charges, and may cause ignition of its vapors.

ALLYL 2,3-EPOXYPROPYL ETHER (106-92-3) Forms explosive mixture with air (flash point 135°F/57°C). Explosive peroxides may be formed in light or air. Unless inhibited, may polymerize on contact with acids or bases. Strong acids, amines, oxidizers may cause fire and explosions. Attacks some forms of plastics, coatings, and rubber.

ALLYL ETHER (557-40-4) Forms explosive mixture with air (flash point 20°F/−7°C). Forms unstable and explosive peroxides. May accumulate static electrical charges, and may cause ignition of its vapors. Incompatible with strong acids, oxidizers.

ALLYLETHYLENE (591-93-5) Forms explosive mixture with air (flash point 40°F/40°C). Strong oxidizers may cause fire and explosions.

ALLYL ETHYL ETHER (557-31-3) Forms explosive mixture with air (flash point less than 70°F/21°C). May form explosive peroxides. May accumulate static electrical charges, and may cause ignition of its vapors. Incompatible with strong acids, oxidizers.

ALLYL GLYCIDYL ETHER (106-92-3) Forms explosive mixture with air (flash point 135°F/57°C). Explosive peroxides may be formed in light or air. Unless inhibited, may polymerize on contact with acids or bases. Strong acids, amines, oxidizers may cause fire and explosions. Attacks some forms of plastics, coatings and rubber.

ALLYLIC ALCOHOL (107-18-6) May form unstable and explosive peroxides. Able to polymerize. May accumulate static electrical charges, and may cause ignition of its vapors. A strong reducing agent; reacts violently with oxidizers. Forms explosive mixture with air (flash point 70°F/21°C). Incompatible with strong acids, amines, isocyanates, carbon tetrachloride, chlorosulfonic acid, diallyl phosphide, oleum, sodium hydroxide, tri-*n*-bromomelamine, metal halides, caustic soda, sodium, magnesium, aluminum, and their alloys. Attacks some coatings, some plastics, and rubber.

1-ALLYLOXY-2,3-EPOXYPROPANE (106-92-3) Forms explosive mixture with air (flash point 135°F/57°C). Explosive peroxides may be formed in light or air. Unless inhibited, may polymerize on contact with acids or bases. Strong acids, amines, oxidizers may cause fire and explosions. Attacks some forms of plastics, coatings, and rubber.

ALLYL PHENYL ETHER (1746-13-0) Forms explosive mixture with air (flash point 143°F/62°C). May form peroxides with air. Incompatible with strong acids, oxidizers.

ALLYLSILICONE TRICHLORIDE (107-37-9) Forms explosive mixture with air (flash point 95°F/35°C). Oxidizers may cause fire and explosions. Reacts violently with water, producing hydrochloric acid. Attacks metal in presence of moisture.

ALLYL TRICHLORIDE (96-18-4) Forms explosive mixture with air (flash point 164°F/74°C). Incompatible with strong oxidizers, strong caustics; chemically active metals may cause fire and explosions. May decompose on contact with aluminum. Attacks some plastics, rubber, or coatings.

ALLYL TRICHLOROSILANE (107-37-9) Forms explosive mixture with air (flash point 95°F/35°C). Reacts violently with water, producing hydrochloric acid. Oxidizers may cause fire and explosions. Attacks metal in presence of moisture.

ALMOND ARTIFICIAL ESSENTIAL OIL (100-52-7) Forms explosive mixture with air (flash point 145°F/63°C). Forms peroxides with oxides of nitrogen. Strong oxidizers may cause fire and explosions. Corrodes metal with moisture.

ALPEROX C (105-74-8) A combustible solid and powerful oxidizer. Reacts violently with reducing agents. May self-ignite if mixed with combustible materials.

ALPHANAPHTHYL THIOUREA (86-88-4) Strong oxidizers may cause fires and explosion. Also reacts with silver nitrate.

ALPHANAPHTYL THIOUREE (French) (86-88-4) Strong oxidizers may cause fires and explosion. Also reacts with silver nitrate.

ALPHASOL OT (119-36-8) Forms explosive mixture with air (flash point 205°F/96°C). Incompatible with strong acids, nitrates, oxidizers.

AL-PHOS (20859-73-8) Contact with moisture (including atmospheric moisture) produces spontaneously combustible phosphine gas.

ALRATO (86-88-4) Strong oxidizers may cause fires and explosion. Also reacts with silver nitrate.

ALROWET D65 (119-36-8) Forms explosive mixture with air (flash point 205°F/96°C). Incompatible with strong acids, nitrates, oxidizers.

ALTOX (309-00-2) Incompatible with concentrated mineral acids, acid catalysts, acid oxidizing agents, phenols, reactive metals.

ALUM (10043-01-3) Forms sulfuric acid with water; strong reaction with bases and many other materials. Dry material is weakly corrosive to carbon steel; aqueous solution attacks metals, forming hydrogen gas.

ALUMINUM ALUM (10043-01-3) Forms sulfuric acid with water; strong reaction with bases and many other materials. Dry material is weakly corrosive to carbon steel; aqueous solution attacks metals, forming hydrogen gas.

ALUMINUMCHLORID (German) (7446-70-0) Contact with air or water produces hydrochloric acid. Water, alcohol, alkenes cause polymerization. Incompatible with nitrobenzene, organic material, and bases. Attacks metal in presence of moisture.

ALUMINUM CHLORIDE or ALUMINUM CHLORIDE, ANHYDROUS (1:3) or ALUMINUM CHLORIDE SOLUTION (7446-70-0) Contact with air or water produces hydrochloric acid. Water, alcohol, alkenes cause polymerization. Incompatible with nitrobenzene, organic material and bases. Attacks metal in presence of moisture.

ALUMINUM ETHYL DICHLORIDE (563-43-9) Forms explosive mixture with air. May ignite spontaneously in air. Reacts violently with water, forming hydrogen chloride and ethane gas. A strong reducing agent; reacts violently with oxidizers. Violent reaction occurs with alcohols, amines, carbon dioxide, phenol, oxides of sulfur or nitrogen, allyl chloride + benzene or toluene. May accumulate static electrical charges, and may cause ignition of its vapors.

ALUMINUM FLUORIDE (7784-18-1) May have a violent reaction with potassium or sodium.

ALUMINUM FOSFIDE (Dutch) (20859-73-8) Contact with moisture (including atmospheric moisture) produces spontaneously combustible phosphine gas.

ALUMINUM LITHIUM HYDRIDE (16853-85-3) Flammable solid. Can ignite spontaneously in moist air or heat above 250°F/121°C. A strong reducing agent. Reacts violently with oxidizers, alcohols, acids, dimethylether, tetrahydrofuran.

ALUMINUM MONOPHOSPHIDE (20859-73-8) Contact with moisture produces spontaneously combustible phosphine gas.

ALUMINUM NITRATE (13473-90-0) A strong oxidizer. Incompatible with combustibles, organics, reducing agents. Aqueous solution is acidic. Attacks metals in the presence of moisture.

ALUMINUM(III) NITRATE (1:3) or ALUMINUM(3+) NITRATE (1:3) (13473-90-0) A strong oxidizer. Incompatible with combustibles, organics, reducing agents. Aqueous solution is acidic. Attacks metals in the presence of moisture.

ALUMINUM NITRATE NONHYDRATE (13473-90-0) A strong oxidizer. Incompatible with combustibles, organics, reducing agents. Aqueous solution is acidic. Attacks metals in the presence of moisture.

ALUMINUM PHOSPHIDE (20859-73-8) Contact with moisture (including atmospheric moisture) produces spontaneously combustible phosphine gas.

ALUMINUM PHOSPHITE (20859-73-8) Contact with moisture (including atmospheric moisture) produces spontaneously combustible phosphine gas.

ALUMINUM POWDER (7429-90-5) A strong reducing agent; reacts violently with many compounds including oxidizers, acids, caustics, halogens, halogenated hydrocarbons, nitrates, sulfates, mercury and mercury compounds, methyl chloride, metal oxides, trichloroethylene, oxygen. May accumulate static electrical charges from friction or stirring, and self-ignite.

ALUMINUM SALT of NITRIC ACID (13473-90-0) A strong oxidizer. Incompatible with combustibles, organics, reducing agents. Aqueous solution is acidic. Attacks metals in the presence of moisture.

ALUMINUM SODIUM FLUORIDE (15096-52-3) Contact with strong acids produces toxic and irritating hydrogen fluoride gas.

ALUMINUM SULFATE or ALUMINUM SULFATE (2:3) or ALU-MINUM SULFATE, LIQUID or ALUMINUM SULFATE SOLU-TION (10043-01-3) Forms sulfuric acid with water; reacts strongly with bases and many other materials. Dry material is weakly corrosive to carbon steel; aqueous solution attacks metals, forming hydrogen gas.

ALUMINUM TRICHLORIDE or ALUMINUM TRICHLORIDE SOLUTION (7446-70-0) Contact with air or water produces hydro-chloric acid. Water, alcohol, alkenes cause polymerization. Incompati-ble with nitrobenzene, organic material, and bases. Attacks metal in presence of moisture.

ALUMINUM TRIETHYL (97-93-8) Poses dangerous explosion hazard. Ignites spontaneously, at all temperatures, in air. A powerful reducing agent; reacts violently with oxidizers, water (forming ethane gas), alcohols, amines, carbon dioxide, carbon tetrachloride, halogen-ated hydrocarbons, oxides of nitrogen or sulfur, phenols, and many other substances. May accumulate static electrical charges, and may cause ignition of its vapors.

ALUMINUM, TRIISOBUTYL- (100-99-2) Ignites spontaneously in air. Water contact produces hydrocarbons. Reacts with oxidizers, carbon dioxide, alcohols, ammonia, halogenated hydrocarbons. At-tacks silicone and urethane rubbers.

ALUMINUM TRINITRATE (13473-90-0) A strong oxidizer. Incom-patible with combustibles, organics, reducing agents. Aqueous solu-tion is acidic. Attacks metals in the presence of moisture.

ALUMINUM TRIS(2-METHYLPROPYL)-(9CI) (100-99-2) Ignites spontaneously in air. Water contact produces hydrocarbons. Reacts with oxidizers, carbon dioxide, alcohols, ammonia, halogenated hydrocarbons. Attacks silicone and urethane rubbers.

ALUMINUM TRISULFATE or ALUMINUM TRISULFATE SOLU-TION (10043-01-3) Forms sulfuric acid with water; with bases and many other materials. Dry material is weakly corrosive to carbon steel; aqueous solution attacks metals, forming hydrogen gas.

ALUNOGENITE (10043-01-3) Forms sulfuric acid with water; reacts strongly with bases and many other materials. Dry material is weakly corrosive to carbon steel; aqueous solution attacks metals, forming hydrogen gas.

ALVIT (60-57-1) Incompatible with concentrated mineral acids, acid catalysts, acid oxidizing agents, phenols, reactive metals.

ALZODEF (156-62-7) Contact with any form of moisture causes decomposition, liberating acetylene and ammonia. Contact with all solvents tested also causes decomposition.

AMATIN (118-74-1) Reacts violently with strong oxidizers, formyldi-methylamine.

AMCHLOR (12125-02-9) Contact with boron trifluoride, boron pen-tafluoride, iodine heptafluoride, or potassium chlorate causes violent reactions. Incompatible with alkalies, acids, salts of lead or silver. At fire temperatures fume corrodes metals.

AMCHLORIDE (12125-02-9) Contact with boron trifluoride, boron pentafluoride, iodine heptafluoride, or potassium chlorate causes violent reactions. Incompatible with alkalies, acids, salts of lead or silver. At fire temperatures fume corrodes metals.

AMCIDE (7773-06-0) Fires and explosions may result from contact with strong oxidizers, hot water, potassium, sodium, sodium nitrite, metal chlorates, hot acid solutions. Corrosive to mild steel.

AMCO (9003-07-0) Incompatible with strong acids, strong oxidizers, chlorine, potassium permanganate.

AMEISENATOD (58-89-9) Not combustible, but may be dissolved in a combustible solvent. If solvent comes in contact with oxidizers, fire and explosions may result.

AMEISENMITTEL MERCK (58-89-9) Not combustible, but may be dissolved in a combustible solvent. If solvent comes in contact with oxidizers, fire and explosions may result.

AMEISENSAEURE (German) (64-18-6) Forms explosive mixture with air (flash point 156°F/69°C). Incompatible with sulfuric acid, bases, ammonia, aliphatic amines, alkanolamines, furfuryl alcohol, hydrogen peroxide, isocyanates, alkylene oxides, epichlorohydrin. Attacks aluminum, cast iron and steel, some plastics, rubber, and coatings.

AMERCIDE (133-06-2) Incompatible with tetraethyl pyrophosphate, parathion.

AMERICAN CYANAMID 4,049 (121-75-5) Incompatible with strong oxidizers, magnesium, alkaline pesticides. Attacks metals, some plastics, rubber, and coatings.

AM-FOL (7664-41-7) **Anhydrous (a flammable gas);** (1336-21-6) **solution in water.** Reacts violently with strong oxidizers, acids. Shock-sensitive compounds may be formed with halogens, mercury oxide, silver oxide. Fire and explosions may be caused by trimethylammonium amide, 1-chloro-2,4-dinitrobenzene, *o*-chloronitrobenzene, platinum, trioxygen difluoride, selenium difluoride dioxide, boron halides, mercury, chlorine, iodine, bromine, hypochlorites, chlorine bleach, amides, organic anhydrides, isocyanates, vinyl acetate, alkylene oxides, epichlorohydrin, aldehydes. Attacks some coatings, plastics, and rubber. Attacks copper, brass, bronze, aluminum, steel, and their alloys.

AMICIDE (7773-06-0) Fires and explosions may result from contact with strong oxidizers, hot water, potassium, sodium, sodium nitrite, metal chlorates, hot acid solutions. Corrosive to mild steel.

AMIDOFOS (299-86-5) Decomposes in strongly alkaline (pH >7) and strongly acidic media. Unstable over long periods of time in water and at temperatures above 140°F/60°C.

AMIDOX (94-75-7) Decomposes in sunlight. Incompatible with strong oxidizers; may cause fire and explosions.

AMILAN (105-60-2) Contact with strong oxidizers may cause fire and explosions.

AMINE (9CI)-O-METHOXYPHENYLAMINE (90-04-0) Incompatible with strong oxidizers, with risk of fire or explosions. Attacks some coatings, and some forms of plastic and rubber.

AMINIC ACID (64-18-6) Forms explosive mixture with air (flash point 156°F/69°C). Incompatible with sulfuric acid, bases, ammonia, aliphatic amines, alkanolamines, furfuryl alcohol, hydrogen peroxide, isocyanates, alkylene oxides, epichlorohydrin. Attacks aluminum, cast iron and steel, some plastics, rubber, and coatings.

4-AMINOANILINE (106-50-3) Incompatible with acids, organic anhydrides, isocyanates, aldehydes, strong oxidizers. Heat and light contribute to instability.

p-**AMINOANILINE or** *para*-**AMINOANILINE** (106-50-3) Incompatible with acids, organic anhydrides, isocyanates, aldehydes, strong oxidizers. Heat and light contribute to instability.

2-AMINOANISOLE (90-04-0) Incompatible with strong oxidizers, with risk of fire or explosions. Attacks some coatings, and some forms of plastic and rubber.

o-**AMINOANISOLE or** *ortho*-**AMINOANISOLE** (90-04-0) Incompatible with strong oxidizers, with risk of fire or explosions. Attacks some coatings, and some forms of plastic and rubber.

AMINOBENZENE (62-53-3) Forms explosive mixture with air (flash point 158°F/70°C). Unless inhibited (usually by methanol), readily able to polymerize. Fires and explosions may result from contact with halogens, strong acids, oxidizers, organic anhydrides, acetic anhydride, isocyanates, aldehydes, sodium peroxide. Reacts with alkali metals and alkali earth metals. Attacks some plastics, rubber, and coatings; incompatible with copper and copper alloys.

4-AMINOBIPHENYL (92-67-1) Strong oxidizers may cause fire and explosions.

p-**AMINOBIPHENYL or** *para*-**AMINOBIPHENYL** (92-67-1) Strong oxidizers may cause fire and explosions.

1-AMINO-BUTAAN (Dutch) (109-73-9) Forms explosive mixture with air (flash point 10°F/−12°C). May accumulate static electrical charges, and may cause ignition of its vapors. Incompatible with nonoxidizing mineral acids, strong acids, organic acids, organic anhydrides, isocyanates, vinyl acetate, acrylates, substituted allyls, alkylene oxides, epichlorohydrin, ketones, aldehydes, alcohols, glycols, phenols, cresols, caprolactam solution, strong oxidizers. In the presence of moisture, corrodes light metals and copper on contact.

1-AMINOBUTAN (German) (109-73-9) Forms explosive mixture with air (flash point 10°F/−12°C). May accumulate static electrical charges, and may cause ignition of its vapors. Incompatible with nonoxidizing mineral acids, strong acids, organic acids, organic anhydrides, isocyanates, vinyl acetate, acrylates, substituted allyls, alkylene oxides, epichlorohydrin, ketones, aldehydes, alcohols, glycols, phenols, cresols, caprolactam solution, strong oxidizers. In the presence of moisture, corrodes light metals and copper on contact.

1-AMINOBUTANE (109-73-9) Forms explosive mixture with air (flash point 10°F/−12°C). May accumulate static electrical charges, and may cause ignition of its vapors. Incompatible with nonoxidizing mineral acids, strong acids, organic acids, organic anhydrides, isocyanates, vinyl acetate, acrylates, substituted allyls, alkylene oxides, epichlorohydrin, ketones, aldehydes, alcohols, glycols, phenols, cresols, caprolactam solution, strong oxidizers. In the presence of moisture, corrodes light metals and copper on contact.

2-AMINOBUTANE (13952-84-6) Forms explosive mixture with air. Incompatible with nonoxidizing mineral acids, strong acids, organic acids, organic anhydrides, isocyanates, vinyl acetate, acrylates, substituted allyls, alkylene oxides, epichlorohydrin, ketones, aldehydes, alcohols, glycols, phenols, cresols, caprolactam solution, strong oxidizers.

AMINOCAPROIC LACTAM (105-60-2) Contact with strong oxidizers may cause fire and explosions.

1-AMINO-4-CHLOROBENZENE (106-47-8) Incompatible with strong acids, oxidizers, acetic anhydride, chlorosulfonic acid, ozone.

1-AMINO-3-CHLORO-6-METHYLBENZENE (95-79-4) Incompatible with acids, organic anhydrides, isocyanates, aldehydes, oxidizers.

2-AMINO-4-CHLOROTOLUENE (95-79-4) Incompatible with acids, organic anhydrides, isocyanates, aldehydes, oxidizers.

2-AMINO-5-CHLOROTOLUENE (95-79-4) Incompatible with acids, organic anhydrides, isocyanates, aldehydes, oxidizers.

AMINOCYCLOHEXANE (108-91-8) Forms explosive mixture with air (flash point 79°F/26°C). Incompatible with nonoxidizing mineral acids, strong acids, organic acids, organic anhydrides, isocyanates, vinyl acetate, acrylates, substituted allyls, alkylene oxides, epichlorohydrin, ketones, aldehydes, alcohols, glycols, phenols, cresols, caprolactam solution, strong oxidizers. Contact with copper alloys, zinc or galvanized steel may cause violent reaction.

4-AMINODIFENIL (Spanish) (92-67-1) Strong oxidizers may cause fire and explosions.

AMINODIMETHYLBENZENE (1300-73-8) Forms explosive mixture with air (flash point 206°F/96.7°C). Contact with strong acids may cause fire and explosions. Contact with hypochlorite bleaches produces explosive chloroamines. Incompatible with strong acids, organic acids and anhydrides, isocyanates, aldehydes.

1-AMINO-2,4-DIMETHYLBENZENE (95-68-1) Forms explosive mixture with air (flash point 206°F/97°C). Contact with strong oxidizers may cause fire and explosions. Hypochlorite bleaches may produce explosive chloramines. Attacks some plastics, rubber, and coatings.

1-AMINO-2,5-DIMETHYLBENZENE (95-78-3) Forms explosive mixture with air (flash point 206°F/97°C). Contact with strong oxidizers may cause fire and explosions. Hypochlorite bleaches may produce explosive chloramines. Attacks some plastics, rubber, and coatings.

3-AMINO-1,4-DIMETHYLBENZENE (95-78-3) Forms explosive mixture with air (flash point 206°F/97°C). Contact with strong oxidizers may cause fire and explosions. Hypochlorite bleaches may produce explosive chloramines. Attacks some plastics, rubber, and coatings.

4-AMINO-1,3-DIMETHYLBENZENE (95-68-1) Forms explosive mixture with air (flash point 206°F/97°C). Contact with strong oxidizers may cause fire and explosions. Hypochlorite bleaches may produce explosive chloramines. Attacks some plastics, rubber, and coatings.

2-AMINODIMETHYLETHANOL beta-AMINOISOBUTANOL
(124-68-5) Forms explosive mixture with air (flash point 153°F/67°C). Incompatible with strong acids, mineral acids, organic acids, organic anhydrides, isocyanates, vinyl acetate, acrylates, substituted allyls, alkylene oxides, epichlorohydrin, aldehydes.

4-AMINODIPHENYL (92-67-1) Strong oxidizers may cause fire and explosions.

p-**AMINODIPHENYL** (92-67-1) Strong oxidizers may cause fire and explosions.

AMINOETHANDIAMINE (111-40-0) Ignites spontaneously with cellulose nitrate. Silver, cobalt, or chromium compounds may cause explosions. Forms explosive mixture with air (flash point 208°F/98°C). Incompatible with nonoxidizing mineral acids, strong acids, organic acids, organic anhydrides, isocyanates, vinyl acetate, acrylates, substituted allyls, alkylene oxides, epichlorohydrin, ketones, aldehydes, alcohols, glycols, mercury, phenols, cresols, caprolactam solution, strong oxidizers. Attacks aluminum, copper, lead, tin, zinc, and alloys.

AMINOETHANE (75-04-7) Forms explosive mixture with air (flash point less than 0°F/−18°C). Incompatible with nonoxidizing mineral acids, strong acids, organic acids, organic anhydrides, isocyanates, vinyl acetate, acrylates, substituted allyls, alkylene oxides, epichlorohydrin, ketones, aldehydes, alcohols, glycols, phenols, cresols, caprolactam solution, strong oxidizers. Attacks aluminum, copper, lead, tin, zinc, and alloys, rubber, and coatings.

1-AMINOETHANE (75-04-7) Forms explosive mixture with air (flash point less than 0°F/−18°C). Incompatible with nonoxidizing mineral acids, strong acids, organic acids, organic anhydrides, isocyanates, vinyl acetate, acrylates, substituted allyls, alkylene oxides, epichlorohydrin, ketones, aldehydes, alcohols, glycols, phenols, cresols, caprolactam solution, strong oxidizers. Attacks aluminum, copper, lead, tin, zinc, and alloys, and some plastics, rubber, and coatings.

2-AMINOETHANOL (141-43-5) Forms explosive mixture with air (flash point 185°F/85°C). Strong oxidizers may cause fire and explosions; incompatible with strong acids (spattering). Attacks plastics.

2-(2-AMINOETHOXY) ETHANOL (929-06-6) Incompatible with acids, organic anhydrides, isocyanates, aldehydes, oxidizers.

n-**(2-AMINOETHYL)** (111-40-0) Ignites spontaneously with cellulose nitrate. Silver, cobalt, or chromium compounds may cause explosions. Forms explosive mixture with air (flash point 208°F/98°C). Incompatible with nonoxidizing mineral acids, strong acids, organic acids, organic anhydrides, isocyanates, vinyl acetate, acrylates, substituted allyls, alkylene oxides, epichlorohydrin, ketones, aldehydes, alcohols, glycols, mercury, phenols, cresols, caprolactam solution, strong oxidizers. Attacks aluminum, copper, lead, tin, zinc, and alloys.

AMINOETHYL ALCOHOL (141-43-5) Forms explosive mixture with air (flash point 185°F/85°C). Strong oxidizers may cause fire and explosions; incompatible with strong acids (spattering). Attacks plastics.

beta-AMINOETHYL ALCOHOL (141-43-5) Forms explosive mixture with air (flash point 185°F/85°C). Strong oxidizers may cause fire and explosions; incompatible with strong acids (spattering). Attacks plastics.

beta-AMINOETHYLAMINE (107-15-3) Forms explosive mixture with air (flash point 104°F/40°C). Incompatible with chlorinated organic compounds, silver perchlorate, 3-propiolactone, mesityl oxide, ethylene dichloride, nonoxidizing mineral acids, strong acids, organic acids, organic anhydrides, isocyanates, vinyl acetate, acrylates, substituted allyls, alkylene oxides, epichlorohydrin, ketones, aldehydes, alcohols, glycols, phenols, cresols, caprolactam solution, strong oxidizers. Attacks aluminum, copper, lead, tin, zinc, and alloys, and some plastics, rubber, and coatings.

2-[(2-AMINOETHYL) AMINO] ETHANOL (929-06-6) Incompatible with acids, organic anhydrides, isocyanates, aldehydes, oxidizers.

N,N'-BIS(2-AMINOETHYL)-1,2-DIAMINOETHANE (112-24-3) Forms explosive mixture with air (flash point 275°F/135°C). Incompatible with nonoxidizing mineral acids, strong acids, organic acids, organic anhydrides, chlorinated hydrocarbons, isocyanates, vinyl acetate, acrylates, substituted allyls, alkylene oxides, epichlorohydrin, ketones, aldehydes, alcohols, glycols, phenols, cresols, caprolactam solution, strong oxidizers. Attacks aluminum, copper, lead, tin, zinc and alloys.

AMINOETHYLENE (151-56-4) Forms explosive mixture with air. Contact with acid, aluminum, carbon dioxide, or silver may cause explosive polymerization.

AMINOETHYLETHANDIAMINE (111-40-0) Ignites spontaneously with cellulose nitrate. Silver, cobalt, or chromium compounds may cause explosions. Forms explosive mixture with air (flash point 208°F/98°C). Incompatible with nonoxidizing mineral acids, strong acids, organic acids, organic anhydrides, isocyanates, vinyl acetate, acrylates, substituted allyls, alkylene oxides, epichlorohydrin, ketones, aldehydes, alcohols, glycols, mercury, phenols, cresols, caprolactam solution, strong oxidizers. Attacks aluminum, copper, lead, tin, zinc and alloys.

AMINO ETHYL-1,2-ETHANEDIAMINE,1,4,7,10,13-PENTAAZA-TRIDECANE-3-AMINOPHENYLMETHANE (108-44-1) Forms explosive mixture with air (flash point 188°F/87°C). Incompatible with strong oxidizers, strong acids. Attacks some plastics, rubber, or coatings.

AMINOETHYLETHANOL AMINE (929-06-6) Incompatible with acids, organic anhydrides, isocyanates, aldehydes, oxidizers.

N-AMINOETHYLETHANOL AMINE (929-06-6) Incompatible with acids, organic anhydrides, isocyanates, aldehydes, oxidizers.

N-(2-AMINOETHYL) ETHANOLAMINE (929-06-6) Incompatible with acids, organic anhydrides, isocyanates, aldehydes, oxidizers.

N,N'-BIS(2-AMINOETHYL)ETHYLENEDIAMINE (112-24-3) Forms explosive mixture with air (flash point 104°F/40°C). Incompatible with nonoxidizing mineral acids, strong acids, organic acids, organic anhydrides, isocyanates, vinyl acetate, acrylates, substituted allyls, alkylene oxides, epichlorohydrin, ketones, aldehydes, alcohols, glycols, phenols, cresols, caprolactam solution, strong oxidizers. Attacks aluminum, copper, lead, tin, zinc, and alloys.

1-AMINO-2-ETHYLHEXAN (Czech) (104-75-6) Forms explosive mixture with air (flash point 140°F/60°C). Incompatible with nonoxidizing mineral acids, strong acids, organic acids, organic anhydrides, isocyanates, vinyl acetate, acrylates, substituted allyls, alkylene oxides, epichlorohydrin, ketones, aldehydes, alcohols, glycols, phenols, cresols, caprolactam solution, strong oxidizers. Attacks aluminum, copper, lead, tin, zinc, and alloys.

1-AMINO-2-ETHYLHEXANE (104-75-6) Forms explosive mixture with air (flash point 140°F/60°C). Incompatible with nonoxidizing mineral acids, strong acids, organic acids, organic anhydrides, isocyanates, vinyl acetate, acrylates, substituted allyls, alkylene oxides, epichlorohydrin, ketones, aldehydes, alcohols, glycols, phenols, cresols, caprolactam solution, strong oxidizers. Attacks aluminum, copper, lead, tin, zinc and alloys.

1-(2-AMINOETHYL) PIPERAZINE (140-31-8) Solution is a strong base. Forms explosive mixture with air (flash point 199°F/93°C). Incompatible with nonoxidizing mineral acids, strong acids, organic acids, organic anhydrides, isocyanates, vinyl acetate, acrylates, substituted allyls, alkylene oxides, epichlorohydrin, ketones, aldehydes, alcohols, glycols, phenols, cresols, caprolactam solution, strong oxidizers. Contact with copper alloys, zinc, or galvanized steel may cause violent reaction.

n-**AMINOETHYLPIPERAZINE** (140-31-8) Solution is a strong base. Forms explosive mixture with air (flash point 199°F/93°C). Incompatible with nonoxidizing mineral acids, strong acids, organic acids, organic anhydrides, isocyanates, vinyl acetate, acrylates, substituted allyls, alkylene oxides, epichlorohydrin, ketones, aldehydes, alcohols, glycols, phenols, cresols, caprolactam solution, strong oxidizers. Contact with copper alloys, zinc or galvanized steel may cause violent reaction.

n-**(2-AMINOETHYL) PIPERAZINE** (140-31-8) Solution is a strong base. Forms explosive mixture with air (flash point 199°F/93°C). Incompatible with nonoxidizing mineral acids, strong acids, organic acids, organic anhydrides, isocyanates, vinyl acetate, acrylates, substituted allyls, alkylene oxides, epichlorohydrin, ketones, aldehydes, alcohols, glycols, phenols, cresols, caprolactam solution, strong oxidizers. Contact with copper alloys, zinc, or galvanized steel may cause violent reaction.

1-AMINO-2-FLUOROBENZENE (348-54-9) Forms explosive mixture with air (flash point 140°F/60°C). Incompatible with strong oxidizers, strong acids.

1-AMINO-4-FLUOROBENZENE (371-40-4) Forms explosive mixture with air (flash point 165°F/74°C). Incompatible with strong acids, strong oxidizers.

AMINOFORM (100-97-0) Incompatible with nonoxidizing mineral acids, strong acids, organic acids, organic anhydrides, isocyanates, vinyl acetate, acrylates, substituted allyls, alkylene oxides, epichlorohydrin, ketones, aldehydes, alcohols, glycols, phenols, cresols, caprolactam solution, sodium peroxide, strong oxidizers. Attacks aluminum, copper, lead, tin, zinc, and alloys.

AMINOHEXAHYDROBENZENE (108-91-8) Forms explosive mixture with air (flash point 79°F/26°C). Incompatible with nonoxidizing mineral acids, strong acids, organic acids, organic anhydrides, isocyanates, vinyl acetate, acrylates, substituted allyls, alkylene oxides, epichlorohydrin, ketones, aldehydes, alcohols, glycols, phenols, cresols, caprolactam solution, strong oxidizers. Contact with copper alloys, zinc, or galvanized steel may cause violent reaction.

1-AMINO-2-HYDROXYPROPANE (75-31-0) Forms explosive mixture with air (flash point −35°F/−37°C). Incompatible with strong acids (explosive spattering), strong oxidizers (fire and explosions). Attacks some plastics, rubber, and coatings.

2-AMINOISOBUTANE (75-64-9) Forms explosive mixture with air. Incompatible with nonoxidizing mineral acids, strong acids, organic acids, organic anhydrides, isocyanates, vinyl acetate, acrylates, substituted allyls, alkylene oxides, epichlorohydrin, ketones, aldehydes, alcohols, glycols, phenols, cresols, caprolactam solution, strong oxidizers.

AMINOMERCURIC CHLORIDE (10124-48-8) Reacts violently with halogens and metal salts of amines.

AMINOMETHANE (74-89-5) Forms explosive gas mixture with air. Incompatible with nonoxidizing mineral acids, strong acids, organic acids, organic anhydrides, isocyanates, vinyl acetate, acrylates, substituted allyls, alkylene oxides, epichlorohydrin, ketones, aldehydes, alcohols, glycols, phenols, cresols, caprolactam solution, strong oxidizers. Attacks aluminum, copper, lead, tin, zinc, and alloys, and some plastics, rubber, and coatings.

AMINOMETHANE (cylinder) (74-89-5) Forms explosive gas mixture with air. Incompatible with nonoxidizing mineral acids, strong acids, organic acids, organic anhydrides, isocyanates, vinyl acetate, acrylates, substituted allyls, alkylene oxides, epichlorohydrin, ketones, aldehydes, alcohols, glycols, phenols, cresols, caprolactam solution, strong oxidizers. Attacks aluminum, copper, lead, tin, zinc and alloys, and some plastics, rubber, and coatings.

1-AMINO-2-METHYLBENZENE (95-53-4) Forms explosive mixture with air (flash point 185°F/85°C). Incompatible with strong acids, mineral acids, organic anhydrides, isocyanates, aldehydes, oxidizers. Attacks some plastics, rubber, or coatings.

2-AMINO-1-METHYLBENZENE (95-53-4) Forms explosive mixture with air (flash point 185°F/85°C). Incompatible with strong acids, mineral acids, organic anhydrides, isocyanates, aldehydes, oxidizers. Attacks some plastics, rubber, or coatings.

3-AMINO-1-METHYLBENZENE (108-44-1) Forms explosive mixture with air (flash point 188°F/87°C). Incompatible with strong oxidizers, strong acids. Attacks some plastics, rubber, or coatings.

4-AMINO-1-METHYLBENZENE (106-49-0) Forms explosive mixture with air (flash point 188°F/87°C). Incompatible with strong oxidizers, strong acids. Attacks some plastics, rubber, or coatings.

1-AMINO-2-METHYLPROPANE (78-81-9) Forms explosive mixture with air (flash point 15°F/−9°C). Incompatible with strong acids, including mineral and organic acids, organic anhydrides, isocyanates, vinyl acetate, acrylates, substituted allyls, alkylene oxides, epichlorohydrin, ketones, aldehydes, alcohols, glycols, phenols, cresols, caprolactam solution, halogens, alkali metals, copper and its alloys.

2-AMINO-2-METHYLPROPANE (75-64-9) Forms explosive mixture with air. Incompatible with nonoxidizing mineral acids, strong acids, organic acids, organic anhydrides, isocyanates, vinyl acetate, acrylates, substituted allyls, alkylene oxides, epichlorohydrin, ketones, aldehydes, alcohols, glycols, mercury, phenols, cresols, caprolactam solution, strong oxidizers.

2-AMINO-2-METHYL-1-PROPANOL (90% or less) (124-68-5) Forms explosive mixture with air (flash point 153°F/67°C). Incompatible with strong acids, mineral acids, organic acids, organic anhydrides, isocyanates, vinyl acetate, acrylates, substituted allyls, alkylene oxides, epichlorohydrin, aldehydes.

4-AMINO-3-METHYL TOLUENE (95-68-1) Forms explosive mixture with air (flash point 206°F/97°C). Contact with strong oxidizers may cause fire and explosions. Hypochlorite bleaches may produce explosive chloramines. Attacks some plastics, rubber, and coatings.

3-AMINOMETHYL-3,5,5-TRIMETHYLCYCLOHEXYLAMINE (2855-13-2) Incompatible with acids, organic anhydrides, isocyanates, vinyl acetate, acrylates, substituted allyls, alkylene oxides, epichlorohydrin, ketones, aldehydes, alcohols, glycols, phenols, cresols, caprolactam solution. Corrodes aluminum and steel in the presence of moisture and carbon dioxide.

1-AMINONAFTALEN (Czech) (134-32-7) Oxidizes in air. Incompatible with nitrous acid, oxidizers, nitrates, organic anhydrides, isocyanates, aldehydes.

2-AMINONAFTHALEN (Czech) (91-59-8) Forms explosive mixture with air (flash point 300°F/149°C). Incompatible strong oxidizers, strong acids. Oxidizes in the presence of air and light.

1-AMINONAPHTHALENE (134-32-7) Oxidizes in air. Incompatible with nitrous acid, oxidizers, nitrates, organic anhydrides, isocyanates, aldehydes.

2-AMINONAPHTHALENE (91-59-8) Forms explosive mixture with air (flash point 300°F/149°C). Incompatible with strong oxidizers, strong acids. Oxidizes in the presence of air and light.

1-AMINO-2-NITROBENZENE (88-74-4) *ortho*; (99-09-2) *meta*. Incompatible with strong acids, combustibles, organics, moisture.

1-AMINO-4-NITROBENZENE (100-01-06) Strong oxidizers and moisture may cause spontaneous heating. Caustics and heat may form explosive mixtures. Attacks some plastics, rubber, and coatings. May accumulate static electrical charges, and may cause ignition of its vapors.

p-**AMINONITROBENZENE** or *para*-**AMINONITROBENZENE** (100-01-06) Strong oxidizers and moisture may cause spontaneous heating. Caustics and heat may form explosive mixtures. Attacks some plastics, rubber, and coatings. May accumulate static electrical charges, and may cause ignition of its vapors.

2-AMINOPENTANE (109-89-7) Forms explosive mixture with air (flash point −9°F/−23°C). Incompatible with nonoxidizing mineral acids, strong acids, organic acids, organic anhydrides, isocyanates, vinyl acetate, acrylates, substituted allyls, alkylene oxides, epichlorohydrin, ketones, aldehydes, alcohols, glycols, mercury, phenols, cresols, caprolactam solution, strong oxidizers. Attacks aluminum, copper, lead, tin, zinc, and alloys.

1-AMINOPROPANE (107-10-8) Forms explosive mixture with air. Incompatible with nonoxidizing mineral acids, strong acids, organic acids, organic anhydrides, halogenated hydrocarbons, isocyanates, nitroparaffins, vinyl acetate, acrylates, substituted allyls, alkylene oxides, epichlorohydrin, ketones, aldehydes, alcohols, glycols, mercury, phenols, cresols, triethyl aluminum, caprolactam solution, strong oxidizers. Attacks aluminum, copper, lead, tin, zinc, and alloys.

2-AMINOPROPANE (75-31-0) Forms explosive mixture with air (flash point −35°F/−37°C). Incompatible with strong acids (explosive spattering), strong oxidizers (fire and explosions). Attacks some plastics, rubber, and coatings.

1-AMINO-PROPANOL-2 (75-31-0) Forms explosive mixture with air (flash point −35°F/−37°C). Incompatible with strong acids (explosive spattering), strong oxidizers (fire and explosions). Attacks some plastics, rubber, and coatings.

1-AMINO-2-PROPANOL (78-96-6) Forms explosive mixture with air (flash point 171°F/77°C). Incompatible with acids, organic anhydrides, isocyanates, aldehydes.

3-AMINO-1-PROPANOL (156-87-6) Forms explosive mixture with air (flash point 175°F/79°C). Incompatible with acids, organic anhydrides, isocyanates, vinyl acetate, acrylates, substituted allyls, alkylene oxides, epichlorohydrin, aldehydes. Corrodes copper and its alloys.

gamma-AMINOPROPANOL (156-87-6) Forms explosive mixture with air (flash point 175°F/79°C). Incompatible with acids, organic anhydrides, isocyanates, vinyl acetate, acrylates, substituted allyls, alkylene oxides, epichlorohydrin, aldehydes. Corrodes copper and its alloys.

3-AMINO-1-PROPENE (107-11-9) Forms explosive mixture with air (flash point −20°F/−29°C). A strong base. Incompatible with acids, strong oxidizers, hypochlorites, nitrosyl perchlorate. Corrodes metals.

3-AMINOPROPYLENE (107-11-9) Forms explosive mixture with air (flash point −20°F/−29°C). A strong base. Incompatible with acids, strong oxidizers, hypochlorites, nitrosyl perchlorate. Corrodes metals.

AMINO-2-PYRIDINE (504-29-0) Reacts with strong oxidizers, strong acids, sodium nitrate.

AMINO-3-PYRIDINE (462-08-8) Forms a strong base with water. Incompatible with strong oxidizers, acids.

2-AMINOPYRIDINE (504-29-0) Reacts with strong oxidizers, strong acids, sodium nitrate.

3-AMINOPYRIDINE (462-08-8) Forms a strong base with water. Incompatible with strong oxidizers, acids.

4-AMINOPYRIDINE (504-24-2) Forms a strong base with water. Incompatible with strong oxidizers, acids.

a-AMINOPYRIDINE or **alpha-AMINOPYRIDINE** (504-29-0) Reacts with strong oxidizers, strong acids, sodium nitrate.

m-**AMINOPYRIDINE** or *meta*-**AMINOPYRIDINE** (462-08-8) Forms a strong base with water. Incompatible with strong oxidizers, acids.

o-**AMINOPYRIDINE** or *ortho*-**AMINOPYRIDINE** (504-29-0) Reacts with strong oxidizers, strong acids, sodium nitrate.

p-**AMINOPYRIDINE** or *para*-**AMINOPYRIDINE** (504-24-2) Forms a strong base with water. Incompatible with strong oxidizers, acids.

3-AMINOTOLUEN (Czech) (108-44-1) Forms explosive mixture with air (flash point 188°F/87°C). Incompatible with strong oxidizers, strong acids. Attacks some plastics, rubber, or coatings.

2-AMINOTOLUENE (95-53-4) Forms explosive mixture with air (flash point 185°F/85°C). Incompatible with strong acids, mineral acids, organic anhydrides, isocyanates, aldehydes, oxidizers. Attacks some plastics, rubber, or coatings.

3-AMINOTOLUENE (108-44-1) Forms explosive mixture with air (flash point 188°F/87°C). Incompatible with strong oxidizers, strong acids. Attacks some plastics, rubber, or coatings.

4-AMINOTOLUENE (106-49-0) Forms explosive mixture with air (flash point 188°F/87°C). Incompatible with strong oxidizers, strong acids. Attacks some plastics, rubber, or coatings.

alpha-AMINOTOLUENE (100-46-9) Forms explosive mixture with air (flash point 168°F/76°C). Incompatible with strong acids, organic anhydrides, isocyanates, aldehydes. Slowly corrodes metals in a moist environment. Attacks some plastics, rubber, or coatings.

m-**AMINOTOLUENE or** *meta*-**AMINOTOLUENE** (108-44-1) Forms explosive mixture with air (flash point 188°F/87°C). Incompatible with strong oxidizers, strong acids. Attacks some plastics, rubber, or coatings.

o-**AMINOTOLUENE or** *ortho*-**AMINOTOLUENE** (95-53-4) Forms explosive mixture with air (flash point 185°F/85°C). Incompatible with strong acids, mineral acids, organic anhydrides, isocyanates, aldehydes, oxidizers. Attacks some plastics, rubber, or coatings.

p-**AMINOTOLUENE or** *para*-**AMINOTOLUENE** (106-49-0) Forms explosive mixture with air (flash point 188°F/87°C). Incompatible with strong oxidizers, strong acids. Attacks some plastics, rubber, or coatings.

2-AMINO-1,4-XYLENE (95-78-3) Forms explosive mixture with air (flash point 206°F/97°C). Contact with strong oxidizers may cause fire and explosions. Hypochlorite bleaches may produce explosive chloramines. Attacks some plastics, rubber, and coatings.

4-AMINO-1,3-XYLENE (95-68-1) Forms explosive mixture with air (flash point 206°F/97°C). Contact with strong oxidizers may cause fire and explosions. Hypochlorite bleaches may produce explosive chloramines. Attacks some plastics, rubber, and coatings.

AMMAT (7773-06-0) Fires and explosions may result from contact with strong oxidizers, hot water, potassium, sodium, sodium nitrite, metal chlorates, hot acid solutions. Corrosive to mild steel.

AMMATE (7773-06-0) Fires and explosions may result from contact with strong oxidizers, hot water, potassium, sodium, sodium nitrite, metal chlorates, hot acid solutions. Corrosive to mild steel.

AMMATE HERBICIDE (7773-06-0) Fires and explosions may result from contact with strong oxidizers, hot water, potassium, sodium, sodium nitrite, metal chlorates, hot acid solutions. Corrosive to mild steel.

AMMONERIC (12125-02-9) Contact with boron trifluoride, boron pentafluoride, iodine heptafluoride, or potassium chlorate causes violent reactions. Incompatible with alkalies, acids, salts of lead or silver. At fire temperatures fume corrodes metals.

AMMONIA (7664-41-7) **anhydrous (a flammable gas);** (1336-21-6) **solution in water.** Reacts violently with strong oxidizers, acids. Shock-sensitive compounds may be formed with halogens, mercury oxide, silver oxide. Fire and explosions may be caused by trimethylammonium amide, 1-chloro-2,4-dinitrobenzene, *o*-chloronitrobenzene, platinum, trioxygen difluoride, selenium difluoride dioxide, boron halides, mercury, chlorine, iodine, bromine, hypochlorites, chlorine bleach, amides, organic anhydrides, isocyanates, vinyl acetate, alkylene oxides, epichlorohydrin, aldehydes. Attacks some coatings, plastics and rubber. Attacks copper, brass, bronze, aluminum, steel and their alloys.

AMMONIA, ANHYDROUS (7664-41-7) **anhydrous (a flammable gas);** (1336-21-6) **solution in water.** Reacts violently with strong oxidizers, acids. Shock-sensitive compounds may be formed with halogens, mercury oxide, silver oxide. Fire and explosions may be caused by trimethylammonium amide, 1-chloro-2,4-dinitrobenzene, *o*-chloronitrobenzene, platinum, trioxygen difluoride, selenium difluoride dioxide, boron halides, mercury, chlorine, iodine, bromine, hypochlorites, chlorine bleach, amides, organic anhydrides, isocyanates, vinyl acetate, alkylene oxides, epichlorohydrin, aldehydes. Attacks some coatings, plastics, and rubber. Attacks copper, brass, bronze, aluminum, steel, and their alloys.

AMMONIA GAS (7664-41-7) **anhydrous (a flammable gas);** (1336-21-6) **solution in water.** Reacts violently with strong oxidizers, acids. Shock-sensitive compounds may be formed with halogens, mercury oxide, silver oxide. Fire and explosions may be caused by trimethylammonium amide, 1-chloro-2,4-dinitrobenzene, *o*-chloronitrobenzene, platinum, trioxygen difluoride, selenium difluoride dioxide, boron halides, mercury, chlorine, iodine, bromine, hypochlorites, chlorine bleach, amides, organic anhydrides, isocyanates, vinyl acetate, alkylene oxides, epichlorohydrin, aldehydes. Attacks some coatings, plastics and rubber. Attacks copper, brass, bronze, aluminum, steel, and their alloys.

AMMONIA SOAP (544-60-5) Incompatible with strong oxidizers, nitric acid.

AMMONIA WATER (1336-21-6) Incompatible with strong acids, mineral acids, organic acids, amides, organic anhydrides, isocyanates, vinyl acetate, epichlorohydrin, aldehydes. Corrosive to copper and aluminum, including their alloys, and galvanized surfaces.

AMMONIAC (French) (7664-41-7) **anhydrous (a flammable gas);** (1336-21-6) **solution in water.** Reacts violently with strong oxidizers, acids. Shock-sensitive compounds may be formed with halogens, mercury oxide, silver oxide. Fire and explosions may be caused by trimethylammonium amide, 1-chloro-2,4-dinitrobenzene, *o*-chloronitrobenzene, platinum, trioxygen difluoride, selenium difluoride dioxide, boron halides, mercury, chlorine, iodine, bromine, hypochlorites, chlorine bleach, amides, organic anhydrides, isocyanates, vinyl acetate, alkylene oxides, epichlorohydrin, aldehydes. Attacks some coatings, plastics, and rubber. Attacks copper, brass, bronze, aluminum, steel, and their alloys.

AMMONIACA (Italian) (7664-41-7) **anhydrous (a flammable gas);** (1336-21-6) **solution in water.** Reacts violently with strong oxidizers, acids. Shock-sensitive compounds may be formed with halogens, mercury oxide, silver oxide. Fire and explosions may be caused by trimethylammonium amide, 1-chloro-2,4-dinitrobenzene, *o*-chloronitrobenzene, platinum, trioxygen difluoride, selenium difluoride dioxide, boron halides, mercury, chlorine, iodine, bromine, hypochlorites, chlorine bleach, amides, organic anhydrides, isocyanates, vinyl acetate, alkylene oxides, epichlorohydrin, aldehydes. Attacks some coatings, plastics, and rubber. Attacks copper, brass, bronze, aluminum, steel, and their alloys.

AMMONIALE (German) (7664-41-7) **anhydrous (a flammable gas);** (1336-21-6) **solution in water.** Reacts violently with strong oxidizers, acids. Shock-sensitive compounds may be formed with halogens, mercury oxide, silver oxide. Fire and explosions may be caused by trimethylammonium amide, 1-chloro-2,4-dinitrobenzene, *o*-chloronitrobenzene, platinum, trioxygen difluoride, selenium difluoride dioxide, boron halides, mercury, chlorine, iodine, bromine, hypochlorites, chlorine bleach, amides, organic anhydrides, isocyanates, vinyl acetate, alkylene oxides, epichlorohydrin, aldehydes. Attacks some coatings, plastics, and rubber. Attacks copper, brass, bronze, aluminum, steel, and their alloys.

AMMONIATED MERCURY (10124-48-8) Reacts violently with halogens and metal salts of amines.

AMMONIOFORMALDEHYDE (100-97-0) Incompatible with nonoxidizing mineral acids, strong acids, organic anhydrides, isocyanates, vinyl acetate, acrylates, substituted allyls, alkylene oxides, epichlorohydrin, ketones, aldehydes, alcohols, glycols, phenols, cresols, caprolactam solution, sodium peroxide, strong oxidizers. Attacks aluminum, copper, lead, tin, zinc, and alloys.

AMMONIUM ACETATE (631-61-8) Incompatible with sodium hypochlorite.

AMMONIUM ACID FLUORIDE (1341-49-7) Incompatible with alkalies, water (forms strong acid). Attacks glass, cement, and most metals.

AMMONIUM AMIDOSULFONATE (7773-06-0) Fires and explosions may result from contact with strong oxidizers, hot water, potassium, sodium, sodium nitrite, metal chlorates, hot acid solutions. Corrosive to mild steel.

AMMONIUM AMIDOSULPHATE (7773-06-0) Fires and explosions may result from contact with strong oxidizers, hot water, potassium, sodium, sodium nitrite, metal chlorates, hot acid solutions. Corrosive to mild steel.

AMMONIUM AMINOSULFONATE (7773-06-0) Fires and explosions may result from contact with strong oxidizers, hot water, potassium, sodium, sodium nitrite, metal chlorates, hot acid solutions. Corrosive to mild steel.

AMMONIUM BICARBONATE (1066-33-7) Incompatible with strong acids. and bases, fluorine, magnesium, sodium hypochlorite.

AMMONIUM BICHROMATE (7789-09-5) Solution is acidic. A strong oxidizer; reacts violently with reducing agents, combustibles, bases. Attacks metals in a moist environment.

AMMONIUM BIFLUORIDE (1341-49-7) Incompatible with alkalies, water (forms strong acid). Attacks glass, cement, and most metals.

AMMONIUM BIPHOSPHATE (7783-28-0) Incompatible with strong oxidizers, strong bases. Contact with air causes substance to give off corrosive anhydrous ammonia fumes.

AMMONIUM BISULFIDE (12124-99-1) Forms explosive mixture with air (flash point 72°F/22°C). Incompatible with strong oxidizers, strong acids. Contact with strong bases produces ammonia. Corrodes aluminum, brass, copper, zinc and related alloys; attacks some plastics, rubber and coatings. May accumulate static electrical charges, and may cause ignition of its vapors.

AMMONIUM BISULFITE (10192-30-0) Incompatible with strong acids, strong oxidizers.

AMMONIUM CARBAZOATE (131-74-8) A self-reactive explosive. Heating it above 250°F/121°C can cause explosions. A powerful oxidizer. Contact with metal, concrete or plaster produces salts that are more shock-sensitive than ammonium picrate. Water increases rate of reactivity with metals.

AMMONIUM CHLORIDE (12125-02-9) Contact with boron trifluoride, boron pentafluoride, iodine heptafluoride, or potassium chlorate causes violent reactions. Incompatible with alkalies, acids, salts of lead or silver. At fire temperatures fume corrodes metals.

AMMONIUM CHROMATE (7788-98-9) Contact with water produces an alkaline solution, evolving free ammonia.

AMMONIUM CITRATE (3012-65-5) Combustible solid. Reacts with strong oxidizers.

AMMONIUM CITRATE, DIBASIC (3012-65-5) Combustible solid. Reacts with strong oxidizers.

AMMONIUM CUPRIC SULFATE (10380-29-7) Water contact produces ammonia. Incompatible with chlorine, fluorine, peroxides, hydroxylamine, magnesium.

AMMONIUM DICHROMATE (7789-09-5) Solution is acidic. A strong oxidizer; reacts violently with reducing agents, combustibles, bases. Attacks metals in a moist environment.

AMMONIUM DIHYDROGEN PHOSPHATE (7783-28-0) Incompatible with strong oxidizers, strong bases. Contact with air causes substance to give off corrosive anhydrous ammonia fumes.

AMMONIUM DISULFATONICKELATE (II) (15699-18-0) Incompatible with strong acids, selenium.

AMMONIUM FERROUS SULFATE (10045-89-3) Reacts with aluminum, magnesium.

AMMONIUM FLUORIDE (12125-01-8) Forms hydrochloric acid with water. Incompatible with strong acids, caustics. Corrosive to glass, cement, and most metals.

AMMONIUM FLUOROSILICATE (16919-19-0) Aqueous solution is corrosive. Keep away from strong oxidizers.

AMMONIUM HYDROGEN CARBONATE (1066-33-7) Incompatible with strong acids and bases, fluorine, magnesium, sodium hypochlorite.

AMMONIUM HYDROGEN DIFLUORIDE (1341-49-7) Incompatible with alkalies, water (forms strong acid). Attacks glass, cement, and most metals.

AMMONIUM HYDROGEN FLUORIDE (1341-49-7) Incompatible with alkalies, water (forms strong acid). Attacks glass, cement, and most metals.

AMMONIUM HYDROGEN SULFIDE (12124-99-1) Forms explosive mixture with air (flash point 72°F/22°C). Incompatible with strong oxidizers, strong acids. Contact with strong bases produces ammonia. Corrodes aluminum, brass, copper, zinc and related alloys; attacks some plastics, rubber, and coatings. May accumulate static electrical charges, and may cause ignition of its vapors.

AMMONIUM HYDROGEN SULFIDE SOLUTION (12124-99-1) Forms explosive mixture with air (flash point 72°F/22°C). Incompatible with strong oxidizers, strong acids. Contact with strong bases produces ammonia. Corrodes aluminum, brass, copper, zinc and related alloys; attacks some plastics, rubber, and coatings. May accumulate static electrical charges, and may cause ignition of its vapors.

AMMONIUM HYDROGEN SULFATE (7783-20-2) Aqueous solution is a strong acid; reacts with bases forming ammonia. Attacks metals. Hot material reacts with nitrates, nitrites, chlorates.

AMMONIUM HYDROGEN SULFITE (10192-30-0) Incompatible with strong acids, strong oxidizers.

AMMONIUM HYDROSULFITE (10192-30-0) Incompatible with strong acids, strong oxidizers.

AMMONIUM HYDROXIDE (7664-41-7) **anhydrous (a flammable gas);** (1336-21-6) **solution in water.** Reacts violently with strong oxidizers, acids. Shock-sensitive compounds may be formed with halogens, mercury oxide, silver oxide. Fire and explosions may be caused by trimethylammonium amide, 1-chloro-2,4-dinitrobenzene, o-chloronitrobenzene, platinum, trioxygen difluoride, selenium difluoride dioxide, boron halides, mercury, chlorine, iodine, bromine, hypochlorites, chlorine bleach, amides, organic anhydrides, isocyanates, vinyl acetate, alkylene oxides, epichlorohydrin, aldehydes. Attacks some coatings, plastics and rubber. Attacks copper, brass, bronze, aluminum, steel, and their alloys.

AMMONIUM HYDROXIDE (less than 28% AQUEOUS AMMONIA) (1336-21-6) Incompatible with strong acids, mineral acids, organic acids, amides, organic anhydrides, isocyanates, vinyl acetate, epichlorohydrin, aldehydes. Corrosive to copper and aluminum, including their alloys, and galvanized surfaces.

AMMONIUM HYPO SOLUTION (7783-18-8) Incompatible with sulfuric acid, isocyanates. Incompatible with strong oxidizers, chlorates, nitrates, nitrites (forms ammonia), hydrogen sulfide, and sulfur trioxide gases.

AMMONIUM HYPOSULFITE SOLUTION (7783-18-8) Incompatible with sulfuric acid, isocyanates. Incompatible with strong oxidizers, chlorates, nitrates, nitrites (forms ammonia), hydrogen sulfide and sulfur trioxide gases.

AMMONIUM IODIDE (12027-06-4) Reacts explosively with bromine trifluoride.

AMMONIUM IRON SULFATE (10045-89-3) Reacts violently with tetranitromethane, ammonium perchlorate, and mercury(II) dinitrate. Sulfates react with aluminum, magnesium.

AMMONIUM MOLYBDATE (13106-76-8) Incompatible with alkali metals, molten magnesium, and strong oxidizers.

AMMONIUM MONOSULFIDE (12124-99-1) Forms explosive mixture with air (flash point 72°F/22°C). Incompatible with strong oxidizers, strong acids. Contact with strong bases produces ammonia. Corrodes aluminum, brass, copper, zinc and related alloys; attacks some plastics, rubber and coatings. May accumulate static electrical charges, and may cause ignition of its vapors.

AMMONIUM MONOSULFITE (10192-30-0) Incompatible with strong acids, strong oxidizers.

AMMONIUM MURIATE (12125-02-9) Contact with boron trifluoride, boron pentafluoride, iodine heptafluoride, or potassium chlorate cause violent reactions. Incompatible with alkalies, acids, salts of lead or silver. At fire temperatures, fume corrodes metals.

AMMONIUM NICKEL(II) SALT (2:2:1) (15699-18-0) Incompatible with strong acids, selenium.

AMMONIUM NICKEL SULFATE (15699-18-0) Incompatible with strong acids, selenium.

AMMONIUM NITRATE (6484-52-2) A strong oxidizer. With reducing agents, combustibles, organic materials, finely divided metals may form explosive mixtures or cause fire and explosions.

AMMONIUM(1) NITRATE(1) (6484-52-2) A strong oxidizer. With reducing agents, combustibles, organic materials, finely divided metals may form explosive mixtures or cause fire and explosions.

AMMONIUM NITRATE-UREA SOLUTION (6484-52-2) Incompatible with sulfuric acid, isocyanates.

AMMONIUM OLEATATE (544-60-5) Incompatible with strong oxidizers, nitric acid.

AMMONIUM PERCHLORATE (7790-98-9) Explosion-sensitive to friction and heat. Contact with organic matter, sulfur, metal powders, ferrocene produces shock-sensitive materials.

AMMONIUM PEROXYDISULFATE (7727-54-0) A strong oxidizer. Reacts violently with reducing agents, aluminum powder, iron. Mixture with sodium peroxide produces a friction-, heat-, and water-sensitive explosive.

AMMONIUM PERSULFATE (7727-54-0) A strong oxidizer. Reacts violently with reducing agents, aluminum powder, iron. Mixture with sodium peroxide produces a friction-, heat-, and water-sensitive explosive.

AMMONIUM PHOSPHATE (7783-28-0) Incompatible with strong oxidizers, strong bases. Contact with air causes substance to give off corrosive anhydrous ammonia fumes.

sec-**AMMONIUM PHOSPHATE** (7783-28-0) Incompatible with strong oxidizers, strong bases. Contact with air causes substance to give off corrosive anhydrous ammonia fumes.

secondary-**AMMONIUM PHOSPHATE** (7783-28-0) Incompatible with strong oxidizers, strong bases. Contact with air causes substance to give off corrosive anhydrous ammonia fumes.

AMMONIUM PHOSPHATE, DIBASIC (7783-28-0) Incompatible with strong oxidizers, strong bases. Contact with air causes substance to give off corrosive anhydrous ammonia fumes.

AMMONIUM PICRATE, DRY (131-74-8) A self-reactive explosive. Heating it above 250°F/121°C can cause explosions. A powerful oxidizer. Contact with metal, concrete, or plaster produces salts that are more shock-sensitive than ammonium picrate. Water increases rate of reactivity with metals.

AMMONIUM PICRATE, WET (131-74-8) A self-reactive explosive. Heating it above 250°F/121°C can cause explosions. A powerful oxidizer. Contact with metal, concrete, or plaster produces salts that are more shock-sensitive than ammonium picrate. Water increases rate of reactivity with metals.

**AMMONIUM PICRATE WETTED WITH MORE THAN 10%
WATER** (131-74-8) A self-reactive explosive. Heating it above
250°F/121°C can cause explosions. A powerful oxidizer. Contact with
metal, concrete, or plaster produces salts that are more shock-
sensitive than ammonium picrate. Water increases rate of reactivity
with metals.

AMMONIUM PICRATE (YELLOW) (131-74-8) A self-reactive explosive. Heating it above 250°F/121°C can cause explosions. A
powerful oxidizer. Contact with metal, concrete, or plaster produces
salts that are more shock-sensitive than ammonium picrate. Water
increases rate of reactivity with metals.

AMMONIUM PICRONITRATE (131-74-8) A self-reactive explosive.
Heating it above 250°F/121°C can cause explosions. A powerful
oxidizer. Contact with metal, concrete, or plaster produces salts that
are more shock-sensitive than ammonium picrate. Water increases
rate of reactivity with metals.

AMMONIUM RHODANIDE (1762-95-4) Reacts violently with
strong oxidizers. Light may cause decomposition.

AMMONIUM SALZ der AMIDOSULFONSAEURE (German)
(7773-06-0) Fires and explosions may result from contact with strong
oxidizers, hot water, potassium, sodium, sodium nitrite, metal chlorates, hot acid solutions. Corrosive to mild steel.

AMMONIUM SILICOFLUORIDE (16919-19-0) Aqueous solution is
corrosive. Keep away from strong oxidizers.

AMMONIUM STEARATE (1002-89-7) Combustible solid (flash
point 140°F/60°C). Strong oxidizers may cause fire and explosions.

AMMONIUM STEARATE DISPERSION (1002-89-7) Combustible
solid (flash point 140°F/60°C). Strong oxidizers may cause fire and
explosions.

AMMONIUM SULFAMATE (7773-06-0) Fires and explosions may
result from contact with strong oxidizers, hot water, potassium,
sodium, sodium nitrite, metal chlorates, hot acid solutions. Corrosive
to mild steel.

AMMONIUM SULFATE (7783-20-2) Aqueous solution is a strong
acid; reacts with bases, forming ammonia. Attacks metals. Hot
material reacts with nitrates, nitrites, chlorates.

AMMONIUM SULFATE (2:1) (7783-20-2) Aqueous solution is a
strong acid; reacts with bases, forming ammonia. Attacks metals. Hot
material reacts with nitrates, nitrites, chlorates.

AMMONIUM SULFHYDRATE SOLUTION (12124-99-1) Forms
explosive mixture with air (flash point 72°F/22°C). Incompatible with
strong oxidizers, strong acids. Contact with strong bases produces
ammonia. Corrodes aluminum, brass, copper, zinc and related alloys;
attacks some plastics, rubber, and coatings. May accumulate static
electrical charges, and may cause ignition of its vapors.

AMMONIUM SULFIDE (12124-99-1) Forms explosive mixture with
air (flash point 72°F/22°C). Incompatible with strong oxidizers, strong
acids. Contact with strong bases produces ammonia. Corrodes
aluminum, brass, copper, zinc, and related alloys; attacks some
plastics, rubber and coatings. May accumulate static electrical
charges, and cause ignition of its vapors.

AMMONIUM SULFITE (10196-04-0) Aqueous solution is a base and evolves hydrogen sulfide or ammonia, depending on pH. Acids form hydrogen sulfide; bases form ammonia. Attacks aluminum, brass, bronze, copper, tin, zinc in the presence of moisture.

AMMONIUM SULPHAMATE (7773-06-0) Fires and explosions may result from contact with strong oxidizers, hot water, potassium, sodium, sodium nitrite, metal chlorates, hot acid solutions. Corrosive to mild steel.

AMMONIUM SULPHATE (7783-20-2) Aqueous solution is a strong acid; reacts with bases forming ammonia. Attacks metals. Hot material reacts with nitrates, nitrites, chlorates.

AMMONIUM THIOSULFATE SOLUTION (60% OR LESS) (7783-18-8) Incompatible with sulfuric acid, isocyanates. Incompatible with strong oxidizers, chlorates, nitrates, nitrites (forms ammonia), hydrogen sulfide and sulfur trioxide gases.

AMOIL (131-18-0) Incompatible with strong acids, nitrates. Attacks some plastics, rubber, or coatings.

AMONIAK (Polish) (7664-41-7) **anhydrous (a flammable gas);** (1336-21-6) **solution in water.** Reacts violently with strong oxidizers, acids. Shock-sensitive compounds may be formed with halogens, mercury oxide, silver oxide. Fire and explosions may be caused by trimethylammonium amide, 1-chloro-2,4-dinitrobenzene, o-chloronitrobenzene, platinum, trioxygen difluoride, selenium difluoride dioxide, boron halides, mercury, chlorine, iodine, bromine, hypochlorites, chlorine bleach, amides, organic anhydrides, isocyanates, vinyl acetate, alkylene oxides, epichlorohydrin, aldehydes. Attacks some coatings, plastics and rubber. Attacks copper, brass, bronze, aluminum, steel, and their alloys.

AMORPHOUS PHOSPHORUS, RED (7723-14-0) Contact with ALL oxidizers may cause fire and explosions or form shock-sensitive compounds. Reacts violently with caustics + heat, chlorosulfonic acid + heat, lead dioxide, performic acid, selenium oxychloride, chlorates, and other materials. Incompatible with many other substances, and forms phosphine gas on contact with moisture and oxygen; opened packages should be stored under inert gas blanket.

AMORPHOUS SILICA (68855-54-9) Contact with fluorine, oxygen difluoride, or chlorine trifluoride may cause fire.

AMOXONE (94-75-7) Decomposes in sunlight. Incompatible with strong oxidizers; may cause fire and explosions.

AMP (124-68-5) Forms explosive mixture with air (flash point 153°F/67°C). Incompatible with strong acids, mineral acids, organic acids, organic anhydrides, isocyanates, vinyl acetate, acrylates, substituted allyls, alkylene oxides, epichlorohydrin, aldehydes.

AMP-95 (124-68-5) Forms explosive mixture with air (flash point 153°F/67°C). Incompatible with strong acids, mineral acids, organic acids, organic anhydrides, isocyanates, vinyl acetate, acrylates, substituted allyls, alkylene oxides, epichlorohydrin, aldehydes.

AMS (98-83-9) Forms explosive mixture with air (flash point 129°F/54°C). Reacts with heat and/or lack of appropriate inhibitor concentration, and/or contact with catalysts for vinyl or ionic polymerization, such as aluminum or iron chloride. Incompatible with strong acids, oxidizers (including peroxides and halogens).

AMS (7773-06-0) Fires and explosions may result from contact with strong oxidizers, hot water, potassium, sodium, sodium nitrite, metal chlorates, hot acid solutions. Corrosive to mild steel.

AMSCO TETRAMER (6842-15-5) Forms explosive mixture with air (flash point 144°F/62°C). Incompatible with strong acids, oxidizers.

AMYAZETAT (German) (628-63-7) Incompatible with strong alkalies, strong acids, nitrates, strong oxidizers. Attacks some plastics, coatings, and rubber.

AMYL ACETATE (626-38-0) Incompatible with strong acids, nitrates, strong alkalies, strong oxidizers, heat. May soften or dissolve plastics.

n-**AMYL ACETATE** (628-63-7) Incompatible with strong alkalies, strong acids, nitrates, strong oxidizers. Attacks some plastics, coatings and rubber.

sec-**AMYL ACETATE** (626-38-0) Incompatible with strong acids, nitrates, strong alkalies, strong oxidizers, heat. May soften or dissolve plastics.

tert-**AMYL ACETATE** (625-16-1) Forms explosive mixture with air (flash point 77°F/25°C). Incompatible with strong acids, nitrates, oxidizers. Attacks some plastics, rubber, or coatings.

AMYL ACETATE, MIXED ISOMERS (628-63-7) Incompatible with strong alkalies, strong acids, nitrates, strong oxidizers. Attacks some plastics, coatings, and rubber.

AMYLACETIC ESTER (626-38-0) Incompatible with strong acids, nitrates, strong alkalies, strong oxidizers, heat. May soften or dissolve plastics.

AMYLACETIC ESTER (123-92-2) Forms explosive mixture with air (flash point 77°F/25°C). Incompatible with strong alkalies, strong acids, nitrates, oxidizers. Reacts violently with reducing agents. Attacks asbestos; softens and dissolves many plastics, rubber, and coatings.

AMYL ACETIC ETHER (628-63-7) Incompatible with strong alkalies, strong acids, nitrates, oxidizers.

AMYL ALCOHOL (71-41-0) Forms explosive mixture with air (flash point 91°F/33°C). Incompatible with strong acids, caustics, aliphatic amines, isocyanates. Incompatible with alkali metals and alkaline earth, forming hydrogen gas.

1-AMYL ALCOHOL (71-41-0) Forms explosive mixture with air (flash point 91°F/33°C). Incompatible with strong acids, caustics, aliphatic amines, isocyanates. Incompatible with alkali metals and alkaline earth, forming hydrogen gas.

n-**AMYL ALCOHOL or** *prim-n*-**AMYL ALCOHOL** (71-41-0) Forms explosive mixture with air (flash point 91°F/33°C). Incompatible with strong acids, caustics, aliphatic amines, isocyanates. Incompatible with alkali metals and alkaline earth, forming hydrogen gas.

AMYL ALCOHOL, NORMAL (71-41-0) Forms explosive mixture with air (flash point 91°F/33°C). Incompatible with strong acids, caustics, aliphatic amines, isocyanates. Incompatible with alkali metals and alkaline earth, forming hydrogen gas.

prim-n-**AMYL ALCOHOL** (71-41-0) Forms explosive mixture with air (flash point 91°F/33°C). Incompatible with strong acids, caustics, aliphatic amines, isocyanates. Incompatible with alkali metals and alkaline earth, forming hydrogen gas.

sec-n-AMYL ALCOHOL or *sec-n*-AMYL ALCOHOL (6032-29-7) Forms explosive mixture with air (flash point 93°F/34°C). Reacts violently with strong oxidizers and alkali metals. May also react with alkaline earth metals.

tert-AMYL ALCOHOL (75-85-4) Forms explosive mixture with air (flash point 77°F/25°C). Incompatible with strong acids, caustics, aliphatic amines, isocyanates, strong oxidizers, alkali metals (i.e., lithium, sodium, potassium, rubidium, cesium, francium).

N-AMYLALKOHOL (Czech) (71-41-0) Forms explosive mixture with air (flash point 91°F/33°C). Incompatible with strong acids, caustics, aliphatic amines, isocyanates. Incompatible with alkali metals and alkaline earth, forming hydrogen gas.

AMYL ALDEHYDE (110-62-3) Forms explosive mixture with air (flash point 54°F/12°C). Incompatible with strong acids, oxidizers, caustics, amines.

AMYL BROMIDE (107-81-3) Forms explosive mixture with air (flash point 90°F/32°C). Incompatible with strong oxidizers, strong acids.

AMYL CARBINOL (111-27-3) Forms explosive mixture with air (flash point 145°F/63°C). Incompatible with strong acids, caustics, aliphatic amines, isocyanates, strong oxidizers.

n-AMYL CARBINOL (111-27-3) Forms explosive mixture with air (flash point 145°F/63°C). Incompatible with strong acids, caustics, aliphatic amines, isocyanates, strong oxidizers.

AMYL CHLORIDE (543-59-9) Forms explosive mixture with air (flash point 34°F/1°C). Strong oxidizers may cause fire and explosions.

n-AMYL CHLORIDE (543-59-9) Forms explosive mixture with air (flash point 34°F/1°C). Strong oxidizers may cause fire and explosions.

AMYLENE (109-67-1) Forms explosive mixture with air (flash point 0°F/−8°C). Incompatible with strong acids, oxidizers.

alpha-*n*-AMYLENE (109-67-1) Forms explosive mixture with air (flash point 0°F/−8°C). Incompatible with strong acids, oxidizers.

AMYLENE DIMER (16736-42-8) Forms explosive mixture with air (flash point 118°F/48°C). Reacts violently with strong oxidizers. May accumulate static electrical charges, and may cause ignition of its vapors.

AMYLENE HYDRATE (75-85-4) Forms explosive mixture with air (flash point 77°F/25°C). Incompatible with strong acids, caustics, aliphatic amines, isocyanates, strong oxidizers, alkali metals (i.e., lithium, sodium, potassium, rubidium, cesium, francium).

AMYL ETHER (693-65-2) Forms explosive mixture with air (flash point 134°F/56°C). Reacts violently with strong oxidizers. Ethers can forms unstable and explosive peroxides.

AMYL ETHYL KETONE (541-85-5) Forms explosive mixture with air (flash point 110°F/43°C). Incompatible with sulfuric acid, nitric acid, aliphatic amines, oxidizers.

AMYL HYDRIDE (109-66-0) Forms explosive mixture with air (flash point −57°F/−49°C). Incompatible with strong oxidizers (fire and explosions). Attacks some plastics, rubber, or coatings.

AMYL HYDROSULFIDE (110-66-7) Forms explosive mixture with air (flash point 65°F/18°C). Incompatible with oxidizers, reducing agents, alkali metals, calcium hypochlorite.

AMYL MERCAPTAN (110-66-7) Forms explosive mixture with air (flash point 65°F/18°C). Incompatible with oxidizers, reducing agents, alkali metals, calcium hypochlorite.

n-**AMYL MERCAPTAN** (110-66-7) Forms explosive mixture with air (flash point 65°F/18°C). Incompatible with oxidizers, reducing agents, alkali metals, calcium hypochlorite.

AMYL METHYL ALCOHOL (108-11-2) Forms explosive mixture with air (flash point 106°F/41°C). Contact with alkali metals produces hydrogen gas. Incompatible with strong acids, caustics, aliphatic amines, isocyanates. Attacks some plastics, rubber, and coatings. May accumulate static electrical charges, and may cause ignition of its vapors.

AMYL-METHYL-CETONE (French) (110-43-0) Forms explosive mixture with air (flash point 102°F/39°C). Incompatible with strong acids, alkalies, aliphatic amines, oxidizers. Attacks some plastics and rubber.

AMYL METHYL KETONE (110-43-0) Forms explosive mixture with air (flash point 102°F/39°C). Incompatible with strong acids, alkalies, aliphatic amines, oxidizers. Attacks some plastics and rubber.

n-**AMYL METHYL KETONE** (110-43-0) Forms explosive mixture with air (flash point 102°F/39°C). Incompatible with strong acids, alkalies, aliphatic amines, oxidizers. Attacks some plastics and rubber.

n-**AMYL NITRATE** (1002-16-0) Forms explosive mixture with air (flash point 120°F/49°C). An oxidizer; reacts strongly with reducing agents, strong acids, esters, combustibles. Attacks some plastics, rubber, or coatings.

AMYL NITRITE (463-04-7) Forms explosive mixture with air (flash point 0°F/−18°C). Decomposes in light, air, or water; forms oxides of nitrogen. Strong oxidizers may cause fire and explosions. Corrodes metal with moisture.

iso-**AMYL NITRITE** (110-46-3) Forms explosive mixture with air (flash point 0°F/−18°C). Decomposes in light, air, or water; forms oxides of nitrogen. A strong oxidizer; reacts with reducing agents, combustibles, organics, other strong oxidizers. Corrodes metal with moisture.

AMYLOL (71-41-0) Forms explosive mixture with air (flash point 91°F/33°C). Incompatible with strong acids, caustics, aliphatic amines, isocyanates. Incompatible with alkali metals and alkaline earth, forming hydrogen gas.

beta-AMYLOSE (9004-34-6) Incompatible with water, bromine pentafluoride, hydrogen peroxide, sodium hypochlorite, sodium nitrate, fluorine, strong oxidizers.

p-tert-**AMYLPHENOL** (80-46-6) Reacts with strong oxidizers. May accumulate static electrical charges, and may cause ignition of its vapors.

AMYL PHTHALATE (131-18-0) Incompatible with strong acids, nitrates. Attacks some plastics, rubber, or coatings.

n-**AMYL PROPIONATE** (624-54-4) Forms explosive mixture with air (flash point 106°F/41°C). Incompatible with sulfuric acid, nitric acid, nitrates, oxidizers.

AMYL SULFHYDRATE (110-66-7) Forms explosive mixture with air (flash point 65°F/18°C). Incompatible with oxidizers, reducing agents, alkali metals, calcium hypochlorite.

AMYL THIOALCOHOL (110-66-7) Forms explosive mixture with air (flash point 65°F/18°C). Incompatible with oxidizers, reducing agents, alkali metals, calcium hypochlorite.

AMYL TRICHLOROSILANE (107-72-2) Forms explosive mixture with air (flash point 145°F/62°C). Water contact produces hydrochloric acid. Incompatible with strong bases. Attacks metal in presence of moisture.

n-**AMYL TRICHLOROSILANE** (107-72-2) Forms explosive mixture with air (flash point 145°F/62°C). Water contact produces hydrochloric acid. Incompatible with strong bases. Attacks metal in presence of moisture.

AN (107-13-1) Forms explosive mixture with air (flash point 32°F/0°C o.c.). Forms explosive peroxides; heat, light, caustics, silver nitrate and peroxides can cause polymerization. Incompatible with strong acids, strong oxidizers, amines, 2-aminoethanol, bromine, chlorosulfonic acid, ethylene diamine, nitric acid, oleum, potassium hydroxide, sodium hydroxide, sulfuric acid. Attacks copper and copper alloys; attacks aluminum in high concentrations. May accumulate static electrical charges, and may cause ignition of its vapors.

ANACARDIC ACID (8001-24-7) Incompatible with sulfuric acid, caustics, ammonia, amines, isocyanates, alkylene oxides, epichlorohydrin.

ANAESTHETIC ETHER (60-29-7) Forms explosive mixture with air (flash point −49°F/−45°C). Incompatible with strong acids, strong oxidizers. Can form peroxides with air or light; may explode when container is unstoppered or otherwise opened. Being a nonconductor, chemical may accumulate static electric charges that may result in ignition of vapor.

ANAMENTH (79-01-6) Contact with caustics produces a toxic and flammable gas. Reacts violently with chemically active metals. Contact with aluminum may produce a violent, self-accelerating polymerization reaction. Incompatible with nonoxidizing mineral acids, strong acids, organic acids, organic anhydrides, isocyanates, alkylene oxides, aldehydes, alcohols, glycols, phenols, cresols, caprolactam solution, epichlorohydrin, nitrogen tetroxide, metal powders, oxygen. May accumulate static electrical charges, and may cause ignition of its vapors.

ANCHRED STANDARD (1309-37-1) Contact with hydrogen peroxide, ethylene oxide, calcium hypochlorite will cause explosion. Reacts violently with powdered aluminum, hydrazine, hydrogen trisulfide.

ANESTHENYL (109-87-5) Forms explosive mixture with air (flash point −4°F/−18°C). Incompatible with strong oxidizers, acids. Attacks some plastics, rubber, and coatings.

ANESTHESIA ETHER (60-29-7) Forms explosive mixture with air (flash point −49°F/−45°C). Incompatible with strong acids, strong oxidizers. Can form peroxides with air or light; may explode when container is unstoppered or otherwise opened. Being a nonconductor, chemical may accumulate static electric charges that may result in ignition of vapor.

ANESTHETIC ETHER (60-29-7) Forms explosive mixture with air (flash point −49°F/−45°C). Incompatible with strong acids, strong oxidizers. Can form peroxides with air or light; may explode when container is unstoppered or otherwise opened. Being a nonconductor, chemical may accumulate static electric charges that may result in ignition of vapor.

ANGININE (55-63-0) Liquid portion of this chemical is an alcohol; reacts with strong acids, caustics, aliphatic amines, isocyanates. Dry chemical is nitroglycerine, a shock- and friction-sensitive explosive. Acids, heat, or mechanical shock may result in explosions.

ANGLISITE (7446-14-2) Reacts with aluminum, magnesium, potassium (violently).

ANHYDRIDE ACETIQUE (French) (108-24-7) Forms explosive mixture with air (flash point 121°F/50°C). Water contact produces acetic acid and considerable heat. Reacts violently with potassium hydroxide, sodium hydroxide, alcohols, amines. Incompatible with acids, ammonia, aniline, chromic anhydride, ethylene diamine, glycerol, hydrogen peroxide, nitrogen tetroxide, oxidizers, permanganates, sodium peroxide. Attacks some coatings, some plastics, and rubber. Corrodes iron and steel. May accumulate static electrical charges, and may cause ignition of its vapors.

ANHYDRIDE ARSENIEUX (French) (1327-53-3) Incompatible with fluorine, fluorides, sodium chlorate.

ANHYDRIDE ARSENIQUE (French) (1303-28-2) Incompatible with acids, aluminum, halogens, rubidium carbide, zinc. Aqueous solutions are acidic.

ANHYDRIDE CARBONIQUE (French) (124-38-9) Incompatible with acrylaldehyde, amines, anhydrous ammonia, cesium monoxide, lithium, metal dusts, potassium, sodium, sodium carbide, sodium–potassium alloy, sodium peroxide, titanium.

ANHYDRIDE CHROMIQUE (French) (1333-82-0) A strong oxidizer. Reacts with acetic acid, acetic anhydride, acetone, anthracene, chromous sulfide, diethyl ether, dimethyl formamide, ethanol, hydrogen sulfide, methanol, naphthalene, camphor, glycerol, potassium ferricyanide, pyridine, turpentine, combustibles, organics. Aqueous solution is strongly acidic. Attacks metals in presence of moisture.

ANHYDRIDE PHTALIQUE (French) (85-44-9) Incompatible with strong acids, caustics, ammonia, amines, strong oxidizers. Attacks some plastics, rubber, or coatings.

ANHYDROFLUORIC ACID (7664-39-3) Incompatible with acetic anhydride, aliphatic amines, alcohols, alkanolamines, alkylene oxides, aromatic amines, amides, 2-aminoethanol, ammonia, ammonium hydroxide, arsenic trioxide, bismuthic acid, calcium oxide, ethylene diamine, ethyleneimine, epichlorohydrin, isocyanates, metal acetylides, nitrogen trifluoride, oleum, organic anhydrides, oxygen difluoride, phosphorus pentoxide, sulfuric acid, sodium hydroxide and other bases, strong oxidizers, vinyl acetate, vinylidene fluoride. Attacks glass, concrete, ceramics, metals, some plastics, rubber, and coatings.

ANHYDROL (64-17-5) Forms explosive mixture with air (flash point 65°F/18°C). May accumulate static electrical charges, and may cause ignition of its vapors. Reactions may be violent with oleum, sulfuric acid, nitric acid, bases, aliphatic amines, isocyanates, oxidizers.

ANHYDRONE (10034-81-8) A powerful oxidizer; reacts violently with reducing agents, organic matter, ethylene oxide, powdered metals, phosphorus, dimethylsulfoxide. Forms explosive material with ethyl alcohol. Incompatible with many materials, including nonoxidizing mineral acids.

ANHYDROUS ALUMINUM CHLORIDE (7446-70-0) Contact with air or water produces hydrochloric acid. Water, alcohol, alkenes cause polymerization. Incompatible with nitrobenzene, organic material, and bases. Attacks metal in presence of moisture.

ANHYDROUS AMMONIA (7664-41-7) **anhydrous (a flammable gas); (1336-21-6) solution in water.** Reacts violently with strong oxidizers, acids. Shock-sensitive compounds may be formed with halogens, mercury oxide, silver oxide. Fire and explosions may be caused by trimethylammonium amide, 1-chloro-2,4-dinitrobenzene, o-chloronitrobenzene, platinum, trioxygen difluoride, selenium difluoride dioxide, boron halides, mercury, chlorine, iodine, bromine, hypochlorites, chlorine bleach, amides, organic anhydrides, isocyanates, vinyl acetate, alkylene oxides, epichlorohydrin, aldehydes. Attacks some coatings, plastics and rubber. Attacks copper, brass, bronze, aluminum, steel, and their alloys.

ANHYDROUS BORIC ACID (1303-86-2) Water contact produces boric acid. Fires and explosion may result from contact with bromine pentafluoride or calcium oxide.

ANHYDROUS CALCIUM SULFATE (7778-18-9) Contact with diazomethane, aluminum, phosphorus may cause explosion.

ANHYDROUS CHLORAL (75-87-6) Contact with oxidizers may cause fire and explosions.

ANHYDROUS GYPSUM (7778-18-9) Contact with diazomethane, aluminum, phosphorus may cause explosion.

ANHYDROUS HYDROBROMIC ACID (10035-10-6) Incompatible with aliphatic amines, alkanolamines, alkylene oxides, aromatic amines, amides, ammonia, ammonium hydroxide, bases, calcium oxide, epichlorohydrin, fluorine, isocyanates, oleum, organic anhydrides, sulfuric acid, sodium tetrahydroborate, strong oxidizers, vinyl acetate, water. Attacks most metals with the formation of flammable hydrogen gas.

ANHYDROUS HYDROCHLORIC ACID (7647-01-0) Incompatible with acetic anhydride, aliphatic amines, alkanolamines, alkylene oxides, aromatic amines, amides, 2-aminoethanol, ammonia, ammonium hydroxide, calcium phosphide, chlorosulfonic acid, ethylene diamine, ethyleneimine, epichlorohydrin, isocyanates, metal acetylides, oleum, organic anhydrides, perchloric acid, 3-propiolactone, uranium phosphide, sulfuric acid, sodium hydroxide and other bases, strong oxidizers, vinyl acetate, vinylidene fluoride. Attacks most metals and some plastics, rubber, and coatings.

ANHYDROUS HYDROGEN FLUORIDE (7664-39-3) Incompatible with acetic anhydride, aliphatic amines, alcohols, alkanolamines, alkylene oxides, aromatic amines, amides, 2-aminoethanol, ammonia, ammonium hydroxide, arsenic trioxide, bismuthic acid, calcium oxide, ethylene diamine, ethyleneimine, epichlorohydrin, isocyanates, metal acetylides, nitrogen trifluoride, oleum, organic anhydrides, oxygen difluoride, phosphorus pentoxide, sulfuric acid, sodium hydroxide and other bases, strong oxidizers, vinyl acetate, vinylidene fluoride. Attacks glass, concrete, ceramics, metals, some plastics, rubber, and coatings.

ANHYDROUS IRON OXIDE (1309-37-1) Contact with hydrogen peroxide, ethylene oxide, calcium hypochlorite will cause explosion. Reacts violently with powdered aluminum, hydrazine, hydrogen trisulfide.

ARMENIAN BOLE (1309-37-1) Contact with hydrogen peroxide, ethylene oxide, calcium hypochlorite will cause explosion. Reacts violently with powdered aluminum, hydrazine, hydrogen trisulfide.

ANHYDROUS SULFATE OF LIME (7778-18-9) Contact with diazomethane, aluminum, phosphorus may cause explosion.

ANIDRIDE ACETICA (Italian) (108-24-7) Forms explosive mixture with air (flash point 121°F/50°C). Water contact produces acetic acid and considerable heat. Reacts violently with potassium hydroxide, sodium hydroxide, alcohols, amines. Incompatible with acids, ammonia, aniline, chromic anhydride, ethylene diamine, ethylenediamine, glycerol, hydrogen peroxide, nitrogen tetroxide, oxidizers, permanganates, sodium peroxide. Attacks some coatings, some plastics, and rubber. Corrodes iron and steel. May accumulate static electrical charges, and may cause ignition of its vapors.

ANIDRIDE CHROMICA (Italian) (1333-82-0) A strong oxidizer. Reacts with acetic acid, acetic anhydride, acetone, anthracene, chromous sulfide, diethyl ether, dimethyl formamide, ethanol, hydrogen sulfide, methanol, naphthalene, camphor, glycerol, potassium ferricyanide, pyridine, turpentine, combustibles, organics. Aqueous solution is strongly acidic. Attacks metals in presence of moisture.

ANIDRIDE FTALICA (Italian) (85-44-9) Incompatible with strong acids, caustics, ammonia, amines, strong oxidizers. Attacks some plastics, rubber, or coatings.

ANILINE (62-53-3) Forms explosive mixture with air (flash point 158°F/70°C). Unless inhibited (usually by methanol), readily able to polymerize. Fires and explosions may result from contact with halogens, strong acids, oxidizers, organic anhydrides, acetic anhydride, isocyanates, aldehydes, sodium peroxide. Reacts with alkali metals and alkali earth metals. Attacks some plastics, rubber, and coatings; copper and copper alloys.

ANILINE, 2,6-DIETHYL (579-66-8) Incompatible with strong acids, mineral acids, organic anhydrides, isocyanates, aldehydes, strong oxidizers.

ANILINE, N,N-DIMETHYL- (121-69-7) Forms explosive mixture with air (flash point 145°F/63°C). Incompatible with strong oxidizers (fire and explosions), strong acids (violent spattering). Contact with benzoyl peroxide, diisopropyl perdicarbonate, and other materials may cause explosions. Attacks some plastics, rubber, and coatings.

ANILINE, 2,6-DIMETHYL (87-62-7) Forms explosive mixture with air (flash point 206°F/96.7°C). Incompatible with strong acids (fire and explosions); hypochlorite bleaches form explosive chloroamines. Incompatible with organic acids and anhydrides, isocyanates, aldehydes. Attacks some plastics, rubber, or coatings.

ANILINE, HEXAHYDRO- (108-91-8) Forms explosive mixture with air (flash point 79°F/26°C). Incompatible with nonoxidizing mineral acids, strong acids, organic acids, organic anhydrides, isocyanates, vinyl acetate, acrylates, substituted allyls, alkylene oxides, epichloro-

hydrin, ketones, aldehydes, alcohols, glycols, phenols, cresols, caprolactam solution, strong oxidizers. Contact with copper alloys, zinc or
galvanized steel may cause violent reaction.

ANILINE, 3-METHYL- (108-44-1) Forms explosive mixture with air
(flash point 188°F/87°C). Incompatible with strong oxidizers, strong
acids. Attacks some plastics, rubber, or coatings.

ANILINE OIL (62-53-3) Forms explosive mixture with air (flash point
158°F/70°C). Unless inhibited (usually by methanol), readily able to
polymerize. Fires and explosions may result from contact with
halogens, strong acids, oxidizers, organic anhydrides, acetic anhydride, isocyanates, aldehydes, sodium peroxide. Reacts with alkali
metals and alkali earth metals. Attacks some plastics, rubber,
coatings, copper and copper alloys.

ANILINE, N-PHENYL (122-39-4) Incompatible with strong acids,
organic anhydrides, isocyanates, ketones, oxidizers, hexachloromelamine, trichloromelamine.

ANILINOBENZENE (122-39-4) Incompatible with strong acids,
organic anhydrides, isocyanates, ketones, oxidizers, hexachloromelamine, trichloromelamine.

ANILINOMETHANE (100-61-8) Incompatible with strong acids,
strong oxidizers. Attacks some plastics, rubber, and coatings. May
accumulate static electrical charges, and may cause ignition of its
vapors.

*o-***ANISIC ACID** (119-36-0) Incompatible with strong acids, nitrates,
strong oxidizers.

ANISOYL CHLORIDE (100-07-2) May be unstable at room temperature. Water contact produces hydrochloric acid. Attacks metals in the
presence of moisture, forming hydrogen gas.

*p-***ANISOYL CHLORIDE** (100-07-2) May be unstable at room
temperature. Water contact produces hydrochloric acid. Attacks
metals in the presence of moisture, forming hydrogen gas.

ANNAMENE (100-42-5) If inhibitor (often *tert*-butylcatechol) is not
present in adequate concentrations, polymerization may occur and
explode containers. Temperatures above 150°F/66°C speed up
polymerization. Incompatible with oxidizers, acids, rust, catalysts for
vinyl polymerization, such as peroxides, strong acids, and aluminum
chloride. Corrodes copper and copper alloys, and dissolves rubber.
Attacks some plastics, rubber, or coatings. May accumulate static
electrical charges, and may cause ignition of its vapors.

ANODYNON (75-00-3) Flammable gas. Forms explosive mixture
with air. Contact with moisture produces hydrochloric acid. May
accumulate static electrical charges, and may cause ignition of its
vapors. Forms explosive mixture with air (flash point −58°F/−50°C).
Contact with chemically active metals may cause fire and explosions.
Attacks some plastics and rubber.

ANOFEX (50-29-3) Incompatible with salts of iron or aluminum, and
bases. Do not store in iron containers.

ANOL (108-94-1) Forms explosive mixture with air (flash point
111°F/44°C). Incompatible with aliphatic amines, sulfuric acid, nitric
acid, oxidizers, strong acids. Dissolves plastics, resins, and rubber.

ANOL (108-93-0) Forms explosive mixture with air (flash point
154°F/68°C). Incompatible with strong acids, caustics, aliphatic
amines, isocyanates. Attacks some plastics, rubber, or coatings.

ANONE (108-94-1) Forms explosive mixture with air (flash point 111°F/44°C). Incompatible with aliphatic amines, sulfuric acid, nitric acid, oxidizers, strong acids. Dissolves plastics, resins, and rubber.

ANOZOL (84-66-2) Reacts violently with strong acids, strong oxidizers, permanganates, water. Attacks some forms of plastic.

ANPROLENE (75-21-8) Forms explosive mixture with air (flash point 20°F/−6°C). Incompatible with alkali metal hydroxides, highly active catalysts (e.g., anhydrous chlorides of iron, tin, or aluminum and oxides of iron or aluminum). Avoid contact with copper. Protect container from physical damage, sun, and heat. Attacks some plastics, rubber, or coatings.

ANSAR (75-60-5) Aqueous solution reacts with chemically active metals. Incompatible with sulfuric acid, caustics, ammonia, amines, isocyanates, alkylene oxides, epichlorohydrin.

ANSAR 160 (124-65-2) Corrodes common metals.

2-ANSIDINE (90-04-0) Incompatible with strong oxidizers, with risk of fire or explosions. Attacks some coatings, and some forms of plastic and rubber.

o-**ANISIDINE or** *ortho*-**ANISIDINE** (90-04-0) Incompatible with strong oxidizers, with risk of fire or explosions. Attacks some coatings, and some forms of plastic and rubber.

o-**ANISYLAMINE** (90-04-0) Incompatible with strong oxidizers, with risk of fire or explosions. Attacks some coatings, and some forms of plastic and rubber.

ANSUL ETHER 121 (110-71-4) Incompatible with sulfuric acid, isocyanates, strong oxidizers.

ANTAK (112-30-1) Forms explosive mixture with air (flash point 180°F/82°C). Incompatible with strong acids, caustics, aliphatic amines, isocyanates, strong oxidizers.

ANTHIUM DIOXCIDE (10049-04-4) Water contact produces perchloric and hydrochloric acids with toxic and corrosive fumes. Explodes on contact with many substances, including carbon monoxide, hydrogen, mercury, potassium, phosphorus, strong caustics, sulfur, granulated sugar. A strong oxidizer. May decompose explosively from shock, friction, or heat: Will detonate if heated rapidly to 212°F (100°C) Contact with reducing agents, dust and other combustible materials, organic matter, sulfur may cause fires and explosions. Attacks some forms of plastics, rubber, and coatings.

ANTHON (52-68-6) Contact with strong oxidizers may cause fire and explosions.

ANTHRACEN (German) (120-12-7) As dust or fine powder, forms an explosive mixture with air. Strong oxidizers may cause fire and explosions.

ANTHRACENE (120-12-7) As dust or fine powder, forms an explosive mixture with air. Strong oxidizers may cause fire and explosions.

ANTHRACIN (120-12-7) As dust or fine powder, forms an explosive mixture with air. Strong oxidizers may cause fire and explosions.

ANTIBULIT (7681-49-4) Water contact produces a corrosive. Reacts with acids.

ANTICARIE (118-74-1) Reacts violently with strong oxidizers, formyldimethylamine.

ANTIFORMIN (7681-52-9) A powerful oxidizer and a strong base. Decomposes in sunlight, forming oxygen and increasing the risk of fire. Stability decreases with concentration, heat, light, decrease in pH, and contamination with metals. Incompatible with strong acids, reducing agents, combustible substances, all causing violent reaction, fire, and explosions. Contact with amines and ammonia salts produces explosive chloroamines. Corrodes many metals: steel, 12% and 17% chrome steel, cast iron, monel, aluminum, nickel, brass, bronze, iconel.

ANTIKNOCK-33 (12108-13-3) Incompatible with strong acids, strong oxidizers.

ANTIMIMONWASSERSTOFFES (German) (7803-52-3) Forms explosive mixture with air. Incompatible with nitric acid, ozone, halogenated hydrocarbons, oxidizers, and moisture.

ANTIMOINE FLUORURE (French) (7783-56-4) Contact with strong oxidizers may cause fire and explosions. Explosive mixture may be formed with perchloric acid.

ANTIMOINE (TRICHLORURE d') **(French)** (10025-91-9) Water contact produces corrosive solution. Air contact produces corrosive vapor.

ANTIMONIC CHLORIDE (7647-18-9) Water contact produces hydrogen chloride gas. Reacts violently with ammonia, caustics. Attacks many metals. Contact with air produces heavier-than-air corrosive vapor.

ANTIMONIO (PENTACLORURO di) (Italian) (7647-18-9) Water contact produces hydrogen chloride gas. Reacts violently with ammonia, caustics. Attacks many metals. Contact with air produces heavier-than-air corrosive vapor.

ANTIMONIO (TRICHLORURO di) (Italian) (10025-91-9) Water contact produces corrosive solution. Air contact produces corrosive vapor.

ANTIMONOUS BROMIDE (7789-61-9) Contact with air, alcohol or water produces corrosive hydrogen bromide.

ANTIMONOUS CHLORIDE (10025-91-9) Water contact produces corrosive solution. Air contact produces corrosive vapor.

ANTIMONOUS FLUORIDE (7783-56-4) Contact with strong oxidizers may cause fire and explosions. Explosive mixture may be formed with perchloric acid.

ANTIMONPENTACHLORID (German) (7647-18-9) Water contact produces hydrogen chloride gas. Reacts violently with ammonia, caustics. Attacks many metals. Contact with air produces heavier-than-air corrosive vapor.

ANTIMONY or ANTIMONY BLACK (7440-36-0) Reacts violently with strong oxidizers, strong acids. Forms explosive mixtures with chloric and perchloric acid. Oxidizers, acids, especially halogenated acids, produce toxic stibine gas (antimony hydride). Mixtures with nitrates or halogenated compounds cause combustion.

ANTIMONY BUTTER (10025-91-9) Water contact produces corrosive solution. Air contact produces corrosive vapor.

ANTIMONY(III) CHLORIDE (10025-91-9) Water contact produces corrosive solution. Air contact produces corrosive vapor.

ANTIMONY(V) CHLORIDE, ANTIMONY PERCHLORIDE (7647-18-9) Water contact produces hydrogen chloride gas. Reacts violently with ammonia, caustics. Attacks many metals. Contact with air produces heavier-than-air corrosive vapor.

ANTIMONY FLUORIDE (7783-70-2) Water contact produces hydrofluoric acid. Incompatible with reducing agents, organics, phosphates, siliceous materials. Severe corrosion of glass or metals (with formation of explosive hydrogen gas) occurs in presence of moisture.

ANTIMONY(III) FLUORIDE (1:3) (7783-56-4) Contact with strong oxidizers may cause fire and explosions. Explosive mixture may be formed with perchloric acid.

ANTIMONY(5+) FLUORIDE (7783-70-2) Water contact produces hydrofluoric acid. Incompatible with reducing agents, organics, phosphates, siliceous materials. Severe corrosion of glass or metals (with formation of explosive hydrogen gas) occurs in presence of moisture.

ANTIMONY(V) FLUORIDE (7783-70-2) Water contact produces hydrofluoric acid. Incompatible with reducing agents, organics, phosphates, siliceous materials. Severe corrosion of glass or metals (with formation of explosive hydrogen gas) occurs in presence of moisture.

ANTIMONY HYDRIDE (7803-52-3) Forms explosive mixture with air. Incompatible with nitric acid, ozone, halogenated hydrocarbons, oxidizers, and moisture.

ANTIMONY PENTACHLORIDE (7647-18-9) Water contact produces hydrogen chloride gas. Reacts violently with ammonia, caustics. Attacks many metals. Contact with air produces heavier-than-air corrosive vapor.

ANTIMONY PENTAFLUORIDE (7783-70-2) Water contact produces hydrofluoric acid. Incompatible with reducing agents, organics, phosphates, siliceous materials. Severe corrosion of glass or metals (with formation of explosive hydrogen gas) occurs in presence of moisture.

ANTIMONY(5+) PENTAFLUORIDE (7783-70-2) Water contact produces hydrofluoric acid. Incompatible with reducing agents, organics, phosphates, siliceous materials. Severe corrosion of glass or metals (with formation of explosive hydrogen gas) occurs in presence of moisture.

ANTIMONY(V) PENTAFLUORIDE (7783-70-2) Water contact produces hydrofluoric acid. Incompatible with reducing agents, organics, phosphates, siliceous materials. Severe corrosion of glass or metals (with formation of explosive hydrogen gas) occurs in presence of moisture.

ANTIMONY, REGULUS (7440-36-0) Reacts violently with strong oxidizers, strong acids. Forms explosive mixtures with chloric and perchloric acid. Oxidizers, acids, especially halogenated acids, produce toxic stibine gas (antimony hydride). Mixtures with nitrates or halogenated compounds cause combustion.

ANTIMONY TRIBROMIDE (7789-61-9) Contact with air, alcohol, or water produces corrosive hydrogen bromide.

ANTIMONY TRICHLORIDE (10025-91-9) Water contact produces corrosive solution. Air contact produces corrosive vapor.

ANTIMONY TRIFLUORIDE (7783-56-4) Contact with strong oxidizers may cause fire and explosions. Explosive mixture may be formed with perchloric acid.

68

ANTIMONY TRIHYDRIDE (7803-52-3) Forms explosive mixture with air. Incompatible with nitric acid, ozone, halogenated hydrocarbons, oxidizers, and moisture.

ANTIMOONPENTACHLORIDE (Dutch) (7647-18-9) Water contact produces hydrogen chloride gas. Reacts violently with ammonia, caustics. Attacks many metals. Contact with air produces heavier-than-air corrosive vapor.

ANTIMOONTRICHLORIDE (Dutch) (10025-91-9) Water contact produces corrosive solution. Air contact produces corrosive vapor.

ANTIREN (110-85-0) Forms explosive mixture with air (flash point 178°F/81°C). Incompatible with nitrogen compounds, carbon tetrachloride, strong oxidizers. Aqueous solutions react with acids. Attacks aluminum, copper, nickel, magnesium and zinc.

ANTI-RUST (7632-00-0) A strong oxidizer; reacts violently with reducing agents and combustibles. Reacts with acids, ammonium salts + heat, butadiene, cyanides, lithium, phthalic acid, thiosulfate, sodium amide, and many other substances. Under certain conditions the chemical can also be a strong reducing agent.

ANTISAL LA (108-88-3) Forms explosive mixture with air (flash point 40°F/4°C). Strong oxidizers may cause fire and explosions. Attacks some plastics, rubber, or coatings. May accumulate static electrical charges, and may cause ignition of its vapors.

ANTU (86-88-4) Strong oxidizers may cause fires and explosion. Also reacts with silver nitrate.

ANTURAT (86-88-4) Strong oxidizers may cause fires and explosion. Also reacts with silver nitrate.

ANTYMON (Polish) (7440-36-0) Reacts violently with strong oxidizers, strong acids. Forms explosive mixtures with chloric and perchloric acid. Oxidizers, acids, especially halogenated acids, produce toxic stibine gas (antimony hydride). Mixtures with nitrates or halogenated compounds cause combustion.

ANTYMONOWODOR (Polish) (7803-52-3) Forms explosive mixture with air. Incompatible with nitric acid, ozone, halogenated hydrocarbons, oxidizers, and moisture.

2 AP (504-29-0) Reacts with strong oxidizers, strong acids, sodium nitrate.

APARASIN (58-89-9) Not combustible, but may be dissolved in a combustible solvent. If solvent comes in contact with oxidizers, fire and explosions may result.

APAVAP (62-73-7) Attacks some plastics, rubber, and coatings.

APAVINPHOS (7786-34-7) Contact with strong oxidizers may cause fire and explosions. Attacks some plastics, rubber, and coatings.

APHAMITE (56-38-2) A combustible liquid. Mixtures with endrin may be explosive. Strong oxidizers may cause fire and explosions. Attacks some plastics, rubber, or coatings.

APHTIRIA (58-89-9) Not combustible, but may be dissolved in a combustible solvent. If solvent comes in contact with oxidizers, fire and explosions may result.

APLIDAL (58-89-9) Not combustible, but may be dissolved in a combustible solvent. If solvent comes in contact with oxidizers, fire and explosions may result.

AQUA AMMONIA (7664-41-7) **anhydrous (a flammable gas)**; (1336-21-6) **solution in water.** Reacts violently with strong oxidizers, acids. Shock-sensitive compounds may be formed with halogens, mercury oxide, silver oxide. Fire and explosions may be caused by trimethylammonium amide, 1-chloro-2,4-dinitrobenzene, *o*-chloronitrobenzene, platinum, trioxygen difluoride, selenium difluoride dioxide, boron halides, mercury, chlorine, iodine, bromine, hypochlorites, chlorine bleach, amides, organic anhydrides, isocyanates, vinyl acetate, alkylene oxides, epichlorohydrin, aldehydes. Attacks some coatings, plastics, and rubber. Attacks copper, brass, bronze, aluminum, steel, and their alloys.

AQUACIDE (85-00-7) Concentrated solution corrodes aluminum.

AQUA FORTIS (7697-37-2) A strong oxidizer that can react violently with reducing agents, combustible materials. Incompatible with many substances including acrylates, aliphatic amines, alcohols, aldehydes, alkanolamines, alkylene oxides, anion exchange resins, aromatic amines, amides, bases, cresols, cyanides, cyclic ketones, epichlorohydrin, glycols, isocyanates, ketones, oleum, organic anhydrides, phenols, substituted allyls, sulfuric acid, strong oxidizers, terpenes. Attacks most metals, and some plastics, rubber and coatings.

AQUA-KLEEN (94-75-7) Decomposes in sunlight. Incompatible with strong oxidizers; may cause fire and explosions.

AQUALIN (107-02-8) Forms explosive mixture with air (flash point −15°F/−26°C). Unstable and very reactive. In storage can form heat- and shock-sensitive compounds. Unless inhibited (usually by hydroquinone), readily forms explosive peroxides. Able to polymerize. A strong reducing agent. Reacts violently with oxidizers, strong acids, caustics, amines, 2-aminoethanol, ammonia, ammonium hydroxide, ethylene diamine, ethyleneimine, hydroxides, metal salts, oxidizers, sulfur dioxide, thiourea. Attacks metals: cadmium and zinc. May accumulate static electrical charges, and may cause ignition of its vapors.

AQUALINE (107-02-8) Unstable and very reactive. In storage can form heat- and shock-sensitive compounds. Unless inhibited (usually by hydroquinone), readily forms explosive peroxides that cause polymerization. Incompatible with strong acids, caustics, amines (aliphatic, alkanol, aromatic), 2-aminoethanol, ammonia, ammonium hydroxide, chlorosulfonic acid, ethylene diamine, ethyleneimine, hydroxides, metal salts, oleum, oxidizers, sulfur dioxide, thiourea. Attacks metals: cadmium and zinc.

AQUA-VEX (93-72-1) Contact with strong oxidizers may cause fire and explosions.

AQUEOUS AMMONIA (1336-21-6) Incompatible with strong acids, mineral acids, organic acids, amides, organic anhydrides, isocyanates, vinyl acetate, epichlorohydrin, aldehydes. Corrosive to copper and aluminum, including their alloys, and galvanized surfaces.

AQUEOUS HYDROGEN CHLORIDE (7647-01-0) Incompatible with acetic anhydride, aliphatic amines, alkanolamines, alkylene oxides, aromatic amines, amides, 2-aminoethanol, ammonia, ammonium hydroxide, calcium phosphide, chlorosulfonic acid, ethylene diamine, ethyleneimine, epichlorohydrin, isocyanates, metal acetylides, oleum, organic anhydrides, perchloric acid, 3-propiolactone, uranium phosphide, sulfuric acid, sodium hydroxide and other bases, strong oxidizers, vinyl acetate, vinylidene fluoride. Attacks most metals and some plastics, rubber, and coatings.

AQUEOUS HYDROGEN FLUORIDE (7664-39-3) Incompatible with acetic anhydride, aliphatic amines, alcohols, alkanolamines, alkylene oxides, aromatic amines, amides, 2-aminoethanol, ammonia, ammonium hydroxide, arsenic trioxide, bismuthic acid, calcium oxide, ethylene diamine, ethyleneimine, epichlorohydrin, isocyanates, metal acetylides, nitrogen trifluoride, oleum, organic anhydrides, oxygen difluoride, phosphorus pentoxide, sulfuric acid, sodium hydroxide and other bases, strong oxidizers, vinyl acetate, vinylidene fluoride. Attacks glass, concrete, ceramics, metals, some plastics, rubber, and coatings.

AR-TOLUENOL (1319-77-3) Incompatible with strong acids, oxidizers, alkalies, aliphatic amines, amides, chlorosulfonic acid, oleum. Liquid attacks some plastics, coatings, and rubber.

ARACHIS OIL (8002-03-7) Incompatible with strong acids, oxidizers.

ARAGONITE (1317-65-3) Incompatible with acids, alum, ammonium salts, fluorine.

ARALO (56-38-2) A combustible liquid. Mixtures with endrin may be explosive. Strong oxidizers may cause fire and explosions. Attacks some plastics, rubber, or coatings.

ARASAN (137-26-8) Combustible solid (flash point 192°F/89°C). Strong oxidizers may cause fire and explosions; contact with strong acid or oxidizable materials produces toxic gases.

ARBITEX (58-89-9) Not combustible, but may be dissolved in a combustible solvent. If solvent comes in contact with oxidizers, fire and explosions may result.

ARBOCEL (9004-34-6) Incompatible with water, bromine pentafluoride, hydrogen peroxide, sodium hypochlorite, sodium nitrate, fluorine, strong oxidizers.

ARCOSOLV (34590-94-8) Ethers, as a class, can form dangerous peroxides on standing. Strong oxidizers may cause fire and explosions. Attacks some plastics, rubber, and coatings.

ARCTON-3 (75-72-9) Reacts with aluminum, magnesium, zinc, and their alloys.

ARCTON-4 (75-45-5) Moisture causes slow decomposition. Attacks some plastics, rubber, or coatings. Thermal decomposition occurs at high temperature with alkalies and alkaline earth metals.

ARCTON-6 (75-71-8) Reacts violently with liquid aluminum. Incompatible with chemically active metals. Attacks some plastics, rubber, and coatings.

ARCTON-7 (75-43-4) Incompatible with chemically active metals. Attacks some plastics, rubber, and coatings.

ARCTON-9 (75-69-4) Incompatible with chemically active metals. Attacks some plastics, rubber, and coatings.

ARCTON-63 (76-13-1) Incompatible with chemically active metals. Contact with alloys containing more than 2% magnesium may cause decomposition (hydrogen chloride, hydrogen fluoride, and carbon monoxide released). Attacks some plastics, rubber, or coatings.

ARCTUVIN (123-31-9) Incompatible with strong oxidizers, caustics. May be oxidized to quinone at room temperatures in the presence of moisture. May explode on contact with oxygen.

ARGENTIC FLUORIDE (7783-95-1) Contact with acetylene produces shock-sensitive material. Ammonia contact produces compounds that are explosive when dry. Hydrogen peroxide causes violent decomposition to oxygen gas. Soluble silver compounds attack some forms of plastics, rubber, and coatings.

ARGENTOUS FLUORIDE (7783-95-1) Contact with acetylene produces shock-sensitive material. Ammonia contact produces compounds that are explosive when dry. Hydrogen peroxide causes violent decomposition to oxygen gas. Soluble silver compounds attack some forms of plastics, rubber and coatings.

ARGENTOUS OXIDE (20667-12-3) A strong oxidizer; reacts with reducing agents, acids, amines, combustible materials. Reacts violently with ammonia.

ARGEZIN (1912-24-9) Incompatible with strong acids.

ARILATE (17804-35-2) Heat, water, strong acids, and strong alkalies can cause decomposition and formation of toxic gas (oxides of nitrogen).

ARIZOLE (8002-09-3) Forms explosive mixture with air (flash point 138°F/59°C). Incompatible with strong acids, strong oxidizers.

ARKLONE P (76-13-1) Incompatible with chemically active metals. Contact with alloys containing more than 2% magnesium may cause decomposition (hydrogen chloride, hydrogen fluoride, and carbon monoxide released). Attacks some plastics, rubber, or coatings.

ARKOTINE (50-29-3) Incompatible with salts of iron or aluminum, and bases. Do not store in iron containers.

AROCLOR (1336-36-3) Incompatible with strong oxidizers, strong acids.

AROCLOR 1016 (12674-11-2) (generic CAS for PCBs 1336-36-3) Incompatible with strong oxidizers, strong acids.

AROCLOR 1221 (1104-28-2) (generic CAS for PCBs 1336-36-3) Incompatible with strong oxidizers, strong acids.

AROCLOR 1232 (11141-16-5) (generic CAS for PCBs 1336-36-3) Incompatible with strong oxidizers, strong acids.

AROCLOR 1242 (53469-21-9) (generic CAS for PCBs 1336-36-3) Incompatible with strong oxidizers, strong acids.

AROCLOR 1248 (12672-29-6) (generic CAS for PCBs 1336-36-3) Incompatible with strong oxidizers, strong acids.

AROCLOR 1254 (11097-69-1) (generic CAS for PCBs 1336-36-3) Incompatible with strong oxidizers, strong acids.

AROCLOR 1260 (11096-82-5) (generic CAS for PCBs 1336-36-3) Incompatible with strong oxidizers, strong acids.

AROCLOR 1262 (37324-23-5) (generic CAS for PCBs 1336-36-3) Incompatible with strong oxidizers, strong acids.

AROCLOR 1268 (11100-14-4) (generic CAS for PCBs 1336-36-3) Incompatible with strong oxidizers, strong acids.

AROCLOR 2565 (37324-24-6) (generic CAS for PCBs 1336-36-3) Incompatible with strong oxidizers, strong acids.

AROCLOR 4465 (11120-29-9) (generic CAS for PCBs 1336-36-3) Incompatible with strong oxidizers, strong acids.

AROMATIC SOLVENT (8032-32-4) Forms explosive mixture with air (flash point −40°F to −86°F/−40°C to −66°C). Incompatible with strong acids, strong oxidizers. Attacks some plastics, rubber, and coatings. May accumulate static electrical charges, and may cause ignition of its vapors.

AROSOL (122-99-6) Incompatible with sulfuric acid, isocyanates, strong oxidizers.

ARSEN (German, Polish) (7440-38-2) Incompatible with strong acids, strong oxidizers, peroxides, bromine pentafluoride, bromine trifluoride, cesium acetylene carbide, chromium trioxide, nitrogen trichloride, silver nitrate.

ARSENATE (7778-39-4) Incompatible with sulfuric acid, caustics, ammonia, amines, isocyanates, alkylene oxides, oxidizers, epichlorohydrin, vinyl acetate, amides. Corrodes metals.

ARSENATE OF LEAD (7784-40-9) Contact with strong oxidizers may cause fire and explosions.

ARSENIC (7440-38-2) Incompatible with strong acids, strong oxidizers, peroxides, bromine pentafluoride, bromine trifluoride, cesium acetylene carbide, chromium trioxide, nitrogen trichloride, silver nitrate.

ARSENIC 75 (7440-38-2) Incompatible with strong acids, strong oxidizers, peroxides, bromine pentafluoride, bromine trifluoride, cesium acetylene carbide, chromium trioxide, nitrogen trichloride, silver nitrate.

ARSENIC ACID (7778-39-4) Incompatible with sulfuric acid, caustics, ammonia, amines, isocyanates, alkylene oxides, oxidizers, epichlorohydrin, vinyl acetate, amides. Corrodes metals.

o-**ARSENIC ACID** (7778-39-4) Incompatible with sulfuric acid, caustics, ammonia, amines, isocyanates, alkylene oxides, oxidizers, epichlorohydrin, vinyl acetate, amides. Corrodes metals.

ARSENIC ACID ANHYDRIDE (1303-28-2) Incompatible with acids, aluminum, halogens, rubidium carbide, zinc. Aqueous solutions are acidic.

ARSENICALS (7440-38-2) Incompatible with strong acids, strong oxidizers, peroxides, bromine pentafluoride, bromine trifluoride, cesium acetylene carbide, chromium trioxide, nitrogen trichloride, silver nitrate.

ARSENIC ANHYDRIDE (1303-28-2) Incompatible with acids, aluminum, halogens, rubidium carbide, zinc. Aqueous solutions are acidic.

ARSENIC BLACK (7440-38-2) Incompatible with strong acids, strong oxidizers, peroxides, bromine pentafluoride, bromine trifluoride, cesium acetylene carbide, chromium trioxide, nitrogen trichloride, silver nitrate.

ARSENIC BLANC (French) (1327-53-3) Incompatible with fluorine, fluorides, sodium chlorate.

ARSENIC CHLORIDE (7784-34-1) Water contact produces corrosive mixture of acids. Light exposure produces toxic gas. Reacts violently with anhydrous ammonia, strong acids, strong oxidizers, halogens. Corrodes metals in the presence of moisture.

ARSENIC HYDRIDE (7784-42-1) Strong oxidizers, nitric acid may cause fire and explosions. Light exposure causes decomposition in the presence of moisture. May accumulate static electrical charges, and may cause ignition of its vapors.

ARSENIC, METALLIC (7440-38-2) Incompatible with strong acids, strong oxidizers, peroxides, bromine pentafluoride, bromine trifluoride, cesium acetylene carbide, chromium trioxide, nitrogen trichloride, silver nitrate.

ARSENIC OXIDE (1303-28-2) Incompatible with acids, aluminum, halogens, rubidium carbide, zinc. Aqueous solutions are acidic.

ARSENIC(III) OXIDE (1327-53-3) Incompatible with acids, fluorine, fluorides, sodium chlorate.

ARSENIC(V) OXIDE (1303-28-2) Incompatible with acids, aluminum, halogens, rubidium carbide, zinc. Aqueous solutions are acidic.

ARSENIC PENTAOXIDE (1303-28-2) Incompatible with acids, aluminum, halogens, rubidium carbide, zinc. Aqueous solutions are acidic.

ARSENIC PENTOXIDE (1303-28-2) Incompatible with acids, aluminum, halogens, rubidium carbide, zinc. Aqueous solutions are acidic.

ARSENIC SESQUIOXIDE (1327-53-3) Incompatible with acids, fluorine, fluorides, sodium chlorate.

ARSENIC SESQUISULFIDE (1303-33-9) Water contact produces hydrogen sulfide. Incompatible with oxidizers, acids, halogens.

ARSENIC, SOLID (7440-38-2) Contact with hydrogen gas produces arsine. Incompatible with strong acids, strong oxidizers, peroxides, bromine pentafluoride, bromine trifluoride, cesium acetylene carbide, chromium trioxide, nitrogen trichloride, silver nitrate.

ARSENIC SULFIDE (1303-33-9) Water contact produces hydrogen sulfide. Incompatible with oxidizers, acids, halogens.

ARSENIC TRICHLORIDE (7784-34-1) Water contact produces corrosive mixture of acids. Light exposure produces toxic gas. Reacts violently with anhydrous ammonia, strong acids, strong oxidizers, halogens. Corrodes metals in the presence of moisture.

ARSENIC TRIHYDRIDE (7784-42-1) Strong oxidizers, nitric acid may cause fire and explosions. Light exposure causes decomposition in the presence of moisture. May accumulate static electrical charges, and may cause ignition of its vapors.

ARSENIC TRIOXIDE (1327-53-3) Incompatible with acids, fluorine, fluorides, sodium chlorate.

ARSENIC TRISULFIDE (1303-33-9) Water contact produces hydrogen sulfide. Incompatible with oxidizers, acids, halogens.

ARSENICUM ALBUM (1327-53-3) Incompatible with acids, fluorine, fluorides, sodium chlorate.

ARSENIC YELLOW (1303-33-9) Water contact produces hydrogen sulfide. Incompatible with oxidizers, acids, halogens.

ARSENIGEN SAURE (German) (1327-53-3) Incompatible with acids, fluorine, fluorides, sodium chlorate.

ARSENIOUS ACID (1327-53-3) Incompatible with acids, fluorine, fluorides, sodium chlorate.

ARSENIOUS OXIDE (1327-53-3) Incompatible with acids, fluorine, fluorides, sodium chlorate.

ARSENIOUS TRIOXIDE (1327-53-3) Incompatible with acids, fluorine, fluorides, sodium chlorate.

ARSENITE (1327-53-3) Incompatible with acids, fluorine, fluorides, sodium chlorate.

ARSENIURETTED HYDROGEN (7784-42-1) Strong oxidizers, nitric acid may cause fire and explosions. Light exposure causes decomposition in the presence of moisture. May accumulate static electrical charges, and may cause ignition of its vapors.

ARSENOLITE (1327-53-3) Incompatible with acids, fluorine, fluorides, sodium chlorate.

ARSENOUS ACID (1327-53-3) Incompatible with acids, fluorine, fluorides, sodium chlorate.

ARSENOUS ACID ANHYDRIDE (1327-53-3) Incompatible with acids, fluorine, fluorides, sodium chlorate.

ARSENOUS ANHYDRIDE (1327-53-3) Incompatible with acids, fluorine, fluorides, sodium chlorate.

ARSENOUS CHLORIDE (7784-34-1) Water contact produces corrosive mixture of acids. Light exposure produces toxic gas. Reacts violently with anhydrous ammonia, strong acids, strong oxidizers, halogens. Corrodes metals in the presence of moisture.

ARSENOUS HYDRIDE (Polish) (7784-42-1) Strong oxidizers, nitric acid may cause fire and explosions. Light exposure causes decomposition in the presence of moisture. May accumulate static electrical charges, and may cause ignition of its vapors.

ARSENOUS OXIDE (1327-53-3) Incompatible with acids, fluorine, fluorides, sodium chlorate.

ARSENOUS OXIDE ANHYDRIDE (1327-53-3) Incompatible with acids, fluorine, fluorides, sodium chlorate.

ARSENOUS TRICHLORIDE (7784-34-1) Water contact produces corrosive mixture of acids. Light exposure produces toxic gas. Reacts violently with anhydrous ammonia, strong acids, strong oxidizers, halogens. Corrodes metals in the presence of moisture.

ARSENTRIOXIDE (1327-53-3) Incompatible with acids, fluorine, fluorides, sodium chlorate.

ARSENWASSERSTOFF (German) (7784-42-1) Strong oxidizers, nitric acid may cause fire and explosions. Light exposure causes decomposition in the presence of moisture. May accumulate static electrical charges, and may cause ignition of its vapors.

ARSICODILE (124-65-2) Corrodes common metals.

ARSINE (7784-42-1) Strong oxidizers, nitric acid may cause fire and explosions. Light exposure causes decomposition in the presence of moisture. May accumulate static electrical charges, and may cause ignition of its vapors.

ARSODENT (1327-53-3) Incompatible with acids, fluorine, fluorides, sodium chlorate.

ARSYCODILE (124-65-2) Corrodes common metals.

ARTHODIBROM (300-76-5) Contact with strong oxidizers may cause fire and explosions. Attacks some plastics, rubber, and coatings.

ARTIC (74-87-3) Incompatible with chemically active metals. Attacks plastics, rubber, and coatings.

ARTIFICIAL ALMOND OIL (100-52-7) Forms explosive mixture with air (flash point 145°F/63°C). A strong reducing agent. Forms peroxides with air. Reacts violently with strong acids, aluminum, caustics, ammonia, amines, iron, strong oxidizers. Attacks some plastics, rubber and coatings.

ARTIFICIAL ANT OIL (98-01-1) Forms explosive mixture with air (flash point 140°F/60°C). Incompatible with strong acids, caustics, ammonia, aliphatic amines, alkanolamines, aromatic amines, oxidizers. Attacks many plastics and coatings.

ARTIFICIAL BARITE (7727-43-7) Explosions may result from contact with aluminum in the presence of heat. Incompatible with potassium, phosphorus.

ARTIFICIAL CINNABAR (1344-48-5) Contact with acids or water evolves flammable hydrogen sulfide, which forms an explosive mixture with air. Contact with strong oxidizers may cause a violent reaction.

ARTIFICIAL HEAVY SPAR (7727-43-7) Explosions may result from contact with aluminum in the presence of heat. Incompatible with potassium, phosphorus.

ASA (50-78-2) Fires and explosions may result from contact with strong oxidizers. Alkali hydroxides or carbonates cause decomposition.

A.S.A. EMPRIN (50-78-2) Fires and explosions may result from contact with strong oxidizers. Alkali hydroxides or carbonates cause decomposition.

ASATARD (50-78-2) Fires and explosions may result from contact with strong oxidizers. Alkali hydroxides or carbonates cause decomposition.

ASEX (7775-09-9) A powerful oxidizer; reacts violently with reducing agents and combustible matter. Explosions may be caused by contact with ammonia salts, carbon, oils, metal sulfides, nitrobenzene, powdered metals, sugar. Contact with strong acids produces carbon dioxide. Forms shock-sensitive mixtures with some organic materials. Solution (50%) decomposes at 300°F(149°C), liberating oxygen.

ASP 47 (3689-24-5) Containers may burst in heat. Strong oxidizers may cause fire and explosions. Attacks some plastics, rubber, or coatings.

ASPHALT (8052-42-4) Incompatible with nitric acid, fluorine, strong oxidizers.

ASPHALT BITUMEN (8052-42-4) Incompatible with nitric acid, fluorine, strong oxidizers.

ASPHALT BLENDING STOCKS: ROOFERS FLUX (8052-42-4) Incompatible with nitric acid, fluorine, strong oxidizers.

ASPHALT BLENDING STOCKS: STRAIGHT RUN RESIDUE (8052-42-4) Incompatible with nitric acid, fluorine, strong oxidizers.

ASPHALT CEMENTS (8052-42-4) Incompatible with nitric acid, fluorine, strong oxidizers.

ASPHALT (CUT BACK) (8052-42-4) Incompatible with nitric acid, fluorine, strong oxidizers.

ASPHALTIC BITUMEN (8052-42-4) Incompatible with nitric acid, fluorine, strong oxidizers.

ASPHALT, PETROLEUM ASPHALTUM (8052-42-4) Incompatible with nitric acid, fluorine, strong oxidizers.

ASPHALTUM (8052-42-4) Incompatible with nitric acid, fluorine, strong oxidizers.

ASPHALTUM OIL (8052-42-4) Incompatible with nitric acid, fluorine, strong oxidizers.

ASPIRIN (50-78-2) Fires and explosions may result from contact with strong oxidizers. Alkali hydroxides or carbonates cause decomposition.

ASPIRINE (50-78-2) Fires and explosions may result from contact with strong oxidizers. Alkali hydroxides or carbonates cause decomposition.

ASPON-CHLORDANE (57-74-9) Contact with strong oxidizers may cause fire and explosions. Attacks some plastics, rubber, and coatings.

ASPRO (50-78-2) Fires and explosions may result from contact with strong oxidizers. Alkali hydroxides or carbonates cause decomposition.

ASTERIC (50-78-2) Fires and explosions may result from contact with strong oxidizers. Alkali hydroxides or carbonates cause decomposition.

ASTROBOT (62-73-7) Attacks some plastics, rubber, and coatings.

ASUNTHOL (56-72-4) Contact with strong oxidizers may cause fire and explosions.

ASUNTOL (56-72-4) Contact with strong oxidizers may cause fire and explosions.

ATAZINAX (1912-24-9) Incompatible with strong acids.

ATE (97-93-8) A dangerous explosion hazard. Ignites spontaneously, at all temperatures, in air. A powerful reducing agent; reacts violently with oxidizers, water (forms ethane gas), alcohols, amines, carbon dioxide, carbon tetrachloride, halogenated hydrocarbons, oxides of nitrogen or sulfur, phenols, and many other substances. May accumulate static electrical charges, and may cause ignition of its vapors.

ATGARD (62-73-7) Attacks some plastics, rubber, and coatings.

ATHYLEN (German) (74-85-1) Incompatible with strong acids, oxidizers, halogen acids, aluminum chloride, bromotrichloromethane, carbon tetrachloride, chlorine, chlorine dioxide, nitrogen dioxide.

ATHYLENGLYKOL-MONOATHYLATHER (German) (110-80-5) Forms explosive mixture with air (flash point 120°F/49°C). Strong oxidizers may cause fire and explosions. Attacks some plastics, rubber, and coatings.

ATLACIDE (7775-09-9) A powerful oxidizer; reacts violently with reducing agents and combustible matter. Explosions may be caused by contact with ammonia salts, carbon, oils, metal sulfides, nitrobenzene, powdered metals, sugar. Contact with strong acids produces carbon dioxide. Forms shock-sensitive mixtures with some organic materials. Solution (50%) decomposes at 300°F(149°C), liberating oxygen.

ATM 2 (NYLON) (105-60-2) Contact with strong oxidizers may cause fire and explosions.

ATOMIT (1317-65-3) Incompatible with acids, alum, ammonium salts, fluorine.

ATRANEX (1912-24-9) Incompatible with strong acids.

ATRASINE (1912-24-9) Incompatible with strong acids.

ATRATOL A (1912-24-9) Incompatible with strong acids.

ATRATOL B-HERBATOX (7775-09-9) A powerful oxidizer; reacts violently with reducing agents and combustible matter. Explosions may be caused by contact with ammonia salts, carbon, oils, metal sulfides, nitrobenzene, powdered metals, sugar. Contact with strong acids produces carbon dioxide. Forms shock-sensitive mixtures with some organic materials. Solution (50%) decomposes at 300°F/(149°C), liberating oxygen.

ATRAZIN (1912-24-9) Incompatible with strong acids.

ATRAZINE (1912-24-9) Incompatible with strong acids.

ATRED (1912-24-9) Incompatible with strong acids.

ATROMBINE-K (81-81-2) Strong oxidizers may cause fire and explosions.

ATTAC-2 (8001-35-2) Reacts with oxidizers, with a risk of fire or explosions. Attacks metals in the presence of moisture.

ATTAC 6 (8001-35-2) Reacts with oxidizers, with a risk of fire or explosions. Attacks metals in the presence of moisture.

ATTAC 6-3 (8001-35-2) Reacts with oxidizers, with a risk of fire or explosions. Attacks metals in the presence of moisture.

AULES (137-26-8) Combustible solid (flash point 192°F/89°C). Strong oxidizers may cause fire and explosions; contact with strong acid or oxidizable materials produces toxic gases.

AUSTRALENE (80-56-8) Incompatible with sulfuric acid, nitric acid.

AVICEL (9004-34-6) Incompatible with water, bromine pentafluoride, hydrogen peroxide, sodium hypochlorite, sodium nitrate, fluorine, strong oxidizers.

AVITROL (504-24-2) Forms a strong base with water. Incompatible with strong oxidizers, acids.

AVLOTHANE (67-72-1) Incompatible with hot iron, zinc, and aluminum; alkalies form spontaneously explosive chloroacetylene. Attacks some plastics, rubber and coatings.

AVOLIN (131-11-3) Strong alkalies, strong acids, nitrates, oxidizers.

AWPA 1 (65996-93-2) Incompatible with oxidizers, strong acids, caustics, aliphatic amines, isocyanates.

9-AZAANTHRACENE (260-94-6) Contact with strong oxidizers may cause fire and explosions.

10-AZAANTHRACENE (260-94-6) Contact with strong oxidizers may cause fire and explosions.

AZABENZENE (110-86-1) Forms explosive mixture with air (flash point 68°F/20°C). Strong oxidizers may cause fire and explosions. Strong acids may cause violent spattering. Attacks some plastics, rubber, or coatings.

AZACYCLOHEPTANE (111-49-9) Forms explosive mixture with air (flash point 99°F/37°C). Incompatible with nonoxidizing mineral acids, strong acids, organic acids, organic anhydrides, isocyanates, vinyl acetate, acrylates, substituted allyls, alkylene oxides, epichlorohydrin, ketones, aldehydes, alcohols, glycols, phenols, cresols, caprolactam solution, strong oxidizers. Attacks aluminum, copper, lead, tin, zinc, and alloys.

1-AZACYCLOHEPTANE (111-49-9) Forms explosive mixture with air (flash point 99°F/37°C). Incompatible with nonoxidizing mineral acids, strong acids, organic acids, organic anhydrides, isocyanates, vinyl acetate, acrylates, substituted allyls, alkylene oxides, epichlorohydrin, ketones, aldehydes, alcohols, glycols, phenols, cresols, caprolactam solution, strong oxidizers. Attacks aluminum, copper, lead, tin, zinc, and alloys.

AZACYCLOPROPANE (151-56-4) Forms explosive mixture with air. Contact with acid, aluminum, carbon dioxide, or silver may cause explosive polymerization. Attacks rubber, coatings, and plastics. Self-reactive with heat or atmospheric carbon dioxide.

1-AZANAPHTHALENE (91-22-5) Contact with strong oxidizers may cause fire and explosions. Attacks some plastics, rubber, and coatings.

3-AZAPENTANE-1,5-DIAMINE (111-40-0) Ignites spontaneously with cellulose nitrate. Silver, cobalt, or chromium compounds may cause explosions. Forms explosive mixture with air (flash point 208°F/98°C). Incompatible with nonoxidizing mineral acids, strong acids, organic acids, organic anhydrides, isocyanates, vinyl acetate, acrylates, substituted allyls, alkylene oxides, epichlorohydrin, ketones, aldehydes, alcohols, glycols, mercury, phenols, cresols, caprolactam solution, strong oxidizers. Attacks aluminum, copper, lead, tin, zinc, and alloys.

AZDEL (9003-07-0) Incompatible with strong acids, strong oxidizers, chlorine, potassium permanganate.

AZIDE (26628-22-8) Incompatible with acids, with some metals (i.e, lead, copper, silver, mercury). Forms explosion-sensitive compounds.

AZIJNZUUR (Dutch) (64-19-7) Vapor forms explosive mixture with air above 105°F/40°C. Reacts violently with oxidizers and bases. Incompatible with strong acids, aliphatic amines, alkanolamines, isocyanates, alkylene oxides, epichlorohydrin, acetaldehyde, 2-aminoethanol, ammonia, ammonium nitrate, chlorosulfonic acid, chromic acid, ethylene diamine, ethyleneimine, perchloric acid, permanganates, phosphorus isocyanate, phosphorus trichloride, potassium *tert*-butoxide, xylene. Attacks cast iron and other metals forming flammable hydrogen gas.

AZIJNZUURANHYDRIDE (Dutch) (108-24-7) Forms explosive mixture with air (flash point 121°F/50°C). Water contact produces acetic acid and considerable heat. Reacts violently with potassium hydroxide, sodium hydroxide, alcohols, amines. Incompatible with acids, ammonia, aniline, chromic anhydride, ethylene diamine, ethylenediamine, glycerol, hydrogen peroxide, nitrogen tetroxide, oxidizers, permanganates, sodium peroxide. Attacks some coatings, some plastics, and rubber. Corrodes iron and steel. May accumulate static electrical charges, and may cause ignition of its vapors.

AZIMETHYLENE (334-88-3) Heat, shock, friction, sunlight or other bright illuminations may cause explosions. Contact with alkali metals, drying agents (e.g., calcium sulfate), or rough edges (e.g., ground glass) will cause explosions.

AZINE (110-86-1) Forms explosive mixture with air (flash point 68°F/20°C). Strong oxidizers may cause fire and explosions. Strong acids may cause violent spattering. Attacks some plastics, rubber, or coatings.

AZINPHOS-METHYL (86-50-0) Strong oxidizers may cause fire and explosions.

AZINPHOSMETILE (Italian) (86-50-0) Strong oxidizers may cause fire and explosions.

AZIRANE (151-56-4) Forms explosive mixture with air. Contact with acids, aluminum, carbon dioxide, or silver may cause explosive polymerization. Attacks rubber, coatings and plastics. Self-reactive with heat or atmospheric carbon dioxide.

AZIRIDINE (151-56-4) Forms explosive mixture with air. Contact with acids, aluminum, carbon dioxide, or silver may cause explosive polymerization. Attacks rubber, coatings and plastics. Self-reactive with heat or atmospheric carbon dioxide.

AZIRIDINE, 2-METHYL- (6CI, 8CI, 9CI) (75-55-8) Forms explosive mixture with air (flash point 25°F/−4°C). Contact with acids or high heat can cause violent polymerization. Strong oxidizers may cause fire and explosions. Attacks some plastics, rubber, or coatings.

AZIRINE (151-56-4) Forms explosive mixture with air. Contact with acids, aluminum, carbon dioxide, or silver may cause explosive polymerization. Attacks rubber, coatings, and plastics. Self-reactive with heat or atmospheric carbon dioxide.

1H-AZIRINE, DIHYDRO- (151-56-4) Forms explosive mixture with air. Contact with acids, aluminum, carbon dioxide, or silver may cause explosive polymerization. Attacks rubber, coatings, and plastics. Self-reactive with heat or atmospheric carbon dioxide.

AZIUM (26628-22-8) Incompatible with acids, some metals (i.e, lead, copper, silver, mercury). Forms explosion-sensitive compounds.

AZOENE FAST ORANGE GR SALT (88-74-4) *ortho;* (99-09-2) *meta.* Incompatible with strong acids, combustibles, organics, moisture.

AZOFOS (298-00-0) Reacts explosively with magnesium.

AZOGEN DEVELOPER-H (95-80-7) Incompatible with oxidizers, acids, organic anhydrides, isocyanates, aldehydes. Attacks aluminum, brass, bronze, copper, zinc.

AZOIC DIAZO COMPONENT 6 (88-74-4) *ortho;* (99-09-2) *meta.* Incompatible with strong acids, combustibles, organics, moisture.

AZOIC DIAZO COMPONENT 37 (100-01-06) Strong oxidizers and moisture may cause spontaneous heating. Caustics and heat may form explosive mixtures. Attacks some plastics, rubber, and coatings. May accumulate static electrical charges, and may cause ignition of its vapors.

AZOPHOS (298-00-0) May react explosively with magnesium.

AZOTIC ACID (7697-37-2) A strong oxidizer that can react violently with reducing agents, combustible materials. Incompatible with many substances, including acrylates, aliphatic amines, alcohols, aldehydes, alkanolamines, alkylene oxides, anion exchange resins, aromatic amines, amides, bases, cresols, cyanides, cyclic ketones, epichlorohydrin, glycols, isocyanates, ketones, oleum, organic anhydrides, phenols, substituted allyls, sulfuric acid, strong oxidizers, terpenes. Attacks most metals, some plastics, rubber, and coatings.

AZOTOWY KWAS (Polish) (7697-37-2) A strong oxidizer that can react violently with reducing agents, combustible materials. Incompatible with many substances, including acrylates, aliphatic amines, alcohols, aldehydes, alkanolamines, alkylene oxides, anion exchange resins, aromatic amines, amides, bases, cresols, cyanides, cyclic ketones, epichlorohydrin, glycols, isocyanates, ketones, oleum, organic anhydrides, phenols, substituted allyls, sulfuric acid, strong oxidizers, terpenes. Attacks most metals, some plastics, rubber, and coatings.

AZOTOX (50-29-3) Incompatible with salts of iron or aluminum, and bases. Do not store in iron containers.

AZOTURE de SODIUM (French) (26628-22-8) Incompatible with acids, some metals (i.e., lead, copper, silver, mercury). Forms explosion-sensitive compounds.

AZUNTHOL (56-72-4) Contact with strong oxidizers may cause fire and explosions.

- B -

B (98-51-1) Forms explosive mixture with air (flash point 155°F/68°C). Incompatible with nitric acid, strong oxidizers.

B 404 (56-38-2) Combustible liquid. Mixtures with endrin may be explosive. Strong oxidizers may cause fire and explosions. Attacks some plastics, rubber, and coatings.

BABULUM OIL (N/A) Incompatible with nitric acid; oxidizers may cause fire and explosions.

BACILLOL (1319-77-3) Incompatible with strong acids, oxidizers, alkalies, aliphatic amines, amides, chlorosulfonic acid, oleum. Liquid attacks some plastics, coatings, and rubber.

BAKER'S P AND S LIQUID or BAKER'S OINTMENT (108-95-2) Forms explosive mixture with air (flash point 174°F/79°C). Incompatible with strong oxidizers, strong acids, caustics, aliphatic amines, amides, oxidizers, formaldehyde, butadiene, calcium hypochlorite. Liquid attacks some plastics, rubber and coatings; hot liquid attacks aluminum, magnesium, lead and zinc metals.

BAKONTAL (7727-43-7) Explosions may result from contact with aluminum in the presence of heat. Incompatible with potassium, phosphorus.

BAN-MITE (121-75-5) Incompatible with strong oxidizers, magnesium, alkaline pesticides. Attacks metals, some plastics, rubber, and coatings.

BANANA OIL (123-92-2) Forms explosive mixture with air (flash point 77°F/25°C). Incompatible with strong alkalies, strong acids, nitrates, oxidizers. Reacts violently with reducing agents. Attacks asbestos; softens and dissolves many plastics, rubber, and coatings.

BANANA OIL (626-38-0) Incompatible with strong acids, nitrates, strong alkalies, strong oxidizers, heat. May soften or dissolve plastics.

BANEX (1918-00-9) Incompatible with sulfuric acid, bases, ammonia, aliphatic amines, alkanolamines, isocyanates, alkylene oxides, epichlorohydrin.

BANGTON (133-06-2) Incompatible with tetraethyl pyrophosphate, parathion.

BANLEN (1918-00-9) Incompatible with sulfuric acid, bases, ammonia, aliphatic amines, alkanolamines, isocyanates, alkylene oxides, epichlorohydrin.

BANTU (86-88-4) Strong oxidizers may cause fires and explosion. Also reacts with silver nitrate.

BANVEL (1918-00-9) Incompatible with sulfuric acid, bases, ammonia, aliphatic amines, alkanolamines, isocyanates, alkylene oxides, epichlorohydrin.

BANVEL D (1918-00-9) Incompatible with sulfuric acid, bases, ammonia, aliphatic amines, alkanolamines, isocyanates, alkylene oxides, epichlorohydrin.

BANVEL HERBICIDE (1918-00-9) Incompatible with sulfuric acid, bases, ammonia, aliphatic amines, alkanolamines, isocyanates, alkylene oxides, epichlorohydrin.

BAP (50-32-8) Strong oxidizers may cause fires and explosions.

BARIDOL (7727-43-7) Explosions may result from contact with aluminum in the presence of heat. Incompatible with potassium, phosphorus.

BARIO (PEROSSIDO di) (Italian) (1304-29-6) Reacts with water, forming oxygen. A strong oxidizer. Incompatible with combustibles, reducing agents, acids, finely divided metals. Contact with organic substances produces shock- and friction-sensitive compounds. Attacks metals in the presence of moisture.

BARITE (7727-43-7) Explosions may result from contact with aluminum in the presence of heat. Incompatible with potassium, phosphorus.

BARITOP (7727-43-7) Explosions may result from contact with aluminum in the presence of heat. Incompatible with potassium, phosphorus.

BARIUM (7440-39-3) Can self-ignite in air. Contact with water causes rapid decomposition and possible ignition. Fire and explosions may be caused by halogenated solvents, carbon tetrachloride, trichlorotrifluoroethane, acids, or oxidizers.

BARIUM AZIDE (18810-58-7) Explodes when heated or shocked. Reacts violently with acids, heavy metals such as lead, carbon disulfide, and shock. Keeping chemical wet greatly reduces its fire and explosion hazard.

BARIUM BINOXIDE (1304-29-6) Reacts with water, forming oxygen. A strong oxidizer. Incompatible with combustibles, reducing agents, acids, finely divided metals. Contact with organic substances produces shock- and friction-sensitive compounds. Attacks metals in the presence of moisture.

BARIUM BROMATE (13967-90-3) Keep away from heat, combustible materials, water, sources of ignition, aluminum, arsenic, carbon, copper, phosphorus, and other oxidizable materials.

BARIUM CARBONATE or BARIUM CARBONATE (1:1) (513-77-9) Incompatible with acids, aluminum, fluorine, magnesium, silicon, oxidizers.

BARIUM CHLORATE (13477-00-4) 1 A powerful oxidizer. Incompatible with sulfuric acid, oleum, combustibles, reducing agents. Contact with ammonium compounds or powdered metals produces shock- and friction-sensitive compounds that ignite or explode. Also reacts with aluminum, arsenic, copper, cyanides, metal sulfides, organic acids, sulfur, zinc.

BARIUM CHLORATE MONOHYDRATE (13477-00-4) A powerful oxidizer. Incompatible with sulfuric acid, oleum, combustibles, reducing agents. Contact with ammonium compounds or powdered metals produces shock- and friction-sensitive compounds that ignite or explode. Also reacts with aluminum, arsenic, copper, cyanides, metal sulfides, organic acids, sulfur, zinc.

BARIUM CYANIDE (542-62-1) Incompatible with nitric acid, nitrates, nitrites, chlorates, magnesium, strong oxidizers. Corrosive to metals in the presence of moisture.

BARIUM DICYANIDE (542-62-1) Incompatible with nitric acid, nitrates, nitrites, chlorates, magnesium, strong oxidizers. Corrosive to metals in the presence of moisture.

BARIUM DINITRATE (10022-31-8) Incompatible with strong acids, aluminum–magnesium alloys, sulfur, combustibles, finely divided metals may form shock-sensitive compounds.

BARIUM DIOXIDE (1304-29-6) Reacts with water, forming oxygen. A strong oxidizer. Incompatible with combustibles, reducing agents, acids, finely divided metals. Contact with organic substances produces shock- and friction-sensitive compounds. Attacks metals in the presence of moisture.

BARIUM HYDROXIDE (17194-00-2) A strong base; reacts violently with acids and chlorine rubber. Attacks chemically active metals.

BARIUM MANGANATE(8+) or BARIUM MANGANATE(VIII) (7787-36-2) May be spontaneously combustible. A strong oxidizer. Incompatible with reducing agents. Mixtures with combustibles may be ignited by friction or acids.

BARIUM METAL (7440-39-3) A strong reducing agent; reacts violently with oxidizers. Can self-ignite in air. Water contact causes rapid decomposition and possible ignition. Fire and explosions may be caused by halogenated solvents, carbon tetrachloride, trichlorotrifluoroethane, acids, or oxidizers.

BARIUM METAL, PROPHORIC (7440-39-3) A strong reducing agent; reacts violently with oxidizers. Can self-ignite in air. Water contact causes rapid decomposition and possible ignition with water. Fire and explosions may be caused by halogenated solvents, carbon tetrachloride, trichlorotrifluoroethane, acids or oxidizers.

BARIUM MONOXIDE (1304-28-5) Reacts with water, hydrogen sulfide, carbon dioxide, hydroxlamine, nitrogen tetroxide, sulfur trioxide, and triuranium.

BARIUM NITRATE (10022-31-8) Incompatible with strong acids, aluminum–magnesium alloys, sulfur, combustibles, finely divided metals; may form shock-sensitive compounds.

BARIUM OXIDE (1304-28-5) Reacts with water, hydrogen sulfide, carbon dioxide, hydroxlamine, nitrogen tetroxide, sulfur trioxide, and triuranium.

BARIUM PERCHLORATE (13465-95-7) A strong oxidizer. Reacts violently with reducing agents and combustibles. Contact with strong acids produces explosive perchloric acid. Contact with ammonium compounds, metal powders, or sulfur produces friction- and impact-sensitive compounds.

BARIUM PERCHLORATE TRIHYDRATE (13465-95-7) A strong oxidizer. Reacts violently with reducing agents and combustibles. With strong acids forms explosive perchloric acid. Contact with ammonium compounds, metal powders or sulfur produces friction- and impact-sensitive compounds.

BARIUM PERMANGANATE (7787-36-2) May be spontaneously combustible. A strong oxidizer. Incompatible with reducing agents. Mixtures with combustibles may be ignited by friction or acids.

BARIUMPEROXID (German) (1304-29-6) Reacts with water, producing oxygen. A strong oxidizer. Reacts violently with combustibles, reducing agents, acids, finely divided metals. Contact with organic substances produces shock- and friction-sensitive compounds. Attacks metals in the presence of moisture.

BARIUM PEROXIDE (1304-29-6) Reacts with water, forming oxygen. A strong oxidizer. Reacts violently with combustibles, reducing agents, acids, finely divided metals. Contact with organic substances produces shock- and friction-sensitive compounds. Attacks metals in the presence of moisture.

BARIUM PEROXYDE (Dutch) (1304-29-6) Reacts with water, producing oxygen. A strong oxidizer. Reacts violently with combustibles, reducing agents, acids, finely divided metals. Contact with organic substances produces shock- and friction-sensitive compounds. Attacks metals in the presence of moisture.

BARIUM PROTOXIDE (1304-28-5) Incompatible with water, hydrogen sulfide, carbon dioxide, hydroxlamine, nitrogen tetroxide, sulfur trioxide, and triuranium.

BARIUM SULFATE or BARIUM SULFATE(1:1) (7727-43-7) Explosions may result from contact with aluminum in the presence of heat. Incompatible with potassium, phosphorus.

BARIUM SULFIDE (21109-95-5) Contact with acids causes decomposition with the formation of toxic hydrogen sulfide gas.

BARIUM SULFITE (7787-39-5) Contact with acids causes decomposition with the formation of toxic hydrogen sulfide gas.

BARIUM SULPHATE (7727-43-7) Explosions may result from contact with aluminum in the presence of heat. Incompatible with potassium, phosphorus.

BARIUM SUPEROXIDE (1304-29-6) Reacts with water, forming oxygen. A strong oxidizer. Reacts violently with combustibles, reducing agents, acids, finely divided metals. Contact with organic substances produces shock- and friction-sensitive compounds. Attacks metals in the presence of moisture.

BARIUM TOXIDE (1304-28-5) Reacts with water, hydrogen sulfide, carbon dioxide, hydroxlamine, nitrogen tetroxide, sulfur trioxide, and triuranium.

BAROSPERSE (7727-43-7) Explosions may result from contact with aluminum in the presence of heat. Incompatible with potassium, phosphorus.

BAROTRAST (7727-43-7) Explosions may result from contact with aluminum in the presence of heat. Incompatible with potassium, phosphorus.

BARSITO (1304-28-5) Reacts with water, hydrogen sulfide, carbon dioxide, hydroxlamine, nitrogen tetroxide, sulfur trioxide, and triuranium.

BARYTA (1304-28-5) Incompatible with water, hydrogen sulfide, carbon dioxide, hydroxlamine, nitrogen tetroxide, sulfur trioxide, and triuranium.

BARYTA WHITE (7727-43-7) Explosions may result from contact with aluminum in the presence of heat. Incompatible with potassium, phosphorus.

BARYTES or BARYTES 22 (7727-43-7) Explosions may result from contact with aluminum in the presence of heat. Incompatible with potassium, phosphorus.

BASFAPON (75-99-0) Corrosive to iron, aluminum, and copper.

BASFAPON B (75-99-0) Corrosive to iron, aluminum, and copper.

BASFAPON/BASFAPON N (75-99-0) Corrosive to iron, aluminum, and copper.

BASINEX (75-99-0) Corrosive to iron, aluminum, and copper.

BASUDIN (333-41-5) Incompatible with water, copper-containing compounds, oxidizers, acids, or bases.

BASUDIN 10 G (333-41-5) Incompatible with water, copper-containing compounds, oxidizers, acids, or bases.

BATTERY ACID (7664-93-9) A strong oxidizer that can react violently with risk of fire and explosion with many substances, including reducing agents, organic and combustible substances, and bases. Incompatible with nonoxidizing mineral acids, organic acids, bases, acrylates, aldehydes, alcohols, alkylene oxides, ammonia, aliphatic amines, alkanolamines, aromatic amines, amides, chlorates, epichlorohydrin, fulminates, glycols, isocyanates, ketones, metals (powdered), organic anhydrides, perchlorates, picrates, substituted allyls, phenols and cresols, water, acetic anhydride, acetone, cyanhydrin, acetonitrile, acrolein, acrylonitrile, allyl alcohol, allyl chloride, 2-aminoethanol, ammonium hydroxide, aniline, bromine pentafluoride, *n*-butyraldehyde, caprolactam solution, carbides, cesium, acetylene carbide, chlorine trifluoride, chlorosulfonic acid, cuprous nitride, diisobutylene, ethylene cyanohydrin, ethylene diamine, ethylene glycol, ethyleneimine, hydrochloric acid, iodine heptafluoride, iron, isoprene, lithium silicide, mercuric nitride, mesityl oxide, nitric acid, *p*-nitrotoluene, perchloric acid, phosphorus, potassium *tert*-butoxide, potassium chlorate, potassium permanganate, 3-propiolactone, propylene oxide, pyridine, rubidium acetylene, silver permanganate, sodium, sodium carbonate, sodium chlorate, sodium hydroxide, styrene monomer, vinyl acetate, Attacks most metals, some plastics, rubber, and coatings.

BAUXITE RESIDUE (1309-37-1) Contact with hydrogen peroxide, ethylene oxide, calcium hypochlorite will cause explosion. Reacts violently with powdered aluminum, hydrazine, hydrogen trisulfide.

BAY 1145 (298-00-0) Mixtures with magnesium may be explosive.

BAY 9026 (2032-65-7) Contact with strong oxidizers may cause fire and explosions.

BAY 9027 (86-50-0) Strong oxidizers may cause fire and explosions.

BAY 10756 (8065-48-3) Forms explosive mixture with air (flash point 113°F/45°C). Incompatible with water, strong oxidizers, caustics.

BAY 15922 (52-68-6) Contact with strong oxidizers may cause fire and explosions.

BAY 19639 (298-04-4) Forms explosive mixture with air (flash point 180°F/82°C). Incompatible with alkalis, strong oxidizers.

BAY 37344 (2032-65-7) Contact with strong oxidizers may cause fire and explosions.

BAY 70143 (1563-66-2) Incompatible with alkaline media, acids or strong oxidizers.

BAY E-601 (298-00-0) Mixtures with magnesium may be explosive.

BAY E-605 (56-38-2) Combustible liquid. Mixtures with endrin may be explosive. Strong oxidizers may cause fire and explosions. Attacks some plastics, rubber, and coatings.

BAYER-E-393 (3689-24-5) Containers may burst in heat. Strong oxidizers may cause fire and explosions. Attacks some plastics, rubber, or coatings.

BAYER E-605 (56-38-2) Combustible liquid. Mixtures with endrin may be explosive. Strong oxidizers may cause fire and explosions. Attacks some plastics, rubber, and coatings.

BAYER 13/59 (52-68-6) Contact with strong oxidizers may cause fire and explosions.

BAYER 21/199 (56-72-4) Contact with strong oxidizers may cause fire and explosions.

BAYER 8169 (8065-48-3) Forms explosive mixture with air (flash point 113°F/45°C). Incompatible with water, strong oxidizers, caustics.

BAYER 17147 (86-50-0) Strong oxidizers may cause fire and explosions.

BAYER 19639 (298-04-4) Forms explosive mixture with air (flash point 180°F/82°C). Incompatible with alkalis, strong oxidizers.

BAYER 37344 (2032-65-7) Contact with strong oxidizers may cause fire and explosions.

BAYMIX (56-72-4) Contact with strong oxidizers may cause fire and explosions.

BAYMIX 50 (56-72-4) Contact with strong oxidizers may cause fire and explosions.

BAYOL F (8012-95-1) Incompatible with nitric acid; oxidizers may cause fire and explosions.

BAYRITES (7727-43-7) Explosions may result from contact with aluminum in the presence of heat. Incompatible with potassium, phosphorus.

BAZUDEN (333-41-5) Incompatible with water, copper-containing compounds, oxidizers, acids, or bases.

BBC (17804-35-2) Heat, water, strong acids and strong alkalies can cause decomposition and formation of toxic oxides of nitrogen.

BBC 12 (96-12-8) Forms explosive mixture with air (flash point 170°F/77°C). Incompatible with oxidizers, chemically active metals (i.e., aluminum, magnesium, and tin alloys). Attacks some rubber materials and coatings.

BBH (58-89-9) Not combustible, but may be dissolved in a combustible solvent. If solvent comes in contact with oxidizers, fire and explosions may result.

BBP (85-68-7) Incompatible with strong acids, nitrates, oxidizers. Destructive to rubber and paint.

BCF-BUSHKILLER (93-76-5) Sealed metal containers may burst in heat above 316°F/158°C. Incompatible with sulfuric acid, bases, ammonia, aliphatic amines, alkanolamines, isocyanates, alkylene oxides, epichlorohydrin.

BCS COPPER FUNGICIDE (7758-98-7) Aqueous solution is an acid. Incompatible with strong bases, hydroxylamine, magnesium.

BEAN SEED PROTECTANT (133-06-2) Incompatible with tetraethyl pyrophosphate, parathion.

BEARING OIL (N/A) Incompatible with nitric acid; oxidizers may cause fire and explosions.

BEET SUGAR (57-50-1) Reacts with potassium hydroxide, strong acids, strong oxidizers.

BEHA (103-23-1) Incompatible with strong acids, nitrates, oxidizers.

BEHP (117-81-7) Incompatible with strong acids, strong alkalies, nitrates, oxidizers.

BELL MINE (1305-62-0) Contact with maleic anhydride, phosphorus, nitroethane, nitromethane, nitroparaffins, or nitropropane may cause explosion.

BELL MINE PULVERIZED LIMESTONE (1317-65-3) Incompatible with acids, alum, ammonium salts, fluorine.

BELT (57-74-9) Contact with strong oxidizers may cause fire and explosions. Attacks some plastics, rubber, and coatings.

1,2-BENENEDICARBOXYLIC ACID, DIHEPTYL ESTER (9CI) (3648-21-3) Incompatible with strong acids, strong alkalies, nitrates, strong oxidizers. Attacks some plastics, rubber, and coatings.

BENFOS (62-73-7) Attacks some plastics, rubber, and coatings.

BEN-HEX (58-89-9) Not combustible, but may be dissolved in a combustible solvent. If solvent comes in contact with oxidizers, fire and explosions may result.

BENLAT or BENLATE (17804-35-2) Heat, water, strong acids and strong alkalies can cause decomposition and formation of toxic oxides of nitrogen.

BENLATE 40 W (17804-35-2) Heat, water, strong acids, and strong alkalies can cause decomposition and formation of toxic oxides of nitrogen.

BENLATE 50 or BENLATE 50 W (17804-35-2) Heat, water, strong acids, and strong alkalies can cause decomposition and formation of toxic oxides of nitrogen.

BENOMYL (17804-35-2) Heat, water, strong acids, and strong alkalies can cause decomposition and formation of toxic oxides of nitrogen.

BENOXYL (94-36-0) Confined storage of dry chemical may lead to decomposition and explosion. A strong oxidant; extremely reactive. Fires and explosion may result from heat or contamination, and from contact with strong acids, combustible materials, oxidizers, acids, bases, alcohols, reducing agents, metals, metal oxides, amines, accelerators, methyl methacrylate, organic matter, lithium aluminum carbide, dimethyl aniline, amines, metallic naphthenates. May attack some plastics, rubber, and coatings. Protect containers from shock and friction.

BENSULFOID (7704-34-9) Combustible solid. Liquid produces sulfur dioxide with air. Reacts violently with strong oxidizers. Forms explosive, shock-sensitive, or pyrophoric mixtures with ammonia, ammonium nitrate, bromates, calcium carbide, charcoal, chlorates, hydrocarbons, iodates, iron. Reacts violently with halogen compounds, sodium, tin, uranium, and other compounds. Attacks steel when moist. May accumulate static electrical charges; may cause ignition of its vapors.

BENTOX 10 (58-89-9) Not combustible, but may be dissolved in a combustible solvent. If solvent comes in contact with oxidizers, fire and explosions may result.

BENZAC (94-36-0) Confined storage of dry chemical may lead to decomposition and explosion. A strong oxidant; extremely reactive. Fires and explosion may result from heat or contamination, and from contact with strong acids, combustible materials, oxidizers, acids, bases, alcohols, reducing agents, metals, metal oxides, amines, accelerators, methyl methacrylate, organic matter, lithium aluminum carbide, dimethyl aniline, amines, metallic naphthenates. May attack some plastics, rubber, and coatings. Protect containers from shock and friction.

BENZAL CHLORIDE (98-87-3) Forms explosive mixture with air (flash point 153°F/67°C). Reacts with acids, bases, strong oxidizers. Forms acid fumes with air. Attacks plastics and coatings.

BENZALDEHYDE (98-88-4) Forms explosive mixture with air (flash point 162°F/72°C). Alcohols, amines above 150°F/66°C speed up polymerization. Incompatible with oxidizers, acids, rust, catalysts for vinyl polymerization, such as peroxides, strong acids, and aluminum chloride. Corrodes copper, copper alloys, and dissolves rubber. Attacks some plastics, rubber, and coatings. May accumulate static electrical charges; may cause ignition of its vapors.

BENZALDEHYDE, alpha-CHLORO- (98-88-4) Forms explosive mixture with air (flash point 162°F/72°C). Water contact may be violent; produces hydrochloric acid. Incompatible with amines, alcohols, alkali metals, dimethylsulfoxide. Attacks metals in the presence of moisture. Attacks some plastics, rubber, and coatings.

1-BENZAZINE (91-22-5) Contact with strong oxidizers may cause fire and explosions. Attacks some plastics, rubber, and coatings.

BENZEEN (Dutch) (71-43-2) Forms explosive mixture with air (flash point 12°F/−11°C).) May accumulate static electrical charges, and may cause ignition of its vapors. Incompatible with strong oxidizers, nitric acid, oxygen, ozone, perchlorates. Attacks some forms of plastics, coatings, and rubber.

BENZELENE (71-43-2) Forms explosive mixture with air (flash point 12°F/−11°C).) May accumulate static electrical charges, and may cause ignition of its vapors. Incompatible with strong oxidizers, nitric acid, oxygen, ozone, perchlorates. Attacks some forms of plastics, coatings, and rubber.

BENZEN (Polish) (71-43-2) Forms explosive mixture with air (flash point 12°F/−11°C).) May accumulate static electrical charges, and may cause ignition of its vapors. Incompatible with strong oxidizers, nitric acid, oxygen, ozone, perchlorates. Attacks some forms of plastics, coatings, and rubber.

BENZENAMINE, 5-CHLORO-2-METHYL- (95-79-4) Incompatible with acids, organic anhydrides, isocyanates, aldehydes, oxidizers.

BENZENAMINE, 4-NITRO- (100-01-06) Strong oxidizers and moisture may cause spontaneous heating. Caustics and heat may produces explosive mixtures. Attacks some plastics, rubber, and coatings. May accumulate static electrical charges, and may cause ignition of its vapors.

BENZENE (71-43-2) Forms explosive mixture with air (flash point 12°F/−11°C).) May accumulate static electrical charges, and may cause ignition of its vapors. Incompatible with strong oxidizers, nitric acid, oxygen, ozone, perchlorates. Attacks some forms of plastics, coatings, and rubber.

BENZENEAMINE (62-53-3) Forms explosive mixture with air (flash point 158°F/70°C). Unless inhibited (usually by methanol), readily able to polymerize. Fires and explosions may result from contact with halogens, strong acids, oxidizers, organic anhydrides, acetic anhydride, isocyanates, aldehydes, sodium peroxide. Reacts with alkali metals and alkali earth metals. Attacks some plastics, rubber, coatings, copper, and copper alloys.

BENZENEAMINE, 2,6-DIETHYL- (9CI) (579-66-8) Incompatible with strong acids, mineral acids, organic anhydrides, isocyanates, aldehydes, strong oxidizers.

BENZENAMINE, N,N-DIMETHYL- (121-69-7) Forms explosive mixture with air (flash point 145°F/63°C). Contact with strong oxidizers may cause fire and explosions; contact with strong acids can cause violent spattering. Contact with benzoyl peroxide, diisopropyl perdicarbonate and other material may cause explosions. Attacks some plastics, rubber, and coatings.

BENZENEAMINE, 4-FLUORO- (9CI) (371-40-4) Forms explosive mixture with air (flash point 165°F/74°C). Incompatible with strong acids, strong oxidizers.

BENZENEAMINE, N-METHYL- (9CI) (100-61-8) Incompatible with strong acids, strong oxidizers. Attacks some plastics, rubber, and coatings. May accumulate static electrical charges, and may cause ignition of its vapors.

BENZENE, ANILINO- (122-39-4) Incompatible with strong acids, organic anhydrides, isocyanates, ketones, oxidizers, hexachloromelamine, trichloromelamine.

BENZENE, (BROMOMETHYL)- (100-39-0) Forms explosive mixture with air (flash point 174°F/79°C). Water contact produces hydrobromic acid. Incompatible with strong oxidizers, bases. Attacks metals, except nickel and lead, in the presence of moisture.

BENZENE CARBALDEHYDE (100-52-7) Forms explosive mixture with air (flash point 145°F/63°C). A strong reducing agent. Forms peroxides with air. Reacts violently with strong acids, aluminum, caustics, ammonia, amines, iron, strong oxidizers. Attacks some plastics, rubber, and coatings.

BENZENECARBINOL or BENZENECARBONAL (100-51-6) Forms explosive mixture with air (flash point 213°F/101°C). Slowly oxidizes in air and oxygen. Incompatible with mineral acids, caustics, aliphatic amines, isocyanates. Reacts violently with strong oxidizers. Corrodes aluminum at high temperature. Attacks some nonfluorinated plastics; may not attack polypropylene.

BENZENECARBONYL CHLORIDE (98-88-4) Forms explosive mixture with air (flash point 162°F/72°C). Water contact may be violent; forms hydrochloric acid. Incompatible with amines, alcohols, alkali metals, dimethylsulfoxide. Attacks metals in the presence of moisture. Attacks some plastics, rubber, and coatings.

BENZENECARBOXYLIC ACID (65-85-0) Incompatible with strong oxidizers, caustics, ammonia, amines, isocyanates.

BENZENE CARCABOXALDEHYDE (100-52-7) Forms explosive mixture with air (flash point 145°F/63°C). A strong reducing agent. Forms peroxides with air. Reacts violently with strong acids, aluminum, caustics, ammonia, amines, iron, strong oxidizers. Attacks some plastics, rubber, and coatings.

BENZENE CHLORIDE (108-90-7) Incompatible with strong oxidizers, dimethyl sulfoxide, powdered sodium, silver perchlorate.

BENZENE, CHLORO- (108-90-7) Incompatible with oxidizers, dimethylsulfoxide, sodium powder, silver perchloride. Attacks some plastics, rubber, and coatings.

BENZENE, CHLOROMETHYL- (100-44-7) Forms explosive mixture with air (flash point 153°F/67°C). Water contact produces hydrogen chloride fume. Strong oxidizers may cause fire and explosions. Violent polymerization may be caused by contact with copper, aluminum, iron, zinc, magnesium, and tin. May accumulate static electrical charges, and may cause ignition of its vapors. Attacks some plastics and rubber.

BENZENE, 1-CHLORO-2-METHYL- (95-49-8) Reacts with oxidizers, with a risk of fire or explosions. (95-49-8) Forms explosive mixture with air (flash point 96°F/36°C). Strong oxidizers may cause fire and explosions. Forms hydrochloric acid with water. Corrodes metal in the presence of moisture. Attacks some plastics, rubber, and coatings.

BENZENE, 1-CHLORO-2-METHYL-BENZENEAMINE, 3-METHYL- (108-44-1) Forms explosive mixture with air (flash point 188°F/87°C). Incompatible with strong oxidizers, strong acids. Attacks some plastics, rubber, and coatings.

BENZENE, CYANO- (100-47-0) Forms explosive mixture with air (flash point 167°F/75°C). Incompatible with strong acids, oxidizers. Attacks some plastics, rubber, and coatings.

BENZENEDIAMINE, AR-METHYL- (95-80-7) Incompatible with oxidizers, acids, organic anhydrides, isocyanates, aldehydes. Attacks aluminum, brass, bronze, copper, zinc.

1,2-BENZENEDICARBOXYLIC ACID (27554-26-3) Incompatible with strong acids, nitrates, oxidizers.

BENZENE-1,3-DICARBOXYLIC ACID (88-99-3) Incompatible with strong acids, nitrates, oxidizers.

1,2-BENZENEDICARBOXYLIC ACID ANHYDRIDE (85-44-9) Incompatible with strong acids, caustics, ammonia, amines, strong oxidizers. Attacks some plastics, rubber, and coatings.

1,2-BENZENEDICARBOXYLIC ACID, DIBUTYL ESTER (84-74-2) Incompatible with strong acids, nitrates, strong oxidizers, alkalies.

O-BENZENEDICARBOXYLIC ACID, DIBUTYL ESTER (84-74-2) Incompatible with strong acids, nitrates, strong oxidizers, alkalies.

BENZENE-O-DICARBOXYLIC ACID DI-N-BUTYL ESTER (84-74-2) Incompatible with strong acids, nitrates, strong oxidizers, strong alkalies.

1,2-BENZENEDICARBOXYLIC ACID, DIETHYL ESTER (84-66-2) Reacts violently with strong acids, strong oxidizers, permanganates, water. Attacks some forms of plastic.

1,2-BENZENEDICARBOXYLIC ACID, DIISOOCTYL ESTER (27554-26-3) Incompatible with strong acids, nitrates, oxidizers.

1,2-BENZENEDICARBOXYLIC ACID, DI-ISONONYL ESTER (88-99-3) Incompatible with strong acids, nitrates, oxidizers.

1,2-BENZENEDICARBOXYLIC ACID, DIMETHYL ESTER (131-11-3) Incompatible with strong alkalies, strong acids, nitrates, oxidizers.

1,2-BENZENE DICARBOXYLIC ACID, DI-(2-METHYLPROPYL)ESTER (84-69-5) Incompatible with strong acids, strong oxidizers, nitrates.

1,2-BENZENEDICARBOXYLIC ACID, DI-*N*-OCTYL ESTER (117-84-0) Water contact causes foaming. Incompatible with strong acids, nitrates.

1,2-BENZENEDICARBOXYLIC ACID, DIPENTYL ESTER (131-18-0) Incompatible with strong acids, nitrates. Attacks some plastics, rubber, and coatings.

1,2-BENZENEDICARBOXYLIC ACID, DI-UNDECYL ESTER (3648-20-2) Incompatible with strong acids, nitrates.

1,4-BENZENE DICARBOXYLIC ACID, METHYL ESTER (9CI) (120-61-6) Incompatible with strong acids, nitrates, strong oxidizers.

1,2-BENZENEDICARBOXYLIC ANHYDRIDE (85-44-9) Incompatible with strong acids, caustics, ammonia, amines, strong oxidizers. Attacks some plastics, rubber, and coatings.

BENZENE, 1,2-DICHLORO- (95-50-1) Forms explosive mixture with air (flash point 151°F/66°C). Incompatible with strong oxidizers, hot aluminum, or aluminum alloy. Attacks some plastics, rubber, and coatings.

BENZENE, 1,4-DICHLORO- (106-46-7) Forms explosive mixture with air (flash point 150°F/66°C). Incompatible with strong oxidizers, metal powders and alkali metals (i.e., lithium, sodium, potassium, rubidium, cesium, francium). Attacks some plastics, rubber, and coatings.

BENZENE, DICHLOROMETHYL- (98-87-3) Forms explosive mixture with air (flash point 153°F/67°C). Reacts with acids, bases, strong oxidizers. Forms acid fumes with air. Attacks plastics and coatings.

BENZENE, *m*-DIHYDROXY- (108-46-3) Reacts violently with strong oxidizers, nitric acid. May accumulate static electrical charges; may cause ignition of its vapors.

BENZENE, O-DIHYDROXY- (120-80-9) Strong oxidizers may cause fire and explosion.

BENZENE,2,4-DIISOCYANATO-1-METHYL- (584-84-9) Incompatible with strong acids, including nonoxidizing mineral and organic acids, caustics, ammonia, amines, amides, alcohols, glycols, caprolactam solution. Water contact causes violent foaming and spattering; produces carbon dioxide and an organic base. Attacks copper and its oxides of nitrogen. A strong oxidizer; reacts with reducing agents, combustibles, organics, and other strong oxidizers. Corrodes metal with moisture.

BENZENE,2,4-DIISOCYANATOMETHYL- (584-84-9) Incompatible with strong acids, including nonoxidizing mineral and organic acids, caustics, ammonia, amines, amides, alcohols, glycols, caprolactam solution. Water contact causes violent foaming and spattering; forms carbon dioxide and an organic base. Attacks copper and its alloys, some plastics including polyethylene and rubber.

BENZENE, DIISOPROPYL (98-51-1) Forms explosive mixture with air (flash point 155°F/68°C). Incompatible with nitric acid, strong oxidizers.

1,2-BENZENEDIOL (120-80-9) Strong oxidizers may cause fire and explosion.

1,3-BENZENEDIOL (108-46-3) Reacts violently with strong oxidizers, nitric acid. Incompatible with acetanilide, albumin, alkalies, antipyrine, camphor, ferric salts, menthol, spirit nitrous ether. Absorbs moisture from air (hygroscopic). May accumulate static electrical charges; may cause ignition of its vapors.

1,4-BENZENEDIOL (123-31-9) Incompatible with strong oxidizers, caustics. May be oxidized to quinone at room temperatures in the presence of moisture. May explode on contact with oxygen.

m-**BENZENEDIOL** or *meta*-**BENZENEDIOL** (108-46-3) Reacts violently with strong oxidizers, nitric acid. Incompatible with acetanilide, albumin, alkalies, antipyrine, camphor, ferric salts, menthol, spirit nitrous ether. Absorbs moisture from air (hygroscopic). May accumulate static electrical charges; may cause ignition of its vapors.

o-**BENZENEDIOL** or *ortho*-**BENZENEDIOL** (120-80-9) Strong oxidizers may cause fire and explosion.

p-**BENZENEDIOL** or *para*-**BENZENEDIOL** (123-31-9) Incompatible with strong oxidizers, caustics. May be oxidized to quinone at room temperatures in the presence of moisture. May explode on contact with oxygen.

BENZENE FLUORIDE (462-06-6) Forms explosive mixture with air (flash point 5°F/−15°C). Incompatible with oxidizers, ammonium nitrate, chromic acid, halogens, hydrogen peroxide, nitric acid. Attacks some plastics, rubber, and coatings.

BENZENE, FLUORO- (462-06-6) Forms explosive mixture with air (flash point 5°F/−15°C). Incompatible with oxidizers, ammonium nitrate, chromic acid, halogens, hydrogen peroxide, nitric acid. Attacks some plastics, rubber, and coatings.

BENZENE HEXACHLORIDE (118-74-1) Reacts violently with strong oxidizers, formyldimethylamine.

BENZENE HEXACHLORIDE, gamma isomer (58-89-9) Not combustible, but may be dissolved in a combustible solvent. If solvent comes in contact with oxidizers, fire and explosions may result.

gamma-BENZENE HEXACHLORIDE (58-89-9) Not combustible, but may be dissolved in a combustible solvent. If solvent comes in contact with oxidizers, fire and explosions may result.

BENZENE, HEXACHLORO- (118-74-1) Reacts violently with strong oxidizers, formyldimethylamine.

BENZENE HEXAHYDRIDE (110-82-7) Forms explosive mixture with air (flash point −4°F/−20°C). Incompatible with oxidizers, nitrogen dioxide.

BENZENE, HEXAHYDRO- (110-82-7) Forms explosive mixture with air (flash point −4°F/−20°C). Incompatible with oxidizers, nitrogen dioxide.

BENZENE, HYDROXY- (108-95-2) Forms explosive mixture with air (flash point 174°F/79°C). Incompatible with strong oxidizers, strong acids, caustics, aliphatic amines, amides, oxidizers, formaldehyde, butadiene, calcium hypochlorite. Liquid attacks some plastics, rubber, and coatings; hot liquid attacks aluminum, magnesium, lead, and zinc metals.

BENZENE, METHYL- (108-88-3) Forms explosive mixture with air (flash point 40°F/4°C). Strong oxidizers may cause fire and explosions. Attacks some plastics, rubber, and coatings. May accumulate static electrical charges; may cause ignition of its vapors.

BENZENE, 1-METHYL-2,4-DINITRO- (121-14-2) Strong oxidizers or caustics may cause fire and explosions. Incompatible with nitric acid; produces an explosive material. Sodium oxide contact causes ignition. Attacks metals.

BENZENE, (1-METHYLETHYL-)- (98-82-8) Forms explosive mixture with air (flash point 99°F/37°C). Incompatible with strong acids, strong oxidizers. Air contact produces cumene hydroperoxide. Attacks rubber. May accumulate static electrical charges; may cause ignition of its vapors.

BENZENENITRILE (100-47-0) Forms explosive mixture with air (flash point 167°F/75°C). Incompatible with strong acids, oxidizers. Attacks some plastics, rubber, and coatings.

BENZENE, NITRO- (98-95-3) Forms explosive mixture with air (flash point 190°F/88°C). Incompatible with nitric acid, nitrogen tetroxide, caustics, ammonia, amines. Attacks some plastics, rubber, and coatings.

BENZENE, 1,1'-OXYBIS- (101-84-8) Incompatible with strong acids, strong oxidizers may cause fire and explosions. Attacks some plastics, rubber, and coatings.

BENZENE PHOSPHORUS DICHLORIDE (644-97-3) Water contact produces hydrochloric acid. Corrodes metals except 316 stainless steel, nickel, and Hastelloy.

BENZENE PHOSPHORUS THIODICHLORIDE (14684-25-4) Water contact produces hydrochloric acid; reacts with bases and attacks metals in the presence of moisture.

BENZENE, PROPYL- (103-65-1) Forms explosive mixture with air (flash point 86°F/30°C). Incompatible with nitric acid.

BENZENE SULFOCHLORIDE (98-09-9) Incompatible with ammonia, aliphatic amines, water solutions. Water contact produces hydrochloric and chlorosulfonic acids; aqueous solutions react violently with bases. Attacks metals in presence of moisture.

BENZENE SULFONE-CHLORIDE or BENZENE SULFONECHLORIDE (98-09-9) Incompatible with ammonia, aliphatic amines, water solutions. Water contact produces hydrochloric and chlorosulfonic acids; aqueous solutions react violently with bases. Attacks metals in presence of moisture.

BENZENESULFONIC (ACID) CHLORIDE or BENZENESULFONIC ACID CHLORIDE (98-09-9) Incompatible with ammonia, aliphatic amines, water solutions. Water contact produces hydrochloric and chlorosulfonic acids; aqueous solutions react violently with bases. Attacks metals in presence of moisture.

BENZENESULFONIC ACID, DODECYL- (27176-87-0) Forms explosive mixture with air (flash point 100°F/38°C). Strong oxidizers may cause fire and explosions. Do not store in carbon steel or aluminum.

BENZENE SULFONIC ACID, DODECYL ESTER (27176-87-0) Forms explosive mixture with air (flash point 100°F/38°C). Strong oxidizers may cause fire and explosions. Do not store in carbon steel or aluminum.

BENZENESULFONYL CHLORIDE (98-09-9) Incompatible with ammonia, aliphatic amines, water solutions. Water contact produces hydrochloric and chlorosulfonic acids; aqueous solutions react violently with bases. Attacks metals in presence of moisture.

BENZENE, 1,2,3,5-TETRAMETHYL- (527-53-7) Forms explosive mixture with air (flash point 160°F/71°C). Strong oxidizers may cause fire and explosions.

BENZENETHIOL (108-98-5) Forms explosive mixture with air (flash point 132°F/56°C). Incompatible with strong acids, caustics, alkali metals (i.e., lithium, sodium, potassium, rubidium, cesium, francium). Oxidizes on contact with air. Corrosive to carbon steel.

BENZENETHIOPHOSPHONYL CHLORIDE (14684-5-4) Water contact produces hydrochloric acid. Corrosive to metals.

BENZENE, 1,2,3-TRICHLORO- (87-61-6) Flash point 210°F/99°C. Strong oxidizers may cause fire and explosions.

BENZENE, 1,2,4-TRICHLORO- (120-82-1) Contact with strong oxidizers may cause fire and explosions. Attacks most rubbers.

BENZENE, 1,1'-(2,2,2-TRICHLOROETHYLIDENE) BIS(4-CHLORO- (50-29-3) Incompatible with salts of iron or aluminum, and bases. Do not store in iron containers.

1,2,3-BENZENETRIOL (87-66-1) Incompatible with strong oxidizers, caustics, ammonia, amines, isocyanates, alkylene oxides, epichlorohydrin.

BENZENE, VINYL- (100-42-5) If inhibitor (often *tert*-butylcatechol) is not present in adequate concentrations, polymerization may occur and explode containers. Temperatures above 150°F/66°C speed up polymerization. Incompatible with oxidizers, acids, rust, catalysts for vinyl polymerization, such as peroxides, strong acids, and aluminum chloride. Corrodes copper, copper alloys, and dissolves rubber. Attacks some plastics, rubber, and coatings. May accumulate static electrical charges; may cause ignition of its vapors.

BENZENOSULFOCHLOREK (Polish) (98-09-9) Incompatible with ammonia, aliphatic amines, water solutions. Water contact produces hydrochloric and chlorosulfonic acids; aqueous solutions react violently with bases. Attacks metals in presence of moisture.

BENZENOSULPHOCHLORIDE (98-09-9) Incompatible with ammonia, aliphatic amines, water solutions. Water contact produces hydrochloric and chlorosulfonic acids; aqueous solutions react violently with bases. Attacks metals in presence of moisture.

BENZENYL CHLORIDE (98-07-7) Reacts with water, forming a corrosive acid mixture. Reacts violently with amines and finely divided light metals (aluminum, magnesium, beryllium, etc.). Attacks some plastics, rubber, and coatings.

BENZENYL TRICHLORIDE (98-07-7) Reacts with water forming a corrosive acid mixture. Reacts violently with amines and finely divided light metals (aluminum, magnesium, beryllium, etc.). Attacks some plastics, rubber, and coatings.

BENZIDIN (Czech) (92-87-5) Oxidizes on exposure to light and air. Strong oxidizers. Red fuming nitric acid may cause fire.

BENZIDINA (Italian) (92-87-5) Oxidizes on exposure to light and air. Reacts violently with strong oxidizers. Red fuming nitric acid may cause fire.

BENZIDINE (92-87-5) Oxidizes on exposure to light and air. Reacts violently with strong oxidizers. Red fuming nitric acid may cause fire.

BENZILE (CLORURO di) (Italian) (100-44-7) Forms explosive mixture with air (flash point 153°F/67°C). Water contact produces hydrogen chloride fume. Strong oxidizers may cause fire and explosions. Violent polymerization may be caused by contact with copper, aluminum, iron, zinc, magnesium and tin. May accumulate static electrical charges, and may cause ignition of its vapors. Attacks some plastics and rubber.

2-BENZIMIDAZOLECARBAMIC ACID, 1-(BUTYLCARBA-MOYL), METHYL ESTER (17804-35-2) Heat, water, strong acids, and strong alkalies can cause decomposition and formation of toxic oxides of nitrogen.

BENZIN (8030-30-6) Forms explosive mixture with air (flash point 100°F/38°C). Incompatible with nitric acid, strong oxidizers. Attacks some plastics, rubber, and coatings.

BENZIN (8032-32-4) Forms explosive mixture with air (flash point −40°F to −86°F/−40°C to −66°C). Incompatible with strong acids, strong oxidizers. Attacks some plastics, rubber, and coatings. May accumulate static electrical charges, and may cause ignition of its vapors.

BENZIN (German) (8006-61-9) Forms explosive mixture with air (flash point 125°F/51°C). Incompatible with nitric acid; oxidizers may cause fire and explosions.

BENZIN (OBS) (71-43-2) Forms explosive mixture with air (flash point 12°F/−11°C).) May accumulate static electrical charges, and may cause ignition of its vapors. Incompatible with strong oxidizers, nitric acid, oxygen, ozone, perchlorates. Attacks some forms of plastics, coatings, and rubber.

BENZINE (OBS) (71-43-2) Forms explosive mixture with air (flash point 12°F/−11°C). May accumulate static electrical charges, and may cause ignition of its vapors. Incompatible with strong oxidizers, nitric acid, oxygen, ozone, perchlorates. Attacks some forms of plastics, coatings, and rubber.

1-BENZINE (91-22-5) Contact with strong oxidizers may cause fire and explosions. Attacks some plastics, rubber, and coatings.

BENZINOFORM (56-23-5) Incompatible with allyl alcohol, zirconium, fluorine gas, alkali metals, aluminum. Attacks some rubbers and plastics.

BENZINOL (79-01-6) Contact with caustics produces a toxic and flammable gas. Reacts violently with chemically active metals. Contact with aluminum may produce a violent, self-accelerating polymerization reaction. Incompatible with nonoxidizing mineral acids, strong acids, organic acids, organic anhydrides, isocyanates, alkylene oxides, aldehydes, alcohols, glycols, phenols, cresols, caprolactam solution, epichlorohydrin, nitrogen tetroxide, metal powders, oxygen. May accumulate static electrical charges; may cause ignition of its vapors.

BENZO-CHINON (German) (106-51-4) Forms explosive mixture with air (flash point 104°F/40°C). Incompatible with strong bases, reducing agents, strong oxidizers. Attacks some plastics, rubber, and coatings.

BENZOEPIN (115-29-7) Hydrolyzed by acids and alkalis. Corrosive to iron.

BENZOFLEX 9-88 or BENZOFLEX 9-88 SG or BENZOFLEX 9-98 (94-51-9) Incompatible with strong acids, nitrates, oxidizers.

BENZOHYDROQUINONE (123-31-9) Incompatible with strong oxidizers, caustics. May be oxidized to quinone at room temperatures in the presence of moisture. May explode on contact with oxygen.

BENZOIC ACID (65-85-0) Incompatible with strong oxidizers, caustics, ammonia, amines, isocyanates.

BENZOIC ACID, 2-(ACETYLOXY)- (50-78-2) Fires and explosions may result from contact with strong oxidizers. Alkali hydroxides or carbonates cause decomposition.

BENZOIC ACID, BENZOPEROXIDE (94-36-0) Confined storage of dry chemical may lead to decomposition and explosion. A strong oxidant; extremely reactive. Fires and explosion may result from heat or contamination, and from contact with strong acids, combustible materials, oxidizers, acids, bases, alcohols, reducing agents, metals, metal oxides, amines, accelerators, methyl methacrylate, organic matter, lithium aluminum carbide, dimethyl aniline, amines, metallic naphthenates. May attack some plastics, rubber, and coatings. Protect containers from shock and friction.

BENZOIC ACID, CHLORIDE (98-88-4) Forms explosive mixture with air (flash point 162°F/72°C). Water contact may be violent; produces hydrochloric acid. Incompatible with amines, alcohols, alkali metals, dimethylsulfoxide. Attacks metals in the presence of moisture. Attacks some plastics, rubber, and coatings.

BENZOIC ACID, 2-METHOXY- (119-36-0) Incompatible with strong acids, nitrates, strong oxidizers.

BENZOIC ACID, METHYL ESTER (93-58-3) Forms explosive mixture with air (flash point 181°F/83°C). Incompatible with strong acids, nitrates, oxidizers.

BENZOIC ACID NITRILE (100-47-0) Forms explosive mixture with air (flash point 167°F/75°C). Incompatible with strong acids, oxidizers. Attacks some plastics, rubber, and coatings.

BENZOIC ACID PEROXIDE (94-36-0) Confined storage of dry chemical may lead to decomposition and explosion. A strong oxidant; extremely reactive. Fires and explosion may result from heat or contamination, and from contact with strong acids, combustible materials, oxidizers, acids, bases, alcohols, reducing agents, metals, metal oxides, amines, accelerators, methyl methacrylate, organic matter, lithium aluminum carbide, dimethyl aniline, amines, metallic naphthenates. May attack some plastics, rubber, and coatings. Protect containers from shock and friction.

BENZOIC ALDEHYDE (100-52-7) Forms explosive mixture with air (flash point 145°F/63°C). A strong reducing agent. Forms peroxides with air. Reacts violently with strong acids, aluminum, caustics, ammonia, amines, iron, strong oxidizers. Attacks some plastics, rubber and coatings.

BENZOL (71-43-2) Forms explosive mixture with air (flash point 12°F/−11°C). May accumulate static electrical charges, and may cause ignition of its vapors. Incompatible with strong oxidizers, nitric acid, oxygen, ozone, perchlorates. Attacks some forms of plastics, coatings, and rubber.

BENZOLE (71-43-2) Forms explosive mixture with air (flash point 12°F/−11°C). May accumulate static electrical charges, and may cause ignition of its vapors. Incompatible with strong oxidizers, nitric acid, oxygen, ozone, perchlorates. Attacks some forms of plastics, coatings, and rubber.

BENZOLENE (71-43-2) Forms explosive mixture with air (flash point 12°F/−11°C).) May accumulate static electrical charges, and may cause ignition of its vapors. Incompatible with strong oxidizers, nitric acid, oxygen, ozone, perchlorates. Attacks some forms of plastics, coatings, and rubber.

BENZOLINE (8032-32-4) Forms explosive mixture with air (flash point −40°F to −86°F/−40°C to −66°C). Incompatible with strong acids, strong oxidizers. Attacks some plastics, rubber, and coatings. May accumulate static electrical charges, and may cause ignition of its vapors.

BENZOLO (Italian) (71-43-2) Forms explosive mixture with air (flash point 12°F/−11°C).) May accumulate static electrical charges, and may cause ignition of its vapors. Incompatible with strong oxidizers, nitric acid, oxygen, ozone, perchlorates. Attacks some forms of plastics, coatings, and rubber.

BENZONITRILE (100-47-0) Forms explosive mixture with air (flash point 167°F/75°C). Incompatible with strong acids, oxidizers. Attacks some plastics, rubber, and coatings.

BENZOPHENONE (119-61-9) Combustible solid. Strong oxidizers may cause fire and explosions. Attacks some plastics, rubber, and coatings.

BENZOPEROXIDE (94-36-0) Confined storage of dry chemical may lead to decomposition and explosion. A strong oxidant; extremely reactive. Fires and explosion may result from heat or contamination, and from contact with strong acids, combustible materials, oxidizers, acids, bases, alcohols, reducing agents, metals, metal oxides, amines, accelerators, methyl methacrylate, organic matter, lithium aluminum carbide, dimethyl aniline, amines, metallic naphthenates. May attack some plastics, rubber, and coatings. Protect containers from shock and friction.

BENZO(a)PYRENE (50-32-8) Strong oxidizers may cause fires and explosions.

3,4-BENZOPYRENE (50-32-8) Strong oxidizers may cause fires and explosions.

BENZO(B) PYRIDINE (91-22-5) Contact with strong oxidizers may cause fire and explosions. Attacks some plastics, rubber, and coatings.

BENZOQUINOL (123-31-9) Incompatible with strong oxidizers, caustics. May be oxidized to quinone at room temperatures in the presence of moisture. May explode on contact with oxygen.

2,3-BENZOQUINOLINE (260-94-6) Contact with strong oxidizers may cause fire and explosions.

BENZO(B) QUINOLINE (260-94-6) Contact with strong oxidizers may cause fire and explosions.

BENZOQUINONE (106-51-4) Forms explosive mixture with air (flash point 104°F/40°C). Incompatible with strong bases, reducing agents, strong oxidizers. Attacks some plastics, rubber, and coatings.

1,4-BENZOQUINONE (106-51-4) Forms explosive mixture with air (flash point 104°F/40°C). Incompatible with strong bases, reducing agents, strong oxidizers. Attacks some plastics, rubber, and coatings.

p-BENZOQUINONE or *para*-BENZOQUINONE (106-51-4) Forms explosive mixture with air (flash point 104°F/40°C). Incompatible with strong bases, reducing agents, strong oxidizers. Attacks some plastics, rubber, and coatings.

2-BENZOTHIAZOLETHIOL, SODIUM SALT (2492-26-4) Concentrated aqueous solutions are corrosive.

2-(3H)-BENZOTHIAZOLETHIONE, SODIUM SALT (2492-26-4) Concentrated aqueous solutions are corrosive.

BENZOTRICHLORIDE (98-07-7) Reacts with water, forming a corrosive acid mixture. Reacts violently with amines and finely divided light metals (e.g., aluminum, magnesium, beryllium). Attacks some plastics, rubber, and coatings.

BENZOYL BENZENE (119-61-9) Combustible solid. Strong oxidizers may cause fire and explosions. Attacks some plastics, rubber, and coatings.

BENZOYL CHLORIDE (98-88-4) Forms explosive mixture with air (flash point 162°F/72°C). Water contact may be violent; produces hydrochloric acid. Incompatible with amines, alcohols, alkali metals, dimethylsulfoxide. Attacks metals in the presence of moisture. Attacks some plastics, rubber, and coatings.

BENZOYL METHIDE HYPNONE (98-86-2) Incompatible with strong acids, aliphatic amines, oxidizers. Forms explosive mixture with air (flash point 170°F/77°C).

BENZOYL PEROXIDE (94-36-0) Confined storage of dry chemical may lead to decomposition and explosion. A strong oxidant; extremely reactive. Fires and explosion may result from heat or contamination, and from contact with strong acids, combustible materials, oxidizers, acids, bases, alcohols, reducing agents, metals, metal oxides, amines, accelerators, methyl methacrylate, organic matter, lithium aluminum carbide, dimethyl aniline, amines, metallic naphthenates. May attack some plastics, rubber, and coatings. Protect containers from shock and friction.

BENZOYL SUPEROXIDE (94-36-0) Confined storage of dry chemical may lead to decomposition and explosion. A strong oxidant; extremely reactive. Fires and explosion may result from heat or contamination, and from contact with strong acids, combustible materials, oxidizers, acids, bases, alcohols, reducing agents, metals, metal oxides, amines, accelerators, methyl methacrylate, organic matter, lithium aluminum carbide, dimethyl aniline, amines, metallic naphthenates. May attack some plastics, rubber, and coatings. Protect containers from shock and friction.

BENZYDYNA (Polish) (92-87-5) Oxidizes on exposure to light and air. Reacts violently with strong oxidizers. Red fuming nitric acid may cause fire.

BENZYL ACETATE (140-11-4) Forms explosive mixture with air (flash point 216°F/102°C). Reacts with strong acids, nitrates, oxidizers.

BENZYL ALCOHOL (100-51-6) Forms explosive mixture with air (flash point 213°F/101°C). Slowly oxidizes in air and oxygen. Incompatible with mineral acids, caustics, aliphatic amines, isocyanates. Reacts violently with strong oxidizers. Corrodes aluminum at high temperature. Attacks some nonfluorinated plastics; may not attack polypropylene.

BENZYL AMINE (100-46-9) Forms explosive mixture with air (flash point 168°F/76°C). Incompatible with strong acids, organic anhydrides, isocyanates, aldehydes. Slowly corrodes metals in a moist environment. Attacks some plastics, rubber, and coatings.

BENZYL BROMIDE (100-39-0) Forms explosive mixture with air (flash point 174°F/79°C). Water contact produces hydrobromic acid. Incompatible with strong oxidizers, bases. Attacks metals, except nickel and lead, in the presence of moisture.

BENZYL, *n*-BUTYL PHTHALATE (85-68-7) Incompatible with strong acids, nitrates, oxidizers. Destructive to rubber and paint. May accumulate static electrical charges, and may cause ignition of its vapors.

BENZYLCARBONYL CHLORIDE (501-53-1) Aqueous solution produces hydrochloric acid; reacts violently with bases. Reacts with nitrates. Attacks metals in the presence of moisture.

BENZYLCHLORID (German) (100-44-7) Forms explosive mixture with air (flash point 153°F/67°C). Water contact produces hydrogen chloride fume. Strong oxidizers may cause fire and explosions. Violent polymerization may be caused by contact with copper, aluminum, iron, zinc, magnesium, and tin. May accumulate static electrical charges, and may cause ignition of its vapors. Attacks some plastics and rubber.

BENZYL CHLORIDE (100-44-7) Forms explosive mixture with air (flash point 153°F/67°C). Water contact produces hydrogen chloride fume. Strong oxidizers may cause fire and explosions. Violent polymerization may be caused by contact with copper, aluminum, iron, zinc, magnesium, and tin. May accumulate static electrical charges, and may cause ignition of its vapors. Attacks some plastics and rubber.

BENZYL CHLOROCARBONATE (501-53-1) Aqueous solution produces hydrochloric acid; reacts violently with bases. Reacts with nitrates. Attacks metals in the presence of moisture.

BENZYL CHLOROFORMATE (501-53-1) Aqueous solution produces hydrochloric acid; reacts violently with bases. Reacts with nitrates. Attacks metals in the presence of moisture.

BENZYL DICHLORIDE (98-87-3) Forms explosive mixture with air (flash point 153°F/67°C). Reacts with acids, bases, strong oxidizers. Produces acid fumes with air. Attacks plastics and coatings.

BENZYL DIMETHYLAMINE (103-83-3) Forms explosive mixture with air (flash point 170°F/77°C). Incompatible with acids, organic anhydrides, isocyanates, aldehydes, strong oxidizers, calcium hypochlorite. Attacks some plastics, rubber, and coatings.

***n*-BENZYLDIMETHYLAMINE** (103-83-3) Forms explosive mixture with air (flash point 170°F/77°C). Incompatible with acids, organic anhydrides, isocyanates, aldehydes, strong oxidizers, calcium hypochlorite. Attacks some plastics, rubber, and coatings.

BENZYL DIMETHYLOCTADECYL AMMONIUM CHLORIDE (959-55-7) Incompatible with strong oxidizers, strong acids.

BENZYL DIMETHYLSTEARYL AMMONIUM CHLORIDE (959-55-7) Incompatible with strong oxidizers, strong acids.

BENZYLE (CHLORURE de) (French) (100-44-7) Forms explosive mixture with air (flash point 153°F/67°C). Water contact produces hydrogen chloride fume. Strong oxidizers may cause fire and explosions. Violent polymerization may be caused by contact with copper, aluminum, iron, zinc, magnesium, and tin. May accumulate static electrical charges, and may cause ignition of its vapors. Attacks some plastics and rubber.

BENZYLENE CHLORIDE (98-87-3) Forms explosive mixture with air (flash point 153°F/67°C). Reacts with acids, bases, strong oxidizers. Produces acid fumes with air. Attacks plastics and coatings.

BENZYL ETHANOATE (140-11-4) Forms explosive mixture with air (flash point 216°F/102°C). Reacts with strong acids, nitrates, oxidizers.

BENZYL ETHER (103-50-4) Forms unstable peroxides. Reacts violently with strong oxidizers. May accumulate static electrical charges; may cause ignition of its vapors.

BENZYLIDENE CHLORIDE (98-87-3) Forms explosive mixture with air (flash point 153°F/67°C). Reacts with acids, bases, strong oxidizers. Produces acid fumes with air. Attacks plastics and coatings.

BENZYL OXIDE (103-50-4) Forms unstable peroxides. Reacts violently with strong oxidizers. May accumulate static electrical charges; may cause ignition of its vapors.

BENZYL TRIMETHYL AMMONIUM CHLORIDE (56-93-9) Incompatible with strong acids, oxidizers.

BEOSIT (115-29-7) Hydrolyzed by acids and alkalis. Corrosive to iron.

BEROL 478 (119-36-8) Forms explosive mixture with air (flash point 205°F/96°C). Incompatible with strong acids, nitrates, oxidizers.

BERTHOLITE (7782-50-5) A strong oxidizer; reacts violently with combustible materials, reducing agents and many other substances. Forms explosive mixtures with gasoline and petroleum products, turpentine, alcohols, acetylene, carbon disulfide, hydrogen, anhydrous ammonia, finely divided metals, organic compounds, phosphorus. Fire, explosion, and the formation of toxic fumes may result from contact with many substances, including alkyphosphines, aluminum, antimony, arsenic compounds, arsine, bismuth, boron, brass, calcium compounds, carbon, diethyl zinc, fluorine, germanium, hydrocarbons, rubber. Attacks some plastics and coatings. Moist substance is extremely corrosive to metals, iron, steel, red metals, zinc. Contact with hot steel may cause a dangerous iron–chlorine fire.

BERTHOLLET'S SALT (3811-04-9) A powerful oxidizer. Reacts violently with reducing agents or combustibles, ammonia gas, ammonium salts, organic matter, hydrogen iodide, organic acids, sulfuric acid. Forms explosive mixtures with metallic powders, ammonium chloride, organic solids, including agricultural materials.

BERYLLIA (1304-56-9) Incompatible with acids, bases, chlorinated hydrocarbons, oxidizers, molten lithium.

BERYLLIUM (7440-41-7) Contact with acids causes evolution of flammable hydrogen gas. Incompatible with alkalies, chlorinated hydrocarbons, oxidizable agents, carbon tetrachloride, trichloroethylene, lithium, phosphorus.

BERYLLIUM CHLORIDE (7787-47-5) Water contact produces heat and hydrochloric acid. Attacks metals in the presence of moisture.

BERYLLIUM DICHLORIDE (7787-47-5) Water contact produces heat and hydrochloric acid. Attacks metals in the presence of moisture.

BERYLLIUM DIFLUORIDE (7787-49-7) Incompatible with acids, caustics, chlorinated hydrocarbons, oxidizers, molten lithium.

BERYLLIUM DINITRATE (13597-99-4); (778-75-5) Produces nitric acid with water. Incompatible with acids, caustics, chlorinated hydrocarbons, oxidizers, molten lithium. Corrodes metals in a moist environment.

BERYLLIUM DUST (7440-41-7) Contact with acids causes evolution of flammable hydrogen gas. Incompatible with alkalies, chlorinated hydrocarbons, oxidizable agents, carbon tetrachloride, trichloroethylene, lithium, phosphorus.

BERYLLIUM FLUORIDE (7787-49-7) Incompatible with acids, caustics, chlorinated hydrocarbons, oxidizers, molten lithium.

BERYLLIUM, METAL or BERYLLIUM, METAL POWDER (7440-41-7) Contact with acids causes evolution of flammable hydrogen gas. Incompatible with alkalies, chlorinated hydrocarbons, oxidizable agents, carbon tetrachloride, trichloroethylene, lithium, phosphorus.

BERYLLIUM MONOXIDE (1304-56-9) Incompatible with acids, bases, chlorinated hydrocarbons, oxidizers, molten lithium.

BERYLLIUM NITRATE (13597-99-4); (778-75-5) Produces nitric acid with water. Incompatible with acids, caustics, chlorinated hydrocarbons, oxidizers, molten lithium. Corrodes metals in a moist environment.

BERYLLIUM NITRATE TRIHYDRATE (13597-99-4); (778-75-5) Produces nitric acid with water. Incompatible with acids, caustics, chlorinated hydrocarbons, oxidizers, molten lithium. Corrodes metals in a moist environment.

BERYLLIUM OXIDE (1304-56-9) Incompatible with acids, bases, chlorinated hydrocarbons, oxidizers, molten lithium.

BERYLLIUM POWDER (7440-41-7) Contact with acids causes evolution of flammable hydrogen gas. Incompatible with alkalies, chlorinated hydrocarbons, oxidizable agents, carbon tetrachloride, trichloroethylene, lithium, phosphorus.

BERYLLIUM SULFATE (13510-49-1) Incompatible with acids, caustics, chlorinated hydrocarbons, oxidizers, lithium.

BERYLLIUM SULFATE TETRAHYDRATE (13510-49-1) Incompatible with acids, caustics, chlorinated hydrocarbons, oxidizers, lithium.

BETANAFTILAMINA (Italian) (91-59-8) Forms explosive mixture with air (flash point 300°F/149°C). Strong oxidizers, strong acids may cause fire and explosions. Oxidizes in the presence of air and light.

BETRAPRONE (57-57-8) Polymerizes in storage. High temperatures cause ruptured containers. Decomposes at room temperature. Hydrolyzes in water.

BETULA OIL (119-36-0) Incompatible with strong acids, nitrates, strong oxidizers.

BEXOL (58-89-9) Not combustible, but may be dissolved in a combustible solvent. If solvent comes in contact with oxidizers, fire and explosions may result.

BFV (50-00-0) Forms explosive mixture with air (flash point 122°F/50°C). Incompatible with strong acids, amines, strong oxidizers, alkaline materials, nitrogen dioxide, performic acid. Reaction with hydrochloric acid forms bis-chloromethyl ether, a carcinogen.

102

BGE (2426-08-6) Forms explosive mixture with air (flash point 130°F/54°C). Air and light produces unstable and explosive peroxides. Contact with strong oxidizers may cause fire and explosions. Strong caustics may cause polymerization. Attacks some plastics and rubber.

n-**BGE** (2426-08-6) Forms explosive mixture with air (flash point 130°F/54°C). Air and light produces unstable and explosive peroxides. Contact with strong oxidizers may cause fire and explosions. Strong caustics may cause polymerization. Attacks some plastics and rubber.

BH 2,4-D (94-75-7) Decomposes in sunlight. Incompatible with strong oxidizers; may cause fire and explosions.

BHC (58-89-9) Not combustible, but may be dissolved in a combustible solvent. If solvent comes in contact with oxidizers, fire and explosions may result.

gamma-BHC (58-89-9) Not combustible, but may be dissolved in a combustible solvent. If solvent comes in contact with oxidizers fire and explosions may result.

BH DALAPON (75-99-0) Corrosive to iron, aluminum, and copper.

BH DOCK KILLER (123-33-1) Contact with strong oxidizers may cause fire and explosions.

BIANISIDINE (95-53-4) Forms explosive mixture with air (flash point 185°F/85°C). Incompatible with strong acids, mineral acids, organic anhydrides, isocyanates, aldehydes, oxidizers. Attacks some plastics, rubber, and coatings.

4,4′-BIANILINE (92-87-5) Oxidizes on exposure to light and air. Reacts violently with strong oxidizers. Red fuming nitric acid may cause fire.

p,p-**BIANILINE** (92-87-5) Oxidizes on exposure to light and air. Reacts violently with strong oxidizers. Red fuming nitric acid may cause fire.

BIBENZENE (92-52-4) Mist forms explosive mixture with air. Strong oxidizers may cause fire and explosions.

BIBESOL (62-73-7) Attacks some plastics, rubber, and coatings.

BICARBURET OF HYDROGEN or BICARBURETTED HYDRO-GEN (74-85-1) Incompatible with strong acids, oxidizers, halogen acids, aluminum chloride, bromotrichloromethane, carbon tetrachloride, chlorine, chlorine dioxide, nitrogen dioxide.

BICHLORIDE of MERCURY (7487-94-7) Incompatible with light metals (aluminum, magnesium, beryllium, etc.), sodium, potassium.

BICHLORURE de MERCURE (French) (7487-94-7) Incompatible with light metals (aluminum, magnesium, beryllium, etc.), sodium, potassium.

BICHLORURE de PROPYLENE (French) (78-87-5) Forms explosive mixture with air (flash point 60°F/15.6°C). Strong oxidizers may cause fire and explosions; incompatible with strong acids (decomposition). Attacks some plastics, rubber, and coatings.

BICHROMATE de SODIUM (French) (10588-01-9) Aqueous solution is corrosive. A strong oxidizer; reacts violently with reducing agents, acids, acetic anhydride, combustibles. Attacks copper, zinc, tin, brass, bronze.

BICHROMATE OF POTASH (7778-50-9) Powdered combustibles will ignite. Strong acids produces toxic vapors.

BICHROMATE OF SODA (10588-01-9) Aqueous solution is corrosive. A strong oxidizer; reacts violently with reducing agents, acids, acetic anhydride, combustibles. Attacks copper, zinc, tin, brass, bronze.

BICHROME (7778-50-9) Powdered combustibles will ignite. Strong acids form toxic vapors.

BICYCLO [4.4.0] DECANE (91-17-8) Forms explosive mixture with air (flash point 134°F/57°C). Strong oxidizers may cause fire and explosions.

BICYCLO 221 HEPT-2-ENE, 5-ETHYLIDENE- (16219-75-3) Reacts violently with oxygen and strong oxidizers. Forms explosive mixture with air. Inhibit peroxide formation with *tert*-butyl catechol.

BICYCLOPENTADIENE (77-73-6) Forms explosive mixture with air (flash point 90°F/3°C). Forms peroxides with air. May polymerize unless inhibited and maintained under inert atmosphere. Reacts with oxidizers.

BIDRIN (141-66-2) Corrosive to cast iron, mild steel, brass, and stainless steel 304.

BIEBERITE (10124-43-3) Incompatible with oxidizers, acetylene.

BIETHYLENE (106-99-0) Self-reactive. Forms explosive peroxides with air. Fires, explosions, or hazardous polymerization may result from contact with air, strong oxidizers, strong acids, ozone, nitrogen dioxide, copper and its alloys, phenol, chlorine dioxide, crotonaldehyde, or a free radical polymerization initiator such as hydroquinone. Add inhibitor (such as *tert*-butyl catechol) and monitor to ensure that effective levels are maintained at all times. May accumulate static electrical charges, and may cause ignition of its vapors.

(1,1'-BIFENYL)-4,4'-DIAMINE (92-87-5) Oxidizes on exposure to light and air. Reacts violently with strong oxidizers. Red fuming nitric acid may cause fire.

BIFLUORIDEN (Dutch) (7782-41-4) A powerful oxidizer and a dangerously reactive gas. Reacts violently with reducing agents and combustible matter. Water contact produces hydrochloric acid, oxygen, and oxygen difluoride. Reacts violently with most oxidizable materials at room temperature, frequently with ignition. Corrodes most metals (except the cylinders in which it is shipped), especially in the presence of moisture. Contact with nitric acid produces explosive gas. Attacks some plastics, rubber, and coatings. Reacts explosively with ammonia, graphite, halocarbons, hydrocarbons, hydrogen, and many other substances. Reacts with nearly every known element.

BIFORMAL (107-22-2) Water contact causes polymerization. Incompatible with strong acids, caustics, ammonia, amines, chlorosulfonic acid, ethylene amine. Corrosive to metals.

BIFORMYL (107-22-2) Water contact causes polymerization. Incompatible with strong acids, caustics, ammonia, amines, chlorosulfonic acid, ethylene amine. Corrosive to metals.

BIG DIPPER (122-39-4) Incompatible with strong acids, organic anhydrides, isocyanates, ketones, oxidizers, hexachloromelamine, trichloromelamine.

B-I-K (57-13-6) Heat forms anhydrous ammonia fumes. Contact with oxidizers, nitrates may cause fire and explosions. Contact with chlorinating agents including hypochlorite bleaches, may produce explosive nitrogen trichloride. Reacts with nitrosyl perchlorate.

BILORIN (64-18-6) Forms explosive mixture with air (flash point 156°F/69°C). Incompatible with sulfuric acid, bases, ammonia, aliphatic amines, alkanolamines, furfuryl alcohol, hydrogen peroxide, isocyanates, alkylene oxides, epichlorohydrin. Attacks aluminum, cast iron, steel, some plastics, rubber, and coatings.

BIMETHYL (74-84-0) Flammable gas; forms explosive mixture with air. Strong oxidizers may cause fire and explosions. May accumulate static electrical charges; may cause ignition of its vapors.

BINITROBENZENE (99-65-0) Explodes: impact- and friction sensitive. Strong oxidizers may cause fire and explosions; avoid contact with caustics, chemically active metals. Attacks some plastics.

BIO 5,462 (115-29-7) Hydrolyzed by acids and alkalis. Corrosive to iron.

BIOCIDE (107-02-8) Forms explosive mixture with air (flash point −15°F/−26°C). Unstable and very reactive. In storage can form heat- and shock-sensitive compounds. Unless inhibited (usually by hydroquinone), readily forms explosive peroxides. Able to polymerize. A strong reducing agent. Reacts violently with oxidizers, strong acids, caustics, amines, 2-aminoethanol, ammonia, ammonium hydroxide, ethylene diamine, ethyleneimine, hydroxides, metal salts, oxidizers, sulfur dioxide, thiourea. Attacks metals: cadmium and zinc. May accumulate static electrical charges, and may cause ignition of its vapors.

BIOFLEX 91 (84-76-4) Incompatible with strong acids, nitrates, oxidizers.

BIOGAS (74-82-8) Flammable gas. Forms an explosive mixture with air. Reacts violently with strong oxidizers.

BIO-SOFT D-40 (25155-30-0) Reacts with acids, including fumes.

BIPHENYL (92-52-4) Mist forms explosive mixture with air. Strong oxidizers may cause fire and explosions.

1,1′-BIPHENYL (92-52-4) Mist forms explosive mixture with air. Strong oxidizers may cause fire and explosions.

4,4′-BIPHENYLDIAMINE (92-87-5) Oxidizes on exposure to light and air. Reacts violently with strong oxidizers. Red fuming nitric acid may cause fire.

(1,1′-BIPHENYL)-4,4′-DIAMINE (92-87-5) Oxidizes on exposure to light and air. Reacts violently with strong oxidizers. Red fuming nitric acid may cause fire.

(1,1′-BIPHENYL)-4,4′-DIAMINE (9CI) (92-87-5) Oxidizes on exposure to light and air. Reacts violently with strong oxidizers. Red fuming nitric acid may cause fire.

1,1′-BIPHENYL-4,4′-DIAMINE, 3,3′-DIMETHYL- (95-53-4) Forms explosive mixture with air (flash point 185°F/85°C). Incompatible with strong acids, mineral acids, organic anhydrides, isocyanates, aldehydes, oxidizers. Attacks some plastics, rubber, and coatings.

BIPHENYL, 4,4′-DIAMINO- (92-87-5) Oxidizes on exposure to light and air. Reacts violently with strong oxidizers. Red fuming nitric acid may cause fire.

4,4′-BIPHENYLENEDIAMINE (92-87-5) Oxidizes on exposure to light and air. Reacts violently with strong oxidizers. Red fuming nitric acid may cause fire.

BIPHENYL ETHER (101-84-8) Incompatible with strong acids; strong oxidizers may cause fire and explosions. Attacks some plastics, rubber, and coatings.

BIPHENYL, MIXED WITH BIPHENYL OXIDE (3:7) (8004-13-5) Contact with strong oxidizers may cause fire and explosions. Attacks some plastics, rubber, and coatings.

BIPHENYL, 4-NITRO- (92-93-3) Contact with strong oxidizers may cause fire and explosions.

1,1'-BIPHENYL, 4-NITRO- (92-93-3) Contact with strong oxidizers may cause fire and explosions.

BIPHENYL OXIDE (101-84-8) Incompatible with strong acids; strong oxidizers may cause fire and explosions. Attacks some plastics, rubber, and coatings.

BIPHENYL, POLYCHLORO- (generic CAS for PCBs 1336-36-3) Incompatible with strong oxidizers, strong acids.

BIPOTASSIUM CHROMATE (7789-00-6) A powerful oxidizer. Contact with combustibles, reducing agents, organic material, or finely divided metals may cause fire and explosions.

BIRNENOEL (628-63-7) Incompatible with strong alkalies, strong acids, nitrates, strong oxidizers. Attacks some plastics, coatings, and rubber.

BIS(ACETO) COBALT (71-48-7) Contact with strong oxidizers may cause fire and explosions. Cobalt compounds may react with acetylene.

BIS(ACETOXY) CADMIUM (543-90-8) Incompatible with strong oxidizers, elemental sulfur, selenium, tellurium, strong acids, nitrates.

BIS(ACETYLOXY) MERCURY (1600-27-7) Heat and light cause decomposition.

BIS(2-AMINOETHYL)AMINE (111-40-0) Ignites spontaneously with cellulose nitrate. Silver, cobalt, or chromium compounds may cause explosions. Forms explosive mixture with air (flash point 208°F/98°C). Incompatible with nonoxidizing mineral acids, strong acids, organic acids, organic anhydrides, isocyanates, vinyl acetate, acrylates, substituted allyls, alkylene oxides, epichlorohydrin, ketones, aldehydes, alcohols, glycols, mercury, phenols, cresols, caprolactam solution, strong oxidizers. Attacks aluminum, copper, lead, tin, zinc, and alloys.

BIS(beta-AMINOETHYL)AMINE (111-40-0) Ignites spontaneously with cellulose nitrate. Silver, cobalt, or chromium compounds may cause explosions. Forms explosive mixture with air (flash point 208°F/98°C). Incompatible with nonoxidizing mineral acids, strong acids, organic acids, organic anhydrides, isocyanates, vinyl acetate, acrylates, substituted allyls, alkylene oxides, epichlorohydrin, ketones, aldehydes, alcohols, glycols, mercury, phenols, cresols, caprolactam solution, strong oxidizers. Attacks aluminum, copper, lead, tin, zinc, and alloys.

BIS(2-BUTOXYETHYL) ETHER (112-73-2) Incompatible with sulfuric acid, isocyanates, strong oxidizers. May accumulate static electrical charges; may cause ignition of its vapors.

BIS(*tert*-BUTYL)CHROMATE (1189-85-1) Incompatible with acids, alcohols, combustible or easily oxidized materials, reducing agents, moisture.

BIS-(*p*-CHLOROBENZOYL) PEROXIDE (94-17-7) A powerful oxidizer and explosion hazard. Reacts violently with reducing agents, combustibles, polymerization initiators, heat, or contaminants.

BIS(2-CHLOROETHYL) ETHER (111-44-4) Forms explosive mixture with air (flash point 131°F/55°C). Strong oxidizers may cause fire and explosions. Attacks some plastics, rubber, and coatings.

BIS(2-CHLOROISOPROPYL) ETHER (108-60-1) Forms explosive mixture with air (flash point 170°F/77°C). Reacts violently with strong oxidizers. Incompatible with aluminum, copper, epoxy coatings.

BIS(2-CHLORO-1-METHYLETHYL ETHER) (108-60-1) Forms explosive mixture with air (flash point 170°F/77°C). Reacts violently with strong oxidizers. Incompatible with aluminum, copper, epoxy coatings.

1,1-BIS(*p*-CHLOROPHENYL)-2,2-DICHLOROETHANE (72-54-8) Contact with strong oxidizers may cause fire and explosions.

1,1-BIS-(*p*-CHLOROPHENYL)-2,2,2-TRICHLOROETHANE (50-29-3) Incompatible with salts of iron or aluminum, and bases. Do not store in iron containers.

2,2-BIS(*p*-CHLOROPHENYL)-1,1-TRICHLOROETHANE (50-29-3) Incompatible with salts of iron or aluminum, and bases. Do not store in iron containers.

2,2-BIS(*p*-CHLOROPHENYL)-1,1,1-TRICHLOROETHANE (50-29-3) Incompatible with salts of iron or aluminum, and bases. Do not store in iron containers.

alpha,alpha-BIS(*p*-CHLOROPHENYL)-beta,beta,beta-TRICHLOR-ETHANE (50-29-3) Incompatible with salts of iron or aluminum, and bases. Do not store in iron containers.

1,1-BIS(*p*-CHLOROPHENYL)-2,2,2-TRICHLOROETHANOL (115-32-2) Incompatible with strong acids, caustics, aliphatic amines, isocyanates

BIS-CYCLOPENTADIENE (77-73-6) Forms explosive mixture with air (flash point 90°F/3°C). Forms peroxides with air. May polymerize unless inhibited and maintained under inert atmosphere. Reacts with oxidizers.

BIS-O,O-DIAETHYLPHOSPHORSAEURE-ANHYDRID (German) (107-49-3) Decomposes above 300°F/150°C, forming flammable ethylene gas. Strong oxidizers may cause fire and explosions. Attacks some plastics, rubber or coatings.

BIS(*s*-(DIETHOXYPHOSPHINOTHIOYL)MERCAPTO)METHANE (563-12-2) Incompatible with alkaline formulations. Mixtures with magnesium may be explosive.

BIS-O,O-DIETHYLPHOSPHORIC ANHYDRIDE (107-49-3) Decomposes above 300°F/150°C, forming flammable ethylene gas. Strong oxidizers may cause fire and explosions. Attacks some plastics, rubber, and coatings.

BIS-O,O-DIETHYLPHOSPHOROTHIONIC ANHYDRIDE (3689-24-5) Containers may burst in heat. Strong oxidizers may cause fire and explosions. Attacks some plastics, rubber, and coatings.

BIS(DIMETHYLAMINO)CARBONOTHIOYL) DISULPHIDE (137-26-8) Combustible solid (flash point 192°F/89°C). Strong oxidizers may cause fire and explosions; contact with strong acid or oxidizable materials form toxic gases.

BIS(DIMETHYL-THIOCARBAMOYL) DISULFID (German) (137-26-8) Combustible solid (flash point 192°F/89°C). Strong oxidizers may cause fire and explosions; contact with strong acid or oxidizable materials produces toxic gases.

BIS(DITHIOPHOSPHATEDE-O,O-DIETHYLE) de S,S'-METH-YLENE (French) (563-12-2) Incompatible with alkaline formulations. Mixtures with magnesium may be explosive.

BIS(2-3-EPOXYPROPYL)ETHER (2238-07-5) Forms explosive mixture with air (flash point 147°F/64°C). Incompatible with strong oxidizers. Ethers, as a class, tend to form peroxides upon contact with air and exposure to light. Attacks some forms of plastics, coatings, and rubber.

2,2-BIS(4-(2,3-EPOXYPROPYLOXY)PHENYL)PROPANE (1675-54-8) Forms explosive mixture with air (flash point 175°F/79°C). Incompatible with strong acids, strong oxidizers. Ethers form peroxides on contact with air and light.

1,2-BIS(ETHOXYCARBONYL)ETHYL (121-75-5) Incompatible with strong oxidizers, magnesium, alkaline pesticides. Attacks metals, some plastics, rubber, and coatings.

S-1,2-BIS(ETHOXYCARBONYL)ETHYL-O,O-DIMETHYL DITHI-OPHOSPHATE OF DIETHYL MERCAPTOSUCCINATE (121-75-5) Incompatible with strong oxidizers, magnesium, alkaline pesticides. Attacks metals, some plastics, rubber, and coatings.

BIS(2-ETHYLHEXYL)ADIPATE (103-23-1) Incompatible with strong acids, nitrates, oxidizers.

BIS(2-ETHYLHEXYL) HYDROGEN PHOSPHATE (298-07-7) Corrosive to metals; forms flammable hydrogen gas.

BIS(2-ETHYLHEXYL)ORTHOPHOSPHORIC ACID (298-07-7) Corrosive to metals; produces flammable hydrogen gas.

BIS(2-ETHYLHEXYL) PHTHALATE (117-81-7) Incompatible with strong acids, strong alkalies, nitrates, oxidizers.

BIS(2-ETHYLHEXYL)PHOSPHORIC ACID (298-07-7) Corrosive to metals; produces flammable hydrogen gas.

BIS(2-ETHYLHEXYL) SODIUM SULFOSUCCINATE (119-36-8) Forms explosive mixture with air (flash point 205°F/96°C). Incompatible with strong acids, nitrates, oxidizers.

BIS FENOL A (80-05-7) Contact with strong oxidizers may cause fire and explosions.

BIS(4-GLYCIDYLOXYPHENYL)DIMETHYAMETHANE (1675-54-8) Forms explosive mixture with air (flash point 175°F/79°C). Incompatible with strong acids, strong oxidizers. Reactions of ethers produces peroxides on contact with air and light.

2,2-BIS(*p*-GLYCIDYLOXYPHENYL)PROPANE (1675-54-8) Forms explosive mixture with air (flash point 175°F/79°C). Incompatible with strong acids, strong oxidizers. Reactions of ethers produces peroxides on contact with air and light.

1,2-BIS(2-HYDROXYETHOXY)ETHANE (112-27-6) Incompatible with sulfuric acid, isocyanates, perchloric acid, strong oxidizers.

BIS-[2-(2-HYDROXYETHOXY)ETHYL] ETHER (112-60-7) Incompatible with sulfuric acid, isocyanates, perchloric acid. Attacks some plastics, rubber, and coatings.

BIS(2-HYDROXYETHYL)AMINE (111-42-2) Incompatible with non-oxidizing mineral acids, strong acids, organic acids, organic anhydrides, isocyanates, vinyl acetate, acrylates, substituted allyls, alkylene oxides, epichlorohydrin, aldehydes, oxidizers. Corrosive to copper, copper alloys, zinc, and galvanized iron.

BIS(2-HYDROXYETHYL)ETHER (111-46-6) Incompatible with sulfuric acid, isocyanates, strong oxidizers.

2,2-BIS(4-HYDROXYFENYL) PROPANE (80-05-7) Contact with strong oxidizers may cause fire and explosions.

BIS(4-HYDROXYPHENYL)DIMETHYLMETHANE DIGLYCIDYL ETHER (1675-54-8) Forms explosive mixture with air (flash point 175°F/79°C). Incompatible with strong acids, strong oxidizers. Ethers form peroxides on contact with air and light.

2,2-BIS(4-HYDROXYPHENYL) PROPANE (80-05-7) Contact with strong oxidizers may cause fire and explosions.

2,2-BIS(4-HYDROXYPHENYL)PROPANE, DIGLYCIDYL ETHER (1675-54-8) Forms explosive mixture with air (flash point 175°F/79°C). Incompatible with strong acids, strong oxidizers. Ethers form peroxides on contact with air and light.

BIS(HYDROXYLAMINE) SULFATE (10039-54-0) Violent, possibly explosive, reaction occurs with caustics or strong oxidizers. Magnesium or aluminum contact may be explosive.

2,2-BIS(HYDROXYMETHYL)-1,3-PROPANEDIOL (115-77-5) Incompatible with organic acids, oxidizers. Forms explosive material with thiophosphoryl chloride + heat.

BIS(2-HYDROXYPROPYL) ETHER (110-98-5) Incompatible with sulfuric acid, perchloric acid, isocyanates, strong oxidizers.

BIS(ISOPROPYL)AMINE (108-18-9) Forms explosive mixture with air (flash point 30°F/−1°C). Incompatible with nonoxidizing mineral acids, strong acids, organic acids, organic anhydrides, isocyanates, vinyl acetate, acrylates, substituted allyls, alkylene oxides, epichlorohydrin, ketones, aldehydes, alcohols, glycols, mercury, phenols, cresols, caprolactam solution, strong oxidizers. Attacks aluminum, copper, lead, tin, zinc and alloys, and some plastics, rubber, and coatings.

BIS(2-METHOXYETHYL)-ETHER (111-96-6) Forms explosive mixture on exposure to air, heat, or light. Reacts violently with oxidizers, metal halides. Incompatible with sulfuric acid, isocyanates, perchloric acid. May accumulate static electrical charges; may cause ignition of its vapors.

2,2-BIS(*p*-METHOXYPHENYL)-1,1,1-TRICHLOROETHANE (72-43-5) Contact with strong oxidizers may cause fire and explosions. Attacks some plastics, rubber, and coatings.

BIS(METHYLCYCLOPENTADIENE) (26472-00-4) Forms explosive mixture with air (flash point 80°F/27°C). Incompatible with strong acids, strong oxidizers.

BIS(1-METHYLETHYL)-BENZENE (98-51-1) Forms explosive mixture with air (flash point 155°F/68°C). Incompatible with nitric acid, strong oxidizers.

BIS(6-METHYLHEPTYL) ESTER OF PHTHALIC ACID (27554-26-3) Incompatible with strong acids, nitrates, oxidizers.

BIS(6-METHYLHEPTYL) PHTHALATE (27554-26-3) Incompatible with strong acids, nitrates, oxidizers.

N,N-BIS(2-METHYLPROPYL) AMINE (110-96-3) Forms explosive mixture with air (flash point 85°F/29°C). Incompatible with nonoxidizing mineral acids, strong acids, organic acids, organic anhydrides, isocyanates, vinyl acetate, acrylates, substituted allyls, alkylene oxides, epichlorohydrin, ketones, aldehydes, alcohols, glycols, mercury, phenols, cresols, caprolactam solution, strong oxidizers. Attacks aluminum, copper, lead, tin, zinc and alloys.

BISMUTH SESQUITELLURIDE (1304-82-1) A violent reaction occurs with strong oxidizers, and a toxic gas may evolve from contact with moisture.

BISMUTH TELLURIDE (1304-82-1) A violent reaction occurs with strong oxidizers, and a toxic gas may evolve from contact with moisture.

BISMUTH TRITELLURIDE (1304-82-1) A violent reaction occurs with strong oxidizers, and a toxic gas may evolve from contact with moisture.

BIS(NITRATO-O,O')DIOXO URANIUM E (10102-06-4) Incompatible with combustible materials (fire and explosions). Water contact produces hydrochloric acid; solution attacks most metals. Reacts with esters. Attacks some plastics, rubber, and coatings.

BISOFLEX-81 (117-81-7) Incompatible with strong acids, strong alkalies, nitrates, oxidizers.

BISOFLEX DOA (103-23-1) Incompatible with strong acids, nitrates, oxidizers.

BISOFLEX DOP (117-81-7) Incompatible with strong acids, strong alkalies, nitrates, oxidizers.

BISPHENOL A (80-05-7) Contact with strong oxidizers may cause fire and explosions.

BISPHENOL A DIGLYCIDYL ETHER (1675-54-8) Forms explosive mixture with air (flash point 175°F/79°C). Incompatible with strong acids, strong oxidizers. Ethers form peroxides on contact with air and light.

BISPHENOL A EPICHLOROHYDRIN CONDENSATE (1675-54-8) Forms explosive mixture with air (flash point 175°F/79°C). Incompatible with strong acids, strong oxidizers. Ethers form peroxides on contact with air and light.

BIS(THIOCYANATO)-MERCURY (592-85-8) Heat can cause violent decomposition.

BISULFITE (7446-09-5) Forms corrosive hydrochloric acid fumes with air. Contact with powdered metals or with alkali metals may cause fire and explosions. Reacts violently with alcohols, caustics, amines, water. Decomposes above 140°F/60°C; forms toxic and corrosive oxides of sulfur. Attacks some plastics, rubber, and coatings.

BISULFITE de SODIUM (French) (7631-90-5) Slowly oxidized to the sulfate on contact with air. Reactions with oxidizers or acids produces sulfur dioxide gas.

4,4′-BI-O-TOLUIDINE (95-53-4) Forms explosive mixture with air (flash point 185°F/85°C). Incompatible with strong acids, mineral acids, organic anhydrides, isocyanates, aldehydes, oxidizers. Attacks some plastics, rubber, and coatings.

BITUMEN (8052-42-4) Incompatible with nitric acid, fluorine, strong oxidizers.

BIVINYL (106-99-0) Self-reactive. Forms explosive peroxides with air. Fires, explosions, or hazardous polymerization may result from contact with air, strong oxidizers, strong acids, ozone, nitrogen dioxide, copper and its alloys, phenol, chlorine dioxide, crotonaldehyde, or a free radical polymerization initiator such as hydroquinone.

Add inhibitor (such as *tert*-butycatechol) and monitor to ensure that effective levels are maintained at all times. May accumulate static electrical charges, and may cause ignition of its vapors.

BK (105-60-2) Contact with strong oxidizers may cause fire and explosions.

B-K LIQUID (7681-52-9) A strong oxidizer and a strong base. Stability decreases with concentration, heat, light, decrease in pH, and contamination with metals. Incompatible with strong acids, reducing agents, combustible substances; all cause violent reaction, fire, and explosions. Contact with amines and ammonia salts produces explosive chloroamines. Corrodes many metals: steels, cast iron, monel, aluminum, brass, iconel.

B-K POWDER (7778-54-3) Decomposes in heat or sunlight. Incompatible with acids, moisture, reducing agents, combustible materials, all other chemicals, especially acetylene, aniline and all other amines, anthracene, carbon tetrachloride, iron oxide, manganese oxide, mercaptans, diethylene glycol monomethyl ether, nitromethane, organic matter, organic sulfides, phenol, 1-propanethiol, propyl mercaptan, sulfur, organic sulfur compounds.

BLACK AND WHITE BLEACHING CREAM (123-31-9) Incompatible with strong oxidizers, caustics. May be oxidized to quinone at room temperatures in the presence of moisture. May explode on contact with oxygen.

BLACK LEAF (54-11-5) Incompatible with strong acids, strong oxidizers. Attacks some plastics, rubber, and coatings. May accumulate static electrical charges, and may cause ignition of its vapors.

BLACK OXIDE OF IRON (1309-37-1) Contact with hydrogen peroxide, ethylene oxide, calcium hypochlorite will cause explosion. Reacts violently with powdered aluminum, hydrazine, hydrogen trisulfide.

BLACOSOLV (79-01-6) Contact with caustics produces a toxic and flammable gas. Reacts violently with chemically active metals. Contact with aluminum may produce a violent, self-accelerating polymerization reaction. Incompatible with nonoxidizing mineral acids, strong acids, organic acids, organic anhydrides, isocyanates, alkylene oxides, aldehydes, alcohols, glycols, phenols, cresols, caprolactam solution, epichlorohydrin, nitrogen tetroxide, metal powders, oxygen. May accumulate static electrical charges; may cause ignition of its vapors.

BLADAFUME (3689-24-5) Containers may burst in heat. Strong oxidizers may cause fire and explosions. Attacks some plastics, rubber, and coatings.

BLADAFUN (3689-24-5) Containers may burst in heat. Strong oxidizers may cause fire and explosions. Attacks some plastics, rubber, and coatings.

BLADAN (56-38-2) Combustible liquid. Mixtures with endrin may be explosive. Strong oxidizers may cause fire and explosions. Attacks some plastics, rubber, and coatings.

BLADAN (107-49-3) Decomposes above 300°F/150°C forming flammable ethylene gas. Strong oxidizers may cause fire and explosions. Attacks some plastics, rubber, and coatings.

BLADAN (563-12-2) Incompatible with alkaline formulations. Mixtures with magnesium may be explosive.

111

BLADAN-F (56-38-2) Combustible liquid. Mixtures with endrin may be explosive. Strong oxidizers may cause fire and explosions. Attacks some plastics, rubber, and coatings.

BLADAN-M (298-00-0) Mixtures with magnesium may be explosive.

BLADON (107-49-3) Decomposes above 300°F/150°C, forming flammable ethylene gas. Strong oxidizers may cause fire and explosions. Attacks some plastics, rubber, and coatings.

BLANC FIXE (7727-43-7) Explosions may result from contact with aluminum in the presence of heat. Incompatible with potassium, phosphorus.

BLANDLUBE (8012-95-1) Incompatible with nitric acid; oxidizers may cause fire and explosions.

BLASTING GELATIN (55-63-0) Liquid portion of this chemical is an alcohol; reacts with strong acids, caustics, aliphatic amines, isocyanates. Dry portion of chemical is nitroglycerin, a shock- and friction-sensitive explosive. Acids, heat, or mechanical shock may result in explosions.

BLASTING OIL (55-63-0) Liquid portion of this chemical is an alcohol; reacts with strong acids, caustics, aliphatic amines, isocyanates. Dry portion of chemical is nitroglycerin, a shock- and friction-sensitive explosive. Acids, heat, or mechanical shock may result in explosions.

BLAUSAEURE (German) (74-90-8) Unless stabilized and maintained, samples stored more than 90 days are hazardous. Samples containing more than 2–5% water are less stable than dry material.Can be self-reactive, forming an explosive mixture with air (flash point 0°F/−18°C). Heat or contact with amines or strong bases can cause polymerization. Incompatible with acetaldehyde. Oxidizers may cause fire and explosions. Attacks some plastics, rubber, and coatings.

BLAUWZUUR (Dutch) (74-90-8) Unless stabilized and maintained, samples stored more than 90 days are hazardous. Samples containing more than 2–5% water are less stable than dry material. Can be self-reactive, forming an explosive mixture with air (flash point 0°F/−18°C). Heat or contact with amines or strong bases can cause polymerization. Incompatible with acetaldehyde. Oxidizers may cause fire and explosions. Attacks some plastics, rubber, and coatings.

BLEACH (7681-52-9) A strong oxidizer and a strong base. Stability decreases with concentration, heat, light, decrease in pH, and contamination with metals. Incompatible with strong acids, reducing agents, combustible substances; all cause violent reaction, fire, and explosions. Contact with amines and ammonia salts produces explosive chloroamines. Corrodes many metals: steels, cast iron, monel, aluminum, brass, iconel.

BLEACHING POWDER (7778-54-3) Decomposes in heat or sunlight. Incompatible with acids, moisture, reducing agents, combustible materials, all other chemicals, especially acetylene, aniline and all other amines, anthracene, carbon tetrachloride, iron oxide, manganese oxide, mercaptans, diethylene glycol monomethyl ether, nitromethane, organic matter, organic sulfides, phenol, 1-propanethiol, propyl mercaptan, sulfur, organic sulfur compounds.

BLEISULFAT (German) (7446-14-2) Reacts with aluminum, magnesium, potassium (violently).

BLENDED RED OXIDES OF IRON (1309-37-1) Contact with hydrogen peroxide, ethylene oxide, calcium hypochlorite will cause explosion. Reacts violently with powdered aluminum, hydrazine, hydrogen trisulfide.

BLUE COPPER (7758-98-7) Aqueous solution is an acid. Incompatible with strong bases, hydroxylamine, magnesium.

BLUE OIL (62-53-3) Forms explosive mixture with air (flash point 158°F/70°C). Unless inhibited (usually by methanol), readily able to polymerize. Fires and explosions may result from contact with halogens, strong acids, oxidizers, organic anhydrides, acetic anhydride, isocyanates, aldehydes, sodium peroxide. Reacts with alkali metals and alkaline earth metals. Attacks some plastics, rubber, coatings, copper, and copper alloys.

BLUE-OX (1314-84-7) Incompatible with oxidizers, acids, water.

BLUE STONE (7758-98-7) Aqueous solution is an acid. Incompatible with strong bases, hydroxylamine, magnesium.

BLUE VITRIOL (7758-98-7) Aqueous solution is an acid. Incompatible with strong bases, hydroxylamine, magnesium.

BOILED LINSEED OIL (8001-26-1) Incompatible with nitric acid; oxidizers may cause fire and explosions.

BOLETIC ACID (110-17-8) Incompatible with sulfuric acid, caustics, ammonia, amines, isocyanates, alkylene oxides, epichlorohydrin.

BONAMID (105-60-2) Contact with strong oxidizers may cause fire and explosions.

BONAZEN (7733-02-0) Incompatible with strong bases.

BONOFORM (79-34-5) Exposure to heat, light, and air produces corrosive and toxic vapors. Reacts with strong caustics to form explosive dichloroacetylene. Reacts violently with chemically active metals or sodium amide. In presence of steam, contact with hot iron, aluminum or zinc may produces toxic vapors. Attacks some plastics, rubber, and coatings.

BORACIC ACID (10043-35-3) Incompatible with bases, potassium. Acetic anhydride produces heat-sensitive explosive.

BORANE, TRIBROMO- (10294-33-4) Incompatible with water, steam, potassium, sodium, alcohol. Attacks metals, rubber.

BORANE, TRIFLUORO- (7637-07-2) Decomposes in heat, water and moist air. Fire and explosions may result from contact with alkali metals, alkaline earth metals (except magnesium), alkyl nitrate, calcium oxide. Incandesces when heated with alkali metals or alkaline earth metals (except magnesium). Corrodes most metals in the presence of moisture.

BORAX (1303-96-4) Reacts with acids, metallic salts.

BORIC ACID (10043-35-3) Incompatible with bases, potassium. Acetic anhydride contact causes heat-sensitive explosive.

BORIC ANHYDRIDE (1303-86-2) Contact with water produces boric acid. Fires and explosion may result from contact with bromine pentafluoride or calcium oxide.

BORIC OXIDE (1303-86-2) Contact with water forms boric acid. Fires and explosion may result from contact with bromine pentafluoride or calcium oxide.

BORON FLUORIDE (7637-07-2) Decomposes in heat, water, and moist air. Fire and explosions may result from contact with alkali metals, alkaline earth metals (except magnesium), alkyl nitrate, calcium oxide. Incandesces when heated with alkali metals or alkaline earth metals (except magnesium). Corrodes most metals in the presence of moisture.

BORON OXIDE (1303-86-2) Contact with water produces boric acid. Fires and explosion may result from contact with bromine pentafluoride or calcium oxide.

BORON SEQUIOXIDE (1303-86-2) Contact with water produces boric acid. Fires and explosion may result from contact with bromine pentafluoride or calcium oxide.

BORON TRIFLUORIDE (7637-07-2) Decomposes in heat, water, and moist air. Fire and explosions may result from contact with alkali metals, alkaline earth metals (except magnesium), alkyl nitrate, calcium oxide. Incandesces when heated with alkali metals or alkaline earth metals (except magnesium). Corrodes most metals in the presence of moisture.

BORON TRIOXIDE (1303-86-2) Contact with water produces boric acid. Fires and explosion may result from contact with bromine pentafluoride or calcium oxide.

BORNANE, 2-OXO- (76-22-2); (8008-51-3) Incompatible with strong oxidizers, especially chromic anhydride, to form explosive mixtures.

2-BORNANONE (76-22-2); (8008-51-3) Incompatible with strong oxidizers especially chromic anhydride, to form explosive mixtures.

BOROETHANE (19287-45-7) Unstable above 0°F/−18°C; may ignite spontaneously in moist air above room temperature. Air, halogenated compounds may cause fire and explosion. Contact with aluminum, lithium, and other active metals produces hydrides, which may ignite spontaneously. Reacts with oxidized surfaces as a strong reducing agent. Attacks some plastics, rubber, and coatings.

BOROFAX (10043-35-3) Incompatible with bases, potassium. Acetic anhydride contact produces heat-sensitive explosive.

BOROHYDRIDE SOLUTION (16940-66-2) Produces hydrogen gas on contact with water or moist air. Reacts violently with oxidizers, dimethyl formamide, aldehydes, acids (forms diborane gas), ketones, metal powders. Polymerization may be caused by contact with bases, acrylonitrile. Corrodes glass slowly.

BOROHYDRURE de SODIUM (French) (16940-66-2) Forms hydrogen gas on contact with water or moist air. Reacts violently with oxidizers, dimethyl formamide, aldehydes, acids (forms diborane gas), ketones, metal powders. Polymerization may be caused by contact with bases, acrylonitrile. Corrodes glass slowly.

BORON BROMIDE (10294-33-4) Incompatible with water, steam, potassium, sodium, alcohol. Attacks metals, rubber.

BORON CHLORIDE (10294-34-5) Water contact produces hydrochloric and boric acids. Incompatible with lead, graphite-impregnated asbestos, potassium, sodium. Vigorously attacks elastomers, packing materials, natural and synthetic rubber, viton, tygon, saran, silastic elastomers. Corrodes metals in the presence of moisture.

BORON FLUORIDE (7787-71-5) A powerful oxidizer. Water contact produces hydrogen fluoride gas. Reacts violently with reducing agents, organic materials, acids, halogens, salts, alkalies, metal oxides, and many other materials. Attacks some plastics, rubber, and coatings.

BORON HYDRIDE (17702-41-9) Incompatible with oxidizers, oxygenated solvents, dimethyl sulfoxide. Carbon tetrachloride, ethers, halocarbons, halogenated compounds form shock-sensitive mixtures. Attacks some plastics, rubber, and coatings.

BORON HYDRIDE (19287-45-7) Unstable above 0°F/−18°C; may ignite spontaneously in moist air above room temperature. Air, halogenated compounds may cause fire and explosion. Contact with aluminum, lithium, and other active metals produces hydrides, which may ignite spontaneously. Reacts with oxidized surfaces as a strong reducing agent. Attacks some plastics, rubber, and coatings.

BORON TRIBROMIDE or BORON TRIBROMIDE 6 (10294-33-4) Incompatible with water, steam, potassium, sodium, alcohol. Attacks metals, rubber.

BORON TRICHLORIDE (10294-34-5) Water contact produces hydrochloric and boric acids. Incompatible with lead, graphite-impregnated asbestos, potassium, sodium. Vigorously attacks elastomers, packing materials, natural and synthetic rubber, viton, tygon, saran, silastic elastomers. Corrodes metals in the presence of moisture.

BORSAEURE (German) (10043-35-3) Incompatible with bases, potassium. Acetic anhydride contact produces heat-sensitive explosive.

BOS MH (123-33-1) Contact with strong oxidizers may cause fire and explosions.

BOSAN SUPRA (50-29-3) Incompatible with salts of iron or aluminum, and bases. Do not store in iron containers.

BOTTLED GAS (74-98-6) Forms explosive gas mixture with air. Strong oxidizers may cause fire and explosions. Liquid attacks some plastics, rubber, and coatings. May accumulate static electrical charges; may cause ignition of its vapors.

BOTTLED GAS (68476-85-7) Forms explosive mixture with air. Strong oxidizers may cause fire and explosions. Attacks some plastics, rubber, and coatings.

BOV (7664-93-9) A strong oxidizer that can react violently with risk of fire and explosion with many substances, including reducing agents, organic and combustible substances, and bases. Incompatible with nonoxidizing mineral acids, organic acids, bases, acrylates, aldehydes, alcohols, alkylene oxides, ammonia, aliphatic amines, alkanolamines, aromatic amines, amides, chlorates, epichlorohydrin, fulminates, glycols, isocyanates, ketones, metals (powdered), organic anhydrides, perchlorates, picrates, substituted allyls, phenols, cresols, water, acetic anhydride, acetone cyanhydrin, acetonitrile, acrolein, acrylonitrile, allyl alcohol, allyl chloride, 2-aminoethanol, ammonium hydroxide, aniline, bromine pentafluoride, *n*-butyraldehyde, caprolactam solution, carbides, cesium acetylene carbide, chlorine trifluoride, chlorosulfonic acid, cuprous nitride, diisobutylene, ethylene cyanohydrin, ethylene diamine, ethylene glycol, ethyleneimine, hydrochloric acid, iodine heptafluoride, iron, isoprene, lithium silicide, mercuric nitride, mesityl oxide, nitric acid, *p*-nitrotoluene, perchloric acid, phosphorus, potassium *tert*-butoxide, potassium chlorate, potassium permanganate, 3-propiolactone, propylene oxide, pyridine, rubidium acetylene, silver permanganate, sodium, sodium carbonate, sodium chlorate, sodium hydroxide, styrene monomer, vinyl acetate. Attacks most metals, and some plastics, rubber and coatings.

BOVIDERMOL (50-29-3) Incompatible with salts of iron or aluminum, and bases. Do not store in iron containers.

BOX TOE GUM (9004-70-0) Forms explosive mixture with air (flash point −64°F/−53°C). Strong oxidizers may cause fire and explosions. May accumulate static electrical charges; may cause ignition of its vapors. Attacks some plastics, rubber, and coatings.

BP (94-36-0) Confined storage of dry chemical may lead to decomposition and explosion. A strong oxidant; extremely reactive. Fires and explosion may result from heat or contamination, and from contact with strong acids, combustible materials, oxidizers, acids, bases, alcohols, reducing agents, metals, metal oxides, amines, accelerators, methyl methacrylate, organic matter, lithium aluminum carbide, dimethyl aniline, amines, metallic naphthenates. May attack some plastics, rubber and coatings. Protect containers from shock and friction.

BPL (57-57-8) Polymerizes in storage. High temperatures cause ruptured containers. Decomposes at room temperature; hydrolyzes in water.

BPO (94-36-0) Confined storage of dry chemical may lead to decomposition and explosion. A strong oxidant; extremely reactive. Fires and explosion may result from heat or contamination, and from contact with strong acids, combustible materials, oxidizers, acids, bases, alcohols, reducing agents, metals, metal oxides, amines, accelerators, methyl methacrylate, organic matter, lithium aluminum carbide, dimethyl aniline, amines, metallic naphthenates. May attack some plastics, rubber and coatings. Protect containers from shock and friction.

BRECOLANE NDG (111-46-6) Incompatible with sulfuric acid, isocyanates, strong oxidizers.

BREVINYL (62-73-7) Attacks some plastics, rubber, and coatings.

BRICK OIL (65996-93-2) Incompatible with oxidizers, strong acids, caustics, aliphatic amines, isocyanates.

BRIMSTONE (7704-34-9) Combustible solid. Liquid forms sulfur dioxide with air. Reacts violently with strong oxidizers. Forms explosive, shock-sensitive or pyrophoric mixtures with ammonia, ammonium nitrate, bromates, calcium carbide, charcoal, chlorates, hydrocarbons, iodates, iron. Reacts violently with halogen compounds, sodium, tin, uranium, and other compounds. Attacks steel when moist. May accumulate static electrical charges; may cause ignition of its vapors.

BRITON (52-68-6) Contact with strong oxidizers may cause fire and explosions.

BRITTEN (52-68-6) Contact with strong oxidizers may cause fire and explosions.

BROCIDE (107-06-2) Forms explosive mixture with air (flash point 55°F/13°C). Incompatible with strong oxidizers, strong caustics, chemically active metals. Corrosive to iron and other metals unless stabilized with alkylamines. Attacks some plastics, rubber, and coatings.

BRODAN (921-88-2) Incompatible with strong acids. Hydrolysis results from acid or alkaline solutions.

BROM (German) (7726-95-6) A powerful oxidizer. May cause fire and explosions in contact with organic or other readily oxidizable materials. Contact with ammonia, acetaldehyde, acetylene, acrylonitrile, or with metals may cause violent reactions. Reacts violently with aluminum, titanium, mercury, or potassium. Also incompatible with alcohols, antimony, alkali hydroxides, arsenites, boron, calcium nitrite, cesium monoxide, carbonyls, dimethyl formamide, ethyl phosphine, fluorine, ferrous and mercurous salts, germanium, hypophosphites, iron carbide, isobutyronphenone, magnesium phosphide, methanol, nickel carbonyl, olefins, ozone, sodium, and many other substances. Attacks some coatings, and some forms of plastic and rubber. Corrodes iron, steel, stainless steels, and copper.

BROMALLYLENE (106-95-1) Forms explosive mixture with air (flash point 30°F/−1°C). Reacts violently with oxidizers. May accumulate static electrical charges, and may cause ignition of its vapors.

BROMCHLOPHOS (300-76-5) Contact with strong oxidizers may cause fire and explosions. Attacks some plastics, rubber, and coatings.

BROME (French) (7726-95-6) A powerful oxidizer. May cause fire and explosions in contact with organic or other readily oxidizable materials. Contact with ammonia, acetaldehyde, acetylene, acrylonitrile, or with metals may cause violent reactions. Reacts violently with aluminum, titanium, mercury or potassium. Also incompatible with alcohols, antimony, alkali hydroxides, arsenites, boron, calcium nitrite, cesium monoxide, carbonyls, dimethyl formamide, ethyl phosphine, fluorine, ferrous and mercurous salts, germanium, hypophosphites, iron carbide, isobutyronphenone, magnesium phosphide, methanol, nickel carbonyl, olefins, ozone, sodium and many other substances. Attacks some coatings, and some forms of plastic and rubber. Corrodes iron, steel, stainless steels and copper.

BROMELITE (1304-56-9) Incompatible with acids, bases, chlorinated hydrocarbons, oxidizers, molten lithium.

BROMELLITE (1304-56-9) Incompatible with acids, bases, chlorinated hydrocarbons, oxidizers, molten lithium.

BROMEX (300-76-5) Contact with strong oxidizers may cause fire and explosions. Attacks some plastics, rubber, and coatings.

BROMIC ACID, BARIUM SALT (13967-90-3) Keep away from heat, combustible materials, water, sources of ignition, aluminum, arsenic, carbon, copper, phosphorus, and other oxidizable materials.

BROMINE (7726-95-6) A powerful oxidizer. May cause fire and explosions in contact with organic or other readily oxidizable materials. Contact with ammonia, acetaldehyde, acetylene, acrylonitrile, or metals may cause violent reactions. Reacts violently with aluminum, titanium, mercury, or potassium. Also incompatible with alcohols, antimony, alkali hydroxides, arsenites, boron, calcium nitrite, cesium monoxide, carbonyls, dimethyl formamide, ethyl phosphine, fluorine, ferrous and mercurous salts, germanium, hypophosphites, iron carbide, isobutyronphenone, magnesium phosphide, methanol, nickel carbonyl, olefins, ozone, sodium, and many other substances. Attacks some coatings, and some forms of plastic and rubber. Corrodes iron, steel, stainless steels, and copper.

BROMINE CYANIDE (506-68-3) May be unstable unless dry and pure. Reacts violently with acids, ammonia, amines. Water contact produces hydrogen cyanide and hydrogen bromide.

BROMINE FLUORIDE (7787-71-5) A powerful oxidizer. Water contact produces hydrogen fluoride gas. Reacts violently with reducing agents, organic materials, acids, halogens, salts, alkalies, metal oxides, and many other materials. Attacks some plastics, rubber, and coatings.

BROMINE FLUORIDE (7789-30-2) Highly reactive with every known element except the inert gases, nitrogen, and oxygen. Incompatible with acids, halogens, arsenic, selenium, alkaline halides, sulfur, iodine, glass, metallic halides, metal oxides, and metals (except copper, stainless steel, nickel and Monel®). Fire may result from contact with combustibles or organic matter at room temperature, and contact of this substance with water produces an explosion. Even under mild conditions this substance attacks organic compounds vigorously, often causing explosion.

BROMINE PENTAFLUORIDE (7789-30-2) Highly reactive with every known element except the inert gases, nitrogen, and oxygen. Incompatible with acids, halogens, arsenic, selenium, alkaline halides, sulfur, iodine, glass, metallic halides, metal oxides, and metals (except copper, stainless steel, nickel, and Monel®). Fire may result from contact with combustibles or organic matter at room temperature, and contact of this substance with water produces an explosion. Even under mild conditions this substance attacks organic compounds vigorously, often causing explosion.

BROMINE TRIFLUORIDE (7787-71-5) A powerful oxidizer. Water contact produces hydrogen fluoride gas. Reacts violently with reducing agents, organic materials, acids, halogens, salts, alkalies, metal oxides, and many other materials. Attacks some plastics, rubber, and coatings.

BROMO (Italian) (7726-95-6) A powerful oxidizer. May cause fire and explosions in contact with organic or other readily oxidizable materials. Contact with ammonia, acetaldehyde, acetylene, acrylonitrile, or metals may cause violent reactions. Reacts violently with aluminum, titanium, mercury, or potassium. Also incompatible with alcohols, antimony, alkali hydroxides, arsenites, boron, calcium nitrite, cesium monoxide, carbonyls, dimethyl formamide, ethyl phosphine, fluorine, ferrous and mercurous salts, germanium, hypophosphites, iron carbide, isobutyronphenone, magnesium phosphide, methanol, nickel carbonyl, olefins, ozone, sodium, and many other substances. Attacks some coatings, and some forms of plastic and rubber. Corrodes iron, steel, stainless steels and copper.

BROMOACETONE (−2) Forms explosive mixture with air (flash point 113°F/45°C). Contact with oxidizers may cause fire and explosions.

BROMOACETYL BROMIDE (598-21-0) Water contact produces hydrogen bromide. Corrodes metals.

BROMOBENZENE (108-86-1) Forms explosive mixture with air (flash point 124°F/51°C). Incompatible with strong oxidizers, alkaline earth metals (barium, calcium, magnesium, strontium, etc.), metallic salts, with risk of violent reactions. May accumulate static electrical charges; may cause ignition of its vapors.

118

BROMOBENZOL (108-86-1) Forms explosive mixture with air (flash point 124°F/51°C). Incompatible with strong oxidizers, alkaline earth metals (barium, calcium, magnesium, strontium, etc.), metallic salts, with risk of violent reactions. May accumulate static electrical charges; may cause ignition of its vapors.

N-BROMOBUTANE (109-65-9) Forms explosive mixture with air (flash point 65°F/18°C). Incompatible with strong oxidizers, strong acids. May accumulate static electrical charges; may cause ignition of its vapors.

1-BROMO BUTANE (109-65-9) Forms explosive mixture with air (flash point 65°F/18°C). Incompatible with strong oxidizers, strong acids. May accumulate static electrical charges; may cause ignition of its vapors.

2-BROMOBUTANE (78-76-2) Forms explosive mixture with air (flash point 70°F/26°C). Reacts with oxidizers.

BROMOCHLOROMETHANE (74-97-5) Incompatible with chemically active metals such as calcium, powdered aluminum, zinc, and magnesium. Liquid attacks some plastics, rubber, and coatings.

BROMOCYAN (506-68-3) May be unstable unless dry and pure. Reacts violently with acids, ammonia, amines. Water contact produces hydrogen cyanide and hydrogen bromide.

BROMOCYANOGEN (506-68-3) May be unstable unless dry and pure. Reacts violently with acids, ammonia, amines. Water contact produces hydrogen cyanide and hydrogen bromide.

BROMOETHANE (74-96-4) Forms explosive mixture with air (flash point less than −4°F/−20°C). Oxidizers may cause fire or explosions. Incompatible with chemically active metals. Attacks some plastic, rubber, and coatings.

BROMOETHANOYL BROMIDE (598-21-0) Water contact produces hydrogen bromide. Corrodes metals.

BROMOFLUOROFORM (75-63-8) Reacts with chemically active metals, powdered aluminum, zinc, magnesium. May attack some plastics, rubber, and coatings.

BROMOFORM (75-25-2) Reacts with chemically active metals, calcium, acetone, strong caustics. Attacks some plastics, rubber, and coatings.

BROMOFORME (French) (75-25-2) Reacts with chemically active metals, calcium, acetone, strong caustics. Attacks some plastics, rubber, and coatings.

BROMOFORMIO (Italian) (75-25-2) Reacts with chemically active metals, calcium, acetone, strong caustics. Attacks some plastics, rubber, and coatings.

BROMOFUME (106-93-4) Reacts with chemically active metals, liquid ammonia, strong oxidizers. Heat and light cause slow decomposition. Attacks some plastics and rubber.

BROM-O-GAS (74-83-9) Incompatible with strong oxidizers, aluminum, dimethylsulfoxide, ethylene oxide, water. Attacks zinc, magnesium, alkali metals, and their alloys, and some plastics, rubber, and coatings.

BROMOMETHANE (74-83-9) Incompatible with strong oxidizers, aluminum, dimethylsulfoxide, ethylene oxide, water. Attacks zinc, magnesium, alkali metals and their alloys.

(BROMOMETHYL)BENZENE (100-39-0) Forms explosive mixture with air (flash point 174°F/79°C). Water contact produces hydrobromic acid. Incompatible with strong oxidizers, bases. Attacks metals, except nickel and lead, in the presence of moisture.

BROMOMETHYL METHYL KETONE (−2) Forms explosive mixture with air (flash point 113°F/45°C). Contact with oxidizers may cause fire and explosions.

p-**(BROMOMETHYL)NITROBENZENE** (100-39-0) Forms explosive mixture with air (flash point 174°F/79°C). Water contact produces hydrobromic acid. Incompatible with strong oxidizers, bases. Attacks metals, except nickel and lead, in the presence of moisture.

2-BROMOPENTANE (107-81-3) Forms explosive mixture with air (flash point 90°F/32°C). Incompatible with strong oxidizers, strong acids.

BROMOPHENYLMETHANE (100-39-0) Forms explosive mixture with air (flash point 174°F/79°C). Water contact produces hydrobromic acid. Incompatible with strong oxidizers, bases. Attacks metals, except nickel and lead, in the presence of moisture.

BROMOPROPANE (−2) Forms explosive mixture with air (flash point 113°F/45°C). Contact with oxidizers may cause fire and explosions.

1-BROMOPROPANE (106-94-5) Forms explosive mixture with air (flash point 78°F/26°C). Strong oxidizers may cause fire and explosions.

BROMO-2-PROPANONE (−2) Forms explosive mixture with air (flash point 113°F/45°C). Contact with oxidizers may cause fire and explosions.

1-BROMO-2-PROPANONE (−2) Forms explosive mixture with air (flash point 113°F/45°C). Contact with oxidizers may cause fire and explosions.

3-BROMO-1-PROPENE (106-95-1) Forms explosive mixture with air (flash point 30°F/−1°C). Reacts violently with oxidizers. May accumulate static electrical charges, and may cause ignition of its vapors.

3-BROMOPROPENE (106-95-1) Forms explosive mixture with air (flash point 30°F/−1°C). Reacts violently with oxidizers. May accumulate static electrical charges, and may cause ignition of its vapors.

3-BROMOPROPYLENE (106-95-1) Forms explosive mixture with air (flash point 30°F/−1°C). Reacts violently with oxidizers. May accumulate static electrical charges, and may cause ignition of its vapors.

a-BROMOTOLUENE or **alpha-BROMOTOLUENE** (100-39-0) Forms explosive mixture with air (flash point 174°F/79°C). Water contact produces hydrobromic acid. Incompatible with strong oxidizers, bases. Attacks metals, except nickel and lead, in the presence of moisture.

omega-BROMOTOLUENE (100-39-0) Forms explosive mixture with air (flash point 174°F/79°C). Water contact produces hydrobromic acid. Incompatible with strong oxidizers, bases. Attacks metals, except nickel and lead, in the presence of moisture.

BROMOTOLUENE, alpha (100-39-0) Forms explosive mixture with air (flash point 174°F/79°C). Water contact produces hydrobromic acid. Incompatible with strong oxidizers, bases. Attacks metals, except nickel and lead, in the presence of moisture.

BROMOTRIFLUOROMETHANE (75-63-8) Reacts with chemically active metals, powdered aluminum, zinc, magnesium. May attack some plastics, rubber, and coatings.

BROMOWODOR (Polish) (10035-10-6) Incompatible with aliphatic amines, alkanolamines, alkylene oxides, aromatic amines, amides, ammonia, ammonium hydroxide, bases, calcium oxide, epichlorohydrin, fluorine, isocyanates, oleum, organic anhydrides, sulfuric acid, sodium tetrahydroborate, strong oxidizers, vinyl acetate, water. Attacks most metals with the formation of flammable hydrogen gas.

BROMURE de CYANOGEN (French) (506-68-3) May be unstable unless dry and pure. Reacts violently with acids, ammonia, amines. Water contact produces hydrogen cyanide and hydrogen bromide.

BROMURE d'ETHYLE (French) (74-96-4) Forms explosive mixture with air (flash point less than −4°F/−20°C). Oxidizers may cause fire or explosions. Incompatible with chemically active metals. Attacks some plastic, rubber, and coatings.

BROMURO di ETILE (Italian) (106-93-4) Reacts with chemically active metals, liquid ammonia, strong oxidizers. Heat and light cause slow decomposition. Attacks some plastics and rubber.

BROMWASSERSTOFF (German) (10035-10-6) Incompatible with aliphatic amines, alkanolamines, alkylene oxides, aromatic amines, amides, ammonia, ammonium hydroxide, bases, calcium oxide, epichlorohydrin, fluorine, isocyanates, oleum, organic anhydrides, sulfuric acid, sodium tetrahydroborate, strong oxidizers, vinyl acetate, water. Attacks most metals with the formation of flammable hydrogen gas.

BROOM (Dutch) (7726-95-6) A powerful oxidizer. May cause fire and explosions in contact with organic or other readily oxidizable materials. Contact with ammonia, acetaldehyde, acetylene, acrylonitrile, or metals may cause violent reactions. Reacts violently with aluminum, titanium, mercury, or potassium. Also incompatible with alcohols, antimony, alkali hydroxides, arsenites, boron, calcium nitrite, cesium monoxide, carbonyls, dimethyl formamide, ethyl phosphine, fluorine, ferrous and mercurous salts, germanium, hypophosphites, iron carbide, isobutyronphenone, magnesium phosphide, methanol, nickel carbonyl, olefins, ozone, sodium, and many other substances. Attacks some coatings, and some forms of plastic and rubber. Corrodes iron, steel, stainless steels, and copper.

BROOMWATERSTOF (Dutch) (10035-10-6) Incompatible with aliphatic amines, alkanolamines, alkylene oxides, aromatic amines, amides, ammonia, ammonium hydroxide, bases, calcium oxide, epichlorohydrin, fluorine, isocyanates, oleum, organic anhydrides, sulfuric acid, sodium tetrahydroborate, strong oxidizers, vinyl acetate, water. Attacks most metals with the formation of flammable hydrogen gas.

11460 BROWN (1300-73-8) Forms explosive mixture with air (flash point 206°F/96.7°C). Contact with strong acids may cause fire and explosions. Contact with hypochlorite bleaches produces explosive chloroamines. Incompatible with strong acids, organic acids and anhydrides, isocyanates, aldehydes.

BRUCIN (German) (357-57-3) Contact with strong oxidizers may cause fire and explosions. May accumulate static electrical charges; may cause ignition of its vapors.

BRUCINA (Italian) (357-57-3) Contact with strong oxidizers may cause fire and explosions. May accumulate static electrical charges; may cause ignition of its vapors.

BRUCINE or (−)BRUCINE or BRUCINE ALKALOID, BRUCINE, SOLID (357-57-3) Contact with strong oxidizers may cause fire and explosions. May accumulate static electrical charges; may cause ignition of its vapors.

(−)BRUCINE DIHYDRATE (357-57-3) Contact with strong oxidizers may cause fire and explosions. May accumulate static electrical charges; may cause ignition of its vapors.

BRUCINE HYDRATE (357-57-3) Contact with strong oxidizers may cause fire and explosions. May accumulate static electrical charges; may cause ignition of its vapors.

BRUMIN (81-81-2) Strong oxidizers may cause fire and explosions.

BRUSH BUSTER (1918-00-9) Incompatible with sulfuric acid, bases, ammonia, aliphatic amines, alkanolamines, isocyanates, alkylene oxides, epichlorohydrin.

BRUSH-OFF 445 LOW VOLATILE BRUSH KILLER (93-76-5) Sealed metal containers may burst in heat above 316°F/158°C. Incompatible with sulfuric acid, bases, ammonia, aliphatic amines, alkanolamines, isocyanates, alkylene oxides, epichlorohydrin.

BRUSH RHAP (93-76-5) Sealed metal containers may burst in heat above 316°F/158°C. Incompatible with sulfuric acid, bases, ammonia, aliphatic amines, alkanolamines, isocyanates, alkylene oxides, epichlorohydrin.

BRUSHTOX (93-76-5) Sealed metal containers may burst in heat above 316°F/158°C. Incompatible with sulfuric acid, bases, ammonia, aliphatic amines, alkanolamines, isocyanates, alkylene oxides, epichlorohydrin.

BSC-REFINED D (98-09-9) Incompatible with ammonia, aliphatic amines, water solutions. Water contact produces hydrochloric and chlorosulfonic acids; aqueous solutions react violently with bases. Attacks metals in presence of moisture.

B-SELEKTONON (94-75-7) Decomposes in sunlight. Incompatible with strong oxidizers; may cause fire and explosions.

BTMAC (56-93-9) Incompatible with strong acids, oxidizers.

BU2AE (102-81-8) Forms explosive mixture with air (flash point 200°F/93°C). Reacts violently with strong oxidizers.

BUCS (111-76-2) Forms explosive mixture with air (flash point 143°F/62°C). Reacts violently with strong caustics and strong oxidizers. Attacks some coatings, plastics and rubber. Attacks metallic aluminum at high temperatures.

BUFOPTO ZINC SULFATE (7733-02-0) Reacts with strong bases.

BUHACH (8003-34-7) Strong oxidizers may cause fire and explosions.

BUNKER C OIL (68476-33-5) Forms explosive mixture with air (flash point 142°F/61°C). Incompatible with nitric acid; oxidizers may cause fire and explosions.

BUNT-CURE (118-74-1) Reacts violently with strong oxidizers, formyldimethylamine.

BUNT-NO-MORE (118-74-1) Reacts violently with strong oxidizers, formyldimethylamine.

122

BURNED LIME or **BURNT LIME** (1305-78-8) Reacts violently with water. Incompatible with ethanol, liquid hydrogen fluoride, some acids, some halogens, some metal halides, some metals, some oxides, boron trifluoride, chlorine trifluoride, liquid hydrofluoric acid, phosphorus pentoxide, boric oxide and calcium chloride mixtures, fluorine, chlorine trifluoride, carbon dioxide.

BURNT ISLAND RED (1309-37-1) Contact with hydrogen peroxide, ethylene oxide, calcium hypochlorite will cause explosion. Reacts violently with powdered aluminum, hydrazine, hydrogen trisulfide.

BURNT SIENNA (1309-37-1) Contact with hydrogen peroxide, ethylene oxide, calcium hypochlorite will cause explosion. Reacts violently with powdered aluminum, hydrazine, hydrogen trisulfide.

BURNT UMBER (1309-37-1) Contact with hydrogen peroxide, ethylene oxide, calcium hypochlorite will cause explosion. Reacts violently with powdered aluminum, hydrazine, hydrogen trisulfide.

BUSH KILLER (94-75-7) Decomposes in sunlight. Incompatible with strong oxidizers; may cause fire and explosions.

BUTADIEEN (Dutch) (106-99-0) Self-reactive. Forms explosive peroxides with air. Fires, explosions, or hazardous polymerization may result from contact with air, strong oxidizers, strong acids, ozone, nitrogen dioxide, copper and its alloys, phenol, chlorine dioxide, crotonaldehyde, or a free radical polymerization initiator such as hydroquinone. Add inhibitor (e.g., *tert*-butylcatechol) and monitor to ensure that effective levels are maintained at all times. May accumulate static electrical charges, and may cause ignition of its vapors.

BUTA-1,3-DIEEN (Dutch) (106-99-0) Self-reactive. Forms explosive peroxides with air. Fires, explosions, or hazardous polymerization may result from contact with air, strong oxidizers, strong acids, ozone, nitrogen dioxide, copper and its alloys, phenol, chlorine dioxide, crotonalde- hyde, or a free radical polymerization initiator such as hydroquinone. Add inhibitor (e.g., *tert*-butylcatechol) and monitor to ensure that effective levels are maintained at all times. May accumulate static electrical charges, and may cause ignition of its vapors.

BUTADIEN (Polish) (106-99-0) Self-reactive. Forms explosive peroxides with air. Fires, explosions, or hazardous polymerization may result from contact with air, strong oxidizers, strong acids, ozone, nitrogen dioxide, copper and its alloys, phenol, chlorine dioxide, crotonaldehyde, or a free radical polymerization initiator such as hydroquinone. Add inhibitor (e.g., *tert*-butylcatechol) and monitor to ensure that effective levels are maintained at all times. May accumulate static electrical charges, and may cause ignition of its vapors.

BUTA-1,3-DIEN (German) (106-99-0) Self-reactive. Forms explosive peroxides with air. Fires, explosions, or hazardous polymerization may result from contact with air, strong oxidizers, strong acids, ozone, nitrogen dioxide, copper and its alloys, phenol, chlorine dioxide, crotonaldehyde, or a free radical polymerization initiator such as hydroquinone. Add inhibitor (e.g., *tert*-butylcatechol) and monitor to ensure that effective levels are maintained at all times. May accumulate static electrical charges, and may cause ignition of its vapors.

BUTADIENE (106-99-0) Self-reactive. Forms explosive peroxides with air. Fires, explosions, or hazardous polymerization may result from contact with air, strong oxidizers, strong acids, ozone, nitrogen dioxide, copper and its alloys, phenol, chlorine dioxide, crotonaldehyde, or a free radical polymerization initiator such as hydroquinone.

Add inhibitor (e.g., *tert*-butylcatechol) and monitor to ensure that effective levels are maintained at all times. May accumulate static electrical charges, and may cause ignition of its vapors.

1,2-BUTADIENE (106-99-0) Self-reactive. Forms explosive peroxides with air. Fires, explosions, or hazardous polymerization may result from contact with air, strong oxidizers, strong acids, ozone, nitrogen dioxide, copper and its alloys, phenol, chlorine dioxide, crotonaldehyde, or a free radical polymerization initiator such as hydroquinone. Add inhibitor (e.g., *tert*-butylcatechol) and monitor to ensure that effective levels are maintained at all times. May accumulate static electrical charges, and may cause ignition of its vapors.

1,3-BUTADIENE (106-99-0) Self-reactive. Forms explosive peroxides with air. Fires, explosions, or hazardous polymerization may result from contact with air, strong oxidizers, strong acids, ozone, nitrogen dioxide, copper and its alloys, phenol, chlorine dioxide, crotonaldehyde, or a free radical polymerization initiator such as hydroquinone. Add inhibitor (e.g., *tert*-butylcatechol) and monitor to ensure that effective levels are maintained at all times. May accumulate static electrical charges, and may cause ignition of its vapors.

alpha-gamma-BUTADIENE (106-99-0) Self-reactive. Forms explosive peroxides with air. Fires, explosions, or hazardous polymerization may result from contact with air, strong oxidizers, strong acids, ozone, nitrogen dioxide, copper and its alloys, phenol, chlorine dioxide, crotonaldehyde, or a free radical polymerization initiator such as hydroquinone. Add inhibitor (e.g., *tert*-butylcatechol) and monitor to ensure that effective levels are maintained at all times. May accumulate static electrical charges, and may cause ignition of its vapors.

BUTADIENE DIOXIDE (111-42-2) Incompatible with nonoxidizing mineral acids, strong acids, organic acids, organic anhydrides, isocyanates, vinyl acetate, acrylates, substituted allyls, alkylene oxides, epichlorohydrin, aldehydes, oxidizers. Corrosive to copper, copper alloys, zinc, and galvanized iron.

1,3-BUTADIENE, 1,1,2,3,4,4-HEXACHLORO- (87-68-3) Forms explosive mixture with air (flash point 195°F/90°C). Reacts strongly with oxidizers, aluminum powder. Attacks aluminum, and some plastics, rubber, and coatings.

BUTAL (123-72-8) May accumulate static electrical charges, and may cause ignition of its vapors. Forms explosive mixture with air (flash point −8°F/−22°C). Incompatible with strong oxidizers, strong acids, caustics, ammonia, aliphatic amines, alkanolamines, aromatic amines.

BUTALDEHYDE (123-72-8) May accumulate static electrical charges, and may cause ignition of its vapors. Forms explosive mixture with air (flash point −8°F/−22°C). Incompatible with strong oxidizers, strong acids, caustics, ammonia, aliphatic amines, alkanolamines, aromatic amines.

n-BUTALDEHYDE (123-72-8) May accumulate static electrical charges, and may cause ignition of its vapors. Forms explosive mixture with air (flash point −8°F/−22°C). Incompatible with strong oxidizers, strong acids, caustics, ammonia, aliphatic amines, alkanolamines, aromatic amines.

BUTALYDE (123-72-8) May accumulate static electrical charges, and may cause ignition of its vapors. Forms explosive mixture with air (flash point −8°F/−22°C). Incompatible with strong oxidizers, strong acids, caustics, ammonia, aliphatic amines, alkanolamines, aromatic amines.

BUTANAL (123-72-8) May accumulate static electrical charges, and may cause ignition of its vapors. Forms explosive mixture with air (flash point −8°F/−22°C). Incompatible with strong oxidizers, strong acids, caustics, ammonia, aliphatic amines, alkanolamines, aromatic amines.

(E)-2-BUTANAL (123-73-9) A strong reducing agent. Forms explosive mixture with air (flash point 55°F/12.8°C). Readily converted by oxygen to peroxides and acids; heat or contact with many other substances may cause polymerization. Incompatible with strong oxidizers, strong acids including nonoxidizing mineral acids, ammonia, aliphatic amines, aromatic amines, 1,3-butadiene, strong bases. Liquid attacks some plastics, rubber, and coatings.

n-**BUTANAL (Czech)** (123-72-8) May accumulate static electrical charges, and may cause ignition of its vapors. Forms explosive mixture with air (flash point −8°F/−22°C). Incompatible with strong oxidizers, strong acids, caustics, ammonia, aliphatic amines, alkanolamines, aromatic amines.

BUTANALDEHYDE (123-72-8) May accumulate static electrical charges, and may cause ignition of its vapors. Forms explosive mixture with air (flash point −8°F/−22°C). Incompatible with strong oxidizers, strong acids, caustics, ammonia, aliphatic amines, alkanolamines, aromatic amines.

1-BUTANAMINE (109-73-9) Forms explosive mixture with air (flash point 10°F/−12°C). May accumulate static electrical charges, and may cause ignition of its vapors. Incompatible with nonoxidizing mineral acids, strong acids, organic acids, organic anhydrides, isocyanates, vinyl acetate, acrylates, substituted allyls, alkylene oxides, epichlorohydrin, ketones, aldehydes, alcohols, glycols, phenols, cresols, caprolactam solution, strong oxidizers. In the presence of moisture, corrodes light metals and copper on contact.

2-BUTANAMINE (13952-84-6) Forms explosive mixture with air. Incompatible with nonoxidizing mineral acids, strong acids, organic acids, organic anhydrides, isocyanates, vinyl acetate, acrylates, substituted allyls, alkylene oxides, epichlorohydrin, ketones, aldehydes, alcohols, glycols, phenols, cresols, caprolactam solution, strong oxidizers.

1-BUTANAMINE, N-BUTYL (111-92-2) Forms explosive mixture with air (flash point 107°F/42°C). Incompatible with nonoxidizing mineral acids, strong acids, organic acids, organic anhydrides, isocyanates, vinyl acetate, acrylates, substituted allyls, alkylene oxides, epichlorohydrin, ketones, aldehydes, alcohols, glycols, phenols, cresols, caprolactam solution, strong oxidizers. Contact with copper alloys, zinc or galvanized steel may cause violent reaction.

1-BUTANECARBOXYLIC ACID (109-52-4) Forms explosive mixture with air (flash point 205°F/96°C). Incompatible with sulfuric acid, caustics, ammonia, amines, isocyanates, alkylene oxides, epichlorohydrin, strong oxidizers.

BUTANE, 1-CHLORO- (109-69-3) Forms explosive mixture with air (flash point 15°F/−9°C). May accumulate static electrical charges, and may cause ignition of its vapors. Water contact slowly forms hydrochloric acid. Incompatible with strong oxidizers, alkaline earth and alkali metals, finely divided metal. Attacks metals in presence of moisture. Attacks some plastics, rubber, and coatings.

1,4-BUTANEDICARBOXYLIC ACID (124-04-9) May accumulate static electrical charges, and may cause ignition of its vapors. Contact with strong oxidizers may cause fire and explosions.

BUTANE, 1,2,3,4-DIEPOXY (111-42-2) Incompatible with nonoxidizing mineral acids, strong acids, organic acids, organic anhydrides, isocyanates, vinyl acetate, acrylates, substituted allyls, alkylene oxides, epichlorohydrin, aldehydes, oxidizers. Corrosive to copper, copper alloys, zinc and galvanized iron.

1,3-BUTANEDIOL (107-88-0) Incompatible with strong acids, caustics, aliphatic amines, isocyanates, oxidizers.

1,4-BUTANEDIOL (110-63-4) Incompatible with strong acids, caustics, aliphatic amines, isocyanates, oxidizers.

2,3-BUTANEDIOL (513-85-9) Forms explosive mixture with air (flash point 185°F/85°C). Incompatible with strong acids, caustics, aliphatic amines, isocyanates, strong oxidizers.

BUTANE-1,3-DIOL (107-88-0) Incompatible with strong acids, caustics, aliphatic amines, isocyanates, oxidizers.

BUTANE-1,4-DIOL (110-63-4) Incompatible with strong acids, caustics, aliphatic amines, isocyanates, oxidizers.

BUTANE, 1,4-EPOXY- (109-99-9) Forms explosive mixture with air (flash point 6°F/−14°C). Unless inhibited, can form unstable and explosive peroxides. Polymerization will occur in the presence of acids, bases, and certain salts. Storage tanks and other equipment should be absolutely dry and free from air, ammonia, acetylene, hydrogen sulfide, rust, and other contaminants. Reacts with strong oxidizers. Attacks some plastics. May accumulate static electric charges that can result in ignition of its vapors.

BUTANETHIOL or *n*-BUTANETHIOL (109-79-5) Forms explosive mixture with air (flash point 35°F/2°C). Incompatible with strong oxidizers and nitric acid. Attacks some plastics and rubber.

BUTANI (Italian) (106-97-8) Forms explosive gas mixture with air. Reacts with oxidizers. Mixed with oxygen this substance will explode on contact with nickel carbonyl in the 68°F to 104°F/20°C to 40°C range.

BUTANIC ACID (107-92-6) Forms explosive mixture with air (flash point 162°F/72°C). Incompatible with sulfuric acid, caustics, ammonia, aliphatic amines, isocyanates, strong oxidizers, alkylene oxides, epichlorohydrin.

BUTANOIC ACID (107-92-6) Forms explosive mixture with air (flash point 162°F/72°C). Incompatible with sulfuric acid, caustics, ammonia, aliphatic amines, isocyanates, strong oxidizers, alkylene oxides, epichlorohydrin.

BUTANOIC ACID, BUTYL ESTER (109-21-7) Forms explosive mixture with air (flash point 48°F/−9°C). Attacks some plastics, rubber, and coatings. Incompatible with strong acids, nitrates; reacts violently with strong oxidizers.

126

BUTANOIC ACID, METHYL ESTER (623-42-7) Incompatible with strong acids, nitrates, oxidizers.

BUTANOIC ACID, 3-OXO-METHYL ESTER (9CI) (105-45-3) Incompatible with oxidizers, strong acids, nitrates.

BUTANOL (71-36-3) Forms explosive mixture with air (flash point 98°F/37°C). May react with aluminum above 120°F/49°C. Attacks some plastics, rubber, and coatings. Incompatible with strong acids, halogens, caustics, alkali metals, aliphatic amines, isocyanates.

BUTAN-1-OL (71-36-3) Forms explosive mixture with air (flash point 98°F/37°C). May react with aluminum above 120°F/49°C. Attacks some plastics, rubber, and coatings. Incompatible with strong acids, halogens, caustics, alkali metals, aliphatic amines, isocyanates.

BUTAN-2-OL (78-92-2) Forms explosive mixture with air (flash point 75°F/4°C). Attacks some plastics, rubber, and coatings. Forms an explosive peroxide in air. Ignites with chromium trioxide. Incompatible with strong oxidizers, strong acids, aliphatic amines, isocyanates, organic peroxides.

1-BUTANOL (71-36-3) Forms explosive mixture with air (flash point 98°F/37°C). May react with aluminum above 120°F/49°C. Attacks some plastics, rubber, and coatings. Incompatible with strong acids, halogens, caustics, alkali metals, aliphatic amines, isocyanates.

2-BUTANOL (78-92-2) Forms explosive mixture with air (flash point 75°F/4°C). Attacks some plastics, rubber, and coatings. Forms an explosive peroxide in air. Ignites with chromium trioxide. Incompatible with strong oxidizers, strong acids, aliphatic amines, isocyanates, organic peroxides.

BUTANOL-2 (78-92-2) Forms explosive mixture with air (flash point 75°F/4°C). Attacks some plastics, rubber, and coatings. Forms an explosive peroxide in air. Ignites with chromium trioxide. Incompatible with strong oxidizers, strong acids, aliphatic amines, isocyanates, organic peroxides.

n-**BUTANOL** or *normal*-**BUTANOL** (71-36-3) Forms explosive mixture with air (flash point 98°F/37°C). May react with aluminum above 120°F/49°C. Attacks some plastics, rubber, and coatings. Incompatible with strong acids, halogens, caustics, alkali metals, aliphatic amines, isocyanates.

sec-**BUTANOL** or *secondary*-**BUTANOL** (78-92-2) Forms explosive mixture with air (flash point 75°F/4°C). Attacks some plastics, rubber, and coatings. Forms an explosive peroxide in air. Ignites with chromium trioxide. Incompatible with strong oxidizers, strong acids, aliphatic amines, isocyanates, organic peroxides.

t-**BUTANOL** or *tert*-**BUTANOL** or *tertiary*-**BUTANOL** (75-65-0) Forms explosive mixture with air (flash point 52°F/11°C). May accumulate static electrical charges, and may cause ignition of its vapors. Incompatible with strong acids, including mineral acids, strong oxidizers or caustics, aliphatic amines, isocyanates, alkali metals (i.e., lithium, sodium, potassium, rubidium, cesium, francium). Attacks many plastics and some coatings.

BUTANOLEN (Dutch) (71-36-3) Forms explosive mixture with air (flash point 98°F/37°C). May react with aluminum above 120°F/49°C. Attacks some plastics, rubber, and coatings. Incompatible with strong acids, halogens, caustics, alkali metals, aliphatic amines, isocyanates.

1-BUTANOL, 3-METHOXY-, ACETATE or 1-BUTANOL, 3-METHOXYACETATE (4435-53-4) Incompatible with strong acids, nitrates, oxidizers.

BUTANOLO (Italian) (71-36-3) Forms explosive mixture with air (flash point 98°F/37°C). May react with aluminum above 120°F/49°C. Attacks some plastics, rubber, and coatings. Incompatible with strong acids, halogens, caustics, alkali metals, aliphatic amines, isocyanates.

BUTANOL TERTIAIRE (French) (75-65-0) Forms explosive mixture with air (flash point 52°F/11°C). May accumulate static electrical charges, and may cause ignition of its vapors. Incompatible with strong acids, including mineral acids, strong oxidizers or caustics, aliphatic amines, isocyanates, alkali metals (i.e., lithium, sodium, potassium, rubidium, cesium, francium). Attacks many plastics and some coatings.

BUTANONE (78-93-3) Forms explosive mixture with air (flash point 16°F/−9°C). Extremely flammable. Incompatible with sulfuric acid, nitric acid, aliphatic amines, strong oxidizers, potassium *tert*-butoxide, 2-propanol, chlorosulfonic acid, oleum (hydrogen peroxide + nitric acid).

2-BUTANONE (78-93-3) Forms explosive mixture with air (flash point 16°F/−9°C). Extremely flammable. Incompatible with sulfuric acid, nitric acid, aliphatic amines, strong oxidizers, potassium *tert*-butoxide, 2-propanol, chlorosulfonic acid, oleum (hydrogen peroxide + nitric acid).

BUTANONE 2 (French) (78-93-3) Forms explosive mixture with air (flash point 16°F/−9°C). Extremely flammable. Incompatible with sulfuric acid, nitric acid, aliphatic amines, strong oxidizers, potassium *tert*-butoxide, 2-propanol, chlorosulfonic acid, oleum (hydrogen peroxide + nitric acid).

2-BUTANONE PEROXIDE (1338-23-4) Forms explosive mixture with air (flash point 125°F/52°C). Explosive decomposition above 176°F/80°C. Pure substance is shock-sensitive. Strong oxidizer. Reacts violently with strong acids, strong bases, reducing agents, combustible substances, organic materials, oxides of heavy metals, salts, trace contaminants, amines. May accumulate static electrical charges; may cause ignition of its vapors.

BUTANOX M50 (1338-23-4) Forms explosive mixture with air (flash point 125°F/52°C). Explosive decomposition above 176°F/80°C. Pure substance is shock-sensitive. Strong oxidizer. Reacts violently with strong acids, strong bases, reducing agents, combustible substances, organic materials, oxides of heavy metals, salts, trace contaminants, amines. May accumulate static electrical charges; may cause ignition of its vapors.

BUTANOYL CHLORIDE (141-75-3) Forms explosive mixture with air (flash point 71°F/22°C). Incompatible with oxidizers, strong bases. Aqueous solution forms hydrochloric acid; corrodes metals.

2-BUTENAL (4170-30-3) A strong reducing agent. Forms explosive mixture with air (flash point 55°F/12.8°C). Readily converted by oxygen to peroxides and acids; heat or contact with many other substances may cause polymerization. Incompatible with strong oxidizers, strong acids including nonoxidizing mineral acids, ammonia, aliphatic amines, aromatic amines, 1,3-butadiene, strong bases. Liquid attacks some plastics, rubber, and coatings.

trans-2-BUTENAL (123-73-9) A strong reducing agent. Forms explosive mixture with air (flash point 55°F/12.8°C). Readily converted by oxygen to peroxides and acids; heat or contact with many other substances may cause polymerization. Incompatible with strong oxidizers, strong acids including nonoxidizing mineral acids, ammonia, aliphatic amines, aromatic amines, 1,3-butadiene, strong bases. Liquid attacks some plastics, rubber, and coatings.

BUTENE (106-98-9) Flammable gas. Reacts violently with strong oxidizers. Can form unstable peroxides; may polymerize. Incompatible with acids, halogens, aluminum borohydride, oxides of nitrogen. May accumulate static electrical charges; may cause ignition of its vapors.

1-BUTENE (106-98-9) Flammable gas. Reacts violently with strong oxidizers. Can form unstable peroxides; may polymerize. Incompatible with acids, halogens, aluminum borohydride, oxides of nitrogen. May accumulate static electrical charges; may cause ignition of its vapors.

n-BUTENE (106-98-9) Flammable gas. Reacts violently with strong oxidizers. Can form unstable peroxides; may polymerize. Incompatible with acids, halogens, aluminum borohydride, oxides of nitrogen. May accumulate static electrical charges; may cause ignition of its vapors.

(E)-BUTENEDIOIC ACID (110-17-8) Incompatible with sulfuric acid, caustics, ammonia, amines, isocyanates, alkylene oxides, epichlorohydrin.

(Z)-BUTENEDIOIC ACID (110-16-7) Aqueous solution is an acid. Incompatible with oxidizers, bases. Corrodes metal if wet.

cis-BUTENEDIOIC ACID (110-16-7) Aqueous solution is an acid. Incompatible with oxidizers, bases. Corrodes metal if wet.

trans-BUTENEDIOIC ACID (110-17-8) Incompatible with sulfuric acid, caustics, ammonia, amines, isocyanates, alkylene oxides, epichlorohydrin.

BUTENEDIOIC ACID, (E)- (110-17-8) Incompatible with sulfuric acid, caustics, ammonia, amines, isocyanates, alkylene oxides, epichlorohydrin.

2-BUTENEDIOIC ACID (E) (110-17-8) Incompatible with sulfuric acid, caustics, ammonia, amines, isocyanates, alkylene oxides, epichlorohydrin.

BUTENEDIOIC ACID, (Z)- (110-16-7) Aqueous solution is an acid. Incompatible with oxidizers, bases. Corrodes metal if wet.

cis-BUTENEDIOIC ANHYDRIDE (108-31-6) Strong oxidizers may cause fire and explosions; contact with amines or alkali metals causes polymerization if temperature is greater than 150°F/66°C. Attacks metals in the presence of moisture.

cis-BUTENEDIOIC ANHYDRIDE (123-33-1) Contact with strong oxidizers may cause fire and explosions.

1,4-BUTENEDIOL (110-64-5) Contact with strong oxidizers may cause fire and explosions.

2-BUTENE-1,4-DIOL (110-64-5) Contact with strong oxidizers may cause fire and explosions.

cis-2-BUTENE-1,4-DIOL (110-64-5) Contact with strong oxidizers may cause fire and explosions.

1-BUTENE OXIDE (106-88-7) Forms explosive mixture with air (flash point −7°F/−22°C). Polymerization will occur in the presence of acids, bases, or certain salts. Storage tanks and other equipment should be absolutely dry and free from air, ammonia, acetylene, hydrogen

129

sulfide, rust, and other contaminants. Reacts violently with oxidizers, acids. May accumulate static electrical charges, and may cause ignition of its vapors.

BUTENE RESINS (9003-29-6) Incompatible with sulfuric acid, nitric acid.

1-BUTEN-3-OL, 3-METHYL (115-18-4) Forms explosive mixture with air (flash point 56°F/13°C). Incompatible with strong acids, caustics, aliphatic amines, isocyanates, oxidizers.

BUTENONE (78-94-4) Heat can cause polymerization. Reacts violently with strong oxidizers.

2-BUTENONE (78-94-4) Heat can cause polymerization. Reacts violently with strong oxidizers.

3-BUTEN-2-ONE (78-94-4) Heat can cause polymerization. Reacts violently with strong oxidizers.

N-BUTILAMINA (Italian) (109-73-9) Forms explosive mixture with air (flash point 10°F/−12°C). May accumulate static electrical charges, and may cause ignition of its vapors. Incompatible with nonoxidizing mineral acids, strong acids, organic acids, organic anhydrides, isocyanates, vinyl acetate, acrylates, substituted allyls, alkylene oxides, epichlorohydrin, ketones, aldehydes, alcohols, glycols, phenols, cresols, caprolactam solution, strong oxidizers. In the presence of moisture, corrodes light metals and copper on contact.

o-(4-*tera*-**BUTIL-2-CLORO-FENIL)-O-METIL-FOSFORAMMIDE** **(Italian)** (299-86-5) Decomposes in strongly alkaline (pH >7) and strongly acidic media. Unstable over long periods of time in water and at temperatures above 140°F/60°C).

BUTILE (ACETATI di) (Italian) (123-86-4) Forms explosive mixture with air (flash point 72°F/22°C). Reacts with water on standing to form acetic acid and *n*-butyl alcohol. Reacts violently with strong oxidizers and potassium-*tert*-butoxide. Incompatible with caustics, strong acids, nitrates. Dissolves rubber, many plastics, resins, and some coatings. May accumulate static electrical charges, and may cause ignition of its vapors.

BUTILMETACRILATO (Italian) (97-88-1) Unless inhibitor is maintained at the proper level, oxidizers, heat, ultraviolet light, or moisture may cause polymerization. May accumulate static electrical charges; may cause ignition of its vapors.

BUTONIC ACID ETHYL ESTER (105-54-4) Incompatible with strong acids, nitrates, oxidizers.

1-BUTOXY BUTANE (142-96-1) Forms explosive mixture with air (flash point 77°F/25°C). May accumulate static electrical charges, and may cause ignition of its vapors. Incompatible with strong acids, oxidizers. Air contact or light may produces unstable and explosive peroxides.

BUTOXYDIETHYLENE GLYCOL (112-34-5) Forms explosive mixture with air (flash point 172°F/78°C). Forms unstable peroxides; may polymerize. Incompatible with strong oxidizers, sulfuric acid, isocyanates, perchloric acid. May accumulate static electrical charges; may cause ignition of its vapors.

BUTOXYDIGLYCOL (112-34-5) Forms explosive mixture with air (flash point 172°F/78°C). Forms unstable peroxides; may polymerize. Incompatible with strong oxidizers, sulfuric acid, isocyanates, perchloric acid. May accumulate static electrical charges; may cause ignition of its vapors.

1-BUTOXY-2,3-EPOXYPROPANE (2426-08-6) Forms explosive mixture with air (flash point 130°F/54°C). Air and light form unstable and explosive peroxides. Contact with strong oxidizers may cause fire and explosions. Strong caustics may cause polymerization. Attacks some plastics and rubber.

2-BUTOXYETHANOL (111-76-2) Forms explosive mixture with air (flash point 143°F/62°C). Reacts violently with strong caustics and strong oxidizers. Attacks some coatings, plastics, and rubber. Attacks metallic aluminum at high temperatures.

2-BUTOXY-1-ETHANOL (111-76-2) Forms explosive mixture with air (flash point 143°F/62°C). Reacts violently with strong caustics and strong oxidizers. Attacks some coatings, plastics, and rubber. Attacks metallic aluminum at high temperatures.

b-BUTOXYETHANOL or beta-BUTOXYETHANOL (111-76-2) Forms explosive mixture with air (flash point 143°F/62°C). Reacts violently with strong caustics and strong oxidizers. Attacks some coatings, plastics and rubber. Attacks metallic aluminum at high temperatures.

2-BUTOXYETHANOL ACETATE (112-07-2) Incompatible with strong acids, perchloric acid, oxidizers, nitrates.

2-(2-BUTOXYETHOXY) ETHANOL (112-34-5) Forms explosive mixture with air (flash point 172°F/78°C). Incompatible with strong oxidizers, strong bases, sulfuric acid, isocyanates. Attacks light metals.

2-(2-BUTOXYETHOXY) ETHANOL ACETATE (124-17-4) Incompatible with strong acids, nitrates, strong alkalies, oxidizers.

2-(2-BUTOXYETHOXY) ETHYL ACETATE (124-17-4) Incompatible with strong acids, nitrates, strong alkalies, oxidizers.

2-BUTOXYETHYL ACETATE (112-07-2) Incompatible with strong acids, perchloric acid, oxidizers, nitrates.

BUTOXYETHYL CELLOSOLVE ACETATE (112-07-2) Incompatible with strong acids, perchloric acid, oxidizers, nitrates.

BUTOXYL (4435-53-4) Incompatible with strong acids, nitrates, oxidizers.

BUTOXYPROPYL TRICHLOROPHENOXYACETATE (8002-26-4) Water contact causes foaming. Strong oxidizers may cause fire and explosions. Attacks some plastics, rubber, and coatings.

BUTTER OF ANTIMONY (7647-18-9) Water contact produces hydrogen chloride gas. Reacts violently with ammonia, caustics. Attacks many metals. Contact with air produces heavier-than-air corrosive vapor.

BUTTER OF ANTIMONY (10025-91-9) Water contact produces corrosive solution. Air contact produces corrosive vapor.

BUTTER OF ARSENIC (7440-38-2) Contact with hydrogen gas produces arsine. Incompatible with strong acids, strong oxidizers, peroxides, bromine pentafluoride, bromine trifluoride, cesium acetylene carbide, chromium trioxide, nitrogen trichloride, silver nitrate.

BUTTER OF ARSENIC (7784-34-1) Water contact produces corrosive mixture of acids. Light exposure produces toxic gas. Reacts violently with anhydrous ammonia, strong acids, strong oxidizers and halogens. Corrodes metals in the presence of moisture.

BUTTER OF ZINC (7646-85-7) Aqueous solution is strongly acidic; reacts with strong bases, and attacks metals in the presence of moisture.

BUTTERSAEURE (German) (107-92-6) Forms explosive mixture with air (flash point 162°F/72°C). Incompatible with sulfuric acid, caustics, ammonia, aliphatic amines, isocyanates, strong oxidizers, alkylene oxides, epichlorohydrin.

BUTYLACETAT (German) (123-86-4) Forms explosive mixture with air (flash point 72°F/22°C). Reacts with water on standing to form acetic acid and *n*-butyl alcohol. Reacts violently with strong oxidizers and potassium-*tert*-butoxide. Incompatible with caustics, strong acids, nitrates. Dissolves rubber, many plastics, resins and some coatings. May accumulate static electrical charges, and may cause ignition of its vapors.

BUTYL ACETATE (123-86-4) Forms explosive mixture with air (flash point 72°F/22°C). Reacts with water on standing to form acetic acid and *n*-butyl alcohol. Reacts violently with strong oxidizers and potassium-*tert*-butoxide. Incompatible with caustics, strong acids, nitrates. Dissolves rubber, many plastics, resins and some coatings. May accumulate static electrical charges, and may cause ignition of its vapors.

1-BUTYL ACETATE (123-86-4) Forms explosive mixture with air (flash point 72°F/22°C). Reacts with water on standing to form acetic acid and *n*-butyl alcohol. Reacts violently with strong oxidizers and potassium-*tert*-butoxide. Incompatible with caustics, strong acids, nitrates. Dissolves rubber, many plastics, resins and some coatings. May accumulate static electrical charges, and may cause ignition of its vapors.

2-BUTYL ACETATE (105-46-4) Forms explosive mixture with air (flash point 70°F/19°C). Reacts violently with oxidizers. Incompatible with strong acids, nitrates, potassium-*tert*-butoxide. Attacks some plastics, rubber, and coatings. May accumulate static electrical charges, and may cause ignition of its vapors.

N-BUTYL ACETATE (123-86-4) Forms explosive mixture with air (flash point 72°F/22°C). Reacts with water on standing to form acetic acid and *n*-butyl alcohol. Reacts violently with strong oxidizers and potassium-*tert*-butoxide. Incompatible with caustics, strong acids, nitrates. Dissolves rubber, many plastics, resins, and some coatings. May accumulate static electrical charges, and may cause ignition of its vapors.

s-**BUTYL ACETATE or** *sec*-**BUTYL ACETATE or** *secondary*-**BUTYL ACETATE** (105-46-4) Forms explosive mixture with air (flash point 70°F/19°C). Reacts violently with oxidizers. Incompatible with strong acids, nitrates, potassium-*tert*-butoxide. Attacks some plastics, rubber and coatings. May accumulate static electrical charges, and may cause ignition of its vapors.

t-**BUTYL ACETATE or** *tert*-**BUTYL ACETATE or** *tertiary*-**BUTYL ACETATE** (540-88-5) Incompatible with strong acids, oxidizers, and alkalies; nitrates. Softens and dissolves plastics.

BUTYL ACETATE ESTER (110-19-0) Forms explosive mixture with air (flash point 64°F/18°C). Reacts with water on standing to form acetic acid and *n*-butyl alcohol. Reacts violently with strong oxidizers. Reacts with caustics, strong acids, nitrates. Dissolves rubber, many plastics, resins and some coatings. May accumulate static electrical charges, and may cause ignition of its vapors.

BUTYLACETATEN (Dutch) (123-86-4) Forms explosive mixture with air (flash point 72°F/22°C). Reacts with water on standing to form acetic acid and n-butyl alcohol. Reacts violently with strong oxidizers and potassium-*tert*-butoxide. Incompatible with caustics, strong acids, nitrates. Dissolves rubber, many plastics, resins and some coatings. May accumulate static electrical charges, and may cause ignition of its vapors.

BUTYLACETIC ACID (142-62-1) Incompatible with sulfuric acid, caustics, ammonia, amines, isocyanates, alkylene oxides, epichlorohydrin. Attacks common metals.

BUTYL ACRYLATE (141-32-2) Forms explosive mixture with air (flash point 103°F/39°C). Peroxides, heat, or light may cause polymerization. Incompatible with strong acids, amines. Reacts violently with strong oxidizers.

BUTYL ACRYLATE (141-32-2) Forms explosive mixture with air (flash point 120°F/49°C). Heat, sparks, open flame, light, or peroxides may cause explosive polymerization. Incompatible with strong acids, amines, halogens, hydrogen compounds, oxidizers, sunlight, or other catalysts.

iso-**BUTYL ACRYLATE** (106-63-8) Forms explosive mixture with air (flash point 94°F/34°C). Incompatible with strong acids, aliphatic amines, alkanolamines.

n-**BUTYL ACRYLATE** (141-32-2) Forms explosive mixture with air (flash point 103°F/39°C). Peroxides, heat, or light may cause polymerization. Incompatible with strong acids, amines. Reacts violently with strong oxidizers.

BUTYLACRYLATE, INHIBITED (141-32-2) Forms explosive mixture with air (flash point 103°F/39°C). Peroxides, heat, or light may cause polymerization. Incompatible with strong acids, amines. Reacts violently with strong oxidizers.

BUTYL ADIPATE (103-23-1) Incompatible with strong acids, strong oxidizers, nitrates.

BUTYL ALCOHOL (71-36-3) Forms explosive mixture with air (flash point 98°F/37°C). May react with aluminum above 120°F/49°C). Attacks some plastics, rubber, and coatings. Incompatible with strong acids, halogens, caustics, alkali metals, aliphatic amines, isocyanates.

BUTYL ALCOHOL or *n*-BUTYL ALCOHOL or *normal*-BUTYL ALCOHOL (71-36-3) Forms explosive mixture with air (flash point 98°F/37°C). May react with aluminum above 120°F/49°C. Attacks some plastics, rubber, and coatings. Incompatible with strong acids, halogens, caustics, alkali metals, aliphatic amines, isocyanates.

2-BUTYL ALCOHOL (78-92-2) Forms explosive mixture with air (flash point 75°F/4°C). Attacks some plastics, rubber, and coatings. Produces an explosive peroxide in air. Ignites with chromium trioxide. Incompatible with strong oxidizers, strong acids, aliphatic amines, isocyanates, organic peroxides.

133

sec-**BUTYL ALCOHOL** or *secondary*-**BUTYL ALCOHOL** (78-92-2) Forms explosive mixture with air (flash point 75°F/4°C). Attacks some plastics, rubber, and coatings. Produces an explosive peroxide in air. Ignites with chromium trioxide. Incompatible with strong oxidizers, strong acids, aliphatic amines, isocyanates, organic peroxides.

t-**BUTYL ALCOHOL** or *tert*-**BUTYL ALCOHOL** (75-65-0) Forms explosive mixture with air (flash point 52°F/11°C). May accumulate static electrical charges, and may cause ignition of its vapors. Incompatible with strong acids, including mineral acids, strong oxidizers or caustics, aliphatic amines, isocyanates, alkali metals (i.e., lithium, sodium, potassium, rubidium, cesium, francium). Attacks many plastics and some coatings.

s-**BUTYL ALCOHOL ACETATE** or *sec*-**BUTYL ALCOHOL ACETATE** or *secondary*-**BUTYL ALCOHOL ACETATE** (105-46-4) Forms explosive mixture with air (flash point 70°F/19°C). Reacts violently with oxidizers. Incompatible with strong acids, nitrates, potassium tert-butoxide. Attacks some plastics, rubber, and coatings. May accumulate static electrical charges, and may cause ignition of its vapors.

BUTYL ALDEHYDE (123-72-8) May accumulate static electrical charges, and may cause ignition of its vapors. Forms explosive mixture with air (flash point −8°F/−22°C). Incompatible with strong oxidizers, strong acids, caustics, ammonia, aliphatic amines, alkanolamines, aromatic amines.

n-**BUTYL ALDEHYDE** or *normal*-**BUTYL ALDEHYDE** (123-72-8) May accumulate static electrical charges, and may cause ignition of its vapors. Forms explosive mixture with air (flash point −8°F/−22°C). Incompatible with strong oxidizers, strong acids, caustics, ammonia, aliphatic amines, alkanolamines, aromatic amines.

N-BUTYL ALPHA-METHYLACRYLATE (97-88-1) Unless inhibitor is maintained at the proper level, oxidizers, heat, ultraviolet light, or moisture may cause polymerization. May accumulate static electrical charges; may cause ignition of its vapors.

N-BUTYLAMIN (German) (109-73-9) Forms explosive mixture with air (flash point 10°F/−12°C). May accumulate static electrical charges, and may cause ignition of its vapors. Incompatible with nonoxidizing mineral acids, strong acids, organic acids, organic anhydrides, isocyanates, vinyl acetate, acrylates, substituted allyls, alkylene oxides, epichlorohydrin, ketones, aldehydes, alcohols, glycols, phenols, cresols, caprolactam solution, strong oxidizers. In the presence of moisture, corrodes light metals and copper on contact.

BUTYLAMINE (109-73-9) Forms explosive mixture with air (flash point 10°F/−12°C). May accumulate static electrical charges, and may cause ignition of its vapors. Incompatible with nonoxidizing mineral acids, strong acids, organic acids, organic anhydrides, isocyanates, vinyl acetate, acrylates, substituted allyls, alkylene oxides, epichlorohydrin, ketones, aldehydes, alcohols, glycols, phenols, cresols, caprolactam solution, strong oxidizers. In the presence of moisture, corrodes light metals and copper on contact.

n-**BUTYLAMINE** or *normal*-**BUTYLAMINE** (109-73-9) Forms explosive mixture with air (flash point 10°F/−12°C). May accumulate static electrical charges, and may cause ignition of its vapors. Incompatible with nonoxidizing mineral acids, strong acids, organic acids, organic anhydrides, isocyanates, vinyl acetate, acrylates, substituted allyls, alkylene oxides, epichlorohydrin, ketones, aldehydes, alcohols, glycols, phenols, cresols, caprolactam solution, strong oxidizers. In the presence of moisture, corrodes light metals and copper on contact.

s-BUTYLAMINE or *sec*-BUTYLAMINE or *secondary*-BUTYLAMINE (13952-84-6) Forms explosive mixture with air. Incompatible with nonoxidizing mineral acids, strong acids, organic acids, organic anhydrides, isocyanates, vinyl acetate, acrylates, substituted allyls, alkylene oxides, epichlorohydrin, ketones, aldehydes, alcohols, glycols, phenols, cresols, caprolactam solution, strong oxidizers.

t-BUTYLAMINE or *tert*-BUTYLAMINE or *tertiary*-BUTYLAMINE (75-64-9) Forms explosive mixture with air (flash point 48°F/−9°C). May accumulate static electrical charges, and may cause ignition of its vapors. Incompatible with nonoxidizing mineral acids, strong acids, organic acids, organic anhydrides, isocyanates, vinyl acetate, acrylates, substituted allyls, alkylene oxides, epichlorohydrin, ketones, aldehydes, alcohols, glycols, phenols, cresols, caprolactam solution, strong oxidizers.

BUTYLAMINE, TERTIARY (75-64-9) Forms explosive mixture with air (Flash point 48°F/−9°C). May accumulate static electrical charges, and may cause ignition of its vapors. Incompatible with nonoxidizing mineral acids, strong acids, organic acids, organic anhydrides, isocyanates, vinyl acetate, acrylates, substituted allyls, alkylene oxides, epichlorohydrin, ketones, aldehydes, alcohols, glycols, phenols, cresols, caprolactam solution, strong oxidizers.

1-(BUTYLAMINO)CARBONYL-1H-BENZIMIDAZOL-2-YL-, METHYL ESTER (17804-35-2) Heat, water, strong acids, and strong alkalies can cause decomposition and formation of toxic oxides of nitrogen.

BUTYL BENZYL PHTHALATE (85-68-7) Incompatible with strong acids, nitrates, oxidizers. Destructive to rubber and paint.

BUTYL BROMIDE (109-65-9) Forms explosive mixture with air (flash point 65°F/18°C). Incompatible with strong oxidizers, strong acids. May accumulate static electrical charges; may cause ignition of its vapors.

n-BUTYL BROMIDE or *normal*-BUTYL BROMIDE (109-65-9) Forms explosive mixture with air (flash point 65°F/18°C). Incompatible with strong oxidizers, strong acids. May accumulate static electrical charges; may cause ignition of its vapors.

s-BUTYL BROMIDE or *sec*-BUTYL BROMIDE or *secondary*-BUTYL BROMIDE (78-76-2) Forms explosive mixture with air (flash point 70°F/26°C). Reacts with oxidizers.

N-BUTYL-1-BUTANAMINE (111-92-2) Forms explosive mixture with air (flash point 107°F/42°C). Incompatible with nonoxidizing mineral acids, strong acids, organic acids, organic anhydrides, isocyanates, vinyl acetate, acrylates, substituted allyls, alkylene oxides, epichlorohydrin, ketones, aldehydes, alcohols, glycols, phenols, cresols, caprolactam solution, strong oxidizers. Contact with copper alloys, zinc, or galvanized steel may cause violent reaction.

BUTYL BUTANOATE (109-21-7) Forms explosive mixture with air (flash point 48°F/−9°C). Attacks some plastics, rubber, and coatings. Incompatible with strong acids, nitrates; reacts violently with strong oxidizers.

n-**BUTYL,** *n*-**BUTANOATE** (109-21-7) Forms explosive mixture with air (flash point 48°F/−9°C). Attacks some plastics, rubber, and coatings. Incompatible with strong acids, nitrates; reacts violently with strong oxidizers.

2-BUTYLBUTANOIC ACID (149-57-5) Reacts with oxidizers, with a risk of fire or explosions. Attacks common metals, especially in a moist environment.

BUTYL BUTYRATE (109-21-7) Forms explosive mixture with air (flash point 48°F/−9°C). Attacks some plastics, rubber, and coatings. Incompatible with strong acids, nitrates; reacts violently with strong oxidizers.

n-**BUTYL,** *n*-**BUTYRATE** (109-21-7) Forms explosive mixture with air (flash point 48°F/−9°C). Attacks some plastics, rubber and coatings. Incompatible with strong acids, nitrates; reacts violently with strong oxidizers.

1-(BUTYLCARBAMOYL)-2-BENZIMIDAZOLEC ARBAMIC ACID, METHYL ESTER (17804-35-2) Heat, water, strong acids, and strong alkalies can cause decomposition and formation of toxic oxides of nitrogen.

1-(n-BUTYLCARBAMOYL)-2-(METHOXY-CARBOXAMIDO)-BENZIMIDAZOL (German) (17804-35-2) Heat, water, strong acids, and strong alkalies can cause decomposition and formation of toxic oxides of nitrogen.

N-BUTYL CARBINOL (71-41-0) Forms explosive mixture with air (flash point 91°F/33°C). Incompatible with strong acids, caustics, aliphatic amines, isocyanates. Alkali metals and alkaline earth contact produces hydrogen gas.

N-BUTYLCARBINYL CHLORIDE (543-59-9) Forms explosive mixture with air (flash point 34°F/1°C). Strong oxidizers may cause fire and explosions.

BUTYL CARBITOL (112-34-5) Forms explosive mixture with air (flash point 172°F/78°C). Incompatible with strong oxidizers, strong bases, sulfuric acid, isocyanates. Attacks light metals.

BUTYL CARBITOL ACETATE (124-17-4) Incompatible with strong acids, nitrates, strong alkalies, oxidizers.

BUTYL CELLOSOLVE (111-76-2) Forms explosive mixture with air (flash point 143°F/62°C). Reacts violently with strong caustics and strong oxidizers. Attacks some coatings, plastics and rubber. Attacks metallic aluminum at high temperatures.

BUTYL CELLOSOLVE ACETATE (112-07-2) Incompatible with strong acids, perchloric acid, oxidizers, nitrates.

o-**(4-*tert*-BUTYL-2-CHLOOR-FENYL)-O-METHYL-FOSFORZUUR-N-METHYL-AMIDE (Dutch)** (299-86-5) Decomposes in strongly alkaline (pH >7) and strongly acidic media. Unstable over long periods of time in water and at temperatures above 140°F/60°C.

BUTYL CHLORIDE (109-69-3) Forms explosive mixture with air (flash point 15°F/−9°C). May accumulate static electrical charges, and may cause ignition of its vapors. Water contact slowly produces hydrochloric acid. Incompatible with strong oxidizers, alkaline earth and alkali metals, finely divided metal. Attacks metals in presence of moisture. Attacks some plastics, rubber, and coatings.

n-BUTYL CHLORIDE or *normal*-BUTYL CHLORIDE (109-69-3) Forms explosive mixture with air (flash point 15°F/−9°C). May accumulate static electrical charges, and may cause ignition of its vapors. Water contact slowly produces hydrochloric acid. Incompatible with strong oxidizers, alkaline earth and alkali metals, finely divided metal. Attacks metals in presence of moisture. Attacks some plastics, rubber or coatings.

n-BUTYL CHLOROFORMATE (592-34-7) Water contact produces hydrochloric acid; corrodes metals.

4-*tert*-BUTYL-2-CHLOROPHENYL METHYL METHYLPHOS-PHORAMIDATE (299-86-5) Decomposes in strongly alkaline (pH >7) and strongly acidic media. Unstable over long periods of time in water and at temperatures above 140°F/60°C.

4-*tert*-BUTYL 2-CHLOROPHENYL METHYLPHOSPHORAMI-DATE de METHYLE (French) (299-86-5) Decomposes in strong alkaline (pH >7) and strongly acidic media. Unstable over long periods of time in water and at temperatures above 140°F/60°C.

o-(4-*tert*-BUTYL-2-CHLOR-PHENYL)-O-METHYL-PHOSPHOR-SAEURE-N-METHYLAMID (German) (299-86-5) Decomposes in strongly alkaline (pH >7) and strongly acidic media. Unstable over long periods of time in water and at temperatures above 140°F/60°C.

t-BUTYL CHROMATE or *tert*-BUTYL CHROMATE or *tertiary*-BUTYL CHROMATE (1189-85-1) Incompatible with acids, alcohols, combustible or easily oxidized materials, reducing agents, moisture.

BUTYL DIGLYME (112-73-2) Incompatible with sulfuric acid, isocyanates, strong oxidizers. May accumulate static electrical charges; may cause ignition of its vapors.

BUTYL 2,4-DIISOPROPYL ESTER (94-11-1) Forms explosive mixture with air (175°F/79°C). Incompatible with strong oxidizers, strong acids, nitrates. Attacks some plastics, rubber, and coatings.

BUTYLE (ACETATE de) (French) (123-86-4) Forms explosive mixture with air (flash point 72°F/22°C). Reacts with water on standing to form acetic acid and *n*-butyl alcohol. Reacts violently with strong oxidizers and potassium-*tert*-butoxide. Incompatible with caustics, strong acids, nitrates. Dissolves rubber, many plastics, resins, and some coatings. May accumulate static electrical charges, and may cause ignition of its vapors.

BUTYLENE (106-98-9) Flammable gas. Reacts violently with strong oxidizers. Can form unstable peroxides; may polymerize. Incompatible with acids, halogens, aluminum borohydride, oxides of nitrogen. May accumulate static electrical charges; may cause ignition of its vapors.

1-BUTYLENE (106-98-9) Flammable gas. Reacts violently with strong oxidizers. Can form unstable peroxides; may polymerize. Incompatible with acids, halogens, aluminum borohydride, oxides of nitrogen. May accumulate static electrical charges; may cause ignition of its vapors.

a-BUTYLENE or alpha-BUTYLENE (106-98-9) Flammable gas. Reacts violently with strong oxidizers. Can form unstable peroxides; may polymerize. Incompatible with acids, halogens, aluminum borohydride, oxides of nitrogen. May accumulate static electrical charges; may cause ignition of its vapors.

gamma-BUTYLENE (115-11-7) Flammable gas. Forms explosive mixture with air. May be able to form unstable peroxides; may cause polymerization. Reacts violently with strong oxidizers, strong acids, oxides of nitrogen. May accumulate static electrical charges; may cause ignition of its vapors.

2-BUTYLENE DICHLORIDE (764-41-0) Forms explosive mixture with air (flash point 126°F/52°C). Water produces hydrochloric acid. Incompatible with strong oxidizers, bases. Corrodes metals in the presence of moisture.

1,3-BUTYLENE GLYCOL (107-88-0) Incompatible with strong acids, caustics, aliphatic amines, isocyanates, oxidizers.

1,4-BUTYLENE GLYCOL (110-63-4) Incompatible with strong acids, caustics, aliphatic amines, isocyanates, oxidizers.

2,3-BUTYLENE GLYCOL (513-85-9) Forms explosive mixture with air (flash point 185°F/85°C). Incompatible with strong acids, caustics, aliphatic amines, isocyanates, strong oxidizers.

beta-BUTYLENE GLYCOL (107-88-0) Incompatible with strong acids, caustics, aliphatic amines, isocyanates, oxidizers.

BUTYLENE GLYCOL (PSEUDO) (513-85-9) Forms explosive mixture with air (flash point 185°F/85°C). Incompatible with strong acids, caustics, aliphatic amines, isocyanates, strong oxidizers.

1,4-BUTYLENEGLYCOL (110-63-4) Incompatible with strong acids, caustics, aliphatic amines, isocyanates, oxidizers.

BUTYLENE HYDRATE (78-92-2) Forms explosive mixture with air (flash point 75°F/4°C). Attacks some plastics, rubber, and coatings. Produces an explosive peroxide in air. Ignites with chromium trioxide. Incompatible with strong oxidizers, strong acids, aliphatic amines, isocyanates, organic peroxides.

BUTYLENE MIXTURES (N/A) Form explosive mixture with air (flash point approximately −24°F/−42°C). React with acids, alkyl halides, strong oxidizers.

BUTYLENE OXIDE (109-99-9) Forms explosive mixture with air (flash point 6°F/−14°C). Unless inhibited, can produce unstable and explosive peroxides. Polymerization will occur in the presence of acids, bases, and certain salts. Storage tanks and other equipment should be absolutely dry and free from air, ammonia, acetylene, hydrogen sulfide, rust, and other contaminants. Reacts with strong oxidizers. Attacks some plastics. May accumulate static electric charges that can result in ignition of its vapors.

1,2-BUTYLENE OXIDE (106-88-7) Forms explosive mixture with air (flash point −7°F/−22°C). Polymerization will occur in the presence of acids, bases or certain salts. Storage tanks and other equipment should be absolutely dry and free from air, ammonia, acetylene, hydrogen sulfide, rust, and other contaminants. Reacts violently with oxidizers, acids. May accumulate static electrical charges, and may cause ignition of its vapors.

1,2-BUTYLENE OXIDE, stabilized (106-88-7) Forms explosive mixture with air (flash point −7°F/−22°C). Polymerization will occur in the presence of acids, bases, or certain salts. Storage tanks and other equipment should be absolutely dry and free from air, ammonia, acetylene, hydrogen sulfide, rust, and other contaminants. Reacts violently with oxidizers, acids. May accumulate static electrical charges, and may cause ignition of its vapors.

N-BUTYL ESTER OF ACETIC ACID (123-86-4) Forms explosive mixture with air (flash point 72°F/22°C). Reacts with water on standing to form acetic acid and *n*-butyl alcohol. Reacts violently with strong oxidizers and potassium-*tert*-butoxide. Incompatible with caustics, strong acids, nitrates. Dissolves rubber, many plastics, resins and some coatings. May accumulate static electrical charges, and may cause ignition of its vapors.

tert-**BUTYL ESTER OF ACETIC ACID** (540-88-5) Incompatible with strong acids, oxidizers, alkalies, nitrates. Softens and dissolves plastics.

BUTYL ETHANOATE (123-86-4) Forms explosive mixture with air (flash point 72°F/22°C). Reacts with water on standing to form acetic acid and *n*-butyl alcohol. Reacts violently with strong oxidizers and potassium-*tert*-butoxide. Incompatible with caustics, strong acids, nitrates. Dissolves rubber, many plastics, resins, and some coatings. May accumulate static electrical charges, and may cause ignition of its vapors.

BUTYL ETHER (142-96-1) Forms explosive mixture with air (flash point 77°F/25°C). May accumulate static electrical charges, and may cause ignition of its vapors. Incompatible with strong acids, oxidizers. Air contact or light may produce unstable and explosive peroxides.

n-**BUTYL ETHER** (142-96-1) Forms explosive mixture with air (flash point 77°F/25°C). May accumulate static electrical charges, and may cause ignition of its vapors. Incompatible with strong acids, oxidizers. Air contact or light may produce unstable and explosive peroxides.

BUTYL ETHYL ACETALDEHYDE (123-05-7) Forms explosive mixture with air (flash point 112°F/44°C). Reacts violently with oxidizers. May ignite spontaneously when spilled on clothing or other absorbent materials. Under certain conditions ignites spontaneously with air. Incompatible with strong acids, caustics, ammonia, amines. Produces unstable peroxides on contact with air.

BUTYLETHYLACETIC ACID (149-57-5) Reacts with oxidizers, with a risk of fire or explosions. Attacks common metals, especially in a moist environment.

BUTYLETHYLAMINE (617-79-8) Incompatible with strong acids, organic anhydrides, isocyanates, aldehydes, oxidizers.

BUTYL ETHYL KETONE (106-35-4) Forms explosive mixture with air (flash point 115°F/46°C). Reacts with oxidizers. Attacks some plastics, rubber, and coatings.

n-**BUTYL ETHYL KETONE** (106-35-4) Forms explosive mixture with air (flash point 115°F/46°C). Reacts with oxidizers. Attacks some plastics, rubber, and coatings.

BUTYL ETHYLENE (592-41-6) Forms explosive mixture with air (flash point −15°F/−26°C). Reacts violently with oxidizers.

o-**BUTYL ETHYLENE GLYCOL** (111-76-2) Forms explosive mixture with air (flash point 143°F/62°C). Reacts violently with strong caustics and strong oxidizers. Attacks some coatings, plastics and rubber. Attacks metallic aluminum at high temperatures.

p-tert-**BUTYLFENOL (Czech)** (98-54-4) Incompatible with strong acids, caustics, aliphatic amines, amides, oxidizers.

BUTYL FORMAL (110-62-3) Forms explosive mixture with air (flash point 54°F/12°C). Incompatible with strong acids, oxidizers, caustics, amines.

BUTYL FORMATE (592-84-7) Reacts with strong acids, nitrates, oxidizers (possibly violently). May accumulate static electrical charges, and may cause ignition of its vapors.

n-**BUTYL FORMATE** (592-84-7) Reacts with strong acids, nitrates, oxidizers (possibly violently). May accumulate static electrical charges, and may cause ignition of its vapors.

n-**BUTYL GLYCIDYL ETHER** (2426-08-6) Forms explosive mixture with air (flash point 130°F/54°C). Air and light form unstable and explosive peroxides. Contact with strong oxidizers may cause fire and explosions. Strong caustics may cause polymerization. Attacks some plastics and rubber.

BUTYL GLYCOL (111-76-2) Forms explosive mixture with air (flash point 143°F/62°C). Reacts violently with strong caustics and strong oxidizers. Attacks some coatings, plastics and rubber. Attacks metallic aluminum at high temperatures.

BUTYL GLYCOL ACETATE (112-07-2) Incompatible with strong acids, perchloric acid, oxidizers, nitrates.

BUTYL GLYCOL MONOBUTYL ETHER (111-76-2) Forms explosive mixture with air (flash point 143°F/62°C). Reacts violently with strong caustics and strong oxidizers. Attacks some coatings, plastics and rubber. Attacks metallic aluminum at high temperatures.

BUTYL HYDRIDE (106-97-8) Forms explosive gas mixture with air. Reacts with oxidizers. Mixed with oxygen this substance will explode on contact with nickel carbonyl in the 68°F to 104°F/20°C to 40°C range.

tert-**BUTYL HYDROPEROXIDE** (75-91-2) Forms explosive mixture with air (flash point 100°F/38°C). A powerful oxidizer; reacts violently with reducing agents, organic materials, ethylene dichloride.

BUTYL HYDROXIDE (71-36-3) Forms explosive mixture with air (flash point 98°F/37°C). May react with aluminum above 120°F/49°C. Attacks some plastics, rubber, and coatings. Incompatible with strong acids, halogens, caustics, alkali metals, aliphatic amines, isocyanates.

tert-**BUTYL HYDROXIDE** (75-65-0) Forms explosive mixture with air (flash point 52°F/11°C). May accumulate static electrical charges, and may cause ignition of its vapors. Incompatible with strong acids including mineral acids, strong oxidizers or caustics, aliphatic amines, isocyanates, alkali metals (i.e., lithium, sodium, potassium, rubidium, cesium, francium). Attacks many plastics and some coatings.

BUTYL-a-HYDROXYPROPIONATE or BUTYL-alpha-HYDROXY-PROPIONATE (138-22-7) Forms explosive mixture with air (flash point 160°F/71°C). Incompatible with strong oxidizers, strong bases.

iso-**BUTYL ISOBUTYRATE** (97-85-8) Forms explosive mixture with air (flash point 99°F/37°C). Incompatible with strong acids, nitrates, strong oxidizers.

BUTYL LACTATE or *n*-BUTYL LACTATE (138-22-7) Forms explosive mixture with air (flash point 160°F/71°C). Incompatible with strong oxidizers, strong bases.

BUTYL MERCAPTAN or *n*-BUTYL MERCAPTAN (109-79-5) Forms explosive mixture with air (flash point 35°F/2°C). Incompatible with strong oxidizers and nitric acid. Attacks some plastics and rubber.

BUTYL METHACRYLAAT (Dutch) (97-88-1) Unless inhibitor is maintained at the proper level, oxidizers, heat, ultraviolet light, or moisture may cause polymerization. May accumulate static electrical charges; may cause ignition of its vapors.

BUTYL METHACRYLATE (97-88-1) Unless inhibitor is maintained at the proper level, oxidizers, heat, ultraviolet light, or moisture may cause polymerization. May accumulate static electrical charges; may cause ignition of its vapors.

BUTYL-2-METHACRYLATE (97-88-1) Unless inhibitor is maintained at the proper level, oxidizers, heat, ultraviolet light, or moisture may cause polymerization. May accumulate static electrical charges; may cause ignition of its vapors.

iso-**BUTYL METHACRYLATE (97-86-9)** Forms explosive mixture with air (flash point 112°F/44°C). Able to form unstable peroxides. Incompatible with strong acids, aliphatic amines, alkanolamines, catalysts, strong oxidizers.

n-**BUTYL METHACRYLATE (97-88-1)** Unless inhibitor is maintained at the proper level, oxidizers, heat, ultraviolet light, or moisture may cause polymerization. May accumulate static electrical charges; may cause ignition of its vapors.

BUTYL METHANOATE (592-84-7) Reacts with strong acids, nitrates, oxidizers (possibly violently). May accumulate static electrical charges, and may cause ignition of its vapors.

tert-**BUTYL METHYL ETHER (1634-04-4)** Forms explosive mixture with air (flash point −14°F/−26°C). May be able to form unstable peroxides. Incompatible with strong acids; reacts violently with strong oxidizers. May accumulate static electrical charges; may cause ignition of its vapors.

BUTYL METHYL KETONE (591-78-6) Forms explosive mixture with air (flash point 77°F/25°C). Strong oxidizers may cause fire and explosions. Dissolves some plastics, resins, and rubber.

N-BUTYL METHYL KETONE (591-78-6) Forms explosive mixture with air (flash point 77°F/25°C). Strong oxidizers may cause fire and explosions. Dissolves some plastics, resins, and rubber.

BUTYL-2-METHYL-2-PROPENOATE (97-88-1) Unless inhibitor is maintained at the proper level, oxidizers, heat, ultraviolet light, or moisture may cause polymerization. May accumulate static electrical charges; may cause ignition of its vapors.

BUTYLOWY ALKOHOL (Polish) (71-36-3) Forms explosive mixture with air (flash point 98°F/37°C). May react with aluminum above 120°F/49°C. Attacks some plastics, rubber, and coatings. Incompatible with strong acids, halogens, caustics, alkali metals, aliphatic amines, isocyanates.

BUTYL OXITOL (111-76-2) Forms explosive mixture with air (flash point 143°F/62°C). Reacts violently with strong caustics and strong oxidizers. Attacks some coatings, plastics and rubber. Attacks metallic aluminum at high temperatures.

BUTYLPHEN (98-54-4) Incompatible with strong acids, caustics, aliphatic amines, amides, oxidizers.

4-*t*-BUTYLPHENOL or 4-*tert*-BUTYLPHENOL (98-54-4) Incompatible with strong acids, caustics, aliphatic amines, amides, oxidizers.

p-tert-**BUTYLPHENOL** or *p-tert*-**BUTYLPHENOL (MAK)** (98-54-4) Incompatible with strong acids, caustics, aliphatic amines, amides, oxidizers.

BUTYL PHOSPHATE, TRI- (126-73-8) Contact with water produces corrosive phosphoric acid. Incompatible with strong acids, strong oxidizers.

BUTYL PHTHALATE or *n*-**BUTYL PHTHALATE** (84-74-2) Incompatible with strong acids, nitrates, strong oxidizers, strong alkalies.

BUTYL PROPANOATE (590-01-2) Forms explosive mixture with air (flash point 90°F/32°C). Incompatible with strong oxidizers, strong acids.

BUTYL 2-PROPENOATE or *n*-**BUTYL 2-PROPENOATE** (141-32-2) Forms explosive mixture with air (flash point 120°F/49°C). Heat, sparks, open flame, light, or peroxides may cause explosive polymerization. Incompatible with strong acids, amines, halogens, hydrogen compounds, oxidizers, sunlight, or other catalysts.

n-**BUTYL PROPIONATE** (590-01-2) Forms explosive mixture with air (flash point 90°F/32°C). Incompatible with strong oxidizers, strong acids.

BUTYL 2,4,5-T (8002-26-4) Water contact causes foaming. Strong oxidizers may cause fire and explosions. Attacks some plastics, rubber, and coatings.

N-BUTYL THIOALCOHOL (109-79-5) Forms explosive mixture with air (flash point 35°F/2°C). Incompatible with strong oxidizers and nitric acid. Attacks some plastics and rubber.

BUTYL TITANATE or **BUTYL TITANATE MONOMER** (5593-70-4) Forms explosive mixture with air (flash point 170°F/77°C). Reacts violently with oxidizers. Water contact produces butanol and titanium dioxide. Reacts with oxidizers.

BUTYL TOLUENE (98-51-1) Forms explosive mixture with air (flash point 155°F/67°C). Reacts with strong oxidizers. May accumulate static electrical charges, and may cause ignition of its vapors.

4-*tert*-BUTYLTOLUENE (98-51-1) Forms explosive mixture with air (flash point 155°F/67°C). Reacts with strong oxidizers. May accumulate static electrical charges, and may cause ignition of its vapors.

p-tert-**BUTYLTOLUENE** (98-51-1) Forms explosive mixture with air (flash point 155°F/67°C). Reacts with strong oxidizers. May accumulate static electrical charges, and may cause ignition of its vapors.

BUTYL 2,4,5-TRICHLOROPHENOXYACETATE (8002-26-4) Water contact causes foaming. Strong oxidizers may cause fire and explosions. Attacks some plastics, rubber, and coatings.

BUTYL TRICHLOROSILANE or *n*-**BUTYL TRICHLOROSILANE** (7521-80-4) Forms explosive mixture with air (flash point 126°F/52°C). Reacts violently with water producing hydrochloric acid and fumes. Strong oxidizers may cause fire and explosions. Attacks metals in a moist environment.

2-BUTYNE (503-17-3) Forms explosive mixture with air (flash point less than −4°F/−20°C). Reacts with oxidizers.

1,4-BUTYNEDIOL (110-65-6) Reacts violently with strong acids, bases, alkali metals and earth.

2-BUTYNE-1,4-DIOL (110-65-6) Reacts violently with strong acids, bases, alkali metals and earth.

BUTYRAL (123-72-8) May accumulate static electrical charges, and may cause ignition of its vapors. Forms explosive mixture with air (flash point −8°F/−22°C). Incompatible with strong oxidizers, strong acids, caustics, ammonia, aliphatic amines, alkanolamines, aromatic amines.

BUTYRAL BUTYRIC ALDEHYDE (123-72-8) May accumulate static electrical charges, and may cause ignition of its vapors. Forms explosive mixture with air (flash point −8°F/−22°C). Incompatible with strong oxidizers, strong acids, caustics, ammonia, aliphatic amines, alkanolamines, aromatic amines.

BUTYRALDEHYD (German) (123-72-8) May accumulate static electrical charges, and may cause ignition of its vapors. Forms explosive mixture with air (flash point −8°F/−22°C). Incompatible with strong oxidizers, strong acids, caustics, ammonia, aliphatic amines, alkanolamines, aromatic amines.

BUTYRALDEHYDE (123-72-8) May accumulate static electrical charges, and may cause ignition of its vapors. Forms explosive mixture with air (flash point −8°F/−22°C). Incompatible with strong oxidizers, strong acids, caustics, ammonia, aliphatic amines, alkanolamines, aromatic amines.

iso-**BUTYRALDEHYDE** (78-84-2) Forms explosive gas mixture with air. Incompatible with strong acids, caustics, aliphatic amines, alkanolamines, aromatic amines, strong oxidizers.

n-**BUTYRALDEHYDE** (123-72-8) May accumulate static electrical charges, and may cause ignition of its vapors. Forms explosive mixture with air (flash point −8°F/−22°C). Incompatible with strong oxidizers, strong acids, caustics, ammonia, aliphatic amines, alkanolamines, aromatic amines.

BUTYRIC ACID or *n*-BUTYRIC ACID (107-92-6) Forms explosive mixture with air (flash point 162°F/72°C). Incompatible with sulfuric acid, caustics, ammonia, aliphatic amines, isocyanates, strong oxidizers, alkylene oxides, epichlorohydrin.

BUTYRIC ACID, BUTYL ESTER (109-21-7) Forms explosive mixture with air (flash point 48°F/−9°C). Attacks some plastics, rubber, and coatings. Incompatible with strong acids, nitrates; reacts violently with strong oxidizers.

BUTYRIC ACID, ETHYL ESTER (105-54-4) Incompatible with strong acids, nitrates, oxidizers.

BUTYRIC ACID, 2-HYDROXY-4-METHYLTHIO- (583-91-5) Incompatible with sulfuric acid, caustics, ammonia, amines, isocyanates, alkylene oxides, epichlorohydrin.

BUTYRIC ACID, METHYL ESTER (623-42-7) Incompatible with strong acids, nitrates, oxidizers.

BUTYRIC ACID NITRILE (109-74-0) Forms explosive mixture with air (flash point 79°F/26°C). Reacts violently with strong acids (forms hydrogen cyanide gas), strong oxidizers (with possible fire and explosions). May accumulate static electrical charges; may cause ignition of its vapors.

BUTYRIC ALDEHYDE (123-72-8) May accumulate static electrical charges, and may cause ignition of its vapors. Forms explosive mixture with air (flash point −8°F/−22°C). Incompatible with strong oxidizers, strong acids, caustics, ammonia, aliphatic amines, alkanolamines, aromatic amines.

BUTYRIC ETHER (105-54-4) Incompatible with strong acids, nitrates, oxidizers.

BUTYRONE (123-19-3) Oxidizers may cause violent reaction.

BUTYRONITRILE (109-74-0) Forms explosive mixture with air (flash point 79°F/26°C). Reacts violently with strong acids (producing hydrogen cyanide gas), strong oxidizers (with possible fire and explosions). May accumulate static electrical charges, and may cause ignition of its vapors.

BUTYROYL CHLORIDE (141-75-3) Forms corrosive hydrogen chloride fumes on contact with air. May accumulate static electrical charges, and may cause ignition of its vapors. Forms explosive mixture with air (flash point 71°F/22°C). Incompatible with oxidizers, strong bases. Aqueous solution produces hydrochloric acid; corrodes metals.

BUTYRYL CHLORIDE (141-75-3) Produces corrosive hydrogen chloride fumes on contact with air. May accumulate static electrical charges, and may cause ignition of its vapors. Forms explosive mixture with air (flash point 71°F/22°C). Incompatible with oxidizers, strong bases. Aqueous solution produces hydrochloric acid; corrodes metals.

N-BUTYRYL CHLORIDE (141-75-3) Produces corrosive hydrogen chloride fumes on contact with air. May accumulate static electrical charges, and may cause ignition of its vapors. Forms explosive mixture with air (flash point 71°F/22°C). Incompatible with oxidizers, strong bases. Aqueous solution produces hydrochloric acid; corrodes metals.

B-W (1344-09-8) Solution is a strong base; reacts with acids, organic anhydrides, alkylene oxides, epichlorohydrin, aldehydes, alcohols, glycols, phenols, cresols, caprolactam solution. Attacks chemically active metals.

- C -

C-56 (77-47-4) Incompatible with water, producing hydrochloric acid. Contact with sodium may be explosive. Corrodes iron and other metals in the presence of moisture.

C-709 (141-66-2) Corrosive to cast iron, mild steel, brass, and stainless steel 304.

C-709 (CIBA-GEIGY) (141-66-2) Corrosive to cast iron, mild steel, brass, and stainless steel 304.

CAA (372-09-8) Incompatible with oxidizers, strong acids, organic acids, caustics, reducing agents.

CABLE OIL (8012-95-1) Incompatible with nitric acid. Strong oxidizers may cause fire and explosions.

CACODYLIC ACID (75-60-5) Aqueous solution reacts with chemically active metals. Incompatible with sulfuric acid, caustics, ammonia, amines, isocyanates, alkylene oxides, epichlorohydrin.

CADDY (10108-64-2) Incompatible with strong oxidizers, elemental sulfur, selenium, tellurium.

CADMIUM (7440-43-9) With air exposure, especially in powdered form, may self-ignite. Reacts violently with strong oxidizers, hydrozoic acid, tellurium. Acid contact produces hydrogen gas. May react with selenium, elemental sulfur, zinc.

CADMIUM ACETATE (543-90-8) Incompatible with strong oxidizers, elemental sulfur, selenium, tellurium, strong acids, nitrates.

CADMIUM(II) ACETATE (543-90-8) Incompatible with strong oxidizers, elemental sulfur, selenium, tellurium, strong acids, nitrates

CADMIUM ACETATE DIHYDRATE (543-90-8) Incompatible with strong oxidizers, elemental sulfur, selenium, tellurium, strong acids, nitrates

CADMIUM BROMIDE (7789-42-6) Incompatible with potassium.

CADMIUM BROMIDE TETRAHYDRATE (7789-42-6) Incompatible with potassium.

CADMIUM CHLORIDE (10108-64-2) Incompatible with strong oxidizers, elemental sulfur, selenium, tellurium.

CADMIUM DIACETATE (543-90-8) Incompatible with strong oxidizers, elemental sulfur, selenium, tellurium, strong acids, nitrates

CADMIUM DICHLORIDE (10108-64-2) Incompatible with strong oxidizers, elemental sulfur, selenium, tellurium.

CADMIUM FLUOBORATE (14486-19-2) May react with strong oxidizers, sulfur, selenium or tellurium.

CADMIUM FLUOROBORATE (14486-19-2) May react with strong oxidizers, sulfur, selenium or tellurium.

CADMIUM FUME (1306-19-0) May produce explosive with magnesium + heat. May react with sulfur, selenium, zinc. Acid contact produces hydrogen gas.

CADMIUM MONOXIDE (1306-19-0) May produce explosive with magnesium + heat. May react with sulfur, selenium, zinc. Acid contact produces hydrogen gas.

CADMIUM(II) NITRATE, TETRAHYDRATE (1:2:4) (10022-68-1) May react with strong oxidizers, selenium, sulfur or tellurium.

CADMIUM OXIDE (1306-19-0) May form an explosive mixture with magnesium + heat. May react with sulfur, selenium, zinc. Acid contact forms hydrogen gas.

CADMIUM SULFATE or CADMIUM SULFATE(1:1) (10124-36-4) May react with strong oxidizers, sulfur, selenium, tellurium, zinc.

CADMIUM SULPHATE (10124-36-4) May react with strong oxidizers, sulfur, selenium, tellurium, zinc.

CADOX HDP (78-18-2) A strong oxidizer. Reacts violently with combustibles, reducing agents, caustics, ammonia.

CADOX PS (94-17-7) A powerful oxidizer and explosion hazard. Reacts violently with reducing agents, combustibles, polymerization initiators, heat or contaminants.

CADOX TBH (75-91-2) Forms explosive mixture with air (flash point 100°F/38°C). A powerful oxidizer; reacts violently with reducing agents, organic materials, ethylene dichloride.

CAIROX (7722-64-7) A strong oxidizer. Produces heat- and shock-sensitive compound with sulfuric acid. Reacts violently with combustibles or reducing agents. Reacts explosively with acetic acid, acetic anhydride, anhydrous ammonia, hydrogen peroxide, glycerol, hydroxylamine, organic matter, powdered sulfur, etc. Incompatible with nitric acid, producing toxic chlorine fumes. A dangerous fire and explosion hazard; isolate from all other materials.

CAJEPUTENE (138-86-3) Forms explosive mixture with air (flash point 115°F/46°C). Strong oxidizers may cause fire and explosions.

CAKE ALUM (10043-01-3) Incompatible with water, producing sulfuric acid; reacts strongly with bases and many other materials. Dry material is weakly corrosive to carbon steel; aqueous solution attacks metals, producing hydrogen gas.

CAKE ALUMINUM (10043-01-3) Incompatible with water, producing sulfuric acid; reacts strongly with bases and many other materials. Dry material is weakly corrosive to carbon steel; aqueous solution attacks metals, producing hydrogen gas.

CALCIA (1305-78-8) Reacts violently with water. Incompatible with ethanol, liquid hydrogen fluoride, acids, halogens, metal halides, light metals, some oxides, boron trifluoride, chlorine trifluoride, liquid hydrofluoric acid, phosphorus pentoxide, boric oxide and calcium chloride mixtures, fluorine, chlorine trifluoride, carbon dioxide.

CALCICAT (7440-70-2) Produces hydrogen gas on contact with air; may self-ignite. A strong reducing agent; reacts violently with water, acids, oxidizers, carbonates, dinitrogen tetroxide, halogenated hydrocarbons, lead chloride, halogens, alkaline hydroxides, oxygen, silicon, sulfur, chlorine, fluorine, chlorine trifluoride, and many other substances.

CALCINED BARSITO (1304-28-5) Incompatible with water, hydrogen sulfide, carbon dioxide, hydroxlamine, nitrogen tetroxide, sulfur trioxide, and triuranium.

CALCINED BARYTA (1304-28-5) Incompatible with water, hydrogen sulfide, carbon dioxide, hydroxlamine, nitrogen tetroxide, sulfur trioxide, and triuranium.

CALCINED BRUCITE (1309-48-4) Reacts violently with chlorine trifluoride, bromine pentalfluoride, phosphorus pentachloride.

CALCINED MAGNESIA (1309-48-4) Reacts violently with chlorine trifluoride, bromine pentalfluoride, phosphorus pentachloride.

CALCINED MAGNESITE (1309-48-4) Reacts violently with chlorine trifluoride, bromine pentalfluoride, phosphorus pentachloride.

CALCITE (1317-65-3) Incompatible with acids, alum, ammonium salts, fluorine.

CALCIUM (7440-70-2) Produces hydrogen gas on contact with air; may self-ignite. A strong reducing agent; reacts violently with water, acids, oxidizers, carbonates, dinitrogen tetroxide, halogenated hydrocarbons, lead chloride, halogens, alkaline hydroxides, oxygen, silicon, sulfur, chlorine, fluorine, chlorine trifluoride, and many other substances.

CALCIUM ABIETATE (9007-13-0) Reacts with oxidizers, with a risk of fire or explosions.

CALCIUM ALKYLAROMATIC SULFONATE (27176-87-0) Forms explosive mixture with air (flash point 100°F/38°C). Strong oxidizers may cause fire and explosions. Do not store in carbon steel or aluminum.

CALCIUM ALKYLBENZENESULFONATE (27176-87-0) Forms explosive mixture with air (flash point 100°F/38°C). Strong oxidizers may cause fire and explosions. Do not store in carbon steel or aluminum.

CALCIUM ALLOYS (7440-70-2) Form hydrogen gas on contact with air; may self-ignite. Strong reducing agents; react violently with water, acids, oxidizers, carbonates, dinitrogen tetroxide, halogenated hydrocarbons, lead chloride, halogens, alkaline hydroxides, oxygen, silicon, sulfur, chlorine, fluorine, chlorine trifluoride and many other substances.

CALCIUM BIPHOSPHATE (10103-46-5) Water contact produces acid; attacks metals.

CALCIUM BROMATE (10102-75-7) A strong oxidizer. Incompatible with combustible materials, reducing agents, strong acids, aluminum, ammonium salts, arsenic, carbon, copper, metal powder and sulfides, organic matter, phosphorus, sulfur.

CALCIUM BROMIDE SOLUTION (7789-41-5) Incompatible with sulfuric acid, isocyanates.

CALCIUM CARBIDE (75-20-7) Water contact or moist air produces explosive acetylene gas. Incompatible with acids, oxidizers, hydrogen chloride, methanol, copper salt solutions, lead fluoride, magnesium, selenium, silver nitrate, iron trichloride, tin dichloride, sodium peroxide, stannous chloride, sulfur.

CALCIUM CARBIMIDE (156-62-7) Contact with any form of moisture causes decomposition, liberating acetylene and ammonia. Contact with all solvents tested also cause decomposition.

CALCIUM CARBONATE (1317-65-3) Incompatible with acids, alum, ammonium salts, fluorine.

CALCIUM(II) CARBONATE (1:1) (1317-65-3) Incompatible with acids, alum, ammonium salts, fluorine.

CALCIUM CHLORATE (10137-74-3) A strong reducing agent. Reacts, possibly violently, with acids (especially organic), reducing agents, aluminum, arsenic, chemically active metals, combustible materials, ammonium compounds, charcoal, copper, cyanides, manganese dioxide, metal sulfides, phosphorus, sulfur.

CALCIUM CHLORIDE (10043-52-4) Incompatible with water, bromine trifluoride, 2-furan, percarboxylic acid. Attack metals.

CALCIUM CHLORIDE, ANHYDROUS (10043-52-4) Incompatible with water, bromine trifluoride, 2-furan, percarboxylic acid. Attacks metals.

CALCIUM CHLORIDE HYDRATES (10043-52-4) Incompatible with water, bromine trifluoride, 2-furan, percarboxylic acid. Attack metals.

CALCIUM CHLORITE (14674-72-7) Incompatible with chlorine.

CALCIUM CHLOROHYDROCHLORITE (7778-54-3) Decomposes in heat or sunlight. Incompatible with acids, moisture, reducing agents, combustible materials, all other chemicals, especially acetylene, aniline and all other amines, anthracene, carbon tetrachloride, iron oxide, manganese oxide, mercaptans, diethylene glycol monomethyl ether, nitromethane, organic matter, organic sulfides, phenol, 1-propanethiol, propyl mercaptan, sulfur, organic sulfur compounds.

CALCIUM CHROMATE (13765-19-0) A strong oxidizer. Incompatible with boron (with violent reaction), combustible material.

CALCIUM CHROMATE(VI) (13765-19-0) A strong oxidizer. Incompatible with boron (with violent reaction), combustible material.

CALCIUM CHROMATE DIHYDRATE (13765-19-0) A strong oxidizer. Incompatible with boron (with violent reaction), combustible material.

CALCIUM CHROME YELLOW (13765-19-0) A strong oxidizer. Incompatible with boron (with violent reaction), combustible material.

CALCIUM CHROMIUM OXIDE (13765-19-0) A strong oxidizer. Incompatible with boron (with violent reaction), combustible material.

CALCIUM CYANAMID (156-62-7) Contact with any form of moisture causes decomposition, liberating acetylene and ammonia. Contact with all solvents tested also causes decomposition.

CALCIUM CYANAMIDE (156-62-7) Contact with any form of moisture causes decomposition, liberating acetylene and ammonia. Contact with all solvents tested also causes decomposition.

CALCIUM CYANIDE (592-01-8) Incompatible with water, strong acids, fluorine, magnesium, nitrates, nitrites.

CALCIUM CYANIDE MIXTURE or CALCIUM CYANIDE MIXTURE, SOLID (592-01-8) Incompatible with water, strong acids, fluorine, magnesium, nitrates, nitrites.

CALCIUM DICARBIDE (75-20-7) Water contact or moist air produces explosive acetylene gas. Incompatible with acids, oxidizers, hydrogen chloride, methanol, copper salt solutions, lead fluoride, magnesium, selenium, silver nitrate, iron trichloride, tin dichloride, sodium peroxide, stannous chloride, sulfur.

CALCIUM DIFLUORIDE (7789-75-5) Incompatible with acids, chemically active metals, reducing agents, water.

CALCIUM DIOXIDE (1305-79-9) A strong alkali, and strong oxidizer. Incompatible with combustible materials, polysulfide polymers.

CALCIUM FLUORIDE (7789-75-5) Incompatible with acids, chemically active metals, reducing agents, water.

CALCIUM HYDRATE (1305-62-0) Contact with maleic anhydride, phosphorus, nitroethane, nitromethane, nitroparaffins, or nitropropane may cause explosion.

CALCIUM HYDROXIDE (1305-62-0) Contact with maleic anhydride, phosphorus, nitroethane, nitromethane, nitroparaffins, or nitropropane may cause explosion.

CALCIUM HYPOCHLORIDE (7778-54-3) Decomposes in heat or sunlight. Incompatible with acids, moisture, reducing agents, combustible materials, all other chemicals, especially acetylene, aniline and all other amines, anthracene, carbon tetrachloride, iron oxide, manganese oxide, mercaptans, diethylene glycol monomethyl ether, nitromethane, organic matter, organic sulfides, phenol, 1-propanethiol, propyl mercaptan, sulfur, organic sulfur compounds.

CALCIUM HYPOCHLORITE (7778-54-3) Decomposes in heat or sunlight. Incompatible with acids, moisture, reducing agents, combustible materials, all other chemicals, especially acetylene, aniline and all other amines, anthracene, carbon tetrachloride, iron oxide, manganese oxide, mercaptans, diethylene glycol monomethyl ether, nitromethane, organic matter, organic sulfides, phenol, 1-propanethiol, propyl mercaptan, sulfur, organic sulfur compounds.

CALCIUM HYPOPHOSPHITE (N/A) Incompatible with strong acids. Shock and heat-sensitive mixture formed with potassium chlorate.

CALCIUM IODATE (N/A) Incompatible with aluminum, arsenic, carbon, copper, metal sulfides, organic matter, phosphorus, sulfur.

CALCIUM LIMED WOOD ROSIN (9007-13-0) Incompatible with oxidizers, strong acids.

CALCIUM METAL or CALCIUM METAL, CRYSTALLINE (7440-70-2) Produces hydrogen gas on contact with air; may self-ignite. A strong reducing agent; reacts violently with water, acids, oxidizers, carbonates, dinitrogen tetroxide, halogenated hydrocarbons, lead chloride, halogens, alkaline hydroxides, oxygen, silicon, sulfur, chlorine, fluorine, chlorine trifluoride and many other substances.

CALCIUM MONOCHROMATE (13765-19-0) A strong oxidizer. Incompatible with boron (with violent reaction), combustible material.

CALCIUM NITRATE (10124-37-5) A strong oxidizer. Incompatible with combustible materials, reducing agents, organics and other oxidizable materials, chemically active metals, aluminum nitrate, ammonium nitrate.

CALCIUM(II) NITRATE (1:2) (10124-37-5) A strong oxidizer. Incompatible with combustible materials, reducing agents, organics and other oxidizable materials, chemically active metals, aluminum nitrate, ammonium nitrate.

CALCIUM NITRATE TETRAHYDRATE (10124-37-5) A strong oxidizer. Incompatible with combustible materials, reducing agents, organics and other oxidizable materials, chemically active metals, aluminum nitrate, ammonium nitrate.

CALCIUM NITRITE (13780-06-8) Rapid heating may cause explosion. A strong oxidizer; reacts violently (with possible fire and explosions) with reducing agents, combustibles, some ammonium compounds, cyanides. Contact with acids may produce toxic fumes.

CALCIUM OXIDE (1305-78-8) Reacts violently with water. Incompatible with ethanol, hydrogen fluoride, acids, halogens, metal halides, light metals, some oxides, boron trifluoride, chlorine trifluoride, liquid hydrofluoric acid, phosphorus pentoxide, boric oxide and calcium chloride mixtures, fluorine, chlorine trifluoride, carbon dioxide.

CALCIUM OXYCHLORIDE (7778-54-3) Decomposes in heat or sunlight. Incompatible with acids, moisture, reducing agents, combustible materials, all other chemicals, especially acetylene, aniline and all other amines, anthracene, carbon tetrachloride, iron oxide, manganese oxide, mercaptans, diethylene glycol monomethyl ether, nitromethane, organic matter, organic sulfides, phenol, 1-propanethiol, propyl mercaptan, sulfur, organic sulfur compounds.

CALCIUM PEROXIDE (1305-79-9) A strong alkali, and strong oxidizer. Incompatible with combustible materials, polysulfide polymers.

CALCIUM PHOSPHATE (10103-46-5) Water contact produces acid; attacks metals.

CALCIUM PHOSPHIDE (1305-99-3) A strong reducing agent. Forms phosphine in air. Water or acids can cause explosions. Incompatible with oxidizers, acids, chlorine, chlorine monoxide, oxygen, sulfur.

CALCIUM PYROPHOSPHATE (10103-46-5) Water contact forms acid; attacks metals.

CALCIUM RESINATE or CALCIUM RESINATE, FUSED (9007-13-0) Incompatible with oxidizers, strong acids.

CALCIUM ROSIN (9007-13-0) Incompatible with oxidizers, strong acids.

CALCIUM SULFATE (7778-18-9) Contact with diazomethane, aluminum, phosphorus may cause explosion.

CALCIUM SULFIDE (20548-54-3) Reacts with air, water, or acids producing toxic and flammable hydrogen sulfide gas. Reacts violently with strong oxidizers.

CALCIUM SUPEROXIDE (1305-79-9) A strong alkali, and strong oxidizer. Incompatible with combustible materials, polysulfide polymers.

CALCIUM SUPERPHOSPHATE (10103-46-5) Water contact produces acid; attacks metals.

CALCOTONE RED (1309-37-1) Contact with hydrogen peroxide, ethylene oxide, calcium hypochlorite will cause explosion. Reacts violently with powdered aluminum, hydrazine, hydrogen trisulfide.

CAL HYPO (7778-54-3) Decomposes in heat or sunlight. Acids, moisture, reducing agents, combustible materials, all other chemicals, especially acetylene, aniline and all other amines, anthracene, carbon tetrachloride, iron oxide, manganese oxide, mercaptans, diethylene glycol monomethyl ether, nitromethane, organic matter, organic sulfides, phenol, 1-propanethiol, propyl mercaptan, sulfur, organic sulfur compounds.

CALMATHION (121-75-5) Incompatible with strong oxidizers, magnesium, alkaline pesticides. Attacks metals, some plastics, rubber, and coatings.

CALOCHLOR (7487-94-7) Incompatible with light metals (aluminum, magnesium, beryllium, etc.), sodium, potassium.

CAL PLUS (10043-52-4) Incompatible with bromine trifluoride, 2-furan, percarboxylic acid. Attacks metals.

CALSOFT F-90 (25155-30-0) Reacts with acids, including fumes

CALTAC (10043-52-4) Incompatible with water, bromine trifluoride, 2-furan, percarboxylic acid. Attacks metals.

150

CALX (1305-78-8) Reacts violently with water. Incompatible with ethanol, hydrogen fluoride, acids, halogens, metal halides, light metals, some oxides, boron trifluoride, chlorine trifluoride, liquid hydrofluoric acid, phosphorus pentoxide, boric oxide and calcium chloride mixtures, fluorine, chlorine trifluoride, carbon dioxide.

2-CAMPHANONE (76-22-2); (8008-51-3) Contact with strong oxidizers, especially chromic anhydride, produce explosive mixtures.

CAMPHECHLOR (8001-35-2) Reacts with oxidizers, with a risk of fire or explosions. Attacks metals in the presence of moisture.

CAMPHENE (79-92-5) Contact with strong oxidizers may cause fire and explosions. Emulsions in xylene may violently decompose on contact with iron or aluminum above 158°F/70°C.

CAMPHENE, OCTACHLORO- (8001-35-2) Reacts with oxidizers, with a risk of fire or explosions. Attacks metals in the presence of moisture.

CAMPHOCLOR (8001-35-2) Reacts with oxidizers, with a risk of fire or explosions. Attacks metals in the presence of moisture.

CAMPHOFENE HUILEUX (8001-35-2) Reacts with oxidizers, with a risk of fire or explosions. Attacks metals in the presence of moisture.

CAMPHOR or CAMPHOR, NATURAL (76-22-2) Forms explosive mixture with air (flash point 155°F/66°C). Violent, possibly explosive, reaction occurs with strong oxidizers. May accumulate static electrical charges, and may cause ignition of its vapors.

CAMPHOR OIL (76-22-2) Forms explosive mixture with air (flash point 155°F/66°C). Violent, possibly explosive, reaction occurs with strong oxidizers. May accumulate static electrical charges, and may cause ignition of its vapors.

2-CAMPHORONE (76-22-2) Forms explosive mixture with air (flash point 155°F/66°C). Violent, possibly explosive, reaction occurs with strong oxidizers. May accumulate static electrical charges, and may cause ignition of its vapors.

CAMPHOR TAR (91-20-3) Forms explosive mixture with air (flash point 174°F/79°C). Incompatible with strong oxidizers, chromium oxide (with violent reaction). Attacks some plastics, rubber, and coatings. May accumulate static electrical charges, and may cause ignition of its vapors.

CAMPILIT (506-68-3) May be unstable unless dry and pure. Reacts violently with acids, ammonia, amines. Water contact produces hydrogen cyanide and hydrogen bromide.

CANDEX (1912-24-9) Incompatible with strong acids.

CANE SUGAR (57-50-1) Reacts with potassium hydroxide, strong acids, strong oxidizers.

CANOGARD (62-73-7) Attacks some plastics, rubber, and coatings.

CAPORIT (7778-54-3) Decomposes in heat or sunlight. Incompatible with acids, moisture, reducing agents, combustible materials, all other chemicals, especially acetylene, aniline and all other amines, anthracene, carbon tetrachloride, iron oxide, manganese oxide, mercaptans, diethylene glycol monomethyl ether, nitromethane, organic matter, organic sulfides, phenol, 1-propanethiol, propyl mercaptan, sulfur, organic sulfur compounds.

CAPRAN 80 (105-60-2) Contact with strong oxidizers may cause fire and explosions.

CAPRALDEHYDE (112-31-2) Incompatible with strong acids, ammonia, caustics, amines. Attacks galvanized steel.

CAPRIC ACID (334-48-5) Incompatible with sulfuric acid, caustics, ammonia, aliphatic amines, alkanolamines, isocyanates, alkylene oxides, epichlorohydrin. Attacks most common metals.

N-CAPRIC ACID (334-48-5) Incompatible with sulfuric acid, caustics, ammonia, aliphatic amines, alkanolamines, isocyanates, alkylene oxides, epichlorohydrin. Attacks most common metals.

CAPRIC ALCOHOL (112-30-1) Forms explosive mixture with air (flash point 180°F/82°C). Incompatible with strong acids, caustics, aliphatic amines, isocyanates, strong oxidizers.

CAPRIC ALDEHYDE (112-31-2) Incompatible with strong acids, ammonia, caustics, amines. Attacks galvanized steel.

CAPRINIC ACID (334-48-5) Incompatible with sulfuric acid, caustics, ammonia, aliphatic amines, alkanolamines, isocyanates, alkylene oxides, epichlorohydrin. Attacks most common metals.

CAPRINIC ALCOHOL (112-30-1) Forms explosive mixture with air (flash point 180°F/82°C). Incompatible with strong acids, caustics, aliphatic amines, isocyanates, strong oxidizers.

CAPROALDEHYDE (66-25-1) Forms explosive mixture with air (flash point 90°F/32°C). Incompatible with strong acids, caustics, ammonia, amines. Attacks some plastics, rubber, and coatings.

CAPROIC ACID or *n*-CAPROIC ACID (142-62-1) Reacts with strong oxidizers. Incompatible with sulfuric acid, caustics, ammonia, amines, isocyanates, alkylene oxides, epichlorohydrin. Attacks common metals.

CAPROLACTAM (105-60-2) Contact with strong oxidizers may cause fire and explosions.

epsilon-CAPROLACTAM (105-60-2) Contact with strong oxidizers may cause fire and explosions.

CAPROLIN (63-25-2) Incompatible with strong oxidizers, strongly alkaline pesticides.

CAPROLON B (105-60-2) Contact with strong oxidizers may cause fire and explosions.

CAPRON (105-60-2) Contact with strong oxidizers may cause fire and explosions.

CAPRONALDEHYDE (66-25-1) Forms explosive mixture with air (flash point 90°F/32°C). Incompatible with strong acids, caustics, ammonia, amines. Attacks some plastics, rubber, and coatings.

CAPRONIC ACID (142-62-1) Reacts with strong oxidizers. Incompatible with sulfuric acid, caustics, ammonia, amines, isocyanates, alkylene oxides, epichlorohydrin. Attacks common metals.

CAPRONIC ALDEHYDE (66-25-1) Forms explosive mixture with air (flash point 90°F/32°C). Incompatible with strong acids, caustics, ammonia, amines. Attacks some plastics, rubber, and coatings.

CAPROYL ALCOHOL (111-27-3) Forms explosive mixture with air (flash point 145°F/63°C). Incompatible with strong acids, caustics, aliphatic amines, isocyanates, strong oxidizers.

N-CAPROYLALDEHYDE (66-25-1) Forms explosive mixture with air (flash point 90°F/32°C). Incompatible with strong acids, caustics, ammonia, amines. Attacks some plastics, rubber, and coatings.

CAPRYL ALCOHOL (111-87-5) Forms explosive mixture with air (flash point 178°F/81°C). Incompatible with strong acids, caustics, aliphatic amines, isocyanates, strong oxidizers.

CAPRYLALDEHYDE (124-13-0) Forms explosive mixture with air (flash point 125°F/52°C). Incompatible with strong oxidizers, caustics, ammonia, aliphatic amines, alkanolamines, aromatic amines, strong oxidizers.

CAPRYLENE (111-66-0) Forms explosive mixture with air (flash point 70°F/21°C). Incompatible with strong acids, strong oxidizers.

N-CAPRYLIC ACID (111-65-4) Forms explosive mixture with air (flash point 56°F/13°C). Incompatible with oxidizers. May accumulate static electrical charges, and may cause ignition of its vapors.

CAPRYLIC ALCOHOL (111-87-5) Forms explosive mixture with air (flash point 178°F/81°C). Incompatible with strong acids, caustics, aliphatic amines, isocyanates, strong oxidizers.

CAPRYLIC ALDEHYDE (124-13-0) Forms explosive mixture with air (flash point 125°F/52°C). Incompatible with strong oxidizers, caustics, ammonia, aliphatic amines, alkanolamines, aromatic amines, strong oxidizers.

CAPRYNIC ACID (334-48-5) Incompatible with sulfuric acid, caustics, ammonia, aliphatic amines, alkanolamines, isocyanates, alkylene oxides, epichlorohydrin. Attacks most common metals.

CAPTAF (133-06-2) Incompatible with tetraethyl pyrophosphate, parathion.

CAPTAF 85W (133-06-2) Incompatible with tetraethyl pyrophosphate, parathion.

CAPTAFOL (2425-06-1) Incompatible with acids or acid vapor. Strong alkaline conditions contribute to instability.

CAPTAN (133-06-2) Incompatible with tetraethyl pyrophosphate, parathion.

CAPTANCAPTENEET 26,538 (133-06-2) Incompatible with tetraethyl pyrophosphate, parathion.

CAPTANE (133-06-2) Incompatible with tetraethyl pyrophosphate, parathion.

CAPTATOL (2425-06-1) Incompatible with acids or acid vapor. Strong alkaline conditions contribute to instability.

CAPTEX (133-06-2) Incompatible with tetraethyl pyrophosphate, parathion.

CAPUT MORTUUM (1309-37-1) Contact with hydrogen peroxide, ethylene oxide, calcium hypochlorite will cause explosion. Reacts violently with powdered aluminum, hydrazine, hydrogen trisulfide.

CARADATE 30 (101-68-8) Incompatible with strong alkalies, acids, alcohol. Unstable above 100°F/37.8°C. Attacks some plastics, rubber, and coatings.

CARBICRIN (141-66-2) Corrosive to cast iron, mild steel, brass, and stainless steel 304.

CARBACRYL (107-13-1) Forms explosive mixture with air (flash point 32°F/0°C o.c.). Forms explosive peroxides; heat, light, caustics, silver nitrate, and peroxides can cause polymerization. Incompatible with strong acids, strong oxidizers, amines, 2-aminoethanol, bromine, chlorosulfonic acid, ethylene diamine, nitric acid, oleum, potassium hydroxide, sodium hydroxide, sulfuric acid. Attacks copper and copper alloys; attacks aluminum in high concentrations. May accumulate static electrical charges, and may cause ignition of its vapors.

CARBAMALDEHYDE (75-12-7) Incompatible with nonoxidizing mineral acids, strong acids, ammonia, cresols, iodine, isocyanates, oleum, phenols, pyridine, sulfur trioxide.

CARBAMIC ACID, 1- (BUTYLAMINO)CARBONYL-1H-BENZI-MIDAZOL-2YL, METHYL ESTER (17804-35-2) Heat, water, strong acids and strong alkalies can cause decomposition and formation of toxic oxides of nitrogen.

CARBAMIC ACID, METHYL-, 2,2-DIMETHYL-2,3-DIHYDRO-BENZOFURAN-7-YL ESTER (1563-66-2) Incompatible with alkaline media, acids or strong oxidizers.

CARBAMIC CHLORIDE, DIMETHYL- (79-44-7) Rapidly hydrolyzed in water. Incompatible with strong acids, oxidizers.

CARBAMIDE (57-13-6) Heat causes evolution of anhydrous ammonia fumes. Contact with oxidizers, nitrates may cause fire and explosions. Contact with chlorinating agents including hypochlorite bleaches may form explosive nitrogen trichloride. Reacts with nitrosyl perchlorate.

CARBAMIDE ACID (57-13-6) Heat forms anhydrous ammonia fumes. Contact with oxidizers, nitrates may cause fire and explosions. Contact with chlorinating agents, including hypochlorite bleaches, may produce explosive nitrogen trichloride. Reacts with nitrosyl perchlorate.

CARBAMIDE PEROXIDE (124-43-6) Combustible solid. At 122°F/50°C reacts with dust, organics.

CARBAMIDE RESIN (57-13-6) Heat causes formation of anhydrous ammonia fumes. Contact with oxidizers, nitrates may cause fire and explosions. Contact with chlorinating agents, including hypochlorite bleaches, may produce explosive nitrogen trichloride. Reacts with nitrosyl perchlorate.

CARBAMIMIDIC ACID (57-13-6) Heat causes formation of anhydrous ammonia fumes. Contact with oxidizers, nitrates may cause fire and explosions. Contact with chlorinating agents, including hypochlorite bleaches may produce explosive nitrogen trichloride. Reacts with nitrosyl perchlorate.

CARBAMOYL CHLORIDE, DIMETHYL- (79-44-7) Rapidly hydrolyzed in water. Incompatible with strong acids, oxidizers.

CARBAPHOS (Russian) (121-75-5) Incompatible with strong oxidizers, magnesium, alkaline pesticides. Attacks metals, some plastics, rubber, and coatings.

CARBARIL (Italian) (63-25-2) Incompatible with strong oxidizers, strongly alkaline pesticides.

CARBARYL (63-25-2) Incompatible with strong oxidizers, strongly alkaline pesticides.

CARBATOX (63-25-2) Incompatible with strong oxidizers, strongly alkaline pesticides.

CARBATOX-60 (63-25-2) Incompatible with strong oxidizers, strongly alkaline pesticides.

CARBATOX 75 (63-25-2) Incompatible with strong oxidizers, strongly alkaline pesticides.

CARBAX (115-32-2) Incompatible with strong acids, caustics, aliphatic amines, isocyanates.

CARBAZOLE (86-74-8) Strong oxidizers may cause fires and explosions.

CARBETHOXY MALATHION (121-75-5) Incompatible with strong oxidizers, magnesium, alkaline pesticides. Attacks metals, some plastics, rubber, and coatings.

CARBETOVUR (121-75-5) Incompatible with strong oxidizers, magnesium, alkaline pesticides. Attacks metals, some plastics, rubber, and coatings.

CARBETOX (121-75-5) Incompatible with strong oxidizers, magnesium, alkaline pesticides. Attacks metals, some plastics, rubber, and coatings.

CARBIDE (75-20-7) Water contact or moist air causes formation of explosive acetylene gas. Incompatible with acids, oxidizers, hydrogen chloride, methanol, copper salt solutions, lead fluoride, magnesium, selenium, silver nitrate, iron trichloride, tin dichloride, sodium peroxide, stannous chloride, sulfur.

CARBINAMINE (74-89-5) Forms explosive gas mixture with air. Incompatible with nonoxidizing mineral acids, strong acids, organic acids, organic anhydrides, isocyanates, vinyl acetate, acrylates, substituted allyls, alkylene oxides, epichlorohydrin, ketones, aldehydes, alcohols, glycols, phenols, cresols, caprolactam solution, strong oxidizers. Attacks aluminum, copper, lead, tin, zinc, and alloys, and some plastics, rubber, and coatings.

CARBINAMINE SOLUTION (74-89-5) Forms explosive gas mixture with air. Incompatible with nonoxidizing mineral acids, strong acids, organic acids, organic anhydrides, isocyanates, vinyl acetate, acrylates, substituted allyls, alkylene oxides, epichlorohydrin, ketones, aldehydes, alcohols, glycols, phenols, cresols, caprolactam solution, strong oxidizers. Attacks aluminum, copper, lead, tin, zinc and alloys, and some plastics, rubber, and coatings.

CARBINOL (67-56-1) Forms explosive mixture with air (flash point 52°F/11°C). Incompatible with strong acids, strong oxidizers, caustics, aliphatic amines, isocyanates, chromic anhydride, lead perchlorate, perchloric acid, phosphorus trioxide. May react with metallic aluminum at high temperature. Attacks some plastics, rubber, and coatings.

CARBITOL (111-90-0) Incompatible with strong acids, isocyanates, strong oxidizers.

CARBITOL ACETATE (112-15-2) Incompatible with sulfuric acid, nitric acid, nitrates, strong oxidizers.

CARBITOL CELLOSOLVE (111-90-0) Incompatible with strong acids, isocyanates, strong oxidizers.

CARBITOL SOLVENT (111-90-0) Incompatible with strong acids, isocyanates, strong oxidizers.

CARBOBENZOXY CHLORIDE (501-53-1) Aqueous solution produces hydrochloric acid; reacts violently with bases. Reacts with nitrates. Attacks metals in the presence of moisture.

CARBOFOS (121-75-5) Incompatible with strong oxidizers, magnesium, alkaline pesticides. Attacks metals, some plastics, rubber, and coatings.

CARBOFURAN (1563-66-2) Incompatible with alkaline media, acids or strong oxidizers.

CARBOLIC ACID (108-95-2) Forms explosive mixture with air (flash point 174°F/79°C). Incompatible with strong oxidizers, strong acids, caustics, aliphatic amines, amides, oxidizers, formaldehyde, butadiene, calcium hypochlorite. Liquid attacks some plastics, rubber, and coatings; hot liquid attacks aluminum, magnesium, lead and zinc metals.

CARBOLSAEURE (German) (108-95-2) Forms explosive mixture with air (flash point 174°F/79°C). Incompatible with strong oxidizers, strong acids, caustics, aliphatic amines, amides, oxidizers, formaldehyde, butadiene, calcium hypochlorite. Liquid attacks some plastics, rubber, and coatings; hot liquid attacks aluminum, magnesium, lead, and zinc, metals.

CARBOMETHANE (463-51-4) Vigorous reaction occurs with water and a wide variety of organic compounds. Forms explosive compound when mixed with hydrogen peroxide. Can dimerize to diketene even at low temperatures. Diketene forms an explosive mixture with air above 90°F/32°C. Reacts violently with oxidizers; forms unstable and explosive peroxides, with a chance of violent polymerization.

CARBOMETHENE (463-51-4) Vigorous reaction occurs with water and a wide variety of organic compounds. Forms explosive compound when mixed with hydrogen peroxide. Can dimerize to diketene even at low temperatures. Diketene forms an explosive mixture with air above 90°F/32°C. Reacts violently with oxidizers; forms unstable and explosive peroxides, with a chance of violent polymerization.

2-CARBOMETHOXY-1-METHYLVINYLDIMETHYL PHOSPHATE (7786-34-7) Contact with strong oxidizers may cause fire and explosions. Attacks some plastics, rubber, and coatings.

alpha-2-CARBOMETHOXY-1-METHYLVINYL DIMETHYL PHOSPHATE (7786-34-7) Contact with strong oxidizers may cause fire and explosions. Attacks some plastics, rubber, and coatings.

1-CARBOMETHOXY-1-PROPEN-2-Y PHOSPHATE (7786-34-7) Contact with strong oxidizers may cause fire and explosions. Attacks some plastics, rubber, and coatings.

CARBONA (56-23-5) Becomes corrosive when in contact with water. Corrosive to metals. Reacts violently with many compounds. Decomposes on contact with chemically active metals such as sodium, potassium, and magnesium. Incompatible with allyl alcohol, fluorine gas, alkali metals, aluminum. Attacks some coatings, plastics, and rubber.

CARBON, ACTIVATED (64365-11-3) activated; (7440-44-0) purified Incompatible with strong oxidizers, strong acids, oxides, unsaturated oils.

CARBON BICHLORIDE (127-18-4) Incompatible with strong oxidizers, finely divided metals, caustics. Stable up to 258°F/126°C; at this temperature product gives off poisonous fumes.

CARBON BISULFIDE or CARBON BISULPHIDE (75-15-0) Highly reactive. Contact with many substances can cause fire and explosions. Forms explosive mixture with air (flash point −22°F/−30°C). Shock can cause explosive decomposition. Incompatible with alkali metals, aliphatic amines, alkanolamines, aluminum, azides, chlorine monoxide, combustible substances, ethylene diamine, ethyleneimine, lead azide, lithium azide, nitric oxide, nitrogen dioxide, potassium, potassium azide, reducing agents, rubidium azide, sodium azide, zinc.

CARBON BLACK (1333-86-4) Dust can form an explosive mixture in air. Oxidizers may cause fire and explosions.

CARBON BROMIDE (558-13-4) Lithium or hexylcyclohexyldilead may cause an explosion.

CARBON CHLORIDE (56-23-5) Becomes corrosive when in contact with water. Corrosive to metals. Incompatible with allyl alcohol, fluorine gas, alkali metals, aluminum, rubbers, and plastics.

CARBON DIFLUORIDE OXIDE (353-50-4) Moisture forms corosive and toxic hydrogen fluoride gas and carbon dioxide.

CARBON DIOXIDE (124-38-9) Incompatible with acrylaldehyde, amines, anhydrous ammonia, cesium monoxide, lithium, metal dusts, potassium, sodium, sodium carbide, sodium–potassium alloy, sodium peroxide, titanium.

CARBON DISULFIDE or CARBON DISULPHIDE (75-15-0) Highly reactive. Contact with many substances can cause fire and explosions. Forms explosive mixture with air (flash point −22°F/−30°C). Shock can cause explosive decomposition. Incompatible with alkali metals, aliphatic amines, alkanolamines, aluminum, azides, chlorine monoxide, combustible substances, ethylene diamine, ethyleneimine, lead azide, lithium azide, nitric oxide, nitrogen dioxide, potassium, potassium azide, reducing agents, rubidium azide, sodium azide, zinc.

CARBONE (OXYCHLORURE de) (French) (75-44-5) Incompatible with water, producing hydrochloric acid. Reacts violently with strong oxidizers, anhydrous ammonia, isopropyl alcohol and chemically active metals. Forms shock-sensitive material with potassium. Attacks most metal in moist conditions.

CARBONE (OXYDE de) (French) (630-08-0) Forms extremely explosive mixture with air. Reacts violently with strong oxidizers.

CARBONE (SUFURE de) (French) (75-15-0) Highly reactive. Contact with many substances can cause fire and explosions. Forms explosive mixture with air (flash point −22°F/−30°C). Shock can cause explosive decomposition. Alkali metals, aliphatic amines, alkanolamines, aluminum, azides, chlorine monoxide, combustible substances, ethylene diamine, ethyleneimine, lead azide, lithium azide, nitric oxide, nitrogen dioxide, potassium, potassium azide, reducing agents, rubidium azide, sodium azide, zinc.

CARBON FLUORIDE OXIDE (353-50-4) Moisture forms hydrogen fluoride gas and carbon dioxide.

CARBON HEXACHLORIDE (67-72-1) Incompatible with hot iron, zinc, and aluminum; alkalies form spontaneously explosive chloroacetylene. Attacks some plastics, rubber, and coatings.

CARBONIC ACID, BARIUM SALT (1:1) (513-77-9) Incompatible with acids, aluminum, fluorine, magnesium, silicon, oxidizers.

CARBONIC ACID, CALCIUM SALT (1:1) (1317-65-3) Incompatible with acids, alum, ammonium salts, fluorine.

CARBONIC ACID, DIETHYL ESTER (105-58-8) Forms explosive mixture with air (flash point 77°F/25°C). Incompatible with oxidizers.

CARBONIC ACID GAS (124-38-9) Incompatible with acrylaldehyde, amines, anhydrous ammonia, cesium monoxide, lithium, metal dusts, potassium, sodium, sodium carbide, sodium–potassium alloy, sodium peroxide, titanium.

CARBONIC ACID, MONOAMMONIUM SALT (1066-33-7) Incompatible with strong acids and bases, fluorine, magnesium, sodium hypochlorite.

CARBONIC ANHYDRIDE (124-38-9) Incompatible with acrylaldehyde, amines, anhydrous ammonia, cesium monoxide, lithium, metal dusts, potassium, sodium, sodium carbide, sodium–potassium alloy, sodium peroxide, titanium.

CARBONIC DICHLORIDE (75-44-5) Incompatible with water, forms hydrochloric acid. Reacts violently with strong oxidizers, anhydrous ammonia, isopropyl alcohol and chemically active metals. Forms shock-sensitive material with potassium. Attacks most metal in moist conditions.

CARBONIC DIFLUORIDE (353-50-4) Moisture forms hydrogen fluoride gas and carbon dioxide.

CARBONIC OXIDE (630-08-0) Forms extremely explosive mixture with air. Reacts violently with strong oxidizers.

CARBONIO (OSSICLORURO di) (Italian) (75-44-5) Incompatible with water, producing hydrochloric acid. Reacts violently with strong oxidizers, anhydrous ammonia, isopropyl alcohol and chemically active metals. Forms shock-sensitive material with potassium. Attacks most metal in moist conditions.

CARBONIO (OSSIDO di) (Italian) (630-08-0) Forms extremely explosive mixture with air. Reacts violently with strong oxidizers.

CARBONIO (SOLFURO di) (Italian) (75-15-0) Highly reactive. Contact with many substances can cause fire and explosions. Forms explosive mixture with air (flash point −22°F/−30°C). Shock can cause explosive decomposition. Incompatible with alkali metals, aliphatic amines, alkanolamines, aluminum, azides, chlorine monoxide, combustible substances, ethylene diamine, ethyleneimine, lead azide, lithium azide, nitric oxide, nitrogen dioxide, potassium, potassium azide, reducing agents, rubidium azide, sodium azide, zinc.

CARBON MONOXIDE (630-08-0) Forms extremely explosive mixture with air. Reacts violently with strong oxidizers.

CARBON NAPHTHA (71-43-2) Forms explosive mixture with air (flash point 12°F/−11°C).) May accumulate static electrical charges, and may cause ignition of its vapors. Incompatible with strong oxidizers, nitric acid, oxygen, ozone, perchlorates. Attacks some forms of plastics, coatings and rubber.

CARBON NITRIDE (460-19-5) Incompatible with acids, water, liquid oxygen, oxidizers; may cause explosions.

CARBONOCHLORIDIC ACID, BUTYL ESTER (592-34-7) Water contact forms hydrochloric acid; corrodes metals.

CARBON OIL (71-43-2) Forms explosive mixture with air (Flash point 12°F/−11°C).) May accumulate static electrical charges, and may cause ignition of its vapors. Incompatible with strong oxidizers, nitric acid, oxygen, ozone, perchlorates. Attacks some forms of plastics, coatings, and rubber.

CARBON OXIDE (630-08-0) Forms extremely explosive mixture with air. Reacts violently with strong oxidizers.

CARBON OXYCHLORIDE (75-44-5) Incompatible with water, forms hydrochloric acid. Reacts violently with strong oxidizers, anhydrous ammonia, isopropyl alcohol and chemically active metals. Forms shock-sensitive material with potassium. Attacks most metal in moist conditions.

CARBON OXYFLUORIDE (353-50-4) Moisture forms hydrogen fluoride gas and carbon dioxide.

CARBON SULFIDE (75-15-0) Highly reactive. Contact with many substances can cause fire and explosions. Forms explosive mixture with air (flash point $-22°F/-30°C$). Shock can cause explosive decomposition. Alkali metals, aliphatic amines, alkanolamines, aluminum, azides, chlorine monoxide, combustible substances, ethylene diamine, ethyleneimine, lead azide, lithium azide, nitric oxide, nitrogen dioxide, potassium, potassium azide, reducing agents, rubidium azide, sodium azide, zinc.

CARBON TET (56-23-5) Becomes corrosive when in contact with water. Corrosive to metals. Reacts violently with many compounds. Decomposes on contact with chemically active metals such as sodium, potassium and magnesium. Allyl alcohol, fluorine gas, alkali metals, aluminum. Attacks some coatings, plastics and rubber.

CARBON TETRACHLORIDE (56-23-5) Becomes corrosive when in contact with water. Corrosive to metals. Reacts violently with many compounds. Decomposes on contact with chemically active metals such as sodium, potassium and magnesium. Allyl alcohol, fluorine gas, alkali metals, aluminum. Attacks some coatings, plastics and rubber.

CARBONYETHANE (79-09-4) Forms explosive mixture with air (flash point 126°F/52°C). Incompatible with sulfuric acid, strong bases, ammonia, aliphatic amines, alkanolamines, isocyanates, alkylene oxides, epichlorohydrin, oxidizers.

CARBONYLCHLORID (German) (75-44-5) Incompatible with water, forms hydrochloric acid. Reacts violently with strong oxidizers, anhydrous ammonia, isopropyl alcohol and chemically active metals. Forms shock-sensitive material with potassium. Attacks most metal in moist conditions.

CARBONYL CHLORIDE (75-44-5) Incompatible with water, producing hydrochloric acid. Reacts violently with strong oxidizers, anhydrous ammonia, isopropyl alcohol and chemically active metals. Forms shock-sensitive material with potassium. Attacks most metal in moist conditions.

CARBONYL DIAMIDE or CARBONYLDIAMIDE (57-13-6) Heat causes formation of anhydrous ammonia fumes. Contact with oxidizers, nitrates may cause fire and explosions. Contact with chlorinating agents, including hypochlorite bleaches may produce explosive nitrogen trichloride. Reacts with nitrosyl perchlorate.

CARBONYL DIAMINE (57-13-6) Heat causes formation of anhydrous ammonia fumes. Contact with oxidizers, nitrates may cause fire and explosions. Contact with chlorinating agents, including hypochlorite bleaches may produce explosive nitrogen trichloride. Reacts with nitrosyl perchlorate.

CARBONYL DIAMINE PEROXIDE (124-43-6) Combustible solid. At 122°F/50°C reacts with dust, organics.

CARBONYL DIFLUORIDE (353-50-4) Moisture causes formation of hydrogen fluoride gas and carbon dioxide.

CARBONYL FLUORIDE (353-50-4) Moisture causes formation of hydrogen fluoride gas and carbon dioxide.

CARBOPHOS (121-75-5) Incompatible with strong oxidizers, magnesium, alkaline pesticides. Attacks metals, some plastics, rubber, and coatings.

CARBORAFFIN activated (64365-11-3); purified (7440-44-0) Incompatible with strong oxidizers, strong acids, oxides, unsaturated oils.

CARBORAFINE activated (64365-11-3); purified (7440-44-0) Incompatible with strong oxidizers, strong acids, oxides, unsaturated oils.

CARBOSIP 5G (1563-66-2) Incompatible with alkaline media, acids, or strong oxidizers.

CARBOXIDE (1305-62-0) Contact with maleic anhydride, phosphorus, nitroethane, nitromethane, nitroparaffins, or nitropropane may cause explosion.

CARBOXYETHANE (79-09-4) Forms explosive mixture with air (flash point 126°F/52°C). Incompatible with sulfuric acid, strong bases, ammonia, aliphatic amines, alkanolamines, isocyanates, alkylene oxides, epichlorohydrin, oxidizers.

CARBOXYLBENZENE (100-47-0) Forms explosive mixture with air (flash point 167°F/75°C). Incompatible with strong acids, oxidizers. Attacks some plastics, rubber, and coatings.

CARFENE (86-50-0) Strong oxidizers may cause fire and explosions.

CARNAUBA WAX Incompatible with sulfuric acid, nitric acid, oxidizers, nitrates.

CAROLID AL (92-52-4) Mist forms explosive mixture with air. Strong oxidizers may cause fire and explosions.

CARPETING MEDIUM (8052-42-4) Incompatible with nitric acid, fluorine, strong oxidizers.

CARPIN (50-78-2) Fires and explosions may result from contact with strong oxidizers. Alkali hydroxides or carbonates cause decomposition.

CARPOLIN (63-25-2) Incompatible with strong oxidizers, strongly alkaline pesticides.

CARREL-DAKINS SOLUTION (7681-52-9) A strong oxidizer and a strong base. Stability decreases with concentration, heat, light, decrease in pH, and contamination with metals. Incompatible with strong acids, reducing agents, combustible substances; all cause violent reaction, fire, and explosions. Contact with amines and ammonia salts produce explosive chloroamines. Corrodes many metals: steels, cast iron, monel, aluminum, brass, iconel.

CARTHAMUS TINCTORIUS OIL (8001-23-8) Incompatible with strong acids, oxidizers.

CARWINATE 125 M (101-68-8) Incompatible with strong alkalies, acids, alcohol. Unstable above 100°F/37.8°C. Attacks some plastics, rubber, and coatings.

CASHEW NUTSHELL LIQUID (8001-24-7) Incompatible with sulfuric acid, caustics, ammonia, amines, isocyanates, alkylene oxides, epichlorohydrin.

CASHEW NUTSHELL OIL (8001-24-7) Incompatible with sulfuric acid, caustics, ammonia, amines, isocyanates, alkylene oxides, epichlorohydrin.

160

CASING HEAD GASOLINE (8006-61-9) Forms explosive mixture with air (flash point 125°F/51°C). Incompatible with nitric acid. Strong oxidizers may cause fire and explosions.

CASTER OIL (8001-79-4) Incompatible with strong acids, oxidizers, nitrates.

CASTOR OIL (8001-79-4) Incompatible with strong acids, oxidizers, nitrates.

CASTOR OIL, HYDROGENATED (8001-79-4) Incompatible with strong acids, oxidizers, nitrates.

CASWELL NO. 112 (776-95-6) A powerful oxidizer. May cause fire and explosions in contact with organic or other readily oxidizable materials. Contact with ammonia, acetaldehyde, acetylene, acryloni-trile, or with metals may cause violent reactions. Reacts violently with aluminum, titanium, mercury or potassium. Also incompatible with alcohols, antimony, alkali hydroxides, arsenites, boron, calcium nitrite, cesium monoxide, carbonyls, dimethyl formamide, ethyl phosphine, fluorine, ferrous and mercurous salts, germanium, hypophosphites, iron carbide, isobutyronphenone, magnesium phosphide, methanol, nickel carbonyl, olefins, ozone, sodium, and many other substances. Attacks some coatings, and some forms of plastic and rubber. Corrodes iron, steel, stainless steels, and copper.

CATALYST 9915 (103-83-3) Forms explosive mixture with air (flash point 170°F/77°C). Incompatible with acids, organic anhydrides, isocyanates, aldehydes, strong oxidizers, calcium hypochlorite. Attacks some plastics, rubber, and coatings.

CATECHIN (120-80-9) Strong oxidizers may cause fire and explosion.

CATECHOL (120-80-9) Strong oxidizers may cause fire and explosion.

CAUSTIC ARSENIC CHLORIDE (7784-34-1) Water contact produces corrosive mixture of acids. Light exposure forms toxic gas. Reacts violently with anhydrous ammonia, strong acids, strong oxidizers, and halogens. Corrodes metals in the presence of moisture.

CAUSTIC POTASH (1310-58-3) Aqueous solution is a strong base. Reacts violently with acids, organic anhydrides, isocyanates, alkylene oxides, epichlorohydrin, aldehydes, alcohols, glycols, phenols, cresols, caprolactam solution. Dissolves in water, producing high heat, caustic fumes and flammable hydrogen gas. Also reacts with halogenated hydrocarbons, maleic anhydride. Attacks metals such as aluminum, tin, lead, and zinc, producing flammable hydrogen gas.

CAUSTIC SODA (1310-73-2) Contact with water, acids, flammable liquids, and organic halogens, especially trichloroethylene, may cause fires and explosions. Contact with metals such as aluminum, tin and zinc causes corrosion and the formation of flammable hydrogen gas. Contact with nitromethane, and similar nitro compounds, metal powders, or sulfur produces friction-/impact-sensitive compounds. Water contact produces heat and corrosive fumes. Attacks some plastics, rubber, and coatings.

CAUSTIC SODA, BEAD, DRY, GRANULAR, SOLID, or FLAKE (1310-73-2) Contact with water, acids, flammable liquids, and organic halogens, especially trichloroethylene, may cause fires and explosions. Contact with metals such as aluminum, tin, and zinc causes corrosion and the formation of flammable hydrogen gas. Contact with nitromethane, and similar nitro compounds produces shock-sensitive salts. Water contact produces heat and corrosive fumes. Attacks some plastics, rubber, and coatings.

CAVI-TROL (7681-49-4) Incompatible with water, forming a corrosive substance. Reacts with acids.

CB (74-97-5) Reacts with chemically active metals such as calcium, powdered aluminum, zinc, and magnesium. Liquid attacks some plastics, rubber, and coatings.

CBD 90 (87-90-1) A powerful oxidizer. Forms explosive material with nitrogen compounds. Contact with organic materials or reducing agents may cause fire.

CBM (74-97-5) Reacts with chemically active metals such as calcium, powdered aluminum, zinc, and magnesium. Liquid attacks some plastics, rubber, and coatings.

CCC (156-62-7) Contact with any form of moisture causes decomposition, liberating acetylene and ammonia. Contact with all solvents tested also causes decomposition.

CCH (7778-54-3) Decomposes in heat or sunlight. Incompatible with acids, moisture, reducing agents, combustible materials, all other chemicals, especially acetylene, aniline and all other amines, anthracene, carbon tetrachloride, iron oxide, manganese oxide, mercaptans, diethylene glycol monomethyl ether, nitromethane, organic matter, organic sulfides, phenol, 1-propanethiol, propyl mercaptan, sulfur, organic sulfur compounds.

CCS 203 (71-36-3) Forms explosive mixture with air (flash point 98°F/37°C. May react with aluminum above 120°F/49°C. Attacks some plastics, rubber, and coatings. Incompatible with strong acids, halogens, caustics, alkali metals, aliphatic amines, isocyanates.

CCS 301 (78-92-2) Forms explosive mixture with air (flash point 75°F/4°C). Attacks some plastics, rubber, and coatings. Forms an explosive peroxide in air. Ignites with chromium trioxide. Incompatible with strong oxidizers, strong acids, aliphatic amines, isocyanates, organic peroxides.

CD 68 (57-74-9) Contact with strong oxidizers may cause fire and explosions. Attacks some plastics, rubber, and coatings.

CECOLENE (79-01-6) Contact with caustics produces a toxic and flammable gas. Reacts violently with chemically active metals. Contact with aluminum may produce a violent, self-accelerating polymerization reaction. Incompatible with nonoxidizing mineral acids, strong acids, organic acids, organic anhydrides, isocyanates, alkylene oxides, aldehydes, alcohols, glycols, phenols, cresols, caprolactam solution, epichlorohydrin, nitrogen tetroxide, metal powders, oxygen. May accumulate static electrical charges, and may cause ignition of its vapors.

CEKIURON (330-54-1) Hydrolyzes in fairly strong acids.

CEKUMETHION (298-00-0) Mixtures with magnesium may be explosive.

CEKUSAN (62-73-7) Attacks some plastics, rubber, and coatings.

CEKUZINA-T (1912-24-9) Incompatible with strong acids.

CELANEX (58-89-9) Not combustible, but may be dissolved in a combustible solvent. If solvent comes in contact with oxidizers, fire and explosions may result.

CELANOL DOS 75 (119-36-8) Forms explosive mixture with air (flash point 205°F/96°C). Incompatible with strong acids, nitrates, oxidizers.

CELLEX MX (9004-34-6) Incompatible with water, bromine pentafluoride, hydrogen peroxide, sodium hypochlorite, sodium nitrate, fluorine, or strong oxidizers.

CELLOIDIN (9004-70-0) If ether solution evaporates, dry nitrocellulose material is a shock-sensitive explosive and fire hazard. Forms explosive mixture with air (flash point −64°F/−53°C). Strong oxidizers may cause fire and explosions. May accumulate static electrical charges, and may cause ignition of its vapors. Attacks some plastics, rubber, and coatings.

CELLON (79-34-5) Exposure to heat, light, and air produces corrosive and toxic vapors. Reacts with strong caustics to form explosive dichloroacetylene. Reacts violently with chemically active metals or sodium amide. In presence of steam, contact with hot iron, aluminum or zinc may produce toxic vapors. Attacks some plastics, rubber, and coatings.

CELLOSOLVE (110-80-5) Forms explosive mixture with air (flash point 120°F/49°C). Strong oxidizers may cause fire and explosions. Attacks some plastics, rubber, and coatings.

CELLOSOLVE ACETATE (111-15-9) Forms explosive mixture with air (flash point 117°F/47°C). Incompatible with strong acids, nitrates, oxidizers. Attacks rubber.

CELLOSOLVE SOLVENT (110-80-5) Forms explosive mixture with air (flash point 120°F/49°C). Strong oxidizers may cause fire and explosions. Attacks some plastics, rubber, and coatings.

CELLULEX DOP (117-84-0) Water contact causes foaming. Incompatible with strong acids, nitrates.

CELLULOSE (9004-34-6) Incompatible with water, bromine pentafluoride, hydrogen peroxide, sodium hypochlorite, sodium nitrate, fluorine, or strong oxidizers.

alpha-CELLULOSE (9004-34-6) Incompatible with water, bromine pentafluoride, hydrogen peroxide, sodium hypochlorite, sodium nitrate, fluorine, or strong oxidizers.

CELLULOSE 248 (9004-34-6) Incompatible with water, bromine pentafluoride, hydrogen peroxide, sodium hypochlorite, sodium nitrate, fluorine or strong oxidizers.

CELLULOSE NITRATE SOLUTION (9004-70-0) If ether solution evaporates, dry nitrocellulose material is a shock-sensitive explosive and fire hazard. Forms explosive mixture with air (flash point −64°F/−53°C). Strong oxidizers may cause fire and explosions. May accumulate static electrical charges, and may cause ignition of its vapors. Attacks some plastics, rubber, and coatings.

CELLUPHOS 4 (126-73-8) Contact with water produces corrosive phosphoric acid. Incompatible with strong acids, strong oxidizers.

CELMIDE (106-93-4) Reacts with chemically active metals, liquid ammonia, strong oxidizers. Heat and light cause slow decomposition. Attacks some plastics and rubber.

CELON A (60-00-4) Incompatible with sulfuric acid, bases, ammonia, aliphatic amines, alkanolamines, isocyanates, alkylene oxides, epichlorohydrin

CELON ATH (60-00-4) Incompatible with sulfuric acid, bases, ammonia, aliphatic amines, alkanolamines, isocyanates, alkylene oxides, epichlorohydrin

CELPHOS (Indian) (20859-73-8) Contact with moisture (including atmospheric moisture) forms spontaneously combustible phosphine gas.

CELTHION (Indian) (121-75-5) Incompatible with strong oxidizers, magnesium, alkaline pesticides. Attacks metals, some plastics, rubber, and coatings.

CENTURY-1240 (57-11-4) Incompatible with strong oxidizers, sulfuric acid, caustics, ammonia, amines, isocyanates, alkylene oxides, epichlorohydrin. Attacks chemically active metals.

CESIUM HYDRATE (21351-79-1) Contact with many organic compounds, many metals (i.e., aluminum, lead, tin, zinc), glass, oxygen, or carbon dioxide causes a violent reaction.

CESIUM HYDROXIDE (21351-79-1) Contact with many organic compounds, many metals (i.e., aluminum, lead, tin, zinc), glass, oxygen, or carbon dioxide causes a violent reaction.

CESIUM HYDROXIDE DIMER (21351-79-1) Contact with many organic compounds, many metals (i.e., aluminum, lead, tin, zinc), glass, oxygen, or carbon dioxide causes a violent reaction.

CETYLENE TETRACHLORIDE (79-34-5) Exposure to heat, light and air produces corrosive and toxic vapors. Reacts with strong caustics to form explosive dichloroacetylene. Reacts violently with chemically active metals or sodium amide. In presence of steam, contact with hot iron, aluminum, or zinc may produces toxic vapors. Attacks some plastics, rubber, and coatings.

CFC-11 (75-69-4) Incompatible with chemically active metals. Attacks some plastics, rubber, and coatings.

CFC-12 (75-71-8) Reacts violently with liquid aluminum. Incompatible with chemically active metals. Attacks some plastics, rubber, and coatings.

CFC-113 (76-13-1) Incompatible with chemically active metals. Contact with alloys containing more than 2% magnesium may cause decomposition (hydrogen chloride, hydrogen fluoride, and carbon monoxide released). Attacks some plastics, rubber, and coatings.

CFC-115 (76-15-3) Thermal decomposition occurs at high temperatures with alkalies and alkaline earth metals.

CG (75-44-5) Incompatible with water, producing hydrochloric acid. Reacts violently with strong oxidizers, anhydrous ammonia, isopropyl alcohol, and chemically active metals. Forms shock-sensitive material with potassium. Attacks most metal in moist conditions.

CGA 24705 (51218-45-0) Incompatible with strong acids, nitrates, oxidizers.

CHA (108-91-8) Forms explosive mixture with air (flash point 79°F/26°C). Incompatible with nonoxidizing mineral acids, strong acids, organic acids, organic anhydrides, isocyanates, vinyl acetate, acrylates, substituted allyls, alkylene oxides, epichlorohydrin, ketones, aldehydes, alcohols, glycols, phenols, cresols, caprolactam solution, strong oxidizers. Contact with copper alloys, zinc, or galvanized steel may cause violent reaction.

CHALK (1317-65-3) Incompatible with acids, alum, ammonium salts, fluorine.

CHALOXYD (1338-23-4) Forms explosive mixture with air (flash point 125°F/52°C). Explosive decomposition occurs above 176°F/80°C. Pure substance is shock-sensitive; strong oxidizer. Reacts violently with strong acids, strong bases, reducing agents, combustible substances, organic materials, oxides of heavy metals, salts, trace contaminants, amines. May accumulate static electrical charges, and may cause ignition of its vapors.

CHAMBER ACID (7664-93-9) A strong oxidizer that can react violently with risk of fire and explosion with many substances, including reducing agents, organic and combustible substances, and bases. Incompatible with nonoxidizing mineral acids, organic acids, bases, acrylates, aldehydes, alcohols, alkylene oxides, ammonia, aliphatic amines, alkanolamines, aromatic amines, amides, chlorates, epichlorohydrin, fulminates, glycols, isocyanates, ketones, metals (powdered), organic anhydrides, perchlorates, picrates, substituted allyls, phenols and cresols, water, acetic anhydride, acetone cyanhydrin, acetonitrile, acrolein, acrylonitrile, allyl alcohol, allyl chloride, 2-aminoethanol, ammonium hydroxide, aniline, bromine pentafluoride, *n*-butyraldehyde, caprolactam solution, carbides, cesium acetylene carbide, chlorine trifluoride, chlorosulfonic acid, cuprous nitride, diisobutylene, ethylene cyanohydrin, ethylene diamine, ethylene glycol, ethyleneimine, hydrochloric acid, iodine heptafluoride, iron, isoprene, lithium silicide, mercuric nitride, mesityl oxide, nitric acid, *p*-nitrotoluene, perchloric acid, phosphorus, potassium-*tert*-butoxide, potassium chlorate, potassium permanganate, 3-propiolactone, propylene oxide, pyridine, rubidium acetylene, silver permanganate, sodium, sodium carbonate, sodium chlorate, sodium hydroxide, styrene monomer, vinyl acetate, Attacks most metals, and some plastics, rubber, and coatings.

CHAMELEON MINERAL (7722-64-7) A strong oxidizer. Forms heat- and shock-sensitive compound with sulfuric acid. Reacts violently with combustibles or reducing agents. Reacts explosively with acetic acid, acetic anhydride, anhydrous ammonia, hydrogen peroxide, glycerol, hydroxylamine, organic matter, powdered sulfur, etc. Incompatible with nitric acid, producing toxic chlorine fumes. A dangerous fire and explosion hazard; isolate from all other materials.

CHANNEL BLACK (1333-86-4) Dust can form an explosive mixture in air. Oxidizers may cause fire and explosions.

CHARCOAL activated (64365-11-3); purified (7440-44-0). Incompatible with strong oxidizers, strong acids, oxides, unsaturated oils.

CHARCOAL, ACTIVATED activated (64365-11-3); purified (7440-44-0). Incompatible with strong oxidizers, strong acids, oxides, unsaturated oils.

CHARCOAL, SHELL activated (64365-11-3); purified (7440-44-0). Incompatible with strong oxidizers, strong acids, oxides, unsaturated oils.

CHEELOX (60-00-4) Incompatible with sulfuric acid, bases, ammonia, aliphatic amines, alkanolamines, isocyanates, alkylene oxides, epichlorohydrin.

CHELEN (75-00-3) Flammable gas. Forms explosive mixture with air. Contact with moisture produces hydrochloric acid. May accumulate static electrical charges, and may cause ignition of its vapors. Forms explosive mixture with air (flash point −58°F/−50°C). Contact with chemically active metals may cause fire and explosions. Attacks some plastics and rubber.

CHEMAID (124-65-2) Corrodes common metals.

CHEMATHION (121-75-5) Incompatible with strong oxidizers, magnesium, alkaline pesticides. Attacks metals, some plastics, rubber, and coatings.

CHEM BAM (142-59-6) Boiling water causes formation of hydrogen sulfide and carbon disulfide vapors.

CHEMCOLOX 340 (60-00-4) Incompatible with sulfuric acid, bases, ammonia, aliphatic amines, alkanolamines, isocyanates, alkylene oxides, epichlorohydrin.

CHEMFORM (72-43-5) Contact with strong oxidizers may cause fire and explosions. Attacks some plastics, rubber, and coatings.

CHEMIFLUOR (7681-49-4) Incompatible with water, producing a corrosive substance. Reacts with acids.

CHEMICAL 109 (86-88-4) Strong oxidizers may cause fires and explosion. Also reacts with silver nitrate.

CHEMLON (105-60-2) Contact with strong oxidizers may cause fire and explosions.

CHEMOX PE (51-28-5) Explosion are caused by heat, friction, or shock. Contact with reducing agents, combustibles may cause fire and explosions. Forms explosive salts with ammonia or strong bases. May accumulate static electrical charges, and may cause ignition of its vapors.

CHEM-PHENE (8001-35-2) Reacts with oxidizers, with a risk of fire or explosions. Attacks metals in the presence of moisture.

CHEM-TOL (87-86-5) Incompatible with strong acids, caustics, aliphatic amines, amides. Decomposes at 375°F/191°C, forming toxic and flammable vapors.

CHILE SALTPETER (7631-99-4) Powerful oxidizer; reacts violently with reducing agents, combustible substances, strong acids, organic materials, powdered metals, bitumens.

CHINESE RED (1344-48-5) Contact with acids or water evolves flammable hydrogen sulfide, which forms an explosive mixture with air. Contact with strong oxidizers may cause a violent reaction.

CHINOLEINE (91-22-5) Contact with strong oxidizers may cause fire and explosions. Attacks some plastics, rubber, and coatings.

CHINOLIN (Czech) (91-22-5) Contact with strong oxidizers may cause fire and explosions. Attacks some plastics, rubber, and coatings.

CHINOLINE (91-22-5) Contact with strong oxidizers may cause fire and explosions. Attacks some plastics, rubber, and coatings.

CHINON (Dutch) (106-51-4) Forms explosive mixture with air (flash point 104°F/40°C). Incompatible with strong bases, reducing agents, strong oxidizers. Attacks some plastics, rubber, and coatings.

CHINON (German) (106-51-4) Forms explosive mixture with air (flash point 104°F/40°C). Incompatible with strong bases, reducing agents, strong oxidizers. Attacks some plastics, rubber, and coatings.

p-**CHINON (German)** (106-51-4) Forms explosive mixture with air (flash point 104°F/40°C). Incompatible with strong bases, reducing agents, strong oxidizers. Attacks some plastics, rubber, and coatings.

166

CHINONE (106-51-4) Forms explosive mixture with air (flash point 104°F/40°C). Incompatible with strong bases, reducing agents, strong oxidizers. Attacks some plastics, rubber, and coatings.

CHINUFUR (1563-66-2) Incompatible with alkaline media, acids, or strong oxidizers.

CHIPCO THIRAM 75 (137-26-8) Combustible solid (flash point 192°F/89°C). Strong oxidizers may cause fire and explosions; contact with strong acid or oxidizable materials produces toxic gases.

CHIPCO TURF HERBICIDE 'D' (94-75-7) Decomposes in sunlight. Incompatible with strong oxidizers; may cause fire and explosions.

CHLOOR (Dutch) (7782-50-5) A powerful oxidizer. Reacts violently with combustible materials, reducing agents and many other substances. Forms explosive mixtures with gasoline and petroleum products, turpentine, alcohols, acetylene, carbon disulfide, hydrogen, anhydrous ammonia, finely divided metals, organic compounds, phosphorus. Fire, explosion, and the formation of toxic fumes may result from contact with many substances, including alkyphosphines, aluminum, antimony, arsenic compounds, arsine, bismuth, boron, brass, calcium compounds, carbon, diethyl zinc, fluorine, germanium, hydrocarbons, rubber. Attacks some plastics and coatings. In the presence of moisture, extremely corrosive to iron, steel, copper, bronze, zinc.

CHLOORBENZENE (Dutch) (108-90-7) Incompatible with oxidizers, dimethylsulfoxide, sodium powder, silver perchloride. Attacks some plastics, rubber, and coatings.

CHLOORDAAN (Dutch) (57-74-9) Contact with strong oxidizers may cause fire and explosions. Attacks some plastics, rubber, and coatings.

CHLOORETHAAN (Dutch) (75-00-3) Flammable gas. Forms explosive mixture with air. Contact with moisture produces hydrochloric acid. May accumulate static electrical charges, and may cause ignition of its vapors. Forms explosive mixture with air (flash point −58°F/−50°C). Contact with chemically active metals may cause fire and explosions. Attacks some plastics and rubber.

CHLOOR-METHAAN (Dutch) (74-87-3) Incompatible with chemically active metals. Attacks plastics, rubber, and coatings.

CHLOORPIKRINE (Dutch) (76-06-2) Can be self-reactive. Fast heating, shock, alkali metals, or alkaline earth may cause explosions. A strong oxidizer; reacts violently with reducing agents, aniline in presence of heat, alcoholic sodium hydroxide, combustible substances, sodium methoxide, propargyl bromide. Liquid attacks some plastics, rubber, and coatings.

CHLOORWATERSTOF (Dutch) (7647-01-0) Incompatible with acetic anhydride, aliphatic amines, alkanolamines, alkylene oxides, aromatic amines, amides, 2-aminoethanol, ammonia, ammonium hydroxide, calcium phosphide, chlorosulfonic acid, ethylene diamine, ethyleneimine, epichlorohydrin, isocyanates, metal acetylides, oleum, organic anhydrides, perchloric acid, 3-propiolactone, uranium phosphide, sulfuric acid, sodium hydroxide and other bases, strong oxidizers, vinyl acetate, vinylidene fluoride. Attacks most metals and some plastics, rubber, and coatings.

CHLOPHEN (generic CAS for PCBs 1336-36-3) Incompatible with strong oxidizers, strong acids.

167

CHLOR (German) (7782-50-5) A powerful oxidizer. Reacts violently with combustible materials, reducing agents, and many other substances. Forms explosive mixtures with gasoline and petroleum products, turpentine, alcohols, acetylene, carbon disulfide, hydrogen, anhydrous ammonia, finely divided metals, organic compounds, phosphorus. Fire, explosion, and the formation of toxic fumes may result from contact with many substances, including alkyphosphines, aluminum, antimony, arsenic compounds, arsine, bismuth, boron, brass, calcium compounds, carbon, diethyl zinc, fluorine, germanium, hydrocarbons, rubber. Attacks some plastics and coatings. In the presence of moisture, extremely corrosive to iron, steel, copper, bronze, zinc.

CHLOR KIL (57-74-9) Contact with strong oxidizers may cause fire and explosions. Attacks some plastics, rubber, and coatings.

CHLOR-METHAN (German) (74-87-3) Incompatible with chemically active metals. Attacks plastics, rubber, and coatings.

CHLORACETIC ACID or CHLORACETIC ACID, LIQUID or CHLORACETIC ACID, SOLID (79-11-8) Aqueous solution is a strong acid. Incompatible with strong oxidizers, bases. Attacks most common metals in the presence of moisture.

a-CHLORACETIC ACID or alpha-CHLORACETIC ACID (79-11-8) Aqueous solution is a strong acid. Incompatible with strong oxidizers, bases. Attacks most common metals in the presence of moisture.

CHLOROACETONITRILE (107-14-2) Forms explosive mixture with air (flash point 133°F/56°C). Contact with moisture produces hydrogen cyanide gas. Reacts with sulfuric acid and strong oxidizers.

CHLORACETYL CHLORIDE (79-04-9) Forms corrosive vapors with air. Reacts strongly with water, producing hydrochloric acid. Incompatible with sulfuric acid, caustics, alkalis, alcohols, aliphatic amines, alkanolamines, ammonia, isocyanates, alkylene oxides, epichlorohydrin.

CHLORAL or CHLORAL, ANHYDROUS (75-87-6) Contact with acids or exposure to light may cause polymerization. Reacts with water produce corrosive and toxic chloral hydrate. Reacts with oxidizers, with a risk of fire or explosions.

CHLORAL HYDRATE (302-17-0) Reacts with strong oxidizers, strong bases. Attacks metals in the presence of moisture.

CHLORALLYLENE (107-05-1) Forms explosive mixture with air (flash point −25°F/−32). Violent polymerization and explosion may occur from heat, light, or contact with acid catalysts, ferric chloride, aluminum chloride, Lewis acids, or Ziegler catalyst (e.g., titanium tetrachloride + triethylaluminum monochloride). Incompatible with strong acids, amines, aluminum chloride, boron trifluoride, chlorosulfonic acid, ethylene diamine, ethyleneimine, ferric chloride, oleum, oxidizers, sodium hydroxide. Slow decomposition occurs with moisture; attacks some coatings, plastics and rubber. Corrosive to steel. May accumulate static electrical charges, and may cause ignition of its vapors.

CHLORATE de CALCIUM (French) (10137-74-3) A strong reducing agent. Reacts, possibly with violence, with acids (especially organic), reducing agents, aluminum, arsenic, chemically active metals, combustible materials, ammonium compounds, charcoal, copper, cyanides, manganese dioxide, metal sulfides, phosphorus, sulfur.

CHLORATE de POTASSIUM (French) (3811-04-9) A powerful oxidizer. Reacts violently with reducing agents or combustibles, ammonia gas, ammonium salts, organic matter, hydrogen iodide, organic acids, sulfuric acid. Forms explosive mixtures with metallic powders, ammonium chloride, organic solids, including agricultural materials.

CHLORATE OF POTASH (3811-04-9) A powerful oxidizer. Reacts violently with reducing agents or combustibles, ammonia gas, ammonium salts, organic matter, hydrogen iodide, organic acids, sulfuric acid. Forms explosive mixtures with metallic powders, ammonium chloride, organic solids including agricultural materials.

CHLORATE OF SODA (7775-09-9) A powerful oxidizer. Reacts violently with reducing agents and combustible matter. Explosions may be caused by contact with ammonia salts, carbon, oils, metal sulfides, nitrobenzene, powdered metals, sugar. Contact with strong acids produces carbon dioxide. Forms shock-sensitive mixtures with some organic materials. Solution (50%) decomposes at 300°F/149°C, liberating oxygen.

CHLORATE SALT OF SODIUM (7775-09-9) A powerful oxidizer. Reacts violently with reducing agents and combustible matter. Explosions may be caused by contact with ammonia salts, carbon, oils, metal sulfides, nitrobenzene, powdered metals, sugar. Contact with strong acids produces carbon dioxide. Forms shock-sensitive mixtures with some organic materials. Solution (50%) decomposes at 300°F/149°C liberating oxygen.

CHLORAX (7775-09-9) A powerful oxidizer. Reacts violently with reducing agents and combustible matter. Explosions may be caused by contact with ammonia salts, carbon, oils, metal sulfides, nitrobenzene, powdered metals, sugar. Contact with strong acids produces carbon dioxide. Forms shock-sensitive mixtures with some organic materials. Solution (50%) decomposes at 300°F/149°C, liberating oxygen.

CHLORBENZEN (108-90-7) Incompatible with oxidizers, dimethylsulfoxide, sodium powder, silver perchloride. Attacks some plastics, rubber, and coatings.

CHLORCYAN (506-77-4) Violent polymerization can be caused by chlorine. Water, steam, alcohols, acids, acid salts, amines, strong alkalis, olefins, strong oxidizers may cause fire and explosion. In crude form chemical trimerizes violently if catalyzed by traces of hydrogen chloride or ammonium chloride. Corrodes brass, copper, bronze.

CHLORDAN (57-74-9) Contact with strong oxidizers may cause fire and explosions. Attacks some plastics, rubber, and coatings.

CHLORDANE (57-74-9) Contact with strong oxidizers may cause fire and explosions. Attacks some plastics, rubber, and coatings.

CHLORDIMETHYLETHER (Czech) (107-30-2) Forms explosive mixture with air (flash point 0°F/−17.8°C). May be able to form unstable and explosive peroxides. Produces hydrochloric acid on contact with moisture. Corrosive to metals in presence of moisture.

CHLORE (French) (7782-50-5) A powerful oxidizer. Reacts violently with combustible materials, reducing agents, and many other substances. Forms explosive mixtures with gasoline and petroleum products, turpentine, alcohols, acetylene, carbon disulfide, hydrogen, anhydrous ammonia, finely divided metals, organic compounds, phosphorus. Fire, explosion, and the formation of toxic fumes may result from contact with many substances, including alkyphosphines, aluminum, antimony, arsenic compounds, arsine, bismuth, boron, brass, calcium compounds, carbon, diethyl zinc, fluorine, germanium, hydrocarbons, rubber. Attacks some plastics and coatings. In the presence of moisture, extremely corrosive to iron, steel, copper, bronze, zinc.

CHLORESENE (58-89-9) Not combustible, but may be dissolved in a combustible solvent. If solvent comes in contact with oxidizers, fire and explosions may result.

2-CHLORETHANOL (107-07-3) Forms explosive mixture with air (flash point 140°F/60°C). Strong oxidizers may cause fire and explosions. Incompatible with strong caustics (with formation of ethylene gas), strong acids, aliphatic amines, isocyanates. Attacks some plastics, rubber, and coatings.

delta-CHLORETHANOL (107-07-3) Forms explosive mixture with air (flash point 140°F/60°C). Strong oxidizers may cause fire and explosions. Incompatible with strong caustics (with formation of ethylene gas), strong acids, aliphatic amines, isocyanates. Attacks some plastics, rubber, and coatings.

CHLORETHYL (75-00-3) Flammable gas. Forms explosive mixture with air. Contact with moisture produces hydrochloric acid. May accumulate static electrical charges, and may cause ignition of its vapors. Forms explosive mixture with air (flash point −58°F/−50°C). Contact with chemically active metals may cause fire and explosions. Attacks some plastics and rubber.

beta-CHLORETHYL ALCOHOL (107-07-3) Forms explosive mixture with air (flash point 140°F/60°C). Strong oxidizers may cause fire and explosions. Incompatible with strong caustics (with formation of ethylene gas), strong acids, aliphatic amines, isocyanates. Attacks some plastics, rubber, and coatings.

CHLOREX (111-44-4) Forms explosive mixture with air (flash point 131°F/55°C). Strong oxidizers may cause fire and explosions. Attacks some plastics, rubber, and coatings.

CHLOREXTOL (generic CAS for PCBs 1336-36-3) Incompatible with strong oxidizers, strong acids.

p-**CHLORFENOL (Czech)** (106-48-9) Forms explosive mixture with air (flash point 147°F/64°C). Reacts with oxidizers, with a risk of fire or explosions. Attacks active metals, aluminum, copper, etc. May accumulate static electrical charges, and may cause ignition of its vapors.

CHLORIC ACID, BARIUM SALT (13477-00-4) Can form explosive mixture with combustibles. Can be ignited by friction. Keep away from combustible materials, sulfur, aluminum, copper, ammonium salts, and other oxidizable materials.

CHLORIC ACID, CALCIUM SALT (10137-74-3) A strong reducing agent. Reacts, possibly with violence, with acids (especially organic), reducing agents, aluminum, arsenic, chemically active metals, combustible materials, ammonium compounds, charcoal, copper, cyanides, manganese dioxide, metal sulfides, phosphorus, sulfur.

CHLORIC ACID, POTASSIUM SALT (3811-04-9) A powerful oxidizer. Reacts violently with reducing agents or combustibles, ammonia gas, ammonium salts, organic matter, hydrogen iodide, organic acids, sulfuric acid. Forms explosive mixtures with metallic powders, ammonium chloride, organic solids, including agricultural materials.

CHLORIC ACID, SODIUM SALT (7775-09-9) A powerful oxidizer; reacts violently with reducing agents and combustible matter. Explosions may be caused by contact with ammonia salts, carbon, oils, metal sulfides, nitrobenzene, powdered metals, sugar. Contact with strong acids produces carbon dioxide. Forms shock-sensitive mixtures with some organic materials. Solution (50%) decomposes at 300°F/149°C liberating oxygen.

CHLORID ANTIMONITY (10025-91-9) Water contact produces corrosive solution. Air contact produces corrosive vapor.

CHLORIDE OF AMYL (543-59-9) Forms explosive mixture with air (flash point 34°F/1°C). Strong oxidizers may cause fire and explosions.

CHLORIDE of LIME (7778-54-3) Decomposes in heat or sunlight. Incompatible with acids, moisture, reducing agents, combustible materials, all other chemicals, especially acetylene, aniline and all other amines, anthracene, carbon tetrachloride, iron oxide, manganese oxide, mercaptans, diethylene glycol monomethyl ether, nitromethane, organic matter, organic sulfides, phenol, 1-propanethiol, propyl mercaptan, sulfur, organic sulfur compounds.

CHLORIDE OF PHOSPHORUS (7719-12-2) Contact with water or alcohol may cause fire and explosions, particularly when there is contact with combustible organic matter. Incompatible with ammonia, caustics, reducing agents, nitric acid, chemically active metals. In the presence of moisture, corrodes most metals. Attacks some plastics, rubber, and coatings.

CHLORIDRTUTNATY (Czech) (7487-94-7) Incompatible with light metals (aluminum, magnesium, beryllium, etc.), sodium, potassium.

CHLORIDUM (75-00-3) Flammable gas. Forms explosive mixture with air. Contact with moisture produces hydrochloric acid. May accumulate static electrical charges, and may cause ignition of its vapors. Forms explosive mixture with air (flash point −58°F/−50°C). Contact with chemically active metals may cause fire and explosions. Attacks some plastics and rubber.

CHLORINATED BIPHENYLS (generic CAS for PCBs 1336-36-3) Incompatible with strong oxidizers, strong acids.

CHLORINATED CAMPHENE (8001-35-2) Reacts with oxidizers, with a risk of fire or explosions. Attacks metals in the presence of moisture.

CHLORINATED DIPHENYL (generic CAS for PCBs 1336-36-3) Incompatible with strong oxidizers, strong acids.

CHLORINATED DIPHENYLENE (generic CAS for PCBs 1336-36-3) Incompatible with strong oxidizers, strong acids.

CHLORINATED HYDROCHLORIC ETHER (75-34-3) Incompatible with strong oxidizers, strong caustics. Attacks plastics and rubber.

CHLORINATED LIME (7778-54-3) Decomposes in heat or sunlight. Incompatible with acids, moisture, reducing agents, combustible materials, all other chemicals, especially acetylene, aniline and all other amines, anthracene, carbon tetrachloride, iron oxide, manganese oxide, mercaptans, diethylene glycol monomethyl ether, nitromethane, organic matter, organic sulfides, phenol, 1-propanethiol, propyl mercaptan, sulfur, organic sulfur compounds.

CHLORINDAN (57-74-9) Contact with strong oxidizers may cause fire and explosions. Attacks some plastics, rubber, and coatings.

CHLORINE (7782-50-5) A powerful oxidizer. Reacts violently with combustible materials, reducing agents, and many other substances. Forms explosive mixtures with gasoline and petroleum products, turpentine, alcohols, acetylene, carbon disulfide, hydrogen, anhydrous ammonia, finely divided metals, organic compounds, phosphorus. Fire, explosion, and the formation of toxic fumes may result from contact with many substances including alkyphosphines, aluminum, antimony, arsenic compounds, arsine, bismuth, boron, brass, calcium compounds, carbon, diethyl zinc, fluorine, germanium, hydrocarbons, rubber. Attacks some plastics and coatings. In the presence of moisture, extremely corrosive to iron, steel, copper, bronze, zinc.

CHLORINE CYANIDE (506-77-4) Violent polymerization can be caused by chlorine. Water, steam, alcohols, acids, acid salts, amines, strong alkalis, olefins, strong oxidizers may cause fire and explosion. In crude form, chemical trimerizes violently if catalyzed by traces of hydrogen chloride or ammonium chloride. Corrodes brass, copper, bronze.

CHLORINE DIOXIDE (10049-04-4) Highly reactive. Explosive reaction can be caused by heat, shock, friction, carbon dioxide, carbon monoxide, mercury, phosphorus, sulfur, and many other materials. Water contact forms perchloric and hydrochloric acid solution. A strong oxidizer; reacts violently with reducing agents, combustibles, organic substances. Attacks metals in the presence of moisture.

CHLORINE FLUORIDE (7790-91-2) Will cause most combustible materials to ignite spontaneously. Dangerously reactive with many materials and substances, including water, acids, aluminum oxide, ammonia, arsenic trioxide, asbestos, benzene, chromic anhydride, chromic oxide, copper, most elements, fuels, organic matter, silicon-containing compounds, sand, glass and glass wool, graphite, oxides. Attacks all forms of plastics, rubber, coatings, and resins except for highly fluorinated polymers such as Teflon and Kel-F.

CHLORINE FLUORIDE OXIDE (7616-94-6) A powerful oxidizer. Incompatible with strong bases, amines, finely divided metals. Contact with reducing agents, combustibles, organic materials and readily oxidizable materials may cause fire and explosions or yield explosive product. Attacks some plastics, rubber, and coatings.

CHLORINE, MOLECULAR (7782-50-5) A powerful oxidizer. Reacts violently with combustible materials, reducing agents and many other substances. Forms explosive mixtures with gasoline and petroleum products, turpentine, alcohols, acetylene, carbon disulfide, hydrogen, anhydrous ammonia, finely divided metals, organic compounds, phosphorus. Fire, explosion, and the formation of toxic fumes may result from contact with many substances including alkyphosphines, aluminum, antimony, arsenic compounds, arsine, bismuth, boron, brass, calcium compounds, carbon, diethyl zinc, fluorine, germanium, hydrocarbons, rubber. Attacks some plastics and coatings. In the presence of moisture, extremely corrosive to iron, steel, copper, bronze, zinc.

CHLORINE OXIDE (10049-04-4) Highly reactive. Explosive reaction can be caused by heat, shock, friction, carbon dioxide, carbon monoxide, mercury, phosphorus, sulfur, and many other materials. Water contact forms perchloric and hydrochloric acid solution. A strong oxidizer; reacts violently with reducing agents, combustibles, organic substances. Attacks metals in the presence of moisture.

CHLORINE(4+) OXIDE or CHLORINE(IV) OXIDE (10049-04-4) Highly reactive. Explosive reaction can be caused by heat, shock, friction, carbon dioxide, carbon monoxide, mercury, phosphorus, sulfur, and many other materials. Water contact forms perchloric and hydrochloric acid solution. A strong oxidizer; reacts violently with reducing agents, combustibles, organic substances. Attacks metals in the presence of moisture.

CHLORINE OXYFLUORIDE (7616-94-6) A powerful oxidizer. Incompatible with strong bases, amines, finely divided metals. Contact with reducing agents, combustibles, organic materials, and readily oxidizable materials may cause fire and explosions or yield explosive product. Attacks some plastics, rubber, and coatings.

CHLORINE PEROXIDE (10049-04-4) Highly reactive. Explosive reaction can be caused by heat, shock, friction, carbon dioxide, carbon monoxide, mercury, phosphorus, sulfur, and many other materials. Water contact forms perchloric and hydrochloric acid solution. A strong oxidizer; reacts violently with reducing agents, combustibles, organic substances. Attacks metals in the presence of moisture.

CHLORINE TRIFLUORIDE (7790-91-2) Will cause most combustible materials to ignite spontaneously. Dangerously reactive with many materials and substances, including water, acids, aluminum oxide, ammonia, arsenic trioxide, asbestos, benzene, chromic anhydride, chromic oxide, copper, most elements, fuels, organic matter, silicon-containing compounds, sand, glass and glass wool, graphite, oxides. Attacks all forms of plastics, rubber, coatings and resins except for highly fluorinated polymers such as Teflon and Kel-F.

CHLOROACETALDEHYDE (107-20-0) Incompatible with oxidizers, acids, water. May form a water-soluble polymer in storage.

2-CHLOROACETALDEHYDE (107-20-0) Incompatible with oxidizers, acids, water. May form a water-soluble polymer in storage.

CHLOROACETALDEHYDE (40% AQUEOUS) (107-20-0) Incompatible with oxidizers, acids, water. May form a water-soluble polymer in storage.

CHLOROACETALDEHYDE MONOMER (107-20-0) Incompatible with oxidizers, acids, water. May form a water-soluble polymer in storage.

CHLOROACETIC ACID (79-11-8) Aqueous solution is a strong acid. Incompatible with strong oxidizers, bases. Attacks most common metals in the presence of moisture.

CHLOROACETIC ACID CHLORIDE (79-04-9) Forms corrosive vapors with air. Reacts strongly with water, producing hydrochloric acid. Incompatible with sulfuric acid, caustics, alkalis, alcohols, aliphatic amines, alkanolamines, ammonia, isocyanates, alkylene oxides, epichlorohydrin.

173

CHLOROACETIC ACID, ETHYL ESTER (105-39-5) Forms explosive mixture with air (flash point 100°F/38°C). Water contact causes formation of toxic and corrosive fumes. Reacts violently with alkaline earth metals (barium, calcium, magnesium, strontium, etc.), alkaline metals, sodium cyanide. Attacks metals in the presence of moisture.

CHLOROACETIC ACID, METHYL ESTER (96-34-4) Forms explosive mixture with air (flash point 135°F/57°C). Strong reaction with oxidizers.

CHLOROACETIC CHLORIDE (79-04-9) Forms corrosive vapors with air. Reacts strongly with water, producing hydrochloric acid. Incompatible with sulfuric acid, caustics, alkalis, alcohols, aliphatic amines, alkanolamines, ammonia, isocyanates, alkylene oxides, epichlorohydrin.

CHLOROACETONE (78-95-5) Forms exposive mixture with air (flash point 40°F/7°C). Strong oxidizers may cause fire and explosions. May accumulate static electrical charges, and may cause ignition of its vapors.

CHLOROACETO PHENONE (532-27-4) Incompatible with water or steam.

2-CHLOROACETOPHENONE (532-27-4) Incompatible with water or steam.

alpha-CHLOROACETOPHENONE (532-27-4) Incompatible with water or steam.

omega-CHLOROACETOPHENONE (532-27-4) Incompatible with water or steam.

CHLOROACETYL CHLORIDE (79-04-9) Forms corrosive vapors with air. Reacts strongly with water, producing hydrochloric acid. Incompatible with sulfuric acid, caustics, alkalis, alcohols, aliphatic amines, alkanolamines, ammonia, isocyanates, alkylene oxides, epichlorohydrin.

CHLOROAETHAN (German) (75-00-3) Flammable gas. Forms explosive mixture with air. Contact with moisture produces hydrochloric acid. May accumulate static electrical charges, and may cause ignition of its vapors. Forms explosive mixture with air (flash point −58°F/−50°C). Contact with chemically active metals may cause fire and explosions. Attacks some plastics and rubber.

2-CHLOROALLYL CHLORIDE (78-88-6) Incompatible with strong acids, oxidizers, aluminum or magnesium compounds, aliphatic amines, alkanolamines, alkaline or corrosive substances.

alpha-CHLOROALLYL CHLORIDE (542-75-6) Incompatible with strong acids, oxidizers, aluminum or magnesium compounds, aliphatic amines, alkanolamines, alkaline or corrosive substances.

gamma-CHLOROALLYL CHLORIDE (542-75-6) Incompatible with strong acids, oxidizers, aluminum or magnesium compounds, aliphatic amines, alkanolamines, alkaline or corrosive substances.

4-CHLORO-1-AMINOBENZENE (106-47-8) Incompatible with strong acids, oxidizers, organic anhydrides, isocyanates, aldehydes, chlorosulfonic acid, ozone.

p-**CHLOROAMINOBENZENE** or *para*-**CHLOROAMINOBENZENE** (106-47-8) Incompatible with strong acids, oxidizers, organic anhydrides, isocyanates, aldehydes, chlorosufonic acid, ozone.

4-CHLORO-2-AMINOTOLUENE (95-79-4) Incompatible with acids, organic anhydrides, isocyanates, aldehydes, oxidizers.

5-CHLORO-2-AMINOTOLUENE (95-79-4) Incompatible with acids, organic anhydrides, isocyanates, aldehydes, oxidizers.

2-CHLOROANILINE (95-51-2) Incompatible with strong acids, organic anhydrides, isocyanates, aldehydes, strong oxidizers.

4-CHLOROANILINE (106-47-8) Incompatible with strong acids, oxidizers, acetic anhydride, chlorosulfonic acid, ozone.

o-**CHLOROANILINE** (95-51-2) Incompatible with strong acids, organic anhydrides, isocyanates, aldehydes, strong oxidizers.

p-**CHLOROANILINE or** *para*-**CHLOROANILINE** (106-47-8) Incompatible with strong acids, oxidizers, organic anhydrides, isocyanates, aldehydes, chlorosufonic acid, ozone.

CHLOROBEN (95-50-1) Forms explosive mixture with air (flash point 151°F/66°C). Incompatible with strong oxidizers, hot aluminum, or aluminum alloy. Attacks some plastics, rubber, and coatings.

CHLOROBENZAL (98-87-3) Forms explosive mixture with air (flash point 153°F/67°C). Reacts with acids, bases, strong oxidizers. Forms acid fumes with air. Attacks plastics and coatings.

(*o*-**CHLOROBENZAL**) **MALONONITRILE** (698-41-1) Incompatible with strong oxidizers, heat.

alpha-CHLOROBENZALDEHYDE (98-88-4) Forms explosive mixture with air (flash point 162°F/72°C). Water contact may be violent; produces hydrochloric acid. Incompatible with amines, alcohols, alkali metals, dimethylsulfoxide. Attacks metals in the presence of moisture. Attacks some plastics, rubber, and coatings.

CHLOROBENZEN (Polish) (108-90-7) Incompatible with oxidizers, dimethylsulfoxide, sodium powder, silver perchloride. Attacks some plastics, rubber, and coatings.

CHLOROBENZENE (108-90-7) Incompatible with oxidizers, dimethylsulfoxide, sodium powder, silver perchloride. Attacks some plastics, rubber, and coatings.

2-CHLOROBENZENEAMINE (95-51-2) Incompatible with strong acids, organic anhydrides, isocyanates, aldehydes, strong oxidizers.

4-CHLOROBENZENEAMINE (106-47-8) Incompatible with strong acids, oxidizers, organic anhydrides, isocyanates, aldehydes, chlorosufonic acid, ozone.

CHLOROBENZOL (108-90-7) Incompatible with oxidizers, dimethylsulfoxide, sodium powder, silver perchloride. Attacks some plastics, rubber, and coatings.

p-**CHLOROBENZOYL PEROXIDE or** *para*-**CHLOROBENZOYL PEROXIDE** (94-17-7) A powerful oxidizer and explosion hazard. Reacts violently with reducing agents, combustibles, polymerization initiators, heat, or contaminants.

p,p'-**CHLOROBENZOYL PEROXIDE** (94-17-7) A powerful oxidizer and explosion hazard. Reacts violently with reducing agents, combustibles, polymerization initiators, heat, or contaminants.

o-**CHLOROBENZYLIDENE MALONONITRILE or** (*o*-**CHLORO-BENZAL**)**MALONONITRILE** (698-41-1) Incompatible with strong oxidizers, heat.

(*o*-**CHLOROBENZAL**)**MALONONITRILE** (698-41-1) Incompatible with strong oxidizers, heat.

CHLORO-1,1-BIPHENYL (generic CAS for PCBs 1336-36-3) Incompatible with strong oxidizers, strong acids.

CHLORO BIPHENYLS (generic CAS for PCBs 1336-36-3) Incompatible with strong oxidizers, strong acids.

CHLOROBROMOMETHANE (74-97-5) Reacts with chemically active metals such as calcium, powdered aluminum, zinc, and magnesium. Liquid attacks some plastics, rubber, and coatings.

CHLOROBUTADIENE (126-99-8) Forms unstable peroxides that cause polymerization. Forms exposive mixture with air (flash point −4°F/−20°C). May accumulate static electrical charges, and may cause ignition of its vapors. Reacts violently with liquid or gaseous fluorine, alkali metals, metal powders, oxidizers. Attacks some plastics, rubber, and coatings.

2-CHLOROBUTADIENE (126-99-8) Forms unstable peroxides that cause polymerization. Forms exposive mixture with air (flash point −4°F/−20°C). May accumulate static electrical charges, and may cause ignition of its vapors. Reacts violently with liquid or gaseous fluorine, alkali metals, metal powders, oxidizers. Attacks some plastics, rubber, and coatings.

2-CHLORO-1,3-BUTADIENE (126-99-8) Forms unstable peroxides that cause polymerization. Forms exposive mixture with air (flash point −4°F/−20°C). May accumulate static electrical charges, and may cause ignition of its vapors. Reacts violently with liquid or gaseous fluorine, alkali metals, metal powders, oxidizers. Attacks some plastics, rubber, and coatings.

1-CHLOROBUTANE (109-69-3) Forms explosive mixture with air (flash point 15°F/−9°C). May accumulate static electrical charges, and may cause ignition of its vapors. Water contact slowly produces hydrochloric acid. Incompatible with strong oxidizers, alkaline earth and alkali metals, finely divided metal. Attacks metals in presence of moisture. Attacks some plastics, rubber, and coatings. May accumulate static electrical charges, and may cause ignition of its vapors.

4-CHLOROBUTYRONITRILE (628-20-6) Reacts with oxidizers, with a risk of fire or explosions. Attacks some plastics, rubber, and coatings.

CHLOROCAMPHENE (8001-35-2) Reacts with oxidizers, with a risk of fire or explosions. Attacks metals in the presence of moisture.

CHLOROCARBONIC ACID, METHYL ESTER (79-22-1) Forms explosive mixture with air (flash point 54°F/12°C). Water contact produces hydrochloric acid. Corrodes metals in the presence of moisture. Attacks some plastics, rubber, and coatings.

CHLOROCARBONIC ACID, *n*-BUTYL ESTER (592-34-7) Water contact forms hydrochloric acid; corrodes metals.

3-CHLOROCHLORDENE (76-44-8) Produces hydrogen chloride gas with iron and rust above 165°F/74°C.

CHLORO(CHLOROMETHOXY)METHANE (542-88-1) Highly volatile. Heat may cause fire. Water contact produces hydrogen chloride and formaldehyde.

CHLOROCHROMIC ANHYDRIDE (14977-61-8) A powerful oxidizer. Water contact produces hydrochloric and chromic acids and chlorine gas. Reacts violently with reducing agents, combustibles, ammonia, halides, phosphorus, sodium azide, elemental sulfur, urea.

176

CHLOROCYAN (506-77-4) Violent polymerization can be caused by chlorine. Water, steam, alcohols, acids, acid salts, amines, strong alkalis, olefins, strong oxidizers may cause fire and explosion. In crude form, chemical trimerizes violently if catalyzed by traces of hydrogen chloride or ammonium chloride. Corrodes brass, copper, bronze.

CHLOROCYANIDE (506-77-4) Violent polymerization can be caused by chlorine. Water, steam, alcohols, acids, acid salts, amines, strong alkalis, olefins, strong oxidizers may cause fire and explosion. In crude form, chemical trimerizes violently if catalyzed by traces of hydrogen chloride or ammonium chloride. Corrodes brass, copper, bronze.

CHLOROCYANOGEN (506-77-4) Violent polymerization can be caused by chlorine. Water, steam, alcohols, acids, acid salts, amines, strong alkalis, olefins, strong oxidizers may cause fire and explosion. In crude form, chemical trimerizes violently if catalyzed by traces of hydrogen chloride or ammonium chloride. Corrodes brass, copper, bronze.

CHLORODANE (57-74-9) Contact with strong oxidizers may cause fire and explosions. Attacks some plastics, rubber, and coatings.

CHLORODEN (95-50-1) Forms explosive mixture with air (flash point 151°F/66°C). Incompatible with strong oxidizers, hot aluminum, or aluminum alloy. Attacks some plastics, rubber, and coatings.

2-CHLORO-1,3-DIENE (126-99-8) Forms unstable peroxides that cause polymerization. Forms exposive mixture with air (flash point −4°F/−20°C). May accumulate static electrical charges, and may cause ignition of its vapors. Reacts violently with liquid or gaseous fluorine, alkali metals, metal powders, oxidizers. Attacks some plastics, rubber, and coatings.

CHLORODIFLUOROMETHANE (75-45-5) Moisture and rust cause slow decomposition, forming toxic gases. Attacks some plastics, rubber, and coatings. Thermal decomposition occurs, at high temperature with alkalies and alkaline earth metals.

CHLORODIPHENYL (54% CHLORINE) (11097-69-1); (generic CAS for PCBs 1336-36-3). Incompatible with strong oxidizers, strong acids.

CHLORO di VINYLE (Italian) (75-01-4) Forms explosive gas mixture with air. Atmospheric oxygen and various contaminants or strong oxidizers cause formation of peroxides, which can initiate a violent polymerization action. Also able to polymerize when heated and exposed, long term, to light. Reacts violently with strong oxidizers or oxides of nitrogen. Contact with copper or other acetylide-forming metals produces explosive compounds. Attacks iron and steel in the presence of moisture.

1-CHLORO-2,3-EPOXYPROPANE (106-89-8) Forms explosive mixture with air (flash point 88°F/31°C). Heat, acids, alkalies and metallic halides can cause explosive polymerization. Aliphatic amines, alkaline earths, alkali metals, alkanolamines, powdered metals, strong oxidizers may cause fire and explosions. Decomposition produces highly toxic phosgene gas. Will pit steel in the presence of moisture. May accumulate static electrical charges, and may cause ignition of its vapors.

3-CHLORO-1,2-EPOXYPROPANE (106-89-8) Forms explosive mixture with air (flash point 88°F/31°C). Heat, acids, alkalies, and metallic halides can cause explosive polymerization. Aliphatic amines, alkaline earths, alkali metals, alkanolamines, powdered metals, strong oxidizers may cause fire and explosions. Decomposition produces highly toxic phosgene gas. Will pit steel in the presence of moisture. May accumulate static electrical charges, and may cause ignition of its vapors.

CHLOROETHANAL (107-20-0) Incompatible with oxidizers, acids, water. May form a water-soluble polymer in storage.

2-CHLOROETHANAL (107-20-0) Incompatible with oxidizers, acids, water. May form a water-soluble polymer in storage.

2-CHLORO-1-ETHANAL (107-20-0) Incompatible with oxidizers, acids, water. May form a water-soluble polymer in storage.

CHLOROETHANE (75-00-3) Flammable gas. Forms explosive mixture with air. Contact with moisture produces hydrochloric acid. May accumulate static electrical charges, and may cause ignition of its vapors. Forms explosive mixture with air (flash point −58°F/−50°C). Contact with chemically active metals may cause fire and explosions. Attacks some plastics and rubber.

CHLOROETHANOL (107-07-3) Forms explosive mixture with air (flash point 140°F/60°C). Strong oxidizers may cause fire and explosions. Incompatible with strong caustics (with formation of ethylene gas), strong acids, aliphatic amines, isocyanates. Attacks some plastics, rubber, and coatings.

2-CHLOROETHANOL (107-07-3) Forms explosive mixture with air (flash point 140°F/60°C). Strong oxidizers may cause fire and explosions. Incompatible with strong caustics (with formation of ethylene gas), strong acids, aliphatic amines, isocyanates. Attacks some plastics, rubber, and coatings.

CHLOROETHENE (75-01-4) Forms explosive gas mixture with air. Atmospheric oxygen and various contaminants or strong oxidizers produce peroxides, which can initiate a violent polymerization action. Also able to polymerize when heated and exposed, long term, to light. Reacts violently with strong oxidizers or oxides of nitrogen. Contact with copper or other acetylide-forming metals produces explosive compounds. Attacks iron and steel in the presence of moisture.

2-CHLOROETHYL ALCOHOL (107-07-3) Forms explosive mixture with air (flash point 140°F/60°C). Strong oxidizers may cause fire and explosions. Incompatible with strong caustics (with formation of ethylene gas), strong acids, aliphatic amines, isocyanates. Attacks some plastics, rubber, and coatings.

2-CHLORO-4-ETHYLAMINEISOPROPYLAMINE-S-TRIAZINE (1912-24-9) Incompatible with strong acids.

1-CHLORO-3-ETHYLAMINO-5-ISOPROPYLAMINO-2,4,6-TRIA-ZINE (1912-24-9) Incompatible with strong acids.

1-CHLORO-3-ETHYLAMINO-5-ISOPROPYLAMINO-S-TRIA-ZINE (1912-24-9) Incompatible with strong acids.

2-CHLORO-4-ETHYLAMINO-6-ISOPROPYLAMINO-1,3,5-TRIA-ZINE (1912-24-9) Incompatible with strong acids.

2-CHLORO-4-ETHYLAMINO-6-ISOPROPYLAMINO-S-TRIA-ZINE (1912-24-9) Incompatible with strong acids.

2-CHLORO-4-ETHYLAMONO-6-ISOPROPYLAMINO- (1912-24-9) Incompatible with strong acids.

CHLOROETHYLENE (75-01-4) Forms explosive gas mixture with air. Atmospheric oxygen and various contaminants or strong oxidizers produce peroxides, which can initiate a violent polymerization action. Also able to polymerize when heated and exposed, long term, to light.

Reacts violently with strong oxidizers or oxides of nitrogen. Contact with copper or other acetylide-forming metals produces explosive compounds. Attacks iron and steel in the presence of moisture.

CHLOROETHYL ETHER (111-44-4) Forms explosive mixture with air (flash point 131°F/55°C). Strong oxidizers may cause fire and explosions. Attacks some plastics, rubber, and coatings.

6-CHLORO-N-ETHYL-N'-(1-METHYLETHYL)-1,3,5-TRIAZINE-2,4-DIAMINE (1912-24-9) Incompatible with strong acids.

CHLOROFORM (67-66-3) Decomposes in the presence of excess water or high temperatures, with formation of phosgene and hydrogen chloride. Can become explosive in the presence of strong alkalies and water. May accumulate static electrical charges, and may cause ignition of its vapors. In contact with water and high temperatures, it becomes corrosive; attacks iron and other metals. Incompatible with acetone, aluminum, strong oxidizers, potassium, sodium, chemically active metals, strong bases. Attacks plastics and rubber.

CHLOROFORME (French) (67-66-3) Decomposes in the presence of excess water or high temperatures, with formation of phosgene and hydrogen chloride. Can become explosive in the presence of strong alkalies and water. Can accumulate static electrical charges. In contact with water and high temperatures, it becomes corrosive; attacks iron and other metals. Incompatible with acetone, aluminum, strong oxidizers, potassium, sodium, chemically active metals, strong bases. Attacks plastics and rubber.

CHLOROFORMIC ACID, BENZYL ESTER (501-53-1) Aqueous solution produces hydrochloric acid; reacts violently with bases. Reacts with nitrates. Attacks metals in the presence of moisture.

CHLOROFORMIC ACID DIMETHYLAMIDE (79-44-7) Rapidly hydrolyzed in water. Incompatible with strong acids, oxidizers.

CHLOROFORMIC ACID, ETHYL ESTER (541-43-3) Forms explosive mixture with air (flash point 61°F/16°C). Strong oxidizers may cause fire and explosions. Attacks metals in the presence of moisture. Water and air contact produces hydrochloric acid.

CHLOROFORMIC ACID, METHYL ESTER (79-22-1) Forms explosive mixture with air (flash point 54°F/12°C). Water contact produces hydrochloric acid. Corrodes metals in the presence of moisture. Attacks some plastics, rubber, and coatings.

CHLOROFORMIC ACID, *n*-BUTYL ESTER (592-34-7) Water contact produces hydrochloric acid; corrodes metals.

CHLOROFORMYL CHLORIDE (75-44-5) Incompatible with water, producing hydrochloric acid. Reacts violently with strong oxidizers, anhydrous ammonia, isopropyl alcohol and chemically active metals. Forms shock-sensitive material with potassium. Attacks most metal in moist conditions.

CHLOROHYDRIC ACID (7647-01-0) Incompatible with acetic anhydride, aliphatic amines, alkanolamines, alkylene oxides, aromatic amines, amides, 2-aminoethanol, ammonia, ammonium hydroxide, calcium phosphide, chlorosulfonic acid, ethylene diamine, ethyleneimine, epichlorohydrin, isocyanates, metal acetylides, oleum, organic anhydrides, perchloric acid, 3-propiolactone, uranium phosphide, sulfuric acid, sodium hydroxide and other bases, strong oxidizers, vinyl acetate, vinylidene fluoride. Attacks most metals and some plastics, rubber, and coatings.

CHLOROHYDRINS (106-89-8) Forms explosive mixture with air (flash point 88°F/31°C). Heat, acids, alkalies and metallic halides can cause explosive polymerization. Aliphatic amines, alkaline earths, alkali metals, alkanolamines, powdered metals, strong oxidizers may cause fire and explosions. Decomposition produces highly toxic phosgene gas. Will pit steel in the presence of moisture. May accumulate static electrical charges, and may cause ignition of its vapors.

1-CHLORO-2-HYDROXYBENZENE (95-57-0) Forms explosive mixture with air (flash point 147°F/64°C) Contact with strong oxidizers may cause fire and explosions. Attacks aluminum and copper.

2-CHLORO-1-HYDROXYBENZENE (95-57-0) Forms explosive mixture with air (flash point 147°F/64°C) Contact with strong oxidizers may cause fire and explosions. Attacks aluminum and copper.

4-CHLORO-1-HYDROXYBENZENE (106-48-9) Forms explosive mixture with air (flash point 147°F/64°C). Reacts with oxidizers, with a risk of fire or explosions. Attacks active metals, aluminum, copper, etc. May accumulate static electrical charges, and may cause ignition of its vapors.

3-CHLORO-7-HYDROXY-4-METHYL-COUMARIN 3-CHLORO-ALLYL CHLORIDE (542-75-6) Incompatible with strong acids, oxidizers, aluminum or magnesium compounds, aliphatic amines, alkanolamines, alkaline or corrosive substances.

3-CHLORO-7-HYDROXY-4-METHYL-COUMARIN-O,O-DIE-THYLPHOSPHOROTHIONATE (56-72-4) Contact with strong oxidizers may cause fire and explosions.

3-CHLORO-7-HYDROXY-4-METHYL-COUMARIN O-ESTER WITH O,O-DIETHYL BICYCLO-(2.2.1)HEPTANE (79-92-5) Contact with strong oxidizers may cause fire and explosions. Emulsions in xylene may violently decompose on contact with iron or aluminum above 158°F/70°C.

gamma-CHLOROISOBUTYLENE (563-47-3) Forms explosive mixture with air (flash point 11°F/−12°C). Strong oxidizers may cause fire and explosions.

1-CHLORO-2-KETOPROPANE (78-95-5) Forms exposive mixture with air (flash point 40°F/7°C). Strong oxidizers may cause fire and explosions. May accumulate static electrical charges, and may cause ignition of its vapors.

CHLOROMETHANE (74-87-3) Incompatible with chemically active metals. Attacks plastics, rubber, and coatings.

3-CHLORO-6-METHYLANILINE (95-79-4) Incompatible with acids, organic anhydrides, isocyanates, aldehydes, oxidizers.

4-CHLORO-2-METHYLANILINE (95-79-4) Incompatible with acids, organic anhydrides, isocyanates, aldehydes, oxidizers.

CHLOROMETHYLBENZENE (100-44-7) Forms explosive mixture with air (flash point 153°F/67°C). Water contact produces hydrogen chloride fume. Strong oxidizers may cause fire and explosions. Violent polymerization may be caused by contact with copper, aluminum, iron, zinc, magnesium and tin. May accumulate static electrical charges, and may cause ignition of its vapors. Attacks some plastics and rubber.

1-CHLORO-3-METHYLBENZENE (108-41-8) Forms explosive mixture with air (flash point 123°F/51°C). Incompatible with strong oxidizers, rare earth metals.

1-CHLORO-4-METHYLBENZENE (106-43-4) Reacts with oxidizers, with a risk of fire or explosions.

2-CHLORO-1-METHYLBENZENE (95-49-8) Reacts with oxidizers, with a risk of fire or explosions.

3-CHLORO-1-METHYLBENZENE (108-41-8) Forms explosive mixture with air (flash point 123°F/51°C). Incompatible with strong oxidizers, rare earth metals.

4-CHLORO-1-METHYLBENZENE (106-43-4) Reacts with oxidizers, with a risk of fire or explosions.

3-CHLORO-4-METHYL-7-COUMARINYLDIETHYLPHOSPHOR-OTHIOATE (56-72-4) Contact with strong oxidizers may cause fire and explosions.

O-CHLORO-4-METHYL-7-COUMARINYL-O,O-DIETHYLPHOS-PHOROTHIOATE (56-72-4) Contact with strong oxidizers may cause fire and explosions.

CHLOROMETHYLCYANIDE (107-14-2) Forms explosive mixture with air (flash point 133°F/56°C). Contact with moisture produces hydrogen cyanide gas. Reacts with sulfuric acid and strong oxidizers.

CHLOROMETHYL ETHER (542-88-1) Highly volatile. Heat may cause fire. Water contact produces hydrogen chloride and formaldehyde.

CHLOROMETHYL METHYL ETHER (107-30-2) Forms explosive mixture with air (flash point 0°F/−17.8°C). May be able to form unstable and explosive peroxides. Produces hydrochloric acid on contact with moisture. Corrosive to metals in presence of moisture.

3-CHLORO-4-METHYL-7-HYDROXYCOUMARINDIETHYL THIOPHOSPHORIC ACID ESTER (56-72-4) Contact with strong oxidizers may cause fire and explosions.

CHLOROMETHYL METHYL KETONE (78-95-5) Forms explosive mixture with air (flash point 40°F/7°C). Strong oxidizers may cause fire and explosions. May accumulate static electrical charges, and may cause ignition of its vapors.

CHLOROMETHYLOXIRANE (106-89-8) Forms explosive mixture with air (flash point 88°F/31°C). Heat, acids, alkalies and metallic halides can cause explosive polymerization. Aliphatic amines, alkaline earths, alkali metals, alkanolamines, powdered metals, strong oxidizers may cause fire and explosions. Decomposition produces highly toxic phosgene gas. Will pit steel in the presence of moisture. May accumulate static electrical charges, and may cause ignition of its vapors.

CHLOROMETHYL OXIRANE (106-89-8) Forms explosive mixture with air (flash point 88°F/31°C). Heat, acids, alkalies, and metallic halides can cause explosive polymerization. Aliphatic amines, alkaline earths, alkali metals, alkanolamines, powdered metals, strong oxidizers may cause fire and explosions. Decomposition produces highly toxic phosgene gas. Will pit steel in the presence of moisture. May accumulate static electrical charges, and may cause ignition of its vapors.

O-(3-CHLORO-4-METHYL-2-OXO-(2H)-1-BENZOPYRAN-7-YL) PHOSPHOROTHIOATE (56-72-4) Contact with strong oxidizers may cause fire and explosions.

CHLOROMETHYL PHENYL KETONE (532-27-4) Incompatible with water, or steam.

3-CHLORO-2-METHYLPROPENE (563-47-3) Forms explosive mixture with air (flash point 11°F/−12°C). Strong oxidizers may cause fire and explosions.

3-CHLORO-4-METHYLUMBELLIFERONEO-ESTER WITH O,O-DIETHYL PHOSPHOROTHIOATE (56-72-4) Contact with strong oxidizers may cause fire and explosions.

CHLORONITROBENZENE or CHLORONITROBENZENE (88-73-3) (*o*-); (121-73-3) (*m*-); (100-00-5) (-*p*). Forms explosive mixture with air. A strong oxidizer; reacts violently with reducing agents, combustibles. Reacts with caustics, ammonia, amines. Attacks some plastics, rubber, and coatings.

1-CHLORO-2-NITROBENZENE (88-73-3) Forms explosive mixture with air. A strong oxidizer; reacts violently with reducing agents, combustibles. Reacts with caustics, ammonia, amines. Attacks some plastics, rubber, and coatings.

1,4-CHLORONITROBENZENE (100-00-5) Forms explosive mixture with air. A strong oxidizer; reacts violently with reducing agents, combustibles. Reacts with caustics, ammonia, amines. Attacks some plastics, rubber, and coatings.

2-CHLORONITROBENZENE (88-73-3) Forms explosive mixture with air. A strong oxidizer; reacts violently with reducing agents, combustibles. Reacts with caustics, ammonia, amines. Attacks some plastics, rubber, and coatings.

2-CHLORO-1-NITROBENZENE (88-73-3) Forms explosive mixture with air. A strong oxidizer; reacts violently with reducing agents, combustibles. Reacts with caustics, ammonia, amines. Attacks some plastics, rubber, and coatings.

m-**CHLORONITROBENZENE, solid (DOT)** or *meta*-**CHLORONITROBENZENE** (121-73-3) Particles form explosive mixture with air. A strong oxidizer; reacts violently with reducing agents, combustibles. Reacts with caustics, ammonia, amines. Attacks some plastics, rubber, and coatings.

o-**CHLORONITROBENZENE (DOT)** or *ortho*-**CHLORONITROBENZENE** (88-73-3) Forms explosive mixture with air. A strong oxidizer; reacts violently with reducing agents, combustibles. Reacts with caustics, ammonia, amines. Attacks some plastics, rubber, and coatings.

p-**CHLORONITROBENZENE (DOT)** or *para*-**CHLORONITROBENZENE** (100-00-5) Forms explosive mixture with air. A strong oxidizer; reacts violently with reducing agents, combustibles. Reacts with caustics, ammonia, amines. Attacks some plastics, rubber, and coatings.

CHLORO-*o*-NITROBENZENE (88-73-3) Forms explosive mixture with air. A strong oxidizer; reacts violently with reducing agents, combustibles. Reacts with caustics, ammonia, amines. Attacks some plastics, rubber, and coatings.

1,1-CHLORONITROETHANE (598-92-5) Shock-sensitive explosive. Forms explosive mixture with air (flash point 133°F/56°C). Reacts violently with strong oxidizers. Quickly elevated temperatures or high heat may cause explosions.

1-CHLORO-2-NITROETHANE (598-92-5) Shock-sensitive explosive. Forms explosive mixture with air (flash point 133°F/56°C). Reacts violently with strong oxidizers. Quickly elevated temperatures or high heat may cause explosions.

1-CHLORO-2-OXOPROPANE (78-95-5) Forms exposive mixture with air (flash point 40°F/7°C). Strong oxidizers may cause fire and explosions. May accumulate static electrical charges, and may cause ignition of its vapors.

CHLOROPENTAFLUOROETHANE (76-15-3) Thermal decomposition occurs at high temperatures with alkalies and alkaline earth metals.

CHLOROPHEN (87-86-5) Incompatible with strong acids, caustics, aliphatic amines, amides. Decomposes at 375°F/191°C, forming toxic and flammable vapors.

CHLOROPHENATE (106-48-9) Forms explosive mixture with air (flash point 147°F/64°C). Reacts with oxidizers, with a risk of fire or explosions. Attacks active metals, aluminum, copper, etc. May accumulate static electrical charges, and may cause ignition of its vapors.

2-CHLOROPHENOL (95-57-0) Forms explosive mixture with air (flash point 147°F/64°C). Contact with strong oxidizers may cause fire and explosions. Attacks aluminum and copper.

4-CHLOROPHENOL (106-48-9) Forms explosive mixture with air (flash point 147°F/64°C). Reacts with oxidizers, with a risk of fire or explosions. Attacks active metals, aluminum, copper, etc. May accumulate static electrical charges, and may cause ignition of its vapors.

o-**CHLOROPHENOL** (95-57-0) Forms explosive mixture with air (flash point 147°F/64°C). Contact with strong oxidizers may cause fire and explosions. Attacks aluminum and copper.

p-**CHLOROPHENOL or** *para*-**CHLOROPHENOL** (106-48-9) Forms explosive mixture with air (flash point 147°F/64°C). Reacts with oxidizers, with a risk of fire or explosions. Attacks active metals, aluminum, copper, etc. May accumulate static electrical charges, and may cause ignition of its vapors.

CHLOROPHENOTHAN (50-29-3) Incompatible with salts of iron or aluminum, and bases. Do not store in iron containers.

CHLOROPHENOTHANE (50-29-3) Incompatible with salts of iron or aluminum, and bases. Do not store in iron containers.

CHLOROPHENOTOXUM (50-29-3) Incompatible with salts of iron or aluminum, and bases. Do not store in iron containers.

2-CHLOROPHENYLAMINE (95-51-2) Incompatible with strong acids, organic anhydrides, isocyanates, aldehydes, strong oxidizers.

4-CHLOROPHENYLAMINE (106-47-8) Incompatible with strong acids, oxidizers, organic anhydrides, isocyanates, aldehydes, chlorosufonic acid, ozone.

183

p-CHLOROPHENYL CHLORIDE or *para*-CHLOROPHENYL CHLORIDE (106-46-7) Forms explosive mixture with air (flash point 150°F/66°C). Incompatible with strong oxidizers, metal powders and alkali metals (i.e., lithium, sodium, potassium, rubidium, cesium, francium). Attacks some plastics, rubber, and coating.

CHLOROPHENYLMETHANE (100-44-7) Forms explosive mixture with air (flash point 153°F/67°C). Water contact produces hydrogen chloride fume. Strong oxidizers may cause fire and explosions. Violent polymerization may be caused by contact with copper, aluminum, iron, zinc, magnesium and tin. May accumulate static electrical charges, and may cause ignition of its vapors. Attacks some plastics and rubber.

CHLOROPHOS (52-68-6) Contact with strong oxidizers may cause fire and explosions.

CHLOR-O-PIC (76-06-2) Can be self-reactive. Fast heating, shock, alkali metals or alkaline earth may cause explosions. A strong oxidizer; reacts violently with reducing agents, aniline in presence of heat, alcoholic sodium hydroxide, combustible substances, sodium methoxide, propargyl bromide. Liquid attacks some plastics, rubber, and coatings.

CHLOROPICRIN or CHLOROPICRIN, ABSORBED (76-06-2) Can be self-reactive. Fast heating, shock, alkali metals, or alkaline earth may cause explosions. A strong oxidizer; reacts violently with reducing agents, aniline in presence of heat, alcoholic sodium hydroxide, combustible substances, sodium methoxide, propargyl bromide. Liquid attacks some plastics, rubber, and coatings.

CHLOROPICRINE (French) (76-06-2) Can be self-reactive. Fast heating, shock, alkali metals, or alkaline earth may cause explosions. A strong oxidizer; reacts violently with reducing agents, aniline in presence of heat, alcoholic sodium hydroxide, combustible substances, sodium methoxide, propargyl bromide. Liquid attacks some plastics, rubber, and coatings.

CHLOROPRENE (126-99-8) Forms unstable peroxides that cause polymerization. Forms exposive mixture with air (flash point −4°F/−20°C). May accumulate static electrical charges, and may cause ignition of its vapors. Reacts violently with liquid or gaseous fluorine, alkali metals, metal powders, oxidizers. Attacks some plastics, rubber, and coatings.

3-CHLOROPRENE (107-05-1) Forms explosive mixture with air (flash point −25°F/−32°C). Violent polymerization and explosion may occur from heat, light, or contact with acid catalysts, ferric chloride, aluminum chloride, Lewis acids, or Ziegler catalyst (e.g., titanium tetrachloride + triethylaluminum monochloride). Incompatible with strong acids, amines, aluminum chloride, boron trifluoride, chlorosulfonic acid, ethylene diamine, ethyleneimine, ferric chloride, oleum, oxidizers, sodium hydroxide. Slow decomposition occurs with moisture; attacks some coatings, plastics, and rubber. Corrosive to steel. May accumulate static electrical charges, and may cause ignition of its vapors.

b-CHLOROPRENE or beta-CHLOROPRENE (126-99-8) Forms unstable peroxides that cause polymerization. Forms exposive mixture with air (flash point−4°F/−20°C). May accumulate static electrical

184

charges, and may cause ignition of its vapors. Reacts violently with liquid or gaseous fluorine, alkali metals, metal powders, oxidizers. Attacks some plastics, rubber, and coatings.

1-CHLOROPROPANE (540-54-5) Forms explosive mixture with air (flash point 0°F/−18°C). Reacts violently with strong oxidizers.

1-CHLORO-2-PROPANONE (78-95-5) Forms exposive mixture with air (flash point 40°F/7°C). Strong oxidizers may cause fire and explosions. May accumulate static electrical charges, and may cause ignition of its vapors.

3-CHLOROPROPANOIC ACID (107-94-8) Reacts with strong oxidizers, caustics, ammonia, amines, isocyanates, alkylene oxide, epichlorohydrin, strong oxidizers. Aqueous solution is acid; attacks metals. Only aluminum or stainless steel with protective lining or coating may contact liquid or vapor.

1-CHLOROPROPENE-2 (107-05-1) Forms explosive mixture with air (flash point −25°F/−32). Violent polymerization and explosion may occur from heat, light, or contact with acid catalysts, ferric chloride, aluminum chloride, Lewis acids, or Ziegler catalyst (e.g., titanium tetrachloride + triethylaluminum monochloride). Incompatible with strong acids, amines, aluminum chloride, boron trifluoride, chlorosulfonic acid, ethylene diamine, ethyleneimine, ferric chloride, oleum, oxidizers, sodium hydroxide. Slow decomposition occurs with moisture; attacks some coatings, plastics, and rubber. Corrosive to steel. May accumulate static electrical charges, and may cause ignition of its vapors.

1-CHLORO-2-PROPENE (107-05-1) Forms explosive mixture with air (flash point −25°F/−32). Violent polymerization and explosion may occur from heat, light, or contact with acid catalysts, ferric chloride, aluminum chloride, Lewis acids, or Ziegler catalyst (e.g., titanium tetrachloride + triethylaluminum monochloride). Incompatible with strong acids, amines, aluminum chloride, boron trifluoride, chlorosulfonic acid, ethylene diamine, ethyleneimine, ferric chloride, oleum, oxidizers, sodium hydroxide. Slow decomposition occurs with moisture; attacks some coatings, plastics, and rubber. Corrosive to steel. May accumulate static electrical charges, and may cause ignition of its vapors.

3-CHLOROPROPENE (107-05-1) Forms explosive mixture with air (flash point −25°F/−32°C). Violent polymerization and explosion may occur from heat, light, or contact with acid catalysts, ferric chloride, aluminum chloride, Lewis acids, or Ziegler catalyst (e.g., titanium tetrachloride + triethylaluminum monochloride). Incompatible with strong acids, amines, aluminum chloride, boron trifluoride, chlorosulfonic acid, ethylene diamine, ethyleneimine, ferric chloride, oleum, oxidizers, sodium hydroxide. Slow decomposition occurs with moisture; attacks some coatings, plastics, and rubber. Corrosive to steel. May accumulate static electrical charges, and may cause ignition of its vapors.

3-CHLORO-1-PROPENE (107-05-1) Forms explosive mixture with air (flash point −25°F/−32). Violent polymerization and explosion may occur from heat, light, or contact with acid catalysts, ferric chloride, aluminum chloride, Lewis acids, or Ziegler catalyst (e.g., titanium tetrachloride + triethylaluminum monochloride). Incompatible with strong acids, amines, aluminum chloride, boron trifluoride, chlorosulfonic acid, ethylene diamine, ethyleneimine, ferric chloride, oleum, oxidizers, sodium hydroxide. Slow decomposition occurs with moisture; attacks some coatings, plastics and rubber. Corrosive to steel. May accumulate static electrical charges, and may cause ignition of its vapors.

3-CHLOROPROPENE-1 (107-05-1) Forms explosive mixture with air (flash point −25°F/−32). Violent polymerization and explosion may occur from heat, light, or contact with acid catalysts, ferric chloride, aluminum chloride, Lewis acids, or Ziegler catalyst (e.g., titanium tetrachloride + triethylaluminum monochloride). Incompatible with strong acids, amines, aluminum chloride, boron trifluoride, chlorosulfonic acid, ethylene diamine, ethyleneimine, ferric chloride, oleum, oxidizers, sodium hydroxide. Slow decomposition occurs with moisture; attacks some coatings, plastics, and rubber. Corrosive to steel. May accumulate static electrical charges, and may cause ignition of its vapors.

3-CHLOROPROPENYL CHLORIDE (542-75-6) Incompatible with strong acids, oxidizers, aluminum or magnesium compounds, aliphatic amines, alkanolamines, alkaline or corrosive substances.

2-CHLOROPROPIONIC ACID (598-78-7) Reacts with strong oxidizers, caustics, ammonia, amines, isocyanates, alkylene oxide, epichlorohydrin, strong oxidizers. Aqueous solution is acid; attacks metals. Only aluminum or stainless steel with protective lining or coating may contact liquid or vapor.

3-CHLOROPROPIONIC ACID (107-94-8) Reacts with strong oxidizers, caustics, ammonia, amines, isocyanates, alkylene oxide, epichlorohydrin, strong oxidizers. Aqueous solution is acid; attacks metals. Only aluminum or stainless steel with protective lining or coating may contact liquid or vapor.

alpha-CHLOROPROPIONIC ACID (598-78-7) Reacts with strong oxidizers, caustics, ammonia, amines, isocyanates, alkylene oxide, epichlorohydrin, strong oxidizers. Aqueous solution is acid; attacks metals. Only aluminum or stainless steel with protective lining or coating may contact liquid or vapor.

beta-CHLOROPROPIONIC ACID (107-94-8) Reacts with strong oxidizers, caustics, ammonia, amines, isocyanates, alkylene oxide, epichlorohydrin, strong oxidizers. Aqueous solution is acid; attacks metals. Only aluminum or stainless steel with protective lining or coating may contact liquid or vapor.

2-CHLORO-4-(2-PROPYLAMINO)-6-ETHYLAMINO-S-TRIAZINE (1912-24-9) Incompatible with strong acids.

3-CHLOROPROPYLENE (107-05-1) Forms explosive mixture with air (flash point −25°F/−32). Violent polymerization and explosion may occur from heat, light, or contact with acid catalysts, ferric chloride, aluminum chloride, Lewis acids, or Ziegler catalyst (e.g., titanium tetrachloride + triethylaluminum monochloride). Incompatible with strong acids, amines, aluminum chloride, boron trifluoride, chlorosulfonic acid, ethylene diamine, ethyleneimine, ferric chloride, oleum, oxidizers, sodium hydroxide. Slow decomposition occurs with moisture; attacks some coatings, plastics and rubber. Corrosive to steel. May accumulate static electrical charges, and may cause ignition of its vapors.

alpha-CHLOROPROPYLENE (107-05-1) Forms explosive mixture with air (flash point −25°F/−32°C). Violent polymerization and explosion may occur from heat, light, or contact with acid catalysts, ferric chloride, aluminum chloride, Lewis acids, or Ziegler catalyst (e.g., titanium tetrachloride + triethylaluminum monochloride). Incompatible with strong acids, amines, aluminum chloride, boron trifluoride, chlorosulfonic acid, ethylene diamine, ethyleneimine, ferric chloride, oleum, oxidizers, sodium hydroxide. Slow decomposition occurs with moisture; attacks some coatings, plastics and rubber. Corrosive to steel. May accumulate static electrical charges, and may cause ignition of its vapors.

gamma-CHLOROPROPYLENE (106-89-8) Forms explosive mixture with air (flash point 88°F/31°C). Heat, acids, alkalies and metallic halides can cause explosive polymerization. Aliphatic amines, alkaline earths, alkali metals, alkanolamines, powdered metals, strong oxidizers may cause fire and explosions. Decomposition causes formation of highly toxic phosgene gas. Will pit steel in the presence of moisture. May accumulate static electrical charges, and may cause ignition of its vapors.

2-CHLOROPROPYLENE OXIDE (106-89-8) Forms explosive mixture with air (flash point 88°F/31°C). Heat, acids, alkalies, and metallic halides can cause explosive polymerization. Aliphatic amines, alkaline earths, alkali metals, alkanolamines, powdered metals, strong oxidizers may cause fire and explosions. Decomposition causes formation of highly toxic phosgene gas. Will pit steel in the presence of moisture. May accumulate static electrical charges, and may cause ignition of its vapors.

3-CHLOROPROPYLENE OXIDE (106-89-8) Forms explosive mixture with air (flash point 88°F/31°C). Heat, acids, alkalies and metallic halides can cause explosive polymerization. Aliphatic amines, alkaline earths, alkali metals, alkanolamines, powdered metals, strong oxidizers may cause fire and explosions. Decomposition causes formation of highly toxic phosgene gas. Will pit steel in the presence of moisture. May accumulate static electrical charges, and may cause ignition of its vapors.

3-CHLORO-1,2-PROPYLENE OXIDE (106-89-8) Forms explosive mixture with air (flash point 88°F/31°C). Heat, acids, alkalies, and metallic halides can cause explosive polymerization. Aliphatic amines, alkaline earths, alkali metals, alkanolamines, powdered metals, strong oxidizers may cause fire and explosions. Decomposition causes formation of highly toxic phosgene gas. Will pit steel in the presence of moisture. May accumulate static electrical charges, and may cause ignition of its vapors.

gamma-CHLOROPROPYLENE OXIDE (106-89-8) Forms explosive mixture with air (flash point 88°F/31°C). Heat, acids, alkalies and metallic halides can cause explosive polymerization. Aliphatic amines, alkaline earths, alkali metals, alkanolamines, powdered metals, strong oxidizers may cause fire and explosions. Decomposition causes formation of highly toxic phosgene gas. Will pit steel in the presence of moisture. May accumulate static electrical charges, and may cause ignition of its vapors.

CHLOROS (7681-52-9) A powerful oxidizer and a strong base. Decomposes in sunlight, producing oxygen and increasing the risk of fire. Stability decreases with concentration, heat, light, decrease in pH, and contamination with metals. Incompatible with strong acids, reducing agents, combustible substances; all cause violent reaction, fire, and explosions. Contact with amines and ammonia salts produces explosive chloroamines. Corrodes many metals: steel, 12% and 17% chrome steel, cast iron, monel, aluminum, nickel, brass, bronze, iconel.

CHLOROSULFONIC ACID (7790-94-5) Reacts violently with water, producing sulfuric and hydrochloric acids and dense fumes. Dangerously reactive; avoid contact with all other material. Strong oxidizer and strong acid; reacts violently with bases, reducing agents, combustibles, acids (especially sulfuric), alcohol, diphenyl ether, silver nitrate, water. Phosphorus contact may cause fire and explosions. Forms explosive material with ethyl alcohol. Incompatible with many materials. Attacks many metals.

CHLOROSULFURIC ACID (7790-94-5) Reacts violently with water, producing sulfuric and hydrochloric acids and dense fumes. Dangerously reactive; avoid contact with all other material. Strong oxidizer and strong acid; reacts violently with bases, reducing agents, combustibles, acids (especially sulfuric), alcohol, diphenyl ether, silver nitrate, water. Phosphorus contact may cause fire and explosions. Forms explosive material with ethyl alcohol. Incompatible with many materials. Attacks many metals.

2-CHLOROSTYRENE Incompatible with strong oxidizers.

o-**CHLOROSTYRENE or** *ortho*-**CHLOROSTYRENE** (2039-87-4) Contact with strong oxidizers may cause fire and explosions.

CHLOROTHENE (71-55-6) Incompatible with strong caustics, strong oxidizers, chemically active metals, sodium, and potassium. Attacks some plastics, rubber, and coatings.

2-CHLOROTOLUENE (95-49-8) Reacts with oxidizers, with a risk of fire or explosions.

3-CHLOROTOLUENE (108-41-8) Forms explosive mixture with air (flash point 123°F/51°C). Incompatible with strong oxidizers, rare earth metals.

4-CHLOROTOLUENE (106-43-4) Reacts with oxidizers, with a risk of fire or explosions.

alpha-CHLOROTOLUENE (100-44-7) Forms explosive mixture with air (flash point 153°F/67°C). Water contact produces hydrogen chloride fume. Strong oxidizers may cause fire and explosions. Violent polymerization may be caused by contact with copper, aluminum, iron, zinc, magnesium, and tin. May accumulate static electrical charges, and may cause ignition of its vapors. Attacks some plastics and rubber.

m-**CHLOROTOLUENE or** *meta*-**CHLOROTOLUENE** (108-41-8) Forms explosive mixture with air (flash point 123°F/51°C). Incompatible with strong oxidizers, rare earth metals.

o-**CHLOROTOLUENE or** *ortho*-**CHLOROTOLUENE** (95-49-8) Reacts with oxidizers, with a risk of fire or explosions.

omega-CHLOROTOLUENE (100-44-7) Forms explosive mixture with air (flash point 153°F/67°C). Water contact produces hydrogen chloride fume. Strong oxidizers may cause fire and explosions. Violent polymerization may be caused by contact with copper, aluminum, iron, zinc, magnesium, and tin. May accumulate static electrical charges, and may cause ignition of its vapors. Attacks some plastics and rubber.

p-**CHLOROTOLUENE or** *para*-**CHLOROTOLUENE** (106-43-4) Reacts with oxidizers, with a risk of fire or explosions.

CHLOROTOLUIDINE (95-79-4) Incompatible with acids, organic anhydrides, isocyanates, aldehydes, oxidizers.

4-CHLORO-*o*-TOLUIDINE (95-79-4) Incompatible with acids, organic anhydrides, isocyanates, aldehydes, oxidizers.

5-CHLORO-*o*-TOLUIDINE (95-79-4) Incompatible with acids, organic anhydrides, isocyanates, aldehydes, oxidizers.

CHLOROTRIFLUORIDE (7790-91-2) Will cause most combustible materials to ignite spontaneously. Dangerously reactive with many materials and substances, including water, acids, aluminum oxide, ammonia, arsenic trioxide, asbestos, benzene, chromic anhydride, chromic oxide, copper, most elements, fuels, organic matter, silicon-containing compounds, sand, glass and glass wool, graphite, oxides. Attacks all forms of plastics, rubber, coatings, and resins except for highly fluorinated polymers such as Teflon and Kel-F.

CHLOROTRIFLUOROETHYLENE (79-38-9) Flammable gas. Forms explosive mixture with air (flash point −18°F/−28°C). Reacts violently with oxidizers. Contact with ethylene may cause explosive polymerization.

CHLOROTRIFLUOROMETHANE (75-72-9) Reacts with aluminum, magnesium, zinc, and their alloys.

CHLOROTRIMETHYLSILANE (75-77-4) Forms explosive mixture with air (flash point −18°F/−28°C). Reacts violently with water. Vigorous reaction occurs with aluminum. Moisture and air contact produces hydrochloric acid.

CHLOROTRIMETHYLSILICANE (75-77-4) Forms explosive mixture with air (flash point −18°F/−28°C). Reacts violently with water. Vigorous reaction occurs with aluminum. Moisture and air contact produces hydrochloric acid.

CHLOROWODOR (Polish) (7647-01-0) Incompatible with acetic anhydride, aliphatic amines, alkanolamines, alkylene oxides, aromatic amines, amides, 2-aminoethanol, ammonia, ammonium hydroxide, calcium phosphide, chlorosulfonic acid, ethylene diamine, ethyleneimine, epichlorohydrin, isocyanates, metal acetylides, oleum, organic anhydrides, perchloric acid, 3-propiolactone, uranium phosphide, sulfuric acid, sodium hydroxide and other bases, strong oxidizers, vinyl acetate, vinylidene fluoride. Attacks most metals and some plastics, rubber, and coatings.

CHLOROZONE (94-75-7) Decomposes in sunlight. Incompatible with strong oxidizers; may cause fire and explosions.

1-CHLORPENTANE (543-59-9) Forms explosive mixture with air (flash point 34°F/1°C). Strong oxidizers may cause fire and explosions.

***o*-CHLORPHENOL (German)** (95-57-0) Forms explosive mixture with air (flash point 147°F/64°C). Contact with strong oxidizers may cause fire and explosions. Attacks aluminum and copper.

CHLORPICRINA (Italian) (76-06-2) Can be self-reactive. Fast heating, shock, alkali metals, or alkaline earth may cause explosions. A strong oxidizer; reacts violently with reducing agents, aniline in presence of heat, alcoholic sodium hydroxide, combustible substances, sodium methoxide, propargyl bromide. Liquid attacks some plastics, rubber, and coatings.

CHLORPIKRIN (German) (76-06-2) Can be self-reactive. Fast heating, shock, alkali metals, or alkaline earth may cause explosions. A strong oxidizer; reacts violently with reducing agents, aniline in presence of heat, alcoholic sodium hydroxide, combustible substances, sodium methoxide, propargyl bromide. Liquid attacks some plastics, rubber, and coatings.

3-CHLORPROPEN (German) (107-05-1) Forms explosive mixture with air (flash point −25°F/−32°C). Violent polymerization and explosion may occur from heat, light, or contact with acid catalysts, ferric chloride, aluminum chloride, Lewis acids, or Ziegler catalyst (e.g., titanium tetrachloride + triethylaluminum monochloride). Incompatible with strong acids, amines, aluminum chloride, boron trifluoride, chlorosulfonic acid, ethylene diamine, ethyleneimine, ferric chloride, oleum, oxidizers, sodium hydroxide. Slow decomposition occurs with moisture; attacks some coatings, plastics, and rubber. Corrosive to steel. May accumulate static electrical charges, and may cause ignition of its vapors.

CHLORPYRIFOS (2921-88-) Incompatible with strong acids. Hydrolysis occurs with acid or alkaline solutions.

CHLORSAURE (German) (7775-09-9) A powerful oxidizer; reacts violently with reducing agents and combustible matter. Explosions may be caused by contact with ammonia salts, carbon, oils, metal sulfides, nitrobenzene, powdered metals, sugar. Contact with strong acids produces carbon dioxide. Forms shock-sensitive mixtures with some organic materials. Solution (50%) decomposes at 300°F/149°C, liberating oxygen.

CHLORTHEPIN (115-29-7) Hydrolyzed by acids and alkalis. Corrosive to iron.

alpha-CHLORTOLUOL (German) (100-44-7) Forms explosive mixture with air (flash point 153°F/67°C). Water contact produces hydrogen chloride fume. Strong oxidizers may cause fire and explosions. Violent polymerization may be caused by contact with copper, aluminum, iron, zinc, magnesium and tin. May accumulate static electrical charges, and may cause ignition of its vapors. Attacks some plastics and rubber.

CHLORURE d'ALUMINUM (French) (7446-70-0) Contact with air or water produces hydrochloric acid. Water, alcohol, alkenes cause polymerization. Incompatible with nitrobenzene, organic material, and bases. Attacks metal in presence of moisture.

CHLORURE de BENZYLE (French) (100-44-7) Forms explosive mixture with air (flash point 153°F/67°C). Water contact produces hydrogen chloride fume. Strong oxidizers may cause fire and explosions. Violent polymerization may be caused by contact with copper, aluminum, iron, zinc, magnesium, and tin. May accumulate static electrical charges, and may cause ignition of its vapors. Attacks some plastics and rubber.

CHLORURE de BENZYLIDENE (French) (98-87-3) Forms explosive mixture with air (flash point 153°F/67°C). Reacts with acids, bases, strong oxidizers. Forms acid fumes with air. Attacks plastics and coatings.

CHLORURE de BUTYLE (French) (109-69-3) Forms explosive mixture with air (flash point 15°F/−9°C). May accumulate static electrical charges, and may cause ignition of its vapors. Water contact slowly produces hydrochloric acid. Incompatible with strong oxidizers, alkaline earth and alkali metals, finely divided metal. Attacks metals in presence of moisture. Attacks some plastics, rubber, and coatings. May accumulate static electrical charges, and may cause ignition of its vapors.

CHLORURE de CHLORACETYLE (French) (79-04-9) Forms corrosive vapors with air. Reacts strongly with water, producing hydrochloric acid. Incompatible with sulfuric acid, caustics, alkalis, alcohols, aliphatic amines, alkanolamines, ammonia, isocyanates, alkylene oxides, epichlorohydrin.

CHLORURE de CYANOGENE (French) (506-77-4) Violent polymerization can be caused by chlorine. Water, steam, alcohols, acids, acid salts, amines, strong alkalis, olefins, strong oxidizers may cause fire and explosion. In crude form, chemical trimerizes violently if catalyzed by traces of hydrogen chloride or ammonium chloride. Corrodes brass, copper, bronze.

CHLORURE d'ETHYLE (French) (75-00-3) Flammable gas. Forms explosive mixture with air. Contact with moisture produces hydrochloric acid. May accumulate static electrical charges, and may cause ignition of its vapors. Forms explosive mixture with air (flash point −58°F/−50°C). Contact with chemically active metals may cause fire and explosions. Attacks some plastics and rubber.

CHLORURE d'ETHYLIDENE (French) (75-34-3) Incompatible with strong oxidizers, strong caustics. Attacks plastics and rubber.

CHLORURE de METHYLE (French) (74-87-3) Incompatible with chemically active metals. Attacks plastics, rubber, and coatings.

CHLORURE de VINYLE (French) (75-01-4) Forms explosive gas mixture with air. Atmospheric oxygen and various contaminants or strong oxidizers causes formation of peroxides, which can initiate a violent polymerization action. Also able to polymerize when heated and exposed, long term, to light. Reacts violently with strong oxidizers or oxides of nitrogen. Contact with copper or other acetylide-forming metals produces explosive compounds. Attacks iron and steel in the presence of moisture.

CHLORURE de VINYLIDENE (French) (75-35-4) Incompatible with nitric acid.

CHLORURE de ZINC (French) (7646-85-7) Aqueous solution is strongly acidic; reacts with strong bases, and attacks metals in the presence of moisture.

CHLORURE MERCURIQUE (French) (7487-94-7) Incompatible with light metals (aluminum, magnesium, beryllium, etc.), sodium, potassium.

CHLORWASSERSTOFF (German) (7647-01-0) Incompatible with acetic anhydride, aliphatic amines, alkanolamines, alkylene oxides, aromatic amines, amides, 2-aminoethanol, ammonia, ammonium hydroxide, calcium phosphide, chlorosulfonic acid, ethylene diamine, ethyleneimine, epichlorohydrin, isocyanates, metal acetylides, oleum, organic anhydrides, perchloric acid, 3-propiolactone, uranium phosphide, sulfuric acid, sodium hydroxide and other bases, strong oxidizers, vinyl acetate, vinylidene fluoride. Attacks most metals and some plastics, rubber, and coatings.

CHLORYL (75-00-3) Flammable gas. Forms explosive mixture with air. Contact with moisture produces hydrochloric acid. May accumulate static electrical charges, and may cause ignition of its vapors.

Forms explosive mixture with air (flash point −58°F/−50°C). Contact with chemically active metals may cause fire and explosions. Attacks some plastics and rubber.

CHLORYL ANESTHETIC (75-00-3) Flammable gas. Forms explosive mixture with air. Contact with moisture produces hydrochloric acid. May accumulate static electrical charges, and may cause ignition of its vapors. Forms explosive mixture with air (flash point −58°F/−50°C). Contact with chemically active metals may cause fire and explosions. Attacks some plastics and rubber.

CHLORYLEA (79-01-6) Contact with caustics produces a toxic and flammable gas. Reacts violently with chemically active metals. Contact with aluminum may produce a violent, self-accelerating polymerization reaction. Incompatible with nonoxidizing mineral acids, strong acids, organic acids, organic anhydrides, isocyanates, alkylene oxides, aldehydes, alcohols, glycols, phenols, cresols, caprolactam solution, epichlorohydrin, nitrogen tetroxide, metal powders, oxygen. May accumulate static electrical charges, and may cause ignition of its vapors.

CHOLAXINE (50-70-4) Incompatible with strong acids, bases, aliphatic amines, isocyanates, strong oxidizers.

CHORYLEN (79-01-6) Contact with caustics produces a toxic and flammable gas. Reacts violently with chemically active metals. Contact with aluminum may produce a violent, self-accelerating polymerization reaction. Incompatible with nonoxidizing mineral acids, strong acids, organic acids, organic anhydrides, isocyanates, alkylene oxides, aldehydes, alcohols, glycols, phenols, cresols, caprolactam solution, epichlorohydrin, nitrogen tetroxide, metal powders, oxygen. May accumulate static electrical charges, and may cause ignition of its vapors.

CHP (80-15-9) A strong oxidizer; reacts violently with reducers, combustibles. Explosive decomposition may occur above 120°F/50°C. Forms explosive mixture with air (flash point 175°F/79°C). Incompatible with acids, bases, amines, metallic salts of cobalt, copper, lead. May accumulate static electrical charges, and may cause ignition of its vapors.

CHROMAR (106-42-3) Forms explosive mixture with air (flash point 81°F/27.2°C). Strong oxidizers may cause fire and explosions. Attacks some plastics, rubber, and coatings. May accumulate static electrical charges, and may cause ignition of its vapors.

CHROMATE OF POTASSIUM (7789-00-6) A powerful oxidizer. Contact with combustibles, reducing agents, organic material or finely divided metals may cause fire and explosions.

CHROMATE OF SODA (7777-11-3) Strong oxidizer; reacts violently with reducing agents, combustibles, strong acids, organic materials.

CHROMIC ACETATE (1066-30-4) Contact with strong oxidizers may cause fire and explosions.

CHROMIC(III) ACETATE (1066-30-4) Contact with strong oxidizers may cause fire and explosions.

CHROMIC ACETATE(III) (1066-30-4) Contact with strong oxidizers may cause fire and explosions.

CHROMIC ACID (7738-94-5); **solution** (1308-14-1). A powerful oxidizer. Reducing agents, combustibles, organic, or other oxidizable materials (e.g., wood, sulfur, aluminum, plastics), may cause fire and explosions. Attacks most forms of metals, cloth, leather, plastics, rubber, and coatings, and may cause spontaneous ignition.

CHROMIC ACID (1333-82-0) A strong oxidizer. Reacts with acetic acid, acetic anhydride, acetone, anthracene, chromous sulfide, diethyl ether, dimethyl formamide, ethanol, hydrogen sulfide, methanol, naphthalene, camphor, glycerol, potassium ferricyanide, pyridine, turpentine, combustibles, organics. Aqueous solution is strongly acidic. Attacks metals in presence of moisture.

CHROMIC(IV) ACID (1333-82-0) A strong oxidizer. Reacts with acetic acid, acetic anhydride, acetone, anthracene, chromous sulfide, diethyl ether, dimethyl formamide, ethanol, hydrogen sulfide, methanol, naphthalene, camphor, glycerol, potassium ferricyanide, pyridine, turpentine, combustibles, organics. Aqueous solution is strongly acidic. Attacks metals in presence of moisture.

CHROMIC(VI) ACID (7738-94-5); **solution** (1308-14-1) A powerful oxidizer. Reducing agents, combustibles, organic, or other oxidizable materials (e.g., wood, sulfur, aluminum, plastics), may cause fire and explosions. Attacks most forms of metals, cloth, leather, plastics, rubber, and coatings and may cause spontaneous ignition.

CHROMIC ACID, CALCIUM SALT (1:1) (13765-19-0) A strong oxidizer. Compatible with boron (with violent reaction), combustible material.

CHROMIC ACID, DI-*tert*-BUTYL ESTER (1189-85-1) Incompatible with acids, alcohols, combustible or easily oxidized materials, reducing agents, moisture.

CHROMIC ACID, DILITHIUM SALT (14307-35-8) Aqueous solution is caustic. An oxidizer; reacts strongly with reducing agents, combustibles, organic materials, acids.

CHROMIC ACID, DIPOTASSIUM SALT (7778-50-9) Powdered combustibles will ignite. Strong acids form toxic vapors.

CHROMIC ACID, DISODIUM SALT (10588-01-9) Aqueous solution is corrosive. A strong oxidizer; reacts violently with reducing agents, acids, acetic anhydride, combustibles. Attacks copper, zinc, tin, brass, bronze.

CHROMIC ACID, STRONTIUM SALT (1:1) (7789-06-2) Incompatible with water, acids, bases.

CHROMIC ANHYDRIDE (1333-82-0) A strong oxidizer. Incompatible with acetic acid, acetic anhydride, acetone, anthracene, chromous sulfide, diethyl ether, dimethyl formamide, ethanol, hydrogen sulfide, methanol, naphthalene, camphor, glycerol, potassium ferricyanide, pyridine, turpentine, combustibles, organics. Aqueous solution is strongly acidic. Attacks metals in presence of moisture.

CHROMIC OXIDE (1333-82-0) A strong oxidizer. Incompatible with acetic acid, acetic anhydride, acetone, anthracene, chromous sulfide, diethyl ether, dimethyl formamide, ethanol, hydrogen sulfide, methanol, naphthalene, camphor, glycerol, potassium ferricyanide, pyridine, turpentine, combustibles, organics. Aqueous solution is strongly acidic. Attacks metals in presence of moisture.

CHROMIC OXYCHLORIDE (14977-61-8) A powerful oxidizer. Water contact produces hydrochloric and chromic acids and chlorine gas. Reacts violently with reducing agents, combustibles, ammonia, halides, phosphorus, sodium azide, elemental sulfur, urea.

CHROMIC TRIOXIDE (1333-82-0) A strong oxidizer. Reacts with acetic acid, acetic anhydride, acetone, anthracene, chromous sulfide, diethyl ether, dimethyl formamide, ethanol, hydrogen sulfide, methanol, naphthalene, camphor, glycerol, potassium ferricyanide, pyridine, turpentine, combustibles, organics. Aqueous solution is strongly acidic. Attacks metals in presence of moisture.

CHROMIUM ACETATE (1066-30-4) Contact with strong oxidizers may cause fire and explosions.

CHROMIUM(III) ACETATE (1066-30-4) Contact with strong oxidizers may cause fire and explosions.

CHROMIUM ANHYDRIDE (7738-94-5); **solution** (1308-14-1). A powerful oxidizer. Reducing agents, combustibles, organic, or other oxidizable materials (e.g., wood, sulfur, aluminum, plastics), may cause fire and explosions. Attacks most forms of metals, cloth, leather, plastics, rubber, and coatings, and may cause spontaneous ignition.

CHROMIUM CHLORIDE OXIDE (14977-61-8) A powerful oxidizer. Water contact produces hydrochloric and chromic acids and chlorine gas. Reacts violently with reducing agents, combustibles, ammonia, halides, phosphorus, sodium azide, elemental sulfur, urea.

CHROMIUM DICHLORIDE (10049-05-5) Very hygroscopic. Water solution produces flammable hydrogen gas.

CHROMIUM DICHLORIDE DIOXIDE (14977-61-8) A powerful oxidizer. Water contact produces hydrochloric and chromic acids and chlorine gas. Reacts violently with reducing agents, combustibles, ammonia, halides, phosphorus, sodium azide, elemental sulfur, urea.

CHROMIUM DIOXIDE DICHLORIDE (14977-61-8) A powerful oxidizer. Water contact produces hydrochloric and chromic acids and chlorine gas. Reacts violently with reducing agents, combustibles, ammonia, halides, phosphorus, sodium azide, elemental sulfur, urea.

CHROMIUM DIOXYCHLORIDE (14977-61-8) A powerful oxidizer. Water contact produces hydrochloric and chromic acids and chlorine gas. Reacts violently with reducing agents, combustibles, ammonia, halides, phosphorus, sodium azide, elemental sulfur, urea.

CHROMIUM DISODIUM OXIDE (7777-11-3) Strong oxidizer; reacts violently with reducing agents, combustibles, strong acids, organic materials.

CHROMIUM LITHIUM OXIDE (14307-35-8) Aqueous solution is caustic. An oxidizer; reacts strongly with reducing agents, combustibles, organic materials, acids.

CHROMIUM OXIDE (1333-82-0) A strong oxidizer. Reacts with acetic acid, acetic anhydride, acetone, anthracene, chromous sulfide, diethyl ether, dimethyl formamide, ethanol, hydrogen sulfide, methanol, naphthalene, camphor, glycerol, potassium ferricyanide, pyridine, turpentine, combustibles, organics. Aqueous solution is strongly acidic. Attacks metals in presence of moisture.

CHROMIUM(IV) OXIDE (1333-82-0) A strong oxidizer. Reacts with acetic acid, acetic anhydride, acetone, anthracene, chromous sulfide, diethyl ether, dimethyl formamide, ethanol, hydrogen sulfide, methanol, naphthalene, camphor, glycerol, potassium ferricyanide, pyridine, turpentine, combustibles, organics. Aqueous solution is strongly acidic. Attacks metals in presence of moisture.

CHROMIUM(IV) OXIDE (1:3) (1333-82-0) A strong oxidizer. Reacts with acetic acid, acetic anhydride, acetone, anthracene, chromous sulfide, diethyl ether, dimethyl formamide, ethanol, hydrogen sulfide, methanol, naphthalene, camphor, glycerol, potassium ferricyanide, pyridine, turpentine, combustibles, organics. Aqueous solution is strongly acidic. Attacks metals in presence of moisture.

CHROMIUM OXYCHLORIDE (14977-61-8) A powerful oxidizer. Water contact produces hydrochloric and chromic acids and chlorine gas. Reacts violently with reducing agents, combustibles, ammonia, halides, phosphorus, sodium azide, elemental sulfur, urea.

CHROMIUM SODIUM OXIDE (10588-01-9) Aqueous solution is corrosive. A strong oxidizer; reacts violently with reducing agents, acids, acetic anhydride, combustibles. Attacks copper, zinc, tin, brass, bronze.

CHROMIUM SODIUM OXIDE (7777-11-3) Strong oxidizer; reacts violently with reducing agents, combustibles, strong acids, organic materials.

CHROMIUM TRIACETATE (1066-30-4) Contact with strong oxidizers may cause fire and explosions.

CHROMIUM TRIOXIDE (7738-94-5); **solution** (1308-14-1). A powerful oxidizer. Reducing agents, combustibles, organic, or other oxidizable materials (e.g., wood, sulfur, aluminum, plastics), may cause fire and explosions. Attacks most forms of metals, cloth, leather, plastics, rubber, and coatings, and may cause spontaneous ignition.

CHROMIUM TRIOXIDE or CHROMIUM TRIOXIDE, ANHYDROUS (1333-82-0) A strong oxidizer. Reacts with acetic acid, acetic anhydride, acetone, anthracene, chromous sulfide, diethyl ether, dimethyl formamide, ethanol, hydrogen sulfide, methanol, naphthalene, camphor, glycerol, potassium ferricyanide, pyridine, turpentine, combustibles, organics. Aqueous solution is strongly acidic. Attacks metals in presence of moisture.

CHROMIUM(6+) TRIOXIDE (1333-82-0) A strong oxidizer. Reacts with acetic acid, acetic anhydride, acetone, anthracene, chromous sulfide, diethyl ether, dimethyl formamide, ethanol, hydrogen sulfide, methanol, naphthalene, camphor, glycerol, potassium ferricyanide, pyridine, turpentine, combustibles, organics. Aqueous solution is strongly acidic. Attacks metals in presence of moisture.

CHROMOCHROMIC ANHYDRIDE (14977-61-8) A powerful oxidizer. Water contact produces hydrochloric and chromic acids and chlorine gas. Reacts violently with reducing agents, combustibles, ammonia, halides, phosphorus, sodium azide, elemental sulfur, urea.

CHROMO (TRIOSSIDO di) (Italian) (1333-82-0) A strong oxidizer. Reacts with acetic acid, acetic anhydride, acetone, anthracene, chromous sulfide, diethyl ether, dimethyl formamide, ethanol, hydrogen sulfide, methanol, naphthalene, camphor, glycerol, potassium ferricyanide, pyridine, turpentine, combustibles, organics. Aqueous solution is strongly acidic. Attacks metals in presence of moisture.

CHROMOUS CHLORIDE (10049-05-5) Very hygroscopic. Water solution produces flammable hydrogen gas.

CHROMSAEUREANHYDRID (German) (1333-82-0) A strong oxidizer. Reacts with acetic acid, acetic anhydride, acetone, anthracene, chromous sulfide, diethyl ether, dimethyl formamide, ethanol, hydrogen sulfide, methanol, naphthalene, camphor, glycerol, potassium ferricyanide, pyridine, turpentine, combustibles, organics. Aqueous solution is strongly acidic. Attacks metals in presence of moisture.

CHROMTRIOXID (German) (1333-82-0) A strong oxidizer. Reacts with acetic acid, acetic anhydride, acetone, anthracene, chromous sulfide, diethyl ether, dimethyl formamide, ethanol, hydrogen sulfide, methanol, naphthalene, camphor, glycerol, potassium ferricyanide, pyridine, turpentine, combustibles, organics. Aqueous solution is strongly acidic. Attacks metals in presence of moisture.

CHROMYL CHLORIDE (14977-61-8) A powerful oxidizer. Water contact produces hydrochloric and chromic acids and chlorine gas. Reacts violently with reducing agents, combustibles, ammonia, halides, phosphorus, sodium azide, elemental sulfur, urea.

CHROOMTRIOXYDE (Dutch) (1333-82-0) A strong oxidizer. Reacts with acetic acid, acetic anhydride, acetone, anthracene, chromous sulfide, diethyl ether, dimethyl formamide, ethanol, hydrogen sulfide, methanol, naphthalene, camphor, glycerol, potassium ferricyanide, pyridine, turpentine, combustibles, organics. Aqueous solution is strongly acidic. Attacks metals in presence of moisture.

CHROOMZUURANHYDRIDE (Dutch) (1333-82-0) A strong oxidizer. Reacts with acetic acid, acetic anhydride, acetone, anthracene, chromous sulfide, diethyl ether, dimethyl formamide, ethanol, hydrogen sulfide, methanol, naphthalene, camphor, glycerol, potassium ferricyanide, pyridine, turpentine, combustibles, organics. Aqueous solution is strongly acidic. Attacks metals in presence of moisture.

C.I. 6000 (62-53-3) Forms explosive mixture with air (flash point 158°F/70°C). Unless inhibited (usually by methanol), readily able to polymerize. Fires and explosions may result from contact with halogens, strong acids, oxidizers, organic anhydrides, acetic anhydride, isocyanates, aldehydes, sodium peroxide. Reacts with alkali metals and alkaline earth metals. Attacks some plastics, rubber, and coatings, copper and copper alloys.

C.I. 6035 (95-80-7) Incompatible with oxidizers, acids, organic anhydrides, isocyanates, aldehydes. Attacks aluminum, brass, bronze, copper, zinc.

C.I. 6500 (120-80-9) Strong oxidizers may cause fire and explosion.

C.I. 6505 (108-46-3) Reacts violently with strong oxidizers, nitric acid. Incompatible with acetanilide, albumin, alkalies, antipyrine, camphor, ferric salts, menthol, spirit nitrous ether. Absorbs moisture from air (hygroscopic). May accumulate static electrical charges, and may cause ignition of its vapors.

C.I. 6515 (87-66-1) Incompatible with strong oxidizers, caustics, ammonia, amines, isocyanates, alkylene oxides, epichlorohydrin.

C.I. 7056 (10025-91-9) Water contact produces corrosive solution. Air contact produces corrosive vapor.

C.I. 7099 (513-77-9) Incompatible with acids, aluminum, fluorine, magnesium, silicon, oxidizers.

C.I. 7185 (543-90-8) Incompatible with strong oxidizers, elemental sulfur, selenium, tellurium, strong acids, nitrates

C.I. 7223 (13765-19-0) A strong oxidizer. Incompatible with boron (with violent reaction), combustible material.

C.I. 7577 (1317-36-8) Reacts violently with aluminum powder, hydrogen peroxide, hydrogen trisulfide, lithium carbide, perchloric acid, strong oxidizers.

C.I. 7630 (7446-14-2) Reacts with aluminum, magnesium, potassium (violently).

C.I. 7640 (1314-87-0) Reacts violently with iodine chloride, hydrogen peroxide.

C.I. 7755 (7722-64-7) A strong oxidizer. Produces heat- and shock-sensitive compound with sulfuric acid. Reacts violently with combustibles or reducing agents. Reacts explosively with acetic acid, acetic anhydride, anhydrous ammonia, hydrogen peroxide, glycerol, hydroxylamine, organic matter, powdered sulfur, etc. Incompatible with nitric acid, producing toxic chlorine fumes. A dangerous fire and explosion hazard; isolate from all other materials.

C.I. 7938 (1314-62-1) Aqueous solution is acidic. Reacts with lithium at high temperature.

C.I. 7940 (27774-13-6) Incompatible with strong oxidizers such as chlorine or fluorine.

C.I. 10355 (122-39-4) Incompatible with strong acids, organic anhydrides, isocyanates, ketones, oxidizers, hexachloromelamine, trichloromelamine.

C.I. 37025 (88-74-4) *ortho-;* (99-09-2) *meta.* Incompatible with strong acids, combustibles, organics, moisture.

C.I. 37107 (106-49-0) Forms explosive mixture with air (flash point 188°F/87°C). Incompatible with strong oxidizers, strong acids. Attacks some plastics, rubber, and coatings.

C.I. 37225 (92-87-5) Oxidizes on exposure to light and air. Reacts violently with strong oxidizers. Red fuming nitric acid may cause fire.

C.I. 37230 (95-53-4) Forms explosive mixture with air (flash point 185°F/85°C). Incompatible with strong acids, mineral acids, organic anhydrides, isocyanates, aldehydes, oxidizers. Attacks some plastics, rubber, and coatings.

C.I. 37270 (91-59-8) Reacts with strong oxidizers, strong acids. Oxidizes slowly in air and light. Incompatible with strong acids, nitrates, oxidizers.

C.I. 76060 (106-50-3) Incompatible with acids, organic anhydrides, isocyanates, aldehydes; strong oxidizers may cause fire and explosions. Heat and light contribute to instability.

C.I. 77050 (7440-36-0) Reacts violently with strong oxidizers, strong acids. Forms explosive mixtures with chloric and perchloric acids. Oxidizers and acids, especially halogenated acids, can produce toxic stibine gas (antimony hydride).

C.I. 77120 (7727-43-7) Explosions may result from contact with aluminum in the presence of heat. Incompatible with potassium, phosphorus.

C.I. 77180 (7440-43-9) Air exposure, especially of powdered form, may cause self-ignition. Reacts violently with strong oxidizers, hyrozoic acid, tellurium. Acid contact produces hydrogen gas. May react with selenium, elemental sulfur, zinc.

C.I. 77491 (1309-37-1) Contact with hydrogen peroxide, ethylene oxide, calcium hypochlorite will cause explosion. Reacts violently with powdered aluminum, hydrazine, hydrogen trisulfide.

C.I. 77775 (7440-02-0) Powder may self-ignite in air. Contact with strong acids produces flammable and explosive hydrogen gas. Sulfur contact evolves heat. Reacts violently with strong oxidizers, nitric acid, hydrazine, and many other chemicals.

CIANURINA (592-04-1) Contact with acidic material produces hydrogen cyanide gas. An impact-sensitive explosive. Reacts strongly with fluorine, hydrogen cyanide, magnesium.

CIANURO di SODIO (Italian) (143-33-9) Incompatible with acids, organic anhydrides, isocyanates, alkylene oxides, epichlorohydrin, aldehydes, alcohols, glycols, phenols, cresols, caprolactam solution, strong oxidizers, nitrates, nitrites.

CIANURO di VINILE (Italian) (107-13-1) Forms explosive mixture with air (flash point 32°F/0°C o.c.). Forms explosive peroxides; heat, light, caustics, silver nitrate, and peroxides can cause polymerization. Incompatible with strong acids, strong oxidizers, amines, 2-amino-ethanol, bromine, chlorosulfonic acid, ethylene diamine, nitric acid, oleum, potassium hydroxide, sodium hydroxide, sulfuric acid. Attacks copper and copper alloys; attacks aluminum in high concentrations. May accumulate static electrical charges, and may cause ignition of its vapors.

C.I. AZOIC COUPLING COMPONENT-107 (106-49-0) Forms explosive mixture with air (flash point 188°F/87°C). Incompatible with strong oxidizers, strong acids. Attacks some plastics, rubber, and coatings.

C.I. AZOIC DIAZO (92-87-5) Oxidizes on exposure to light and air. Reacts violently with strong oxidizers. Red fuming nitric acid may cause fire.

C.I. AZOIC DIAZO COMPONENT 112 (92-87-5) Oxidizes on exposure to light and air. Reacts violently with strong oxidizers. Red fuming nitric acid may cause fire.

C.I. AZOIC DIAZO COMPONENT 113 (95-53-4) Forms explosive mixture with air (flash point 185°F/85°C). Incompatible with strong acids, mineral acids, organic anhydrides, isocyanates, aldehydes, oxidizers. Attacks some plastics, rubber, and coatings.

C.I. AZOIC DIAZO COMPONENT 114 (134-32-7) Oxidizes in air. Incompatible with nitrous acid, oxidizers, nitrates, organic anhydrides, isocyanates, aldehydes.

CICHLORODIOXO CHROMIUM (14977-61-8) A powerful oxidizer. Water contact produces hydrochloric and chromic acids and chlorine gas. Reacts violently with reducing agents, combustibles, ammonia, halides, phosphorus, sodium azide, elemental sulfur, urea.

CICLOESANO (Italian) (110-82-7) Forms explosive mixture with air (flash point −4°F/−20°C). Incompatible with oxidizers, nitrogen dioxide.

CICLOESANOLO (Italian) (108-93-0) Forms explosive mixture with air (flash point 154°F/68°C). Incompatible with strong acids, caustics, aliphatic amines, isocyanates. Attacks some plastics, rubber, or coatings.

CICLOESANONE (Italian) (108-94-1) Forms explosive mixture with air (flash point 111°F/44°C). May accumulate static electrical charges, and may cause ignition of its vapors. Attacks red metals and lead. Incompatible with amines, sulfuric acid, nitric acid, oxidizers, strong acids, aliphatic amines. Dissolves plastics, resins, and rubber.

CICN (506-77-4) Violent polymerization can be caused by chlorine. Water, steam, alcohols, acids, acid salts, amines, strong alkalis, olefins, strong oxidizers may cause fire and explosion. In crude form, chemical trimerizes violently if catalyzed by traces of hydrogen chloride or ammonium chloride. Corrodes brass, copper, bronze.

C.I. DEVELOPER 4 (108-46-3) Reacts violently with strong oxidizers, nitric acid. Incompatible with acetanilide, albumin, alkalies, antipyrine, camphor, ferric salts, menthol, spirit nitrous ether. Absorbs moisture from air (hygroscopic). May accumulate static electrical charges, and may cause ignition of its vapors.

C.I. DEVELOPER 13 (106-50-3) Incompatible with acids, organic anhydrides, isocyanates, aldehydes; strong oxidizers may cause fire and explosions. Heat and light contribute to instability.

CIDEX (111-30-8) Water contact forms a polymer solution. Incompatible with strong acids, caustics, ammonia, amines, strong oxidizers.

CIMEXAN (121-75-5) Incompatible with strong oxidizers, magnesium, alkaline pesticides. Attacks metals, some plastics, rubber, and coatings.

CINENE (138-86-3) Forms explosive mixture with air (flash point 115°F/46°C). Strong oxidizers may cause fire and explosions.

CINERIN I (8003-34-7) Strong oxidizers may cause fire and explosions.

CINERIN II (8003-34-7) Strong oxidizers may cause fire and explosions.

CINNAMENE (100-42-5) If inhibitor (often *tert*-butylcatechol) is not present in adequate concentrations, polymerization may occur and explode containers. Temperatures above 150°F/66°C speed up polymerization. Incompatible with oxidizers, acids, rust, catalysts for vinyl polymerization, such as peroxides, strong acids, and aluminum chloride. Corrodes copper and copper alloys, and dissolves rubber. Attacks some plastics, rubber, and coatings. May accumulate static electrical charges, and may cause ignition of its vapors.

CINNAMENOL (100-42-5) If inhibitor (often *tert*-butylcatechol) is not present in adequate concentrations, polymerization may occur and explode containers. Temperatures above 150°F/66°C speed up polymerization. Incompatible with oxidizers, acids, rust, catalysts for vinyl polymerization, such as peroxides, strong acids, and aluminum chloride. Corrodes copper and copper alloys, and dissolves rubber. Attacks some plastics, rubber, and coatings. May accumulate static electrical charges, and may cause ignition of its vapors.

CINNAMIC ALDEHYDE (104-55-2) A strong reducing agent; reacts violently with oxidizers, acids, bases, alkaline earth and alkali metals; high temperatures increase reaction. Attacks some metals, plastics, and coatings.

CINNAMOL (100-42-5) If inhibitor (often *tert*-butylcatechol) is not present in adequate concentrations, polymerization may occur and explode containers. Temperatures above 150°F/66°C speed up polymerization. Incompatible with oxidizers, acids, rust, catalysts for vinyl polymerization, such as peroxides, strong acids, and aluminum chloride. Corrodes copper and copper alloys, and dissolves rubber. Attacks some plastics, rubber, and coatings. May accumulate static electrical charges, and may cause ignition of its vapors.

CINNAMYL ALDEHYDE (104-55-2) A strong reducing agent; reacts violently with oxidizers, acids, bases, alkaline earth and alkali metals; high temperatures increase reaction. Attacks some metals, plastics, and coatings.

C.I. OXIDATION BASE (95-80-7) Incompatible with oxidizers, acids, organic anhydrides, isocyanates, aldehydes. Attacks aluminum, brass, bronze, copper, zinc.

C.I. OXIDATION BASE 10 (106-50-3) Incompatible with acids, organic anhydrides, isocyanates, aldehydes; strong oxidizers may cause fire and explosions. Heat and light contribute to instability.

C.I. OXIDATION BASE 26 (120-80-9) Strong oxidizers may cause fire and explosion.

C.I. OXIDATION BASE 31 (108-46-3) Reacts violently with strong oxidizers, nitric acid. Incompatible with acetanilide, albumin, alkalies, antipyrine, camphor, ferric salts, menthol, spirit nitrous ether. Absorbs moisture from air (hygroscopic). May accumulate static electrical charges, and may cause ignition of its vapors.

C.I. OXIDATION BASE 32 (87-66-1) Incompatible with strong oxidizers, caustics, ammonia, amines, isocyanates, alkylene oxides, epichlorohydrin.

C.I. PIGMENT BLACK 7 (1333-86-4) Dust can form an explosive mixture in air. Oxidizers may cause fire and explosions.

C.I. PIGMENT RED 101 MORTUUM (1309-37-1) Contact with hydrogen peroxide, ethylene oxide, calcium hypochlorite will cause explosion. Reacts violently with powdered aluminum, hydrazine, hydrogen trisulfide.

C.I. PIGMENT WHITE 3 (7446-14-2) Reacts with aluminum, magnesium, potassium (violent).

C.I. PIGMENT WHITE 10 (513-77-9) Incompatible with acids, aluminum, fluorine, magnesium, silicon, oxidizers.

C.I. PIGMENT WHITE 21 (7727-43-7) Explosions may result from contact with aluminum in the presence of heat. Incompatible with potassium, phosphorus.

C.I. PIGMENT YELLOW 32 (7789-06-2) Incompatible with water, acids, bases.

C.I. PIGMENT YELLOW 33 (13765-19-0) A strong oxidizer. Incompatible with boron (with violent reaction), combustible material.

C.I. PIGMENT YELLOW 46 (1317-36-8) Reacts violently with aluminum powder, hydrogen peroxide, hydrogen trisulfide, lithium carbide, perchloric acid, strong oxidizers.

CIRCOSOLV (79-01-6) Contact with caustics produces a toxic and flammable gas. Reacts violently with chemically active metals. Contact with aluminum may produce a violent, self-accelerating polymerization reaction. Incompatible with nonoxidizing mineral acids, strong acids, organic acids, organic anhydrides, isocyanates, alkylene oxides, aldehydes, alcohols, glycols, phenols, cresols, caprolactam solution, epichlorohydrin, nitrogen tetroxide, metal powders, oxygen. May accumulate static electrical charges, and may cause ignition of its vapors.

cis-1,4-DICHLORO-2-BUTENE (764-41-0) Forms explosive mixture with air (flash point 126°F/52°C). Water contact produces hydrochloric acid. Incompatible with strong oxidizers, bases. Corrodes metals in the presence of moisture.

CITOBARYUM (7727-43-7) Explosions may result from contact with aluminum in the presence of heat. Incompatible with potassium, phosphorus.

CITOX (50-29-3) Incompatible with salts of iron or aluminum, and bases. Do not store in iron containers.

CITRIC ACID (77-92-9) Incompatible with bases, strong oxidizers. Contact with metal nitrates may be explosive. Attacks aluminum, copper, zinc and their alloys, when wet.

CITRIC ACID, AMMONIUM SALT (3012-65-5) Combustible solid. Reacts with strong oxidizers.

CITRIC ACID, DIAMMONIUM SALT (3012-65-5) Combustible solid. Reacts with strong oxidizers.

CITRONELLA or CITRONELLA OIL (8000-29-1) Reacts vigorously with strong oxidizers.

CLAIRSIT (594-42-3) Incompatible with caustics or amines (cause rapid decomposition), hot iron (produces toxic carbon tetrachloride fumes), hot water produces hydrochloric acid), sulfur, and carbon dioxide. Attacks some plastics, rubber, and coatings.

CLARIFIED OILS (PETROLEUM), CATALYTIC CRACKED (64741-62-4) Incompatible with nitric acid. Strong oxidizers may cause fire and explosions.

CLAUDELITE (1327-53-3) Incompatible with acids, fluorine, fluorides, sodium chlorate.

CLAUDETITE (1327-53-3) Incompatible with acids, fluorine, fluorides, sodium chlorate.

CLESTOL (119-36-8) Forms explosive mixture with air (flash point 205°F/96°C). Incompatible with strong acids, nitrates, oxidizers.

CLOFENOTANE (50-29-3) Incompatible with salts of iron or aluminum, and bases. Do not store in iron containers.

CLOPHEN (generic CAS for PCBs 1336-36-3) Incompatible with strong oxidizers, strong acids.

CLOR CHEM T-590 (8001-35-2) Reacts with oxidizers, with a risk of fire or explosions. Attacks metals in the presence of moisture.

CLORDANO (Italian) (57-74-9) Contact with strong oxidizers may cause fire and explosions. Attacks some plastics, rubber, and coatings.

CLORO (Italian) (7782-50-5) A powerful oxidizer. Reacts violently with combustible materials, reducing agents and many other substances. Forms explosive mixtures with gasoline and petroleum products, turpentine, alcohols, acetylene, carbon disulfide, hydrogen, anhydrous ammonia, finely divided metals, organic compounds, phosphorus. Fire, explosion, and the formation of toxic fumes may result from contact with many substances, including alkyphosphines, aluminum, antimony, arsenic compounds, arsine, bismuth, boron, brass, calcium compounds, carbon, diethyl zinc, fluorine, germanium, hydrocarbons, rubber. Attacks some plastics and coatings. In the presence of moisture, extremely corrosive to iron, steel, copper, bronze, zinc.

CLOROBEN (95-50-1) Forms explosive mixture with air (flash point 151°F/66°C). Incompatible with strong oxidizers, hot aluminum or aluminum alloy. Attacks some plastics, rubber, and coatings.

CLOROETANO (Italian) (75-00-3) Flammable gas. Forms explosive mixture with air. Contact with moisture produces hydrochloric acid. May accumulate static electrical charges, and may cause ignition of its vapors. Forms explosive mixture with air (flash point −58°F/−50°C). Contact with chemically active metals may cause fire and explosions. Attacks some plastics and rubber.

CLOROFORMIO (Italian) (67-66-3) Decomposes in the presence of excess water or high temperatures, producing phosgene and hydrogen chloride. Can become explosive in the presence of strong alkalies and water. Can accumulate static electrical charges. With water contact and high temperatures, it becomes corrosive; attacks iron and other metals. Incompatible with acetone, aluminum, strong oxidizers, potassium, sodium, chemically active metals, strong bases. Attacks plastics and rubber.

CLOROMETANO (Italian) (74-87-3) Incompatible with chemically active metals. Attacks plastics, rubber, and coatings.

CLOROX (7681-52-9) A strong oxidizer and a strong base. Stability decreases with concentration, heat, light, decrease in pH, and contamination with metals. Incompatible with strong acids, reducing agents, combustible substances; all cause violent reaction, fire, and explosions. Contact with amines and ammonia salts produces explosive chloroamines. Corrodes many metals: steels, cast iron, monel, aluminum, brass, iconel.

CLORURO di ETILE (Italian) (75-00-3) Flammable gas. Forms explosive mixture with air. Contact with moisture produces hydrochloric acid. May accumulate static electrical charges, and may cause ignition of its vapors. Forms explosive mixture with air (flash point −58°F/−50°C). Contact with chemically active metals may cause fire and explosions. Attacks some plastics and rubber.

CLORURO di ETILIDENE (Italian) (75-34-3) Incompatible with strong oxidizers, strong caustics. Attacks plastics and rubber.

CLORURO di MERCURIO (Italian) (7487-94-7) Incompatible with light metals (aluminum, magnesium, beryllium, etc.), sodium, potassium.

CLORURO di METILE (Italian) (74-87-3) Incompatible with chemically active metals. Attacks plastics, rubber, and coatings.

CLYSAR (9003-07-0) Incompatible with strong acids, strong oxidizers, chlorine, potassium permanganate.

CMDP (7786-34-7) Contact with strong oxidizers may cause fire and explosions. Attacks some plastics, rubber, and coatings.

CMME (107-30-2) Forms explosive mixture with air (flash point 0°F/−17.8°C). May be able to form unstable and explosive peroxides. Produces hydrochloric acid on contact with moisture. Corrosive to metals in presence of moisture.

CNCI (506-77-4) Violent polymerization can be caused by chlorine. Water, steam, alcohols, acids, acid salts, amines, strong alkalis, olefins, strong oxidizers may cause fire and explosion. In crude form chemical trimerizes violently if catalyzed by traces of hydrogen chloride or ammonium chloride. Corrodes brass, copper, bronze.

CO (630-08-0) Forms extremely explosive mixture with air. Reacts violently with strong oxidizers.

CO-12 (112-53-8) Incompatible with strong acids, caustics, aliphatic amines, isocyanates, oxidizers.

COALITE NTP (25155-23-1) Incompatible with sulfuric acid, nitric acid, nitrates, strong oxidizers.

COAL NAPHTHA (71-43-2) Forms explosive mixture with air (flash point 12°F/−11°C).) May accumulate static electrical charges, and may cause ignition of its vapors. Incompatible with strong oxidizers, nitric acid, oxygen, ozone, perchlorates. Attacks some forms of plastics, coatings, and rubber.

COAL NAPHTHA, PHENYL HYDRIDE (71-43-2) Forms explosive mixture with air (Flash point 12°F/−11°C).) May accumulate static electrical charges, and may cause ignition of its vapors. Incompatible with strong oxidizers, nitric acid, oxygen, ozone, perchlorates. Attacks some forms of plastics, coatings and rubber.

COAL OIL (8002-05-9) Incompatible with nitric acid. Strong oxidizers may cause fire and explosions.

COAL OIL (8008-20-6) Incompatible with nitric acid. Strong oxidizers may cause fire and explosions.

COAL TAR CREOSOTE (65996-93-2) Incompatible with oxidizers, strong acids, caustics, aliphatic amines, isocyanates. Swells and softens rubber.

COAL TAR NAPHTHA (8030-30-6) Forms explosive mixture with air (flash point 100°F/38°C). Incompatible with nitric acid, strong oxidizers. Attacks some plastics, rubber, and coatings.

COAL TAR OIL (8007-45-2) Forms explosive mixture with air (flash point > 60°F/15°C). Incompatible with strong acids, nitrates.

COAL TAR OIL (65996-93-2) Incompatible with oxidizers, strong acids, caustics, aliphatic amines, isocyanates. Swells and softens rubber.

COAL TAR PITCH VOLATILES (50-32-8); (86-74-8); (120-12-7); (129-00-0) Strong oxidizers may cause fires and explosions.

COBALT (7440-48-4) An extreme fire hazard. Pyrophoric; particles can self-ignite in air.

COBALT ACETATE (71-48-7) Contact with strong oxidizers may cause fire and explosions. Cobalt compounds may react with acetylene.

COBALT(2+) ACETATE or COBALT(II) ACETATE (71-48-7) Contact with strong oxidizers may cause fire and explosions. Cobalt compounds may react with acetylene.

COBALT ACETATE TETRAHYDRATE (71-48-7) Contact with strong oxidizers may cause fire and explosions. Cobalt compounds may react with acetylene.

COBALT AMINO SULFONATE (16107-41-3) An oxidizer; reacts with reducing agents and combustibles.

COBALT BROMIDE (95-63-6) Forms explosive mixture with air (flash point 112°F/44°C). Incompatible with nitric acid, strong oxidizers.

COBALT(2+) BROMIDE or COBALT(II) BROMIDE (95-63-6) Forms explosive mixture with air (flash point 112°F/44°C). Incompatible with nitric acid, strong oxidizers.

COBALT CHLORIDE (7646-79-9) Incompatible with oxidizers, metals (especially sodium or potassium), acetylene.

COBALT(2+) CHLORIDE or COBALT(II) CHLORIDE (7646-79-9) Incompatible with oxidizers, metals (especially sodium or potassium), acetylene.

COBALT DIACETATE TETRAHYDRATE (71-48-7) Contact with strong oxidizers may cause fire and explosions. Cobalt compounds may react with acetylene.

COBALT DIBROMIDE (95-63-6) Forms explosive mixture with air (flash point 112°F/44°C). Incompatible with nitric acid, strong oxidizers.

COBALT DIFLUORIDE (10026-17-2) Incompatible with strong oxidizers, ammonia, nitrates, acetylene.

COBALT DIFORMATE (544-18-3) Incompatible with ammonia, acetylene, nitrates, strong oxidizers.

COBALT FLUORIDE (10026-17-2) Incompatible with strong oxidizers, ammonia, nitrates, acetylene.

COBALT(2+) FLUORIDE or COBALT(II) FLUORIDE (10026-17-2) Incompatible with strong oxidizers, ammonia, nitrates, acetylene.

COBALT FORMATE (544-18-3) Incompatible with ammonia, acetylene, nitrates, strong oxidizers.

COBALT MURIATE (7646-79-9) Incompatible with oxidizers, metals (especially sodium or potassium), acetylene.

COBALT NITRATE (10141-05-6) A strong oxidizer; reacts with reducing agents, combustibles.

COBALT(2+) NITRATE or COBALT(II) NITRATE (10141-05-6) A strong oxidizer; reacts with reducing agents, combustibles.

COBALTOUS ACETATE TETRAHYDRATE (71-48-7) Contact with strong oxidizers may cause fire and explosions. Cobalt compounds may react with acetylene.

COBALTOUS BROMIDE (95-63-6) Forms explosive mixture with air (flash point 112°F/44°C). Incompatible with nitric acid, strong oxidizers.

COBALTOUS CHLORIDE (7646-79-9) Incompatible with oxidizers, metals (especially sodium or potassium), acetylene.

COBALTOUS CHLORIDE DIHYDRATE (7646-79-9) Incompatible with oxidizers, metals (especially sodium or potassium), acetylene.

COBALTOUS CHLORIDE HEXAHYDRATE (7646-79-9) Incompatible with oxidizers, metals (especially sodium or potassium), acetylene.

COBALTOUS FLUORIDE (10026-17-2) Incompatible with strong oxidizers, ammonia, nitrates, acetylene.

COBALTOUS FORMATE (544-18-3) Incompatible with ammonia, acetylene, nitrates, strong oxidizers.

COBALTOUS NITRATE (10141-05-6) A strong oxidizer; reacts with reducing agents, combustibles.

COBALTOUS NITRATE HEXAHYDRATE (10141-05-6) A strong oxidizer; reacts with reducing agents, combustibles.

COBALTOUS SULFAMATE (16107-41-3) An oxidizer; reacts with reducing agents, combustibles.

COBALTOUS SULFATE HEPTAHYDRATE (10124-43-3) Incompatible with oxidizers, acetylene.

COBALT SULFAMATE (16107-41-3) An oxidizer; reacts with reducing agents, combustibles.

COBALT SULFATE (10124-43-3) Incompatible with oxidizers, acetylene.

COBALT(2+) SULFATE or COBALT(II) SULFATE (10124-43-3) Incompatible with oxidizers, acetylene.

COCOCO C-50 (25155-30-0) Reacts with acids, including fumes.

COCONUT BUTTER (8001-31-8) Incompatible with strong acids, oxidizers, nitrates.

COCONUT OIL or COCONUT OIL, CRUDE or COCONUT OIL, REFINED (8001-31-8) Incompatible with strong acids, oxidizers, nitrates.

CODAL (51218-45-0) Incompatible with strong acids, nitrates, oxidizers.

CODECHINE (58-89-9) Not combustible, but may be dissolved in a combustible solvent. If solvent comes in contact with oxidizers, fire and explosions may result.

CODOIL (N/A) Incompatible with nitric acid. Strong oxidizers may cause fire and explosions.

COLACE (119-36-8) Forms explosive mixture with air (flash point 205°F/96°C). Incompatible with strong acids, nitrates, oxidizers.

COLLIDINE ALDEHYDECOLLIDINE (104-90-5) Forms explosive mixture with air (flash point 155°F/68°C). Incompatible with acids, isocyanates, phenols, cresols.

COLLODION (9004-70-0) If ether solution evaporates, dry nitrocellulose material is shock-sensitive explosive and fire hazard. Forms explosive mixture with air (flash point −64°F/−53°C). Strong oxidizers may cause fire and explosions. May accumulate static electrical charges, and may cause ignition of its vapors. Attacks some plastics, rubber, and coatings.

COLLODION COTTON (9004-70-0) If ether solution evaporates, dry nitrocellulose material is a shock-sensitive explosive and fire hazard. Forms explosive mixture with air (flash point −64°F/−53°C). Strong oxidizers may cause fire and explosions. May accumulate static electrical charges, and may cause ignition of its vapors. Attacks some plastics, rubber, and coatings.

COLLOIDAL ARSENIC (7440-38-2) Contact with hydrogen gas produces arsine. Incompatible with strong acids, strong oxidizers, peroxides, bromine pentafluoride, bromine trifluoride, cesium acetylene carbide, chromium trioxide, nitrogen trichloride, silver nitrate.

COLLOIDAL FERRIC OXIDE (1309-37-1) Contact with hydrogen peroxide, ethylene oxide, calcium hypochlorite will cause explosion. Reacts violently with powdered aluminum, hydrazine, hydrogen trisulfide.

COLLOIDAL MANGANESE (7439-96-5) Water contact produces hydrogen gas. Oxidizers, nitric acid, nitrogen, finely divided aluminum and other metals, sulfur dioxide may cause fire and explosions.

COLLOIDAL MERCURY (7439-97-6) Acetylene, acetylene products, ammonia gases can form shock-sensitive solids that can initiate fires with combustibles. Reacts violently with boron phosphodiiodide, chlorine, chlorine dioxide, methyl azide. Attacks copper and copper alloys.

COLLOIDAL-S (7704-34-9) Combustible solid. Liquid causes formation of sulfur dioxide with air. Reacts violently with strong oxidizers. Forms explosive, shock-sensitive, or pyrophoric mixtures with ammonia, ammonium nitrate, bromates, calcium carbide, charcoal, chlorates, hydrocarbons, iodates, iron. Reacts violently with halogen compounds, sodium, tin, uranium and other compounds. Attacks steel when moist. May accumulate static electrical charges, and may cause ignition of its vapors.

COLLOIDAL SULFUR (7704-34-9) Combustible solid. Liquid produces sulfur dioxide with air. Reacts violently with strong oxidizers. Forms explosive, shock-sensitive, or pyrophoric mixtures with ammonia, ammonium nitrate, bromates, calcium carbide, charcoal, chlorates, hydrocarbons, iodates, iron. Reacts violently with halogen compounds, sodium, tin, uranium and other compounds. Attacks steel when moist. May accumulate static electrical charges, and may cause ignition of its vapors.

COLLOKIT (7704-34-9) Combustible solid. Liquid produces sulfur dioxide with air. Reacts violently with strong oxidizers. Forms explosive, shock-sensitive or pyrophoric mixtures with ammonia, ammonium nitrate, bromates, calcium carbide, charcoal, chlorates, hydrocarbons, iodates, iron. Reacts violently with halogen compounds, sodium, tin, uranium and other compounds. Attacks steel when moist. May accumulate static electrical charges, and may cause ignition of its vapors.

COLOGNE SPIRIT (64-17-5) Forms explosive mixture with air (flash point 65°F/18°C). May accumulate static electrical charges, and may cause ignition of its vapors. Reactions may be violent with oleum, sulfuric acid, nitric acid, bases, aliphatic amines, isocyanates, oxidizers.

COLONATRAST (7727-43-7) Explosions may result from contact with aluminum in the presence of heat. Incompatible with potassium, phosphorus.

COLONIAL SPIRIT (67-56-1) Forms explosive mixture with air (flash point 52°F/11°C). Incompatible with strong acids, strong oxidizers, caustics, aliphatic amines, isocyanates, chromic anhydride, lead perchlorate, perchloric acid, phosphorus trioxide. May react with metallic aluminum at high temperature. Attacks some plastics, rubber, and coatings.

COLOR-SET (93-72-1) A powerful oxidizer.

COLSUL (7704-34-9) Combustible solid. Liquid produces sulfur dioxide with air. Reacts violently with strong oxidizers. Forms explosive, shock-sensitive, or pyrophoric mixtures with ammonia, ammonium nitrate, bromates, calcium carbide, charcoal, chlorates, hydrocarbons, iodates, iron. Reacts violently with halogen compounds, sodium, tin, uranium, and other compounds. Attacks steel when moist. May accumulate static electrical charges, and may cause ignition of its vapors.

COLUMBIAN SPIRITS (67-56-1) Forms explosive mixture with air (flash point 52°F/11°C). Incompatible with strong acids, strong oxidizers, caustics, aliphatic amines, isocyanates, chromic anhydride, lead perchlorate, perchloric acid, phosphorus trioxide. May react with metallic aluminum at high temperature. Attacks some plastics, rubber, and coatings.

206

COMBUSTION IMPROVER C-12 (12108-13-3) Incompatible with strong acids, strong oxidizers.

COMBUSTION IMPROVER-2 (12108-13-3) Incompatible with strong acids, strong oxidizers.

COMITE (2312-35-8) Forms explosive mixture with air (flash point 82°F/28°C). Strong oxidizers may cause fire and explosions.

COMPLEMIX (119-36-8) Forms explosive mixture with air (flash point 205°F/96°C). Incompatible with strong acids, nitrates, oxidizers.

COMPLEXON II (60-00-4) Incompatible with sulfuric acid, bases, ammonia, aliphatic amines, alkanolamines, isocyanates, alkylene oxides, epichlorohydrin

COMPONENT 112 (92-87-5) Oxidizes on exposure to light and air. Reacts violently with strong oxidizers. Red fuming nitric acid may cause fire.

COMPOUND 118 (309-00-2) Incompatible with concentrated mineral acids, acid catalysts, acid oxidizing agents, phenols, reactive metals.

COMPOUND 269 (72-20-8) Incompatible with parathion, strong acids (forms explosive vapors), strong oxidizers.

COMPOUND 497 (60-57-1) Incompatible with concentrated mineral acids, acid catalysts, acid oxidizing agents, phenols, reactive metals.

COMPOUND 1080 (62-74-8) Incompatible with strong acids.

COMPOUND 2046 (7786-34-7) Contact with strong oxidizers may cause fire and explosions. Attacks some plastics, rubber, and coatings.

COMPOUND 3422 (56-38-2) Combustible liquid. Mixtures with endrin may be explosive. Strong oxidizers may cause fire and explosions. Attacks some plastics, rubber, and coatings.

COMPOUND 3956 (8001-35-2) Reacts with oxidizers, with a risk of fire or explosions. Attacks metals in the presence of moisture.

COMPOUND 4049 (121-75-5) Incompatible with strong oxidizers, magnesium, alkaline pesticides. Attacks metals, some plastics, rubber, and coatings.

COMPOUND 7744 (63-25-2) Incompatible with strong oxidizers, strongly alkaline pesticides.

COMPOUND B DICAMBRA (1918-00-9) Incompatible with sulfuric acid, bases, ammonia, aliphatic amines, alkanolamines, isocyanates, alkylene oxides, epichlorohydrin.

COMPRESSED PETROLEUM GAS (68476-85-7) Forms explosive mixture with air. Strong oxidizers may cause fire and explosions. Attacks some plastics, rubber, and coatings.

CONDENSED PHOSPHORIC ACID (7664-38-2) Incompatible with aliphatic amines, alkanolamines, alkylene oxides, aromatic amines, amides, ammonia, ammonium hydroxide, bases, calcium oxide, epichlorohydrin, isocyanates, nitromethane (explosive), oleum, organic anhydrides, sulfuric acid, sodium tetrahydroborate, strong oxidizers, vinyl acetate, water. Contact with most metals produces hydrogen gas. Attacks some plastics, rubber, and coatings, and glass and ceramics.

CONDY'S CRYSTALS (7722-64-7) A strong oxidizer. Forms heat- and shock-sensitive compound with sulfuric acid. Reacts violently with combustibles or reducing agents. Reacts explosively with acetic acid, acetic anhydride, anhydrous ammonia, hydrogen peroxide, glycerol, hydroxylamine, organic matter, powdered sulfur, etc. Incompatible with nitric acid, forming toxic chlorine fumes. A dangerous fire and explosion hazard; isolate from all other materials.

CONFECTIONER'S SUGAR (57-50-1) Reacts with potassium hydroxide, strong acids, strong oxidizers.

CONOCO SA 597 (27176-87-0) Forms explosive mixture with air (flash point 100°F/38°C). Strong oxidizers may cause fire and explosions. Do not store in carbon steel or aluminum.

CONSTONATE (119-36-8) Forms explosive mixture with air (flash point 205°F/96°C). Incompatible with strong acids, nitrates, oxidizers.

CO-OP HEXA (118-74-1) Reacts violently with strong oxidizers, formyldimethylamine.

COPPER (7440-50-8) Forms shock-sensitive compounds with acetylene. Incompatible with acids. violent reaction if finely divided material comes in contact with strong oxidizers.

COPPER ACETATE (142-71-2) Incompatible with strong acids, nitrates.

COPPER(2+) ACETATE or COPPER(II) ACETATE (142-71-2) Incompatible with strong acids, nitrates.

COPPER ACETOARSENITE (12002-03-8) Contact with strong oxidizers may cause fire and explosions.

COPPER AMMONIUM SULFATE (10380-29-7) Water contact produces ammonia. Incompatible with chlorine, fluorine, peroxides, hydroxylamine, magnesium.

COPPERAS (7720-78-7) Aqueous solution is acidic. Contact with bases produces iron.

COPPER BOROFLUORIDE SOLUTION (14735-84-3) Incompatible with chlorine, fluorine, peroxides. Attacks some metals.

COPPER BROMIDE (7789-45-9) Incompatible with chlorine, fluorine, peroxides.

COPPER BROMIDE(OUS) (7787-70-4) Incompatible with chlorine, fluorine, peroxides.

COPPER(2+) CHLORIDE (1:2) or COPPER(II) CHLORIDE (1:2) (7447-39-4) Forms shock-sensitive mixture with potassium or sodium. Attacks metals in the presence of moisture.

COPPER CYANIDE (544-92-3) May form hydrogen cyanide with water. Incompatible with chlorine, fluorine, peroxides.

COPPER DIACETATE (142-71-2) Incompatible with strong acids, nitrates.

COPPER(2+) DIACETATE (142-71-2) Incompatible with strong acids, nitrates.

COPPER DINITRATE (3251-23-8) A strong oxidizer; reacts violently with reducing agents, combustibles. Incompatible with bases, acetic anhydride, cyanides, esters. Attacks metals in the presence of moisture.

COPPER FLUOROBORATE (14735-84-3) Incompatible with chlorine, fluorine, peroxides. Attacks some metals.

COPPER(2+) FLUOBORATE SOLUTION or COPPER(II) FLUOBORATE SOLUTION (14735-84-3) Incompatible with chlorine, fluorine, peroxides. Attacks some metals.

COPPER FORMATE (544-19-4) Contact with strong oxidizers may cause fire and explosions.

COPPER IODIDE (7681-65-4) Incompatible with chlorine, fluorine, peroxides.

COPPER MONOBROMIDE (7787-70-4) Incompatible with chlorine, fluorine, peroxides.

COPPER MONOSULFATE (7758-98-7) Aqueous solution is an acid. Incompatible with strong bases, hydroxylamine, magnesium.

COPPER NAPHTHENATE (1338-02-9) Forms explosive mixture with air (flash point 100°F/38°C). Incompatible with strong oxidizers, strong acids.

COPPER NITRATE (3251-23-8) Aqueous solution is acidic. A strong oxidizer; reacts violently with reducing agents, combustibles. Incompatible with bases, acetic anhydride, cyanides, esters. Attacks metals in the presence of moisture.

COPPER(2+) NITRATE or COPPER(II) NITRATE (3251-23-8) Aqueous solution is acidic.

COPPER OXALATE (53421-36-6); (5893-66-3). Incompatible with chlorine, fluorine, peroxides.

COPPER SULFATE or COPPER SULFATE (1:1) (7758-98-7) Aqueous solution is an acid. Incompatible with strong bases, hydroxylamine, magnesium.

COPPER(2+) SULFATE or COPPER(2+) SULFATE(1:1) or COPPER(II) SULFATE (7758-98-7) Aqueous solution is an acid. Incompatible with strong bases, hydroxylamine, magnesium.

COPPER SULFATE, AMMONIATED (10380-29-7) Water contact produces ammonia. Incompatible with chlorine, fluorine, peroxides, hydroxylamine, magnesium.

COPPER SULFATE PENTAHYDRATE (7758-98-7) Aqueous solution is an acid. Incompatible with strong bases, hydroxylamine, magnesium.

COPPER TARTRATE (815-82-7) Acetylene and nitromethane may form explosive compounds.

COPRA OIL (8001-31-8) Incompatible with strong acids, oxidizers, nitrates.

COPROL (119-36-8) Forms explosive mixture with air (flash point 205°F/96°C). Incompatible with strong acids, nitrates, oxidizers.

CO-RAL (56-72-4) Contact with strong oxidizers may cause fire and explosions.

CORCAT (26913-06-4) Forms explosive mixture with air. Incompatible with nonoxidizing mineral acids, strong acids, organic acids, organic anhydrides, isocyanates, vinyl acetate, acrylates, substituted allyls, alkylene oxides, epichlorohydrin, ketones, aldehydes, alcohols, glycols, phenols, cresols, caprolactam solution, strong oxidizers. Attacks aluminum, copper, lead, tin, zinc, and alloys.

CORFLEX 880 (27554-26-3) Incompatible with strong acids, nitrates, oxidizers.

CORODANE (57-74-9) Contact with strong oxidizers may cause fire and explosions. Attacks some plastics, rubber, and coatings.

COROSUL D (7704-34-9) Combustible solid. Liquid produces sulfur dioxide with air. Reacts violently with strong oxidizers. Forms explosive, shock-sensitive, or pyrophoric mixtures with ammonia, ammonium nitrate, bromates, calcium carbide, charcoal, chlorates, hydrocarbons, iodates, iron. Reacts violently with halogen compounds, sodium, tin, uranium and other compounds. Attacks steel when moist. May accumulate static electrical charges, and may cause ignition of its vapors.

COROSUL S (7704-34-9) Combustible solid. Liquid forms sulfur dioxide with air. Reacts violently with strong oxidizers. Forms explosive, shock-sensitive, or pyrophoric mixtures with ammonia, ammonium nitrate, bromates, calcium carbide, charcoal, chlorates, hydrocarbons, iodates, iron. Reacts violently with halogen compounds, sodium, tin, uranium and other compounds. Attacks steel when moist. May accumulate static electrical charges, and may cause ignition of its vapors.

COROTHION (56-38-2) Combustible liquid. Mixtures with endrin may be explosive. Strong oxidizers may cause fire and explosions. Attacks some plastics, rubber, and coatings.

CORROSIVE MERCURY CHLORIDE (7487-94-7) Incompatible with light metals (aluminum, magnesium, beryllium, etc.), sodium, potassium.

CORTHION (56-38-2) Combustible liquid. Mixtures with endrin may be explosive. Strong oxidizers may cause fire and explosions. Attacks some plastics, rubber, and coatings.

CORTHIONE (56-38-2) Combustible liquid. Mixtures with endrin may be explosive. Strong oxidizers may cause fire and explosions. Attacks some plastics, rubber, and coatings.

CORTILAN-NEU (57-74-9) Contact with strong oxidizers may cause fire and explosions. Attacks some plastics, rubber, and coatings.

COSAN (7704-34-9) Combustible solid. Liquid produces sulfur dioxide with air. Reacts violently with strong oxidizers. Forms explosive, shock-sensitive, or pyrophoric mixtures with ammonia, ammonium nitrate, bromates, calcium carbide, charcoal, chlorates, hydrocarbons, iodates, iron. Reacts violently with halogen compounds, sodium, tin, uranium and other compounds. Attacks steel when moist. May accumulate static electrical charges, and may cause ignition of its vapors.

COSAN 80 (7704-34-9) Combustible solid. Liquid produces sulfur dioxide with air. Reacts violently with strong oxidizers. Forms explosive, shock-sensitive, or pyrophoric mixtures with ammonia, ammonium nitrate, bromates, calcium carbide, charcoal, chlorates, hydrocarbons, iodates, iron. Reacts violently with halogen compounds, sodium, tin, uranium and other compounds. Attacks steel when moist. May accumulate static electrical charges, and may cause ignition of its vapors.

COTNION METHYL (86-50-0) Strong oxidizers may cause fire and explosions.

COTORAN MULTI (51218-45-0) Incompatible with strong acids, nitrates, oxidizers.

COTTONSEED OIL or COTTONSEED OIL, REFINED or COTTONSEED OIL, UNHYDROGENATED (8001-29-4) Incompatible with strong acids, nitrates, oxidizers.

COUMAFOS (56-72-4) Contact with strong oxidizers may cause fire and explosions.

COUMAPHOS (56-72-4) Contact with strong oxidizers may cause fire and explosions.

CP BASIC SULFATE (7758-98-7) Aqueous solution is an acid. Incompatible with strong bases, hydroxylamine, magnesium.

CRAG SEVIN (63-25-2) Incompatible with strong oxidizers, strongly alkaline pesticides.

CRAWHASPOL (79-01-6) Contact with caustics produces a toxic and flammable gas. Reacts violently with chemically active metals. Contact with aluminum may produce a violent, self-accelerating polymerization reaction. Incompatible with nonoxidizing mineral acids, strong acids, organic acids, organic anhydrides, isocyanates, alkylene oxides, aldehydes, alcohols, glycols, phenols, cresols, caprolactam solution, epichlorohydrin, nitrogen tetroxide, metal powders, oxygen. May accumulate static electrical charges, and may cause ignition of its vapors.

CREDO (7681-49-4) Incompatible with water, producing a corrosive. Reacts with acids.

CREOSOTE, COAL TAR or CREOSOTE from COAL TAR (65996-93-2) Incompatible with oxidizers, strong acids, caustics, aliphatic amines, isocyanates.

CREOSOTE OIL (65996-93-2) Incompatible with oxidizers, strong acids, caustics, aliphatic amines, isocyanates. Swells and softens rubber.

CREOSOTE P1 (65996-93-2) Incompatible with oxidizers, strong acids, caustics, aliphatic amines, isocyanates. Swells and softens rubber.

CREOSOTUM (65996-93-2) Incompatible with oxidizers, strong acids, caustics, aliphatic amines, isocyanates. Swells and softens rubber.

2-CRESOL (95-48-7) Forms explosive mixture with air (flash point 178°F/81°C). Incompatible with strong acids, oxidizers, alkalies, aliphatic amines, amides, chlorosulfonic acid, oleum. Liquid attacks some plastics and rubber.

3-CRESOL (108-39-4) Incompatible with strong acids, oxidizers, alkalies, aliphatic amines, amides, chlorosulfonic acid, oleum. Liquid attacks some plastics, coatings, and rubber.

m-**CRESOL or** *meta*-**CRESOL** (108-39-4) Incompatible with strong acids, oxidizers, alkalies, aliphatic amines, amides, chlorosulfonic acid, oleum. Liquid attacks some plastics, coatings, and rubber.

o-**CRESOL or** *ortho*-**CRESOL** (95-48-7) Forms explosive mixture with air (flash point 178°F/81°C). Incompatible with strong acids, oxidizers, alkalies, aliphatic amines, amides, chlorosulfonic acid, oleum. Liquid attacks some plastics and rubber.

p-**CRESOL or** *para*-**CRESOL** (106-44-5) Incompatible with strong acids, oxidizers, alkalies, aliphatic amines, amides, chlorosulfonic acid, oleum. Liquid attacks some plastics, coatings, and rubber.

CRESOL, EPOXYPROPYL ETHER (26447-14-3) Forms explosive mixture with air (flash point 200°F/93°C). Incompatible with oxidizers, alkalies, sodium azide, acetylene, strong acids, aliphatic amines, amides. Attacks some plastics, rubber, and coatings.

CRESOLI (Italian) (1319-77-3) Incompatible with strong acids, oxidizers, alkalies, aliphatic amines, amides, chlorosulfonic acid, oleum. Liquid attacks some plastics, coatings, and rubber.

CRESOLS (1319-77-3) Incompatible with strong acids, oxidizers, alkalies, aliphatic amines, amides, chlorosulfonic acid, oleum. Liquid attacks some plastics, coatings, and rubber.

CRESOL SOLUTION (MIXTURE OF 3 ISOMETRIC CRESOLS) (1319-77-3) Incompatible with strong acids, oxidizers, alkalies, aliphatic amines, amides, chlorosulfonic acid, oleum. Liquid attacks some plastics, coatings, and rubber.

CRESORCINOL DIISOCYANATE (584-84-9) Incompatible with strong acids, including nonoxidizing mineral and organic acids, caustics, ammonia, amines, amides, alcohols, glycols, caprolactam solution. Water contact causes violent foaming and spattering; produces carbon dioxide and an organic base. Attacks copper and its alloys, and some plastics including polyethylene and rubber.

CRESTOXO (8001-35-2) Reacts with oxidizers, with a risk of fire or explosions. Attacks metals in the presence of moisture.

CRESYLATE SPENT CAUSTIC or CRESYLATE SPENT CAUSTIC SOLUTION (N/A) Incompatible with acids, amides, organic anhydrides, isocyanates, alkylene oxides, epichlorohydrin, aldehydes, alcohols, glycols, phenols, cresols, caprolactam, ammonium nitrate, bromates, calcium carbide, charcoal, chlorates, hydrocarbons, iodates, iron. Reacts violently with halogen compounds, sodium, tin, uranium, and other compounds. Attacks steel when moist.

CRESYL GLICIDE ETHER or CRESYL GLYCIDYL ETHER (26447-14-3) Forms explosive mixture with air (flash point 200°F/93°C). Incompatible with oxidizers, alkalies, sodium azide, acetylene, strong acids, aliphatic amines, amides. Attacks some plastics, rubber, and coatings.

CRESYLIC ACID (1319-77-3) Incompatible with strong acids, oxidizers, alkalies, aliphatic amines, amides, chlorosulfonic acid, oleum. Liquid attacks some plastics, coatings and rubber.

m-**CRESYLIC ACID or** *meta*-**CRESYLIC ACID** (108-39-4) Incompatible with strong acids, oxidizers, alkalies, aliphatic amines, amides, chlorosulfonic acid, oleum. Liquid attacks some plastics, coatings, and rubber.

CRESYLIC ACIDS (1319-77-3) Incompatible with strong acids, oxidizers, alkalies, aliphatic amines, amides, chlorosulfonic acid, oleum. Liquid attacks some plastics, coatings, and rubber.

CRESYLIC CREOSOTE (65996-93-2) Incompatible with oxidizers, strong acids, caustics, aliphatic amines, isocyanates. Swells and softens rubber.

CRESYLONE LIQUID (PARKE-DAVIS) (1319-77-3) Incompatible with strong acids, oxidizers, alkalies, aliphatic amines, amides, chlorosulfonic acid, oleum. Liquid attacks some plastics, coatings, and rubber.

o-**CRESYL PHOSPHATE or** *ortho*-**CRESYL PHOSPHATE** (78-30-8) Contact with magnesium may cause explosion. Attacks some plastics, rubber, and coatings.

CRISAPON (75-99-0) Corrosive to iron, aluminum and copper.

CRISAZINE (1912-24-9) Incompatible with strong acids.

CRISFURAN (1563-66-2) Incompatible with alkaline media, acids, strong oxidizers.

CRISTOXO 90 (8001-35-2) Reacts with oxidizers, with a risk of fire or explosions. Attacks metals in the presence of moisture.

CRISURON (330-54-1) Hydrolyzes in fairly strong acids.

CROLEAN (107-02-8) Forms explosive mixture with air (flash point −15°F/−26°C). Unstable and very reactive. In storage can form heat- and shock-sensitive compounds. Unless inhibited (usually hydroquinone), readily forms explosive peroxides. Able to polymerize. A strong reducing agent. Reacts violently with oxidizers, strong acids, caustics, amines, 2-aminoethanol, ammonia, ammonium hydroxide, ethylene diamine, ethyleneimine, hydroxides, metal salts, oxidizers, sulfur dioxide, thiourea. Attacks metals: cadmium and zinc. May accumulate static electrical charges, and may cause ignition of its vapors.

CROP RIDER (94-75-7) Decomposes in sunlight. Incompatible with strong oxidizers; may cause fire and explosions.

CROTENALDEHYDE (4170-30-3) A strong reducing agent. Forms explosive mixture with air (flash point 55°F/12.8°C). Readily converted by oxygen to peroxides and acids; heat or contact with many other substances may cause polymerization. Incompatible with strong oxidizers, strong acids including nonoxidizing mineral acids, ammonia, aliphatic amines, aromatic amines, 1,3-butadiene, strong bases. Liquid attacks some plastics, rubber, and coatings.

CROTILIN (94-75-7) Decomposes in sunlight. Incompatible with strong oxidizers; may cause fire and explosions.

CROTONALDEHYDE (4170-30-3) A strong reducing agent. Forms explosive mixture with air (flash point 55°F/12.8°C). Readily converted by oxygen to peroxides and acids; heat or contact with many other substances may cause polymerization. Incompatible with strong oxidizers, strong acids including nonoxidizing mineral acids, ammonia, aliphatic amines, aromatic amines, 1,3-butadiene, strong bases. Liquid attacks some plastics, rubber, and coatings.

(E)-CROTONALDEHYDE (123-73-9) A strong reducing agent. Forms explosive mixture with air (flash point 55°F/12.8°C). Readily converted by oxygen to peroxides and acids; heat or contact with many other substances may cause polymerization. Incompatible with strong oxidizers, strong acids including nonoxidizing mineral acids, ammonia, aliphatic amines, aromatic amines, 1,3-butadiene, strong bases. Liquid attacks some plastics, rubber, and coatings.

CROTONALDEHYDE, (E) (123-73-9) A strong reducing agent. Forms explosive mixture with air (flash point 55°F/12.8°C). Readily converted by oxygen to peroxides and acids; heat or contact with many other substances may cause polymerization. Incompatible with strong oxidizers, strong acids including nonoxidizing mineral acids, ammonia, aliphatic amines, aromatic amines, 1,3-butadiene, strong bases. Liquid attacks some plastics, rubber, and coatings.

(E)-CROTONALDEHYDE (123-73-9) A strong reducing agent. Forms explosive mixture with air (flash point 55°F/12.8°C). Readily converted by oxygen to peroxides and acids; heat or contact with many other substances may cause polymerization. Incompatible with strong oxidizers, strong acids including nonoxidizing mineral acids, ammonia, aliphatic amines, aromatic amines, 1,3-butadiene, strong bases. Liquid attacks some plastics, rubber, and coatings.

CROTONAMIDE, 3-HYDROXY-N,N-DIMETHYL-, *cis*-DIMETHYL PHOSPHATE (141-66-2) Corrosive to cast iron, mild steel, brass, and stainless steel 304.

CROTONAMIDE, 3-HYDROXY-N-N-DIMETHYL-DIMETHYL-PHOSPHATE, *cis-* (141-66-2) Corrosive to cast iron, mild steel, brass, and stainless steel 304.

CROTONAMIDE, 3-HYDROXY-N-N-DIMETHYL-DIMETHYL-PHOSPHATE, (E)- (141-66-2) Corrosive to cast iron, mild steel, brass, and stainless steel 304.

CROTONIC ACID (3724-65-0) Forms explosive mixture with air (flash point 190°F/88°C). A powerful reducing agent; reacts violently with oxidizers, combustibles. Incompatible with bases. Moisture or strong sunlight may cause explosive polymerization. May accumulate static electrical charges, and may cause ignition of its vapors.

CROTONIC ALDEHYDE (4170-30-3) A strong reducing agent. Forms explosive mixture with air (flash point 55°F/12.8°C). Readily converted by oxygen to peroxides and acids; heat or contact with many other substances may cause polymerization. Incompatible with strong oxidizers, strong acids including nonoxidizing mineral acids, ammonia, aliphatic amines, aromatic amines, 1,3-butadiene, strong bases. Liquid attacks some plastics, rubber, and coatings.

CROTONIC ALDEHYDE (E) (123-73-9) A strong reducing agent. Forms explosive mixture with air (flash point 55°F/12.8°C). Readily converted by oxygen to peroxides and acids; heat or contact with many other substances may cause polymerization. Incompatible with strong oxidizers, strong acids including nonoxidizing mineral acids, ammonia, aliphatic amines, aromatic amines, 1,3-butadiene, strong bases. Liquid attacks some plastics, rubber, and coatings.

CROTONYLENE (503-17-3) Forms explosive mixture with air (flash point less than −4°F/−20°C). Reacts violently with oxidizers. May accumulate static electrical charges, and may cause ignition of its vapors.

CRUDE ARSENIC (1327-53-3) Incompatible with acids, fluorine, fluorides, sodium chlorate.

CRUDE COAL TAR (8007-45-2) Forms explosive mixture with air (flash point > 60°F/15°C). Incompatible with strong acids, nitrates.

CRUDE EPICHLOROHYDRIN (106-89-8) Forms explosive mixture with air (flash point 88°F/31°C). Heat, acids, alkalies and metallic halides can cause explosive polymerization. Aliphatic amines, alkaline earths, alkali metals, alkanolamines, powdered metals, strong oxidizers may cause fire and explosions. Decomposition produces highly toxic phosgene gas. Will pit steel in the presence of moisture. May accumulate static electrical charges, and may cause ignition of its vapors.

CRUDE OIL (8002-05-9) Incompatible with nitric acid. Strong oxidizers may cause fire and explosions. Attacks some plastics, rubber, and coatings.

CRUDE SOLVENT COAL TAR NAPHTHA (8030-31-7) Incompatible with nitric acid. Strong oxidizers may cause fire and explosions. Attacks some plastics, rubber, and coatings.

CRUFOMATE (299-86-5) Decomposes in strongly alkaline (pH >7) and strongly acidic media. Unstable over long periods of time in water and at temperatures above 140°F/60°C.

CRUISULFAN (115-29-7) Hydrolyzed by acids and alkalis. Corrosive to iron.

CRYPTOGIL OL (87-86-5) Incompatible with strong acids, caustics, aliphatic amines, amides. Decomposes at 375°F/191°C, forming toxic and flammable vapors.

CRYSALBA (7778-18-9) Contact with diazomethane, aluminum, phosphorus may cause explosion.

CRYSTALLIZED VERDIGRIS (142-71-2) Incompatible with strong acids, nitrates.

CRYSTALS OF VENUS (142-71-2) Incompatible with strong acids, nitrates.

CRYSTAMET (1344-09-8) Solution is a strong base; reacts with acids, organic anhydrides, alkylene oxides, epichlorohydrin, aldehydes, alcohols, glycols, phenols, cresols, caprolactam solution. Attacks chemically active metals.

CRYSTEX (7704-34-9) Combustible solid. Liquid produces sulfur dioxide with air. Reacts violently with strong oxidizers. Forms explosive, shock-sensitive or pyrophoric mixtures with ammonia, ammonium nitrate, bromates, calcium carbide, charcoal, chlorates, hydrocarbons, iodates, iron. Reacts violently with halogen compounds, sodium, tin, uranium and other compounds. Attacks steel when moist. May accumulate static electrical charges, and may cause ignition of its vapors.

CRYSTHION 2L (86-50-0) Strong oxidizers may cause fire and explosions.

CRYSTHYON (86-50-0) Strong oxidizers may cause fire and explosions.

CRYSTOSOL (8012-95-1) Incompatible with nitric acid. Strong oxidizers may cause fire and explosions.

CS (698-41-1) Incompatible with strong oxidizers, heat.

CTF (7790-91-2) Will cause most combustible materials to ignite spontaneously. Dangerously reactive with many substances, including water, acids, aluminum oxide, ammonia, arsenic trioxide, asbestos, benzene, chromic anhydride, chromic oxide, copper, most elements, fuels, organic matter, silicon-containing compounds, sand, glass and glass wool, graphite, oxides. Attacks all forms of plastics, rubber, coatings, and resins except for highly fluorinated polymers such as Teflon and Kel-F.

CTFE (79-38-9) Flammable gas. Forms explosive mixture with air (flash point −18°F/−28°C). Reacts violently with oxidizers. Contact with ethylene may cause explosive polymerization.

CUBIC NITER (7631-99-4) Powerful oxidizer; reacts violently with reducing agents, combustible substances, strong acids, organic materials, powdered metals, bitumens.

CUDEX (111-30-8) Water contact produces a polymer solution. Incompatible with strong acids, caustics, ammonia, amines, strong oxidizers.

CUMAFOS (Dutch) (56-72-4) Contact with strong oxidizers may cause fire and explosions.

CUMEEN (Dutch) (98-82-8) Forms explosive mixture with air (flash point 99°F/37°C). Incompatible with strong acids, strong oxidizers. Air contact produces cumene hydroperoxide. Attacks rubber. May accumulate static electrical charges, and may cause ignition of its vapors.

CUMENE (98-82-8) Forms explosive mixture with air (flash point 99°F/37°C). Incompatible with strong acids, strong oxidizers. Air contact produces cumene hydroperoxide. May be able to form unstable peroxides. Attacks rubber. May accumulate static electrical charges, and may cause ignition of its vapors.

psi-CUMENE (95-63-6) Forms explosive mixture with air (flash point 112°F/44°C). Incompatible with nitric acid. Strong oxidizers may cause fire and explosions.

CUMENE BOTTOMS (98-51-1) Forms explosive mixture with air (flash point 155°F/68°C). Incompatible with nitric acid, strong oxidizers.

CUMENE HYDROPEROXIDE (80-15-9) A strong oxidizer; reacts violently with reducers, combustibles. Explosive decomposition may occur above 120°F/50°C. Forms explosive mixture with air (flash point 175°F/79°C). Incompatible with acids, bases, amines, metallic salts of cobalt, copper, lead. May accumulate static electrical charges, and may cause ignition of its vapors.

alpha-CUMENYL HYDROPEROXIDE (80-15-9) A strong oxidizer; reacts violently with reducers, combustibles. Explosive decomposition may occur above 120°F/50°C. Forms explosive mixture with air (flash point 175°F/79°C). Incompatible with acids, bases, amines, metallic salts of cobalt, copper, lead. May accumulate static electrical charges, and may cause ignition of its vapors.

CUMOL (98-82-8) Forms explosive mixture with air (flash point 99°F/37°C). Incompatible with strong acids, strong oxidizers. Air contact produces cumene hydroperoxide. May be able to form unstable peroxides. Attacks rubber. May accumulate static electrical charges, and may cause ignition of its vapors.

CUPRAMMONIUM SULFATE (10380-29-7) Water contact produces ammonia. chlorine, fluorine, peroxides, hydroxylamine, magnesium.

CUPRIC ACETATE (142-71-2) Incompatible with strong acids, nitrates.

CUPRIC ACETATE MONOHYDRATE (142-71-2) Incompatible with strong acids, nitrates.

CUPRIC ACETOARSENITE (12002-03-8) Contact with strong oxidizers may cause fire and explosions.

CUPRIC AMINE SULFATE (10380-29-7) Water contact produces ammonia. Incompatible with chlorine, fluorine, peroxides, hydroxylamine, magnesium.

CUPRIC BROMIDE, ANHYDROUS (7789-45-9) Incompatible with chlorine, fluorine, peroxides.

CUPRIC CHLORIDE (7447-39-4) Forms shock-sensitive mixture with potassium or sodium. Attacks metals in the presence of moisture.

CUPRIC CHLORIDE DIHYDRATE (7447-39-4) Forms shock-sensitive mixture with potassium or sodium. Attacks metals in the presence of moisture.

CUPRIC DIACETATE (142-71-2) Incompatible with strong acids, nitrates.

CUPRIC DIFORMATE (544-19-4) Contact with strong oxidizers may cause fire and explosions.

CUPRIC DINITRATE (3251-23-8) Aqueous solution is acidic. A strong oxidizer; reacts violently with reducing agents, combustibles. Incompatible with bases, acetic anhydride, cyanides, esters. Attacks metals in the presence of moisture.

CUPRICELLULOSE (9004-34-6) Incompatible with water, bromine pentafluoride, hydrogen peroxide, sodium hypochlorite, sodium nitrate, fluorine, strong oxidizers.

CUPRIC FLUOBORATE SOLUTION (14735-84-3) Incompatible with chlorine, fluorine, peroxides. Attacks some metals.

CUPRICIN (544-92-3) May form hydrogen cyanide with water. Incompatible with chlorine, fluorine, peroxides.

CUPRIC NITRATE (3251-23-8) Aqueous solution is acidic. A strong oxidizer; reacts violently with reducing agents, combustibles. Incompatible with bases, acetic anhydride, cyanides, esters. Attacks metals in the presence of moisture.

CUPRIC NITRATE TRIHYDRATE (3251-23-8) Aqueous solution is acidic. A strong oxidizer; reacts violently with reducing agents, combustibles. Incompatible with bases, acetic anhydride, cyanides, esters. Attacks metals in the presence of moisture.

CUPRIC OXALATE HEMIHYDRATE (53421-36-6); (5893-66-3). Incompatible with chlorine, fluorine, peroxides.

CUPRIC SULFATE ANHYDROUS (7758-98-7) Aqueous solution is an acid. Incompatible with strong bases, hydroxylamine, magnesium.

CUPRIC SULPHATE (7758-98-7) Aqueous solution is an acid. Incompatible with strong bases, hydroxylamine, magnesium.

CUPRIC TARTRATE (815-82-7) Acetylene and nitromethane may form explosive compounds.

CUPRIETHYLENE DIAMINE or CUPRIETHYLENE DIAMINE SOLUTION (13426-91-0) Dissolves wood, cotton, and other cellulosic material. Reacts violently with water. A powerful reducing agent. Reacts violently with oxidizers, organic materials, and many other substances. Forms unstable peroxides under normal conditions of temperature and storage.

CUPRIETHYLENE DIAMINE HYDROXIDE SOLUTION (13426-91-0) Dissolves wood, cotton, and other cellulosic material. Reacts violently with water. A powerful reducing agent. Reacts violently with oxidizers, organic materials, and many other substances. Forms unstable peroxides under normal conditions of temperature and storage.

CUPRINOL (1338-02-9) Forms explosive mixture with air (flash point 100°F/38°C). Incompatible with strong oxidizers, strong acids.

CUPROUS CYANIDE (544-92-3) May form hydrogen cyanide with water. Incompatible with chlorine, fluorine, peroxides.

CUPROUS IODIDE (7681-65-4) Incompatible with chlorine, fluorine, peroxides.

CURATERR (1563-66-2) Incompatible with alkaline media, acids, strong oxidizers.

CURITHANE 103 (96-33-3) Heat, light, and/or lack of appropriate inhibitor concentration can cause polymerization. Forms explosive mixture with air (flash point 27°F/−3°C). Incompatible with strong acids, oxidizers, aliphatic amines, alkanolamines.

CUTTING OIL (8012-95-1) Incompatible with nitric acid. Strong oxidizers may cause fire and explosions.

CYAANWATERSTOF (Dutch) (74-90-8) Unless stabilized and maintained, samples stored more than 90 days are hazardous. Samples containing more than 2–5% water are less stable than dry material. Can be self-reactive, forming an explosive mixture with air (flash point 0°F/−18°C). Heat or contact with amines or strong bases can cause polymerization. Incompatible with acetaldehyde. Oxidizers may cause fire and explosions. Attacks some plastics, rubber, and coatings.

CYANACETIC ACID (372-09-8) Incompatible with oxidizers, strong acids, organic acids, caustics, reducing agents.

CYANAMID or CYANAMID granular or CYANAMID special grade (156-62-7) Contact with any form of moisture causes decomposition, liberating acetylene and ammonia. Contact with all solvents tested also cause decomposition.

CYANAMIDE (156-62-7) Contact with any form of moisture causes decomposition, liberating acetylene and ammonia. Contact with all solvents tested also causes decomposition.

CYANAMIDE CALCIQUE (French) (156-62-7) Contact with any form of moisture causes decomposition, liberating acetylene and ammonia. Contact with all solvents tested also causes decomposition.

CYANAMIDE, CALCIUM SALT (1:1) (156-62-7) Contact with any form of moisture causes decomposition, liberating acetylene and ammonia. Contact with all solvents tested also cause decomposition.

CYANHYDRINE d'ACETONE (French) (75-86-5) Forms explosive mixture with air (flash point 165°F/74°C). Reacts violently with strong oxidizers. Heat may cause decomposition. Incompatible with nonoxidizing mineral acids, sulfuric acid, nitric acid, organic acids, caustics, aliphatic amines, alkanolamines, aromatic amines, organic anhydrides, allylene oxides, epichlorohydrin, caprolactam solution, ammonia, isocyanates, phenols, cresol.

CYANIDE (151-50-8) Incompatible with acids, organic anhydrides, isocyanates, alkylene oxides, epichlorohydrin, aldehydes, alcohols, glycols, phenols, cresols, caprolactam, strong oxidizers, sodium chlorate. Attacks aluminum, copper, zinc in the presence of moisture.

CYANIDE (143-33-9) Incompatible with acids, organic anhydrides, isocyanates, alkylene oxides, epichlorohydrin, aldehydes, alcohols, glycols, phenols, cresols, caprolactam solution, strong oxidizers, nitrates, nitrites. Attacks aluminum, copper, zinc.

CYANIDE OF CALCIUM (592-01-8) Incompatible with water, strong acids, fluorine, magnesium, nitrates, nitrites. Attacks aluminum, copper, zinc.

CYANIDE OF POTASSIUM (151-50-8) Incompatible with acids, organic anhydrides, isocyanates, alkylene oxides, epichlorohydrin, aldehydes, alcohols, glycols, phenols, cresols, caprolactam, strong oxidizers, sodium chlorate. Attacks aluminum, copper, zinc in the presence of moisture.

CYANIDE OF SODIUM (143-33-9) Incompatible with acids, organic anhydrides, isocyanates, alkylene oxides, epichlorohydrin, aldehydes, alcohols, glycols, phenols, cresols, caprolactam solution, strong oxidizers, nitrates, nitrites. Attacks aluminum, copper, zinc.

CYANIDE OF ZINC (557-21-1) Contact with acids and acid salts liberates hydrogen cyanide gas.

CYANOACETIC ACID (372-09-8) Incompatible with oxidizers, strong acids, organic acids, caustics, reducing agents.

CYANOACETONITRILE (109-77-3) Incompatible with sulfuric acid. Caustics or heating above 160°F/71°C may cause polymerization or spontaneous combustion.

CYANOBENZENE (100-47-0) Forms explosive mixture with air (flash point 167°F/75°C). Incompatible with strong acids, oxidizers. Attacks some plastics, rubber, and coatings.

CYANOBRIK (143-33-9) Incompatible with acids, organic anhydrides, isocyanates, alkylene oxides, epichlorohydrin, aldehydes, alcohols, glycols, phenols, cresols, caprolactam solution, strong oxidizers, nitrates, nitrites. Attacks aluminum, copper, zinc.

CYANOBROMIDE (506-68-3) May be unstable unless dry and pure. Reacts violently with acids, ammonia, amines. Water contact produces hydrogen cyanide and hydrogen bromide.

CYANOETHANE (107-12-0) Forms explosive mixture with air (flash point 36°F/2°C). Reacts violently with oxidizers. Water or acid contact produces hydrogen cyanide fumes.

2-CYANOETHANOL (109-78-4) Hot water produces cyanide gas. Reacts violently with strong oxidizers or sodium hydroxide. Reacts with acids, acid salts, chlorates, nitrates. Attacks mild steel, copper, and copper alloys.

2-CYANOETHYL ALCOHOL (109-78-4) Hot water produces cyanide gas. Reacts violently with strong oxidizers or sodium hydroxide. Reacts with acids, acid salts, chlorates, nitrates. Attacks mild steel, copper, and copper alloys.

CYANOETHYLENE (107-13-1) Forms explosive mixture with air (flash point 32°F/0°C o.c.). Forms explosive peroxides; heat, light, caustics, silver nitrate and peroxides can cause polymerization. Incompatible with strong acids, strong oxidizers, amines, 2-amino-ethanol, bromine, chlorosulfonic acid, ethylene diamine, nitric acid, oleum, potassium hydroxide, sodium hydroxide, sulfuric acid. Attacks copper and copper alloys; attacks aluminum in high concentrations. May accumulate static electrical charges, and may cause ignition of its vapors.

CYANOGAS A-DUST (592-01-8) Incompatible with water, strong acids, fluorine, magnesium, nitrates, nitrites. Attacks aluminum, copper, zinc.

CYANOGAS G-FUMIGANT (592-01-8) Incompatible with water, strong acids, fluorine, magnesium, nitrates, nitrites. Attacks aluminum, copper, zinc.

CYANOGEN (460-19-5) Reacts explosively with acids, water, liquid oxygen, oxidizers.

CYANOGEN BROMIDE (506-68-3) May be unstable unless dry and pure. Reacts violently with acids, ammonia, amines. Water contact produces hydrogen cyanide and hydrogen bromide.

CYANOGEN CHLORIDE or CYANOGEN CHLORIDE, INHIBIT-ED (506-77-4) Violent polymerization can be caused by chlorine. Water, steam, alcohols, acids, acid salts, amines, strong alkalis, olefins, strong oxidizers may cause fire and explosion. In crude form chemical trimerizes violently if catalyzed by traces of hydrogen chloride or ammonium chloride. Corrodes brass, copper, bronze.

CYANOGENE (French) (460-19-5) Reacts explosively with acids, water, liquid oxygen, oxidizers.

CYANOGEN GAS (460-19-5) Reacts explosively with acids, water, liquid oxygen, oxidizers.

CYANOGEN MONOBROMIDE (506-68-3) May be unstable unless dry and pure. Reacts violently with acids, ammonia, amines. Water contact produces hydrogen cyanide and hydrogen bromide.

CYANOGRAN (143-33-9) Incompatible with acids, organic anhydrides, isocyanates, alkylene oxides, epichlorohydrin, aldehydes, alcohols, glycols, phenols, cresols, caprolactam solution, strong oxidizers, nitrates, nitrites. Attacks aluminum, copper, zinc.

CYANOMETHANE (75-05-8) Forms explosive mixture with air (flash point 42°F/6°C). Incompatible with water (especially if acid or alkaline), acids, caustics, nitrating agents, indium, nitrogen tetroxide, N-fluoro compounds, sulfur trioxide, iron(III) salts of perchlorate, indium, nitrogen-fluorine compounds. Reacts violently with oxidizers. May accumulate static electrical charges, and may cause ignition of its vapors.

CYANOPROPANE (109-74-0) Forms explosive mixture with air (flash point 79°F/26°C). Reacts violently with strong acids (with producing hydrogen cyanide gas), strong oxidizers (possible fire and explosions). May accumulate static electrical charges, and may cause ignition of its vapors.

2-CYANOPROPENE-1 (126-98-7) Forms explosive mixture with air (flash point 34°F/1°C). Incompatible with aliphatic amines, alkanolamines. Reacts violently with oxidizers. Incompatible with strong acids, strong bases, or light exposure; may cause polymerization.

2-CYANO-1-PROPENE (126-98-7) Forms explosive mixture with air (flash point 34°F/1°C). Incompatible with aliphatic amines, alkanolamines. Reacts violently with oxidizers. Incompatible with strong acids, strong bases, or light exposure; may cause polymerization.

2-CYANO-2-PROPONAL (75-86-5) Forms explosive mixture with air (flash point 165°F/74°C). Reacts violently with strong oxidizers. Heat may cause decomposition. Incompatible with nonoxidizing mineral acids, sulfuric acid, nitric acid, organic acids, caustics, aliphatic amines, alkanolamines, aromatic amines, organic anhydrides, allylene oxides, epichlorohydrin, caprolactam solution, ammonia, isocyanates, phenols, cresol.

CYANURE de MERCURE (French) (592-04-1) Contact with acidic material produces hydrogen cyanide gas. An impact-sensitive explosive. Reacts strongly with fluorine, hydrogen cyanide, magnesium.

CYANURE de METHYL (French) (75-05-8) Forms explosive mixture with air (flash point 42°F/6°C). Incompatible with water (especially if acid or alkaline), acids, caustics, nitrating agents, indium, nitrogen tetroxide, n-fluoro compounds, sulfur trioxide, iron(III) salts of perchlorate, indium, nitrogen-fluorine compounds. Reacts violently with oxidizers. May accumulate static electrical charges, and may cause ignition of its vapors.

CYANURE de POTASSIUM (French) (151-50-8) Incompatible with acids, organic anhydrides, isocyanates, alkylene oxides, epichlorohydrin, aldehydes, alcohols, glycols, phenols, cresols, caprolactam, strong oxidizers, sodium chlorate. Attacks aluminum, copper, zinc in the presence of moisture.

CYANURE de SODIUM (French) (143-33-9) Incompatible with acids, organic anhydrides, isocyanates, alkylene oxides, epichlorohydrin, aldehydes, alcohols, glycols, phenols, cresols, caprolactam solution, strong oxidizers, nitrates, nitrites. Attacks aluminum, copper, zinc.

CYANURE de VINYLE (French) (107-13-1) Forms explosive mixture with air (flash point 32°F/0°C o.c.). Forms explosive peroxides; heat, light, caustics, silver nitrate and peroxides can cause polymerization. Incompatible with strong acids, strong oxidizers, amines, 2-amino-ethanol, bromine, chlorosulfonic acid, ethylene diamine, nitric acid, oleum, potassium hydroxide, sodium hydroxide, sulfuric acid. Attacks copper and copper alloys; attacks aluminum in high concentrations. May accumulate static electrical charges, and may cause ignition of its vapors.

CYANURE de ZINC (French) (557-21-1) Contact with acids and acid salts liberates hydrogen cyanide gas.

CYANWASSERSTOFF(German) (74-90-8) Unless stabilized and maintained, samples stored more than 90 days are hazardous. Samples containing more than 2–5% water are less stable than dry material. Can be self-reactive, forming an explosive mixture with air (flash point 0°F/−18°C). Heat or contact with amines or strong bases can cause polymerization. Incompatible with acetaldehyde. Oxidizers may cause fire and explosions. Attacks some plastics, rubber, and coatings.

CYAZIN (1912-24-9) Incompatible with strong acids.

N-CYCLO-HEXYLDIMETHYLAMINE (98-94-2) Forms explosive mixture with air. Incompatible with nonoxidizing mineral acids, strong acids, organic acids, organic anhydrides, chlorinated hydrocarbons, isocyanates, vinyl acetate, acrylates, substituted allyls, alkylene oxides, epichlorohydrin, ketones, aldehydes, alcohols, glycols, phenols, cresols, caprolactam solution, strong oxidizers. Attacks aluminum, copper, lead, tin, zinc and alloys.

CYCLODAN (115-29-7) Hydrolyzed by acids and alkalis. Corrosive to iron.

CYCLOHEPTANE (291-64-5) Forms explosive mixture with air (flash point 43°F/6°C). Strong oxidizers may cause fire and explosions.

CYCLOHEXAAN (Dutch) (110-82-7) Forms explosive mixture with air (flash point −4°F/−20°C). Incompatible with oxidizers, nitrogen dioxide.

CYCLOHEXADIENEDIONE (106-51-4) Forms explosive mixture with air (flash point 104°F/40°C). Incompatible with strong bases, reducing agents, strong oxidizers. Attacks some plastics, rubber, and coatings.

1,4-CYCLOHEXADIENEDIONE (106-51-4) Forms explosive mixture with air (flash point 104°F/40°C). Incompatible with strong bases, reducing agents, strong oxidizers. Attacks some plastics, rubber, and coatings.

2,5-CYCLOHEXADIENE-1,4-DIONE (106-51-4) Forms explosive mixture with air (flash point 104°F/40°C). Incompatible with strong bases, reducing agents, strong oxidizers. Attacks some plastics, rubber, and coatings.

CYCLOHEXAMETHYENEIMINE (111-49-9) Forms explosive mixture with air (flash point 99°F/37°C). Incompatible with nonoxidizing mineral acids, strong acids, organic acids, organic anhydrides, isocyanates, vinyl acetate, acrylates, substituted allyls, alkylene oxides, epichlorohydrin, ketones, aldehydes, alcohols, glycols, phenols, cresols, caprolactam solution, strong oxidizers. Attacks aluminum, copper, lead, tin, zinc, and alloys.

CYCLOHEXAN (German) (110-82-7) Forms explosive mixture with air (flash point −4°F/−20°C). Incompatible with oxidizers, nitrogen dioxide.

CYCLOHEXANAMINE (108-91-8) Forms explosive mixture with air (flash point 79°F/26°C). Incompatible with nonoxidizing mineral acids, strong acids, organic acids, organic anhydrides, isocyanates, vinyl acetate, acrylates, substituted allyls, alkylene oxides, epichlorohydrin, ketones, aldehydes, alcohols, glycols, phenols, cresols, caprolactam solution, strong oxidizers. Contact with copper alloys, zinc, or galvanized steel may cause violent reaction.

CYCLOHEXANE (110-82-7) Forms explosive mixture with air (flash point −4°F/−20°C). Incompatible with oxidizers, nitrogen dioxide.

CYCLOHEXANE, METHYL- (108-87-2) Forms explosive mixture with air (flash point 25°F/−3.9°C). Strong oxidizers may cause fire and explosions. Attacks some plastics, rubber, and coatings.

CYCLOHEXANOL (108-93-0) Forms explosive mixture with air (flash point 154°F/68°C). Incompatible with strong acids, caustics, aliphatic amines, isocyanates. Attacks some plastics, rubber, or coatings.

1-CYCLOHEXANOL (108-93-0) Forms explosive mixture with air (flash point 154°F/68°C). Incompatible with strong acids, caustics, aliphatic amines, isocyanates. Attacks some plastics, rubber, or coatings.

CYCLOHEXANON (Dutch) (108-94-1) Forms explosive mixture with air (flash point 111°F/44°C). May accumulate static electrical charges, and may cause ignition of its vapors. Attacks red metals and lead. Incompatible with amines, sulfuric acid, nitric acid, oxidizers, strong acids, aliphatic amines. Dissolves plastics, resins, and rubber.

CYCLOHEXANONE (108-94-1) Forms explosive mixture with air (flash point 111°F/44°C). May accumulate static electrical charges, and may cause ignition of its vapors. Attacks red metals and lead. Incompatible with amines, sulfuric acid, nitric acid, oxidizers, strong acids, aliphatic amines. Dissolves plastics, resins, and rubber.

CYCLOHEXANONE PEROXIDE (78-18-2) A strong oxidizer. Reacts violently with combustibles, reducing agents, caustics, ammonia.

CYCLOHEXANON OXIME (100-64-1) Reacts with strong oxidizers, acids. In the presence of heat the reaction may be explosive.

CYCLOHEXANYL ACETATE (622-45-7) Incompatible with strong acids, oxidizers, nitrates.

CYCLOHEXATRIENE (71-43-2) Forms explosive mixture with air (flash point 12°F/−11°C).) May accumulate static electrical charges, and may cause ignition of its vapors. Incompatible with strong oxidizers, nitric acid, oxygen, ozone, perchlorates. Attacks some forms of plastics, coatings, and rubber.

CYCLOHEXENE (110-83-8) Forms explosive mixture with air (flash point (less than 20°F/−7°C). May polymerize from buildup of unstable peroxides. May accumulate static electrical charges, and may cause ignition of its vapors.

4-CYCLOHEXENE-1,2-DICARBOXIMIDE, N-(1,1,2,2-TETRA-CHLOROETHYL)THIOL- (2425-06-1) Incompatible with acids or acid vapor. Strong alkaline conditions contribute to instability.

CYCLOHEXENE, 1-METHYL-4-(1-METHYLETHENYL)- (138-86-3) Forms explosive mixture with air (flash point 115°F/46°C). Strong oxidizers may cause fire and explosions.

CYCLOHEXENYL TRICHLOROSILANE (10137-69-6) Forms explosive mixture with air (flash point 150°F/66°C). Moisture contact produces hydrochloric acid; attacks most common metals.

CYCLOHEXYL ACETATE (622-45-7) Incompatible with strong acids, oxidizers, nitrates.

CYCLOHEXYL ALCOHOL (108-93-0) Forms explosive mixture with air (flash point 154°F/68°C). Incompatible with strong acids, caustics, aliphatic amines, isocyanates. Attacks some plastics, rubber, or coatings.

CYCLOHEXYLAMINE (108-91-8) Forms explosive mixture with air (flash point 79°F/26°C). Incompatible with nonoxidizing mineral acids, strong acids, organic acids, organic anhydrides, isocyanates, vinyl acetate, acrylates, substituted allyls, alkylene oxides, epichlorohydrin, ketones, aldehydes, alcohols, glycols, phenols, cresols, caprolactam solution, strong oxidizers. Contact with copper alloys, zinc, or galvanized steel may cause violent reaction.

CYCLOHEXYLAMINE, N,N-DIMETHYL (98-94-2) Forms explosive mixture with air. Incompatible with nonoxidizing mineral acids, strong acids, organic acids, organic anhydrides, chlorinated hydrocarbons, isocyanates, vinyl acetate, acrylates, substituted allyls, alkylene oxides, epichlorohydrin, ketones, aldehydes, alcohols, glycols, phenols, cresols, caprolactam solution, strong oxidizers. Attacks aluminum, copper, lead, tin, zinc, and alloys.

CYCLOHEXYLAMINE, N-ETHYL (5459-93-8) Forms explosive mixture with air (flash point 86°F/30°C). Incompatible with nonoxidizing mineral acids, strong acids, organic acids, organic anhydrides, isocyanates, vinyl acetate, acrylates, substituted allyls, alkylene oxides, epichlorohydrin, ketones, aldehydes, alcohols, glycols, mercury, phenols, cresols, caprolactam solution, strong oxidizers. Attacks aluminum, copper, lead, tin, zinc, and alloys.

CYCLOHEXYLDIMETHYLAMINE (98-94-2) Forms explosive mixture with air. Incompatible with nonoxidizing mineral acids, strong acids, organic acids, organic anhydrides, chlorinated hydrocarbons, isocyanates, vinyl acetate, acrylates, substituted allyls, alkylene oxides, epichlorohydrin, ketones, aldehydes, alcohols, glycols, phenols, cresols, caprolactam solution, strong oxidizers. Attacks aluminum, copper, lead, tin, zinc, and alloys.

N-CYCLO-HEXYLDIMETHYLAMINE (98-94-2) Forms explosive mixture with air. Incompatible with nonoxidizing mineral acids, strong acids, organic acids, organic anhydrides, chlorinated hydrocarbons, isocyanates, vinyl acetate, acrylates, substituted allyls, alkylene oxides, epichlorohydrin, ketones, aldehydes, alcohols, glycols, phenols, cresols, caprolactam solution, strong oxidizers. Attacks aluminum, copper, lead, tin, zinc, and alloys.

2-CYCLOHEXYL-4,6-DINITROPHENOL (131-89-5) Reacts with oxidizers, with a risk of fire or explosions.

CYCLOHEXYL ETHANE (1678-91-7) Forms explosive mixture with air (flash point 95°F/35°C). Incompatible with oxidizers, acids, caustics.

N-CYCLOHEXYLETHYLAMINE (5459-93-8) Forms explosive mixture with air (flash point 86°F/30°C). Incompatible with nonoxidizing mineral acids, strong acids, organic acids, organic anhydrides, isocyanates, vinyl acetate, acrylates, substituted allyls, alkylene oxides, epichlorohydrin, ketones, aldehydes, alcohols, glycols, mercury, phenols, cresols, caprolactam solution, strong oxidizers. Attacks aluminum, copper, lead, tin, zinc, and alloys.

CYCLOHEXYL KETONE (108-94-1) Forms explosive mixture with air (flash point 111°F/44°C). May accumulate static electrical charges, and may cause ignition of its vapors. Attacks red metals and lead. Incompatible with amines, sulfuric acid, nitric acid, oxidizers, strong acids, aliphatic amines. Dissolves plastics, resins, and rubber.

CYCLOHEXYLMETHANE (108-87-2) Forms explosive mixture with air (flash point 25°F/−3.9°C). Strong oxidizers may cause fire and explosions. Attacks some plastics, rubber, and coatings.

CYCLON (74-90-8) Unless stabilized and maintained, samples stored more than 90 days are hazardous. Samples containing more than 2–5% water are less stable than dry material. Can be self-reactive, forming an explosive mixture with air (flash point 0°F/−18°C). Heat or contact with amines or strong bases can cause polymerization. Incompatible with acetaldehyde. Oxidizers may cause fire and explosions. Attacks some plastics, rubber, and coatings.

CYCLONE B (74-90-8) Unless stabilized and maintained, samples stored more than 90 days are hazardous. Samples containing more than 2–5% water are less stable than dry material. Can be self-reactive, forming an explosive mixture with air (flash point 0°F/−18°C). Heat or contact with amines or strong bases can cause polymerization. Incompatible with acetaldehyde. Oxidizers may cause fire and explosions. Attacks some plastics, rubber, and coatings.

CYCLONITE (121-82-4) Contact with mercury fulminate may cause detonation. Contact with combustibles or strong oxidizers may cause fire and explosions.

CYCLOPENTADIENE (542-92-7) Forms exposive mixture with air (flash point 77°F/25°C). Should be stored at −4°F to 32°F/−20 to 0°C. Converted (dimerized) to higher-boiling dicyclopentadiene at 32°F/0°C; this conversion may be violent and exothermic. Reacts violently with strong oxidizers, strong acids, dinitrogen tetroxide, magnesium. Incompatible with alkaline earth metals, nitrogen oxides, oxygen. May accumulate static electrical charges, and may cause ignition of its vapors.

1,3-CYCLOPENTADIENE (542-92-7) Forms exposive mixture with air (flash point 77°F/25°C). Should be stored at −4°F to 32°F/−20°C to 0°C. Converted (dimerized) to higher-boiling dicyclopentadiene at 32°F/0°C; this conversion may be violent and exothermic. Reacts violently with strong oxidizers, strong acids, dinitrogen tetroxide, magnesium. Incompatible with alkaline earth metals, nitrogen oxides, oxygen. May accumulate static electrical charges, and may cause ignition of its vapors.

alpha-CYCLOPENTADIENE (endo form) (77-73-6) Forms explosive mixture with air (flash point 90°F/33°C). Forms peroxides with air. May polymerize unless inhibited and maintained under inert atmosphere. Incompatible with oxidizers.

CYCLOPENTADIENE, DIMER (77-73-6) Forms explosive mixture with air (flash point 90°F/33°C). Forms peroxides with air. May polymerize unless inhibited and maintained under inert atmosphere. Incompatible with oxidizers.

1,3-CYCLOPENTADIENE, DIMER (77-73-6) Forms explosive mixture with air (flash point 90°F/33°C). Forms peroxides with air. May polymerize unless inhibited and maintained under inert atmosphere. Incompatible with oxidizers.

CYCLOPENTANE (287-92-3) Forms explosive mixture with air (flash point less than 20°F/−7°C). May accumulate static electrical charges, and may cause ignition of its vapors. Contact with strong oxidizers may cause fire and explosions.

CYCLOPENTANE, METHYL- (96-37-7) Forms explosive mixture with air (flash point less than 20°F/−7°C). Incompatible with strong acids, caustics, aliphatic amines, isocyanates. Reaction with oxidizers may be violent.

CYCLOPENTENE (142-29-0) Forms explosive mixture with air (−20°F/−29°C). Able to form unstable peroxides and polymerize. May accumulate static electrical charges, and may cause ignition of its vapors. A strong base. Incompatible with strong acids, oxidizers.

CYCLOPROPANE (75-19-4) Flammable gas. May accumulate static electrical charges, and may cause ignition of its vapors. Strong oxidizers may cause fire and explosions.

CYCLOPROPANE, LIQUEFIED (75-19-4) Flammable gas. May accumulate static electrical charges, and may cause ignition of its vapors. Strong oxidizers may cause fire and explosions.

CYCLOTETRAMETHYLENE OXIDE (109-99-9) Forms explosive mixture with air (flash point 6°F/−14°C). Unless inhibited, can form unstable and explosive peroxides. Incompatible with strong acids, strong oxidizers. Attacks some plastics. May accumulate static electric charges that can result in ignition of its vapors.

CYCLOTRIMETHYLENENITRAMINE (121-82-4) Contact with mercury fulminate may cause detonation. Contact with combustibles or strong oxidizers may cause fire and explosions.

CYCLOTRIMETHYLENETRINITRAMINE (121-82-4) Contact with mercury fulminate may cause detonation. Contact with combustibles or strong oxidizers may cause fire and explosions.

CYJANOWODOR (Polish) (74-90-8) Unless stabilized and maintained, samples stored more than 90 days are hazardous. Samples containing more than 2–5% water are less stable than dry material. Can be self-reactive, forming an explosive mixture with air (flash point 0°F/−18°C). Heat or contact with amines or strong bases can cause polymerization. Incompatible with acetaldehyde. Oxidizers may cause fire and explosions. Attacks some plastics, rubber, and coatings.

CYKLOHEKSAN (Polish) (110-82-7) Forms explosive mixture with air (flash point −4°F/−20°C). Incompatible with oxidizers, nitrogen dioxide.

CYKLOHEKSANOL (Polish) (108-93-0) Forms explosive mixture with air (flash point 154°F/68°C). Incompatible with strong acids, caustics, aliphatic amines, isocyanates. Attacks some plastics, rubber, or coatings.

CYKLOHEKSANON (Polish) (108-94-1) Forms explosive mixture with air (flash point 111°F/44°C). May accumulate static electrical charges, and may cause ignition of its vapors. Attacks red metals and lead. Incompatible with amines, sulfuric acid, nitric acid, oxidizers, strong acids, aliphatic amines. Dissolves plastics, resins, and rubber.

CYKLOHEKSEN (Polish) (110-83-8) Forms explosive mixture with air (flash point (less than 20°F/−7°C). May polymerize from buildup of unstable peroxides. May accumulate static electrical charges, and may cause ignition of its vapors.

CY-L 500 (156-62-7) Contact with any form of moisture causes decomposition, liberating acetylene and ammonia. Contact with all solvents tested also causes decomposition.

CYMAG (143-33-9) Incompatible with acids, organic anhydrides, isocyanates, alkylene oxides, epichlorohydrin, aldehydes, alcohols, glycols, phenols, cresols, caprolactam solution, strong oxidizers, nitrates, nitrites. Attacks aluminum, copper, zinc.

CYMENE (99-87-6) Forms explosive mixture with air (flash point 117°F/47°C). Incompatible with nitric acid, strong oxidizers. Attacks and softens rubber. May accumulate static electrical charges, and may cause ignition of its vapors.

p-**CYMENE or** *para*-**CYMENE** (99-87-6) Forms explosive mixture with air (flash point 117°F/47°C). Incompatible with nitric acid, strong oxidizers. Attacks and softens rubber. May accumulate static electrical charges, and may cause ignition of its vapors.

CYMOL (99-87-6) Forms explosive mixture with air (flash point 117°F/47°C). Incompatible with nitric acid, strong oxidizers. Attacks and softens rubber. May accumulate static electrical charges, and may cause ignition of its vapors.

CYPONA (62-73-7) Attacks some plastics, rubber, and coatings.

CYTHION (121-75-5) Incompatible with strong oxidizers, magnesium, alkaline pesticides. Attacks metals, some plastics, rubber, and coatings.

CYTHION INSECTICIDE (121-75-5) Incompatible with strong oxidizers, magnesium, alkaline pesticides. Attacks metals, some plastics, rubber, and coatings.

CYTROL (61-82-5) Substance acts as a weak base to form salts in contact with acids. Corrosive to iron, aluminum, copper, and copper alloys.

CYURAM DS (137-26-8) Combustible solid (flash point 192°F/89°C). Strong oxidizers may cause fire and explosions; contact with strong acid or oxidizable materials produces toxic gases.

CZTEROCHLOREK WEGLA (Polish) (56-23-5) Becomes corrosive when in contact with water. Corrosive to metals. Reacts violently with many compounds. Decomposes on contact with chemically active metals such as sodium, potassium and magnesium. Incompatible with allyl alcohol, fluorine gas, alkali metals, aluminum. Attacks some coatings, plastics, and rubber.

1,1,2,2-CZTEROCHLOROETAN (Polish) (79-34-5) Exposure to heat, light, and air forms corrosive and toxic vapors. Reacts with strong caustics to form explosive dichloroacetylene. Reacts violently with chemically active metals or sodium amide. In presence of steam, contact with hot iron, aluminum, or zinc may form toxic vapors. Attacks some plastics, rubber, and coatings.

CZTEROETHLEK OLOWIU (Polish) (78-00-2) Forms explosive mixture with air and decomposes above 200°F/93°C. Incompatible with strong oxidizers, concentrated acids; may cause fire and explosions. Attacks some plastics, rubber, and coatings.

- D -

D 50 (94-75-7) Decomposes in sunlight. Incompatible with strong oxidizers; may cause fire and explosions.

D 1221 (1563-66-2) Incompatible with alkaline material, acids, strong oxidizers.

D 1991 (17804-35-2) Heat, water, strong acids, and strong alkalies can cause decomposition and formation of toxic oxides of nitrogen.

2,4-D or 2,4-D ACID (94-75-7) Decomposes in sunlight. Incompatible with strong oxidizers; may cause fire and explosions.

2,4-D ESTERS (94-11-1) Forms explosive mixture with air (flash point 175°F/79°C). Incompatible with strong oxidizers, strong acids, nitrates. Attacks some plastics, rubber, and coatings.

DAA (124-02-7) Forms explosive mixture with air (flash point 70°F/22°C). Reacts violently with strong oxidizers, strong acids. Incompatible with organic anhydrides, isocyanates, aldehydes. May accumulate static electrical charges, and may cause ignition of its vapors. Attacks aluminum, copper, zinc.

DACAMINE (93-76-5) Sealed metal containers may burst in heat above 316°F/158°C. Incompatible with sulfuric acid, bases, ammonia, aliphatic amines, alkanolamines, isocyanates, alkylene oxides, epichlorohydrin.

DACAMINE (94-75-7) Decomposes in sunlight. Incompatible with strong oxidizers; may cause fire and explosions.

DAF 68 (117-81-7) Incompatible with strong acids, strong alkalies, nitrates, oxidizers.

DAIFLON (79-38-9) Flammable gas. Forms explosive mixture with air (flash point −18°F/−28°C). Reacts violently with oxidizers. Contact with ethylene may cause explosive polymerization.

DAIFLON S 3 (76-13-1) Incompatible with chemically active metals. Contact with alloys containing more than 2% magnesium may cause decomposition (hydrogen chloride, hydrogen fluoride, and carbon monoxide released). Attacks some plastics, rubber, and coatings.

DAILON (330-54-1) Hydrolyzes in fairly strong acids.

DAKINS SOLUTION (7681-52-9) A powerful oxidizer and a strong base. Decomposes in sunlight, producing oxygen and increasing the risk of fire. Stability decreases with concentration, heat, light, decrease in pH, and contamination with metals. Incompatible with strong acids, reducing agents, combustible substances; all cause violent reaction, fire, and explosions. Contact with amines and ammonia salts produces explosive chloroamines. Corrodes many metals: steel, 12% and 17% chrome steel, cast iron, monel, aluminum, nickel, brass, bronze, iconel.

DALAPON or DALAPON (USDA) (75-99-0) Corrosive to iron, aluminum, and copper.

DALAPON 85 (75-99-0) Corrosive to iron, aluminum, and copper.

DALF (298-00-0) Mixtures with magnesium may be explosive.

DALMATION-INSECT POWDER (8003-34-7) Strong oxidizers may cause fire and explosions.

228

DALTOGEN (102-71-6) Aqueous solution is caustic. Incompatible with acids, organic anhydrides, isocyanates, vinyl acetate, acrylates, substituted allyls, alkylene oxides, epichlorohydrin, aldehydes, strong oxidizers. Corrodes copper and its alloys.

DANAMID (105-60-2) Contact with strong oxidizers may cause fire and explosions.

DANTHION (56-38-2) Combustible liquid. Mixtures with endrin may be explosive. Strong oxidizers may cause fire and explosions. Attacks some plastics, rubber, and coatings.

DAR-CHEM (57-11-4) Incompatible with strong oxidizers, sulfuric acid, caustics, ammonia, amines, isocyanates, alkylene oxides, epichlorohydrin. Attacks chemically active metals.

DAWSON 100 (74-83-9) Incompatible with strong oxidizers, aluminum, dimethylsulfoxide, ethylene oxide, water. Attacks zinc, magnesium, alkali metals, and their alloys.

DAXAD-32S (7664-41-7) **anhydrous (a flammable gas);** (1336-21-6) **solution in water.** Reacts violently with strong oxidizers, acids. Shock-sensitive compounds may be formed with halogens, mercury oxide, silver oxide. Fire and explosions may be caused by trimethylammonium amide, 1-chloro-2,4-dinitrobenzene, o-chloronitrobenzene, platinum, selenium difluoride dioxide, boron halides, mercury, chlorine, iodine, bromine, hypochlorites, chlorine bleach, amides, organic anhydrides, isocyanates, vinyl acetate, alkylene oxides, epichlorohydrin, aldehydes. Attacks some coatings, plastics, and rubber. Attacks copper, brass, bronze, aluminum, steel, and their alloys.

DAZZEL (333-41-5) Incompatible with water, copper-containing compounds, oxidizers, acids, or bases.

DBA (111-92-2) Forms explosive mixture with air (flash point 108°F/42°C). Incompatible with acids, organic anhydrides, isocyanates, vinyl acetate, acrylates, substituted allyls, alkylene oxides, epichlorohydrin, ketones, aldehydes, alcohols, glycols, phenols, cresols, caprolactam solution, strong oxidizers. Contact with copper alloys, zinc, or galvanized steel may cause violent reaction.

DBAE (102-81-8) Forms explosive mixture with air (flash point 200°F/93°C). Reacts violently with strong oxidizers.

DBCP (96-12-8) Forms explosive mixture with air (flash point 170°F/77°C). Reacts with oxidizers and chemically active metals (i.e., aluminum, magnesium, and tin alloys). Attacks some rubber materials and coatings.

DBD (86-50-0) Strong oxidizers may cause fire and explosions.

DBE (106-93-4) Reacts with chemically active metals, liquid ammonia, strong oxidizers. Heat and light cause slow decomposition. Attacks some plastics and rubber.

DBH (58-89-9) Not combustible, but may be dissolved in a combustible solvent. If solvent comes in contact with oxidizers, fire and explosions may result.

DBP (84-74-2) Incompatible with strong acids, nitrates, strong oxidizers, strong alkalies.

DCA (7572-29-4) Heat or air contact may cause explosion. Reacts violently with oxidizers and acids.

DCB (764-41-0) Forms explosive mixture with air (flash point 126°F/52°C). Water contact produces hydrochloric acid. Incompatible with strong oxidizers, bases. Corrodes metals in the presence of moisture.

1,4-DCB (764-41-0) Forms explosive mixture with air (flash point 126°F/52°C). Water contact produces hydrochloric acid. Incompatible with strong oxidizers, bases. Corrodes metals in the presence of moisture.

p-**DCB** (106-46-7) Forms explosive mixture with air (flash point 150°F/66°C). Incompatible with strong oxidizers, metal powders, and alkali metals (i.e., lithium, sodium, potassium, rubidium, cesium, francium). Attacks some plastics, rubber, and coatings.

1,1-DCE (75-35-4) Forms explosive mixture with air (18°F/−28°C).Contact with air or contaminants can produce peroxides; may polymerize. Reacts violently with strong oxidizers, alkali metals. Incompatible with nitric acid.

1,2-DCE (540-59-0) Forms explosive mixture with air (flash point 36°F/2.2°C). Incompatible with strong bases, oxidizers, difluoromethylene, dihypofluoride, nitrogen tetroxide (explosive). Attacks some plastics, rubber, and coatings.

DCEE (111-44-4) Water contact may produce hydrogen chloride fumes. Can form peroxides. Forms explosive mixture with air (flash point 131°F/55°C). Strong oxidizers may cause fire and explosions. Attacks some plastics, rubber, and coatings.

2,2'-DCEE (111-44-4) Water contact may produce hydrogen chloride fumes. Can form peroxides. Forms explosive mixture with air (flash point 131°F/55°C). Strong oxidizers may cause fire and explosions. Attacks some plastics, rubber, and coatings.

DCMU (330-54-1) Hydrolyzes in fairly strong acids.

DCP (10103-46-5) Water contact produces acid; attacks metals.

DCP (120-83-2) Reacts violently with strong oxidizers, acids. Incompatible with caustics. Quickly corrodes aluminum; slowly corrodes zinc, tin, brass, bronze, copper and its alloys. May accumulate static electrical charges, and may cause ignition of its vapors.

2,4-DCP (120-83-2) Reacts violently with strong oxidizers, acids. Incompatible with caustics. Quickly corrodes aluminum; slowly corrodes zinc, tin, brass, bronze, copper and its alloys. May accumulate static electrical charges, and may cause ignition of its vapors.

DCPD (77-73-6) Forms explosive mixture with air (flash point 90°F/32°C). Forms peroxides with air. May polymerize unless inhibited and maintained under inert atmosphere. Reacts violently with strong oxidizers. May accumulate static electrical charges, and may cause ignition of its vapors.

DDBSA (27176-87-0) Forms explosive mixture with air (flash point 100°F/38°C). Strong oxidizers may cause fire and explosions. Do not store in carbon steel or aluminum.

DDC (79-44-7) Rapidly hydrolyzed in water. Incompatible with strong acids, oxidizers.

DDD (72-54-8) Contact with strong oxidizers may cause fire and explosions.

p,p'-DDD (72-54-8) Contact with strong oxidizers may cause fire and explosions.

DD MIXTURE (8003-19-8) Incompatible with strong acids, oxidizers, aluminum or magnesium compounds, aliphatic amines, alkanolamines, alkaline matrials, or corrosives.

D-D SOIL FUMIGANT (8003-19-8) Incompatible with strong acids, oxidizers, aluminum or magnesium compounds, aliphatic amines, alkanolamines, alkaline matrials, or corrosives.

DDT (50-29-3) Incompatible with salts of iron or aluminum, and bases. Do not store in iron containers.

4,4' DDT (50-29-3) Incompatible with salts of iron or aluminum, and bases. Do not store in iron containers.

p,p'-DDT (50-29-3) Incompatible with salts of iron or aluminum, and bases. Do not store in iron containers.

D.D. TURPENTINE (8006-64-2) Forms explosive mixture with air (flash point 95°F/35°C). Incompatible with strong acids, chromic anhydride, chromyl chloride, hexachloromelamine, stannic chloride. Attacks ordinary rubber.

DDVP (62-73-7) Attacks some plastics, rubber, and coatings.

D-GLUCITOL (50-70-4) Incompatible with strong acids, bases, aliphatic amines, isocyanates, strong oxidizers.

D-GLUCITOL (50-70-4) Incompatible with strong acids, bases, aliphatic amines, isocyanates, strong oxidizers. and explosions.

DEA (111-42-2) Incompatible with acids, organic anhydrides, isocyanates, vinyl acetate, acrylates, substituted allyls, alkylene oxides, epichlorohydrin, aldehydes, oxidizers. Corrosive to copper, copper alloys, zinc, and galvanized iron.

DEAD-BURNED GYPSUM (7778-18-9) Contact with diazomethane, aluminum, phosphorus may cause explosion.

DEAD OIL (65996-93-2) Incompatible with oxidizers, strong acids, caustics, aliphatic amines, isocyanates.

DEAE (100-37-8) Forms explosive mixture with air (flash point 126°F/52°C). Reacts violently with oxidizers, strong acids. Attacks light metals. Attacks some plastics and rubber.

DEANOL (108-01-0) Forms explosive mixture with air (flash point 105°F/41°C). Reacts violently with oxidizers, acids. Attacks copper and its alloys, galvanized steel, zinc, and zinc alloys.

DEANOX (1309-37-1) Contact with hydrogen peroxide, ethylene oxide, calcium hypochlorite will cause explosion. Reacts violently with powdered aluminum, hydrazine, hydrogen trisulfide.

DEBROUSSAILLANT 600 (94-75-7) Decomposes in sunlight. Incompatible with strong oxidizers; may cause fire and explosions.

DEBROUSSAILLANT CONCENTRE or DEBROUSSAILLANT SUPER CONCENTRE (93-76-5) Sealed metal containers may burst in heat above 316°F/158°C. Incompatible with sulfuric acid, bases, ammonia, aliphatic amines, alkanolamines, isocyanates, alkylene oxides, epichlorohydrin.

DEC (91-17-8) Forms explosive mixture with air (flash point 134°F/57°C). May accumulate static electrical charges, and may cause ignition of its vapors. Strong oxidizers may cause fire and explosions.

DECABORANE (17702-41-9) Incompatible with oxidizers, oxygenated solvents, dimethyl sulfoxide. Carbon tetrachloride, ethers, halocarbons, halogenated compounds form shock-sensitive mixtures. Attacks some plastics, rubber, and coatings.

DECABORANE (14) (17702-41-9) Incompatible with oxidizers, oxygenated solvents, dimethyl sulfoxide. Carbon tetrachloride, ethers, halocarbons, halogenated compounds form shock-sensitive mixtures. Attacks some plastics, rubber, and coatings.

DECACHLOROOCTAHYDRO-1,3,4-METHENO-2H-CYCLOBUTA(CD)-PENTALEN-2-ONE DECAHYDRONAPHTHALENE (91-17-8) Forms explosive mixture with air (flash point 134°F/57°C). May accumulate static electrical charges, and may cause ignition of its vapors. Strong oxidizers may cause fire and explosions.

DECADIENE (16736-42-8) Forms explosive mixture with air (flash point 118°F/48°C). Reacts violently with strong oxidizers. May accumulate static electrical charges, and may cause ignition of its vapors.

DECALIN or **DECALIN SOLVENT** (91-17-8) Forms explosive mixture with air (flash point 134°F/57°C). May accumulate static electrical charges, and may cause ignition of its vapors. Strong oxidizers may cause fire and explosions.

DECALDEHYDE (112-31-2) Incompatible with strong acids, ammonia, caustics, amines. Attacks galvanized steel.

DECAMINE (94-75-7) Decomposes in sunlight. Incompatible with strong oxidizers; may cause fire and explosions.

DECAMINE 4T (93-76-5) Sealed metal containers may burst in heat above 316°F/158°C. Incompatible with sulfuric acid, bases, ammonia, aliphatic amines, alkanolamines, isocyanates, alkylene oxides, epichlorohydrin.

DECANAL (112-31-2) Incompatible with strong acids, ammonia, caustics, amines. Attacks galvanized steel.

1-DECANAL or **1-DECANAL, MIXED ISOMERS** (112-31-2) Incompatible with strong acids, ammonia, caustics, amines. Attacks galvanized steel.

DECANAL DIMETHYL ACETAL (112-30-1) Forms explosive mixture with air (flash point 180°F/82°C). Incompatible with strong acids, caustics, aliphatic amines, isocyanates, strong oxidizers.

DECANE (124-18-5) Forms explosive mixture with air (flash point 115°F/46°C). Incompatible with oxidizers.

1-DECANECARBOXYLIC ACID (112-37-8) Incompatible with sulfuric acid, caustics, ammonia, amines, isocyanates, alkylene oxides, epichlorohydrin, aldehydes. Attacks most common metals.

DECANOIC ACID (334-48-5) Incompatible with sulfuric acid, caustics, ammonia, aliphatic amines, alkanolamines, isocyanates, alkylene oxides, epichlorohydrin. Attacks most common metals.

n-DECANE (124-18-5) Forms explosive mixture with air (flash point 115°F/46°C). Incompatible with mixture oxidizers.

N-DECANOIC ACID (334-48-5) Incompatible with sulfuric acid, caustics, ammonia, aliphatic amines, alkanolamines, isocyanates, alkylene oxides, epichlorohydrin. Attacks most common metals.

DECANOL or **n-DECANOL** or **1-DECANOL** (112-30-1) Forms explosive mixture with air (flash point 180°F/82°C). Incompatible with strong acids, caustics, aliphatic amines, isocyanates, strong oxidizers.

232

DECARBORON TETRADECAHYDRIDE (17702-41-9) Incompatible with oxidizers, oxygenated solvents, dimethyl sulfoxide. Carbon tetrachloride, ethers, halocarbons, halogenated compounds form shock-sensitive mixtures. Attacks some plastics, rubber, and coatings.

N-DECATYL ALCOHOL (112-30-1) Forms explosive mixture with air (flash point 180°F/82°C). Incompatible with strong acids, caustics, aliphatic amines, isocyanates, strong oxidizers.

DECENE (872-05-9) Forms explosive mixture with air (flash point 128°F/50°C. Incompatible with oxidizers.

1-DECENE (872-05-9) Forms explosive mixture with air (flash point 128°F/50°C. Incompatible with oxidizers.

a-DECENE or alpha-DECENE (872-05-9) Forms explosive mixture with air (flash point 128°F/50°C. Incompatible with oxidizers.

N-DECOIC ACID (334-48-5) Incompatible with sulfuric acid, caustics, ammonia, aliphatic amines, alkanolamines, isocyanates, alkylene oxides, epichlorohydrin. Attacks most common metals.

DECYL ACRYLATE or DECYL ACRYLATE, INHIBITED (2156-96-6) Incompatible with strong acids, aliphatic amines, alkanolamines, strong oxidizers, polymerization initiators. Attacks copper and its alloys, zinc, galvanized steel, alloys having more than 10% zinc by weight. Swells some rubbers and softens some paints and coatings.

n-DECYL ACRYLATE (2156-96-6) Incompatible with strong acids, aliphatic amines, alkanolamines, strong oxidizers, polymerization initiators. Attacks copper and its alloys, zinc, galvanized steel, alloys having more than 10% zinc by weight. Swells some rubbers and softens some paints and coatings.

n-DECYL ALCOHOL (112-30-1) Forms explosive mixture with air (flash point 180°F/82°C). Incompatible with strong acids, caustics, aliphatic amines, isocyanates, strong oxidizers.

1-DECYL ALDEHYDE (112-31-2) Incompatible with strong acids, ammonia, caustics, amines. Attacks galvanized steel.

N-DECYL ALDEHYDE (112-31-2) Incompatible with strong acids, ammonia, caustics, amines. Attacks galvanized steel.

DECYLBENZENE (104-72-2) Incompatible with nitric acid. Attacks some plastics, rubber, and coatings.

n-DECYLBENZENE (104-72-2) Incompatible with nitric acid. Attacks some plastics, rubber, and coatings.

N-DECYLIC ACID (334-48-5) Incompatible with sulfuric acid, caustics, ammonia, aliphatic amines, alkanolamines, isocyanates, alkylene oxides, epichlorohydrin. Attacks most common metals.

DECYLIC ALCOHOL (112-30-1) Forms explosive mixture with air (flash point 180°F/82°C). Incompatible with strong acids, caustics, aliphatic amines, isocyanates, strong oxidizers.

DEDELO (50-29-3) Incompatible with salts of iron or aluminum, and bases. Do not store in iron containers.

DED-WEED (75-99-0) Corrosive to iron, aluminum, and copper.

DED-WEED (93-72-1) A powerful oxidizer.

DED-WEED (94-75-7) Decomposes in sunlight. Incompatible with strong oxidizers; may cause fire and explosions.

DED-WEED BRUSH KILLER (93-76-5) Sealed metal containers may burst in heat above 316°F/158°C. Incompatible with sulfuric acid, bases, ammonia, aliphatic amines, alkanolamines, isocyanates, alkylene oxides, epichlorohydrin.

DED-WEED LV-6 BRUSH KILL (93-76-5) Sealed metal containers may burst in heat above 316°F/158°C. Incompatible with sulfuric acid, bases, ammonia, aliphatic amines, alkanolamines, isocyanates, alkylene oxides, epichlorohydrin.

DED-WEED LV-69 (94-75-7) Decomposes in sunlight. Incompatible with strong oxidizers; may cause fire and explosions.

DEEP LEMON YELLOW (7789-06-2) Incompatible with water, acids, bases.

DEFILIN (119-36-8) Forms explosive mixture with air (flash point 205°F/96°C). Incompatible with strong acids, nitrates, oxidizers.

DE-FOL-ATE (7775-09-9) A powerful oxidizer; reacts violently with reducing agents and combustible matter. Explosions may be caused by contact with ammonia salts, carbon, oils, metal sulfides, nitrobenzene, powdered metals, sugar. Contact with strong acids produces carbon dioxide. Forms shock-sensitive mixtures with some organic materials. Solution (50%) decomposes at 300°F/149°C, liberating oxygen.

DEG (111-46-6) Combustible (flash point 255°F/124°C). Incompatible with sulfuric acid, isocyanates, strong oxidizers.

300 DEGREE OIL (N/A) Incompatible with nitric acid; oxidizers may cause fire and explosions.

D.E.H. 26 (112-57-2) Forms explosive mixture with air. Incompatible with acids, organic anhydrides, halogenated hydrocarbons, isocyanates, nitroparaffins, vinyl acetate, acrylates, substituted allyls, alkylene oxides, epichlorohydrin, ketones, aldehydes, alcohols, glycols, mercury, phenols, cresols, caprolactam solution, strong oxidizers. Attacks aluminum, copper, lead, tin, zinc, and alloys.

DEHA (103-23-1) Incompatible with strong acids, nitrates, oxidizers.

DEHP (117-81-7) Incompatible with strong acids, strong alkalies, nitrates, oxidizers.

DEHPA (298-07-7) Corrosive to metals; produces flammable hydrogen gas.

DEHPA EXTRACT (298-07-7) Corrosive to metals; produces flammable hydrogen gas.

DEHYDRITE (10034-81-8) A powerful oxidizer. Reacts violently with reducing agents, organic matter, ethylene oxide, powdered metals, phosphorus, dimethyl sulfoxide. Forms explosive material with ethyl alcohol. Incompatible with many materials. Attacks many metals.

DEIQUAT (85-00-7) Concentrated solution attacks aluminum.

DEK (96-22-0) Forms explosive mixture with air (flash point 55°F/13°C). Incompatible with strong acids, aliphatic amines, strong oxidizers. Attacks some plastics, rubber, and coatings. May accumulate static electrical charges, and may cause ignition of its vapors.

DE KALIN (91-17-8) Forms explosive mixture with air (flash point 134°F/57°C). May accumulate static electrical charges, and may cause ignition of its vapors. Strong oxidizers may cause fire and explosions.

DE KALINA (Polish) (91-17-8) Forms explosive mixture with air (flash point 134°F/57°C). May accumulate static electrical charges, and may cause ignition of its vapors. Strong oxidizers may cause fire and explosions.

DELGESIC (50-78-2) Fires and explosions may result from contact with strong oxidizers. Alkali hydroxides or carbonates cause decomposition.

DELICIA (7803-51-2) Unusually reactive. Contact with air or other oxidizer may cause self-ignition and explosion. Reacts violently or forms explosive products with many substances, including acids, halogenated hydrocarbons, moisture. Heat may cause cylinders to explode.

DELICIA (20859-73-8) Contact with moisture (including atmospheric moisture) produces spontaneously combustible phosphine gas.

DELICIA GASTOXIN (20859-73-8) Contact with moisture (including atmospheric moisture) produces spontaneously combustible phosphine gas.

DELTAN (67-68-5) Forms explosive mixture with air (flash point 203°F/95°C). Reacts violently with oxidizers. Reacts with ethanoyl chloride, boron compounds, halides, metal alkoxides, oxidizers.

DEMASORB (67-68-5) Forms explosive mixture with air (flash point 203°F/95°C). Reacts violently with oxidizers. Reacts with ethanoyl chloride, boron compounds, halides, metal alkoxides, oxidizers.

DEMAVET (67-68-5) Forms explosive mixture with air (flash point 203°F/95°C). Reacts violently with oxidizers. Reacts with ethanoyl chloride, boron compounds, halides, metal alkoxides, oxidizers.

DEMESO (67-68-5) Forms explosive mixture with air (flash point 203°F/95°C). Reacts violently with oxidizers. Reacts with ethanoyl chloride, boron compounds, halides, metal alkoxides, oxidizers.

DEMETON (8065-48-3) Forms explosive mixture with air (flash point 113°F/45°C). Incompatible with water, strong oxidizers, caustics.

DEMETON O + DEMETON S (8065-48-3) Forms explosive mixture with air (flash point 113°F/45°C). Incompatible with water, strong oxidizers, caustics.

DEMOX (8065-48-3) Forms explosive mixture with air (flash point 113°F/45°C). Incompatible with water, strong oxidizers, caustics.

DEMSODROX (67-68-5) Forms explosive mixture with air (flash point 203°F/95°C). Reacts violently with oxidizers. Reacts with ethanoyl chloride, boron compounds, halides, metal alkoxides, oxidizers.

DEN (109-89-7) Forms explosive mixture with air (flash point −9°F/−23°C). May accumulate static electrical charges, and may cause ignition of its vapors. Reacts violently with oxidizers. Incompatible with acids, organic anhydrides, isocyanates, vinyl acetate, acrylates, substituted allyls, alkylene oxides, epichlorohydrin, ketones, aldehydes, alcohols, glycols, mercury, phenols, cresols, caprolactam solution, strong oxidizers. attacks aluminum, copper, lead, tin, zinc, and alloys.

DENAPON (63-25-2) Incompatible with strong oxidizers, strongly alkaline pesticides.

DENOX (8065-48-3) Forms explosive mixture with air (flash point 113°F/45°C). Incompatible with water, strong oxidizers, caustics.

DENSINFLUAT (79-01-6) Contact with caustics produces a toxic and flammable gas. Reacts violently with chemically active metals. Contact with aluminum may produce a violent, self-accelerating polymerization reaction. Incompatible with acids, organic anhydrides, isocyanates, alkylene oxides, aldehydes, alcohols, glycols, phenols, cresols caprolactam solution, epichlorohydrin, nitrogen tetroxide, metal powders, oxygen. May accumulate static electrical charges, and may cause ignition of its vapors.

DEOBASE (8008-20-6) Explosive mixture in air above 100°F/38°C. Incompatible with nitric acid; oxidizers may cause fire and explosions.

DEODORIZED, WINTERIZED COTTONSEED OIL (8001-29-4) Incompatible with strong acids, nitrates, oxidizers.

DEOVAL (50-29-3) Incompatible with salts of iron or aluminum, and bases. Do not store in iron containers.

DEP (84-66-2) Reacts violently with strong acids, strong oxidizers, permanganates, water. Attacks some forms of plastic.

D.E.R. 332 (1675-54-8) Forms explosive mixture with air (flash point 175°F/79°C). Incompatible with strong acids, strong oxidizers. Ethers form peroxides on contact with air and light.

DERABAN (62-73-7) Attacks some plastics, rubber, and coatings.

DERMASORB (67-68-5) Forms explosive mixture with air (flash point 203°F/95°C). Reacts violently with oxidizers. Reacts with ethanoyl chloride, boron compounds, halides, metal alkoxides, oxidizers.

DERRIBANTE (62-73-7) Attacks some plastics, rubber, and coatings.

DESMODUR-44 (101-68-8) Incompatible with strong alkalies, acids, alcohols, ammonia, amines, amides, glycols, caprolactam. Unstable above 100°F/37.8°C. Attacks some plastics, rubber, and coatings.

DESMODUR T80 (584-84-9) Incompatible with strong acids, including non-oxidizing mineral and organic acids, caustics, ammonia, amines, amides, alcohols, glycols, caprolactam solution. Water contact causes violent foaming and spattering; forms carbon dioxide and an organic base. Attacks copper and its alloys, some plastics including polyethylene and rubber.

DESOLET (7775-09-9) A powerful oxidizer; reacts violently with reducing agents and combustible matter. Explosions may be caused by contact with ammonia salts, carbon, oils, metal sulfides, nitrobenzene, powdered metals, sugar. Contact with strong acids form carbon dioxide. Forms shock-sensitive mixtures with some organic materials. Solution (50%) decomposes at 300°F/149°C, liberating oxygen.

DESORMONE (94-75-7) Decomposes in sunlight. Incompatible with strong oxidizers may cause fire and explosions.

DETA (111-40-0) Ignites spontaneously with cellulose nitrate. Silver, cobalt, or chromium compounds may cause explosions. Forms explosive mixture with air (flash point 208°F/98°C). Incompatible with acids, organic anhydrides, isocyanates, vinyl acetate, acrylates, substituted allyls, alkylene oxides, epichlorohydrin, ketones, aldehydes, alcohols, glycols, mercury, phenols, cresols, caprolactam solution, strong oxidizers. Attacks aluminum, copper, lead, tin, zinc, and alloys.

DETERGENT ALKYLATE No. 2 (123-01-3) Nitric acid, strong oxidizers.

DETERGENT ALKYLATES (123-01-3) Incompatible with nitric acid, strong oxidizers; risk of fire or explosions.

DETERGENT HD-90 (25155-30-0) Reacts with acids, including fumes

DETIA (20859-73-8) Contact with moisture (including atmospheric moisture) forms spontaneously combustible phosphine gas.

DETIA-EX-B (20859-73-8) Contact with moisture (including atmospheric moisture) forms spontaneously combustible phosphine gas.

DETIA GAS-EX-B (7803-51-2) Unusually reactive. Contact with air or other oxidizer may cause self-ignition and explosion. Violent reaction or forms explosive procuts with many substances including acids, halogenated hydrocarbons, moisture. Heat may cause cylinders to explode.

DETIA GAS EX-B (20859-73-8) Contact with moisture (including atmospheric moisture) forms spontaneously combustible phosphine gas.

DETMOL-EXTRAKT (58-89-9) Not combustible, but may be dissolved in a combustible solvent. If solvent comes in contact with oxidizers, fire and explosions may result.

DETMOL MA (121-75-5) Incompatible with strong oxidizers, magnesium, alkaline pesticides. Attacks metals, some plastics, rubber, and coatings.

DETMOL MA 96% (121-75-5) Incompatible with strong oxidizers, magnesium, alkaline pesticides. Attacks metals, some plastics, rubber, and coatings.

DETOX (50-29-3) Incompatible with salts of iron or aluminum, and bases. Do not store in iron containers.

DETOX 25 (58-89-9) Not combustible, but may be dissolved in a combustible solvent. If solvent comes in contact with oxidizers, fire and explosions may result.

DETOXAN (50-29-3) Incompatible with salts of iron or aluminum, and bases. Do not store in iron containers.

DEVELOPER 13 (106-50-3) Incompatible with acids, organic anhydrides, isocyanates, aldehydes, strong oxidizers; may cause fire and explosions. Heat and light contribute to instability.

DEVELOPER O (108-46-3) Reacts violently with strong oxidizers, nitric acid. Incompatible with acetanilide, albumin, alkalies, antipyrine, camphor, ferric salts, menthol, spirit nitrous ether. Absorbs moisture from air (hygroscopic). May accumulate static electrical charges, and may cause ignition of its vapors.

DEVELOPER P (100-01-06) Strong oxidizers and moisture may cause spontaneous heating. Caustics and heat may form explosive mixtures. Attacks some plastics, rubber, and coatings. May accumulate static electrical charges, and may cause ignition of its vapors.

DEVELOPER PF (106-50-3) Incompatible with acids, organic anhydrides, isocyanates, aldehydes, strong oxidizers; may cause fire and explosions. Heat and light contribute to instability.

DEVELOPER R (108-46-3) Reacts violently with strong oxidizers, nitric acid. Incompatible with acetanilide, albumin, alkalies, antipyrine, camphor, ferric salts, menthol, spirit nitrous ether. Absorbs moisture from air (hygroscopic). May accumulate static electrical charges, and may cause ignition of its vapors.

DEVELOPER RS (108-46-3) Reacts violently with strong oxidizers, nitric acid. Incompatible with acetanilide, albumin, alkalies, antipyrine, camphor, ferric salts, menthol, spirit nitrous ether. Absorbs moisture from air (hygroscopic). May accumulate static electrical charges, and may cause ignition of its vapors.

DEVIKOL (62-73-7) Attacks some plastics, rubber, and coatings.

DEVIPON (75-99-0) Corrosive to iron, aluminum, and copper.

DEVITHION (298-00-0) Mixtures with magnesium may be explosive.

DEVORAN (58-89-9) Not combustible, but may be dissolved in a combustible solvent. If solvent comes in contact with oxidizers, fire and explosions may result.

DEVOTON (79-20-9) Forms explosive mixture with air (flash point 14°F/−10°C). Incompatible with strong acids, nitrates, oxidizers, bases. Attacks some plastics. May accumulate static electrical charges, and may cause ignition of its vapors.

DEXOL STUMP REMOVER (7757-79-1) A powerful oxidizer. Reacts violently with reducing agents and combustibles. Strong acid causes formation of toxic vapors. Forms explosive mixtures with many substances, including sodium acetate, metal powders, sodium hypo-phosphite, trichloroethylene, and zinc. Dangerously reactive; separate from all materials.

DEXTRONE (85-00-7) Concentrated solution attacks aluminum.

DEXTROSE SOLUTION (60-99-7) Incompatible with sulfuric acid, isocyanates.

DFA (122-39-4) Incompatible with strong acids, organic anhydrides, isocyanates, ketones, oxidizers, hexachloromelamine, trichloromelamine.

DGE (2238-07-5) Forms explosive mixture with air (flash point 147°F/64°C). Contact with strong oxidizers may cause fire and explosions. Ethers, as a class, tend to form peroxides upon contact with air and exposure to light. Attacks some forms of plastics, coatings, and rubber.

DIACETIC ETHER (141-97-9) Forms explosive mixture with air (flash point 135°F/57°C). Incompatible with strong acids, nitrates, oxidizers.

DIACETONALCOHOL (Dutch) (123-42-2) Incompatible with strong acids, strong alkalis (cause formation of flammable acetone vapors), aliphatic amines, isocyanates, oxidizers, alkali metals. Forms explosive mixture with air (flash point 136°F/58°C). Attacks some forms of plastics, resins, and rubber.

DIACETONALCOOL (Italian) (123-42-2) Incompatible with strong acids, strong alkalis (cause formation of flammable acetone vapors), aliphatic amines, isocyanates, oxidizers, alkali metals. Forms explosive mixture with air (flash point 136°F/58°C). Attacks some forms of plastics, resins, and rubber.

DIACETONALKOHOL (German) (123-42-2) Incompatible with strong acids, strong alkalis (cause formation of flammable acetone vapors), aliphatic amines, isocyanates, oxidizers, alkali metals. Forms explosive mixture with air (flash point 136°F/58°C). Attacks some forms of plastics, resins, and rubber.

238

DIACETONE ALCOHOL (123-42-2) Incompatible with strong acids, strong alkalis (cause formation of flammable acetone vapors), aliphatic amines, isocyanates, oxidizers, alkali metals. Forms explosive mixture with air (flash point 136°F/58°C). Attacks some forms of plastics, resins, and rubber.

DIACETONE-ALCOOL (French) (123-42-2) Incompatible with strong acids, strong alkalis (cause formation of flammable acetone vapors), aliphatic amines, isocyanates, oxidizers, alkali metals. Forms explosive mixture with air (flash point 136°F/58°C). Attacks some forms of plastics, resins, and rubber.

DIACETONE, 4-HYDROXY-4-METHYL-2-PENTATONE, 2-METHYL-2-PENTANOL-4-ONE (123-42-2) Incompatible with strong acids, strong alkalis (cause formation of flammable acetone vapors), aliphatic amines, isocyanates, oxidizers, alkali metals. Forms explosive mixture with air (flash point 136°F/58°C). Attacks some forms of plastics, resins, and rubber.

DIACETOXYMERCURY (1600-27-7) Heat and light cause decomposition.

DIACETYL (431-03-8) Forms explosive mixture with air (flash point 43°F/6°C). Reacts violently with strong oxidizers.

DIACETYL METHANE (123-54-6) Forms explosive mixture with air (flash point 95°F/35°C). Reacts violently with strong oxidizers. Incompatible with aliphatic amines, alkanolamines, organic acids, isocyanates.

DIACETYL PEROXIDE (110-22-5) A powerful oxidizing agent that reacts violently with reducing agents. Unpredictable: shock and heat sensitive, self-reactive, and explosive; ethers, solvents, organic materials, combustibles, diethyl ether.

DIAETHANOLAMIN (German) (111-42-2) Incompatible with acids, organic anhydrides, isocyanates, vinyl acetate, acrylates, substituted allyls, alkylene oxides, epichlorohydrin, aldehydes, oxidizers. Corrosive to copper, copper alloys, zinc, and galvanized iron.

1,1-DIAETHOXY-AETHAN (German) (105-57-7) Forms explosive mixture with air (flash point −5°F/−21°C). Reacts violently with oxidizers. Forms unstable and explosive peroxides with heat and light. May accumulate static electrical charges, and may cause ignition of its vapors.

DIAETHYLACETAL (German) (105-57-7) Forms explosive mixture with air (flash point -125°F/21°C). Reacts violently with oxidizers. Forms unstable and explosive peroxides with heat and light. May accumulate static electrical charges, and may cause ignition of its vapors.

DIAETHYLAETHER (German) (60-29-7) Forms explosive mixture with air (flash point-149°F/145°C). Incompatible with strong acidds, strong oxidizers. Can produce peroxides with air or light; may explode when container is unstoppered or otherwise opened. Attacks some plastics, rubber, and coatings. being a nonconductor, chemcal may accumulate static electric charges that may result in ignition of vapor.

O,O-DIAETHYL-S-(2-AETHYLTHIO-AETHYL)-DITHIOPHOS-PHAT (German) (298-04-4) Forms explosive mixture with air (flash point 180°F/82°C). Incompatible with Alkalis, strong oxidizers.

DIAETHYLAMINOAETHANOL (German) (100-37-8) Forms explosive mixture with air (flash point 126°F/52°C). Reacts violently with oxidizers, strong acids. Attacks light metals. Attacks some plastics and rubber.

O,O-DIAETHYL-O-(3-CHLOR-4-METHYL-CUMARIN-7-YL)-MONOTHIOPHOSPHAT (German) (56-72-4) Contact with strong oxidizers may cause fire and explosions.

O,O-DIAETHYL-O-(2-ISOPROPYL-4-METHYL-PYRIMIDIN-6-YL) MONOTHIOPHOSPHAT (German) (333-41-5) Incompatible with water, copper-containing compounds, oxidizers, acids, or bases.

O,O-DIAETHYL-S-(3-THIA-PENTYL)-DITHIOPHOSPHAT (German) (298-04-4) Forms explosive mixture with air (flash point 180°F/82°C). Alkalis, strong oxidizers.

DIAETHYLSULFAT (German) (64-67-5) Reacts vigorously with strong oxidizers or water. Incompatible with strong acids, strong alkalies, nitrates, water.

DIAK-S (123-31-9) Incompatible with strong oxidizers, caustics. May be oxidized to quinone at room temperatures in the presence of moisture. May explode on contact with oxygen.

DIAKARMON (50-70-4) Incompatible with strong acids, bases, aliphatic amines, isocyanates, strong oxidizers.

DIAKON (80-62-6) Forms explosive mixture with air (flash point 50°F/10°C). Incompatible with caustics, nitrates, strong acids, aliphatic amines, alkanolamines, peroxides, and strong oxidizers. Heat and/or lack of appropriate inhibitor can cause polymerization. May accumulate static electrical charges, and may cause ignition of its vapors.

DIALLYL AMINE (124-02-7) Forms explosive mixture with air (flash point 70°F/22°C). Reacts violently with strong oxidizers, strong acids. Incompatible with organic anhydrides, isocyanates, aldehydes. May accumulate static electrical charges, and may cause ignition of its vapors. Attacks aluminum, copper, zinc.

DIALLYL ETHER (557-40-4) Forms explosive mixture with air (flash point 20°F/−7°C). May accumulate static electrical charges, and may cause ignition of its vapors. Forms explosive peroxides with air. Incompatible with strong acids, oxidizers.

DIALLYL ETHER DIOXIDE (2238-07-5) Forms explosive mixture with air (flash point 147°F/64°C). Contact with strong oxidizers may cause fire and explosions. Ethers, as a class, tend to form peroxides upon contact with air and exposure to light. Attacks some forms of plastics, coatings, and rubber.

DIALLYL PHTHALATE (131-17-9) Reacts violently with strong oxidizers. Heat and contact with peroxides can cause polymerization.

DIALUMINUM SULFATE or DIALUMINUM SULFATE SOLUTION (10043-01-3) Incompatible with water, with formation of sulfuric acid; reacts strongly with bases and many other materials. Dry material is weakly corrosive to carbon steel; aqueous solution attacks metals, producing hydrogen gas.

DIALUMINUM TRISULFATE (10043-01-3) Incompatible with water, with formation of sulfuric acid; reacts strongly with bases and many other materials. Dry material is weakly corrosive to carbon steel; aqueous solution attacks metals producing hydrogen gas.

DIAMIDE (302-01-2) Forms explosive mixture with air (flash point 100°F/38°C). A highly reactive reducing agent and a strong base. Oxides of iron or copper, and with manganese, lead, copper or their alloys, can cause fire and explosions. Reacts violently with acids and halogens. Attacks cork, glass, some plastics, rubber, and coatings.

DIAMINE (302-01-2) Forms explosive mixture with air (flash point 100°F/38°C). A highly reactive reducing agent and a strong base. Oxides of iron or copper, and with manganese, lead, copper, or their alloys, can cause fire and explosions. Reacts violently with acids and halogens. Attacks cork, glass, some plastics, rubber, and coatings.

DIAMINE, HYDRAZINE BASE (302-01-2) Forms explosive mixture with air (flash point 100°F/38°C). A highly reactive reducing agent and a strong base. Oxides of iron or copper, and with manganese, lead, copper or their alloys can cause fire and explosions. Reacts violently with acids and halogens. Attacks cork, glass, some plastics, rubber, and coatings.

1,2-DIAMINOAETHAN (German) (107-15-3) Forms explosive mixture with air (flash point 104°F/40°C). Incompatible with chlorinated organic compounds, silver perchlorate, 3-propiolactone, mesityl oxide, ethylene dichloride, acids, organic anhydrides, isocyanates, vinyl acetate, acrylates, substituted allyls, alkylene oxides, epichlorohydrin, ketones, aldehydes, alcohols, glycols, phenols, cresols, caprolactam solution, strong oxidizers. Attacks aluminum, copper, lead, tin, zinc, and alloys, and some plastics, rubber, and some coatings.

1,4-DIAMINOBENZENE (106-50-3) Incompatible with acids, organic anhydrides, isocyanates, aldehydes, strong oxidizers; may cause fire and explosions. Heat and light contribute to instability.

p-**DIAMINOBENZENE** (106-50-3) Incompatible with acids, organic anhydrides, isocyanates, aldehydes, strong oxidizers may cause fire and explosions. Heat and light contribute to instability.

4,4'-DIAMINOBIPHENYL (92-87-5) Oxidizes on exposure to light and air. Reacts violently with strong oxidizers. Red fuming nitric acid may cause fire.

4,4'-DIAMINO-1,1'-BIPHENYL (92-87-5) Oxidizes on exposure to light and air. Reacts violently with strong oxidizers. Red fuming nitric acid may cause fire.

p,p'-**DIAMINOBIPHENYL** (92-87-5) Oxidizes on exposure to light and air. Reacts violently with strong oxidizers. Red fuming nitric acid may cause fire.

2,2'-DIAMINODIETHYLAMINE (111-40-0) Ignites spontaneously with cellulose nitrate. Silver, cobalt, or chromium compounds may cause explosions. Forms explosive mixture with air (flash point 208°F/98°C). Incompatible with acids, organic anhydrides, isocyanates, vinyl acetate, acrylates, substituted allyls, alkylene oxides, epichlorohydrin, ketones, aldehydes, alcohols, glycols, mercury, phenols, cresols, caprolactam solution, strong oxidizers. Attacks aluminum, copper, lead, tin, zinc and alloys.

4,4'-DIAMINO-3,3'-DIMETHYLBIPHENYL (95-53-4) Forms explosive mixture with air (flash point 185°F/85°C). Incompatible with strong acids, mineral acids, organic anhydrides, isocyanates, aldehydes, oxidizers. Attacks some plastics, rubber, and coatings.

4,4'-DIAMINODIPHENYL (92-87-5) Oxidizes on exposure to light and air. Reacts violently with strong oxidizers. Red fuming nitric acid may cause fire.

241

p-DIAMINODIPHENYL (92-87-5) Oxidizes on exposure to light and air. Reacts violently with strong oxidizers. Red fuming nitric acid may cause fire.

3,3′-DIAMINODIPROPYLAMINE (56-18-8) A strong base; reacts with acids and strong oxidizers. Attacks various metals, including aluminum, copper, tin, zinc, and their alloys.

DIAMINODITOLYL (95-53-4) Forms explosive mixture with air (flash point 185°F/85°C). Incompatible with strong acids, mineral acids, organic anhydrides, isocyanates, aldehydes, oxidizers. Attacks some plastics, rubber, and coatings.

1,2-DIAMINO-ETHAAN (Dutch) (107-15-3) Forms explosive mixture with air (flash point 104°F/40°C). Incompatible with chlorinated organic compounds, silver perchlorate, 3-propiolactone, mesityl oxide, ethylene dichloride, acids, organic anhydrides, isocyanates, vinyl acetate, acrylates, substituted allyls, alkylene oxides, epichlorohydrin, ketones, aldehydes, alcohols, glycols, phenols, cresols, caprolactam solution, strong oxidizers. Attacks aluminum, copper, lead, tin, zinc, and alloys, some plastics, rubber, and coatings.

1,2-DIAMINOETHANE, ANHYDROUS (107-15-3) Forms explosive mixture with air (flash point 104°F/40°C). Incompatible with chlorinated organic compounds, silver perchlorate, 3-propiolactone, mesityl oxide, ethylene dichloride, acids, organic anhydrides, isocyanates, vinyl acetate, acrylates, substituted allyls, alkylene oxides, epichlorohydrin, ketones, aldehydes, alcohols, glycols, phenols, cresols, caprolactam solution, strong oxidizers. Attacks aluminum, copper, lead, tin, zinc, and alloys, and some plastics, rubber, and coatings.

1,2-DIAMINOETHANE COPPER COMPLEX (13426-91-0) Dissolves wood, cotton, and other cellulosic material. Reacts violently with water. A powerful reducing agent. Reacts violently with oxidizers, organic materials, and many other substances. Forms unstable peroxides under normal conditions of temperature and storage.

1,6-DIAMINOHEXANE (124-09-4) Forms explosive mixture with air (flash point 80°C). (176°F/C) Ethylene dichloride, acids, organic anhydrides, isocyanates, vinyl acetate, acrylates, substituted allyls, alkylene oxides, epichlorohydrin, ketones, aldehydes, alcohols, glycols, phenols, cresols, caprolactam solution, strong oxidizers. Attacks aluminum, copper, lead, tin, zinc, and alloys.

1,5-DIAMINOPENTANE (462-94-2) Forms explosive mixture with air (flash point 144°F/62°C). Reacts violently with strong oxidizers. Reacts with acids. Attacks metals, some plastics, rubber, and coatings.

2,4-DIAMINOTOLUENE (95-80-7) Incompatible with oxidizers, acids, organic anhydrides, isocyanates, aldehydes. Attacks aluminum, brass, bronze, copper, zinc.

1,11-DIAMINO-3,6,9-TRIAZAUNDECANE (112-57-2) Forms explosive mixture with air. Incompatible with acids, organic anhydrides, halogenated hydrocarbons, isocyanates, nitroparaffins, vinyl acetate, acrylates, substituted allyls, alkylene oxides, epichlorohydrin, ketones, aldehydes, alcohols, glycols, mercury, phenols, cresols, caprolactam solution, strong oxidizers. Attacks aluminum, copper, lead, tin, zinc, and alloys.

1,6-DIAMINO-2,2,4 TRIMETHYLHEXANE (25513-64-8) Incompatible with nitric, sulfuric, mineral, and organic acids, organic anhydrides, isocyanates, vinyl acetate, acrylates, substituted allyls, alkylene oxides, ketones, aldehydes, alcohols, glycols, phenols, cresols, caprolactam solution.

DIAMMONIUM CHROMATE (7788-98-9) Contact with water produces an alkaline solution, with evolution of free ammonia.

DIAMMONIUM CITRATE (3012-65-5) Combustible solid. Reacts with strong oxidizers.

DIAMMONIUM HYDROGEN PHOSPHATE (7783-28-0) Incompatible with strong oxidizers, strong bases. Contact with air causes substance to give off corrosive anhydrous ammonia fumes.

DIAMMONIUM ORTHOPHOSPHATE (7783-28-0) Incompatible with strong oxidizers, strong bases. Contact with air causes substance to give off corrosive anhydrous ammonia fumes.

DIAMMONIUM PHOSPHATE (7783-28-0) Incompatible with strong oxidizers, strong bases. Contact with air causes substance to give off corrosive anhydrous ammonia fumes.

DIAMMONIUM SALT OF ZINC EDTA (12519-36-7) Incompatible with sulfuric acid, isocyanates.

DIAMMONIUM SULFATE (7783-20-2) Aqueous solution is a strong acid; reacts with bases, producing ammonia. Attacks metals. Hot material reacts with nitrates, nitrites, chlorates.

DIAMMONIUM SULFIDE SOLUTION (12124-99-1) Forms explosive mixture with air (flash point 72°F/22°C). Incompatible with strong oxidizers, strong acids. Contact with strong bases produces ammonia. Corrodes aluminum, brass, copper, zinc and related alloys; attacks some plastics, rubber, and coatings. May accumulate static electrical charges, and may cause ignition of its vapors.

DIAMYL ETHER (693-65-2) Has flash point of 134°F/56°C. Reacts violently with strong oxidizers. Ethers can forms unstable and explosive peroxides.

DIAMYL PHTHALATE (131-18-0) Incompatible with strong acids, nitrates. Attacks some plastics, rubber, and coatings.

DI-*n*-AMYL PHTHALATE (3648-21-3) Incompatible with strong acids, strong alkalies, nitrates, strong oxidizers. Attacks some plastics, rubber, and coatings.

DIANAT (Russian) (1918-00-9) Incompatible with sulfuric acid, bases, ammonia, aliphatic amines, alkanolamines, isocyanates, alkylene oxides, epichlorohydrin.

DIANATE (1918-00-9) Incompatible with sulfuric acid, bases, ammonia, aliphatic amines, alkanolamines, isocyanates, alkylene oxides, epichlorohydrin.

p,p'-**DIANILINE** (92-87-5) Oxidizes on exposure to light and air. Reacts violently with strong oxidizers. Red fuming nitric acid may cause fire.

DIANON (333-41-5) Incompatible with water, copper-containing compounds, oxidizers, acids, bases.

DIAPADRIN (141-66-2) Corrosive to cast iron, mild steel, brass, and stainless steel 304.

DIAREX HF 77 (100-42-5) If inhibitor (often *tert*-butyl catechol) is not present in adequate concentrations, polymerization may occur and explode containers. Temperatures above 150°F/66°C speed up polymerization. Incompatible with oxidizers, acids, rust, catalysts for vinyl polymerization, such as peroxides, strong acids, and aluminum chloride. Corrodes copper and copper alloys, and dissolves rubber. Attacks some plastics, rubber, and coatings. May accumulate static electrical charges, and may cause ignition of its vapors.

DIARSENIC PENTOXIDE (1303-28-2) Incompatible with acids, aluminum, halogens, rubidium carbide, zinc. Aqueous solutions are acidic.

DIARSENIC TRIOXIDE (1327-53-3) Incompatible with acids, fluorine, fluorides, sodium chlorate.

DIARSENIC TRISULFIDE (1303-33-9) Water contact produces hydrogen sulfide. Incompatible with oxidizers, acids, halogens.

DIATER (330-54-1) Hydrolyzes in fairly strong acids.

DIATERR-FOS (333-41-5) Incompatible with water, copper-containing compounds, oxidizers, acids, bases.

1,4-DIAZACYCLOHEXANE (110-85-0) Forms explosive mixture with air (flash point 178°F/81°C). Incompatible with nitrogen compounds, carbon tetrachloride, strong oxidizers. Aqueous solutions react with acids. Attacks aluminum, copper, nickel, magnesium, and zinc.

DIAZAJET (333-41-5) Incompatible with water, copper-containing compounds, oxidizers, acids, bases.

3,6-DIAZAOCTANE-1,8-DIAMINE (112-24-3) Forms explosive mixture with air (flash point 275°F/135°C). Incompatible with acids, organic anhydrides, chlorinated hydrocarbons, isocyanates, vinyl acetate, acrylates, substituted allyls, alkylene oxides, epichlorohydrin, ketones, aldehydes, alcohols, glycols, phenols, cresols, caprolactam solution, strong oxidizers. Attacks aluminum, copper, lead, tin, zinc, and alloys.

3,6-DIAZAOCTANEDIOIC ACID, 3,6-BIS(CARBOXYMETHYL)- (60-00-4) Incompatible with sulfuric acid, bases, ammonia, aliphatic amines, alkanolamines, isocyanates, alkylene oxides, epichlorohydrin.

DIAZATOL (333-41-5) Incompatible with water, copper-containing compounds, oxidizers, acids, bases.

DIAZIDE (333-41-5) Incompatible with water, copper-containing compounds, oxidizers, acids, bases.

DIAZINON or DIAZINONE (333-41-5) Incompatible with water, copper-containing compounds, oxidizers, acids, bases.

DIAZIRINE Heat, shock, friction, sunlight or other bright illuminations may cause explosions. Contact with alkali metals, drying agents such as calcium sulfate, or rough edges (e.g., ground glass) will cause explosions.

DIAZITOL (333-41-5) Incompatible with water, copper-containing compounds, oxidizers, acids, bases.

DIAZOL (333-41-5) Incompatible with water, copper-containing compounds, oxidizers, acids, bases.

DIAZOMETHANE (334-88-3) Heat, shock, friction, sunlight, other bright illuminations may cause explosions. Contact with alkali metals, drying agents (e.g., calcium sulfate), or rough edges (e.g., ground glass) will cause explosions.

DIAZONIUM Heat, shock, friction, sunlight, or other bright illuminations may cause explosions. Contact with alkali metals, drying agents such as calcium sulfate, or rough edges (e.g., ground glass) may cause explosions.

DIBAL H (1191-15-7) Ignites spontaneously in air. A powerful reducing agent; reacts violently with oxidizers, combustibles. Highly reactive; contact with many substances, including water, alcohols, amines, carbon dioxide, cresols, glycols, halogenated hydrocarbons, phenols, can cause fire and explosions.

DIBASIC LEAD ARSENATE (7784-40-9) Contact with strong oxidizers may cause fire and explosions.

DIBENZO (B,E) PYRIDINE (260-94-6) Contact with strong oxidizers may cause fire and explosions.

DIBENZOL DIPROPYLENE GLYCOL ESTER (94-51-9) Incompatible with strong acids, nitrates, oxidizers.

DIBENZOYL PEROXIDE (94-36-0) Confined storage of dry chemical may lead to decomposition and explosion. A strong oxidant; extremely reactive. Fires and explosion may result from heat or contamination, and from contact with strong acids, combustible materials, oxidizers, acids, bases, alcohols, reducing agents, metals, metal oxides, amines, accelerators, methyl methacrylate, organic matter, lithium aluminum carbide, dimethyl aniline, amines, metallic naphthenates. May attack some plastics, rubber, and coatings. Protect containers from shock and friction.

DIBENZYL ETHER (103-50-4) Forms unstable peroxides. Reacts violently with strong oxidizers. May accumulate static electrical charges, and may cause ignition of its vapors.

DIBK (108-83-8) Incompatible with strong acids, aliphatic amines, strong oxidizers. Attacks plastics, coatings, and rubber. Forms explosive mixture with air (flash point 140°F/60°C).

DIBORANE (19287-45-7) Unstable above 0°F/−18°C . May ignite spontaneously in moist air (flash point: room temperature). Contact with air, halogenated compounds may cause fire and explosion. Contact with aluminum, lithium, and other active metals produces hydrides, which may ignite spontaneously. Incompatible with aluminum, carbon tetrachloride, nitric acid, nitrogen trifluoride, and many other chemicals. Reacts with oxidized surfaces as a strong reducing agent. Attacks some plastics, rubber, or coatings.

DIBORANE (6) (19287-45-7) Unstable above 0°F/−18°C. May ignite spontaneously in moist air (flash point: room temperature). Contact with air, halogenated compounds may cause fire and explosion. Contact with aluminum, lithium and other active metals produces hydrides, which may ignite spontaneously. Incompatible with aluminum, carbon tetrachloride, nitric acid, nitrogen trifluoride, and many other chemicals. Reacts with oxidized surfaces as a strong reducing agent. Attacks some plastics, rubber, or coatings.

DIBORANE HEXANHYDRIDE (19287-45-7) Unstable above 0°F/−18°C. May ignite spontaneously in moist air (flash point: room temperature). Contact with air, halogenated compounds may cause fire and explosion. Contact with aluminum, lithium, and other active metals produces hydrides, which may ignite spontaneously. Incompatible with aluminum, carbon tetrachloride, nitric acid, nitrogen trifluoride and many other chemicals. Reacts with oxidized surfaces as a strong reducing agent. Attacks some plastics, rubber, or coatings.

DIBORON TRIOXIDE (1303-86-2) Water contact produces boric acid. Fires and explosion may result from contact with bromine pentafluoride or calcium oxide.

DIBOVAN (50-29-3) Incompatible with salts of iron or aluminum, and bases. Do not store in iron containers.

DIBROM (300-76-5) Contact with strong oxidizers may cause fire and explosions. Attacks some plastics, rubber, and coatings.

1,2-DIBROMAETHAN (German) (106-93-4) Reacts with chemically active metals, liquid ammonia, strong oxidizers. Heat and light cause slow decomposition. Attacks some plastics and rubber.

O-(1,2-DIBROM-2,2-DICHLOR-AETHYL)-O,O-DIMETHYL-PHOSPHAT (German) (300-76-5) Contact with strong oxidizers may cause fire and explosions. Attacks some plastics, rubber, and coatings.

1,2-DIBROMO-3-CHLORO- (96-12-8) Forms explosive mixture with air (flash point 170°F/77°C). Reacts with oxidizers and chemically active metals (i.e., aluminum, magnesium, and tin alloys). Attacks some rubber materials and coatings.

DIBROMOCHLOROPROPANE (96-12-8) Forms explosive mixture with air (flash point 170°F/77°C). Reacts with oxidizers and chemically active metals (i.e., aluminum, magnesium, and tin alloys). Attacks some rubber materials and coatings.

1,2-DIBROMO-3-CHLOROPROPANE (96-12-8) Forms explosive mixture with air (flash point 170°F/77°C). Reacts with oxidizers and chemically active metals (i.e., aluminum, magnesium, and tin alloys. Attacks some rubber materials and coatings.

2-DIBROMO-3-CHLORO-PROPANE (96-12-8) Forms explosive mixture with air (flash point 170°F/77°C). Reacts with oxidizers and chemically active metals (i.e., aluminum, magnesium, and tin alloys). Attacks some rubber materials and coatings.

O-(1,2-DIBROMO-2,2-DICLORO-ETIL)-O,O-DIMETIL-FOSFA-TO (Italian) (300-76-5) Contact with strong oxidizers may cause fire and explosions. Attacks some plastics, rubber, and coatings.

1,2-DIBROMO-2,2-DICHLOROETHYL DIMETHYL PHOSPHATE (300-76-5) Contact with strong oxidizers may cause fire and explosions. Attacks some plastics, rubber, and coatings.

DIBROMODIFLUOROMETHANE (75-61-6) Reacts with chemically active metals and oxidizers. Attacks some plastics, rubber, and coatings.

DIBROMODIFLUORO-METHANE (75-61-6) Reacts with chemically active metals and oxidizers. Attacks some plastics, rubber, and coatings.

1,2-DIBROMOETANO (Italian) (106-93-4) Reacts with chemically active metals, liquid ammonia, strong oxidizers. Heat and light cause slow decomposition. Attacks some plastics and rubber.

DIBROMOETHANE (106-93-4) Reacts with chemically active metals, liquid ammonia, strong oxidizers. Heat and light cause slow decomposition. Attacks some plastics and rubber.

1,2-DIBROMOETHANE (106-93-4) Reacts with chemically active metals, liquid ammonia, strong oxidizers. Heat and light cause slow decomposition. Attacks some plastics and rubber.

alpha, beta-DIBROMOETHANE (106-93-4) Reacts with chemically active metals, liquid ammonia, strong oxidizers. Heat and light cause slow decomposition. Attacks some plastics and rubber.

sym-DIBROMOETHANE (106-93-4) Reacts with chemically active metals, liquid ammonia, strong oxidizers. Heat and light cause slow decomposition. Attacks some plastics and rubber.

DIBROMOMETHANE (74-95-3) Mixture with potassium forms a shock-sensitive explosive.

DIBROMURE d'ETHYLENE (French) (106-93-4) Reacts with chemically active metals, liquid ammonia, strong oxidizers. Heat and light cause slow decomposition. Attacks some plastics and rubber.

O-(1,2-DIBROOM-2,2-DICHLOOR-ETHYL)-O,O-DIMETHYL-FOSFAAT (Dutch) (300-76-5) Contact with strong oxidizers may cause fire and explosions. Attacks some plastics, rubber, and coatings.

1,2-DIBROOMETHAAN (Dutch) (106-93-4) Reacts with chemically active metals, liquid ammonia, strong oxidizers. Heat and light cause slow decomposition. Attacks some plastics and rubber.

1,2-DIBUTOXYETHANE (112-48-1) Forms explosive mixture with air (flash point 185°F/85°C). Incompatible with sulfuric acid, isocyanates, perchloric acid.

2,2'-DIBUTOXYETHYL ETHER (112-73-2) Incompatible with sulfuric acid, isocyanates, strong oxidizers. May accumulate static electrical charges, and may cause ignition of its vapors.

DIBUTYL ACID PHOSPHATE (107-66-4) Incompatible with strong oxidizers. Attacks some plastics, rubber, and coatings.

DIBUTYL ACID O-PHOSPHATE (107-66-4) Incompatible with strong oxidizers. Attacks some plastics, rubber, and coatings.

DIBUTYL ADIPINATE (103-23-1) Incompatible with strong acids, strong oxidizers, nitrates.

DIBUTYL ALUMINUM HYDRIDE (1191-15-7) Ignites spontaneously in air. A powerful reducing agent; reacts violently with oxidizers, combustibles. Highly reactive; contact with many substances, including water, alcohols, amines, carbon dioxide, cresols, glycols, halogenated hydrocarbons, phenols, can cause fire and explosions.

DIBUTYLAMINE (111-92-2) Forms explosive mixture with air (flash point 117°F/47°C). Incompatible with acids, organic anhydrides, isocyanates, vinyl acetate, acrylates, substituted allyls, alkylene oxides, epichlorohydrin, ketones, aldehydes, alcohols, glycols, phenols, cresols, caprolactam solution, strong oxidizers. Contact with copper alloys, zinc, or galvanized steel may cause violent reaction.

n-**DIBUTYLAMINE** (111-92-2) Forms explosive mixture with air (flash point 117°F/47°C). Incompatible with acids, organic anhydrides, isocyanates, vinyl acetate, acrylates, substituted allyls, alkylene oxides, epichlorohydrin, ketones, aldehydes, alcohols, glycols, phenols, cresols, caprolactam solution, strong oxidizers. Contact with copper alloys, zinc, or galvanized steel may cause violent reaction.

DI-*n*-BUTYLAMINE (111-92-2) Forms explosive mixture with air (flash point 117°F/47°C). Incompatible with acids, organic anhydrides, isocyanates, vinyl acetate, acrylates, substituted allyls, alkylene oxides, epichlorohydrin, ketones, aldehydes, alcohols, glycols, phenols, cresols, caprolactam solution, strong oxidizers. Contact with copper alloys, zinc,or galvanized steel may cause violent reaction.

DIBUTYLAMINOETHANOL (102-81-8) Forms explosive mixture with air (flash point 200°F/93°C). Reacts violently with strong oxidizers.

N,N-DI-N-BUTYLAMINOETHANOL (102-81-8) Forms explosive mixture with air (flash point 200°F/93°C). Reacts violently with strong oxidizers.

2-DIBUTYLAMINOETHANOL (102-81-8) Forms explosive mixture with air (flash point 200°F/93°C). Reacts violently with strong oxidizers.

2-(DIBUTYLAMINO)ETHANOL 2-N-DIBUTYLAMINOETHA-NOL (102-81-8) Forms explosive mixture with air (flash point 200°F/93°C). Reacts violently with strong oxidizers.

beta-n-DIBUTYLAMINOETHYL ALCOHOL (102-81-8) Forms explosive mixture with air (flash point 200°F/93°C). Reacts violently with strong oxidizers.

DIBUTYL-1,2-BENZENEDICARBOXYLATE (84-74-2) Incompatible with strong acids, nitrates, strong oxidizers, strong alkalies.

DIBUTYL CARBITOL (112-73-2) Incompatible with sulfuric acid, isocyanates, strong oxidizers. May accumulate static electrical charges, and may cause ignition of its vapors.

DIBUTYL CELLOSOLVE (112-48-1) Forms explosive mixture with air (flash point 185°F/85°C). Incompatible with sulfuric acid, isocyanates, perchloric acid.

N,N-DIBUTYLETHANOLAMINE (102-81-8) Forms explosive mixture with air (flash point 200°F/93°C). Reacts violently with strong oxidizers.

DIBUTYL ETHER (142-96-1) Forms explosive mixture with air (flash point 77°F/25°C). May accumulate static electrical charges, and may cause ignition of its vapors. Incompatible with strong acids, oxidizers. Air contact or light may produce unstable and explosive peroxides.

n-**DIBUTYL ETHER** (142-96-1) Forms explosive mixture with air (flash point 77°F/25°C). May accumulate static electrical charges, and may cause ignition of its vapors. Incompatible with strong acids, oxidizers. Air contact or light may produce unstable and explosive peroxides.

DI-*n*-BUTYL ETHER (142-96-1) Forms explosive mixture with air (flash point 77°F/25°C). May accumulate static electrical charges, and may cause ignition of its vapors. Incompatible with strong acids, oxidizers. Air contact or light may produce unstable and explosive peroxides.

DIBUTYL HEXANEDIOATE (103-23-1) Incompatible with strong acids, strong oxidizers, nitrates.

DI-*n*-BUTYL HYDROGEN PHOSPHATE (107-66-4) Incompatible with strong oxidizers. Attacks some plastics, rubber, and coatings.

N,N-DIBUTYL N-(2-HYDROXYETHYL)AMINE (102-81-8) Forms explosive mixture with air (flash point 200°F/93°C). Reacts violently with strong oxidizers.

DIBUTYL OXIDE (142-96-1) Forms explosive mixture with air (flash point 77°F/25°C). May accumulate static electrical charges, and may cause ignition of its vapors. Incompatible with strong acids, oxidizers. Air contact or light may produce unstable and explosive peroxides.

DIBUTYLPHENOL (26746-38-3) Reacts with oxidizers, with a risk of fire or explosions.

2,6-DI-*tert*-BUTYLPHENOL (26746-38-3) Reacts with oxidizers, with a risk of fire or explosions.

DIBUTYL PHOSPHORIC ACID (107-66-4) Incompatible with strong oxidizers. Attacks some plastics, rubber, and coatings.

DIBUTYL PHOSPHATE (107-66-4) Incompatible with strong oxidizers. Attacks some plastics, rubber, and coatings.

DIBUTYL PHTHALATE (84-74-2) Incompatible with strong acids, nitrates, strong oxidizers, strong alkalies.

DI-*n*-BUTYL PHTHALATE (84-74-2) Incompatible with strong acids, nitrates, strong oxidizers, strong alkalies.

DICALCIUM PHOSPHATE (10103-46-5) Water contact forms acid; attacks metals.

DICAMBA (1918-00-9) Incompatible with sulfuric acid, bases, ammonia, aliphatic amines, alkanolamines, isocyanates, alkylene oxides, epichlorohydrin.

DICAPROATE (112-27-6) Incompatible with sulfuric acid, isocyanates, perchloric acid, strong oxidizers.

DICARBAM (63-25-2) Incompatible with strong oxidizers, strongly alkaline pesticides.

DICARBOMETHOXY ZINC (557-34-6) Incompatible with strong acids, strong bases, nitrates.

o-**DICARBOXYBENZENE** (88-99-3) Incompatible with strong acids, nitrates, oxidizers.

DICARBURRETTED HYDROGEN (74-85-1) Incompatible with strong acids, oxidizers, halogens, nitric acids, aluminum chloride, bromotrichloromethane, carbon tetrachloride, chlorine, chlorine dioxide, nitrogen dioxide.

1,4-DICHLOORBENZENE (Dutch) (106-46-7) Forms explosive mixture with air (flash point 150°F/66°C). Incompatible with strong oxidizers, metal powders and alkali metals (i.e., lithium, sodium, potassium, rubidium, cesium, francium). Attacks some plastics, rubber, and coating.

p-**DICHLOORBENZENE (Dutch)** (106-46-7) Forms explosive mixture with air (flash point 150°F/66°C). Incompatible with strong oxidizers, metal powders and alkali metals (i.e., lithium, sodium, potassium, rubidium, cesium, francium). Attacks some plastics, rubber, and coating.

1,1-DICHLOORETHAAN (Dutch) (75-34-3) Forms explosive mixture with air (flash point 2°F/−17°C). Incompatible with strong oxidizers, strong caustics. Attacks plastics and rubber.

(2,4-DICHLOOR-FENOXY)-AZIJNZUUR (Dutch) (94-75-7) Decomposes in sunlight. Incompatible with strong oxidizers; may cause fire and explosions.

3-(3,4-DICHLOOR-FENYL)-1,1-DIMETHYLUREUM (Dutch) (330-54-1) Hydrolyzes in fairly strong acids.

1,1-DICHLORAETHAN (German) (75-34-3) Forms explosive mixture with air (flash point 2°F/−17°C). Incompatible with strong oxidizers, strong caustics. Attacks plastics and rubber.

1,2-DICHLOR-AETHEN (German) (540-59-0) Forms explosive mixture with air (flash point 36°F/2°C). Incompatible with strong bases, oxidizers, difluoromethylene, dihypofluoride, nitrogen tetroxide (explosive). Attacks some plastics, rubber, and coatings.

o-**DICHLORBENZENE** (95-50-1) Forms explosive mixture with air (flash point 151°F/66°C). Incompatible with strong oxidizers, hot aluminum, or aluminum alloy. Attacks some plastics, rubber, and coatings.

1,4-DICHLOR-BENZOL (German) (106-46-7) Forms explosive mixture with air (flash point 150°F/66°C). Incompatible with strong oxidizers, metal powders, and alkali metals (i.e., lithium, sodium, potassium, rubidium, cesium, francium). Attacks some plastics, rubber, and coatings.

o-**DICHLOR BENZOL** (95-50-1) Forms explosive mixture with air (flash point 151°F/66°C). Incompatible with strong oxidizers, hot aluminum, or aluminum, alloy. Attacks some plastics, rubber, and coatings.

p-**DICHLORBENZOL (German)** (106-46-7) Forms explosive mixture with air (flash point 150°F/66°C). Incompatible with strong oxidizers, metal powders and alkali metals (i.e., lithium, sodium, potassium, rubidium, cesium, francium). Attacks some plastics, rubber, and coating.

DICHLORDIMETHYLAETHER (German) (542-88-1) Highly volatile. Heat and light may causes formation of unstable peroxides. Heat may cause fire. Contact with water produces hydrogen chloride and formaldehyde.

1,1-DICHLORETHANE (75-34-3) Forms explosive mixture with air (flash point 2°F/−17°C). Incompatible with strong oxidizers, strong caustics. Attacks plastics and rubber.

DICHLORFENIDIM (330-54-1) Hydrolyzes in fairly strong acids.

DICHLORICIDE (95-50-1) Forms explosive mixture with air (flash point 151°F/66°C). Incompatible with strong oxidizers, hot aluminum, or aluminum alloy. Attacks some plastics, rubber, and coatings.

DI-CHLORICIDE (106-46-7) Forms explosive mixture with air (flash point 150°F/66°C). Incompatible with strong oxidizers, metal powders, and alkali metals (i.e., lithium, sodium, potassium, rubidium, cesium, francium). Attacks some plastics, rubber, and coatings.

DICHLORINE (7782-50-5) A powerful oxidizer. Reacts violently with combustible materials, reducing agents, and many other substances. Forms explosive mixtures with gasoline and petroleum products, turpentine, alcohols, acetylene, carbon disulfide, hydrogen, anhydrous ammonia, finely divided metals, organic compounds, phosphorus. Fire, explosion, and the formation of toxic fumes may result from contact with many substances including alkyphosphines, aluminum, antimony, arsenic compounds, arsine, bismuth, boron, brass, calcium compounds, carbon, diethyl zinc, fluorine, germanium, hydrocarbons, rubber. Attacks some plastics and coatings. In the presence of moisture, extremely corrosive to iron, steel, copper, bronze, zinc.

o-**DICHLORO-** (95-50-1) Forms explosive mixture with air (flash point 151°F/66°C). Incompatible with strong oxidizers, hot aluminum, or aluminum alloy. Attacks some plastics, rubber, and coatings.

DICHLOROACETIC ACID, METHYL ESTER (116-54-1) Forms explosive mixture with air (flash point 176°F/80°C). Water solution is corrosive. Strong oxidizers may cause fire and explosions.

DICHLOROACETYLENE (7572-29-4) Heat or air contact may cause explosion. Reacts violently with oxidizers and acids.

250

3,6-DICHLORO-O-ANISIC ACID (1918-00-9) Incompatible with sulfuric acid, bases, ammonia, aliphatic amines, alkanolamines, isocyanates, alkylene oxides, epichlorohydrin.

1,2-DICHLOROBENZENE (95-50-1) Forms explosive mixture with air (flash point 151°F/66°C). Incompatible with strong oxidizers, hot aluminum, or aluminum alloy. Attacks some plastics, rubber, and coatings.

1,3-DICHLOROBENZENE (541-73-1) Forms explosive mixture with air (flash point 151°F/66°C). Incompatible with strong oxidizers, alkali metals (i.e., lithium, sodium, potassium, rubidium, cesium, francium).

1,4-DICHLOROBENZENE (106-46-7) Forms explosive mixture with air (flash point 150°F/66°C). Incompatible with strong oxidizers, metal powders and alkali metals (i.e., lithium, sodium, potassium, rubidium, cesium, francium). Attacks some plastics, rubber, and coatings.

m-**DICHLOROBENZENE** or *meta*-**DICHLOROBENZENE** (541-73-1) Forms explosive mixture with air (flash point 151°F/66°C). Incompatible with strong oxidizers, alkali metals (i.e., lithium, sodium, potassium, rubidium, cesium, francium).

o-**DICHLOROBENZENE** or *ortho*-**DICHLOROBENZENE** (95-50-1) Forms explosive mixture with air (flash point 151°F/66°C). Incompatible with strong oxidizers, hot aluminum, or aluminum alloy. Attacks some plastics, rubber, and coatings.

p-**DICHLOROBENZENE** or *para*-**DICHLOROBENZENE** (106-46-7) Forms explosive mixture with air (flash point 150°F/66°C). Incompatible with strong oxidizers, metal powders, and alkali metals (i.e., lithium, sodium, potassium, rubidium, cesium, francium). Attacks some plastics, rubber, and coatings.

DICHLOROBENZENE, *ortho*, **LIQUID** (95-50-1) Forms explosive mixture with air (flash point 151°F/66°C). Incompatible with strong oxidizers, hot aluminum or aluminum alloy. Attacks some plastics, rubber, and coatings.

DICHLOROBENZENE, *para*, **SOLID** (106-46-7) Forms explosive mixture with air (flash point 150°F/66°C). Incompatible with strong oxidizers, metal powders, and alkali metals (i.e., lithium, sodium, potassium, rubidium, cesium, francium). Attacks some plastics, rubber, and coatings.

o-**DICHLOROBENZOL** or *ortho*-**DICHLOROBENZOL** (95-50-1) Forms explosive mixture with air (flash point 151°F/66°C). Incompatible with strong oxidizers, hot aluminum, or aluminum alloy. Attacks some plastics, rubber, and coatings.

p-**DICHLOROBENZOL** or *para*-**DICHLOROBENZOL** (106-46-7) Forms explosive mixture with air (flash point 150°F/66°C). Incompatible with strong oxidizers, metal powders, and alkali metals (i.e., lithium, sodium, potassium, rubidium, cesium, francium). Attacks some plastics, rubber, and coatings.

DI-(4-CHLOROBENZOYL) PEROXIDE (94-17-7) A powerful oxidizer and explosion hazard. Reacts violently with reducing agents, combustibles, polymerization initiators, heat, or contaminants.

DI-(p-CHLOROBENZOYL) PEROXIDE (94-17-7) A powerful oxidizer and explosion hazard. Reacts violently with reducing agents, combustibles, polymerization initiators, heat, or contaminants.

p,p'-DICHLOROBENZOYL PEROXIDE (94-17-7) A powerful oxidizer and explosion hazard. Reacts violently with reducing agents, combustibles, polymerization initiators, heat, or contaminants.

DICHLOROBUTENE (764-41-0) Forms explosive mixture with air (flash point 126°F/52°C). Water contact produces hydrochloric acid. Incompatible with strong oxidizers, bases. Corrodes metals in the presence of moisture.

1,4-DICHLORO-2-BUTYLENE (764-41-0) Forms explosive mixture with air (flash point 126°F/52°C). Water contact produces hydrochloric acid. Incompatible with strong oxidizers, bases. Corrodes metals in the presence of moisture.

DICHLOROCHLORDENE (57-74-9) Contact with strong oxidizers may cause fire and explosions. Attacks some plastics, rubber, and coatings.

1,1-DICHLORO-2,2-BIS(P-CHLORO-PHENYL) ETHANE (72-54-8) Contact with strong oxidizers may cause fire and explosions.

1,1-DICHLORO-2,2-DICHLOROETHANE (127-18-4) Reacts violently with strong oxidizers, and with powders from light metals and zinc. Attacks some plastics, rubber, and coatings.

DICHLOROCIDE (106-46-7) Forms explosive mixture with air (flash point 150°F/66°C). Incompatible with strong oxidizers, metal powders, and alkali metals (i.e., lithium, sodium, potassium, rubidium, cesium, francium). Attacks some plastics, rubber, and coating.

DICHLORODIETHYL ETHER (111-44-4) Water contact may produce hydrogen chloride fumes. Can form peroxides. Forms explosive mixture with air (flash point 131°F/55°C). Strong oxidizers may cause fire and explosions. Attacks some plastics, rubber, and coatings.

2,2'-DICHLORODIETHYL ETHER (111-44-4) Water contact may produce hydrogen chloride fumes. Can form peroxides. Forms explosive mixture with air (flash point 131°F/55°C). Strong oxidizers may cause fire and explosions. Attacks some plastics, rubber, and coatings.

DICHLORODIETHYL OXIDE (111-44-4) Water contact may produce hydrogen chloride fumes. Can form peroxides. Forms explosive mixture with air (flash point 131°F/55°C). Strong oxidizers may cause fire and explosions. Attacks some plastics, rubber, and coatings.

DICHLORODIFLUOROMETHANE (75-71-8) Reacts violently with liquid aluminum. Incompatible with chemically active metals. Attacks some plastics, rubber, and coatings.

DICHLORODIISOPROPYL ETHER (108-60-1) Forms explosive mixture with air (flash point 170°F/77°C). Reacts violently with strong oxidizers. Incompatible with aluminum, copper, epoxy coatings.

DICHLORODIMETHYL ETHER (542-88-1) Highly volatile. Heat and light may cause formation of unstable peroxides. Heat may cause fire. Contact with water produces hydrogen chloride and formaldehyde.

sym-DICHLORODIMETHYL ETHER (542-88-1) Highly volatile. Heat and light may cause formation of unstable peroxides. Heat may cause fire. Contact with water produces hydrogen chloride and formaldehyde.

DICHLORODIMETHYLSILANE (75-78-5) Forms explosive gas mixture with air (flash point 15°F/−9°C). Water (vapor) or air form hydrochloric acid. Incompatible with acetone, amines, ammonia, alcohols, strong oxidizers, caustics. Attacks most metals.

DICHLORO-DIOXOCHROMIUM (14977-61-8) A powerful oxidizer. Water contact produces hydrochloric and chromic acids and chlorine gas. Reacts violently with reducing agents, combustibles, ammonia, halides, phosphorus, sodium azide, elemental sulfur, urea.

DICHLORODIOXOCHROMIUM (14977-61-8) A powerful oxidizer. Water contact produces hydrochloric and chromic acids and chlorine gas. Reacts violently with reducing agents, combustibles, ammonia, halides, phosphorus, sodium azide, elemental sulfur, urea.

DICHLORODIPHENYLDICHLORO ETHANE (72-54-8) Contact with strong oxidizers may cause fire and explosions.

DICHLORODIPHENYLSILANE (80-10-4) Water contact produces hydrochloric acid. Strong oxidizers may cause fire and explosions. Corrodes metals in the presence of moisture.

DICHLORODIPHENYLSILICANE (80-10-4) Water contact produces hydrochloric acid. Strong oxidizers may cause fire and explosions. Corrodes metals in the presence of moisture.

DICHLORODIPHENYL TRICHLOROETHANE (50-29-3) Incompatible with salts of iron or aluminum, and bases. Do not store in iron containers.

4,4'-DICHLORODIPHENYLTRICHLOROETHANE (50-29-3) Incompatible with salts of iron or aluminum, and bases. Do not store in iron containers.

P,P'-DICHLORODIPHENYLTRICHLOROETHANE (50-29-3) Incompatible with salts of iron or aluminum, and bases. Do not store in iron containers.

1,1-DICHLOROETHANE (75-34-3) Forms explosive mixture with air (flash point 2°F/−17°C). Incompatible with strong oxidizers, strong caustics. Attacks plastics and rubber.

1,2-DICHLOROETHANE (107-06-2) May accumulate static electrical charges, and may cause ignition of its vapors. Contact with hot water may produce hydrochloric acid. Forms explosive mixture with air (flash point 55°F/13°C). Incompatible with strong oxidizers, strong caustics, chemically active metals. Corrosive to iron and other metals. Attacks some plastics, rubber, or coatings.

asym-**DICHLOROETHANE or** *asymmetrical* **DICHLOROETHANE** (75-34-3) Forms explosive mixture with air (flash point 2°F/−17°C). Incompatible with strong oxidizers, strong caustics. Attacks plastics and rubber.

1,1-DICHLOROETHENE (75-35-4) Forms explosive mixture with air (18°F/−28°C). Air or contaminants can cause formation of peroxides; may polymerize. Reacts violently with strong oxidizers, alkali metals. Incompatible with nitric acid.

DICHLOROETHER (111-44-4) Water contact may produce hydrogen chloride fumes. Can form peroxides. Forms explosive mixture with air (flash point 131°F/55°C). Strong oxidizers may cause fire and explosions. Attacks some plastics, rubber, and coatings.

DICHLOROETHYLENE (540-59-0) Forms explosive mixture with air (flash point 36°F/2.2°C). Incompatible with strong bases, oxidizers, difluoromethylene, dihypofluoride, nitrogen tetroxide (explosive). Attacks some plastics, rubber, and coatings.

1,1-DICHLOROETHYLENE (75-35-4) Forms explosive mixture with air (18°F/−28°C). Air or contaminants can cause formation of peroxides; may polymerize. Reacts violently with strong oxidizers, alkali metals. Incompatible with nitric acid.

DICHLORO-1,2-ETHYLENE (French) (540-59-0) Forms explosive mixture with air (flash point 36°F/2.2°C). Incompatible with strong bases, oxidizers, difluoromethylene, dihypofluoride, nitrogen tetroxide (explosive). Attacks some plastics, rubber, and coatings.

1,2-DICHLOROETHYLENE (540-59-0) Forms explosive mixture with air (flash point 36°F/2.2°C). Incompatible with strong bases, oxidizers, difluoromethylene, dihypofluoride, nitrogen tetroxide (explosive). Attacks some plastics, rubber, and coatings.

cis-**1,2-DICHLOROETHYLENE** (540-59-0) Forms explosive mixture with air (flash point 36°F/2.2°C). Incompatible with strong bases, oxidizers, difluoromethylene, dihypofluoride, nitrogen tetroxide (explosive). Attacks some plastics, rubber, and coatings.

trans-**1,2-DICHLOROETHYLENE** (540-59-0) Forms explosive mixture with air (flash point 36°F/2.2°C). Incompatible with strong bases, oxidizers, difluoromethylene, dihypofluoride, nitrogen tetroxide (explosive). Attacks some plastics, rubber, and coatings.

unsym-**DICHLOROETHYLENE** or *unsymmetrical*-**DICHLOROE-THYLENE** (75-35-4) Forms explosive mixture with air (18°F/−28°C). Air or contaminants can cause peroxides to form; may polymerize. Reacts violently with strong oxidizers, alkali metals. Incompatible with nitric acid.

DI-(2-CHLOROETHYL) ETHER (111-44-4) Water contact may produce hydrogen chloride fumes. Can form peroxides. Forms explosive mixture with air (flash point 131°F/55°C). Strong oxidizers may cause fire and explosions. Attacks some plastics, rubber, and coatings.

2,2′-DICHLOROETHYL ETHER (111-44-4) Water contact may produce hydrogen chloride fumes. Can form peroxides. Forms explosive mixture with air (flash point 131°F/55°C). Strong oxidizers may cause fire and explosions. Attacks some plastics, rubber, and coatings.

DICHLOROETHYLPHENYLSILANE (1125-27-5) Forms explosive mixture with air (flash point 150°F/66°C). Incompatible with strong bases, water (forms hydrochloric acid). Corrodes most metals. Attacks some plastics, rubber, and coatings.

DICHLOROETHYLSILANE (1789-58-8) Forms explosive mixture with air (flash point 30°F/1°C). Water contact produces hydrochloric acid. Incompatible with strong bases. Attacks most common metals.

DICHLOROETHYNE (7572-29-4) Heat or air contact may cause explosion. Reacts violently with oxidizers and acids.

DICHLOROFLUOROMETHANE (75-43-4) Incompatible with chemically active metals. Attacks some plastics, rubber, and coatings.

1,6-DICHLOROHEXANE (2163-00-0) Forms explosive mixture with air 165°F/74°C). Strong oxidizers may cause fire and explosions.

DICHLOROISOCYANURIC ACID POTASSIUM SALT (2244-21-5) Incompatible with water; forms a bleach solution. A strong oxidizer; reacts violently with reducing agents, combustibles, organics, easily chlorinated materials.

DICHLOROISOPROPYL ETHER (108-60-1) Forms explosive mixture with air (flash point 170°F/77°C). Reacts violently with strong oxidizers. Incompatible with aluminum, copper, epoxy coatings.

2,2'-DICHLOROISOPROPYL ETHER (108-60-1) Forms explosive mixture witn air (flash point 170°F/77°C). Reacts violently with strong oxidizers. Incompatible with aluminum, copper, epoxy coatings.

DICHLOROMETHANE (75-09-2) Incompatible with strong oxidizers, strong caustics, chemically active metals. Attacks some plastics, rubber, and coatings. May accumulate static electrical charges, and may cause ignition of its vapors.

DICHLOROMETHYLETHANE (75-34-3) Forms explosive mixture with air (flash point 2°F/−17°C). Incompatible with strong oxidizers, strong caustics. Attacks plastics and rubber.

sym-**DICHLOROMETHYL ETHER** (542-88-1) Highly volatile. Heat and light may cause formation of unstable peroxides. Heat may cause fire. Contact with water produces hydrogen chloride and formaldehyde.

DICHLOROMETHYLSILANE (75-54-7) Forms explosive mixture with air (flash point 15°F/−9°C). Water and air vapor form corrosive fumes. Incompatible with strong bases and oxidizers. Corrodes most metals. Attacks some plastics, rubber, and coatings.

DICHLOROMONOFLUOROMETHANE (75-43-4) Reacts with water and chemically active metals. Attacks some plastics, rubber, and coatings.

DICHLORONITROETHANE (594-72-9) Contact with strong oxidizers may cause fire and explosions. Attacks some plastics, rubber, and coatings.

1,1-DICHLORO-1-NITROETHANE (594-72-9) Contact with strong oxidizers may cause fire and explosions. Attacks some plastics, rubber, and coatings.

2,4-DICHLOROPHENOL (120-83-2) Reacts violently with strong oxidizers, acids. Incompatible with caustics. Quickly corrodes aluminum; slowly corrodes zinc, tin, brass, bronze, copper and its alloys. May accumulate static electrical charges, and may cause ignition of its vapors.

3-(3,4-DICHLOROPHENOL)-1,1-DIMETHYL UREA (330-54-1) Hydrolyzes in fairly strong acids.

(2,4-DICHLOROPHENOXY)ACETIC ACID, ISOPROPYL ESTER (94-11-1) Forms explosive mixture with air 175°F/79°C). Incompatible with strong oxidizers, strong acids, nitrates. Attacks some plastics, rubber, and coatings.

DICHLOROPHENOXYACETIC ACID (94-75-7) Decomposes in sunlight. Incompatible with strong oxidizers; may cause fire and explosions.

2,4-DICHLOROPHENOXYACETATE (94-11-1) Forms explosive mixture with air 175°F/79°C). Incompatible with strong oxidizers, strong acids, nitrates. Attacks some plastics, rubber, and coatings.

2,4-DICHLOROPHENOXYACETIC ACID, BUTOXYETHYL ESTER (94-11-1) Forms explosive mixture with air 175°F/79°C). Incompatible with strong oxidizers, strong acids, nitrates. Attacks some plastics, rubber, and coatings.

2,4-DICHLOROPHENOXYACETIC ACID, SALTS AND ESTERS (94-75-7) Decomposes in sunlight. Incompatible with strong oxidizers; may cause fire and explosions.

3-(3,4-DICHLOROPHENYL)-1,1-DIMETHYLUREA (330-54-1) Hydrolyzes in fairly strong acids.

N′-(3,4-DICHLOROPHENYL)-N,N-DIMETHYLUREA (330-54-1) Hydrolyzes in fairly strong acids.

1-(3,4-DICHLOROPHENYL)-3,3-DIMETHYLUREE (French) (330-54-1) Hydrolyzes in fairly strong acids.

DICHLOROPHENYLARSINE (696-28-6) Forms explosive mixture with air (flash point 60°F/−16°C). Water contact produces hydrochloric acid. Attacks metals in the presence of moisture.

DICHLOROPHENYLPHOSPHINE (644-97-3) Water contact produces hydrochloric acid. Corrodes metals except 316 stainless steel, nickel, and Hastelloy.

DI-(p-CHLOROPHENYL)TRICHLOROMETHYLCARBINOL (115-32-2) Incompatible with strong acids, caustics, aliphatic amines, isocyanates.

DICHLOROPHOS (62-73-7) Attacks some plastics, rubber, and coatings.

DICHLOROPHOSPHORIC ACID, ETHYL ESTER (1498-51-7) Water contact produces hydrochloric acid. Attacks metals when wet.

DICHLOROPROPANE (78-87-5) Forms explosive mixture with air (flash point 60°F/16°C). May accumulate static electrical charges, and may cause ignition of its vapors. Strong oxidizers may cause fire and explosions. Strong acids can cause decomposition and the formation of hydrogen chloride vapors. Corrodes aluminum. Attacks some plastics, rubber, and coatings.

1,1-DICHLOROPROPANE (78-99-9) Forms explosive mixture with air (flash point about 60°F). Strong oxidizers may cause fire and explosions. Corrodes aluminum. Attacks some plastics, rubber, and coatings.

DICHLORO-1,2-PROPANE (78-87-5) Forms explosive mixture with air (flash point 60°F/16°C). May accumulate static electrical charges, and may cause ignition of its vapors. Strong oxidizers may cause fire and explosions. Strong acids can cause decomposition and the formation of hydrogen chloride vapors. Corrodes aluminum. Attacks some plastics, rubber, and coatings.

1,2-DICHLOROPROPANE (78-87-5) Forms explosive mixture with air (flash point 60°F/16°C). May accumulate static electrical charges, and may cause ignition of its vapors. Strong oxidizers may cause fire and explosions. Strong acids can cause decomposition and the formation of hydrogen chloride vapors. Corrodes aluminum. Attacks some plastics, rubber, and coatings.

1,3-DICHLOROPROPANE (142-28-9) Forms explosive mixture with air (flash point 70°F/21°C). Incompatible with oxidizers, light metals. Attacks some plastics, rubber, and coatings. May accumulate static electrical charges, and may cause ignition of its vapors.

2,3-DICHLORO-1-PROPANE (78-88-6) Incompatible with strong acids, oxidizers, aluminum or magnesium compounds, aliphatic amines, alkanolamines, alkaline matrials, or corrosives.

alpha, beta-DICHLOROPROPANE (78-87-5) Forms explosive mixture with air (flash point 60°F/16°C). May accumulate static electrical charges, and may cause ignition of its vapors. Strong oxidizers may cause fire and explosions. Strong acids can cause decomposition and the formation of hydrogen chloride vapors. Corrodes aluminum. Attacks some plastics, rubber, and coatings.

2,2-DICHLOROPROPANOIC ACID (75-99-0) Corrosive to iron, aluminum, and copper.

DICHLOROPROPENE (542-75-6) Forms explosive mixture with air (flash point 95°F/35°C). Reacts violently with strong oxidizers. May accumulate static electrical charges, and may cause ignition of its vapors. Incompatible with strong acids, oxidizers, aluminum or magnesium compounds, aliphatic amines, alkanolamines, alkaline matrials, or corrosives.

1,3-DICHLOROPROPENE (542-75-6) Forms explosive mixture with air (flash point 95°F/35°C). Reacts violently with strong oxidizers. May accumulate static electrical charges, and may cause ignition of its vapors. Incompatible with strong acids, oxidizers, aluminum or magnesium compounds, aliphatic amines, alkanolamines, alkaline matrials, or corrosives.

1,3-DICHLORO-1-PROPENE (542-75-6) Forms explosive mixture with air (flash point 95°F/35°C). Reacts violently with strong oxidizers. May accumulate static electrical charges, and may cause ignition of its vapors. Incompatible with strong acids, oxidizers, aluminum or magnesium compounds, aliphatic amines, alkanolamines, alkaline matrials, or corrosives.

1,3-DICHLOROPROPENE-1 (542-75-6) Forms explosive mixture with air (flash point 95°F/35°C). Reacts violently with strong oxidizers. May accumulate static electrical charges, and may cause ignition of its vapors. Incompatible with strong acids, oxidizers, aluminum or magnesium compounds, aliphatic amines, alkanolamines, alkaline matrials, or corrosives.

2,3-DICHLOROPROPENE (78-88-6) Incompatible with strong acids, oxidizers, aluminum or magnesium compounds, aliphatic amines, alkanolamines, alkaline matrials, or corrosives.

2,3-DICHLORO-1-PROPENE (78-88-6) Incompatible with strong acids, oxidizers, aluminum or magnesium compounds, aliphatic amines, alkanolamines, alkaline matrials, or corrosives.

DICHLOROPROPENE, DICHLOROPROPANE MIXTURE (8003-19-8) Incompatible with strong acids, oxidizers, aluminum or magnesium compounds, aliphatic amines, alkanolamines, alkaline matrials, or corrosives.

1,3-DICHLOROPROPENE AND 1,2-DICHLOROPROPANE MIXTURE (8003-19-8) Incompatible with strong acids, oxidizers, aluminum or magnesium compounds, aliphatic amines, alkanolamines, alkaline matrials or corrosives.

2,2-DICHLOROPROPIONIC ACID (75-99-0) Corrosive to iron, aluminum, and copper.

alpha-DICHLOROPROPIONIC ACID (75-99-0) Corrosive to iron, aluminum, and copper.

alpha,alpha-DICHLOROPROPIONIC ACID (75-99-0) Corrosive to iron, aluminum, and copper.

1,3-DICHLOROPROPYLENE (542-75-6) Forms explosive mixture with air (flash point 95°F/35°C). Reacts violently with strong oxidizers. May accumulate static electrical charges, and may cause ignition of its vapors. Incompatible with strong acids, oxidizers, aluminum or magnesium compounds, aliphatic amines, alkanolamines, alkaline matrials, or corrosives.

2,3-DICHLOROPROPYLENE (78-88-6) Incompatible with strong acids, oxidizers, aluminum or magnesium compounds, aliphatic amines, alkanolamines, alkaline matrials, or corrosives.

alpha,gamma-DICHLOROPROPYLENE (542-75-6) Forms explosive mixture with air (flash point 95°F/35°C). Reacts violently with strong oxidizers. May accumulate static electrical charges, and may cause ignition of its vapors. Incompatible with strong acids, oxidizers, aluminum or magnesium compounds, aliphatic amines, alkanolamines, alkaline matrials, or corrosives.

DICHLOROTETRAFLUOROETHANE (76-14-2) Incompatible with chemically active metals. Attacks some plastics, rubber, and coatings.

1,2-DICHLOROTETRAFLUOROETHANE (76-14-2) Incompatible with chemically active metals. Attacks some plastics, rubber, and coatings.

DICHLOROTHIOCARBONYL (463-71-8) Water contact produces hydrochloric acid, carbon disulfide and carbon dioxide. Corrodes most metals.

alpha,alpha-DICHLOROTOLUENE (98-87-3) Forms explosive mixture with air (flash point 153°F/67°C). Reacts with acids, bases, strong oxidizers. Forms acid fumes with air. Attacks plastics and coatings.

DICHLORO-S-TRIAZINE-2,4,6(1H,3H,5H)-TRIONE POTASSIUM DERIV (2244-21-5) Incompatible with water; forms a bleach solution. A strong oxidizer; reacts violently with reducing agents, combustibles, organics, easily chlorinated materials.

1,3-DICHLORO-S-TRIAZINE-2,4,6(1H,3H,5H)TRIONE POTASSIUM SALT (2244-21-5) Incompatible with water, forms a bleach solution. A strong oxidizer; reacts violently with reducing agents, combustibles, organics, easily chlorinated materials.

4,4'-DICHLORO-alpha-TRICHLOROMETHYLBENZHYDROL (115-32-2) Incompatible with strong acids, caustics, aliphatic amines, isocyanates.

2,2-DICHLOROVINYL DIMETHYL PHOSPHATE (62-73-7) Attacks some plastics, rubber, and coatings.

2,2-DICHLOROVINYL O,O-DIMETHYL PHOSPHATE (62-73-7) Attacks some plastics, rubber, and coatings.

2,4-DICHLORPHENOXYACETIC ACID (94-75-7) Decomposes in sunlight. Incompatible with strong oxidizers; may cause fire and explosions.

(2,4-DICHLOR-PHENOXY)-ESSIG SAEURE (German) (94-75-7) Decomposes in sunlight. Incompatible with strong oxidizers; may cause fire and explosions.

3-(3,4-DICHLOR-PHENYL)-1,1-DIMETHYL-HARNSTOFF (German) (330-54-1) Hydrolyzes in fairly strong acids.

DICHLORPROPAN-DICHLORPROPEN GEMISCH (German) (8003-19-8) Incompatible with strong acids, oxidizers, aluminum or magnesium compounds, aliphatic amines, alkanolamines, alkaline matrials or corrosives.

p-DICHLOR-S-TRIAZIN-2,4,6 (1H,3H,5H) TRIONE POTASSIUM (2244-21-5) Incompatible with water; forms a bleach solution. A strong oxidizer; reacts violently with refucing agents, combustibles, organics, easily chlorinated materials.

DICHLORVOS (62-73-7) Attacks some plastics, rubber, and coatings.

DICHROMIC ACID, DIPOTASSIUM SALT (7778-50-9) Powdered combustibles will ignite. Strong acids form toxic vapors.

1,4-DICLOROBENZENE (Italian) (106-46-7) Forms explosive mixture with air (flash point 150°F/66°C). Incompatible with strong oxidizers, metal powders and alkali metals (i.e., lithium, sodium, potassium, rubidium, cesium, francium). Attacks some plastics, rubber, and coating.

p-DICLOROBENZENE (Italian) (106-46-7) Forms explosive mixture with air (flash point 150°F/66°C). Incompatible with strong oxidizers, metal powders, and alkali metals (i.e., lithium, sodium, potassium, rubidium, cesium, francium). Attacks some plastics, rubber, and coatings.

1,1-DICLOROETANO (Italian) (75-34-3) Forms explosive mixture with air (flash point 2°F/−17°C). Incompatible with strong oxidizers, strong caustics. Attacks plastics and rubber.

3-(3,4-DICLORO-FENYL)-1,1-DIMETIL-UREA (Italian) (330-54-1) Hydrolyzes in fairly strong acids.

DICOFOL (115-32-2) Incompatible with strong acids, caustics, aliphatic amines, isocyanates.

DICOL (111-46-6) Combustible (flash point 255°F/124°C). Incompatible with sulfuric acid, isocyanates, strong oxidizers.

DICOPHANE (50-29-3) Incompatible with salts of iron or aluminum, and bases. Do not store in iron containers.

DICOPUR (94-75-7) Decomposes in sunlight. Incompatible with strong oxidizers; may cause fire and explosions.

DICOTOX (94-75-7) Decomposes in sunlight. Incompatible with strong oxidizers; may cause fire and explosions.

DICROTOPHOS (141-66-2) Corrosive to cast iron, mild steel, brass, and stainless steel 304.

DICROTOFOS (Dutch) (141-66-2) Corrosive to cast iron, mild steel, brass, and stainless steel 304.

DICYAN (460-19-5) Explosive reaction with acids, water, liquid oxygen, oxidizers.

DICYANAMIDE (461-58-5) Mixtures with oxidizers may be explosive.

1,4-DICYANOBUTANE (111-69-3) May accumulate static electrical charges, and may cause ignition of its vapors. Decomposes at its flash point, forming poisonous cyanide gas. Forms explosive mixture with air (flash point 199°F/93°C). Incompatible with sulfuric acid, oxidizers.

beta, beta-DICYANO-O-CHLOROSTYRENE (698-41-1) Incompatible with strong oxidizers, heat.

DICYANOGEN (460-19-5) Explosive reaction with acids, water, liquid oxygen, oxidizers.

DICYCLOHEXANONE DIPEROXIDE (78-18-2) A strong oxidizer. Reacts violently with combustibles, reducing agents, caustics, ammonia.

DICYCLOPENTADIENE (77-73-6) Forms explosive mixture with air (flash point 90°F/32°C). Forms peroxides with air. May polymerize unless inhibited and maintained under inert atmosphere. reacts violently with strong oxidizers. May accumulate static electrical charges, and may cause ignition of its vapors.

1,3-DICYCLOPENTADIENE DIMER (77-73-6) Forms explosive mixture with air (flash point 90°F/32°C). Forms peroxides with air. May polymerize unless inhibited and maintained under inert atmosphere. reacts violently with strong oxidizers. May accumulate static electrical charges, and may cause ignition of its vapors.

DICYKLOPENTADIEN (Czech) (77-73-6) Forms explosive mixture with air (flash point 90°F/32°C). Forms peroxides with air. May polymerize unless inhibited and maintained under inert atmosphere. reacts violently with strong oxidizers. May accumulate static electrical charges, and may cause ignition of its vapors.

DIDIGAM (50-29-3) Incompatible with salts of iron or aluminum, and bases. Do not store in iron containers.

DIDIMAC (50-29-3) Incompatible with salts of iron or aluminum, and bases. Do not store in iron containers.

DIELDREX (60-57-1) Incompatible with concentrated mineral acids, acid catalysts, acid oxidizing agents, phenols, reactive metals.

DIELDRIN (60-57-1) Incompatible with concentrated mineral acids, acid catalysts, acid oxidizing agents, phenols, reactive metals.

DIELDRINE (French) (60-57-1) Incompatible with concentrated mineral acids, acid catalysts, acid oxidizing agents, phenols, reactive metals.

DIELDRITE (60-57-1) Incompatible with concentrated mineral acids, acid catalysts, acid oxidizing agents, phenols, reactive metals.

DI(EPOXYPROPYL)ETHER (2238-07-5) Forms explosive mixture with air (flash point 147°F/64°C). Contact with strong oxidizers may cause fire and explosions. Ethers, as a class, tend to form peroxides upon contact with air and exposure to light. Attacks some forms of plastics, coatings, and rubber.

DI(2,3-EPOXY)PROPYL ETHER (2238-07-5) Forms explosive mixture with air (flash point 147°F/64°C). Contact with strong oxidizers may cause fire and explosions. Ethers, as a class, tend to form peroxides upon contact with air and exposure to light. Attacks some forms of plastics, coatings, and rubber.

DIESEL FUEL 1-D (68334-30-5) Incompatible with nitric acid; oxidizers may cause fire and explosions.

DIESEL IGNITION IMPROVER (1002-16-0) Forms explosive mixture with air (flash point 120°F/49°C). An oxidizer. Reacts strongly reaction with reducing agents, strong acids, esters, combustibles. Attacks some plastics, rubber, and coatings.

DIESEL OIL or DIESEL OIL (LIGHT) or DIESEL OIL, MEDIUM (68334-30-5) Incompatible with nitric acid; oxidizers may cause fire and explosions.

DIETHAMINE (109-89-7) Forms explosive mixture with air (flash point −9°F/−23°C). May accumulate static electrical charges, and may cause ignition of its vapors. Reacts violently with oxidizers. Incompatible with acids, organic anhydrides, isocyanates, vinyl acetate, acrylates, substituted allyls, alkylene oxides, epichlorohydrin,

ketones, aldehydes, alcohols, glycols, mercury, phenols, cresols, caprolactam solution, strong oxidizers. Attacks aluminum, copper, lead, tin, zinc, and alloys.

DIETHANOLAMIN (Czech) (111-42-2) Incompatible with acids, organic anhydrides, isocyanates, vinyl acetate, acrylates, substituted allyls, alkylene oxides, epichlorohydrin, aldehydes, oxidizers. Corrosive to copper, copper alloys, zinc, and galvanized iron.

DIETHANOLAMINE (111-42-2) Incompatible with acids, organic anhydrides, isocyanates, vinyl acetate, acrylates, substituted allyls, alkylene oxides, epichlorohydrin, aldehydes, oxidizers. Corrosive to copper, copper alloys, zinc, and galvanized iron.

N,N-DIETHANOLAMINE (111-42-2) Incompatible with acids, organic anhydrides, isocyanates, vinyl acetate, acrylates, substituted allyls, alkylene oxides, epichlorohydrin, aldehydes, oxidizers. Corrosive to copper, copper alloys, zinc, and galvanized iron.

DIETHION (563-12-2) Incompatible with alkaline formulations. Mixtures with magnesium may be explosive.

1,1-DIETHOXY-ETHAAN (Dutch) (105-57-7) Forms explosive mixture with air (flash point −5°F/−21°C). Reacts violently with oxidizers. Forms unstable and explosive peroxides with heat and light. May accumulate static electrical charges, and may cause ignition of its vapors.

1,1-DIETHOXYETHANE (105-57-7) Forms explosive mixture with air (flash point −5°F/−21°C). Reacts violently with oxidizers. Forms unstable and explosive peroxides with heat and light. May accumulate static electrical charges, and may cause ignition of its vapors.

1,2-DIETHOXYETHANE (629-14-1) Forms explosive mixture with air (flash point 95°F/35°C) Able to form unstable peroxides. Incompatible with sulfuric acid, isocyanates, strong oxidizers.

DIETHYL (106-97-8) Forms explosive gas mixture with air. Reacts violently with strong oxidizers. Mixed with oxygen this substance will explode on contact with nickel carbonyl within the 68°F to 104°F/20°C to 40°C range.

O,O-DIETHYL (56-72-4) Contact with strong oxidizers may cause fire and explosions.

DIETHYL ACETAL (105-57-7) Forms explosive mixture with air (flash point −5°F/−21°C). Reacts violently with oxidizers. Forms unstable and explosive peroxides with heat and light. May accumulate static electrical charges, and may cause ignition of its vapors.

DIETHYLAMINE (109-89-7) Forms explosive mixture with air (flash point −9°F/−23°C). May accumulate static electrical charges, and may cause ignition of its vapors. Reacts violently with oxidizers. Incompatible with acids, organic anhydrides, isocyanates, vinyl acetate, acrylates, substituted allyls, alkylene oxides, epichlorohydrin, ketones, aldehydes, alcohols, glycols, mercury, phenols, cresols, caprolactam solution, strong oxidizers. Attacks aluminum, copper, lead, tin, zinc, and alloys.

N,N-DIETHYLAMINE (109-89-7) Forms explosive mixture with air (flash point −9°F/−23°C). May accumulate static electrical charges, and may cause ignition of its vapors. Reacts violently with oxidizers. Incompatible with acids, organic anhydrides, isocyanates, vinyl acetate, acrylates, substituted allyls, alkylene oxides, epichlorohydrin,

ketones, aldehydes, alcohols, glycols, mercury, phenols, cresols, caprolactam solution, strong oxidizers. Attacks aluminum, copper, lead, tin, zinc, and alloys.

DIETHYLAMINE, 2,2′-DIHYDROXY- (111-42-2) Incompatible with acids, organic anhydrides, isocyanates, vinyl acetate, acrylates, substituted allyls, alkylene oxides, epichlorohydrin, aldehydes, oxidizers. Corrosive to copper, copper alloys, zinc, and galvanized iron.

2-DIETHYLAMINO- (100-37-8) Forms explosive mixture with air (flash point 126°F/52°C). Reacts violently with oxidizers, strong acids. Attacks light metals. Attacks some plastics and rubber.

(DIETHYLAMINO)ETHANE (121-44-8) Forms explosive mixture with air (flash point 20°F/−6.7°C). Incompatible with strong acids, mineral acids, organic acids (acid contact) may cause violent spattering), organic anhydrides, halogenated hydrocarbons, isocyanates, nitroparaffins, nitrogen tetroxide, vinyl acetate, acrylates, substituted allyls, alkylene oxides, epichlorohydrin, ketones, aldehydes, alcohols, glycols, phenols, cresols, caprolactam solution, strong oxidizers. Attacks aluminum, copper, lead, tin, zinc, and alloys, and some plastics, rubber, and coatings.

2-(DIETHYLAMINO)ETHANOL (100-37-8) Forms explosive mixture with air (flash point 126°F/52°C). Reacts violently with oxidizers, strong acids. Attacks light metals. Attacks some plastics and rubber.

2-N-DIETHYLAMINOETHANOL (100-37-8) Forms explosive mixture with air (flash point 126°F/52°C). Reacts violently with oxidizers, strong acids. Attacks light metals. Attacks some plastics and rubber.

N-DIETHYLAMINOETHANOL (100-37-8) Forms explosive mixture with air (flash point 126°F/52°C). Reacts violently with oxidizers, strong acids. Attacks light metals. Attacks some plastics and rubber.

beta-DIETHYLAMINOETHANOL (100-37-8) Forms explosive mixture with air (flash point 126°F/52°C). Reacts violently with oxidizers, strong acids. Attacks light metals. Attacks some plastics and rubber.

2-(DIETHYLAMINO) ETHYL ALCOHOL (100-37-8) Forms explosive mixture with air (flash point 126°F/52°C). Reacts violently with oxidizers, strong acids. Attacks light metals. Attacks some plastics and rubber.

beta-DIETHYLAMINOETHYL ALCOHOL (100-37-8) Forms explosive mixture with air (flash point 126°F/52°C). Reacts violently with oxidizers, strong acids. Attacks light metals. Attacks some plastics and rubber.

N,N-DIETHYLANILINE (91-66-7) Forms explosive mixture with air (flash point 185°F/85°C). Reacts violently with strong oxidizers. Incompatible with acids, organic anhydrides, isocyanates, aldehydes.

2,6-DIETHYL ANILINE (579-66-8) Incompatible with strong acids, mineral acids, organic anhydrides, isocyanates, aldehydes, strong oxidizers.

DIETHYLBENZENE (25340-17-4); (1300-82-9) Forms explosive mixture with air (flash point 135°F/57°C). May accumulate static electrical charges, and may cause ignition of its vapors. Strong oxidizers may cause fire and explosions. Attacks rubber.

DIETHYLCARBONAT (German) (105-58-8) Forms explosive mixture with air (flash point 77°F/25°C) Reacts violently with strong oxidizers.

262

DIETHYL CARBONATE (105-58-8) Forms explosive mixture with air (flash point 77°F/25°C) Reacts violently with strong oxidizers.

DIETHYL CARBINOL (112-36-7) Forms explosive mixture with air (flash point 180°F/82°C). Reacts with strong oxidizers. Heat or atmospheric air may form unstable peroxides.

DIETHYL CELLOSOLVE (629-14-1) Forms explosive mixture with air (flash point 95°F/35°C) Able to form unstable peroxides. Incompatible with sulfuric acid, isocyanates, strong oxidizers.

DIETHYLCETONE (French) (96-22-0) Forms explosive mixture with air (flash point 55°F/13°C). Incompatible with strong acids, aliphatic amines, strong oxidizers. Attacks some plastics, rubber, and coatings. May accumulate static electrical charges, and may cause ignition of its vapors.

O,O-DIETHYL-O-(3-CHLOOR-4-METHYL-CUMARIN-7-YL) MONOTHIOFOSFAAT (Dutch) (56-72-4) Contact with strong oxidizers may cause fire and explosions.

O,O-DIETHYL-O-(3-CHLORO-4-METHYLCOUMARINYL-7) THIOPHOSPHATE (56-72-4) Contact with strong oxidizers may cause fire and explosions.

O,O-DIETHYL-O-(3-CHLORO-4-METHYL-7-COUMARINYL) PHOSPHOROTHIOATE (56-72-4) Contact with strong oxidizers may cause fire and explosions.

O,O-DIETHYL-O-(3-CHLORO-4-METHYL-2-OXO-2H-BENZO-PYRAN-7-YL) PHOSPHOROTHIOATE (56-72-4) Contact with strong oxidizers may cause fire and explosions.

DIETHYL-3-CHLORO-4-METHYLUMBELLIFERYL THIONO-PHOSPHATE (56-72-4) Contact with strong oxidizers may cause fire and explosions.

O,O-DIETHYL-3-CHLORO-4-METHYL-7-UMBELLIFERONE THIOPHOSPHATE (56-72-4) Contact with strong oxidizers may cause fire and explosions.

O,O-DIETHYL-O-(3-CHLORO-4-METHYLUMBELLIFERYL) PHOSPHOROTHIOATE (56-72-4) Contact with strong oxidizers may cause fire and explosions.

1,4-DIETHYLENEDIAMINE (110-85-0) Forms explosive mixture with air (flash point 178°F/81°C). Incompatible with nitrogen compounds, carbon tetrachloride, strong oxidizers. Aqueous solutions react with acids. Attacks aluminum, copper, nickel, magnesium, and zinc.

DIETHYLENE DIOXIDE (123-91-1) Forms explosive mixture with air (flash point 54°F/12°C). Moisture causes formation of unstable peroxides. May accumulate static electrical charges, and may cause ignition of its vapors. Incompatible with strong oxidizers, strong acids. Attacks many plastics.

DIETHYLENE-1,4-DIOXIDE (123-91-1) Forms explosive mixture with air (flash point 54°F/12°C). Moisture causes formation of unstable peroxides. May accumulate static electrical charges, and may cause ignition of its vapors. Incompatible with strong oxidizers, strong acids. Attacks many plastics.

1,4-DIETHYLENEDIOXIDE (123-91-1) Forms explosive mixture with air (flash point 54°F/12°C). Moisture causes formation of unstable peroxides. May accumulate static electrical charges, and may cause ignition of its vapors. Incompatible with strong oxidizers, strong acids. Attacks many plastics.

DIETHYLENE ETHER (123-91-1) Forms explosive mixture with air (flash point 54°F/12°C). Moisture causes formation of unstable peroxides. May accumulate static electrical charges, and may cause ignition of its vapors. Incompatible with strong oxidizers, strong acids. Attacks many plastics.

DIETHYLENE GLYCOL (111-46-6) Combustible (flash point 255°F/124°C). Incompatible with sulfuric acid, isocyanates, strong oxidizers.

DIETHYLENE GLYCOL BUTYL ETHER (112-34-5) Heat or contact with atmospheric air may cause formation of unstable peroxides. Forms explosive mixture with air (flash point 172°F/78°C). Incompatible with strong oxidizers, strong bases, sulfuric acid, isocyanates. Attacks light metals.

DIETHYLENE GLYCOL N-BUTYL ETHER (112-34-5) Heat or contact with atmospheric air may cause formation of unstable peroxides. Forms explosive mixture with air (flash point 172°F/78°C). Incompatible with strong oxidizers, strong bases, sulfuric acid, isocyanates. Attacks light metals.

DIETHYLENE GLYCOL BUTYL ETHER ACETATE (124-17-4) Incompatible with strong acids, nitrates, strong alkalies, oxidizers.

DIETHYLENE GLYCOL DI-N-BUTYL ETHER (112-73-2) Incompatible with sulfuric acid, isocyanates, strong oxidizers. May accumulate static electrical charges, and may cause ignition of its vapors.

DIETHYLENE GLYCOL DIBUTYL ETHER (112-73-2) Incompatible with sulfuric acid, isocyanates, strong oxidizers. May accumulate static electrical charges, and may cause ignition of its vapors.

DIETHYLENE GLYCOL DIETHYL ETHER (112-36-7) Forms explosive mixture with air (flash point 180°F/82°C). Reacts with strong oxidizers. Heat or atmospheric air may cause formation of unstable peroxides.

DIETHYLENE GLYCOL DIMETHYL ETHER (111-96-6) Forms explosive mixture on exposure to air, heat or light. Reacts violently with oxidizers, metal halides. Incompatible with sulfuric acid, isocyanates, perchloric acid. May accumulate static electrical charges, and may cause ignition of its vapors.

DIETHYLENE GLYCOL ETHYL ETHER (111-90-0) Forms explosive mixture with air (flash point 201°F/94°C). Heat or contact with atmospheric air may cause formation of unstable peroxides. Incompatible with strong acids, isocyanates, strong oxidizers.

DIETHYLENE GLYCOL ETHYL ETHER ACETATE (112-15-2) Incompatible with sulfuric acid, nitric acid, nitrates, strong oxidizers.

DIETHYLENE GLYCOL N-HEXYL ETHER (112-59-4) Incompatible with sulfuric acid, isocyanates, perchloric acid.

DIETHYLENE GLYCOL METHYL ETHER (111-77-3) Incompatible with sulfuric acid, isocyanates, perchloric acid.

DIETHYLENE GLYCOL METHYL ETHER ACETATE (629-38-9) Forms explosive mixture with air (flash point 180°F/98°C). Incompatible with acids, organic anhydrides, isocyanates, vinyl acetates,

264

acrylates, substituted allyls, alkylene oxides, epichlorohydrin, ketones, aldehydes, alcohols, glycols, phenols, cresols, caprolactam solution. May form shock-sensitive mixtures with silver, cobalt, chlorinated hydrocarbons, nitromethane, chromium compounds. Attacks copper, brass, zinc.

DIETHYLENE GLYCOL MONOBUTYL ETHER (112-34-5) Heat or contact with atmospheric air may produce unstable peroxides. Forms explosive mixture with air (flash point 172°F/78°C). Incompatible with strong oxidizers, strong bases, sulfuric acid, isocyanates. Attacks light metals.

DIETHYLENE GLYCOL MONOBUTYL ETHER ACETATE (124-17-4) Incompatible with strong acids, nitrates, strong alkalies, oxidizers.

DIETHYLENE GLYCOL MONOMETHYL ETHER (111-77-3) Incompatible with sulfuric acid, isocyanates, perchloric acid.

DIETHYLENE GLYCOL MONOETHYL ETHER (111-90-0) Forms explosive mixture with air (flash point 201°F/94°C). Heat or contact with atmospheric air may produce unstable peroxides. Incompatible with strong acids, isocyanates, strong oxidizers.

DIETHYLENE GLYCOL MONOETHYL ETHER (111-46-6) Combustible (flash point 255°F/124°C). Incompatible with sulfuric acid, isocyanates, strong oxidizers.

DIETHYLENE GLYCOL PHTHALATE (629-38-9) Incompatible with sulfuric acids, nitric acids, nitrates.

DIETHYLENEIMIDE OXIDE (110-91-8) Forms explosive mixture with air (flash point 98°F/37°C). Incompatible with acids, organic anhydrides, isocyanates, vinyl acetate, acrylates, substituted allyls, alkylene oxides, epichlorohydrin, ketones, aldehydes, alcohols, glycols, phenols, cresols, caprolactam solution, strong oxidizers. Attacks aluminum, copper, lead, tin, zinc, and alloys, and some plastics, rubber, and coatings.

DIETHYLENIMIDE OXIDE (110-91-8) Forms explosive mixture with air (flash point 98°F/37°C). Incompatible with acids, organic anhydrides, isocyanates, vinyl acetate, acrylates, substituted allyls, alkylene oxides, epichlorohydrin, ketones, aldehydes, alcohols, glycols, phenols, cresols, caprolactam solution, strong oxidizers. Attacks aluminum, copper, lead, tin, zinc, and alloys, and some plastics, rubber, and coatings.

DIETHYLENE IMIDOXIDE (110-91-8) Forms explosive mixture with air (flash point 98°F/37°C). Incompatible with acids, organic anhydrides, isocyanates, vinyl acetate, acrylates, substituted allyls, alkylene oxides, epichlorohydrin, ketones, aldehydes, alcohols, glycols, phenols, cresols, caprolactam solution, strong oxidizers. Attacks aluminum, copper, lead, tin, zinc, and alloys, and some plastics, rubber, and coatings.

DIETHYLENEIMINE (110-85-0) Forms explosive mixture with air (flash point 178°F/81°C). Incompatible with nitrogen compounds, carbon tetrachloride, strong oxidizers. Aqueous solutions react with acids. Attacks aluminum, copper, nickel, magnesium, and zinc.

DIETHYLENE OXIDE (109-99-9) Forms explosive mixture with air(flash point 6°F/−14°C). Unless inhibited, can form unstable and explosive peroxides. Incompatible with strong acids, strong oxidizers. Attacks some plastics. May accumulate static electric charges that can result in ignition of its vapors.

DI-(ETHYLENE OXIDE) (123-91-1) Forms explosive mixture with air (flash point 54°F/12°C). Moisture causes formation of unstable peroxides. May accumulate static electrical charges, and may cause ignition of its vapors. Incompatible with strong oxidizers, strong acids. Attacks many plastics.

DIETHYLENE OXIMIDE (110-91-8) Forms explosive mixture with air (flash point 98°F/37°C). Incompatible with acids, organic anhydrides, isocyanates, vinyl acetate, acrylates, substituted allyls, alkylene oxides, epichlorohydrin, ketones, aldehydes, alcohols, glycols, phenols, cresols, caprolactam solution, strong oxidizers. Attacks aluminum, copper, lead, tin, zinc, and alloys, and some plastics, rubber, and coatings.

DIETHYLENE TRIAMINE (111-40-0) Ignites spontaneously with cellulose nitrate. Silver, cobalt, or chromium compounds may cause explosions. Forms explosive mixture with air (flash point 208°F/98°C). Incompatible with acids, organic anhydrides, isocyanates, vinyl acetate, acrylates, substituted allyls, alkylene oxides, epichlorohydrin, ketones, aldehydes, alcohols, glycols, mercury, phenols, cresols, caprolactam solution, strong oxidizers. Attacks aluminum, copper, lead, tin, zinc, and alloys.

DIETHYLENETRIAMINE (111-40-0) Ignites spontaneously with cellulose nitrate. Silver, cobalt, or chromium compounds may cause explosions. Forms explosive mixture with air (flash point 208°F/98°C). Incompatible with acids, organic anhydrides, isocyanates, vinyl acetate, acrylates, substituted allyls, alkylene oxides, epichlorohydrin, ketones, aldehydes, alcohols, glycols, mercury, phenols, cresols, caprolactam solution, strong oxidizers. Attacks aluminum, copper, lead, tin, zinc, and alloys.

DIETHYL ESTER OF PHTHALIC ACID (84-66-2) Reacts violently with strong acids, strong oxidizers, permanganates, water. Attacks some forms of plastic.

DIETHYL ESTER SULFURIC ACID (64-67-5) Reacts vigorously with strong oxidizers or water. Incompatible with strong acids, strong alkalies, nitrates, water.

N,N-DIETHYLETHANEAMINE (121-44-8) Forms explosive mixture with air (flash point 20°F/−6.7°C). Incompatible with strong acids, mineral acids, organic acids (acids may cause violent spattering), organic anhydrides, halogenated hydrocarbons, isocyanates, nitroparaffins, nitrogen tetroxide, vinyl acetate, acrylates, substituted allyls, alkylene oxides, epichlorohydrin, ketones, aldehydes, alcohols, glycols, phenols, cresols, caprolactam solution, strong oxidizers. Attacks aluminum, copper, lead, tin, zinc, and alloys, and some plastics, rubber, and coatings.

DIETHYLETHANOLAMINE (100-37-8) Forms explosive mixture with air (flash point 126°F/52°C). Reacts violently with oxidizers, strong acids. Attacks light metals. Attacks some plastics and rubber.

N,N-DIETHYLETHANOLAMINE (100-37-8) Forms explosive mixture with air (flash point 126°F/52°C). Reacts violently with oxidizers, strong acids. Attacks light metals, some plastics, and rubber.

DIETHYL ETHER (60-29-7) Forms explosive mixture with air (flash point −49°F/−45°C). Incompatible with strong acids, strong oxidizers. Can form peroxides with air or light; may explode when container is unstoppered or otherwise opened. Attacks some plastics, rubber, and coatings. Being a nonconductor, chemical may accumulate static electric charges that may result in ignition of vapor.

O,O-DIETHYL-S-(2-ETHTHIOETHYL) PHOSPHORODITHIO-ATE (298-04-4) Forms explosive mixture with air (flash point 180°F/82°C). Incompatible with alkalis, strong oxidizers.

O,O-DIETHYL-S-(2-ETHTHIOETHYL)THIOTHIONOPHOS-PHATE (298-04-4) Forms explosive mixture with air (flash point 180°F/82°C). Incompatible with alkalis, strong oxidizers.

O,O-DIETHYL S-(2-ETHYLMERCAPTOETHYL)DITHIO-PHOSPHATE (298-04-4) Forms explosive mixture with air (flash point 180°F/82°C). Incompatible with alkalis, strong oxidizers.

O,O-DIETHYL-2-ETHYLMERCAPTOETHYL THIOPHOSPHATE, DIETHOXYTHIOPHOSPHORIC ACID (8065-48-3) Forms explosive mixture with air (flash point 113°F/45°C). Incompatible with water, strong oxidizers, caustics.

O,O-DIETHYL-S-(2-ETHYLTHIO-ETHYL)-DITHIOFOSFAAT (Dutch) (298-04-4) Forms explosive mixture with air (flash point 180°F/82°C). Incompatible with alkalis, strong oxidizers.

O,O-DIETHYL-S-2-(ETHYLTHIO)ETHYL PHOSPHODITHIOATE (298-04-4) Forms explosive mixture with air (flash point 180°F/82°C). Incompatible with alkalis, strong oxidizers.

O,O-DIETHYL-2-ETHYLTHIOETHYLPHOSPHORODITHIOATE (298-04-4) Forms explosive mixture with air (flash point 180°F/82°C). Incompatible with alkalis, strong oxidizers.

O,O-DIETHYL-S-2-(ETHYLTHIO)ETHYLPHOSPHORODI-THIOATE (298-04-4) Forms explosive mixture with air (flash point 180°F/82°C). Incompatible with alkalis, strong oxidizers.

O,O[DIETHYL-O (and 5)-]2-(ETHYLTHIO)ETHYLPHOSPHORO-THIOATES (8065-48-3) Forms explosive mixture with air (flash point 113°F/45°C). Incompatible with water, strong oxidizers, caustics.

DIETHYL GLYCOL DIMETHYL ETHER (111-96-6) Forms explosive mixture on exposure to air, heat, or light. Reacts violently with oxidizers, metal halides. Incompatible with sulfuric acid, isocyanates, perchloric acid. May accumulate static electrical charges, and may cause ignition of its vapors.

DI-(2-ETHYLHEXYL) ADIPATE (103-23-1) Incompatible with strong acids, nitrates, oxidizers.

DI-(2-ETHYLHEXYL) PHOSPHATE (298-07-7) Corrosive to metals; forms flammable hydrogen gas.

DI-(2-ETHYLHEXYL) PHOSPHORIC ACID (298-07-7) Corrosive to metals; produces flammable hydrogen gas.

DI-(2-ETHYLHEXYL) PHTHALATE (117-81-7) Incompatible with strong acids, strong alkalies, nitrates, oxidizers.

DI-(2-ETHYLHEXYL) SULFOSUCCINATE, SODIUM SALT (119-36-8) Forms explosive mixture with air (flash point 205°F/96°C). Incompatible with strong acids, nitrates, oxidizers.

N,N-DIETHYL-2-HYDROXYETHYLAMINE (100-37-8) Forms explosive mixture with air (flash point 126°F/52°C). Reacts violently with oxidizers, strong acids. Attacks light metals. Attacks some plastics and rubber.

N,N-DIETHYL-N-(8-HYDROXYETHYL)AMINE (100-37-8) Forms explosive mixture with air (flash point 126°F/52°C). Reacts violently with oxidizers, strong acids. Attacks light metals. Attacks some plastics and rubber.

N,N-DIETHYL-N-(beta-HYDROXYETHYL)AMINE (100-37-8) Forms explosive mixture with air (flash point 126°F/52°C). Reacts violently with oxidizers, strong acids. Attacks light metals. Attacks some plastics and rubber.

O,O-DIETHYL-O-(2-ISOPROPYL-6-METHYL-4-PYRIMIDIN-YL)PHOSPHOROTHIOATE (333-41-5) Incompatible with water, copper-containing compounds, oxidizers, acids, or bases.

DIETHYL KETONE (96-22-0) Forms explosive mixture with air (flash point 55°F/13°C). Incompatible with strong acids, aliphatic amines, strong oxidizers. Attacks some plastics, rubber, and coatings. May accumulate static electrical charges, and may cause ignition of its vapors.

DIETHYL MONOSULFATE (64-67-5) Reacts vigorously with strong oxidizers or water. Incompatible with strong acids, strong alkalies, nitrates, water.

O,O-DIETHYL-O,P-NITROPHENYL PHOSPHOROTHIOATE (56-38-2) Combustible liquid. Mixtures with endrin may be explosive. Strong oxidizers may cause fire and explosions. Attacks some plastics, rubber, and coatings.

DIETHYL 4-NITROPHENYL PHOSPHOROTHIONATE (56-38-2) Combustible liquid. Mixtures with endrin may be explosive. Strong oxidizers may cause fire and explosions. Attacks some plastics, rubber, and coatings.

DIETHYL P-NITROPHENYL THIONOPHOSPHATE (56-38-2) Combustible liquid. Mixtures with endrin may be explosive. Strong oxidizers may cause fire and explosions. Attacks some plastics, rubber, and coatings.

DIETHYL OXIDE (60-29-7) Forms explosive mixture with air (flash point −49°F/−45°C). Incompatible with strong acids, strong oxidizers. Can form peroxides with air or light; may explode when container is unstoppered or otherwise opened. Attacks some plastics, rubber, and coatings. Being a nonconductor, chemical may accumulate static electric charges that may result in ignition of vapor.

DIETHYLPARATHION (56-38-2) Combustible liquid. Mixtures with endrin may be explosive. Strong oxidizers may cause fire and explosions. Attacks some plastics, rubber, and coatings.

DIETHYLPHENYLAMINE (91-66-7) Forms explosive mixture with air (flash point 185°F/85°C). Reacts violently with strong oxidizers. Incompatible with acids, organic anhydrides, isocyanates, aldehydes.

DIETHYL PHTHALATE (84-66-2) Reacts violently with strong acids, strong oxidizers, permanganates, water. Attacks some forms of plastic.

DIETHYL SULFATE (64-67-5) Reacts vigorously with strong oxidizers or water. Incompatible with strong acids, strong alkalies, nitrates, water.

DIETHYL SULFIDE (352-93-2) Forms explosive mixture with air (flash point 15°F/−10°C). May be able to form unstable peroxides. Incompatible with water, strong acids, strong oxidizers. May accumulate static electrical charges, and may cause ignition of its vapors.

DIETHYL SULPHATE (64-67-5) Reacts vigorously with strong oxidizers or water. Incompatible with strong acids, strong alkalies, nitrates, water.

DIETHYLTHIOPHOSPHORIC ACID ESTER OF 3-CHLORO-4-METHYL-7-HYDROXYCOUMARIN (56-72-4) Contact with strong oxidizers may cause fire and explosions.

DIETHYL ZINC (557-20-0) Ignites spontaneously on contact with air or oxidizers. Explosive decomposition occurs at 245°F/120°C. Reacts violently with hydrazine, sulfur dioxide, some alcohols, ozone; with possible fire and explosions. Water forms ethane gas.

O,O-DIETIL-O-(3-CLORO-4-METIL-CUMARIN-7-IL-MONOTI-OFOSFATO) (Italian) (56-72-4) Contact with strong oxidizers may cause fire and explosions.

O,O-DIETIL-S-(2-ETILTIO-ETIL)-DITIOFOSFATO (Italian) (298-04-4) Forms explosive mixture with air (flash point 180°F/82°C). Incompatible with alkalis, strong oxidizers.

1,1-DIETOSSIETANO (Italian) (105-57-7) Forms explosive mixture with air (flash point −5°F/−21°C). Reacts violently with oxidizers. Forms unstable and explosive peroxides with heat and light. May accumulate static electrical charges, and may cause ignition of its vapors.

DIEYANOMETHANE (109-77-3) Incompatible with sulfuric acid. Caustics or heating above 160°F/71°C may cause polymerization or spontaneous combustion.

DIFENYLOL PROPANE (80-05-7) Contact with strong oxidizers may cause fire and explosions.

DIFLUORINE MONOXIDE (7783-41-7) Forms explosive mixture with water. A powerful oxidizer. Contact with all reducing agents, organic and combustible materials, ammonia, platinum and many other metals, metal oxides, and moist air may cause fire and explosions. Attacks some forms of plastics, rubber, and coatings. See 29 CFR 1910.101 for specific regulations on storage of compressed gas cylinders.

DIFLUOROCHLOROMETHANE (75-45-5) Moisture and rust cause slow decomposition, producing toxic gases. Attacks some plastics, rubber, and coatings. Thermal decomposition occurs at high temperature with alkalies and alkaline earth metals.

DIFLUORODIBROMOMETHANE (75-61-6) Reacts with chemically active metals and oxidizers. Attacks some plastics, rubber, and coatings.

DIFLUORODICHLOROMETHANE (75-71-8) Reacts violently with liquid aluminum. Incompatible with chemically active metals. Attacks some plastics, rubber, and coatings.

DIFLUOROETHANE (75-37-6) Flammable gas. May accumulate static electrical charges, and may cause ignition of its vapors. Reacts violently with strong oxidizers. Attacks metals in presence of moisture.

1,1-DIFLUOROETHANE (75-37-6) Flammable gas. May accumulate static electrical charges, and may cause ignition of its vapors. Reacts violently with strong oxidizers. Attacks metals in presence of moisture.

1,1-DIFLUOROETHENE (75-38-7) Flammable gas. Forms peroxides; may polymerize, causing explosion. Reacts violently with strong oxidizers. May accumulate static electrical charges, and may cause ignition of its vapors.

DIFLUORO-1,1-ETHENE (75-38-7) Flammable gas. Forms peroxides; may polymerize, causing explosion. Reacts violently with strong oxidizers. May accumulate static electrical charges, and may cause ignition of its vapors.

1,1-DIFLUOROETHYLENE (75-38-7) Flammable gas. Forms peroxides; may polymerize, causing explosion. Reacts violently with strong oxidizers. May accumulate static electrical charges, and may cause ignition of its vapors.

DIFLUORO-1,1-ETHYLENE (75-38-7) Flammable gas. Forms peroxides; may polymerize, causing explosion. Reacts violently with strong oxidizers. May accumulate static electrical charges, and may cause ignition of its vapors.

DIFLUOROPHOSPHORIC ACID (13779-41-4) Water contact produces hydrofluoric acid. Attacks glass and other siliceous materials. Corrodes metals.

DIFLUOROPHOSPHORUS ACID (ANHYDROUS) (13779-41-4) Water contact produces hydrofluoric acid. Attacks glass and other siliceous materials. Corrodes metals.

DIFOLATAN (2425-06-1) Incompatible with acids or acid vapor. Strong alkaline conditions contribute to instability.

1,3-DIFORMAL PROPANE (111-30-8) Water contact produces a polymer solution. Incompatible with strong acids, caustics, ammonia, amines, strong oxidizers.

DIFORMYL (107-22-2) Water contact causes polymerization. Incompatible with strong acids, caustics, ammonia, amines, chlorosulfonic acid, ethylene amine. Corrosive to metals.

DIFOSAN (2425-06-1) Incompatible with acids or acid vapor. Strong alkaline conditions contribute to instability.

DIGGE (111-46-6) Combustible (flash point 255°F/124°C). Incompatible with sulfuric acid, isocyanates, strong oxidizers.

DIGLYCIDYL ETHER (2238-07-5) Forms explosive mixture with air (flash point 147°F/64°C). Contact with strong oxidizers may cause fire and explosions. Ethers, as a class, tend to form peroxides upon contact with air and exposure to light. Attacks some forms of plastics, coatings, and rubber.

DIGLYCIDYL ETHER OF BISPHENOL A (1675-54-8) Forms explosive mixture with air (flash point 175°F/79°C). Incompatible with strong acids, strong oxidizers. Ethers form peroxides on contact with air and light.

DIGLYCOL (111-46-6) Combustible (flash point 255°F/124°C). Incompatible with sulfuric acid, isocyanates, strong oxidizers.

DIGLYCOL MONOBUTYL ETHER (112-34-5) Heat or contact with atmospheric air may produce unstable peroxides. Forms explosive mixture with air (flash point 172°F/78°C). Incompatible with strong oxidizers, strong bases, sulfuric acid, isocyanates. Attacks light metals.

DIGLYCOL MONOBUTYL ETHER ACETATE (124-17-4) Incompatible with strong acids, nitrates, strong alkalies, oxidizers.

DIGLYCOL MONOETHYL ETHER (111-77-3) Incompatible with sulfuric acid, isocyanates, perchloric acid.

DIGLYCOL MONOETHYL ETHER ACETATE (112-15-2) Incompatible with sulfuric acid, nitric acid, nitrates, strong oxidizers.

270

DIGLYME (111-96-6) Forms explosive mixture on exposure to air, heat, or light. Reacts violently with oxidizers, metal halides. Incompatible with sulfuric acid, isocyanates, perchloric acid. May accumulate static electrical charges, and may cause ignition of its vapors.

DIHEPTYL PHTHALATE (3648-21-3) Incompatible with strong acids, strong alkalies, nitrates, strong oxidizers. Attacks some plastics, rubber, and coatings.

DI-n-HEPTL PHTHALATE (3648-21-3) Incompatible with strong acids, strong alkalies, nitrates, strong oxidizers. Attacks some plastics, rubber, and coatings.

DI-n-HEXYL ADIPATE (N/A) Incompatible with strong acids, nitrates, strong oxidizers.

1,4-DIHIDROBENZENE (Italian) (123-31-9) Incompatible with strong oxidizers, caustics. May be oxidized to quinone at room temperatures in the presence of moisture. May explode on contact with oxygen.

DIHYDROAZIRINE (151-56-4) Forms explosive mixture with air. Contact with acid, aluminum, carbon dioxide, or silver may cause explosive polymerization. Attacks rubber, coatings, and plastics. Self-reactive with heat or atmospheric carbon dioxide.

DIHYDRO-1-AZIRINE (151-56-4) Forms explosive mixture with air. Contact with acid, aluminum, carbon dioxide, or silver may cause explosive polymerization. Attacks rubber, coatings, and plastics. Self-reactive with heat or atmospheric carbon dioxide.

9,10-DIHYDRO-8a,10,-DIAZONIAPHENANTHRENE DIBRO-MIDE (85-00-7) Concentrated solution attacks aluminum.

9,10-DIHYDRO-8a,10a-DIAZONIAPHENANTHRENE(1,1′-ETH-YLENE-2,2′-BIPYRIDYLIUM)DIBROMIDE (85-00-7) Concentrated solution attacks aluminum.

DIHYDRO-2,5-DIOXOFURAN (108-31-6) Strong oxidizers may cause fire and explosions; contact with amines, alkali metals causes polymerization if temperature is greater than 150°F/66°C. Attacks metals in the presence of moisture.

5,6-DIHYDRO-DIPYRIDO(1,2a; 2,1c)PYRAZINIUM DIBROMIDE (85-00-7) Concentrated solution attacks aluminum.

DIHYDROOXIRENE (75-21-8) Forms explosive mixture with air (flash point 20°F/–6°C). Incompatible with alkali metal hydroxides,highly active catalysts (e.g., anhydrous chlorides of iron, tin, or aluminum and oxides of iron or aluminum). Avoid contact with copper. Protect container from physical damage, sun, and heat. Attacks some plastics, rubber, and coatings.

DIHYDROPENTABORANE(9) (19624-22-7) Forms explosive mixture with air (flash point 95°F/35°C). May ignite spontaneously in moist air. Water contact produces boric acid and explosive hydrogen gas. Oxidizers cause formation of highly explosive mixtures. Halogen and halogenated compounds may cause explosions. Attacks some plastics, rubber, and coatings.

2,5-DIHYDROPEROXY-2,5-DIMETHYLHEXANE (3025-88-5) Contact with acids and many metals may cause decomposition.

2,3-DIHYDROPYRAN (110-87-2) Forms explosive mixture with air (flash point 0°F/–18°C). Reacts violently with strong oxidizers. May accumulate static electrical charges, and may cause ignition of its vapors.

1,2-DIHYDRO-3,6-PYRIDAZINEDIONE (123-33-1) Contact with strong oxidizers may cause fire and explosions.

1,2-DIHYDROPYRIDAZINE-3,6-DIONE (123-33-1) Contact with strong oxidizers may cause fire and explosions.

6,7-DIHYDROPYRIDO(1,2a; 2′,1′-c)PYRAZINEDIUM DIBRO-MIDE (85-00-7) Concentrated solution attacks aluminum.

1,4-DIHYDROXYBENZEN (Czech) (123-31-9) Incompatible with strong oxidizers, caustics. May be oxidized to quinone at room temperatures in the presence of moisture. May explode on contact with oxygen.

DIHYDROXYBENZENE (123-31-9) Incompatible with strong oxidizers, caustics. May be oxidized to quinone at room temperatures in the presence of moisture. May explode on contact with oxygen.

1,2-DIHYDROXYBENZENE (120-80-9) Strong oxidizers may cause fire and explosion.

1,3-DIHYDROXYBENZENE (108-46-3) Reacts violently with strong oxidizers, nitric acid. Incompatible with acetanilide, albumin, alkalies, antipyrine, camphor, ferric salts, menthol, spirit nitrous ether. Absorbs moisture from air (hygroscopic). May accumulate static electrical charges, and may cause ignition of its vapors.

1,4-DIHYDROXY-BENZENE (Dutch) (123-31-9) Incompatible with strong oxidizers, caustics. May be oxidized to quinone at room temperatures in the presence of moisture. May explode on contact with oxygen.

1,4-DIHYDROXYBENZENE (123-31-9) Incompatible with strong oxidizers, caustics. May be oxidized to quinone at room temperatures in the presence of moisture. May explode on contact with oxygen.

m-**DIHYDROXYBENZENE** or *meta*-**DIHYDROXYBENZENE** (108-46-3) Reacts violently with strong oxidizers, nitric acid. Incompatible with acetanilide, albumin, alkalies, antipyrine, camphor, ferric salts, menthol, spirit nitrous ether. Absorbs moisture from air (hygroscopic). May accumulate static electrical charges, and may cause ignition of its vapors.

o-**DIHYDROXYBENZENE** or *ortho*-**DIHYDROXYBENZENE** (120-80-9) Strong oxidizers may cause fire and explosion.

p-**DIHYDROXYBENZENE** or *para*-**DIHYDROXYBENZENE** (123-31-9) Incompatible with strong oxidizers, caustics. May be oxidized to quinone at room temperatures in the presence of moisture. May explode on contact with oxygen.

1,4-DIHYDROXY-BENZOL (German) (123-31-9) Incompatible with strong oxidizers, caustics. May be oxidized to quinone at room temperatures in the presence of moisture. May explode on contact with oxygen.

m-**DIHYDROXYBENZOL** or *meta*-**DIHYDROXYBENZOL** (108-46-3) Reacts violently with strong oxidizers, nitric acid. Incompatible with acetanilide, albumin, alkalies, antipyrine, camphor, ferric salts, menthol, spirit nitrous ether. Absorbs moisture from air (hygroscopic). May accumulate static electrical charges, and may cause ignition of its vapors.

1,3-DIHYDROXYBUTANE (107-88-0) Incompatible with strong acids, caustics, aliphatic amines, isocyanates, oxidizers.

1,4-DIHYDROXYBUTANE (110-63-4) Incompatible with strong acids, caustics, aliphatic amines, isocyanates, oxidizers.

2,3-DIHYDROXYBUTANE (513-85-9) Forms explosive mixture with air (flash point 185°F/85°C). Incompatible with strong acids, caustics, aliphatic amines, isocyanates, strong oxidizers.

1,4-DIHYDROXY-2-BUTENE (110-64-5) Contact with strong oxidizers may cause fire and explosions.

1,4-DIHYDROXY-2-BUTYNE (110-65-6) Reacts violently with strong acids, bases, alkali metals and earth.

2,2′-DIHYDROXYDIETHYLAMINE (111-42-2) Incompatible with acids, organic anhydrides, isocyanates, vinyl acetate, acrylates, substituted allyls, alkylene oxides, epichlorohydrin, aldehydes, oxidizers. Corrosive to copper, copper alloys, zinc, and galvanized iron.

beta, beta-DIHYDROXYDIETHYL ETHER (111-46-6) Combustible (flash point 255°F/124°C). Incompatible with sulfuric acid, isocyanates, strong oxidizers.

l′,P′-DIHYDROXYDIPHENYL-DIMETHYLMETHANE (80-05-7) Contact with strong oxidizers may cause fire and explosions.

2,2′-DIHYDROXYDIPROPYLAMINE (110-97-4) Incompatible with sulfuric, nitric, mineral, and organic acids, and organic anhydrides, isocyanates, vinyl acetate, acrylates, substituted allyls, alkylene oxides, epichlorohydrin, aldehydes, oxidizers. Attacks some plastics, rubber, and coatings, and copper and copper alloys.

2,2′-DIHYDROXYDIPROPYLAMINE (110-98-5) Incompatible with sulfuric acid, perchloric acid, isocyanates, strong oxidizers.

1,2-DIHYDROXYETHANE (107-21-1) Incompatible with strong acids, caustics, aliphatic amines, isocyanates, chlorosulfonic acid, oleum, strong oxidizers.

DI-beta-HYDROXYETHOXYETHANE (112-27-6) Incompatible with sulfuric acid, isocyanates, perchloric acid, strong oxidizers.

DI-(2-HYDROXYETHYL) AMINE (111-42-2) Incompatible with acids, organic anhydrides, isocyanates, vinyl acetate, acrylates, substituted allyls, alkylene oxides, epichlorohydrin, aldehydes, oxidizers. Corrosive to copper, copper alloys, zinc, and galvanized iron.

2,2′-DIHYDROXYETHYLAMINE (111-42-2) Incompatible with acids, organic anhydrides, isocyanates, vinyl acetate, acrylates, substituted allyls, alkylene oxides, epichlorohydrin, aldehydes, oxidizers. Corrosive to copper, copper alloys, zinc, and galvanized iron.

2,4-DIHYDROXY-2-METHYLPENTANE (107-41-5) Forms explosive mixture with air (flash point 209°F/98°C). Incompatible with strong acids, caustics, aliphatic amines, isocyanates, strong oxidizers.

1,2-DIHYDROXYPROPANE (57-55-7) Incompatible with strong acids, caustics, aliphatic amines, isocyanates, strong oxidizers (with violent reaction).

DIIRONTRISULFATE (10028-22-5) Corrosive to copper and its alloys, and to mild and galvanized steel.

DIISIPROPYL-ACETONE (108-83-8) Incompatible with strong acids, aliphatic amines, strong oxidizers. Attacks plastics, coatings and rubber. Forms explosive mixture with air (flash point 140°F/60°C).

DIISOBUTILCHETONE (Italian) (108-83-8) Incompatible with strong acids, aliphatic amines, strong oxidizers. Attacks plastics, coatings, and rubber. Forms explosive mixture with air (flash point 140°F/60°C).

DIISOBUTYLAMINE (110-96-3) Forms explosive mixture with air (flash point 85°F/29°C). Incompatible with acids, organic anhydrides, isocyanates, vinyl acetate, acrylates, substituted allyls, alkylene oxides, epichlorohydrin, ketones, aldehydes, alcohols, glycols, mercury, phenols, cresols, caprolactam solution, strong oxidizers. Attacks aluminum, copper, lead, tin, zinc, and alloys.

DI-ISOBUTYLCETONE (French) (108-83-8) Incompatible with strong acids, aliphatic amines, strong oxidizers. Attacks plastics, coatings, and rubber. Forms explosive mixture with air (flash point 140°F/60°C).

DIISOBUTYLENE (25167-70-8) Forms explosive mixture with air (flash point 23°F/−5°C). May be able to form unstable and explosive peroxides. Incompatible with strong acids, strong oxidizers. May accumulate static electrical charges, and may cause ignition of its vapors.

DIISOBUTYLKETON (Dutch, German) (108-83-8) Incompatible with strong acids, aliphatic amines, strong oxidizers. Attacks plastics, coatings and rubber. Forms explosive mixture with air (flash point 140°F/60°C).

DIISOBUTYL CARBINOL (108-82-7) Forms explosive mixture with air (flash point 165°F/74°C). Incompatible with strong acids, caustics, aliphatic amines, isocyanates, strong oxidizers.

DIISOBUTYL KETONE (108-83-8) Incompatible with strong acids, aliphatic amines, strong oxidizers. Attacks plastics, coatings, and rubber. Forms explosive mixture with air (flash point 140°F/60°C).

DIISOBUTYL PHTHALATE (84-69-5) Incompatible with strong acids, strong oxidizers, nitrates.

N,N-DI-ISO-1-CYANATOLUENE (584-84-9) Incompatible with strong acids, including nonoxidizing mineral and organic acids, caustics, ammonia, amines, amides, alcohols, glycols, caprolactam solution. Water contact causes violent foaming and spattering; produces carbon dioxide and an organic base. Attacks copper and its alloys, some plastics including polyethylene, and rubber.

DI-ISOCYANATE de TOLUYLENE (French) (584-84-9) Incompatible with strong acids, including nonoxidizing mineral and organic acids, caustics, ammonia, amines, amides, alcohols, glycols, caprolactam solution. Water contact causes violent foaming and spattering; produces carbon dioxide and an organic base. Attacks copper and its alloys, some plastics including polyethylene, and rubber.

DIISOCYANAT-TOLUOL (German) (584-84-9) Incompatible with strong acids including nonoxidizing mineral and organic acids, caustics, ammonia, amines, amides, alcohols, glycols, caprolactam solution. Water contact causes violent foaming and spattering; produces carbon dioxide and an organic base. Attacks copper and its alloys, some plastics including polyethylene, and rubber.

2,4-DIISOCYANATO-1-METHYLBENZENE (9CI) (584-84-9) Incompatible with strong acids, including nonoxidizing mineral and organic acids, caustics, ammonia, amines, amides, alcohols, glycols, caprolactam solution. Water contact causes violent foaming and spattering; produces carbon dioxide and an organic base. Attacks copper and its alloys, some plastics including polyethylene, and rubber.

2,4-DIISOCYANATOTOLUENE (584-84-9) Incompatible with strong acids including nonoxidizing mineral and organic acids, caustics, ammonia, amines, amides, alcohols, glycols, caprolactam solution. Water contact causes violent foaming and spattering; produces carbon dioxide and an organic base. Attacks copper and its alloys, some plastics including polyethylene, and rubber.

DIISODECYL PHTHALATE (26761-40-0) Incompatible with strong acids, strong alkalies, nitrates, oxidizers.

DIISONONYL ADIPATE (N/A) Incompatible with strong acids, nitrates, oxidizers.

DIISONONYL PHTHALATE (88-99-3) Incompatible with strong acids, nitrates, oxidizers.

DIISOOCTYL ACID PHOSPHATE (298-07-7) Corrosive to metals; forms flammable hydrogen gas.

DIISOOCTYL PHTHALATE (27554-26-3) May accumulate static electrical charges, and may cause ignition of its vapors. Incompatible with strong acids, nitrates, oxidizers.

DIISOPROPANOLAMINE (110-97-4) Incompatible with sulfuric, nitric, mineral, and organic acids, and organic anhydrides, isocyanates, vinyl acetate, acrylates, substituted allyls, alkylene oxides, epichlorohydrin, aldehydes, oxidizers. Attacks some plastics, rubber, and coatings, copper, and copper alloys.

DIISOPROPYLACETONE (108-83-8) Incompatible with strong acids, aliphatic amines, strong oxidizers. Attacks plastics, coatings and rubber. Forms explosive mixture with air (flash point 140°F/60°C).

5-DIISOPROPYLACETONE (108-83-8) Incompatible with strong acids, aliphatic amines, strong oxidizers. Attacks plastics, coatings and rubber. Forms explosive mixture with air (flash point 140°F/60°C).

DIISOPROPYLAMINE (108-18-9) Forms explosive mixture with air (flash point 30°F/−1°C). Incompatible with acids, organic anhydrides, isocyanates, vinyl acetate, acrylates, substituted allyls, alkylene oxides, epichlorohydrin, ketones, aldehydes, alcohols, glycols, mercury, phenols, cresols, caprolactam solution, strong oxidizers. Attacks aluminum, copper, lead, tin, zinc and alloys, and some plastics, rubber, and coatings.

DIISOPROPYLBENZENE HYDROPEROXIDE, not more than 72% in solution (26762-93-6) Forms explosive mixture with air (flash point 175°F/79°C). Rapid decomposition may be caused by contact with oxidizers, reducing agents, minerals acids; incompatible with aluminum, copper, brass, lead, zinc salts.

DIISOPROPYLBENZENE HYDROPEROXIDE (26762-93-6) Forms explosive mixture with air (flash point 175°F/79°C). Rapid decomposition may be caused by contact with oxidizers, reducing agents, minerals acids; incompatible with aluminum, copper, brass, lead, zinc salts.

DIISOPROPYLBENZENE (all isomers) (98-51-1) Forms explosive mixture with air (flash point 155°F/68°C). Incompatible with nitric acid, strong oxidizers.

DIISOPROPYL CARBINOL (600-36-2) Forms explosive mixture with air (flash point 118°F/48°C). Reacts violently with strong oxidizers.

DIISOPROPYL ETHER (108-20-3) Forms explosive mixture with air (flash point −18°F/−28°C). Air contact produces explosive peroxides that may detonate with heat or shock. Strong oxidizers may cause fire and explosions. Attacks some plastics, rubber, and coatings.

DIISOPROPYL NAPHTHALENE (24157-81-1) Incompatible with nitric acid; oxidizers may cause fire and explosions.

2,6-DIISOPROPYL NAPHTHALENE (24157-81-1) Incompatible with nitric acid; oxidizers may cause fire and explosions.

DIISOPROPYL OXIDE (108-20-3) Forms explosive mixture with air (flash point −18°F/−28°C). Air contact produces explosive peroxides that may detonate with heat or shock. Strong oxidizers may cause fire and explosions. Attacks some plastics, rubber, and coatings.

DIISOPROPYL PERCARBONATE (105-64-6) Combustible solid. Contact with some metals may cause decomposition.

DIISOPROPYL PEROXYDICARBONATE (105-64-6) Combustible solid. Contact with some metals may cause decomposition.

DIKETENE (674-82-8) Reacts violently with water. Incompatible with alcohols, strong acids, amines, caustics.

DIKETONE ALCOHOL (123-42-2) Incompatible with strong acids, strong alkalis (cause formation of flammable acetone vapors), aliphatic amines, isocyanates, oxidizers, alkali metals. Forms explosive mixture with air (flash point 136°F/58°C). Attacks some forms of plastics, resins, and rubber.

DILANTIN DB (95-50-1) Forms explosive mixture with air (flash point 151°F/66°C). Incompatible with strong oxidizers, hot aluminum, or aluminum alloy. Attacks some plastics, rubber, and coatings.

DILATIN DB (95-50-1) Forms explosive mixture with air (flash point 151°F/66°C). Incompatible with strong oxidizers, hot aluminum, or aluminum alloy. Attacks some plastics, rubber, and coatings.

DILAUROYL PEROXIDE (105-74-8) A combustible solid and powerful oxidizer. Reacts violently with reducing agents. May self-ignite if mixed with combustible materials.

DILENE (72-54-8) Contact with strong oxidizers may cause fire and explosions.

DILITHIUM CHROMATE (14307-35-8) Aqueous solution is caustic. An oxidizer; reacts strongly with reducing agents, combustibles, organic materials, acids.

DIMAZ (298-04-4) Forms explosive mixture with air (flash point 180°F/82°C). Incompatible with alkalis, strong oxidizers.

DIMAZINE (57-14-7) Forms explosive mixture with air (flash point 5°F/−15°C). A strong reducing agent. Incompatible with strong acids, halogens, metallic mercury, strong oxidizers (with possible spontaneous ignition). Attacks some plastics, rubber, and coatings. May accumulate static electrical charges, and may cause ignition of its vapors.

DI-p-MENTHA-1,8-DIENE (138-86-3) Forms explosive mixture with air (flash point 115°F/46°C). Strong oxidizers may cause fire and explosions.

DIMER CYKLOPENTADIENU (Czech) (77-73-6) Forms explosive mixture with air (flash point 90°F/32°C). Forms peroxides with air. May polymerize unless inhibited and maintained under inert atmosphere. Reacts violently with strong oxidizers. May accumulate static electrical charges, and may cause ignition of its vapors.

276

DIMER OF NITROGEN DIOXIDE (10544-72-6); (10102-44-0 CAS for nitrogen dioxide) A powerful oxidizer. Water contact produces nitric acid and nitric oxide. Incompatible with nitric oxide; reacts with air to produce more nitrogen tetroxide. Explosive reaction with alcohols, fuels and petroleum, nitrobenzene, organic matter, anhydrous ammonia, chlorinated hydrocarbons, toluene. Vigorous reaction occurs with combustibles, reducing agents. Attacks metals in presence of moisture.

2,7:3,6-DIMETHANONAPTH 2,3-B OXIRENE, 3,4,5,6,9,9-HEXA-CHLORO-1A,2,2A,3,6,6A,7,7A-OCTAHYDRO-(1A alpha, 2 beta, 2A alpha, 3 beta, 6 beta, 6A alpha, 7 beta, 7A alpha)- (60-57-1) Incompatible with concentrated mineral acids, acid catalysts, acid oxidizing agents, phenols, reactive metals.

DIMETHICONE 350 (9016-00-6) Incompatible with sulfuric acid, isocyanates.

P,P'-DIMETHOXYDIPHENYLTRICHLOROETHANE (72-43-5) Contact with strong oxidizers may cause fire and explosions. Attacks some plastics, rubber, and coatings.

1,1-DIMETHOXYETHANE (534-15-6) Forms explosive mixture with air (flash point −15°F/−26°C). Reacts violently with strong oxidizers. Incompatible with acids. May accumulate static electrical charges, and may cause ignition of its vapors.

1,2-DIMETHOXYETHANE (110-71-4) Incompatible with sulfuric acid, isocyanates, strong oxidizers.

DIMETHOXYMETHANE (109-87-5) Forms explosive mixture with air (flash point −4°F/−18°C). Incompatible with strong oxidizers, acids. Attacks some plastics, rubber, and coatings.

3-(DIMETHOXYPHOSPHINYLOXY)-N,N-DIMETHYL-cis-CRO-TONAMIDE (141-66-2) Corrosive to cast iron, mild steel, brass, and stainless steel 304.

3-(DIMETHOXYPHOSPHINYLOXY)-N,N-DIMETHYLISOCRO-TONAMIDE (141-66-2) Corrosive to cast iron, mild steel, brass, and stainless steel 304.

2,3-DIMETHOXYSTRICHNIDIN-10-ONE (357-57-3) Contact with strong oxidizers may cause fire and explosions. May accumulate static electrical charges, and may cause ignition of its vapors.

DIMETHOXY STRYCHNINE (357-57-3) Contact with strong oxidizers may cause fire and explosions. May accumulate static electrical charges, and may cause ignition of its vapors.

1,1-DIMETHOXYSTRYCHNINE (357-57-3) Contact with strong oxidizers may cause fire and explosions. May accumulate static electrical charges, and may cause ignition of its vapors.

2,3-DIMETHOXYSTRYCHNINE (357-57-3) Contact with strong oxidizers may cause fire and explosions. May accumulate static electrical charges, and may cause ignition of its vapors.

DIMETHYL ACETAL (534-15-6) Forms explosive mixture with air (flash point −15°F/−26°C). Reacts violently with strong oxidizers. Incompatible with acids. May accumulate static electrical charges, and may cause ignition of its vapors.

DIMETHYLACETAL FORMALDEHYDE (109-87-5) Forms explosive mixture with air (flash point −4°F/−18°C). Incompatible with strong oxidizers, acids. Attacks some plastics, rubber, and coatings.

DIMETHYLACETAMIDE (127-19-5) Forms explosive mixture with air (flash point 158°F/70°C). Incompatible with nonoxidizing mineral acids, strong acids, ammonia, isocyanates, phenols, cresols, halogenated compounds above 185°F/85°C. Attacks some plastics, rubber, and coatings.

N,N-DIMETHYL ACETAMIDE or N,N-DIMETHYL ACETAMIDE SOLUTION (127-19-5) Forms explosive mixture with air (flash point 158°F/70°C). Incompatible with nonoxidizing mineral acids, strong acids, ammonia, isocyanates, phenols, cresols, halogenated compounds above 185°F/85°C. Attacks some plastics, rubber, and coatings.

DIMETHYLACETIC ACID (79-31-2) Forms explosive mixture with air (flash point 132°F/56°C). Incompatible with sulfuric acid, caustics, ammonia, amines, isocyanates, alkylene oxides, epichlorohydrin. Attacks aluminum and other metals.

N,N-DIMETHYLACETIMIDE (127-19-5) Forms explosive mixture with air (flash point 158°F/70°C). Incompatible with nonoxidizing mineral acids, strong acids, ammonia, isocyanates, phenols, cresols, halogenated compounds above 185°F/85°C. Attacks some plastics, rubber, and coatings.

DIMETHYLACETONE (96-22-0) Forms explosive mixture with air (flash point 55°F/13°C). Incompatible with strong acids, aliphatic amines, strong oxidizers. Attacks some plastics, rubber, and coatings. May accumulate static electrical charges, and may cause ignition of its vapors.

DIMETHYLACETONE AMIDE (127-19-5) Forms explosive mixture with air (flash point 158°F/70°C). Incompatible with nonoxidizing mineral acids, strong acids, ammonia, isocyanates, phenols, cresols, halogenated compounds above 185°F/85°C. Attacks some plastics, rubber, and coatings.

DIMETHYLACETONYLCARBINOL (123-42-2) Incompatible with strong acids, strong alkalis (with formation of flammable acetone vapors), aliphatic amines, isocyanates, oxidizers, alkali metals. Forms explosive mixture with air (flash point 136°F/58°C). Attacks some forms of plastics, resins, and rubber.

DIMETHYLACETYLENE (503-17-3) Forms explosive mixture with air (flash point less than −4°F/−20°C). Reacts violently with oxidizers. May accumulate static electrical charges, and may cause ignition of its vapors.

DIMETHYLACETYLENECARBINOL (75-85-4) Forms explosive mixture with air (flash point 77°F/25°C). May accumulate static electrical charges, and may cause ignition of its vapors. Incompatible with strong acids, caustics, aliphatic amines, isocyanates, strong oxidizers, alkali metals (i.e., lithium, sodium, potassium, rubidium, cesium, francium).

DIMETHYLADIPATE (627-93-0) Incompatible with strong acids, nitrates, oxidizers.

DIMETHYLAMIDE ACETATE (127-19-5) Forms explosive mixture with air (flash point 158°F/70°C). Incompatible with nonoxidizing mineral acids, strong acids, ammonia, isocyanates, phenols, cresols, halogenated compounds above 185°F/85°C. Attacks some plastics, rubber, and coatings.

DIMETHYLAMIDE ACETATE SOLUTION (127-19-5) Forms explosive mixture with air (flash point 158°F/70°C. Incompatible with nonoxidizing mineral acids, strong acids, ammonia, isocyanates, phenols, cresols, halogenated compounds above 185°F/85°C. Attacks some plastics, rubber, and coatings.

DIMETHYLAMINE (124-40-3) A gas. Incompatible with acids, organic anhydrides, isocyanates, vinyl acetate, acrylates, substituted allyls, alkylene oxides, epichlorohydrin, ketones, aldehydes, alcohols, glycols, mercury, phenols, cresols, caprolactam solution, strong oxidizers. Attacks aluminum, copper, lead, tin, zinc, and alloys, and some plastics, rubbers, and coatings.

N,N-DIMETHYLAMINE (124-40-3) A gas. Incompatible with acids, organic anhydrides, isocyanates, vinyl acetate, acrylates, substituted allyls, alkylene oxides, epichlorohydrin, ketones, aldehydes, alcohols, glycols, mercury, phenols, cresols, caprolactam solution, strong oxidizers. Attacks aluminum, copper, lead, tin, zinc and alloys, and some plastics, rubbers, and coatings.

DIMETHYLAMINE, ANHYDROUS (124-40-3) A gas. Incompatible with acids, organic anhydrides, isocyanates, vinyl acetate, acrylates, substituted allyls, alkylene oxides, epichlorohydrin, ketones, aldehydes, alcohols, glycols, mercury, phenols, cresols, caprolactam solution, strong oxidizers. Attacks aluminum, copper, lead, tin, zinc, and alloys, and some plastics, rubbers, and coatings.

DIMETHYLAMINE, AQUEOUS SOLUTION or DIMETHYL-AMINE SOLUTION (124-40-3) A gas. Incompatible with acids, organic anhydrides, isocyanates, vinyl acetate, acrylates, substituted allyls, alkylene oxides, epichlorohydrin, ketones, aldehydes, alcohols, glycols, mercury, phenols, cresols, caprolactam solution, strong oxidizers. Attacks aluminum, copper, lead, tin, zinc and alloys, and some plastics, rubbers, and coatings.

DIMETHYLAMINOBENZENE or (DIMETHYLAMINO)BENZENE (121-69-7) Forms explosive mixture with air (flash point 145°F/63°C). Contact with strong acids may cause fire and explosions. Contact with hypochlorite bleaches produces explosive chloroamines. Incompatible with anhydrides, isocyanates, aldehydes.

DIMETHYLAMINO CARBONYL CHLORIDE (79-44-7) Rapidly hydrolyzed in water. Incompatible with strong acids, oxidizers.

N,N-DIMETHYLAMINOCYCLOHEXANE (98-94-2) Forms explosive mixture with air. Incompatible with acids, organic anhydrides, chlorinated hydrocarbons, isocyanates, vinyl acetate, acrylates, substituted allyls, alkylene oxides, epichlorohydrin, ketones, aldehydes, alcohols, glycols, phenols, cresols, caprolactam solution, strong oxidizers. Attacks aluminum, copper, lead, tin, zinc, and alloys.

b-DIMETHYLAMINOETHYL ALCOHOL (108-01-0) Forms explosive mixture with air (flash point 105°F/41°C). Reacts violently with oxidizers, acids. Attacks copper and its alloys, galvanized steel, zinc, and zinc alloys.

2-(DIMETHYLAMINO)ETHANOL (108-01-0) Forms explosive mixture with air (flash point 105°F/41°C). Reacts violently with oxidizers, acids. Attacks copper and its alloys, galvanized steel, zinc, and zinc alloys.

3-(DIMETHYLAMINO)-1-METHYL-3-OXO-1-PROPENYL DI-METHYL PHOSPHATE (141-66-2) Corrosive to cast iron, mild steel, brass, and stainless steel 304.

3-DIMETHYLAMINOPROPYLAMINE (109-55-7) Forms explosive mixture with air (flash point 100°F/38°C). reacts violently with strong oxidizers. Incompatible with and strong acids, organic anhydrides, isocyanates, aldehydes.

a-(DIMETHYLAMINO)TOLUENE (103-83-3) Forms explosive mixture with air (flash point 170°F/77°C). Incompatible with acids, organic anhydrides, isocyanates, aldehydes, strong oxidizers, calcium hypochlorite. Attacks some plastics, rubber, and coatings.

DIMETHYLANILINE (121-69-7) Forms explosive mixture with air (flash point 145F/63°C). Contact with strong acids may cause fire and explosions. Contact with hypochlorite bleaches produces explosive chloroamines. Incompatible with anhydrides, isocyanates, aldehydes.

DIMETHYLANILINE (1300-73-8) Forms explosive mixture with air (flash point 206°F/96.7°C). Contact with strong acids may cause fire and explosions. Contact with hypochlorite bleaches produces explosive chloroamines. Incompatible with organic acids and anhydrides, isocyanates, aldehydes.

2,6-DIMETHYLANILINE (87-62-7) Forms explosive mixture with air (flash point 206°F/96.7°C). Contact with strong acids may cause fire and explosions. Contact with hypochlorite bleaches produces explosive chloroamines. Incompatible with organic acids and anhydrides, isocyanates, aldehydes.

N,N-DIMETHYLANILINE (121-69-7) Forms explosive mixture with air (flash point 145°F/63°C). Contact with strong oxidizers may cause fire and explosions; incompatible with strong acids (with violent spattering). Contact with benzoyl peroxide, diisopropyl perdicarbonate, and other material may cause explosions. Attacks some plastics, rubber, and coatings.

DIMETHYLARSENIC ACID (75-60-5) Aqueous solution reacts with chemically active metals. Incompatible with sulfuric acid, caustics, ammonia, amines, isocyanates, alkylene oxides, epichlorohydrin.

2,6-DIMETHYLBENZENAMINE (87-62-7) Forms explosive mixture with air (flash point 206°F/96.7°C). Contact with strong acids may cause fire and explosions. Contact with hypochlorite bleaches produces explosive chloroamines. Incompatible with strong acids, organic acids and anhydrides, isocyanates, aldehydes.

N,N-DIMETHYLBENZENEAMINE (121-69-7) Forms explosive mixture with air (flash point 145F/63°C). Contact with strong acids may cause fire and explosions. Contact with hypochlorite bleaches produces explosive chloroamines. Incompatible with anhydrides, isocyanates, aldehydes.

1,2-DIMETHYLBENZENE (95-47-6) Forms explosive mixture with air (flash point 90°F/32°C). Contact with strong oxidizers may cause fire and explosions. Attacks some plastics, rubber, and coatings. May accumulate static electrical charges, and may cause ignition of its vapors.

1,3-DIMETHYLBENZENE (108-38-3) Forms explosive mixture with air (flash point 84°F/28.9°). Contact with strong oxidizers may cause fire and explosions. Attacks some plastics, rubber, and coatings. May accumulate static electrical charges, and may cause ignition of its vapors.

1,4-DIMETHYLBENZENE (106-42-3) Forms explosive mixture with air (flash point 81°F/27.2°C). Contact with strong oxidizers may cause fire and explosions. Attacks some plastics, rubber, and coatings. May accumulate static electrical charges, and may cause ignition of its vapors.

m-DIMETHYLBENZENE or *meta*-DIMETHYLBENZENE (108-38-3) Forms explosive mixture with air (flash point 84°F/28.9°). Contact with strong oxidizers may cause fire and explosions. Attacks some plastics, rubber, and coatings. May accumulate static electrical charges, and may cause ignition of its vapors.

o-DIMETHYLBENZENE or *ortho*-DIMETHYLBENZENE (95-47-6) Forms explosive mixture with air (flash point 90°F/32°C). Strong oxidizers may cause fire and explosions. Attacks some plastics, rubber, and coatings. May accumulate static electrical charges, and may cause ignition of its vapors.

p-DIMETHYLBENZENE or *para*-DIMETHYLBENZENE (106-42-3) Forms explosive mixture with air (flash point 81°F/27.2°C). Contact with strong oxidizers may cause fire and explosions. Attacks some plastics, rubber, and coatings. May accumulate static electrical charges, and may cause ignition of its vapors.

N,N-DIMETHYLBENZENEAMINE (121-69-7) Forms explosive mixture with air (flash point 145°F/63°C). Contact with strong oxidizers may cause fire and explosions; incompatible with strong acids (with violent spattering). Contact with benzoyl peroxide, diisopropyl perdicarbonate and other material may cause explosions. Attacks some plastics, rubber, and coatings.

DIMETHYL-1,2-BENZENEDICARBOXYLATE (131-11-3) Incompatible with strong alkalies, strong acids, nitrates, oxidizers.

alpha, alpha-DIMETHYLBENZENE HYDROPEROXIDE (80-15-9) A strong oxidizer; reacts violently with reducers, combustibles. Explosive decomposition may occur above 120°F/50°C. Forms explosive mixture with air (flash point 175°F/79°C). Incompatible with acids, bases, amines, metallic salts of cobalt, copper, lead. May accumulate static electrical charges, and may cause ignition of its vapors.

N,N-DIMETHYL BENZENE METHANAMINE (103-83-3) Forms explosive mixture with air (flash point 170°F/77°C). Incompatible with acids, organic anhydrides, isocyanates, aldehydes, strong oxidizers, calcium hypochlorite. Attacks some plastics, rubber, and coatings.

DIMETHYL BENZENEORTHODICARBOXYLATE (131-11-3) Incompatible with strong alkalies, strong acids, nitrates, oxidizers.

3,3′-DIMETHYLBENZIDIN (95-53-4) Forms explosive mixture with air (flash point 185°F/85°C). Incompatible with strong acids, mineral acids, organic anhydrides, isocyanates, aldehydes, oxidizers. Attacks some plastics, rubber, and coatings.

3,3′-DIMETHYLBENZIDINE (95-53-4) Forms explosive mixture with air (flash point 185°F/85°C). Incompatible with strong acids, mineral acids, organic anhydrides, isocyanates, aldehydes, oxidizers. Attacks some plastics, rubber, and coatings.

DIMETHYLBENZYL HYDROPEROXIDE (80-15-9) A strong oxidizer; reacts violently with reducers, combustibles. Explosive decomposition may occur above 120°F/50°C. Forms explosive mixture with air (flash point 175°F/79°C). Incompatible with acids, bases, amines, metallic salts of cobalt, copper, lead. May accumulate static electrical charges, and may cause ignition of its vapors.

alpha,alpha-DIMETHYLBENZYL HYDROPEROXIDE (80-15-9) A strong oxidizer; reacts violently with reducers, combustibles. Explosive decomposition may occur above 120°F/50°C. Forms explosive mixture with air (flash point 175°F/79°C). Incompatible with acids, bases, amines, metallic salts of cobalt, copper, lead. May accumulate static electrical charges, and may cause ignition of its vapors.

3,3'-DIMETHYLBIPHENYL-4,4'-DIAMINE (95-53-4) Forms explosive mixture with air (flash point 185°F/85°C). Incompatible with strong acids, mineral acids, organic anhydrides, isocyanates, aldehydes, oxidizers. Attacks some plastics, rubber, and coatings.

3,3'-DIMETHYL-4,4'-BIPHENYLDIAMINE (95-53-4) Forms explosive mixture with air (flash point 185°F/85°C). Incompatible with strong acids, mineral acids, organic anhydrides, isocyanates, aldehydes, oxidizers. Attacks some plastics, rubber, and coatings.

2,2-DIMETHYLBUTANE (75-83-2) Forms explosive mixture with air (flash point −54°F/−48°C). Reacts violently with strong oxidizers.

1,3-DIMETHYLBUTANOL (108-11-2) Forms explosive mixture with air (flash point 106°F/41°C). Contact with alkali metals produces hydrogen gas. Incompatible with strong acids, caustics, aliphatic amines, isocyanates. Attacks some plastics, rubber, and coatings. May accumulate static electrical charges, and may cause ignition of its vapors.

1,3-DIMETHYLBUTYL ACETATE (108-84-9) Forms explosive mixture with air (flash point 113°F/45°C). Incompatible with strong acids, strong alkalies, nitrates, strong oxidizers.

2,2-DIMETHYLCAPRYLIC ACID (N/A) Incompatible with sulfuric acid, caustics, ammonia, amines, isocyanates, alkylene oxides, epichlorohydrin.

DIMETHYLCARBAMIC ACID CHLORIDE (79-44-7) Rapidly hydrolyzed in water. Incompatible with strong acids, oxidizers.

DIMETHYLCARBAMIC CHLORIDE (79-44-7) Rapidly hydrolyzed in water. Incompatible with strong acids, oxidizers.

DIMETHYLCARBAMIDOYL CHLORIDE (79-44-7) Rapidly hydrolyzed in water. Incompatible with strong acids, oxidizers.

DIMETHYL CARBAMOYL CHLORIDE (79-44-7) Rapidly hydrolyzed in water. Incompatible with strong acids, oxidizers.

N,N-DIMETHYLCARBAMOYL CHLORIDE (79-44-7) Rapidly hydrolyzed in water. Incompatible with strong acids, oxidizers.

cis-**2-DIMETHYLCARBAMOYL-1-METHYLVINYL DIMETHYLPHOSPHATE** (141-66-2) Corrosive to cast iron, mild steel, brass, and stainless steel 304.

DIMETHYLCARBAMYL CHLORIDE (79-44-7) Rapidly hydrolyzed in water. Incompatible with strong acids, oxidizers.

N,N-DIMETHYLCARBAMYL CHLORIDE (79-44-7) Rapidly hydrolyzed in water. Incompatible with strong acids, oxidizers.

DIMETHYLCARBINOL (67-63-0) Forms explosive mixture with air (flash point 53°F/12°C). Incompatible with alkaline earth and alkali metals, crotonaldehyde, phosgene, strong acids, amines, ammonia, caustics, strong oxidizers. Attacks some plastics, rubber, and coatings. Reacts with metallic aluminum at high temperatures.

DIMETHYL CARBONATE (616-38-6) Forms explosive mixture with air (flash point 66°F/19°C). Reacts violently with strong oxidizers. Incompatible with aluminum, magnesium, silicone. May accumulate static electrical charges, and may cause ignition of its vapors.

DIMETHYL CELLOSOLVE (110-71-4) Incompatible with sulfuric acid, isocyanates, strong oxidizers.

DIMETHYLCHLOROETHER (107-30-2) Forms explosive mixture with air (flash point 0°F/−17.8°C). May be able to form unstable and explosive peroxides. Forms hydrochloric acid on contact with moisture. Corrosive to metals in presence of moisture.

N,N-DIMETHYLCHLOROFORMAMIDE (79-44-7) Rapidly hydrolyzed in water. Incompatible with strong acids, oxidizers.

N-DIMETHYLCYCLOHEXANAMINE (98-94-2) Forms explosive mixture with air. Incompatible with acids, organic anhydrides, chlorinated hydrocarbons, isocyanates, vinyl acetate, acrylates, substituted allyls, alkylene oxides, epichlorohydrin, ketones, aldehydes, alcohols, glycols, phenols, cresols, caprolactam solution, strong oxidizers. Attacks aluminum, copper, lead, tin, zinc, and alloys.

1,4-DIMETHYLCYCLOHEXANE-*trans* (589-90-2) Forms explosive mixture with air (flash point 50°F/10°C). Reacts violently with strong oxidizers. May accumulate static electrical charges, and may cause ignition of its vapors.

N,N-DIMETHYLCYCLOHEXYLAMINE (98-94-2) Forms explosive mixture with air. Incompatible with acids, organic anhydrides, chlorinated hydrocarbons, isocyanates, vinyl acetate, acrylates, substituted allyls, alkylene oxides, epichlorohydrin, ketones, aldehydes, alcohols, glycols, phenols, cresols, caprolactam solution, strong oxidizers. Attacks aluminum, copper, lead, tin, zinc, and alloys.

DIMETHYL-1,2-DIBROMO-2,2-DICHLOROETHYLPHOSPHATE (300-76-5) Contact with strong oxidizers may cause fire and explosions. Attacks some plastics, rubber, and coatings.

O,O-DIMETHYL-O-(1,2-DIBROMO-2,2-DICHLOROETHYL) PHOSPHATE (300-76-5) Contact with strong oxidizers may cause fire and explosions. Attacks some plastics, rubber, and coatings.

O,O-DIMETHYL-(1,2-DICARBETHOXY-ETHYL)PHOSPHORO-DITHIOCITE (121-75-5) Incompatible with strong oxidizers, magnesium, alkaline pesticides. Attacks metals, some plastics, rubber, and coatings.

O,O-DIMETHYL-O-2,2-DICHLORO-1,2-DIBROMOETHYL PHOSPHATE (300-76-5) Contact with strong oxidizers may cause fire and explosions. Attacks some plastics, rubber, and coatings.

DIMETHYL-1,1'-DICHLOROETHER (542-88-1) Highly volatile. Heat and light may form unstable peroxides. Heat may cause fire. Contact with water forms hydrogen chloride and formaldehyde.

1,1-DIMETHYL-3-(3,4-DICHLOROPHENYL)UREA (330-54-1) Hydrolyzes in fairly strong acids.

DIMETHYLDICHLOROSILAN or DIMETHYLDICHLOROSI-LANE (75-78-5) Forms explosive gas mixture with air (flash point 15°F/−9°C). Water (vapor) or air form hydrochloric acid. Incompatible with acetone, amines, ammonia, alcohols, strong oxidizers, caustics. Attacks most metals.

DIMETHYL-DICHLORSILAN (Czech) or **DIMETHYL-DICHLO-ROSILAN (Czech)** (75-78-5) Forms explosive gas mixture with air (flash point 15°F/−9°C). Water (vapor) or air form hydrochloric acid. Incompatible with acetone, amines, ammonia, alcohols, strong oxidizers, caustics. Attacks most metals.

1,1'-DIMETHYLDIETHYLENE GLYCOL (110-98-5) Incompatible with sulfuric acid, perchloric acid, isocyanates, strong oxidizers.

2,2-DIMETHYL-2,2-DIHYDROBENZOFURANYL-7 N-METHYL-CARBAMATE (1563-66-2) Incompatible with alkaline material, acids, strong oxidizers.

O,O-DIMETHYL-O-(2-DIMETHYL-CARBAMOYL-1-METHYL-VINYL)PHOSPHAT (German) (141-66-2) Corrosive to cast iron, mild steel, brass, and stainless steel 304.

O,O-DIMETHYL-O-(N,N-DIMETHYLCARBAMOYL-1-METHYL-VINYL) PHOSPHATE (141-66-2) Corrosive to cast iron, mild steel, brass, and stainless steel 304.

O,O-DIMETHYL-O-(1,4-DIMETHYL-3-OXO-4-AZA-PENT-1-ENYL)FOSFAAT (Dutch) (141-66-2) Corrosive to cast iron, mild steel, brass, and stainless steel 304.

O,O-DIMETHYL-O-(1,4-DIMETHYL-3-OXO-4-AZA-PENT-1-ENYL)PHOSPHATE (141-66-2) Corrosive to cast iron, mild steel, brass, and stainless steel 304.

3,3'-DIMETHYL-4,4'-DIPHENYLDIAMINE (95-53-4) Forms explosive mixture with air (flash point 185°F/85°C). Incompatible with strong acids, mineral acids, organic anhydrides, isocyanates, aldehydes, oxidizers. Attacks some plastics, rubber, and coatings.

O,O-DIMETHYL DITHIOPHOSPHATE OF DIETHYL MERCAPTOSUCCINATE (121-75-5) Incompatible with strong oxidizers, magnesium, alkaline pesticides. Attacks metals, some plastics, rubber, and coatings.

DIMETHYLENEDIAMINE (107-15-3) Forms explosive mixture with air (flash point 104°F/40°C). Incompatible with chlorinated organic compounds, silver perchlorate, 3-propiolactone, mesityl oxide, ethylene dichloride, acids, organic anhydrides, isocyanates, vinyl acetate, acrylates, substituted allyls, alkylene oxides, epichlorohydrin, ketones, aldehydes, alcohols, glycols, phenols, cresols, caprolactam solution, strong oxidizers. Attacks aluminum, copper, lead, tin, zinc, and alloys, and some plastics, rubber, and coatings.

DIMETHYLENE GLYCOL (513-85-9) Forms explosive mixture with air (flash point 185°F/85°C). Incompatible with strong acids, caustics, aliphatic amines, isocyanates, strong oxidizers.

DIMETHYLENEIMINE (151-56-4) Forms explosive mixture with air. Contact with acid, aluminum, carbon dioxide or silver may cause explosive polymerization. Attacks rubber, coatings and plastics. Self-reactive with heat or atmospheric carbon dioxide.

3,3-DIMETHYLENENORCAMPHENE (79-92-5) Contact with strong oxidizers may cause fire and explosions. Emulsions in xylene may violently decompose on contact with iron or aluminum above 158°F/70°C.

DIMETHYLENE OXIDE (75-21-8) Forms explosive mixture with air (flash point 20°F/−6°C). Incompatible with alkali metal hydroxides, highly active catalysts (e.g., anhydrous chlorides of iron, tin, or aluminum and oxides of iron or aluminum). Avoid contact with copper. Protect container from physical damage, sun, and heat. Attacks some plastics, rubber, or coatings.

DIMETHYLENIMINE (151-56-4) Forms explosive mixture with air. Contact with acid, aluminum, carbon dioxide or silver may cause explosive polymerization. Attacks rubber, coatings, and plastics. Self-reactive with heat or atmospheric carbon dioxide.

DIMETHYLESTER KYSELINY SIROVE (Czech) (77-78-1) Forms explosive mixture with air (flash point 182°F/83°C). Incompatible with strong oxidizers, strong acids, strong alkalies, strong ammonia solutions. Attacks some plastics, rubber, and coatings.

DIMETHYL ESTER OF SULFURIC ACID (77-78-1) Forms explosive mixture with air (flash point 182°F/83°C). Incompatible with strong oxidizers, strong acids, strong alkalies, strong ammonia solutions. Attacks some plastics, rubber, and coatings.

1,1-DIMETHYLETHANE (75-28-5) Forms explosive gas mixture with air. Strong oxidizers may cause fire and explosions. May accumulate static electrical charges, and may cause ignition of its vapors.

1,1-DIMETHYLETHANOL (75-65-0) Forms explosive mixture with air (flash point 52°F/11°C). May accumulate static electrical charges, and may cause ignition of its vapors. Incompatible with strong acids, including mineral acids, strong oxidizers or caustics, aliphatic amines, isocyanates, alkali metals (i.e., lithium, sodium, potassium, rubidium, cesium, francium). Attacks many plastics and some coatings.

DIMETHYLETHANOLAMINE (108-01-0) Forms explosive mixture with air (flash point 105°F/41°C). Reacts violently with oxidizers, acids. Attacks copper and its alloys, galvanized steel, zinc, and zinc alloys.

N,N-DIMETHYLETHANOLAMINE (108-01-0) Forms explosive mixture with air (flash point 105°F/41°C). Reacts violently with oxidizers, acids. Attacks copper and its alloys, galvanized steel, zinc, and zinc alloys.

DIMETHYL ETHER (115-10-6) Flammable gas. Forms explosive mixture with air. Forms unstable peroxides in storage. Reacts violently with strong oxidizers.

1,1-DIMETHYLETHYLAMINE (75-64-9) Forms explosive mixture with air (flash point 48°F/−9°C). May accumulate static electrical charges, and may cause ignition of its vapors. Incompatible with acids, organic anhydrides, isocyanates, vinyl acetate, acrylates, substituted allyls, alkylene oxides, epichlorohydrin, ketones, aldehydes, alcohols, glycols, phenols, cresols, caprolactam solution, strong oxidizers.

DIMETHYLETHYLCARBINOL (75-85-4) Forms explosive mixture with air (flash point 77°F/25°C). May accumulate static electrical charges, and may cause ignition of its vapors. Incompatible with strong acids, caustics, aliphatic amines, isocyanates, strong oxidizers, alkali metals (i.e., lithium, sodium, potassium, rubidium, cesium, francium).

1,1-DIMETHYLETHYL HYDROPEROXIDE (75-91-2) Forms explosive mixture with air (flash point 100°F/38°C). A powerful oxidizer; reacts violently with reducing agents, organic materials, ethylene dichloride.

285

1-(1,1-DIMETYLETHYL)-4-METHYLBENZENE (98-51-1) Forms explosive mixture with air (flash point 155°F/67°C). Reacts with strong oxidizers. May accumulate static electrical charges, and may cause ignition of its vapors.

4-(1,1-DIMETHYLETHYL) PHENOL (98-54-4) Incompatible with strong acids, caustics, aliphatic amines, amides, oxidizers.

DIMETHYL FORMAL (109-87-5) Forms explosive mixture with air (flash point −4°F/−18°C). Incompatible with strong oxidizers, acids. Attacks some plastics, rubber, and coatings.

DIMETHYLFORMALDEHYDE (67-64-1) Forms explosive mixture with air (flash point −4°F/−20°C). Reacts violently with chloroform. Incompatible with strong acids, aliphatic amines, chloroform, chromic anhydride, chromyl chloride, hexachloromelamine, hydrogen peroxide, nitrosyl chloride, nitrosyl perchlorate, nitryl perchlorate, permonosulfuric acid, potassium *tert*-butoxide. Unstable and explosive peroxides are formed with strong oxidizers. May accumulate static electrical charges, and may cause ignition of its vapors. Dissolves most rubber, resins, and plastics.

DIMETHYLFORMAMID (German) (68-12-2) Forms explosive mixture with air (flash point 136°F/58°C). Contact with carbon tetrachloride and other halogenated compounds, particularly in presence of iron or strong oxidizers may cause fire and explosions. Reacts vigorously with alkylaluminums. Incompatible with nonoxidizing mineral acids, strong acids, chlorinated hydrocarbons, isocyanates, nitrates, organic nitrates, phenols, cresols, ammonia, bromine, chromic anhydride, magnesium nitrate, methylene diisocyanate, phosphorus trioxide, triethyaluminum. Attacks some plastics, rubber, and coatings.

DIMETHYL FORMAMIDE (68-12-2) Forms explosive mixture with air (flash point 136°F/58°C). Contact with carbon tetrachloride and other halogenated compounds, particularly in presence of iron or strong oxidizers, may cause fire and explosions. Reacts vigorously with alkylaluminums. Incompatible with nonoxidizing mineral acids, strong acids, chlorinated hydrocarbons, isocyanates, nitrates, organic nitrates, phenols, cresols, ammonia, bromine, chromic anhydride, magnesium nitrate, methylene diisocyanate, phosphorus trioxide, triethylaluminum. Attacks some plastics, rubber, and coatings.

N,N-DIMETHYLFORMAMIDE (68-12-2) Forms explosive mixture with air (flash point 136°F/58°C). Contact with carbon tetrachloride and other halogenated compounds, particularly in presence of iron or strong oxidizers, may cause fire and explosions. Reacts vigorously with alkylaluminums. Incompatible with nonoxidizing mineral acids, strong acids, chlorinated hydrocarbons, isocyanates, nitrates, organic nitrates, phenols, cresols, ammonia, chromic anhydride, magnesium nitrate, methylene diisocyanate, phosphorus trioxide, triethylaluminum. Attacks some plastics, rubber, and coatings.

DIMETYLFORMAMIDU (Czech) (68-12-2) Forms explosive mixture with air (flash point 136°F/58°C). Contact with carbon tetrachloride and other halogenated compounds, particularly in presence of with iron or strong oxidizers may cause fire and explosions. Reacts vigorously with alkylaluminums. Incompatible with nonoxidizing mineral acids, strong acids, chlorinated hydrocarbons, isocyanates, nitrates, organic nitrates, phenols, cresols, ammonia, chromic anhydride, magnesium nitrate, methylene diisocyanate, phosphorus trioxide, triethylaluminum. Attacks some plastics, rubber, and coatings.

DIMETHYLFORMEHYDE (67-64-1) Forms explosive mixture with air (flash point −4°F/−20°C). Reacts violently with chloroform. Incompatible with strong acids, aliphatic amines, chloroform, chromic anhydride, chromyl chloride, hexachloromelamine, hydrogen peroxide, nitrosyl chloride, nitrosyl perchlorate, nitryl perchlorate, permonosulfuric acid, potassium *tert*-butoxide. Unstable and explosive peroxides are formed with strong oxidizers. May accumulate static electrical charges, and may cause ignition of its vapors. Dissolves most rubber, resins, and plastics.

para-**DIMETHYLFTALAAT (Dutch)** (120-61-6) Incompatible with strong acids, nitrates, strong oxidizers.

DIMETHYL GLUTARATE (1119-40-0) Incompatible with strong acids, nitrates, oxidizers.

2,6-DIMETHYL-4-HEPTANE (108-83-8) Incompatible with strong acids, aliphatic amines, strong oxidizers. Attacks plastics, coatings, and rubber. Forms explosive mixture with air (flash point 140°F/60°C).

2,6-DIMETHYL-4-HEPTANOL (108-82-7) Forms explosive mixture with air (flash point 165°F/74°C). Incompatible with strong acids, caustics, aliphatic amines, isocyanates, strong oxidizers.

2,6-DIMETHYL-HEPTAN-4-ON (Dutch or German) (108-83-8) Incompatible with strong acids, aliphatic amines, strong oxidizers. Attacks plastics, coatings, and rubber. Forms explosive mixture with air (flash point 140°F/60°C).

2,6-DIMETHYLHEPTANONE (108-83-8) Incompatible with strong acids, aliphatic amines, strong oxidizers. Attacks plastics, coatings, and rubber. Forms explosive mixture with air (flash point 140°F/60°C).

2,6-DIMETHYLHEPTAN-4-ONE (108-83-8) Incompatible with strong acids, aliphatic amines, strong oxidizers. Attacks plastics, coatings, and rubber. Forms explosive mixture with air (flash point 140°F/60°C).

2,6-DIMETHYL-4-HEPTANONE (108-83-8) Incompatible with strong acids, aliphatic amines, strong oxidizers. Attacks plastics, coatings, and rubber. Forms explosive mixture with air (flash point 140°F/60°C).

DI (6-METHYLHEPTYL) PHTHALATE (27554-26-3) May accumulate static electrical charges, and may cause ignition of its vapors. Incompatible with strong acids, nitrates, oxidizers.

DI-(6-METHYLHEPTYL) PHTHALATE (27554-26-3) May accumulate static electrical charges, and may cause ignition of its vapors. Incompatible with strong acids, nitrates, oxidizers.

DIMETHYLHEXANALS (N/A) Forms explosive mixture with air (flash point 104°F/40°C). Incompatible with strong acids, caustics, ammonia, amines.

2,5-DIMETHYLHEXANE-2,5-DIHY-DROPEROXIDE (3025-88-5) Contact with acids and many metals may cause decomposition.

DIMETHYLHEXANE DIHYDROPEROXIDE (3025-88-5) An organic peroxide. Contact with acids and many metals may cause decomposition.

DIMETHYLHEXANE DIHYDROPEROXIDE (DRY) or DIMETHYLHEXANE DIHYDROPEROXIDE (WITH 18% OR MORE WATER) (3025-88-5) Contact with acids and many metals may cause decomposition.

DIMETHYL HEXANEDIOATE (627-93-0) Incompatible with strong acids, nitrates, oxidizers.

DIMETHYL-1-HEXANOLS (26952-21-6) Forms explosive mixture with air (flash point 180°F/82°C). Incompatible with strong acids, caustics, amines, isocyanates.

1,1-DIMETHYLHYDRAZIN (German) (57-14-7) Forms explosive mixture with air (flash point 5°F/−15°C). A strong reducing agent. Incompatible with strong acids, halogens, metallic mercury, strong oxidizers (with possible spontaneous ignition). Attacks some plastics, rubber, and coatings. May accumulate static electrical charges, and may cause ignition of its vapors.

1,2-DIMETHYLHYDRAZIN (German) (540-73-8) Forms explosive mixture with air (flash point 73°F/23°C). A strong reducing agent. Reacts violently with strong oxidizers, strong acids. Attacks some plastics, rubber, and coatings. May accumulate static electrical charges, and may cause ignition of its vapors.

DIMETHYLHYDRAZINE (57-14-7) Forms explosive mixture with air (flash point 5°F/−15°C). A strong reducing agent. Incompatible with strong acids, halogens, metallic mercury, strong oxidizers (with possible spontaneous ignition). Attacks some plastics, rubber, and coatings. May accumulate static electrical charges, and may cause ignition of its vapors.

1,1-DIMETHYLHYDRAZINE (57-14-7) Forms explosive mixture with air (flash point 5°F/−15°C). A strong reducing agent. Incompatible with strong acids, halogens, metallic mercury, strong oxidizers (with possible spontaneous ignition). Attacks some plastics, rubber, and coatings. May accumulate static electrical charges, and may cause ignition of its vapors.

1,2-DIMETHYLHYDRAZINE (540-73-8) Forms explosive mixture with air (flash point 73°F/23°C). A strong reducing agent. Reacts violently with strong oxidizers, strong acids. Attacks some plastics, rubber, and coatings. May accumulate static electrical charges, and may cause ignition of its vapors.

N,N-DIMETHYLHYDRAZINE (540-73-8) Forms explosive mixture with air (flash point 73°F/23°C). A strong reducing agent. Reacts violently with strong oxidizers, strong acids. Attacks some plastics, rubber, and coatings. May accumulate static electrical charges, and may cause ignition of its vapors.

asym-**DIMETHYLHYDRAZINE** (57-14-7) Forms explosive mixture with air (flash point 5°F/−15°C). A strong reducing agent. Incompatible with strong acids, halogens, metallic mercury, strong oxidizers (with possible spontaneous ignition). Attacks some plastics, rubber, and coatings. May accumulate static electrical charges, and may cause ignition of its vapors.

sym-**DIMETHYLHYDRAZINE** (540-73-8) Forms explosive mixture with air (flash point 73°F/23°C). A strong reducing agent. Reacts violently with strong oxidizers, strong acids. Attacks some plastics, rubber, and coatings. May accumulate static electrical charges, and may cause ignition of its vapors.

unsym-**DIMETHYLHYDRAZINE or** *unsymmetrical*-**DIMETHYLHY-DRAZINE** (57-14-7) Forms explosive mixture with air (flash point 5°F/−15°C). A strong reducing agent. Incompatible with strong acids, halogens, metallic mercury, strong oxidizers (with possible spontaneous ignition). Attacks some plastics, rubber, and coatings. May accumulate static electrical charges, and may cause ignition of its vapors.

unsym-DIMETHYLHYDRAZINE (57-14-7) Forms explosive mixture with air (flash point 5°F/−15°C). A strong reducing agent. Incompatible with strong acids, halogens, metallic mercury, strong oxidizers (possible spontaneous ignition). Attacks some plastics, rubber, and coatings. May accumulate static electrical charges, and may cause ignition of its vapors.

DIMETHYLHYDRAZINE, SYMMETRICAL (540-73-8) Forms explosive mixture with air (flash point 73°F/23°C). A strong reducing agent. Reacts violently with strong oxidizers, strong acids. Attacks some plastics, rubber, and coatings. May accumulate static electrical charges, and may cause ignition of its vapors.

DIMETHYLHYDRAZINE, UNSYMMETRICAL (57-14-7) Forms explosive mixture with air (flash point 5°F/−15°C). A strong reducing agent. Incompatible with strong acids, halogens, metallic mercury, strong oxidizers (with possible spontaneous ignition). Attacks some plastics, rubber, and coatings. May accumulate static electrical charges, and may cause ignition of its vapors.

DIMETHYL HYDROGEN PHOSPHITE (868-85-9) Incompatible with acids (nitric, sulfuric, nonoxidizing mineral, organic), nitrates, oxidizers.

N,N-DIMETHYL-N-(2-HYDROXYETHYL) AMINE (108-01-0) Forms explosive mixture with air (flash point 105°F/41°C). Reacts violently with oxidizers, acids. Attacks copper and its alloys, galvanized steel, zinc, and zinc alloys.

O,O-DIMETHYL-(1-HYDROXY-2,2,2-TRICHLORO)ETHYL PHOSPHATE (52-68-6) Contact with strong oxidizers may cause fire and explosions.

DIMETHYLKETAL (67-64-1) Forms explosive mixture with air (flash point −4°F/−20°C). Reacts violently with chloroform. Incompatible with strong acids, aliphatic amines, chloroform, chromic anhydride, chromyl chloride, hexachloromelamine, hydrogen peroxide, nitrosyl chloride, nitrosyl perchlorate, nitryl perchlorate, permonosulfuric acid, potassium *tert*-butoxide. Unstable and explosive peroxides formed with strong oxidizers. May accumulate static electrical charges, and may cause ignition of its vapors. Dissolves most rubber, resins, and plastics.

N,N-DIMETHYLMETHANAMINE (75-50-3) Forms explosive mixture with air (flash point 10°F/−12°C). Incompatible with oxidizers, acids. Attacks chemically active metals: aluminum, copper, zinc, and their alloys.

DIMETHYLMETHANE (74-98-6) Forms explosive gas mixture with air. Strong oxidizers may cause fire and explosions. Liquid attacks some plastics, rubber, and coatings. May accumulate static electrical charges, and may cause ignition of its vapors.

N,N-DIMETHYLMETHANIDE (68-12-2) Forms explosive mixture with air (flash point 136°F/58°C). Contact with carbon tetrachloride and other halogenated compounds, particularly in presence of iron or strong oxidizers, may cause fire and explosions. Reacts vigorously reaction with alkylaluminums. Incompatible with nonoxidizing mineral acids, strong acids, chlorinated hydrocarbons, isocyanates, nitrates, organic nitrates, phenols, cresols, ammonia, chromic anhydride, magnesium nitrate, methylene diisocyanate, phosphorus trioxide, triethylaluminum. Attacks some plastics, rubber, and coatings.

2,2-DIMETHYL-3-METHYLENE- (79-92-5) Contact with strong oxidizers may cause fire and explosions. Emulsions in xylene may violently decompose on contact with iron or aluminum above 158°F/70°C.

2-2-DIMETHYL-3-METHYLENE NORBORANE (79-92-5) Contact with strong oxidizers may cause fire and explosions. Emulsions in xylene may violently decompose on contact with iron or aluminum above 158°F/70°C.

3,3-DIMETHYL-2-METHYLENE NORCAMPHANE (79-92-5) Contact with strong oxidizers may cause fire and explosions. Emulsions in xylene may violently decompose on contact with iron or aluminum above 158°F/70°C.

3,5-DIMETHYL-4-(METHYLTHIO)METHYLCARBAMATE (2032-65-7) Contact with strong oxidizers may cause fire and explosions.

3,5-DIMETHYL-4-(METHYLTHIO)PHENOLMETHYLCARBA-MATE (2032-65-7) Contact with strong oxidizers may cause fire and explosions.

3,5-DIMETHYL-4-METHYLTHIOPHENYL N-METHYLCARBA-MATE (2032-65-7) Contact with strong oxidizers may cause fire and explosions.

DIMETHYL MONOSULFATE (77-78-1) Forms explosive mixture with air (flash point 182°F/83°C). Incompatible with strong oxidizers, strong acids, strong alkalies, strong ammonia solutions. Attacks some plastics, rubber, and coatings.

DIMETHYLNITROMETHANE (79-46-9) Forms explosive mixture with air (flash point 74°F/24°C). Incompatible with amines, strong acids, alkalies, strong oxidizers. Contact with some metal oxides may cause decomposition. Mixtures with hydrocarbons are extremely flammable. Attacks some plastics, rubber, and coatings.

O,O-DIMETHYL-O-4-NITROPHENYL PHOSPHOROTHIOATE (298-00-0) Mixtures with magnesium may be explosive.

DIMETHYLNITROSAMINE (62-75-9) Ultraviolet light, strong oxidizers can cause reactions. Store in dark bottles.

N,N-DIMETHYLNITROSAMINE (62-75-9) Ultraviolet light, strong oxidizers can cause reactions. Store in dark bottles.

DIMETHYLNITROSOAMINE (62-75-9) Ultraviolet light, strong oxidizers can cause reactions. Store in dark bottles.

2,2-DIMETHYLOCTANOIC ACID (N/A) Incompatible with sulfuric acid, caustics, ammonia, amines, isocyanates, alkylene oxides, epichlorohydrin, oxidizers.

N,N-DIMETHYL-N-OCTYLBENZENEMETHANAMINIUM-CHLORIDE (959-55-7) Incompatible with strong oxidizers, strong acids.

DI- (7-METHYLOCTYL) PHTHALATE (88-99-3) Incompatible with strong acids, nitrates, oxidizers.

DIMETHYLOL PROPANE (126-30-7) Forms explosive mixture with air (flash point 225°F/107°C). Incompatible with strong acids, caustics, aliphatic amines, isocyanates, oxidizers.

O,O-DIMETHYLO-(2,4,5-TRICHLOROPHENYL) PHOSPHORO-THIOATE (299-84-3) Temperatures above 300°F/150°C may cause explosive decomposition. Contact with strong oxidizers may cause fire and explosions. Attacks some plastics, rubber, and coatings.

O,O-DIMETHYL-S-(4-OXO-1,2,3-BEZOTRIAZIN-3(4H)-YL METHYL)PHOSPHORO DIMETHYLPHENOL (1300-71-6) Forms explosive mixture with air (flash point 186°F/86°C) Reacts violently with strong oxidizers.

DIMETHYLPHENOL PHOSPHATE (3:1) (25155-23-1) Incompatible with sulfuric acid, nitric acid, nitrates, strong oxidizers.

DIMETHYLPHENYLAMINE (1300-73-8) Forms explosive mixture with air (flash point 206°F/96.7°C). Contact with strong acids may cause fire and explosions. Contact with hypochlorite bleaches produces explosive chloroamines. Incompatible with strong acids, organic acids and anhydrides, isocyanates, aldehydes. Attacks some plastics, rubber, and coatings.

DIMETHYLPHENYLAMINE (121-69-7) Forms explosive mixture with air (flash point 145°F/63°C). Contact with strong acids may cause fire and explosions. Contact with hypochlorite bleaches produces explosive chloroamines. Incompatible with anhydrides, isocyanates, aldehydes.

DIMETHYL PHOSPHATE OF 3-HYDROXY-N,N-DIMETHYL-cis-CROTONAMIDE (141-66-2) Corrosive to cast iron, mild steel, brass, and stainless steel 304.

3-((DIMETHYLPHOSPHINYL)OXY)-2-BUTENOIC ACID METH-YL ESTHER (7786-34-7) Contact with strong oxidizers may cause fire and explosions. Attacks some plastics, rubber, and coatings.

DIMETHYL PHOSPHITE (868-85-9) Incompatible with acids (nitric, sulfuric, nonoxidizing mineral, organic), nitrates, oxidizers.

DIMETHYL PHOSPHONATE (868-85-9) Incompatible with acids (nitric, sulfuric, nonoxidizing mineral, organic), nitrates, oxidizers.

O,O-DIMETHYL-PHOSPHORODITHIOATE (121-75-5) Incompatible with strong oxidizers, magnesium, alkaline pesticides. Attacks metals, some plastics, rubber, and coatings.

O,O-DIMETHYL PHOSTENE (7786-34-7) Contact with strong oxidizers may cause fire and explosions. Attacks some plastics, rubber, and coatings.

DIMETHYL PHTHALATE (131-11-3) Incompatible with strong alkalies, strong acids, nitrates, oxidizers.

DIMETHYLPOLYSILOXANE (N/A) Incompatible with strong acids, nitrates, oxidizers.

2,2-DIMETHYL-1,3-PROPANEDIOL (126-30-7) Forms explosive mixture with air (flash point 225°F/107°C). Incompatible with strong acids, caustics, aliphatic amines, isocyanates, oxidizers.

2,2-DIMETHYLPROPANE-1,3-DIOL (126-30-7) Forms explosive mixture with air (flash point 225°F/107°C). Incompatible with strong acids, caustics, aliphatic amines, isocyanates, oxidizers.

2,2-DIMETHYLPROPIONIC ACID (75-98-9) Forms explosive mixture with air (flash point 147°F/64°C). Incompatible with sulfuric acid, caustics, ammonia, amines, isocyanates, alkylene oxides, epichlorohydrin, oxidizers.

alpha,alpha-DIMETHYLPROPIONIC ACID (75-98-9) Forms explosive mixture with air (flash point 147°F/64°C). Incompatible with sulfuric acid, caustics, ammonia, amines, isocyanates, alkylene oxides, epichlorohydrin, oxidizers.

(DIMETHYL-1,1-PROPYL)-4-PHENOL (80-46-6) Reacts with strong oxidizers. May accumulate static electrical charges, and may cause ignition of its vapors.

4-(1,1-DIMETHYLPROPYL)PHENOL (80-46-6) Reacts with strong oxidizers. May accumulate static electrical charges, and may cause ignition of its vapors.

DIMETHYL SILICONE or **DIMETHYL SILICONE FLUIDS** or **DIMETHYL SILICONE OIL** (N/A) Incompatible with strong acids, nitrates, oxidizers.

10,11-DIMETHYLSTRYCHNINE (357-57-3) Contact with strong oxidizers may cause fire and explosions. May accumulate static electrical charges, and may cause ignition of its vapors.

DIMETHYL SUCCINATE (106-65-0) Incompatible with strong acids, nitrates, oxidizers.

DIMETHYLSULFAAT (Dutch) (77-78-1) Forms explosive mixture with air (flash point 182°F/83°C). Incompatible with strong oxidizers, strong acids, strong alkalies, strong ammonia solutions. Attacks some plastics, rubber, and coatings.

DIMETHYLSULFAT (Czech) (77-78-1) Forms explosive mixture with air (flash point 182°F/83°C). Incompatible with strong oxidizers, strong acids, strong alkalies, strong ammonia solutions. Attacks some plastics, rubber, and coatings.

DIMETHYLSULFATE (77-78-1) Forms explosive mixture with air (flash point 182°F/83°C). Incompatible with strong oxidizers, strong acids, strong alkalies, strong ammonia solutions. Attacks some plastics, rubber, and coatings.

DIMETHYLSULFID (Czech) (75-18-3) Forms explosive mixture with air (flash point −36°F/−38°C) Reacts violently with strong oxidizers. May accumulate static electrical charges, and may cause ignition of its vapors.

DIMETHYL SULFIDE (75-18-3) Forms explosive mixture with air (flash point −36°F/−38°C) Reacts violently with strong oxidizers. May accumulate static electrical charges, and may cause ignition of its vapors.

DIMETHYL SULFOXIDE (67-68-5) Forms explosive mixture with air (flash point 203°F/95°C). Reacts violently with oxidizers. Reacts with ethanoyl chloride, boron compounds, halides, metal alkoxides, oxidizers.

DIMETHYL SULPHATE (77-78-1) Forms explosive mixture with air (flash point 182°F/83°C). Incompatible with strong oxidizers, strong acids, strong alkalies, strong ammonia solutions. Attacks some plastics, rubber, and coatings.

DIMETHYL SULPHIDE (75-18-3) Forms explosive mixture with air (flash point −36°F/−38°C) Reacts violently with strong oxidizers. May accumulate static electrical charges, and may cause ignition of its vapors.

DIMETHYL SULPHOXIDE (67-68-5) Forms explosive mixture with air (flash point 203°F/95°C). Reacts violently with oxidizers. Reacts with ethanoyl chloride, boron compounds, halides, metal alkoxides, oxidizers.

DIMETHYL TEREPHTHALATE (120-61-6) Incompatible with strong acids, nitrates, strong oxidizers.

N,N-(DIMETHYL) a-TOLUENEAMINE (103-83-3) Forms explosive mixture with air (flash point 170°F/77°C). Incompatible with acids, organic anhydrides, isocyanates, aldehydes, strong oxidizers, calcium hypochlorite. Attacks some plastics, rubber, and coatings.

DIMETHYL TRICHLOROPHENYL THIOPHOSPHOROTHIOATE (299-84-3) Temperatures above 300°F/150°C may cause explosive decomposition. Contact with strong oxidizers may cause fire and explosions. Attacks some plastics, rubber, and coatings.

O,O-DIMETHYL-O-(2,4,5-TRICHLOROPHENYL)THIOPHOS-PHAT (German) (299-84-3) Temperatures above 300°F/150°C may cause explosive decomposition. Contact with strong oxidizers may cause fire and explosions. Attacks some plastics, rubber, and coatings.

O,O-DIMETHYL-O-(2,4,5-TRICHLOROPHENYL)THIO-PHOSPHATE (299-84-3) Temperatures above 300°F/150°C may cause explosive decomposition. Contact with strong oxidizers may cause fire and explosions. Attacks some plastics, rubber, and coatings.

DIMETHYLTRIMETHYLENE GLYCOL (126-30-7) Forms explosive mixture with air (flash point 225°F/107°C). Incompatible with strong acids, caustics, aliphatic amines, isocyanates, oxidizers.

DIMETHYLZINC (544-97-8) Ignites spontaneously with air. Water contact produces methane gas.

O,O-DIMETIL-O-(1,4-DIMETIL-3-OXO-4-AZA-PENT-1-ENIL)-FOSFATO (Italian) (141-66-2) Corrosive to cast iron, mild steel, brass, and stainless steel 304.

2,6-DIMETIL-EPTAN-4-ONE (Italian) (108-83-8) Incompatible with strong acids, aliphatic amines, strong oxidizers. Attacks plastics, coatings, and rubber. Forms explosive mixture with air (flash point 140°F/60°C).

DIMETILFORMAMIDE (Italian) (68-12-2) Forms explosive mixture with air (flash point 136°F/58°C). Contact with carbon tetrachloride and other halogenated compounds, particularly in contact with iron or strong oxidizers, may cause fire and explosions. Vigorous reaction with alkylaluminums. Incompatible with nonoxidizing mineral acids, strong acids, chlorinated hydrocarbons, isocyanates, nitrates, organic nitrates, phenols, cresols, ammonia, chromic anhydride, magnesium nitrate, methylene diisocyanate, phosphorus trioxide, triethylaluminum. Attacks some plastics, rubber, and coatings.

DIMETILSOLFATO (Italian) (77-78-1) Forms explosive mixture with air (flash point 182°F/83°C). Incompatible with strong oxidizers, strong acids, strong alkalies, strong ammonia solutions. Attacks some plastics, rubber, and coatings.

DIMETOX (52-68-6) Contact with strong oxidizers may cause fire and explosions.

DIMEXIDE (67-68-5) Forms explosive mixture with air (flash point 203°F/95°C). Reacts violently with oxidizers. Reacts with ethanoyl chloride, boron compounds, halides, metal alkoxides, oxidizers.

DIMPYLATE (333-41-5) Incompatible with water, copper-containing compounds, oxidizers, acids, or bases.

DINIL (8004-13-5) Contact with strong oxidizers may cause fire and explosions. Attacks some plastics, rubber, and coatings.

2,4-DINITRANILINE (97-02-9) Reacts violently with strong oxidizers, strong acids.

DINITROANILINE (97-02-9) Reacts violently with strong oxidizers, strong acids.

2,4-DINITROANILINE (97-02-9) Reacts violently with strong oxidizers, strong acids.

2,4-DINITROBENZENAMIME (97-02-9) Reacts violently with strong oxidizers, strong acids.

1,2-DINITROBENZENE (528-29-0) Explodes: is impact- and friction-sensitive. Strong oxidizers may cause fire and explosions; avoid contact with caustics, chemically active metals. Attacks some plastics.

1,3-DINITROBENZENE (99-65-0) Explodes: is impact- and friction-sensitive. Strong oxidizers may cause fire and explosions; avoid contact with caustics, chemically active metals. Attacks some plastics.

1,4-DINITROBENZENE (25154-54-8) Explodes: is impact- and friction- sensitive. Strong oxidizers may cause fire and explosions; avoid contact with caustics, chemically active metals. Attacks some plastics.

2,4-DINITROBENZENE (99-65-0) Explodes: is impact- and friction-sensitive. Strong oxidizers may cause fire and explosions; avoid contact with caustics, chemically active metals. Attacks some plastics.

m-**DINITROBENZENE** or *meta*-**DINITROBENZENE** (99-65-0) Explodes: is impact- and friction-sensitive. Strong oxidizers may cause fire and explosions; avoid contact with caustics, chemically active metals. Attacks some plastics.

o-**DINITROBENZENE** or *ortho*-**DINITROBENZENE** (528-29-0) Explodes: is impact- and friction-sensitive. Strong oxidizers may cause fire and explosions; avoid contact with caustics, chemically active metals. Attacks some plastics.

p-**DINITROBENZENE** or *para*-**DINITROBENZENE** (25154-54-8) Explodes: is impact- and friction-sensitive. Strong oxidizers may cause fire and explosions; avoid contact with caustics, chemically active metals. Attacks some plastics.

DINITROBENZINE (25154-54-8) Explodes: is impact- and friction-sensitive. Strong oxidizers may cause fire and explosions; avoid contact with caustics, chemically active metals. Attacks some plastics.

DINITROBENZOL, SOLID (DOT) or DINITROBENZOL, SOLUTION (DOT) (25154-54-8) Explodes: is impact- and friction-sensitive. Strong oxidizers may cause fire and explosions; avoid contact with caustics, chemically active metals. Attacks some plastics.

1,3-DINITROBENZOL (99-65-0) Explodes: is impact- and friction-sensitive. Strong oxidizers may cause fire and explosions; avoid contact with caustics, chemically active metals. Attacks some plastics.

o-**DINITROBENZOL** or *ortho*-**DINITROBENZOL** (528-29-0) Explodes: is impact- and friction-sensitive. Strong oxidizers may cause fire and explosions; avoid contact with caustics, chemically active metals. Attacks some plastics.

DINITROCRESOL (534-52-1) Incompatible with heat and strong oxidizers. Dust contact can produce explosive mixtures with air.

DINITRO-O-CRESOL (534-52-1) Incompatible with heat and strong oxidizers. Dust contact can produce explosive mixtures with air.

2,6-DINITRO-O-CRESOL (534-52-1) Incompatible with heat and strong oxidizers. Dust contact can produce explosive mixtures with air.

3,5-DINITRO-O-CRESOL (534-52-1) Heat and strong oxidizers. Dust can form explosive mixtures with air.

4,6-DINITRO-O-CRESOL (534-52-1) Heat and strong oxidizers. Dust can form explosive mixtures with air.

DINITROCYCLOHEXYL PHENOL (131-89-5) Reacts with oxidizers, with a risk of fire or explosions.

DINITRO-O-CYCLOHEXYLPHENOL (131-89-5) Reacts with oxidizers, with a risk of fire or explosions.

2,4-DINITRO-6-CYCLOHEXYLPHENOL (131-89-5) Reacts with oxidizers, with a risk of fire or explosions.

4,6-DINITRO-O-CYCLOHEXYL PHENOL (131-89-5) Reacts with oxidizers, with a risk of fire or explosions.

2,6-DINITRO-N,N-DI-N-PROPYL-alpha,alpha,alpha,-TRI-FLUORO-P-TOLUIDINE (112-57-2) Forms explosive mixture with air. Incompatible with acids, organic anhydrides, halogenated hydrocarbons, isocyanates, nitroparaffins, vinyl acetate, acrylates, substituted allyls, alkylene oxides, epichlorohydrin, ketones, aldehydes, alcohols, glycols, mercury, phenols, cresols, caprolactam solution, strong oxidizers. Attacks aluminum, copper, lead, tin, zinc, and alloys.

DINITROGEN MONOXIDE (10024-97-2) May form explosive mixture with air. Incompatible with nonflammable gas (supports combustion), aluminum, anhydrous ammonia (may be explosive), boron, hydrazine lithium hydride, phosphine, sodium.

DINITROGEN TETROXIDE (10544-72-6); (10102-44-0 CAS for nitrogen dioxide). A powerful oxidizer. Water contact produces nitric acid and nitric oxide. Incompatible with nitric oxide reacts with air, forming more nitrogen tetroxide. Reacts explosively with alcohols, fuels and petroleum, nitrobenzene, organic matter, anhydrous ammonia, chlorinated hydrocarbons, toluene. Reacts vigorously reaction with combustibles, reducing agents. Attacks metals in presence of moisture.

DINITROGLYCOL (628-96-6) Acids, heat, or mechanical shock may result in explosions.

3,5-DINITRO-2-HYDROXYTOLUENE (534-52-1) Incompatible with heat and strong oxidizers. Dust contact can produce explosive mixtures with air.

4,6-DINITRO-2-METHYL PHENOL (534-52-1) Incompatible with heat and strong oxidizers. Dust contact can produce explosive mixtures with air.

2,4-DINITRONILIN (German) (97-02-9) Reacts violently with strong oxidizers. May detonate when heated under confinement.

2,4-DINITRONILINA (Italian) (97-02-9) Reacts violently with strong oxidizers. May detonate when heated under confinement.

2,4-DINITROPHENOL (51-28-5) Explosion is caused by heat, friction, or shock. Contact with reducing agents, combustibles may cause fire and explosions. Forms explosive salts with ammonia or strong bases. May accumulate static electrical charges, and may cause ignition of its vapors.

2,5-DINITROPHENOL (392-71-5) Incompatible with sulfuric acid, nitric acid, caustics, aliphatic amines, isocyanates.

2,6-DINITROPHENOL (573-56-8) Incompatible with sulfuric acid, nitric acid, caustics, aliphatic amines, isocyanates.

a-DINITROPHENOL or alpha-DINITROPHENOL (51-28-5) Explosion is caused by heat, friction, or shock. Contact with reducing agents, combustibles may cause fire and explosions. Forms explosive salts with ammonia or strong bases. May accumulate static electrical charges, and may cause ignition of its vapors.

beta-DINITROPHENOL (573-56-8) Incompatible with sulfuric acid, nitric acid, caustics, aliphatic amines, isocyanates.

gamma-DINITROPHENOL (392-71-5) Incompatible with sulfuric acid, nitric acid, caustics, aliphatic amines, isocyanates.

O,O-DINITROPHENOL (573-56-8) Incompatible with sulfuric acid, nitric acid, caustics, aliphatic amines, isocyanates.

DINITROTOLUENE (5321-14-6) Incompatible with strong oxidizers; caustics may cause fire and explosions. Attacks some metals.

2,4-DINITROTOLUENE (121-14-2) Strong oxidizers or caustics may cause fire and explosions. Incompatible with nitric acid, forming an explosive material. Sodium oxide contact causes ignition. Attacks metals.

2,6-DINITROTOLUENE Strong oxidizers or caustics may cause fire and explosions. Contact with nitric acid produces an explosive material. Sodium oxide contact causes ignition. Attacks some metals.

3,4-DINITROTOLUENE (610-39-9) Strong oxidizers or caustics may cause fire and explosions. Contact with nitric acid produces an explosive material. Sodium oxide contact causes ignition. Attacks some metals.

ar,ar-DINITROTOLUENE (5321-14-6) Strong oxidizers or caustics may cause fire and explosions. Attacks some metals.

DINITROTOLUOL (121-14-2) Strong oxidizers or caustics may cause fire and explosions. Incompatible with nitric acid, forming an explosive material. Sodium oxide contact causes ignition. Attacks metals.

2,4-DINITROTOLUOL (121-14-2) Strong oxidizers or caustics may cause fire and explosions. Incompatible with nitric acid, forming an explosive material. Sodium oxide contact causes ignition. Attacks metals.

DINONYL 1,2-BENZENEDICARBOXYLATE (84-76-4) Incompatible with strong acids, nitrates, oxidizers.

DINONYL PHTHALATE (84-76-4) Incompatible with strong acids, nitrates, oxidizers.

DI-N-NONYL PHTHALATE (84-76-4) Incompatible with strong acids, nitrates, oxidizers.

DINOPOL NOP (117-84-0) Water contact causes foaming. Incompatible with strong acids, nitrates.

DINOXOL (93-76-5) Sealed metal containers may burst at temperatures above 316°F/158°C. Incompatible with sulfuric acid, bases, ammonia, aliphatic amines, alkanolamines, isocyanates, alkylene oxides, epichlorohydrin.

DINOXOL (94-75-7) Decomposes in sunlight. Incompatible with strong oxidizers; may cause fire and explosions.

DINYL (8004-13-5) Contact with strong oxidizers may cause fire and explosions. Attacks some plastics, rubber, and coatings.

DIOCTYAL (119-36-8) Forms explosive mixture with air (flash point 205°F/96°C). Incompatible with strong acids, nitrates, oxidizers.

DIOCTYL ADIPATE (103-23-1) Incompatible with strong acids, nitrates, oxidizers.

DIOCTYL PHTHALATE (117-84-0) Water contact causes foaming. Incompatible with strong acids, nitrates.

DI-*sec*-OCTYL PHTHALATE (117-81-7) Incompatible with strong acids, strong alkalies, nitrates, oxidizers.

DIOCTYL SODIUM SULFOSUCCINATE (119-36-8) Forms explosive mixture with air (flash point 205°F/96°C). Incompatible with strong acids, nitrates, oxidizers.

DIOCTYN (119-36-8) Forms explosive mixture with air (flash point 205°F/96°C). Incompatible with strong acids, nitrates, oxidizers.

DIOFORM (540-59-0) Forms explosive mixture with air (flash point 36°F/2.2°C). Incompatible with strong bases, oxidizers, difluoromethylene, dihypofluoride, nitrogen tetroxide (explosive). Attacks some plastics, rubber, and coatings.

DIOKAN (123-91-1) Forms explosive mixture with air (flash point 54°F/12°C). Moisture contact produces unstable peroxides. May accumulate static electrical charges, and may cause ignition of its vapors. Incompatible with strong oxidizers, strong acids. Attacks many plastics.

DIOKSAN (Polish) (123-91-1) Forms explosive mixture with air (flash point 54°F/12°C). Moisture contact produces unstable peroxides. May accumulate static electrical charges, and may cause ignition of its vapors. Incompatible with strong oxidizers, strong acids. Attacks many plastics.

DIOLAMINE (111-42-2) Incompatible with acids, organic anhydrides, isocyanates, vinyl acetate, acrylates, substituted allyls, alkylene oxides, epichlorohydrin, aldehydes, oxidizers. Corrosive to copper, copper alloys, zinc, and galvanized iron.

DIOLANE (107-41-5) Forms explosive mixture with air (flash point 209°F/98°C). Incompatible with strong acids, caustics, aliphatic amines, isocyanates, strong oxidizers.

DIOMEDICONE (119-36-8) Forms explosive mixture with air (flash point 205°F/96°C). Incompatible with strong acids, nitrates, oxidizers.

DI-ON (330-54-1) Hydrolyzes in fairly strong acids.

DIOP (27554-26-3) May accumulate static electrical charges, and may cause ignition of its vapors. Incompatible with strong acids, nitrates, oxidizers.

DIOSSANO-1,4 (Italian) (123-91-1) Forms explosive mixture with air (flash point 54°F/12°C). Moisture contact produces unstable peroxides. May accumulate static electrical charges, and may cause ignition of its vapors. Incompatible with strong oxidizers, strong acids. Attacks many plastics.

DIOSUCCIN (119-36-8) Forms explosive mixture with air (flash point 205°F/96°C). Incompatible with strong acids, nitrates, oxidizers.

DIOTILAN (119-36-8) Forms explosive mixture with air (flash point 205°F/96°C). Incompatible with strong acids, nitrates, oxidizers.

DIOVAC (119-36-8) Forms explosive mixture with air (flash point 205°F/96°C). Incompatible with strong acids, nitrates, oxidizers.

DIOXAAN-1,4 (Dutch) (123-91-1) Forms explosive mixture with air (flash point 54°F/12°C). Moisture contact produces unstable peroxides. May accumulate static electrical charges, and may cause ignition of its vapors. Incompatible with strong oxidizers, strong acids. Attacks many plastics.

1,4-DIOXACYCLOHEXANE (123-91-1) Forms explosive mixture with air (flash point 54°F/12°C). Moisture contact produces unstable peroxides. May accumulate static electrical charges, and may cause ignition of its vapors. Incompatible with strong oxidizers, strong acids. Attacks many plastics.

DIOXAN (123-91-1) Forms explosive mixture with air (flash point 54°F/12°C). Moisture contact produces unstable peroxides. May accumulate static electrical charges, and may cause ignition of its vapors. Incompatible with strong oxidizers, strong acids. Attacks many plastics.

DIOXAN-1,4 (German) (123-91-1) Forms explosive mixture with air (flash point 54°F/12°C). Moisture contact produces unstable peroxides. May accumulate static electrical charges, and may cause ignition of its vapors. Incompatible with strong oxidizers, strong acids. Attacks many plastics.

DIOXANE (123-91-1) Forms explosive mixture with air (flash point 54°F/12°C). Moisture contact produces unstable peroxides. May accumulate static electrical charges, and may cause ignition of its vapors. Incompatible with strong oxidizers, strong acids. Attacks many plastics.

1,4-DIOXANE (123-91-1) Forms explosive mixture with air (flash point 54°F/12°C). Moisture contact produces unstable peroxides. May accumulate static electrical charges, and may cause ignition of its vapors. Incompatible with strong oxidizers, strong acids. Attacks many plastics.

DIOXANE-1,4 (123-91-1) Forms explosive mixture with air (flash point 54°F/12°C). Moisture contact produces unstable peroxides. May accumulate static electrical charges, and may cause ignition of its vapors. Incompatible with strong oxidizers, strong acids. Attacks many plastics.

P-DIOXANE (123-91-1) Forms explosive mixture with air (flash point 54°F/12°C). Moisture contact produces unstable peroxides. May accumulate static electrical charges, and may cause ignition of its vapors. Incompatible with strong oxidizers, strong acids. Attacks many plastics.

DIOXANNE (French) (123-91-1) Forms explosive mixture with air (flash point 54°F/12°C). Moisture contact produces unstable peroxides. May accumulate static electrical charges, and may cause ignition of its vapors. Incompatible with strong oxidizers, strong acids. Attacks many plastics.

P-DIOXAN, TETRAHYDRO- (123-91-1) Forms explosive mixture with air (flash point 54°F/12°C). Moisture contact produces unstable peroxides. May accumulate static electrical charges, and may cause ignition of its vapors. Incompatible with strong oxidizers, strong acids. Attacks many plastics.

3,6-DIOXAOCTANE-1,8-DIOL (112-27-6) Incompatible with sulfuric acid, isocyanates, perchloric acid, strong oxidizers.

DIOXITOL (111-46-6) Combustible (flash point 255°F/124°C). Incompatible with sulfuric acid, isocyanates, strong oxidizers.

P-DIOXOBENZENE (123-31-9) Incompatible with strong oxidizers, caustics. May be oxidized to quinone at room temperatures in the presence of moisture. May explode on contact with oxygen.

DIOXODICHLOROCHROMIUM (14977-61-8) A powerful oxidizer. Water contact produces hydrochloric and chromic acids and chlorine gas. Reacts violently with reducing agents, combustibles, ammonia, halides, phosphorus, sodium azide, elemental sulfur, urea.

DIOXONIUM PERCHLORATE SOLUTION (7601-90-3) Aqueous solution is a strong acid. Anhydrous material decomposes explosively at atmospheric pressure. Temperature above 165°F/74°C may cause explosion. A strong oxidizer. Reacts violently with reducing agents, combustible materials, acetic acid, alcohols, glycols, hypophosphites, ketones, metal powders, strong acids, sulfoxides with risk of fire and explosions. Contact with common materials (wood, fabric) can cause spontaneous ignition.

1,3-DIOXOPHTHALAN (85-44-9) Incompatible with strong acids, caustics, ammonia, amines, strong oxidizers. Attacks some plastics, rubber, and coatings.

1,4-DIOXYBENZENE (106-51-4) Forms explosive mixture with air (flash point 104°F/40°C). Incompatible with strong bases, reducing agents, strong oxidizers. Attacks some plastics, rubber, and coatings.

m-**DIOXYBENZENE or** *meta*-**DIOXYBENZENE** (108-46-3) Reacts violently with strong oxidizers, nitric acid. Incompatible with acetanilide, albumin, alkalies, antipyrine, camphor, ferric salts, menthol, spirit nitrous ether. Absorbs moisture from air (hygroscopic). May accumulate static electrical charges, and may cause ignition of its vapors.

o-**DIOXYBENZENE or** *ortho*-**DIOXYBENZENE** (120-80-9) Strong oxidizers may cause fire and explosion.

DIOXYDE de BARYUM (French) (1304-29-6) Reacts with water, forming oxygen. A strong oxidizer. Reacts violently with combustibles, reducing agents, acids, finely divided metals. Contact with organic substances form shock- and friction-sensitive compounds. Attacks metals in the presence of moisture.

DIOXYETHYLENE ETHER (123-91-1) Forms explosive mixture with air (flash point 54°F/12°C). Moisture contact produces unstable peroxides. May accumulate static electrical charges, and may cause ignition of its vapors. Incompatible with strong oxidizers, strong acids. Attacks many plastics.

4,8-DIOXY-UNDECANE DIOL,-1,11 (24800-44-0) Incompatible with sulfuric acid, isocyanates, perchloric acid, strong oxidizers. Attacks some plastics, rubber, and coatings.

DIPA (108-18-9) Forms explosive mixture with air (flash point 30°F/−1°C). Incompatible with acids, organic anhydrides, isocyanates, vinyl acetate, acrylates, substituted allyls, alkylene oxides, epichlorohydrin, ketones, aldehydes, alcohols, glycols, mercury, phenols, cresols, caprolactam solution, strong oxidizers. Attacks aluminum, copper, lead, tin, zinc, and alloys, and some plastics, rubber, and coatings.

DIPANOL (138-86-3) Forms explosive mixture with air (flash point 115°F/46°C). Strong oxidizers may cause fire and explosions.

DIPENTENE (138-86-3) Forms explosive mixture with air (flash point 115°F/46°C). Strong oxidizers may cause fire and explosions.

DIPENTYL PHTHALATE (131-18-0) Incompatible with strong acids, nitrates. Attacks some plastics, rubber, and coatings.

O-DIPHENOL (120-80-9) Strong oxidizers may cause fire and explosions.

DIPHENYL (92-52-4) Mist forms explosive mixture with air. Strong oxidizers may cause fire and explosions.

DIPHENYLAMINE (122-39-4) Incompatible with strong acids, aldehydes, organic anhydrides, isocyanates, oxidizers, hexachloromelamine, trichloromelamine.

N,N-DIPHENYLAMINE (122-39-4) Incompatible with strong acids, aldehydes, organic anhydrides, isocyanates, oxidizers, hexachloromelamine, trichloromelamine.

DIPHENYLDICHLOROSILANE (80-10-4) Water contact produces hydrochloric acid. Strong oxidizers may cause fire and explosions. Corrodes metals in the presence of moisture.

DIPHENYL–DIPHENYL ETHER MIXTURE (8004-13-5) Contact with strong oxidizers may cause fire and explosions. Attacks some plastics, rubber, and coatings.

4,4'-DIPHENYLENEDIAMINE (92-87-5) Oxidizes on exposure to light and air. Reacts violently with strong oxidizers. Red fuming nitric acid may cause fire.

DIPHENYL ETHER (101-84-8) Incompatible with strong acids, strong oxidizers; may cause fire and explosions. Attacks some plastics, rubber, and coatings.

DIPHENYLGLYOXAL (94-36-0) Confined storage of dry chemical may lead to decomposition and explosion. A strong oxidant; extremely reactive. Fires and explosion may result from heat or contamination, and from contact with strong acids, combustible materials, oxidizers, acids, bases, alcohols, reducing agents, metals, metal oxides, amines, accelerators, methyl methacrylate, organic matter, lithium aluminum carbide, dimethyl aniline, amines, metallic naphthenates. May attack some plastics, rubber, and coatings. Protect containers from shock and friction.

DIPHENYL KETONE (119-61-9) Combustible solid. Strong oxidizers may cause fire and explosions. Attacks some plastics, rubber, and coatings.

DIPHENYLMETHANE DIISOCYANATE (101-68-8) Incompatible with strong alkalies, acids, alcohols, ammonia, amines, amides, glycols, caprolactam. Unstable above 100°F/37.8°C. Attacks some plastics, rubber, and coatings.

DIPHENYLMETHANE-4,4'-DIISOCYANATE (101-68-8) Incompatible with strong alkalies, acids, alcohols, ammonia, amines, amides, glycols, caprolactam. Unstable above 100°F/37.8°C. Attacks some plastics, rubber, and coatings.

4,4'-DIPHENYLMETHANE DIISOCYANATE (101-68-8) Incompatible with strong alkalies, acids, alcohols, ammonia, amines, amides, glycols, caprolactam. Unstable above 100°F/37.8°C. Attacks some plastics, rubber, and coatings.

DIPHENYL METHANONE (119-61-9) Combustible solid. Strong oxidizers may cause fire and explosions. Attacks some plastics, rubber, and coatings.

DIPHENYL MIXED WITH DIPHENYL ETHER (8004-13-5) Contact with strong oxidizers may cause fire and explosions. Attacks some plastics, rubber, and coatings.

DIPHENYL OXIDE (101-84-8) Incompatible with strong acids, strong oxidizers; may cause fire and explosions. Attacks some plastics, rubber, and coatings.

DIPHENYLSILICON DICHLORIDE (80-10-4) Water contact produces hydrochloric acid. Strong oxidizers may cause fire and explosions. Corrodes metals in the presence of moisture.

DIPHENYLTRICHLOROETHANE (50-29-3) Incompatible with salts of iron or aluminum, and bases. Do not store in iron containers.

DIPHOSGENE (75-44-5) Incompatible with water, forming hydrochloric acid. Reacts violently with strong oxidizers, anhydrous ammonia, isopropyl alcohol, and chemically active metals. Forms shock-sensitive material with potassium. Attacks most metal in moist conditions.

DIPHOSPHORIC ACID, TETRAETHYL ESTER (107-49-3) Decomposes above 300°F/150°C, forming flammable ethylene gas. Strong oxidizers may cause fire and explosions. Attacks some plastics, rubber, and coatings.

DIPHYL (8004-13-5) Contact with strong oxidizers may cause fire and explosions. Attacks some plastics, rubber, and coatings.

DIPIRATRIL-TROPICO (67-68-5) Forms explosive mixture with air (flash point 203°F/95°C). Reacts violently with oxidizers. Reacts with ethanoyl chloride, boron compounds, halides, metal alkoxides, oxidizers.

DIPOFENE (333-41-5) Incompatible with water, copper-containing compounds, oxidizers, acids, or bases.

DIPOTASSIUM CHROMATE (7789-00-6) A powerful oxidizer. Contact with combustibles, reducing agents, organic material, or finely divided metals may cause fire and explosions.

DIPOTASSIUM DICHROMATE (7778-50-9) Powdered combustibles will cause ignition. Strong acids cause formation of toxic vapors.

DIPOTASSIUM MONOCHROMATE (7789-00-6) A powerful oxidizer. Contact with combustibles, reducing agents, organic material, or finely divided metals may cause fire and explosions.

DIPPING ACID (7664-93-9) A strong oxidizer that can react violently, with risk of fire and explosion, with many substances, including reducing agents, organic and combustible substances and bases. Incompatible with nonoxidizing mineral acids, organic acids, bases, acrylates, aldehydes, alcohols, alkylene oxides, ammonia, aliphatic amines, alkanolamines, aromatic amines, amides, chlorates, epichlorohydrin, fulminates, glycols, isocyanates, ketones, metals (powdered), organic anhydrides, perchlorates, picrates, substituted allyls, phenols and cresols, water, acetic anhydride, acetone cyanhydrin, acetonitrile, acrolein, acrylonitrile, allyl alcohol, allyl chloride, 2-aminoethanol, ammonium hydroxide, aniline, bromine pentafluoride, N-butyraldehyde, caprolactam solution, carbides, cesium acetylene carbide, chlorine trifluoride, chlorosulfonic acid, cuprous nitride, diisobutylene, ethylene cyanohydrin, ethylene diamine, ethylene glycol, ethyleneimine, hydrochloric acid, iodine heptafluoride, iron, isoprene, lithium silicide, mercuric nitride, mesityl oxide, nitric acid, p-nitrotoluene, perchloric acid, phosphorus, potassium *tert*-butoxide, potassium chlorate, potassium permanganate, propiolactone (beta-), propylene oxide, pyridine, rubidium acetylene, silver permanganate, sodium, sodium carbonate, sodium chlorate, sodium hydroxide, styrene monomer, vinyl acetate. Attacks most metals and some plastics, rubber, and coatings.

DIPROPAL METHANE (142-82-5) Forms explosive mixture with air (flash point 25°F/−4°C). Strong oxidizers may cause fire and explosions. Attacks some plastics, rubber, and coatings. May accumulate static electric charges that can ignite its vapors.

DIPROPANEDIOL DIBENZOATE (94-51-9) Incompatible with strong acids, nitrates, oxidizers.

DI-N-PROPENYLAMINE (124-02-7) Forms explosive mixture with air (flash point 70°F/22°C). Reacts violently with strong oxidizers, strong acids. Incompatible with organic anhydrides, isocyanates, aldehydes. May accumulate static electrical charges, and may cause ignition of its vapors. Attacks aluminum, copper, zinc.

DIPROPYLAMINE (142-84-7) Forms explosive mixture with air (flash point 63°F/17°C). Incompatible with acids, organic anhydrides, isocyanates, vinyl acetate, acrylates, substituted allyls, alkylene oxides, epichlorohydrin, ketones, aldehydes, alcohols, glycols, mercury, phenols, cresols, caprolactam solution, strong oxidizers. Attacks aluminum, copper, lead, tin, zinc, and alloys.

DI-*n*-PROPYLAMINE (142-84-7) Forms explosive mixture with air (flash point 63°F/17°C). Incompatible with acids, organic anhydrides, isocyanates, vinyl acetate, acrylates, substituted allyls, alkylene oxides, epichlorohydrin, ketones, aldehydes, alcohols, glycols, mercury, phenols, cresols, caprolactam solution, strong oxidizers. Attacks aluminum, copper, lead, tin, zinc, and alloys.

DIPROPYL ETHER (111-43-3) Forms explosive mixture with air (flash point 70°F/21°C). Forms explosive peroxides, especially when anhydrous. Incompatible with strong acids, strong oxidizers.

DIPROPYL KETONE (123-19-3) Oxidizers may cause violent reaction.

DIPROPYLENE GLYCOL (110-98-5) Incompatible with sulfuric acid, perchloric acid, isocyanates, strong oxidizers.

DIPROPYLENE GLYCOL DIBENZOATE (94-51-9) Incompatible with strong acids, nitrates, oxidizers.

DIPROPYLENE GLYCOL METHYL ETHER (34590-94-8) Ethers, as a class, can form dangerous peroxides on standing. Strong oxidizers may cause fire and explosions. Attacks some plastics, rubber, and coatings.

DIPROPYLENE GLYCOL MONOMETHYL ETHER (34590-94-8) Ethers, as a class, can form dangerous peroxides on standing. Strong oxidizers may cause fire and explosions. Attacks some plastics, rubber, and coatings.

DIPROPYLENE TRIAMINE (56-18-8) A strong base; reacts with acids and strong oxidizers. Attacks various metals, including aluminum, copper, tin, zinc, and their alloys.

DIPROPYL METHANE (142-82-5) Forms explosive mixture with air (flash point 25°F/−4°C). Strong oxidizers may cause fire and explosions. Attacks some plastics, rubber, and coatings. May accumulate static electric charges that can ignite its vapors.

DIPROPYL OXIDE (111-43-3) Forms explosive mixture with air (flash point 70°F/21°C). Forms explosive peroxides, especially when anhydrous. Incompatible with strong acids, strong oxidizers.

N,N-DIPROPYL-1-PROPANAMINE (102-69-2) Incompatible with strong acids, strong oxidizers.

DIPTEREX (52-68-6) Contact with strong oxidizers may cause fire and explosions.

DIQUAT (85-00-7) Concentrated solution attacks aluminum.

DIQUAT DIBROMIDE (85-00-7) Concentrated solution attacks aluminum.

DISB (84-69-5) Incompatible with strong acids, strong oxidizers, nitrates.

DISILVER OXIDE (20667-12-3) A strong oxidizer; reacts with reducing agents, acids, amines, combustible materials. Reacts violently with ammonia.

DISODIUM CHROMATE (7777-11-3) A strong oxidizer; reacts violently with reducing agents, combustibles, strong acids, organic materials.

DISODIUM DICHROMATE (10588-01-9) Aqueous solution is corrosive. A strong oxidizer; reacts violently with reducing agents, acids, acetic anhydride, combustibles. Attacks copper, zinc, tin, brass, bronze.

DISODIUM DIFLUORIDE (7681-49-4) Incompatible with water, forming a corrosive. Reacts with acids.

DISODIUM ETHYLENEBIS [DITHIOCARBAMATE] (142-59-6) Boiling water cause formation of hydrogen sulfide and carbon disulfide vapors.

DISODIUM METASILICATE (1344-09-8) Solution is a strong base; reacts with acids, organic anhydrides, alkylene oxides, epichlorohydrin, aldehydes, alcohols, glycols, phenols, cresols, caprolactam solution. Attacks chemically active metals.

DISODIUM MONOSILICATE (1344-09-8) Solution is a strong base; reacts with acids, organic anhydrides, alkylene oxides, epichlorohydrin, aldehydes, alcohols, glycols, phenols, cresols, caprolactam solution. Attacks chemically active metals.

DISODIUM MONOSULFIDE (1313-82-2) Once moist, can self-ignite upon drying in air. Incompatible with carbon, oxidizers, diazonium salts. Attacks metals.

DISODIUM SULFIDE (1313-82-2) Once moist, can self-ignite upon drying in air. Incompatible with acids, carbon, oxidizers, diazonium salts. Attacks metals.

DISODIUM SULFITE (7757-83-7) A reducing agent. Incompatible with oxidizers, combustibles, acids, organic materials.

DISOLFURO di TETRAMETILTIOURAME (Italian) (137-26-8) Combustible solid (flash point 192°F/89°C). Strong oxidizers may cause fire and explosions; contact with strong acid or oxidizable materials produces toxic gases.

DISPERMINE (110-85-0) Forms explosive mixture with air (flash point 178°F/81°C). Incompatible with nitrogen compounds, carbon tetrachloride, strong oxidizers. Aqueous solutions react with acids. Attacks aluminum, copper, nickel, magnesium, and zinc.

DISTILLATES: FLASHED FEED STOCKS (8002-05-9) Flash point −40°F to −86°F/−40°C to −66°C. Incompatible with nitric acid; oxidizers may cause fire and explosions.

DISTILLATES: STRAIGHT RUN (64741-43-1) Incompatible with nitric acid; oxidizers may cause fire and explosions.

DISTOKAL (67-72-1) Incompatible with hot iron, zinc, and aluminum; alkalies cause formation of spontaneously explosive chloroacetylene. Attacks some plastics, rubber, and coatings.

DISTOPAN (67-72-1) Incompatible with hot iron, zinc and aluminum; alkalies cause formation of spontaneously explosive chloroacetylene. Attacks some plastics, rubber, and coatings.

DISULFATON (298-04-4) Forms explosive mixture with air (flash point 180°F/82°C). Incompatible with alkalis, strong oxidizers.

DISULFOTON (298-04-4) Forms explosive mixture with air (flash point 180°F/82°C). Incompatible with alkalis, strong oxidizers.

DISULFUR DICHLORIDE (10025-67-9) Water contact produces hydrochloric acid, sulfur dioxide, sulfur, sulfur thiosulfite, hydrogen sulfide; solution is a strong acid. Oxidizers, strong bases, oxides of phosphorus, organics may cause fires. Attacks some plastics, rubber, and coatings.

DISULFURE de TETRAMETHYLTHIOURAME (French) (137-26-8) Combustible solid (flash point 192°F/89°C). Strong oxidizers may cause fire and explosions; contact with strong acid or oxidizable materials produces toxic gases.

DISULFURIC ACID (8014-95-7) A powerful oxidizer and strong acid. Reacts violently with reducing agents, organic materials. Reacts with air, forming corrosive fumes. Contact with cast iron may cause a violent reaction. Extremely hazardous on contact with many materials including chlorates, carbides, fulminates. Reacts vigorously with metals (powders will ignite) releasing hydrogen. Attacks some plastics, rubber, and coatings on brief contact.

DI-SYSTON (298-04-4) Forms explosive mixture with air (flash point 180°F/82°C). Incompatible with Alkalis, strong oxidizers.

DISYSTOX (298-04-4) Forms explosive mixture with air (flash point 180°F/82°C). Incompatible with alkalis, strong oxidizers.

DITHANE (142-59-6) Boiling water causes formation of hydrogen sulfide and carbon disulfide vapors.

DITHANE A-40 (142-59-6) Boiling water causes formation of hydrogen sulfide and carbon disulfide vapors.

DITHANE D-14 (142-59-6) Boiling water causes formation of hydrogen sulfide and carbon disulfide vapors.

DITHIO (3689-24-5) Containers may burst in heat. Strong oxidizers may cause fire and explosions. Attacks some plastics, rubber, and coatings.

DITHIOATE (86-50-0) Strong oxidizers may cause fire and explosions.

alpha,alpha'-DITHIOBIS(DIMETHYLTHIO)FORMAMIDE (137-26-8) Combustible solid (flash point 192°F/89°C). Strong oxidizers may cause fire and explosions; contact with strong acid or oxidizable materials produces toxic gases.

DITHIOCARBONIC ANHYDRIDE (75-15-0) Highly reactive. Contact with many substances can cause fire and explosions. Forms explosive mixture with air (flash point −22°F/−30°C). Shock can cause explosive decomposition. Incompatible with alkali metals, aliphatic amines, alkanolamines, aluminum, azides, chlorine monoxide, combustible substances, ethylene diamine, ethyleneimine, lead azide, lithium azide, nitric oxide, nitrogen dioxide, potassium, potassium azide, reducing agents, rubidium azide, sodium azide, zinc.

DITHIODEMETON (298-04-4) Forms explosive mixture with air (flash point 180°F/82°C). Incompatible with alkalis, strong oxidizers.

N,N'-(DITHIODICARBOROTHIOYL)BIS(N-METHYL-METHAN-AMINE) (137-26-8) Combustible solid (flash point 192°F/89°C). Strong oxidizers may cause fire and explosions; contact with strong acid or oxidizable materials produces toxic gases.

DITHIODIPHOSPHORIC ACID, TETRAETHYL ESTER (3689-24-5) Containers may burst in heat. Strong oxidizers may cause fire and explosions. Attacks some plastics, rubber, and coatings.

DITHIOFOS (3689-24-5) Containers may burst in heat. Strong oxidizers may cause fire and explosions. Attacks some plastics, rubber, and coatings.

DITHION (3689-24-5) Containers may burst in heat. Strong oxidizers may cause fire and explosions. Attacks some plastics, rubber, and coatings.

DITHIONE (3689-24-5) Containers may burst in heat. Strong oxidizers may cause fire and explosions. Attacks some plastics, rubber, and coatings.

DITHIONIC ACID (8014-95-7) A powerful oxidizer and strong acid. Reacts violently with reducing agents, organic materials. Reacts with air, forming corrosive fumes. Contact with cast iron may cause a violent reaction. Extremely hazardous on contact with many materials including chlorates, carbides, fulminates. Reacts vigorously reaction with metals (powders will ignite), releasing hydrogen. Attacks some plastics, rubber, and coatings on brief contact.

DITHIOPHOS (3689-24-5) Containers may burst in heat. Strong oxidizers may cause fire and explosions. Attacks some plastics, rubber, and coatings.

DITHIOPHOSPHATE de O,O-DIETHYLE ETDE S-(2-ETHYL-THIO-ETHYLE) (French) (298-04-4) Forms explosive mixture with air (flash point 180°F/82°C). Incompatible with alkalis, strong oxidizers.

DI(THIOPHOSPHORIC) ACID, TETRAETHYL ESTER (3689-24-5) Containers may burst in heat. Strong oxidizers may cause fire and explosions. Attacks some plastics, rubber, and coatings.

DITHIOPYROPHOSPHATE de TETRAETHYLE (French) (3689-24-5) Containers may burst in heat. Strong oxidizers may cause fire and explosions. Attacks some plastics, rubber, and coatings.

DITHIOPYROPHOSPHORIC ACID, O,O,O,O-TETRAETHYL ESTER (3689-24-5) Containers may burst in heat. Strong oxidizers may cause fire and explosions. Attacks some plastics, rubber, and coatings.

DITHIOSYSTOX (298-04-4) Forms explosive mixture with air (flash point 180°F/82°C). Incompatible with alkalis, strong oxidizers.

DITHIOTEP (3689-24-5) Containers may burst in heat. Strong oxidizers may cause fire and explosions. Attacks some plastics, rubber, and coatings.

4,4'-DI-O-TOLUIDINE (95-53-4) Forms explosive mixture with air (flash point 185°F/85°C). Incompatible with strong acids, mineral acids, organic anhydrides, isocyanates, aldehydes, oxidizers. Attacks some plastics, rubber, and coatings.

DITRIDECYL PHTHALATE (119-06-2) Incompatible with strong acids, nitrates, oxidizers.

DITTHIONOUS ACID, ZINC SALT(1:1) (7779-86-4) Water contact produces oxides of sulfur. Incompatible with oxidizers, acids.

DIUNDECYL PHTHALATE (3648-20-2) Incompatible with strong acids, nitrates.

DIUREX (330-54-1) Hydrolyzes in fairly strong acids.

DIUROL (61-82-5) Substance acts as a weak base to form salts in contact with acids. Corrosive to iron, aluminum, copper, and copper alloys.

DIUROL (330-54-1) Hydrolyzes in fairly strong acids.

DIURON (330-54-1) Hydrolyzes in fairly strong acids.

DIURON 4L (330-54-1) Hydrolyzes in fairly strong acids.

DIVINYL (106-99-0) Self-reactive. Forms explosive peroxides with air. Fires, explosions, or hazardous polymerization may result from contact with air, strong oxidizers, strong acids, ozone, nitrogen dioxide, copper and its alloys, phenol, chlorine dioxide, crotonaldehyde, or a free radical polymerization initiator such as hydroquinone. Add inhibitor (such as *tert*-butyl catechol) and monitor to ensure that effective levels are maintained at all times. May accumulate static electrical charges, and may cause ignition of its vapors.

1,4-DIVINYLBENZENE (1321-74-0) Forms explosive mixture with air (flash point 165°F/74°C). Reacts violently with strong oxidizers. Can polymerize; add inhibitor and monitor to ensure that effective levels are maintained at all times. May accumulate static electrical charges, and may cause ignition of its vapors.

DIVINYLENE OXIDE (110-00-9) Forms explosive mixture with air (flash point −58°F/−50°C). Reacts violently with acids, oxidizers. Unless chemical is stabilized with an inhibitor, air exposure produces unstable peroxides.

DIVINYLMETHANE (591-93-5) Forms explosive mixture with air (flash point 40°F/40°C). Strong oxidizers may cause fire and explosions.

DIVIPAN (62-73-7) Attacks some plastics, rubber, and coatings.

DIZENE (95-50-1) Forms explosive mixture with air (flash point 151°F/66°C). Incompatible with strong oxidizers, hot aluminum, or aluminum alloy. Attacks some plastics, rubber, and coatings.

DIZINON (333-41-5) Incompatible with water, copper-containing compounds, oxidizers, acids, or bases.

DMA (121-69-7) Forms explosive mixture with air (flash point 145°F/63°C). Contact with strong oxidizers may cause fire and explosions; incompatible with strong acids (cause violent spattering). Contact with benzoyl peroxide, diisopropyl perdicarbonate, and other material may cause explosions. Attacks some plastics, rubber, and coatings.

DMA (124-40-3) A gas. Incompatible with acids, organic anhydrides, isocyanates, vinyl acetate, acrylates, substituted allyls, alkylene oxides, epichlorohydrin, ketones, aldehydes, alcohols, glycols, mercury, phenols, cresols, caprolactam solution, strong oxidizers. Attacks aluminum, copper, lead, tin, zinc and alloys, and some plastics, rubbers, and coatings.

DMA (127-19-5) Forms explosive mixture with air (flash point 158°F/70°C). Incompatible with nonoxidizing mineral acids, strong acids, ammonia, isocyanates, phenols, cresols, halogenated compounds above 185°F/85°C). Attacks some plastics, rubber, and coatings.

DMA-4 (94-75-7) Decomposes in sunlight. Incompatible with strong oxidizers; may cause fire and explosions.

DMAC (127-19-5) Forms explosive mixture with air (flash point 158°F/70°C). Incompatible with nonoxidizing mineral acids, strong acids, ammonia, isocyanates, phenols, cresols, halogenated compounds, above 185°F/85°C. Attacks some plastics, rubber, and coatings.

DMB (95-53-4) Forms explosive mixture with air (flash point 185°F/85°C). Incompatible with strong acids, mineral acids, organic anhydrides, isocyanates, aldehydes, oxidizers. Attacks some plastics, rubber, and coatings.

DMCC (79-44-7) Rapidly hydrolyzed in water. Incompatible with strong acids, oxidizers. Attacks some plastics, rubber, and coatings.

DMDT (72-43-5) Contact with strong oxidizers may cause fire and explosions. Attacks some plastics, rubber, and coatings.

DME (115-10-6) Flammable gas. Forms explosive mixture with air. Forms unstable peroxides in storage. Reacts violently with strong oxidizers.

DMF (68-12-2) Forms explosive mixture with air (flash point 136°F/58°C). Contact with carbon tetrachloride and other halogenated compounds, particularly in presence of iron or strong oxidizers may cause fire and explosions. Reacts vigorously with alkylaluminums. Incompatible with nonoxidizing mineral acids, strong acids, chlorinated hydrocarbons, isocyanates, nitrates, organic nitrates, phenols, cresols, ammonia, bromine, chromic anhydride, magnesium nitrate, methylene diisocyanate, phosphorus trioxide, triethylaluminum. Attacks some plastics, rubber, and coatings.

DMFA (68-12-2) Forms explosive mixture with air (flash point 136°F/58°C). Contact with carbon tetrachloride and other halogenated compounds, particularly in presence of iron or strong oxidizers may cause fire and explosions. Reacts vigorously reaction with alkylaluminums. Incompatible with nonoxidizing mineral acids, strong acids, chlorinated hydrocarbons, isocyanates, nitrates, organic nitrates, phenols, cresols, ammonia, chromic anhydride, magnesium nitrate, methylene diisocyanate, phosphorus trioxide, triethylaluminum. Attacks some plastics, rubber, and coatings.

DMH (540-73-8) Forms explosive mixture with air (flash point 73°F/23°C). A strong reducing agent. Reacts violently with strong oxidizers, strong acids. Attacks some plastics, rubber, and coatings. May accumulate static electrical charges, and may cause ignition of its vapors.

DMH (57-14-7) Forms explosive mixture with air (flash point 5°F/−15°C). A strong reducing agent. Incompatible with strong acids, halogens, metallic mercury, strong oxidizers (with possible spontaneous ignition). Attacks some plastics, rubber, and coatings. May accumulate static electrical charges, and may cause ignition of its vapors.

DMN or DMNA (62-75-9) Ultraviolet light, strong oxidizers can cause reactions. Store in dark bottles.

DMP (131-11-3) Incompatible with strong alkalies, strong acids, nitrates, oxidizers.

DMS (75-18-3) Forms explosive mixture with air (flash point −36°F/−38°C) Reacts violently with strong oxidizers. May accumulate static electrical charges, and may cause ignition of its vapors.

DMS (77-78-1) Forms explosive mixture with air (flash point 182°F/83°C). Incompatible with strong oxidizers, strong acids, strong alkalies, strong ammonia solutions. Attacks some plastics, rubber, and coatings.

DMS-70 (67-68-5) Forms explosive mixture with air (flash point 203°F/95°C). Reacts violently with oxidizers. Reacts with ethanoyl chloride, boron compounds, halides, metal alkoxides, oxidizers.

DMS-90 (67-68-5) Forms explosive mixture with air (flash point 203°F/95°C). Reacts violently with oxidizers. Reacts with ethanoyl chloride, boron compounds, halides, metal alkoxides, oxidizers.

DMS (METHYL SULFATE) (77-78-1) Forms explosive mixture with air (flash point 182°F/83°C). Incompatible with strong oxidizers, strong acids, strong alkalies, strong ammonia solutions. Attacks some plastics, rubber, and coatings.

DMSO (67-68-5) Forms explosive mixture with air (flash point 203°F/95°C). Reacts violently with oxidizers. Reacts with ethanoyl chloride, boron compounds, halides, metal alkoxides, oxidizers.

DMT (120-61-6) Incompatible with strong acids, nitrates, strong oxidizers.

DMU (330-54-1) Hydrolyzes in fairly strong acids.

DN (534-52-1) Reacts with heat and strong oxidizers. Dust contact can produce explosive mixtures with air.

DNA (97-02-9) Reacts violently with strong oxidizers. May detonate when heated under confinement.

DNBA (111-92-2) Forms explosive mixture with air (flash point 117°F/47°C). Incompatible with acids, organic anhydrides, isocyanates, vinyl acetate, acrylates, substituted allyls, alkylene oxides, epichlorohydrin, ketones, aldehydes, alcohols, glycols, phenols, cresols, caprolactam solution, strong oxidizers. Contact with copper alloys, zinc, or galvanized steel may cause violent reaction.

DNC (534-52-1) Reacts with heat and strong oxidizers. Dust contact can produce explosive mixtures with air.

DNOC (534-52-1) Reacts with heat and strong oxidizers. Dust contact can produce explosive mixtures with air.

DNOP (117-84-0) Water contact causes foaming. Incompatible with strong acids, nitrates.

DNP (573-56-8) Incompatible with sulfuric acid, nitric acid, caustics, aliphatic amines, isocyanates.

2,4-DNP (51-28-5) Explosion is caused by heat, friction or shock. Contact with reducing agents, combustibles may cause fire and explosions. Forms explosive salts with ammonia or strong bases. May accumulate static electrical charges, and may cause ignition of its vapors.

2,5-DNP (392-71-5) Incompatible with sulfuric acid, nitric acid, caustics, aliphatic amines, isocyanates.

2,4-DNT (121-14-2) Strong oxidizers or caustics may cause fire and explosions. Incompatible with nitric acid, forming an explosive material. Sodium oxide contact causes ignition. Attacks metals.

2,6-DNT (606-20-2) Strong oxidizers or caustics may cause fire and explosions. Contact with nitric acid produces an explosive material. Sodium oxide contact causes ignition. Attacks some metals.

3,4-DNT (610-39-9) Strong oxidizers or caustics may cause fire and explosions. Contact with nitric acid produces an explosive material. Sodium oxide contact causes ignition. Attacks some metals.

DNTP (56-38-2) Combustible liquid. Mixtures with endrin may be explosive. Strong oxidizers may cause fire and explosions. Attacks some plastics, rubber, and coatings.

DO 14 (2312-35-8) Forms explosive mixture with air (flash point 82°F/28°C). Strong oxidizers may cause fire and explosions.

DOCUSATE SODIUM (119-36-8) Forms explosive mixture with air (flash point 205°F/96°C). Incompatible with strong acids, nitrates, oxidizers.

DODAT (50-29-3) Incompatible with salts of iron or aluminum, and bases. Do not store in iron containers.

DODECANE or 1-DODECANE (112-41-4) Forms explosive mixture with air (flash point 165°F/74°C). Reacts with acids and oxidizers.

1-DODECANETHIOL (112-55-0) Contact with strong oxidizers may cause fire and explosions.

DODECONIC ACID (143-07-7) Incompatible with sulfuric acid, bases, ammonia, amines, alkylene oxide, epichlorohydrin.

DODECANOIC ACID (143-07-7) Incompatible with sulfuric acid, bases, ammonia, amines, alkylene oxide, epichlorohydrin.

N-DODECANOIC ACID (143-07-7) Incompatible with sulfuric acid, bases, ammonia, amines, alkylene oxide, epichlorohydrin.

DODECANOL (112-53-8) Incompatible with strong acids, caustics, aliphatic amines, isocyanates, oxidizers.

1-DODECANOL (112-53-8) Incompatible with strong acids, caustics, aliphatic amines, isocyanates, oxidizers.

N-DODECANOL (112-53-8) Incompatible with strong acids, caustics, aliphatic amines, isocyanates, oxidizers.

DODECANOYL PEROXIDE (105-74-8) A powerful oxidizer. May ignite or explode spontaneously when mixed with combustible materials or rapidly heated.

DODECENE (6842-15-5) Forms explosive mixture with air (flash point 144°F/62°C). Incompatible with strong acids, oxidizers.

1-DODECENE (6842-15-5) Forms explosive mixture with air (flash point 144°F/62°C). Incompatible with strong acids, oxidizers.

DODECENE (NONLINEAR) (6842-15-5) Forms explosive mixture with air (flash point 144°F/62°C). Incompatible with strong acids, oxidizers.

DODECYL ALCOHOL (112-53-8) Incompatible with strong acids, caustics, aliphatic amines, isocyanates, oxidizers.

N-DODECYL ALCOHOL (112-53-8) Incompatible with strong acids, caustics, aliphatic amines, isocyanates, oxidizers.

DODECYLAMINE, TETRADECYLAMINE MIXTURE (CAS unknown) Forms explosive mixture with air. Incompatible with acids, organic anhydrides, chlorinated hydrocarbons, isocyanates, vinyl acetate, acrylates, substituted allyls, alkylene oxides, epichlorohydrin, ketones, aldehydes, alcohols, glycols, phenols, cresols, caprolactam solution, strong oxidizers. Attacks aluminum, copper, lead, tin, zinc, and alloys.

DODECYLBENZENE (123-01-3) Incompatible with nitric acid, strong oxidizers; presents risk of fire or explosions.

N-DODECYLBENZENE (123-01-3) Incompatible with nitric acid, strong oxidizers; presents risk of fire or explosions.

DODECYL BENZENESULFONATE (27176-87-0) Forms explosive mixture with air (flash point 100°F/38°C). Strong oxidizers may cause fire and explosions. Do not store in carbon steel or aluminum.

DODECYLBENZENESULFONATE SODIUM SALT (25155-30-0) Reacts with acids, including fumes.

DODECYL BENZENESULFONIC ACID (27176-87-0) Contact with strong oxidizers may cause fire and explosions. Do not store in carbon steel or aluminum.

DODECYLBENZENESULFONIC ACID, CALCIUM SALT (27176-87-0) Forms explosive mixture with air (flash point 100°F/38°C). Strong oxidizers may cause fire and explosions. Do not store in carbon steel or aluminum.

DODECYLBENZENESULFONIC ACID, ISOPROPYLAMINE SALT (42504-46-1) Incompatible with oxidizers, strong acids.

DODECYL BENZENE SULFONIC ACID, SODIUM SALT (25155-30-0) Reacts with acids, including fumes.

DODECYL DIPHENYL ETHER DISULFONATE SOLUTION (25167-32-2) Incompatible with sulfuric acid, isocyanates.

DODECYL DIPHENYL ETHER SULFONATE, DISODIUM SALT, AQUEOUS SOLUTION (25167-32-2) Incompatible with sulfuric acid, isocyanates.

DODECYLENE (6842-15-5) Forms explosive mixture with air (flash point 144°F/62°C). Incompatible with strong acids, oxidizers.

DODECYLENE (alpha) (6842-15-5) Forms explosive mixture with air (flash point 144°F/62°C). Incompatible with strong acids, oxidizers.

a-DODECYLENE or alpha-DODECYLENE (112-41-4) Forms explosive mixture with air (flash point 165°F/74°C). Reacts with acids and oxidizers.

alpha-DODECYLENE (6842-15-5) Forms explosive mixture with air (flash point 144°F/62°C). Incompatible with strong acids, oxidizers.

DODECYL AND TETRADECYL AMINE MIXTURE (CAS unknown) Forms explosive mixture with air. Incompatible with acids, organic anhydrides, chlorinated hydrocarbons, isocyanates, vinyl acetate, acrylates, substituted allyls, alkylene oxides, epichlorohydrin, ketones, aldehydes, alcohols, glycols, phenols, cresols, caprolactam solution, strong oxidizers. Attacks aluminum, copper, lead, tin, zinc and alloys.

DODECYL TRICHLOROSILANE (4484-7-4) Forms explosive mixture with air (flash point 150°F/66°C). Water and air vapor form hydrochloric acid fumes. Strong bases and oxidizers. Corrodes metals.

DODECYLETHYLENE (N/A) Forms explosive mixture with air (flash point 230°F/110°C). Incompatible with sulfuric acid, nitric acid, strong oxidizers.

DODECYL MERCAPTAN (112-55-0) Contact with strong oxidizers may cause fire and explosions.

tert-**DODECYL MERCAPTAN** (2466-19-1) Reacts vigorously reaction with strong oxidizers.

DODECYLMETHACRYLATE (142-90-5) Forms explosive mixture with air (flash point 230°F/110°C). Oxidizers or reducing agents may cause polymerization.

DODECYL-2-METHYL-2-PROPENOATE (142-90-5) Forms explosive mixture with air (flash point 230°F/110°C). Oxidizers or reducing agents may cause polymerization.

DODECYL/PENTADECYL METHACRYLATE (N/A) Incompatible with sulfuric acid, nitric acid, aliphatic amines, alkanolamines.

DODECYL PHENOL (27193-86-8) Incompatible with strong acids, caustics, aliphatic amines, amides.

DODECYLTRICHLOROSILANE (4484-7-4) Forms explosive mixture with air (flash point 150°F/66°C). Water and air vapor contact produce hydrochloric acid fumes. Incompatible with strong bases and oxidizers. Corrodes metals.

DOLCO MOUSE CEREAL (57-24-9) Contact with strong oxidizers may cause fire and explosions.

DOLCYMENE (99-87-6) Forms explosive mixture with air (flash point 117°F/47°C). Incompatible with nitric acid, strong oxidizers. Attacks and softens rubber. May accumulate static electrical charges, and may cause ignition of its vapors.

DOLEN-PUR (87-68-3) Forms explosive mixture with air (flash point 195°F/90°C). Reacts strongly with oxidizers, aluminum powder. Attacks aluminum. Attacks some plastics, rubber, and coatings.

DOL GRANULE (58-89-9) Not combustible, but may be dissolved in a combustible solvent. If solvent comes in contact with oxidizers fire and explosions may result.

DOLICUR (67-68-5) Forms explosive mixture with air (flash point 203°F/95°C). Reacts violently with oxidizers. Reacts with ethanoyl chloride, boron compounds, halides, metal alkoxides, oxidizers.

DOMATOL (61-82-5) Substance acts as a weak base to form salts in contact with acids. Corrosive to iron, aluminum, copper, and copper alloys.

DOMOLITE (1317-65-3) Incompatible with acids, alum, ammonium salts, fluorine.

DOMOSO (67-68-5) Forms explosive mixture with air (flash point 203°F/95°C). Reacts violently with oxidizers. Reacts with ethanoyl chloride, boron compounds, halides, metal alkoxides, oxidizers.

DOP (117-84-0) Water contact causes foaming. Incompatible with strong acids, nitrates.

DOP (117-81-7) Incompatible with strong acids, strong alkalies, nitrates, oxidizers.

DORMANT FUNGICIDE (DOW) (131-52-2) Contact with strong oxidizers may cause fire and explosions.

DORMANT OIL (N/A) Incompatible with nitric acid; contact with oxidizers may cause fire and explosions.

DORMONE (94-75-7) Decomposes in sunlight. Incompatible with strong oxidizers; may cause fire and explosions.

DOWANOL (111-90-0) Forms explosive mixture with air (flash point 201°F/94°C). Heat or contact with atmospheric air may cause formation of unstable peroxides. Incompatible with strong acids, isocyanates, strong oxidizers.

DOWANOL-50B (34590-94-8) Ethers, as a class, can form dangerous peroxides on standing. Strong oxidizers may cause fire and explosions. Attacks some plastics, rubber, and coatings.

DOWANOL DB (112-34-5) Heat or contact with atmospheric air may cause formation of unstable peroxides. Forms explosive mixture with air (flash point 172°F/78°C). Incompatible with strong oxidizers, strong bases, sulfuric acid, isocyanates. Attacks light metals.

DOWANOL DE (111-90-0) Forms explosive mixture with air (flash point 201°F/94°C). Heat or contact with atmospheric air may cause formation unstable peroxides. Incompatible with strong acids, isocyanates, strong oxidizers.

DOWANOL DM (111-77-3) Incompatible with sulfuric acid, isocyanates, perchloric acid.

DOWANOL DPM (34590-94-8) Ethers, as a class, can form dangerous peroxides on standing. Strong oxidizers may cause fire and explosions. Attacks some plastics, rubber, and coatings.

DOWANOL E (110-80-5) Forms explosive mixture with air (flash point 120°F/49°C). Strong oxidizers may cause fire and explosions. Attacks some plastics, rubber, and coatings.

DOWANOL EB (111-76-2) Forms explosive mixture with air (flash point 143°F/62°C). Reacts violently with strong caustics and strong oxidizers. Attacks some coatings, plastics, and rubber. Attacks metallic aluminum at high temperatures.

DOWANOL EE (110-80-5) Forms explosive mixture with air (flash point 120°F/49°C). Strong oxidizers may cause fire and explosions. Attacks some plastics, rubber, and coatings.

DOWANOL EIPAT (109-59-1) Forms explosive mixture with air 92°F/33°C). Incompatible with sulfuric acid, perchloric acid, isocyanates, oxidizers.

DOWANOL EM (109-86-4) Forms explosive mixture with air (flash point 103°F/39°C). Heat or oxidizers may cause formation of unstable peroxides. Attacks many metals. Strong oxidizers cause fire and explosions. Strong bases cause decomposition. Attacks some plastics, rubber, and coatings. May accumulate static electrical charges, and may cause ignition of its vapors.

DOWANOL EP (122-99-6) Incompatible with sulfuric acid, isocyanates, strong oxidizers.

DOWANOL EPH (122-99-6) Incompatible with sulfuric acid, isocyanates, strong oxidizers.

DOWANOL PM (107-98-2) Forms explosive mixture with air (flash point 90°F/32°C). Incompatible with sulfuric acid, isocyanates, perchloric acid.

DOWANOL TE (112-50-5) Incompatible with sulfuric acid, perchloric acid, isocyanates, strong oxidizers.

DOWANOL-TE (112-50-5) Incompatible with sulfuric acid, isocyanates, perchloric acid, oxidizers.

DOWANOL TPM (25498-49-1) Forms explosive mixture with air (flash point 250°F/121°C). Incompatible with sulfuric acid, perchloric acid, isocyanates, oxidizers.

DOWCHLOR (57-74-9) Contact with strong oxidizers may cause fire and explosions. Attacks some plastics, rubber, and coatings.

DOWCIDE 7 (87-86-5) Hot water causes decomposition, producing hydrochloric acid. Strong oxidizers may cause fire and explosions.

DOWCIDE G-ST (131-52-2) Contact with strong oxidizers may cause fire and explosions.

DOWCO 132 (299-86-5) Decomposes in strong alkaline (pH >7) and strongly acidic media. Unstable over long periods of time in water and at temperatures above 140°F/60°C.

DOWCO 139 (315-18-4) Usually dissolved in a combustible liquid; avoid oxidizer contact.

DOWCO-163 (1929-82-4) Reacts with aluminum, magnesium, or their alloys.

DOWCO 179 (2921-88-) Incompatible with strong acids. Hydrolyzes from acid or alkaline solutions.

DOW CORNING 346 (9016-00-6) Incompatible with sulfuric acid, isocyanates.

DOWFAX 2A1 (25167-32-2) Incompatible with sulfuric acid, isocyanates.

DOWFLAKE (10043-52-4) Incompatible with water, bromine trifluoride, 2-furan, percarboxylic acid. Attacks metals.

DOWFROST (57-55-7) Incompatible with strong acids, caustics, aliphatic amines, isocyanates, strong oxidizers (violent reaction).

DOWFUME (74-83-9) Incompatible with strong oxidizers, aluminum, dimethylsulfoxide, ethylene oxide, water. Attacks zinc, magnesium, alkali metals, and their alloys.

DOWFUME 40 (106-93-4) Reacts with chemically active metals, liquid ammonia, strong oxidizers. Heat and light cause slow decomposition. Attacks some plastics and rubber.

DOWFUME EDB (106-93-4) Reacts with chemically active metals, liquid ammonia, strong oxidizers. Heat and light cause slow decomposition. Attacks some plastics and rubber.

DOWFUME N (8003-19-8) Incompatible with strong acids, oxidizers, aluminum or magnesium compounds, aliphatic amines, alkanolamines, alkaline matrials, or corrosives.

DOWFUME W-8 (106-93-4) Reacts with chemically active metals, liquid ammonia, strong oxidizers. Heat and light cause slow decomposition. Attacks some plastics and rubber.

DOWFUME W-10 (106-93-4) Reacts with chemically active metals, liquid ammonia, strong oxidizers. Heat and light cause slow decomposition. Attacks some plastics and rubber.

DOWFUME W-15 (106-93-4) Reacts with chemically active metals, liquid ammonia, strong oxidizers. Heat and light cause slow decomposition. Attacks some plastics and rubber.

DOWFUME W-40 (106-93-4) Reacts with chemically active metals, liquid ammonia, strong oxidizers. Heat and light cause slow decomposition. Attacks some plastics and rubber.

DOWFUME W-85 (106-93-4) Reacts with chemically active metals, liquid ammonia, strong oxidizers. Heat and light cause slow decomposition. Attacks some plastics and rubber.

DOWICIDE 2 (99-95-4) Contact with strong oxidizers may cause fire and explosions.

DOWICIDE 7 (87-86-5) Hot water causes decomposition, producing hydrochloric acid. Strong oxidizers may cause fire and explosions.

DOWICIDE B (99-95-4) Contact with strong oxidizers may cause fire and explosions.

DOWICIDE EC-7 (87-86-5) Hot water causes decomposition, producing hydrochloric acid. Strong oxidizers may cause fire and explosions.

DOWICIDE G (87-86-5) Hot water causes decomposition, producing hydrochloric acid. Strong oxidizers may cause fire and explosions.

DOW PENTACHLOROPHENOL DP-2 ANTIMICROBIAL (87-86-5) Hot water causes decomposition, producing hydrochloric acid. Strong oxidizers may cause fire and explosions.

DOW-PER (127-18-4) Incompatible with strong oxidizers, finely divided metals, caustics. Stable up to 258°F/126°C; at this temperature product gives off poisonous fumes.

DOWPON (75-99-0) Corrosive to iron, aluminum, and copper.

DOWPON M (75-99-0) Corrosive to iron, aluminum, and copper.

DOWTHERM (8004-13-5) Contact with strong oxidizers may cause fire and explosions. Attacks some plastics, rubber, and coatings.

DOWTHERM (101-84-8) Incompatible with strong acids, strong oxidizers; may cause fire and explosions. Attacks some plastics, rubber, and coatings.

DOWTHERM 209 (107-98-2) Forms explosive mixture with air (flash point 90°F/32°C). Incompatible with sulfuric acid, isocyanates, perchloric acid.

DOWTHERM A (8004-13-5) Contact with strong oxidizers may cause fire and explosions. Attacks some plastics, rubber, and coatings.

DOWTHERM E (95-50-1) Forms explosive mixture with air (flash point 151°F/66°C). Incompatible with strong oxidizers, hot aluminum, or aluminum alloy. Attacks some plastics, rubber, and coatings.

314

DOWTHERM SR-1 (107-21-1) Incompatible with strong acids, caustics, aliphatic amines, isocyanates, chlorosulfonic acid, oleum, strong oxidizers.

DOW-TRI (79-01-6) Contact with caustics produces a toxic and flammable gas. Reacts violently with chemically active metals. Contact with aluminum may produce a violent, self-accelerating polymerization reaction. Incompatible with acids, organic anhydrides, isocyanates, alkylene oxides, aldehydes, alcohols, glycols, phenols, cresols, caprolactam solution, epichlorohydrin, nitrogen tetroxide, metal powders, oxygen. May accumulate static electrical charges, and may cause ignition of its vapors.

DOXINATE (119-36-8) Forms explosive mixture with air (flash point 205°F/96°C). Incompatible with strong acids, nitrates, oxidizers.

DOXOL (119-36-8) Forms explosive mixture with air (flash point 205°F/96°C). Incompatible with strong acids, nitrates, oxidizers.

DPA (122-39-4) Incompatible with strong acids, aldehydes, organic anhydrides, isocyanates, oxidizers, hexachloromelamine, trichloromelamine.

DPK (123-19-3) Oxidizers may cause violent reaction.

DPP (131-18-0) Incompatible with strong acids, nitrates. Attacks some plastics, rubber, and coatings.

DPP (56-38-2) Combustible liquid. Mixtures with endrin may be explosive. Strong oxidizers may cause fire and explosions. Attacks some plastics, rubber, and coatings.

DPY-97 F (105-74-8) A powerful oxidizer. May ignite or explode spontaneously when mixed with combustible materials or rapidly heated.

DRACYCLIC ACID (65-85-0) Incompatible with strong oxidizers, caustics, ammonia, amines, isocyanates.

DRAKEOL (8012-95-1) Incompatible with nitric acid; oxidizers may cause fire and explosions.

DRAZA (2032-65-7) Contact with strong oxidizers may cause fire and explosions.

DREXEL (330-54-1) Hydrolyzes in fairly strong acids.

DREXEL DEFOL (7775-09-9) A powerful oxidizer; reacts violently with reducing agents and combustible matter. Explosions may be caused by contact with ammonia salts, carbon, oils, metal sulfides, nitrobenzene, powdered metals, sugar. Contact with strong acids produces carbon dioxide. Forms shock-sensitive mixtures with some organic materials. Solution (50%) decomposes at 300°F/149°C, liberating oxygen.

DREXEL PARATHION 8E (56-38-2) Combustible liquid. Mixtures with endrin may be explosive. Strong oxidizers may cause fire and explosions. Attacks some plastics, rubber, and coatings.

DRIERITE (7778-18-9) Contact with diazomethane, aluminum, phosphorus may cause explosion.

DRILL TOX-SPEZIAL (58-89-9) Not combustible, but may be dissolved in a combustible solvent. If solvent comes in contact with oxidizers, fire and explosions may result.

DRINOX (309-00-2) Incompatible with concentrated mineral acids, acid catalysts, acid oxidizing agents, phenols, reactive metals.

315

DRINOX (76-44-8) Produces hydrogen chloride gas with iron and rust above 165°F/74°C.

DRINOX H-34 (76-44-8) Produces hydrogen chloride gas with iron and rust above 165°F/74°C.

DROP LEAF (7775-09-9) A powerful oxidizer; reacts violently with reducing agents and combustible matter. Explosions may be caused by contact with ammonia salts, carbon, oils, metal sulfides, nitrobenzene, powdered metals, sugar. Contact with strong acids produces carbon dioxide. Forms shock-sensitive mixtures with some organic materials. Solution (50%) decomposes at 300°F/149°C, liberating oxygen.

DRY ICE (SOLID) (124-38-9) Incompatible with acrylaldehyde, amines, anhydrous ammonia, cesium monoxide, lithium, metal dusts, potassium, sodium, sodium carbide, sodium–potassium alloy, sodium peroxide, titanium.

DRYCLEANER NAPHTHA (8052-41-3) Forms explosive mixture with air (flash point 102°F to 140°F/38.7°C to 60°C). Incompatible with nitric acid, strong oxidizers. Attacks some plastics, rubber, and coatings.

DRYCLEANING SAFETY SOLVENT (8052-41-3) Forms explosive mixture with air (flash point 102°F to 140°F/38.7°C to 60°C). Incompatible with nitric acid, strong oxidizers. Attacks some plastics, rubber, and coatings.

DS (64-67-5) Reacts vigorously reaction with strong oxidizers or water. Incompatible with strong acids, strong alkalies, nitrates, water.

DSE (142-59-6) Boiling water contact produces hydrogen sulfide and carbon disulfide vapors.

DSS (119-36-8) Forms explosive mixture with air (flash point 205°F/96°C). Incompatible with strong acids, nitrates, oxidizers.

DTDP (119-06-2) Incompatible with strong acids, nitrates, oxidizers.

DTMC (115-32-2) Incompatible with strong acids, caustics, aliphatic amines, isocyanates.

DUAL (51218-45-0) Incompatible with strong acids, nitrates, oxidizers.

DUKERON (79-01-6) Contact with caustics produces a toxic and flammable gas. Reacts violently with chemically active metals. Contact with aluminum may produce a violent, self-accelerating polymerization reaction. Incompatible with acids, organic anhydrides, isocyanates, alkylene oxides, aldehydes, alcohols, glycols, phenols, cresols, caprolactam solution, epichlorohydrin, nitrogen tetroxide, metal powders, oxygen. May accumulate static electrical charges, and may cause ignition of its vapors.

DULSIVAC (119-36-8) Forms explosive mixture with air (flash point 205°F/96°C). Incompatible with strong acids, nitrates, oxidizers.

DUODECYLIC ACID (143-07-7) Incompatible with sulfuric acid, bases, ammonia, amines, alkylene oxide, epichlorohydrin.

DUODEX (2492-26-4) Incompatible with concentrated aqueous solutions (are corrosive).

DUO-KILL (62-73-7) Attacks some plastics, rubber, and coatings.

DUOSOL (119-36-8) Forms explosive mixture with air (flash point 205°F/96°C). Incompatible with strong acids, nitrates, oxidizers.

DURAFUR BLACK R (106-50-3) Incompatible with acids, organic anhydrides, isocyanates, aldehydes; strong oxidizers may cause fire and explosions. Heat and light contribute to instability.

DURAFUR DEVELOPER C (120-80-9) Strong oxidizers may cause fire and explosion.

DURAFUR DEVELOPER G (108-46-3) Reacts violently with strong oxidizers, nitric acid. Incompatible with acetanilide, albumin, alkalies, antipyrine, camphor, ferric salts, menthol, spirit nitrous ether. Absorbs moisture from air (hygroscopic). May accumulate static electrical charges, and may cause ignition of its vapors.

DURALYN (137-26-8) Combustible solid (flash point 192°F/89°C). Strong oxidizers may cause fire and explosions; contact with strong acid or oxidizable materials produces toxic gases.

DURAMAX (50-78-2) Fires and explosions may result from contact with strong oxidizers. Alkali hydroxides or carbonates cause decomposition.

DURAMITEX (121-75-5) Incompatible with strong oxidizers, magnesium, alkaline pesticides. Attacks metals, some plastics, rubber, and coatings.

DURAN (330-54-1) Hydrolzses in fairly strong acids.

DURAPHOS (7786-34-7) Contact with strong oxidizers may cause fire and explosions. Attacks some plastics, rubber, and coatings.

DURASORB (67-68-5) Forms explosive mixture with air (flash point 203°F/95°C). Reacts violently with oxidizers. Reacts with ethanoyl chloride, boron compounds, halides, metal alkoxides, oxidizers.

DURAVOS (62-73-7) Attacks some plastics, rubber, and coatings.

DURETHAN (105-60-2) Contact with strong oxidizers may cause fire and explosions.

DURETTER (7720-78-7) Aqueous solution is acidic. Contact with bases produces iron.

DUROFERON (7720-78-7) Aqueous solution is acidic. Contact with bases produces iron.

DUROTOX (87-86-5) Hot water causes decomposition, producing hydrochloric acid. Strong oxidizers may cause fire and explosions.

DURSBAN (2921-88-) Incompatible with strong acids. Hydrolyzis from acid or alkaline solutions.

DURSBAN F (2921-88-) Incompatible with strong acids. Hydrolysis from acid or alkaline solutions.

DUSICNAN BARNATY (Czech) (10022-31-8) Incompatible with strong acids, aluminum–magnesium alloys, sulfur, combustibles, finely divided metals; may form shock-sensitive compounds.

DUST-LAYING OIL (8052-42-4) Incompatible with nitric acid, fluorine, strong oxidizers.

DUTCH LIQUID (107-06-2) May accumulate static electrical charges, and may cause ignition of its vapors. Contact with hot water may produce hydrochloric acid. Forms explosive mixture with air (flash point 55°F/13°C). Incompatible with strong oxidizers, strong caustics, chemically active metals. Corrosive to iron and other metals. Attacks some plastics, rubber, or coatings.

DUTCH TREAT (124-65-2) Corrodes common metals.

DWUBROMOETAN (Polish) (106-93-4) Reacts with chemically active metals, liquid ammonia, strong oxidizers. Heat and light cause slow decomposition. Attacks some plastics and rubber.

DWUCHLORODWUFLUOROMETAN (Polish) (75-71-8) Reacts violently with liquid aluminum. Incompatible with chemically active metals. Attacks some plastics, rubber, and coatings.

2,4-DWUCHLOROFENOKSYOCTOWY KWAS (Polish) (94-75-7) Decomposes in sunlight. Incompatible with strong oxidizers; may cause fire and explosions.

DWUCHLOROPROPAN (Polish) (78-87-5) Forms explosive mixture with air (flash point 60°F/16°C). May accumulate static electrical charges, and may cause ignition of its vapors. Strong oxidizers may cause fire and explosions. Strong acids can cause decomposition and the formation of hydrogen chloride vapors. Corrodes aluminum. Attacks some plastics, rubber, and coatings.

DWUETYLOWYETER (Polish) (60-29-7) Forms explosive mixture with air (flash point −49°F/−45°C). Incompatible with strong acids, strong oxidizers. Can form peroxides from contact with air or light; may explode when container is unstoppered or otherwise opened. Attacks some plastics, rubber, and coatings. Being a nonconductor, chemical may accumulate static electric charges that may result in ignition of vapor.

DWUMETYLOANILINA (Polish) (121-69-7) Forms explosive mixture with air (flash point 145°F/63°C). Contact with strong oxidizers may cause fire and explosions; incompatible with strong acids (with violent spattering). Contact with benzoyl peroxide, diisopropyl perdicarbonate and other material may cause explosions. Attacks some plastics, rubber, and coatings.

DWUMETYLOFORMAMID (Polish) (68-12-2) Forms explosive mixture with air (flash point 136°F/58°C). Contact with carbon tetrachloride and other halogenated compounds, particularly in presence of iron or strong oxidizers, may cause fire and explosions. Reacts vigorously reaction with alkylaluminums. Incompatible with nonoxidizing mineral acids, strong acids, chlorinated hydrocarbons, isocyanates, nitrates, organic nitrates, phenols, cresols, ammonia, chromic anhydride, magnesium nitrate, methylene diisocyanate, phosphorus trioxide, triethylaluminum. Attacks some plastics, rubber, and coatings.

DWUMETYLOWY SIARCZAN (Polish) (77-78-1) Forms explosive mixture with air (flash point 182°F/83°C). Incompatible with strong oxidizers, strong acids, strong alkalies, strong ammonia solutions. Attacks some plastics, rubber, and coatings.

DWUNITROBENZEN (Polish) (99-65-0) Explodes: is impact- and friction-sensitive. Strong oxidizers may cause fire and explosions; avoid contact with caustics, chemically active metals. Attacks some plastics.

DYKANOL (generic CAS for PCBs 1336-36-3) Incompatible with strong oxidizers, strong acids.

DYKOL (50-29-3) Incompatible with salts of iron or aluminum, and bases. Do not store in iron containers.

DYLOX (52-68-6) Contact with strong oxidizers may cause fire and explosions.

DYNEX (330-54-1) Hydrolyzes in fairly strong acids.

DYREX (52-68-6) Contact with strong oxidizers may cause fire and explosions.

DYTOL M-83 (111-87-5) Forms explosive mixture with air (flash point 178°F/81°C). Incompatible with strong acids, caustics, aliphatic amines, isocyanates, strong oxidizers.

DYTOL S-91 (112-30-1) Forms explosive mixture with air (flash point 180°F/82°C). Incompatible with strong acids, caustics, aliphatic amines, isocyanates, strong oxidizers.

DYVON (52-68-6) Contact with strong oxidizers may cause fire and explosions.

DYZOL (333-41-5) Incompatible with water, copper-containing compounds, oxidizers, acids, or bases.

- E -

E-1 (151-56-4) Forms explosive mixture with air (flash point 12°F/−11°C). May accumulate static electrical charges, and may cause ignition of its vapors. Contact with acid, aluminum, carbon dioxide or silver may cause explosive polymerization. Attacks rubber, coatings, plastics, and chemically active metals. Self-reactive with heat or atmospheric carbon dioxide.

E 393 (3689-24-5) Containers may burst in heat. Strong oxidizers may cause fire and explosions. Attacks some plastics, rubber, and coatings.

E 605 (56-38-2) Combustible liquid. Mixtures with endrin may be explosive. Strong oxidizers may cause fire and explosions. Attacks some plastics, rubber, and coatings.

E 1059 (8065-48-3) Forms explosive mixture with air (flash point 113°F/45°C). Incompatible with water, strong oxidizers, caustics.

E 3314 (76-44-8) Forms hydrogen chloride gas with iron and rust above 165°F/74°C.

EA (75-04-7) Forms explosive mixture with air (flash point less than −0°F/−18°C). Incompatible with acids, organic anhydrides, isocyanates, vinyl acetate, acrylates, substituted allyls, alkylene oxides, epichlorohydrin, ketones, aldehydes, alcohols, glycols, phenols, cresols, caprolactam solution, strong oxidizers. Attacks aluminum, copper, lead, tin, zinc, and alloys, and some plastics, rubber, and coatings.

EAA (141-97-9) Forms explosive mixture with air (flash point 135°F/57°C). Incompatible with strong acids, nitrates, oxidizers.

EADC (563-43-9) Forms explosive mixture with air. May ignite spontaneously in air. Reacts violently with water, producing hydrogen chloride and ethane gas. A strong reducing agent; reacts violently with oxidizers. Reacts violently with alcohols, amines, carbon dioxide, phenol, oxides of sulfur or nitrogen. May accumulate static electrical charges, and may cause ignition of its vapors.

EAK (541-85-5) Forms explosive mixture with air (flash point 110°F/43°C). Incompatible with sulfuric acid, nitric acid, aliphatic amines, oxidizers.

EARTHNUT OIL (8002-03-7) Incompatible with strong acids, oxidizers.

EASC (12075-68-2) Ignites spontaneously with air. A strong reducing agent; reacts violently with oxidizers, alcohols, amines, carbon dioxide, carbon tetrachloride, cresols, oxides of nitrogen, oxides of sulfur, phenols and other substances. May accumulate static electrical charges, and may cause ignition of its vapors. Attacks metals in a moist environment.

EB (100-41-4) May accumulate static electrical charges, and may cause ignition of its vapors. Forms explosive mixture with air (flash point 59°F/15°C). Incompatible with strong oxidizers, nitric acid.

EBDC, SODIUM SALT (142-59-6) Boiling water contact produces hydrogen sulfide and carbon disulfide vapors.

EBRAGOM TB (137-26-8) Combustible solid (flash point 192°F/89°C). Strong oxidizers may cause fire and explosions; contact with strong acid or oxidizable materials produces toxic gases.

EC 300 (123-33-1) Contact with strong oxidizers may cause fire and explosions.

ECATOX (56-38-2) Combustible liquid. Mixtures with endrin may be explosive. Strong oxidizers may cause fire and explosions. Attacks some plastics, rubber, and coatings.

ECF (541-43-3) Forms explosive mixture with air (flash point 61°F/16°C). Strong oxidizers may cause fire and explosions. Attacks metals in the presence of moisture. Water and air contact produce hydrochloric acid.

ECH (106-89-8) Forms explosive mixture with air (flash point 88°F/31°C). Heat, acids, alkalies and metallic halides can cause explosive polymerization. Aliphatic amines, alkaline earths, alkali metals, alkanolamines, powdered metals, strong oxidizers may cause fire and explosions. Decomposition produces highly toxic phosgene gas. Will pit steel in the presence of moisture. May accumulate static electrical charges, and may cause ignition of its vapors.

ECM (50-78-2) Fires and explosions may result from contact with strong oxidizers. Alkali hydroxides or carbonates cause decomposition.

ECOTRIN (50-78-2) Fires and explosions may result from contact with strong oxidizers. Alkali hydroxides or carbonates cause decomposition.

ECRINITRIT (7632-00-0) A strong oxidizer; reacts violently with reducing agents and combustibles. Reacts with acids, ammonium salts + heat, butadiene, cyanides, lithium, phthalic acid, thiosulfate, sodium amide, and many other substances. Under certain conditions the chemical can also be a strong reducing agent.

EDATHAMIL (60-00-4) Incompatible with sulfuric acid, bases, ammonia, aliphatic amines, alkanolamines, isocyanates, alkylene oxides, epichlorohydrin.

EDB or E-D-BEE (106-93-4) Reacts with chemically active metals, liquid ammonia, strong oxidizers. Heat and light cause slow decomposition. Attacks some plastics and rubber.

EDB-85 (106-93-4) Reacts with chemically active metals, liquid ammonia, strong oxidizers. Heat and light cause slow decomposition. Attacks some plastics and rubber.

EDC (107-06-2) May accumulate static electrical charges, and may cause ignition of its vapors. Contact with hot water may produce hydrochloric acid. Forms explosive mixture with air (flash point 55°F/13°C). Incompatible with strong oxidizers, strong caustics, chemically active metals. Corrosive to iron and other metals. Attacks some plastics, rubber, or coatings.

EDCO (74-83-9) Incompatible with strong oxidizers, aluminum, dimethylsulfoxide, ethylene oxide, water. Attacks zinc, magnesium, alkali metals and their alloys.

EDETIC ACID (60-00-4) Incompatible with sulfuric acid, bases, ammonia, aliphatic amines, alkanolamines, isocyanates, alkylene oxides, epichlorohydrin.

EDETIC (60-00-4) Incompatible with sulfuric acid, bases, ammonia, aliphatic amines, alkanolamines, isocyanates, alkylene oxides, epichlorohydrin.

EDIBLE TALLOW (N/A) Incompatible with strong acids, oxidizers.

EDTA (60-00-4) Incompatible with sulfuric acid, bases, ammonia, aliphatic amines, alkanolamines, isocyanates, alkylene oxides, epichlorohydrin.

EDTA ACID (60-00-4) Incompatible with sulfuric acid, bases, ammonia, aliphatic amines, alkanolamines, isocyanates, alkylene oxides, epichlorohydrin.

EDTA-ZINC or EDTA-ZINC COMPLEX or EDTA-ZINC SALT (12519-36-7) Incompatible with sulfuric acid, isocyanates.

EFFEMOLL DOA (103-23-1) Incompatible with strong acids, nitrates, oxidizers.

EG (107-21-1) Incompatible with strong acids, caustics, aliphatic amines, isocyanates, chlorosulfonic acid, oleum, strong oxidizers.

EGBE (111-76-2) Forms explosive mixture with air (flash point 143°F/62°C). Reacts violently with strong caustics and strong oxidizers. Attacks some coatings, plastics, and rubber. Attacks metallic aluminum at high temperatures.

EGBEA (112-07-2) Forms explosive mixture with air (flash point 160°F/71°C). Ethers, as a class, can form peroxides. Incompatible with strong acids, perchloric acid, oxidizers, nitrates.

EGDME (110-71-4) Forms explosive mixture with air (flash point 29°F/−2°C). May accumulate static electrical charges, and may cause ignition of its vapors. Incompatible with sulfuric acid, isocyanates, strong oxidizers.

EGDN (628-96-6) Contact with acids, heat, or mechanical shock may cause explosions.

EGEE (110-80-5) Forms explosive mixture with air (flash point 120°F/49°C). Strong oxidizers may cause fire and explosions. Attacks some plastics, rubber, and coatings. Able to form peroxides. Incompatible with strong acids, aluminum and its alloys.

EGEEA (111-15-9) Forms explosive mixture with air (flash point 117°F/47°C). Incompatible with strong acids, nitrates. Reacts violently with oxidizers. May form unstable peroxides. Softens many plastics. Attacks some plastics, rubber, and coatings.

EGITOL (67-72-1) Incompatible with hot iron, zinc and aluminum; contact with alkalies produces spontaneously explosive chloroacetylene. Attacks some plastics, rubber, and coatings.

EGM (109-86-4) Forms explosive mixture with air (flash point 103°F/39°C). Heat or oxidizers may cause formation of form unstable peroxides. Attacks many metals. Strong oxidizers cause fire and explosions. Strong bases cause decomposition. Attacks some plastics, rubber, and coatings. May accumulate static electrical charges, and may cause ignition of its vapors.

EGME (109-86-4) Forms explosive mixture with air (flash point 103°F/39°C). Heat or oxidizers may cause formation of unstable peroxides. Attacks many metals. Strong oxidizers cause fire and explosions. Strong bases cause decomposition. Attacks some plastics, rubber, and coatings. May accumulate static electrical charges, and may cause ignition of its vapors.

322

EGMEA (110-49-6) Incompatible with nitrates, strong oxidizers, strong alkalies, strong acids.

EHYDE ACETIQUE (French) (75-07-0) Slowly polymerizes to paraldehyde. Explodes when mixed with iodine. Contact with strong bases can cause explosive polymerization. A strong reducing agent; reacts violently with combustibles, strong acids, caustics, ammonia, aliphatic amines, alkanolamines, aromatic amines, organic substances, halogens, oxidizers. Forms explosive peroxides with air. May dissolve rubber. May accumulate static electrical charges, and may cause ignition of its vapors.

EHYDE ACRYLIQUE (French) (107-02-8) Forms explosive mixture with air (flash point −15°F/−26°C). Unstable and very reactive. In storage can form heat- and shock-sensitive compounds. Unless inhibited (usually by hydroquinone), readily forms explosive peroxides. Able to polymerize. A strong reducing agent; reacts violently with oxidizers, strong acids, caustics, amines, 2-aminoethanol, ammonia, ammonium hydroxide, ethylene diamine, ethyleneimine, hydroxides, metal salts, oxidizers, sulfur dioxide, thiourea. Attacks metals: cadmium and zinc. May accumulate static electrical charges, and may cause ignition of its vapors.

EIDE ACETICA (Italian) (75-07-0) Slowly polymerizes to paraldehyde. Explodes when mixed with iodine. Contact with strong bases can cause explosive polymerization. A strong reducing agent; reacts violently with combustibles, strong acids, caustics, ammonia, aliphatic amines, alkanolamines, aromatic amines, organic substances, halogens, oxidizers. Forms explosive peroxides with air. May dissolve rubber. May accumulate static electrical charges, and may cause ignition of its vapors.

EIDE ACRILICA (Italian) (107-02-8) Forms explosive mixture with air (flash point −15°F/−26°C). Unstable and very reactive. In storage can form heat- and shock-sensitive compounds. Unless inhibited (usually by hydroquinone), readily forms explosive peroxides. Able to polymerize. A strong reducing agent; reacts violently with oxidizers, strong acids, caustics, amines, 2-aminoethanol, ammonia, ammonium hydroxide, ethylene diamine, ethyleneimine, hydroxides, metal salts, oxidizers, sulfur dioxide, thiourea. Attacks metals: cadmium and zinc. May accumulate static electrical charges, and may cause ignition of its vapors.

EINECS NO. 231-635-3 (7664-41-7) anhydrous (a flammable gas); (1336-21-6) solution in water. Reacts violently with strong oxidizers, acids. Shock-sensitive compounds may be formed with halogens, mercury oxide, silver oxide. Fire and explosions may be caused by trimethylammonium amide, 1-chloro-2,4-dinitrobenzene, o-chloronitrobenzene, platinum, selenium difluoride dioxide, boron halides, mercury, chlorine, iodine, bromine, hypochlorites, chlorine bleach, amides, organic anhydrides, isocyanates, vinyl acetate, alkylene oxides, epichlorohydrin, aldehydes. Attacks some coatings, plastics, and rubber. Attacks copper, brass, bronze, aluminum, steel, and their alloys.

EKTAFOS (141-66-2) Corrosive to cast iron, mild steel, brass, and stainless steel 304.

EKAGOM TB (137-26-8) Combustible solid (flash point 192°F/89°C). Strong oxidizers may cause fire and explosions; contact with strong acid or oxidizable materials produces toxic gases.

EKATIN WF & WF ULV (56-38-2) Combustible liquid. Mixtures with endrin may be explosive. Strong oxidizers may cause fire and explosions. Attacks some plastics, rubber, and coatings.

EKATOX (56-38-2) Combustible liquid. Mixtures with endrin may be explosive. Strong oxidizers may cause fire and explosions. Attacks some plastics, rubber, and coatings.

EKTASOLVE DB ACETATE (124-17-4) Incompatible with strong acids, nitrates, strong alkalies, oxidizers.

EKTASOLVE EB ACETATE (112-07-2) Forms explosive mixture with air (flash point 160°F/71°C). Ethers, as a class, can form peroxides. Incompatible with strong acids, perchloric acid, oxidizers, nitrates.

EKTASOLVE EB SOLVENT (111-76-2) Forms explosive mixture with air (flash point 143°F/62°C). Reacts violently with strong caustics and strong oxidizers. Attacks some coatings, plastics, and rubber. Attacks metallic aluminum at high temperatures.

EKTASOLVE EE (110-80-5) Forms explosive mixture with air (flash point 120°F/49°C). Strong oxidizers may cause fire and explosions. Attacks some plastics, rubber, and coatings. Able to form peroxides. Incompatible with strong acids, aluminum and its alloys.

EKTASOLVE EP (2807-30-9) Incompatible with sulfuric acid, isocyanates, strong oxidizers.

EL 4049 (121-75-5) Incompatible with strong oxidizers, magnesium, alkaline pesticides. Attacks metals and some plastics, rubber, and coatings.

ELAIC ACID (112-80-1) Incompatible with strong oxidizers; contact with perchloric acid and aluminum powder may cause explosions. Corrodes aluminum.

ELALDEHYDE (123-63-7) Reacts with strong acids, caustics, ammonia, amines, oxidizers. Contact with acids produces acetaldehyde.

ELAOL (84-74-2) Incompatible with strong acids, nitrates, strong oxidizers, strong alkalies.

ELAYL (74-85-1) A flammable gas. May accumulate static electrical charges, and may cause ignition of its vapors. Oxidizers may cause explosive polymerization and fire. Incompatible with acids, halogens, nitrogen oxides, hydrogen bromide, aluminum chloride, bromotrichloromethane, carbon tetrachloride, chlorine, chlorine dioxide, nitrogen dioxide.

ELDOPOQUE (123-31-9) Incompatible with strong oxidizers, caustics. May be oxidized to quinone at room temperatures in the presence of moisture. May explode on contact with oxygen.

ELDOQUIN (123-31-9) Incompatible with strong oxidizers, caustics. May be oxidized to quinone at room temperatures in the presence of moisture. May explode on contact with oxygen.

ELECTRICAL INSULATING OIL (N/A) Incompatible with nitric acid; contact with oxidizers may cause fire and explosions. May accumulate static electrical charges, and may cause ignition of its vapors.

ELECTRO-CF 11 (75-69-4) Reacts with barium, lithium, sodium, magnesium, titanium. Attacks some plastics, rubber, and coatings.

ELECTRO-CF 12 (75-71-8) Reacts violently with liquid aluminum. Incompatible with chemically active metals. Attacks some plastics, rubber, and coatings.

ELEMENTAL WHITE PHOSPHORUS (7723-14-0) A strong reducing agent and fire risk; store under water away from heat. Ignites spontaneously on contact with air (flash point at or above 86°F/30°C). Contact with all oxidizers (including elemental sulfur) will cause fire and explosions. Contact with strong caustics produces poisonous and flammable phosphine gas. Liquid attacks some plastics, rubber, and coatings.

ELMASIL (61-82-5) Substance acts as a weak base to form salts in contact with acids. Corrosive to iron, aluminum, copper, and copper alloys.

ELPON (9003-07-0) Incompatible with strong acids, strong oxidizers, chlorine, potassium permanganate.

EMBAFUME (74-83-9) Incompatible with strong oxidizers, aluminum, dimethylsulfoxide, ethylene oxide, water. Attacks zinc, magnesium, alkali metals, and their alloys.

EMBATHION (563-12-2) Incompatible with alkaline formulations. Mixtures with magnesium may be explosive.

EMERALD GREEN (12002-03-8) Contact with strong oxidizers may cause fire and explosions.

EMERSOL-120 (57-11-4) Incompatible with strong oxidizers, sulfuric acid, caustics, ammonia, amines, isocyanates, alkylene oxides, epichlorohydrin. Attacks chemically active metals.

EMERSOL 210 (112-80-1) Incompatible with strong oxidizers; contact with perchloric acid and aluminum powder may cause explosions. Corrodes aluminum.

EMERSSENCE 1160 (122-99-6) Incompatible with sulfuric acid, isocyanates, strong oxidizers.

EMERY 6705 (122-99-6) Incompatible with sulfuric acid, isocyanates, strong oxidizers.

EMISOL (61-82-5) Substance acts as a weak base to form salts in contact with acids. Corrosive to iron, aluminum, copper, and copper alloys.

EMMATOS (121-75-5) Incompatible with strong oxidizers, magnesium, alkaline pesticides. Attacks metals, some plastics, rubber, and coatings.

EMO-NIK (54-11-5) Incompatible with strong acids, strong oxidizers. Attacks some plastics, rubber, and coatings. May accumulate static electrical charges, and may cause ignition of its vapors.

EMOTOS EXTRA (121-75-5) Incompatible with strong oxidizers, magnesium, alkaline pesticides. Attacks metals, some plastics, rubber, and coatings.

EMULSAMINE BK (94-75-7) Decomposes in sunlight. Incompatible with strong oxidizers; may cause fire and explosions.

EMULSAMINE E-3 (94-75-7) Decomposes in sunlight. Incompatible with strong oxidizers; may cause fire and explosions.

ENAMEL WHITE (7727-43-7) Explosions may result from contact with aluminum in the presence of heat. Incompatible with potassium, phosphorus.

ENANTHIC ACID (111-14-8) Incompatible with sulfuric acid, caustics, ammonia, aliphatic amines, alkanolamines, isocyanates, alkylene oxides, epichlorohydrin.

325

ENANTHIC ALCOHOL (111-70-6) Forms explosive mixture with air above 170°F/77°C. Incompatible with strong acids, caustics, aliphatic amines, isocyanates, oxidizers.

ENB (16219-75-3) Reacts violently with oxygen and strong oxidizers. Forms explosive mixture with air (flash point 100°F/38°C). May accumulate static electrical charges, and may cause ignition of its vapors. Inhibit peroxide formation with *tert*-butyl catechol.

ENDOCEL (115-29-7) Hydrolyzed by acids and alkalis. Corrosive to iron.

1,4-ENDO-5,8-DIMETHANONAPHTHALENE (309-00-2) Incompatible with concentrated mineral acids, acid catalysts, acid oxidizing agents, phenols, reactive metals.

ENDOSULFAN (115-29-7) Hydrolyzed by acids and alkalis. Corrosive to iron.

ENDOSULPHAN (115-29-7) Hydrolyzed by acids and alkalis. Corrosive to iron.

ENDRATE (60-00-4) Incompatible with sulfuric acid, bases, ammonia, aliphatic amines, alkanolamines, isocyanates, alkylene oxides, epichlorohydrin.

ENDREX (72-20-8) Incompatible with parathion, strong acids (forms explosive vapors), strong oxidizers.

ENDRIN (72-20-8) Incompatible with parathion, strong acids (forms explosive vapors), strong oxidizers.

ENDRINE (French) (72-20-8) Parathion, strong acids (forms explosive vapors), strong oxidizers.

ENDYDOL (50-78-2) Fires and explosions may result from contact with strong oxidizers. Alkali hydroxides or carbonates cause decomposition.

ENGLISH RED (1309-37-1) Contact with hydrogen peroxide, ethylene oxide, calcium hypochlorite will cause explosion. Reacts violently with powdered aluminum, hydrazine, hydrogen trisulfide.

ENGRAVERS ACID (7697-37-2) A strong oxidizer that can react violently with reducing agents, combustible materials. Incompatible with many substances, including acrylates, aliphatic amines, alcohols, aldehydes, alkanolamines, alkylene oxides, anion exchange resins, aromatic amines, amides, bases, cresols, cyanides, cyclic ketones, epichlorohydrin, glycols, isocyanates, ketones, oleum, organic anhydrides, phenols, substituted allyls, sulfuric acid, strong oxidizers, terpenes. Attacks most metals, and some plastics, rubber, and coatings.

ENSURE (115-29-7) Hydrolyzed by acids and alkalis. Corrosive to iron.

ENT 54 (107-13-1) Forms explosive mixture with air (flash point 32°F/0°C o.c.). Forms explosive peroxides; heat, light, caustics, silver nitrate, and peroxides can cause polymerization. Incompatible with strong acids, strong oxidizers, amines, 2-aminoethanol, bromine, chlorosulfonic acid, ethylene diamine, nitric acid, oleum, potassium hydroxide, sodium hydroxide, sulfuric acid. Attacks copper and copper alloys; attacks aluminum in high concentrations. May accumulate static electrical charges, and may cause ignition of its vapors.

ENT 262 (131-11-3) Incompatible with strong alkalies, strong acids, nitrates, oxidizers.

ENT 987 (137-26-8) Combustible solid (flash point 192°F/89°C). Strong oxidizers may cause fire and explosions; contact with strong acid or oxidizable materials produces toxic gases.

ENT 1,506 (50-29-3) Incompatible with salts of iron or aluminum, and bases. Do not store in iron containers.

ENT 1,656 (107-06-2) May accumulate static electrical charges, and may cause ignition of its vapors. Contact with hot water may produce hydrochloric acid. Forms explosive mixture with air (flash point 55°F/13°C). Incompatible with strong oxidizers, strong caustics, chemically active metals. Corrosive to iron and other metals. Attacks some plastics, rubber, or coatings.

ENT 1,716 (72-43-5) Contact with strong oxidizers may cause fire and explosions. Attacks some plastics, rubber, and coatings.

ENT 1,860 (127-18-4) Incompatible with strong oxidizers, finely divided metals, caustics. Stable up to 258°F/126°C; at this temperature product gives off poisonous fumes.

ENT 3,424 (54-11-5) Incompatible with strong acids, strong oxidizers. Attacks some plastics, rubber, and coatings. May accumulate static electrical charges, and may cause ignition of its vapors.

ENT 4,225 (72-54-8) Contact with strong oxidizers may cause fire and explosions.

ENT 4,705 (56-23-5) Becomes corrosive when in contact with water. Corrosive to metals. Reacts violently with many compounds. Decomposes on contact with chemically active metals such as sodium, potassium, and magnesium. Incompatible with allyl alcohol, fluorine gas, alkali metals, aluminum. Attacks some coatings, plastics, and rubber.

ENT 7,796 (58-89-9) Not combustible, but may be dissolved in a combustible solvent. If solvent comes in contact with oxidizers, fire and explosions may result.

ENT 8,420 (8003-19-8) Incompatible with strong acids, oxidizers, aluminum or magnesium compounds, aliphatic amines, alkanolamines, alkaline materials, or corrosives.

ENT 9,735 (8001-35-2) Reacts with oxidizers, with a risk of fire or explosions. Attacks metals in the presence of moisture.

ENT 9,932 (57-74-9) Contact with strong oxidizers may cause fire and explosions. Attacks some plastics, rubber, and coatings.

ENT 15,108 (56-38-2) Combustible liquid. Mixtures with endrin may be explosive. Strong oxidizers may cause fire and explosions. Attacks some plastics, rubber, and coatings.

ENT 15,152 (76-44-8) Produces hydrogen chloride gas with iron and rust above 165°F/74°C.

ENT 15,349 (106-93-4) Reacts with chemically active metals, liquid ammonia, strong oxidizers. Heat and light cause slow decomposition. Attacks some plastics and rubber.

ENT 15,406 (78-87-5) Forms explosive mixture with air (flash point 60°F/16°C). May accumulate static electrical charges, and may cause ignition of its vapors. Strong oxidizers may cause fire and explosions. Strong acids can cause decomposition and the formation of hydrogen chloride vapors. Corrodes aluminum. Attacks some plastics, rubber, and coatings.

ENT 15,949 (309-00-2) Incompatible with concentrated mineral acids, acid catalysts, acid oxidizing agents, phenols, reactive metals.

ENT 16,225 (60-57-1) Incompatible with concentrated mineral acids, acid catalysts, acid oxidizing agents, phenols, reactive metals.

ENT 16,273 (3689-24-5) Containers may burst in heat. Strong oxidizers may cause fire and explosions. Attacks some plastics, rubber, and coatings.

ENT 17,034 (121-75-5) Incompatible with strong oxidizers, magnesium, alkaline pesticides. Attacks metals and some plastics, rubber, and coatings.

ENT 17,251 (72-20-8) Incompatible with parathion, strong acids (forms explosive vapors), strong oxidizers.

ENT 17,292 (298-00-0) Mixtures with magnesium may be explosive.

ENT 17,295 (8065-48-3) Forms explosive mixture with air (flash point 113°F/45°C). Incompatible with water, strong oxidizers, caustics.

ENT 17,957 (56-72-4) Contact with strong oxidizers may cause fire and explosions.

ENT 18,771 (107-49-3) Decomposes above 300°F/150°C, producing flammable ethylene gas. Strong oxidizers may cause fire and explosions. Attacks some plastics, rubber, and coatings.

ENT 18,870 (123-33-1) Contact with strong oxidizers may cause fire and explosions.

ENT 19,507 (333-41-5) Incompatible with water, copper-containing compounds, oxidizers, acids, or bases.

ENT 20,738 (62-73-7) Attacks some plastics, rubber, and coatings.

ENT 22,374 (7786-34-7) Contact with strong oxidizers may cause fire and explosions. Attacks some plastics, rubber, and coatings.

ENT 23,233 (86-50-0) Strong oxidizers may cause fire and explosions.

ENT 23,437 (298-04-4) Forms explosive mixture with air above 180°F/82°C. Incompatible with alkalis, strong oxidizers.

ENT 23,969 (63-25-2) Incompatible with strong oxidizers, strongly alkaline pesticides.

ENT 23,979 (115-29-7) Hydrolyzed by acids and alkalis. Corrosive to iron.

ENT 24,105 (563-12-2) Incompatible with alkaline formulations. Mixtures with magnesium may be explosive.

ENT 24,482 (141-66-2) Corrosive to cast iron, mild steel, brass, and stainless steel 304.

ENT 24,984 (15096-52-3) Contact with strong acids may produce toxic and corrosive hydrogen fluoride gas.

ENT 24,988 (300-76-5) Contact with strong oxidizers may cause fire and explosions. Attacks some plastics, rubber, and coatings.

ENT 25,445 (61-82-5) Substance acts as a weak base to form salts in contact with acids. Corrosive to iron, aluminum, copper, and copper alloys.

ENT 25,552 (57-74-9) Contact with strong oxidizers may cause fire and explosions. Attacks some plastics, rubber, and coatings.

ENT 25,602-X (299-86-5) Decomposes in strongly alkaline (pH >7) and strongly acidic media. Unstable over long periods of time in water and at temperatures above 140°F/60°C.

ENT 25,726 (2032-65-7) Contact with strong oxidizers may cause fire and explosions.

ENT 26,538 (133-06-2) Incompatible with tetraethyl pyrophosphate, parathion.

ENT 27,164 (1563-66-2) Incompatible with alkaline material, acids, strong oxidizers.

ENT 27,226 (2312-35-8) Forms explosive mixture with air (flash point 82°F/28°C). Strong oxidizers may cause fire and explosions.

ENT 27,311 (2921-88-) Incompatible with strong acids. Hydrolyzes from acid or alkaline solutions.

ENTEROSARINE (50-78-2) Fires and explosions may result from contact with strong oxidizers. Alkali hydroxides or carbonates cause decomposition.

ENTOMOXAN (58-89-9) Not combustible, but may be dissolved in a combustible solvent. If solvent comes in contact with oxidizers, fire and explosions may result.

ENTROPHEN (50-78-2) Fires and explosions may result from contact with strong oxidizers. Alkali hydroxides or carbonates cause decomposition.

ENVERT 171 (94-75-7) Decomposes in sunlight. Incompatible with strong oxidizers; may cause fire and explosions.

ENVERT DT (94-75-7) Decomposes in sunlight. Incompatible with strong oxidizers; may cause fire and explosions.

ENVERT-T (93-76-5) Sealed metal containers may burst in heat above 316°F/158°C). Incompatible with sulfuric acid, bases, ammonia, aliphatic amines, alkanolamines, isocyanates, alkylene oxides, epichlorohydrin.

E.O. (75-21-8) Forms explosive mixture with air (flash point 20°F/−6°C). Incompatible with alkali metal hydroxides, highly active catalysts (e.g., anhydrous chlorides of iron, tin, or aluminum and oxides of iron or aluminum). Avoid contact with copper. Protect container from physical damage, sun and heat. Attacks some plastics, rubber, and coatings.

EP 30 (87-86-5) Hot water causes decomposition forming hydrochloric acid. Strong oxidizers may cause fire and explosions.

EP-161E (556-61-6) Forms explosive mixture with air (flash point 90°F/32°C). Incompatible with strong acids, caustics, ammonia, amines, amides, alcohols, glycols, caprolactam solution, strong oxidizers.

EPAL-6 (111-27-3) Forms explosive mixture with air above 145°F/63°C). Incompatible with strong acids, caustics, aliphatic amines, isocyanates, strong oxidizers.

EPAL 10 (112-30-1) Forms explosive mixture with air above 180°F/82°C). Incompatible with strong acids, caustics, aliphatic amines, isocyanates, strong oxidizers.

EPAL 12 (112-53-8) Incompatible with strong acids, caustics, aliphatic amines, isocyanates, oxidizers.

EPICHLOROHYDRIN (106-89-8) Forms explosive mixture with air (flash point 88°F/31°C). Heat, acids, alkalies, and metallic halides can cause explosive polymerization. Aliphatic amines, alkaline earths, alkali metals, alkanolamines, powdered metals, strong oxidizers may cause fire and explosions. Decomposition produces highly toxic phosgene gas. Will pit steel in the presence of moisture. May accumulate static electrical charges, and may cause ignition of its vapors.

a-EPICHLOROHYDRIN or alpha-EPICHLOROHYDRIN (106-89-8) Forms explosive mixture with air (flash point 88°F/31°C). Heat, acids, alkalies, and metallic halides can cause explosive polymerization. Aliphatic amines, alkaline earths, alkali metals, alkanolamines, powdered metals, strong oxidizers may cause fire and explosions. Decomposition produces highly toxic phosgene gas. Will pit steel in the presence of moisture. May accumulate static electrical charges, and may cause ignition of its vapors.

EPICLEAR (94-36-0) Confined storage of dry chemical may lead to decomposition and explosion. A strong oxidant; extremely reactive. Fires and explosion may result from heat or contamination, and from contact with strong acids, combustible materials, oxidizers, acids, bases, alcohols, reducing agents, metals, metal oxides, amines, accelerators, methyl methacrylate, organic matter, lithium aluminum carbide, dimethyl aniline, amines, metallic naphthenates. May attack some plastics, rubber, and coatings. Protect containers from shock and friction.

EPIHYDRIN ALCOHOL (556-52-5) Forms explosive mixture with air (flash point 158°F/70°C). Reacts violently with strong oxidizers. Contact with caustics, acids, barium, lithium, sodium, magnesium, titanium can cause polymerization. Attacks some plastics, rubber, and coatings.

EPI-REZ 508 (1675-54-8) Forms explosive mixture with air (flash point 175°F/79°C). Incompatible with strong acids, strong oxidizers. Ethers form peroxides on contact with air and light.

EPI-REZ 510 (1675-54-8) Forms explosive mixture with air (flash point 175°F/79°C). Incompatible with strong acids, strong oxidizers. Ethers form peroxides on contact with air and light.

EPON 828 (1675-54-8) Forms explosive mixture with air (flash point 175°F/79°C). Incompatible with strong acids, strong oxidizers. Ethers form peroxides on contact with air and light.

EPOXIDE A (1675-54-8) Forms explosive mixture with air (flash point 175°F/79°C). Incompatible with strong acids, strong oxidizers. Ethers form peroxides on contact with air and light.

EPOXIDIZED TALL OIL, OCTYL ESTER (N/A) Incompatible with strong acids, nitrates, oxidizers.

1,2-EPOXYAETHAN (German) (75-21-8) Forms explosive mixture with air (flash point 20°F/−6°C). Incompatible with alkali metal hydroxides, highly active catalysts (e.g., anhydrous chlorides of iron, tin, or aluminum and oxides of iron or aluminum). Avoid contact with copper. Protect container from physical damage, sun, and heat. Attacks some plastics, rubber, or coatings.

1,2-EPOXYBUTANE (106-88-7) Forms explosive mixture with air. (flash point −7°F/−22°C). Strong bases or metal chlorides cause polymerization. Reacts violently with oxidizers, acids. May accumulate static electrical charges, and may cause ignition of its vapors.

1,4-EPOXYBUTANE (109-99-9) Forms explosive mixture with air (flash point 6°F/−14°C). Unless inhibited, can form unstable and explosive peroxides. Incompatible with strong acids, strong oxidizers. Attacks some plastics. May accumulate static electric charges that can result in ignition of its vapors.

330

1,2-EPOXY-3-BUTOXY PROPANE (2426-08-6) Forms explosive mixture with air (flash point 130°F/54°C). Air and light cause formation of unstable and explosive peroxides. Contact with strong oxidizers may cause fire and explosions. Strong caustics may cause polymerization. Attacks some plastics and rubber.

1,2-EPOXY-3-CHLOROPROPANE (106-89-8) Forms explosive mixture with air (flash point 88°F/31°C). Heat, acids, alkalies, metallic halides can cause explosive polymerization. Aliphatic amines, alkaline earths, alkali metals, alkanolamines, powdered metals, strong oxidizers may cause fire and explosions. Decomposition produces highly toxic phosgene gas. Will pit steel in the presence of moisture. May accumulate static electrical charges, and may cause ignition of its vapors.

EPOXYETHANE (French) (75-21-8) Forms explosive mixture with air (flash point 20°F/−6°C). Incompatible with alkali metal hydroxides, highly active catalysts (e.g., anhydrous chlorides of iron, tin, or aluminum and oxides of iron or aluminum). Avoid contact with copper. Protect container from physical damage, sun, and heat. Attacks some plastics, rubber, or coatings.

1,2-EPOXY ETHANE (75-21-8) Forms explosive mixture with air (flash point 20°F/−6°C). Incompatible with alkali metal hydroxides,highly active catalysts (e.g., anhydrous chlorides of iron, tin, or aluminum and oxides of iron or aluminum). Avoid contact with copper. Protect container from physical damage, sun, and heat. Attacks some plastics, rubber, or coatings.

1,2-EPOXYETHANE (75-21-8) Forms explosive mixture with air (flash point 20°F/−6°C). Incompatible with alkali metal hydroxides, highly active catalysts (e.g., anhydrous chlorides of iron, tin, or aluminum and oxides of iron or aluminum). Avoid contact with copper. Protect container from physical damage, sun, and heat. Attacks some plastics, rubber, or coatings.

1,2-EPOXY-3-HYDROXY PROPANE (556-52-5) Forms explosive mixture with air (flash point 158°F/70°C). Reacts violently with strong oxidizers. Contact with caustics, acids, barium, lithium, sodium, magnesium, titanium can cause polymerization. Attacks some plastics, rubber, and coatings.

1,2-EPOXY-3-ISOPROPOXYPROPANE (4016-08-6) Forms explosive mixture with air (flash point 92°F/33°C). Incompatible with strong oxidizers (cause fire), strong caustics (cause polymerization). Air and light cause formation of unstable and explosive peroxides. Attacks some rubbers and plastics.

EPOXYPROPANE (75-56-9) Forms explosive gas mixture with air (flash point −35°F/−37°C). Incompatible with anhydrous metal chlorides. Strong acids, caustics, peroxides cause polymerization. Reacts with ammonia, amines, acetylene-forming metals. Attacks some plastics, rubber, and coatings.

1,2-EPOXYPROPANE (75-56-9) Forms explosive gas mixture with air (flash point −35°F/−37°C). Incompatible with anhydrous metal chlorides. Strong acids, caustics, peroxides cause polymerization. Reacts with ammonia, amines, acetylene-forming metals. Attacks some plastics, rubber, and coatings.

2,3-EPOXYPROPANE (75-56-9) Forms explosive gas mixture with air (flash point −35°F/−37°C). Incompatible with anhydrous metal chlorides. Strong acids, caustics, peroxides cause polymerization. Reacts with ammonia, amines, acetylene-forming metals. Attacks some plastics, rubber, and coatings.

1,2-EPOXY-1-PROPANOL (556-52-5) Forms explosive mixture with air (flash point 158°F/70°C). Reacts violently with strong oxidizers. Contact with caustics, acids, barium, lithium, sodium, magnesium, titanium can cause polymerization. Attacks some plastics, rubber, and coatings.

2,3-EPOXYPROPANOL (556-52-5) Forms explosive mixture with air (flash point 158°F/70°C). Reacts violently with strong oxidizers. Contact with caustics, acids, barium, lithium, sodium, magnesium, titanium can cause polymerization. Attacks some plastics, rubber, and coatings.

2,3-EPOXYPROPYL BUTYL ETHER (2426-08-6) Forms explosive mixture with air (flash point 130°F/54°C). Air and light cause formation of unstable and explosive peroxides. Contact with strong oxidizers may cause fire and explosions. Strong caustics may cause polymerization. Attacks some plastics and rubber.

2-EPOXYPROPYL ETHER (2238-07-5) Forms explosive mixture with air (flash point 147°F/64°C). Contact with strong oxidizers may cause fire and explosions. Ethers, as a class, tend to form peroxides upon contact with air and exposure to light. Attacks some forms of plastics, coatings, and rubber.

EPTACLORO (Italian) (76-44-8) Forms hydrogen chloride gas with iron and rust above 165°F/74°C.

1,4,5,6,7,8,8-EPTACLORO-3A,4,7,7A-TETRAIDRO-4,7-ENDO-METANO-INDENE (Italian) (76-44-8) Forms hydrogen chloride gas with iron and rust above 165°F/74°C.

EPTANI (Italian) (142-82-5) Forms explosive mixture with air (flash point 25°F/−4°C). Strong oxidizers may cause fire and explosions. Attacks some plastics, rubber, and coatings. May accumulate static electric charges that can ignite its vapors.

EQUIGEL (62-73-7) Attacks some plastics, rubber, and coatings.

ERADEX (2921-88-) Incompatible with strong acids. Hydrolyzes from acid or alkaline solutions.

ERGOPLAST AdDO (103-23-1) Incompatible with strong acids, nitrates, oxidizers.

ERL-27774 (1675-54-8) Forms explosive mixture with air (flash point 175°F/79°C). Incompatible with strong acids, strong oxidizers. Ethers form peroxides on contact with air and light.

ERYTHRENE (106-99-0) Self-reactive. Forms explosive peroxides with air. Fires, explosions, or hazardous polymerization may result from contact with air, strong oxidizers, strong acids, ozone, nitrogen dioxide, copper and its alloys, phenol, chlorine dioxide, crotonaldehyde, or a free radical polymerization initiator such as hydroquinone. Add inhibitor (such as *tert*-butylcatechol) and monitor to ensure that effective levels are maintained at all times. May accumulate static electrical charges, and may cause ignition of its vapors.

ESAIDRO-1,3,5-TRINITRO-1,3,5-TRIAZINA (Italian) (121-82-4) Contact with mercury fulminate may cause detonation. Contact with combustibles or strong oxidizers may cause fire and explosions.

ESANI (Italian) (110-54-3) Forms explosive mixture with air (flash point −7°F/−22°C). Strong oxidizers may cause fire and explosions. Attacks some plastics, rubber, and coatings. May accumulate static electrical charges, and may cause ignition of its vapors.

ESEN (85-44-9) Incompatible with strong acids, caustics, ammonia, amines, strong oxidizers. Attacks some plastics, rubber, and coatings.

ESKIMON 11 (75-69-4) Reacts with barium, lithium, sodium, magnesium, titanium. Attacks some plastics, rubber, and coatings.

ESKIMON 12 (75-71-8) Reacts violently with liquid aluminum. Incompatible with chemically active metals. Attacks some plastics, rubber, and coatings.

ESKIMON 22 (75-45-5) Moisture and rust cause slow decomposition, producing toxic gases. Attacks some plastics, rubber, and coatings. Thermal decomposition occurs at high temperature with alkalies and alkaline earth metals.

ESOPHOTRAST (7727-43-7) Explosions may result from contact with aluminum in the presence of heat. Incompatible with potassium, phosphorus.

ESSENCE (French) (8006-61-9) Forms explosive mixture with air (flash point 125°F/51°C). Incompatible with nitric acid; oxidizers may cause fire and explosions.

ESSENCE OF MIRBANE (98-95-3) Forms explosive mixture with air (flash point 190°F/88°C). Incompatible with nitric acid, nitrogen tetroxide, caustics, ammonia, amines. Attacks some plastics, rubber, and coatings.

ESSENCE OF MYRBANE (98-95-3) Forms explosive mixture with air (flash point 190°F/88°C). Incompatible with nitric acid, nitrogen tetroxide, caustics, ammonia, amines. Attacks some plastics, rubber, and coatings.

ESSENCE OF NIOBE (93-58-3) Forms explosive mixture with air (flash point 181°F/83°C). Incompatible with strong acids, nitrates, oxidizers.

ESSIGESTER (German) (141-78-6) Will hydrolyze on standing, producing acetic acid and ethyl alcohol. This reaction is greatly accelerated by bases (alkalies). Forms explosive mixture with air (flash point 24°F/−4.4°C). Incompatible with strong acids, nitrates, oxidizers, chlorosulfonic acid, lithium aluminum hydride, oleum.

ESSIGSAEURE (German) (64-19-7) Vapor forms explosive mixture with air (flash point 105°F/40°C). Reacts violently with oxidizers and bases. Incompatible with strong acids, aliphatic amines, alkanolamines, isocyanates, alkylene oxides, epichlorohydrin, acetaldehyde, 2-aminoethanol, ammonia, ammonium nitrate, chlorosulfonic acid, chromic acid, ethylene diamine, ethyleneimine, perchloric acid, permanganates, phosphorus isocyanate, phosphorus trichloride, potassium *tert*-butoxide, xylene. Attacks cast iron and other metals, producing flammable hydrogen gas.

ESSIGSAEUREANHYDRID (German) (108-24-7) Forms explosive mixture with air (flash point 121°F/50°C). Water contact produces acetic acid and considerable heat. Reacts violently with potassium hydroxide, sodium hydroxide, alcohols amines. Incompatible with acids, ammonia, aniline, chromic anhydride, ethylene diamine, ethylenediamine, glycerol, hydrogen peroxide, nitrogen tetroxide, oxidizers, permanganates, sodium peroxide. Attacks some coatings, some plastics, and rubber. Corrodes iron and steel. May accumulate static electrical charges, and may cause ignition of its vapors.

ESSO FUNGICIDE 406 (133-06-2) Incompatible with tetraethyl pyrophosphate, parathion.

ESTAR (8007-45-2) Forms explosive mixture with air (flash point 60°F/15°C). Incompatible with strong acids, nitrates.

ESTERCIDE T-2 (93-76-5) Sealed metal containers may burst at temperatures above 316°F/158°C. Incompatible with sulfuric acid, bases, ammonia, aliphatic amines, alkanolamines, isocyanates, alkylene oxides, epichlorohydrin.

ESTERCIDE T-245 (93-76-5) Sealed metal containers may burst at temperatures above 316°F/158°C. Incompatible with sulfuric acid, bases, ammonia, aliphatic amines, alkanolamines, isocyanates, alkylene oxides, epichlorohydrin.

ESTERON (93-76-5) Sealed metal containers may burst at temperatures above 316°F/158°C. Incompatible with sulfuric acid, bases, ammonia, aliphatic amines, alkanolamines, isocyanates, alkylene oxides, epichlorohydrin.

ESTERON (94-75-7) Decomposes in sunlight. Incompatible with strong oxidizers; may cause fire and explosions.

ESTERON 44 or ESTERON 44 WEED KILLER (94-11-1) Forms explosive mixture with air (flash point 175°F/79°C). Incompatible with strong oxidizers, strong acids, nitrates. Attacks some plastics, rubber, and coatings.

ESTERON 76 BE (94-75-7) Decomposes in sunlight. Incompatible with strong oxidizers; may cause fire and explosions.

ESTERON 99 or ESTERON 99 CONCENTRATE (94-75-7) Decomposes in sunlight. Incompatible with strong oxidizers; may cause fire and explosions.

ESTERON 245 (93-76-5) Sealed metal containers may burst at temperatures above 316°F/158°C. Incompatible with sulfuric acid, bases, ammonia, aliphatic amines, alkanolamines, isocyanates, alkylene oxides, epichlorohydrin.

ESTERON BRUSH KILLER (93-76-5) Sealed metal containers may burst at temperatures above 316°F/158°C. Incompatible with sulfuric acid, bases, ammonia, aliphatic amines, alkanolamines, isocyanates, alkylene oxides, epichlorohydrin.

ESTERONE FOUR (94-75-7) Decomposes in sunlight. Incompatible with strong oxidizers; may cause fire and explosions.

ESTOL-1550 (84-66-2) Reacts violently with strong acids, strong oxidizers, permanganates, water. Attacks some forms of plastic.

ESTONATE (50-29-3) Incompatible with salts of iron or aluminum, and bases. Do not store in iron containers.

ESTONE (94-75-7) Decomposes in sunlight. Incompatible with strong oxidizers; may cause fire and explosions.

ESTONOX (8001-35-2) Reacts with oxidizers, with a risk of fire or explosions. Attacks metals in the presence of moisture.

ESTROSEL (62-73-7) Attacks some plastics, rubber, and coatings.

ESTROSOL (62-73-7) Attacks some plastics, rubber, and coatings.

ETAHOLO (Italian) (64-17-5) Forms explosive mixture with air (flash point 65°F/18°C). May accumulate static electrical charges, and may cause ignition of its vapors. Reactions may be violent with oleum, sulfuric acid, nitric acid, bases, aliphatic amines, isocyanates, oxidizers.

ETAIN (TETRACHLORURE d') (French) (7646-78-8) Water contact produces hydrogen chloride. Contact with alcohols, amines, potassium, sodium, turpentine may cause fire and explosions. Attacks some plastics, rubber, and coatings.

ETERE ETILICO (Italian) (60-29-7) Forms explosive mixture with air (flash point −49°F/−45°C). Incompatible with strong acids, strong oxidizers. Can form peroxides with air or light; may explode when container is unstoppered or otherwise opened. Attacks some plastics, rubber, and coatings. Being a nonconductor, chemical may accumulate static electric charges that may result in ignition of its vapor.

ETHANAL (75-07-0) Slowly polymerizes to paraldehyde. Explodes when mixed with iodine. Contact with strong bases can cause explosive polymerization. A strong reducing agent; reacts violently with combustibles, strong acids, caustics, ammonia, aliphatic amines, alkanolamines, aromatic amines, organic substances, halogens, oxidizers. Forms explosive peroxides with air. May dissolve rubber. May accumulate static electrical charges, and may cause ignition of its vapors.

ETHANAL, TRICHLORO- (75-87-6) Contact with acids or exposure to light may cause polymerization. Reacts with water, producing forming corrosive and toxic chloral hydrate. Contact with oxidizers may cause fire and explosions.

ETHANAMINE (75-04-7) Forms explosive mixture with air (flash point less than 0°F/−18°C). Incompatible with acids, organic anhydrides, isocyanates, vinyl acetate, acrylates, substituted allyls, alkylene oxides, epichlorohydrin, ketones, aldehydes, alcohols, glycols, phenols, cresols, caprolactam solution, strong oxidizers. Attacks aluminum, copper, lead, tin, zinc, and alloys, and some plastics, rubber, and coatings.

ETHANAMINE, N,N-DIETHYL- (121-44-8) Forms explosive mixture with air (flash point 20°F/−6.7°C). Incompatible with strong acids, mineral acids, organic acids (acids may cause violent spattering), organic anhydrides, halogenated hydrocarbons, isocyanates, nitroparaffins, nitrogen tetroxide, vinyl acetate, acrylates, substituted allyls, alkylene oxides, epichlorohydrin, ketones, aldehydes, alcohols, glycols, phenols, cresols, caprolactam solution, strong oxidizers. Attacks aluminum, copper, lead, tin, zinc, and alloys, and some plastics, rubber, and coatings.

ETHANDIAL (107-22-2) Water contact causes polymerization. Incompatible with strong acids, caustics, ammonia, amines, chlorosulfonic acid, ethylene amine. Corrosive to metals.

ETHANE (74-84-0) Flammable gas; forms explosive mixture with air. Strong oxidizers may cause fire and explosions. May accumulate static electrical charges, and may cause ignition of its vapors.

ETHANE CARBOXYLIC ACID (79-09-4) Forms explosive mixture with air (flash point 126°F/52°C). Incompatible with sulfuric acid, strong bases, ammonia, aliphatic amines, alkanolamines, isocyanates, alkylene oxides, epichlorohydrin, oxidizers.

ETHANE, CHLOROPENTAFLUORO- (76-15-3) Thermal decomposition occurs at high temperatures with alkalies and alkaline earth metals.

ETHANEDIAL (107-22-2) Water contact causes polymerization. Incompatible with strong acids, caustics, ammonia, amines, chlorosulfonic acid, ethylene amine. Corrosive to metals.

1,2-ETHANEDIAMINE (107-15-3) Forms explosive mixture with air (flash point 104°F/40°C). Incompatible with chlorinated organic compounds, silver perchlorate, 3-propiolactone, mesityl oxide, ethylene dichloride, acids, organic anhydrides, isocyanates, vinyl acetate, acrylates, substituted allyls, alkylene oxides, epichlorohydrin, ketones, aldehydes, alcohols, glycols, phenols, cresols, caprolactam solution, strong oxidizers. Attacks aluminum, copper, lead, tin, zinc, and alloys, and some plastics, rubber, and coatings.

1,2-ETHANEDIAMINE, N-(2-AMINOETHYL)- (111-40-0) Ignites spontaneously with cellulose nitrate. Silver, cobalt, or chromium compounds may cause explosions. Forms explosive mixture with air (flash point 208°F/98°C). Incompatible with acids, organic anhydrides, isocyanates, vinyl acetate, acrylates, substituted allyls, alkylene oxides, epichlorohydrin, ketones, aldehydes, alcohols, glycols, mercury, phenols, cresols, caprolactam solution, strong oxidizers. Attacks aluminum, copper, lead, tin, zinc, and alloys.

1,2-ETHANEDIAMINE, N-(2-AMINOETHYL)-N'-(2-AMINOETHYL)AMINOETHYL- (112-57-2) Forms explosive mixture with air. Incompatible with acids, organic anhydrides, halogenated hydrocarbons, isocyanates, nitroparaffins, vinyl acetate, acrylates, substituted allyls, alkylene oxides, epichlorohydrin, ketones, aldehydes, alcohols, glycols, mercury, phenols, cresols, caprolactam solution, strong oxidizers. Attacks aluminum, copper, lead, tin, zinc, and alloys.

1,2-ETHANEDIAMINE, N-(2-AMINOETHYL)-N'-(2-(2-AMINOETHYL)ETHYL)- (112-57-2) Forms explosive mixture with air. Incompatible with acids, organic anhydrides, halogenated hydrocarbons, isocyanates, nitroparaffins, vinyl acetate, acrylates, substituted allyls, alkylene oxides, epichlorohydrin, ketones, aldehydes, alcohols, glycols, mercury, phenols, cresols, caprolactam solution, strong oxidizers. Attacks aluminum, copper, lead, tin, zinc, and alloys.

ETHANE, 1,2-DIBUTOXY (112-48-1) Forms explosive mixture with air above 185°F/85°C. Oxidizers may cause fire and explosions. Incompatible with sulfuric acid, isocyanates, perchloric acid.

ETHANE DICHLORIDE (107-06-2) May accumulate static electrical charges, and may cause ignition of its vapors. Contact with hot water may produce hydrochloric acid. Forms explosive mixture with air (flash point 55°F/13°C). Incompatible with strong oxidizers, strong caustics, chemically active metals. Corrosive to iron and other metals. Attacks some plastics, rubber, or coatings.

ETHANE, 1,1-DICHLORO- (75-34-3) Forms explosive mixture with air (flash point 2°F/−17°C). Incompatible with strong oxidizers, strong caustics. Attacks plastics and rubber.

ETHANE DINITRILE (460-19-5) Reacts explosively with acids, water, liquid oxygen, oxidizers.

ETHANEDIOIC ACID (144-62-7) Contact with silver compounds may produce explosive salt. Incompatible with caustics, furfuryl alcohol, oxidizers, mercury, silver, sodium chlorite, sodium hypochlorite.

1,2-ETHANEDIOL (107-21-1) Incompatible with strong acids, caustics, aliphatic amines, isocyanates, chlorosulfonic acid, oleum, strong oxidizers.

1,2-ETHANEDIOL DIACETATE (111-55-7) Forms explosive mixture with air above 191°F/88°C. Strong oxidizers may cause fire and explosions.

ETHANEDIOL DINITRATE (628-96-6) Acids, heat, or mechanical shock may result in explosions.

1,2-ETHANEDIOL DIPROPANOATE (123-73-9) A strong reducing agent. Forms explosive mixture with air (flash point 55°F/12.8°C). Readily converted by oxygen to peroxides and acids; heat or contact with many other substances may cause polymerization. Incompatible with strong oxidizers, strong acids including nonoxidizing mineral acids, ammonia, aliphatic amines, aromatic amines, 1,3-butadiene, strong bases. Liquid attacks some plastics, rubber, and coatings.

1,2-ETHANEDIOL, MONOACETATE (542-59-6) Forms explosive mixture with air above 191°F/88°C. Incompatible with sulfuric acid, nitric acid, nitrates.

1,2-ETHANEDIONE (107-22-2) Water contact causes polymerization. Incompatible with strong acids, caustics, ammonia, amines, chlorosulfonic acid, ethylene amine. Corrosive to metals.

2,2'-(1,2-ETHANEDIYLBIS(OXY)) BISEHANOL (112-27-6) Incompatible with sulfuric acid, isocyanates, perchloric acid, strong oxidizers.

ETHANE HEXACHLORIDE (67-72-1) Incompatible with hot iron, zinc, and aluminum; alkalies cause formation of spontaneously explosive chloroacetylene. Attacks some plastics, rubber, and coatings.

ETHANE ISOCYANATE (109-90-0) Forms explosive mixture with air (flash point 12°F/−6°C). Reacts violently with water and strong oxidizers. Incompatible with acids, bases, ammonia, amines, amides, alcohols, glycols, caprolactam solution. May accumulate static electrical charges, and may cause ignition of its vapors.

ETHANENITRILE (75-05-8) Forms explosive mixture with air (flash point 42°F/6°C). Incompatible with water (especially if acid or alkaline), acids, caustics, nitrating agents, indium, nitrogen tetroxide, n-fluoro compounds, sulfur trioxide, iron(III) salts of perchlorate, indium, nitrogen-fluorine compounds. Reacts violently with oxidizers. May accumulate static electrical charges, and may cause ignition of its vapors.

ETHANE, 1,1'-OXYBIS- (60-29-7) Forms explosive mixture with air (flash point −49°F/−45°C). Incompatible with strong acids, strong oxidizers. Can form peroxides with air or light; may explode when container is unstoppered or otherwise opened. Attacks some plastics, rubber, and coatings. Being a nonconductor, chemical may accumulate static electric charges that may result in ignition of vapor.

ETHANE, 1,1'-OXYBIS (2-CHLORO-) (111-44-4) Water contact may produce hydrogen chloride fumes. Can form peroxides. Forms explosive mixture with air (flash point 131°F/55°C). Strong oxidizers may cause fire and explosions. Attacks some plastics, rubber, and coatings.

ETHANE PENTACHLORIDE (76-01-7) Incompatible with water, producing dichloroacetic acid. May self-ignite. Reacts violently with alkali metals (i.e., lithium, sodium, potassium, rubidium, cesium, francium). Forms spontaneously explosive materials with alkalies, metals. Shock- and friction-sensitive material is formed by mixing with potassium.

ETHANE, PENTACHLORO- (76-01-7) Incompatible with water, producing dichloroacetic acid. May self-ignite. Reacts violently with alkali metals (i.e., lithium, sodium, potassium, rubidium, cesium, francium). Forms spontaneously explosive materials with alkalies, metals. Shock- and friction-sensitive material is formed by mixing with potassium.

ETHANEPEROXIC ACID (79-21-0) A powerful oxidizer and extremely sensitive explosive. Forms explosive mixture with air (flash point 100°F/38°C). Reacts violently with many substances, including acetic anhydride, combustibles, strong bases, ether solvents, olefins, organic matter, magnesium, metal oxides, heavy metals, metal chloride solutions, reducing agents, nickel, phosphorus, sodium nitride, zinc.

ETHANE, 1,1,2,2-TETRABROMO- (79-27-6) Reacts with chemically active metals, caustics. Hot iron, aluminum, or zinc in the presence of steam may produce toxic vapors. Softens or destroys most plastics and rubbers.

ETHANE, 1,1,2,2-TETRACHLORO- (79-34-5) Exposure to heat, light, and air produces corrosive and toxic vapors. Reacts with strong caustics to form explosive dichloroacetylene. Reacts violently with chemically active metals or sodium amide. In presence of steam, contact with hot iron, aluminum, or zinc may form toxic vapors. Attacks some plastics, rubber, and coatings.

ETHANE, 1,1,2,2-TETRACHLORO-1,2-DIFLUORO- (76-12-0) Reacts with chemically active metals, powdered aluminum, zinc, and magnesium. Attacks some forms of plastics, rubber, and coatings.

ETHANE, 1,1,1,2-TETRACHLORO-2,2-DIFLUORO- (76-11-9) Reacts with chemically active metals, oxidizers. Attacks some forms of plastics, rubber, and coatings.

ETHANETHIOL (75-08-1) Forms explosive mixture with air (flash point less than 0°F/−18°C). May accumulate static electrical charges, and may cause ignition of its vapors. Contact with strong oxidizers may cause fire and explosions. Attacks some forms of plastics, coatings, and rubber.

ETHANE TRICHLORIDE (79-00-5) Incompatible with strong oxidizers, strong caustics and chemically active metals (especially powders), or sodium amide; may cause fire and explosions. Attacks some plastics, rubber, and coatings.

ETHANE, 1,1,2-TRICHLORO- (79-00-5) Incompatible with strong oxidizers, strong caustics and chemically active metals (especially powders), or sodium amide; may cause fire and explosions. Attacks some plastics, rubber, and coatings.

ETHANOIC ACID (64-19-7) Vapor forms explosive mixture with air (flash point 105°F/40°C). Reacts violently with oxidizers and bases. Incompatible with strong acids, aliphatic amines, alkanolamines, isocyanates, alkylene oxides, epichlorohydrin, acetaldehyde, 2-aminoethanol, ammonia, ammonium nitrate, chlorosulfonic acid, chromic acid, ethylene diamine, ethyleneimine, perchloric acid, permanganates, phosphorus isocyanate, phosphorus trichloride, potassium *tert*-butoxide, xylene. Attacks cast iron and other metals, producing flammable hydrogen gas.

ETHANOIC ANHYDRATE (108-24-7) Forms explosive mixture with air (flash point 121°F/50°C). Water contact produces acetic acid and considerable heat. Reacts violently with potassium hydroxide, sodium hydroxide, alcohols, amines. Incompatible with acids, ammonia, aniline, chromic anhydride, ethylene diamine, ethylenediamine, glycerol, hydrogen peroxide, nitrogen tetroxide, oxidizers, permanganates, sodium peroxide. Attacks some coatings, some plastics, and rubber. Corrodes iron and steel. May accumulate static electrical charges, and may cause ignition of its vapors.

ETHANOL (64-17-5) Forms explosive mixture with air (flash point 65°F/18°C). May accumulate static electrical charges, and may cause ignition of its vapors. Reactions may be violent with oleum, sulfuric acid, nitric acid, bases, aliphatic amines, isocyanates, oxidizers.

ETHANOLAMINE (141-43-5) Will oxidize in air; often shipped under a pad of inert gas. Forms explosive mixture with air (flash point 185°F/85°C). Strong oxidizers may cause fire and explosions. A strong base. Incompatible with acids, organic anhydrides, isocyanates, alkylene oxides, epichlorohydrin, aldehydes, alcohols, glycols, phenols, cresols, caprolactam solution. Contact with strong acids may cause spattering. Attacks plastics, rubber, and aluminum, copper, tin, and their alloys.

ETHANOL, 1,2-DIBROMO-2,2-DICHLORO-, DIMETHYL PHOSPHATE (300-76-5) Contact with strong oxidizers may cause fire and explosions. Attacks some plastics, rubber, and coatings.

ETHANOL, 2-(DIBUTYLAMINO)- (102-81-8) Forms explosive mixture with air (flash point 200°F/93°C). Reacts violently with strong oxidizers.

ETHANOL, 2-(DIETHYLAMINO)- (100-37-8) Forms explosive mixture with air (flash point 126°F/52°C). Reacts violently with oxidizers, strong acids. Attacks light metals. Attacks some plastics and rubber.

ETHANOL, 2-ETHOXY- (110-80-5) Forms explosive mixture with air (flash point 120°F/49°C). Strong oxidizers may cause fire and explosions. Attacks some plastics, rubber, and coatings. Able to form peroxides. Incompatible with strong acids, aluminum and its alloys.

ETHANOL, 2,2′-(ETHYLENEDIOXY)DI- (112-27-6) Incompatible with sulfuric acid, isocyanates, perchloric acid, strong oxidizers.

ETHANOL, 2,2′-IMINOBIS- (111-42-2) Incompatible with acids, organic anhydrides, isocyanates, vinyl acetate, acrylates, substituted allyls, alkylene oxides, epichlorohydrin, aldehydes, oxidizers. Corrosive to copper, copper alloys, zinc, and galvanized iron.

ETHANOL, 2-ISOPROPOXY (109-59-1) Forms explosive mixture with air. 92°F/33°C). Incompatible with sulfuric acid, perchloric acid, isocyanates, oxidizers.

ETHANOL,1-PHENYL- (98-85-1) Forms explosive mixture with air above 205°F/96°C. Incompatible with strong acids, caustics, aliphatic amines, isocyanates, oxidizers.

ETHANOL 200 PROOF (64-17-5) Forms explosive mixture with air (flash point 65°F/18°C). May accumulate static electrical charges, and may cause ignition of its vapors. Reactions may be violent with oleum, sulfuric acid, nitric acid, bases, aliphatic amines, isocyanates, oxidizers.

ETHANOYL CHLORIDE (75-36-5) Forms explosive mixture with air (flash point 40°F/4°C). May accumulate static electrical charges, and may cause ignition of its vapors. Reacts violently with water, oxidizers, strong bases, alcohols (especially ethanol), dimethyl sulfoxide, phosphorus trichloride; produces corrosive hydrochloric acid with air.

ETHENE (74-85-1) A flammable gas. May accumulate static electrical charges, and may cause ignition of its vapors. Oxidizers may cause explosive polymerization and fire. Incompatible with acids, halogens, nitrogen oxides, hydrogen bromide, aluminum chloride, bromotrichloromethane, carbon tetrachloride, chlorine, chlorine dioxide, nitrogen dioxide.

1,2-ETHENEDICARBOXYLIC ACID, *trans-* (110-17-8) Incompatible with sulfuric acid, caustics, ammonia, amines, isocyanates, alkylene oxides, epichlorohydrin.

ETHENE, 1,1-DICHLORO- (75-35-4) Forms explosive mixture with air (18°F/−28°C). Air or contaminants can cause formation of peroxides; may polymerize. Reacts violently with strong oxidizers, alkali metals. Incompatible with nitric acid.

ETHENE, 1,2-DICHLORO- (540-59-0) Forms explosive mixture with air (flash point 36°F/2.2°C). Incompatible with strong bases, oxidizers, difluoromethylene, dihypofluoride, nitrogen tetroxide (explosive). Attacks some plastics, rubber, and coatings.

ETHENOID (463-51-4) Reacts vigorously with water and a wide variety of organic compounds. Forms explosive compound when mixed with hydrogen peroxide. Can dimerize to diketene even at low temperatures. Diketene forms an explosive mixture with air (flash point 90°F/32°C.) Reacts violently with oxidizers; forms unstable and explosive peroxides, with a chance of violent polymerization.

ETHENONE (463-51-4) Reacts vigorously with water and a wide variety of organic compounds. Forms explosive compound when mixed with hydrogen peroxide. Can dimerize to diketene even at low temperatures. Diketene forms an explosive mixture with air (flash point 90°F/32°C.) Reacts violently with oxidizers; forms unstable and explosive peroxides, with a chance of violent polymerization.

ETHENYL ACETATE (108-05-4) Forms explosive mixture with air (flash point 18°F/−8°C). Polymerizes readily if not inhibited; heat can initiate reaction. Reacts violently with oxidizers. Incompatible with nonoxidizing mineral acids, strong acids, ammonia, aliphatic amines, alkanolamines. Also reacts with 2-aminoethanol, chlorosulfonic acid, ethylene diamine, ethyeneimine, ethyleneimine, ozone, oleum, peroxides. May accumulate static electrical charges, and may cause ignition of its vapors.

ETHENYL ETHANOATE (108-05-4) Forms explosive mixture with air (flash point 18°F/−8°C). Polymerizes readily if not inhibited; heat can initiate reaction. Reacts violently with oxidizers. Incompatible with non-oxidizing mineral acids, strong acids, ammonia, aliphatic amines, alkanolamines. Also reacts with 2-aminoethanol, chlorosulfonic acid, ethylene diamine, ethyeneimine, ethyleneimine, ozone, oleum, peroxides. May accumulate static electrical charges, and may cause ignition of its vapors.

ETHENYLETHANOATE (108-05-4) Forms explosive mixture with air (flash point 18°F/−8°C). Polymerizes readily if not inhibited; heat can initiate reaction. Reacts violently with oxidizers. Incompatible with nonoxidizing mineral acids, strong acids, ammonia, aliphatic amines,

340

alkanolamines. Also reacts with, 2-aminoethanol, chlorosulfonic acid, ethylene diamine, ethyeneimine, ethyleneimine, ozone, oleum, peroxides. May accumulate static electrical charges, and may cause ignition of its vapors.

ETHER (60-29-7) Forms explosive mixture with air (flash point −49°F/−45°C). Incompatible with strong acids, strong oxidizers. Can form peroxides with air or light; may explode when container is unstoppered or otherwise opened. Attacks some plastics, rubber, and coatings. Being a nonconductor, chemical may accumulate static electric charges that may result in ignition of vapor.

ETHER, BIS(2-CHLORO-1-METHYLETHYL) (108-60-1) Forms explosive mixture with air above 170°F/77°C. Reacts violently with strong oxidizers. Incompatible with aluminum, copper, epoxy coatings.

ETHER, BIS(2,3-EPOXYPROPYL)- (2238-07-5) Forms explosive mixture with air (flash point 147°F/64°C). Contact with strong oxidizers may cause fire and explosions. Ethers, as a class, tend to form peroxides upon contact with air and exposure to light. Attacks some forms of plastics, coatings, and rubber.

ETHER BUTYLIQUE (French) (142-96-1) Forms explosive mixture with air. (flash point 77°F/25°C). May accumulate static electrical charges, and may cause ignition of its vapors. Incompatible with strong acids, oxidizers. Air contact or light may cause formation of unstable and explosive peroxides.

ETHER CHLORATUS (75-00-3) Flammable gas. Reacts slowly with water; produces hydrogen chloride gas. Contact with moisture produces hydrochloric acid. May accumulate static electrical charges, and may cause ignition of its vapors. Forms explosive mixture with air (flash point −58°F/−55°C). Contact with aluminum, lithium, magnesium, sodium, potassium, zinc may cause fire and explosions. Attacks some plastics and rubber.

ETHER CYANATUS (107-12-0) Forms explosive mixture with air (flash point 36°F/2°C). Reacts violently with oxidizers. Water or acid contact produces hydrogen cyanide fumes.

ETHER CYANIDE (107-12-0) Forms explosive mixture with air (flash point 36°F/2°C). Reacts violently with oxidizers. Water or acid contact produces hydrogen cyanide fumes.

ETHER, DIGLYCIDYL (2238-07-5) Forms explosive mixture with air (flash point 147°F/64°C). Contact with strong oxidizers may cause fire and explosions. Ethers, as a class, tend to form peroxides upon contact with air and exposure to light. Attacks some forms of plastics, coatings, and rubber.

ETHER, DIMETHYL CHLORO (107-30-2) Forms explosive mixture with air (flash point 0°F/−17.8°C). May be able to form unstable and explosive peroxides. Produces hydrochloric acid on contact with moisture. Corrosive to metals in presence of moisture.

ETHER, ETHYL (60-29-7) Forms explosive mixture with air (flash point −49°F/−45°C). Incompatible with strong acids, strong oxidizers. Can form peroxides with air or light; may explode when container is unstoppered or otherwise opened. Attacks some plastics, rubber, and coatings. Being a nonconductor, chemical may accumulate static electric charges that may result in ignition of vapor.

341

ETHER ETHYLENE GLYCOL DIBUTYL (112-48-1) Forms explosive mixture with air above 185°F/85°C. Oxidizers may cause fire and explosions. Incompatible with sulfuric acid, isocyanates, perchloric acid.

ETHER ETHYLIQUE (French) (60-29-7) Forms explosive mixture with air (flash point −49°F/−45°C). Incompatible with strong acids, strong oxidizers. Can form peroxides with air or light; may explode when container is unstoppered or otherwise opened. Attacks some plastics, rubber, and coatings. Being a nonconductor, chemical may accumulate static electric charges that may result in ignition of vapor.

ETHER HYDROCHLORIC (75-00-3) Flammable gas. Reacts slowly with water; produces hydrogen chloride gas. Contact with moisture produces hydrochloric acid. May accumulate static electrical charges, and may cause ignition of its vapors. Forms explosive mixture with air (flash point −58°F/−55°C). Contact with aluminum, lithium, magnesium, sodium, potassium, zinc may cause fire and explosions. Attacks some plastics and rubber.

ETHER METHYLIQUE MONOCHLORE (French) (107-30-2) Forms explosive mixture with air (flash point 0°F/−17.8°C). May be able to form unstable and explosive peroxides. Produces hydrochloric acid on contact with moisture. Corrosive to metals in presence of moisture.

ETHER MONOETHYLIQUE de L'ETHYLENE-GLYCOL (French) (110-80-5) Forms explosive mixture with air (flash point 120°F/49°C). Strong oxidizers may cause fire and explosions. Attacks some plastics, rubber, and coatings. Able to form peroxides. Incompatible with strong acids, aluminum and its alloys.

ETHER MONOMETHYLIQUE de L'ETHYLENE-GLYCIL (French) (109-86-4) Forms explosive mixture with air (flash point 103°F/39°C). Heat or oxidizers may cause formation of unstable peroxides. Attacks many metals. Strong oxidizers cause fire and explosions. Strong bases cause decomposition. Attacks some plastics, rubber, and coatings. May accumulate static electrical charges, and may cause ignition of its vapors.

ETHER MURIATIC (75-00-3) Flammable gas. Reacts slowly with water; produces hydrogen chloride gas. Contact with moisture produces hydrochloric acid. May accumulate static electrical charges, and may cause ignition of its vapors. Forms explosive mixture with air (flash point −58°F/−55°C). Contact with aluminum, lithium, magnesium, sodium, potassium, zinc may cause fire and explosions. Attacks some plastics and rubber.

ETHER, VINYL ETHYL (109-92-2) Incompatible with acids (nonoxidizing, sulfuric, nitric), ammonia, aliphatic amines, alkanolamines.

ETHIDE (594-72-9) Contact with strong oxidizers may cause fire and explosions. Attacks some plastics, rubber, and coatings.

ETHINE (74-86-2) A strong reducing agent that reacts violently with oxidizers. Forms explosive mixture with air (flash point 0°F/−18°C). Forms shock-sensitive mixture with copper and copper salts, mercury and mercury salts, silver and silver salts. Reacts with brass, bromine, cesium hydride, chlorine, cobalt, cuprous acetylise, fluorine, iodine, mercuric nitrate, nitric acid, potassium, rubidium hydride, trifluoromethyl hypofluorite, sodium hydride.

ETHINYL TRICHLORIDE (79-01-6) Contact with caustics produces a toxic and flammable gas. Reacts violently with chemically active metals. Contact with aluminum may produce a violent, self-accelerat-

342

ing polymerization reaction. Incompatible with acids, organic anhydrides, isocyanates, alkylene oxides, aldehydes, alcohols, glycols, phenols, cresols, caprolactam solution, epichlorohydrin, nitrogen tetroxide, metal powders, oxygen. May accumulate static electrical charges, and may cause ignition of its vapors.

ETHIOL (563-12-2) Incompatible with alkaline formulations. Mixtures with magnesium may be explosive.

ETHIOLACAR (121-75-5) Incompatible with strong oxidizers, magnesium, alkaline pesticides. Attacks metals, some plastics, rubber, and coatings.

ETHION (563-12-2) Incompatible with alkaline formulations. Mixtures with magnesium may be explosive.

ETHIOPS MINERAL (1344-48-5) Contact with acids or water causes evolution of flammable hydrogen sulfide, which forms an explosive mixture with air. Contact with strong oxidizers may cause a violent reaction.

ETHLON (56-38-2) Combustible liquid. Mixtures with endrin may be explosive. Strong oxidizers may cause fire and explosions. Attacks some plastics, rubber, and coatings.

ETHODAN (563-12-2) Incompatible with alkaline formulations. Mixtures with magnesium may be explosive.

2-ETHOXYBUTANE (628-81-9) Forms explosive mixture with air (flash point 40°F/4°C). Heat or air contact may produce unstable peroxides. Reacts violently with strong oxidizers. Attacks some plastics, rubber, and coatings. May accumulate static electrical charges, and may cause ignition of its vapors.

ETHOXYCARBONYLETHYLENE (140-88-5) Forms explosive mixture with air (flash point 48°F/9°C). Atmospheric moisture and strong alkalies may cause fire and explosions. Unless properly inhibited (note: inert gas blanket not recommended), heat, light, or peroxides can cause polymerization. Incompatible with oxidizers (reaction may be violent), strong acids, amines. May accumulate static electrical charges, and may cause ignition of its vapors.

ETHOXY DIGLYCOL (111-90-0) Forms explosive mixture with air (flash point 201°F/94°C). Heat or atmospheric air contact may produce unstable peroxides. Incompatible with strong acids, isocyanates, strong oxidizers.

ETHOXY DIHYDROPYRAN (103-75-3) Forms explosive mixture with air above 108°F/42°C. Oxidizers may cause fire and explosions.

2-ETHOXY DIHYDROPYRAN (103-75-3) Forms explosive mixture with air above 108°F/42°C. Oxidizers may cause fire and explosions.

2-ETHOXY-2,3-DIHYDRO-*gamma*-PYRAN (103-75-3) Forms explosive mixture with air above 108°F/42°C. Oxidizers may cause fire and explosions.

2-ETHOXY-3,4-DIHYDRO-2-PYRAN (103-75-3) Forms explosive mixture with air above 108°F/42°C. Oxidizers may cause fire and explosions.

2-ETHOXY-3,4-DIHYDRO-1,2-PYRAN (103-75-3) Forms explosive mixture with air above 108°F/42°C. Oxidizers may cause fire and explosions.

2-ETHOXY-3,4-DIHYDRO-2H-PYRAN (103-75-3) Forms explosive mixture with air above 108°F/42°C. Oxidizers may cause fire and explosions.

343

ETHOXYETHANE (60-29-7) Forms explosive mixture with air (flash point −49°F/−45°C). Incompatible with strong acids, strong oxidizers. Can form peroxides with air or light; may explode when container is unstoppered or otherwise opened. Attacks some plastics, rubber, and coatings. Being a nonconductor, chemical may accumulate static electric charges that may result in ignition of vapor.

2-ETHOXYETHANOL (110-80-5) Forms explosive mixture with air (flash point 120°F/49°C). Strong oxidizers may cause fire and explosions. Attacks some plastics, rubber, and coatings. Able to form peroxides. Incompatible with strong acids, aluminum and its alloys.

2-(2-ETHOXYETHOXY) ETHANOL (111-90-0) Forms explosive mixture with air (flash point 201°F/94°C). Heat or atmospheric air contact may produce unstable peroxides. Incompatible with strong acids, isocyanates, strong oxidizers.

2-(2-ETHOXYETHOXY)ETHANOL (111-46-6) Combustible (flash point 255°F/124°C). Incompatible with sulfuric acid, isocyanates, strong oxidizers.

2-(2-ETHOXYETHOXY)ETHANOL ACETATE (112-15-2) Incompatible with sulfuric acid, nitric acid, nitrates, strong oxidizers.

ETHOXYETHYL ACETATE (111-15-9) Forms explosive mixture with air (flash point 117°F/47°C). Incompatible with strong acids, nitrates. Reacts violently with oxidizers. May form unstable peroxides. Softens many plastics. Attacks some plastics, rubber, and coatings.

2-ETHOXYETHYL ACETATE (111-15-9) Forms explosive mixture with air (flash point 117°F/47°C). Incompatible with strong acids, nitrates. Reacts violently with oxidizers. May form unstable peroxides. Softens many plastics. Attacks some plastics, rubber, and coatings.

ETHOXYFORMIC ANHYDRIDE (105-58-8) Forms explosive mixture with air (flash point 77°F/25°C). Reacts violently with strong oxidizers.

ETHOXYLATED DODECANOL (9008-57-5) Incompatible with strong acids, caustics, aliphatic amines, isocyanates.

ETHOXYLATED DODECYL ALCOHOL (9008-57-5) Incompatible with strong acids, caustics, aliphatic amines, isocyanates.

ETHOXYLATED LAURYL ALCOHOL (9008-57-5) Incompatible with strong acids, caustics, aliphatic amines, isocyanates.

ETHOXYLATED MYRISTYL ALCOHOL (N/A) Incompatible with strong acids, caustics, aliphatic amines, isocyanates.

ETHOXYLATED PENTADECANOL (N/A) Incompatible with sulfuric acid, nitric acid, caustics, aliphatic amines, isocyanates.

ETHOXYLATED PENTADE-CYLALCOHOL (N/A) Incompatible with sulfuric acid, nitric acid, caustics, aliphatic amines, isocyanates.

ETHOXYLATED TETRADECANOL (N/A) Incompatible with strong acids, caustics, aliphatic amines, isocyanates.

ETHOXYLATED TETRADECYL ALCOHOL (N/A) Incompatible with strong acids, caustics, aliphatic amines, isocyanates.

ETHOXYLATED TRIDECANOL (N/A) Incompatible with strong acids, caustics, aliphatic amines, isocyanates.

ETHOXYLATED TRIDECYL ALCOHOL (N/A) Incompatible with strong acids, caustics, aliphatic amines, isocyanates.

1-ETHOXY-2-PROPANOL (1569-02-4) Forms explosive mixture with air above 109°F/43°C. Incompatible with sulfuric acid, isocyanates, perchlorates.

ETHOXY PROPIONIC ACID, ETHYL ESTER (763-69-9) Forms explosive mixture with air above 138°F/59°C. Incompatible with strong acids, nitrates, oxidizers.

ETHOXYTRIETHYLENE GLYCOL (112-50-5) Incompatible with sulfuric acid, perchloric acid, isocyanates, strong oxidizers.

ETHOXYTRIGLYCOL (112-50-5) Incompatible with sulfuric acid, isocyanates, perchloric acid, oxidizers.

ETHYLACETAAT (Dutch) (141-78-6) Will hydrolyze on standing, forming acetic acid and ethyl alcohol. This reaction is greatly accelerated by bases (alkalies). Forms explosive mixture with air (flash point 24°F/−4.4°C). Incompatible with strong acids, nitrates, oxidizers, chlorosulfonic acid, lithium aluminum hydride, oleum.

ETHYL ACETATE (141-78-6) Will hydrolyze on standing, forming acetic acid and ethyl alcohol. This reaction is greatly accelerated by bases (alkalies). Forms explosive mixture with air (flash point 24°F/−4.4°C). Incompatible with strong acids, nitrates, oxidizers, chlorosulfonic acid, lithium aluminum hydride, oleum.

ETHYLACETIC ACID (107-92-6) Forms explosive mixture with air (flash point 162°F/72°C). Incompatible with sulfuric acid, caustics, ammonia, aliphatic amines, isocyanates, strong oxidizers, alkylene oxides, epichlorohydrin.

ETHYL ACETIC ESTER (141-78-6) Will hydrolyze on standing, forming acetic acid and ethyl alcohol. This reaction is greatly accelerated by bases (alkalies). Forms explosive mixture with air (flash point 24°F/−4.4°C). Incompatible with strong acids, nitrates, oxidizers, chlorosulfonic acid, lithium aluminum hydride, oleum.

ETHYL ACETOACETATE (141-97-9) Forms explosive mixture with air (flash point 135°F/57°C). Incompatible with strong acids, nitrates, oxidizers.

ETHYL ACETONE (107-87-9) Forms explosive mixture with air (flash point 45°F/7°C). Reacts violently with strong oxidizers. Attacks some plastics, rubber, and coatings.

ETHYL ACETYL ACETATE (141-97-9) Forms explosive mixture with air (flash point 135°F/57°C). Incompatible with strong acids, nitrates, oxidizers.

ETHYLACRYLAAT (Dutch) (140-88-5) Forms explosive mixture with air (flash point 48°F/9°C). Atmospheric moisture and strong alkalies may cause fire and explosions. Unless properly inhibited (note: inert gas blanket not recommended), heat, light, or peroxides can cause polymerization. Incompatible with oxidizers (reaction may be violent), strong acids, amines. May accumulate static electrical charges, and may cause ignition of its vapors.

ETHYL ACRYLATE or ETHYL ACRYLATE, INHIBITED (140-88-5) Forms explosive mixture with air (flash point 48°F/9°C). Atmospheric moisture and strong alkalies may cause fire and explosions. Unless properly inhibited (note: inert gas blanket not recommended), heat, light, or peroxides can cause polymerization. Incompatible with oxidizers (reaction may be violent), strong acids, amines. May accumulate static electrical charges, and may cause ignition of its vapors.

ETHYLAKRYLAT (Czech) (140-88-5) Forms explosive mixture with air (flash point 48°F/9°C). Atmospheric moisture and strong alkalies may cause fire and explosions. Unless properly inhibited (note: inert gas blanket not recommended), heat, light, or peroxides can cause polymerization. Incompatible with oxidizers (reaction may be violent), strong acids, amines. May accumulate static electrical charges, and may cause ignition of its vapors.

ETHYLALCOHOL (Dutch) (64-17-5) Forms explosive mixture with air (flash point 65°F/18°C). May accumulate static electrical charges, and may cause ignition of its vapors. Reactions may be violent with oleum, sulfuric acid, nitric acid, bases, aliphatic amines, isocyanates, oxidizers.

ETHYL ALCOHOL (64-17-5) Forms explosive mixture with air (flash point 65°F/18°C). May accumulate static electrical charges, and may cause ignition of its vapors. Reactions may be violent with oleum, sulfuric acid, nitric acid, bases, aliphatic amines, isocyanates, oxidizers.

ETHYL ALCOHOL ANHYDRO-S (64-17-5) Forms explosive mixture with air (flash point 65°F/18°C). May accumulate static electrical charges, and may cause ignition of its vapors. Reactions may be violent with oleum, sulfuric acid, nitric acid, bases, aliphatic amines, isocyanates, oxidizers.

ETHYL ALDEHYDE (75-07-0) Slowly polymerizes to paraldehyde. Explodes when mixed with iodine. Contact with strong bases can cause explosive polymerization. A strong reducing agent; reacts violently with combustibles, strong acids, caustics, ammonia, aliphatic amines, alkanolamines, aromatic amines, organic substances, halogens, oxidizers. Forms explosive peroxides with air. May dissolve rubber. May accumulate static electrical charges, and may cause ignition of its vapors.

ETHYL ALUMINUM DICHLORIDE (563-43-9) Forms explosive mixture with air. May ignite spontaneously in air. Reacts violently with water, producing hydrogen chloride and ethane gas. A strong reducing agent; reacts violently with oxidizers. Reacts violently with alcohols, amines, carbon dioxide, phenol, oxides of sulfur or nitrogen. May accumulate static electrical charges, and may cause ignition of its vapors.

ETHYLALUMINUM SESQUICHLORIDE (12075-68-2) Ignites spontaneously with air. A strong reducing agent; reacts violently with oxidizers, alcohols, amines, carbon dioxide, carbon tetrachloride, cresols, oxides of nitrogen, oxides of sulfur, phenols, and other substances. May accumulate static electrical charges, and may cause ignition of its vapors. Attacks metals in a moist environment.

ETHYLAMINE (75-04-7) Forms explosive mixture with air (flash point less than 0°F/−18°C). Incompatible with acids, organic anhydrides, isocyanates, vinyl acetate, acrylates, substituted allyls, alkylene oxides, epichlorohydrin, ketones, aldehydes, alcohols, glycols, phenols, cresols, caprolactam solution, strong oxidizers. Attacks aluminum, copper, lead, tin, zinc, and alloys, and some plastics, rubber, and coatings.

ETHYL AMYL KETONE (541-85-5) Forms explosive mixture with air (flash point 110°F/43°C). Incompatible with sulfuric acid, nitric acid, aliphatic amines, oxidizers.

ETHYLANILINE or n-ETHYLANILINE (103-69-5) Forms explosive mixture with air (flash point 185°F/85°C). Reacts violently with strong oxidizers and strong acids, with risk of fire and explosions.

ETHYLBENZOATE (93-89-0) Forms explosive mixture with air (flash point 190°F/88°C). Incompatible with strong oxidizers. May accumulate static electrical charges, and may cause ignition of its vapors.

ETHYLBENZEEN (Dutch) (100-41-4) May accumulate static electrical charges, and may cause ignition of its vapors. Forms explosive mixture with air (flash point 59°F/15°C). Incompatible with strong oxidizers, nitric acid.

ETHYLBENZENE (100-41-4) May accumulate static electrical charges, and may cause ignition of its vapors. Forms explosive mixture with air (flash point 59°F/15°C). Incompatible with strong oxidizers, nitric acid.

n-ETHYLBENZENEAMINE (103-69-5) Forms explosive mixture with air (flash point 185°F/85°C). Reacts violently with strong oxidizers and strong acids, with risk of fire and explosions.

ETHYLBENZOL (100-41-4) May accumulate static electrical charges, and may cause ignition of its vapors. Forms explosive mixture with air (flash point 59°F/15°C). Incompatible with strong oxidizers, nitric acid.

ETHYL BROMIDE (74-96-4) Forms explosive mixture with air (flash point less than −4°F/−20°C)). Hydrolyzes in water, producing hydrogen bromide. Oxidizers may cause fire or explosions. Fire and explosions may be caused by contact with aluminum, magnesium, or zinc powders, or lithium, potassium, or sodium. Attacks some plastic, rubber, and coatings.

ETHYL BUTANOATE (105-54-4) Incompatible with strong acids, nitrates, oxidizers.

ETHYLBUTANOL (97-95-0) Forms explosive mixture with air (flash point 70°F/21°C). Incompatible with strong acids, caustics, isocyanates, amines, isocyanates.

2-ETHYL BUTANOL (97-95-0) Forms explosive mixture with air. (flash point 70°F/21°C). Incompatible with strong acids, caustics, isocyanates, amines, isocyanates.

2-ETHYL BUTANOL-1 (97-95-0) Forms explosive mixture with air. (flash point 70°F/21°C). Incompatible with strong acids, caustics, isocyanates, amines, isocyanates.

2-ETHYL-1-BUTANOL (97-95-0) Forms explosive mixture with air. (flash point 70°F/21°C). Incompatible with strong acids, caustics, isocyanates, amines, isocyanates.

2-ETHYLBUTYL ALCOHOL (97-95-0) Forms explosive mixture with air. (flash point 70°F/21°C). Incompatible with strong acids, caustics, isocyanates, amines, isocyanates.

ETHYLBUTYLAMINE (617-79-8) Incompatible with strong acids, organic anhydrides, isocyanates, aldehydes, oxidizers.

N-ETHYL-n-BUTYLAMINE (617-79-8) Incompatible with strong acids, organic anhydrides, isocyanates, aldehydes, oxidizers.

ETHYLBUTYLETHER or ETHYL-N-BUTYL ETHER (628-81-9) Forms explosive mixture with air (flash point 40°F/4°C). Heat or air contact may produce unstable peroxides. Reacts violently with strong oxidizers. Attacks some plastics, rubber, and coatings. May accumulate static electrical charges, and may cause ignition of its vapors.

ETHYL BUTYL KETONE (106-35-4) Forms explosive mixture with air (flash point 115°F/46°C) Reacts violently with strong oxidizers. Attacks some plastics, rubber, and coatings.

ETHYL BUTYLACETALDEHYDE (123-05-7) Forms explosive mixture with air (flash point 112°F/44°C). Reacts violently with oxidizers. May ignite spontaneously when spilled on clothing or other absorbent materials. Under certain conditions, ignites spontaneously with air. Incompatible with strong acids, caustics, ammonia, amines. Forms unstable peroxides on contact with air.

ETHYL BUTYRATE (105-54-4) Incompatible with strong acids, nitrates, oxidizers.

ETHYL-N-BUTYRATE (105-54-4) Incompatible with strong acids, nitrates, oxidizers.

2-ETHYLCAPROALDEHYDE (123-05-7) Forms explosive mixture with air (flash point 112°F/44°C). Reacts violently with oxidizers. May ignite spontaneously when spilled on clothing or other absorbent materials. Under certain conditions, ignites spontaneously with air. Incompatible with strong acids, caustics, ammonia, amines. Forms unstable peroxides on contact with air.

alpha-ETHYLCAPROALDEHYDE (123-05-7) Forms explosive mixture with air (flash point 112°F/44°C). Reacts violently with oxidizers. May ignite spontaneously when spilled on clothing or other absorbent materials. Under certain conditions, ignites spontaneously with air. Incompatible with strong acids, caustics, ammonia, amines. Forms unstable peroxides on contact with air.

2-ETHYLCAPROIC ACID (149-57-5) May accumulate static electrical charges, and may cause ignition of its vapors. Reacts with oxidizers, with a risk of fire or explosions. Attacks common metals, especially in a moist environment.

alpha-ETHYLCAPROIC ACID (149-57-5) May accumulate static electrical charges, and may cause ignition of its vapors. Reacts with oxidizers, with a risk of fire or explosions. Attacks common metals, especially in a moist environment.

ETHYL CARBONATE (105-58-8) Forms explosive mixture with air (flash point 77°F/25°C). Reacts violently with strong oxidizers.

ETHYL CARBINOL (71-23-8) Forms explosive mixture with air (flash point 74°F/23°C). Attacks some plastics, rubber, and coatings. Incompatible with strong acids, caustics, aliphatic amines, isocyanates.

ETHYL CARBITOL (111-90-0) Forms explosive mixture with air (flash point 201°F/94°C). Heat or contact with atmospheric air may produce unstable peroxides. Incompatible with strong acids, isocyanates, strong oxidizers.

ETHYL CELLOSOLVE (110-80-5) Forms explosive mixture with air (flash point 120°F/49°C). Strong oxidizers may cause fire and explosions. Attacks some plastics, rubber, and coatings. Able to form peroxides. Incompatible with strong acids, aluminum and its alloys.

ETHYL CHLORIDE (75-00-3) Flammable gas. Reacts slowly with water; produces hydrogen chloride gas. Contact with moisture produces hydrochloric acid. May accumulate static electrical charges, and may cause ignition of its vapors. Forms explosive mixture with air (flash point −58°F/−55°C). Contact with aluminum, lithium, magnesium, sodium, potassium, zinc may cause fire and explosions. Attacks some plastics and rubber.

ETHYL CHLOROACETATE (105-39-5) Forms explosive mixture with air above 100°F/38°C. Water contact produces toxic and corrosive fumes. Reacts violently with alkaline earth metals (barium, calcium, magnesium, strontium, etc.), alkaline metals, sodium cyanide. Attacks metals in the presence of moisture.

ETHYL-alpha-CHLOROACETATE (105-39-5) Forms explosive mixture with air above 100°F/38°C. Water contact produces toxic and corrosive fumes. Reacts violently with alkaline earth metals (barium, calcium, magnesium, strontium, etc.), alkaline metals, sodium cyanide. Attacks metals in the presence of moisture.

ETHYL CHLOROCARBONATE (541-43-3) Forms explosive mixture with air (flash point 61°F/16°C). Strong oxidizers may cause fire and explosions. Attacks metals in the presence of moisture. Water and air contact produce hydrochloric acid.

ETHYL CHLOROETHANOATE (105-39-5) Forms explosive mixture with air above 100°F/38°C. Water contact produces toxic and corrosive fumes. Reacts violently with alkaline earth metals (barium, calcium, magnesium, strontium, etc.), alkaline metals, sodium cyanide. Attacks metals in the presence of moisture.

ETHYL-2-CHLOROETHANOATE (105-39-5) Forms explosive mixture with air above 100°F/38°C. Water contact produces toxic and corrosive fumes. Reacts violently with alkaline earth metals (barium, calcium, magnesium, strontium, etc.), alkaline metals, sodium cyanide. Attacks metals in the presence of moisture.

ETHYL CHLOROFORMATE (541-43-3) Forms explosive mixture with air. (flash point 61°F/16°C). Strong oxidizers may cause fire and explosions. Attacks metals in the presence of moisture. Water and air contact produce hydrochloric acid.

ETHYL CHLOROTHIOFORMATE (2941-64-2) Forms explosive mixture with air (flash point 86°F/19°C). Oxidizers may cause fire or explosions.

ETHYL CHLOROTHIOLFORMATE (2941-64-2) Forms explosive mixture with air (flash point 86°F/19°C). Oxidizers may cause fire or explosions.

ETHYL CYANIDE (107-12-0) Forms explosive mixture with air (flash point 36°F/2°C). Reacts violently with oxidizers. Water or acid contact produces hydrogen cyanide fumes.

N-ETHYLCYCLOHEXANAMINE (5459-93-8) Forms explosive mixture with air (flash point 86°F/30°C). Incompatible with acids, organic anhydrides, isocyanates, vinyl acetate, acrylates, substituted allyls, alkylene oxides, epichlorohydrin, ketones, aldehydes, alcohols, glycols, mercury, phenols, cresols, caprolactam solution, strong oxidizers. Attacks aluminum, copper, lead, tin, zinc, and alloys.

ETHYL CYCLOHEXANE (1678-91-7) Forms explosive mixture with air above 95°F/35°C. Incompatible with oxidizers, acids, caustics.

N-ETHYL(CYCLO)HEXYLAMINE (5459-93-8) Forms explosive mixture with air (flash point 86°F/30°C). Incompatible with acids, organic anhydrides, isocyanates, vinyl acetate, acrylates, substituted allyls, alkylene oxides, epichlorohydrin, ketones, aldehydes, alcohols, glycols, mercury, phenols, cresols, caprolactam solution, strong oxidizers. Attacks aluminum, copper, lead, tin, zinc, and alloys.

ETHYL DICHLOROPHOSPHATE (1498-51-7) Water contact produces hydrochloric acid. Attacks metals when wet.

ETHYLDICHLOROSILANE (1789-58-8) Forms explosive mixture with air (flash point 30°F/1°C). Water contact produces hydrochloric acid. Incompatible with strong bases. Attacks most common metals.

ETHYL DIETHYLENE GLYCOL (111-90-0) Forms explosive mixture with air (flash point 201°F/94°C). Heat or contact with atmospheric air may produce unstable peroxides. Incompatible with strong acids, isocyanates, strong oxidizers.

ETHYL DIMETHYL METHANE (78-78-4) Highly volatile liquid (flash point less than −60°F). Mixture with air may explode. Attacks some plastics, rubber, and coatings. May accumulate static electrical charges, and may cause ignition of its vapors.

ETHYLE (ACETATE d') (French) (141-78-6) Will hydrolyze on standing, forming acetic acid and ethyl alcohol. This reaction is greatly accelerated by bases (alkalies). Forms explosive mixture with air (flash point 24°F/−4.4°C). Incompatible with strong acids, nitrates, oxidizers, chlorosulfonic acid, lithium aluminum hydride, oleum.

ETHYLEENOXIDE (Dutch) (75-21-8) Forms explosive mixture with air (flash point 20°F/−6°C). Incompatible with alkali metal hydroxides, highly active catalysts (e.g., anhydrous chlorides of iron, tin, or aluminum and oxides of iron or aluminum). Avoid contact with copper. Protect container from physical damage, sun, and heat. Attacks some plastics, rubber, and coatings.

ETHYL EHYDE (75-07-0) Slowly polymerizes to paraldehyde. Explodes when mixed with iodine. Contact with strong bases can cause explosive polymerization. A strong reducing agent; reacts violently with combustibles, strong acids, caustics, ammonia, aliphatic amines, alkanolamines, aromatic amines, organic substances, halogens, oxidizers. Forms explosive peroxides with air. May dissolve rubber. May accumulate static electrical charges, and may cause ignition of its vapors.

ETHYLENE (74-85-1) A flammable gas. May accumulate static electrical charges, and may cause ignition of its vapors. Oxidizers may cause explosive polymerization and fire. Incompatible with acids, halogens, nitrogen oxides, hydrogen bromide, aluminum chloride, bromotrichloromethane, carbon tetrachloride, chlorine, chlorine dioxide, nitrogen dioxide.

ETHYLENE ACETATE (111-55-7) Forms explosive mixture with air above 191°F/88°C. Strong oxidizers may cause fire and explosions.

ETHYLENE ALCOHOL (107-21-1) Incompatible with strong acids, caustics, aliphatic amines, isocyanates, chlorosulfonic acid, oleum, strong oxidizers.

ETHYLENE ALDEHYDE (107-02-8) Forms explosive mixture with air (flash point −15°F/−26°C). Unstable and very reactive. In storage can form heat- and shock-sensitive compounds. Unless inhibited (usually by hydroquinone), readily forms explosive peroxides. Able to polymerize. A strong reducing agent. Reacts violently with oxidizers, strong acids, caustics, amines, 2-aminoethanol, ammonia, ammonium hydroxide, ethylene diamine, ethyleneimine, hydroxides, metal salts, oxidizers, sulfur dioxide, thiourea. Attacks metals: cadmium and zinc. May accumulate static electrical charges, and may cause ignition of its vapors.

1,1'-ETHYLENE-2,2'-BIPYRIDYLIUMDIBROMIDE (85-00-7) Concentrated solution attacks aluminum.

ETHYLENE BIS(IMINODIACETIC ACID) (60-00-4) Incompatible with sulfuric acid, bases, ammonia, aliphatic amines, alkanolamines, isocyanates, alkylene oxides, epichlorohydrin.

ETHYLENEBIS (DITHIOCARBAMIC ACID), DISODIUM SALT (142-59-6) Boiling water causes formation of hydrogen sulfide and carbon disulfide vapors.

ETHYLENE BROMIDE (106-93-4) Reacts with chemically active metals, liquid ammonia, strong oxidizers. Heat and light cause slow decomposition. Attacks some plastics and rubber.

ETHYLENECARBOXAMIDE (79-06-1) Unless inhibited, ultraviolet light and heat (above 184°F/85°C) can cause polymerization. Incompatible with nonoxidizing mineral acids, strong acids, ammonia, oleum, oxidizers, isocyanates.

ETHYLENE CARBOXYLIC ACID (79-10-7) Forms explosive mixture with air (flash point 122°F/50°C). Forms explosive peroxides; light, heat, and peroxides can cause polymerization. Incompatible with sulfuric acid, caustics, ammonia, amines,isocyanates, alkylene oxides, epichlorohydrin, oxidizers, toluenediamine, pyridine, methyl pyridine, n-methyl pyrrolidone, 2-methyl-6-ethyl aniline, aniline, ethylene diamine, ethyleneimine, 2-aminoethanol. Severely corrodes carbon steel and iron; attacks other metals. May accumulate static electrical charges, and may cause ignition of its vapors.

ETHYLENE CHLORHYDRIN (107-07-3) Forms explosive mixture with air (flash point 140°F/60°C). Strong oxidizers may cause fire and explosions; incompatible with strong caustics (with formation of ethylene gas), strong acids, aliphatic amines, isocyanates. Attacks some plastics, rubber, and coatings.

ETHYLENE CHLORIDE (107-06-2) May accumulate static electrical charges, and may cause ignition of its vapors. Contact with hot water may form hydrochloric acid. Forms explosive mixture with air (flash point 55°F/13°C). Incompatible with strong oxidizers, strong caustics, chemically active metals. Corrosive to iron and other metals. Attacks some plastics, rubber, or coatings.

ETHYLENE CHLOROHYDRIN (107-07-3) Forms explosive mixture with air (flash point 140°F/60°C). Strong oxidizers may cause fire and explosions, strong caustics (formation of ethylene gas), strong acids, aliphatic amines, isocyanates. Attacks some plastics, rubber, and coatings.

ETHYLENE CYANOHYDRIN (109-78-4) Forms explosive mixture with air (flash point 140°F/60°C). Avoid basic contamination to prevent polymerization. A powerful reducing agent. Hot water causes formation of cyanide gas. Reacts violently with strong oxidizers or sodium hydroxide. Reacts with acids, acid salts, chlorates, nitrates. Attacks mild steel, copper, and copper alloys.

ETHYLENE DIACETATE (111-55-7) Forms explosive mixture with air above 191°F/88°C. Strong oxidizers may cause fire and explosions.

ETHYLENE DIAMINE (107-15-3) Forms explosive mixture with air (flash point 104°F/40°C). Incompatible with chlorinated organic compounds, silver perchlorate, 3-propiolactone, mesityl oxide, ethylene dichloride, acids, organic anhydrides, isocyanates, vinyl acetate, acrylates, substituted allyls, alkylene oxides, epichlorohydrin, ketones, aldehydes, alcohols, glycols, phenols, cresols, caprolactam solution, strong oxidizers. Attacks aluminum, copper, lead, tin, zinc, and alloys, and some plastics, rubber, and coatings.

ETHYLENE-DIAMINE (French) (107-15-3) Forms explosive mixture with air (flash point 104°F/40°C). Incompatible with chlorinated organic compounds, silver perchlorate, 3-propiolactone, mesityl oxide, ethylene dichloride, acids, organic anhydrides, isocyanates, vinyl acetate, acrylates, substituted allyls, alkylene oxides, epichlorohydrin, ketones, aldehydes, alcohols, glycols, phenols, cresols, caprolactam solution, strong oxidizers. Attacks aluminum, copper, lead, tin, zinc, and alloys, and some plastics, rubber, and coatings.

ETHYLENEDIAMINE (111-40-0) Ignites spontaneously with cellulose nitrate. Silver, cobalt, or chromium compounds may cause explosions. Forms explosive mixture with air (flash point 208°F/98°C). Incompatible with acids, organic anhydrides, isocyanates, vinyl acetate, acrylates, substituted allyls, alkylene oxides, epichlorohydrin, ketones, aldehydes, alcohols, glycols, mercury, phenols, cresols, caprolactam solution, strong oxidizers. Attacks aluminum, copper, lead, tin, zinc, and alloys.

1,2-ETHYLENEDIAMINE (112-24-30) Forms explosive mixture with air (flash point 104°F/40°C). Incompatible with acids, organic anhydrides, isocyanates, vinyl acetate, acrylates, substituted allyls, alkylene oxides, epichlorohydrin, ketones, aldehydes, alcohols, glycols, phenols, cresols, caprolactam solution, strong oxidizers. Attacks aluminum, copper, lead, tin, zinc, and alloys.

ETHYLENEDIAMINETETRAACETATE (60-00-4) Incompatible with sulfuric acid, bases, ammonia, aliphatic amines, alkanolamines, isocyanates, alkylene oxides, epichlorohydrin.

ETHYLENEDIAMINE-N,N,N′,N′-TETRAACETIC ACID (60-00-4) Incompatible with sulfuric acid, bases, ammonia, aliphatic amines, alkanolamines, isocyanates, alkylene oxides, epichlorohydrin.

ETHYLENEDIAMINETETRAACETIC ACID (60-00-4) Incompatible with sulfuric acid, bases, ammonia, aliphatic amines, alkanolamines, isocyanates, alkylene oxides, epichlorohydrin.

ETHYLENE DIBROMIDE (106-93-4) Reacts with chemically active metals, liquid ammonia, strong oxidizers. Heat and light cause slow decomposition. Attacks some plastics and rubber.

1,2-ETHYLENE DIBROMIDE (106-93-4) Reacts with chemically active metals, liquid ammonia, strong oxidizers. Heat and light cause slow decomposition. Attacks some plastics and rubber.

1,2-ETHYLENEDICARBOXYLIC ACID, (Z) (110-16-7) Aqueous solution is an acid. Incompatible with oxidizers, bases. Corrodes metal if wet.

1,2-ETHYLENEDICARBOXYLIC ACID, (E) (110-17-8) Incompatible with sulfuric acid, caustics, ammonia, amines, isocyanates, alkylene oxides, epichlorohydrin.

cis-**1,2-ETHYLENEDICARBOXYLIC ACID** (110-16-7) Aqueous solution is an acid. Incompatible with oxidizers, bases. Corrodes metal if wet.

trans-**1,2-ETHYLENEDICARBOXYLIC ACID** (110-17-8) Incompatible with sulfuric acid, caustics, ammonia, amines, isocyanates, alkylene oxides, epichlorohydrin.

ETHYLENE DICHLORIDE (107-06-2) May accumulate static electrical charges, and may cause ignition of its vapors. Contact with hot water may produce hydrochloric acid. Forms explosive mixture with air (flash point 55°F/13°C). Incompatible with strong oxidizers, strong caustics, chemically active metals. Corrosive to iron and other metals. Attacks some plastics, rubber, or coatings.

1,2-ETHYLENE DICHLORIDE (107-06-2) May accumulate static electrical charges, and may cause ignition of its vapors. Contact with hot water may produce hydrochloric acid. Forms explosive mixture with air (flash point 55°F/13°C). Incompatible with strong oxidizers, strong caustics, chemically active metals. Corrosive to iron and other metals. Attacks some plastics, rubber, or coatings.

ETHYLENE DIGLYCOL (111-46-6) Combustible (flash point 255°F/124°C). Incompatible with sulfuric acid, isocyanates, strong oxidizers.

ETHYLENE DIGLYCOL MONOMETHYL ETHER (111-77-3) Incompatible with sulfuric acid, isocyanates, perchloric acid.

ETHYLENE DIHYDRATE (107-21-1) Incompatible with strong acids, caustics, aliphatic amines, isocyanates, chlorosulfonic acid, oleum, strong oxidizers.

ETHYLENE DIMETHYL ETHER (110-71-4) Forms explosive mixture with air (flash point 29°F/−2°C). May accumulate static electrical charges, and may cause ignition of its vapors. Incompatible with sulfuric acid, isocyanates, strong oxidizers.

ETHYLENE DINITRATE (628-96-6) Acids, heat, or mechanical shock may result in explosions.

(ETHYLENEDINITRILO)TETRAACETIC ACID (60-00-4) Incompatible with sulfuric acid, bases, ammonia, aliphatic amines, alkanolamines, isocyanates, alkylene oxides, epichlorohydrin.

ETHYLENEDINITRILOTETRAACETIC ACID (60-00-4) Incompatible with sulfuric acid, bases, ammonia, aliphatic amines, alkanolamines, isocyanates, alkylene oxides, epichlorohydrin.

2,2′-ETHYLENEDIOXYDIETHANOL (112-27-6) Incompatible with sulfuric acid, isocyanates, perchloric acid, strong oxidizers.

2,2′-ETHYLENEDIOXYETHANOL (112-27-6) Incompatible with sulfuric acid, isocyanates, perchloric acid, strong oxidizers.

ETHYLENE DIPYRIDYLIUM DIBROMIDE (85-00-7) Concentrated solution attacks aluminum.

1,1′-ETHYLENE-2,2′-DIPYRIDYLIUMDIBROMIDE (85-00-7) Concentrated solution attacks aluminum.

1,1-ETHYLENE 2,2-DIPYRIDYLIUM DIBROMIDE (85-00-7) Concentrated solution attacks aluminum.

ETHYLENE EHYDE (107-02-8) Forms explosive mixture with air (flash point −15°F/−26°C). Unstable and very reactive. In storage can form heat- and shock-sensitive compounds. Unless inhibited (usually by hydroquinone), readily forms explosive peroxides. Able to polymerize. A strong reducing agent. Reacts violently with oxidizers, strong acids, caustics, amines, 2-aminoethanol, ammonia, ammonium hydroxide, ethylene diamine, ethyleneimine, hydroxides, metal salts, oxidizers, sulfur dioxide, thiourea. Attacks metals: cadmium and zinc. May accumulate static electrical charges, and may cause ignition of its vapors.

ETHYLENE FLUORIDE (75-37-6) Flammable gas. May accumulate static electrical charges, and may cause ignition of its vapors. Reacts violently with strong oxidizers. Attacks metals in presence of moisture.

ETHYLENE GLYCOL (107-21-1) Incompatible with strong acids, caustics, aliphatic amines, isocyanates, chlorosulfonic acid, oleum, strong oxidizers.

ETHYLENE GLYCOL ACETATE (111-55-7) Forms explosive mixture with air above 191°F/88°C. Strong oxidizers may cause fire and explosions.

ETHYLENE GLYCOL-BIS-(2-HYDROXYETHYL ETHER) (112-27-6) Incompatible with sulfuric acid, isocyanates, perchloric acid, strong oxidizers.

ETHYLENE GLYCOL BUTYL ETHER or ETHYLENE GLYCOL n-BUTYL ETHER (111-76-2) Forms explosive mixture with air (flash point 143°F/62°C). Reacts violently with strong caustics and strong oxidizers. Attacks some coatings, plastics, and rubber. Attacks metallic aluminum at high temperatures.

ETHYLENE GLYCOL DIACETATE (111-55-7) Forms explosive mixture with air above 191°F/88°C. Strong oxidizers may cause fire and explosions.

ETHYLENE GLYCOL DIBUTYL ETHER (112-48-1) Forms explosive mixture with air above 185°F/85°C. Oxidizers may cause fire and explosions. Incompatible with sulfuric acid, isocyanates, perchloric acid.

ETHYLENE GLYCOL DIETHYL ETHER (629-14-1) Forms explosive mixture with air (flash point 95°F/35°C). Able to form unstable peroxides. Incompatible with sulfuric acid, isocyanates, strong oxidizers.

ETHYLENE GLYCOL DIHYDROXYDIETHYL ETHER (112-27-6) Incompatible with sulfuric acid, isocyanates, perchloric acid, strong oxidizers.

ETHYLENE GLYCOL DIMETHYL ETHER (110-71-4) Forms explosive mixture with air (flash point 29°F/−2°C). May accumulate static electrical charges, and may cause ignition of its vapors. Incompatible with sulfuric acid, isocyanates, strong oxidizers.

ETHYLENE GLYCOL DINITRATE (628-96-6) Acids, heat, or mechanical shock may result in explosions.

ETHYLENE GLYCOL ETHYL ETHER (110-80-5) Forms explosive mixture with air (flash point 120°F/49°C). Strong oxidizers may cause fire and explosions. Attacks some plastics, rubber, and coatings. Able to form peroxides. Incompatible with strong acids, aluminum and its alloys.

ETHYLENE GLYCOL ETHYL ETHER ACETATE (111-15-9) Forms explosive mixture with air (flash point 117°F/47°C). Incompatible with strong acids, nitrates. Reacts violently with oxidizers. May form unstable peroxides. Softens many plastics. Attacks some plastics, rubber, and coatings.

ETHYLENE GLYCOL ISOPROPYL ETHER (109-59-1) Forms explosive mixture with air (flash point 92°F/33°C). Incompatible with sulfuric acid, perchloric acid, isocyanates, oxidizers.

ETHYLENE GLYCOL METHYL ETHER (109-86-4) Forms explosive mixture with air (flash point 103°F/39°C). Heat or oxidizers may cause formation of unstable peroxides. Attacks many metals. Strong oxidizers cause fire and explosions. Strong bases cause decomposition. Attacks some plastics, rubber, and coatings. May accumulate static electrical charges, and may cause ignition of its vapors.

ETHYLENE GLYCOL METHYL ETHER ACETATE (110-49-6) Incompatible with nitrates, strong oxidizers, strong alkalies, strong acids.

ETHYLENE GLYCOL MONOACETATE (542-59-6) Forms explosive mixture with air above 191°F/88°C. Incompatible with sulfuric acid, nitric acid, nitrates.

ETHYLENE GLYCOL MONOBUTYL ETHER or ETHYLENE GLYCOL MONOBUTYL ETHER (DOT) (111-76-2) Forms explosive mixture with air (flash point 143°F/62°C). Reacts violently with strong caustics and strong oxidizers. Attacks some coatings, plastics, and rubber. Attacks metallic aluminum at high temperatures.

ETHYLENE GLYCOL MONOBUTYL ETHER (111-76-2) Forms explosive mixture with air (flash point 143°F/62°C). Reacts violently with strong caustics and strong oxidizers. Attacks some coatings, plastics, and rubber. Attacks metallic aluminum at high temperatures.

ETHYLENE GLYCOL MONOBUTYL ETHER ACETATE (112-07-2) Forms explosive mixture with air (flash point 160°F/71°C). Ethers, as a class, can form peroxides. Incompatible with strong acids, perchloric acid, oxidizers, nitrates.

ETHYLENE GLYCOL MONOETHYL ETHER (110-80-5) Forms explosive mixture with air (flash point 120°F/49°C). Strong oxidizers may cause fire and explosions. Attacks some plastics, rubber, and coatings. Able to form peroxides. Incompatible with strong acids, aluminum and its alloys.

ETHYLENE GLYCOL MONOETHYL ETHER ACETATE (111-15-9) Forms explosive mixture with air (flash point 117°F/47°C). Incompatible with strong acids, nitrates. Reacts violently with oxidizers. May form unstable peroxides. Softens many plastics. Attacks some plastics, rubber, and coatings.

ETHYLENE GLYCOL MONOETHYL ETHER MONOACETATE (111-15-9) Forms explosive mixture with air (flash point 117°F/47°C). Incompatible with strong acids, nitrates. Reacts violently with oxidizers. May form unstable peroxides. Softens many plastics. Attacks some plastics, rubber, and coatings.

ETHYLENE GLYCOL MONOISOPROPYL ETHER (109-59-1) Forms explosive mixture with air. 92°F/33°C). Incompatible with sulfuric acid, perchloric acid, isocyanates, oxidizers.

ETHYLENE GLYCOL MONOMETHYL ETHER (109-86-4) Forms explosive mixture with air (flash point 103°F/39°C). Heat or oxidizers may cause formation of unstable peroxides. Attacks many metals. Strong oxidizers cause fire and explosions. Strong bases cause decomposition. Attacks some plastics, rubber, and coatings. May accumulate static electrical charges, and may cause ignition of its vapors.

ETHYLENE GLYCOL MONOMETHYL ETHER ACETATE (110-49-6) Incompatible with nitrates, strong oxidizers, strong alkalies, strong acids.

ETHYLENE GLYCOL MONOPHENYL ETHER (122-99-6) Incompatible with sulfuric acid, isocyanates, strong oxidizers.

ETHYLENE GLYCOL MONOPROPYL ETHER (2807-30-9) Incompatible with sulfuric acid, isocyanates, strong oxidizers.

ETHYLENE GLYCOL PHENYL ETHER (122-99-6) Incompatible with sulfuric acid, isocyanates, strong oxidizers.

ETHYLENE GLYCOL PROPYL ETHER (2807-30-9) Incompatible with sulfuric acid, isocyanates, strong oxidizers.

ETHYLENE HEXACHLORIDE (67-72-1) Incompatible with hot iron, zinc, and aluminum; contact with alkalies produces spontaneously explosive chloroacetylene. Attacks some plastics, rubber, and coatings.

ETHYLENE IMINE (151-56-4) Forms explosive mixture with air (flash point 12°F/−11°C). May accumulate static electrical charges, and may cause ignition of its vapors. Contact with acid, aluminum, carbon dioxide, or silver may cause explosive polymerization. Attacks rubber, coatings, plastics, and chemically active metals. Self-reactive with heat or atmospheric carbon dioxide.

ETHYLENE MONOCHLORIDE (75-01-4) Forms explosive gas mixture with air. Atmospheric oxygen and various contaminants or strong oxidizers cause formation of peroxides, which can initiate a violent polymerization action. Also able to polymerize when heated and exposed, long term, to light. Reacts violently with strong oxidizers or oxides of nitrogen. Contact with copper or other acetylide-forming metals produces explosive compounds. Attacks iron and steel in the presence of moisture.

ETHYLENE MONOCLINIC TABLETS CARBOXAMIDE (79-06-1) Unless inhibited, ultraviolet light and heat (above 184°F/85°C) can cause polymerization. Incompatible with nonoxidizing mineral acids, strong acids, ammonia, oleum, oxidizers, isocyanates.

ETHYLENE NITRATE (628-96-6) Acids, heat, or mechanical shock may result in explosions.

ETHYLENE OXIDE (75-21-8) Forms explosive mixture with air (flash point 20°F/−6°C). Incompatible with alkali metal hydroxides, highly active catalysts (e.g., anhydrous chlorides of iron, tin, or aluminum and oxides of iron or aluminum). Avoid contact with copper. Protect container from physical damage, sun, and heat. Attacks some plastics, rubber, or coatings.

ETHYLENE OXIDE, METHYL- (75-56-9) Forms explosive gas mixture with air (flash point −35°F/−37°C). Incompatible with anhydrous metal chlorides. Strong acids, caustics, peroxides cause polymerization. Reacts with ammonia, amines, acetylene-forming metals. Attacks some plastics, rubber, and coatings.

ETHYLENE (OXYDE D′) (French) (75-21-8) Forms explosive mixture with air (flash point 20°F/−6°C). Incompatible with alkali metal hydroxides, highly active catalysts (e.g., anhydrous chlorides of iron, tin, or aluminum and oxides of iron or aluminum). Avoid contact with copper. Protect container from physical damage, sun and heat. Attacks some plastics, rubber, or coatings.

ETHYLENE, PHENYL- (100-42-5) If inhibitor (often *tert*-butyl catechol) is not present in adequate concentrations, polymerization may occur and explode containers. Temperatures above 150°F/66°C speed up polymerization. Incompatible with oxidizers, acids, rust, catalysts for vinyl polymerization such as peroxides, strong acids, and

356

aluminum chloride. Corrodes copper and copper alloys, and dissolves rubber. Attacks some plastics, rubber, and coatings. May accumulate static electrical charges, and may cause ignition of its vapors.

ETHYLENE TETRACHLORIDE (127-18-4) Incompatible with strong oxidizers, finely divided metals, caustics. Stable up to 258°F/126°C; at this temperature product gives off poisonous fumes.

ETHYLENE, TETRAFLUORO- (INHIBITED) (116-14-3) A highly reactive flammable gas. Able to form unstable peroxides; if inhibitor is not present in adequate concentrations, explosive polymerization may occur. Reacts violently with air, oxygen, oxidizers, sulfur trioxide.

ETHYLENE TRICHLORIDE (79-01-6) Contact with caustics produces a toxic and flammable gas. Reacts violently with chemically active metals. Contact with aluminum may produce a violent, self-accelerating polymerization reaction. Incompatible with acids, organic anhydrides, isocyanates, alkylene oxides, aldehydes, alcohols, glycols, phenols, cresols, caprolactam solution, epichlorohydrin, nitrogen tetroxide, metal powders, oxygen. May accumulate static electrical charges, and may cause ignition of its vapors.

ETHYLENIMINE (151-56-4) Forms explosive mixture with air (flash point 12°F/−11°C). May accumulate static electrical charges, and may cause ignition of its vapors. Contact with acid, aluminum, carbon dioxide, or silver may cause explosive polymerization. Attacks rubber, coatings, plastics, and chemically active metals. Self-reactive with heat or atmospheric carbon dioxide.

ETHYL ESTER OF FORMIC ACID (109-94-4) Forms explosive mixture with air (flash point −4°F/−20°C). May accumulate static electrical charges, and may cause ignition of its vapors. Incompatible with nitrates, strong oxidizers, strong alkalies, strong acids.

n-ETHYLETHANAMINE (109-89-7) Forms explosive mixture with air (flash point −9°F/−23°C). May accumulate static electrical charges, and may cause ignition of its vapors. Reacts violently with oxidizers. Incompatible with acids, organic anhydrides, isocyanates, vinyl acetate, acrylates, substituted allyls, alkylene oxides, epichlorohydrin, ketones, aldehydes, alcohols, glycols, mercury, phenols, cresols, caprolactam solution, strong oxidizers. Attacks aluminum, copper, lead, tin, zinc, and alloys.

ETHYL ETHANOATE (141-78-6) Will hydrolyze on standing, forming acetic acid and ethyl alcohol. This reaction is greatly accelerated by bases (alkalies). Forms explosive mixture with air (flash point 24°F/−4.4°C). Incompatible with strong acids, nitrates, oxidizers, chlorosulfonic acid, lithium aluminum hydride, oleum

ETHYL ETHER (60-29-7) Forms explosive mixture with air (flash point −49°F/−45°C). Incompatible with strong acids, strong oxidizers. Can form peroxides from air or light; may explode when container is unstoppered or otherwise opened. Attacks some plastics, rubber, and coatings. Being a nonconductor, chemical may accumulate static electric charges that may result in ignition of vapor.

ETHYL-3-ETHOXYPROPIONATE (763-69-9) Forms explosive mixture with air above 138°F/59°C. Incompatible with strong acids, nitrates, oxidizers.

ETHYL beta-ETHOXYPROPIONATE (763-69-9) Forms explosive mixture with air above 138°F/59°C. Incompatible with strong acids, nitrates, oxidizers.

ETHYLETHYLENE (106-98-9) Flammable gas. Reacts violently with strong oxidizers. Can form unstable peroxides; may polymerize. Incompatible with acids, halogens, aluminum borohydride, oxides of nitrogen. May accumulate static electrical charges, and may cause ignition of its vapors.

ETHYL FORMATE (109-94-4) Forms explosive mixture with air (flash point −4°F/−20°C). May accumulate static electrical charges, and may cause ignition of its vapors. Incompatible with nitrates, strong oxidizers, strong alkalies, strong acids.

ETHYLFORMIC ACID (79-09-4) Forms explosive mixture with air (flash point 126°F/52°C). Incompatible with sulfuric acid, strong bases, ammonia, aliphatic amines, alkanolamines, isocyanates, alkylene oxides, epichlorohydrin, oxidizers.

ETHYL FORMIC ESTER (109-94-4) Forms explosive mixture with air (flash point −4°F/−20°C). May accumulate static electrical charges, and may cause ignition of its vapors. Incompatible with nitrates, strong oxidizers, strong alkalies, strong acids.

ETHYL GLYME (629-14-1) Forms explosive mixture with air (flash point 95°F/35°C). Able to form unstable peroxides. Incompatible with sulfuric acid, isocyanates, strong oxidizers.

ETHYLHEXALDEHYDE (123-05-7) Forms explosive mixture with air (flash point 112°F/44°C). Reacts violently with oxidizers. May ignite spontaneously when spilled on clothing or other absorbent materials. Under certain conditions, ignites spontaneously with air. Incompatible with strong acids, caustics, ammonia, amines. Forms unstable peroxides on contact with air.

2-ETHYLHEXALDEHYDE (123-05-7) Forms explosive mixture with air (flash point 112°F/44°C). Reacts violently with oxidizers. May ignite spontaneously when spilled on clothing or other absorbent materials. Under certain conditions, ignites spontaneously with air. Incompatible with strong acids, caustics, ammonia, amines. Forms unstable peroxides on contact with air.

2-ETHYLHEXANAL (123-05-7) Forms explosive mixture with air (flash point 112°F/44°C). Reacts violently with oxidizers. May ignite spontaneously when spilled on clothing or other absorbent materials. Under certain conditions, ignites spontaneously with air. Incompatible with strong acids, caustics, ammonia, amines. Forms unstable peroxides on contact with air.

2-ETHYL-1-HEXANAL (123-05-7) Forms explosive mixture with air (flash point 112°F/44°C). Reacts violently with oxidizers. May ignite spontaneously when spilled on clothing or other absorbent materials. Under certain conditions, ignites spontaneously with air. Incompatible with strong acids, caustics, ammonia, amines. Forms unstable peroxides on contact with air.

2-ETHYLHEXANOIC ACID (149-57-5) May accumulate static electrical charges, and may cause ignition of its vapors. Reacts with oxidizers, with a risk of fire or explosions. Attacks common metals, especially in a moist environment.

2-ETHYL HEXANOL (104-76-7) Forms explosive mixture with air (flash point 175°F/80°C). Incompatible with strong acids, caustics, aliphatic amines, isocyanates, oxidizers. May accumulate static electrical charges, and may cause ignition of its vapors.

2-ETHYL-1-HEXANOL (104-76-7) Forms explosive mixture with air (flash point 175°F/80°C). Incompatible with strong acids, caustics, aliphatic amines, isocyanates, oxidizers. May accumulate static electrical charges, and may cause ignition of its vapors.

2-ETHYL-1-HEXANOL HYDROGEN PHOSPHATE (298-07-7) Corrosive to metals; forms flammable hydrogen gas.

2-ETHYL-2-HEXENAL (645-62-5) Forms explosive mixture with air above 155°F/68°C. Incompatible with acids, caustics, ammonia, amines, oxidizers.

2-ETHYLHEXOIC ACID (149-57-5) May accumulate static electrical charges, and may cause ignition of its vapors. Reacts with oxidizers, with a risk of fire or explosions. Attacks common metals, especially in a moist environment.

2-ETHYLHEXYL ACETATE (103-09-3) Forms explosive mixture with air above 160°F/71°C. Strong oxidizers may cause fire or explosions.

2-ETHYLHEXYL ACRYLATE (103-11-7) Sunlight, heat, contaminants, or peroxides can cause polymerization. Reacts violently with strong oxidizers, with risk of fire and explosions. Forms explosive mixture with air (flash point 180°F/82°C). Incompatible with strong acids, aliphatic amines, alkanolamines.

2-ETHYLHEXYL ALCOHOL (104-76-7) Forms explosive mixture with air (flash point 175°F/80°C). Incompatible with strong acids, caustics, aliphatic amines, isocyanates, oxidizers. May accumulate static electrical charges, and may cause ignition of its vapors.

2-ETHYL HEXYLAMINE (104-75-6) Forms explosive mixture with air (flash point 140°F/60°C). A strong base. Incompatible with acids, organic anhydrides, isocyanates, vinyl acetate, acrylates, substituted allyls, alkylene oxides, epichlorohydrin, ketones, aldehydes, alcohols, glycols, phenols, cresols, caprolactam solution, strong oxidizers. Attacks aluminum, copper, lead, tin, zinc, and alloys.

2-ETHYLHEXYLAMINE-1 (104-75-6) Forms explosive mixture with air (flash point 140°F/60°C). A strong base. Incompatible with acids, organic anhydrides, isocyanates, vinyl acetate, acrylates, substituted allyls, alkylene oxides, epichlorohydrin, ketones, aldehydes, alcohols, glycols, phenols, cresols, caprolactam solution, strong oxidizers. Attacks aluminum, copper, lead, tin, zinc and alloys.

2-ETHYL-1-HEXYLAMINE (104-75-6) Forms explosive mixture with air (flash point 140°F/60°C). A strong base. Incompatible with acids, organic anhydrides, isocyanates, vinyl acetate, acrylates, substituted allyls, alkylene oxides, epichlorohydrin, ketones, aldehydes, alcohols, glycols, phenols, cresols, caprolactam solution, strong oxidizers. Attacks aluminum, copper, lead, tin, zinc, and alloys.

beta-ETHYLHEXYLAMINE (104-75-6) Forms explosive mixture with air (flash point 140°F/60°C). A strong base. Incompatible with acids, organic anhydrides, isocyanates, vinyl acetate, acrylates, substituted allyls, alkylene oxides, epichlorohydrin, ketones, aldehydes, alcohols, glycols, phenols, cresols, caprolactam solution, strong oxidizers. Attacks aluminum, copper, lead, tin, zinc, and alloys.

2-ETHYLHEXYL, 2-PROPENOATE (103-11-7) Sunlight, heat, contaminants, or peroxides can cause polymerization. Reacts violently with strong oxidizers, with risk of fire and explosions. Forms explosive mixture with air (flash point 180°F/82°C). Incompatible with strong acids, aliphatic amines, alkanolamines.

ETHYL HYDRATE (64-17-5) Forms explosive mixture with air (flash point 65°F/18°C). May accumulate static electrical charges, and may cause ignition of its vapors. Reactions may be violent with oleum, sulfuric acid, nitric acid, bases, aliphatic amines, isocyanates, oxidizers.

ETHYL HYDRIDE (74-84-0) Flammable gas; forms explosive mixture with air. Strong oxidizers may cause fire and explosions. May accumulate static electrical charges, and may cause ignition of its vapors.

ETHYL HYDROSULFIDE (75-08-1) Forms explosive mixture with air (flash point less than 0°F/−18°C). May accumulate static electrical charges, and may cause ignition of its vapors. Contact with strong oxidizers may cause fire and explosions. Attacks some forms of plastics, coatings, and rubber.

ETHYL HYDROXIDE (64-17-5) Forms explosive mixture with air (flash point 65°F/18°C). May accumulate static electrical charges, and may cause ignition of its vapors. Reactions may be violent with oleum, sulfuric acid, nitric acid, bases, aliphatic amines, isocyanates, oxidizers.

ETHYL-2-HYDROXYPROPANOATE (97-64-3) Fire or explosion hazard (flash point 115°F/46°C). Strong oxidizers may cause fire and explosions.

ETHYL-2-HYDROXYPROPIONATE (97-64-3) Fire or explosion hazard (flash point 115°F/46°C). Strong oxidizers may cause fire and explosions.

ETHYL alpha-HYDROXYPROPIONATE (97-64-3) Fire or explosion hazard (flash point 115°F/46°C). Strong oxidizers may cause fire and explosions.

ETHYLIC ACID (64-19-7) Vapor forms explosive mixture with air (flash point 105°F/40°C). Reacts violently with oxidizers and bases. Incompatible with strong acids, aliphatic amines, alkanolamines, isocyanates, alkylene oxides, epichlorohydrin, acetaldehyde, 2-aminoethanol, ammonia, ammonium nitrate, chlorosulfonic acid, chromic acid, ethylene diamine, ethyleneimine, perchloric acid, permanganates, phosphorus isocyanate, phosphorus trichloride, potassium *tert*-butoxide, xylene. Attacks cast iron and other metals, producing flammable hydrogen gas.

5-ETHYLIDENEBICYCLO(2.2.1)HEPT-2-ENE (16219-75-3) Reacts violently with oxygen and strong oxidizers. Forms explosive mixture with air (flash point 100°F/38°C). May accumulate static electrical charges, and may cause ignition of its vapors. Inhibit peroxide formation with *tert*-butyl catechol.

ETHYLIDENE CHLORIDE (75-34-3) Forms explosive mixture with air (flash point 2°F/−17°C). Incompatible with strong oxidizers, strong caustics. Attacks plastics and rubber.

ETHYLIDENE DICHLORIDE (75-34-3) Forms explosive mixture with air (flash point 2°F/−17°C). Incompatible with strong oxidizers, strong caustics. Attacks plastics and rubber.

1,1-ETHYLIDENE DICHLORIDE (75-34-3) Forms explosive mixture with air (flash point 2°F/−17°C). Incompatible with strong oxidizers, strong caustics. Attacks plastics and rubber.

ETHYLIDENE DIETHYL ETHER (105-57-7) Forms explosive mixture with air (flash point −5°F/−21°C). Reacts violently with oxidizers. Forms unstable and explosive peroxides with heat and light. May accumulate static electrical charges, and may cause ignition of its vapors.

ETHYLIDENE DIFLUORIDE (75-37-6) Flammable gas. May accumulate static electrical charges, and may cause ignition of its vapors. Reacts violently with strong oxidizers. Attacks metals in presence of moisture.

ETHYLIDENE FLUORIDE (75-37-6) Flammable gas. May accumulate static electrical charges, and may cause ignition of its vapors. Reacts violently with strong oxidizers. Attacks metals in presence of moisture.

ETHYLIDENELACTIC ACID (598-8-3) Incompatible with strong acids.

ETHYLIDENE NORBORNENE (16219-75-3) Reacts violently with oxygen and strong oxidizers. Forms explosive mixture with air (flash point 100°F/38°C). May accumulate static electrical charges, and may cause ignition of its vapors. Inhibit peroxide formation with *tert*-butyl catechol.

5-ETHYLIDENE-2-NORBORNENE (16219-75-3) Reacts violently with oxygen and strong oxidizers. Forms explosive mixture with air (flash point 100°F/38°C). May accumulate static electrical charges, and may cause ignition of its vapors. Inhibit peroxide formation with *tert*-butyl catechol.

ETHYLIDENENORBORNYLENE (16219-75-3) Reacts violently with oxygen and strong oxidizers. Forms explosive mixture with air (flash point 100°F/38°C). May accumulate static electrical charges, and may cause ignition of its vapors. Inhibit peroxide formation with *tert*-butyl catechol.

ETHYLIDENENORCAMPHENE (16219-75-3) Reacts violently with oxygen and strong oxidizers. Forms explosive mixture with air (flash point 100°F/38°C). May accumulate static electrical charges, and may cause ignition of its vapors. Inhibit peroxide formation with *tert*-butyl catechol.

ETHYLIMENE (151-56-4) Forms explosive mixture with air (flash point 12°F/−11°C). May accumulate static electrical charges, and may cause ignition of its vapors. Contact with acid, aluminum, carbon dioxide or silver may cause explosive polymerization. Attacks rubber, coatings, plastics, and chemically active metals. Self-reactive with heat or atmospheric carbon dioxide.

ETHYLIMINE (151-56-4) Forms explosive mixture with air (flash point 12°F/−11°C). May accumulate static electrical charges, and may cause ignition of its vapors. Contact with acid, aluminum, carbon dioxide, or silver may cause explosive polymerization. Attacks rubber, coatings, plastics, and chemically active metals. Self-reactive with heat or atmospheric carbon dioxide.

ETHYL ISOCYANATE (109-90-0) Forms explosive mixture with air (flash point 12°F/−6°C). Reacts violently with water and strong oxidizers. Incompatible with acids, bases, ammonia, amines, amides, alcohols, glycols, caprolactam solution. May accumulate static electrical charges, and may cause ignition of its vapors.

ETHYL KETONE (96-22-0) Forms explosive mixture with air (flash point 55°F/13°C). Incompatible with strong acids, aliphatic amines, strong oxidizers. Attacks some plastics, rubber, and coatings. May accumulate static electrical charges, and may cause ignition of its vapors.

ETHYL LACTATE (97-64-3) Fire or explosion hazard (flash point 115°F/46°C). Strong oxidizers may cause fire and explosions.

ETHYL DL-LACTATE (97-64-3) Fire or explosion hazard (flash point 115°F/46°C). Strong oxidizers may cause fire and explosions.

ETHYL MERCAPTAN (75-08-1) Forms explosive mixture with air (flash point less than 0°F/−18°C). May accumulate static electrical charges, and may cause ignition of its vapors. Contact with strong oxidizers may cause fire and explosions. Attacks some forms of plastics, coatings and rubber.

ETHYL METHACRYLATE (97-63-2) Forms explosive mixture with air (flash point 80°F/27°C). Incompatible with strong acids, amines, oxidizers. Corrodes some metals.

ETHYL-1-2-METHACRYLATE (97-63-2) Forms explosive mixture with air (flash point 80°F/27°C). Incompatible with strong acids, amines, oxidizers. Corrodes some metals.

ETHYL-2-METHACRYLATE (97-63-2) Forms explosive mixture with air (flash point 80°F/27°C). Incompatible with strong acids, amines, oxidizers. Corrodes some metals.

ETHYL METHACRYLATE—INHIBITED (97-63-2) Forms explosive mixture with air (flash point 80°F/27°C). Incompatible with strong acids, amines, oxidizers. Corrodes some metals.

ETHYL METHANOATE (109-94-4) Forms explosive mixture with air (flash point −4°F/−20°C). May accumulate static electrical charges, and may cause ignition of its vapors. Incompatible with nitrates, strong oxidizers, strong alkalies, strong acids.

ETHYLMETHANOL (107-19-7) Incompatible with strong acids, aliphatic amines, phosphorus pentoxide, caustics, isocyanates, strong oxidizers.

ETHYL METHYL ACRYLATE (97-63-2) Forms explosive mixture with air (flash point 80°F/27°C). Incompatible with strong acids, amines, oxidizers. Corrodes some metals.

ETHYL-alpha-METHYLACRYLATE (97-63-2) Forms explosive mixture with air (flash point 80°F/27°C). Incompatible with strong acids, amines, oxidizers. Corrodes some metals.

1-ETHYL-2-METHYLBENZENE (611-14-3) Forms explosive mixture with air above 103°F/39°C. Incompatible with nitric acid, oxidizers (with fire and explosions).

O-ETHYLMETHYLBENZENE (611-14-3) Forms explosive mixture with air above 103°F/39°C. Incompatible with nitric acid, oxidizers (with fire and explosions).

ETHYL METHYL CARBINOL (78-92-2) Forms explosive mixture with air (flash point 75°F/4°C). Attacks some plastics, rubber, and coatings. Forms an explosive peroxide in air. Ignites with chromium trioxide. Incompatible with strong oxidizers, strong acids, aliphatic amines, isocyanates, organic peroxides.

ETHYL METHYL CETONE (French) (78-93-3) Forms explosive mixture with air (flash point 16°F/−9°C). Extremely flammable. Incompatible with sulfuric acid, nitric acid, aliphatic amines, strong oxidizers, potassium *tert*-butoxide, 2-propanol, chlorosulfonic acid, oleum, (hydrogen peroxide + nitric acid).

ETHYL METHYLENE PHOSPHORODITHIOATE (563-12-2) Incompatible with alkaline formulations. Mixtures with magnesium may be explosive.

ETHYLMETHYLKETON (Dutch) (78-93-3) Forms explosive mixture with air (flash point 16°F/−9°C). Extremely flammable. Incompatible with sulfuric acid, nitric acid, aliphatic amines, strong oxidizers, potassium *tert*-butoxide, 2-propanol, chlorosulfonic acid, oleum (hydrogen peroxide + nitric acid).

ETHYL METHYL KETONE (78-93-3) Forms explosive mixture with air (flash point 16°F/−9°C). Extremely flammable. Incompatible with sulfuric acid, nitric acid, aliphatic amines, strong oxidizers, potassium *tert*-butoxide, 2-propanol, chlorosulfonic acid, oleum (hydrogen peroxide + nitric acid).

ETHYL METHYL KETONE PEROXIDE (1338-23-4) Forms explosive mixture with air (flash point 125°F/52°C). Explosive decomposition occurs above 176°F/80°C. Pure substance is shock-sensitive. Strong oxidizer. Reacts violently with strong acids, strong bases, reducing agents, combustible substances, organic materials, oxides of heavy metals, salts, trace contaminants, amines. May accumulate static electrical charges, and may cause ignition of its vapors.

2-ETHYL-2-METHYL-2-PROPENOATE (97-63-2) Forms explosive mixture with air (flash point 80°F/27°C). Incompatible with strong acids, amines, oxidizers. Corrodes some metals.

5-ETHYL-2-METHYLPYRIDINE (104-90-5) Forms explosive mixture with air above 155°F/68°C. Incompatible with acids, isocyanates, phenols, cresols.

ETHYL MONOCHLOROACETATE (105-39-5) Forms explosive mixture with air above 100°F/38°C. Water contact produces toxic and corrosive fumes. Reacts violently with alkaline earth metals (barium, calcium, magnesium, strontium, etc.), alkaline metals, sodium cyanide. Attacks metals in the presence of moisture.

4-ETHYLMORPHOLINE (100-74-3) Forms explosive mixture with air (flash point 90°F/32.2°C). Incompatible with strong acids, strong oxidizers. Attacks some plastics, rubber, and coatings.

n-ETHYL MORPHOLINE (100-74-3) Forms explosive mixture with air (flash point 90°F/32.2°C). Incompatible with strong acids, strong oxidizers. Attacks some plastics, rubber, and coatings.

ETHYL NITRIL (75-05-8) Forms explosive mixture with air (flash point 42°F/6°C). Incompatible with water (especially if acid or alkaline), acids, caustics, nitrating agents, indium, nitrogen tetroxide, n-fluoro compounds, sulfur trioxide, iron(III) salts of perchlorate, indium, nitrogen-fluorine compounds. Reacts violently with oxidizers. May accumulate static electrical charges, and may cause ignition of its vapors.

ETHYL NITRILE (75-05-8) Forms explosive mixture with air (flash point 42°F/6°C). Incompatible with water (especially if acid or alkaline), acids, caustics, nitrating agents, indium, nitrogen tetroxide, n-fluoro compounds, sulfur trioxide, iron(III) salts of perchlorate, indium, nitrogen-fluorine compounds. Reacts violently with oxidizers. May accumulate static electrical charges, and may cause ignition of its vapors.

ETHYL NITRITE or ETHYL NITRITE SOLUTION (109-95-5) Forms explosive mixture with air (flash point −31°F/−35°C). A powerful oxidizer. Heat above 190°F/88°C can cause explosive decomposition. Incompatible with nitric acid; reducing agents can cause a violent reaction, fire, or explosions.

ETHYL ORTHOSILICATE (78-10-4) Forms explosive mixture with air (flash point 99°F/37°C). Incompatible with strong oxidizers, strong acids, water (producing volatile ethyl alcohol). Attacks some plastics and rubber.

ETHYLOXITOL ACETATE (111-15-9) Forms explosive mixture with air (flash point 117°F/47°C). Incompatible with strong acids, nitrates. Reacts violently with oxidizers. May form unstable peroxides. Softens many plastics. Attacks some plastics, rubber, and coatings.

ETHYL-3-OXOBUTANOATE (141-97-9) Forms explosive mixture with air (flash point 135°F/57°C). Incompatible with strong acids, nitrates, oxidizers.

ETHYL PARATHION (56-38-2) Combustible liquid. Mixtures with endrin may be explosive. Strong oxidizers may cause fire and explosions. Attacks some plastics, rubber, and coatings.

ETHYLPHENOL (90-00-6) Forms explosive mixture with air above 173°F/78°C. Incompatible with strong acids, caustics, aliphatic amines, amides.

2-ETHYLPHENOL (90-00-6) Forms explosive mixture with air above 173°F/78°C. Incompatible with strong acids, caustics, aliphatic amines, amides.

o-**ETHYLPHENOL** (90-00-6) Forms explosive mixture with air above 173°F/78°C. Incompatible with strong acids, caustics, aliphatic amines, amides.

ETHYL PHENYL DICHLOROSILANE (1125-27-5) Forms explosive mixture with air (flash point 150°F/66°C). Incompatible with strong bases, water (forms hydrochloric acid). Corrodes most metals. Attacks some plastics, rubber, and coatings.

ETHYL PHOSPHATE (78-40-0) Incompatible with strong acids, nitrates, oxidizers.

ETHYL PHOSPHONOTHIOIC DICHLORIDE (993-43-1) Forms explosive mixture with air above 203°F/95°C. Water contact produces hydrochloric acid. Reacts violently with strong oxidizers. Attacks metals in the presence of moisture.

ETHYL PHOSPHORODICHLORIDATE (1498-51-7) Water contact produces hydrochloric acid. Attacks metals when wet.

ETHYL PHOSPHORODICHLORIDOTHIONATE (993-43-1) Forms explosive mixture with air above 203°F/95°C. Water contact produces hydrochloric acid. Reacts violently with strong oxidizers. Attacks metals in the presence of moisture.

ETHYL PHTHALATE (84-66-2) Reacts violently with strong acids, strong oxidizers, permanganates, water. Attacks some forms of plastic.

5-ETHYL-2-PICOLINE (104-90-5) Forms explosive mixture with air above 155°F/68°C. Incompatible with acids, isocyanates, phenols, cresols.

ETHYL PROPANOATE (105-37-3) Forms explosive mixture with air (flash point 54°F/12°C). May accumulate static electrical charges, and may cause ignition of its vapors. Incompatible with strong acids, nitrates, strong oxidizers.

ETHYL PROPENOATE (140-88-5) Forms explosive mixture with air (flash point 48°F/9°C). Atmospheric moisture and strong alkalies may cause fire and explosions. Unless properly inhibited (note: inert gas blanket not recommended), heat, light, or peroxides can cause polymerization. Incompatible with oxidizers (reaction may be violent), strong acids, amines. May accumulate static electrical charges, and may cause ignition of its vapors.

ETHYL-2-PROPENOATE (140-88-5) Forms explosive mixture with air (flash point 48°F/9°C). Atmospheric moisture and strong alkalies may cause fire and explosions. Unless properly inhibited (note: inert gas blanket not recommended), heat, light, or peroxides can cause polymerization. Incompatible with oxidizers (reactions may be violent), strong acids, amines. May accumulate static electrical charges, and may cause ignition of its vapors.

ETHYL PROPIONATE (105-37-3) Forms explosive mixture with air (flash point 54°F/12°C). May accumulate static electrical charges, and may cause ignition of its vapors. Incompatible with strong acids, nitrates, strong oxidizers.

ETHYL PROPIONYL (96-22-0) Forms explosive mixture with air (flash point 55°F/13°C). Incompatible with strong acids, aliphatic amines, strong oxidizers. Attacks some plastics, rubber, and coatings. May accumulate static electrical charges, and may cause ignition of its vapors.

2-ETHYL-3-PROPYLACROLEIN (645-62-5) Forms explosive mixture with air above 155°F/68°C. Incompatible with acids, caustics, ammonia, amines, oxidizers.

2-ETHYL-3-PROPYL ACRYLALDEHYDE (1645-62-5) Forms explosive mixture with air above 155°F/68°C. Incompatible with acids, caustics, ammonia, amines, oxidizers.

ETHYL PYROPHOSPHATE, TETRA- (107-49-3) Decomposes above 300°F/150°C, forming flammable ethylene gas. Strong oxidizers may cause fire and explosions. Attacks some plastics, rubber, and coatings.

ETHYL SILICATE (78-10-4) Forms explosive mixture with air (flash point 99°F/37°C). Incompatible with strong oxidizers, strong acids, water (producing volatile ethyl alcohol). Attacks some plastics and rubber.

ETHYL SILICATE 40 (78-10-4) Forms explosive mixture with air (flash point 99°F/37°C). Incompatible with strong oxidizers, strong acids, water (producing volatile ethyl alcohol). Attacks some plastics and rubber.

ETHYL SILICATE, CONDENSED (78-10-4) Forms explosive mixture with air (flash point 99°F/37°C). Incompatible with strong oxidizers, strong acids, water (producing volatile ethyl alcohol). Attacks some plastics and rubber.

ETHYL SILICON TRICHLORIDE (115-1-9) Forms explosive mixture with air (flash point 72°F/22°C). Water contact produces hydrochloric acid. Strong oxidizers may cause fire and explosions. Corrodes most common metals.

ETHYL SULFATE (64-67-5) Reacts vigorously with strong oxidizers or water. Incompatible with strong acids, strong alkalies, nitrates, water.

ETHYL SULFHYDRATE (75-08-1) Forms explosive mixture with air (flash point less than 0°F/−18°C). May accumulate static electrical charges, and may cause ignition of its vapors. Contact with strong oxidizers may cause fire and explosions. Attacks some forms of plastics, coatings, and rubber.

ETHYL THIOALCOHOL (75-08-1) Forms explosive mixture with air (flash point less than 0°F/−18°C). May accumulate static electrical charges, and may cause ignition of its vapors. Contact with strong oxidizers may cause fire and explosions. Attacks some forms of plastics, coatings, and rubber.

S-2-(ETHYLTHIO)ETHYL O,O-DIETHYL ESTER OF PHOSPHO-RODITHIOIC ACID (298-04-4) Forms explosive mixture with air above 180°F/82°C. Incompatible with alkalis, strong oxidizers.

O,O-ETHYL S-2(ETHYLTHIO)ETHYL PHOSPHORODITHIOATE (298-04-4) Forms explosive mixture with air above 180°F/82°C. Incompatible with alkalis, strong oxidizers.

ETHYL THIONOPHOSPHORYL DICHLORIDE (993-43-1) Forms explosive mixture with air above 203°F/95°C. Water contact produces hydrochloric acid. Reacts violently with strong oxidizers. Attacks metals in the presence of moisture.

ETHYL THIOPYROPHOSPHATE (3689-24-5) Containers may burst in heat. Strong oxidizers may cause fire and explosions. Attacks some plastics, rubber, and coatings.

2-ETHYL TOLUENE (611-14-3) Forms explosive mixture with air above 103°F/39°C. Incompatible with nitric acid, oxidizers (with fire and explosions).

O-ETHYLTOLUENE (611-14-3) Forms explosive mixture with air above 103°F/39°C. Incompatible with nitric acid, oxidizers (with fire and explosions).

ETHYLTRICHLORO SILANE (115-1-9) Forms explosive mixture with air (flash point 72°F/22°C). Water contact produces hydrochloric acid. Strong oxidizers may cause fire and explosions. Corrodes most common metals.

ETHYL TRICHLOROSILANE (115-1-9) Forms explosive mixture with air (flash point 72°F/22°C). Water contact produces hydrochloric acid. Strong oxidizers may cause fire and explosions. Corrodes most common metals.

ETHYL VINYL ETHER (109-92-2) Incompatible with acids (nonoxidizing, sulfuric, nitric), ammonia, aliphatic amines, alkanolamines.

ETHYL VINYL ETHER, INHIBITED (109-92-2) Acids (non-oxidizing, sulfuric, nitric), ammonia, aliphatic amines, alkanolamines.

ETHYLZINC (557-20-0) Ignites spontaneously on contact with air or oxidizers. Explosive decomposition occurs at 245°F/120°C. Reacts violently with hydrazine, sulfur dioxide, some alcohols, ozone; with possible fire and explosions. Water contact produces ethane gas.

ETHYNE (74-86-2) A strong reducing agent that reacts violently with oxidizers. Forms explosive mixture with air (flash point 0°F/−18°C). Forms shock-sensitive mixture with copper and copper salts, mercury and mercury salts, silver and silver salts. Reacts with brass, bromine, cesium hydride, chlorine, cobalt, cuprous acetylise, fluorine, iodine, mercuric nitrate, nitric acid, potassium, rubidium hydride, trifluoro-methyl hypofluorite, sodium hydride.

ETHYNE, DICHLORO- (9CI) (7572-29-4) Heat or air contact may cause explosion. Reacts violently with oxidizers and acids.

ETHYNYL CARBINOL (107-19-7) Incompatible with strong acids, aliphatic amines, phosphorus pentoxide, caustics, isocyanates, strong oxidizers.

ETHYNYLCARBINOL (107-19-7) Incompatible with strong acids, aliphatic amines, phosphorus pentoxide, caustics, isocyanates, strong oxidizers.

ETHYNYLDIMETHYLCARBINOL (115-19-5) Forms explosive mix-ture with air (flash point less than 70°F/21°C). Incompatible with strong acids, caustics, aliphatic amines, isocyanates, oxidizers.

ETIL ACRILATO (Italian) (140-88-5) Forms explosive mixture with air (flash point 48°F/9°C). Atmospheric moisture and strong alkalies may cause fire and explosions. Unless properly inhibited (note: inert gas blanket not recommended), heat, light, or peroxides can cause polymerization. Incompatible with oxidizers (reaction may be violent), strong acids, amines. May accumulate static electrical charges, and may cause ignition of its vapors.

ETILACRILATULUI (Rumanian) (140-88-5) Forms explosive mix-ture with air (flash point 48°F/9°C). Atmospheric moisture and strong alkalies may cause fire and explosions. Unless properly inhibited (note: inert gas blanket not recommended), heat, light, or peroxides can cause polymerization. Incompatible with oxidizers (reaction may be violent), strong acids, amines. May accumulate static electrical charges, and may cause ignition of its vapors.

ETILAMINA (Italian) (75-04-7) Forms explosive mixture with air (flash point less than −0°F/−18°C). Incompatible with acids, organic anhydrides, isocyanates, vinyl acetate, acrylates, substituted allyls, alkylene oxides, epichlorohydrin, ketones, aldehydes, alcohols, gly-cols, phenols, cresols, caprolactam solution, strong oxidizers. Attacks aluminum, copper, lead, tin, zinc and alloys, and some plastics, rubber, and coatings.

ETILBENZENE (Italian) (100-41-4) May accumulate static electrical charges, and may cause ignition of its vapors. Forms explosive mixture with air (flash point 59°F/15°C). Incompatible with strong oxidizers, nitric acid.

ETILE (ACETATO di) (Italian) (141-78-6) Will hydrolyze on standing, producing acetic acid and ethyl alcohol. This reaction is greatly accelerated by bases (alkalies). Forms explosive mixture with air (flash point 24°F/−4.4°C). Incompatible with strong acids, nitrates, oxidiz-ers, chlorosulfonic acid, lithium aluminum hydride, oleum.

ETILENE (OSSIDO di) (Italian) (75-21-8) Forms explosive mixture with air (flash point 20°F/−6°C). Incompatible with alkali metal hydroxides, highly active catalysts (e.g., anhydrous chlorides of iron, tin, or aluminum and oxides of iron or aluminum). Avoid contact with copper. Protect container from physical damage, sun and heat. Attacks some plastics, rubber, or coatings.

ETILENIMINA (Italian) (151-56-4) Forms explosive mixture with air (flash point 12°F/−11°C). May accumulate static electrical charges, and may cause ignition of its vapors. Contact with acid, aluminum,

367

carbon dioxide, or silver may cause explosive polymerization. Attacks rubber, coatings, plastics, and chemically active metals. Self-reactive with heat or atmospheric carbon dioxide.

ETIOL (121-75-5) Incompatible with strong oxidizers, magnesium, alkaline pesticides. Attacks metals, some plastics, rubber, and coatings.

ETO (75-21-8) Forms explosive mixture with air (flash point 20°F/−6°C). Incompatible with alkali metal hydroxides, highly active catalysts (e.g., anhydrous chlorides of iron, tin, or aluminum and oxides of iron or aluminum). Avoid contact with copper. Protect container from physical damage, sun, and heat. Attacks some plastics, rubber, or coatings.

ETOKSYETYLOWY ALKOHOL (Polish) (110-80-5) Forms explosive mixture with air (flash point 120°F/49°C). Strong oxidizers may cause fire and explosions. Attacks some plastics, rubber, and coatings.

ETOXYETHENE (109-92-2) Incompatible with acids (nonoxidizing, sulfuric, nitric), ammonia, aliphatic amines, alkanolamines. Incompatible with aluminum and its alloys.

ETROLENE (299-84-3) Temperatures above 300°F/150°C may cause explosive decomposition. Contact with strong oxidizers may cause fire and explosions. Attacks some plastics, rubber, and coatings.

ETYLENU TLENEK (Polish) (75-21-8) Forms explosive mixture with air (flash point 20°F/−6°C). Incompatible with alkali metal hydroxides, highly active catalysts (e.g., anhydrous chlorides of iron, tin, or aluminum and oxides of iron or aluminum). Avoid contact with copper. Protect container from physical damage, sun, and heat. Attacks some plastics, rubber, or coatings.

ETYLOAMINA (Polish) (75-04-7) Forms explosive mixture with air (flash point less than 0°F/−18°C). Incompatible with acids, organic anhydrides, isocyanates, vinyl acetate, acrylates, substituted allyls, alkylene oxides, epichlorohydrin, ketones, aldehydes, alcohols, glycols, phenols, cresols, caprolactam solution, strong oxidizers. Attacks aluminum, copper, lead, tin, zinc, and alloys, and some plastics, rubber, and coatings.

ETYLOBENZEN (Polish) (100-41-4) May accumulate static electrical charges, and may cause ignition of its vapors. Forms explosive mixture with air (flash point 59°F/15°C). Incompatible with strong oxidizers, nitric acid.

ETYLOWY ALKOHOL (Polish) (64-17-5) Forms explosive mixture with air (flash point 65°F/18°C). May accumulate static electrical charges, and may cause ignition of its vapors. Reactions may be violent with oleum, sulfuric acid, nitric acid, bases, aliphatic amines, isocyanates, oxidizers.

ETYLU BROMEK (Polish) (74-96-4) Forms explosive mixture with air (flash point less than −4°F/−20°C)). Hydrolyzes in water, producing hydrogen bromide. Oxidizers may cause fire or explosions. Fire and explosions may be caused by contact with aluminum, magnesium, or zinc powders; incompatible with lithium, potassium, sodium. Attacks some plastic, rubber, and coatings.

ETYLU CHLOREK (Polish) (75-00-3) Flammable gas. Reacts slowly with water; producing hydrogen chloride gas. Contact with moisture forms hydrochloric acid. May accumulate static electrical charges, and may cause ignition of its vapors. Forms explosive mixture with air (flash point −58°F/−55°C). Contact with aluminum, lithium, magnesium, sodium, potassium, zinc may cause fire and explosions. Attacks some plastics and rubber.

ETYLU KRZEMIAN (Polish) (78-10-4) Forms explosive mixture with air (flash point 99°F/37°C). Incompatible with strong oxidizers, strong acids, water (produces volatile ethyl alcohol). Attacks some plastics and rubber.

EUFIN (105-58-8) Forms explosive mixture with air (flash point 77°F/25°C). Reacts violently with strong oxidizers.

EUNATROL (143-19-1) Combustible solid. Strong oxidizers may cause fires or explosions.

EVAU-SUPERFALL (7775-09-9) A powerful oxidizer; reacts violently with reducing agents and combustible matter. Explosions may be caused by contact with ammonia salts, carbon, oils, metal sulfides, nitrobenzene, powdered metals, sugar. Contact with strong acids produces carbon dioxide. Forms shock-sensitive mixtures with some organic materials. Solution (50%) decomposes at 300°F/149°C, liberating oxygen.

EVE (109-92-2) Incompatible with acids (nonoxidizing, sulfuric, nitric), ammonia, aliphatic amines, alkanolamines.

EVOLA (106-46-7) Forms explosive mixture with air (flash point 150°F/66°C). Incompatible with strong oxidizers, metal powders and alkali metals (i.e., lithium, sodium, potassium, rubidium, cesium, francium). Attacks some plastics, rubber, and coating.

EWEISS (7727-43-7) Explosions may result from contact with aluminum in the presence of heat. Incompatible with potassium, phosphorus.

EXAGAMA (58-89-9) Not combustible, but may be dissolved in a combustible solvent. If solvent comes in contact with oxidizers, fire and explosions may result.

EXHAUST GAS (630-08-0) Forms extremely explosive mixture with air. Reacts violently with strong oxidizers.

EXPERIMENTAL INSECTICIDE 7744 (63-25-2) Incompatible with strong oxidizers, strongly alkaline pesticides.

EXPLOSIVE D (131-74-8) A self-reactive explosive. Heat above 250°F/121°C can cause explosions. A powerful oxidizer. Contact with metal, concrete, or plaster produces salts that are more shock-sensitive than ammonium picrate. Water increases rate of reactivity with metals.

EXSICCATED FERROUS SULFATE (7720-78-7) Aqueous solution is acidic. Contact with bases produces iron.

EXTERMATHION (121-75-5) Incompatible with strong oxidizers, magnesium, alkaline pesticides. Attacks metals, some plastics, rubber, and coatings.

EXTRACT-S (75-18-3) Forms explosive mixture with air (flash point −36°F/−38°C). Reacts violently with strong oxidizers. May accumulate static electrical charges, and may cause ignition of its vapors.

EXTREMA (10026-04-7) Water contact produces hydrochloric acid. Reacts violently with potassium or sodium. Attacks metals in the presence of moisture.

EXTREMA (78-10-4) Forms explosive mixture with air (flash point 99°F/37°C). Incompatible with strong oxidizers, strong acids, water (produces volatile ethyl alcohol). Attacks some plastics and rubber.

E-Z-PAQUE (7727-43-7) Explosions may result from contact with aluminum in the presence of heat. Incompatible with potassium, phosphorus.

- F -

F 11 (75-69-4) Reacts with barium, lithium, sodium, magnesium, titanium. Attacks some plastics, rubber, and coatings.

F 12 (75-71-8) Reacts violently with liquid aluminum. Incompatible with chemically active metals. Attacks some plastics, rubber, and coatings.

F 13 (75-72-9) Reacts with aluminum, magnesium, zinc, and their alloys.

F-13B1 (75-63-8) Reacts with chemically active metals, powdered aluminum, zinc, magnesium. May attack some plastics, rubber, and coatings.

F 21 (75-43-4) Reacts with water and chemically active metals. Attacks some plastics, rubber, and coatings.

F 22 (75-45-5) Moisture and rust cause slow decomposition, producing toxic gases. Attacks some plastics, rubber, and coatings. Thermal decomposition occurs at high temperature with alkalies and alkaline earth metals.

F 112 (76-12-0) Reacts with chemically active metals, powdered aluminum, zinc, and magnesium. Attacks some forms of plastics, rubber, and coatings.

F 114 (76-14-2) Reacts with barium, lithium, sodium, magnesium, titanium. Attacks some plastics, rubber, and coatings.

F 115 (76-15-3) Thermal decomposition occurs at high temperatures with alkalies and alkaline earth metals.

F 1991 (17804-35-2) Heat, water, strong acids and strong alkalies can cause decomposition and formation of toxic oxides of nitrogen.

FA (50-00-0) May polymerize unless properly inhibited (usually with methanol). Forms explosive mixture with air (flash point 122°F/50°C). Incompatible with strong acids, amines, strong oxidizers, alkaline materials, nitrogen dioxide, performic acid. Reaction with hydrochloric acid produces bis-chloromethyl ether, a carcinogen.

FAA (53-96-3) Incompatible with cyanides.

2-FAA (53-96-3) Incompatible with cyanides.

N-FAA (53-96-3) Incompatible with cyanides.

FACTITIOUS AIR (10024-97-2) May form explosive mixture with air. Incompatible with nonflammable gas; supports combustion. Incompatible with aluminum, anhydrous ammonia (may be explosive), boron, hydrazine lithium hydride, phosphine, sodium.

FALITIRAM (137-26-8) Combustible solid (flash point 192°F/89°C). Strong oxidizers may cause fire and explosions; contact with strong acid or oxidizable materials produces toxic gases.

FALKITOL, FASCIOLIN (67-72-1) Incompatible with hot iron, zinc, and aluminum; alkalies cause formation of spontaneously explosive chloroacetylene. Attacks some plastics, rubber, and coatings.

FANNOFORM (50-00-0) May polymerize unless properly inhibited (usually with methanol). Forms explosive mixture with air (flash point 122°F/50°C). Incompatible with strong acids, amines, strong oxidizers, alkaline materials, nitrogen dioxide, performic acid. Reaction with hydrochloric acid produces bis-chloromethyl ether, a carcinogen.

FASCIOLIN (56-23-5) Becomes corrosive when in contact with water. Corrosive to metals. Reacts violently with many compounds. Decomposes on contact with chemically active metals such as sodium, potassium, and magnesium. Incompatible with allyl alcohol, fluorine gas, alkali metals, aluminum. Attacks some coatings, plastics, and rubber.

FASCO-TERPENE (8001-35-2) Reacts with oxidizers, with a risk of fire or explosions. Attacks metals in the presence of moisture.

FAST CORINTH BASE B (92-87-5) Oxidizes on exposure to light and air. Reacts violently with strong oxidizers. Red fuming nitric acid may cause fire.

FAST DARK BLUE BASE R (95-53-4) Forms explosive mixture with air (flash point 185°F/85°C). Incompatible with strong acids, mineral acids, organic anhydrides, isocyanates, aldehydes, oxidizers. Attacks some plastics, rubber, and coatings.

FAST GARNET B BASE or FAST GARNET BASE B (134-32-7) Oxidizes in air. Incompatible with nitrous acid, oxidizers, nitrates, organic anhydrides, isocyanates, aldehydes.

FAST RED BASE (100-01-06) Strong oxidizers and moisture may cause spontaneous heating. Caustics and heat may produce explosive mixtures. Attacks some plastics, rubber, and coatings. May accumulate static electrical charges, and may cause ignition of its vapors.

FAST RED GG BASE (100-01-06) Strong oxidizers and moisture may cause spontaneous heating. Caustics and heat may produce explosive mixtures. Attacks some plastics, rubber, and coatings. May accumulate static electrical charges, and may cause ignition of its vapors.

FAST RED IG BASE (100-01-06) Strong oxidizers and moisture may cause spontaneous heating. Caustics and heat may produce explosive mixtures. Attacks some plastics, rubber, and coatings. May accumulate static electrical charges, and may cause ignition of its vapors.

FAST RED 2G BASE (100-01-06) Strong oxidizers and moisture may cause spontaneous heating. Caustics and heat may produce explosive mixtures. Attacks some plastics, rubber, and coatings. May accumulate static electrical charges, and may cause ignition of its vapors.

FAST RED TR BASE (95-79-4) Incompatible with acids, organic anhydrides, isocyanates, aldehydes, oxidizers.

FAST SCARLET BASE B (91-59-8) Forms explosive mixture with air (flash point 300°F/190°C). Strong oxidizers, strong acids may cause fire and explosions. Oxidizes in the presence of air and light.

FAST WHITE (7446-14-2) Reacts violently with potassium. Incompatible with aluminum, magnesium.

FB/2 (85-00-7) Concentrated solution attacks aluminum.

FC-11 (75-69-4) Reacts with barium, lithium, sodium, magnesium, titanium. Attacks some plastics, rubber, and coatings.

FC 12 (75-71-8) Reacts violently with liquid aluminum. Incompatible with chemically active metals. Attacks some plastics, rubber, and coatings.

FDA (696-28-6) Forms explosive mixture with air (flash point 60°F/−16°C). Water contact produces hydrochloric acid. Attacks metals in the presence of moisture.

FECAMA (62-73-7) Attacks some plastics, rubber, and coatings.

FEGLOX (85-00-7) Concentrated solution attacks aluminum.

FEKABIT (3811-04-9) A powerful oxidizer. Reacts violently with reducing agents or combustibles, ammonia gas, ammonium salts, organic matter, hydrogen iodide, organic acids, sulfuric acid. Forms explosive mixtures with metallic powders, ammonium chloride, organic solids, including agricultural materials.

FEMA NO. 2003 (75-07-0) Slowly polymerizes to paraldehyde. Explodes when mixed with iodine. Contact with strong bases can cause explosive polymerization. A strong reducing agent. Reacts violently with combustibles, strong acids, caustics, ammonia, aliphatic amines, alkanolamines, aromatic amines, organic substances, halogens, oxidizers; forms explosive peroxides with air. May dissolve rubber. May accumulate static electrical charges, and may cause ignition of its vapors.

FEMA NO. 2006 (64-19-7) Vapor forms explosive mixture with air (flash point 105°F/40°C). Reacts violently with oxidizers and bases. Incompatible with strong acids, aliphatic amines, alkanolamines, isocyanates, alkylene oxides, epichlorohydrin, acetaldehyde, 2-aminoethanol, ammonia, ammonium nitrate, chlorosulfonic acid, chromic acid, ethylene diamine, ethyleneimine, perchloric acid, permanganates, phosphorus isocyanate, phosphorus trichloride, potassium *tert*-butoxide, xylene. Attacks cast iron and other metals, producing flammable hydrogen gas.

FEMA NO. 2009 (98-86-2) Incompatible with strong acids, aliphatic amines, oxidizers. Forms explosive mixture with air (flash point 170°F/77°C).

FEMA NO. 2011 (124-04-9) May accumulate static electrical charges, and may cause ignition of its vapors. Contact with strong oxidizers may cause fire and explosions.

FEMA NO. 2134 (119-61-9) Combustible solid. Strong oxidizers may cause fire and explosions. Attacks some plastics, rubber, and coatings.

FEMA NO. 2135 (140-11-4) Forms explosive mixture with air (flash point 216°F/102°C). Reacts with strong acids, nitrates, oxidizers.

FEMA NO. 2137 (100-51-6) Forms explosive mixture with air (flash point 213°F/101°C). Slowly oxidizes in air and oxygen. Incompatible with mineral acids, caustics, aliphatic amines, isocyanates. Reacts violently with strong oxidizers. Corrodes aluminum at high temperature. Attacks some nonfluorinated plastics; may not attack polypropylene.

FEMA NO. 2174 (123-86-4) Forms explosive mixture with air (flash point 72°F/22°C). Reacts with water on standing, to produce acetic acid and n-butyl alcohol. Reacts violently with strong oxidizers and potassium-*tert*-butoxide. Incompatible with caustics, strong acids, nitrates. Dissolves rubber, many plastics, resins and some coatings. May accumulate static electrical charges, and may cause ignition of its vapors.

FEMA NO. 2178 (71-36-3) Forms explosive mixture with air (flash point 98°F/37°C). May react with aluminum above 120°F/49°C. Attacks some plastics, rubber, and coatings. Incompatible with strong acids, halogens, caustics, alkali metals, aliphatic amines, isocyanates.

FEMA NO. 2186 (109-21-7) Forms explosive mixture with air (flash point 48°F/−9°C). Attacks some plastics, rubber, and coatings. Incompatible with strong acids, nitrates; reacts violently with strong oxidizers.

FEMA NO. 2219 (123-72-8) May accumulate static electrical charges, and may cause ignition of its vapors. Forms explosive mixture with air (flash point −8°F/−22°C). Incompatible with strong oxidizers, strong acids, caustics, ammonia, aliphatic amines, alkanolamines, aromatic amines.

FEMA NO. 2220 (78-84-2) Forms explosive gas mixture with air. Incompatible with strong acids, caustics, aliphatic amines, alkanolamines, aromatic amines, strong oxidizers.

FEMA NO. 2221 (107-92-6) Forms explosive mixture with air (flash point 162°F/72°C). Incompatible with sulfuric acid, caustics, ammonia, aliphatic amines, isocyanates, strong oxidizers, alkylene oxides, epichlorohydrin.

FEMA NO. 2229 (79-92-5) Contact with strong oxidizers may cause fire and explosions. Emulsions in xylene may violently decompose on contact with iron or aluminum above 158°F/70°C.

FEMA NO. 2356 (99-87-6) Forms explosive mixture with air (flash point 117°F/47°C). Incompatible with nitric acid, strong oxidizers. Attacks and softens rubber. May accumulate static electrical charges, and may cause ignition of its vapors.

FEMA NO. 2362 (112-31-2) Incompatible with strong acids, ammonia, caustics, amines. Attacks galvanized steel.

FEMA NO. 2365 (112-30-1) Forms explosive mixture with air (flash point 180°F/82°C). Incompatible with strong acids, caustics, aliphatic amines, isocyanates, strong oxidizers.

FEMA NO. 2371 (103-50-4) Forms unstable peroxides. Reacts violently with strong oxidizers. May accumulate static electrical charges, and may cause ignition of its vapors.

FEMA NO. 2414 (141-78-6) Will hydrolyze on standing, producing acetic acid and ethyl alcohol. This reaction is greatly accelerated by bases (alkalies). Forms explosive mixture with air (flash point 24°F/−4.4°C). Incompatible with strong acids, nitrates, oxidizers, chlorosulfonic acid, lithium aluminum hydride, oleum.

FEMA NO. 2415 (141-97-9) Forms explosive mixture with air (flash point 135°F/57°C). Incompatible with strong acids, nitrates, oxidizers.

FEMA NO. 2418 (140-88-5) Forms explosive mixture with air (flash point 48°F/9°C). Atmospheric moisture and strong alkalies may cause fire and explosions. Unless properly inhibited (note: inert gas blanket not recommended), heat, light, or peroxides can cause polymerization. Incompatible with oxidizers (reaction may be violent), strong acids, amines. May accumulate static electrical charges, and may cause ignition of its vapors.

FEMA NO. 2427 (105-54-4) Incompatible with strong acids, nitrates, oxidizers.

374

FEMA NO. 2434 (109-94-4) Forms explosive mixture with air (flash point −4°F/−20°C). May accumulate static electrical charges, and may cause ignition of its vapors. Incompatible with nitrates, strong oxidizers, strong alkalies, strong acids.

FEMA NO. 2440 (97-64-3) Fire or explosion hazard (flash point 115°F/46°C). Strong oxidizers may cause fire and explosions.

FEMA NO. 2456 (105-37-3) Forms explosive mixture with air (flash point 54°F/12°C). May accumulate static electrical charges, and may cause ignition of its vapors. Incompatible with strong acids, nitrates, strong oxidizers.

FEMA NO. 2489 (98-01-1) Forms explosive mixture with air (flash point 140°F/60°C). Incompatible with strong acids, caustics, ammonia, aliphatic amines, alkanolamines, aromatic amines, oxidizers. Attacks many plastics and coatings.

FEMA NO. 2544 (110-43-0) Forms explosive mixture with air (flash point 102°F/39°C). Incompatible with strong acids, alkalies, aliphatic amines, oxidizers. Attacks some plastics and rubber.

FEMA NO. 2545 (106-35-4) Forms explosive mixture with air (flash point 115°F/46°C). Reacts violently with strong oxidizers. Attacks some plastics, rubber, and coatings.

FEMA NO. 2548 (111-70-6) Forms explosive mixture with air (flash point 170°F/77°C). Incompatible with strong acids, caustics, aliphatic amines, isocyanates, oxidizers.

FEMA NO. 2557 (66-25-1) Forms explosive mixture with air (flash point 90°F/32°C). Incompatible with strong acids, caustics, ammonia, amines. Attacks some plastics, rubber, and coatings.

FEMA NO. 2559 (142-62-1) Reacts with strong oxidizers. Incompatible with sulfuric acid, caustics, ammonia, amines, isocyanates, alkylene oxides, epichlorohydrin. Attacks common metals.

FEMA NO. 2567 (111-27-3) Forms explosive mixture with air (flash point 145°F/63°C). Incompatible with strong acids, caustics, aliphatic amines, isocyanates, strong oxidizers.

FEMA NO. 2617 (112-53-8) Incompatible with strong acids, caustics, aliphatic amines, isocyanates, oxidizers.

FEMA NO. 2683 (93-58-3) Forms explosive mixture with air (flash point 181°F/83°C). Incompatible with strong acids, nitrates, oxidizers.

FEMA NO. 2685 (98-85-1) Forms explosive mixture with air (flash point 205°F/96°C). Incompatible with strong acids, caustics, aliphatic amines, isocyanates, oxidizers.

FEMA No. 2731 (108-10-1) Forms explosive mixture with air (flash point 64°F/18°C). Incompatible with oxidizers, strong acids, aliphatic amines. Dissolves some plastics, resins and rubber.

FEMA NO. 2788 (143-13-5) Incompatible with strong acids, nitrates

FEMA NO. 2789 (143-08-8) Forms explosive mixture with air (flash point 165°F/74°C). Incompatible with strong acids, caustics, aliphatic amines, isocyanates, strong oxidizers.

FEMA NO. 2797 (124-13-0) Forms explosive mixture with air (flash point 125°F/52°C). Incompatible with strong oxidizers, caustics, ammonia, aliphatic amines, alkanolamines, aromatic amines, strong oxidizers.

FEMA NO. 2800 (111-87-5) Forms explosive mixture with air (flash point 178°F/81°C). Incompatible with strong acids, caustics, aliphatic amines, isocyanates, strong oxidizers.

FEMA NO. 2841 (123-54-6) Forms explosive mixture with air (flash point 95°F/35°C). Reacts violently with strong oxidizers. Incompatible with aliphatic amines, alkanolamines, organic acids, isocyanates.

FEMA NO. 2842 (107-87-9) Forms explosive mixture with air (flash point 45°F/7°C). Reacts violently with strong oxidizers. Attacks some plastics, rubber, and coatings.

FEMA NO. 2902 (80-56-8) Incompatible with sulfuric acid, nitric acid.

FEMA No. 2923 (123-38-6) Forms explosive mixture with air (flash point −2°F/−30°C). Incompatible with strong acids, caustics, amines. Reacts violently with strong oxidizers. Can self-ignite if finely dispersed on porous or combustible material. Heat or ultraviolet can cause decomposition. May accumulate static electrical charges, and may cause ignition of its vapors.

FEMA NO. 3101 (109-52-4) Forms explosive mixture with air (flash point 205°F/96°C). Incompatible with sulfuric acid, caustics, ammonia, amines, isocyanates, alkylene oxides, epichlorohydrin, strong oxidizers.

FEMA NO. 3326 (67-64-1) Forms explosive mixture with air (flash point −4°F/−20°C). Reacts violently with chloroform. Incompatible with strong acids, aliphatic amines, chloroform, chromic anhydride, chromyl chloride, hexachloromelamine, hydrogen peroxide, nitrosyl chloride, nitrosyl perchlorate, nitryl perchlorate, permonosulfuric acid, potassium *tert*-butoxide. Unstable and explosive peroxides are formed with strong oxidizers. May accumulate static electrical charges, and may cause ignition of its vapors. Dissolves most rubber, resins, and plastics.

FENAMIN or FENAMINE (1912-24-9) Incompatible with strong acids.

FENAMINE (61-82-5) Substance acts as a weak base to produce salts in contact with acids. Corrosive to iron, aluminum, copper, and copper alloys.

FENATROL (1912-24-9) Incompatible with strong acids.

FENAVAR (61-82-5) Substance acts as a weak base to form salts in contact with acids. Corrosive to iron, aluminum, copper, and copper alloys.

FENCE RIDER (93-76-5) Sealed metal containers may burst in heat above 316°F/158°C. Incompatible with sulfuric acid, bases, ammonia, aliphatic amines, alkanolamines, isocyanates, alkylene oxides, epichlorohydrin.

FENCHLOORFOS (Dutch) (299-84-3) Temperatures above 300°F/150°C may cause explosive decomposition. Contact with strong oxidizers may cause fire and explosions. Attacks some plastics, rubber, and coatings.

FENCHLOROPHOS (299-84-3) Temperatures above 300°F/150°C may cause explosive decomposition. Contact with strong oxidizers may cause fire and explosions. Attacks some plastics, rubber, and coatings.

FENCHLORPHOS (299-84-3) Temperatures above 300°F/150°C may cause explosive decomposition. Contact with strong oxidizers may cause fire and explosions. Attacks some plastics, rubber, and coatings.

FENCLOR (generic CAS for PCBs 1336-36-3) Incompatible with strong oxidizers, strong acids.

2-FENILPROPANO (Italian) (98-82-8) Forms explosive mixture with air (flash point 99°F/37°C). Incompatible with strong acids, strong oxidizers. Air contact produces cumene hydroperoxide. May be able to form unstable peroxides. Attacks rubber. May accumulate static electrical charges, and may cause ignition of its vapors.

FENOL (Dutch or Polish) (108-95-2) Forms explosive mixture with air (flash point 174°F/79°C). Incompatible with strong oxidizers, strong acids, caustics, aliphatic amines, amides, oxidizers, formaldehyde, butadiene, calcium hypochlorite. Liquid attacks some plastics, rubber, and coatings; hot liquid attacks aluminum, magnesium, lead, and zinc metals.

FENOLO (Italian) (108-95-2) Forms explosive mixture with air (flash point 174°F/79°C). Incompatible with strong oxidizers, strong acids, caustics, aliphatic amines, amides, oxidizers, formaldehyde, butadiene, calcium hypochlorite. Liquid attacks some plastics, rubber, and coatings; hot liquid attacks aluminum, magnesium, lead, and zinc metals.

FENOMORE (93-72-1) A powerful oxidizer; reacts violently with reducing agents, combustibles.

FENOPROP (93-72-1) A powerful oxidizer; reacts violently with reducing agents, combustibles.

FENOXYL CARBON N (51-28-5) Explosion may be caused by heat, friction, or shock. Contact with reducing agents, combustibles may cause fire and explosions. Forms explosive salts with ammonia or strong bases. May accumulate static electrical charges, and may cause ignition of its vapors.

FENYLENODWUAMINA (Polish) (106-50-3) Incompatible with acids, organic anhydrides, isocyanates, aldehydes, strong oxidizers; may cause fire and explosions. Heat and light contribute to instability.

2-FENYL-PROPAN (Dutch) (98-82-8) Forms explosive mixture with air (flash point 99°F/37°C). Incompatible with strong acids, strong oxidizers. Air contact produces cumene hydroperoxide. May be able to form unstable peroxides. Attacks rubber. May accumulate static electrical charges, and may cause ignition of its vapors.

FEOSOL (7720-78-7) Aqueous solution is acidic. Contact with bases produces iron.

FEOSPAN (7720-78-7) Aqueous solution is acidic. Contact with bases produces iron.

FER-IN-SOL (7720-78-7) Aqueous solution is acidic. Contact with bases produces iron.

FERMENICIDE (7446-09-5) Produces corrosive hydrochloric acid fumes with air. Contact with copper, bronze, or alkali metals may cause fire and explosions. Reacts violently with alcohols, caustics, amines, water. Decomposes above 140°F/60°C; produces toxic and corrosive oxides of sulfur. Attacks some plastics, rubber, and coatings.

FERMENTATION ALCOHOL (64-17-5) Forms explosive mixture with air (flash point 65°F/18°C). May accumulate static electrical charges, and may cause ignition of its vapors. Reactions may be violent with oleum, sulfuric acid, nitric acid, bases, aliphatic amines, isocyanates, oxidizers.

FERMENTATION AMYL ALCOHOL (123-51-3) Forms explosive mixture with air (flash point 109°F/43°C). Strong oxidizers may cause fire and explosions. Attacks some plastics, rubber, and coatings.

FERMENTATION BUTYL ALCOHOL (78-83-1) Forms explosive mixture with air (flash point 82°F/28°C). Incompatible with strong acids, strong oxidizers, caustics, aliphatic amines, isocyanates, alkali metals, and alkali earth. Attacks some plastics, rubber, and coatings. May react with aluminum at high temperatures.

FERMINE (131-11-3) Incompatible with strong alkalies, strong acids, nitrates, oxidizers.

FERNACOL (137-26-8) Combustible solid (flash point 192°F/89°C). Strong oxidizers may cause fire and explosions; contact with strong acid or oxidizable materials produces toxic gases.

FERNASAN (137-26-8) Combustible solid (flash point 192°F/89°C). Strong oxidizers may cause fire and explosions; contact with strong acid or oxidizable materials produces toxic gases.

FERNESTA (94-75-7) Decomposes in sunlight. Incompatible with strong oxidizers; may cause fire and explosions.

FERNIDE (137-26-8) Combustible solid (flash point 192°F/89°C). Strong oxidizers may cause fire and explosions; contact with strong acid or oxidizable materials produces toxic gases.

FERNIMINE (94-75-7) Decomposes in sunlight. Incompatible with strong oxidizers; may cause fire and explosions.

FERNOXONE (94-75-7) Decomposes in sunlight. Incompatible with strong oxidizers; may cause fire and explosions.

FERNOZONE (94-75-7) Decomposes in sunlight. Incompatible with strong oxidizers; may cause fire and explosions.

FERRALYN (7720-78-7) Aqueous solution is acidic. Contact with bases produces iron.

FERRIC CHLORIDE or FERRIC CHLORIDE, ANHYDROUS or FERRIC CHLORIDE, SOLID (7705-08-0) Incompatible with water. Solution is very acidic; reacts violently with bases. Shock- and friction-sensitive explosive material is formed with potassium, sodium and other active metals. Strong reaction occurs with allyl chloride. Attacks metals when wet.

FERRIC CHLORIDE, HEXAHYDRATE (7705-08-0) Incompatible with water. Solution is very acidic; reacts violently with bases. Shock- and friction-sensitive explosive material is formed with potassium, sodium and other active metals. Strong reaction occurs with allyl chloride. Attacks metals when wet.

FERRIC FLUORIDE (7783-50-8) Contact with strong oxidizers may cause fire and explosions.

FERRIC NITRATE or FERRIC NITRATE, NONHYDRATE (10421-48-4) Solution is corrosive to metals. Contact with combustibles may cause fire.

FERRIC OXIDE (1309-37-1) Contact with hydrogen peroxide, ethylene oxide, calcium hypochlorite will cause explosion. Reacts violently with powdered aluminum, hydrazine, hydrogen trisulfide.

FERRIC SULFATE (10028-22-5) Corrosive to copper and its alloys, mild and galvanized steel.

FERRO-GRADUMET (7720-78-7) Aqueous solution is acidic. Contact with bases produces iron.

FERROSILICON (8049-17-0) Water contact forms toxic and explosive gases. Reacts with acids and sodium hydroxide.

FERROSILICON, CONTAINING MORE THAN 30% BUT LESS THAN 90% SILICON (8049-17-0) Water contact forms toxic and explosive gases. Reacts with acids and sodium hydroxide.

FERROSULFAT (German) (7720-78-7) Aqueous solution is acidic. Contact with bases produces iron.

FERROSULFATE (7720-78-7) Aqueous solution is acidic. Contact with bases produces iron.

FERRO-THERON (7720-78-7) Aqueous solution is acidic. Contact with bases produces iron.

FERROUS AMMONIUM SULFATE (10045-89-3) Reacts violently with tetranitromethane, ammonium perchlorate, and mercury(II) dinitrate. Sulfates react with aluminum, magnesium.

FERROUS AMMONIUM SULFATE HEXAHYDRATE (10045-89-3) Reacts violently with tetranitromethane, ammonium perchlorate and mercury(II) dinitrate. Sulfates react with aluminum, magnesium.

FERROUS CHLORIDE (7758-94-3) Solution attacks metals. Contact with ethylene oxide may initiate polymerization. Potassium or sodium contact produces impact-sensitive materials.

FERROUS(III) CHLORIDE (7705-08-0) Incompatible with water. Solution is very acidic; reacts violently with bases. Shock- and friction-sensitive explosive material is formed with potassium, sodium and other active metals. Strong reaction occurs with allyl chloride. Attacks metals when wet.

FERROUS CHLORIDE TETRAHYDRATE (7758-94-3) Solution attacks metals. Contact with ethylene oxide may initiate polymerization. Potassium or sodium contact produces impact-sensitive materials.

FERROUS SULFATE (7720-78-7) Aqueous solution is acidic. Contact with bases produces iron.

FERROUS SULFATE (1:1) (7720-78-7) Aqueous solution is acidic. Contact with bases form iron.

FERROUS(II) SULFATE (7782-63-0) Aqueous solution is strongly acidic; reacts violently with strong acids. Incompatible with oxidizers and oxygen in the air.

FERROUS SULPHATE (7720-78-7) Aqueous solution is acidic. Contact with bases produces iron.

FERRUGO (1309-37-1) Contact with hydrogen peroxide, ethylene oxide, calcium hypochlorite will cause explosion. Reacts violently with powdered aluminum, hydrazine, hydrogen trisulfide.

FERSOLATE (7720-78-7) Aqueous solution is acidic. Contact with bases produces iron.

FERTILIZER ACID (7664-93-9) A strong oxidizer that can react violently with risk of fire and explosion with many substances, including reducing agents, organic and combustible substances, and bases. Incompatible with nonoxidizing mineral acids, organic acids, bases, acrylates, aldehydes, alcohols, alkylene oxides, ammonia, aliphatic amines, alkanolamines, aromatic amines, amides, chlorates, epichlorohydrin, fulminates, glycols, isocyanates, ketones, metals (powdered), organic anhydrides, perchlorates, picrates, substituted allyls, phenols and cresols, water, acetic anhydride, acetone cyanhydrin, acetonitrile, acrolein, acrylonitrile, allyl alcohol, allyl chloride, 2-aminoethanol, ammonium hydroxide, aniline, bromine pentafluoride, n-butyraldehyde, caprolactam solution, carbides, cesium acetylene carbide, chlorine trifluoride, chlorosulfonic acid, cuprous nitride, diiso-butylene, ethylene cyanohydrin, ethylene diamine, ethylene glycol, ethyleneimine, hydrochloric acid, iodine heptafluoride, iron, isoprene, lithium silicide, mercuric nitride, mesityl oxide, nitric acid, p-nitrotolu-ene, perchloric acid, phosphorus, potassium *tert*-butoxide, potassium chlorate, potassium permanganate, propiolactone (beta-), propylene oxide, pyridine, rubidium acetylene, silver permanganate, sodium, sodium carbonate, sodium chlorate, sodium hydroxide, styrene mono-mer, vinyl acetate. Attacks most metals, and some plastics, rubber, and coatings.

FICLOR 91 (87-90-1) A powerful oxidizer. Forms explosive material with nitrogen compounds. Contact with organic materials or reducing agents may cause fire.

FILMERINE (7632-00-0) A strong oxidizer; reacts violently with reducing agents and combustibles. Reacts with acids, ammonium salts + heat, butadiene, cyanides, lithium, phthalic acid, thiosulfate, sodium amide, and many other substances. Under certain conditions the chemical can also be a strong reducing agent.

FILTER ALUM (10043-01-3) Aqueous solution is a strong acid. Incompatible with caustics.

FINEMEAL (7727-43-7) Explosions may result from contact with aluminum in the presence of heat. Incompatible with potassium, phosphorus.

FIRE DAMP (74-82-8) Forms an explosive mixture with air. Reacts violently with strong oxidizers.

FIRMOTOX (8003-34-7) Strong oxidizers may cause fire and explo-sions.

FISH OIL (N/A) Incompatible with strong acids, oxidizers.

FLAXSEED OIL (8001-26-1) Incompatible with nitric acid; oxidizers may cause fire and explosions.

FLECK-FLIP (79-01-6) Contact with caustics produces a toxic and flammable gas. Reacts violently with chemically active metals. Contact with aluminum may produce a violent, self-accelerating polymerization reaction. Incompatible with acids, organic anhydrides, isocyanates, alkylene oxides, aldehydes, alcohols, glycols, phenols, cresols, capro-lactam solution, epichlorohydrin, nitrogen tetroxide, metal powders, oxygen. May accumulate static electrical charges, and may cause ignition of its vapors.

FLEXOL A-26 (103-23-1) Incompatible with strong acids, nitrates, oxidizers.

FLEXOL PLASTICIZER P (27554-26-3) May accumulate static electrical charges, and may cause ignition of its vapors. Incompatible with strong acids, nitrates, oxidizers.

FLIT 406 (133-06-2) Incompatible with tetraethyl pyrophosphate, parathion.

FLO-MORE (30525-89-4) Forms explosive mixture with air (flash point 160°F/71°C). Aqueous solution causes formation of formaldehyde. Reacts violently with strong oxidizers or liquid oxygen. May accumulate static electrical charges, and may cause ignition of its vapors.

FLO PRO T SEED PROTECTANT (137-26-8) Combustible solid (flash point 192°F/89°C). Strong oxidizers may cause fire and explosions; contact with strong acid or oxidizable materials produces toxic gases.

FLORES MARTIS (7705-08-0) Incompatible with water. Solution is very acidic; reacts violently with bases. Shock- and friction-sensitive explosive material is formed with potassium, sodium, and other active metals. Strong reaction occurs with allyl chloride. Attacks metals when wet.

FLORIDINE (7681-49-4) Incompatible with water; forms a corrosive. Reacts with acids.

FLOROCID (7681-49-4) Incompatible with water; forms a corrosive. Reacts with acids.

FLOUR SULPHUR (7704-34-9) Combustible solid. Liquid contact produces sulfur dioxide with air. Reacts violently with strong oxidizers. Forms explosive, shock-sensitive, or pyrophoric mixtures with ammonia, ammonium nitrate, bromates, calcium carbide, charcoal, chlorates, hydrocarbons, iodates, iron. Reacts violently with halogen compounds, sodium, tin, uranium, and other compounds. Attacks steel when moist. May accumulate static electrical charges, and may cause ignition of its vapors.

FLOWERS OF SULPHUR (7704-34-9) Combustible solid. Liquid contact produces sulfur dioxide with air. Reacts violently with strong oxidizers. Forms explosive, shock-sensitive, or pyrophoric mixtures with ammonia, ammonium nitrate, bromates, calcium carbide, charcoal, chlorates, hydrocarbons, iodates, iron. Reacts violently with halogen compounds, sodium, tin, uranium, and other compounds. Attacks steel when moist. May accumulate static electrical charges, and may cause ignition of its vapors.

FLOZENGES (7681-49-4) Incompatible with water; forms a corrosive. Reacts with acids.

FLUATE (79-01-6) Contact with caustics produces a toxic and flammable gas. Reacts violently with chemically active metals. Contact with aluminum may produce a violent, self-accelerating polymerization reaction. Incompatible with acids, organic anhydrides, isocyanates, alkylene oxides, aldehydes, alcohols, glycols, phenols, cresols, caprolactam solution, epichlorohydrin, nitrogen tetroxide, metal powders, oxygen.

FLUBORIC ACID (16872-11-0) A strong acid; reacts violently with strong alkali materials. Incompatible with strong oxidizers. Attacks metals.

FLUE GAS (630-08-0) Forms extremely explosive mixture with air. Reacts violently with strong oxidizers.

FLUKOIDS (56-23-5) Becomes corrosive when in contact with water. Corrosive to metals. Reacts violently with many compounds. Decomposes on contact with chemically active metals such as sodium, potassium, and magnesium. Incommpatible with allyl alcohol, fluorine gas, alkali metals, aluminum. Attacks some coatings, plastics, and rubber.

FLUOPHOSGENE (353-50-4) Moisture contact produces hydrogen fluoride gas and carbon dioxide.

FLUOR (Dutch) (7782-41-4) A powerful oxidizer and a dangerously reactive gas. Reacts violently with reducing agents and combustible materials. Contact with water can produce corrosive hydrogen fluoride, hydrochloric acid, oxygen, and oxygen difluoride. Vigorous reaction with most oxidizable materials at room temperature, frequently with ignition. Corrodes most metals (except the cylinders in which it is shipped), especially in the presence of moisture. Contact with nitric acid produces explosive gas. Attacks some plastics, rubber, and coatings. Reacts explosively reaction with ammonia, graphite, halocarbons, hydrocarbons, hydrogen, and many other substances. Reacts with nearly every known element.

FLUOR (French) (7782-41-4) A powerful oxidizer and a dangerously reactive gas. Reacts violently with reducing agents and combustible materials. Contact with water can produce corrosive hydrogen fluoride, hydrochloric acid, oxygen and oxygen difluoride. Reacts vigorously with most oxidizable materials at room temperature, frequently with ignition. Corrodes most metals (except the cylinders in which it is shipped), especially in the presence of moisture. Contact with nitric acid produces explosive gas. Attacks some plastics, rubber, and coatings. Reacts explosively with ammonia, graphite, halocarbons, hydrocarbons, hydrogen, and many other substances. Reacts with nearly every known element.

FLUOR (German or Polish) (7782-41-4) A powerful oxidizer and a dangerously reactive gas. Reacts violently with reducing agents and combustible materials. Contact with water can produce corrosive hydrogen fluoride, hydrochloric acid, oxygen, and oxygen difluoride. Reacts vigorously with most oxidizable materials at room temperature, frequently with ignition. Corrodes most metals (except the cylinders in which it is shipped), especially in the presence of moisture. Contact with nitric acid produces explosive gas. Attacks some plastics, rubber, and coatings. Reacts explosively with ammonia, graphite, halocarbons, hydrocarbons, hydrogen, and many other substances. Reacts with nearly every known element.

FLUORAL (7681-49-4) Incompatible with water; forms a corrosive. Reacts with acids.

FLUORANE 114 (76-14-2) Reacts with barium, lithium, sodium, magnesium, titanium. Attacks some plastics, rubber, and coatings.

2-FLUORENYLACETAMIDE (53-96-3) Incompatible with cyanides.

FLUORETHYLENE (75-02-5) Flammable gas. May be able to polymerize. Reacts violently with oxidizers. May accumulate static electrical charges, and may cause ignition of its vapors.

FLUORINE (7782-41-4) A powerful oxidizer and a dangerously reactive gas. Reacts violently with reducing agents and combustible materials. Contact with water can produce corrosive hydrogen fluoride, hydrochloric acid, oxygen, and oxygen difluoride. Reacts vigorously with most oxidizable materials at room temperature, frequently with ignition. Corrodes most metals (except the cylinders in which it is shipped), especially in the presence of moisture. Contact with nitric acid produces explosive gas. Attacks some plastics, rubber, and coatings. Reacts explosively with ammonia, graphite, halocarbons, hydrocarbons, hydrogen and many other substances. Reacts with nearly every known element.

FLUORINE-19 (7782-41-4) A powerful oxidizer and a dangerously reactive gas. Reacts violently with reducing agents and combustible materials. Contact with water can produce corrosive hydrogen fluoride, hydrochloric acid, oxygen and oxygen difluoride. Reacts vigorously with most oxidizable materials at room temperature, frequently with ignition. Corrodes most metals (except the cylinders in which it is shipped), especially in the presence of moisture. Contact with nitric acid produces explosive gas. Attacks some plastics, rubber, and coatings. Reacts explosively with ammonia, graphite, halocarbons, hydrocarbons, hydrogen, and many other substances. Reacts with nearly every known element.

FLUORINE MONOXIDE (7783-41-7) Forms explosive mixture with water. A powerful oxidizer. Contact with all reducing agents, organic and combustible materials, ammonia, platinum and many other metals, metal oxides, and moist air may cause fire and explosions. Attacks some forms of plastics, rubber, and coatings. See 29 CFR 1910.101 for specific regulations on storage of compressed gas cylinders.

FLUORINE OXIDE (7783-41-7) Forms explosive mixture with water. A powerful oxidizer. Contact with all reducing agents, organic and combustible materials, ammonia, platinum and many other metals, metal oxides, and moist air may cause fire and explosions. Attacks some forms of plastics, rubber, and coatings. See 29 CFR 1910.101 for specific regulations on storage of compressed gas cylinders.

FLUORISTAN (7783-47-3) Acid contact produces hydrogen fluoride fumes.

FLUORO (Italian) (7782-41-4) A powerful oxidizer and a dangerously reactive gas. Reacts violently with reducing agents and combustible materials. Contact with water can produce corrosive hydrogen fluoride, hydrochloric acid, oxygen and oxygen difluoride. Reacts vigorously with most oxidizable materials at room temperature, frequently with ignition. Corrodes most metals (except the cylinders in which it is shipped), especially in the presence of moisture. Contact with nitric acid produces explosive gas. Attacks some plastics, rubber, and coatings. Reacts explosively with ammonia, graphite, halocarbons, hydrocarbons, hydrogen, and many other substances. Reacts with nearly every known element.

FLUOROACETIC ACID, SODIUM SALT (62-74-8) Incompatible with strong acids.

2-FLUOROANILINE (348-54-9) Forms explosive mixture with air (flash point 140°F/60°C). Incompatible with strong oxidizers, strong acids.

4-FLUOROANILINE (371-40-4) Forms explosive mixture with air (flash point 165°F/74°C). Incompatible with strong acids, strong oxidizers.

O-FLUOROANILINE (348-54-9) Forms explosive mixture with air (flash point 140°F/60°C). Incompatible with strong oxidizers, strong acids.

P-FLUOROANILINE (371-40-4) Forms explosive mixture with air (flash point 165°F/74°C). Incompatible with strong acids, strong oxidizers.

2-FLUOROBENZENAMINE (348-54-9) Forms explosive mixture with air (flash point 140°F/60°C). Incompatible with strong oxidizers, strong acids.

4-FLUOROBENZENAMINE (371-40-4) Forms explosive mixture with air (flash point 165°F/74°C). Incompatible with strong acids, strong oxidizers.

FLUOROBENZENE (462-06-6) Forms explosive mixture with air (flash point 5°F/−15°C). Incompatible with oxidizers, ammonium nitrate, chromic acid, halogens, hydrogen peroxide, nitric acid. Attacks some plastics, rubber, and coatings.

FLUOROCARBON 11 (75-69-4) Reacts with barium, lithium, sodium, magnesium, titanium. Attacks some plastics, rubber, and coatings.

FLUOROCARBON 12 (75-71-8) Reacts violently with liquid aluminum. Incompatible with chemically active metals. Attacks some plastics, rubber, and coatings.

FLUOROCARBON 115 (76-15-3) Thermal decomposition occurs at high temperatures with alkalies and alkaline earth metals.

FLUORODICHLOROMETHANE (75-43-4) Reacts with water and chemically active metals. Attacks some plastics, rubber, and coatings.

FLUOROETHENE (75-02-5) Flammable gas. May be able to polymerize. Reacts violently with oxidizers. May accumulate static electrical charges, and may cause ignition of its vapors.

FLUOROFORMYL FLUORIDE (353-50-4) Moisture produces hydrogen fluoride gas and carbon dioxide.

1-FLUORO-2-METHYLBENZENE (95-52-3) Forms explosive mixture with air (flash point 55°F/13°C) Reacts violently with strong oxidizers.

1-FLUORO-3-METHYLBENZENE (352-70-5) Forms explosive mixture with air (flash point 49°F/9°C). Reacts violently with strong oxidizers.

1-FLUORO-4-METHYLBENZENE (352-32-9) Forms explosive mixture with air (flash point 105°F/49°C). Reacts violently with oxidizers.

2-FLUORO-1-METHYLBENZENE (95-52-3) Forms explosive mixture with air (flash point 55°F/13°C). Reacts violently with strong oxidizers.

4-FLUORO-1-METHYLBENZENE (352-32-9) Forms explosive mixture with air (flash point 105°F/49°C). Reacts violently with oxidizers.

2-FLUOROPHENYLAMINE (348-54-9) Forms explosive mixture with air (flash point 140°F/60°C). Incompatible with strong oxidizers, strong acids.

4-FLUOROPHENYLAMINE (371-40-4) Forms explosive mixture with air (flash point 165°F/74°C). Incompatible with strong acids, strong oxidizers.

FLUOROPHOSGENE (353-50-4) Moisture contact produces hydrogen fluoride gas and carbon dioxide.

FLUOROPLAST-3 (79-38-9) Flammable gas. Forms explosive mixture with air (flash point −18°F/−28°C). Reacts violently with oxidizers. Contact with ethylene may cause explosive polymerization.

FLUOROSILICIC ACID (16961-83-4) Incompatible with aliphatic amines, alkanolamines, alkylene oxides, aromatic amines, amides, ammonia, ammonium hydroxide, bases, calcium oxide, epichlorohydrin, isocyanates, oleum, organic anhydrides, sulfuric acid, strong oxidizers, vinyl acetate, water.

FLUOROSULFONIC ACID (7789-21-1) Water contact produces hydrogen fluoride and sulfuric acid mist. Incompatible with acids, caustics, ammonia, amines, amides, organic anhydrides, isocyanates, vinyl acetate, acrylates, substituted allyls, alkylene oxides, epichlorohydrin, ketones, aldehydes, alcohols, glycols, phenols, cresols, caprolactam solution.

FLUOROSULFURIC ACID (7789-21-1) Water contact produces hydrogen fluoride and sulfuric acid mist. Incompatible with acids, caustics, ammonia, amines, amides, organic anhydrides, isocyanates, vinyl acetate, acrylates, substituted allyls, alkylene oxides, epichlorohydrin, ketones, aldehydes, alcohols, glycols, phenols, cresols, caprolactam solution.

2-FLUOROTOLUENE (95-52-3) Forms explosive mixture with air (flash point 55°F/13°C). Reacts violently with strong oxidizers.

3-FLUOROTOLUENE (352-70-5) Forms explosive mixture with air (flash point 49°F/9°C). Reacts violently with strong oxidizers.

4-FLUOROTOLUENE (352-32-9) Forms explosive mixture with air (flash point 105°F/49°C). Reacts violently with oxidizers.

m-**FLUOROTOLUENE** (352-70-5) Forms explosive mixture with air (flash point 49°F/9°C). Reacts violently with strong oxidizers.

o-**FLUOROTOLUENE** or *ortho*-**FLUOROTOLUENE** (95-52-3) Forms explosive mixture with air (flash point 55°F/13°C). Reacts violently with strong oxidizers.

p-**FLUOROTOLUENE** or *para*-**FLUOROTOLUENE** (352-32-9) Forms explosive mixture with air (flash point 105°F/49°C). Reacts violently with oxidizers.

FLUOROTRICHLOROMETHANE (75-69-4) Reacts with barium, lithium, sodium, magnesium, titanium. Attacks some plastics, rubber, and coatings.

FLUORSPAR (7789-75-5) Incompatible with acids, chemically active metals, reducing agents, water.

FLUORURE de BORE (French) (7637-07-2) Decomposes in heat, water and moist air. Fire and explosions may result from contact with alkali metals, alkaline earth metals (except magnesium), alkyl nitrate, calcium oxide. Incandesces when heated with alkali metals or alkaline earth metals (except magnesium). Corrodes most metals in the presence of moisture.

FLUORURE de SULFURYLE (French) (2699-79-8) Reacts with moisture. A compressed gas. High temperatures may cause cylinders to burst. See 29 CFR 1910.101 for specific storage regulations.

FLUORURES ACIDE (French) (7782-41-4) A powerful oxidizer and a dangerously reactive gas. Reacts violently with reducing agents and combustible materials. Contact with water can produce corrosive hydrogen fluoride, hydrochloric acid, oxygen, and oxygen difluoride. Reacts vigorously with most oxidizable materials at room temperature, frequently with ignition. Corrodes most metals (except the cylinders in which it is shipped), especially in the presence of moisture. Contact with nitric acid produces explosive gas. Attacks some plastics, rubber, and coatings. Reacts explosively with ammonia, graphite, halocarbons, hydrocarbons, hydrogen, and many other substances. Reacts with nearly every known element.

FLUORURI ACIDI (Italian) (7782-41-4) A powerful oxidizer and a dangerously reactive gas. Reacts violently with reducing agents and combustible materials. Contact with water can form corrosive hydrogen fluoride, hydrochloric acid, oxygen, and oxygen difluoride. Reacts vigorously with most oxidizable materials at room temperature, frequently with ignition. Corrodes most metals (except the cylinders in which it is shipped) especially in the presence of moisture. Contact with nitric acid forms explosive gas. Attacks some plastics, rubber, and coatings. Reacts explosively with ammonia, graphite, halocarbons, hydrocarbons, hydrogen, and many other substances. Reacts with nearly every known element.

FLUOSILIC ACID (16961-83-4) Incompatible with aliphatic amines, alkanolamines, alkylene oxides, aromatic amines, amides, ammonia, ammonium hydroxide, bases, calcium oxide, epichlorohydrin, isocyanates, oleum, organic anhydrides, sulfuric acid, strong oxidizers, vinyl acetate, water.

FLUOSILICIC ACID (16961-83-4) Incompatible with aliphatic amines, alkanolamines, alkylene oxides, aromatic amines, amides, ammonia, ammonium hydroxide, bases, calcium oxide, epichlorohydrin, isocyanates, oleum, organic anhydrides, sulfuric acid, strong oxidizers, vinyl acetate, water.

FLUOSPAR (7789-75-5) Incompatible with acids, chemically active metals, reducing agents, water.

FLUOSULFONIC ACID (7789-21-1) Water contact produces hydrogen fluoride and sulfuric acid mist. Incompatible with acids, caustics, ammonia, amines, amides, organic anhydrides, isocyanates, vinyl acetate, acrylates, substituted allyls, alkylene oxides, epichlorohydrin, ketones, aldehydes, alcohols, glycols, phenols, cresols, caprolactam solution.

FLUROPLAST 4 (116-14-3) A highly reactive flammable gas. Able to form unstable peroxides; if inhibitor is not present in adequate concentrations, explosive polymerization may occur. Reacts violently with air, oxygen, oxidizers, sulfur trioxide.

FLUXING OIL (8052-42-4) Incompatible with nitric acid, fluorine, strong oxidizers.

FLUX MAAY (54-11-5) Incompatible with strong acids, strong oxidizers. Attacks some plastics, rubber, and coatings. May accumulate static electrical charges, and may cause ignition of its vapors.

FLY-DIE (62-73-7) Attacks some plastics, rubber, and coatings.

FLY FIGHTER (62-73-7) Attacks some plastics, rubber, and coatings.

FMC-1240 (563-12-2) Incompatible with alkaline formulations. Mixtures with magnesium may be explosive.

FMC-5462 (115-29-7) Hydrolyzed by acids and alkalis. Corrosive to iron.

FMC-10242 (1563-66-2) Incompatible with alkaline material, acids, strong oxidizers.

FOLCID (2425-06-1) Incompatible with acids or acid vapor. Strong alkaline conditions contribute to instability.

FOLIAGE OIL (N/A) Incompatible with nitric acid; oxidizers may cause fire and explosions.

FOLIDOL (56-38-2) Combustible liquid. Mixtures with endrin may be explosive. Strong oxidizers may cause fire and explosions. Attacks some plastics, rubber, and coatings.

FOLIDOL E 605 or FOLIDOL E&E 605 (56-38-2) Combustible liquid. Mixtures with endrin may be explosive. Strong oxidizers may cause fire and explosions. Attacks some plastics, rubber, and coatings.

FOLIDOL M (298-00-0) Mixtures with magnesium may be explosive.

FONOLINE (8012-95-1) Incompatible with nitric acid; oxidizers may cause fire and explosions.

FOREDEX 75 (94-75-7) Decomposes in sunlight. Incompatible with strong oxidizers; may cause fire and explosions.

FORLIN (58-89-9) Not combustible, but may be dissolved in a combustible solvent. If solvent comes in contact with oxidizers, fire and explosions may result.

FORMAGENE (30525-89-4) Forms explosive mixture with air (flash point 160°F/71°C). Aqueous solution causes formation of formaldehyde. Reacts violently with strong oxidizers or liquid oxygen. May accumulate static electrical charges, and may cause ignition of its vapors.

FORMAL (109-87-5) Forms explosive mixture with air (flash point −4°F/−18°C). Incompatible with strong oxidizers, acids. Attacks some plastics, rubber, and coatings.

FORMAL (121-75-5) Incompatible with strong oxidizers, magnesium, alkaline pesticides. Attacks metals, some plastics, rubber, and coatings.

FORMALDEHYD (Czech or Polish) (50-00-0) May polymerize unless properly inhibited (usually with methanol). Forms explosive mixture with air (flash point 122°F/50°C). Incompatible with strong acids, amines, strong oxidizers, alkaline materials, nitrogen dioxide, performic acid. Reaction with hydrochloric acid produces bis-chloromethyl ether, a carcinogen.

FORMALDEHYDE DIMETHYLACETAL (109-87-5) Forms explosive mixture with air (flash point −4°F/−18°C). Incompatible with strong oxidizers, acids. Attacks some plastics, rubber, and coatings.

FORMALDEHYDE POLYMER (30525-89-4) Forms explosive mixture with air (flash point 160°F/71°C). Aqueous solution causes formation of formaldehyde. Reacts violently with strong oxidizers or liquid oxygen. May accumulate static electrical charges, and may cause ignition of its vapors.

FORMALDEHYDE SOLUTION (50-00-0) May polymerize unless properly inhibited (usually with methanol). Forms explosive mixture with air (flash point 122°F/50°C). Incompatible with strong acids, amines, strong oxidizers, alkaline materials, nitrogen dioxide, performic acid. Reaction with hydrochloric acid produces bis-chloromethyl ether, a carcinogen.

FORMALIN (50-00-0) May polymerize unless properly inhibited (usually with methanol). Forms explosive mixture with air (flash point 122°F/50°C). Incompatible with strong acids, amines, strong oxidizers, alkaline materials, nitrogen dioxide, performic acid. Reaction with hydrochloric acid produces bis-chloromethyl ether, a carcinogen.

FORMALIN 40 (50-00-0) May polymerize unless properly inhibited (usually with methanol). Forms explosive mixture with air (flash point 122°F/50°C). Incompatible with strong acids, amines, strong oxidizers, alkaline materials, nitrogen dioxide, performic acid. Reaction with hydrochloric acid produces bis-chloromethyl ether, a carcinogen.

FORMALINA (Italian) (50-00-0) May polymerize unless properly inhibited (usually with methanol). Forms explosive mixture with air (flash point 122°F/50°C). Incompatible with strong acids, amines, strong oxidizers, alkaline materials, nitrogen dioxide, performic acid. Reaction with hydrochloric acid produces bis-chloromethyl ether, a carcinogen.

FORMALINE (German) (50-00-0) May polymerize unless properly inhibited (usually with methanol). Forms explosive mixture with air (flash point 122°F/50°C). Incompatible with strong acids, amines, strong oxidizers, alkaline materials, nitrogen dioxide, performic acid. Reaction with hydrochloric acid produces bis-chloromethyl ether, a carcinogen.

FORMALIN LOESUNGEN (German) (50-00-0) May polymerize unless properly inhibited (usually with methanol). Forms explosive mixture with air (flash point 122°F/50°C). Incompatible with strong acids, amines, strong oxidizers, alkaline materials, nitrogen dioxide, performic acid. Reaction with hydrochloric acid produces bis-chloromethyl ether, a carcinogen.

FORMALITH (50-00-0) May polymerize unless properly inhibited (usually with methanol). Forms explosive mixture with air (flash point 122°F/50°C). Incompatible with strong acids, amines, strong oxidizers, alkaline materials, nitrogen dioxide, performic acid. Reaction with hydrochloric acid produces bis-chloromethyl ether, a carcinogen.

FORMAMIDE (75-12-7) Forms hydrocyanic acid with water solutions. Incompatible with nonoxidizing mineral acids, strong acids, ammonia, cresols, iodine, isocyanates, oleum, phenols, pyridine, sulfur trioxide.

FORMAMIDE, N,N-DIMETHYL- (68-12-2) Forms explosive mixture with air (flash point 136°F/58°C). Contact with carbon tetrachloride and other halogenated compounds, particularly in presence of iron or strong oxidizers, may cause fire and explosions. Reacts vigorously with alkylaluminums. Incompatible with nonoxidizing mineral acids, strong acids, chlorinated hydrocarbons, isocyanates, nitrates, organic nitrates, phenols, cresols, ammonia, chromic anhydride, magnesium nitrate, methylene diisocyanate, phosphorus trioxide, triethylaluminum. Attacks some plastics, rubber, and coatings.

FORMIATE de METHYLE (French) (107-31-3) Incompatible with strong acids, nitrates. Reacts violently with strong oxidizers. Attacks some plastics and coatings.

FORMIC ACID or FORMIC ACID, 85–95% IN AQUEOUS SOLUTION (64-18-6) Forms explosive mixture with air (flash point 156°F/(69°C). A strong reducing agent; reacts violently with oxidizers. Incompatible with sulfuric acid, bases, ammonia, aliphatic amines, alkanolamines, furfuryl alcohol, hydrogen peroxide, isocyanates, alkylene oxides, epichlorohydrin. Attacks aluminum, cast iron, and steel, and some plastics, rubber, and coatings.

FORMIC ACID, AMIDE (75-12-7) Forms hydrocyanic acid with water solutions. Incompatible with nonoxidizing mineral acids, strong acids, ammonia, cresols, iodine, isocyanates, oleum, phenols, pyridine, sulfur trioxide.

FORMIC ACID, BUTYL ESTER (592-84-7) Reacts with strong acids, nitrates, oxidizers (possibly violently). May accumulate static electrical charges, and may cause ignition of its vapors.

FORMIC ACID, ETHYL ESTER (109-94-4) Forms explosive mixture with air (flash point −4°F/−20°C). May accumulate static electrical charges, and may cause ignition of its vapors. Incompatible with nitrates, strong oxidizers, strong alkalies, strong acids.

FORMIC ACID, METHYL ESTER (107-31-3) Incompatible with strong acids, nitrates. Reacts violently with strong oxidizers. Attacks some plastics and coatings.

FORMIC ALDEHYDE (50-00-0) May polymerize unless properly inhibited (usually with methanol). Forms explosive mixture with air (flash point 122°F/50°C). Incompatible with strong acids, amines, strong oxidizers, alkaline materials, nitrogen dioxide, performic acid. Reaction with hydrochloric acid produces bis-chloromethyl ether, a carcinogen.

FORMIC ETHER (109-94-4) Forms explosive mixture with air (flash point −4°F/−20°C). May accumulate static electrical charges, and may cause ignition of its vapors. Incompatible with nitrates, strong oxidizers, strong alkalies, strong acids.

FORMIMIDIC ACID (75-12-7) Forms hydrocyanic acid with water solutions. Incompatible with nonoxidizing mineral acids, strong acids, ammonia, cresols, iodine, isocyanates, oleum, phenols, pyridine, sulfur trioxide.

FORMOL (50-00-0) May polymerize unless properly inhibited (usually with methanol). Forms explosive mixture with air (flash point 122°F/50°C). Incompatible with strong acids, amines, strong oxidizers, alkaline materials, nitrogen dioxide, performic acid. Reaction with hydrochloric acid produces bis-chloromethyl ether, a carcinogen.

FORMONITRILE (74-90-8) Unless stabilized and maintained, samples stored more than 90 days are hazardous. Samples containing more than 2–5% water are less stable than dry material. Can be self-reactive, forming an explosive mixture with air (flash point 0°F/−18°C). Heat or contact with amines or strong bases can cause polymerization. Incompatible with acetaldehyde. Oxidizers may cause fire and explosions. Attacks some plastics, rubber, and coatings.

FORMOSA CAMPHOR (76-22-2) Forms explosive mixture with air (flash point 155°F/66°C). Violent, possibly explosive, reaction occurs with strong oxidizers. May accumulate static electrical charges, and may cause ignition of its vapors.

FORMULA 40 (94-75-7) Decomposes in sunlight. Incompatible with strong oxidizers; may cause fire and explosions.

FORMYL TRIBROMIDE (75-25-2) Reacts with chemically active metals, calcium, acetone, strong caustics. Attacks some plastics, rubber, and coatings.

FORMYL TRICHLORIDE (67-66-3) Decomposes in the presence of excess water or high temperatures, producing phosgene and hydrogen chloride. Can become explosive in the presence of strong alkalies and water. Can accumulate static electrical charges. In contact with water and high temperatures, it becomes corrosive; attacks iron and other metals. Incompatible with acetone, aluminum, strong oxidizers, potassium, sodium, chemically active metals, strong bases. Attacks plastics and rubber.

N-FORMYLDIMETHYLAMINE (68-12-2) Forms explosive mixture with air (flash point 136°F/58°C). Contact with carbon tetrachloride and other halogenated compounds, particularly in presence of iron or strong oxidizers, may cause fire and explosions. Reacts vigorously with alkylaluminums. Incompatible with nonoxidizing mineral acids, strong acids, chlorinated hydrocarbons, isocyanates, nitrates, organic nitrates, phenols, cresols, ammonia, chromic anhydride, magnesium nitrate, methylene diisocyanate, phosphorus trioxide, triethyaluminum. Attacks some plastics, rubber, and coatings.

FORMYLFORMIC ACID (298-12-4) Reacts with alkali metals; releases flammable hydrogen gas.

FORMYLIC ACID (64-18-6) Forms explosive mixture with air (flash point 156°F/69°C). A strong reducing agent; reacts violently with oxidizers. Incompatible with sulfuric acid, bases, ammonia, aliphatic amines, alkanolamines, furfuryl alcohol, hydrogen peroxide, isocyanates, alkylene oxides, epichlorohydrin. Attacks aluminum, cast iron and steel, and some plastics, rubber, and coatings.

2-FORMYLPHENOL (90-02-8) Forms explosive mixture with air (flash point 172°F/78°C). Reacts violently with strong oxidizers.

alpha-FORMYLPHENOL (90-02-8) Forms explosive mixture with air (flash point 172°F/78°C). Reacts violently with strong oxidizers.

FOROTOX (52-68-6) Contact with strong oxidizers may cause fire and explosions.

FORRON (93-76-5) Sealed metal containers may burst in heat above 316°F/158°C. Incompatible with sulfuric acid, bases, ammonia, aliphatic amines, alkanolamines, isocyanates, alkylene oxides, epichlorohydrin.

FORSTU 46 (93-76-5) Sealed metal containers may burst in heat above 316°F/158°C. Incompatible with sulfuric acid, bases, ammonia, aliphatic amines, alkanolamines, isocyanates, alkylene oxides, epichlorohydrin.

FORTEX (93-76-5) Sealed metal containers may burst in heat above 316°F/158°C. Incompatible with sulfuric acid, bases, ammonia, aliphatic amines, alkanolamines, isocyanates, alkylene oxides, epichlorohydrin.

FORTHION (121-75-5) Incompatible with strong oxidizers, magnesium, alkaline pesticides. Attacks metals, some plastics, rubber, and coatings.

FOSCHLOR (52-68-6) Contact with strong oxidizers may cause fire and explosions.

FOSDRIN (7786-34-7) Contact with strong oxidizers may cause fire and explosions. Attacks some plastics, rubber, and coatings.

FOSFERMO (56-38-2) Combustible liquid. Mixtures with endrin may be explosive. Strong oxidizers may cause fire and explosions. Attacks some plastics, rubber, and coatings.

FOSFEX (56-38-2) Combustible liquid. Mixtures with endrin may be explosive. Strong oxidizers may cause fire and explosions. Attacks some plastics, rubber, and coatings.

FOSFIVE (56-38-2) Combustible liquid. Mixtures with endrin may be explosive. Strong oxidizers may cause fire and explosions. Attacks some plastics, rubber, and coatings.

FOSFONO 50 (563-12-2) Incompatible with alkaline formulations. Mixtures with magnesium may be explosive.

FOSFORO (PENTACHLORURO di) (Italian) (10026-13-8) Moisture will cause development of pressure in a closed container. Water causes violent spattering and formation of toxic and corrosive

390

hydrogen chloride gas and phosphoric acid mist. Reacts violently with magnesium oxide, chemically active metals, alkalies. Attacks some plastics, rubber, and coatings.

FOSFORO (TRICLORURO di) (Italian) (7719-12-2) Contact with water or alcohol may cause fire and explosions, particularly in presence of combustible organic matter. Incompatible with ammonia, caustics, reducing agents, nitric acid, chemically active metals. In the presence of moisture, corrodes most metals. Attacks some plastics, rubber, and coatings.

FOSFOROWODOR (Polish) (7803-51-2) Unusually reactive. Contact with air or other oxidizer may cause self-ignition and explosion. Reacts violently or forms explosive procuts with many substances, including acids, halogenated hydrocarbons, moisture. Heat may cause cylinders to explode.

FOSFORPENTACHLORIDE (Dutch) (10026-13-8) Moisture will cause development of pressure in a closed container. Water causes violent spattering and formation of toxic and corrosive hydrogen chloride gas and phosphoric acid mist. Reacts violently with magnesium oxide, chemically active metals, alkalies. Attacks some plastics, rubber, and coatings.

FOSFORTRICHLORIDE (Dutch) (7719-12-2) Contact with water or alcohol may cause fire and explosions, particularly in presence of combustible organic matter. Incompatible with ammonia, caustics, reducing agents, nitric acid, chemically active metals. In the presence of moisture, corrodes most metals. Attacks some plastics, rubber, and coatings.

FOSFORYN TROJMETYLOWY (Czech) (121-45-9) Forms explosive mixture with air (flash point 82°F/28°C). Reacts violently with strong acids, nitrates, oxidizers, magnesium salts of perchloric acid.

FOSFORZUUROPLOSSINGEN (Dutch) (7664-38-2) Incompatible with aliphatic amines, alkanolamines, alkylene oxides, aromatic amines, amides, ammonia, ammonium hydroxide, bases, calcium oxide, epichlorohydrin, isocyanates, (explosive with) nitromethane, oleum, organic anhydrides, sulfuric acid, sodium tetrahydroborate, strong oxidizers, vinyl acetate, water. Contact with most metals produces hydrogen gas. Attacks some plastics, rubber, and coatings, and glass and ceramics.

FOSFOTHION (121-75-5) Incompatible with strong oxidizers, magnesium, alkaline pesticides. Attacks metals, some plastics, rubber, and coatings.

FOSFURI di ALLUMINIO (Italian) (20859-73-8) Contact with moisture (including atmospheric moisture) produces spontaneously combustible phosphine gas.

FOSGEEN (Dutch) (75-44-5) Incompatible with water, producing hydrochloric acid. Reacts violently with strong oxidizers, anhydrous ammonia, isopropyl alcohol, and chemically active metals. Forms shock-sensitive material with potassium. Attacks most metal in moist conditions.

FOSGEN (Polish) (75-44-5) Incompatible with water, producing hydrochloric acid. Reacts violently with strong oxidizers, anhydrous ammonia, isopropyl alcohol, and chemically active metals. Forms shock-sensitive material with potassium. Attacks most metal in moist conditions.

FOSGENE (Italian) (75-44-5) Incompatible with water, producing hydrochloric acid. Reacts violently with strong oxidizers, anhydrous ammonia, isopropyl alcohol, and chemically active metals. Forms shock-sensitive material with potassium. Attacks most metal in moist conditions.

FOSOVA (56-38-2) Combustible liquid. Mixtures with endrin may be explosive. Strong oxidizers may cause fire and explosions. Attacks some plastics, rubber, and coatings.

FOSTERN (56-38-2) Combustible liquid. Mixtures with endrin may be explosive. Strong oxidizers may cause fire and explosions. Attacks some plastics, rubber, and coatings.

FOSTEX (94-36-0) Confined storage of dry chemical may lead to decomposition and explosion. A strong oxidant; extremely reactive. Fires and explosion may result from heat or contamination, and from contact with strong acids, combustible materials, oxidizers, acids, bases, alcohols, reducing agents, metals, metal oxides, amines, accelerators, methyl methacrylate, organic matter, lithium aluminum carbide, dimethyl aniline, amines, metallic naphthenates. May attack some plastics, rubber, and coatings. Protect containers from shock and friction.

FOSTOX (56-38-2) Combustible liquid. Mixtures with endrin may be explosive. Strong oxidizers may cause fire and explosions. Attacks some plastics, rubber, and coatings.

FOSVEX (107-49-3) Decomposes above 300°F/150°C, producing flammable ethylene gas. Strong oxidizers may cause fire and explosions. Attacks some plastics, rubber, and coatings.

FOTOX (1303-28-2) Incompatible with acids, aluminum, halogens, rubidium carbide, zinc. Aqueous solutions are acidic.

FOURAMINE D (106-50-3) Incompatible with acids, organic anhydrides, isocyanates, aldehydes, strong oxidizers; may cause fire and explosions. Heat and light contribute to instability.

FOURAMINE PCH (120-80-9) Strong oxidizers may cause fire and explosion.

FOURAMINE RS (108-46-3) Reacts violently with strong oxidizers, nitric acid. Incompatible with acetanilide, albumin, alkalies, antipyrine, camphor, ferric salts, menthol, spirit nitrous ether. Absorbs moisture from air (hygroscopic). May accumulate static electrical charges, and may cause ignition of its vapors.

FOURRINE 1 (106-50-3) Incompatible with acids, organic anhydrides, isocyanates, aldehydes, strong oxidizers; may cause fire and explosions. Heat and light contribute to instability.

FOURRINE 68 (120-80-9) Strong oxidizers may cause fire and explosion.

FOURRINE 79 (108-46-3) Reacts violently with strong oxidizers, nitric acid. Incompatible with acetanilide, albumin, alkalies, antipyrine, camphor, ferric salts, menthol, spirit nitrous ether. Absorbs moisture from air (hygroscopic). May accumulate static electrical charges, and may cause ignition of its vapors.

FOURRINE D (106-50-3) Incompatible with acids, organic anhydrides, isocyanates, aldehydes, strong oxidizers; may cause fire and explosions. Heat and light contribute to instability.

FOURRINE EW (108-46-3) Reacts violently with strong oxidizers, nitric acid. Incompatible with acetanilide, albumin, alkalies, antipyrine, camphor, ferric salts, menthol, spirit nitrous ether. Absorbs moisture from air (hygroscopic). May accumulate static electrical charges, and may cause ignition of its vapors.

FOURRINE PG (87-66-1) Incompatible with strong oxidizers, caustics, ammonia, amines, isocyanates, alkylene oxides, epichlorohydrin.

FRANKLIN (1317-65-3) Incompatible with acids, alum, ammonium salts, fluorine.

FRATOL (62-74-8) Incompatible with strong acids.

FREEMANS WHITE LEAD (7446-14-2) Reacts violently with potassium. Incompatible with aluminum, magnesium.

FREON 10 (56-23-5) Becomes corrosive when in contact with water. Corrosive to metals. Reacts violently with many compounds. Decomposes on contact with chemically active metals such as sodium, potassium and magnesium. Incompatible with allyl alcohol, fluorine gas, alkali metals, aluminum. Attacks some coatings, plastics, and rubber.

FREON 11 (75-69-4) Reacts with barium, lithium, sodium, magnesium, titanium. Attacks some plastics, rubber, and coatings.

FREON 12 (75-71-8) Reacts violently with liquid aluminum. Incompatible with chemically active metals. Attacks some plastics, rubber, and coatings.

FREON 12-B2 (75-61-6) Reacts with chemically active metals and oxidizers. Attacks some plastics, rubber, and coatings.

FREON 13 (75-72-9) Reacts with aluminum, magnesium, zinc, and their alloys.

FREON 13B1 (75-63-8) Reacts with chemically active metals, powdered aluminum, zinc, magnesium. May attack some plastics, rubber, and coatings.

FREON 20 (67-66-3) Decomposes in the presence of excess water or at high temperatures, producing phosgene and hydrogen chloride. Can become explosive in the presence of strong alkalies and water. Can accumulate static electrical charges. In contact with water and at high temperatures, it becomes corrosive; attacks iron and other metals. Incompatible with acetone, aluminum, strong oxidizers, potassium, sodium, chemically active metals, strong bases. Attacks plastics and rubber.

FREON 21 (75-43-4) Reacts with water and chemically active metals. Attacks some plastics, rubber, and coatings.

FREON 22 (75-45-5) Moisture and rust cause slow decomposition, producing toxic gases. Attacks some plastics, rubber, and coatings. Thermal decomposition occurs at high temperature with alkalies and alkaline earth metals.

FREON 112 (76-12-0) Reacts with chemically active metals, powdered aluminum, zinc, and magnesium. Attacks some forms of plastics, rubber, and coatings.

FREON 113 (76-13-1) Reacts with barium, lithium, sodium, magnesium, titanium. Contact with alloys containing more than 2% magnesium may cause decomposition (with hydrogen chloride, hydrogen fluoride, and carbon monoxide released). Attacks some plastics, rubber, and coatings.

FREON 114 (76-14-2) Reacts with barium, lithium, sodium, magnesium, titanium. Attacks some plastics, rubber, and coatings.

FREON 115 (76-15-3) Thermal decomposition occurs at high temperatures with alkalies and alkaline earth metals.

FREON 152 (75-37-6) Flammable gas. May accumulate static electrical charges, and may cause ignition of its vapors. Reacts violently with strong oxidizers. Attacks metals in presence of moisture.

FREON 1301 (75-63-8) Reacts with chemically active metals, powdered aluminum, zinc, magnesium. May attack some plastics, rubber, and coatings.

FREON F-12 (75-71-8) Reacts violently with liquid aluminum. Incompatible with chemically active metals. Attacks some plastics, rubber, and coatings.

FREON MF (75-69-4) Reacts with barium, lithium, sodium, magnesium, titanium. Attacks some plastics, rubber, and coatings.

FREON TF (76-13-1) Reacts with barium, lithium, sodium, magnesium, titanium. Contact with alloys containing more than 2% magnesium may cause decomposition (with hydrogen chloride, hydrogen fluoride, and carbon monoxide released). Attacks some plastics, rubber, and coatings.

FRIDEX (107-21-1) Incompatible with strong acids, caustics, aliphatic amines, isocyanates, chlorosulfonic acid, oleum, strong oxidizers.

FRIGEN 11 (75-69-4) Reacts with barium, lithium, sodium, magnesium, titanium. Attacks some plastics, rubber, and coatings.

FRIGEN 12 (75-71-8) Reacts violently with liquid aluminum. Incompatible with chemically active metals. Attacks some plastics, rubber, and coatings.

FRIGEN 113A (76-13-1) Reacts with barium, lithium, sodium, magnesium, titanium. Contact with alloys containing more than 2% magnesium may cause decomposition (with hydrogen chloride, hydrogen fluoride, and carbon monoxide released). Attacks some plastics, rubber, and coatings.

FRUIT-O-NET (93-72-1) A powerful oxidizer; reacts violently with reducing agents, combustibles.

FRUITONE A (93-76-5) Sealed metal containers may burst in heat above 316°F/158°C. Incompatible with sulfuric acid, bases, ammonia, aliphatic amines, alkanolamines, isocyanates, alkylene oxides, epichlorohydrin.

FRUMIN AL (298-04-4) Forms explosive mixture with air (flash point 180°F/82°C). Incompatible with alkalis, strong oxidizers.

FRUMIN G (298-04-4) Forms explosive mixture with air (flash point 180°F/82°C). Incompatible with alkalis, strong oxidizers.

FTAALZUURANHYDRIDE (Dutch) (85-44-9) Incompatible with strong acids, caustics, ammonia, amines, strong oxidizers. Attacks some plastics, rubber, and coatings.

FTALOWY BEZWODNIK (Polish) (85-44-9) Incompatible with strong acids, caustics, ammonia, amines, strong oxidizers. Attacks some plastics, rubber, and coatings.

FUEL OIL NO. 1 (8008-20-6) Explosive mixture in air above 100°F/38°C. Incompatible with nitric acid; oxidizers may cause fire and explosions.

FUEL OIL NO. 1 (68476-31-3) Explosive mixture in air above 100°F/38°C. Incompatible with nitric acid; oxidizers may cause fire and explosions.

FUEL OIL NO. 1-D (64741-44-2) (68334-30-5) Incompatible with nitric acid; oxidizers may cause fire and explosions.

FUEL OIL NO. 2 (8006-61-9) Forms explosive mixture with air (flash point 125°F/51°C). Incompatible with nitric acid; oxidizers may cause fire and explosions.

FUEL OIL NO. 2-D (64741-44-2) (68334-30-5) Incompatible with nitric acid; oxidizers may cause fire and explosions.

FUEL OIL NO. 4 (68476-33-5) Forms explosive mixture with air (flash point 142°F/61°C). Incompatible with nitric acid; oxidizers may cause fire and explosions.

FUEL OIL NO. 5 (68476-33-5) Forms explosive mixture with air (flash point 156°F/69°C). Incompatible with nitric acid; oxidizers may cause fire and explosions.

FUEL OIL NO. 6 (68476-33-5) Forms explosive mixture with air (flash point 150°F/66°C). Incompatible with nitric acid; oxidizers may cause fire and explosions.

FUMAGON (96-12-8) Forms explosive mixture with air (flash point 170°F/77°C). Reacts with oxidizers and chemically active metals (i.e., aluminum, magnesium, and tin alloys). Attacks some rubber materials and coatings.

FUMARIC ACID (110-17-8) Reacts with strong oxidizers. Incompatible with sulfuric acid, caustics, ammonia, amines, isocyanates, alkylene oxides, epichlorohydrin.

FUMAZONE (96-12-8) Forms explosive mixture with air (flash point 170°F/77°C). Reacts with oxidizers and chemically active metals (i.e., aluminum, magnesium, and tin alloys). Attacks some rubber materials and coatings.

FUMIGRAIN (107-13-1) Forms explosive mixture with air (flash point 32°F/0°C o.c.). Forms explosive peroxides; heat, light, caustics, silver nitrate, and peroxides can cause polymerization. Incompatible with strong acids, strong oxidizers, amines, 2-aminoethanol, bromine, chlorosulfonic acid, ethylene diamine, nitric acid, oleum, potassium hydroxide, sodium hydroxide, sulfuric acid. Attacks copper and copper alloys; attacks aluminum in high concentrations. May accumulate static electrical charges, and may cause ignition of its vapors.

FUMING LIQUID ARSENIC (7784-34-1) Water contact produces corrosive mixture of acids. Light exposure produces toxic gas. Reacts violently with anhydrous ammonia, strong acids, strong oxidizers, and halogens. Corrodes metals in the presence of moisture.

FUMING SULFURIC ACID (8014-95-7) A powerful oxidizer and strong acid. Reacts violently with reducing agents, organic materials. Reacts with air, producing corrosive fumes. Contact with cast iron may cause a violent reaction. Extremely hazardous on contact with many materials including chlorates, carbides, fulminates. Reacts vigorously with metals (powders will ignite), releasing hydrogen. Attacks some plastics, rubber, and coatings on brief contact.

FUMO-GAS (106-93-4) Reacts with chemically active metals, liquid ammonia, strong oxidizers. Heat and light cause slow decomposition. Attacks some plastics and rubber.

FUNGCHEX (7487-94-7) Incompatible with light metals (aluminum, magnesium, beryllium, etc.), sodium, potassium.

FUNGICIDE 1991 (17804-35-2) Heat, water, strong acids, and strong alkalies can cause decomposition and formation of toxic oxides of nitrogen.

FUNGIFEN (87-86-5) Hot water causes decomposition, producing hydrochloric acid. Strong oxidizers may cause fire and explosions.

FUNGUS BAN TYPE II (133-06-2) Incompatible with tetraethyl pyrophosphate, parathion.

FURADAN (1563-66-2) Incompatible with alkaline material, acids, strong oxidizers.

FURAL (98-01-1) Forms explosive mixture with air (flash point 140°F/60°C). Incompatible with strong acids, caustics, ammonia, aliphatic amines, alkanolamines, aromatic amines, oxidizers. Attacks many plastics and coatings.

2-FURALDEHYDE (98-01-1) Forms explosive mixture with air (flash point 140°F/60°C). Incompatible with strong acids, caustics, ammonia, aliphatic amines, alkanolamines, aromatic amines, oxidizers. Attacks many plastics and coatings.

FURALE (98-01-1) Forms explosive mixture with air (flash point 140°F/60°C). Incompatible with strong acids, caustics, ammonia, aliphatic amines, alkanolamines, aromatic amines, oxidizers. Attacks many plastics and coatings.

FURAL/PYROMUCIC ALDEHYDE (98-01-1) Forms explosive mixture with air (flash point 140°F/60°C). Incompatible with strong acids, caustics, ammonia, aliphatic amines, alkanolamines, aromatic amines, oxidizers. Attacks many plastics and coatings.

FURAN (110-00-9) Forms explosive mixture with air (flash point −58°F/−50°C). Reacts violently with acids, oxidizers. Unless stabilized with an inhibitor, air exposure causes formation of unstable peroxides.

2-FURANALDEHYDE (98-01-1) Forms explosive mixture with air (flash point 140°F/60°C). Incompatible with strong acids, caustics, ammonia, aliphatic amines, alkanolamines, aromatic amines, oxidizers. Attacks many plastics and coatings.

2-FURANCARBINOL (98-00-0) Forms explosive mixture with air (flash point 170°F/77°C). Strong acids (including some organic acids) or acid catalysts may cause polymerization. Reacts strongly with oxidizers and acids. Incompatible with caustics, aliphatic amines, isocyanates. Attacks some plastics, coatings, and rubber.

2-FURANCARBONAL (98-01-1) Forms explosive mixture with air (flash point 140°F/60°C). Incompatible with strong acids, caustics, ammonia, aliphatic amines, alkanolamines, aromatic amines, oxidizers. Attacks many plastics and coatings.

2-FURANCARBOXALDEHYDE (98-01-1) Forms explosive mixture with air (flash point 140°F/60°C). Incompatible with strong acids, caustics, ammonia, aliphatic amines, alkanolamines, aromatic amines, oxidizers. Attacks many plastics and coatings.

2,5-FURANEDIONE (108-31-6) Strong oxidizers may cause fire and explosions. Contact with amines or alkali metals may cause polymerization if temperature is greater than 150°F/66°C. Attacks some plastics, rubber, and coatings.

FURANIDINE (109-99-9) Forms explosive mixture with air (flash point 6°F/−14°C). Unless inhibited, can form unstable and explosive peroxides. Incompatible with strong acids, strong oxidizers. Attacks some plastics. May accumulate static electric charges that can result in ignition of its vapors.

2-FURANMETHANOL (98-00-0) Forms explosive mixture with air (flash point 170°F/77°C). Strong acids (including some organic acids) or acid catalysts may cause polymerization. Reacts strongly with oxidizers and acids. Incompatible with caustics, aliphatic amines, isocyanates. Attacks some plastics, coatings, and rubber.

FURAN, TETRAHYDRO- (109-99-9) Forms explosive mixture with air (flash point 6°F/−14°C). Unless inhibited, can form unstable and explosive peroxides. Incompatible with strong acids, strong oxidizers. Attacks some plastics. May accumulate static electric charges that can result in ignition of its vapors.

FURATOL (62-74-8) Incompatible with strong acids.

FUR BLACK 41867 (106-50-3) Incompatible with acids, organic anhydrides, isocyanates, aldehydes, strong oxidizers; may cause fire and explosions. Heat and light contribute to instability.

FUR BROWN 41866 (106-50-3) Incompatible with acids, organic anhydrides, isocyanates, aldehydes, strong oxidizers; may cause fire and explosions. Heat and light contribute to instability.

FURFURAL (98-01-1) Forms explosive mixture with air (flash point 140°F/60°C). Incompatible with strong acids, caustics, ammonia, aliphatic amines, alkanolamines, aromatic amines, oxidizers. Attacks many plastics and coatings.

FURFURAL ALCOHOL (98-00-0) Forms explosive mixture with air (flash point 170°F/77°C). Strong acids (including some organic acids) or acid catalysts may cause polymerization. Reacts strongly with oxidizers and acids. Incompatible with caustics, aliphatic amines, isocyanates. Attacks some plastics, coatings, and rubber.

FURFURALDEHYDE (98-01-1) Forms explosive mixture with air (flash point 140°F/60°C). Incompatible with strong acids, caustics, ammonia, aliphatic amines, alkanolamines, aromatic amines, oxidizers. Attacks many plastics and coatings.

FURFURAN (110-00-9) Forms explosive mixture with air (flash point −58°F/−50°C). Reacts violently with acids, oxidizers. Unless stabilized with an inhibitor, air exposure causes formation of unstable peroxides.

FURFUROL (98-01-1) Forms explosive mixture with air (flash point 140°F/60°C). Incompatible with strong acids, caustics, ammonia, aliphatic amines, alkanolamines, aromatic amines, oxidizers. Attacks many plastics and coatings.

FURFUROLE (98-01-1) Forms explosive mixture with air (flash point 140°F/60°C). Incompatible with strong acids, caustics, ammonia, aliphatic amines, alkanolamines, aromatic amines, oxidizers. Attacks many plastics and coatings.

FURFURYL ALCOHOL (98-00-0) Forms explosive mixture with air (flash point 170°F/77°C). Strong acids (including some organic acids) or acid catalysts may cause polymerization. Reacts strongly with oxidizers and acids. Incompatible with caustics, aliphatic amines, isocyanates. Attacks some plastics, coatings, and rubber.

FURNACE BLACK (1333-86-4) Dust contact can produce an explosive mixture in air. Oxidizers may cause fire and explosions.

FURODAN (1563-66-2) Incompatible with alkaline material, acids, strong oxidizers.

FUROLE (98-01-1) Forms explosive mixture with air (flash point 140°F/60°C). Incompatible with strong acids, caustics, ammonia, aliphatic amines, alkanolamines, aromatic amines, oxidizers. Attacks many plastics and coatings.

alpha-FUROLE (98-01-1) Forms explosive mixture with air (flash point 140°F/60°C). Incompatible with strong acids, caustics, ammonia, aliphatic amines, alkanolamines, aromatic amines, oxidizers. Attacks many plastics and coatings.

FURRO D (106-50-3) Incompatible with acids, organic anhydrides, isocyanates, aldehydes, strong oxidizers; may cause fire and explosions. Heat and light contribute to instability.

FUR YELLOW (106-50-3) Incompatible with acids, organic anhydrides, isocyanates, aldehydes, strong oxidizers; may cause fire and explosions. Heat and light contribute to instability.

FURYLALCOHOL (98-00-0) Forms explosive mixture with air (flash point 170°F/77°C). Strong acids (including some organic acids) or acid catalysts may cause polymerization. Reacts strongly with oxidizers and acids. Incompatible with caustics, aliphatic amines, isocyanates. Attacks some plastics, coatings, and rubber.

2-FURYLCARBINOL (98-00-0) Forms explosive mixture with air (flash point 170°F/77°C). Strong acids (including some organic acids) or acid catalysts may cause polymerization. Reacts strongly with oxidizers and acids. Incompatible with caustics, aliphatic amines, isocyanates. Attacks some plastics, coatings, and rubber.

2-FURYL-METHANAL (98-01-1) Forms explosive mixture with air (flash point 140°F/60°C). Incompatible with strong acids, caustics, ammonia, aliphatic amines, alkanolamines, aromatic amines, oxidizers. Attacks many plastics and coatings.

FUSED BORIC ACID (1303-86-2) Water contact produces boric acid. Fires and explosion may result from contact with bromine pentafluoride or calcium oxide.

FUSEL OIL (123-51-3) Forms explosive mixture with air (flash point 109°F/43°C). Strong oxidizers may cause fire and explosions. Attacks some plastics, rubber, and coatings.

FUTRAMINE D (106-50-3) Incompatible with acids, organic anhydrides, isocyanates, aldehydes, strong oxidizers; may cause fire and explosions. Heat and light contribute to instability.

FYDE (50-00-0) May polymerize unless properly inhibited (usually with methanol). Forms explosive mixture with air (flash point 122°F/50°C). Incompatible with strong acids, amines, strong oxidizers, alkaline materials, nitrogen dioxide, performic acid. Reaction with hydrochloric acid produces bis-chloromethyl ether, a carcinogen.

FYFANON (Danish) (121-75-5) Incompatible with strong oxidizers, magnesium, alkaline pesticides. Attacks metals, some plastics, rubber, and coatings.

- G -

G-301 (333-41-5) Incompatible with water, copper-containing compounds, oxidizers, acids, or bases.

G-24480 (333-41-5) Incompatible with water, copper-containing compounds, oxidizers, acids, or bases.

G-30027 (1912-24-9) Incompatible with strong acids.

GAFCOL EB (111-76-2) Forms explosive mixture with air (flash point 143°F/62°C). Reacts violently with strong caustics and strong oxidizers. Attacks some coatings, plastics, and rubber. Attacks metallic aluminum at high temperatures.

GALENA (1314-87-0) Reacts violently with iodine monochloride or hydrogen peroxide.

GALLIC ACID or GALLIC ACID MONOHYDRATE (149-91-7) Incompatible with sulfuric acid, caustics, ammonia, aliphatic and alkanol amines, isocyanates, alkylene oxides, epichlorohydrin, strong oxidizers.

GALLOGAMA (58-89-9) Not combustible, but may be dissolved in a combustible solvent. If solvent comes in contact with oxidizers, fire and explosions may result.

GAMACID (58-89-9) Not combustible, but may be dissolved in a combustible solvent. If solvent comes in contact with oxidizers, fire and explosions may result.

GAMAPHEX (58-89-9) Not combustible, but may be dissolved in a combustible solvent. If solvent comes in contact with oxidizers, fire and explosions may result.

GAMASOL-90 (67-68-5) Forms explosive mixture with air (flash point 203°F/95°C). Reacts violently with oxidizers. Reacts with ethanoyl chloride, boron compounds, halides, metal alkoxides, oxidizers.

GAMENE (58-89-9) Not combustible, but may be dissolved in a combustible solvent. If solvent comes in contact with oxidizers, fire and explosions may result.

GAMMAHEXA (58-89-9) Not combustible, but may be dissolved in a combustible solvent. If solvent comes in contact with oxidizers, fire and explosions may result.

GAMMALIN (58-89-9) Not combustible, but may be dissolved in a combustible solvent. If solvent comes in contact with oxidizers, fire and explosions may result.

GAMMATERR (58-89-9) Not combustible, but may be dissolved in a combustible solvent. If solvent comes in contact with oxidizers, fire and explosions may result.

GAMMEX (58-89-9) Not combustible, but may be dissolved in a combustible solvent. If solvent comes in contact with oxidizers, fire and explosions may result.

GAMMEXANE (58-89-9) Not combustible, but may be dissolved in a combustible solvent. If solvent comes in contact with oxidizers, fire and explosions may result.

GAMMOPAZ (58-89-9) Not combustible, but may be dissolved in a combustible solvent. If solvent comes in contact with oxidizers, fire and explosions may result.

GAMONIL (63-25-2) Incompatible with strong oxidizers, strongly alkaline pesticides.

GARDENTOX (333-41-5) Incompatible with water, copper-containing compounds, oxidizers, acids, or bases.

GAS, NATURAL (74-82-8) Flammable gas. Reacts violently with strong oxidizers.

GAS OIL or GAS OIL, CRACKED (64741-44-2) Forms explosive mixture with air (flash point approximately 151°F/55°C) Incompatible with nitric acid. Strong oxidizers may cause fire and explosions. May accumulate static electrical charges, and may cause ignition of its vapors.

GAS OILS (PETROLEUM), LIGHT VACUUM (64741-44-2) Forms explosive mixture with air (flash point approximately 151°F/55°C) Incompatible with nitric acid. Strong oxidizers may cause fire and explosions. May accumulate static electrical charges, and may cause ignition of its vapors.

GAS OILS (PETROLEUM), STRAIGHT RUN (64741-43-1) Oxidizers may cause fire and explosions. Incompatible with nitric acid. May accumulate static electrical charges, and may cause ignition of its vapors.

GASOLINE, AVIATION (less than 4.86 g lead/gal) (8006-61-9) Forms explosive mixture with air (flash point 125°F/51°C). Oxidizers may cause fire and explosions. Incompatible with nitric acid. May accumulate static electrical charges, and may cause ignition of its vapors.

GASOLINE, AVIATION GRADE (100–130 OCTANE) (8006-61-9) Forms explosive mixture with air (flash point 125°F/51°C). Oxidizers may cause fire and explosions. Incompatible with nitric acid. May accumulate static electrical charges, and may cause ignition of its vapors.

GASOLINE, AVIATION GRADE (115–145 OCTANE) (8006-61-9) Forms explosive mixture with air (flash point 125°F/51°C). Oxidizers may cause fire and explosions. Incompatible with nitric acid. May accumulate static electrical charges, and may cause ignition of its vapors.

GASOLINE BLENDING STOCKS: ALKYLATES (N/A) Incompatible with strong oxidizers, nitric acid, perchlorates.

GASOLINE BLENDING STOCKS: REFORMATES (N/A) Forms explosive mixture with air (flash point 0°F to 73°F/−18°C to 23°C). Incompatible with strong oxidizers, nitric acid, perchlorates. May accumulate static electrical charges, and may cause ignition of its vapors.

GASOLINE, NATURAL (8006-61-9) Forms explosive mixture with air (flash point 125°F/51°C). Oxidizers may cause fire and explosions. Incompatible with nitric acid. May accumulate static electrical charges, and may cause ignition of its vapors.

GASOLINES: AUTOMOTIVE (less than 4.23 g lead/gal) (8006-61-9) Forms explosive mixture with air (flash point 125°F/51°C). Oxidizers may cause fire and explosions. Incompatible with nitric acid. May accumulate static electrical charges, and may cause ignition of its vapors.

GASOLINES: AVIATION (8006-61-9) Forms explosive mixture with air (flash point 125°F/51°C). Oxidizers may cause fire and explosions. Incompatible with nitric acid. May accumulate static electrical charges, and may cause ignition of its vapors.

GASOLINES: CASINGHEAD (8006-61-9) Forms explosive mixture with air (flash point 125°F/51°C). Oxidizers may cause fire and explosions. Incompatible with nitric acid. May accumulate static electrical charges, and may cause ignition of its vapors.

GASOLINES: POLYMER (N/A) Forms explosive mixture with air (flash point 0°F to 73°F/−18°C to 3°C). Incompatible with acids, oxidizers, perchlorates.

GASOLINE, STRAIGHT RUN, TOPPING-PLANT (68606-11-1) Oxidizers may cause fire and explosions. Incompatible with nitric acid. May accumulate static electrical charges, and may cause ignition of its vapors.

GASOLINES: STRAIGHT RUN (68606-11-1) Oxidizers may cause fire and explosions. Incompatible with nitric acid. May accumulate static electrical charges, and may cause ignition of its vapors.

GAULTHERIA OIL (119-36-0) Incompatible with strong acids, nitrates, strong oxidizers.

GBL (123-19-3) Oxidizers may cause violent reaction.

GEARPHOS (298-00-0) Mixtures with magnesium may be explosive.

GEARPHOS (56-38-2) Combustible liquid. Mixtures with endrin may be explosive. Strong oxidizers may cause fire and explosions. Attacks some plastics, rubber, and coatings.

GEIGY 24,480 (333-41-5) Incompatible with water, copper-containing compounds, oxidizers, acids, or bases.

GEIGY 30,027 (1912-24-9) Incompatible with strong acids.

GELBIN YELLOW ULTRAMARINE (13765-19-0) A strong oxidizer. Incompatible with boron (with violent reaction), combustible material.

GENETRON 11 (75-69-4) Reacts with barium, lithium, sodium, magnesium, titanium. Attacks some plastics, rubber, and coatings.

GENETRON 12 (75-71-8) Reacts violently with liquid aluminum. Incompatible with chemically active metals. Attacks some plastics, rubber, and coatings.

GENETRON 13 (75-72-9) Reacts with aluminum, magnesium, zinc, and their alloys.

GENETRON 21 (75-43-4) Reacts with water and chemically active metals. Attacks some plastics, rubber, and coatings.

GENETRON 22 (75-45-5) Moisture and rust cause slow decomposition, producing toxic gases. Attacks some plastics, rubber, and coatings. Thermal decomposition occurs at high temperature with alkalies and alkaline earth metals.

GENETRON 100 (75-37-6) Flammable gas. May accumulate static electrical charges, and may cause ignition of its vapors. Reacts violently with strong oxidizers. Attacks metals in presence of moisture.

GENETRON 112 (76-12-0) Reacts with chemically active metals, powdered aluminum, zinc, and magnesium. Attacks some forms of plastics, rubber, and coatings.

GENETRON 113 (76-13-1) Reacts with barium, lithium, sodium, magnesium, titanium. Contact with alloys containing more than 2% magnesium may cause decomposition (with hydrogen chloride, hydrogen fluoride, and carbon monoxide released). Attacks some plastics, rubber, and coatings.

GENETRON 114 (76-14-2) Reacts with barium, lithium, sodium, magnesium, titanium. Attacks some plastics, rubber, and coatings.

GENETRON 115 (76-15-3) Thermal decomposition occurs at high temperatures with alkalies and alkaline earth metals.

GENETRON 1113 (79-38-9) Flammable gas. Forms explosive mixture with air (flash point $-18°F/-28°C$). Reacts violently with oxidizers. Contact with ethylene may cause explosive polymerization.

GENIPHENE (8001-35-2) Reacts with oxidizers, with a risk of fire or explosions. Attacks metals in the presence of moisture.

GENITHION (56-38-2) Combustible liquid. Mixtures with endrin may be explosive. Strong oxidizers may cause fire and explosions. Attacks some plastics, rubber, and coatings.

GENITOX (50-29-3) Incompatible with salts of iron or aluminum, and bases. Do not store in iron containers.

GEON (9016-00-6) Incompatible with sulfuric acid, isocyanates.

GERANIUM CRYSTALS (101-84-8) Incompatible with strong acids, strong oxidizers; may cause fire and explosions. Attacks some plastics, rubber, and coatings.

GERFIL (9003-07-0) Incompatible with strong acids, strong oxidizers, chlorine, potassium permanganate.

GERMAIN'S (63-25-2) Incompatible with strong oxidizers, strongly alkaline pesticides.

GERMALGENE (79-01-6) Contact with caustics produces a toxic and flammable gas. Reacts violently with chemically active metals. Contact with aluminum may produce a violent, self-accelerating polymerization reaction. Incompatible with acids, organic anhydrides, isocyanates, alkylene oxides, aldehydes, alcohols, glycols, phenols, cresols, caprolactam solution, epichlorohydrin, nitrogen tetroxide, metal powders, oxygen. May accumulate static electrical charges, and may cause ignition of its vapors.

GESAFID (50-29-3) Incompatible with salts of iron or aluminum, and bases. Do not store in iron containers.

GESAPON (50-29-3) Incompatible with salts of iron or aluminum, and bases. Do not store in iron containers.

GESAPRIM (1912-24-9) Incompatible with strong acids.

GESAREX (50-29-3) Incompatible with salts of iron or aluminum, and bases. Do not store in iron containers.

GESAROL (50-29-3) Incompatible with salts of iron or aluminum, and bases. Do not store in iron containers.

GESFID (7786-34-7) Contact with strong oxidizers may cause fire and explosions. Attacks some plastics, rubber, and coatings.

GESOPRIM (1912-24-9) Incompatible with strong acids.

GESTID (7786-34-7) Contact with strong oxidizers may cause fire and explosions. Attacks some plastics, rubber, and coatings.

GEXANE (58-89-9) Not combustible, but may be dissolved in a combustible solvent. If solvent comes in contact with oxidizers, fire and explosions may result.

GIBS (7778-18-9) Contact with diazomethane, aluminum, phosphorus may cause explosion.

GLACIAL ACETIC ACID (64-19-7) Vapor forms explosive mixture with air (flash point 105°F/40°C). Reacts violently with oxidizers and bases. Incompatible with strong acids, aliphatic amines, alkanolamines, isocyanates, alkylene oxides, epichlorohydrin, acetaldehyde, 2-aminoethanol, ammonia, ammonium nitrate, chlorosulfonic acid, chromic acid, ethylene diamine, ethyleneimine, perchloric acid, permanganates, phosphorus isocyanate, phosphorus trichloride, potassium *tert*-butoxide, xylene. Attacks cast iron and other metals, producing flammable hydrogen gas.

GLACIAL ACRYLIC ACID (79-10-7) Forms explosive mixture with air (flash point 122°F/50°C). Forms explosive peroxides; light, heat and peroxides can cause polymerization. Incompatible with sulfuric acid, caustics, ammonia, amines, isocyanates, alkylene oxides, epichlorohydrin, oxidizers, toluenediamine, pyridine, methyl pyridine, n-methyl pyrrolidone, 2-methyl-6-ethyl aniline, aniline, ethylene diamine, ethyleneimine, 2-aminoethanol. Severely corrodes carbon steel and iron; attacks other metals. May accumulate static electrical charges, and may cause ignition of its vapors.

GLAZED PENTA (87-86-5) Hot water causes decomposition forming hydrochloric acid. Strong oxidizers may cause fire and explosions.

GLONOIN (55-63-0) Liquid portion of this chemical is an alcohol; reacts with strong acids, caustics, aliphatic amines, isocyanates. Dry portion of chemical is nitroglycerine, a shock- and friction-sensitive explosive. Acids, heat, or mechanical shock may result in explosions.

GLUCINIUM (7440-41-7) Contact with acids causes evolution of flammable hydrogen gas. Incompatible with alkalies, chlorinated hydrocarbons, oxidizable agents, carbon tetrachloride, trichloroethylene, lithium, phosphorus.

GLUCITOL (50-70-4) Incompatible with strong acids, bases, aliphatic amines, isocyanates, strong oxidizers.

D-GLUCITOL (50-70-4) Incompatible with strong acids, bases, aliphatic amines, isocyanates, strong oxidizers.

GLUCOSE SOLUTION (60-99-7) Incompatible with sulfuric acid, isocyanates.

GLUTAMIC DIALDEHYDE (111-30-8) Water contact produces a polymer solution. A strong reducing agent. Incompatible with strong acids, caustics, ammonia, amines, strong oxidizers.

GLUTARAL (111-30-8) Water contact produces a polymer solution. A strong reducing agent. Incompatible with strong acids, caustics, ammonia, amines, strong oxidizers.

GLUTARALDEHYD (Czech) (111-30-8) Water contact forms a polymer solution. A strong reducing agent. Incompatible with strong acids, caustics, ammonia, amines, strong oxidizers.

GLUTARALDEHYDE or GLUTARALDEHYDE SOLUTION (111-30-8) Water contact produces a polymer solution. A strong reducing agent. Incompatible with strong acids, caustics, ammonia, amines, strong oxidizers.

GLUTARD DIALDEHYDE (111-30-8) Water contact produces a polymer solution. A strong reducing agent. Incompatible with strong acids, caustics, ammonia, amines, strong oxidizers.

GLUTARIC ACID (110-94-1) Reacts with oxidizers and strong bases.

GLUTARIC ACID DIALDEHYSE (111-30-8) Water contact produces a polymer solution. A strong reducing agent. Incompatible with strong acids, caustics, ammonia, amines, strong oxidizers.

GLUTARIC DIALDEHYDE (111-30-8) Water contact produces a polymer solution. A strong reducing agent. Incompatible with strong acids, caustics, ammonia, amines, strong oxidizers.

GLYCERIN (90 TECHNICAL) (56-81-5) Polymerizes at approximately 300°F/150°C. Incompatible with acetic anhydride, potassium permanganate, strong acids, caustics, aliphatic amines, isocyanates, oxidizers.

GLYCERINE or GLYCERIN, ANHYDROUS or GLYCERIN, SYNTHETIC (56-81-5) Polymerizes at approximately 300°F/150°C. Incompatible with acetic anhydride, potassium permanganate, strong acids, caustics, aliphatic amines, isocyanates, oxidizers.

GLYCERITOL (56-81-5) Polymerizes at approximately 300°F/150°C. Incompatible with acetic anhydride, potassium permanganate, strong acids, caustics, aliphatic amines, isocyanates, oxidizers.

GLYCEROL (56-81-5) Polymerizes at approximately 300°F/150°C. Incompatible with acetic anhydride, potassium permanganate, strong acids, caustics, aliphatic amines, isocyanates, oxidizers.

GLYCEROL EPICHLOROHYDRIN (106-89-8) Forms explosive mixture with air (flash point 88°F/31°C). Heat, acids, alkalies, and metallic halides can cause explosive polymerization. Aliphatic amines, alkaline earths, alkali metals, alkanolamines, powdered metals, strong oxidizers may cause fire and explosions. Decomposition produces highly toxic phosgene gas. Will pit steel in the presence of moisture. May accumulate static electrical charges, and may cause ignition of its vapors.

GLYCEROL TRINITRATE (55-63-0) Liquid portion of this chemical is an alcohol; reacts with strong acids, caustics, aliphatic amines, isocyanates. Dry portion of chemical is nitroglycerine, a shock- and friction-sensitive explosive. Acids, heat, or mechanical shock may result in explosions.

GLYCERYL NITRATE (55-63-0) Liquid portion of this chemical is an alcohol; reacts with strong acids, caustics, aliphatic amines, isocyanates. Dry portion of chemical is nitroglycerine, a shock- and friction-sensitive explosive. Acids, heat, or mechanical shock may result in explosions.

GLYCERYL TRICHLORHYDRIN (96-18-4) Forms explosive mixture with air (flash point 164°F/74°C). Incompatible with strong oxidizers, strong caustics, chemically active metals; may cause fire and explosions. May decompose on contact with aluminum. Attacks some plastics, rubber and coatings.

GLYCERYL TRINITRATE (55-63-0) Liquid portion of this chemical is an alcohol; reacts with strong acids, caustics, aliphatic amines, isocyanates. Dry portion of chemical is nitroglycerine, a shock- and friction-sensitive explosive. Acids, heat, or mechanical shock may result in explosions.

GLYCIDE (556-52-5) Forms explosive mixture with air (flash point 158°F/70°C). Reacts violently with strong oxidizers. Contact with caustics, acids, barium, lithium, sodium, magnesium, titanium can cause polymerization. Incompatible with nitrates. Attacks some plastics, rubber, and coatings.

GLYCIDOL (556-52-5) Forms explosive mixture with air (flash point 158°F/70°C). Reacts violently with strong oxidizers. Contact with caustics, acids, barium, lithium, sodium, magnesium, titanium can cause polymerization. Incompatible with nitrates. Attacks some plastics, rubber, and coatings.

GLYCIDYL ALCOHOL (556-52-5) Forms explosive mixture with air (flash point 158°F/70°C). Reacts violently with strong oxidizers. Contact with caustics, acids, barium, lithium, sodium, magnesium, titanium can cause polymerization. Incompatible with nitrates. Attacks some plastics, rubber, and coatings.

GLYCIDYL ALPHA-METHYL ACRYLATE (106-99-2) Forms explosive mixture with air (flash point −105°F/−76°C). Air exposure causes formation of unstable peroxides; heat, caustics, or shock may cause explosive polymerization. Reacts violently with aluminum borohydride, sodium nitrite, strong oxidizers.

GLYCIDYL ISOPROPYL ETHER (4016-08-6) Forms explosive mixture with air (flash point 92°F/33°C). Incompatible with strong oxidizers (fire), strong caustics (polymerization). Air and light cause formation of unstable and explosive peroxides. Attacks some rubbers and plastics.

GLYCIDYL METHACRYLATE (106-99-2) Forms explosive mixture with air (flash point −105°F/−76°C). Air exposure causes formation of unstable peroxides; heat, caustics, or shock may cause explosive polymerization. Reacts violently with aluminum borohydride, sodium nitrite, strong oxidizers.

GLYCIDYL METHYLPHENYL ETHER (26447-14-3) Forms explosive mixture with air (flash point 200°F/93°C). Incompatible with oxidizers, alkalies, sodium azide, acetylene, strong acids, aliphatic amines, amides. Attacks some plastics, rubber, and coatings.

GLYCINE, N,N'-1,2-ETHANEDIYLBIS(N-(CARBOXYMETHYL)- (9CI) (60-00-4) Incompatible with sulfuric acid, bases, ammonia, aliphatic amines, alkanolamines, isocyanates, alkylene oxides, epichlorohydrin.

GLYCINOL (141-43-5) Will oxidize in air; often shipped under a pad of inert gas. Forms explosive mixture with air (flash point 185°F/85°C). Strong oxidizers may cause fire and explosions. A strong base. Incompatible with acids, organic anhydrides, isocyanates, alkylene oxides, epichlorohydrin, aldehydes, alcohols, glycols, phenols, cresols, caprolactam solution. Contact with strong acids may cause spattering. Attacks plastics, rubber, and aluminum, copper, tin, and their alloys.

GLYCOL (107-21-1) Incompatible with strong acids, caustics, aliphatic amines, isocyanates, chlorosulfonic acid, oleum, strong oxidizers.

GLYCOL ALCOHOL (107-21-1) Incompatible with strong acids, caustics, aliphatic amines, isocyanates, chlorosulfonic acid, oleum, strong oxidizers.

GLYCOL BIS(HYDROXYETHYL) ETHER (112-27-6) Incompatible with sulfuric acid, isocyanates, perchloric acid, strong oxidizers.

GLYCOL BROMIDE (106-93-4) Reacts with chemically active metals, liquid ammonia, strong oxidizers. Heat and light cause slow decomposition. Attacks some plastics and rubber.

GLYCOL BUTYL ETHER (111-76-2) Forms explosive mixture with air (flash point 143°F/62°C). Reacts violently with strong caustics and strong oxidizers. Attacks some coatings, plastics, and rubber. Attacks metallic aluminum at high temperatures.

GLYCOL CHLOROHYDRIN (107-07-3) Forms explosive mixture with air (flash point 140°F/60°C). Incompatible with strong oxidizers (may cause fire and explosions), strong caustics (cause formation of ethylene gas), strong acids, aliphatic amines, isocyanates. Attacks some plastics, rubber, and coatings.

GLYCOL CYANOHYDRIN (109-78-4) Forms explosive mixture with air (flash point 140°F/60°C). Avoid basic contamination to prevent polymerization. A powerful reducing agent. Hot water forms cyanide gas. Reacts violently with strong oxidizers or sodium hydroxide. Reacts with acids, acid salts, chlorates, nitrates. Attacks mild steel, copper, and copper alloys.

GLYCOL DIACETATE (111-55-7) Forms explosive mixture with air (flash point 191°F/88°C). Strong oxidizers may cause fire and explosions.

GLYCOL DIBROMIDE (106-93-4) Reacts with chemically active metals, liquid ammonia, strong oxidizers. Heat and light cause slow decomposition. Attacks some plastics and rubber.

GLYCOL DICHLORIDE (107-06-2) May accumulate static electrical charges, and may cause ignition of its vapors. Contact with hot water may produce hydrochloric acid. Forms explosive mixture with air (flash point 55°F/13°C). Incompatible with strong oxidizers, strong caustics, chemically active metals. Corrosive to iron and other metals. Attacks some plastics, rubber, or coatings.

GLYCOL DIMETHYL ETHER (110-71-4) Forms explosive mixture with air (flash point 29°F/−2°C). May accumulate static electrical charges, and may cause ignition of its vapors. Incompatible with sulfuric acid, isocyanates, strong oxidizers.

GLYCOL DINITRATE, NITROGLYCOL (628-96-6) Acids, heat, or mechanical shock may result in explosions.

GLYCOL ETHER (111-46-6) Combustible (flash point 255°F/124°C). Incompatible with sulfuric acid, isocyanates, strong oxidizers.

GLYCOL ETHER DB ACETATE (124-17-4) Incompatible with strong acids, nitrates, strong alkalies, oxidizers.

GLYCOL ETHER DE (111-46-6) Combustible (flash point 255°F/124°C). Incompatible with sulfuric acid, isocyanates, strong oxidizers.

GLYCOL ETHER EE (110-80-5) Forms explosive mixture with air (flash point 120°F/49°C). Strong oxidizers may cause fire and explosions. Attacks some plastics, rubber, and coatings. Able to form peroxides. Incompatible with strong acids, aluminum and its alloys.

GLYCOL ETHER EM (109-86-4) Forms explosive mixture with air (flash point 103°F/39°C). Heat or oxidizers may cause formation of unstable peroxides. Attacks many metals. Strong oxidizers cause fire and explosions. Strong bases cause decomposition. Attacks some plastics, rubber, and coatings. May accumulate static electrical charges, and may cause ignition of its vapors.

GLYCOL ETHYL ETHER (110-80-5) Forms explosive mixture with air (flash point 120°F/49°C). Strong oxidizers may cause fire and explosions. Attacks some plastics, rubber, and coatings. Able to form peroxides. Incompatible with strong acids, aluminum and its alloys.

GLYCOL ETHYL ETHER (111-46-6) Combustible (flash point 255°F/124°C). Incompatible with sulfuric acid, isocyanates, strong oxidizers.

GLYCOL ETHYLENE ETHER (123-91-1) Forms explosive mixture with air (flash point 54°F/12°C). Moisture causes formation of unstable peroxides. May accumulate static electrical charges, and may cause ignition of its vapors. Incompatible with strong oxidizers, strong acids. Attacks many plastics.

GLYCOLIC ACID (79-14-1) Incompatible with sulfuric acid, caustics, ammonia, aliphatic amines, isocyanates, alkylene oxides, epichlorohydrin, strong oxidizers.

GLYCOL METHYL ETHER (109-86-4) Forms explosive mixture with air (flash point 103°F/39°C). Heat or oxidizers may cause formation of unstable peroxides. Attacks many metals. Strong oxidizers cause fire and explosions. Strong bases cause decomposition. Attacks some plastics, rubber, and coatings. May accumulate static electrical charges, and may cause ignition of its vapors.

GLYCOL MONOACETATE (542-59-6) Forms explosive mixture with air (flash point 191°F/88°C). Incompatible with sulfuric acid, nitric acid, nitrates.

GLYCOL-MONOACETIN (542-59-6) Forms explosive mixture with air (flash point 191°F/88°C). Incompatible with sulfuric acid, nitric acid, nitrates.

GLYCOL MONOBUTYL ETHER ACETATE (112-07-2) Forms explosive mixture with air (flash point 160°F/71°C). Ethers, as a class, can form peroxides. Incompatible with strong acids, perchloric acid, oxidizers, nitrates

GLYCOL MONOETHYL ETHER (110-80-5) Forms explosive mixture with air (flash point 120°F/49°C). Strong oxidizers may cause fire and explosions. Attacks some plastics, rubber, and coatings. Able to form peroxides. Incompatible with strong acids, aluminum and its alloys.

GLYCOL MONOETHYL ETHER ACETATE (111-15-9) Forms explosive mixture with air (flash point 117°F/47°C). Incompatible with strong acids, nitrates. Reacts violently with oxidizers. May form unstable peroxides. Softens many plastics.

GLYCOL MONOMETHYL ETHER (109-86-4) Forms explosive mixture with air (flash point 103°F/39°C). Heat or oxidizers may cause formation of unstable peroxides. Attacks many metals. Strong oxidizers cause fire and explosions. Strong bases cause decomposition. Attacks some plastics, rubber, and coatings. May accumulate static electrical charges, and may cause ignition of its vapors.

GLYCOL MONOMETHYL ETHER ACETATE (110-49-6) Incompatible with nitrates, strong oxidizers, strong alkalies, strong acids.

GLYCON DP (57-11-4) Incompatible with strong oxidizers, sulfuric acid, caustics, ammonia, amines, isocyanates, alkylene oxides, epichlorohydrin. Attacks chemically active metals.

GLYCON RO (112-80-1) Incompatible with strong oxidizers, perchloric acid, and aluminum powder; may cause explosions. Corrodes aluminum.

GLYCON TP (57-11-4) Incompatible with strong oxidizers, sulfuric acid, caustics, ammonia, amines, isocyanates, alkylene oxides, epichlorohydrin. Attacks chemically active metals.

GLYCON WO (112-80-1) Incompatible with strong oxidizers, perchloric acid, and aluminum powder; may cause explosions. Corrodes aluminum.

GLYCYL ALCOHOL (56-81-5) Polymerizes at approximately 300°F/150°C. Incompatible with acetic anhydride, potassium permanganate, strong acids, caustics, aliphatic amines, isocyanates, oxidizers.

GLYCIDYL PHENYL ETHER (122-60-1) Strong oxidizers may cause fire and explosions. Contact with amines, strong acids, and strong bases may cause polymerization with the liberation of heat and spattering. Exposure to light and air may cause formation of explosive peroxides. Attacks some forms of plastics, coatings, and rubber.

GLYME (110-71-4) Forms explosive mixture with air (flash point 29°F/−2°C). May accumulate static electrical charges, and may cause ignition of its vapors. Incompatible with sulfuric acid, isocyanates, strong oxidizers.

GLYMOL (8012-95-1) Oxidizers may cause fire and explosions. Incompatible with nitric acid. May accumulate static electrical charges, and may cause ignition of its vapors.

GLYODEX 3722 (133-06-2) Incompatible with tetraethyl pyrophosphate, parathion.

GLYOXAL (107-22-2) Forms explosive mixture with air. A strong reducing agent; reacts violently with oxidizers. Heat or contact with water, strong bases causes polymerization. Incompatible with strong acids, caustics, ammonia, amines, chlorosulfonic acid, ethylene amine. Corrosive to metals.

GLY-OXIDE (124-43-6) Combustible solid. At 122°F/50°C reacts with dust, organics.

GLYOXYALDEHYDE (107-22-2) Forms explosive mixture with air. A strong reducing agent; reacts violently with oxidizers. Heat or contact with water, strong bases causes polymerization. Incompatible with strong acids, caustics, ammonia, amines, chlorosulfonic acid, ethylene amine. Corrosive to metals.

GLYOXYLIC ACID (50% OR LESS) (298-12-4) Reacts with alkali metals; releases flammable hydrogen gas.

GOOD-RITE (9016-00-6) Incompatible with sulfuric acid, isocyanates.

GP-40-66:120 (87-68-3) Forms explosive mixture with air (flash point 195°F/90°C). Reacts strongly with oxidizers, aluminum powder. Attacks aluminum. Attacks some plastics, rubber, and coatings.

GPKH (76-44-8) Produces hydrogen chloride gas with iron and rust above 165°F/74°C.

GRAIN ALCOHOL (64-17-5) Forms explosive mixture with air (flash point 65°F/18°C). May accumulate static electrical charges, and may cause ignition of its vapors. Reactions may be violent with oleum, sulfuric acid, nitric acid, bases, aliphatic amines, isocyanates, oxidizers.

408

GRAIN SORGHUM HARVEST AID (7775-09-9) A powerful oxidizer; reacts violently with reducing agents and combustible matter. Explosions may be caused by contact with ammonia salts, carbon, oils, metal sulfides, nitrobenzene, powdered metals, sugar. Contact with strong acids produce carbon dioxide. Forms shock-sensitive mixtures with some organic materials. Solution (50%) decomposes at 300°F/149°C, liberating oxygen.

GRAMEVIN (75-99-0) Corrosive to iron, aluminum, and copper.

GRANEX OK (7775-09-9) A powerful oxidizer; reacts violently with reducing agents and combustible matter. Explosions may be caused by contact with ammonia salts, carbon, oils, metal sulfides, nitrobenzene, powdered metals, sugar. Contact with strong acids produces carbon dioxide. Forms shock-sensitive mixtures with some organic materials. Solution (50%) decomposes at 300°F/149°C, liberating oxygen.

GRANMAG (1309-48-4) Reacts violently with chlorine trifluoride, bromine pentafluoride, phosphorus pentachloride.

GRANOX MN (118-74-1) Reacts violently with strong oxidizers, formyldimethylamine.

GRANULATED SUGAR (57-50-1) Reacts with potassium hydroxide, strong acids, strong oxidizers.

GRAPE SUGAR SOLUTION (60-99-7) Incompatible with sulfuric acid, isocyanates.

GREEN NICKEL OXIDE (300-76-5) Unstable in the presence of iron. Reacts with acids, strong oxidizers in sunlight. Corrosive to metals. Attacks some plastics, rubber, and coatings.

GREEN OIL (120-12-7) Dust or fine powder causes formation of an explosive mixture with air. Strong oxidizers may cause fire and explosions.

GREEN VITRIOL IRON MONOSULFATE (7720-78-7) Aqueous solution is acidic. Contact with bases produces iron.

GREY ARSENIC (7440-38-2) Contact with hydrogen gas produces arsine. Incompatible with strong acids, strong oxidizers, peroxides, bromine pentafluoride, bromine trifluoride, cesium acetylene carbide, chromium trioxide, nitrogen trichloride, silver nitrate.

GRILON (105-60-2) Contact with strong oxidizers may cause fire and explosions.

GRISOL (107-49-3) Decomposes above 300°F/150°C, producing flammable ethylene gas. Strong oxidizers may cause fire and explosions. Attacks some plastics, rubber, and coatings.

GROCO (8001-26-1) Oxidizers may cause fire and explosions. Incompatible with nitric acid. May accumulate static electrical charges, and may cause ignition of its vapors.

GROCO-54 (57-11-4) Incompatible with strong oxidizers, sulfuric acid, caustics, ammonia, amines, isocyanates, alkylene oxides, epichlorohydrin. Attacks chemically active metals.

GROCOLENE (56-81-5) Polymerizes at approximately 300°F/150°C. Incompatible with acetic anhydride, potassium permanganate, strong acids, caustics, aliphatic amines, isocyanates, oxidizers.

GROUNDNUT OIL (8002-03-7) Incompatible with strong acids, oxidizers.

GROUND VOCLE SULPHUR (7704-34-9) Combustible solid. Liquid contact causes formation of sulfur dioxide with air. Reacts violently with strong oxidizers. Forms explosive, shock-sensitive, or pyrophoric mixtures with ammonia, ammonium nitrate, bromates, calcium carbide, charcoal, chlorates, hydrocarbons, iodates, iron. Reacts violently with halogen compounds and sodium, tin, uranium, and other compounds. Attacks steel when moist. May accumulate static electrical charges, and may cause ignition of its vapors.

GRUNDIER ARBEZOL (87-86-5) Hot water causes decomposition, producing hydrochloric acid. Strong oxidizers may cause fire and explosions.

GTN (55-63-0) Liquid portion of this chemical is an alcohol; reacts with strong acids, caustics, aliphatic amines, isocyanates. Dry portion of chemical is nitroglycerine, a shock- and friction-sensitive explosive. Acids, heat, or mechanical shock may result in explosions.

GUAIACOL (90-05-1) Forms explosive mixture with air (flash point 82°F/180°C). Reacts violently with strong oxidizers.

GUESAROL (50-29-3) Incompatible with salts of iron or aluminum, and bases. Do not store in iron containers.

GUM (9016-00-6) Incompatible with sulfuric acid, isocyanates.

GUM CAMPHOR (76-22-2) Forms explosive mixture with air (flash point 155°F/66°C). Violent, possibly explosive, reaction occurs with strong oxidizers. May accumulate static electrical charges, and may cause ignition of its vapors.

GUM SPIRITS (8006-64-2) Forms explosive mixture with air (flash point 95°F/35°C). Incompatible with strong oxidizers, especially chlorine; may cause fire and explosions. Incompatible with strong acids, chromic anhydride, chromyl chloride, hexachloromelamine, stannic chloride. Attacks ordinary rubber.

GUM TURPENTINE (8006-64-2) Forms explosive mixture with air (flash point 95°F/35°C). Incompatible with strong oxidizers, especially chlorine; may cause fire and explosions. Incompatible with strong acids, chromic anhydride, chromyl chloride, hexachloromelamine, stannic chloride. Attacks ordinary rubber.

GUN COTTON (9004-70-0) If ether solution evaporates, dry nitrocellulose material is a shock-sensitive explosive and fire hazard. Forms explosive mixture with air (flash point −64°F/−53°C). Strong oxidizers may cause fire and explosions. May accumulate static electrical charges, and may cause ignition of its vapors. Attacks some plastics, rubber, and coatings.

GUSATHION or GUSATHION INSECTICIDE (86-50-0) Strong oxidizers may cause fire and explosions.

GUSATHION M (86-50-0) Strong oxidizers may cause fire and explosions.

GUTHION or GUTHION INSECTICIDE (86-50-0) Strong oxidizers may cause fire and explosions.

GY-PHENE (8001-35-2) Reacts with oxidizers, with a risk of fire or explosions. Attacks metals in the presence of moisture.

GYLCIDY BUTYL ETHER (2426-08-6) Forms explosive mixture with air (flash point 130°F/54°C). Air and light cause formation of unstable and explosive peroxides. Contact with strong oxidizers may cause fire and explosions. Strong caustics may cause polymerization. Attacks some plastics and rubber.

GYRON (50-29-3) Incompatible with salts of iron or aluminum, and bases. Do not store in iron containers.

- H -

H 34 (76-44-8) Produces hydrogen chloride gas with iron and rust above 165°F/74°C.

H 321 (2032-65-7) Contact with strong oxidizers may cause fire and explosions.

HAFNIUM or HAFNIUM, ELEMENTAL or HAFNIUM, POWDER or HAFNIUM, METALLIC (7440-58-6) Moisture may cause self-ignition or explosions. A powerful oxidizer. Contact with air or nitric acid, or heat, shock, friction, strong oxidizers, or ignition sources may cause explosions of powdered form. Being an oxidizer, substance may have a violent reaction with many substances.

HAIMASED (540-72-7) Incompatible with acids, bases, ammonia, amines, amides, alcohols, glycols, caprolactam.

HAIPEN 50 (2425-06-1) Incompatible with acids or acid vapor. Strongly alkaline conditions contribute to instability.

HALOCARBON 11 (75-69-4) Reacts with barium, lithium, sodium, magnesium, titanium. Attacks some plastics, rubber, and coatings.

HALOCARBON 12 (75-71-8) Reacts violently with liquid aluminum. Incompatible with chemically active metals. Attacks some plastics, rubber, and coatings.

HALOCARBON 13 (75-72-9) Reacts with aluminum, magnesium, zinc, and their alloys.

HALOCARBON 13B1 (75-63-8) Reacts with chemically active metals, powdered aluminum, zinc, magnesium. May attack some plastics, rubber, and coatings.

HALOCARBON 21 (75-43-4) Reacts with water and chemically active metals. Attacks some plastics, rubber, and coatings.

HALOCARBON 22 (75-45-6) Thermal decomposition occurs at high temperature with alkalies and alkaline earth metals.

HALOCARBON 112 (76-12-0) Reacts with chemically active metals, powdered aluminum, zinc, and magnesium. Attacks some forms of plastics, rubber, and coatings.

HALOCARBON 112a (76-11-9) Reacts with chemically active metals, oxidizers. Attacks some forms of plastics, rubber, and coatings.

HALOCARBON 113 (76-13-1) Reacts with barium, lithium, sodium, magnesium, titanium. Contact with alloys containing more than 2% magnesium may cause decomposition (hydrogen chloride, hydrogen fluoride, and carbon monoxide released). Attacks some plastics, rubber, and coatings.

HALOCARBON 115 (76-15-3) Thermal decomposition occurs at high temperatures with alkalies and alkaline earth metals.

HALOCARBON 152a (75-37-6) Flammable gas. May accumulate static electrical charges, and may cause ignition of its vapors. Reacts violently with strong oxidizers. Attacks metals in presence of moisture.

HALOCARBON 1301 (75-63-8) Reacts with chemically active metals, powdered aluminum, zinc, magnesium. May attack some plastics, rubber, and coatings.

HALOGENATED WAXES (generic CAS for PCBs 1336-36-3) Incompatible with strong oxidizers, strong acids.

HALON (75-71-8) Reacts violently with liquid aluminum. Incompatible with chemically active metals. Attacks some plastics, rubber, and coatings.

HALON 104 (56-23-5) Becomes corrosive when in contact with water. Corrosive to metals. Reacts violently with many compounds. Decomposes on contact with chemically active metals such as sodium, potassium and magnesium. Incompatible with allyl alcohol, fluorine gas, alkali metals, aluminum. Attacks some coatings, plastics, and rubber.

HALON 112 (75-43-4) Reacts with water and chemically active metals. Attacks some plastics, rubber, and coatings.

HALON 122 (75-71-8) Reacts violently with liquid aluminum. Incompatible with chemically active metals. Attacks some plastics, rubber, and coatings.

HALON 242 (76-14-2) Reacts with barium, lithium, sodium, magnesium, titanium. Attacks some plastics, rubber, and coatings.

HALON 1001 (74-83-9) Incompatible with strong oxidizers, aluminum, dimethylsulfoxide, ethylene oxide, water. Attacks zinc, magnesium, alkali metals and their alloys.

HALON 1011 (74-97-5) Reacts with chemically active metals such as calcium, powdered aluminum, zinc, and magnesium. Liquid attacks some plastics, rubber, and coatings.

HALON 1202 (75-61-6) Reacts with chemically active metals and oxidizers. Attacks some plastics, rubber, and coatings.

HALON 1301 (75-63-8) Reacts with chemically active metals, powdered aluminum, zinc, magnesium. May attack some plastics, rubber, and coatings.

HALON 2001 (74-96-4) Forms explosive mixture with air (flash point less than $-4°F/-20°C$). Oxidizers may cause fire or explosions. Incompatible with chemically active metals. Attacks some plastic, rubber, and coatings.

HALON 10001 (74-88-4) Contact with strong oxidizers may cause fire and explosions. Heat, light, and moisture contribute to instability.

HALOWAX 1013 (1321-64-8) Strong oxidizers may cause fire and explosions. Heat contributes to instability.

HALOWAX 1014 (1335-87-1) Oxidizers may cause fire and explosions.

HALOWAX 1051 (2234-13-1) Strong oxidizers may cause fire and explosions.

HARD WAX (8002-74-2) Reacts with oxidizers, with a risk of fire or explosions.

HARVEST-AID (7775-09-9) A powerful oxidizer; reacts violently with reducing agents and combustible matter. Explosions may be caused by contact with ammonia salts, carbon, oils, metal sulfides, nitrobenzene, powdered metals, sugar. Contact with strong acids produces carbon dioxide. Forms shock-sensitive mixtures with some organic materials. Solution (50%) decomposes at $300°F/149°C$, liberating oxygen.

HATCOL XP (6196-95-8) Incompatible with nitric acid, strong oxidizers; poses risk of fire or explosions.

HAVERO-EXTRA (50-29-3) Incompatible with salts of iron or aluminum, and bases. Do not store in iron containers.

HAVIDOTE (60-00-4) Incompatible with sulfuric acid, bases, ammonia, aliphatic amines, alkanolamines, isocyanates, alkylene oxides, epichlorohydrin.

HCB (118-74-1) Reacts violently with strong oxidizers, formyldimethylamine.

HCBD (87-68-3) Forms explosive mixture with air (flash point 195°F/90°C). Reacts strongly with oxidizers, aluminum powder. Attacks aluminum. Attacks some plastics, rubber, and coatings.

HCCH (58-89-9) Not combustible, but may be dissolved in a combustible solvent. If solvent comes in contact with oxidizers, fire and explosions may result.

HCCPD (77-47-4) Incompatible with water, producing hydrochloric acid. Contact with sodium may be explosive. Corrodes iron and other metals in the presence of moisture.

HCDB (137-26-8) Combustible solid (flash point 192°F/89°C). Strong oxidizers may cause fire and explosions; contact with strong acid or oxidizable materials produces toxic gases.

HCFC-115 (76-15-3) Thermal decomposition occurs at high temperatures with alkalies and alkaline earth metals.

HCH (58-89-9) Not combustible, but may be dissolved in a combustible solvent. If solvent comes in contact with oxidizers, fire and explosions may result.

gamma-HCH (58-89-9) Not combustible, but may be dissolved in a combustible solvent. If solvent comes in contact with oxidizers, fire and explosions may result.

HCL (7647-01-0) Incompatible with acetic anhydride, aliphatic amines, alkanolamines, alkylene oxides, aromatic amines, amides, 2-aminoethanol, ammonia, ammonium hydroxide, calcium phosphide, chlorosulfonic acid, ethylene diamine, ethyleneimine, epichlorohydrin, isocyanates, metal acetylides, oleum, organic anhydrides, perchloric acid, 3-propiolactone, uranium phosphide, sulfuric acid, sodium hydroxide and other bases, strong oxidizers, vinyl acetate, vinylidene fluoride. Attacks most metals (can produce flammable hydrogen gas), and some plastics, rubber, and coatings.

HCN (74-90-8) Unless stabilized and maintained, samples stored more than 90 days are hazardous. Samples containing more than 2–5% water are less stable than dry material. Can be self-reactive, forming an explosive mixture with air (flash point 0°F/−18°C). Heat or contact with amines or strong bases can cause polymerization. Incompatible with acetaldehyde. Oxidizers may cause fire and explosions. Attacks some plastics, rubber, and coatings.

HCS 3260 (57-74-9) Contact with strong oxidizers may cause fire and explosions. Attacks some plastics, rubber, and coatings.

HDEHP (298-07-7) Corrosive to metals; produces flammable hydrogen gas.

HEA (818-61-6) Incompatible with strong acids, aliphatic amines, alkanolamines. Attacks mild steel and tin plate. Swells rubber; removes paint.

HEAT TREATING OIL (8012-95-1) Oxidizers may cause fire and explosions. Incompatible with nitric acid. May accumulate static electrical charges, and may cause ignition of its vapors.

HEAVY INDUSTRIAL FUEL OIL (68476-33-5) Forms explosive mixture with air (flash point 142°F/61°C). Oxidizers may cause fire and explosions. Incompatible with nitric acid. May accumulate static electrical charges, and may cause ignition of its vapors.

HEAVY OIL (65996-93-2) Incompatible with oxidizers, strong acids, caustics, aliphatic amines, isocyanates.

HECLOTOX (58-89-9) Not combustible, but may be dissolved in a combustible solvent. If solvent comes in contact with oxidizers, fire and explosions may result.

HEDONAL or HEDONAL (THE HERBICIDE) (94-75-7) Decomposes in sunlight. Incompatible with strong oxidizers; may cause fire and explosions.

HEKSAN (Polish) (110-54-3) Forms explosive mixture with air (flash point −7°F/−22°C). Strong oxidizers may cause fire and explosions. Attacks some plastics, rubber, and coatings. May accumulate static electrical charges, and may cause ignition of its vapors.

HEKSOGEN (Polish) (121-82-4) Contact with mercury fulminate may cause detonation. Contact with combustibles or strong oxidizers may cause fire and explosions.

HELICON (50-78-2) Fires and explosions may result from contact with strong oxidizers. Alkali hydroxides or carbonates cause decomposition.

HENDECANOIC ACID (112-37-8) Incompatible with sulfuric acid, caustics, ammonia, amines, isocyanates, alkylene oxides, epichlorohydrin, aldehydes. Attacks most common metals.

HENDECANOIC ALCOHOL (103-08-2) Incompatible with strong acids, caustics, aliphatic amines, isocyanates, acetaldehyde, ethylene oxide, nitrogen tetroxide, tri-isobutyl aluminum.

1-HENDECANOL (103-08-2) Incompatible with strong acids, caustics, aliphatic amines, isocyanates, acetaldehyde, ethylene oxide, nitrogen tetroxide, tri-isobutyl aluminum.

HEOD (60-57-1) Incompatible with concentrated mineral acids, acid catalysts, acid oxidizing agents, phenols, reactive metals.

HEOD ENDO,EXO-1,2,3,4,10- (60-57-1) Incompatible with concentrated mineral acids, acid catalysts, acid oxidizing agents, phenols, reactive metals.

HEPATIC GAS (7783-06-4) A highly flammable and reactive gas. Incompatible with acetaldehyde, barium pentafluoride, chlorine monoxide, chlorine trifluoride, chromic anhydride, copper, lead dioxide, nitric acid, nitrogen iodide, nitrogen trichloride, nitrogen trifluoride, oxygen difluoride, oxidizers, phenyl diazonium chloride, sodium, sodium peroxide. Reacts with alkali metals and alkali earth metals. Attacks metals.

HEPT (107-49-3) Decomposes above 300°F/150°C, forming flammable ethylene gas. Strong oxidizers may cause fire and explosions. Attacks some plastics, rubber, and coatings.

HEPTACHLOOR (Dutch) (76-44-8) Produces hydrogen chloride gas with iron and rust above 165°F/74°C.

1,4,5,6,7,8,8-HEPTACHLOOR-3A,4,7,7A-TETRAHYDRO-4,7-ENDO-METHANO-INDEEN (Dutch) (76-44-8) Produces hydrogen chloride gas with iron and rust above 165°F/74°C.

HEPTACHLOR (76-44-8) Produces hydrogen chloride gas with iron and rust above 165°F/74°C.

HEPTACHLORE (French) (76-44-8) Produces hydrogen chloride gas with iron and rust above 165°F/74°C.

3,4,5,6,7,8,8-HEPTACHLORODICYCLOPENTADIENE (76-44-8) Forms hydrogen chloride gas with iron and rust above 165°F/74°C.

1,4,5,6,7,8,8-HEPTACHLORO-3A,4,7,7A-TETRAHYDRO-4,7-EN-DOMETHANOINDENE (76-44-8) Produces hydrogen chloride gas with iron and rust above 165°F/74°C.

1,4,5,6,7,10,10-HEPTACHLORO-4,7,8,9-TETRAHYDRO-4,7-EN-DOMETHYLENEINDENE (76-44-8) Produces hydrogen chloride gas with iron and rust above 165°F/74°C.

1(3A),4,5,6,7,8,8-HEPTACHLORO-3A,4,7,7A-TETRAHYDRO-4,7-METHANOINDENE (76-44-8) Produces hydrogen chloride gas with iron and rust above 165°F/74°C.

1,4,5,6,7,8,8-HEPTACHLORO-3A,4,7,7A-TETRAHYDRO-4,7-METHANOINDENE (76-44-8) Produces hydrogen chloride gas with iron and rust above 165°F/74°C.

1,4,5,6,7,8,8-HEPTACHLORO-3A,4,7,7A-TETRAHYDRO-4,7-METHANOL-1H-INDENE (76-44-8) Produces hydrogen chloride gas with iron and rust above 165°F/74°C.

1,4,5,6,7,8,8A-HEPTACHLORO-3A,4,7,7A-TETRAHYDRO-4,7-METHANOINDANE (76-44-8) Produces hydrogen chloride gas with iron and rust above 165°F/74°C.

1,4,5,6,7,10,10-HEPTACHLORO-4,7,8,9-TETRAHYDRO-4,7-METHYLENEINDENE (76-44-8) Produces hydrogen chloride gas with iron and rust above 165°F/74°C.

1,4,5,6,7,8,8-HEPTACHLOR-3A,4,7,7,A-TETRAHYDRO-4,7-ENDO-METHANO-INDEN (German) (76-44-8) Produces hydrogen chloride gas with iron and rust above 165°F/74°C.

1-HEPTADECANECARBOXYLIC ACID (57-11-4) Incompatible with strong oxidizers, sulfuric acid, caustics, ammonia, amines, isocyanates, alkylene oxides, epichlorohydrin. Attacks chemically active metals.

cis-8-HEPTADECYLENECARBOXYLIC ACID (112-80-1) Incompatible with strong oxidizers, perchloric acid, and aluminum powder; may cause explosions. Corrodes aluminum.

HEPTAGRAN (76-44-8) Produces hydrogen chloride gas with iron and rust above 165°F/74°C.

HEPTAMUL (76-44-8) Produces hydrogen chloride gas with iron and rust above 165°F/74°C.

HEPTAN (Polish) (142-82-5) Forms explosive mixture with air (flash point 25°F/−4°C). Strong oxidizers may cause fire and explosions. Attacks some plastics, rubber, and coatings. May accumulate static electric charges that can ignite its vapors.

HEPTANAPHTHENE (108-87-2) Forms explosive mixture with air (flash point 25°F/−3.9°C). Strong oxidizers may cause fire and explosions. Attacks some plastics, rubber, and coatings.

HEPTANE (142-82-5) Forms explosive mixture with air (flash point 25°F/−4°C). Strong oxidizers may cause fire and explosions. Attacks some plastics, rubber, and coatings. May accumulate static electric charges that can ignite its vapors.

n-HEPTANE or *normal*-HEPTANE (142-82-5) Forms explosive mixture with air (flash point 25°F/−4°C). Strong oxidizers may cause fire and explosions. Attacks some plastics, rubber, and coatings. May accumulate static electric charges that can ignite its vapors.

1-HEPTANECARBOXYLIC ACID (124-07-2) Forms explosive mixture with air (flash point 56°F/13°C) Reacts violently with strong oxidizers. May accumulate static electrical charges, and may cause ignition of its vapors.

3-HEPTANECARBOXYLIC ACID (149-57-5) May accumulate static electrical charges, and may cause ignition of its vapors. Reacts with oxidizers, with a risk of fire or explosions. Attacks common metals, especially in a moist environment.

HEPTANEN (Dutch) (142-82-5) Forms explosive mixture with air (flash point 25°F/−4°C). Strong oxidizers may cause fire and explosions. Attacks some plastics, rubber, and coatings. May accumulate static electric charges that can ignite its vapors.

HEPTANOIC ACID (111-14-8) Forms explosive mixture with air (flash point greater than 221°F/105°C). Incompatible with sulfuric acid, caustics, ammonia, aliphatic amines, alkanolamines, isocyanates, alkylene oxides, epichlorohydrin.

HEPTANOL (111-70-6) or 1-HEPTANOL or HEPTANOL-1 Forms explosive mixture with air (flash point 170°F/77°C). Incompatible with strong acids, caustics, aliphatic amines, isocyanates, oxidizers.

2-HEPTANOL or HEPTANOL-2 (543-49-7) Forms explosive mixture with air (flash point 160°F/71°C). Reacts violently with strong oxidizers. May accumulate static electrical charges, and may cause ignition of its vapors.

3-HEPTANOL or HEPTANOL-3 (589-82-2) Forms explosive mixture with air (flash point 140°F/60°C). Reacts violently with strong oxidizers. May accumulate static electrical charges, and may cause ignition of its vapors.

n-HEPTANOL (111-70-6) Forms explosive mixture with air (flash point 170°F/77°C). Incompatible with strong acids, caustics, aliphatic amines, isocyanates, oxidizers.

2-HEPTANONE (110-43-0) Forms explosive mixture with air (flash point 102°F/39°C). Incompatible with strong acids, alkalies, aliphatic amines, oxidizers. Attacks some plastics and rubber.

3-HEPTANONE (106-35-4) Forms explosive mixture with air (flash point 115°F/46°C). Reacts violently with strong oxidizers. Attacks some plastics, rubber, and coatings.

HEPTAN-3-ONE (106-35-4) Forms explosive mixture with air (flash point 115°F/46°C) Reacts violently with strong oxidizers. Attacks some plastics, rubber, and coatings.

HEPTAN-4-ONE (123-19-3) Oxidizers may cause violent reaction.

4-HEPTANONE (123-19-3) Oxidizers may cause violent reaction.

4-HEPTANONE, 2,6-DIMETHYL- (108-83-8) Incompatible with strong acids, aliphatic amines, strong oxidizers. Attacks plastics, coatings and rubber. Forms explosive mixture with air (flash point 140°F/60°C).

HEPTANYL ACETATE (112-06-1) Incompatible with strong acids, nitrates.

HEPTENE (592-76-7) Forms explosive mixture with air (flash point less than 25°F/−4°C). Reacts violently with strong oxidizers.

417

1-HEPTENE (592-76-7) Forms explosive mixture with air (flash point less than 25°F/−4°C). Reacts violently with strong oxidizers.

n-HEPTENE (592-76-7) Forms explosive mixture with air (flash point less than 25°F/−4°C). Reacts violently with strong oxidizers.

HEPTHLIC ACID (111-14-8) Forms explosive mixture with air (flash point greater than 221°F/105°C). Incompatible with sulfuric acid, caustics, ammonia, aliphatic amines, alkanolamines, isocyanates, alkylene oxides, epichlorohydrin.

HEPTOIC ACID or n-HEPTOIC ACID (111-14-8) Forms explosive mixture with air (flash point greater than 221°F/105°C). Incompatible with sulfuric acid, caustics, ammonia, aliphatic amines, alkanolamines, isocyanates, alkylene oxides, epichlorohydrin.

HEPTYL ACETATE (112-06-1) Incompatible with strong acids, nitrates.

1-HEPTYL ACETATE (112-06-1) Incompatible with strong acids, nitrates.

N-HEPTYL ACETATE (112-06-1) Incompatible with strong acids, nitrates.

HEPTYL ALCOHOL (111-70-6) Forms explosive mixture with air (flash point 170°F/77°C). Incompatible with strong acids, caustics, aliphatic amines, isocyanates, oxidizers.

sec-**HEPTYL ALCOHOL** (543-49-7) Forms explosive mixture with air (flash point 25°F/−4°C). Reacts violently with strong oxidizers. May accumulate static electrical charges, and may cause ignition of its vapors.

HEPTYL CARBINOL (111-87-5) Forms explosive mixture with air (flash point 178°F/81°C). Incompatible with strong acids, caustics, aliphatic amines, isocyanates, strong oxidizers.

n-HEPTYLIC ACID (111-14-8) Forms explosive mixture with air (flash point greater than 221°F/105°C). Incompatible with sulfuric acid, caustics, ammonia, aliphatic amines, alkanolamines, isocyanates, alkylene oxides, epichlorohydrin.

N-HEPTYLETHYLENE (27214-95-8) Forms explosive mixture with air (flash point 78°F/26°C). Incompatible with sulfuric acid, nitric acid, oxidizers. Attacks some plastics, rubber, and coatings. May accumulate static electrical charges, and may cause ignition of its vapors.

N-HEPTYLETHYLENE (27214-95-8) Forms explosive mixture with air (flash point 78°F/26°C). Incompatible with sulfuric acid, nitric acid, oxidizers. Attacks some plastics, rubber, and coatings. May accumulate static electrical charges, and may cause ignition of its vapors.

HEPTYL HYDRIDE (142-82-5) Forms explosive mixture with air (flash point 25°F/−4°C). Strong oxidizers may cause fire and explosions. Attacks some plastics, rubber, and coatings. May accumulate static electric charges that can ignite its vapors.

HEPTYLENE (592-76-7) Forms explosive mixture with air (flash point less than 25°F/−4°C). Reacts violently with strong oxidizers.

HEPTYL PHTHALATE (3648-21-3) Incompatible with strong acids, strong alkalies, nitrates, strong oxidizers. Attacks some plastics, rubber, and coatings.

1-HEPTYLENE (592-76-7) Forms explosive mixture with air (flash point less than 25°F/−4°C). Reacts violently with strong oxidizers.

HERBATOX (330-54-1) Hydrolyzes in fairly strong acids.

HERBICIDE TOTAL (61-82-5) Substance acts as a weak base to form salts in contact with acids. Corrosive to iron, aluminum, copper, and copper alloys.

HERBIDAL (94-75-7) Decomposes in sunlight. Incompatible with strong oxidizers; may cause fire and explosions.

HERBIZOLE (61-82-5) Substance acts as a weak base to form salts in contact with acids. Corrosive to iron, aluminum, copper, and copper alloys.

HERCULES 3956 (8001-35-2) Reacts with oxidizers, with a risk of fire or explosions. Attacks metals in the presence of moisture.

HERCULES TOXAPHENE (8001-35-2) Reacts with oxidizers, with a risk of fire or explosions. Attacks metals in the presence of moisture.

HERKAL (62-73-7) Attacks some plastics, rubber, and coatings.

HERMAL (137-26-8) Combustible solid (flash point 192°F/89°C). Strong oxidizers may cause fire and explosions; contact with strong acid or oxidizable materials produces toxic gases.

HERMAT TMT (137-26-8) Combustible solid (flash point 192°F/89°C). Strong oxidizers may cause fire and explosions; contact with strong acid or oxidizable materials produce toxic gases.

HERYL (137-26-8) Combustible solid (flash point 192°F/89°C). Strong oxidizers may cause fire and explosions; contact with strong acid or oxidizable materials produces toxic gases.

n-HETANOL-1 (111-70-6) Forms explosive mixture with air (flash point 170°F/77°C). Incompatible with strong acids, caustics, aliphatic amines, isocyanates, oxidizers.

HEXA (58-89-9) Not combustible, but may be dissolved in a combustible solvent. If solvent comes in contact with oxidizers, fire and explosions may result.

HEXA (100-97-0) Incompatible with acids, organic anhydrides, isocyanates, vinyl acetate, acrylates, substituted allyls, alkylene oxides, epichlorohydrin, ketones, aldehydes, alcohols, glycols, phenols, cresols, caprolactam solution, sodium peroxide, strong oxidizers. Attacks aluminum, copper, lead, tin, zinc, and alloys.

HEXAANDIONE-2,5 (110-13-4) Forms explosive mixture with air (flash point 174°F/79°C). Strong oxidizers may cause fire and explosions.

1,2,3,4,5,6-HEXACHLOR-CYCLOHEXANE (58-89-9) Not combustible, but may be dissolved in a combustible solvent. If solvent comes in contact with oxidizers, fire and explosions may result.

HEXA C. B. (118-74-1) Reacts violently with strong oxidizers, formyldimethylamine.

HEXACHLORAN (58-89-9) Not combustible, but may be dissolved in a combustible solvent. If solvent comes in contact with oxidizers, fire and explosions may result.

gamma-HEXACHLORAN (58-89-9) Not combustible, but may be dissolved in a combustible solvent. If solvent comes in contact with oxidizers, fire and explosions may result.

HEXACHLORANE (58-89-9) Not combustible, but may be dissolved in a combustible solvent. If solvent comes in contact with oxidizers, fire and explosions may result.

gamma-HEXACHLORANE (58-89-9) Not combustible, but may be dissolved in a combustible solvent. If solvent comes in contact with oxidizers, fire and explosions may result.

HEXACHLOROBENZENE (118-74-1) Reacts violently with strong oxidizers, formyldimethylamine.

gamma-HEXACHLOROBENZENE (58-89-9) Not combustible, but may be dissolved in a combustible solvent. If solvent comes in contact with oxidizers, fire and explosions may result.

HEXACHLOROBENZOL (German) (118-74-1) Reacts violently with strong oxidizers, formyldimethylamine.

1,2,3,4,7,7-HEXACHLOROBICYCLO(2,2,1)HEPTEN-5,6-BIOXY-METHYLENESULFITE (115-29-7) Hydrolyzed by acids and alkalis. Corrosive to iron.

HEXACHLOROBUTADIENE (87-68-3) Forms explosive mixture with air (flash point 195°F/90°C). Reacts strongly with oxidizers, aluminum powder. Attacks aluminum. Attacks some plastics, rubber, and coatings.

HEXACHLORO-1,3-BUTADIENE (87-68-3) Forms explosive mixture with air (flash point 195°F/90°C). Reacts strongly with oxidizers, aluminum powder. Attacks aluminum. Attacks some plastics, rubber, and coatings.

1-alpha,2-alpha,3-beta,4-alpha,5-alpha,6-beta-HEXACHLOROCYCLOHEXANE (58-89-9) Not combustible, but may be dissolved in a combustible solvent. If solvent comes in contact with oxidizers, fire and explosions may result.

gamma-HEXACHLOROCYCLOHEXANE (58-89-9) Not combustible, but may be dissolved in a combustible solvent. If solvent comes in contact with oxidizers, fire and explosions may result.

gamma-1,2,3,4,5,6-HEXACHLOROCYCLOHEXANE (58-89-9) Not combustible, but may be dissolved in a combustible solvent. If solvent comes in contact with oxidizers, fire and explosions may result.

HEXACHLOROCYCLOHEXANE, GAMMA ISOMER (58-89-9) Not combustible, but may be dissolved in a combustible solvent. If solvent comes in contact with oxidizers, fire and explosions may result.

HEXACHLOROCYCLOHEXANE, GAMMA-ISOMER (58-89-9) Not combustible, but may be dissolved in a combustible solvent. If solvent comes in contact with oxidizers, fire and explosions may result.

1,2,3,4,5,6-HEXACHLOROCYCLOHEXANE, GAMMA-ISOMER (58-89-9) Not combustible, but may be dissolved in a combustible solvent. If solvent comes in contact with oxidizers, fire and explosions may result.

HEXACHLOROCYCLOPENTADIENE (77-47-4) Incompatible with water, producing hydrochloric acid. Contact with sodium may be explosive. Corrodes iron and other metals in the presence of moisture.

HEXACHLORO-1,3-CYCLOPENTADIENE (77-47-4) Incompatible with water, producing hydrochloric acid. Contact with sodium may be explosive. Corrodes iron and other metals in the presence of moisture.

1,2,3,4,5,5-HEXACHLORO-1,3-CYCLOPENTADIENE (77-47-4) Incompatible with water, producing hydrochloric acid. Contact with sodium may be explosive. Corrodes iron and other metals in the presence of moisture.

HEXACHLORODIPHENYL OXIDE (55720-99-5) Contact with strong oxidizers may cause fire and explosions.

HEXACHLOROEPOXYOCTAHYDRO-ENDO,ENDO-DIMETHA-NONAPTHALENE (72-20-8) Incompatible with parathion, strong acids (forms explosive vapors), strong oxidizers.

HEXACHLOROEPOXYOCTAHYDRO-ENDO,EXO-DIMETH-ANONAPHTHALENE (60-57-1) Incompatible with concentrated mineral acids, acid catalysts, acid oxidizing agents, phenols, reactive metals.

HEXACHLORO-6,7-EPOXY-1,4,4A,5,6,7,8,8A-OCTAHYDRO-1,4:5,8-DIMETHANONAPHTHALENE (60-57-1) Incompatible with concentrated mineral acids, acid catalysts, acid oxidizing agents, phenols, reactive metals.

1,2,3,4,10,10-HEXACHLORO-6,7-EPOXY-1,4,4A,5,6,7,8,8A-OCTA-HYDRO-1,4-ENDO-ENDO-1,4,5,8-DIMETHANONAPHTHA-LENE (72-20-8) Incompatible with parathion, strong acids (forms explosive vapors), strong oxidizers.

1,2,3,4,10,10-HEXACHLORO-6,7-EPOXY-1,4,4A,5,6,7,8,8A-OCTA-HYDRO-1,4-ENDO-EXO-5,8-DI-METHANONAPHTHALENE (60-57-1) Incompatible with concentrated mineral acids, acid catalysts, acid oxidizing agents, phenols, reactive metals.

HEXACHLOROETHANE (67-72-1) Incompatible with aluminum or zinc. Contact with alkalies produces spontaneously explosive chloroacetylene. Attacks some plastics, rubber, and coatings.

1,1,1,2,2,2-HEXACHLOROETHANE (67-72-1) Incompatible with aluminum or zinc. Contact with alkalies form spontaneously explosive chloroacetylene. Attacks some plastics, rubber, and coatings.

1,2,3,4,10,10-HEXACHLORO-1,4,4A,5,8,8A-HEXAHYDRO-1,4,5,8-DIMETHANONAPHTHALENE (309-00-2) Incompatible with concentrated mineral acids, acid catalysts, acid oxidizing agents, phenols, reactive metals.

HEXACHLOROHEXAHYDRO-ENDO-EXO-DIMETHANO-NAPHTHALENE (309-00-2) Incompatible with concentrated mineral acids, acid catalysts, acid oxidizing agents, phenols, reactive metals.

1,2,3,4,10,10-HEXACHLORO-1,4,4A,5,8,8A-HEXAHYDRO-1,4,5,8-ENDO,EXO-DIMETHANONAPHTHALENE (309-00-2) Incompatible with concentrated mineral acids, acid catalysts, acid oxidizing agents, phenols, reactive metals.

1,2,3,4,10,10-HEXACHLORO-1,4,4A,5,8,8A-HEXAHYDRO-1,4-ENDO-EXO-5,8-DIMETHANONAPHTHALENE (309-00-2) Incompatible with concentrated mineral acids, acid catalysts, acid oxidizing agents, phenols, reactive metals.

6,7,8,9,10-HEXACHLORO-1,5,5a,6,9,9a,-HEXAHYDRO-6,9-METH-ANO-2,4,3-BENZODIOXATHIEPIN-3-OXIDE (115-29-7) Hydrolyzed by acids and alkalis. Corrosive to iron.

HEXACHLORONAPHTHALENE (1335-87-1) Oxidizers may cause fire and explosions.

3,4,5,6,9,9-HEXACHLORO-1A,2,2A,3,6,6A,7,7A-OCTAHYDRO-2,7:3,6-DIMETHANO (60-57-1) Incompatible with concentrated mineral acids, acid catalysts, acid oxidizing agents, phenols, reactive metals.

HEXACID 698 (142-62-1) Reacts with strong oxidizers. Incompatible with sulfuric acid, caustics, ammonia, amines, isocyanates, alkylene oxides, epichlorohydrin. Attacks common metals.

HEXACID 898 (124-07-2) Forms explosive mixture with air (flash point 56°F/13°C). Reacts violently with strong oxidizers. May accumulate static electrical charges, and may cause ignition of its vapors.

HEXACID 1095 (334-48-5) Incompatible with sulfuric acid, caustics, ammonia, aliphatic amines, alkanolamines, isocyanates, alkylene oxides, epichlorohydrin. Attacks most common metals.

HEXADRIN (72-20-8) Incompatible with parathion, strong acids (forms explosive vapors), strong oxidizers.

HEXAFLUOSILICIC ACID (16961-83-4) Incompatible with aliphatic amines, alkanolamines, alkylene oxides, aromatic amines, amides, ammonia, ammonium hydroxide, bases, calcium oxide, epichlorohydrin, isocyanates, oleum, organic anhydrides, sulfuric acid, strong oxidizers, vinyl acetate, water.

HEXAHYDRATE (10377-60-3) A powerful oxidizer. Reacts violently with dimethylformamide, reducing agents, combustibles, organic and easily oxidizable matter.

HEXAHYDRIC ALCOHOL (50-70-4) Incompatible with strong acids, bases, aliphatic amines, isocyanates, strong oxidizers.

HEXAHYDROANILINE (108-91-8) Forms explosive mixture with air (flash point 79°F/26°C). Incompatible with acids, organic anhydrides, isocyanates, vinyl acetate, acrylates, substituted allyls, alkylene oxides, epichlorohydrin, ketones, aldehydes, alcohols, glycols, phenols, cresols, caprolactam solution, strong oxidizers. Contact with copper alloys, zinc, or galvanized steel may cause violent reaction.

HEXAHYDROAZEPINE (111-49-9) Forms explosive mixture with air (flash point 99°F/37°C). Incompatible with acids, organic anhydrides, isocyanates, vinyl acetate, acrylates, substituted allyls, alkylene oxides, epichlorohydrin, ketones, aldehydes, alcohols, glycols, phenols, cresols, caprolactam solution, strong oxidizers. Attacks aluminum, copper, lead, tin, zinc, and alloys.

HEXAHYDRO-2H-AZEPINE-2-ONE (105-60-2) Contact with strong oxidizers may cause fire and explosions.

HEXAHYDROBENZENAMINE (108-91-8) Forms explosive mixture with air (flash point 79°F/26°C). Incompatible with acids, organic anhydrides, isocyanates, vinyl acetate, acrylates, substituted allyls, alkylene oxides, epichlorohydrin, ketones, aldehydes, alcohols, glycols, phenols, cresols, caprolactam solution, strong oxidizers. Contact with copper alloys, zinc, or galvanized steel may cause violent reaction.

HEXAHYDROBENZENE (110-82-7) Forms explosive mixture with air (flash point −4°F/−20°C). Incompatible with oxidizers, nitrogen dioxide.

HEXAHYDROBENZENE, HEXAMETHYLENE (110-82-7) Forms explosive mixture with air (flash point −4°F/−20°C). Incompatible with oxidizers, nitrogen dioxide.

HEXAHYDROCRESOL (25639-42-3) Forms explosive mixture with air (flash point 149°F/65°C). Strong oxidizers may cause fire and explosions. Attacks some plastics, rubber, and coatings.

HEXAHYDROCUMENE (696-29-7) Forms explosive mixture with air (flash point 96°F/36°C). Incompatible with strong oxidizers, acids.

HEXAHYDRO-1,4-DIAZINE (110-85-0) Forms explosive mixture with air (flash point 178°F/81°C). Incompatible with nitrogen compounds, carbon tetrachloride, strong oxidizers. Aqueous solutions react with acids. Attacks aluminum, copper, nickel, magnesium, and zinc.

HEXAHYDROMETHYLPHENOL (25639-42-3) Forms explosive mixture with air (flash point 149°F/65°C). Strong oxidizers may cause fire and explosions. Attacks some plastics, rubber, and coatings.

HEXAHYDROPHENOL (108-93-0) Forms explosive mixture with air (flash point 154°F/68°C). Incompatible with strong acids, caustics, aliphatic amines, isocyanates. Attacks some plastics, rubber, or coatings.

HEXAHYDROPHENOL (108-94-1) Forms explosive mixture with air (flash point 111°F/44°C). May accumulate static electrical charges, and may cause ignition of its vapors. Attacks red metals and lead. Incompatible with amines, sulfuric acid, nitric acid, oxidizers, strong acids, aliphatic amines. Dissolves plastics, resins, and rubber.

HEXAHYDROPYRAZINE (110-85-0) Forms explosive mixture with air (flash point 178°F/81°C). Incompatible with nitrogen compounds, carbon tetrachloride, strong oxidizers. Aqueous solutions react with acids. Attacks aluminum, copper, nickel, magnesium, and zinc.

HEXAHYDROTOLUENE (108-87-2) Forms explosive mixture with air (flash point 25°F/−3.9°C). Strong oxidizers may cause fire and explosions. Attacks some plastics, rubber, and coatings.

HEXAHYDRO-1,3,5-TRINITRO-1,3,5-TRIAZIN (German) (121-82-4) Contact with mercury fulminate may cause detonation. Contact with combustibles or strong oxidizers may cause fire and explosions.

HEXAHYDRO-1,3,5-TRINITRO-1,3,5-TRIAZINE (121-82-4) Contact with mercury fulminate amy cause detonation. Contact with combustibles or strong oxidizers may cause fire and explosions.

HEXAHYDRO-1,3,5-TRINITRO-S-TRIAZINE (121-82-4) Contact with mercury fulminate amy cause detonation. Contact with combustibles or strong oxidizers may cause fire and explosions.

HEXALDEHYDE (66-25-1) Forms explosive mixture with air (flash point 90°F/32°C). Incompatible with strong acids, caustics, ammonia, amines. Attacks some plastics, rubber, and coatings.

n-HEXALDEHYDE (66-25-1) Forms explosive mixture with air (flash point 90°F/32°C). Incompatible with strong acids, caustics, ammonia, amines. Attacks some plastics, rubber, and coatings.

HEXALIN (108-94-1) Forms explosive mixture with air (flash point 111°F/44°C). May accumulate static electrical charges, and may cause ignition of its vapors. Attacks red metals and lead. Incompatible with amines, sulfuric acid, nitric acid, oxidizers, strong acids, aliphatic amines. Dissolves plastics, resins, and rubber.

HEXALIN (108-93-0) Forms explosive mixture with air (flash point 154°F/68°C). Incompatible with strong acids, caustics, aliphatic amines, isocyanates. Attacks some plastics, rubber, or coatings.

HEXAMETHYLENE (110-82-7) Forms explosive mixture with air (flash point −4°F/−20°C). Incompatible with oxidizers, nitrogen dioxide.

HEXAMETHYLENEDIAMINE or HEXAMETHYLENEDIAMINE, SOLID (124-09-4) Forms explosive mixture with air (flash point 176°F/80°C). A strong reducing agent. Reacts with ethylene dichloride, acids, organic anhydrides, isocyanates, vinyl acetate, acrylates, substituted allyls, alkylene oxides, epichlorohydrin, ketones, aldehydes, alcohols, glycols, phenols, cresols, caprolactam solution, strong oxidizers. Attacks aluminum, copper, lead, tin, zinc, and alloys.

HEXAMETHYLENEDIAMINE, SOLUTION (124-09-4) Forms explosive mixture with air (flash point 176°F/80°C). A strong reducing agent. Reacts with ethylene dichloride, acids, organic anhydrides, isocyanates, vinyl acetate, acrylates, substituted allyls, alkylene oxides, epichlorohydrin, ketones, aldehydes, alcohols, glycols, phenols, cresols, caprolactam solution, strong oxidizers. Attacks aluminum, copper, lead, tin, zinc, and alloys.

HEXAMETHYLENEIMINE (111-49-9) Forms explosive mixture with air (flash point 99°F/37°C). Incompatible with acids, organic anhydrides, isocyanates, vinyl acetate, acrylates, substituted allyls, alkylene oxides, epichlorohydrin, ketones, aldehydes, alcohols, glycols, phenols, cresols, caprolactam solution, strong oxidizers. Attacks aluminum, copper, lead, tin, zinc, and alloys.

HEXAMETHYLENETETRAMINE (100-97-0) Incompatible with acids, organic anhydrides, isocyanates, vinyl acetate, acrylates, substituted allyls, alkylene oxides, epichlorohydrin, ketones, aldehydes, alcohols, glycols, phenols, cresols, caprolactam solution, sodium peroxide, strong oxidizers. Attacks aluminum, copper, lead, tin, zinc, and alloys.

HEXAMINE (100-97-0) Incompatible with acids, organic anhydrides, isocyanates, vinyl acetate, acrylates, substituted allyls, alkylene oxides, epichlorohydrin, ketones, aldehydes, alcohols, glycols, phenols, cresols, caprolactam solution, sodium peroxide, strong oxidizers. Attacks aluminum, copper, lead, tin, zinc, and alloys.

HEXAMITE (107-49-3) Decomposes above 300°F/150°C, producing flammable ethylene gas. Strong oxidizers may cause fire and explosions. Attacks some plastics, rubber, and coatings.

HEXANAL (66-25-1) Forms explosive mixture with air (flash point 90°F/32°C). Incompatible with strong acids, caustics, ammonia, amines. Attacks some plastics, rubber, and coatings.

1-HEXANAL (66-25-1) Forms explosive mixture with air (flash point 90°F/32°C). Incompatible with strong acids, caustics, ammonia, amines. Attacks some plastics, rubber, and coatings.

HEXANAL, 2-ETHYL (123-05-7) Forms explosive mixture with air (flash point 112°F/44°C). Reacts violently with oxidizers. May ignite spontaneously when spilled on clothing or other absorbent materials. Under certain conditions ignites spontaneously with air. Incompatible with strong acids, caustics, ammonia, amines. Forms unstable peroxides on contact with air.

HEXANAPHTHENE (110-82-7) Forms explosive mixture with air (flash point −4°F/−20°C). Incompatible with oxidizers, nitrogen dioxide.

HEXANAPHTHYLENE (110-83-8) Forms explosive mixture with air (flash point less than 20°F/−7°C). May polymerize from buildup of unstable peroxides. May accumulate static electrical charges, and may cause ignition of its vapors.

HEXANE (110-54-3) Forms explosive mixture with air (flash point −7°F/−22°C). Strong oxidizers may cause fire and explosions. Attacks some plastics, rubber, and coatings. May accumulate static electrical charges, and may cause ignition of its vapors.

n-HEXANE or *normal*-HEXANE (110-54-3) Forms explosive mixture with air (flash point $-7°F/-22°C$). Strong oxidizers may cause fire and explosions. Attacks some plastics, rubber, and coatings. May accumulate static electrical charges, and may cause ignition of its vapors.

HEXANE CARBOXYLIC ACID (111-14-8) Forms explosive mixture with air (flash point greater than 221°F/105°C). Incompatible with sulfuric acid, caustics, ammonia, aliphatic amines, alkanolamines, isocyanates, alkylene oxides, epichlorohydrin.

1-HEXANECARBOXYLIC ACID (111-14-8) Forms explosive mixture with air (flash point greater than 221°F/105°C). Incompatible with sulfuric acid, caustics, ammonia, aliphatic amines, alkanolamines, isocyanates, alkylene oxides, epichlorohydrin.

1,6-HEXANEDIAMINE (124-09-4) Forms explosive mixture with air (flash point 176°F/80°C). A strong reducing agent. Reacts with ethylene dichloride, acids, organic anhydrides, isocyanates, vinyl acetate, acrylates, substituted allyls, alkylene oxides, epichlorohydrin, ketones, aldehydes, alcohols, glycols, phenols, cresols, caprolactam solution, strong oxidizers. Attacks aluminum, copper, lead, tin, zinc, and alloys.

1,6-HEXANEDIAMINE, SOLUTION (124-09-4) Forms explosive mixture with air (flash point 176°F/80°C). A strong reducing agent. Reacts with ethylene dichloride, acids, organic anhydrides, isocyanates, vinyl acetate, acrylates, substituted allyls, alkylene oxides, epichlorohydrin, ketones, aldehydes, alcohols, glycols, phenols, cresols, caprolactam solution, strong oxidizers. Attacks aluminum, copper, lead, tin, zinc, and alloys.

1,6-HEXANEDIAMINE, 2,2,4 TRIMETHYL- (25513-64-8) Incompatible with nitric, sulfuric, mineral, and organic acids and organic anhydrides, isocyanates, vinyl acetate, acrylates, substituted allyls, alkylene oxides, ketones, aldehydes, alcohols, glycols, phenols, cresols, caprolactam solution.

HEXANEDINITRILE (111-69-3) May accumulate static electrical charges, and may cause ignition of its vapors. Decomposes at its flash point, forming poisonous cyanide gas. Forms explosive mixture with air (flash point 199°F/93°C). Incompatible with sulfuric acid, oxidizers.

HEXANEDIOIC ACID (124-04-9) May accumulate static electrical charges, and may cause ignition of its vapors. Contact with strong oxidizers may cause fire and explosions.

1,6-HEXANEDIOIC ACID (124-04-9) May accumulate static electrical charges, and may cause ignition of its vapors. Contact with strong oxidizers may cause fire and explosions.

HEXANEDIOIC ACID, DIBUTYL ESTER (103-23-1) Incompatible with strong acids, strong oxidizers, nitrates.

HEXANEDIOIC ACID, DIMETHYL ESTER (627-93-0) Incompatible with strong acids, nitrates, oxidizers.

HEXANEDIOIC ACID, DINITRILE (111-69-3) May accumulate static electrical charges, and may cause ignition of its vapors. Decomposes at its flash point, forming poisonous cyanide gas. Forms explosive mixture with air (flash point 199°F/93°C). Incompatible with sulfuric acid, oxidizers

1,2-HEXANEDIOL (107-41-5) Forms explosive mixture with air (flash point 209°F/98°C). Incompatible with strong acids, caustics, aliphatic amines, isocyanates, strong oxidizers.

2,5-HEXANEDIOL (2935-44-6) Forms explosive mixture with air (flash point 206°F/102°C). Incompatible with strong oxidizers, strong acids, caustics, aliphatic amines, isocyanates.

HEXANEN (Dutch) (110-54-3) Forms explosive mixture with air (flash point −7°F/−22°C). Strong oxidizers may cause fire and explosions. Attacks some plastics, rubber, and coatings. May accumulate static electrical charges, and may cause ignition of its vapors.

1,2,3,4,5,6-HEXANNEHEXOL (50-70-4) Incompatible with strong acids, bases, aliphatic amines, isocyanates, strong oxidizers.

HEXANOIC ACID (142-62-1) Reacts with strong oxidizers. Incompatible with sulfuric acid, caustics, ammonia, amines, isocyanates, alkylene oxides, epichlorohydrin. Attacks common metals.

N-HEXANOIC ACID (142-62-1) Reacts with strong oxidizers. Incompatible with sulfuric acid, caustics, ammonia, amines, isocyanates, alkylene oxides, epichlorohydrin. Attacks common metals.

HEXANOIC ACID, 2-ETHYL- (149-57-5) May accumulate static electrical charges, and may cause ignition of its vapors. Reacts with oxidizers, with a risk of fire or explosions. Attacks common metals, especially in a moist environment.

HEXANOL (111-27-3) Forms explosive mixture with air (flash point 145°F/63°C). Incompatible with strong acids, caustics, aliphatic amines, isocyanates, strong oxidizers.

1-HEXANOL (111-27-3) Forms explosive mixture with air (flash point 145°F/63°C). Incompatible with strong acids, caustics, aliphatic amines, isocyanates, strong oxidizers.

n-**HEXANOL** or *normal*-**HEXANOL** (111-27-3) Forms explosive mixture with air (flash point 145°F/63°C). Incompatible with strong acids, caustics, aliphatic amines, isocyanates, strong oxidizers.

sec-**HEXANOL** or *secondary*-**HEXANOL** (97-95-0) Forms explosive mixture with air (flash point 70°F/21°C). Incompatible with strong acids, caustics, isocyanates, amines, isocyanates.

HEXANON (108-94-1) Forms explosive mixture with air (flash point 111°F/44°C). May accumulate static electrical charges, and may cause ignition of its vapors. Attacks red metals and lead. Incompatible with amines, sulfuric acid, nitric acid, oxidizers, strong acids, aliphatic amines. Dissolves plastics, resins, and rubber.

HEXANONE-2 (591-78-6) Forms explosive mixture with air (flash point 77°F/25°C). Strong oxidizers may cause fire and explosions. Dissolves some plastics, resins, and rubber.

2-HEXANONE (591-78-6) Forms explosive mixture with air (flash point 77°F/25°C). Strong oxidizers may cause fire and explosions. Dissolves some plastics, resins, and rubber.

HEXAPLAS DIOP (27554-26-3) May accumulate static electrical charges, and may cause ignition of its vapors. Incompatible with strong acids, nitrates, oxidizers.

HEXAPLAS M/B (84-74-2) Incompatible with strong acids, nitrates, strong oxidizers, strong alkalies.

HEXAPLAS M/1B (84-69-5) Incompatible with strong acids, strong oxidizers, nitrates.

HEXAPLAS M/O (27554-26-3) May accumulate static electrical charges, and may cause ignition of its vapors. Incompatible with strong acids, nitrates, oxidizers.

HEXASUL (7704-34-9) Combustible solid. Liquid contact produces sulfur dioxide with air. Reacts violently with strong oxidizers. Forms explosive, shock-sensitive, or pyrophoric mixtures with ammonia, ammonium nitrate, bromates, calcium carbide, charcoal, chlorates, hydrocarbons, iodates, iron. Reacts violently with halogen compounds and sodium, tin, uranium, and other compounds. Attacks steel when moist. May accumulate static electrical charges, and may cause ignition of its vapors.

HEXATHIR (137-26-8) Combustible solid (flash point 192°F/89°C). Strong oxidizers may cause fire and explosions; contact with strong acid or oxidizable materials produces toxic gases.

HEXATOX (58-89-9) Not combustible, but may be dissolved in a combustible solvent. If solvent comes in contact with oxidizers, fire and explosions may result.

HEXAVERM (58-89-9) Not combustible, but may be dissolved in a combustible solvent. If solvent comes in contact with oxidizers, fire and explosions may result.

HEXAVIN (63-25-2) Incompatible with strong oxidizers, strongly alkaline pesticides.

HEXENE (592-41-6) Forms explosive mixture with air (flash point −15°F(-26°C). Reacts violently with oxidizers.

1-HEXENE (592-41-6) Forms explosive mixture with air (flash point −15°F(-26°C). Reacts violently with oxidizers.

alpha-HEXENE (592-41-6) Forms explosive mixture with air (flash point −15°F(-26°C). Reacts violently with oxidizers.

HEXICIDE (58-89-9) Not combustible, but may be dissolved in a combustible solvent. If solvent comes in contact with oxidizers, fire and explosions may result.

HEXOGEEN (Dutch) (121-82-4) Contact with mercury fulminate may cause detonation. Contact with combustibles or strong oxidizers may cause fire and explosions.

HEXOGEN (121-82-4) Contact with mercury fulminate may cause detonation. Contact with combustibles or strong oxidizers may cause fire and explosions.

HEXOGEN 5W (121-82-4) Contact with mercury fulminate may cause detonation. Contact with combustibles or strong oxidizers may cause fire and explosions.

n-HEXOIC ACID (142-62-1) Reacts with strong oxidizers. Incompatible with sulfuric acid, caustics, ammonia, amines, isocyanates, alkylene oxides, epichlorohydrin. Attacks common metals.

HEXOLITE (121-82-4) Contact with mercury fulminate may cause detonation. Contact with combustibles or strong oxidizers may cause fire and explosions.

HEXON (Czech) (108-10-1) Forms explosive mixture with air (flash point 64°F/18°C). Incompatible with oxidizers, strong acids, aliphatic amines. Dissolves some plastics, resins, and rubber.

HEXONE (108-10-1) Forms explosive mixture with air (flash point 64°F/18°C). Incompatible with oxidizers, strong acids, aliphatic amines. Dissolves some plastics, resins, and rubber.

HEXYCLAN (58-89-9) Not combustible, but may be dissolved in a combustible solvent. If solvent comes in contact with oxidizers, fire and explosions may result.

HEXYL ACETATE (142-92-7) Incompatible with strong acids, nitrates, oxidizers.

HEXYL ACETATE (108-84-9) Forms explosive mixture with air (flash point 113°F/45°C). Incompatible with strong acids, strong alkalies, nitrates, strong oxidizers.

1-HEXYL ACETATE (142-92-7) Incompatible with strong acids, nitrates, oxidizers.

n-**HEXYL ACETATE** or *normal*-**HEXYL ACETATE** (142-92-7) Incompatible with strong acids, nitrates, oxidizers.

sec-**HEXYL ACETATE** or *secondary*-**HEXYL ACETATE** (108-84-9) Forms explosive mixture with air (flash point 113°F/45°C). Incompatible with strong acids, strong alkalies, nitrates, strong oxidizers.

HEXYL ALCOHOL (111-27-3) Forms explosive mixture with air (flash point 145°F/63°C). Incompatible with strong acids, caustics, aliphatic amines, isocyanates, strong oxidizers.

n-**HEXYL ALCOHOL** or *normal*-**HEXYL ALCOHOL** (111-27-3) Forms explosive mixture with air (flash point 145°F/63°C). Incompatible with strong acids, caustics, aliphatic amines, isocyanates, strong oxidizers.

sec-**HEXYL ALCOHOL** or *secondary*-**HEXYL ALCOHOL** (97-95-0) Forms explosive mixture with air (flash point 70°F/21°C). Incompatible with strong acids, caustics, isocyanates, amines, isocyanates.

HEXYL ALCOHOL, ACETATE (142-92-7) Incompatible with strong acids, nitrates, oxidizers.

HEXYL CARBITOL (112-59-4) Incompatible with sulfuric acid, isocyanates, perchloric acid.

HEXYLENE (592-41-6) Forms explosive mixture with air (flash point −15°F/-26°C). Reacts violently with oxidizers.

HEXYLENE GLYCOL (107-41-5) Forms explosive mixture with air (flash point 209°F/98°C). Incompatible with strong acids, caustics, aliphatic amines, isocyanates, strong oxidizers.

2,5-HEXYLENE GLYCOL (2935-44-6) Forms explosive mixture with air (flash point 206°F/102°C). Incompatible with strong oxidizers, strong acids, caustics, aliphatic amines, isocyanates.

HEXYL ETHANOATE (142-92-7) Incompatible with strong acids, nitrates, oxidizers

HEXYL HYDRIDE (110-54-3) Forms explosive mixture with air (flash point −7°F/−22°C). Strong oxidizers may cause fire and explosions. Attacks some plastics, rubber, and coatings. May accumulate static electrical charges, and may cause ignition of its vapors.

HF (7664-39-3) A strong acid. Reacts—possibly violently–with acetic anhydride, aliphatic amines, alcohols, alkanolamines, alkylene oxides, aromatic amines, amides, 2-aminoethanol, ammonia, ammonium hydroxide, arsenic trioxide, bismuthic acid, calcium oxide, ethylene diamine, ethyleneimine, epichlorohydrin, isocyanates, metal acetylides, nitrogen trifluoride, oleum, organic anhydrides, oxygen difluoride, phosphorus pentoxide, sulfuric acid, sodium hydroxide and other bases, strong oxidizers, vinyl acetate, vinylidene fluoride. Attacks glass, concrete, ceramics, metals, some plastics, rubber, and coatings.

HF-A (7664-39-3) A strong acid. Reacts—possibly violently—with acetic anhydride, aliphatic amines, alcohols, alkanolamines, alkylene oxides, aromatic amines, amides, 2-aminoethanol, ammonia, ammonium hydroxide, arsenic trioxide, bismuthic acid, calcium oxide, ethylene diamine, ethyleneimine, epichlorohydrin, isocyanates, metal acetylides, nitrogen trifluoride, oleum, organic anhydrides, oxygen difluoride, phosphorus pentoxide, sulfuric acid, sodium hydroxide and other bases, strong oxidizers, vinyl acetate, vinylidene fluoride. Attacks glass, concrete, ceramics, metals, some plastics, rubber, and coatings.

HGI (58-89-9) Not combustible, but may be dissolved in a combustible solvent. If solvent comes in contact with oxidizers, fire and explosions may result.

HHDN (309-00-2) Incompatible with concentrated mineral acids, acid catalysts, acid oxidizing agents, phenols, reactive metals.

HI-DRY (112-60-7) Incompatible with sulfuric acid, isocyanates, perchloric acid. Attacks some plastics, rubber, and coatings.

HIGHER FATTY ALCOHOL (112-92-5) Incompatible with strong acids, caustics, aliphatic amines, isocyanates.

HIGH SOLVENT NAPHTHA (8030-31-7) Oxidizers may cause fire and explosions. Incompatible with nitric acid. May accumulate static electrical charges, and may cause ignition of its vapors.

HIGH SPEED BEARING OIL (n/a) Oxidizers may cause fire and explosions. Incompatible with nitric acid. May accumulate static electrical charges, and may cause ignition of its vapors.

HIGH STRENGTH HYDROGEN PEROXIDE (7722-84-1) A powerful oxidizer; attacks many substances. Contact with most organic, readily oxidizable materials, reducing agents, and combustibles causes fire and explosions. Contact with iron, copper, brass, bronze, chromium, zinc, lead, manganese, silver, and other catalytic metals (and their salts), especially in a basic (pH7 or above) environment, causes rapid decomposition with evolution of oxygen gas. Attacks, and may ignite, some plastics, rubber, and coatings. Decomposes slowly at ordinary temperatures and builds up pressure in a closed container. The rate of decomposition doubles for each 50°F/10°C rise (1.5 times 10°C rise) in temperature and becomes self-sustaining at 285°F/141°C.

HILDAN (115-29-7) Hydrolyzed by acids and alkalis. Corrosive to iron.

HILDIT (50-29-3) Incompatible with salts of iron or aluminum, and bases. Do not store in iron containers.

HILTHION (121-75-5) Incompatible with strong oxidizers, magnesium, alkaline pesticides. Attacks metals, some plastics, rubber, and coatings.

HILTHION 25WDP (121-75-5) Incompatible with strong oxidizers, magnesium, alkaline pesticides. Attacks metals, some plastics, rubber, and coatings.

HI-POINT 90 (1338-23-4) Forms explosive mixture with air (flash point 125°F/52°C). Explosive decomposition occurs above 176°F/80°C. Pure substance is shock-sensitive. Strong oxidizer. Reacts violently with strong acids, strong bases, reducing agents, combustible substances, organic materials, oxides of heavy metals, salts, trace contaminants, amines. May accumulate static electrical charges, and may cause ignition of its vapors.

HITTORF'S PHOSPHORUS (7723-14-0) Contact with all oxidizers may cause fire and explosions or produce shock-sensitive compounds. Reacts violently with caustics + heat, chlorosulfonic acid + heat, lead dioxide, performic acid, selenium oxychloride, chlorates, and other materials. Incompatible with many other substances, and produces phosphine gas on contact with moisture and oxygen; opened packages should be stored under inert gas blanket.

HMDA (124-09-4) Forms explosive mixture with air (flash point 176°F/80°C). A strong reducing agent. Reacts with ethylene dichloride, acids, organic anhydrides, isocyanates, vinyl acetate, acrylates, substituted allyls, alkylene oxides, epichlorohydrin, ketones, aldehydes, alcohols, glycols, phenols, cresols, caprolactam solution, strong oxidizers. Attacks aluminum, copper, lead, tin, zinc, and alloys.

HOE-2671 (115-29-7) Hydrolyzed by acids and alkalis. Corrosive to iron.

HOME-HEATING OIL (68476-30-2) Oxidizers may cause fire and explosions. Incompatible with nitric acid. May accumulate static electrical charges, and may cause ignition of its vapors.

HOMOPIPERIDINE (111-49-9) Forms explosive mixture with air (flash point 99°F/37°C). Incompatible with acids, organic anhydrides, isocyanates, vinyl acetate, acrylates, substituted allyls, alkylene oxides, epichlorohydrin, ketones, aldehydes, alcohols, glycols, phenols, cresols, caprolactam solution, strong oxidizers. Attacks aluminum, copper, lead, tin, zinc, and alloys.

HORTEX (58-89-9) Not combustible, but may be dissolved in a combustible solvent. If solvent comes in contact with oxidizers, fire and explosions may result.

HOUSEHOLD AMMONIA (1336-21-6) Incompatible with strong acids, mineral acids, organic acids, amides, organic anhydrides, isocyanates, vinyl acetate, epichlorohydrin, aldehydes. Corrosive to copper and aluminum, including their alloys, and galvanized surfaces.

HPA (999-61-1) Incompatible with strong acids, nitrates.

beta-HPN (109-78-4) Forms explosive mixture with air (flash point 140°F/60°C). Avoid basic contamination to prevent polymerization. A powerful reducing agent. Hot water contact produces cyanide gas. Reacts violently with strong oxidizers or sodium hydroxide. Reacts with acids, acid salts, chlorates, nitrates. Attacks mild steel, copper, and copper alloys.

HS (10039-54-0) A strong oxidizer. Heat or friction may cause explosions. Violent, possibly explosive, reaction occurs with caustics, reducing agents, or combustibles. Incompatible with strong acids, organic anhydrides, isocyanates, aldehydes. Magnesium or aluminum contact may be explosive.

HSDB 5700 (583-91-5) Incompatible with sulfuric acid, caustics, ammonia, amines, isocyanates, alkylene oxides, epichlorohydrin.

HTH or HTH DRY CHLORINE (7778-54-3) Decomposes in heat or sunlight. Incompatible with acids, moisture, reducing agents, combustible materials, all other chemicals, especially acetylene, aniline and all other amines, anthracene, carbon tetrachloride, iron oxide, manganese oxide, mercaptans, diethylene glycol monomethyl ether, nitromethane, organic matter, organic sulfides, phenol, 1-propanethiol, propyl mercaptan, sulfur, organic sulfur compounds.

430

HUILE de CAMPHRE (French) (76-22-2) Forms explosive mixture with air (flash point 155°F/66°C). Violent, possibly explosive, reaction occurs with strong oxidizers. May accumulate static electrical charges, and may cause ignition of its vapors.

HUMIFEN WT-27G (119-36-8) Forms explosive mixture with air (flash point 205°F/96°C). Incompatible with strong acids, nitrates, oxidizers.

HUNGAZIN (1912-24-9) Incompatible with strong acids.

HUNGAZIN PK (1912-24-9) Incompatible with strong acids.

HW 920 (330-54-1) Hydrolyzes in fairly strong acids.

HYADUR (67-68-5) Forms explosive mixture with air (flash point 203°F/95°C). Reacts violently with oxidizers. Reacts with ethanoyl chloride, boron compounds, halides, metal alkoxides, oxidizers.

HYCAR (9016-00-6) Incompatible with sulfuric acid, isocyanates.

HY-CHLOR (7778-54-3) Decomposes in heat or sunlight. Incompatible with acids, moisture, reducing agents, combustible materials, all other chemicals, especially acetylene, aniline and all other amines, anthracene, carbon tetrachloride, iron oxide, manganese oxide, mercaptans, diethylene glycol monomethyl ether, nitromethane, organic matter, organic sulfides, phenol, 1-propanethiol, propyl mercaptan, sulfur, organic sulfur compounds.

HYCLORITE (7681-52-9) A powerful oxidizer and a strong base. Decomposes in sunlight, producing oxygen and increasing the risk of fire. Stability decreases with concentration, heat, light, decrease in pH, and contamination with metals. Incompatible with strong acids, reducing agents, combustible substances; all cause violent reaction, fire, and explosions. Contact with amines and ammonia salts produces explosive chloroamines. Corrodes many metals: steel, 12% and 17% chrome steel, cast iron, monel, aluminum, nickel, brass, bronze, iconel.

HYDRACRYLIC ACID, beta-LACTONE (57-57-8) Polymerizes in storage. High temperatures cause containers to rupture. Decomposes at room temperature; hydrolyzes in water.

HYDRACRYLONITRILE (109-78-4) Forms explosive mixture with air (flash point 140°F/60°C). Avoid basic contamination to prevent polymerization. A powerful reducing agent. Hot water contact produces cyanide gas. Reacts violently with strong oxidizers or sodium hydroxide. Reacts with acids, acid salts, chlorates, nitrates. Attacks mild steel, copper, and copper alloys.

HYDRALIN (108-93-0) Forms explosive mixture with air (flash point 154°F/68°C). Incompatible with strong acids, caustics, aliphatic amines, isocyanates. Attacks some plastics, rubber, or coatings.

HYDRARGYRUM BIJODATUM (German) (7774-29-0) Reacts violently with strong oxidizers, chlorine, fluorine.

HYDRATED KEMIKAL (1305-62-0) Contact with maleic anhydride, phosphorus, nitroethane, nitromethane, nitroparaffins, nitropropane, phosphorus may cause explosion.

HYDRAULIC OIL (8012-95-1) Oxidizers may cause fire and explosions. Incompatible with nitric acid. May accumulate static electrical charges, and may cause ignition of its vapors.

HYDRAZINE or HYDRAZINE, ANHYDROUS (302-01-2) Forms explosive mixture with air (flash point 100°F/38°C). A highly reactive reducing agent and a strong base. Incompatible with oxides of iron or copper, and with manganese, lead, copper, or their alloys; can cause fire and explosions. Reacts violently with acids and halogens. Attacks cork, glass, some plastics, rubber, and coatings.

HYDRAZINE BASE (302-01-2) Forms explosive mixture with air (flash point 100°F/38°C). A highly reactive reducing agent and a strong base. Incompatible with oxides of iron or copper, and with manganese, lead, copper or their alloys; can cause fire and explosions. Reacts violently with acids and halogens. Attacks cork, glass, some plastics, rubber, and coatings.

HYDRAZINE-BENZENE (100-63-0) A highly reactive reducing agent. Forms explosive mixture with air (flash point 190°F/88°C). Incompatible with oxides of iron or copper, and with manganese, lead, copper, or their alloys, and methyl iodide, oxidizers, organic compounds. Attacks some plastics, rubber, and coatings and cork.

HYDRAZINE, 1,1-DIMETHYL- (57-14-7) Forms explosive mixture with air (flash point 5°F/−15°C). A strong reducing agent. Incompatible with strong acids, halogens, metallic mercury, strong oxidizers (with possible spontaneous ignition). Attacks some plastics, rubber, and coatings. May accumulate static electrical charges, and may cause ignition of its vapors.

HYDRAZINE, 1,2-DIMETHYL- (540-73-8) Forms explosive mixture with air (flash point 73°F/23°C). A strong reducing agent. Reacts violently with strong oxidizers, strong acids. Attacks some plastics, rubber, and coatings. May accumulate static electrical charges, and may cause ignition of its vapors.

HYDRAZINE, METHYL- (60-34-4) A highly reactive reducing agent. Forms explosive mixture with air (flash point 17°F/−8°C). Contact with oxides of iron or copper, and manganese, lead, copper, or their alloys can lead to fire and explosions. Attacks cork, some plastics, coatings, and rubber.

HYDRAZINOBENZENE (100-63-0) A highly reactive reducing agent. Forms explosive mixture with air (flash point 190°F/88°C). Incompatible with oxides of iron or copper, and with manganese, lead, copper, or their alloys, and methyl iodide, oxidizers, organic compounds. Attacks some plastics, rubber, and coatings and cork.

HYDRAZOIC ACID, SODIUM SALT (26628-22-8) Incompatible with acids, with some metals (i.e, lead, copper, silver, mercury). Forms explosion-sensitive compounds.

HYDRAZOMETHANE (60-34-4) Highly reactive reducing agent. Forms explosive mixture with air (flash point 17°F/−8°C). Contact with oxides of iron or copper, and manganese, lead, copper, or their alloys can lead to fire and explosions. Attacks cork, some plastics, coatings, and rubber.

HYDRAZYNA (Polish) (302-01-2) Forms explosive mixture with air (flash point 100°F/38°C). A highly reactive reducing agent and a strong base. Incompatible with oxides of iron or copper, and with manganese, lead, copper or their alloys; can cause fire and explosions. Reacts violently with acids and halogens. Attacks cork, glass, some plastics, rubber, and coatings.

HYDROBROMIC ACID or HYDROBROMIC ACID, ANHYDROUS (10035-10-6) Incompatible with aliphatic amines, alkanolamines, alkylene oxides, aromatic amines, amides, ammonia, ammonium hydroxide, bases, calcium oxide, epichlorohydrin, fluorine, isocyanates, oleum, organic anhydrides, sulfuric acid, sodium tetrahydroborate, strong oxidizers, vinyl acetate, water. Attacks most metals with the formation of flammable hydrogen gas.

HYDROCHINON (Czech and Polish) (123-31-9) Incompatible with strong oxidizers, caustics. May be oxidized to quinone at room temperatures in the presence of moisture. May explode on contact with oxygen.

HYDROCHLORIC ACID or HYDROCHLORIC ACID, ANHYDROUS (7647-01-0) Incompatible with acetic anhydride, aliphatic amines, alkanolamines, alkylene oxides, aromatic amines, amides, 2-aminoethanol, ammonia, ammonium hydroxide, calcium phosphide, chlorosulfonic acid, ethylene diamine, ethyleneimine, epichlorohydrin, isocyanates, metal acetylides, oleum, organic anhydrides, perchloric acid, 3-propiolactone, uranium phosphide, sulfuric acid, sodium hydroxide and other bases, strong oxidizers, vinyl acetate, vinylidene fluoride. Attacks most metals (can produce flammable hydrogen gas), and some plastics, rubber, and coatings.

HYDROCHLORIC ETHER (75-00-3) Flammable gas. Reacts slowly with water; produces hydrogen chloride gas. Contact with moisture produces hydrochloric acid. May accumulate static electrical charges, and may cause ignition of its vapors. Forms explosive mixture with air (flash point −58°F/−55°C). Contact with aluminum, lithium, magnesium, sodium, potassium, zinc may cause fire and explosions. Attacks some plastics and rubber.

HYDROCHLORIDE (7647-01-0) Reacts with acetic anhydride, aliphatic amines, alkanolamines, alkylene oxides, aromatic amines, amides, 2-aminoethanol, ammonia, ammonium hydroxide, calcium phosphide, chlorosulfonic acid, ethylene diamine, ethyleneimine, epichlorohydrin, isocyanates, metal acetylides, oleum, organic anhydrides, perchloric acid, 3-propiolactone, uranium phosphide, sulfuric acid, sodium hydroxide and other bases, strong oxidizers, vinyl acetate, vinylidene fluoride. Attacks most metals (can produce flammable hydrogen gas), and some plastics, rubber, and coatings.

HYDROCYANIC ACID (74-90-8) Unless stabilized and maintained, samples stored more than 90 days are hazardous. Samples containing more than 2–5% water are less stable than dry material. Can be self-reactive, forming an explosive mixture with air (flash point 0°F/−18°C). Heat or contact with amines or strong bases can cause polymerization. Incompatible with acetaldehyde. Oxidizers may cause fire and explosions. Attacks some plastics, rubber, and coatings.

HYDROCYANIC ACID, POTASSIUM SALT (151-50-8) Incompatible with acids, organic anhydrides, isocyanates, alkylene oxides, epichlorohydrin, aldehydes, alcohols, glycols, phenols, cresols, caprolactam, strong oxidizers, sodium chlorate. Attacks aluminum, copper, zinc, in the presence of moisture.

HYDROCYANIC ACID, SODIUM SALT (143-33-9) Incompatible with acids, organic anhydrides, isocyanates, alkylene oxides, epichlorohydrin, aldehydes, alcohols, glycols, phenols, cresols, caprolactam solution, strong oxidizers, nitrates, nitrites. Attacks aluminum, copper, zinc.

HYDROCYANIC ETHER (107-12-0) Forms explosive mixture with air (flash point 36°F/2°C). Reacts violently with oxidizers. Water or acid contact produces hydrogen cyanide fumes.

433

HYDROFLUORIC ACID (7664-39-3) A strong acid. Reacts—possibly violently—with acetic anhydride, aliphatic amines, alcohols, alkanolamines, alkylene oxides, aromatic amines, amides, 2-aminoethanol, ammonia, ammonium hydroxide, arsenic trioxide, bismuthic acid, calcium oxide, ethylene diamine, ethyleneimine, epichlorohydrin, isocyanates, metal acetylides, nitrogen trifluoride, oleum, organic anhydrides, oxygen difluoride, phosphorus pentoxide, sulfuric acid, sodium hydroxide and other bases, strong oxidizers, vinyl acetate, vinylidene fluoride. Attacks glass, concrete, ceramics, metals, some plastics, rubber, and coatings.

HYDROFLUORIDE (7664-39-3) A strong acid. Reacts—possibly violently—with acetic anhydride, aliphatic amines, alcohols, alkanolamines, alkylene oxides, aromatic amines, amides, 2-aminoethanol, ammonia, ammonium hydroxide, arsenic trioxide, bismuthic acid, calcium oxide, ethylene diamine, ethyleneimine, epichlorohydrin, isocyanates, metal acetylides, nitrogen trifluoride, oleum, organic anhydrides, oxygen difluoride, phosphorus pentoxide, sulfuric acid, sodium hydroxide and other bases, strong oxidizers, vinyl acetate, vinylidene fluoride. Attacks glass, concrete, ceramics, metals, some plastics, rubber, and coatings.

HYDROFLUOSILIC ACID (16961-83-4) Incompatible with aliphatic amines, alkanolamines, alkylene oxides, aromatic amines, amides, ammonia, ammonium hydroxide, bases, calcium oxide, epichlorohydrin, isocyanates, oleum, organic anhydrides, sulfuric acid, strong oxidizers, vinyl acetate, water.

HYDROFOL ACID 1255 (143-07-7) Incompatible with sulfuric acid, bases, ammonia, amines, alkylene oxide, epichlorohydrin.

HYDROFURAN (109-99-9) Forms explosive mixture with air (flash point 6°F/−14°C). Unless inhibited, can form unstable and explosive peroxides. Incompatible with strong acids, strong oxidizers. Attacks some plastics. May accumulate static electric charges that can result in ignition of its vapors.

HYDROGEN (1333-74-0) Vapors form explosive or combustible mixture with air over a wide range of concentrations. Ignites easily with oxygen. Reacts violently with oxidizers, halogens, acetylene, bromine, chlorine, fluorine, nitrous oxide, and other gases. Mild steel and most iron alloys become brittle at liquid hydrogen temperatures.

HYDROGEN ANTIMONIDE (7803-52-3) Forms explosive mixture with air. Reacts violently with nitric acid, ozone, halogenated hydrocarbons, oxidizers, ozone, and moisture.

HYDROGEN ARSENIDE (7784-42-1) Strong oxidizers, nitric acid may cause fire and explosions. Light exposure causes decomposition in the presence of moisture. May accumulate static electrical charges, and may cause ignition of its vapors.

HYDROGEN BROMIDE or HYDROGEN BROMIDE, ANHYDROUS (10035-10-6) Incompatible with aliphatic amines, alkanolamines, alkylene oxides, aromatic amines, amides, ammonia, ammonium hydroxide, bases, calcium oxide, epichlorohydrin, fluorine, isocyanates, oleum, organic anhydrides, sulfuric acid, sodium tetrahydroborate, strong oxidizers, vinyl acetate, water. Attacks most metals with the formation of flammable hydrogen gas.

HYDROGEN CARBOXYLIC ACID (64-18-6) Forms explosive mixture with air (flash point 156°F/(69°C). A strong reducing agent; reacts violently with oxidizers. Incompatible with sulfuric acid, bases, ammonia, aliphatic amines, alkanolamines, furfuryl alcohol, hydrogen peroxide, isocyanates, alkylene oxides, epichlorohydrin. Attacks aluminum, cast iron, and steel, and some plastics, rubber, and coatings.

HYDROGEN CHLORIDE (7647-01-0) Reacts with acetic anhydride, aliphatic amines, alkanolamines, alkylene oxides, aromatic amines, amides, 2-aminoethanol, ammonia, ammonium hydroxide, calcium phosphide, chlorosulfonic acid, ethylene diamine, ethyleneimine, epichlorohydrin, isocyanates, metal acetylides, oleum, organic anhydrides, perchloric acid, 3-propiolactone, uranium phosphide, sulfuric acid, sodium hydroxide and other bases, strong oxidizers, vinyl acetate, vinylidene fluoride. Attacks most metals (can produce flammable hydrogen gas), and some plastics, rubber, and coatings.

HYDROGEN, COMPRESSED (1333-74-0) Vapors form explosive or combustible mixture with air over a wide range of concentrations. Ignites easily with oxygen. Reacts violently with oxidizers, halogens, acetylene, bromine, chlorine, fluorine, nitrous oxide, and other gases. Mild steel and most iron alloys become brittle at liquid hydrogen temperatures.

HYDROGEN CYANIDE (74-90-8) Unless stabilized and maintained, samples stored more than 90 days are hazardous. Samples containing more than 2–5% water are less stable than dry material. Can be self-reactive, forming an explosive mixture with air (flash point 0°F/−18°C). Heat or contact with amines or strong bases can cause polymerization. Incompatible with acetaldehyde. Oxidizers may cause fire and explosions. Attacks some plastics, rubber, and coatings.

HYDROGEN DIOXIDE (7722-84-1) A powerful oxidizer; attacks many substances. Contact with most organic, readily oxidizable materials, reducing agents, and combustibles causes fire and explosions. Contact with iron, copper, brass, bronze, chromium, zinc, lead, manganese, silver, and other catalytic metals (and their salts), especially in a basic (pH7 or above) environment, causes rapid decomposition with evolution of oxygen gas. Attacks, and may ignite, some plastics, rubber, and coatings. Decomposes slowly at ordinary temperatures and builds up pressure in a closed container. The rate of decomposition doubles for each 50°F/10°C rise (1.5 times 10°C rise) in temperature and becomes self-sustaining at 285°F/141°C.

HYDROGENE SULFURE (French) (7783-06-4) A highly flammable and reactive gas. Incompatible with acetaldehyde, barium pentafluoride, chlorine monoxide, chlorine trifluoride, chromic anhydride, copper, lead dioxide, nitric acid, nitrogen iodide, nitrogen trichloride, nitrogen trifluoride, oxygen difluoride, oxidizers, phenyl diazonium chloride, sodium, sodium peroxide. Reacts with alkali metals and alkaline earth metals. Attacks metals.

HYDROGEN FLUORIDE (7664-39-3) A strong acid. Reacts—possibly with violence—with acetic anhydride, aliphatic amines, alcohols, alkanolamines, alkylene oxides, aromatic amines, amides, 2-aminoethanol, ammonia, ammonium hydroxide, arsenic trioxide, bismuthic acid, calcium oxide, ethylene diamine, ethyleneimine, epichlorohydrin, isocyanates, metal acetylides, nitrogen trifluoride, oleum, organic anhydrides, oxygen difluoride, phosphorus pentoxide, sulfuric acid, sodium hydroxide and other bases, strong oxidizers, vinyl acetate, vinylidene fluoride. Attacks glass, concrete, ceramics, metals, some plastics, rubber, and coatings.

HYDROGEN HEXAFLUOROSILICATE (16961-83-4) Incompatible with aliphatic amines, alkanolamines, alkylene oxides, aromatic amines, amides, ammonia, ammonium hydroxide, bases, calcium oxide, epichlorohydrin, isocyanates, oleum, organic anhydrides, sulfuric acid, strong oxidizers, vinyl acetate, water.

HYDROGEN NITRATE (7697-37-2) A strong oxidizer that can react violently with reducing agents, combustible materials. Incompatible with many substances, including acrylates, aliphatic amines, alcohols, aldehydes, alkanolamines, alkylene oxides, anion exchange resins, aromatic amines, amides, bases, cresols, cyanides, cyclic ketones, epichlorohydrin, glycols, isocyanates, ketones, oleum, organic anhydrides, phenols, substituted allyls, sulfuric acid, strong oxidizers, terpenes. Attacks most metals, and some plastics, rubber, and coatings.

HYDROGEN PEROXIDE (7722-84-1) A powerful oxidizer; attacks many substances. Contact with most organic, readily oxidizable materials, reducing agents, and combustibles causes fire and explosions. Contact with iron, copper, brass, bronze, chromium, zinc, lead, manganese, silver, and other catalytic metals (and their salts), especially in a basic (pH7 or above) environment, causes rapid decomposition with evolution of oxygen gas. Attacks, and may ignite, some plastics, rubber, and coatings. Decomposes slowly at ordinary temperatures and builds up pressure in a closed container. The rate of decomposition doubles for each 50°F/10°C rise (1.5 times 10°C rise) in temperature and becomes self-sustaining at 285°F/141°C.

HYDROGEN PEROXIDE CARBAMIDE (124-43-6) Combustible solid. At 122°F/50°C reacts with dust, organics.

HYDROGEN PEROXIDE WITH UREA (124-43-6) Combustible solid. At 122°F/50°C reacts with dust, organics.

HYDROGEN PHOSPHIDE (7803-51-2) Unusually reactive. Contact with air or other oxidizer may cause self-ignition and explosion. Reacts violently or forms explosive products with many substances, including acids, halogenated hydrocarbons, moisture. Heat may cause cylinders to explode.

HYDROGEN, REFRIGERATED LIQUID (1333-74-0) Vapors form explosive or combustible mixture with air over a wide range of concentrations. Ignites easily with oxygen. Reacts violently with oxidizers, halogens, acetylene, bromine, chlorine, fluorine, nitrous oxide and other gases. Mild steel and most iron alloys become brittle at liquid hydrogen temperatures.

HYDROGEN SELINIDE (7783-07-5) Oxidizers, acids, water, halogenated hydrocarbons may cause fire and explosions.

HYDROGEN SULFIDE (7783-06-4) A highly flammable and reactive gas. Incompatible with acetaldehyde, barium pentafluoride, chlorine monoxide, chlorine trifluoride, chromic anhydride, copper, lead dioxide, nitric acid, nitrogen iodide, nitrogen trichloride, nitrogen trifluoride, oxygen difluoride, oxidizers, phenyl diazonium chloride, sodium, sodium peroxide. Reacts with alkali metals and alkaline earth metals. Attacks metals.

HYDROGEN SULFITE SODIUM (7631-90-5) Slowly oxidized to the sulfate on contact with air. Oxidizers or acids cause formation of sulfur dioxide gas.

HYDROGEN TETRAFLUOROBORATE (16872-11-0) A strong acid; reacts violently with strong alkali materials. Incompatible with strong oxidizers. Attacks metals.

HYDROOXYCYCLOHEXANE (108-94-1) Forms explosive mixture with air (flash point 111°F/44°C). May accumulate static electrical charges, and may cause ignition of its vapors. Attacks red metals and lead. Incompatible with amines, sulfuric acid, nitric acid, oxidizers, strong acids, aliphatic amines. Dissolves plastics, resins, and rubber.

HYDROPEROXIDE (7722-84-1) A powerful oxidizer; attacks many substances. Contact with most organic, readily oxidizable materials, reducing agents, and combustibles causes fire and explosions. Contact with iron, copper, brass, bronze, chromium, zinc, lead, manganese, silver, and other catalytic metals (and their salts), especially in a basic (pH7 or above) environment, causes rapid decomposition with evolution of oxygen gas. Attacks, and may ignite, some plastics, rubber, and coatings. Decomposes slowly at ordinary temperatures and builds up pressure in a closed container. The rate of decomposition doubles for each 50°F/10°C rise (1.5 times 10°C rise) in temperature and becomes self-sustaining at 285°F/141°C.

HYDROPEROXIDE, ACETYL (79-21-0) A powerful oxidizer and extremely sensitive explosive. Forms explosive mixture with air (flash point 100°F/38°C). Reacts violently with many substances including acetic anhydride, combustibles, strong bases, ether solvents, olefins, organic matter, magnesium, metal oxides, heavy metals, metal chloride solutions, reducing agents, nickel, phosphorus, sodium nitride, zinc.

HYDROPEROXIDE, 1-METHYL-1-PHENYLETHYL- (80-15-9) A strong oxidizer; reacts violently with reducers, combustibles. Explosive decomposition may occur above 120°F/50°C. Forms explosive mixture with air (flash point 175°F/79°C). Incompatible with acids, bases, amines, metallic salts of cobalt, copper, lead. May accumulate static electrical charges, and may cause ignition of its vapors.

1-HYDROPEROXYCYCLOHEXYL 1-HYDROXYCYCLOHEXYL PEROXIDE (78-18-2) A strong oxidizer. Reacts violently with combustibles, reducing agents, caustics, ammonia.

HYDROPEROXY-2-METHYL PROPANE (75-91-2) Forms explosive mixture with air (flash point 100°F/38°C). A powerful oxidizer; reacts violently with reducing agents, organic materials, ethylene dichloride.

HYDROPHENOL (108-93-0) Forms explosive mixture with air (flash point 154°F/68°C). Incompatible with strong acids, caustics, aliphatic amines, isocyanates. Attacks some plastics, rubber, or coatings.

HYDROQUINOL (123-31-9) Incompatible with strong oxidizers, caustics. May be oxidized to quinone at room temperatures in the presence of moisture. May explode on contact with oxygen.

HYDROQUINONE (123-31-9) Incompatible with strong oxidizers, caustics. May be oxidized to quinone at room temperatures in the presence of moisture. May explode on contact with oxygen.

a-HYDROQUINONE or alpha-HYDROQUINONE (123-31-9) Incompatible with strong oxidizers, caustics. May be oxidized to quinone at room temperatures in the presence of moisture. May explode on contact with oxygen.

m-**HYDROQUINONE or** *meta*-**HYDROQUINONE** (108-46-3) Reacts violently with strong oxidizers, nitric acid. Incompatible with acetanilide, albumin, alkalies, antipyrine, camphor, ferric salts, menthol, spirit nitrous ether. Absorbs moisture from air (hygroscopic). May accumulate static electrical charges, and may cause ignition of its vapors.

o-**HYDROQUINONE** or *ortho*-**HYDROQUINONE** (120-80-9) Strong oxidizers may cause fire and explosion.

p-**HYDROQUINONE or *para*-HYDROQUINONE** (123-31-9) Incompatible with strong oxidizers, caustics. May be oxidized to quinone at room temperatures in the presence of moisture. May explode on contact with oxygen.

HYDROSULFURIC ACID (7783-06-4) A highly flammable and reactive gas. Incompatible with acetaldehyde, barium pentafluoride, chlorine monoxide, chlorine trifluoride, chromic anhydride, copper, lead dioxide, nitric acid, nitrogen iodide, nitrogen trichloride, nitrogen trifluoride, oxygen difluoride, oxidizers, phenyl diazonium chloride, sodium, sodium peroxide. Reacts with alkali metals and alkali earth metals. Attacks metals.

N-HYDROXETHYL-1,2-ETHANEDIAMINE (929-06-6) Incompatible with acids, organic anhydrides, isocyanates, aldehydes, oxidizers.

HYDROXIDE de POTASSIUM (French) (1310-58-3) Aqueous solution is a strong base. Reacts violently with acids, organic anhydrides, isocyanates, alkylene oxides, epichlorohydrin, aldehydes, alcohols, glycols, phenols, cresols, caprolactam solution. Dissolves in water, producing high heat, caustic fumes, and flammable hydrogen gas. Also reacts with halogenated hydrocarbons, maleic anhydride. Attacks metals such as aluminum, tin, lead, and zinc, producing flammable hydrogen gas.

HYDROXYACETIC ACID (79-14-1) Incompatible with sulfuric acid, caustics, ammonia, aliphatic amines, isocyanates, alkylene oxides, epichlorohydrin, strong oxidizers.

o-**HYDROXYANISOLE** (90-05-1) Forms explosive mixture with air (flash point 82°F/180°C). Reacts violently with strong oxidizers.

2-HYDROXYBENZALDEHYDE (90-02-8) Forms explosive mixture with air (flash point 172°F/78°C). Reacts violently with strong oxidizers.

o-**HYDROXYBENZALDEHYDE** (90-02-8) Forms explosive mixture with air (flash point 172°F/78°C). Reacts violently with strong oxidizers.

HYDROXYBENZENE (108-95-2) Forms explosive mixture with air (flash point 174°F/79°C). Incompatible with strong oxidizers, strong acids, caustics, aliphatic amines, amides, oxidizers, formaldehyde, butadiene, calcium hypochlorite. Liquid attacks some plastics, rubber, and coatings; hot liquid attacks aluminum, magnesium, lead, and zinc metals.

2-HYDROXYBENZOIC ACID (69-72-7) Reacts with strong oxidizers, ethyl nitrite, iodine, iron salts, lead acetate.

o-**HYDROXYBENZOIC ACID** (69-72-7) Reacts with strong oxidizers, ethyl nitrite, iodine, iron salts, lead acetate.

1-HYDROXYBUTANE (71-36-3) Forms explosive mixture with air (flash point 98°F/37°C). May react with aluminum above 120°F/49°C. Attacks some plastics, rubber, and coatings. Incompatible with strong acids, halogens, caustics, alkali metals, aliphatic amines, isocyanates.

2-HYDROXYBUTANE (78-92-2) Forms explosive mixture with air (flash point 75°F/4°C). Attacks some plastics, rubber, and coatings. Forms an explosive peroxide in air. Ignites with chromium trioxide. Incompatible with strong oxidizers, strong acids, aliphatic amines, isocyanates, organic peroxides.

1-HYDROXY-4-*tert*-BUTYLBENZENE (98-54-4) Incompatible with strong acids, caustics, aliphatic amines, amides, oxidizers.

HYDROXYCELLULOSE (9004-34-6) Incompatible with water, bromine pentafluoride, hydrogen peroxide, sodium hypochlorite, sodium nitrate, fluorine or strong oxidizers.

2-HYDROXYCHLOROBENZENE (95-57-0) Forms explosive mixture with air (flash point 147°F/64°C) Contact with strong oxidizers may cause fire and explosions. Attacks aluminum and copper.

4-HYDROXYCHLOROBENZENE (106-48-9) Forms explosive mixture with air (flash point 147°F/64°C). Reacts with oxidizers, with a risk of fire or explosions. Attacks active metals, aluminum, copper, etc. May accumulate static electrical charges, and may cause ignition of its vapors.

3-HYDROXYCROTONIC ACID METHYL ESTER DIMETHYL PHOSPHATE (7786-34-7) Contact with strong oxidizers may cause fire and explosions. Attacks some plastics, rubber, and coatings.

1-HYDROXY-2-CYANOETHANE (109-78-4) Forms explosive mixture with air (flash point 140°F/60°C). Avoid basic contamination to prevent polymerization. A powerful reducing agent. Hot water contact produces cyanide gas. Reacts violently with strong oxidizers or sodium hydroxide. Reacts with acids, acid salts, chlorates, nitrates. Attacks mild steel, copper, and copper alloys.

3-HYDROXYCYCLOHEXADIEN-1-ONE (108-46-3) Reacts violently with strong oxidizers, nitric acid. Incompatible with acetanilide, albumin, alkalies, antipyrine, camphor, ferric salts, menthol, spirit nitrous ether. Absorbs moisture from air (hygroscopic). May accumulate static electrical charges, and may cause ignition of its vapors.

HYDROXYCYCLOHEXANE (108-93-0) Forms explosive mixture with air (flash point 154°F/68°C). Incompatible with strong acids, caustics, aliphatic amines, isocyanates. Attacks some plastics, rubber, or coatings.

HYDROXYDE de SODIUM (French) (1310-73-2) Contact with water, acids, flammable liquids, and organic halogens, especially trichloroethylene, may cause fires and explosions. Contact with metals such as aluminum, tin, and zinc causes corrosion and the formation of flammable hydrogen gas. Contact with nitromethane and similar nitro compounds produces shock-sensitive salts. Water contact produces heat and corrosive fumes. Attacks some plastics, rubber, and coatings.

HYDROXYDIMETHYLARSINE OXIDE (75-60-5) Aqueous solution reacts with chemically active metals. Incompatible with sulfuric acid, caustics, ammonia, amines, isocyanates, alkylene oxides, epichlorohydrin.

HYDROXYDIMETHYLARSINE OXIDE, SODIUM SALT (124-65-2) Corrodes common metals.

cis-**3-HYDROXY-N,N-DIMETHYLCROTONAMIDE** (141-66-2) Corrosive to cast iron, mild steel, brass, and stainless steel 304.

3-HYDROXYDIMETHYLCROTONAMIDE DIMETHYL PHOSPHATE (141-66-2) Corrosive to cast iron, mild steel, brass, and stainless steel 304.

3-HYDROXY-N,N-DIMETHYL-cis-CROTONAMIDE DIMETHYL PHOSPHATE (141-66-2) Corrosive to cast iron, mild steel, brass, and stainless steel 304.

1-HYDROXY-2,4-DINITRO-BENZENE (51-28-5) Explosion may be caused by heat, friction, or shock. Contact with reducing agents, combustibles may cause fire and explosions. Forms explosive salts with ammonia or strong bases. May accumulate static electrical charges, and may cause ignition of its vapors.

HYDROXYETHANOIC ACID (79-14-1) Incompatible with sulfuric acid, caustics, ammonia, aliphatic amines, isocyanates, alkylene oxides, epichlorohydrin, strong oxidizers.

2-HYDROXYETHANOL (107-21-1) Incompatible with strong acids, caustics, aliphatic amines, isocyanates, chlorosulfonic acid, oleum, strong oxidizers.

HYDROXY ETHER (111-15-9) Forms explosive mixture with air (flash point 117°F/47°C). Incompatible with strong acids, nitrates. Reacts violently with oxidizers. May form unstable peroxides. Softens many plastics. Attacks some plastics, rubber, and coatings.

HYDROXY ETHER (110-80-5) Forms explosive mixture with air (flash point 120°F/49°C). Strong oxidizers may cause fire and explosions. Attacks some plastics, rubber, and coatings. Able to form peroxides. Incompatible with strong acids, aluminum and its alloys.

2-HYDROXYETHYL ACETATE (542-59-6) Forms explosive mixture with air (flash point 191°F/88°C). Incompatible with sulfuric acid, nitric acid, nitrates.

2-HYDROXYETHYL ACRYLATE (818-61-6) Able to polymerize. Incompatible with strong acids, aliphatic amines, alkanolamines. Attacks mild steel and tin plate. Swells rubber; removes paint.

b-HYDROXYETHYL ACRYLATE or beta-HYDROXYETHYL ACRYLATE (818-61-6) Able to polymerize. Incompatible with strong acids, aliphatic amines, alkanolamines. Attacks mild steel and tin plate. Swells rubber; removes paint.

2-HYDROXYETHYLAMINE (141-43-5) Will oxidize in air; often shipped under a pad of inert gas. Forms explosive mixture with air (flash point 185°F/85°C). Strong oxidizers may cause fire and explosions. A strong base. Incompatible with acids, organic anhydrides, isocyanates, alkylene oxides, epichlorohydrin, aldehydes, alcohols, glycols, phenols, cresols, caprolactam solution. Contact with strong acids may cause spattering. Attacks plastics, rubber, and aluminum, copper, tin, and their alloys.

1-(2-HYDROXYETHYLAMINO)-2-AMINOETHANE (929-06-6) Incompatible with acids, organic anhydrides, isocyanates, aldehydes, oxidizers.

2-(2-HYDROXYETHYLAMINO)ETHYLAMINE (929-06-6) Incompatible with acids, organic anhydrides, isocyanates, aldehydes, oxidizers.

(1-HYDROXYETHYL)BENZENE (98-85-1) Forms explosive mixture with air (flash point 205°F/96°C). Incompatible with strong acids, caustics, aliphatic amines, isocyanates, oxidizers.

440

b-HYDROXYETHYLDIMETHYLAMINE or beta-HYDROXY-ETHYLDIMETHYLAMINE (108-01-0) Forms explosive mixture with air (flash point 105°F/41°C). Reacts violently with oxidizers, acids. Attacks copper, copper alloys, galvanized steel, zinc, and zinc, alloys.

n-b-HYDROXYETHYLETHYLENEDIAMINE (929-06-6) Incompatible with acids, organic anhydrides, isocyanates, aldehydes, oxidizers.

HYDROXYETHYLETHYLENEDIAMINE, N-B- (929-06-6) Incompatible with acids, organic anhydrides, isocyanates, aldehydes, oxidizers.

b-HYDROXYETHYL ISOPROPYL ETHER or beta-HYDROXY-ETHYL ISOPROPYL ETHER (109-59-1) Forms explosive mixture with air 92°F/33°C). Incompatible with sulfuric acid, perchloric acid, isocyanates, oxidizers.

2-(HYDROXYETHYL) METHYLAMINE (109-83-1) Forms explosive mixture with air (flash point 165°F/74°C). Incompatible with strong oxidizers (may cause fire). A strong base; reacts with acids. Attacks aluminum, copper, zinc, and alloys.

2-HYDROXYETHYL 2-PROPENOATE (818-61-6) Able to polymerize. Incompatible with strong acids, aliphatic amines, alkanolamines. Attacks mild steel and tin plate. Swells rubber; removes paint.

1-HYDROXYHEPTANE (111-70-6) Forms explosive mixture with air (flash point 170°F/77°C). Incompatible with strong acids, caustics, aliphatic amines, isocyanates, oxidizers.

2-HYDROXYHEPTANE (543-49-7) Forms explosive mixture with air (flash point 160°F/71°C). Reacts violently with strong oxidizers. May accumulate static electrical charges, and may cause ignition of its vapors.

3-HYDROXYHEPTANE (589-82-2) Forms explosive mixture with air (flash point 140°F/60°C). Reacts violently with strong oxidizers. May accumulate static electrical charges, and may cause ignition of its vapors.

1-HYDROXYHEXANE (111-27-3) Forms explosive mixture with air (flash point 145°F/63°C). Incompatible with strong acids, caustics, aliphatic amines, isocyanates, strong oxidizers.

2-HYDROXYISOBUTYRONITRILE (75-86-5) Forms explosive mixture with air (flash point 165°F/74°C). Reacts violently with strong oxidizers. Heat may cause decomposition. Incompatible with nonoxidizing mineral acids, sulfuric acid, nitric acid, organic acids, caustics, aliphatic amines, alkanolamines, aromatic amines, organic anhydrides, allylene oxides, epichlorohydrin, caprolactam solution, ammonia, isocyanates, phenols, cresol.

a-HYDROXYISOBUTYRONITRILE or alpha-HYDROXYISOBU-TYRONITRILE (75-86-5) Forms explosive mixture with air (flash point 165°F/74°C). Reacts violently with strong oxidizers. Heat may cause decomposition. Incompatible with nonoxidizing mineral acids, sulfuric acid, nitric acid, organic acids, caustics, aliphatic amines, alkanolamines, aromatic amines, organic anhydrides, allylene oxides, epichlorohydrin, caprolactam solution, ammonia, isocyanates, phenols, cresol.

HYDROXYISOBUTYRONITRITE (75-86-5) Forms explosive mixture with air (flash point 165°F/74°C). Reacts violently with strong oxidizers. Heat may cause decomposition. Incompatible with nonoxidizing mineral acids, sulfuric acid, nitric acid, organic acids, caustics, aliphatic amines, alkanolamines, aromatic amines, organic anhydrides, allylene oxides, epichlorohydrin, caprolactam solution, ammonia, isocyanates, phenols, cresol.

4-HYDROXY-2-KETO-4-METHYLPENTANE (123-42-2) Incompatible with strong acids, strong alkalis (cause formation of flammable acetone vapors), aliphatic amines, isocyanates, oxidizers, alkali metals. Forms explosive mixture with air (flash point 136°F/58°C). Attacks some forms of plastics, resins, and rubber.

HYDROXYLAMINE (7803-49-8) Self-reactive; contaminants or temperatures above 150°F/65°C may cause explosive decomposition. Incompatible with strong acids, organic anhydrides, isocyanates, aldehydes, sodium, zinc. Aqueous solution is a base. Strong oxidizers may cause fire and explosions. Attacks some metals.

HYDROXYLAMINE HYDROSULFATE (10046-00-1) A strong reducing agent; reacts violently with oxidizers. Incompatible with strong acids, organic anhydrides, isocyanates, aldehydes. Aqueous solution is highly acidic; reacts with bases. Attacks some metals and coatings.

HYDROXYLAMINE NEUTRAL (10039-54-0) A strong oxidizer. Heat or friction may cause explosions. Violent, possibly explosive, reaction occurs with caustics, reducing agents, or combustibles. Incompatible with strong acids, organic anhydrides, isocyanates, aldehydes. Magnesium or aluminum contact may be explosive.

HYDROXYLAMINE SULFATE or HYDROXYLAMINE SULFATE (2:1) (10039-54-0) A strong oxidizer. Heat or friction may cause explosions. Violent, possibly explosive, reaction occurs with caustics, reducing agents, or combustibles. Incompatible with strong acids, organic anhydrides, isocyanates, aldehydes. Magnesium or aluminum contact may be explosive.

HYDROXYLAMMONIUM SULFATE (10039-54-0) A strong oxidizer. Heat or friction may cause explosions. Violent, possibly explosive, reaction occurs with caustics, reducing agents, or combustibles. Incompatible with strong acids, organic anhydrides, isocyanates, aldehydes. Magnesium or aluminum contact may be explosive.

4-HYDROXYL-2-KETO-4-METHYLPENTANE (123-42-2) Incompatible with strong acids, strong alkalis (cause formation of flammable acetone vapors), aliphatic amines, isocyanates, oxidizers, alkali metals. Forms explosive mixture with air (flash point 136°F/58°C). Attacks some forms of plastics, resins, and rubber.

1-HYDROXY-2-METHOXYBENZENE (90-05-1) Forms explosive mixture with air (flash point 82°F/180°C). Reacts violently with strong oxidizers.

2-HYDROXY-2-METHYL-3-BUTYNE (115-19-5) Forms explosive mixture with air (flash point less than 70°F/21°C). Incompatible with strong acids, caustics, aliphatic amines, isocyanates, oxidizers.

2-HYDROXYMETHYLFURAN (98-00-0) Forms explosive mixture with air (flash point 170°F/77°C). Strong acids (including some organic acids) or acid catalysts may cause polymerization. Reacts strongly with oxidizers and acids. Incompatible with caustics, aliphatic amines, isocyanates. Attacks some plastics, coatings, and rubber.

4-HYDROXY-4-METHYL-PENTAN-2-ON (German or Dutch) (123-42-2) Incompatible with strong acids, strong alkalis (cause formation of flammable acetone vapors), aliphatic amines, isocyanates, oxidizers, alkali metals. Forms explosive mixture with air (flash point 136°F/58°C). Attacks some forms of plastics, resins, and rubber.

4-HYDROXY-4-METHYL-2-PENTANONE (123-42-2) Incompatible with strong acids, strong alkalis (cause formation of flammable acetone vapors), aliphatic amines, isocyanates, oxidizers, alkali metals. Forms explosive mixture with air (flash point 136°F/58°C). Attacks some forms of plastics, resins, and rubber.

4-HYDROXY-4-METHYLPENTAN-2-ONE (123-42-2) Incompatible with strong acids, strong alkalis (cause formation of flammable acetone vapors), aliphatic amines, isocyanates, oxidizers, alkali metals. Forms explosive mixture with air (flash point 136°F/58°C). Attacks some forms of plastics, resins, and rubber.

1-HYDROXYMETHYLPROPANE (78-83-1) Forms explosive mixture with air (flash point 82°F/28°C). Incompatible with strong acids, strong oxidizers, caustics, aliphatic amines, isocyanates, alkali metals and alkaline earth. Attacks some plastics, rubber, and coatings. May react with aluminum at high temperatures.

2-HYDROXY-2-METHYLPROPIONITRILE (75-86-5) Forms explosive mixture with air (flash point 165°F/74°C). Reacts violently with strong oxidizers. Heat may cause decomposition. Incompatible with nonoxidizing mineral acids, sulfuric acid, nitric acid, organic acids, caustics, aliphatic amines, alkanolamines, aromatic amines, organic anhydrides, allylene oxides, epichlorohydrin, caprolactam solution, ammonia, isocyanates, phenols, cresol.

2-HYDROXY-4-(METHYLTHIO)-BUTANOIC ACID (583-91-5) Incompatible with sulfuric acid, caustics, ammonia, amines, isocyanates, alkylene oxides, epichlorohydrin.

2-HYDROXYNITROBENZENE (88-75-5) A strong oxidizer; reacts violently with reducing agents, combustibles, organic matter, chlorosulfonic acid, potassium hydroxide.

3-HYDROXYNITROBENZENE (554-84-7) A strong oxidizers; reacts with combustibles, reducing agents, organics, and other easily oxidizable materials.

4-HYDROXYNITROBENZENE (100-02-7) Combustible solid. A strong oxidizer; reacts with reducing agents, combustibles, organic, and other easily oxidizable materials. Incompatible with strong acids, caustics, aliphatic amines, amides.

m-HYDROXYNITROBENZENE (554-84-7) A strong oxidizers; reacts with combustibles, reducing agents, organics, and other easily oxidizable materials.

1-HYDROXYOCTANE (111-87-5) Forms explosive mixture with air (flash point 178°F/81°C). Incompatible with strong acids, caustics, aliphatic amines, isocyanates, strong oxidizers.

2-HYDROXYPHENOL (120-80-9) Strong oxidizers may cause fire and explosion.

3-HYDROXYPHENOL (108-46-3) Reacts violently with strong oxidizers, nitric acid. Incompatible with acetanilide, albumin, alkalies, antipyrine, camphor, ferric salts, menthol, spirit nitrous ether. Absorbs moisture from air (hygroscopic). May accumulate static electrical charges, and may cause ignition of its vapors.

m-HYDROXYPHENOL or *meta*-HYDROXYPHENOL (108-46-3) Reacts violently with strong oxidizers, nitric acid. Incompatible with acetanilide, albumin, alkalies, antipyrine, camphor, ferric salts, menthol, spirit nitrous ether. Absorbs moisture from air (hygroscopic). May accumulate static electrical charges, and may cause ignition of its vapors.

o-HYDROXYPHENOL or *ortho*-HYDROXYPHENOL (120-80-9) Strong oxidizers may cause fire and explosion.

p-HYDROXYPHENOL or *para*-HYDROXYPHENOL (123-31-9) Incompatible with strong oxidizers, caustics. May be oxidized to quinone at room temperatures in the presence of moisture. May explode on contact with oxygen.

1-HYDROXYPROPANE (71-23-8) Forms explosive mixture with air (flash point 74°F/23°C). Attacks some plastics, rubber, and coatings. Incompatible with strong acids, caustics, aliphatic amines, isocyanates.

3-HYDROXYPROPANENITRILE (109-78-4) Forms explosive mixture with air (flash point 140°F/60°C). Avoid basic contamination to prevent polymerization. A powerful reducing agent. Hot water contact produces cyanide gas. Reacts violently with strong oxidizers or sodium hydroxide. Reacts with acids, acid salts, chlorates, nitrates. Attacks mild steel, copper, and copper alloys.

2-HYDROXY-1,2,3-PROPANE-TRICARBOXYLIC ACID (77-92-9) Incompatible with bases, strong oxidizers, amines. Contact with metal nitrates may be explosive. Attacks aluminum, copper, zinc, and their alloys, when wet.

2-HYDROXYPROPANOIC ACID (598-82-3) Incompatible with strong acids. Aqueous solution reacts with bases.

a-HYDROXYPROPIONIC ACID or alpha-HYDROXYPROPIONIC ACID (598-82-3) Incompatible with strong acids. Aqueous solution reacts with bases.

2-HYDROXYPROPANOL (57-55-7) Incompatible with strong acids, caustics, aliphatic amines, isocyanates, strong oxidizers (reacts violently).

3-HYDROXYPROPENE (107-18-6) May form unstable and explosive peroxides. Able to polymerize. May accumulate static electrical charges, and may cause ignition of its vapors. A strong reducing agent; reacts violently with oxidizers. Forms explosive mixture with air (flash point 70°F/21°C). Incompatible with strong acids, amines, isocyanates, carbon tetrachloride, chlorosulfonic acid, diallyl phosphide, oleum, sodium hydroxide, tri-n-bromomelamine, metal halides, caustic soda, sodium, magnesium, aluminum, and their alloys. Attacks some coatings, some plastics, and rubber.

3-HYDROXY-beta-PROPIOLACTONE (57-57-8) Polymerizes in storage. High temperatures cause containers to rupture. Decomposes at room temperature; hydrolyzes in water.

HYDROXYPROPYL ACRYLATE (999-61-1) Incompatible with strong acids, nitrates.

b-HYDROXYPROPYL ACRYLATE or beta-HYDROXYPROPYL ACRYLATE (999-61-1) Incompatible with strong acids, nitrates.

2-HYDROXYPROPYLAMINE (78-96-6) Forms explosive mixture with air (flash point 171°F/77°C). Incompatible with acids, organic anhydrides, isocyanates, aldehydes.

3-HYDROXYPROPYLAMINE (156-87-6) Forms explosive mixture with air (flash point 175°F/79°C). Incompatible with acids, organic anhydrides, isocyanates, vinyl acetate, acrylates, substituted allyls, alkylene oxides, epichlorohydrin, aldehydes. Corrodes copper and its alloys.

444

HYDROXYPROPYL METHACRYLATE (N/A) Forms explosive mixture with air (flash point 250°F/121°C). Strong oxidizers may cause fire and explosions. May polymerize when hot or when exposed to ultraviolet light or free-radical catalysts.

6-HYDROXY-3(2H)-PYRIDAZINONE (123-33-1) Contact with strong oxidizers may cause fire and explosions.

3-HYDROXYTOLUENE (108-39-4) Incompatible with strong acids, oxidizers, alkalies, aliphatic amines, amides, chlorosulfonic acid, oleum. Liquid attacks some plastics, coatings, and rubber.

4-HYDROXYTOLUENE (106-44-5) Incompatible with strong acids, oxidizers, alkalies, aliphatic amines, amides, chlorosulfonic acid, oleum. Liquid attacks some plastics, coatings, and rubber.

a-HYDROXYTOLUENE or alpha-HYDROXYTOLUENE (100-51-6) Forms explosive mixture with air (flash point 213°F/101°C). Slowly oxidizes in air and oxygen. Incompatible with mineral acids, caustics, aliphatic amines, isocyanates. Reacts violently with strong oxidizers. Corrodes aluminum at high temperature. Attacks some nonfluorinated plastics; may not attack polypropylene.

o-**HYDROXYTOLUENE** (95-48-7) Forms explosive mixture with air (flash point 178°F/81°C). Incompatible with strong acids, oxidizers, alkalies, aliphatic amines, amides, chlorosulfonic acid, oleum. Liquid attacks some plastics and rubber.

HYDROXYTOLUENES (1319-77-3) Incompatible with strong acids, oxidizers, alkalies, aliphatic amines, amides, chlorosulfonic acid, oleum. Liquid attacks some plastics, coatings, and rubber.

HYDROXYTOLUOLE (German) (1319-77-3) Incompatible with strong acids, oxidizers, alkalies, aliphatic amines, amides, chlorosulfonic acid, oleum. Liquid attacks some plastics, coatings, and rubber.

b-HYDROXYTRICARBALLYLIC ACID or beta-HYDROXYTRICARBALLYLIC ACID (77-92-9) Incompatible with bases, strong oxidizers, amines. Contact with metal nitrates may be explosive. Attacks aluminum, copper, zinc, and their alloys, when wet.

b-HYDROXYTRICARBOXYLIC ACID or beta-HYDROXYTRICARBOXYLIC ACID or beta-HYDROXY-TRICARBOXYLIC ACID (77-92-9) Incompatible with bases, strong oxidizers, amines. Contact with metal nitrates may be explosive. Attacks aluminum, copper, zinc, and their alloys, when wet.

2-HYDROXYTRIETHYLAMINE (100-37-8) Forms explosive mixture with air (flash point 126°F/52°C). Reacts violently with oxidizers, strong acids. Attacks light metals. Attacks some plastics and rubber.

2-HYDROXY-M-XYLENE (1300-71-6) Forms explosive mixture with air (flash point 186°F/86°C) Reacts violently with strong oxidizers.

HYDRURE de LITHIUM (French) (7580-67-8) Incompatible with oxidizers, halogenated hydrocarbons, acids; can cause fire and explosion. Contact with water produces hydrogen gas; reaction may cause ignition.

HYLEMOX (563-12-2) Incompatible with alkaline formulations. Mixtures with magnesium may be explosive.

HYLENE M50 (101-68-8) Incompatible with strong alkalies, acids, alcohols, ammonia, amines, amides, glycols, caprolactam. Unstable above 100°F/37.8°C. Attacks some plastics, rubber, and coatings.

HYLENE T (584-84-9) Incompatible with strong acids, including nonoxidizing mineral and organic acids, caustics, ammonia, amines, amides, alcohols, glycols, caprolactam solution. Water contact causes violent foaming and spattering; produces carbon dioxide and an organic base. Attacks copper and its alloys, some plastics including polyethylene, and rubber.

HYLENE TCPA (584-84-9) Incompatible with strong acids, including nonoxidizing mineral and organic acids, caustics, ammonia, amines, amides, alcohols, glycols, caprolactam solution. Water contact causes violent foaming and spattering; produces carbon dioxide and an organic base. Attacks copper and its alloys, some plastics including polyethylene, and rubber.

HYLENE TLC (584-84-9) Incompatible with strong acids, including nonoxidizing mineral and organic acids, caustics, ammonia, amines, amides, alcohols, glycols, caprolactam solution. Water contact causes violent foaming and spattering; produces carbon dioxide and an organic base. Attacks copper and its alloys, some plastics including polyethylene, and rubber.

HYLENE TM (584-84-9) Incompatible with strong acids, including nonoxidizing mineral and organic acids, caustics, ammonia, amines, amides, alcohols, glycols, caprolactam solution. Water contact causes violent foaming and spattering; produces carbon dioxide and an organic base. Attacks copper and its alloys, some plastics including polyethylene, and rubber.

HYLENE TM-65 (584-84-9) Incompatible with strong acids, including nonoxidizing mineral and organic acids, caustics, ammonia, amines, amides, alcohols, glycols, caprolactam solution. Water contact causes violent foaming and spattering; produces carbon dioxide and an organic base. Attacks copper and its alloys, some plastics including polyethylene, and rubber.

HYLENE TRF (584-84-9) Incompatible with strong acids, including nonoxidizing mineral and organic acids, caustics, ammonia, amines, amides, alcohols, glycols, caprolactam solution. Water contact causes violent foaming and spattering; produces carbon dioxide and an organic base. Attacks copper and its alloys, some plastics including polyethylene, and rubber.

HYOXYL (7722-84-1) A powerful oxidizer; attacks many substances. Contact with most organic, readily oxidizable materials, reducing agents, and combustibles causes fire and explosions. Contact with iron, copper, brass, bronze, chromium, zinc, lead, manganese, silver, and other catalytic metals (and their salts), especially in a basic (pH7 or above) environment, causes rapid decomposition with evolution of oxygen gas. Attacks, and may ignite, some plastics, rubber, and coatings. Decomposes slowly at ordinary temperatures and builds up pressure in a closed container. The rate of decomposition doubles for each 50°F/10°C rise (1.5 times 10°C rise) in temperature and becomes self-sustaining at 285°F/141°C.

HYPEROL (124-43-6) Combustible solid. At 122°F/50°C reacts with dust, organics.

HY-PHI 2066 (112-80-1) Incompatible with strong oxidizers, perchloric acid, and aluminum powder; may cause explosions. Corrodes aluminum.

HY-PHY 1055 (112-80-1) Incompatible with strong oxidizers, perchloric acid, and aluminum powder; may cause explosions. Corrodes aluminum.

HYPNONE (98-86-2) Incompatible with strong acids, aliphatic amines, oxidizers. Forms explosive mixture with air (flash point 170°F/77°C).

HYPOCHLOROUS ACID, CALCIUM (7778-54-3) Decomposes in heat or sunlight. Incompatible with acids, moisture, reducing agents, combustible materials, all other chemicals, especially acetylene, aniline and all other amines, anthracene, carbon tetrachloride, iron oxide, manganese oxide, mercaptans, diethylene glycol monomethyl ether, nitromethane, organic matter, organic sulfides, phenol, 1-propanethiol, propyl mercaptan, sulfur, organic sulfur compounds.

HYPOCHLOROUS ACID, SODIUM SALT (7681-52-9) A powerful oxidizer and a strong base. Decomposes in sunlight, forming oxygen and increasing the risk of fire. Stability decreases with concentration, heat, light, decrease in pH, and contamination with metals. Incompatible with strong acids, reducing agents, combustible substances; all cause violent reaction, fire and explosions. Contact with amines and ammonia salts produces explosive chloroamines. Corrodes many metals: steel, 12% and 17% chrome steel, cast iron, monel, aluminum, nickel, brass, bronze, iconel.

HYPODERMACIDE (52-68-6) Contact with strong oxidizers may cause fire and explosions.

HYPONITROUS ACID ANHYDRIDE (10024-97-2) May form explosive mixture with air. Incompatible with nonflammable gas; supports combustion. Incompatible with aluminum, anhydrous ammonia (may be explosive), boron, hydrazine lithium hydride, phosphine, sodium.

HYSTRENE 9512 (143-07-7) Incompatible with sulfuric acid, bases, ammonia, amines, alkylene oxide, epichlorohydrin.

HYTROL O (108-94-1) Forms explosive mixture with air (flash point 111°F/44°C). May accumulate static electrical charges, and may cause ignition of its vapors. Attacks red metals and lead. Incompatible with amines, sulfuric acid, nitric acid, oxidizers, strong acids, aliphatic amines. Dissolves plastics, resins, and rubber.

IBA (78-83-1) Forms explosive mixture with air (flash point 82°F/28°C). Incompatible with strong acids, strong oxidizers, caustics, aliphatic amines, isocyanates, alkali metals, and alkali earth. Attacks some plastics, rubber, and coatings. May react with aluminum at high temperatures.

IBN (78-82-0) Forms explosive mixture with air (flash point 47°F/8°C). Reacts violently with strong oxidizers.

ICEON 22 (75-45-5) Moisture and rust cause slow decomposition, producing toxic gases. Attacks some paints, rubber, and coatings. Thermal decomposition occurs at high temperature with alkalies and alkaline earth metals.

IDROCHINONE (Italian) (123-31-9) Incompatible with strong oxidizers, caustics. May be oxidized to quinone at room temperatures in the presence of moisture. May explode on contact with oxygen.

IDROGENOSOLFORATO (Italian) (7783-06-4) A highly flammable and reactive gas. Incompatible with acetaldehyde, barium pentafluoride, chlorine monoxide, chlorine trifluoride, chromic anhydride, copper, lead dioxide, nitric acid, nitrogen iodide, nitrogen trichloride, nitrogen trifluoride, oxygen difluoride, oxidizers, phenyl diazonium chloride, sodium, sodium peroxide. Reacts with alkaline metals and alkali earth metals. Attacks metals.

4-IDROSSI-4-METIL-PENTAN-2-ONE (Italian) (123-42-2) Incompatible with strong acids, strong alkalies (cause formation of flammable acetone vapors), aliphatic amines, isocyanates, oxidizers, alkali metals. Forms explosive mixture with air (flash point 136°F/58°C). Attacks some forms of plastics, resins, and rubber.

IGE (4016-08-6) Forms explosive mixture with air (flash point 92°F/33°C). Incompatible with strong oxidizers (fire), strong caustics (polymerization). Air and light cause formation of unstable and explosive peroxides. Attacks some rubbers and plastics.

IKURIN (7773-06-0) Fires and explosions may result from contact with strong oxidizers, hot water, potassium, sodium, sodium nitrite, metal chlorates, hot acid solutions. Corrosive to mild steel.

ILLOXOL (60-57-1) Incompatible with concentrated mineral acids, acid catalysts, acid oxidizing agents, phenols, reactive metals.

ILLUMINATING OIL (8008-20-6) Forms explosive mixture with air above 100°F/38°C. Oxidizers may cause fire and explosions. Incompatible with nitric acid. May accumulate static electrical charges, and may cause ignition of its vapors.

2,2′-IMINOBISETHANOL (111-42-2) Incompatible with acids, organic anhydrides, isocyanates, vinyl acetate, acrylates, substituted allyls, alkylene oxides, epichlorohydrin, aldehydes, oxidizers. Corrosive to copper, copper alloys, zinc, and galvanized iron.

2,2′-IMINOBISETHYLAMINE (111-40-0) Ignites spontaneously with cellulose nitrate. Silver, cobalt, or chromium compounds may cause explosions. Forms explosive mixture with air (flash point 208°F/98°C). Incompatible with acids, organic anhydrides, isocyanates, vinyl acetate, acrylates, substituted allyls, alkylene oxides, epichlorohydrin,

ketones, aldehydes, alcohols, glycols, mercury, phenols, cresols, caprolactam solution, strong oxidizers. Attacks aluminum, copper, lead, tin, zinc, and alloys.

2,2'-IMINODIETHANOL (111-42-2) Incompatible with acids, organic anhydrides, isocyanates, vinyl acetate, acrylates, substituted allyls, alkylene oxides, epichlorohydrin, aldehydes, oxidizers. Corrosive to copper, copper alloys, zinc, and galvanized iron.

1,1'-IMINODI-2-PROPANOL (110-97-4) Incompatible with sulfuric, nitric, mineral and organic acids, organic anhydrides, isocyanates, vinyl acetate, acrylates, substituted allyls, alkylene oxides, epichlorohydrin, aldehydes, oxidizers. Attacks some paints, rubber, and coatings, and copper and copper alloys.

IMPERIAL GREEN (12002-03-8) Contact with strong oxidizers may cause fire and explosions.

INACTIVE LIMONENE (138-86-3) Forms explosive mixture with air (flash point 115°F/46°C). Strong oxidizers may cause fire and explosions.

INAKOR (1912-24-9) Incompatible with strong acids.

INCIDOL (94-36-0) Confined storage of dry chemical may lead to decomposition and explosion. A strong oxidant; extremely reactive. Fires and explosion may result from heat or contamination, and from contact with strong acids, combustible materials, oxidizers, acids, bases, alcohols, reducing agents, metals, metal oxides, amines, accelerators, methyl methacrylate, organic matter, lithium aluminum carbide, dimethyl aniline, amines, metallic naphthenates. May attack some plastics, rubber, and coatings. Protect containers from shock and friction.

INDIAN RED (1309-37-1) Contact with hydrogen peroxide, ethylene oxide, calcium hypochlorite will cause explosion. Reacts violently with powdered aluminum, hydrazine, hydrogen trisulfide.

INDOPOL L 50 (9003-29-6) Incompatible with sulfuric acid, nitric acid.

INERTEEN (generic CAS for PCBs 1336-36-3) Incompatible with strong oxidizers, strong acids.

INEXIT (58-89-9) Not combustible, but may be dissolved in a combustible solvent. If solvent comes in contact with oxidizers, fire and explosions may result.

INFILTRINA (67-68-5) Forms explosive mixture with air (flash point 203°F/95°C). Reacts violently with oxidizers. Reacts with ethanoyl chloride, boron compounds, halides, metal alkoxides, oxidizers.

INHIBINE (7722-84-1) A powerful oxidizer; attacks many substances. Contact with most organic, readily oxidizable materials, reducing agents, and combustibles causes fire and explosions. Contact with iron, copper, brass, bronze, chromium, zinc, lead, manganese, silver, and other catalytic metals (and their salts), especially in a basic (pH 7 or above) environment, causes rapid decomposition with evolution of oxygen gas. Attacks, and may ignite, some plastics, rubber, and coatings. Decomposes slowly at ordinary temperatures and builds up pressure in a closed container. The rate of decomposition doubles for each 50°F/10°C rise (1.5 times 10°C rise) in temperature and becomes self-sustaining at 285°F/141°C.

INSULATING OIL (N/A) Oxidizers may cause fire and explosions. Incompatible with nitric acid. May accumulate static electrical charges, and may cause ignition of its vapors.

INVERTON 245 (93-76-5) Sealed metal containers may burst in heat above 316°F/158°C. Incompatible with sulfuric acid, bases, ammonia, aliphatic amines, alkanolamines, isocyanates, alkylene oxides, epichlorohydrin.

IODE (French) (7553-56-2) Contact with ammonia produces a shock-sensitive compound, which will explode, causing fires. A powerful oxidizer. Reacts violently with reducing agents, combustibles, acetylene, acetaldehyde, antimony, cesium, francium, lithium, potassium, powdered metals, rubidium, phosphorus, sodium, sulfur.

IODINE or IODINE CRYSTALS or IODINE, ELEMENTAL or IODINE SUBLIMED (7553-56-2) Contact with ammonia produces a shock-sensitive compound, which will explode, causing fires. A powerful oxidizer. Reacts violently with reducing agents, combustibles, acetylene, acetaldehyde, antimony, cesium, francium, lithium, potassium, powdered metals, rubidium, phosphorus, sodium, sulfur.

IODINE CHLORIDE (7790-99-0) Heat may cause explosions. Reacts violently with water, organic matter, aluminum foil, cadmium sulfide, lead sulfide, phosphorus, phosphorus trichloride, potassium, rubber, silver sulfide, sodium, zinc sulfide, and other metals.

IODINE MONOCHLORIDE (7790-99-0) Heat may cause explosions. Reacts violently with water, organic matter, aluminum foil, cadmium sulfide, lead sulfide, phosphorus, phosphorus trichloride, potassium, rubber, silver sulfide, sodium, zinc sulfide, and other metals.

IODIO (Italian) (7553-56-2) Contact with ammonia produces a shock-sensitive compound, which will explode, causing fires. A powerful oxidizer. Reacts violently with reducing agents, combustibles, acetylene, acetaldehyde, antimony, cesium, francium, lithium, potassium, powdered metals, rubidium, phosphorus, sodium, sulfur.

IODOMETANO (Italian) (74-88-4) Contact with strong oxidizers may cause fire and explosions. Heat, light, and moisture contribute to instability.

IODOMETHANE (74-88-4) Contact with strong oxidizers may cause fire and explosions. Heat, light, and moisture contribute to instability.

IODURE de METHYLE (French) (74-88-4) Contact with strong oxidizers may cause fire and explosions. Heat, light, and moisture contribute to instability.

IOPEZITE (7778-50-9) Powdered combustibles will ignite. Strong acids cause formation of toxic vapors.

IPA (67-63-0) Forms explosive mixture with air (flash point 53°F/12°C). Incompatible with alkaline earth and alkali metals, crotonaldehyde, phosgene, strong acids, amines, ammonia, caustics, strong oxidizers. Attacks some plastics, rubber, and coatings. Reacts with metallic aluminum at high temperatures.

IPANER (94-75-7) Decomposes in sunlight. Incompatible with strong oxidizers; may cause fire and explosions.

IPDI (4098-71-9) Forms explosive mixture with air (flash point 212°F/100°C). Water contact produces a corrosive. Incompatible with all acids, caustics, ammonia, aliphatic amines, alkanolamines, aromatic amines, amides, alcohols, glycols, caprolactam, strong oxidizers.

IPE (109-59-1) Forms explosive mixture with air flash point 92°F/33°C). Incompatible with sulfuric acid, perchloric acid, isocyanates, oxidizers.

450

IRON AMMONIUM SULFATE (10045-89-3) Reacts violently with tetranitromethane, ammonium perchlorate, and mercury(II) dinitrate. Sulfates react with aluminum, magnesium.

IRONATE (7782-63-0) Aqueous solution is strongly acidic; reacts violently with strong acids. Incompatible with oxidizers and oxygen in the air.

IRON(II) CHLORIDE (1:2) (7758-94-3) Solution attacks metals. Contact with ethylene oxide may initiate polymerization. Potassium or sodium contact produces impact-sensitive materials.

IRON(III) CHLORIDE (7705-08-0) Incompatible with water. Solution is very acidic; reacts violently with bases. Forms shock- and friction-sensitive explosive material with potassium, sodium and other active metals. Reacts strongly with allyl chloride. Attacks metals when wet.

IRON DICHLORIDE (7758-94-3) Solution attacks metals. Contact with ethylene oxide may initiate polymerization. Potassium or sodium contact produces impact-sensitive materials.

IRON FLUORIDE (7783-50-8) Contact with strong oxidizers may cause fire and explosions.

IRON NITRATE (10421-48-4) Solution is corrosive to metals. Contact with combustibles may cause fire.

IRON(III) NITRATE, ANHYDROUS (10421-48-4) Solution is corrosive to metals. Contact with combustibles may cause fire.

IRON (OUS) SULFATE (7720-78-7) Aqueous solution is acidic. Contact with bases produces iron.

IRON OXIDE (1309-37-1) Contact with hydrogen peroxide, ethylene oxide, calcium hypochlorite will cause explosion. Reacts violently with powdered aluminum, hydrazine, hydrogen trisulfide.

IRON(III) OXIDE (1309-37-1) Contact with hydrogen peroxide, ethylene oxide, calcium hypochlorite will cause explosion. Reacts violently with powdered aluminum, hydrazine, hydrogen trisulfide.

IRON OXIDE RED (1309-37-1) Contact with hydrogen peroxide, ethylene oxide, calcium hypochlorite will cause explosion. Reacts violently with powdered aluminum, hydrazine, hydrogen trisulfide.

IRON PERCHLORIDE (7705-08-0) Incompatible with water. Solution is very acidic; reacts violently with bases. Forms shock- and friction-sensitive explosive material with potassium, sodium, and other active metals. Reacts strongly with allyl chloride. Attacks metals when wet.

IRON PERSULFATE (10028-22-5) Corrosive to copper and its alloys, mild and galvanized steel.

IRON PROTOCHLORIDE (7758-94-3) Solution attacks metals. Contact with ethylene oxide may initiate polymerization. Potassium or sodium contact produces impact-sensitive materials.

IRON PROTOSULFATE (7720-78-7) Aqueous solution is acidic. Contact with bases produces iron.

IRON SESQUIOXIDE (1309-37-1) Contact with hydrogen peroxide, ethylene oxide, calcium hypochlorite will cause explosion. Reacts violently with powdered aluminum, hydrazine, hydrogen trisulfide.

IRON SESQUISULFATE (10028-22-5) Corrosive to copper and its alloys, mild and galvanized steel.

IRON–SILICON ALLOY (8049-17-0) Water contact produces toxic and explosive gases. Reacts with acids and sodium hydroxide.

IRON(2+) SULFATE (7720-78-7) Aqueous solution is acidic. Contact with bases produces iron.

IRON(2+) SULFATE (1:1) (7720-78-7) Aqueous solution is acidic. Contact with bases produces iron.

IRON(3+) SULFATE (10028-22-5) Corrosive to copper and its alloys, mild and galvanized steel.

IRON(II) SULFATE (7720-78-7) Aqueous solution is acidic. Contact with bases produces iron.

IRON(III) SULFATE (10028-22-5) Corrosive to copper and its alloys, mild and galvanized steel.

IRON SULFATE (1:1) (7720-78-7) Aqueous solution is acidic. Contact with bases produces iron.

IRON SULFATE (2:3) (10028-22-5) Corrosive to copper and its alloys, mild and galvanized steel.

IRON TERSULFATE (10028-22-5) Corrosive to copper and its alloys, mild and galvanized steel.

IRON(II) TERSULFATE (7782-63-0) Aqueous solution is strongly acidic; reacts violently with strong acids. Incompatible with oxidizers and oxygen in the air.

IRON TRICHLORIDE (7705-08-0) Incompatible with water. Solution is very acidic; reacts violently with bases. Forms shock- and friction-sensitive explosive material with potassium, sodium, and other active metals. Reacts strongly with allyl chloride. Attacks metals when wet.

IRON TRINITRATE (10421-48-4) Solution is corrosive to metals. Contact with combustibles may cause fire.

IRON VITRIOL (7720-78-7) Aqueous solution is acidic. Contact with bases produces iron.

IRON VITRIOL (7782-63-0) Aqueous solution is strongly acidic; reacts violently with strong acids. Incompatible with oxidizers and oxygen in the air.

IROSPAN (7720-78-7) Aqueous solution is acidic. Contact with bases produces iron.

ISCEON 113 (76-13-1) Reacts with barium, lithium, sodium, magnesium, titanium. Contact with alloys containing more than 2% magnesium may cause decomposition (hydrogen chloride, hydrogen fluoride, and carbon monoxide released). Attacks some paints, rubber, and coatings.

ISCEON 122 (75-71-8) Reacts violently with liquid aluminum. Incompatible with chemically active metals. Attacks some plastics, rubber, and coatings.

ISCEON 131 (75-69-4) Reacts with barium, lithium, sodium, magnesium, titanium. Attacks some plastics, rubber, and coatings.

ISCOBROME (74-83-9) Incompatible with strong oxidizers, aluminum, dimethylsulfoxide, ethylene oxide, water. Attacks zinc, magnesium, alkali metals, and their alloys.

ISCOBROME D (106-93-4) Reacts with chemically active metals, liquid ammonia, strong oxidizers. Heat and light cause slow decomposition. Attacks some plastics and rubber.

ISOACETOPHORONE (78-59-1) Forms explosive mixture with air (flash point 184°F/84°C). Incompatible with strong oxidizers, aliphatic amines. Strong oxidizers may cause fire and explosions. Attacks some plastics, resins, and rubber.

ISOAMYL ACETATE (123-92-2) Forms explosive mixture with air (flash point 77°F/25°C). Incompatible with strong alkalies, strong acids, nitrates, oxidizers. Reacts violently with reducing agents. Attacks asbestos; softens and dissolves many plastics, rubber, and coatings.

ISOAMYL ALCOHOL or ISOAMYL ALCOHOL, PRIMARY (123-51-3) Forms explosive mixture with air (flash point 109°F/43°C). Strong oxidizers may cause fire and explosions. Attacks some plastics, rubber, and coatings.

sec-**ISOAMYL ALCOHOL or ISOAMYL ALCOHOL, SECONDARY** (582-75-4) Forms explosive mixture with air (flash point 95°F/35°C). Incompatible with strong acids, caustics, aliphatic amines, isocyanates. May accumulate static electrical charges, and may cause ignition of its vapors.

ISOAMYL ETHANOATE (123-92-2) Forms explosive mixture with air (flash point 77°F/25°C). Incompatible with strong alkalies, strong acids, nitrates, oxidizers. Reacts violently with reducing agents. Attacks asbestos; softens and dissolves many plastics, rubber, and coatings.

ISOAMYL FORMATE (110-45-2) Forms explosive mixture with air (flash point 72°F/22°C). Reacts violently with strong oxidizers. May accumulate static electrical charges, and may cause ignition of its vapors.

ISOAMYL HYDRIDE (78-78-4) Highly volatile liquid (flash point −70°F/−57°C). Mixture with air may explode. Attacks some paints, rubber, and coatings. May accumulate static electrical charges, and may cause ignition of its vapors.

ISOAMYOL (123-51-3) Forms explosive mixture with air (flash point 109°F/43°C). Strong oxidizers may cause fire and explosions. Attacks some plastics, rubber, and coatings.

ISOBENZOFURAN, 1,3-DIHYDRO-1,3-DIOXO- (85-44-9) Incompatible with strong acids, caustics, ammonia, amines, strong oxidizers. Attacks some paints, rubber, and coatings.

1,3-ISOBENZOFURANDIONE (85-44-9) Incompatible with strong acids, caustics, ammonia, amines, strong oxidizers. Attacks some paints, rubber, and coatings.

ISOBUTANAL (78-84-2) Forms explosive gas mixture with air. Incompatible with strong acids, caustics, aliphatic amines, alkanolamines, aromatic amines, strong oxidizers.

ISOBUTANE (75-28-5) Forms explosive gas mixture with air. Strong oxidizers may cause fire and explosions. May accumulate static electrical charges, and may cause ignition of its vapors.

ISOBUTANOL (78-83-1) Forms explosive mixture with air (flash point 82°F/28°C). Incompatible with strong acids, strong oxidizers, caustics, aliphatic amines, isocyanates, alkali metals, and alkaline earth. Attacks some plastics, rubber, and coatings. May react with aluminum at high temperatures.

ISOBUTANOL AMINE (124-68-5) Forms explosive mixture with air (flash point 153°F/67°C). Incompatible with strong acids, mineral acids, organic acids, organic anhydrides, isocyanates, vinyl acetate, acrylates, substituted allyls, alkylene oxides, epichlorohydrin, aldehydes.

ISOBUTANOL-2-AMINE (124-68-5) Forms explosive mixture with air (flash point 153°F/67°C). Incompatible with strong acids, mineral acids, organic acids, organic anhydrides, isocyanates, vinyl acetate, acrylates, substituted allyls, alkylene oxides, epichlorohydrin, aldehydes.

ISOBUTENE (115-11-7) Flammable gas; forms explosive mixture with air. May be able to form unstable peroxides; may cause polymerization. Reacts violently with strong oxidizers, strong acids, oxides of nitrogen. May accumulate static electrical charges, and may cause ignition of its vapors.

ISOBUTENE TRIMER (7756-94-7) Incompatible with sulfuric or nitric acids.

ISOBUTENYL METHYL KETONE (141-79-7) Forms explosive mixture with air (flash point 87°F/31°C). Forms peroxides. Incompatible with strong acids, aliphatic amines, alkanolamines, 2-aminoethanol, ethylene diamine, chlorosulfonic acid, oleum, oxidizers. Dissolves some forms of plastics, resins, and rubber. Attacks copper.

ISOBUTYL ACETATE (110-19-0) Forms explosive mixture with air (flash point 64°F/18°C). Reacts with water on standing, producing acetic acid and n-butyl alcohol. Reacts violently with strong oxidizers. Reacts with caustics, strong acids, nitrates. Dissolves rubber, many plastics, resins, and some coatings. May accumulate static electrical charges, and may cause ignition of its vapors.

ISOBUTYL ACRYLATE, INHIBITED (106-63-8) Forms explosive mixture with air (flash point 94°F/34°C). Incompatible with strong acids, aliphatic amines, alkanolamines.

ISOBUTYL ALCOHOL (78-83-1) Forms explosive mixture with air (flash point 82°F/28°C). Incompatible with strong acids, strong oxidizers, caustics, aliphatic amines, isocyanates, alkali metals and alkali earth. Attacks some plastics, rubber, and coatings. May react with aluminum at high temperatures.

ISOBUTYL ALDEHYDE (78-84-2) Forms explosive gas mixture with air. Incompatible with strong acids, caustics, aliphatic amines, alkanolamines, aromatic amines, strong oxidizers.

ISOBUTYL ALUMINUM DICHLORIDE (1888-87-5) May self-ignite on contact with air. Water or other moisture contact may cause explosion. A strong reducing agent; reacts violently with oxidizers, combustible materials. Incompatible with amines, carbon dioxide, halogens. May accumulate static electrical charges, and may cause ignition of its vapors.

ISOBUTYLAMINE or ISO-BUTYLAMINE (78-81-9) Forms explosive mixture with air (flash point 15°F/−9°C). Incompatible with strong acids, including mineral and organic acids, organic anhydrides, isocyanates, vinyl acetate, acrylates, substituted allyls, alkylene oxides, epichlorohydrin, ketones, aldehydes, alcohols, glycols, phenols, cresols, caprolactam solution, halogens, alkali metals, copper and its alloys.

ISOBUTYL CARBINOL (78-83-1) Forms explosive mixture with air (flash point 82°F/28°C). Incompatible with strong acids, strong oxidizers, caustics, aliphatic amines, isocyanates, alkali metals, and alkali earth. Attacks some plastics, rubber, and coatings. May react with aluminum at high temperatures.

ISOBUTYL CARBINOL (123-51-3) Forms explosive mixture with air (flash point 109°F/43°C). Strong oxidizers may cause fire and explosions. Attacks some plastics, rubber, and coatings.

ISOBUTYL CHLORIDE (513-36-0) Forms explosive mixture with air (flash point less than 70°F/21°C). Reacts violently with strong oxidizers. May accumulate static electrical charges, and may cause ignition of its vapors.

ISOBUTYLENE (115-11-7) Flammable gas; forms explosive mixture with air. May be able to form unstable peroxides; inappropriate level of inhibitor may cause polymerization. Reacts violently with strong oxidizers, strong acids, oxides of nitrogen. May accumulate static electrical charges, and may cause ignition of its vapors.

ISOBUTYL KETONE (108-83-8) Incompatible with strong acids, aliphatic amines, strong oxidizers. Attacks plastics, coatings, and rubber. Forms explosive mixture with air (flash point 140°F/60°C).

ISOBUTYL alpha-METHACRYLATE (97-86-9) Forms explosive mixture with air (flash point 112°F/44°C). Able to form unstable peroxides. Incompatible with strong acids, aliphatic amines, alkanolamines, catalysts, strong oxidizers.

ISOBUTYL METHYL CARBINOL (108-11-2) Forms explosive mixture with air (flash point 106°F/41°C). Contact with alkali metals produces hydrogen gas. Incompatible with strong acids, caustics, aliphatic amines, isocyanates. Attacks some plastics, rubber, and coatings. May accumulate static electrical charges, and may cause ignition of its vapors.

ISOBUTYL METHYL KETONE (108-10-1) Forms explosive mixture with air (flash point 64°F/18°C). Incompatible with oxidizers, strong acids, aliphatic amines. Dissolves some plastics, resins, and rubber.

ISOBUTYL METHYL METHANOL (108-11-2) Forms explosive mixture with air (flash point 106°F/41°C). Contact with alkali metals produces hydrogen gas. Incompatible with strong acids, caustics, aliphatic amines, isocyanates. Attacks some plastics, rubber, and coatings. May accumulate static electrical charges, and may cause ignition of its vapors.

ISOBUTYL 2-METHYL-2-PROPENOATE (97-86-9) Forms explosive mixture with air (flash point 112°F/44°C). Able to form unstable peroxides. Incompatible with strong acids, aliphatic amines, alkanolamines, catalysts, strong oxidizers.

ISOBUTYL PHTHALATE (84-69-5) Incompatible with strong acids, strong oxidizers, nitrates.

ISOBUTYL PROPENOATE (106-63-8) Forms explosive mixture with air (flash point 94°F/34°C). Incompatible with strong acids, aliphatic amines, alkanolamines.

ISOBUTYL-2-PROPENOATE (106-63-8) Forms explosive mixture with air (flash point 94°F/34°C). Incompatible with strong acids, aliphatic amines, alkanolamines.

ISOBUTYRALDEHYDE (78-84-2) Forms explosive gas mixture with air. Incompatible with strong acids, caustics, aliphatic amines, alkanolamines, aromatic amines, strong oxidizers.

ISOBUTYRIC ACID (79-31-2) Forms explosive mixture with air (flash point 132°F/56°C). Incompatible with sulfuric acid, caustics, ammonia, amines, isocyanates, alkylene oxides, epichlorohydrin. Attacks aluminum and other metals.

ISOBUTYRIC ACID, ISOBUTYL ESTER (97-85-8) Forms explosive mixture with air (flash point 99°F/37°C). Incompatible with strong acids, nitrates, strong oxidizers.

ISOBUTYRIC ALDEHYDE (78-84-2) Forms explosive gas mixture with air. Incompatible with strong acids, caustics, aliphatic amines, alkanolamines, aromatic amines, strong oxidizers.

ISOBUTYRONITRILE (78-82-0) Forms explosive mixture with air (flash point 47°F/8°C) Reacts violently with strong oxidizers.

ISOCTYL TRICHLOROPHENOXYACETATE (8002-26-4) Water contact causes foaming. Strong oxidizers may cause fire and explosions. Attacks some paints, rubber, and coatings.

ISOCUMENE (103-65-1) Forms explosive mixture with air (flash point 86°F/30°C). Incompatible with nitric acid.

ISOCYANATE de METHYLE (French) (624-83-9) Forms explosive mixture with air (flash point 0°F/−18°C). Contact with iron, tin, copper (or salts of these elements) and with certain other catalysts (e.g., triphenylarsenic oxide, triethylphosphine, and tributyltin oxide) or elevated temperatures may cause polymerization. Reacts violently with warm water. Incompatible with strong acids, alcohols, glycols, amines, amides, ammonia, caprolactam, caustics, strong oxidizers. Attacks some plastics, rubber, or coatings.

ISOCYANATE METHANE (624-83-9) Forms explosive mixture with air (flash point 0°F/−18°C). Contact with iron, tin, copper (or salts of these elements) and with certain other catalysts (e.g., triphenylarsenic oxide, triethylphosphine, and tributyltin oxide) or elevated temperatures may cause polymerization. Reacts violently with warm water. Incompatible with strong acids, alcohols, glycols, amines, amides, ammonia, caprolactam, caustics, strong oxidizers. Attacks some plastics, rubber, or coatings.

ISO-CYANATOMETHANE (624-83-9) Forms explosive mixture with air (flash point 0°F/−18°C). Contact with iron, tin, copper (or salts of these elements), and with certain other catalysts (e.g., triphenylarsenic oxide, triethylphosphine, and tributyltin oxide) or elevated temperatures may cause polymerization. Reacts violently with warm water. Incompatible with strong acids, alcohols, glycols, amines, amides, ammonia, caprolactam, caustics, strong oxidizers. Attacks some plastics, rubber, or coatings.

3-ISOCYANATOMETHYL-3,5,5-TRIMETHYLCYCLOHEXYL-ISOCYATE (4098-71-9) Forms explosive mixture with air (flash point 212°F/100°C). Water contact produces a corrosive. Incompatible with all acids, caustics, ammonia, aliphatic amines, alkanolamines, aromatic amines, amides, alcohols, glycols, caprolactam, strong oxidizers.

ISOCYANIC ACID, METHYL ESTER (624-83-9) Forms explosive mixture with air (flash point 0°F/−18°C). Contact with iron, tin, copper (or salts of these elements) and with certain other catalysts (such as triphenylarsenic oxide, triethylphosphine, and tributyltin oxide) or elevated temperatures may cause polymerization. Reacts violently with warm water. Incompatible with strong acids, alcohols, glycols, amines, amides, ammonia, caprolactam, caustics, strong oxidizers. Attacks some plastics, rubber, or coatings.

ISOCYANIC ACID, METHYLPHENYLENE ESTER (584-84-9) Incompatible with strong acids, including nonoxidizing mineral and organic acids, caustics, ammonia, amines, amides, alcohols, glycols, caprolactam solution. Water contact causes violent foaming and spattering; produces carbon dioxide and an organic base. Attacks copper and its alloys, some plastics including polyethylene, and rubber.

ISOCYANIC ACID, 4-METHYL-M-PHENYLENE ESTER (584-84-9) Incompatible with strong acids, including nonoxidizing mineral and organic acids, caustics, ammonia, amines, amides, alcohols, glycols, caprolactam solution. Water contact causes violent foaming and spattering; produces carbon dioxide and an organic base. Attacks copper and its alloys, some plastics, including polyethylene, and rubber.

ISOCYANURIC ACID, DICHLORO-, POTASSIUM SALT (2244-21-5) Incompatible with water; forms a bleach solution. A strong oxidizer; reacts violently with reducing agents, combustibles, organics, easily chlorinated materials.

ISOCYANURIC CHLORIDE (87-90-1) A powerful oxidizer. Forms explosive material with nitrogen compounds. Contact with organic materials or reducing agents may cause fire.

ISODECALDEHYDE (N/A) Forms explosive mixture with air (flash point 185°F/85°C). Incompatible with strong acids, caustics, ammonia, amines.

ISODECALDEHYDE, MIXED ISOMERS (N/A) Forms explosive mixture with air (flash point 185°F/85°C). Incompatible with strong acids, caustics, ammonia, amines.

ISODECANOL (25339-17-7) Forms explosive mixture with air (flash point 220°F/104°C). Incompatible with strong acids, caustics, amines, isocyanates.

ISODECYL ACRYLATE (1330-61-6) Incompatible with strong acids, aliphatic amines, alkanolamines, oxidizers.

ISODECYL ALCOHOL (25339-17-7) Forms explosive mixture with air (flash point 220°F/104°C). Strong oxidizers may cause fire and explosions. Incompatible with strong acids, caustics, amines, isocyanates.

ISODECYL PROPENOATE (1330-61-6) Incompatible with strong acids, aliphatic amines, alkanolamines, oxidizers.

ISODURENE (527-53-7) Forms explosive mixture with air (flash point 160°F/71°C). Strong oxidizers may cause fire and explosions.

ISOHEXANE (107-83-5) Forms explosive mixture with air (flash point −20°F/−9°C). Reacts vigorously with strong oxidizers. May accumulate static electrical charges, and may cause ignition of its vapors.

ISOHEXENE (763-29-1) Forms explosive mixture with air (flash point less than 20°F/−7°C) Reacts violently with strong oxidizers.

ISOHEXYL ALCOHOL (108-11-2) Forms explosive mixture with air (flash point 106°F/41°C). Contact with alkali metals produces hydrogen gas. Incompatible with strong acids, caustics, aliphatic amines, isocyanates. Attacks some plastics, rubber, and coatings. May accumulate static electrical charges, and may cause ignition of its vapors.

1H-ISOINDOLE-1,3(2H)-DIONE, 3A,4,7,7A-TETRAHYDRO-2-(1,1,2,2-TETRACHLOROETHYL)THIO- (2425-06-1) Incompatible with acids or acid vapor. Strongly alkaline conditions contribute to instability.

ISOL (107-41-5) Forms explosive mixture with air (flash point 209°F/98°C). Incompatible with strong acids, caustics, aliphatic amines, isocyanates, strong oxidizers.

ISONATE (101-68-8) Incompatible with strong alkalies, acids, alcohols, ammonia, amines, amides, glycols, caprolactam. Unstable above 100°F/37.8°C. Attacks some plastics, rubber, and coatings.

ISONITROPROPANE (79-46-9) Forms explosive mixture with air (flash point 74°F/24°C). Incompatible with amines, strong acids, alkalies, strong oxidizers. Contact with some metal oxides may cause decomposition. Mixtures with hydrocarbons are extremely flammable. Attacks some plastics, rubber, and coatings.

ISOOCTALDEHYDE (N/A) Forms explosive mixture with air (flash point 104°F/40°C). Incompatible with strong acids, caustics, ammonia, amines.

ISOOCTANE (540-84-1) Forms explosive mixture with air (flash point 40°F/5°C). Reacts with oxidizers. May accumulate static electrical charges, and may cause ignition of its vapors.

ISOOCTANOL (26952-21-6) Forms explosive mixture with air (flash point 180°F/82°C). Incompatible with strong acids, caustics, amines, isocyanates.

ISOOCTENE (11071-47-9) Forms explosive mixture with air (flash point 23°F/−5°C). May be able to form unstable and explosive peroxides. Incompatible with strong acids, strong oxidizers. May accumulate static electrical charges, and may cause ignition of its vapors.

ISOOCTYL ALCOHOL (26952-21-6) Forms explosive mixture with air (flash point 180°F/82°C). Incompatible with strong acids, caustics, amines, isocyanates.

ISOOCTYLALDEHYDE (N/A) Forms explosive mixture with air (flash point 104°F/40°C). Incompatible with strong acids, caustics, ammonia, amines.

ISOOCTYL PHTHALATE (27554-26-3) May accumulate static electrical charges, and may cause ignition of its vapors. Incompatible with strong acids, nitrates, oxidizers.

ISOPENTALDEHYDE (107-83-5) Forms explosive mixture with air (flash point −20°F/−29°C). Reacts violently with strong oxidizers. May accumulate static electrical charges, and may cause ignition of its vapors.

ISOPENTANE (78-78-4) Highly volatile liquid (flash point −70°F/−57°C). Mixture with air may explode. Attacks some paints, rubber, and coatings. May accumulate static electrical charges, and may cause ignition of its vapors.

ISOPENTYL ACETATE (123-92-2) Forms explosive mixture with air (flash point 77°F/25°C). Incompatible with strong alkalies, strong acids, nitrates, oxidizers. Reacts violently with reducing agents. Attacks asbestos; softens and dissolves many plastics, rubber, and coatings.

ISOPENTYL ALCOHOL (123-51-3) Forms explosive mixture with air (flash point 109°F/43°C). Strong oxidizers may cause fire and explosions. Attacks some plastics, rubber, and coatings.

ISOPENTYL ALCOHOL NITRITE or ISOPENTYL NITRITE (110-46-3) Forms explosive mixture with air (flash point 0°F/−18°C). Decomposes in light, air, or water; forms oxides of nitrogen. A strong oxidizer; reacts with reducing agents, combustibles, organics, and other strong oxidizers. Corrodes metal with moisture.

ISOPHORONE (78-59-1) Forms explosive mixture with air (flash point 184°F/84°C). Incompatible with strong oxidizers, aliphatic amines. Strong oxidizers may cause fire and explosions. Attacks some plastics, resins, and rubber.

ISOPHORONE DIAMINE (2855-13-2) Incompatible with acids, organic anhydrides, isocyanates, vinyl acetate, acrylates, substituted allyls, alkylene oxides, epichlorohydrin, ketones, aldehydes, alcohols, glycols, phenols, cresols, caprolactam solution. Corrodes aluminum and steel in the presence of moisture and carbon dioxide.

ISOPHORONE DIAMINE DIISOCYANATE (4098-71-9) Forms explosive mixture with air (flash point 212°F/100°C). Water contact produces a corrosive. Incompatible with all acids, caustics, ammonia, aliphatic amines, alkanolamines, aromatic amines, amides, alcohols, glycols, caprolactam, strong oxidizers.

ISOPHORONE DIISOCYANATE (4098-71-9) Forms explosive mixture with air (flash point 212°F/100°C). Water contact produces a corrosive. Incompatible with all acids, caustics, ammonia, aliphatic amines, alkanolamines, aromatic amines, amides, alcohols, glycols, caprolactam, strong oxidizers.

ISOPHTHALIC ACID (88-99-3) Incompatible with strong acids, nitrates, oxidizers.

ISOPRENE (78-79-5) Store under inert atmosphere (preferably nitrogen) with at least 50 ppm *tert*-butyl catechol present as an inhibitor. Forms explosive mixture with air (flash point −65°F/−54°C). Air contact causes formation of unstable peroxides; heat or buildup of peroxides may cause violent polymerization. Reacts violently with oxidizers, acids, reducing agents. Attacks some paints, rubber, and coatings. May accumulate static electrical charges, and may cause ignition of its vapors.

ISOPROPANOL (67-63-0) Forms explosive mixture with air (flash point 53°F/12°C). Incompatible with alkaline earth and alkali metals, crotonaldehyde, phosgene, strong acids, amines, ammonia, caustics, strong oxidizers. Attacks some plastics, rubber, and coatings. Reacts with metallic aluminum at high temperatures.

ISOPROPANOLAMINE (78-96-6) Forms explosive mixture with air (flash point 171°F/77°C). Strong oxidizers may cause fire and explosions. Incompatible with acids, organic anhydrides, isocyanates, aldehydes.

ISOPROPENE CYANIDE (126-98-7) Forms explosive mixture with air (flash point 34°F/1°C). Incompatible with aliphatic amines, alkanolamines. Reacts violently with oxidizers. Incompatible with strong acids; strong bases or light exposure may cause polymerization.

ISOPROPENYL BENZENE (98-83-9) Forms explosive mixture with air (flash point 129°F/54°C). Avoid heat and/or inappropriate inhibitor concentration, and/or contact with catalysts for vinyl or ionic polymerization, such as aluminum or iron chloride. Incompatible with strong acids, oxidizers (including peroxides and halogens).

ISOPROPENYL CARBINOL (513-42-8) Forms explosive mixture with air (flash point 92°F/33°C). Incompatible with strong acids, caustics, aliphatic amines, isocyanates, oxidizers.

ISOPROPENYL METHYL KETONE or ISOPROPENYL METHYL KETONE, INHIBITED (814-78-8) Forms explosive mixture with air (flash point 73°F/23°C) Reacts violently with strong oxidizers. Heat and/or inappropriate level of inhibitor may cause polymerization.

ISOPROPENYLNITRILE (126-98-7) Forms explosive mixture with air (flash point 34°F/1°C). Incompatible with aliphatic amines, alkanolamines. Reacts violently with oxidizers. Incompatible with strong acids; strong bases or light exposure may cause polymerization.

ISOPROPILBENZENE (Italian) (98-82-8) Forms explosive mixture with air (flash point 99°F/37°C). Incompatible with strong acids, strong oxidizers. Air contact causes formation of cumene hydroperoxide. May be able to form unstable peroxides. Attacks rubber. May accumulate static electrical charges, and may cause ignition of its vapors.

2-ISOPROPOXY PROPANE (108-20-3) Forms explosive mixture with air (flash point −18°F/−28°C). Air contact causes formation of explosive peroxides that may detonate with heat or shock. Strong oxidizers may cause fire and explosions. Attacks some plastics, rubber, and coatings.

2-ISOPROPOXYETHANOL (109-59-1) Forms explosive mixture with air (flash point 92°F/33°C). Incompatible with sulfuric acid, perchloric acid, isocyanates, oxidizers.

ISOPROPYL ACETATE (108-21-4) Forms explosive mixture with air (flash point 40°F/4.4°C). Incompatible with strong alkalies, strong acids, nitrates, strong oxidizers. Dissolves rubber and many plastic materials. Contact with steel may cause slow decomposition. May accumulate static electrical charges, and may cause ignition of its vapors.

ISOPROPYLACETONE (108-10-1) Forms explosive mixture with air (flash point 64°F/18°C). Incompatible with oxidizers, strong acids, aliphatic amines. Dissolves some plastics, resins, and rubber.

ISOPROPYL ALCOHOL (67-63-0) Forms explosive mixture with air (flash point 53°F/12°C). Incompatible with alkaline earth and alkali metals, crotonaldehyde, phosgene, strong acids, amines, ammonia, caustics, strong oxidizers. Attacks some plastics, rubber, and coatings. Reacts with metallic aluminum at high temperatures.

ISOPROPYLAMINE (75-31-0) Forms explosive mixture with air (flash point −35°F/−37°C). A strong base. Reacts with alkali metals, alkaline earth, strong acids (explosive spattering), strong oxidizers (fire and explosions). Attacks some plastics, rubber, and coatings.

ISOPROPYLAMINEDODECYLBENZENESULFONATE (42504-46-1) Incompatible with oxidizers, strong acids.

ISOPROPYLBENZEEN (Dutch) (98-82-8) Forms explosive mixture with air (flash point 99°F/37°C). Incompatible with strong acids, strong oxidizers. Air contact causes formation of cumene hydroperoxide. May be able to form unstable peroxides. Attacks rubber. May accumulate static electrical charges, and may cause ignition of its vapors.

ISOPROPYL BENZENE (98-82-8) Forms explosive mixture with air (flash point 99°F/37°C). Incompatible with strong acids, strong oxidizers. Air contact causes formation of cumene hydroperoxide.

May be able to form unstable peroxides. Attacks rubber. May accumulate static electrical charges, and may cause ignition of its vapors.

ISOPROPYLBENZENE HYDROPEROXIDE (80-15-9) A strong oxidizer; reacts violently with reducers, combustibles. Explosive decomposition may occur above 120°F/50°C. Forms explosive mixture with air (flash point 175°F/79°C). Incompatible with acids, bases, amines, metallic salts of cobalt, copper, lead. May accumulate static electrical charges, and may cause ignition of its vapors.

ISOPROPYLBENZOL (98-82-8) Forms explosive mixture with air (flash point 99°F/37°C). Incompatible with strong acids, strong oxidizers. Air contact causes formation of cumene hydroperoxide. May be able to form unstable peroxides. Attacks rubber. May accumulate static electrical charges, and may cause ignition of its vapors.

ISOPROPYL BENZOL (German) (98-82-8) Forms explosive mixture with air (flash point 99°F/37°C). Incompatible with strong acids, strong oxidizers. Air contact causes formation of cumene hydroperoxide. May be able to form unstable peroxides. Attacks rubber. May accumulate static electrical charges, and may cause ignition of its vapors.

ISOPROPYLCARBINOL (78-83-1) Forms explosive mixture with air (flash point 82°F/28°C). Incompatible with strong acids, strong oxidizers, caustics, aliphatic amines, isocyanates, alkali metals, and alkaline earth. Attacks some plastics, rubber, and coatings. May react with aluminum at high temperatures.

ISOPROPYL CELLOSOLVE (109-59-1) Forms explosive mixture with air (flash point 92°F/33°C). Incompatible with sulfuric acid, perchloric acid, isocyanates, oxidizers.

ISOPROPYLCUMYLHYDROPEROXIDE (26762-93-6) Forms explosive mixture with air (flash point 175°F/79°C). Rapid decomposition may be caused by active metals: aluminum, copper, zinc, and their alloys.

ISOPROPYL CYANIDE (78-82-0) Forms explosive mixture with air (flash point 47°F/8°C). Reacts violently with strong oxidizers.

ISOPROPYL CYCLOHEXANE (696-29-7) Forms explosive mixture with air (flash point 96°F/36°C). Incompatible with strong oxidizers, acids.

ISOPROPYL 2,4-DICHLOROPHENOXY ACETATE (94-11-1) Forms explosive mixture with air (flash point 175°F/79°C). Incompatible with strong oxidizers, strong acids, nitrates. Attacks some paints, rubber, and coatings.

ISOPROPYL EPOXYPROPYL ETHER (4016-08-6) Forms explosive mixture with air (flash point 92°F/33°C). Incompatible with strong oxidizers (fire), strong caustics (polymerization). Air and light causes formation of unstable and explosive peroxides. Attacks some rubbers and plastics.

ISOPROPYL ESTER OF ACETIC ACID (108-21-4) Forms explosive mixture with air (flash point 40°F/4.4°C). Incompatible with strong alkalies, strong acids, nitrates, strong oxidizers. Dissolves rubber and many plastic materials. Contact with steel may cause slow decomposition. May accumulate static electrical charges, and may cause ignition of its vapors.

ISOPROPYL ETHER (108-20-3) Forms explosive mixture with air (flash point −18°F/−28°C). May accumulate static electrical charges, and may cause ignition of its vapors. Air contact causes formation of explosive peroxides that may detonate with heat or shock. Strong oxidizers may cause fire and explosions. Attacks some plastics, rubber, and coatings.

ISOPROPYLFORMIC ACID (79-31-2) Forms explosive mixture with air (flash point 132°F/56°C). Incompatible with sulfuric acid, caustics, ammonia, amines, isocyanates, alkylene oxides, epichlorohydrin. Attacks aluminum and other metals.

ISOPROPYL GLYCIDYL ETHER (4016-08-6) Forms explosive mixture with air (flash point 92°F/33°C). Incompatible with strong oxidizers (fire), strong caustics (polymerization). Air and light causes formation of unstable and explosive peroxides. Attacks some rubbers and plastics.

ISOPROPYL GLYCOL (109-59-1) Forms explosive mixture with air 92°F/33°C). Incompatible with sulfuric acid, perchloric acid, isocyanates, oxidizers.

ISOPROPYLIDENEACETONE (141-79-7) Forms explosive mixture with air (flash point 87°F/31°C). Forms peroxides. Incompatible with strong acids aliphatic amines, alkanolamines, 2-aminoethanol, ethylene diamine, chlorosulfonic acid, oleum, oxidizers. Dissolves some forms of plastics, resins, and rubber. Attacks copper.

4,4′-ISOPROPYLIDENEDIPHENO EPICHLOROHYDRIN RESIN (1675-54-8) Forms explosive mixture with air (flash point 175°F/79°C). Incompatible with strong acids, strong oxidizers. Ethers form peroxides on contact with air and light.

4,4′-ISOPROPYLIDENEDIPHENOL (80-05-7) Combustible. Fine dust can form an explosive mixture with air. Contact with strong oxidizers may cause fire and explosions.

4,4′-ISOPROPYLIDENEDIPHENOL DIGLYCIDYL ETHER (1675-54-8) Forms explosive mixture with air (flash point 175°F/79°C). Incompatible with strong acids, strong oxidizers. Ethers form peroxides on contact with air and light.

ISOPROPYL MERCAPTAN (75-33-2) Forms explosive mixture with air (flash point −30°F/−34°C). Reacts violently with strong oxidizers.

4-ISOPROPYL-1-METHYL BENZENE (99-87-6) Forms explosive mixture with air (flash point 117°F/47°C). Incompatible with nitric acid, strong oxidizers. Attacks and softens rubber. May accumulate static electrical charges, and may cause ignition of its vapors.

ISOPROPYL METHYL KETONE (563-80-4) Forms explosive mixture with air at 43°F/6°C. Reacts violently with strong oxidizers.

ISOPROPYLOXITOL (109-59-1) Forms explosive mixture with air (flash point 92°F/33°C). Incompatible with sulfuric acid, perchloric acid, isocyanates, oxidizers.

ISOPROPYL PERCARBONATE (105-64-6) Combustible solid. Contact with some metals may cause decomposition.

ISOPROPYL PEROXYDICARBONATE (105-64-6) Combustible solid. Contact with some metals may cause decomposition.

2-ISOPROPYL PHENOL (88-69-7) Forms explosive mixture with air (flash point 220°F/104°C). Incompatible with oxidizers, acids.

o-**ISOPROPYL PHENOL** (88-69-7) Forms explosive mixture with air (flash point 220°F/104°C). Incompatible with oxidizers, acids.

ISOPROPYLTHIOL (75-33-2) Forms explosive mixture with air (flash point −30°F/−34°C) Reacts violently with strong oxidizers.

4-ISOPROPYL TOLUENE (99-87-6) Forms explosive mixture with air (flash point 117°F/47°C). Incompatible with nitric acid, strong oxidizers. Attacks and softens rubber. May accumulate static electrical charges, and may cause ignition of its vapors.

p-**ISOPROPYLTOLUENE** (99-87-6) Forms explosive mixture with air (flash point 117°F/47°C). Incompatible with nitric acid, strong oxidizers. Attacks and softens rubber. May accumulate static electrical charges, and may cause ignition of its vapors.

ISOPROPYLTOLUOL (99-87-6) Forms explosive mixture with air (flash point 117°F/47°C). Incompatible with nitric acid, strong oxidizers. Attacks and softens rubber. May accumulate static electrical charges, and may cause ignition of its vapors.

(ISOPROXYMETHYL)OXIRANE (4016-08-6) Forms explosive mixture with air (flash point 92°F/33°C). Incompatible with strong oxidizers (fire), strong caustics (polymerization). Air and light cause formation of unstable and explosive peroxides. Attacks some rubbers and plastics.

ISOTHIOCYANATE de METHYLE (French) (556-61-6) Forms explosive mixture with air (flash point 90°F/32°C). Incompatible with strong acids, caustics, ammonia, amines, amides, alcohols, glycols, caprolactam solution, strong oxidizers.

ISOTHIOCYANIC ACID, METHYL ESTER (556-61-6) Forms explosive mixture with air (flash point 90°F/32°C). Incompatible with strong acids, caustics, ammonia, amines, amides, alcohols, glycols, caprolactam solution, strong oxidizers.

ISOTHIOCYANOMETHANE (556-61-6) Forms explosive mixture with air (flash point 90°F/32°C). Incompatible with strong acids, caustics, ammonia, amines, amides, alcohols, glycols, caprolactam solution, strong oxidizers.

ISOTHIOUREA (62-56-6) Aqueous solution is a base; reacts strongly with acids. Incompatible with nitric acid, hydrogen peroxide, acrolein, metals.

ISOTOX (58-89-9) Not combustible, but may be dissolved in a combustible solvent. If solvent comes in contact with oxidizers, fire and explosions may result.

ISOTRIDECANOL (112-70-9) Incompatible with strong acids, caustics, aliphatic amines, isocyanates, strong oxidizers. Attacks aluminum. May accumulate static electrical charges, and may cause ignition of its vapors.

ISOTRIDECYL ALCOHOL (112-70-9) Incompatible with strong acids, caustics, aliphatic amines, isocyanates, strong oxidizers. Attacks aluminum. May accumulate static electrical charges, and may cause ignition of its vapors.

ISOTRON-11 (75-69-4) Reacts with barium, lithium, sodium, magnesium, titanium. Attacks some plastics, rubber, and coatings.

ISOTRON-12 (75-71-8) Reacts violently with liquid aluminum. Incompatible with chemically active metals. Attacks some plastics, rubber, and coatings.

ISOTRON-22 (75-45-5) Moisture and rust cause slow decomposition, forming toxic gases. Attacks some paints, rubber, and coatings. Thermal decomposition occurs at high temperature with alkalies and alkaline earth metals.

ISOUREA (57-13-6) Heat causes formation of anhydrous ammonia fumes. Contact with oxidizers, nitrates may cause fire and explosions. Contact with chlorinating agents including hypochlorite bleaches may produce explosive nitrogen trichloride. Reacts with nitrosyl perchlorate.

ISOVALERAL (107-83-5) Forms explosive mixture with air (flash point −20°F/−29°C). Reacts violently with strong oxidizers. May accumulate static electrical charges, and may cause ignition of its vapors.

ISOVALERALDEHYDE (107-83-5) Forms explosive mixture with air (flash point −20°F/−29°C). Reacts violently with strong oxidizers. May accumulate static electrical charges, and may cause ignition of its vapors.

ISOVALERIC ALDEHYDE (107-83-5) Forms explosive mixture with air (flash point −20°F/−29°C). Reacts violently with strong oxidizers. May accumulate static electrical charges, and may cause ignition of its vapors.

ISOVALERONE (108-83-8) Incompatible with strong acids, aliphatic amines, strong oxidizers. Attacks plastics, coatings, and rubber. Forms explosive mixture with air (flash point 140°F/60°C).

p-**ISOXAZINE, TETRAHYDRO-** (110-91-8) Forms explosive mixture with air (flash point 98°F/37°C). Incompatible with acids, organic anhydrides, isocyanates, vinyl acetate, acrylates, substituted allyls, alkylene oxides, epichlorohydrin, ketones, aldehydes, alcohols, glycols, phenols, cresols, caprolactam solution, strong oxidizers. Attacks aluminum, copper, lead, tin, zinc, and alloys, and some plastics, rubber, and coatings.

ITAMID (105-60-2) Contact with strong oxidizers may cause fire and explosions.

ITOPAZ (563-12-2) Incompatible with alkaline formulations. Mixtures with magnesium may be explosive.

IVALON (50-00-0) May polymerize unless properly inhibited (usually with methanol). Forms explosive mixture with air (flash point 122°F/50°C). Incompatible with strong acids, amines, strong oxidizers, alkaline materials, nitrogen dioxide, performic acid. Reaction with hydrochloric acid produces bis-chloromethyl ether, a carcinogen.

IVORAN (50-29-3) Incompatible with salts of iron or aluminum, and bases. Do not store in iron containers.

IXODEX (50-29-3) Incompatible with salts of iron or aluminum, and bases. Do not store in iron containers.

- J -

JACUTIN (58-89-9) Not combustible, but may be dissolved in a combustible solvent. If solvent comes in contact with oxidizers, fire and explosions may result.

JAPAN CAMPHOR (76-22-2) Forms explosive mixture with air (flash point 155°F/66°C). Violent, possibly explosive, reaction occurs with strong oxidizers. May accumulate static electrical charges, and may cause ignition of its vapors.

JASMOLIN I or JASMOLIN II (8003-34-7) Strong oxidizers may cause fire and explosions.

JAVELLE WATER (7681-52-9) A strong oxidizer and a strong base. Stability decreases with concentration, heat, light, decrease in pH, and contamination with metals. Incompatible with strong acids, reducing agents, combustible substances; all cause violent reaction, fire and explosions. Contact with amines and ammonia salts produce explosive chloroamines. Corrodes many metals: steels, cast iron, monel, aluminum, brass, iconel.

JAYFLEX DTDP (119-06-2) Incompatible with strong acids, nitrates, oxidizers.

JAYSOL S (64-17-5) Forms explosive mixture with air (flash point 65°F/18°C). May accumulate static electrical charges, and may cause ignition of its vapors. Reactions may be violent with oleum, sulfuric acid, nitric acid, bases, aliphatic amines, isocyanates, oxidizers

JDB-50-T (78-18-2) A strong oxidizer. Reacts violently with combustibles, reducing agents, caustics, ammonia.

JEFFERSOL EB (111-76-2) Forms explosive mixture with air (flash point 143°F/62°C). Reacts violently with strong caustics and strong oxidizers. Attacks some coatings, plastics, and rubber. Attacks metallic aluminum at high temperatures.

JEFFERSOL EE (110-80-5) Forms explosive mixture with air (flash point 120°F/49°C). Strong oxidizers may cause fire and explosions. Attacks some plastics, rubber, and coatings. Able to form peroxides. Incompatible with strong acids, aluminum and its alloys.

JEFFERSOL EM (109-86-4) Forms explosive mixture with air (flash point 103°F/39°C). Heat or oxidizers may cause formation of unstable peroxides. Attacks many metals. Strong oxidizers cause fire and explosions. Strong bases cause decomposition. Attacks some plastics, rubber, and coatings. May accumulate static electrical charges, and may cause ignition of its vapors.

JEFFOX (25322-69-4) Incompatible with sulfuric acid, isocyanates, perchloric acid, oxidizers.

JET A or JET A-1 (8008-20-6) Forms explosive mixture with air (flash point 100°F/43°C). Incompatible with nitric acid, oxidizers; may cause fire and explosions. May rust steel. May accumulate static electrical charges, and may cause ignition of its vapors.

JET FUEL (68476-31-3) Forms explosive mixture in air above 100°F/38°C. Oxidizers may cause fire and explosions. Incompatible with nitric acid. May accumulate static electrical charges, and may cause ignition of its vapors.

JET FUEL: JP-1 (8008-20-6) Forms explosive mixture in air above 100°F/38°C. Oxidizers may cause fire and explosions. Incompatible with nitric acid. May accumulate static electrical charges, and may cause ignition of its vapors.

JET FUEL: JP-3 (8008-20-6) Forms explosive mixture with air (flash point 110°F/43°C). Oxidizers may cause fire and explosions. Incompatible with nitric acid. May accumulate static electrical charges, and may cause ignition of its vapors. May rust steel.

JET FUEL: JP-4 (8008-20-6) Forms explosive mixture with air (flash point −10°F to 30°F/−23°C to −1°C). Oxidizers may cause fire and explosions. Incompatible with nitric acid. May accumulate static electrical charges, and may cause ignition of its vapors. May rust steel.

JET FUEL: JP-5 (8008-20-6) Forms explosive mixture with air (flash point 95–145°F/35–63°C). Oxidizers may cause fire and explosions. Incompatible with nitric acid. May accumulate static electrical charges, and may cause ignition of its vapors. May rust steel.

JEWELER'S ROUGE (1309-37-1) Contact with hydrogen peroxide, ethylene oxide, calcium hypochlorite will cause explosion. Reacts violently with powdered aluminum, hydrazine, hydrogen trisulfide.

JOD (German or Polish) (7553-56-2) Contact with ammonia produces a shock-sensitive compound, which will explode, causing fires. A powerful oxidizer. Reacts violently with reducing agents, combustibles, acetylene, acetaldehyde, antimony, cesium, francium, lithium, potassium, powdered metals, rubidium, phosphorus, powdered metals, sodium, sulfur.

JOOD (Dutch) (7553-56-2) Contact with ammonia produces a shock-sensitive compound, which will explode, causing fires. A powerful oxidizer. Reacts violently with reducing agents, combustibles, acetylene, acetaldehyde, antimony, cesium, francium, lithium, potassium, powdered metals, rubidium, phosphorus, powdered metals, sodium, sulfur.

JOODMETHAAN (Dutch) (74-88-4) Contact with strong oxidizers may cause fire and explosions. Heat light and moisture contribute to instability.

JP-1 (68476-31-3) Forms explosive mixture in air above 100°F/38°C. Oxidizers may cause fire and explosions. Incompatible with nitric acid. May accumulate static electrical charges, and may cause ignition of its vapors. May rust steel.

JP-3 or JP-3 JET FUEL (8008-20-6) Forms explosive mixture with air (flash point 110°F/43°C). Oxidizers may cause fire and explosions. Incompatible with nitric acid. May accumulate static electrical charges, and may cause ignition of its vapors. May rust steel.

JP-4 (8008-20-6) Forms explosive mixture with air (flash point −10°F to 30°F/−23°C to −1°C). Oxidizers may cause fire and explosions. Incompatible with nitric acid. May accumulate static electrical charges, and may cause ignition of its vapors. May rust steel.

JP-5 NAVY FUEL (8008-20-6) Forms explosive mixture with air (flash point 95–145°F/35–63°C). Oxidizers may cause fire and explosions. Incompatible with nitric acid. May accumulate static electrical charges, and may cause ignition of its vapors. May rust steel.

JUDEAN PITCH (8052-42-4) Incompatible with nitric acid, fluorine, strong oxidizers.

JULIN'S CARBON CHLORIDE (118-74-1) Reacts violently with strong oxidizers, formyldimethylamine.

- K -

K-52 (112-80-1) Incompatible with strong oxidizers, perchloric acid, and aluminum powder; may cause explosions. Corrodes aluminum.

KADMIUM (German) (7440-43-9) Air exposure, especially of powdered form, may cause chemical to self-ignite. Reacts violently with strong oxidizers, hydrozoic acid, tellurium. Acid contact produces hydrogen gas. May react with selenium, elemental sulfur, zinc.

KADMIUM CHLORID (German) (10108-64-2) Incompatible with strong oxidizers, elemental sulfur, selenium, tellurium.

KALIUM (7440-09-7) A dangerous explosion hazard. Air contact causes spontaneous ignition. Reacts violently with water. A powerful reducing agent. Reacts violently with oxidizers, organic materials, and many other substances. Forms unstable peroxides under normal conditions of temperature and storage.

KALIUM-CYANID (German) (151-50-8) Incompatible with acids, organic anhydrides, isocyanates, alkylene oxides, epichlorohydrin, aldehydes, alcohols, glycols, phenols, cresols, caprolactam, strong oxidizers, sodium chlorate. Attacks aluminum, copper, zinc in the presence of moisture.

KALIUMCHLORAAT (Dutch) (3811-04-9) A powerful oxidizer. Reacts violently with reducing agents or combustibles, ammonia gas, ammonium salts, organic matter, hydrogen iodide, organic acids, sulfuric acid. Forms explosive mixtures with metallic powders, ammonium chloride, organic solids including agricultural materials.

KALIUMCHLORAT (German) (3811-04-9) A powerful oxidizer. Reacts violently with reducing agents or combustibles, ammonia gas, ammonium salts, organic matter, hydrogen iodide, organic acids, sulfuric acid. Forms explosive mixtures with metallic powders, ammonium chloride, organic solids, including agricultural materials.

KALIUMDICHROMAT (German) (7778-50-9) Powdered combustibles will ignite. Strong acids produce toxic vapors.

KALIUMHYDROXID (German) (1310-58-3) Aqueous solution is a strong base. Reacts violently with acids, organic anhydrides, isocyanates, alkylene oxides, epichlorohydrin, aldehydes, alcohols, glycols, phenols, cresols, caprolactam solution. Dissolves in water, producing high heat, caustic fumes, and flammable hydrogen gas. Also reacts with halogenated hydrocarbons, maleic anhydride. Attacks metals such as aluminum, tin, lead, and zinc, producing flammable hydrogen gas.

KALIUMHYDROXYDE (Dutch) (1310-58-3) Aqueous solution is a strong base. Reacts violently with acids, organic anhydrides, isocyanates, alkylene oxides, epichlorohydrin, aldehydes, alcohols, glycols, phenols, cresols, caprolactam solution. Dissolves in water, producing high heat, caustic fumes, and flammable hydrogen gas. Also reacts with halogenated hydrocarbons, maleic anhydride. Attacks metals such as aluminum, tin, lead, and zinc, producing flammable hydrogen gas.

KALIUMPERMANGANAAT (Dutch) (7722-64-7) A strong oxidizer. Forms heat- and shock-sensitive compound with sulfuric acid. Reacts violently with combustibles or reducing agents. Reacts explosively with acetic acid, acetic anhydride, anhydrous ammonia, hydrogen peroxide, glycerol, hydroxylamine, organic matter, powdered sulfur, etc. Incompatible with nitric acid, producing toxic chlorine fumes. A dangerous fire and explosion hazard; isolate from all other materials.

KALIUMPERMANGANAT (German) (7722-64-7) A strong oxidizer. Forms heat- and shock-sensitive compound with sulfuric acid. Reacts violently with combustibles or reducing agents. Reacts explosively with acetic acid, acetic anhydride, anhydrous ammonia, hydrogen peroxide, glycerol, hydroxylamine, organic matter, powdered sulfur, etc. Incompatible with nitric acid, producing toxic chlorine fumes. A dangerous fire and explosion hazard; isolate from all other materials.

KALPHOS (56-38-2) Combustible liquid. Mixtures with endrin may be explosive. Strong oxidizers may cause fire and explosions. Attacks some plastics, rubber, and coatings.

KAM 1000 (57-11-4) Incompatible with strong oxidizers, sulfuric acid, caustics, ammonia, amines, isocyanates, alkylene oxides, epichlorohydrin. Attacks chemically active metals.

KAM 2000 (57-11-4) Incompatible with strong oxidizers, sulfuric acid, caustics, ammonia, amines, isocyanates, alkylene oxides, epichlorohydrin. Attacks chemically active metals.

KAMFOCHLOR (8001-35-2) Reacts with oxidizers, with a risk of fire or explosions. Attacks metals in the presence of moisture.

KAMPFER (German) (76-22-2) Forms explosive mixture with air (flash point 155°F/66°C). Violent, possibly explosive, reaction occurs with strong oxidizers. May accumulate static electrical charges, and may cause ignition of its vapors.

KANECHLOR (1336-36-3) Incompatible with strong oxidizers, strong acids.

KANECHLOR 300 (37353-63-2) (generic CAS for PCBs 1336-36-3). Incompatible with strong oxidizers, strong acids.

KANECHLOR 400 (12737-87-0) (generic CAS for PCBs 1336-36-3). Incompatible with strong oxidizers, strong acids.

KANECHLOR 500 (37317-412) (generic CAS for PCBs 1336-36-3). Incompatible with strong oxidizers, strong acids.

KAPROLIT (105-60-2) Contact with strong oxidizers may cause fire and explosions.

KARBARYL (Polish) (63-25-2) Incompatible with strong oxidizers, strongly alkaline pesticides.

KARBASPRAY (63-25-2) Incompatible with strong oxidizers, strongly alkaline pesticides.

KARBATOX (63-25-2) Incompatible with strong oxidizers, strongly alkaline pesticides.

KARBOFOS (121-75-5) Incompatible with strong oxidizers, magnesium, alkaline pesticides. Attacks metals, some plastics, rubber, and coatings.

KARBOSEP (63-25-2) Incompatible with strong oxidizers, strongly alkaline pesticides.

KARLAN (299-84-3) Temperatures above 300°F/150°C may cause explosive decomposition. Contact with strong oxidizers may cause fire and explosions. Attacks some plastics, rubber, and coatings.

KARMEX (330-54-1) Hydrolyzes in fairly strong acids.

KARMEX DIURON HERBICIDE (330-54-1) Hydrolyzes in fairly strong acids.

KARMEX DW (330-54-1) Hydrolyzes in fairly strong acids.

KARSAN (50-00-0) May polymerize unless properly inhibited (usually with methanol). Forms explosive mixture with air (flash point 122°F/50°C). Incompatible with strong acids, amines, strong oxidizers, alkaline materials, nitrogen dioxide, performic acid. Reaction with hydrochloric acid produces bis-chloromethyl ether, a carcinogen.

KARSTENITE (7778-18-9) Contact with diazomethane, aluminum, phosphorus may cause explosion.

KATCHUNG OIL (8002-03-7) Incompatible with strong acids, oxidizers.

KAUTSCHIN (138-86-3) Forms explosive mixture with air (flash point 115°F/46°C). Strong oxidizers may cause fire and explosions.

KAYAFUME (74-83-9) Incompatible with strong oxidizers, aluminum, dimethylsulfoxide, ethylene oxide, water. Attacks zinc, magnesium, alkali metals, and their alloys.

KAYAZINON (333-41-5) Incompatible with water, copper-containing compounds, oxidizers, acids, or bases.

KAYAZOL (333-41-5) Incompatible with water, copper-containing compounds, oxidizers, acids, or bases.

KAZOE (26628-22-8) Incompatible with acids, with some metals (i.e, lead, copper, silver, mercury). Forms explosion-sensitive compounds.

KCN (151-50-8) Incompatible with acids, organic anhydrides, isocyanates, alkylene oxides, epichlorohydrin, aldehydes, alcohols, glycols, phenols, cresols, caprolactam, strong oxidizers, sodium chlorate. Attacks aluminum, copper, zinc in the presence of moisture.

KELENE (75-00-3) Flammable gas. Reacts slowly with water, producing hydrogen chloride gas. Contact with moisture produces hydrochloric acid. May accumulate static electrical charges, and may cause ignition of its vapors. Forms explosive mixture with air (flash point −58°F/−55°C). Contact with aluminum, lithium, magnesium, sodium, potassium, zinc may cause fire and explosions. Attacks some plastics and rubber.

KEL F MONOMER (79-38-9) Flammable gas. Forms explosive mixture with air (flash point −18°F/−28°C). Reacts violently with oxidizers. Contact with ethylene may cause explosive polymerization.

KELTHANE (115-32-2) Incompatible with strong acids, caustics, aliphatic amines, isocyanates.

p,p-KELTHANE (115-32-2) Incompatible with strong acids, caustics, aliphatic amines, isocyanates.

KELTHANETHANOL (115-32-2) Incompatible with strong acids, caustics, aliphatic amines, isocyanates.

KENAPON (75-99-0) Corrosive to iron, aluminum, and copper.

KERALYT (69-72-7) Reacts with strong oxidizers, ethyl nitrite, iodine, iron salts, lead acetate.

KEROSENE (8008-20-6) Forms explosive mixture in air above 100°F/38°C. Oxidizers may cause fire and explosions. Incompatible with nitric acid. May accumulate static electrical charges, and may cause ignition of its vapors. May rust steel.

KEROSENE, HEAVY (8008-20-6) Forms explosive mixture with air (flash point 95–145°F/35–63°C). Oxidizers may cause fire and explosions. Incompatible with nitric acid. May accumulate static electrical charges, and may cause ignition of its vapors. May rust steel.

KEROSINE (8008-20-6) Forms explosive mixture in air above 100°F/38°C. Oxidizers may cause fire and explosions. Incompatible with nitric acid. May accumulate static electrical charges, and may cause ignition of its vapors. May rust steel.

KETEN or KETENE (463-51-4) Reacts vigorously with water and a wide variety of organic compounds. Forms explosive compound when mixed with hydrogen peroxide. Can dimerize to diketene even at low temperatures. Diketene forms an explosive mixture with air (flash point 90°F/32°C). Reacts violently with oxidizers; forms unstable and explosive peroxides, with a chance of violent polymerization.

KETOETHYLENE (463-51-4) Reacts vigorously with water and a wide variety of organic compounds. Forms explosive compound when mixed with hydrogen peroxide. Can dimerize to diketene even at low temperatures. Diketene forms an explosive mixture with air (flash point 90°F/32°C). Reacts violently with oxidizers; forms unstable and explosive peroxides, with a chance of violent polymerization.

2-KETOHEPTANE (110-43-0) Forms explosive mixture with air (flash point 102°F/39°C). Incompatible with strong acids, alkalies, aliphatic amines. Attacks some plastics and rubber.

KETOHEXAMETHYLENE (108-94-1) Forms explosive mixture with air (flash point 111°F/44°C). May accumulate static electrical charges, and may cause ignition of its vapors. Attacks red metals and lead. Incompatible with amines, sulfuric acid, nitric acid, oxidizers, strong acids, aliphatic amines. Dissolves plastics, resins, and rubber.

2-KETOHEXAMETHYLENIMINE (105-60-2) Contact with strong oxidizers may cause fire and explosions.

KETONE (67-64-1) Forms explosive mixture with air (flash point −4°F/−20°C). Reacts violently with chloroform. Incompatible with strong acids, aliphatic amines, chloroform, chromic anhydride, chromyl chloride, hexachloromelamine, hydrogen peroxide, nitrosyl chloride, nitrosyl perchlorate, nitryl perchlorate, permonosulfuric acid, potassium *tert*-butoxide. Forms unstable and explosive peroxides with strong oxidizers. May accumulate static electrical charges, and may cause ignition of its vapors. Dissolves most rubber, resins, and plastics.

KETONE, BUTYL METHYL (591-78-6) Forms explosive mixture with air (flash point 77°F/25°C). Strong oxidizers may cause fire and explosions. Dissolves some plastics, resins, and rubber.

KETONE, DIMETHYL (67-64-1) Forms explosive mixture with air (flash point −4°F/−20°C). Reacts violently with chloroform. Incompatible with strong acids, aliphatic amines, chloroform, chromic anhydride, chromyl chloride, hexachloromelamine, hydrogen peroxide, nitrosyl chloride, nitrosyl perchlorate, nitryl perchlorate, permonosulfuric acid, potassium *tert*-butoxide. Forms unstable and explosive peroxides with strong oxidizers. May accumulate static electrical charges, and may cause ignition of its vapors. Dissolves most rubber, resins, and plastics.

KETONE, ETHYL METHYL (78-93-3) Forms explosive mixture with air (flash point 16°F/−9°C). Extremely flammable. Incompatible with sulfuric acid, nitric acid, aliphatic amines, strong oxidizers, potassium *tert*-butoxide, 2-propanol, chlorosulfonic acid, oleum (hydrogen peroxide + nitric acid).

KETONE, HEPTYL METHYL (821-55-6) Forms explosive mixture with air (flash point 140°F/60°C). Incompatible with strong acids, aliphatic amines, oxidizers.

KETONE METHYL PHENYL (98-86-2) Incompatible with strong acids, aliphatic amines, oxidizers. Forms explosive mixture with air (flash point 170°F/77°C).

KETONE PROPANE (67-64-1) Forms explosive mixture with air (flash point −4°F/−20°C). Reacts violently with chloroform. Incompatible with strong acids, aliphatic amines, chloroform, chromic anhydride, chromyl chloride, hexachloromelamine, hydrogen peroxide, nitrosyl chloride, nitrosyl perchlorate, nitryl perchlorate, permonosulfuric acid, potassium *tert*-butoxide. Forms unstable and explosive peroxides with strong oxidizers. May accumulate static electrical charges, and may cause ignition of its vapors. Dissolves most rubber, resins, and plastics.

KETONOX (1338-23-4) Forms explosive mixture with air (flash point 125°F/52°C). Explosive decomposition occurs above 176°F/80°C. Pure substance is shock-sensitive. Strong oxidizer. Reacts violently with strong acids, strong bases, reducing agents, combustible substances, organic materials, oxides of heavy metals, salts, trace contaminants, amines. May accumulate static electrical charges, and may cause ignition of its vapors.

beta-KETOPROPANE (67-64-1) Forms explosive mixture with air (flash point −4°F/−20°C). Reacts violently with chloroform. Incompatible with strong acids, aliphatic amines, chloroform, chromic anhydride, chromyl chloride, hexachloromelamine, hydrogen peroxide, nitrosyl chloride, nitrosyl perchlorate, nitryl perchlorate, permonosulfuric acid, potassium *tert*-butoxide. Forms unstable and explosive peroxides with strong oxidizers. May accumulate static electrical charges, and may cause ignition of its vapors. Dissolves most rubber, resins, and plastics.

2-KETO-1,7,7-TRIMETHYLNORCAMPHANE (76-22-2) Forms explosive mixture with air (flash point 155°F/66°C). Violent, possibly explosive, reaction occurs with strong oxidizers. May accumulate static electrical charges, and may cause ignition of its vapors.

KETTLE RENDERED LARD (N/A) Incompatible with strong acids, oxidizers.

K-FLEX DP (94-51-9) Incompatible with strong acids, nitrates, oxidizers.

KIEZELFLUORWATERSTOFZUUR (Dutch) (95-52-3) Forms explosive mixture with air (flash point 55°F/13°C). Reacts violently with strong oxidizers.

KILEX-3 (8002-26-4) Water contact causes foaming. Strong oxidizers may cause fire and explosions. Attacks some plastics, rubber, and coatings.

KILLAX (107-49-3) Decomposes above 300°F/150°C, producing flammable ethylene gas. Strong oxidizers may cause fire and explosions. Attacks some plastics, rubber, and coatings.

KILLGERM DETHLAC INSECTICIDAL LACQUER (60-57-1) Incompatible with concentrated mineral acids, acid catalysts, acid oxidizing agents, phenols, reactive metals.

KILL KANTZ (86-88-4) Strong oxidizers may cause fires and explosion. Also reacts with silver nitrate.

KILLMASTER (2921-88-) Incompatible with strong acids. Hydrolyzes from acid or alkaline solutions.

KILMITE 40 (107-49-3) Decomposes above 300°F/150°C, producing flammable ethylene gas. Strong oxidizers may cause fire and explosions. Attacks some plastics, rubber, and coatings.

KILRAT (1314-84-7) Incompatible with oxidizers, acids, water.

KING'S GOLD (1303-33-9) Water contact produces hydrogen sulfide. Incompatible with oxidizers, acids, halogens.

KING'S GREEN (12002-03-8) Contact with strong oxidizers may cause fire and explosions.

KING'S YELLOW (1303-33-9) Water contact produces hydrogen sulfide. Incompatible with oxidizers, acids, halogens.

KLEER-LOT (61-82-5) Substance acts as a weak base to produce salts in contact with acids. Corrosive to iron, aluminum, copper, and copper alloys.

KLOREX (7775-09-9) A powerful oxidizer; reacts violently with reducing agents and combustible matter. Explosions may be caused by contact with ammonia salts, carbon, oils, metal sulfides, nitrobenzene, powdered metals, sugar. Contact with strong acids produces carbon dioxide. Forms shock-sensitive mixtures with some organic materials. Solution (50%) decomposes at 300°F/149°C, liberating oxygen.

KOBALT CHLORID (German) (7646-79-9) Incompatible with oxidizers, metals (especially sodium or potassium), acetylene.

KODAFLEX DOA (103-23-1) Incompatible with strong acids, nitrates, oxidizers.

K-O-H (1310-58-3) Aqueous solution is a strong base. Reacts violently with acids, organic anhydrides, isocyanates, alkylene oxides, epichlorohydrin, aldehydes, alcohols, glycols, phenols, cresols, caprolactam solution. Dissolves in water, producing high heat, caustic fumes, and flammable hydrogen gas. Also reacts with halogenated hydrocarbons, maleic anhydride. Attacks metals such as aluminum, tin, lead, and zinc, producing flammable hydrogen gas.

KOHLENDIOXYD (German) (124-38-9) Incompatible with acrylaldehyde, amines, anhydrous ammonia, cesium monoxide, lithium, metal dusts, potassium, sodium, sodium carbide, sodium–potassium alloy, sodium peroxide, titanium.

KOHLENDISULFID (SCHWEFELKOHLENSTOFF) (German) (75-15-0) Highly reactive. Contact with many substances can cause fire and explosions. Forms explosive mixture with air (Flash point −22°F/−30°C). Shock can cause explosive decomposition. Incompatible with alkali metals, aliphatic amines, alkanolamines, aluminum, azides, chlorine monoxide, combustible substances, ethylene diamine, ethyleneimine, lead azide, lithium azide, nitric oxide, nitrogen dioxide, potassium, potassium azide, reducing agents, rubidium azide, sodium azide, zinc.

KOHLENMONOXID (German) (630-08-0) Forms extremely explosive mixture with air. Reacts violently with strong oxidizers.

KOHLENSAEURE (German) (124-38-9) Incompatible with acrylaldehyde, amines, anhydrous ammonia, cesium monoxide, lithium, metal dusts, potassium, sodium, sodium carbide, sodium–potassium alloy, sodium peroxide, titanium.

KOKOTINE (58-89-9) Not combustible, but may be dissolved in a combustible solvent. If solvent comes in contact with oxidizers, fire and explosions may result.

KOLOFOG (7704-34-9) Combustible solid. Liquid contact produces sulfur dioxide with air. Reacts violently with strong oxidizers. Forms explosive, shock-sensitive, or pyrophoric mixtures with ammonia, ammonium nitrate, bromates, calcium carbide, charcoal, chlorates, hydrocarbons, iodates, iron. Reacts violently with halogen compounds, sodium, tin, uranium, and other compounds. Attacks steel when moist. May accumulate static electrical charges, and may cause ignition of its vapors.

KOLOSPRAY (7704-34-9) Combustible solid. Liquid contact produces sulfur dioxide with air. Reacts violently with strong oxidizers. Forms explosive, shock-sensitive, or pyrophoric mixtures with ammonia, ammonium nitrate, bromates, calcium carbide, charcoal, chlorates, hydrocarbons, iodates, iron. Reacts violently with halogen compounds, sodium, tin, uranium, and other compounds. Attacks steel when moist. May accumulate static electrical charges, and may cause ignition of its vapors.

KONLAX (119-36-8) Forms explosive mixture with air (flash point 205°F/96°C). Incompatible with strong acids, nitrates, oxidizers.

KOOLMONOXYDE (Dutch) (630-08-0) Forms extremely explosive mixture with air. Reacts violently with strong oxidizers.

KOOLSTOFDISULFIDE (ZWAVELKOOLSTOF) (Dutch) (75-15-0) Highly reactive. Contact with many substances can cause fire and explosions. Forms explosive mixture with air (flash point −22°F/−30°C). Shock can cause explosive decomposition. Incompatible with alkali metals, aliphatic amines, alkanolamines, aluminum, azides, chlorine monoxide, combustible substances, ethylene diamine, ethyleneimine, lead azide, lithium azide, nitric oxide, nitrogen dioxide, potassium, potassium azide, reducing agents, rubidium azide, sodium azide, zinc.

KOOLSTOFOXYCHLORIDE (Dutch) (75-44-5) Incompatible with water, producing hydrochloric acid. Reacts violently with strong oxidizers, anhydrous ammonia, isopropyl alcohol, and chemically active metals. Forms shock-sensitive material with potassium. Attacks most metal in moist conditions.

KOP-THIODAN (115-29-7) Hydrolyzed by acids and alkalies. Corrosive to iron.

KOP-THION (121-75-5) Incompatible with strong oxidizers, magnesium, alkaline pesticides. Attacks metals, some plastics, rubber, and coatings.

KOPFUME (106-93-4) Reacts with chemically active metals, liquid ammonia, strong oxidizers. Heat and light cause slow decomposition. Attacks some plastics and rubber.

KOPSOL (50-29-3) Incompatible with salts of iron or aluminum, and bases. Do not store in iron containers.

KORAX (600-25-9) Contact with strong oxidizers may cause fire and explosions. Attacks some plastics, rubber, and coatings.

KORAX 6 (600-25-9) Contact with strong oxidizers may cause fire and explosions. Attacks some plastics, rubber, and coatings.

KORLAN (299-84-3) Temperatures above 300°F/150°C may cause explosive decomposition. Contact with strong oxidizers may cause fire and explosions. Attacks some plastics, rubber, and coatings.

KOSATE (119-36-8) Forms explosive mixture with air (flash point 205°F/96°C). Incompatible with strong acids, nitrates, oxidizers.

KOYOSIDE (15096-52-3) Contact with strong acids may produce toxic and corrosive hydrogen fluoride gas.

KRECALVIN (62-73-7) Attacks some plastics, rubber, and coatings.

KREGASAN (137-26-8) Combustible solid (flash point 192°F/89°C). Strong oxidizers may cause fire and explosions; contact with strong acid or oxidizable materials produces toxic gases.

KRESOLE (German) (1319-77-3) Incompatible with strong acids, oxidizers, alkalies, aliphatic amines, amides, chlorosulfonic acid, oleum. Liquid attacks some plastics, coatings, and rubber.

KRESOLEN (Dutch) (1319-77-3) Incompatible with strong acids, oxidizers, alkalies, aliphatic amines, amides, chlorosulfonic acid, oleum. Liquid attacks some plastics, coatings, and rubber.

KREZOL (Polish) (1319-77-3) Incompatible with strong acids, oxidizers, alkalies, aliphatic amines, amides, chlorosulfonic acid, oleum. Liquid attacks some plastics, coatings, and rubber.

KROTILINE (94-75-7) Decomposes in sunlight. Incompatible with strong oxidizers; may cause fire and explosions.

KROTONALDEHYD (Czech) (4170-30-3) A strong reducing agent. Forms explosive mixture with air (flash point 55°F/12.8°C). Readily converted by oxygen to peroxides and acids; heat or contact with many other substances may cause polymerization. Incompatible with strong oxidizers, strong acids including nonoxidizing mineral acids, ammonia, aliphatic amines, aromatic amines, 1,3-butadiene, strong bases. Liquid attacks some plastics, rubber, and coatings.

KRYOCIDE (15096-52-3) Contact with strong acids may produce toxic and corrosive hydrogen fluoride gas.

KRYOLITH (German) (15096-52-3) Contact with strong acids may produce toxic and corrosive hydrogen fluoride gas.

KRYSID (86-88-4) Strong oxidizers may cause fires and explosion. Also reacts with silver nitrate.

KRYSID PI (86-88-4) Strong oxidizers may cause fires and explosion. Also reacts with silver nitrate.

KUMANDER (81-81-2) Strong oxidizers may cause fire and explosions.

KUMULUS (7704-34-9) Combustible solid. Liquid contact produces sulfur dioxide with air. Reacts violently with strong oxidizers. Forms explosive, shock-sensitive, or pyrophoric mixtures with ammonia, ammonium nitrate, bromates, calcium carbide, charcoal, chlorates, hydrocarbons, iodates, iron. Reacts violently with halogen compounds, sodium, tin, uranium, and other compounds. Attacks steel when moist. May accumulate static electrical charges, and may cause ignition of its vapors.

KUPPERSULFAT (German) (7758-98-7) Aqueous solution is an acid. Incompatible with strong bases, hydroxylamine, magnesium.

KURAN (93-72-1) A powerful oxidizer; reacts violently with reducing agents, combustibles.

KURON (93-72-1) A powerful oxidizer; reacts violently with reducing agents, combustibles.

KUROSAL or KUROSAL G (93-72-1) A powerful oxidizer; reacts violently with reducing agents, combustibles.

KUSA-TOHRUKUSATOL (7775-09-9) A powerful oxidizer; reacts violently with reducing agents and combustible matter. Explosions may be caused by contact with ammonia salts, carbon, oils, metal sulfides, nitrobenzene, powdered metals, sugar. Contact with strong acids produces carbon dioxide. Forms shock-sensitive mixtures with some organic materials. Solution (50%) decomposes at 300°F/149°C, liberating oxygen.

KWELL (58-89-9) Not combustible, but may be dissolved in a combustible solvent. If solvent comes in contact with oxidizers, fire and explosions may result.

KWIK (Dutch) (7439-97-6) Contact with acetylene, acetylene products, ammonia gases can produce shock-sensitive solids that can initiate fires with combustibles. Reacts violently with boron phosphodiiodide, chlorine, chlorine dioxide, methyl azide. Attacks copper and copper alloys.

KWIK-KIL (57-24-9) Contact with strong oxidizers may cause fire and explosions.

KWIT (563-12-2) Incompatible with alkaline formulations. Mixtures with magnesium may be explosive.

KYANID SODNY (Czech) (143-33-9) Incompatible with acids, organic anhydrides, isocyanates, alkylene oxides, epichlorohydrin, aldehydes, alcohols, glycols, phenols, cresols, caprolactam solution, strong oxidizers, nitrates, nitrites. Attacks aluminum, copper, zinc.

KYPCHLOR (57-74-9) Contact with strong oxidizers may cause fire and explosions. Attacks some plastics, rubber, and coatings.

KYPFOS (121-75-5) Incompatible with strong oxidizers, magnesium, alkaline pesticides. Attacks metals, some plastics, rubber, and coatings.

KYPTHION (56-38-2) Combustible liquid. Mixtures with endrin may be explosive. Strong oxidizers may cause fire and explosions. Attacks some plastics, rubber, and coatings.

KYSELINA ADIPOVA (Czech), MOLTEN ADIPIC ACID (124-04-9) May accumulate static electrical charges, and may cause ignition of its vapors. Contact with strong oxidizers may cause fire and explosions.

KYSELINA FUMAROVA (Czech) (110-17-8) Reacts with strong oxidizers. Incompatible with sulfuric acid, caustics, ammonia, amines, isocyanates, alkylene oxides, epichlorohydrin.

KYSELINA STAVELOVA (Czech) (144-62-7) Silver compounds may form explosive salt. Incompatible with caustics, furfuryl alcohol, oxidizers, mercury, silver, sodium chlorite, sodium hypochlorite.

KYSELINA TRICHLOISOKYANUROVA (Czech) (87-90-1) A powerful oxidizer. Forms explosive material with nitrogen compounds. Contact with organic materials or reducing agents may cause fire.

- L -

L-310 (8001-26-1) Oxidizers may cause fire and explosions. Incompatible with nitric acid. May accumulate static electrical charges, and may cause ignition of its vapors.

LACTATE d'ETHYLE (French) (97-64-3) Fire or explosion hazard (flash point 115°F/46°C). Strong oxidizers may cause fire and explosions.

LACTIC ACID (598-82-3) Incompatible with strong acids. Aqueous solution reacts with bases.

LD-LACTIC ACID (598-82-3) Incompatible with strong acids. Aqueous solution reacts with bases.

LACTIC ACID, BUTYL ESTER (138-22-7) Forms explosive mixture with air (flash point 160°F/71°C). Incompatible with strong oxidizers, strong bases, acids.

LACTIC ACID, ETHYL ESTER (97-64-3) Fire or explosion hazard (flash point 115°F/46°C). Strong oxidizers may cause fire and explosions.

LACTOBARYT (7727-43-7) Explosions may result from contact with aluminum in the presence of heat. Incompatible with potassium, phosphorus.

LAH (16853-85-3) Flammable solid. Can ignite spontaneously in moist air or heat above 250°F/121°C. A strong reducing agent. Reacts violently with oxidizers, alcohols, acids, dimethylether, tetrahydrofuran.

LAMBETH (9003-07-0) Incompatible with strong acids, strong oxidizers, chlorine, potassium permanganate.

LAMP BLACK (1333-86-4) Dust contact can produce an explosive mixture in air. Oxidizers may cause fire and explosions.

LANADIN (79-01-6) Contact with caustics produces a toxic and flammable gas. Reacts violently with chemically active metals. Contact with aluminum may produce a violent, self-accelerating polymerization reaction. Incompatible with acids, organic anhydrides, isocyanates, alkylene oxides, aldehydes, alcohols, glycols, phenols, cresols, caprolactam solution, epichlorohydrin, nitrogen tetroxide, metal powders, oxygen. May accumulate static electrical charges, and may cause ignition of its vapors.

LANARKITE (7446-14-2) Reacts violently with potassium. Incompatible with aluminum, magnesium.

LANSTAN (600-25-9) Contact with strong oxidizers may cause fire and explosions. Attacks some plastics, rubber, and coatings.

LAPIS INFERNALIS (7761-88-8) Contact with acetylene produces shock-sensitive material. Ammonia contact produces compounds that are explosive when dry. Hydrogen peroxide causes violent decomposition to oxygen gas. Attacks some plastics, rubber, and coatings.

LARD (N/A) Incompatible with strong acids, oxidizers.

LARD OIL, ANIMAL or LARD OIL, COMMERCIAL or LARD OIL, MINERAL or LARD OIL, PURE (N/A) Incompatible with strong acids, oxidizers.

LARVACIDE (76-06-2) Can be self-reactive. Fast heating, shock, alkali metals or alkaline earth may cause explosions. A strong oxidizer; reacts violently with reducing agents, aniline in presence of heat, alcoholic sodium hydroxide, combustible substances, sodium methoxide, propargyl bromide. Liquid attacks some plastics, rubber, and coatings.

LATEX or LATEX, LIQUID SYNTHETIC (9016-00-6) Incompatible with sulfuric acid, isocyanates.

LAUGHING GAS (10024-97-2) May form explosive mixture with flammable gases. Nonflammable but supports combustion; as temperature increases it becomes both a strong oxidizer and self-reactive. High temperatures may cause explosive decomposition. Incompatible with aluminum, anhydrous ammonia (may cause explosion), boron, carbon monoxide, hydrazine, hydrogen, hydrogen sulfide, lithium hydride, phosphine, sodium.

LAUREL CAMPHOR (76-22-2) Forms explosive mixture with air (flash point 155°F/66°C). Violent, possibly explosive, reaction occurs with strong oxidizers. May accumulate static electrical charges, and may cause ignition of its vapors.

LAURIC ACID (143-07-7) Incompatible with sulfuric acid, bases, ammonia, amines, alkylene oxide, epichlorohydrin.

LAURIC ALCOHOL (112-53-8) Incompatible with strong acids, caustics, aliphatic amines, isocyanates, oxidizers.

LAUROSTEARIC ACID (143-07-7) Incompatible with sulfuric acid, bases, ammonia, amines, alkylene oxide, epichlorohydrin.

LAUROX (105-74-8) A powerful oxidizer. May ignite or explode spontaneously when mixed with combustible materials or rapidly heated.

LAUROYL PEROXIDE (105-74-8) A powerful oxidizer. May ignite or explode spontaneously when mixed with combustible materials or rapidly heated.

LAURYDOL (105-74-8) A powerful oxidizer. May ignite or explode spontaneously when mixed with combustible materials or rapidly heated.

LAURYL ALCOHOL (112-53-8) Incompatible with strong acids, caustics, aliphatic amines, isocyanates, oxidizers.

n-LAURYL ALCOHOL (112-53-8) Incompatible with strong acids, caustics, aliphatic amines, isocyanates, oxidizers.

LAURYLBENZENE (123-01-3) Incompatible with nitric acid, strong oxidizers; poses risk of fire or explosions.

LAURYLBENZENESULFONIC ACID (27176-87-0) Forms explosive mixture with air (flash point 100°F/38°C). Strong oxidizers may cause fire and explosions. Do not store in carbon steel or aluminum.

LAURYL MERCAPTAN (112-55-0) Contact with strong oxidizers may cause fire and explosions.

LAURYL METHACRYLATE (142-90-5) Forms explosive mixture with air (flash point 230°F/110°C). Oxidizers or reducing agents may cause polymerization.

LAUXTOL (87-86-5) Hot water causes decomposition, producing hydrochloric acid. Strong oxidizers may cause fire and explosions.

LAUXTOL A (87-86-5) Hot water causes decomposition, producing hydrochloric acid. Strong oxidizers may cause fire and explosions.

LAVATAR (8007-45-2) Forms explosive mixture with air (flash point > 60°F/15°C). Incompatible with strong acids, nitrates.

LAWN-KEEP (94-75-7) Decomposes in sunlight. Incompatible with strong oxidizers; may cause fire and explosions.

LAWRENCITE (7758-94-3) Solution attacks metals. Contact with ethylene oxide may initiate polymerization. Reacts with potassium or sodium mixtures produces impact-sensitive materials.

LAXINATE (119-36-8) Forms explosive mixture with air (flash point 205°F/96°C). Incompatible with strong acids, nitrates, oxidizers.

LEAD (7439-92-1) Powder is incompatible with ammonium nitrate.

LEAD ACETATE (301-04-2) Contact with strong acids produces acetic acid. Incompatible with bases, ammonia, amines, cresols, isocyanates, alkylene oxides, epichlorohydrin, phenols, sulfites. Reacts violently with bromates, strong oxidizers.

LEAD(2+) ACETATE (301-04-2) Contact with strong acids produces acetic acid. Incompatible with bases, ammonia, amines, cresols, isocyanates, alkylene oxides, epichlorohydrin, phenols, sulfites. Reacts violently with bromates, strong oxidizers.

LEAD(4+) ACETATE (546-67-8) Incompatible with water, produces lead dioxide and acetic acid. Corrodes metal in the presence of moisture.

LEAD(II) ACETATE (301-04-2) Contact with strong acids produces acetic acid. Incompatible with bases, ammonia, amines, cresols, isocyanates, alkylene oxides, epichlorohydrin, phenols, sulfites. Reacts violently with bromates, strong oxidizers.

LEAD(IV) ACETATE or LEAD(4+) ACETATE (546-67-8) Incompatible with water, producing lead dioxide and acetic acid. Corrodes metal in the presence of moisture.

LEAD ACETATE TRIHYDRATE (301-04-2) Contact with strong acids produces acetic acid. Incompatible with bases, ammonia, amines, cresols, isocyanates, alkylene oxides, epichlorohydrin, phenols, sulfites. Reacts violently with bromates, strong oxidizers.

LEAD ACETATE(II), TRIHYDRATE (301-04-2) Contact with strong acids produces acetic acid. Incompatible with bases, ammonia, amines, cresols, isocyanates, alkylene oxides, epichlorohydrin, phenols, sulfites. Reacts violently with bromates, strong oxidizers.

LEAD ACID ARSENATE (7784-40-9) Contact with strong oxidizers may cause fire and explosions.

LEAD ARSENATE (7784-40-9) Contact with strong oxidizers may cause fire and explosions.

LEAD BOTTOMS (7446-14-2) Reacts violently with potassium. Incompatible with aluminum, magnesium.

LEAD CARBONATE (598-63-0) Incompatible with acids, fluorine, magnesium, hot water.

LEAD CHLORIDE (7758-95-4) Reacts violently with calcium when heated.

LEAD(2+) CHLORIDE or LEAD(II) CHLORIDE (7758-95-4) Reacts violently with calcium when heated.

LEAD CHROMATE (7758-97-6) Incompatible with combustible materials, hydrazine, sulfur.

LEAD(2+) CHROMATE or LEAD(II) CHROMATE (7758-97-6) Incompatible with combustible materials, hydrazine, sulfur.

478

LEAD DIACETATE (301-04-2) Contact with strong acids produces acetic acid. Incompatible with bases, ammonia, amines, cresols, isocyanates, alkylene oxides, epichlorohydrin, phenols, sulfites. Reacts violently with bromates, strong oxidizers.

LEAD DICHLORIDE (7758-95-4) Reacts violently with calcium when heated.

LEAD DIFLUORIDE (7783-46-2) Reacts violently with fluorine, calcium carbide.

LEAD DINITRATE (10099-74-8) A strong oxidizer; reacts violently with reducing agents and combustible materials. Incompatible with carbon, lead hypophosphite, ammonium thiocyanate, potassium acetate.

LEAD FLUORIDE (7783-46-2) Reacts violently with fluorine, calcium carbide.

LEAD(2+) FLUORIDE or LEAD(II) FLUORIDE (7783-46-2) Reacts violently with fluorine, calcium carbide.

LEAD FLUOROBORATE or LEAD FLUOROBORATE SOLUTION (13814-96-5) Aqueous solution is acidic; corrodes many metals.

LEAD HYPOSULFITE (13478-50-7) Aqueous solution is acidic; corrodes many metals.

LEAD IOSULFATE (13478-50-7) Aqueous solution is acidic; corrodes many metals.

LEAD MONOXIDE (1317-36-8) Reacts violently with aluminum powder, hydrogen peroxide, hydrogen trisulfide, lithium carbide, perchloric acid, strong oxidizers.

LEAD NITRATE (10099-74-8) A strong oxidizer; reacts violently with reducing agents and combustible materials. Incompatible with carbon, lead hypophosphite, ammonium thiocyanate, potassium acetate.

LEAD(2+) NITRATE(1:2) (10099-74-8) A strong oxidizer; reacts violently with reducing agents and combustible materials. Incompatible with carbon, lead hypophosphite, ammonium thiocyanate, potassium acetate.

LEAD OXIDE (1317-36-8) Reacts violently with aluminum powder, hydrogen peroxide, hydrogen trisulfide, lithium carbide, perchloric acid, strong oxidizers.

LEAD(2+) OXIDE (1317-36-8) Reacts violently with aluminum powder, hydrogen peroxide, hydrogen trisulfide, lithium carbide, perchloric acid, strong oxidizers.

LEAD(II) OXIDE (1317-36-8) Reacts violently with aluminum powder, hydrogen peroxide, hydrogen trisulfide, lithium carbide, perchloric acid, strong oxidizers.

LEAD OXIDE YELLOW (1317-36-8) Reacts violently with aluminum powder, hydrogen peroxide, hydrogen trisulfide, lithium carbide, perchloric acid, strong oxidizers.

LEAD PEROXIDE (1309-60-0) A strong oxidizer; reacts violently with reducing agents and combustible materials. Incompatible with acids.

LEAD PROTOXIDE (1317-36-8) Reacts violently with aluminum powder, hydrogen peroxide, hydrogen trisulfide, lithium carbide, perchloric acid, strong oxidizers.

LEAD SULFATE (7446-14-2) Reacts violently with potassium. Incompatible with aluminum, magnesium.

LEAD(II) SULFATE (1:1) (7446-14-2) Reacts violently with potassium. Incompatible with aluminum, magnesium.

LEAD SULFIDE (1314-87-0) Reacts violently with iodine monochloride or hydrogen peroxide.

LEAD SULFOCYANATE (592-87-0) Incompatible with strong oxidizers, strong acids.

LEAD TETRAACETATE or LEAD(4+) ACETATE (546-67-8) Incompatible with water, producing lead dioxide and acetic acid. Corrodes metal in the presence of moisture.

LEAD TETRAETHYL (78-00-2) Forms explosive mixture with air and decomposes above 200°F/93°C. Incompatible with strong oxidizers, concentrated acids; may cause fire and explosions. Attacks some plastics, rubber, and coatings.

LEAD, TETRAMETHYL- (75-74-1) Forms explosive mixture with air (flash point 100°F/38°C). Decomposes above 212°F/100°C. Strong oxidizers and strong acids cause fire and explosions. Attacks some plastics, rubber, and coatings.

LEAD THIOCYANATE (592-87-0) Incompatible with strong oxidizers, strong acids.

LEAF LARD (N/A) Incompatible with strong acids, oxidizers.

LE CAPTANE (French) (133-06-2) Incompatible with tetraethyl pyrophosphate, parathion.

LEDON 11 (75-69-4) Reacts with barium, lithium, sodium, magnesium, titanium. Attacks some plastics, rubber, and coatings.

LEDON 12 (75-71-8) Reacts violently with liquid aluminum. Incompatible with chemically active metals. Attacks some plastics, rubber, and coatings.

LEIVASOM (52-68-6) Contact with strong oxidizers may cause fire and explosions.

LEMONENE (92-52-4) Mist forms explosive mixture with air. Strong oxidizers may cause fire and explosions.

LENDINE (58-89-9) Not combustible, but may be dissolved in a combustible solvent. If solvent comes in contact with oxidizers, fire and explosions may result.

LENTOX (58-89-9) Not combustible, but may be dissolved in a combustible solvent. If solvent comes in contact with oxidizers, fire and explosions may result.

LETHALAIRE G-52 (107-49-3) Decomposes above 300°F/150°C, producing flammable ethylene gas. Strong oxidizers may cause fire and explosions. Attacks some plastics, rubber, and coatings.

LETHALAIRE G-54 (56-38-2) Combustible liquid. Mixtures with endrin may be explosive. Strong oxidizers may cause fire and explosions. Attacks some plastics, rubber, and coatings.

LETHALAIRE G-57 (3689-24-5) Containers may burst in heat. Strong oxidizers may cause fire and explosions. Attacks some plastics, rubber, and coatings.

LETHURIN (79-01-6) Contact with caustics produces a toxic and flammable gas. Reacts violently with chemically active metals. Contact with aluminum may produce a violent, self-accelerating polymerization reaction. Incompatible with acids, organic anhydrides, isocyanates, alkylene oxides, aldehydes, alcohols, glycols, phenols, cresols, caprolactam solution, epichlorohydrin, nitrogen tetroxide, metal powders, oxygen. May accumulate static electrical charges, and may cause ignition of its vapors.

LEUCOL (91-22-5) Contact with strong oxidizers may cause fire and explosions. Attacks some plastics, rubber, and coatings.

LEUCOLINE (91-22-5) Contact with strong oxidizers may cause fire and explosions. Attacks some plastics, rubber, and coatings.

LEUKOL (91-22-5) Contact with strong oxidizers may cause fire and explosions. Attacks some plastics, rubber, and coatings.

LEVANOX RED 130A (1309-37-1) Contact with hydrogen peroxide, ethylene oxide, calcium hypochlorite will cause explosion. Reacts violently with powdered aluminum, hydrazine, hydrogen trisulfide.

LEWIS-RED DEVIL LYE (1310-73-2) Contact with water, acids, flammable liquids, and organic halogens, especially trichloroethylene, may cause fires and explosions. Contact with metals such as aluminum, tin and zinc causes corrosion and the formation of flammable hydrogen gas. Contact with nitromethane and similar nitro compounds produces shock-sensitive salts. Water contact produces heat and corrosive fumes. Attacks some plastics, rubber, and coatings.

LICHENIC ACID (110-17-8) Reacts with strong oxidizers. Incompatible with sulfuric acid, caustics, ammonia, amines, isocyanates, alkylene oxides, epichlorohydrin.

LIDENAL (58-89-9) Not combustible, but may be dissolved in a combustible solvent. If solvent comes in contact with oxidizers, fire and explosions may result.

LIGHT NAPHTHA (8002-05-9) Forms explosive mixture with air (flash point 100°F/38°C). Incompatible with nitric acid, strong oxidizers. Attacks some plastics, rubber, and coatings. May accumulate static electrical charges, and may cause ignition of its vapors.

LIGHT OIL (8007-45-2) Forms explosive mixture with air (flash point > 60°F/15°C). Incompatible with strong acids, nitrates.

LIGHT RED (1309-37-1) Contact with hydrogen peroxide, ethylene oxide, calcium hypochlorite will cause explosion. Reacts violently with powdered aluminum, hydrazine, hydrogen trisulfide.

LIGROIN (8032-32-4) Forms explosive mixture with air (flash point −40°F to −86°F/−40°C to −66°C). Incompatible with strong acids, strong oxidizers. Attacks some plastics, rubber, and coatings. May accumulate static electrical charges, and may cause ignition of its vapors.

LIME (1305-78-8) Reacts violently with water. Incompatible with ethanol, hydrogen fluoride, acids, halogens, metal halides, light metals, some oxides, boron trifluoride, chlorine trifluoride, liquid hydrofluoric acid, phosphorus pentoxide, boric oxide and calcium chloride mixtures, fluorine, chlorine trifluoride, carbon dioxide.

LIME, BURNED or LIME, BURNT (1305-78-8) Reacts violently with water. Incompatible with ethanol, hydrogen fluoride, acids, halogens, metal halides, light metals, some oxides, boron trifluoride, chlorine trifluoride, liquid hydrofluoric acid, phosphorus pentoxide, boric oxide and calcium chloride mixtures, fluorine, chlorine trifluoride, carbon dioxide.

481

LIME CHLORIDE (7778-54-3) Decomposes in heat or sunlight. Incompatible with acids, moisture, reducing agents, combustible materials, all other chemicals, especially acetylene, aniline and all other amines, anthracene, carbon tetrachloride, iron oxide, manganese oxide, mercaptans, diethylene glycol monomethyl ether, nitromethane, organic matter, organic sulfides, phenol, 1-propanethiol, propyl mercaptan, sulfur, organic sulfur compounds.

LIMED ROSIN (9007-13-0) Incompatible with oxidizers, strong acids.

LIME NITROGEN (156-62-7) Contact with any form of moisture causes decomposition, liberating acetylene and ammonia. Contact with all solvents tested also causes decomposition.

LIME SALTPETER (10124-37-5) A strong oxidizer. Incompatible with combustible materials, reducing agents, organics and other oxidizable materials, chemically active metals, aluminum nitrate, ammonium nitrate.

LIMESTONE (1317-65-3) Incompatible with acids, alum, ammonium salts, fluorine.

LIME, UNSLAKED (1305-78-8) Reacts violently with water. Incompatible with ethanol, hydrogen fluoride, acids, halogens, metal halides, light metals, some oxides, boron trifluoride, chlorine trifluoride, liquid hydrofluoric acid, phosphorus pentoxide, boric oxide and calcium chloride mixtures, fluorine, chlorine trifluoride, carbon dioxide.

LIME WATER (1305-62-0) Contact with maleic anhydride, phosphorus, nitroethane, nitromethane, nitroparaffins, or nitropropane may cause explosion.

LIMONENE (138-86-3) Forms explosive mixture with air (flash point 115°F/46°C). Strong oxidizers may cause fire and explosions.

DL-LIMONENE (138-86-3) Forms explosive mixture with air (flash point 115°F/46°C). Strong oxidizers may cause fire and explosions.

LINDAFOR (58-89-9) Not combustible, but may be dissolved in a combustible solvent. If solvent comes in contact with oxidizers, fire and explosions may result.

LINDAGAM (58-89-9) Not combustible, but may be dissolved in a combustible solvent. If solvent comes in contact with oxidizers, fire and explosions may result.

LINDAGRAIN (58-89-9) Not combustible, but may be dissolved in a combustible solvent. If solvent comes in contact with oxidizers, fire and explosions may result.

LINDAGRANOX (58-89-9) Not combustible, but may be dissolved in a combustible solvent. If solvent comes in contact with oxidizers, fire and explosions may result.

LINDANE (58-89-9) Not combustible, but may be dissolved in a combustible solvent. If solvent comes in contact with oxidizers, fire and explosions may result.

gamma-LINDANE (58-89-9) Not combustible, but may be dissolved in a combustible solvent. If solvent comes in contact with oxidizers, fire and explosions may result.

LINDAPOUDRE (58-89-9) Not combustible, but may be dissolved in a combustible solvent. If solvent comes in contact with oxidizers, fire and explosions may result.

LINDATOX (58-89-9) Not combustible, but may be dissolved in a combustible solvent. If solvent comes in contact with oxidizers, fire and explosions may result.

LINEAR ALCOHOLS (27196-00-5) Incompatible with strong acids, caustics, aliphatic amines, isocyanates, strong oxidizers.

LINDOSEP (58-89-9) Not combustible, but may be dissolved in a combustible solvent. If solvent comes in contact with oxidizers, fire and explosions may result.

LINE RIDER (93-76-5) Sealed metal containers may burst in heat above 316°F/158°C. Incompatible with sulfuric acid, bases, ammonia, aliphatic amines, alkanolamines, isocyanates, alkylene oxides, epichlorohydrin.

LINSEED OIL (8001-26-1) Oxidizers may cause fire and explosions. Incompatible with nitric acid. May accumulate static electrical charges, and may cause ignition of its vapors.

LINTOX (58-89-9) Not combustible, but may be dissolved in a combustible solvent. If solvent comes in contact with oxidizers, fire and explosions may result.

LIQUAMON 28 (15978-77-5) Incompatible with oxidizers, combustibles, organics, acids, amides, organic anhydrides, isocyanates, vinyl acetate, alkylene oxides, epichlorohydrin, aldehydes, sulfur. Reacts violently with copper and copper alloys. Attacks lead, zinc.

LIQUA-TOX (81-81-2) Strong oxidizers may cause fire and explosions.

LIQUEFIED HYDROCARBON GAS (68476-85-7) Forms explosive mixture with air. Strong oxidizers may cause fire and explosions. Attacks some plastics, rubber, and coatings.

LIQUEFIED NATURAL GAS (74-82-8) Flammable gas. Reacts violently with strong oxidizers.

LIQUEFIED PETROLEUM GAS (68476-85-7) Forms explosive mixture with air. Strong oxidizers may cause fire and explosions. Attacks some plastics, rubber, and coatings.

LIQUIBARINE (7727-43-7) Explosions may result from contact with aluminum in the presence of heat. Incompatible with potassium, phosphorus.

LIQUID AMMONIA (7664-41-7) anhydrous (a flammable gas); (1336-21-6) solution in water. Reacts violently with strong oxidizers, acids. Shock-sensitive compounds may be produced with halogens, mercury oxide, silver oxide. Fire and explosions may be caused by trimethylammonium amide, 1-chloro-2,4-dinitrobenzene, o-chloronitrobenzene, platinum, selenium difluoride dioxide, boron halides, mercury, chlorine, iodine, bromine, hypochlorites, chlorine bleach, amides, organic anhydrides, isocyanates, vinyl acetate, alkylene oxides, epichlorohydrin, aldehydes. Attacks some coatings, plastics, and rubber. Attacks copper, brass, bronze, aluminum, steel, and their alloys.

LIQUID ASPHALT or LIQUID ASPHALTUM (8052-42-4) Incompatible with nitric acid, fluorine, strong oxidizers.

LIQUID BLEACH (7681-52-9) A strong oxidizer and a strong base. Stability decreases with concentration, heat, light, decrease in pH, and contamination with metals. Incompatible with strong acids, reducing agents, combustible substances; all cause violent reaction, fire and explosions. Contact with amines and ammonia salts produce explosive chloroamines. Corrodes many metals: steels, cast iron, monel, aluminum, brass, iconel.

LIQUID CAMPHOR (76-22-2) Forms explosive mixture with air (flash point 155°F/66°C). Violent, possibly explosive, reaction occurs with strong oxidizers. May accumulate static electrical charges, and may cause ignition of its vapors.

LIQUID ETHELYNE (74-85-1) A flammable gas. May accumulate static electrical charges, and may cause ignition of its vapors. Oxidizers may cause explosive polymerization and fire. Incompatible with acids, halogens, nitrogen oxides, hydrogen bromide, aluminum chloride, bromotrichloromethane, carbon tetrachloride, chlorine, chlorine dioxide, nitrogen dioxide.

LIQUID ETHYENE (74-85-1) A flammable gas. May accumulate static electrical charges, and may cause ignition of its vapors. Oxidizers may cause explosive polymerization and fire. Incompatible with acids, halogens, nitrogen oxides, hydrogen bromide, aluminum chloride, bromotrichloromethane, carbon tetrachloride, chlorine, chlorine dioxide, nitrogen dioxide.

LIQUID GUM CAMPHOR (76-22-2) Forms explosive mixture with air (flash point 155°F/66°C). Violent, possibly explosive, reaction occurs with strong oxidizers. May accumulate static electrical charges, and may cause ignition of its vapors.

LIQUID HYDROGEN (1333-74-0) Vapors form explosive or combustible mixture with air over a wide range of concentrations. Ignites easily with oxygen. Reacts violently with oxidizers, halogens, acetylene, bromine, chlorine, fluorine, nitrous oxide, and other gases. Mild steel and most iron alloys become brittle at liquid hydrogen temperatures.

LIQUID IMPURE CAMPHOR (76-22-2) Forms explosive mixture with air (flash point 155°F/66°C). Violent, possibly explosive, reaction occurs with strong oxidizers. May accumulate static electrical charges, and may cause ignition of its vapors.

LIQUID NITROGEN (79-24-3) Forms explosive mixture with air (flash point 82°F/28°C). Reacts violently with oxidizers, caustics, combustibles. Can form shock-sensitive salts with calcium hydroxide and water. Incompatible with amines, strong acids, heavy metal oxides. Organic contamination may cause substance to become shock-sensitive. Attacks some plastics, rubber, and coatings. May accumulate static electrical charges, and may cause ignition of its vapors.

LIQUIDOW (10043-52-4) Incompatible with water, bromine trifluoride, 2-furan, percarboxylic acid. Attacks metals.

LIQUID OXYGEN (7782-44-7) Heat of water contact will vigorously vaporize liquid oxygen. A strong oxidizer; reacts violently with reducing agents, combustibles, organic and easily oxidizable materials; contact may cause fire and explosions. The low temperature may cause brittleness on contact with some materials.

LIQUID PETROLATUM (8012-95-1) Oxidizers may cause fire and explosions. Incompatible with nitric acid. May accumulate static electrical charges, and may cause ignition of its vapors.

LIQUID PITCH OIL (65996-93-2) Incompatible with oxidizers, strong acids, caustics, aliphatic amines, isocyanates.

LIQUID ROSIN (8002-26-4) Incompatible with strong acids, oxidizers, nitrates.

LIROHEX (107-49-3) Decomposes above 300°F/150°C, producing flammable ethylene gas. Strong oxidizers may cause fire and explosions. Attacks some plastics, rubber, and coatings.

LIROPON (75-99-0) Corrosive to iron, aluminum, and copper.

LIROPREM (87-86-5) Hot water causes decomposition, producing hydrochloric acid. Strong oxidizers may cause fire and explosions.

LIROTHION (56-38-2) Combustible liquid. Mixtures with endrin may be explosive. Strong oxidizers may cause fire and explosions. Attacks some plastics, rubber, and coatings.

LITHARGE (1317-36-8) Reacts violently with aluminum powder, hydrogen peroxide, hydrogen trisulfide, lithium carbide, perchloric acid, strong oxidizers.

LITHARGE YELLOW L-28 (1317-36-8) Reacts violently with aluminum powder, hydrogen peroxide, hydrogen trisulfide, lithium carbide, perchloric acid, strong oxidizers.

LITHIUM (7439-93-2) Reacts violently with water, producing flammable hydrogen gas, corrosive fumes, and a strong caustic solution. Finely divided particles or powdered form may ignite spontaneously in air. Contact with air produces corrosive fumes of lithium hydroxide. Reacts violently with oxidizers, acetonitrile, nitric acid, arsenic, bromobenzene, carbon tetrachloride, sulfur, and many other substances. Forms impact- and friction-sensitive mixtures with bromobenzene, carbon tetrabromide, chloroform (weak explosion), iodoform, halogens, halocarbons, methyl dichloride, methyl diiodide and other substances. Attacks plastics, rubber, ceramic materials, concrete, sand, and metal alloys: cobalt, iron, manganese, nickel.

LITHIUM ALUMINUM HYDRIDE (16853-85-3) Flammable solid. Can ignite spontaneously in moist air or heat above 250°F/121°C. A strong reducing agent. Reacts violently with oxidizers, alcohols, acids, dimethylether, tetrahydrofuran.

LITHIUM BICHROMATE (13843-81-7) Aqueous solution is caustic. An oxidizer; reacts violently with reducing agents, combustibles, organics, and easily oxidizable materials.

LITHIUM BICHROMATE DIHYDRATE (13843-81-7) Aqueous solution is caustic. An oxidizer; reacts violently with reducing agents, combustibles, organics, and easily oxidizable materials.

LITHIUM CARBONATE (554-13-2) Incompatible with lithium, acids. Corrodes aluminum, copper, zinc.

LITHIUM CHROMATE (14307-35-8) Aqueous solution is caustic. An oxidizer; reacts strongly with reducing agents, combustibles, organic materials, acids.

LITHIUM DICHROMATE (13843-81-7) Aqueous solution is caustic. An oxidizer; reacts violently with reducing agents, combustibles, organics, and easily oxidizable materials.

LITHIUM, ELEMENTAL (7439-93-2) Reacts violently with water, producing flammable hydrogen gas, corrosive fumes, and a strong caustic solution. Finely divided particles or powdered form may ignite spontaneously in air. Contact with air produces corrosive fumes of lithium hydroxide. Reacts violently with oxidizers, acetonitrile, nitric acid, arsenic, bromobenzene, carbon tetrachloride, sulfur, and many other substances. Forms impact- and friction-sensitive mixtures with bromobenzene, carbon tetrabromide, chloroform (weak explosion), iodoform, halogens, halocarbons, methyl dichloride, methyl diiodide, and other substances. Attacks plastics, rubber, ceramic materials, concrete, sand, and metal alloys: cobalt, iron, manganese, nickel.

LITHIUM HYDRIDE (7580-67-8) Incompatible with oxidizers, halogenated hydrocarbons, acids; can cause fire and explosion. Contact with water produces hydrogen gas; reaction may cause ignition.

LITHIUM METAL (7439-93-2) Reacts violently with water, producing flammable hydrogen gas, corrosive fumes, and a strong caustic solution. Finely divided particles or powdered form may ignite spontaneously in air. Contact with air produces corrosive fumes of lithium hydroxide. Reacts violently with oxidizers, acetonitrile, nitric acid, arsenic, bromobenzene, carbon tetrachloride, sulfur, and many other substances. Forms impact- and friction-sensitive mixtures with bromobenzene, carbon tetrabromide, chloroform (weak explosion), iodoform, halogens, halocarbons, methyl dichloride, methyl diiodide, and other substances. Attacks plastics, rubber, ceramic materials, concrete, sand, and metal alloys: cobalt, iron, manganese, nickel.

LITHIUM MONOHYDRIDE (7439-93-2) Reacts violently with water, producing flammable hydrogen gas, corrosive fumes, and a strong caustic solution. Finely divided particles or powdered form may ignite spontaneously in air. Contact with air produces corrosive fumes of lithium hydroxide. Reacts violently with oxidizers, acetonitrile, nitric acid, arsenic, bromobenzene, carbon tetrachloride, sulfur, and many other substances. Forms impact- and friction-sensitive mixtures with bromobenzene, carbon tetrabromide, chloroform (weak explosion), iodoform, halogens, halocarbons, methyl dichloride, methyl diiodide, and other substances. Attacks plastics, rubber, ceramic materials, concrete, sand, and metal alloys: cobalt, iron, manganese, nickel.

LITHIUM MONOHYDRIDE (7580-67-8) Incompatible with oxidizers, halogenated hydrocarbons, acids; can cause fire and explosion. Contact with water produces hydrogen gas; reaction may cause ignition.

LITHIUM TETRAHYDROALUMINATE (16853-85-3) Flammable solid. Can ignite spontaneously in moist air or heat above 250°F/121°C. A strong reducing agent. Reacts violently with oxidizers, alcohols, acids, dimethylether, tetrahydrofuran.

LITHOGRAPHIC STONE (1317-65-3) Incompatible with acids, alum, ammonium salts, fluorine.

LO-BAX (7778-54-3) Decomposes in heat or sunlight. Incompatible with acids, moisture, reducing agents, combustible materials, all other chemicals, especially acetylene, aniline and all other amines, anthracene, carbon tetrachloride, iron oxide, manganese oxide, mercaptans, diethylene glycol monomethyl ether, nitromethane, organic matter, organic sulfides, phenol, 1-propanethiol, propyl mercaptan, sulfur, organic sulfur compounds.

LOISOL (52-68-6) Contact with strong oxidizers may cause fire and explosions.

LOREX (7775-09-9) A powerful oxidizer; reacts violently with reducing agents and combustible matter. Explosions may be caused by contact with ammonia salts, carbon, oils, metal sulfides, nitrobenzene, powdered metals, sugar. Contact with strong acids produces carbon dioxide. Forms shock-sensitive mixtures with some organic materials. Solution (50%) decomposes at 300°F/149°C, liberating oxygen.

LOREXANE (58-89-9) Not combustible, but may be dissolved in a combustible solvent. If solvent comes in contact with oxidizers, fire and explosions may result.

LOROL (112-53-8) Incompatible with strong acids, caustics, aliphatic amines, isocyanates, oxidizers.

LOROL-20 (111-87-5) Forms explosive mixture with air (flash point 178°F/81°C). Incompatible with strong acids, caustics, aliphatic amines, isocyanates, strong oxidizers.

LOROL-22 (112-30-1) Forms explosive mixture with air (flash point 180°F/82°C). Incompatible with strong acids, caustics, aliphatic amines, isocyanates, strong oxidizers.

LORSBAN (2921-88-) Incompatible with strong acids. Hydrolyzes from acid or alkaline solutions.

LOSANTIN (7778-54-3) Decomposes in heat or sunlight. Incompatible with acids, moisture, reducing agents, combustible materials, all other chemicals, especially acetylene, aniline and all other amines, anthracene, carbon tetrachloride, iron oxide, manganese oxide, mercaptans, diethylene glycol monomethyl ether, nitromethane, organic matter, organic sulfides, phenol, 1-propanethiol, propyl mercaptan, sulfur, organic sulfur compounds.

LOSUNGSMITTEL APV (111-90-0) Forms explosive mixture with air (flash point 201°F/94°C). Heat or contact with atmospheric air may cause formation of unstable peroxides. Incompatible with strong acids, isocyanates, strong oxidizers.

LOX (7782-44-7) Heat of water contact will vigorously vaporize liquid oxygen. A strong oxidizer; reacts violently with reducing agents, combustibles, organic and easily oxidizable materials; contact may cause fire and explosions. The low temperature may cause brittleness on contact with some materials.

LNG (74-82-8) Flammable gas. Reacts violently with strong oxidizers.

LPG or L.P.G. (68476-85-7) Forms explosive mixture with air. Strong oxidizers may cause fire and explosions. Attacks some plastics, rubber, and coatings.

LPG ETHYL MERCAPTAN 1010 (75-08-1) Forms explosive mixture with air (flash point less than 0°F/−18°C). May accumulate static electrical charges, and may cause ignition of its vapors. Contact with strong oxidizers may cause fire and explosions. Attacks some forms of plastics, coatings, and rubber.

LPT (1338-23-4) Forms explosive mixture with air (flash point 125°F/52°C). Explosive decomposition occurs above 176°F/80°C. Pure substance is shock-sensitive. Strong oxidizer. Reacts violently with strong acids, strong bases, reducing agents, combustible substances, organic materials, oxides of heavy metals, salts, trace contaminants, amines. May accumulate static electrical charges, and may cause ignition of its vapors.

LUBRICATING OIL (8012-95-1) Oxidizers may cause fire and explosions. Incompatible with nitric acid. May accumulate static electrical charges, and may cause ignition of its vapors.

LUBRICATING OIL, TURBINE (N/A) Incompatible with nitric acid, strong oxidizers; poses risk of fire or explosions.

LUCIDOL (94-36-0) Confined storage of dry chemical may lead to decomposition and explosion. A strong oxidant; extremely reactive. Fires and explosion may result from heat or contamination, and from

487

contact with strong acids, combustible materials, oxidizers, acids, bases, alcohols, reducing agents, metals, metal oxides, amines, accelerators, methyl methacrylate, organic matter, lithium aluminum carbide, dimethyl aniline, amines, metallic naphthenates. May attack some plastics, rubber, and coatings. Protect containers from shock and friction.

LUMBRICAL (110-85-0) Forms explosive mixture with air (flash point 178°F/81°C). Incompatible with nitrogen compounds, carbon tetrachloride, strong oxidizers. Aqueous solutions react with acids. Attacks aluminum, copper, nickel, magnesium, and zinc.

LUMBUCAL (110-85-0) Forms explosive mixture with air (flash point 178°F/81°C). Incompatible with nitrogen compounds, carbon tetrachloride, strong oxidizers. Aqueous solutions react with acids. Attacks aluminum, copper, nickel, magnesium, and zinc.

LUNAR CAUSTIC (7761-88-8) Contact with acetylene produces shock-sensitive material. Ammonia contact produces compounds that are explosive when dry. Hydrogen peroxide causes violent decomposition to oxygen gas. Attacks some plastics, rubber, and coatings.

LUPERCO (78-18-2) A strong oxidizer. Reacts violently with combustibles, reducing agents, caustics, ammonia.

LUPERCO AA (94-36-0) Confined storage of dry chemical may lead to decomposition and explosion. A strong oxidant; extremely reactive. Fires and explosion may result from heat or contamination, and from contact with strong acids, combustible materials, oxidizers, acids, bases, alcohols, reducing agents, metals, metal oxides, amines, accelerators, methyl methacrylate, organic matter, lithium aluminum carbide, dimethyl aniline, amines, metallic naphthenates. May attack some plastics, rubber, and coatings. Protect containers from shock and friction.

LUPERSOL (1338-23-4) Forms explosive mixture with air (flash point 125°F/52°C). Explosive decomposition occurs above 176°F/80°C. Pure substance is shock-sensitive. Strong oxidizer. Reacts violently with strong acids, strong bases, reducing agents, combustible substances, organic materials, oxides of heavy metals, salts, trace contaminants, amines. May accumulate static electrical charges, and may cause ignition of its vapors.

2,6-LUTIDINE (108-48-5) Forms explosive mixture with air (flash point 100°F/38°C). Strong oxidizers may cause fire and explosions.

LUTROL-9 (107-21-1) Incompatible with strong acids, caustics, aliphatic amines, isocyanates, chlorosulfonic acid, oleum, strong oxidizers.

LYDDITE (88-89-1) Dry material is explosive. Impact-sensitive; protect from shock. Contact with copper, lead, zinc, and other metals, or their salts, can produce salts that are initiators and much more sensitive to shock than this chemical. Shock-sensitive salts include ammonium salts and calcium salts; the calcium salt may form when picric acid comes into contact with plaster and concrete. May accumulate static electrical charges, and may cause explosion. Aqueous solution is a strong oxidizer and a strong acid; reacts violently with reducing agents, combustibles, organics and easily oxidized materials, caustics, aluminum, and other metal powders. Attacks metals.

488

LYE (1310-73-2) Contact with water, acids, flammable liquids, and organic halogens, especially trichloroethylene, may cause fires and explosions. Contact with metals such as aluminum, tin, and zinc causes corrosion and the formation of flammable hydrogen gas. Contact with nitromethane, and similar nitro compounds, produce shock-sensitive salts. Water contact produces heat and corrosive fumes. Attacks some plastics, rubber, and coatings.

LYE (1310-58-3) Aqueous solution is a strong base. Reacts violently with acids, organic anhydrides, isocyanates, alkylene oxides, epichlorohydrin, aldehydes, alcohols, glycols, phenols, cresols, caprolactam solution. Dissolves in water, producing high heat, caustic fumes and flammable hydrogen gas. Also reacts with halogenated hydrocarbons, maleic anhydride. Attacks metals such as aluminum, tin, lead, and zinc, producing flammable hydrogen gas.

LYP (105-74-8) A powerful oxidizer. May ignite or explode spontaneously when mixed with combustible materials or rapidly heated.

LYP 97 (105-74-8) A powerful oxidizer. May ignite or explode spontaneously when mixed with combustible materials or rapidly heated.

LYSOFORM (50-00-0) May polymerize unless properly inhibited (usually with methanol). Forms explosive mixture with air (flash point 122°F/50°C). Incompatible with strong acids, amines, strong oxidizers, alkaline materials, nitrogen dioxide, performic acid. Reaction with hydrochloric acid produces bis-chloromethyl ether, a carcinogen.

- M -

M-74 (298-04-4) Forms explosive mixture with air (flash point 180°F/82°C). Incompatible with alkalies, strong oxidizers.

M 105 (1338-23-4) Forms explosive mixture with air (flash point 125°F/52°C). Explosive decomposition occurs above 176°F/80°C. Pure substance is shock-sensitive. Strong oxidizer. Reacts violently with strong acids, strong bases, reducing agents, combustible substances, organic materials, oxides of heavy metals, salts, trace contaminants, amines. May accumulate static electrical charges, and may cause ignition of its vapors.

M 140 (57-74-9) Contact with strong oxidizers may cause fire and explosions. Attacks some plastics, rubber, and coatings.

M-176 (67-68-5) Forms explosive mixture with air (flash point 203°F/95°C). Reacts violently with oxidizers. Reacts with ethanoyl chloride, boron compounds, halides, metal alkoxides, oxidizers.

M 410 (57-74-9) Contact with strong oxidizers may cause fire and explosions. Attacks some plastics, rubber, and coatings.

M 5055 (8001-35-2) Reacts with oxidizers, with a risk of fire or explosions. Attacks metals in the presence of moisture.

MA (100-61-8) Incompatible with strong acids, strong oxidizers. Attacks some plastics, rubber, and coatings. May accumulate static electrical charges, and may cause ignition of its vapors.

MA 1214 (112-53-8) Incompatible with strong acids, caustics, aliphatic amines, isocyanates, oxidizers.

MAA (108-11-2) Forms explosive mixture with air (flash point 106°F/41°C). Contact with alkali metals produces hydrogen gas. Incompatible with strong acids, caustics, aliphatic amines, isocyanates. Attacks some plastics, rubber, and coatings. May accumulate static electrical charges, and may cause ignition of its vapors.

MAA (108-31-7) Aqueous solution is a strong acid. Contact with strong oxidizers may cause fire and explosions. Contact with amines or alkali metals causes polymerization if temperature is greater than 150°F/66°C. Attacks some plastics, rubber, and coatings.

MAAC (108-84-9) Forms explosive mixture with air (flash point 113°F/45°C). Incompatible with strong acids, strong alkalies, nitrates, strong oxidizers.

MACE (532-27-4) Incompatible with water or steam.

MACH-NIC (54-11-5) Incompatible with strong acids, strong oxidizers. Attacks some plastics, rubber, and coatings. May accumulate static electrical charges, and may cause ignition of its vapors.

MACROGOL 400 BPC (107-21-1) Incompatible with strong acids, caustics, aliphatic amines, isocyanates, chlorosulfonic acid, oleum, strong oxidizers.

MACRONDRAY (94-75-7) Decomposes in sunlight. Incompatible with strong oxidizers; may cause fire and explosions.

MACROPAQUE (7727-43-7) Explosions may result from contact with aluminum in the presence of heat. Incompatible with potassium, phosphorus.

MAFU (62-73-7) Attacks some plastics, rubber, and coatings.

MAGCAL (1309-48-4) Reacts violently with chlorine trifluoride, bromine pentafluoride, phosphorus pentachloride. Incompatible with acids.

MAGLITE (1309-48-4) Reacts violently with chlorine trifluoride, bromine pentafluoride, phosphorus pentachloride. Incompatible with acids.

MAGNACAT (7439-96-5) Water contact provides hydrogen gas. Oxidizers, nitric acid, nitrogen, finely divided aluminum and other metals, sulfur dioxide may cause fire and explosions.

MAGNACIDE (107-02-8) Forms explosive mixture with air (flash point −15°F/−26°C). Unstable and very reactive. In storage can form heat- and shock-sensitive compounds. Unless inhibited (usually with hydroquinone), readily forms explosive peroxides. Able to polymerize. A strong reducing agent. Reacts violently with oxidizers, strong acids, caustics, amines, 2-aminoethanol, ammonia, ammonium hydroxide, ethylene diamine, ethyleneimine, hydroxides, metal salts, oxidizers, sulfur dioxide, thiourea. Attacks metals: cadmium and zinc. May accumulate static electrical charges, and may cause ignition of its vapors.

MAGNESIA (1309-48-4) Incompatible with acids, phosphorus pentachloride.

MAGNESIO (Italian) (7439-95-4) Strong reducing agent; reacts violently with oxidizers, strong acids, carbon tetrachloride, carbonates, chloroform, phosphates, silver nitrate, sodium peroxide, sulfates, trichloroethylene, and many other substances.

MAGNESIUM or MAGNESIUM, ELEMENTAL or MAGNESIUM METAL or MAGNESIUM PELLETS or MAGNESIUM POWDER or MAGNESIUM RIBBONS or MAGNESIUM SCALPINGS or MAGNESIUM SHAVINGS or MAGNESIUM SHEET or MAGNESIUM TURNINGS (7439-95-4) Strong reducing agent; reacts violently with oxidizers, strong acids, carbon tetrachloride, carbonates, chloroform, phosphates, silver nitrate, sodium peroxide, sulfates, trichloroethylene, and many other substances.

MAGNESIUM CHLORIDE (7786-30-3) Incompatible with water. Attacks steel in the presence of moisture.

MAGNESIUM HYDROXIDE (1309-42-8) Incompatible with acids, phosphorus.

MAGNESIUM NITRATE (10377-60-3) A powerful oxidizer. Reacts violently with dimethylformamide, reducing agents, combustibles, organic and easily oxidizable matter.

MAGNESIUM(2+) NITRATE(1:2) or MAGNESIUM(II) NITRATE(1:2) (10377-60-3) A powerful oxidizer. Reacts violently with dimethylformamide, reducing agents, combustibles, organic and easily oxidizable matter.

MAGNESIUM OXIDE (1309-48-4) Reacts violently with chlorine trifluoride, bromine pentafluoride, phosphorus pentachloride. Incompatible with acids.

MAGNESIUM PERCHLORATE or MAGNESIUM PERCHLORATE, ANHYDROUS (10034-81-8) A powerful oxidizer. Reacts violently with reducing agents, organic matter, ethylene oxide, powdered metals, phosphorus, dimethylsulfoxide. Forms explosive material with ethyl alcohol. Incompatible with many materials. Attacks many metals.

MAGNESIUM PERCHLORATE HEXAHYDRATE (10034-81-8) A powerful oxidizer. Reacts violently with reducing agents, organic matter, ethylene oxide, powdered metals, phosphorus, dimethylsulfoxide. Forms explosive material with ethyl alcohol. Incompatible with many materials. Attacks many metals.

MAGNETIC 70 (7704-34-9) Combustible solid. Liquid contact produces sulfur dioxide with air. Reacts violently with strong oxidizers. Forms explosive, shock-sensitive or pyrophoric mixtures with ammonia, ammonium nitrate, bromates, calcium carbide, charcoal, chlorates, hydrocarbons, iodates, iron. Reacts violently with halogen compounds, sodium, tin, uranium, and other compounds. Attacks steel when moist. May accumulate static electrical charges, and may cause ignition of its vapors.

MAGNETIC 90 (7704-34-9) Combustible solid. Liquid contact produces sulfur dioxide with air. Reacts violently with strong oxidizers. Forms explosive, shock-sensitive, or pyrophoric mixtures with ammonia, ammonium nitrate, bromates, calcium carbide, charcoal, chlorates, hydrocarbons, iodates, iron. Reacts violently with halogen compounds, sodium, tin, uranium, and other compounds. Attacks steel when moist. May accumulate static electrical charges, and may cause ignition of its vapors.

MAGNETIC 95 (7704-34-9) Combustible solid. Liquid contact produces sulfur dioxide with air. Reacts violently with strong oxidizers. Forms explosive, shock-sensitive, or pyrophoric mixtures with ammonia, ammonium nitrate, bromates, calcium carbide, charcoal, chlorates, hydrocarbons, iodates, iron. Reacts violently with halogen compounds, sodium, tin, uranium, and other compounds. Attacks steel when moist. May accumulate static electrical charges, and may cause ignition of its vapors.

MAGNEZU TLENEK (Polish) (1309-48-4) Reacts violently with chlorine trifluoride, bromine pentafluoride, phosphorus pentachloride. Incompatible with acids.

MAGOX (1309-48-4) Reacts violently with chlorine trifluoride, bromine pentafluoride, phosphorus pentachloride. Incompatible with acids.

MALACIDE (121-75-5) Incompatible with strong oxidizers, magnesium, alkaline pesticides. Attacks metals, some plastics, rubber, and coatings.

MALAFOR (121-75-5) Incompatible with strong oxidizers, magnesium, alkaline pesticides. Attacks metals, some plastics, rubber, and coatings.

MALAGRAN (121-75-5) Incompatible with strong oxidizers, magnesium, alkaline pesticides. Attacks metals, some plastics, rubber, and coatings.

MALAKILL (121-75-5) Incompatible with strong oxidizers, magnesium, alkaline pesticides. Attacks metals, some plastics, rubber, and coatings.

MALAMAR (121-75-5) Incompatible with strong oxidizers, magnesium, alkaline pesticides. Attacks metals, some plastics, rubber, and coatings.

MALAMAR 50 (121-75-5) Incompatible with strong oxidizers, magnesium, alkaline pesticides. Attacks metals, some plastics, rubber, and coatings.

MALAPHELE (121-75-5) Incompatible with strong oxidizers, magnesium, alkaline pesticides. Attacks metals, some plastics, rubber, and coatings.

MALAPHOS (121-75-5) Incompatible with strong oxidizers, magnesium, alkaline pesticides. Attacks metals, some plastics, rubber, and coatings.

MALASOL (121-75-5) Incompatible with strong oxidizers, magnesium, alkaline pesticides. Attacks metals, some plastics, rubber, and coatings.

MALASPRAY (121-75-5) Incompatible with strong oxidizers, magnesium, alkaline pesticides. Attacks metals, some plastics, rubber, and coatings.

MALATHION (121-75-5) Incompatible with strong oxidizers, magnesium, alkaline pesticides. Attacks metals, some plastics, rubber, and coatings.

MALATHIOZOO (121-75-5) Incompatible with strong oxidizers, magnesium, alkaline pesticides. Attacks metals, some plastics, rubber, and coatings.

MALATHON (121-75-5) Incompatible with strong oxidizers, magnesium, alkaline pesticides. Attacks metals, some plastics, rubber, and coatings.

MALATION (Polish) (121-75-5) Incompatible with strong oxidizers, magnesium, alkaline pesticides. Attacks metals, some plastics, rubber, and coatings.

MALATOL (121-75-5) Incompatible with strong oxidizers, magnesium, alkaline pesticides. Attacks metals, some plastics, rubber, and coatings.

MALATOX (121-75-5) Incompatible with strong oxidizers, magnesium, alkaline pesticides. Attacks metals, some plastics, rubber, and coatings.

MALAZIDE (123-33-1) Contact with strong oxidizers may cause fire and explosions.

MALDISON (121-75-5) Incompatible with strong oxidizers, magnesium, alkaline pesticides. Attacks metals, some plastics, rubber, and coatings.

MALEIC ACID (110-16-7) May accumulate static electrical charges, and may cause ignition of its vapors. Aqueous solution is an acid. Incompatible with strong oxidizers, bases. Corrodes metal if wet.

MALEIC ACID ANHYDRIDE (108-31-7) Aqueous solution is a strong acid. Contact with strong oxidizers may cause fire and explosions. Contact with amines or alkali metals cause, polymerization if temperature is greater than 150°F/66°C. Attacks some plastics, rubber, and coatings.

MALEIC ACID HYDRAZIDE (123-33-1) Contact with strong oxidizers may cause fire and explosions.

MALEIC ANHYDRIDE (108-31-7) Aqueous solution is a strong acid. Contact with strong oxidizers may cause fire and explosions. Contact with amines or alkali metals cause, polymerization if temperature is greater than 150°F/66°C. Attacks some plastics, rubber, and coatings.

MALEIC HYDRAZIDE (123-33-1) Contact with strong oxidizers may cause fire and explosions.

MALEINIC ACID (110-16-7) May accumulate static electrical charges, and may cause ignition of its vapors. Aqueous solution is an acid. Incompatible with strong oxidizers, bases. Corrodes metal if wet.

MALENIC ACID (110-16-7) May accumulate static electrical charges, and may cause ignition of its vapors. Aqueous solution is an acid. Incompatible with strong oxidizers, bases. Corrodes metal if wet.

N,N-MALEOXYHYDRAZINE (123-33-1) Contact with strong oxidizers may cause fire and explosions.

MALIX (115-29-7) Hydrolyzed by acids and alkalies. Corrosive to iron.

MALMED (121-75-5) Incompatible with strong oxidizers, magnesium, alkaline pesticides. Attacks metals, some plastics, rubber, and coatings.

MALONIC DINITRILE (109-77-3) Incompatible with sulfuric acid. Caustics or heat above 160°F/71°C may cause polymerization or spontaneous combustion.

MALONIC MONONITRILE (372-09-8) Incompatible with oxidizers, strong acids, organic acids, caustics, reducing agents.

MALONONITRILE (109-77-3) Incompatible with sulfuric acid. Caustics or heat above 160°F/71°C may cause polymerization or spontaneous combustion.

MALPHOS (121-75-5) Incompatible with strong oxidizers, magnesium, alkaline pesticides. Attacks metals, some plastics, rubber, and coatings.

MALTOX (121-75-5) Incompatible with strong oxidizers, magnesium, alkaline pesticides. Attacks metals, some plastics, rubber, and coatings.

MANGAN (Polish) (7439-96-5) Water contact produces hydrogen gas. Oxidizers, nitric acid, nitrogen, finely divided aluminum and other metals, sulfur dioxide may cause fire and explosions.

MANGANESE (7439-96-5) Water contact produces hydrogen gas. Oxidizers, nitric acid, nitrogen, finely divided aluminum and other metals, sulfur dioxide may cause fire and explosions.

MANGANESE BINOXIDE (1313-13-9) A strong oxidizer; reacts violently with reducing agents, combustible materials, hydrochloric acid, chlorates, hydrogen peroxide, sodium peroxide, and many other substances.

MANGANESE BLACK (1313-13-9) A strong oxidizer; reacts violently with reducing agents, combustible materials, hydrochloric acid, chlorates, hydrogen peroxide, sodium peroxide, and many other substances.

MANGANESE DIOXIDE (1313-13-9) A strong oxidizer; reacts violently with reducing agents, combustible materials, hydrochloric acid, chlorates, hydrogen peroxide, sodium peroxide, and many other substances.

MANGANESE TRICARBONYL METHYLCYCLOPENTADIENYL (12108-13-3) Incompatible with strong acids, strong oxidizers.

MANGAN NITRIDOVANY (Czech) (7439-96-5) Water contact produces hydrogen gas. Oxidizers, nitric acid, nitrogen, finely divided aluminum and other metals, sulfur dioxide may cause fire and explosions.

MANOXAL OT (119-36-8) Forms explosive mixture with air (flash point 205°F/96°C). Incompatible with strong acids, nitrates, oxidizers.

MANUFACTURED IRON OXIDES (1309-37-1) Contact with hydrogen peroxide, ethylene oxide, calcium hypochlorite will cause explosion. Reacts violently with powdered aluminum, hydrazine, hydrogen trisulfide.

MAOH (108-11-2) Forms explosive mixture with air (flash point 106°F/41°C). Contact with alkali metals produces hydrogen gas. Incompatible with strong acids, caustics, aliphatic amines, isocyanates. Attacks some plastics, rubber, and coatings. May accumulate static electrical charges, and may cause ignition of its vapors.

MAPP or MAPP GAS (59355-75-8) Forms explosive gas mixture with air. Incompatible with strong oxidizers, copper alloys containing more than 6% copper; may cause formation of explosive compounds. Attacks some plastics, rubber, and coatings. May accumulate static electrical charges, and may cause ignition of its vapors.

MARANYL (105-60-2) Contact with strong oxidizers may cause fire and explosions.

MARBLE (1317-65-3) Incompatible with acids, alum, ammonium salts, fluorine.

MARLATE (72-43-5) Contact with strong oxidizers may cause fire and explosions. Attacks some plastics, rubber, and coatings.

MARLATE 50 (72-43-5) Contact with strong oxidizers may cause fire and explosions. Attacks some plastics, rubber, and coatings.

MARMER (330-54-1) Hydrolyzes in fairly strong acids.

MAROXOL-50 (51-28-5) Explosion may be caused by heat, friction, or shock. Contact with reducing agents, combustibles may cause fire and explosions. Produces explosive salts with ammonia or strong bases. May accumulate static electrical charges, and may cause ignition of its vapors.

MARS BROWN (1309-37-1) Contact with hydrogen peroxide, ethylene oxide, calcium hypochlorite will cause explosion. Reacts violently with powdered aluminum, hydrazine, hydrogen trisulfide.

MARSH GAS (74-82-8) Forms an explosive mixture with air. Reacts violently with strong oxidizers.

MARSHITE (7681-65-4) Incompatible with chlorine, fluorine, peroxides.

MARS RED (1309-37-1) Contact with hydrogen peroxide, ethylene oxide, calcium hypochlorite will cause explosion. Reacts violently with powdered aluminum, hydrazine, hydrogen trisulfide.

MARVEX (62-73-7) Attacks some plastics, rubber, and coatings.

MASSICOT (1317-36-8) Reacts violently with aluminum powder, hydrogen peroxide, hydrogen trisulfide, lithium carbide, perchloric acid, strong oxidizers.

MATRICARIA CAMPHOR (76-22-2) Forms explosive mixture with air (flash point 155°F/66°C). Violent, possibly explosive, reaction occurs with strong oxidizers. May accumulate static electrical charges, and may cause ignition of its vapors.

MATTING ACID (7664-93-9) A strong oxidizer that can react violently with risk of fire and explosion with many substances, including reducing agents, organic and combustible substances, and bases. Incompatible with nonoxidizing mineral acids, organic acids, bases, acrylates, aldehydes, alcohols, alkylene oxides, ammonia, aliphatic amines, alkanolamines, aromatic amines, amides, chlorates, epichlorohydrin, fulminates, glycols, isocyanates, ketones, metals

495

(powdered), organic anhydrides, perchlorates, picrates, substituted allyls, phenols and cresols, water, acetic anhydride, acetone cyanhydrin, acetonitrile, acrolein, acrylonitrile, allyl alcohol, allyl chloride, 2-aminoethanol, ammonium hydroxide, aniline, bromine pentafluoride, n-butyraldehyde, caprolactam solution, carbides, cesium acetylene carbide, chlorine trifluoride, chlorosulfonic acid, cuprous nitride, diisobutylene, ethylene cyanohydrin, ethylene diamine, ethylene glycol, ethyleneimine, hydrochloric acid, iodine heptafluoride, iron, isoprene, lithium silicide, mercuric nitride, mesityl oxide, nitric acid, p-nitrotoluene, perchloric acid, phosphorus, potassium $tert$-butoxide, potassium chlorate, potassium permanganate, propiolactone (beta-), propylene oxide, pyridine, rubidium acetylene, silver permanganate, sodium, sodium carbonate, sodium chlorate, sodium hydroxide, styrene monomer, vinyl acetate, Attacks most metals, and some plastics, rubber, and coatings.

MAZIDE (123-33-1) Contact with strong oxidizers may cause fire and explosions.

MAZOTEN (52-68-6) Contact with strong oxidizers may cause fire and explosions.

MBC (17804-35-2) Heat, water, strong acids and strong alkalies can cause decomposition and formation of toxic oxides of nitrogen.

M-B-C FUMIGANT (74-83-9) Incompatible with strong oxidizers, aluminum, dimethylsulfoxide, ethylene oxide, water. Attacks zinc, magnesium, alkali metals, and their alloys. Forms aluminum alkyls in presence of aluminum; aluminum alkyls are spontaneously ignitable materials.

MBDA (1918-00-9) Incompatible with sulfuric acid, bases, ammonia, aliphatic amines, alkanolamines, isocyanates, alkylene oxides, epichlorohydrin.

MBI (101-68-8) Incompatible with strong alkalies, acids, alcohols, ammonia, amines, amides, glycols, caprolactam. Unstable above 100°F/37.8°C. Attacks some plastics, rubber, and coatings.

MBK (591-78-6) Forms explosive mixture with air (flash point 77°F/25°C). Strong oxidizers may cause fire and explosions. Dissolves some plastics, resins, and rubber.

MBT (149-30-4) Incompatible with oxidizers. May accumulate static electrical charges, and may cause ignition of its vapors.

MC (7487-94-7) Incompatible with light metals (aluminum, magnesium, beryllium, etc.), sodium, potassium.

MCA (79-11-8) Aqueous solution is a strong acid. Incompatible with strong oxidizers, bases. Attacks most common metals in the presence of moisture.

MCB (108-90-7) Forms explosive mixture with air (flash point 82°F/28°C). May accumulate static electrical charges, and may cause ignition of its vapors. Incompatible with oxidizers, dimethylsulfoxide, sodium powder, silver perchloride. Attacks some plastics, rubber, and coatings.

MCF (79-22-1) Forms explosive mixture with air (flash point 54°F/12°C). Water contact produces hydrochloric acid. Corrodes metals in the presence of moisture. Attacks some plastics, rubber, and coatings.

MCP (10103-46-5) Water contact forms acid; attacks metals.

MDB (541-73-1) Forms explosive mixture with air (flash point 151°F/66°C). Incompatible with strong oxidizers, alkali metals (i.e., lithium, sodium, potassium, rubidium, cesium, francium).

MDCB (541-73-1) Forms explosive mixture with air (flash point 151°F/66°C). Incompatible with strong oxidizers, alkali metals (i.e., lithium, sodium, potassium, rubidium, cesium, francium).

ME-1700 (72-54-8) Contact with strong oxidizers may cause fire and explosions.

MEADOW GREEN (12002-03-8) Contact with strong oxidizers may cause fire and explosions.

MEASURIN (50-78-2) Fires and explosions may result from contact with strong oxidizers. Alkali hydroxides or carbonates cause decomposition.

MECB (111-77-3) Incompatible with sulfuric acid, isocyanates, perchloric acid.

MECS (109-86-4) Forms explosive mixture with air (flash point 103°F/39°C). Heat or oxidizers may cause formation of unstable peroxides. Attacks many metals. Strong oxidizers cause fire and explosions. Strong bases cause decomposition. Attacks some plastics, rubber, and coatings. May accumulate static electrical charges, and may cause ignition of its vapors.

MEDIBEN (1918-00-9) Incompatible with sulfuric acid, bases, ammonia, aliphatic amines, alkanolamines, isocyanates, alkylene oxides, epichlorohydrin.

MEETCO (78-93-3) Forms explosive mixture with air (flash point 16°F/−9°C). Extremely flammable. Incompatible with sulfuric acid, nitric acid, aliphatic amines, strong oxidizers, potassium *tert*-butoxide, 2-propanol, chlorosulfonic acid, oleum.

MEG (107-21-1) Incompatible with strong acids, caustics, aliphatic amines, isocyanates, chlorosulfonic acid, oleum, strong oxidizers.

MEK (78-93-3) Forms explosive mixture with air (flash point 16°F/−9°C). Extremely flammable. Incompatible with sulfuric acid, nitric acid, aliphatic amines, strong oxidizers, potassium *tert*-butoxide, 2-propanol, chlorosulfonic acid, oleum.

MEK-OXIME (96-29-7) Forms explosive mixture with air (flash point above 155°F/68°C). Strong oxidizers may cause fire and explosions.

MEKP (1338-23-4) Forms explosive mixture with air (flash point 125°F/52°C). Explosive decomposition occurs above 176°F/80°C. Pure substance is shock-sensitive. Strong oxidizer. Reacts violently with strong acids, strong bases, reducing agents, combustible substances, organic materials, oxides of heavy metals, salts, trace contaminants, amines. May accumulate static electrical charges, and may cause ignition of its vapors.

MEK PEROXIDE (1338-23-4) Forms explosive mixture with air (flash point 125°F/52°C). Explosive decomposition occurs above 176°F/80°C. Pure substance is shock-sensitive. Strong oxidizer. Reacts violently with strong acids, strong bases, reducing agents, combustible substances, organic materials, oxides of heavy metals, salts, trace contaminants, amines. May accumulate static electrical charges, and may cause ignition of its vapors.

MEKP-HA 1 (1338-23-4) Forms explosive mixture with air (flash point 125°F/52°C). Explosive decomposition occurs above 176°F/80°C. Pure substance is shock-sensitive. Strong oxidizer. Reacts violently with strong acids, strong bases, reducing agents, combustible substances, organic materials, oxides of heavy metals, salts, trace contaminants, amines. May accumulate static electrical charges, and may cause ignition of its vapors.

MEKP-LA 1 (1338-23-4) Forms explosive mixture with air (flash point 125°F/52°C). Explosive decomposition occurs above 176°F/80°C. Pure substance is shock-sensitive. Strong oxidizer. Reacts violently with strong acids, strong bases, reducing agents, combustible substances, organic materials, oxides of heavy metals, salts, trace contaminants, amines. May accumulate static electrical charges, and cause ignition of its vapors.

MELDANE (56-72-4) Contact with strong oxidizers may cause fire and explosions.

MELDONE (56-72-4) Contact with strong oxidizers may cause fire and explosions.

MELINITE (88-89-1) Dry material is explosive. Impact-sensitive; protect from shock. Copper, lead, zinc and other metals, or their salts can form salts that are initiators and much more sensitive to shock than this chemical. Shock-sensitive salts include ammonium salts and calcium salts; the calcium salt may form when picric acid comes into contact with plaster and concrete. May accumulate static electrical charges, and may cause explosion. Aqueous solution is a strong oxidizer and a strong acid; reacts violently with reducing agents, combustibles, organics and easily oxidized materials, caustics, aluminum and other metal powders. Attacks metals.

MELIPAX (8001-35-2) Reacts with oxidizers, with a risk of fire or explosions. Attacks metals in the presence of moisture.

MENDRIN (72-20-8) Incompatible with parathion, strong acids (forms explosive vapors), strong oxidizers.

MENIPHOS (7786-34-7) Contact with strong oxidizers may cause fire and explosions. Attacks some plastics, rubber, and coatings.

MENITE (7786-34-7) Contact with strong oxidizers may cause fire and explosions. Attacks some plastics, rubber, and coatings.

p-**MENTHA-1,8-DIENE or** *p*-**MENTHA-1,8-DIENE, DL-** (138-86-3) Forms explosive mixture with air (flash point 115°F/46°C). Strong oxidizers may cause fire and explosions.

1,8(9)-P-MENTHADIENE (138-86-3) Forms explosive mixture with air (flash point 115°F/46°C). Strong oxidizers may cause fire and explosions.

MEP (104-90-5) Forms explosive mixture with air (flash point 155°F/68°C). Incompatible with acids, isocyanates, phenols, cresols.

ME-PARATHION (298-00-0) Mixtures with magnesium may be explosive.

MEPATON (298-00-0) Mixtures with magnesium may be explosive.

MEPOX (298-00-0) Mixtures with magnesium may be explosive.

MER (80-62-6) Forms explosive mixture with air (flash point 50°F/10°C). Incompatible with caustics, nitrates, strong acids, aliphatic amines, alkanolamines, peroxides, and strong oxidizers. Heat and/or lack of appropriate inhibitor can cause polymerization. May accumulate static electrical charges, and may cause ignition of its vapors.

MERCAPTAN METHYLIQUE (French) (74-93-1) May accumulate static electrical charges, and may cause ignition of its vapors. Forms explosive air–gas mixture (flash point 0°F/−18°C). Strong oxidizers may cause fire and explosions. Attacks some plastics, coatings, and rubber.

MERCAPTOBENZENE (108-98-5) Forms explosive mixture with air (flash point 132°F/56°C). Incompatible with strong acids, caustics, alkali metals (i.e., lithium, sodium, potassium, rubidium, cesium, francium). Oxidizes on contact with air. Corrosive to carbon steel.

2-MERCAPTOBENZOTHIAZOL (149-30-4) Incompatible with oxidizers. May accumulate static electrical charges, and may cause ignition of its vapors.

2-MERCAPTOBENZOTHIAZOL SODIUM SALT (2492-26-4) Incompatible with oxidizers. Concentrated aqueous solutions are corrosive.

1-MERCAPTOBUTANE (109-79-5) Forms explosive mixture with air (flash point 35°F/2°C). Incompatible with strong oxidizers and nitric acid. Attacks some plastics and rubber.

MERCAPTODIMETHUR (2032-65-7) Contact with strong oxidizers may cause fire and explosions.

1-MERCAPTODODECANE (112-55-0) Contact with strong oxidizers may cause fire and explosions.

MERCAPTOETHANE (75-08-1) Forms explosive mixture with air (flash point less than 0°F/−18°C). May accumulate static electrical charges, and may cause ignition of its vapors. Contact with strong oxidizers may cause fire and explosions. Attacks some forms of plastics, coatings, and rubber.

MERCAPTOETHANOL (60-24-2) Forms explosive mixture with air (flash point 165°F/74°C). Incompatible with oxidizers, strong acids, caustics, aliphatic amines, isocyanates.

MERCAPTOMETHANE (74-93-1) May accumulate static electrical charges, and may cause ignition of its vapors. Forms explosive air–gas mixture (flash point 0°F/−18°C). Strong oxidizers may cause fire and explosions. Attacks some plastics, coatings, and rubber.

MERCAPTOPHOS (8065-48-3) Forms explosive mixture with air (flash point 113°F/45°C). Incompatible with water, strong oxidizers, caustics.

2-MERCAPTOPROPANE (75-33-2) Forms explosive mixture with air (flash point −30°F/−34°C). Reacts violently with strong oxidizers.

3-MERCAPTOPROPANE (107-03-9) Incompatible with oxidizers, reducing agents, strong acids, caustics, alkali metal, calcium hypochlorite (reacts violently).

3-MERCAPTOPROPANOL (107-03-9) Incompatible with oxidizers, reducing agents, strong acids, caustics, alkali metal, calcium hypochlorite (reacts violently).

MERCAPTOTHION (121-75-5) Incompatible with strong oxidizers, magnesium, alkaline pesticides. Attacks metals, some plastics, rubber, and coatings.

MERCOL 25 (25155-30-0) Reacts with acids, including fumes.

MERCURAM (137-26-8) Combustible solid (flash point 192°F/89°C). Strong oxidizers may cause fire and explosions; contact with strong acid or oxidizable materials produces toxic gases.

MERCURE (French) (7439-97-6) Acetylene, acetylene products, ammonia gases can produce shock-sensitive solids that can initiate fires of combustibles. Reacts violently with boron phosphodiiodide, chlorine, chlorine dioxide, methyl azide. Attacks copper and copper alloys.

MERCURIACETATE (1600-27-7) Heat and light cause decomposition.

MERCURIALIN (74-89-5) Forms explosive gas mixture with air. Incompatible with acids, organic anhydrides, isocyanates, vinyl acetate, acrylates, substituted allyls, alkylene oxides, epichlorohydrin, ketones, aldehydes, alcohols, glycols, phenols, cresols, caprolactam solution, strong oxidizers. Attacks aluminum, ,copper, lead, tin, zinc, and alloys, and some plastics, rubber, and coatings.

MERCURIALIN SOLUTION (74-89-5) Forms explosive gas mixture with air. Incompatible with acids, organic anhydrides, isocyanates, vinyl acetate, acrylates, substituted allyls, alkylene oxides, epichlorohydrin, ketones, aldehydes, alcohols, glycols, phenols, cresols, caprolactam solution, strong oxidizers. Attacks aluminum, copper, lead, tin, zinc, and alloys.

MERCURIC ACETATE (1600-27-7) Heat and light cause decomposition.

MERCURIC AMMONIUM CHLORIDE (10124-48-8) Reacts violently with halogens and metal salts of amines.

MERCURIC BICHLORIDE (7487-94-7) Incompatible with light metals (aluminum, magnesium, beryllium, etc.), sodium, potassium.

MERCURIC CHLORIDE (7487-94-7) Incompatible with light metals (aluminum, magnesium, beryllium, etc.), sodium, potassium.

MERCURIC CHLORIDE, AMMONIATED (10124-48-8) Reacts violently with halogens and metal salts of amines.

MERCURIC CYANIDE (592-04-1) Contact with acidic material produces hydrogen cyanide gas. An impact-sensitive explosive. Reacts strongly with fluorine, hydrogen cyanide, magnesium.

MERCURIC DIACETATE (1600-27-7) Heat and light cause decomposition.

MERCURIC IODIDE (7774-29-0) Reacts violently with strong oxidizers, chlorine, fluorine.

MERCURIC IODIDE, RED (7774-29-0) Reacts violently with strong oxidizers, chlorine, fluorine.

MERCURIC NITRATE (10045-94-0) A powerful oxidizer. Incompatible with reducing agents, combustibles; forms shock-sensitive compounds with acetylene, ethanol, isobutene, phosphine, sulfur. Forms heat-sensitive compound with potassium cyanide. Incompatible with strong acids, hydrocarbons. Aqueous solution corrodes metals.

MERCURIC OXIDE (21908-53-2) A strong oxidizer; reacts violently with reducing agents and combustible materials. Incompatible with alcohols, chlorine, phosphorus, sulfur, alkali metals (i.e., lithium, sodium, potassium, rubidium, cesium, francium).

MERCURIC OXIDE, RED (21908-53-2) A strong oxidizer; reacts violently with reducing agents and combustible materials. Incompatible with alcohols, chlorine, phosphorus, sulfur, alkali metals (i.e., lithium, sodium, potassium, rubidium, cesium, francium).

MERCURIC OXIDE, YELLOW (21908-53-2) A strong oxidizer; reacts violently with reducing agents and combustible materials. Incompatible with alcohols, chlorine, phosphorus, sulfur, alkali metals (i.e., lithium, sodium, potassium, rubidium, cesium, francium).

MERCURIC RHODANIDE (592-85-8) Heat can cause violent decomposition.

MERCURIC SULFATE (7783-35-9) Water contact produces sulfuric acid. Light may cause decomposition. Incompatible with aluminum or magnesium. Reacts violently with hydrogen chloride above 250°F/121°C. Attacks metals when wet.

MERCURIC SULFIDE, RED or BLACK (1344-48-5) Contact with acids or water causes evolution of flammable hydrogen sulfide, which forms an explosive mixture with air. Contact with strong oxidizers may cause a violent reaction.

MERCURIC SULFOCYANATE or MERCURIC SULFO CYANATE, SOLID (592-85-8) Heat can cause violent decomposition.

MERCURIC SULFOCYANIDE (592-85-8) Heat can cause violent decomposition.

MERCURIC THIOCYANATE (592-85-8) Heat can cause violent decomposition.

MERCURIO (Italian) (7439-97-6) Acetylene, acetylene products, ammonia gases can produce shock-sensitive solids that can initiate fires of combustibles. Reacts violently with boron phosphodiiodide, chlorine, chlorine dioxide, methyl azide. Attacks copper and copper alloys.

MERCUROUS NITRATE or MERCUROUS NITRATE MONOHYDRATE (10415-75-5) A powerful oxidizer. Incompatible with reducing agents, combustible materials. Forms shock-sensitive compounds with phosphorus. Aqueous solutions are corrosive.

MERCURY (7439-97-6) Acetylene, acetylene products, ammonia gases can produce shock-sensitive solids that can initiate fires of combustibles. Reacts violently with boron phosphodiiodide, chlorine, chlorine dioxide, methyl azide. Attacks copper and copper alloys.

MERCURY ACETATE (1600-27-7) Heat and light cause decomposition.

MERCURY(2+) ACETATE or MERCURY(II) ACETATE (1600-27-7) Heat and light cause decomposition.

MERCURY AMIDE CHLORIDE (10124-48-8) Reacts violently with halogens and metal salts of amines.

MERCURY AMINE CHLORIDE (10124-48-8) Reacts violently with halogens and metal salts of amines.

MERCURY AMMONIUM CHLORIDE (10124-48-8) Reacts violently with halogens and metal salts of amines.

MERCURY BICHLORIDE (7487-94-7) Incompatible with light metals (aluminum, magnesium, beryllium, etc.), sodium, potassium.

MERCURY BINIODIDE (7774-29-0) Reacts violently with strong oxidizers, chlorine, fluorine.

MERCURY BISULFATE (7783-35-9) Water contact produces sulfuric acid. Light may cause decomposition. Incompatible with aluminum or magnesium. Reacts violently with hydrogen chloride above 250°F/121°C. Attacks metals when wet.

501

MERCURY(2+) CHLORIDE or MERCURY(II) CHLORIDE (7487-94-7) Incompatible with light metals (aluminum, magnesium, beryllium, etc.), sodium, potassium.

MERCURY CYANIDE (592-04-1) Contact with acidic material produces hydrogen cyanide gas. An impact-sensitive explosive. Reacts strongly with fluorine, hydrogen cyanide, magnesium.

MERCURY(2+) CYANIDE or MERCURY(II) CYANIDE (592-04-1) Contact with acidic material produces hydrogen cyanide gas. An impact-sensitive explosive. Reacts stongly with fluorine, hydrogen cyanide, magnesium.

MERCURY DIACETATE (1600-27-7) Heat and light cause decomposition.

MERCURY, ELEMENTAL (7439-97-6) Acetylene, acetylene products, ammonia gases can produce shock-sensitive solids that can initiate fires of combustibles. Reacts violently with boron phosphodiiodide, chlorine, chlorine dioxide, methyl azide. Attacks copper and copper alloys.

MERCURY(2+) IODIDE or MERCURY(II) IODIDE (7774-29-0) Reacts violently with strong oxidizers, chlorine, fluorine.

MERCURYL ACETATE (1600-27-7) Heat and light cause decomposition.

MERCURY, METALLIC (7439-97-6) Acetylene, acetylene products, ammonia gases can produce shock-sensitive solids that can initiate fires of combustibles. Reacts violently with boron phosphodiiodide, chlorine, chlorine dioxide, methyl azide. Attacks copper and copper alloys.

MERCURY(2+) NITRATE (1:2) or MERCURY(II) NITRATE (1:2) (10045-94-0) A powerful oxidizer. Incompatible with reducing agents, combustibles; forms shock-sensitive compounds with acetylene, ethanol, isobutene, phosphine, sulfur. Forms heat-sensitive compound with potassium cyanide. Incompatible with strong acids, hydrocarbons. Aqueous solution corrodes metals.

MERCURY NITRATE (10045-94-0) A powerful oxidizer. Incompatible with reducing agents, combustibles; forms shock-sensitive compounds with acetylene, ethanol, isobutene, phosphine, sulfur. Forms heat-sensitive compound with potassium cyanide. Incompatible with strong acids, hydrocarbons. Aqueous solution corrodes metals.

MERCURY(I) NITRATE (1:1) (10415-75-5) A powerful oxidizer. Incompatible with reducing agents, combustible materials. Forms shock-sensitive compounds with phosphorus. Aqueous solutions are corrosive.

MERCURY NITRATE MONOHYDRATE (10045-94-0) A powerful oxidizer. Incompatible with reducing agents, combustibles; forms shock-sensitive compounds with acetylene, ethanol, isobutene, phosphine, sulfur. Forms heat-sensitive compound with potassium cyanide. Incompatible with strong acids, hydrocarbons. Aqueous solution corrodes metals.

MERCURY OXIDE (21908-53-2) A strong oxidizer; reacts violently with reducing agents and combustible materials. Incompatible with alcohols, chlorine, phosphorus, sulfur, alkali metals (i.e., lithium, sodium, potassium, rubidium, cesium, francium).

MERCURY PERCHLORIDE (7487-94-7) Incompatible with light metals (aluminum, magnesium, beryllium, etc.), sodium, potassium.

MERCURY PERNITRATE (10045-94-0) A powerful oxidizer. Incompatible with reducing agents, combustibles; forms shock-sensitive compounds with acetylene, ethanol, isobutene, phosphine, sulfur. Forms heat-sensitive compound with potassium cyanide. Incompatible with strong acids, hydrocarbons. Aqueous solution corrodes metals.

MERCURY PERSULFATE (7783-35-9) Water contact produces sulfuric acid. Light may cause decomposition. Incompatible with aluminum or magnesium. Reacts violently with hydrogen chloride above 250°F/121°C. Attacks metals when wet.

MERCURY PROTONITRATE (10415-75-5) A powerful oxidizer. Incompatible with reducing agents, combustible materials. Forms shock-sensitive compounds with phosphorus. Aqueous solutions are corrosive.

MERCURY RHODANIDE (592-85-8) Heat can cause violent decomposition.

MERCURY(2+) SULFATE (1:1) or **MERCURY(II) SULFATE** (7783-35-9) Water contact produces sulfuric acid. Light may cause decomposition. Incompatible with aluminum or magnesium. Reacts violently with hydrogen chloride above 250°F/121°C. Attacks metals when wet.

MERCURY(2+) THIOCYANATE or **MERCURY(II) THIOCYANATE** (592-85-8) Heat can cause violent decomposition.

MERPAN (133-06-2) Incompatible with tetraethyl pyrophosphate, parathion.

MERVAMINE (119-36-8) Forms explosive mixture with air (flash point 205°F/96°C). Incompatible with strong acids, nitrates, oxidizers.

MESITYLENE (108-67-8) Forms explosive mixture with air (flash point 117°F/47°C). Reacts violently with strong oxidizers. May accumulate static electrical charges, and may cause ignition of its vapors.

MESITYL OXIDE (141-79-7) Forms explosive mixture with air (flash point 87°F/31°C). Forms peroxides. Incompatible with strong acids aliphatic amines, alkanolamines, 2-aminoethanol, ethylene diamine, chlorosulfonic acid, oleum, oxidizers. Dissolves some forms of plastics, resins and rubber. Attacks copper.

MESUROL (2032-65-7) Contact with strong oxidizers may cause fire and explosions.

MESYL CHLORIDE (124-63-0) Forms explosive mixture with air. Incompatible with water, caustics, ammonia, amines.

METACETONE (96-22-0) Forms explosive mixture with air (flash point 55°F/13°C). Incompatible with strong acids, aliphatic amines, strong oxidizers. Attacks some plastics, rubber, and coatings. May accumulate static electrical charges, and may cause ignition of its vapors.

METACETONIC ACID (79-09-4) Forms explosive mixture with air (flash point 126°F/52°C). Incompatible with sulfuric acid, strong bases, ammonia, aliphatic amines, alkanolamines, isocyanates, alkylene oxides, epichlorohydrin, oxidizers.

METACIDE (298-00-0) Mixtures with magnesium may be explosive.

METAFOS (298-00-0) Mixtures with magnesium may be explosive.

METAKRYLAN METYLU (Polish) (80-62-6) Forms explosive mixture with air (flash point 50°F/10°C). Incompatible with caustics, nitrates, strong acids, aliphatic amines, alkanolamines, peroxides and

503

strong oxidizers. Heat and/or lack of appropriate inhibitor can cause polymerization. May accumulate static electrical charges, and may cause ignition of its vapors.

METALLIC ARSENIC (7440-38-2) Contact with hydrogen gas produces arsine. Incompatible with strong acids, strong oxidizers, peroxides, bromine pentafluoride, bromine trifluoride, cesium acetylene carbide, chromium trioxide, nitrogen trichloride, silver nitrate.

METALLIC MERCURY (7439-97-6) Acetylene, acetylene products, ammonia gases can produce shock-sensitive solids that can initiate fires of combustibles. Reacts violently with boron phosphodiiodide, chlorine, chlorine dioxide, methyl azide. Attacks copper and copper alloys.

METALLIC RESINATE (9007-13-0) Incompatible with oxidizers, strong acids.

METANOLO (Italian) (67-56-1) May accumulate static electrical charges, and may cause ignition of its vapors. Forms explosive mixture with air (flash point 52°F/11°C). Incompatible with strong acids, strong oxidizers, caustics, aliphatic amines, isocyanates, chromic anhydride, lead perchlorate, perchloric acid, phosphorus trioxide. May react with metallic aluminum at high temperature. Attacks some plastics, rubber, and coatings.

METAPHOR (298-00-0) Mixtures with magnesium may be explosive.

METAPHOS (298-00-0) Mixtures with magnesium may be explosive.

METAPHOSPHORIC ACID (7664-38-2) Incompatible with aliphatic amines, alkanolamines, alkylene oxides, aromatic amines, amides, ammonia, ammonium hydroxide, bases, calcium oxide, epichlorohydrin, isocyanates, nitromethane (explosive), oleum, organic anhydrides, sulfuric acid, sodium tetrahydroborate, strong oxidizers, vinyl acetate, water. Contact with most metals produces hydrogen gas. Attacks some plastics, rubber, and coatings, and glass and ceramics.

METAQUEST A (60-00-4) Incompatible with sulfuric acid, bases, ammonia, aliphatic amines, alkanolamines, isocyanates, alkylene oxides, epichlorohydrin.

METAUPON (112-80-1) Incompatible with strong oxidizers, perchloric acid, and aluminum powder; may cause explosions. Corrodes aluminum.

METELILACHLOR (51218-45-0) Incompatible with strong acids, nitrates, oxidizers.

METHAANTHIOL (Dutch) (74-93-1) May accumulate static electrical charges, and may cause ignition of its vapors. Forms explosive air–gas mixture (flash point 0°F/−18°C). Strong oxidizers may cause fire and explosions. Attacks some plastics, coatings, and rubber.

METHACETALDEHYDE (9002-91-9) Forms explosive mixture with air (flash point 97°F/36°C). Reacts violently with strong oxidizers. Incompatible with strong acids, caustics, ammonia, amines. May accumulate static electrical charges, and may cause ignition of its vapors.

METHACETONE (96-22-0) Forms explosive mixture with air (flash point 55°F/13°C). Incompatible with strong acids, aliphatic amines, strong oxidizers. Attacks some plastics, rubber, and coatings. May accumulate static electrical charges, and may cause ignition of its vapors.

METHACIDE (108-88-3) Forms explosive mixture with air (flash point 40°F/4°C). Strong oxidizers may cause fire and explosions. Attacks some plastics, rubber, and coatings. May accumulate static electrical charges, and may cause ignition of its vapors.

METHACRYLATE de BUTYLE (French) (97-88-1) Unless inhibitor is maintained at the proper level, oxidizers, heat, ultraviolet light, or moisture may cause polymerization. May accumulate static electrical charges, and may cause ignition of its vapors.

METHACRYLATE de METHYLE (French) (80-62-6) Forms explosive mixture with air (flash point 50°F/10°C). Incompatible with caustics, nitrates, strong acids, aliphatic amines, alkanolamines, peroxides, and strong oxidizers. Heat and/or lack of appropriate inhibitor can cause polymerization. May accumulate static electrical charges, and may cause ignition of its vapors.

METHACRYLIC ACID (79-41-4) Forms explosive mixture with air (flash point 152°F/67°C). A reducing agent; reacts with oxidizers. Forms unstable peroxides; can polymerize violently, especially if stored above 120°F/49°C. Incompatible with strong acids, caustics, ammonia, amines, isocyanates, alkylene oxides, epichlorohydrin.

2-METHACRYLIC ACID (79-41-4) Forms explosive mixture with air (flash point 152°F/67°C). A reducing agent; reacts with oxidizers. Forms unstable peroxides; can polymerize violently, especially if stored above 120°F/49°C. Incompatible with strong acids, caustics, ammonia, amines, isocyanates, alkylene oxides, epichlorohydrin.

a-METHACRYLIC ACID or alpha-METHACRYLIC ACID (79-41-4) Forms explosive mixture with air (flash point 152°F/67°C). A reducing agent; reacts with oxidizers. Forms unstable peroxides; can polymerize violently, especially if stored above 120°F/49°C. Incompatible with strong acids, caustics, ammonia, amines, isocyanates, alkylene oxides, epichlorohydrin.

METHACRYLIC ACID, BUTYL ESTER (97-88-1) Unless inhibitor is maintained at the proper level, oxidizers, heat, ultraviolet light, or moisture may cause polymerization. May accumulate static electrical charges, and may cause ignition of its vapors.

METHACRYLIC ACID, DODECYL ESTER (142-90-5) Forms explosive mixture with air (flash point 230°F/110°C). Oxidizers or reducing agents may cause polymerization.

METHACRYLIC ACID, DODECYL AND PENTADECYL ESTER MIX (N/A) Incompatible with sulfuric acid, nitric acid, aliphatic amines, alkanolamines.

METHACRYLIC ACID, 2,3-EPOXY PROPYL ESTER (106-99-0) Self-reactive. Forms explosive peroxides with air. Fires, explosions, or hazardous polymerization may result from contact with air, strong oxidizers, strong acids, ozone, nitrogen dioxide, copper and its alloys, phenol, chlorine dioxide, crotonaldehyde, or a free radical polymerization initiator such as hydroquinone. Add inhibitor (such as *tert*-butyl catechol) and monitor to ensure that effective levels are maintained at all times. May accumulate static electrical charges, and may cause ignition of its vapors.

1-2-METHACRYLIC ACID, ETHYL ESTER (97-63-2) Forms explosive mixture with air (flash point 80°F/27°C). Incompatible with strong acids, amines, oxidizers. Corrodes some metals.

METHACRYLIC ACID, GLACIAL (79-41-4) Forms explosive mixture with air (flash point 152°F/67°C). A reducing agent; reacts with oxidizers. Forms unstable peroxides; can polymerize violently, especially if stored above 120°F/49°C. Incompatible with strong acids, caustics, ammonia, amines, isocyanates, alkylene oxides, epichlorohydrin.

METHACRYLIC ACID, INHIBITED (79-41-4) Forms explosive mixture with air (flash point 152°F/67°C). A reducing agent; reacts with oxidizers. Forms unstable peroxides; can polymerize violently, especially if stored above 120°F/49°C. Incompatible with strong acids, caustics, ammonia, amines, isocyanates, alkylene oxides, epichlorohydrin.

METHACRYLIC ACID, ISOBUTY ESTER (97-86-9) Forms explosive mixture with air (flash point 112°F/44°C). Able to form unstable peroxides. Incompatible with strong acids, aliphatic amines, alkanolamines, catalysts, strong oxidizers.

METHACRYLIC ACID, METHYL ESTER (80-62-6) Forms explosive mixture with air (flash point 50°F/10°C). Incompatible with caustics, nitrates, strong acids, aliphatic amines, alkanolamines, peroxides and strong oxidizers. Heat and/or lack of appropriate inhibitor can cause polymerization. May accumulate static electrical charges, and may cause ignition of its vapors.

METHACRYLONITRILE (126-98-7) Forms explosive mixture with air (flash point 34°F/1°C). Incompatible with aliphatic amines, alkanolamines. Reacts violently with oxidizers. Incompatible with strong acids, strong bases, or light exposure; may cause polymerization.

METHACRYLSAEURE BUTYL ESTER (German) (97-88-1) Unless inhibitor is maintained at the proper level, oxidizers, heat, ultraviolet light, or moisture may cause polymerization. May accumulate static electrical charges, and may cause ignition of its vapors.

METHACRYLSAEUREMETHYL ESTER (German) (80-62-6) Forms explosive mixture with air (flash point 50°F/10°C). Incompatible with caustics, nitrates, strong acids, aliphatic amines, alkanolamines, peroxides, and strong oxidizers. Heat and/or lack of appropriate inhibitor can cause polymerization. May accumulate static electrical charges, and may cause ignition of its vapors.

METHALDEHYDE (9002-91-9) Forms explosive mixture with air (flash point 97°F/36°C). Reacts violently with strong oxidizers. Incompatible with strong acids, caustics, ammonia, amines. May accumulate static electrical charges, and may cause ignition of its vapors.

METHALLYL ALCOHOL (513-42-8) Forms explosive mixture with air (flash point 92°F/33°C). Incompatible with strong acids, caustics, aliphatic amines, isocyanates, oxidizers.

METHALLYL CHLORIDE (563-47-3) Forms explosive mixture with air (flash point 11°F/−12°C). Strong oxidizers may cause fire and explosions.

2-METHALLYL CHLORIDE (563-47-3) Forms explosive mixture with air (flash point 11°F/−12°C). Strong oxidizers may cause fire and explosions.

a-METHALLYL CHLORIDE or alpha-METHALLYL CHLORIDE (563-47-3) Forms explosive mixture with air (flash point 11°F/−12°C). Strong oxidizers may cause fire and explosions.

b-METHALLYL CHLORIDE or beta-METHALLYL CHLORIDE (563-47-3) Forms explosive mixture with air (flash point 11°F/−12°C). Strong oxidizers may cause fire and explosions.

METHANAL or METHANAL SOLUTION (50-00-0) May polymerize unless properly inhibited (usually with methanol). Forms explosive mixture with air (flash point 122°F/50°C). Incompatible with strong acids, amines, strong oxidizers, alkaline materials, nitrogen dioxide, performic acid. Reaction with hydrochloric acid produces bis-chloromethyl ether, a carcinogen.

METHANAMIDE (75-12-7) Produces hydrocyanic acid with water solutions. Incompatible with nonoxidizing mineral acids, strong acids, ammonia, cresols, iodine, isocyanates, oleum, phenols, pyridine, sulfur trioxide.

METHANAMINE or METHANAMINE (9CI) or METHANAMINE SOLUTION (74-89-5) Forms explosive gas mixture with air. Incompatible with acids, organic anhydrides, isocyanates, vinyl acetate, acrylates, substituted allyls, alkylene oxides, epichlorohydrin, ketones, aldehydes, alcohols, glycols, phenols, cresols, caprolactam solution, strong oxidizers. Attacks aluminum, copper, lead, tin, zinc, and alloys, and some plastics, rubber, and coatings.

METHANAMINE, N-METHYL- (9CI) (124-40-3) A gas. Incompatible with acids, organic anhydrides, isocyanates, vinyl acetate, acrylates, substituted allyls, alkylene oxides, epichlorohydrin, ketones, aldehydes, alcohols, glycols, mercury, phenols, cresols, caprolactam solution, strong oxidizers. Attacks aluminum, copper, lead, tin, zinc, and alloys, and some plastics, rubber, and coatings.

METHANAMINE, N-METHYL-N-NITROSO- (62-75-9) Ultraviolet light, strong oxidizers can cause reactions. Store in dark bottles.

METHANE (74-82-8) Forms an explosive mixture with air. Reacts violently with strong oxidizers.

METHANE, BROMOCHLORO- (74-97-5) Reacts with chemically active metals such as calcium, powdered aluminum, zinc, and magnesium. Liquid attacks some plastics, rubber, and coatings.

METHANE, BROMOTRIFLUORO (75-63-8) Reacts with chemically active metals, powdered aluminum, zinc, magnesium. May attack some plastics, rubber, and coatings.

METHANECARBONITRILE (75-05-8) Forms explosive mixture with air (flash point 42°F/6°C). Incompatible with water (especially if acid or alkaline), acids, caustics, nitrating agents, indium, nitrogen tetroxide, n-fluoro compounds, sulfur trioxide, iron(III) salts of perchlorate, indium, nitrogen-fluorine compounds. Reacts violently with oxidizers. May accumulate static electrical charges, and may cause ignition of its vapors.

METHANE CARBOXYLIC ACID (64-19-7) Vapor forms explosive mixture with air (flash point 105°F/40°C). Reacts violently with oxidizers and bases. Incompatible with strong acids, aliphatic amines, alkanolamines, isocyanates, alkylene oxides, epichlorohydrin, acetaldehyde, 2-aminoethanol, ammonia, ammonium nitrate, chlorosulfonic acid, chromic acid, ethylene diamine, ethyleneimine, perchloric acid, permanganates, phosphorus isocyanate, phosphorus trichloride, potassium *tert*-butoxide, xylene. Attacks cast iron and other metals, producing flammable hydrogen gas.

METHANE, CHLORO- (74-87-3) Moisture causes decomposition. Contact with aluminum causes formation of a product that may ignite spontaneously in air. Reacts with barium, lithium, sodium, magnesium, titanium. Attacks plastics, rubber, and coatings.

METHANE, CHLOROMETHOXY- (107-30-2) Forms explosive mixture with air (flash point 0°F/−17.8°C). May be able to form unstable and explosive peroxides. Produces hydrochloric acid on contact with moisture. Corrosive to metals in presence of moisture.

METHANE, CYANO- (75-05-8) Forms explosive mixture with air (flash point 42°F/6°C). Incompatible with water (especially if acid or alkaline), acids, caustics, nitrating agents, indium, nitrogen tetroxide, n-fluoro compounds, sulfur trioxide, iron(III) salts of perchlorate, indium, nitrogen-fluorine compounds. Reacts violently with oxidizers. May accumulate static electrical charges, and may cause ignition of its vapors.

METHANE, DIBROMO- (74-95-3) Mixing with potassium produces a shock-sensitive explosive.

METHANE, DICHLORODIFLUORO- (75-71-8) Reacts violently with liquid aluminum. Incompatible with chemically active metals. Attacks some plastics, rubber, and coatings.

METHANE, DIMETHOXY- (109-87-5) May be able to form unstable and explosive peroxides. Forms explosive mixture with air (flash point −4°F/−18°C). Incompatible with strong oxidizers, acids. Attacks some plastics, rubber, and coatings.

METHANEDITHIOL, S,S-DIESTER WITH O,O-DIETHYL PHOS-PHORODITHIOATE ACID (563-12-2) Incompatible with alkaline formulations. Mixtures with magnesium may be explosive.

METHANEETHIOL (74-93-1) May accumulate static electrical charges, and may cause ignition of its vapors. Forms explosive air–gas mixture (flash point 0°F/−18°C). Strong oxidizers may cause fire and explosions. Attacks some plastics, coatings, and rubber.

METHANE GAS (74-82-8) Forms an explosive mixture with air. Reacts violently with strong oxidizers.

METHANE, IODO- (74-88-4) Contact with strong oxidizers may cause fire and explosions. Heat, light and moisture contribute to instability.

METHANE, ISOCYANATO- (624-83-9) Forms explosive mixture with air (flash point 0°F/−18°C). Contact with iron, tin, copper (or salts of these elements) and with certain other catalysts (such as triphenylarsenic oxide, triethylphosphine, and tributyltin oxide) or elevated temperatures may cause polymerization. Reacts violently with warm water. Incompatible with strong acids, alcohols, glycols, amines, amides, ammonia, caprolactam, caustics, strong oxidizers. Attacks some plastics, rubber, or coatings.

METHANE, ISOTHIOCYANATO- (556-61-6) Forms explosive mixture with air (flash point 90°F/32°C). Incompatible with strong acids, caustics, ammonia, amines, amides, alcohols, glycols, caprolactam solution, strong oxidizers.

METHANE, NITRO- (75-52-5) Forms explosive mixture with air (flash point 95°F/35°C). Caustics, ammonia, strong acids, amines can cause sensitization, which can cause explosions. Strong oxidizers may cause fire and explosions. Mixtures with hydrocarbons (and other

508

combustible materials) are highly flammable. Contact with metallic oxides may cause decomposition and development of pressure. Attacks some plastics, rubber, and coatings.

METHANE OXYBIS (CHLORO-) or METHANE, OXY-BIS[CHLORO] (542-88-1) Highly volatile. Heat and light may cause formation of unstable peroxides. Heat may cause fire. Contact with water produces hydrogen chloride and formaldehyde.

METHANE, PHENYL- (108-88-3) Forms explosive mixture with air (flash point 40°F/4°C). Strong oxidizers may cause fire and explosions. Attacks some plastics, rubber, and coatings. May accumulate static electrical charges, and may cause ignition of its vapors.

METHANE, REFRIGERATED LIQUID (74-82-8) Flammable gas. Reacts violently with strong oxidizers.

METHANESULFONYL CHLORIDE (124-63-0) Forms explosive mixture with air. Incompatible with water, caustics, ammonia, amines.

METHANE TETRABROMIDE (558-13-4) Lithium or hexylcyclohexyldilead may cause an explosion.

METHANE TETRACHLORIDE (56-23-5) Becomes corrosive when in contact with water. Corrosive to metals. Reacts violently with many compounds. Decomposes on contact with chemically active metals such as sodium, potassium and magnesium. Incompatible with allyl alcohol, fluorine gas, alkali metals, aluminum. Attacks some coatings, plastics, and rubber.

METHANE, TETRACHLORO- (56-23-5) Becomes corrosive when in contact with water. Corrosive to metals. Reacts violently with many compounds. Decomposes on contact with chemically active metals such as sodium, potassium and magnesium. Incompatible with allyl alcohol, fluorine gas, alkali metals, aluminum. Attacks some coatings, plastics, and rubber.

METHANE, TETRANITRO- (509-14-8) Combustible organic matter wet with this chemical may be highly explosive, and susceptible to mild shock. Contact with hydrocarbons, alkalies, or metals produces explosive mixtures. Attacks some plastics, rubber, and coatings.

METHANETHIOL (74-93-1) May accumulate static electrical charges, and may cause ignition of its vapors. Forms explosive air-gas mixture (flash point 0°F/−18°C). Strong oxidizers may cause fire and explosions. Attacks some plastics, coatings, and rubber.

METHANETHIOMETHANE (75-18-3) Forms explosive mixture with air (flash point −36°F/−38°C) Reacts violently with strong oxidizers. May accumulate static electrical charges, and may cause ignition of its vapors.

METHANE, TRIBROMO- (75-25-2) Reacts with chemically active metals, calcium, acetone, strong caustics. Attacks some plastics, rubber, and coatings.

METHANE TRICHLORIDE (67-66-3) Decomposes in the presence of excess water or high temperatures, producing phosgene and hydrogen chloride. Can become explosive in the presence of strong alkalies and water. Can accumulate static electrical charges. In contact with water and high temperatures, it becomes corrosive; attacks iron and other metals. Incompatible with acetone, aluminum, strong oxidizers, potassium, sodium, chemically active metals, strong bases. Attacks plastics and rubber.

METHANE, TRICHLORO- (67-66-3) Decomposes in the presence of excess water or high temperatures, producing phosgene and hydrogen chloride. Can become explosive in the presence of strong alkalies and water. Can accumulate static electrical charges. In contact with water and high temperatures, it becomes corrosive; attacks iron and other metals. Incompatible with acetone, aluminum, strong oxidizers, potassium, sodium, chemically active metals, strong bases. Attacks plastics and rubber.

METHANE, TRICHLOROFLUORO- (75-69-4) Reacts with barium, lithium, sodium, magnesium, titanium. Attacks some plastics, rubber, and coatings.

METHANE, TRICHLORONITRO- (76-06-2) Can be self-reactive. Fast heating, shock, alkali metals or alkaline earth may cause explosions. A strong oxidizer; reacts violently with reducing agents, aniline in presence of heat, alcoholic sodium hydroxide, combustible substances, sodium methoxide, propargyl bromide. Liquid attacks some plastics, rubber, and coatings.

METHANOIC ACID (64-18-6) Forms explosive mixture with air (flash point 156°F/69°C). A strong reducing agent; reacts violently with oxidizers. Incompatible with sulfuric acid, bases, ammonia, aliphatic amines, alkanolamines, furfuryl alcohol, hydrogen peroxide, isocyanates, alkylene oxides, epichlorohydrin. Attacks aluminum, cast iron and steel, some plastics, rubber, and coatings.

METHANOIC ACID, AMIDE (75-12-7) Produces hydrocyanic acid with water solutions. Incompatible with nonoxidizing mineral acids, strong acids, ammonia, cresols, iodine, isocyanates, oleum, phenols, pyridine, sulfur trioxide.

4,7-METHANO-1H-INDENE (77-73-6) Forms explosive mixture with air (flash point 90°F/32°C). Forms peroxides with air. May polymerize unless inhibited and maintained under inert atmosphere. Reacts violently with strong oxidizers. May accumulate static electrical charges, and may cause ignition of its vapors.

4,7-METHANO-1H-INDENE, 1,2,4,5,6,7,8,8-OCTACHLORO-2,3,3A,4,7,7A-HEXAHYDRO- (57-74-9) Contact with strong oxidizers may cause fire and explosions. Attacks some plastics, rubber, and coatings.

4,7-METHANOINDENE, 3A,4,7,7A TETRAHYDRODIMETHYL (26472-00-4) Forms explosive mixture with air (flash point 80°F/27°C). Incompatible with strong acids, strong oxidizers.

METHANOL (67-56-1) May accumulate static electrical charges, and may cause ignition of its vapors. Forms explosive mixture with air (flash point 52°F/11°C). Incompatible with strong acids, strong oxidizers, caustics, aliphatic amines, isocyanates, chromic anhydride, lead perchlorate, perchloric acid, phosphorus trioxide. May react with metallic aluminum at high temperature. Attacks some plastics, rubber, and coatings.

METHANOLACETONITRILE (109-78-4) Forms explosive mixture with air (flash point 140°F/60°C). Avoid basic contamination to prevent polymerization. A powerful reducing agent. Hot water contact produces cyanide gas. Reacts violently with strong oxidizers or sodium hydroxide. Reacts with acids, acid salts, chlorates, nitrates. Attacks mild steel, copper, and copper alloys.

METHANOL, ETHYNYL (107-19-7) Incompatible with strong acids, aliphatic amines, phosphorus pentoxide, caustics, isocyanates, strong oxidizers.

METHANOL, SODIUM SALT (124-41-4) Flammable solid; may ignite spontaneously in moist air. A strong reducing agent; reacts violently with oxidizers, water (ignition), chloroform, chlorine oxyflour-ide, aluminum, magnesium, beryllium. Attacks some plastics, rubber, and coatings.

METHANOL, TRIMETHYL- (75-65-0) Forms explosive mixture with air (flash point 52°F/11°C). May accumulate static electrical charges, and may cause ignition of its vapors. Incompatible with strong acids including mineral acids, strong oxidizers, caustics, aliphatic amines, isocyanates, alkali metals (i.e., lithium, sodium, potassium, rubidium, cesium, francium). Attacks many plastics and some coatings.

METHANTHIOL (German) (74-93-1) May accumulate static electrical charges, and may cause ignition of its vapors. Forms explosive air–gas mixture (flash point 0°F/−18°C). Strong oxidizers may cause fire and explosions. Attacks some plastics, coatings, and rubber.

METHENEAMINE (100-97-0) Incompatible with acids, organic anhydrides, isocyanates, vinyl acetate, acrylates, substituted allyls, alkylene oxides, epichlorohydrin, ketones, aldehydes, alcohols, glycols, phenols, cresols, caprolactam solution, sodium peroxide, strong oxidizers. Attacks aluminum, copper, lead, tin, zinc, and alloys.

METHENYL TRIBROMIDE (75-25-2) Reacts with chemically active metals, calcium, acetone, strong caustics. Attacks some plastics, rubber, and coatings.

METHENYL TRICHLORIDE (67-66-3) Decomposes in the presence of excess water or high temperatures, producing phosgene and hydrogen chloride. Can become explosive in the presence of strong alkalies and water. Can accumulate static electrical charges. In contact with water and high temperatures, it becomes corrosive; attacks iron and other metals. Incompatible with acetone, aluminum, strong oxidizers, potassium, sodium, chemically active metals, strong bases. Attacks plastics and rubber.

METHIONINE HYDROXY ANALOG (583-91-5) Incompatible with sulfuric acid, caustics, ammonia, amines, isocyanates, alkylene oxides, epichlorohydrin.

METHMERCAPTURON (2032-65-7) Contact with strong oxidizers may cause fire and explosions.

METHOGAS (74-83-9) Incompatible with strong oxidizers, aluminum, dimethylsulfoxide, ethylene oxide, water. Attacks zinc, magnesium, alkali metals, and their alloys. Produces aluminum alkyls in presence of aluminum; aluminum alkyls are spontaneously ignitable materials.

O-METHOXYBENZOIC ACID (119-36-0) Incompatible with strong acids, nitrates, strong oxidizers.

METHOXYBENZOYL CHLORIDE (100-07-2) May be unstable at room temperature. Water contact produces hydrochloric acid. Attacks metals in the presence of moisture, producing hydrogen gas.

1-METHOXYBUTANE-1,3-DIONE (105-45-3) Incompatible with oxidizers, strong acids, nitrates.

3-METHOXY-1-BUTANOL ACETATE (4435-53-4) Incompatible with strong acids, nitrates, oxidizers.

3-METHOXYBUTYL ACETATE (4435-53-4) Incompatible with strong acids, nitrates, oxidizers.

METHOXYCARBONYL CHLORIDE (79-22-1) Forms explosive mixture with air (flash point 54°F/12°C). Water contact produces hydrochloric acid. Corrodes metals in the presence of moisture. Attacks some plastics, rubber, and coatings.

METHOXYCARBONYLETHYLENE (96-33-3) Heat, light, and/or lack of appropriate inhibitor concentration can cause polymerization. Forms explosive mixture with air (flash point 27°F/−3°C). Incompatible with strong acids, oxidizers, aliphatic amines, alkanolamines.

1-METHOXYCARBONYL-1-PROPEN-2-YLDIMETHYL PHOS-PHATE (7786-34-7) Contact with strong oxidizers may cause fire and explosions. Attacks some plastics, rubber, and coatings.

METHOXYCHLOR (72-43-5) Contact with strong oxidizers may cause fire and explosions. Attacks some plastics, rubber, and coatings.

METHOXY DDT (72-43-5) Contact with strong oxidizers may cause fire and explosions. Attacks some plastics, rubber, and coatings.

METHOXYDIGLYCOL (111-77-3) Incompatible with sulfuric acid, isocyanates, perchloric acid.

METHOXYETHANE (540-67-0) Flammable gas. Reacts violently with strong oxidizers, sulfuric and nitric acids. May be able to polymerize. Attacks some plastics, rubber, and coatings.

2-METHOXYETHANOL (109-86-4) Forms explosive mixture with air (flash point 103°F/39°C). Heat or oxidizers may cause formation of unstable peroxides. Attacks many metals. Strong oxidizers cause fire and explosions. Strong bases cause decomposition. Attacks some plastics, rubber, and coatings. May accumulate static electrical charges, and may cause ignition of its vapors.

METHOXY ETHER OF PROPYLENE GLYCOL (107-98-2) Forms explosive mixture with air (flash point 90°F/32°C). Incompatible with sulfuric acid, isocyanates, perchloric acid.

2-(2-METHOXYETHOXY)-ETHANOL (111-77-3) Incompatible with sulfuric acid, isocyanates, perchloric acid.

2-METHOXYETHYL ACETATE (110-49-6) Forms explosive mixture with air (flash point 117°F/47°C). May form unstable and explosive peroxides. Incompatible with nitrates, strong oxidizers, strong alkalies, strong acids.

2-METHOXYETHYL ACRYLATE (110-49-6) Forms explosive mixture with air (flash point 117°F/47°C). May form unstable and explosive peroxides. Incompatible with nitrates, strong oxidizers, strong alkalies, strong acids.

METHOXYETHYLENE (107-25-5) Forms explosive mixture with air (flash point −69°F/−56°C). Water contact slowly produces acetaldehyde and methyl alcohol. Forms unstable peroxides; able to polymerize. Acids, oxidizers may cause fire and explosions. May accumulate static electrical charges, and may cause ignition of its vapors.

1-METHOXY-2-HYDROXYPROPANE (107-98-2) Forms explosive mixture with air (flash point 90°F/32°C). Incompatible with sulfuric acid, isocyanates, perchloric acid.

METHOXYMETHYL CHLORIDE (107-30-2) Forms explosive mixture with air (flash point 0°F/−17.8°C). May be able to form unstable and explosive peroxides. Produces hydrochloric acid on contact with moisture. Corrosive to metals in presence of moisture.

2-METHOXY-1-METHYLETHANOL (107-98-2) Forms explosive mixture with air (flash point 90°F/32°C). Incompatible with sulfuric acid, isocyanates, perchloric acid.

METHOXYMETHYL ETHER (109-87-5) May be able to form unstable and explosive peroxides. Forms explosive mixture with air (flash point −4°F/−18°C). Incompatible with strong oxidizers, acids. Attacks some plastics, rubber, and coatings.

2-METHOXY-2-METHYL PROPANE (1634-04-4) Forms explosive mixture with air (flash point −14°F/−26°C). May be able to form unstable peroxides. Incompatible with strong acids. Reacts violently with strong oxidizers. May accumulate static electrical charges, and may cause ignition of its vapors.

2-METHOXYPHENOL (90-05-1) Forms explosive mixture with air (flash point 82°F/180°C). Reacts violently with strong oxidizers.

1-METHOXY-2-PROPANOL (107-98-2) Forms explosive mixture with air (flash point 90°F/32°C). Incompatible with sulfuric acid, isocyanates, perchloric acid.

1-METHOXY-2-PROPANOL ACETATE (108-65-6) Forms explosive mixture with air (flash point 108°F/42°C). Incompatible with strong acids, nitrates, oxidizers.

METHVTIOLO (Italian) (74-93-1) May accumulate static electrical charges, and may cause ignition of its vapors. Forms explosive air–gas mixture (flash point 0°F/−18°C). Strong oxidizers may cause fire and explosions. Attacks some plastics, coatings, and rubber.

METHYLACETAAT (Dutch) (79-20-9) Forms explosive mixture with air (flash point 14°F/−10°C). Incompatible with strong acids, nitrates, oxidizers, bases. Attacks some plastics. May accumulate static electrical charges, and may cause ignition of its vapors.

METHYLACETALDEHYDE (123-38-6) Forms explosive mixture with air (flash point −2°F/−30°C). Incompatible with strong acids, caustics, amines. Reacts violently with strong oxidizers. Can self-ignite if finely dispersed on porous or combustible material. Heat or ultraviolet can cause decomposition. May accumulate static electrical charges, and may cause ignition of its vapors.

METHYLACETAT (German) (79-20-9) Forms explosive mixture with air (flash point 14°F/−10°C). Incompatible with strong acids, nitrates, oxidizers, bases. Attacks some plastics. May accumulate static electrical charges, and may cause ignition of its vapors.

METHYL ACETATE (79-20-9) Forms explosive mixture with air (flash point 14°F/−10°C). Incompatible with strong acids, nitrates, oxidizers, bases. Attacks some plastics. May accumulate static electrical charges, and may cause ignition of its vapors.

METHYLACETIC ACID (79-09-4) Forms explosive mixture with air (flash point 126°F/52°C). Incompatible with sulfuric acid, strong bases, ammonia, aliphatic amines, alkanolamines, isocyanates, alkylene oxides, epichlorohydrin, oxidizers.

METHYLACETIC ACID ANHYDRIDE (123-62-6) Forms explosive mixture with air (flash point 145°F/63°C). Water contact produces propionic acid. Incompatible with strong oxidizers, acids, caustics; may cause fire and explosions. Incompatible with ammonia, amines.

METHYL ACETIC ESTER (79-20-9) Forms explosive mixture with air (flash point 14°F/−10°C). Incompatible with strong acids, nitrates, oxidizers, bases. Attacks some plastics. May accumulate static electrical charges, and may cause ignition of its vapors.

METHYL ACETOACETATE (105-45-3) Incompatible with oxidizers, strong acids, nitrates.

METHYL ACETONE (78-93-3) Forms explosive mixture with air (flash point 16°F/−9°C). Extremely flammable. Incompatible with sulfuric acid, nitric acid, aliphatic amines, strong oxidizers, potassium *tert*-butoxide, 2-propanol, chlorosulfonic acid, oleum.

METHYL ACETYLACETATE (105-45-3) Incompatible with oxidizers, strong acids, nitrates.

METHYL ACETYLACETONATE (105-45-3) Incompatible with oxidizers, strong acids, nitrates.

METHYL ACETYLENE (74-99-7) Forms explosive gas mixture with air. Can form explosive peroxide. Strong oxidizers may cause fire and explosions. Forms shock-sensitive compounds with copper, magnesium, silver, and their alloys. Copper or copper alloys containing more than 67% copper should not be used in handling equipment. Attacks some plastics, rubber, and coatings. May accumulate static electrical charges, and may cause ignition of its vapors.

METHYL ACETYLENE–ALLENE MIXTURE (59355-75-8) Forms explosive gas mixture with air. Incompatible with strong oxidizers. Copper alloys may form explosive compounds. Attacks some plastics, rubber, and coatings. May accumulate static electrical charges, and may cause ignition of its vapors.

METHYL ACETYLENE–PROPADIENE MIXTURE (59355-75-8) Forms explosive gas mixture with air. Incompatible with strong oxidizers. Copper alloys may form explosive compounds. Attacks some plastics, rubber, and coatings. May accumulate static electrical charges, and may cause ignition of its vapors.

b-METHYL ACROLEIN or beta-METHYL ACROLEIN (4170-30-3) A strong reducing agent. Forms explosive mixture with air (flash point 55°F/12.8°C). Readily converted by oxygen to peroxides and acids; heat or contact with many other substances may cause polymerization. Incompatible with strong oxidizers, strong acids including nonoxidizing mineral acids, ammonia, aliphatic amines, aromatic amines, 1,3-butadiene, strong bases. Liquid attacks some plastics, rubber, and coatings.

METHYL ACRYLAAT (Dutch) (96-33-3) Heat, light, and/or lack of appropriate inhibitor concentration can cause polymerization. Forms explosive mixture with air (flash point 27°F/−3°C). Incompatible with strong acids, oxidizers, aliphatic amines, alkanolamines.

METHYL ACRYLAT (German) (96-33-3) Heat, light, and/or lack of appropriate inhibitor concentration can cause polymerization. Forms explosive mixture with air (flash point 27°F/−3°C). Incompatible with strong acids, oxidizers, aliphatic amines, alkanolamines.

METHYL ACRYLATE (96-33-3) Heat, light, and/or lack of appropriate inhibitor concentration can cause polymerization. Forms explosive mixture with air (flash point 27°F/−3°C). Incompatible with strong acids, oxidizers, aliphatic amines, alkanolamines.

2-METHYLACRYLONITRILE (126-98-7) Forms explosive mixture with air (flash point 34°F/1°C). Incompatible with aliphatic amines, alkanolamines. Reacts violently with oxidizers. Incompatible with strong acids, strong bases, or light exposure; may cause polymerization.

a-METHYLACRYLONITRILE or alpha-METHYLACRYLONITRILE (126-98-7) Forms explosive mixture with air (flash point 34°F/1°C). Incompatible with aliphatic amines, alkanolamines. Reacts violently with oxidizers. Incompatible with strong acids, strong bases, or light exposure; may cause polymerization.

2-METHYLACTONITRILE (75-86-5) Forms explosive mixture with air (flash point 165°F/74°C). Reacts violently with strong oxidizers. Heat may cause decomposition. Incompatible with nonoxidizing mineral acids, sulfuric acid, nitric acid, organic acids, caustics, aliphatic amines, alkanolamines, aromatic amines, organic anhydrides, allylene oxides, epichlorohydrin, caprolactam solution, ammonia, isocyanates, phenols, cresol.

METHYL ADIPATE (627-93-0) Incompatible with strong acids, nitrates, oxidizers.

METHYLAL (109-87-5) May be able to form unstable and explosive peroxides. Forms explosive mixture with air (flash point −4°F/−18°C). Incompatible with strong oxidizers, acids. Attacks some plastics, rubber, and coatings.

METHYL ALCOHOL (67-56-1) May accumulate static electrical charges, and may cause ignition of its vapors. Forms explosive mixture with air (flash point 52°F/11°C). Incompatible with strong acids, strong oxidizers, caustics, aliphatic amines, isocyanates, chromic anhydride, lead perchlorate, perchloric acid, phosphorus trioxide. May react with metallic aluminum at high temperature. Attacks some plastics, rubber, and coatings.

METHYL ALCOHOL, SODIUM SALT (124-41-4) Flammable solid; may ignite spontaneously in moist air. A strong reducing agent; reacts violently with oxidizers, water (ignition), chloroform, chlorine oxyfluoride, aluminum, magnesium, beryllium. Attacks some plastics, rubber, and coatings.

METHYL ALDEHYDE (50-00-0) May polymerize unless properly inhibited (usually with methanol). Forms explosive mixture with air (flash point 122°F/50°C). Incompatible with strong acids, amines, strong oxidizers, alkaline materials, nitrogen dioxide, performic acid. Reaction with hydrochloric acid produces bis-chloromethyl ether, a carcinogen.

METHYLALKOHOL (German) (67-56-1) May accumulate static electrical charges, and may cause ignition of its vapors. Forms explosive mixture with air (flash point 52°F/11°C). Incompatible with strong acids, strong oxidizers, caustics, aliphatic amines, isocyanates, chromic anhydride, lead perchlorate, perchloric acid, phosphorus trioxide. May react with metallic aluminum at high temperature. Attacks some plastics, rubber, and coatings.

METHYL ALLYL ALCOHOL (513-42-8) Forms explosive mixture with air (flash point 92°F/33°C). Incompatible with strong acids, caustics, aliphatic amines, isocyanates, oxidizers.

b-METHYLALLYL CHLORIDE or beta-METHYLALLYL CHLORIDE (563-47-3) Forms explosive mixture with air (flash point 11°F/−12°C). Strong oxidizers may cause fire and explosions.

METHYLALUMINUM DICHLORIDE (917-65-7) Air contact may cause self-ignition. Reacts with water. A strong reducing agent; reacts violently with oxidizers, combustible materials, highly reactive substances. Incompatible with alcohols, amines, glycols, phenols and many other substances. May accumulate static electrical charges, and may cause ignition of its vapors.

METHYLALUMINUM SESQUICHLORIDE or METHYLALUMINUM SESQUICHLORIDE (DOT) (12542-85-7) Ignites instantly in air. A strong reducing agent; reacts violently with water, alcohols, amines, glycols, phenols, and many other substances. May accumulate static electrical charges, and may cause ignition of its vapors.

METHYLAMINE (74-89-5) Forms explosive gas mixture with air. Incompatible with acids, organic anhydrides, isocyanates, vinyl acetate, acrylates, substituted allyls, alkylene oxides, epichlorohydrin, ketones, aldehydes, alcohols, glycols, phenols, cresols, caprolactam solution, strong oxidizers. Attacks aluminum, copper, lead, tin, zinc, and alloys, and some plastics, rubber, and coatings.

METHYLAMINEN (Dutch) (74-89-5) Forms explosive gas mixture with air. Incompatible with acids, organic anhydrides, isocyanates, vinyl acetate, acrylates, substituted allyls, alkylene oxides, epichlorohydrin, ketones, aldehydes, alcohols, glycols, phenols, cresols, caprolactam solution, strong oxidizers. Attacks aluminum, copper, lead, tin, zinc and alloys, and some plastics, rubber, and coatings.

METHYLAMINE SOLUTION (40%) (74-89-5) Forms explosive mixture with air (flash point 10°F/−12°C). Incompatible with acids, organic anhydrides, isocyanates, vinyl acetate, acrylates, substituted allyls, alkylene oxides, epichlorohydrin, ketones, aldehydes, alcohols, glycols, phenols, cresols, caprolactam solution, strong oxidizers. Attacks aluminum, copper, lead, magnesium, tin, zinc, and alloys, and some plastics, rubber, and coatings.

(METHYLAMINO)BENZENE (100-61-8) Incompatible with strong acids, strong oxidizers. Attacks some plastics, rubber, and coatings. May accumulate static electrical charges, and may cause ignition of its vapors.

2-(METHYLAMINO)ETHANOL (109-83-1) Forms explosive mixture with air (flash point 165°F/74°C). Contact with strong oxidizers may cause fire and explosions. A strong base; reacts with acids. Attacks aluminum, copper, zinc, and alloys.

N-METHYLAMINOBENZENE (100-61-8) Incompatible with strong acids, strong oxidizers. Attacks some plastics, rubber, and coatings. May accumulate static electrical charges, and may cause ignition of its vapors.

1-METHYL-1,2-AMINO-BENZENE (95-53-4) Forms explosive mixture with air (flash point 185°F/85°C). Incompatible with strong acids, mineral acids, organic anhydrides, isocyanates, aldehydes, oxidizers. Attacks some plastics, rubber, and coatings.

n-METHYLAMINOETHANOL (109-83-1) Forms explosive mixture with air (flash point 165°F/74°C). Contact with strong oxidizers may cause fire and explosions. A strong base; reacts with acids. Attacks aluminum, copper, zinc, and alloys.

METHYLAMYL ACETATE (142-92-7) Incompatible with strong acids, nitrates, oxidizers.

METHYL AMYL ACETATE (108-84-9) Forms explosive mixture with air (flash point 113°F/45°C). Incompatible with strong acids, strong alkalies, nitrates, strong oxidizers. Will swell rubber, and can dissolve certain coatings.

METHYL AMYL ALCOHOL (108-11-2) Forms explosive mixture with air (flash point 106°F/41°C). Contact with alkali metals produces hydrogen gas. Incompatible with strong acids, caustics, aliphatic amines, isocyanates. Attacks some plastics, rubber, and coatings. May accumulate static electrical charges, and may cause ignition of its vapors.

METHYL AMYL-CETONE (French) (110-43-0) Forms explosive mixture with air (flash point 102°F/39°C). Incompatible with strong acids, alkalies, aliphatic amines, oxidizers. Attacks some plastics and rubber.

METHYL AMYL KETONE (110-43-0) Forms explosive mixture with air (flash point 102°F/39°C). Incompatible with strong acids, alkalies, aliphatic amines, oxidizers. Attacks some plastics and rubber.

METHYL n-AMYL KETONE (110-43-0) Forms explosive mixture with air (flash point 102°F/39°C). Incompatible with strong acids, alkalies, aliphatic amines, oxidizers. Attacks some plastics and rubber.

METHYL ANILINE or *n*-METHYLANILINE or *normal*-METHYLA-NILINE (100-61-8) Forms explosive mixture with air (flash point 174°F/79°C). Incompatible with strong acids, strong oxidizers. Attacks some plastics, rubber, and coatings. May accumulate static electrical charges, and may cause ignition of its vapors.

2-METHYLANILINE (95-53-4) Forms explosive mixture with air (flash point 185°F/85°C). Incompatible with strong acids, mineral acids, organic anhydrides, isocyanates, aldehydes, oxidizers. Attacks some plastics, rubber, and coatings.

3-METHYLANILINE (108-44-1) Forms explosive mixture with air (flash point 188°F/87°C). Incompatible with strong oxidizers, strong acids. Attacks some plastics, rubber, and coatings.

4-METHYLANILINE (106-49-0) Forms explosive mixture with air (flash point 188°F/87°C). Incompatible with strong oxidizers, strong acids. Attacks some plastics, rubber, and coatings.

m-**METHYLANILINE or *meta*-METHYLANILINE** (108-44-1) Forms explosive mixture with air (flash point 188°F/87°C). Incompatible with strong oxidizers, strong acids. Attacks some plastics, rubber, and coatings.

o-**METHYLANILINE or *ortho*-METHYLANILINE** (95-53-4) Forms explosive mixture with air (flash point 185°F/85°C). Incompatible with strong acids, mineral acids, organic anhydrides, isocyanates, aldehydes, oxidizers. Attacks some plastics, rubber, and coatings.

p-**METHYLANILINE or *para*-METHYLANILINE** (106-49-0) Forms explosive mixture with air (flash point 188°F/87°C). Incompatible with strong oxidizers, strong acids. Attacks some plastics, rubber, and coatings.

METHYLANILINE (MONO) (100-61-8) Forms explosive mixture with air (flash point 174°F/79°C). Incompatible with strong acids, strong oxidizers. Attacks some plastics, rubber, and coatings. May accumulate static electrical charges, and may cause ignition of its vapors.

2-METHYLAZACLYCLOPROPANE (75-55-8) Forms explosive mixture with air (flash point 25°F/−4°C). Contact with acids or high heat can cause violent polymerization. Strong oxidizers may cause fire and explosions. Attacks some plastics, rubber, and coatings.

METHYL AZINPHOS (86-50-0) Strong oxidizers may cause fire and explosions.

2-METHYLAZIRIDINE (75-55-8) Forms explosive mixture with air (flash point 25°F/−4°C). Contact with acids or high heat can cause violent polymerization. Strong oxidizers may cause fire and explosions. Attacks some plastics, rubber, and coatings.

N-METHYLBENZENAMINE (100-61-8) Forms explosive mixture with air (flash point 174°F/79°C). Incompatible with strong acids, strong oxidizers. Attacks some plastics, rubber, and coatings. May accumulate static electrical charges, and may cause ignition of its vapors.

2-METHYLBENZENAMINE (95-53-4) Forms explosive mixture with air (flash point 185°F/85°C). Incompatible with strong acids, mineral acids, organic anhydrides, isocyanates, aldehydes, oxidizers. Attacks some plastics, rubber, and coatings.

3-METHYLBENZENAMINE (108-44-1) Forms explosive mixture with air (flash point 188°F/87°C). Incompatible with strong oxidizers, strong acids. Attacks some plastics, rubber, and coatings.

M-METHYLBENZENAMINE (108-44-1) Forms explosive mixture with air (flash point 188°F/87°C). Incompatible with strong oxidizers, strong acids. Attacks some plastics, rubber, and coatings.

METHYL BENZENE (108-88-3) Forms explosive mixture with air (flash point 40°F/4°C). Strong oxidizers may cause fire and explosions. Attacks some plastics, rubber, and coatings. May accumulate static electrical charges, and may cause ignition of its vapors.

4-METHYLBENZENEAMINE (106-49-0) Forms explosive mixture with air (flash point 188°F/87°C). Incompatible with strong oxidizers, strong acids. Attacks some plastics, rubber, and coatings.

METHYL BENZENECARBOXYLATE (93-58-3) Forms explosive mixture with air (flash point 181°F/83°C). Incompatible with strong acids, nitrates, oxidizers.

alpha-METHYL BENZENE METHANOL (98-85-1) Forms explosive mixture with air (flash point 205°F/96°C). Incompatible with strong acids, caustics, aliphatic amines, isocyanates, oxidizers.

METHYL BENZOATE (93-58-3) Forms explosive mixture with air (flash point 181°F/83°C). Incompatible with strong acids, nitrates, oxidizers.

METHYLBENZOL (108-88-3) Forms explosive mixture with air (flash point 40°F/4°C). Strong oxidizers may cause fire and explosions. Attacks some plastics, rubber, and coatings. May accumulate static electrical charges, and may cause ignition of its vapors.

a-METHYLBENZYL ALCOHOL or alpha-METHYLBENZYL AL-COHOL (98-85-1) Forms explosive mixture with air (flash point 205°F/96°C). Incompatible with strong acids, caustics, aliphatic amines, isocyanates, oxidizers.

a-METHYLBIVINYL or alpha-METHYLBIVINYL (504-60-9) Forms explosive mixture with air (flash point −20°F/−29°C). Reacts violently with oxidizers. Attacks some plastics, rubber, and coatings. May accumulate static electrical charges, and may cause ignition of its vapors.

b-METHYLBIVINYL or beta-METHYLBIVINYL (78-79-5) Store under inert atmosphere (preferably nitrogen) with at least 50 ppm *tert*-butyl catechol present as an inhibitor. Forms explosive mixture with air (flash point −65°F/−54°C). Air contact produces unstable peroxides; heat or buildup of peroxides may cause violent polymerization. Reacts violently with oxidizers, acids, reducing agents. Attacks some plastics, rubber, and coatings. May accumulate static electrical charges, and may cause ignition of its vapors.

METHYL BROMIDE (74-83-9) Incompatible with strong oxidizers, aluminum, dimethylsulfoxide, ethylene oxide, water. Attacks zinc, magnesium, alkali metals, and their alloys. Forms aluminum alkyls in presence of aluminum; aluminum alkyls are spontaneously ignitable materials.

1-METHYLBUTADIENE (504-60-9) Forms explosive mixture with air (flash point −20°F/−29°C). Reacts violently with oxidizers. Attacks some plastics, rubber, and coatings. May accumulate static electrical charges, and may cause ignition of its vapors.

2-METHYL-1,3-BUTADIENE (78-79-5) Store under inert atmosphere (preferably nitrogen) with at least 50 ppm *tert*-butyl catechol present as an inhibitor. Forms explosive mixture with air (flash point −65°F/−54°C). Air contact produces unstable peroxides; heat or buildup of peroxides may cause violent polymerization. Reacts violently with oxidizers, acids, reducing agents. Attacks some plastics, rubber, and coatings. May accumulate static electrical charges, and may cause ignition of its vapors.

3-METHYL-1,3-BUTADIENE (78-79-5) Store under inert atmosphere (preferably nitrogen) with at least 50 ppm *tert*-butyl catechol present as an inhibitor. Forms explosive mixture with air (flash point −65°F/−54°C). Air contact produces unstable peroxides; heat or buildup of peroxides may cause violent polymerization. Reacts violently with oxidizers, acids, reducing agents. Attacks some plastics, rubber, and coatings. May accumulate static electrical charges, and may cause ignition of its vapors.

3-METHYLBUTANAL (107-83-5) Forms explosive mixture with air (flash point −20°F/−29°C). Reacts violently with strong oxidizers. May accumulate static electrical charges, and may cause ignition of its vapors.

2-METHYLBUTANE (78-78-4) Highly volatile liquid (flash point −70°F/−57°C). Mixture with air may explode. Attacks some plastics, rubber, and coatings. May accumulate static electrical charges, and may cause ignition of its vapors.

METHYL-N-BUTANOATE (623-42-7) Incompatible with strong acids, nitrates, oxidizers.

2-METHYL-2-BUTANOL (75-85-4) Forms explosive mixture with air (flash point 77°F/25°C). May accumulate static electrical charges, and may cause ignition of its vapors. Incompatible with strong acids, caustics, aliphatic amines, isocyanates, strong oxidizers, alkali metals (i.e., lithium, sodium, potassium, rubidium, cesium, francium).

3-METHYL-1-BUTANOL (primary) (123-51-3) Forms explosive mixture with air (flash point 109°F/43°C). Strong oxidizers may cause fire and explosions. Attacks some plastics, rubber, and coatings.

3-METHYL-1-BUTANOL ACETATE (626-38-0) Incompatible with strong acids, nitrates, strong alkalies, strong oxidizers, heat. May soften or dissolve plastics.

3-METHYL-1-BUTANOL ACETATE (123-92-2) Forms explosive mixture with air (flash point 77°F/25°C). Incompatible with strong alkalies, strong acids, nitrates, oxidizers. Reacts violently with reducing agents. Attacks asbestos; softens and dissolves many plastics, rubber, and coatings.

3-METHYL-2-BUTANOL SECONDARY ISOAMYL ALCOHOL (582-75-4) Forms explosive mixture with air (flash point 95°F/35°C). Incompatible with strong acids, caustics, aliphatic amines, isocyanates. May accumulate static electrical charges, and may cause ignition of its vapors.

3-METHYLBUTANOL NITRITE (110-46-3) Forms explosive mixture with air (flash point 0°F/−18°C). Decomposes in light, air or water; produces oxides of nitrogen. A strong oxidizer; reacts with reducing agents, combustibles, organics, and other strong oxidizers. Corrodes metal with moisture.

3-METHYL-2-BUTANONE (563-80-4) Forms explosive mixture with air at 43°F/6°C). Reacts violently with strong oxidizers.

3-METHYL BUTAN-2-ONE (563-80-4) Forms explosive mixture with air at 43°F/6°C). Reacts violently with strong oxidizers.

2-METHYL-1-BUTENE-3-ONE (814-78-8) Forms explosive mixture with air (flash point 73°F/23°C). Reacts violently with strong oxidizers. Heat and/or inappropriate level of inhibitor may cause polymerization.

2-METHYL-3-BUTEN-2-OL (115-18-4) Forms explosive mixture with air (flash point 56°F/13°C). Incompatible with strong acids, caustics, aliphatic amines, isocyanates, oxidizers.

3-METHYL-BUTEN-OL-(3) (115-18-4) Forms explosive mixture with air (flash point 56°F/13°C). Incompatible with strong acids, caustics, aliphatic amines, isocyanates, oxidizers.

3-METHYL-1-BUTEN-3-OL (115-18-4) Forms explosive mixture with air (flash point 56°F/13°C). Incompatible with strong acids, caustics, aliphatic amines, isocyanates, oxidizers.

METHYL BUTENOL (115-18-4) Forms explosive mixture with air (flash point 56°F/13°C). Incompatible with strong acids, caustics, aliphatic amines, isocyanates, oxidizers.

3-METHYL-3-BUTEN-2-ONE (German) (814-78-8) Forms explosive mixture with air (flash point 73°F/23°C). Reacts violently with strong oxidizers. Heat and/or inappropriate level of inhibitor may cause polymerization.

1-METHYLBUTYL ACETATE (626-38-0) Incompatible with strong acids, nitrates, strong alkalies, strong oxidizers, heat. May soften or dissolve plastics.

3-METHYL-1-BUTYL ACETATE (123-92-2) Forms explosive mixture with air (flash point 77°F/25°C). Incompatible with strong acids, nitrates, oxidizers. Reacts violently with reducing agents. Attacks asbestos; softens and dissolves many plastics, rubber, and coatings.

2-METHYL-BUTYLACRYLATE (97-88-1) Unless inhibitor is maintained at the proper level, oxidizers, heat, ultraviolet light, or moisture may cause polymerization. May accumulate static electrical charges, and may cause ignition of its vapors.

1-METHYL-4-*tert*-BUTYLBENZENE (98-51-1) Forms explosive mixture with air (flash point 155°F/67°C). Reacts with strong oxidizers. May accumulate static electrical charges, and may cause ignition of its vapors.

p-**METHYL-*tert*-BUTYLBENZENE** (98-51-1) Forms explosive mixture with air (flash point 155°F/67°C). Reacts with strong oxidizers. May accumulate static electrical charges, and may cause ignition of its vapors.

METHYL 1-(BUTYLCARBAMOYL)-2-BENZIMIDAZOYLYLCAR-BAMATE (17804-35-2) Heat, water, strong acids, and strong alkalies can cause decomposition and formation of toxic oxides of nitrogen.

METHYL-1,3-BUTYLENE GLYCOL ACETATE (4435-53-4) Incompatible with strong acids, nitrates, oxidizers.

3-METHYLBUTYL ESTER OF ACETIC ACID (123-92-2) Forms explosive mixture with air (flash point 77°F/25°C). Incompatible with strong alkalies, strong acids, nitrates, oxidizers. Reacts violently with reducing agents. Attacks asbestos; softens and dissolves many plastics, rubber, and coatings.

METHYLBUTYL ETHANOATE (123-92-2) Forms explosive mixture with air (flash point 77°F/25°C). Incompatible with strong alkalies, strong acids, nitrates, oxidizers. Reacts violently with reducing agents. Attacks asbestos; softens and dissolves many plastics, rubber, and coatings.

1-METHYLBUTYL ETHANOATE (626-38-0) Incompatible with strong acids, nitrates, strong alkalies, strong oxidizers, heat. May soften or dissolve plastics.

3-METHYLBUTYL ETHANOATE (123-92-2) Forms explosive mixture with air (flash point 77°F/25°C). Incompatible with strong alkalies, strong acids, nitrates, oxidizers. Reacts violently with reducing agents. Attacks asbestos; softens and dissolves many plastics, rubber, and coatings.

METHYL *tert*-BUTYL ETHER (1634-04-4) Forms explosive mixture with air (flash point −14°F/−26°C). May be able to form unstable peroxides. Incompatible with strong acids. Reacts violently with strong oxidizers. May accumulate static electrical charges, and may cause ignition of its vapors.

METHYL BUTYL KETONE (591-78-6) Forms explosive mixture with air (flash point 77°F/25°C). Strong oxidizers may cause fire and explosions. Dissolves some plastics, resins, and rubber.

METHYL n-BUTYL KETONE (591-78-6) Forms explosive mixture with air (flash point 77°F/25°C). Strong oxidizers may cause fire and explosions. Dissolves some plastics, resins, and rubber.

3-METHYLBUTYL NITRITE (110-46-3) Forms explosive mixture with air (flash point 0°F/−18°C). Decomposes in light, air or water; produces oxides of nitrogen. A strong oxidizer; reacts with reducing agents, combustibles, organics, and other strong oxidizers. Corrodes metal with moisture.

METHYL BUTYNOL (115-19-5) Forms explosive mixture with air (flash point less than 70°F/21°C). Incompatible with strong acids, caustics, aliphatic amines, isocyanates, oxidizers.

METHYL BUTYNOL (75-85-4) Forms explosive mixture with air (flash point 77°F/25°C). May accumulate static electrical charges, and may cause ignition of its vapors. Incompatible with strong acids, caustics, aliphatic amines, isocyanates, strong oxidizers, alkali metals (i.e., lithium, sodium, potassium, rubidium, cesium, francium).

2-METHYL-2-BUTYNOL (75-85-4) Forms explosive mixture with air (flash point 77°F/25°C). May accumulate static electrical charges, and may cause ignition of its vapors. Incompatible with strong acids, caustics, aliphatic amines, isocyanates, strong oxidizers, alkali metals (i.e., lithium, sodium, potassium, rubidium, cesium, francium).

2-METHYL-3-BUTYN-2-OL (115-19-5) Forms explosive mixture with air (flash point less than 70°F/21°C). Incompatible with strong acids, caustics, aliphatic amines, isocyanates, oxidizers.

3-METHYLBUTYRALDEHYDE (107-83-5) Forms explosive mixture with air (flash point −20°F/−29°C). Reacts violently with strong oxidizers. May accumulate static electrical charges, and may cause ignition of its vapors.

METHYL BUTYRATE (623-42-7) Incompatible with strong acids, nitrates, oxidizers.

METHYL N-BUTYRATE (623-42-7) Incompatible with strong acids, nitrates, oxidizers.

METHYLCARBAMATE (1563-66-2) Incompatible with alkaline material, acids, strong oxidizers.

METHYLCARBAMATE 1-NAPHTHALENOL (63-25-2) Incompatible with strong oxidizers, strongly alkaline pesticides.

METHYLCARBAMATE 1-NAPHTHALENOL, METHYCARBAMATE (63-25-2) Incompatible with strong oxidizers, strongly alkaline pesticides.

N-METHYLCARBAMATE de 1-NAPHTYLE (French) (63-25-2) Incompatible with strong oxidizers, strongly alkaline pesticides.

METHYL CARBAMIC ACID 4-(METHYLTHIO)-3,5-XYLYL ESTER 4-METHYLMERCAPTO-3,5-DIMETHYLPHENYL N-METHYLCARBAMATE (2032-65-7) Contact with strong oxidizers may cause fire and explosions.

METHYLCARBAMIC ACID, 1-NAPHTHYL ESTER (63-25-2) Incompatible with strong oxidizers, strongly alkaline pesticides.

METHYLCARBAMYL AMINE (624-83-9) Forms explosive mixture with air (flash point 0°F/−18°C). Contact with iron, tin, copper (or salts of these elements) and with certain other catalysts (such as triphenyl-arsenic oxide, triethylphosphine, and tributyltin oxide) or elevated temperatures may cause polymerization. Reacts violently with warm water. Incompatible with strong acids, alcohols, glycols, amines, amides, ammonia, caprolactam, caustics, strong oxidizers. Attacks some plastics, rubber, or coatings.

METHYL CARBINOL (64-17-5) Forms explosive mixture with air (flash point 65°F/18°C). May accumulate static electrical charges, and may cause ignition of its vapors. Reactions may be violent with oleum, sulfuric acid, nitric acid, bases, aliphatic amines, isocyanates, oxidizers.

METHYL CARBITOL (111-77-3) Incompatible with sulfuric acid, isocyanates, perchloric acid.

METHYL CARBITOL ACETATE (629-38-9) Forms explosive mixture with air (flash point 180°F/98°C). Incompatible with acids, organic anhydrides, isocyanates, vinyl acetates, acrylates, substituted allyls, alkylene oxides, epichlorohydrin, ketones, aldehydes, alcohols, glycols, phenols, cresols, caprolactam solution. May form shock-sensitive mixtures with silver, cobalt, chlorinated hydrocarbons, nitromethane, chromium compounds. Attacks copper, brass, zinc.

METHYL-4-CARBOMETHOXY BENZOATE (120-61-6) Incompatible with strong acids, nitrates, strong oxidizers.

METHYL CARBONIMIDE (624-83-9) Forms explosive mixture with air (flash point 0°F/−18°C). Contact with iron, tin, copper (or salts of these elements) and with certain other catalysts (such as triphenylarsenic oxide, triethylphosphine, and tributyltin oxide) or elevated temperatures may cause polymerization. Reacts violently with warm water. Incompatible with strong acids, alcohols, glycols, amines, amides, ammonia, caprolactam, caustics, strong oxidizers. Attacks some plastics, rubber, or coatings.

METHYLCATECHOL (90-05-1) Forms explosive mixture with air (flash point 82°F/180°C). Reacts violently with strong oxidizers.

METHYL CELLOSOLVE (109-86-4) Forms explosive mixture with air (flash point 103°F/39°C). Heat or oxidizers may cause formation of unstable peroxides. Attacks many metals. Strong oxidizers cause fire and explosions. Strong bases cause decomposition. Attacks some plastics, rubber, and coatings. May accumulate static electrical charges, and may cause ignition of its vapors.

METHYL CELLOSOLVE ACETATE (110-49-6) Forms explosive mixture with air (flash point 117°F/47°C). May form unstable and explosive peroxides. Incompatible with nitrates, strong oxidizers, strong alkalies, strong acids.

METHYLCHLORID (German) (74-87-3) Moisture causes decomposition. Contact with aluminum results in a product that may ignite spontaneously in air. Reacts with barium, lithium, sodium, magnesium, titanium. Attacks plastics, rubber, and coatings.

METHYL CHLORIDE (74-87-3) Moisture causes decomposition. Contact with aluminum results in a product that may ignite spontaneously in air. Reacts with barium, lithium, sodium, magnesium, titanium. Attacks plastics, rubber, and coatings.

METHYL CHLOROACETATE (96-34-4) Forms explosive mixture with air (flash point 135°F/57°C). Reacts strongly with oxidizers.

1-METHYL-2-CHLOROBENZENE (95-49-8) Forms explosive mixture with air (flash point 96°F/36°C). Strong oxidizers may cause fire and explosions. Produces hydrochloric acid with water. Corrodes metal in the presence of moisture. Attacks some plastics, rubber, and coatings.

2-METHYLCHLOROBENZENE (95-49-8) Forms explosive mixture with air (flash point 96°F/36°C). Strong oxidizers may cause fire and explosions. Produces hydrochloric acid with water. Corrodes metal in the presence of moisture. Attacks some plastics, rubber, and coatings.

o-**METHYL-O-2-CHLORO-4-*tert*-BUTYLPHENYL N-METHYL-AMIDOPHOSPHATE** (299-86-5) Decomposes in strongly alkaline (pH >7) and strongly acidic media. Unstable over long periods of time in water and at temperatures above 140°F/60°C.

METHYL CHLOROCARBONATE (79-22-1) Forms explosive mixture with air (flash point 54°F/12°C). Water contact produces hydrochloric acid. Corrodes metals in the presence of moisture. Attacks some plastics, rubber, and coatings.

METHYLCHLOROFORM (71-55-6) Reacts with acetone, strong caustics, alkaline earth, alkali metals, dinitrogen tetroxide, sodium hydroxide, amides, strong oxidizers, sodium, potassium. Reacts violently with copper, bronze and other metal powders. Uninhibited grade is corrosive to aluminum. Attacks some plastics, rubber, and coatings.

METHYL CHLOROFORMATE (79-22-1) Forms explosive mixture with air (flash point 54°F/12°C). Water contact produces hydrochloric acid. Corrodes metals in the presence of moisture. Attacks some plastics, rubber, and coatings.

METHYLCHLOROMETHYL ETHER (107-30-2) Forms explosive mixture with air (flash point 0°F/−17.8°C). May be able to form unstable and explosive peroxides. Produces hydrochloric acid on contact with moisture. Corrosive to metals in presence of moisture.

METHYLCHLOROMETHYL ETHER, ANHYDROUS (107-30-2) Forms explosive mixture with air (flash point 0°F/−17.8°C). May be able to form unstable and explosive peroxides. Produces hydrochloric acid on contact with moisture. Corrosive to metals in presence of moisture.

METHYL CHLOROPHOS (52-68-6) Contact with strong oxidizers may cause fire and explosions.

METHYL CHLOROSILANE (75-79-6) Forms explosive gas mixture with air (flash point 15°F/−9°C). Water or air contact produces hydrochloric acid. Strong oxidizers may cause fire and explosions. Incompatible with bases. Corrodes most metals. Attacks some plastics, rubber, and coatings.

METHYL CYANIDE (75-05-8) Forms explosive mixture with air (flash point 42°F/6°C). Incompatible with water (especially if acid or alkaline), acids, caustics, nitrating agents, indium, nitrogen tetroxide, n-fluoro compounds, sulfur trioxide, iron(III) salts of perchlorate, indium, nitrogen-fluorine compounds. Reacts violently with oxidizers. May accumulate static electrical charges, and may cause ignition of its vapors.

METHYLCYCLOHEXANE (108-87-2) Forms explosive mixture with air (flash point 25°F/−3.9°C). Strong oxidizers may cause fire and explosions. Attacks some plastics, rubber, and coatings.

METHYLCYCLOHEXANOL (25639-42-3) Forms explosive mixture with air (flash point 149°F/65°C). Strong oxidizers may cause fire and explosions. Attacks some plastics, rubber, and coatings.

2-METHYLCYCLOHEXANOL (25639-42-3) Forms explosive mixture with air (flash point 149°F/65°C). Strong oxidizers may cause fire and explosions. Attacks some plastics, rubber, and coatings.

1-METHYLCYCLOHEXAN-2-ONE (583-60-8) Incompatible with oxidizers (cause fire and explosions). Attacks some plastics, rubber, and coatings.

2-METHYLCYCLOHEXANONE (583-60-8) Incompatible with oxidizers (cause fire and explosions). Attacks some plastics, rubber, and coatings.

o-**METHYLCYCLOHEXANONE** (583-60-8) Incompatible with oxidizers (cause fire and explosions). Attacks some plastics, rubber, and coatings.

METHYLCYCLOPENTADIENE DIMER (26472-00-4) Forms explosive mixture with air (flash point 80°F/27°C). Incompatible with strong acids, strong oxidizers.

METHYLCYCLOPENTADIENYL MANGANESE TRICARBONYL (12108-13-3) Incompatible with strong acids, strong oxidizers.

METHYLCYCLOPENTANE (96-37-7) Forms explosive mixture with air (flash point less than 20°F/7°C). Incompatible with strong acids, caustics, aliphatic amines, isocyanates. Reaction with oxidizers may be violent.

METHYL DICHLOROACETATE (116-54-1) Forms explosive mixture with air (flash point 176°F/80°C). Water solution is corrosive. Strong oxidizers may cause fire and explosions.

METHYL DICHLOROETHANOATE (116-54-1) Forms explosive mixture with air (flash point 176°F/80°C). Water solution is corrosive. Strong oxidizers may cause fire and explosions.

METHYL DICHLOROSILANE (75-54-7) Forms explosive mixture with air (flash point 15°F/−9°C). Water and air vapor cause formation of corrosive fumes. Incompatible with strong bases and oxidizers. Corrodes most metals. Attacks some plastics, rubber, and coatings.

METHYL-DICHLORSILAN (Czech) (75-54-7) Forms explosive mixture with air (flash point 15°F/−9°C). Water and air vapor cause formation of corrosive fumes. Incompatible with strong bases and oxidizers. Corrodes most metals. Attacks some plastics, rubber, and coatings.

METHYL DIETHANOLAMINE (105-59-9) Incompatible with mineral acids, organic acids, sulfuric acid, nitric acid, organic anhydrides, isocyanates, vinyl acetate, acrylates, substituted allyls, alkylene oxides, epichlorohydrin, aldehydes, oxidizers.

METHYL 1,1-DIMETHYLETHYL ETHER (1634-04-4) Forms explosive mixture with air (flash point −14°F/−26°C). May be able to form unstable peroxides. Incompatible with strong acids. Reacts violently with strong oxidizers. May accumulate static electrical charges, and may cause ignition of its vapors.

METHYL 3-DIMETHOXYPHOSPHINYLOXYCROTONATE (7786-34-7) Contact with strong oxidizers may cause fire and explosions. Attacks some plastics, rubber, and coatings.

1-METHYL-2,4-DINITROBENZENE (121-14-2) Strong oxidizers or caustics may cause fire and explosions. Incompatible with nitric acid; produces an explosive material. Sodium oxide contact causes ignition. Attacks metals.

2-METHYL-1,3-DINIROBENZENE (606-20-2) Strong oxidizers or caustics may cause fire and explosions. Contact with nitric acid produces an explosive material. Sodium oxide contact causes ignition. Attacks some metals.

4-METHYL-1,2-DINITROBENZENE (610-39-9) Strong oxidizers or caustics may cause fire and explosions. Contact with nitric acid produces an explosive material. Sodium oxide contact causes ignition. Attacks some metals.

METHYL-E 605 (298-00-0) Mixtures with magnesium may be explosive.

METHYLE (ACETATE de) (French) (79-20-9) Forms explosive mixture with air (flash point 14°F/−10°C). Incompatible with strong acids, nitrates, oxidizers, bases. Attacks some plastics. May accumulate static electrical charges, and may cause ignition of its vapors.

METHYLEEN-S,S′-BIS(O,O-DIETHYL-DITHIOFOSFAAT) (Dutch) (563-12-2) Incompatible with alkaline formulations. Mixtures with magnesium may be explosive.

METHYLE (FORMIATE de) (French) (107-31-3) Forms explosive mixture with air (flash point −2°F/−19°C). Incompatible with strong acids, caustics, water, nitrates. Reacts violently with strong oxidizers. Attacks some plastics and coatings. May accumulate static electrical charges, and may cause ignition of its vapors.

METHYLENE ACETONE (78-94-4) Forms explosive mixture with air (flash point 20°F/−7°C). May accumulate static electrical charges, and may cause ignition of its vapors. Heat can cause polymerization. Reacts violently with strong oxidizers.

METHYLENE-S,S′-BIS(O,O-DIAETHYL-DITHIOPHOSPHAT) (German) (563-12-2) Incompatible with alkaline formulations. Mixtures with magnesium may be explosive.

METHYLENE BIS(PHENYLISOCYANATE) (101-68-8) Incompatible with strong alkalies, acids, alcohols, ammonia, amines, amides, glycols, caprolactam. Unstable above 100°F/37.8°C. Attacks some plastics, rubber, and coatings.

METHYLENE BISPHENYL ISOCYANATE (101-68-8) Incompatible with strong alkalies, acids, alcohols, ammonia, amines, amides, glycols, caprolactam. Unstable above 100°F/37.8°C. Attacks some plastics, rubber, and coatings.

METHYLENE BIS(4-PHENYL ISOCYANATE) (101-68-8) Incompatible with strong alkalies, acids, alcohols, ammonia, amines, amides, glycols, caprolactam. Unstable above 100°F/37.8°C. Attacks some plastics, rubber, and coatings.

METHYLENE BROMIDE (74-95-3) Mixting with potassium produces a shock-sensitive explosive.

METHYLENE CHLORIDE (75-09-2) Incompatible with strong oxidizers, strong caustics, chemically active metals. Attacks some plastics, rubber and coatings. May accumulate static electrical charges, and may cause ignition of its vapors.

METHYLENE CHLOROBROMIDE (74-97-5) Reacts with chemically active metals such as calcium, powdered aluminum, zinc and magnesium. Liquid attacks some plastics, rubber, and coatings.

METHYLENE CYANIDE (109-77-3) Incompatible with sulfuric acid. Caustics or heat above 160°F/71°C may cause polymerization or spontaneous combustion.

METHYLENE DIBROMIDE (74-95-3) Mixing with potassium produces a shock-sensitive explosive.

METHYLENE DIMETHYL ETHER (109-87-5) May be able to form unstable and explosive peroxides. Forms explosive mixture with air (flash point −4°F/−18°C). Incompatible with strong oxidizers, acids. Attacks some plastics, rubber, and coatings.

METHYLENE DICHLORIDE (75-09-2) Incompatible with strong oxidizers, strong caustics, chemically active metals. Attacks some plastics, rubber, and coatings. May accumulate static electrical charges, and cause ignition of its vapors.

4,4-METHYLENEDIPHENYL DIISOCYANATE (101-68-8) Incompatible with strong alkalies, acids, alcohols, ammonia, amines, amides, glycols, caprolactam. Unstable above 100°F/37.8°C. Attacks some plastics, rubber, and coatings.

METHYLENE DI-p-PHENYLENE ESTER OF ISOCYANIC ACID (101-68-8) Incompatible with strong alkalies, acids, alcohols, ammonia, amines, amides, glycols, caprolactam. Unstable above 100°F/37.8°C. Attacks some plastics, rubber, and coatings.

METHYLENE GLYCOL (50-00-0) May polymerize unless properly inhibited (usually with methanol). Forms explosive mixture with air (flash point 122°F/50°C). Incompatible with strong acids, amines, strong oxidizers, alkaline materials, nitrogen dioxide, performic acid. Reaction with hydrochloric acid produces bis-chloromethyl ether, a carcinogen.

3-METHYLENE-7-METHYL 1,6-OCTADIENE (123-35-3) Forms explosive mixture with air (flash point 103°F/40°C). Incompatible with strong oxidizers, sulfuric acid, nitric acid.

METHYLENE OXIDE (50-00-0) May polymerize unless properly inhibited (usually with methanol). Forms explosive mixture with air (flash point 122°F/50°C). Incompatible with strong acids, amines, strong oxidizers, alkaline materials, nitrogen dioxide, performic acid. Reaction with hydrochloric acid produces bis-chloromethyl ether, a carcinogen.

S,S'-METHYLENE O,O,O',O'-TETRAETHYL ESTER PHOSPHORODITHIOIC ACID (563-12-2) Incompatible with alkaline formulations. Mixtures with magnesium may be explosive.

S,S'-METHYLENE O,O,O',O'-TETRAETHYL PHOSPHORODITHIOATE (563-12-2) Incompatible with alkaline formulations. Mixtures with magnesium may be explosive.

METHYLENE TRIBROMIDE (75-25-2) Reacts with chemically active metals, calcium, acetone, strong caustics. Attacks some plastics, rubber, and coatings.

METHYLESTER KISELINY OCTOVE (Czech) (79-20-9) Forms explosive mixture with air (flash point 14°F/−10°C). Incompatible with strong acids, nitrates, oxidizers, bases. Attacks some plastics. May accumulate static electrical charges, and may cause ignition of its vapors.

METHYL ESTER OF ACETIC ACID (79-20-9) Forms explosive mixture with air (flash point 14°F/−10°C). Incompatible with strong acids, nitrates, oxidizers, bases. Attacks some plastics. May accumulate static electrical charges, and may cause ignition of its vapors.

METHYL ESTER OF ACRYLIC ACID (96-33-3) Heat, light and/or lack of appropriate inhibitor concentration can cause polymerization. Forms explosive mixture with air (flash point 27°F/−3°C). Incompatible with strong acids, oxidizers, aliphatic amines, alkanolamines.

METHYL ESTER OF FORMIC ACID (107-31-3) Forms explosive mixture with air (flash point −2°F/−19°C). Incompatible with strong acids, caustics, water, nitrates. Reacts violently with strong oxidizers. Attacks some plastics and coatings. May accumulate static electrical charges, and may cause ignition of its vapors.

METHYL ESTER OF ISOCYANIC ACID (624-83-9) Forms explosive mixture with air (flash point 0°F/−18°C). Contact with iron, tin, copper (or salts of these elements) and with certain other catalysts (such as triphenylarsenic oxide, triethylphosphine, and tributyltin oxide) or elevated temperatures may cause polymerization. Reacts violently with warm water. Incompatible with strong acids, alcohols, glycols, amines, amides, ammonia, caprolactam, caustics, strong oxidizers. Attacks some plastics, rubber, or coatings.

METHYL ESTER OF SULFURIC ACID (77-78-1) Forms explosive mixture with air (flash point 182°F/83°C). Incompatible with strong oxidizers, strong acids, strong alkalies, strong ammonia solutions. Attacks some plastics, rubber, and coatings.

METHYLE (SULFATE de) (French) (77-78-1) Forms explosive mixture with air (flash point 182°F/83°C). Incompatible with strong oxidizers, strong acids, strong alkalies, strong ammonia solutions. Attacks some plastics, rubber, and coatings.

METHYL ETHANOATE (79-20-9) Forms explosive mixture with air (flash point 14°F/−10°C). Incompatible with strong acids, nitrates, oxidizers, bases. Attacks some plastics. May accumulate static electrical charges, and may cause ignition of its vapors.

n-METHYL ETHANOLAMINE (109-83-1) Forms explosive mixture with air (flash point 165°F/74°C). Contact with strong oxidizers may cause fire and explosions. A strong base; reacts with acids. Attacks aluminum, copper, zinc, and alloys.

METHYLETHENE (115-07-1) Forms explosive mixture with air (flash point −162°F/−108°C). Incompatible with strong oxidizers, strong acids.

METHYL ETHER (115-10-6) Flammable gas. Forms explosive mixture with air. Forms unstable peroxides in storage. Reacts violently with strong oxidizers.

METHYL ETHOXOL (109-86-4) Forms explosive mixture with air (flash point 103°F/39°C). Heat or oxidizers may cause formation of unstable peroxides. Attacks many metals. Strong oxidizers cause fire and explosions. Strong bases cause decomposition. Attacks some plastics, rubber, and coatings. May accumulate static electrical charges, and may cause ignition of its vapors.

1-METHYLETHYL ACETATE (108-21-4) Forms explosive mixture with air (flash point 40°F/4.4°C). Incompatible with strong alkalies, strong acids, nitrates, strong oxidizers. Dissolves rubber and many plastic materials. Contact with steel may cause slow decomposition. May accumulate static electrical charges, and may cause ignition of its vapors.

(1-METHYLETHYL) BENZENE or 1-(METHYLETHYL) BENZENE (98-82-8) Forms explosive mixture with air (flash point 99°F/37°C). Incompatible with strong acids, strong oxidizers. Air contact produces cumene hydroperoxide. May be able to form unstable peroxides. Attacks rubber. May accumulate static electrical charges, and may cause ignition of its vapors.

O-METHYLETHYLBENZENE (611-14-3) Forms explosive mixture with air (flash point 103°F/39°C). Incompatible with nitric acid, oxidizers (cause fire and explosions).

METHYL ETHYL BROMO-METHANE (78-76-2) Forms explosive mixture with air (flash point 70°F/26°C) Reacts violently with strong oxidizers.

METHYL ETHYL CARBINOL (78-92-2) Forms explosive mixture with air (flash point 75°F/4°C). Attacks some plastics, rubber, and coatings. Forms an explosive peroxide in air. Ignites with chromium trioxide. Incompatible with strong oxidizers, strong acids, aliphatic amines, isocyanates, organic peroxides.

METHYLETHYLCARBINOL (78-92-2) Forms explosive mixture with air (flash point 75°F/4°C). Attacks some plastics, rubber, and coatings. Forms an explosive peroxide in air. Ignites with chromium trioxide. Incompatible with strong oxidizers, strong acids, aliphatic amines, isocyanates, organic peroxides.

1-METHYLETHYLCYCLOHEXANE (696-29-7) Forms explosive mixture with air (flash point 96°F/36°C). Incompatible with strong oxidizers, acids.

METHYLETHYLENE (115-07-1) Forms explosive mixture with air (flash point −162°F/−108°C). Incompatible with strong oxidizers, strong acids.

METHYLETHYLENE GLYCOL (57-55-7) Incompatible with strong acids, caustics, aliphatic amines, isocyanates, strong oxidizers (with violent reaction).

METHYL ETHYLENE OXIDE (75-56-9) Forms explosive gas mixture with air (flash point −35°F/−37°C). Incompatible with anhydrous metal chlorides. Strong acids, caustics, peroxides cause polymerization. Reacts with ammonia, amines, acetylene-forming metals. Attacks some plastics, rubber, and coatings.

2-METHYLETHYLEN IMINE (75-55-8) Forms explosive mixture with air (flash point 25°F/−4°C). Contact with acids or high heat can cause violent polymerization. Strong oxidizers may cause fire and explosions. Attacks some plastics, rubber, and coatings.

2-METHYLETHYLENIMINE (75-55-8) Forms explosive mixture with air (flash point 25°F/−4°C). Contact with acids or high heat can cause violent polymerization. Strong oxidizers may cause fire and explosions. Attacks some plastics, rubber, and coatings.

METHYL ETHYL ETHER or METHYL ETHYL ETHER (DOT) (540-67-0) Flammable gas. Reacts violently with strong oxidizers, sulfuric and nitric acids. May be able to polymerize. Attacks some plastics, rubber, and coatings.

METHYL ETHYL KETONE (78-93-3) Forms explosive mixture with air (flash point 16°F/−9°C). Extremely flammable. Incompatible with sulfuric acid, nitric acid, aliphatic amines, strong oxidizers, potassium *tert*-butoxide, 2-propanol, chlorosulfonic acid, oleum.

METHYL ETHYL KETONE HYDROPEROXIDE (1338-23-4) Forms explosive mixture with air (flash point 125°F/52°C). Explosive decomposition occurs above 176°F/80°C. Pure substance is shock-sensitive. Strong oxidizer. Reacts violently with strong acids, strong bases, reducing agents, combustible substances, organic materials, oxides of heavy metals, salts, trace contaminants, amines. May accumulate static electrical charges, and may cause ignition of its vapors.

METHYL ETHYL KETONE PEROXIDE (1338-23-4) Forms explosive mixture with air (flash point 125°F/52°C). Explosive decomposition occurs above 176°F/80°C. Pure substance is shock-sensitive. Strong oxidizer. Reacts violently with strong acids, strong bases, reducing agents, combustible substances, organic materials, oxides of heavy metals, salts, trace contaminants, amines. May accumulate static electrical charges, and may cause ignition of its vapors.

METHYL ETHYL KETONOXIME (96-29-7) Forms explosive mixture with air (flash point above 155°F/68°C). Strong oxidizers may cause fire and explosions. Keep cool.

METHYL ETHYL METHANE (106-97-8) May accumulate static electrical charges, and may cause ignition of its vapors. Forms explosive gas mixture with air. Reacts violently with strong oxidizers. Mixed with oxygen this substance will explode on contact with nickel carbonyl in the 68°F to 104°F/20°C to 40°C range.

METHYLETHYLMETHANE (106-97-8) May accumulate static electrical charges, and may cause ignition of its vapors. Forms explosive gas mixture with air. Reacts violently with strong oxidizers. Mixed with oxygen this substance will explode on contact with nickel carbonyl in the 68° to 104°F/20° to 40°C range.

N-(1-METHYLETHYL)-2-PROPANAMINE (108-18-9) Forms explosive mixture with air (flash point 30°F/−1°C). Incompatible with acids, organic anhydrides, isocyanates, vinyl acetate, acrylates, substituted allyls, alkylene oxides, epichlorohydrin, ketones, aldehydes, alcohols, glycols, mercury, phenols, cresols, caprolactam solution, strong oxidizers. Attacks aluminum, copper, lead, tin, zinc, and alloys, and some plastics, rubber, and coatings.

METHYLETHYLPYRIDINE (104-90-5) Forms explosive mixture with air (flash point 155°F/68°C). Incompatible with acids, isocyanates, phenols, cresols.

2-METHYL-5-ETHYLPYRIDINE (104-90-5) Forms explosive mixture with air (flash point 155°F/68°C). Incompatible with acids, isocyanates, phenols, cresols.

1-METHYL-2-FLUOROBENZENE (95-52-3) Forms explosive mixture with air (flash point 55°F/13°C). Reacts violently with strong oxidizers.

1-METHYL-3-FLUOROBENZENE (352-70-5) Forms explosive mixture with air (flash point 49°F/9°C). Reacts violently with strong oxidizers.

METHYL FORMAL (109-87-5) May be able to form unstable and explosive peroxides. Forms explosive mixture with air (flash point −4°F/−18°C). Incompatible with strong oxidizers, acids. Attacks some plastics, rubber, and coatings.

METHYL FORMATE (107-31-3) Forms explosive mixture with air (flash point −2°F/−19°C). Incompatible with strong acids, caustics, water, nitrates. Reacts violently with strong oxidizers. Attacks some plastics and coatings. May accumulate static electrical charges, and may cause ignition of its vapors.

METHYLFORMIAAT (Dutch) (107-31-3) Forms explosive mixture with air (flash point −2°F/−19°C). Incompatible with strong acids, caustics, water, nitrates. Reacts violently with strong oxidizers. Attacks some plastics and coatings. May accumulate static electrical charges, and may cause ignition of its vapors.

METHYLFORMIAT (German) (107-31-3) Forms explosive mixture with air (flash point −2°F/−19°C). Incompatible with strong acids, caustics, water, nitrates. Reacts violently with strong oxidizers. Attacks some plastics and coatings. May accumulate static electrical charges, and may cause ignition of its vapors.

METHYL FOSFERNO (298-00-0) Mixtures with magnesium may be explosive.

METHYL GLYCOL (109-86-4) Forms explosive mixture with air (flash point 103°F/39°C). Heat or oxidizers may cause formation of unstable peroxides. Attacks many metals. Strong oxidizers cause fire and explosions. Strong bases cause decomposition. Attacks some plastics, rubber, and coatings. May accumulate static electrical charges, and may cause ignition of its vapors.

METHYL GLYCOL (57-55-7) Incompatible with strong acids, caustics, aliphatic amines, isocyanates, strong oxidizers (with violent reaction).

METHYL GLYCOL ACETATE (110-49-6) Forms explosive mixture with air (flash point 117°F/47°C). May form unstable and explosive peroxides. Incompatible with nitrates, strong oxidizers, strong alkalies, strong acids.

METHYL GUTHION (86-50-0) Strong oxidizers may cause fire and explosions.

6-METHYL-1-HEPTANAL (N/A) Forms explosive mixture with air (flash point 104°F/40°C). Incompatible with strong acids, caustics, ammonia, amines.

6-METHYL-1-HEPTANOL (26952-21-6) Forms explosive mixture with air (flash point 180°F/82°C). Incompatible with strong acids, caustics, amines, isocyanates.

5-METHYL-3-HEPTANONE (541-85-5) Forms explosive mixture with air (flash point 110°F/43°C). Incompatible with sulfuric acid, nitric acid, aliphatic amines, oxidizers.

METHYL HEPTYL KETONE (821-55-6) Forms explosive mixture with air (flash point 140°F/60°C) Incompatible with strong acids, aliphatic amines, oxidizers.

METHYL HYDRATE (67-56-1) May accumulate static electrical charges, and may cause ignition of its vapors. Forms explosive mixture with air (flash point 52°F/11°C). Incompatible with strong acids, strong oxidizers, caustics, aliphatic amines, isocyanates, chromic anhydride, lead perchlorate, perchloric acid, phosphorus trioxide. May react with metallic aluminum at high temperature. Attacks some plastics, rubber, and coatings.

METHYLHYDRAZINE (60-34-4) Highly reactive reducing agent and a strong base. Forms explosive mixture with air (flash point 17°F/−8°C). Reacts violently with oxidizers, combustibles, acids. Incompatible with alcohols, glycols, isocyanates, phenols, cresols. Contact with oxides of iron or copper, manganese, lead, copper, or their alloys can lead to fire and explosions. Attacks cork, some plastics, coatings, and rubber.

531

1-METHYLHYDRAZINE (60-34-4) Highly reactive reducing agent and a strong base. Forms explosive mixture with air (flash point 17°F/−8°C). Reacts violently with oxidizers, combustibles, acids. Incompatible with alcohols, glycols, isocyanates, phenols, cresols. Contact with oxides of iron or copper, manganese, lead, copper or their alloys can lead to fire and explosions. Attacks cork, some plastics, coatings, and rubber.

METHYL HYDRIDE (74-82-8) Forms an explosive mixture with air. Reacts violently with strong oxidizers.

METHYL HYDROXIDE (67-56-1) May accumulate static electrical charges, and may cause ignition of its vapors. Forms explosive mixture with air (flash point 52°F/11°C). Incompatible with strong acids, strong oxidizers, caustics, aliphatic amines, isocyanates, chromic anhydride, lead perchlorate, perchloric acid, phosphorus trioxide. May react with metallic aluminum at high temperature. Attacks some plastics, rubber, and coatings.

P-METHYLHYDROXYBENZENE (106-44-5) Incompatible with strong acids, oxidizers, alkalies, aliphatic amines, amides, chlorosulfonic acid, oleum. Liquid attacks some plastics, coatings, and rubber.

2-METHYL-2-HYDROXY-3-BUTYNE (75-85-4) Forms explosive mixture with air (flash point 77°F/25°C). May accumulate static electrical charges, and may cause ignition of its vapors. Incompatible with strong acids, caustics, aliphatic amines, isocyanates, strong oxidizers, alkali metals (i.e., lithium, sodium, potassium, rubidium, cesium, francium).

1-METHYL-2-HYDROXY-ETHYLAMINE (78-91-1) Forms explosive mixture with air (flash point 145°F/63°C). Incompatible with acids, organic anhydrides, isocyanates, vinyl acetate, acrylates, substituted allyls, alkylene oxides, epichlorohydrin, aldehydes.

4-METHYL-4-HYDROXY-2-PENTANONE (123-42-2) Incompatible with strong acids, strong alkalies (cause formation of flammable acetone vapors), aliphatic amines, isocyanates, oxidizers, alkali metals. Forms explosive mixture with air (flash point 136°F/58°C). Attacks some forms of plastics, resins, and rubber.

METHYLIDE (334-88-3) Heat, shock, friction, sunlight or other bright illuminations may cause explosions. Contact with alkali metals, drying agents such as calcium sulfate, or rough edges (such as ground glass) will cause explosions.

2,2′-METHYLIMINODIETHANOL (105-59-9) Incompatible with mineral acids, organic acids, sulfuric acid, nitric acid, organic anhydrides, isocyanates, vinyl acetate, acrylates, substituted allyls, alkylene oxides, epichlorohydrin, aldehydes, oxidizers.

METHYL IODIDE (74-88-4) Contact with strong oxidizers may cause fire and explosions. Heat, light and moisture contribute to instability.

METHYL ISOAMYL ACETATE (108-84-9) Forms explosive mixture with air (flash point 113°F/45°C). Incompatible with strong acids, strong alkalies, nitrates, strong oxidizers. Will swell rubber and can dissolve certain coatings.

METHYL ISOBUTENYL KETONE (141-79-7) Forms explosive mixture with air (flash point 87°F/31°C). Forms peroxides. Incompatible with strong acids aliphatic amines, alkanolamines, 2-aminoethanol, ethylene diamine, chlorosulfonic acid, oleum, oxidizers. Dissolves some forms of plastics, resins, and rubber. Attacks copper.

METHYL ISOBUTYL CARBINOL (108-11-2) Forms explosive mixture with air (flash point 106°F/41°C). Contact with alkali metals produces hydrogen gas. Incompatible with strong acids, caustics, aliphatic amines, isocyanates. Attacks some plastics, rubber, and coatings. May accumulate static electrical charges, and may cause ignition of its vapors.

METHYLISOBUTYLCARBINOL ACETATE (108-84-9) Forms explosive mixture with air (flash point 113°F/45°C). Incompatible with strong acids, strong alkalies, nitrates, strong oxidizers. Will swell rubber and can dissolve certain coatings.

METHYLISOBUTYLCARBINYL ACETATE (108-84-9) Forms explosive mixture with air (flash point 113°F/45°C). Incompatible with strong acids, strong alkalies, nitrates, strong oxidizers.

METHYL-ISOBUTYL-CETONE (French) (108-10-1) Forms explosive mixture with air (flash point 64°F/18°C). Incompatible with strong oxidizers, strong acids, aliphatic amines. Dissolves some plastics, resins, and rubber.

METHYL ISOBUTYL KETONE (108-10-1) Forms explosive mixture with air (flash point 64°F/18°C). Incompatible with strong oxidizers, strong acids, aliphatic amines. Dissolves some plastics, resins, and rubber.

METHYLISOCYANAAT (Dutch) (624-83-9) Forms explosive mixture with air (flash point 0°F/−18°C). Contact with iron, tin, copper (or salts of these elements) and with certain other catalysts (such as triphenylarsenic oxide, triethylphosphine, and tributyltin oxide) or elevated temperatures may cause polymerization. Reacts violently with warm water. Incompatible with strong acids, alcohols, glycols, amines, amides, ammonia, caprolactam, caustics, strong oxidizers. Attacks some plastics, rubber, or coatings.

METHYL ISOCYANAT (German) (624-83-9) Forms explosive mixture with air (flash point 0°F/−18°C). Contact with iron, tin, copper (or salts of these elements) and with certain other catalysts (such as triphenylarsenic oxide, triethylphosphine, and tributyltin oxide) or elevated temperatures may cause polymerization. Reacts violently with warm water. Incompatible with strong acids, alcohols, glycols, amines, amides, ammonia, caprolactam, caustics, strong oxidizers. Attacks some plastics, rubber, or coatings.

METHYL ISOCYANATE (624-83-9) Forms explosive mixture with air (flash point 0°F/−18°C). Contact with iron, tin, copper (or salts of these elements) and with certain other catalysts (such as triphenylarsenic oxide, triethylphosphine, and tributyltin oxide) or elevated temperatures may cause polymerization. Reacts violently with warm water. Incompatible with strong acids, alcohols, glycols, amines, amides, ammonia, caprolactam, caustics, strong oxidizers. Attacks some plastics, rubber, or coatings.

1-METHYL-4-ISOPROPYLBENZENE (99-87-6) Forms explosive mixture with air (flash point 117°F/47°C). Incompatible with nitric acid, strong oxidizers. Attacks and softens rubber. May accumulate static electrical charges, and may cause ignition of its vapors.

1-METHYL-4-ISOPROPENYL-1-CYCLOHEXENE (138-86-3) Forms explosive mixture with air (flash point 115°F/46°C). Strong oxidizers may cause fire and explosions.

METHYL ISOPROPENYL KETONE (814-78-8) Forms explosive mixture with air (flash point 73°F/23°C) Reacts violently with strong oxidizers. Heat and/or inappropriate level of inhibitor may cause polymerization.

METHYL ISOPROPYL KETONE (563-80-4) Forms explosive mixture with air at 43°F/6°C. Reacts violently with strong oxidizers.

METHYL ISOTHIOCYANATE (556-61-6) Forms explosive mixture with air (flash point 90°F/32°C). Incompatible with strong acids, caustics, ammonia, amines, amides, alcohols, glycols, caprolactam solution, strong oxidizers.

METHYLJODID (German) (74-88-4) Contact with strong oxidizers may cause fire and explosions. Heat, light and moisture contribute to instability.

METHYL KETONE (67-64-1) Forms explosive mixture with air (flash point −4°F/−20°C). Reacts violently with chloroform. Incompatible with strong acids, aliphatic amines, chloroform, chromic anhydride, chromyl chloride, hexachloromelamine, hydrogen peroxide, nitrosyl chloride, nitrosyl perchlorate, nitryl perchlorate, permonosulfuric acid, potassium *tert*-butoxide. Unstable and explosive peroxides may be formed with strong oxidizers. May accumulate static electrical charges, and may cause ignition of its vapors. Dissolves most rubber, resins, and plastics.

METHYL KETONE (MEK) (78-93-3) Forms explosive mixture with air (flash point 16°F/−9°C). Extremely flammable. Incompatible with sulfuric acid, nitric acid, aliphatic amines, strong oxidizers, potassium *tert*-butoxide, 2-propanol, chlorosulfonic acid, oleum.

METHYL MERCAPTAN (74-93-1) May accumulate static electrical charges, and may cause ignition of its vapors. Forms explosive air–gas mixture (flash point 0°F/−18°C). Strong oxidizers may cause fire and explosions. Attacks some plastics, coatings, and rubber.

METHYL MERCAPTANE (74-93-1) May accumulate static electrical charges, and may cause ignition of its vapors. Forms explosive air–gas mixture (flash point 0°F/−18°C). Strong oxidizers may cause fire and explosions. Attacks some plastics, coatings, and rubber.

METHYLMERCAPTLAAN (Dutch) (74-93-1) May accumulate static electrical charges, and may cause ignition of its vapors. Forms explosive air–gas mixture (flash point 0°F/−18°C). Strong oxidizers may cause fire and explosions. Attacks some plastics, coatings, and rubber.

4-METHYLMERCAPTO-3,5-XYLYLMETHYLCARBAMATE (2032-65-7) Contact with strong oxidizers may cause fire and explosions.

METHYLMETHACRYLAAT (Dutch) (80-62-6) Forms explosive mixture with air (flash point 50°F/10°C). Incompatible with caustics, nitrates, strong acids, aliphatic amines, alkanolamines, peroxides and strong oxidizers. Heat and/or lack of appropriate inhibitor can cause polymerization. May accumulate static electrical charges, and may cause ignition of its vapors.

METHYL alpha-METHYLACRYLATE (80-62-6) Forms explosive mixture with air (flash point 50°F/10°C). Incompatible with caustics, nitrates, strong acids, aliphatic amines, alkanolamines, peroxides, and strong oxidizers. Heat and/or lack of appropriate inhibitor can cause polymerization. May accumulate static electrical charges, and may cause ignition of its vapors.

534

METHYL-METHACRYLAT (German) (80-62-6) Forms explosive mixture with air (flash point 50°F/10°C). Incompatible with caustics, nitrates, strong acids, aliphatic amines, alkanolamines, peroxides, and strong oxidizers. Heat and/or lack of appropriate inhibitor can cause polymerization. May accumulate static electrical charges, and may cause ignition of its vapors.

METHYL METHACRYLATE or METHYL METHACRYLATE MONOMER or METHYL METHACRYLATE MONOMER, INHIBITED (80-62-6) Forms explosive mixture with air (flash point 50°F/10°C). Incompatible with caustics, nitrates, strong acids, aliphatic amines, alkanolamines, peroxides and strong oxidizers. Heat and/or lack of appropriate inhibitor can cause polymerization. May accumulate static electrical charges, and may cause ignition of its vapors.

N-METHYLMETHANAMINE (124-40-3) A gas. Incompatible with acids, organic anhydrides, isocyanates, vinyl acetate, acrylates, substituted allyls, alkylene oxides, epichlorohydrin, ketones, aldehydes, alcohols, glycols, mercury, phenols, cresols, caprolactam solution, strong oxidizers. Attacks aluminum, copper, lead, tin, zinc, and alloys, and some plastics, rubbers, and coatings.

METHYLMETHANE (74-84-0) Flammable gas; forms explosive mixture with air. Strong oxidizers may cause fire and explosions. May accumulate static electrical charges, and may cause ignition of its vapors.

METHYL METHANOATE (107-31-3) Forms explosive mixture with air (flash point −2°F/−19°C). Incompatible with strong acids, caustics, water, nitrates. Reacts violently with strong oxidizers. Attacks some plastics and coatings. May accumulate static electrical charges, and may cause ignition of its vapors.

2-METHYL-2-METHOXY PROPANE (1634-04-4) Forms explosive mixture with air (flash point −14°F/−26°C). May be able to form unstable peroxides. Incompatible with strong acids. Reacts violently with strong oxidizers. May accumulate static electrical charges, and may cause ignition of its vapors.

METHYL a-METHYLACRYLATE (80-62-6) Forms explosive mixture with air (flash point 50°F/10°C). Incompatible with caustics, nitrates, strong acids, aliphatic amines, alkanolamines, peroxides, and strong oxidizers. Heat and/or lack of appropriate inhibitor can cause polymerization. May accumulate static electrical charges, and may cause ignition of its vapors.

2-METHYL-6-METHYLENE-2,7-OCTADIENE (123-35-3) Forms explosive mixture with air (flash point 103°F/40°C). Incompatible with strong oxidizers, sulfuric acid, nitric acid.

METHYL-2-METHYLPROPENOATE (80-62-6) Forms explosive mixture with air (flash point 50°F/10°C). Incompatible with caustics, nitrates, strong acids, aliphatic amines, alkanolamines, peroxides and strong oxidizers. Heat and/or lack of appropriate inhibitor can cause polymerization. May accumulate static electrical charges, and may cause ignition of its vapors.

METHYL 2-METHYL-2-PROPENOATE (80-62-6) Forms explosive mixture with air (flash point 50°F/10°C). Incompatible with caustics, nitrates, strong acids, aliphatic amines, alkanolamines, peroxides and strong oxidizers. Heat and/or lack of appropriate inhibitor can cause polymerization.May accumulate static electrical charges, and may cause ignition of its vapors.

2-METHYL-n-(2-METHYLPROPYL)-1-PROPANAMINE (110-96-3) Forms explosive mixture with air (flash point 85°F/29°C). Incompatible with acids, organic anhydrides, isocyanates, vinyl acetate, acrylates, substituted allyls, alkylene oxides, epichlorohydrin, ketones, aldehydes, alcohols, glycols, mercury, phenols, cresols, caprolactam solution, strong oxidizers. Attacks aluminum, copper, lead, tin, zinc, and alloys.

METHYL MONOCHLOROACETATE (96-34-4) Forms explosive mixture with air (flash point 135°F/57°C). Reacts strongly with oxidizers.

METHYL MORPHOLINE or n-METHYL MORPHOLINE (109-02-4) Forms explosive mixture with air (75°F/24°C). Reacts strongly with oxidizers.

4-METHYL MORPHOLINE (109-02-4) Forms explosive mixture with air (75°F/24°C). Reacts strongly with oxidizers.

METHYL MUSTARD OIL (556-61-6) Forms explosive mixture with air (flash point 90°F/32°C). Incompatible with strong acids, caustics, ammonia, amines, amides, alcohols, glycols, caprolactam solution, strong oxidizers.

N-METHYL-1-NAFTYL-CARBAMAAT (Dutch) (63-25-2) Incompatible with strong oxidizers, strongly alkaline pesticides.

1-METHYLNAPHTHALENE (90-12-0) Forms explosive mixture with air (flash point 180°F/82°C). Oxidizers may cause fire and explosions. Incompatible with nitric acid. May accumulate static electrical charges, and may cause ignition of its vapors.

a-METHYLNAPHTHALENE or alpha-METHYLNAPHTHALENE (90-12-0) Forms explosive mixture with air (flash point 180°F/82°C). Oxidizers may cause fire and explosions. Incompatible with nitric acid. May accumulate static electrical charges, and may cause ignition of its vapors.

a-METHYL NAPHTHALENE or alpha-METHYL NAPHTHALENE (90-12-0) Forms explosive mixture with air (flash point 180°F/82°C). Oxidizers may cause fire and explosions. Incompatible with nitric acid. May accumulate static electrical charges, and may cause ignition of its vapors.

N-METHYL-1-NAPHTHYL-CARBAMAT (German) (63-25-2) Incompatible with strong oxidizers, strongly alkaline pesticides.

N-METHYL-1-NAPHTHYL CARBAMATE (63-25-2) Incompatible with strong oxidizers, strongly alkaline pesticides.

N-METHYL-alpha-NAPHTHYLCARBAMATE (63-25-2) Incompatible with strong oxidizers, strongly alkaline pesticides.

N-METHYL-alpha-NAPHTHYLURETHAN (63-25-2) Incompatible with strong oxidizers, strongly alkaline pesticides.

METHYL NIRAN (298-00-0) Mixtures with magnesium may be explosive.

METHYL NITROBENZENE (88-72-2) Incompatible with strong acids, caustics, ammonia, amines, reducing agents, oxidizers. Attacks some plastics, rubber, and coatings.

METHYL NITROBENZENE (99-99-0) Incompatible with strong acids, caustics, ammonia, amines, reducing agents, oxidizers. Attacks some plastics, rubber, and coatings.

2-METHYLNITROBENZENE (88-72-2) Incompatible with strong acids, caustics, ammonia, amines, reducing agents, oxidizers. Attacks some plastics, rubber, and coatings.

3-METHYLNITROBENZENE (99-08-1) Incompatible with strong acids, caustics, ammonia, amines, reducing agents, oxidizers. Attacks some plastics, rubber, and coatings.

3-METHYLNITROBENZENE (554-84-7) A strong oxidizers; reacts with combustibles, reducing agents, organics and other easily oxidizable materials.

4-METHYLNITROBENZENE (99-99-0) Incompatible with strong acids, caustics, ammonia, amines, reducing agents, oxidizers. Attacks some plastics, rubber, and coatings.

m-**METHYLNITROBENZENE or** *meta*-**METHYLNITROBENZENE** (99-08-1) Incompatible with strong acids, caustics, ammonia, amines, reducing agents, oxidizers. Attacks some plastics, rubber, and coatings.

o-**METHYLNITROBENZENE or** *ortho*-**METHYLNITROBENZENE** (88-72-2) Incompatible with strong acids, caustics, ammonia, amines, reducing agents, oxidizers. Attacks some plastics, rubber, and coatings.

p-**METHYL NITROBENZENE or** *para*-**METHYL NITROBENZENE** (99-99-0) Incompatible with strong acids, caustics, ammonia, amines, reducing agents, oxidizers. Attacks some plastics, rubber, and coatings.

N-METHYL-N-NITROSOMETHANAMINE (62-75-9) Ultraviolet light, strong oxidizers can cause reactions. Store in dark bottles.

METHYLOCYKLOHEKSAN (Polish) (108-87-2) Forms explosive mixture with air (flash point 25°F/−3.9°C). Strong oxidizers may cause fire and explosions. Attacks some plastics, rubber, and coatings.

METHYLOL (67-56-1) May accumulate static electrical charges, and may cause ignition of its vapors. Forms explosive mixture with air (flash point 52°F/11°C). Incompatible with strong acids, strong oxidizers, caustics, aliphatic amines, isocyanates, chromic anhydride, lead perchlorate, perchloric acid, phosphorus trioxide. May react with metallic aluminum at high temperature. Attacks some plastics, rubber, and coatings.

2-METHYLOLPENTANE (97-95-0) Forms explosive mixture with air (flash point 70°F/21°C). Incompatible with strong acids, caustics, isocyanates, amines, isocyanates.

METHYLOLPROPANE (71-36-3) Forms explosive mixture with air (flash point 98°F/37°C). May react with aluminum above 120°F/49°C. Attacks some plastics, rubber, and coatings. Incompatible with strong acids, halogens, caustics, alkali metals, aliphatic amines, isocyanates.

METHYL OXIRANE (75-56-9) Forms explosive gas mixture with air (flash point −35°F/−37°C). Incompatible with anhydrous metal chlorides. Strong acids, caustics, peroxides cause polymerization. Reacts with ammonia, amines, acetylene-forming metals. Attacks some plastics, rubber, and coatings.

METHYL OXITOL (109-86-4) Forms explosive mixture with air (flash point 103°F/39°C). Heat or oxidizers may cause formation of unstable peroxides. Attacks many metals. Strong oxidizers cause fire and explosions. Strong bases cause decomposition. Attacks some plastics, rubber, and coatings. May accumulate static electrical charges, and may cause ignition of its vapors.

METHYL 3-OXOBUTYRATE (105-45-3) Incompatible with oxidizers, strong acids, nitrates.

METHYL PARATHION (298-00-0) Mixtures with magnesium may be explosive.

2-METHYL PENTANE (107-83-5) Forms explosive mixture with air (flash point $-20°F/-9°C$). Reacts vigorously with strong oxidizers. May accumulate static electrical charges, and may cause ignition of its vapors.

2-METHYLPENTANE-2,4-DIOL (107-41-5) Forms explosive mixture with air (flash point 209°F/98°C). Incompatible with strong acids, caustics, aliphatic amines, isocyanates, strong oxidizers.

2-METHYL-2,4-PENTANEDIOL (107-41-5) Forms explosive mixture with air (flash point 209°F/98°C). Incompatible with strong acids, caustics, aliphatic amines, isocyanates, strong oxidizers.

4-METHYL-2,4-PENTANEDIOL (107-41-5) Forms explosive mixture with air (flash point 209°F/98°C). Incompatible with strong acids, caustics, aliphatic amines, isocyanates, strong oxidizers.

2-METHYL-4-PENTANOL (108-11-2) Forms explosive mixture with air (flash point 106°F/41°C). Contact with alkali metals produces hydrogen gas. Incompatible with strong acids, caustics, aliphatic amines, isocyanates. Attacks some plastics, rubber, and coatings. May accumulate static electrical charges, and may cause ignition of its vapors.

4-METHYL-2-PENTANOL (108-11-2) Forms explosive mixture with air (flash point 106°F/41°C). Contact with alkali metals produces hydrogen gas. Incompatible with strong acids, caustics, aliphatic amines, isocyanates. Attacks some plastics, rubber, and coatings. May accumulate static electrical charges, and may cause ignition of its vapors.

4-METHYLPENTANOL-2 (108-11-2) Forms explosive mixture with air (flash point 106°F/41°C). Contact with alkali metals produces hydrogen gas. Incompatible with strong acids, caustics, aliphatic amines, isocyanates. Attacks some plastics, rubber, and coatings. May accumulate static electrical charges, and may cause ignition of its vapors.

4-METHYL-2-PENTANOL ACETATE (108-84-9) Forms explosive mixture with air (flash point 113°F/45°C). Incompatible with strong acids, strong alkalies, nitrates, strong oxidizers. Will swell rubber, and can dissolve certain coatings.

2-METHYL-2-PENTANOL-4-ONE (123-42-2) Incompatible with strong acids, strong alkalis (cause formation of flammable acetone vapors), aliphatic amines, isocyanates, oxidizers, alkali metals. Forms explosive mixture with air (flash point 136°F/58°C). Attacks some forms of plastics, resins, and rubber.

4-METHYL-2-PENTANON (Czech) (108-10-1) Forms explosive mixture with air (flash point 64°F/18°C). Incompatible with strong oxidizers, strong acids, aliphatic amines. Dissolves some plastics, resins, and rubber.

538

2-METHYL-4-PENTANONE (108-10-1) Forms explosive mixture with air (flash point 64°F/18°C). Incompatible with strong oxidizers, strong acids, aliphatic amines. Dissolves some plastics, resins, and rubber.

4-METHYL-2-PENTANONE (108-10-1) Forms explosive mixture with air (flash point 64°F/18°C). Incompatible with strong oxidizers, strong acids, aliphatic amines. Dissolves some plastics, resins, and rubber.

2-METHYL-1-PENTENE (763-29-1) Forms explosive mixture with air (flash point less than 20°F/−7°C). Reacts violently with strong oxidizers.

2-METHYL PENTENE-1 (763-29-1) Forms explosive mixture with air (flash point less than 20°F/−7°C). Reacts violently with strong oxidizers.

4-METHYL-1-PENTENE (691-37-2) Forms explosive mixture with air (flash point −25°F/−32°C). Incompatible with sulfuric acid, nitric acid, oxidizers.

4-METHYL-4-PENTENE (763-29-1) Forms explosive mixture with air (flash point less than 20°F/−7°C). Reacts violently with strong oxidizers.

4-METHYL-3-PENTENE-2-ONE (141-79-7) Forms explosive mixture with air (flash point 87°F/31°C). Forms peroxides. Incompatible with strong acids aliphatic amines, alkanolamines, 2-aminoethanol, ethylene diamine, chlorosulfonic acid, oleum, oxidizers. Dissolves some forms of plastics, resins, and rubber. Attacks copper.

4-METHYL-2-PENTYL ACETATE (108-84-9) Forms explosive mixture with air (flash point 113°F/45°C). Incompatible with strong acids, strong alkalies, nitrates, strong oxidizers. Will swell rubber, and can dissolve certain coatings.

4-METHYL-2-PENTYL ALCOHOL (108-11-2) Forms explosive mixture with air (flash point 106°F/41°C). Contact with alkali metals produces hydrogen gas. Incompatible with strong acids, caustics, aliphatic amines, isocyanates. Attacks some plastics, rubber, and coatings. May accumulate static electrical charges, and may cause ignition of its vapors.

METHYL PENTYL KETONE (110-43-0) Forms explosive mixture with air (flash point 102°F/39°C). Incompatible with strong acids, alkalies, aliphatic amines, oxidizers. Attacks some plastics and rubber.

METHYLPHENOL (1319-77-3) Incompatible with strong acids, oxidizers, alkalies, aliphatic amines, amides, chlorosulfonic acid, oleum. Liquid attacks some plastics, coatings, and rubber.

2-METHYL PHENOL (1319-77-3) Incompatible with strong acids, oxidizers, alkalies, aliphatic amines, amides, chlorosulfonic acid, oleum. Liquid attacks some plastics, coatings, and rubber.

2-METHYLPHENOL (95-48-7) Forms explosive mixture with air (flash point 178°F/81°C). Incompatible with strong acids, oxidizers, alkalies, aliphatic amines, amides, chlorosulfonic acid, oleum. Liquid attacks some plastics and rubber.

3-METHYL PHENOL (1319-77-3) Incompatible with strong acids, oxidizers, alkalies, aliphatic amines, amides, chlorosulfonic acid, oleum. Liquid attacks some plastics, coatings, and rubber.

4-METHYL PHENOL (1319-77-3) Incompatible with strong acids, oxidizers, alkalies, aliphatic amines, amides, chlorosulfonic acid, oleum. Liquid attacks some plastics, coatings, and rubber.

m-METHYLPHENOL or *meta*-METHYLPHENOL (108-39-4) Incompatible with strong acids, oxidizers, alkalies, aliphatic amines, amides, chlorosulfonic acid, oleum. Liquid attacks some plastics, coatings, and rubber.

p-METHYLPHENOL or *para*-METHYLPHENOL (106-44-5) Incompatible with strong acids, oxidizers, alkalies, aliphatic amines, amides, chlorosulfonic acid, oleum. Liquid attacks some plastics, coatings, and rubber.

METHYLPHENOLS (1319-77-3) Incompatible with strong acids, oxidizers, alkalies, aliphatic amines, amides, chlorosulfonic acid, oleum. Liquid attacks some plastics, coatings, and rubber.

METHYLPHENYL AMINE (100-61-8) Forms explosive mixture with air (flash point 174°F/79°C). Incompatible with strong acids, strong oxidizers. Attacks some plastics, rubber, and coatings. May accumulate static electrical charges, and may cause ignition of its vapors.

N-METHYLPHENYLAMINE (100-61-8) Forms explosive mixture with air (flash point 174°F/79°C). Incompatible with strong acids, strong oxidizers. Attacks some plastics, rubber, and coatings. May accumulate static electrical charges, and may cause ignition of its vapors.

4-METHYL-PHENYLENE DIISOCYANATE (584-84-9) Incompatible with strong acids including nonoxidizing mineral and organic acids, caustics, ammonia, amines, amides, alcohols, glycols, caprolactam solution. Water contact causes violent foaming and spattering; produces carbon dioxide and an organic base. Attacks copper and its alloys, some plastics including polyethylene, and rubber.

4-METHYL-PHENYLENE ISOCYANATE (584-84-9) Incompatible with strong acids including nonoxidizing mineral and organic acids, caustics, ammonia, amines, amides, alcohols, glycols, caprolactam solution. Water contact causes violent foaming and spattering; produces carbon dioxide and an organic base. Attacks copper and its alloys, some plastics including polyethylene, and rubber.

1-METHYL-1-PHENYL-ETHENE (98-83-9) Forms explosive mixture with air (flash point 129°F/54°C). Avoid heat and/or lack of appropriate inhibitor concentration, and/or contact with catalysts for vinyl or ionic polymerization such as aluminum or iron chloride. Incompatible with strong acids, oxidizers (including peroxides and halogens).

1-METHYL-1-PHENYL-ETHYLENE (98-83-9) Forms explosive mixture with air (flash point 129°F/54°C). Avoid heat and/or lack of appropriate inhibitor concentration, and/or contact with catalysts for vinyl or ionic polymerization such as aluminum or iron chloride. Incompatible with strong acids, oxidizers (including peroxides and halogens).

METHYL PHENYL KETONE (98-86-2) Incompatible with strong acids, aliphatic amines, oxidizers. Forms explosive mixture with air (flash point 170°F/77°C).

METHYLPHENYL METHANOL (98-85-1) Forms explosive mixture with air (flash point 205°F/96°C). Incompatible with strong acids, caustics, aliphatic amines, isocyanates, oxidizers.

METHYL PHOSPHITE (121-45-9) Forms explosive mixture with air (flash point 82°F/28°C). Reacts violently with strong acids, nitrates, oxidizers, magnesium salts of perchloric acid.

METHYL PHOSPHONOTHIOIC DICHLORIDE (676-97-1) Reacts violently with water, producing hydrochloric acid and hydrogen chloride fumes. Corrodes metals.

METHYLPHOSPHORAMIDIC ACID, 4-t-BUTYL-2-CHLORO-PHENYL METHYL ESTER (299-86-5) Decomposes in strongly alkaline (pH >7) and strongly acidic media. Unstable over long periods of time in water and at temperatures above 140°F/60°C.

METHYL PHTHALATE (131-11-3) Incompatible with strong alkalies, strong acids, nitrates, oxidizers.

2-METHYLPROPANAL (78-84-2) Forms explosive gas mixture with air. Incompatible with strong acids, caustics, aliphatic amines, alkanol-amines, aromatic amines, strong oxidizers.

2-METHYL-2-PROPANAMINE (75-64-9) Forms explosive mixture with air (flash point 48°F/−9°C). May accumulate static electrical charges, and may cause ignition of its vapors. Incompatible with acids, organic anhydrides, isocyanates, vinyl acetate, acrylates, substituted allyls, alkylene oxides, epichlorohydrin, ketones, aldehydes, alcohols, glycols, phenols, cresols, caprolactam solution, strong oxidizers.

2-METHYLPROPANE (75-28-5) Forms explosive gas mixture with air. Strong oxidizers may cause fire and explosions. May accumulate static electrical charges, and may cause ignition of its vapors.

2-METHYLPROPANENITRILE (78-82-0) Forms explosive mixture with air (flash point 47°F/8°C). Reacts violently with strong oxidizers.

2-METHYLPROPANOIC ACID (79-31-2) Forms explosive mixture with air (flash point 132°F/56°C). Incompatible with sulfuric acid, caustics, ammonia, amines, isocyanates, alkylene oxides, epichlorohydrin. Attacks aluminum and other metals.

2-METHYL-2-PROPANOIC ACID, ETHYL ESTER (97-63-2) Forms explosive mixture with air (flash point 80°F/27°C). Incompatible with strong acids, amines, oxidizers. Corrodes some metals.

2-METHYL-1-PROPANOL (78-83-1) Forms explosive mixture with air (flash point 82°F/28°C). Incompatible with strong acids, strong oxidizers, caustics, aliphatic amines, isocyanates, alkaline metals, and alkali earth. Attacks some plastics, rubber, and coatings. May react with aluminum at high temperatures.

2-METHYL-2-PROPANOL (75-65-0) Forms explosive mixture with air (flash point 52°F/11°C). May accumulate static electrical charges, and may cause ignition of its vapors. Incompatible with strong acids including mineral acids, strong oxidizers or caustics, aliphatic amines, isocyanates, alkali metals (i.e., lithium, sodium, potassium, rubidium, cesium, francium). Attacks many plastics and some coatings.

METHYL PROPENATE (96-33-3) Heat, light, and/or lack of appropriate inhibitor concentration can cause polymerization. Forms explosive mixture with air (flash point 27°F/−3°C). Incompatible with strong acids, oxidizers, aliphatic amines, alkanolamines.

2-METHYLPROPENE (115-11-7) Flammable gas; forms explosive mixture with air. May be able to form unstable peroxides; inappropriate level of inhibitor may cause polymerization. Reacts violently with strong oxidizers, strong acids, oxides of nitrogen. May accumulate static electrical charges, and may cause ignition of its vapors.

2-METHYLPROPENENITRILE (126-98-7) Forms explosive mixture with air (flash point 34°F/1°C). Incompatible with aliphatic amines, alkanolamines. Reacts violently with oxidizers. Incompatible with strong acids, strong bases, or light exposure; may cause polymerization.

2-METHYL-2-PROPENENITRILE (126-98-7) Forms explosive mixture with air (flash point 34°F/1°C). Incompatible with aliphatic amines, alkanolamines. Reacts violently with oxidizers. Incompatible with strong acids, strong bases, or light exposure; may cause polymerization.

2-METHYL PROPENIC ACID (79-41-4) Forms explosive mixture with air (flash point 152°F/67°C). A reducing agent; reacts with oxidizers. Forms unstable peroxides; can polymerize violently, especially if stored above 120°F/49°C. Incompatible with strong acids, caustics, ammonia, amines, isocyanates, alkylene oxides, epichlorohydrin.

2-METHYL PROPENIC ACID, METHYL ESTER (80-62-6) Forms explosive mixture with air (flash point 50°F/10°C). Incompatible with caustics, nitrates, strong acids, aliphatic amines, alkanolamines, peroxides, and strong oxidizers. Heat and/or lack of appropriate inhibitor can cause polymerization. May accumulate static electrical charges, and may cause ignition of its vapors.

METHYL-2-PROPENOATE (96-33-3) Heat, light, and/or lack of appropriate inhibitor concentration can cause polymerization. Forms explosive mixture with air (flash point 27°F/−3°C). Incompatible with strong acids, oxidizers, aliphatic amines, alkanolamines.

2-METHYL-2-PROPEN-1-OL (513-42-8) Forms explosive mixture with air (flash point 92°F/33°C). Incompatible with strong acids, caustics, aliphatic amines, isocyanates, oxidizers.

METHYLPROPIONATE (554-12-1) Forms explosive mixture with air (28°F/−2°C). Strong oxidizers may cause fire and explosions. Incompatible with strong acids and nitrates. May accumulate static electrical charges, and may cause ignition of its vapors.

2-METHYLPROPIONIC ACID (79-41-4) Forms explosive mixture with air (flash point 152°F/67°C). A reducing agent; reacts with oxidizers. Forms unstable peroxides; can polymerize violently, especially if stored above 120°F/49°C. Incompatible with strong acids, caustics, ammonia, amines, isocyanates, alkylene oxides, epichlorohydrin.

alpha-METHYLPROPIONIC ACID (79-31-2) Forms explosive mixture with air (flash point 132°F/56°C). Incompatible with sulfuric acid, caustics, ammonia, amines, isocyanates, alkylene oxides, epichlorohydrin. Attacks aluminum and other metals.

2-METHYLPROPIONITRILE (78-82-0) Forms explosive mixture with air (flash point 47°F/8°C). Reacts violently with strong oxidizers.

1-METHYL PROPYL ACETATE (105-46-4) Forms explosive mixture with air (flash point 70°F/19°C). Reacts violently with oxidizers. Incompatible with strong acids, nitrates, potassium *tert*-butoxide. Attacks some plastics, rubber, and coatings. May accumulate static electrical charges, and may cause ignition of its vapors.

2-METHYL PROPYL ACETATE (110-19-0) Forms explosive mixture with air (flash point 64°F/18°C). Reacts with water on standing to produce acetic acid and n-butyl alcohol. Reacts violently with strong oxidizers. Reacts with caustics, strong acids, nitrates. Dissolves rubber, many plastics, resins, and some coatings. May accumulate static electrical charges, and may cause ignition of its vapors.

2-METHYL-1-PROPYL ACETATE (110-19-0) Forms explosive mixture with air (flash point 64°F/18°C). Reacts with water on standing to produce acetic acid and n-butyl alcohol. Reacts violently with strong oxidizers. Reacts with caustics, strong acids, nitrates. Dissolves rubber, many plastics, resins, and some coatings. May accumulate static electrical charges, and may cause ignition of its vapors.

2-METHYL-1-PROPYL ACRYLATE (106-63-8) Forms explosive mixture with air (flash point 94°F/34°C). Incompatible with strong acids, aliphatic amines, alkanolamines.

1-METHYLPROPYLAMINE (13952-84-6) Forms explosive mixture with air. Incompatible with acids, organic anhydrides, isocyanates, vinyl acetate, acrylates, substituted allyls, alkylene oxides, epichlorohydrin, ketones, aldehydes, alcohols, glycols, phenols, cresols, caprolactam solution, strong oxidizers.

2-METHYLPROPYLAMINE (78-81-9) Forms explosive mixture with air (flash point 15°F/−9°C). Incompatible with strong acids including mineral and organic acids, organic anhydrides, isocyanates, vinyl acetate, acrylates, substituted allyls, alkylene oxides, epichlorohydrin, ketones, aldehydes, alcohols, glycols, phenols, cresols, caprolactam solution, halogens, alkali metals, copper and its alloys.

METHYL PROPYL BENZENE (99-87-6) Forms explosive mixture with air (flash point 117°F/47°C). Incompatible with nitric acid, strong oxidizers. Attacks and softens rubber. May accumulate static electrical charges, and may cause ignition of its vapors.

METHYL PROPYL CARBINOL (6032-29-7) Forms explosive mixture with air (flash point 93°F/34°C). Reacts violently with strong oxidizers and alkali metals. May also react with alkali earth metals.

2-METHYLPROPYL ESTER (9CI) (97-85-8) Forms explosive mixture with air (flash point 99°F/37°C). Incompatible with strong acids, nitrates, strong oxidizers.

b-METHYLPROPYL ETHANOATE or beta-METHYLPROPYL ETHANOATE (110-19-0) Forms explosive mixture with air (flash point 64°F/18°C). Reacts with water on standing to produce acetic acid and n-butyl alcohol. Reacts violently with strong oxidizers. Reacts with caustics, strong acids, nitrates. Dissolves rubber, many plastics, resins, and some coatings. May accumulate static electrical charges, and may cause ignition of its vapors.

2-METHYL-2-PROPYLETHANOL (108-11-2) Forms explosive mixture with air (flash point 106°F/41°C). Contact with alkali metals produces hydrogen gas. Incompatible with strong acids, caustics, aliphatic amines, isocyanates. Attacks some plastics, rubber, and coatings. May accumulate static electrical charges, and may cause ignition of its vapors.

METHYL PROPYL ETHER (557-17-5) Forms explosive mixture with air (flash point less than −4°F/−20°C). May be able to form unstable and explosive peroxides. Reacts violently with strong oxidizers. Incompatible with sulfuric or nitric acids. May accumulate static electrical charges, and may cause ignition of its vapors.

1-METHYL-1-PROPYLETHYLENE (763-29-1) Forms explosive mixture with air (flash point less than 20°F/−7°C). Reacts violently with strong oxidizers.

2-METHYLPROPYLISOBUTYRATE (97-85-8) Forms explosive mixture with air (flash point 99°F/37°C). Incompatible with strong acids, nitrates, strong oxidizers.

METHYL PROPYL KETONE or METHYL-n-PROPYL KETONE (107-87-9) May accumulate static electrical charges, and may cause ignition of its vapors. Forms explosive mixture with air (flash point 45°F/7°C) Reacts violently with strong oxidizers. Attacks some plastics, rubber, and coatings.

2-METHYLPYRIDINE (109-06-8) Forms explosive mixture with air (flash point 79°F/26°C). Strong oxidizers may cause fire and explosions. Attacks copper and its alloys.

3-METHYLPYRIDINE (108-99-6) Forms explosive mixture with air (flash point 97°F/36°C). Incompatible with acids, organic anhydrides, isocyanates, aldehydes, oxidizers.

4-METHYLPYRIDINE (108-89-4) Forms explosive mixture with air (flash point 134°F/57°C). Incompatibel with acids, organic anhydrides, isocyanates, aldehydes, oxidizers.

a-METHYLPYRIDINE or alpha-METHYLPYRIDINE (109-06-8) Forms explosive mixture with air (flash point 79°F/26°C). Strong oxidizers may cause fire and explosions. Attacks copper and its alloys.

1-METHYL-2-(3-PYRIDYL)PYRROLIDINE (54-11-5) Incompatible with strong acids, strong oxidizers. Attacks some plastics, rubber, and coatings. May accumulate static electrical charges, and may cause ignition of its vapors.

1-METHYLPYRROLIDINONE (872-50-4) Forms explosive mixture with air (flash point 7°F/−14°C). A strong base. Incompatible with strong acids, organic anhydrides, isocyanates, aldehydes, light metals, flammable and porous materials. Attacks some plastics, rubber, and coatings.

1-METHYL-2-PYRROLIDINONE (872-50-4) Forms explosive mixture with air (flash point 7°F/−14°C). A strong base. Incompatible with strong acids, organic anhydrides, isocyanates, aldehydes, light metals, flammable and porous materials. Attacks some plastics, rubber, and coatings.

N-METHYLPYRROLIDINONE (872-50-4) Forms explosive mixture with air (flash point 7°F/−14°C). A strong base. Incompatible with strong acids, organic anhydrides, isocyanates, aldehydes, light metals, flammable and porous materials. Attacks some plastics, rubber, and coatings.

1-METHYLPYRROLIDONE (872-50-4) Forms explosive mixture with air (flash point 7°F/−14°C). A strong base. Incompatible with strong acids, organic anhydrides, isocyanates, aldehydes, light metals, flammable and porous materials. Attacks some plastics, rubber, and coatings.

n-METHYL-alpha-PYRROLIDONE (872-50-4) Forms explosive mixture with air (flash point 7°F/−14°C). A strong base. Incompatible with strong acids, organic anhydrides, isocyanates, aldehydes, light metals, flammable and porous materials. Attacks some plastics, rubber, and coatings.

3-(1-METHYL-2-PYRROLIDYL)PYRIDINE (54-11-5) Incompatible with strong acids, strong oxidizers. Attacks some plastics, rubber, and coatings. May accumulate static electrical charges, and may cause ignition of its vapors.

METHYL SALICYLATE (119-36-0) Incompatible with strong acids, nitrates, strong oxidizers.

METHYLSENFOEL (German) (556-61-6) Forms explosive mixture with air (flash point 90°F/32°C). Incompatible with strong acids, caustics, ammonia, amines, amides, alcohols, glycols, caprolactam solution, strong oxidizers.

METHYL SILICONE (9016-00-6) Incompatible with sulfuric acid, isocyanates.

METHYL STYRENE (25013-15-4) Forms explosive mixture with air (flash point 17°F/53°C). Polymerization inhibitor (usually 10 to 50 ppm of *tert*-butyl catechol) must be present in adequate concentrations to avoid explosive polymerization. Oxidizers may cause fire and explosions. Also, the following should be avoided: peroxides, aluminum chloride, acids (nonoxidizing mineral, sulfuric, nitric), caustics, ammonia, aliphatic amines, alkanolamines.

3-METHYL STYRENE (25013-15-4) Forms explosive mixture with air (flash point 17°F/53°C). Polymerization inhibitor (usually 10 to 50 ppm of *tert*-butyl catechol) must be present in adequate concentrations to avoid explosive polymerization. Oxidizers may cause fire and explosions. Also, the following should be avoided: peroxides, aluminum chloride, acids (nonoxidizing mineral, sulfuric, nitric), caustics, ammonia, aliphatic amines, alkanolamines.

4-METHYL STYRENE (25013-15-4) Forms explosive mixture with air (flash point 17°F/53°C). Polymerization inhibitor (usually 10 to 50 ppm of *tert*-butyl catechol) must be present in adequate concentrations to avoid explosive polymerization. Oxidizers may cause fire and explosions. Also, the following should be avoided: peroxides, aluminum chloride, acids (nonoxidizing mineral, sulfuric, nitric), caustics, ammonia, aliphatic amines, alkanolamines.

a-METHYL STYRENE or alpha-METHYL STYRENE (98-83-9) Forms explosive mixture with air (flash point 129°F/54°C). Avoid heat and/or lack of appropriate inhibitor concentration, and/or contact with catalysts for vinyl or ionic polymerization such as aluminum or iron chloride. Incompatible with strong acids, oxidizers (including peroxides and halogens).

m-**METHYL STYRENE or** *meta*-**METHYL STYRENE** (25013-15-4) Forms explosive mixture with air (flash point 17°F/53°C). Polymerization inhibitor (usually 10 to 50 ppm of *tert*-butyl catechol) must be present in adequate concentrations to avoid explosive polymerization. Oxidizers may cause fire and explosions. Also, the following should be avoided: peroxides, aluminum chloride, acids (nonoxidizing mineral, sulfuric, nitric), caustics, ammonia, aliphatic amines, alkanolamines.

p-**METHYL STYRENE or** *para*-**METHYL STYRENE** (25013-15-4) Forms explosive mixture with air (flash point 17°F/53°C). Polymerization inhibitor (usually 10 to 50 ppm of *tert*-butyl catechol) must be present in adequate concentrations to avoid explosive polymerization. Oxidizers may cause fire and explosions. Also, the following should be avoided: peroxides, aluminum chloride, acids (nonoxidizing mineral, sulfuric, nitric), caustics, ammonia, aliphatic amines, alkanolamines.

METHYL SULFATE (77-78-1) Forms explosive mixture with air (flash point 182°F/83°C). Incompatible with strong oxidizers, strong acids, strong alkalies, strong ammonia solutions. Attacks some plastics, rubber, and coatings.

METHYL SULFHYDRATE (74-93-1) May accumulate static electrical charges, and may cause ignition of its vapors. Forms explosive air–gas mixture (flash point 0°F/−18°C). Strong oxidizers may cause fire and explosions. Attacks some plastics, coatings, and rubber.

METHYL SULFIDE (75-18-3) Forms explosive mixture with air (flash point −36°F/−38°C). Reacts violently with strong oxidizers. May accumulate static electrical charges, and may cause ignition of its vapors.

METHYL SULFOXIDE (67-68-5) Forms explosive mixture with air (flash point 203°F/95°C). Reacts violently with oxidizers. Reacts with ethanoyl chloride, boron compounds, halides, metal alkoxides, oxidizers.

METHYL SULPHIDE (75-18-3) Forms explosive mixture with air (flash point −36°F/−38°C) Reacts violently with strong oxidizers. May accumulate static electrical charges, and may cause ignition of its vapors.

n-METHYL-N-2,4,6-TETRANITROANILINE (479-45-8) A powerful oxidizer and highly sensitive explosive. Protect containers from shock, friction, heat, oxidizers, and hydrazine.

METHYLTHIOALCOHOL (74-93-1) May accumulate static electrical charges, and may cause ignition of its vapors. Forms explosive air–gas mixture (flash point 0°F/−18°C). Strong oxidizers may cause fire and explosions. Attacks some plastics, coatings, and rubber.

4-METHYLTHIO-3,5-DIMETHYLPHENYLMETHYLCARBA-MATE (2032-65-7) Contact with strong oxidizers may cause fire and explosions.

4-(METHYLTHIO)-3,5-XYLYLMETHYLCARBAMATE (2032-65-7) Contact with strong oxidizers may cause fire and explosions.

METHYLTHIOPHOS (298-00-0) Mixtures with magnesium may be explosive.

METHYL THIRAM (137-26-8) Combustible solid (flash point 192°F/89°C). Strong oxidizers may cause fire and explosions; contact with strong acid or oxidizable materials produces toxic gases.

METHYL THIURAM DISULFIDE (137-26-8) Combustible solid (flash point 192°F/89°C). Strong oxidizers may cause fire and explosions; contact with strong acid or oxidizable materials produces toxic gases.

m-**METHYLTOLUENE** or *meta*-**METHYLTOLUENE** (108-38-3) Forms explosive mixture with air (flash point 90°F/32°C). Strong oxidizers may cause fire and explosions. Attacks some plastics, rubber, and coatings. May accumulate static electrical charges, and may cause ignition of its vapors.

o-**METHYLTOLUENE** or *ortho*-**METHYLTOLUENE** (95-47-6) Forms explosive mixture with air (flash point 90°F/32°C). Strong oxidizers may cause fire and explosions. Attacks some plastics, rubber, and coatings. May accumulate static electrical charges, and may cause ignition of its vapors.

p-**METHYLTOLUENE** or *para*-**METHYLTOLUENE** (106-42-3) Forms explosive mixture with air (flash point 81°F/27.2°C). Strong oxidizers may cause fire and explosions. Attacks some plastics, rubber, and coatings. May accumulate static electrical charges, and may cause ignition of its vapors.

2-METHYL-*p*-TOLUIDINE (95-68-1) Forms explosive mixture with air (flash point 206°F/97°C). Contact with strong oxidizers may cause fire and explosions. Hypochlorite bleaches may cause formation of explosive chloramines. Attacks some plastics, rubber, and coatings.

4-METHYL-*o*-TOLUIDINE (95-68-1) Forms explosive mixture with air (flash point 206°F/97°C). Contact with strong oxidizers may cause fire and explosions. Hypochlorite bleaches may cause formation of explosive chloramines. Attacks some plastics, rubber, and coatings.

METHYL TRIBROMIDE (75-25-2) Reacts with chemically active metals, calcium, acetone, strong caustics. Attacks some plastics, rubber, and coatings.

METHYL TRICHLORIDE (67-66-3) Decomposes in the presence of excess water or high temperatures, producing phosgene and hydrogen chloride. Can become explosive in the presence of strong alkalies and water. Can accumulate static electrical charges. In contact with water and at high temperatures, it becomes corrosive; attacks iron and other metals. Incompatible with acetone, aluminum, strong oxidizers, potassium, sodium, chemically active metals, strong bases. Attacks plastics and rubber.

METHYLTRICHLOROMETHANE (71-55-6) Reacts with acetone, strong caustics, alkaline earth, alkali metals, dinitrogen tetroxide, sodium hydroxide, amides, strong oxidizers, sodium, potassium. Reacts violently with copper, bronze, and other metal powders. The uninhibited grade is corrosive to aluminum. Attacks some plastics, rubber, and coatings.

METHYL TRICHLOROSILANE (75-79-6) Forms explosive gas mixture with air (flash point 15°F/−9°C). Water or air contact produces hydrochloric acid. Strong oxidizers may cause fire and explosions. Incompatible with bases. Corrodes most metals. Attacks some plastics, rubber, and coatings.

METHYL TRICHLORSILAN (Czech) (75-79-6) Forms explosive gas mixture with air (flash point 15°F/−9°C). Water or air contact produces hydrochloric acid. Strong oxidizers may cause fire and explosions. Incompatible with bases. Corrodes most metals. Attacks some plastics, rubber, and coatings.

METHYLTRIMETHYLENE GLYCOL (107-88-0) Incompatible with strong acids, caustics, aliphatic amines, isocyanates, oxidizers.

METHYL TUADS (137-26-8) Combustible solid (flash point 192°F/89°C). Strong oxidizers may cause fire and explosions; contact with strong acid or oxidizable materials produces toxic gases.

METHYL VINYL ETHER (107-25-5) Forms explosive mixture with air (flash point −69°F/−56°C). Water contact slowly produces acetaldehyde and methyl alcohol. Forms unstable peroxides; able to polymerize. Acids, oxidizers may cause fire and explosions. May accumulate static electrical charges, and may cause ignition of its vapors.

METHYL VINYL KETONE (78-94-4) Forms explosive mixture with air (flash point 20°F/−7°C). May accumulate static electrical charges, and may cause ignition of its vapors. Heat can cause polymerization. Reacts violently with strong oxidizers.

METHYLZINC (544-97-8) Ignites spontaneously with air. Water contact produces methane gas.

1-METHYPROPYL ALCOHOL (78-92-2) Forms explosive mixture with air (flash point 75°F/4°C). Attacks some plastics, rubber, and coatings. Forms an explosive peroxide in air. Ignites with chromium trioxide. Incompatible with strong oxidizers, strong acids, aliphatic amines, isocyanates, organic peroxides.

METILACRILATO (Italian) (96-33-3) Heat, light and/or lack of appropriate inhibitor concentration can cause polymerization. Forms explosive mixture with air (flash point 27°F/−3°C). Incompatible with strong acids, oxidizers, aliphatic amines, alkanolamines.

METILAMIL ALCOHOL (Italian) (108-11-2) Forms explosive mixture with air (flash point 106°F/41°C). Contact with alkali metals produces hydrogen gas. Incompatible with strong acids, caustics, aliphatic amines, isocyanates. Attacks some plastics, rubber, and coatings. May accumulate static electrical charges, and may cause ignition of its vapors.

METILAMINE (Italian) (74-89-5) Forms explosive gas mixture with air. Incompatible with acids, organic anhydrides, isocyanates, vinyl acetate, acrylates, substituted allyls, alkylene oxides, epichlorohydrin, ketones, aldehydes, alcohols, glycols, phenols, cresols, caprolactam solution, strong oxidizers. Attacks aluminum, copper, lead, tin, zinc, and alloys, and some plastics, rubber, and coatings.

METIL CELLOSOLVE (Italian) (109-86-4) Forms explosive mixture with air (flash point 103°F/39°C). Heat or oxidizers may produce unstable peroxides. Attacks many metals. Strong oxidizers cause fire and explosions. Strong bases cause decomposition. Attacks some plastics, rubber, and coatings. May accumulate static electrical charges, and may cause ignition of its vapors.

METILE (ACETATO di) (Italian) (79-20-9) Forms explosive mixture with air (flash point 14°F/−10°C). Incompatible with strong acids, nitrates, oxidizers, bases. Attacks some plastics. May accumulate static electrical charges, and may cause ignition of its vapors.

METILEN-S,S′-BIS(O,O-DIETIL-DITIOFOSFATO) (Italian) (563-12-2) Incompatible with alkaline formulations. Mixtures with magnesium may be explosive.

METILETILCHETONE (Italian) (78-93-3) Forms explosive mixture with air (flash point 16°F/−9°C). Extremely flammable. Incompatible with sulfuric acid, nitric acid, aliphatic amines, strong oxidizers, potassium *tert*-butoxide, 2-propanol, chlorosulfonic acid, oleum.

METIL (FORMIATO di) (Italian) (107-31-3) Forms explosive mixture with air (flash point −2°F/−19°C). Incompatible with strong acids, caustics, water, nitrates. Reacts violently with strong oxidizers. Attacks some plastics and coatings. May accumulate static electrical charges, and may cause ignition of its vapors.

METIL ISOCIANATO (Italian) (624-83-9) Forms explosive mixture with air (flash point 0°F/−18°C). Contact with iron, tin, copper (or salts of these elements), and with certain other catalysts (such as triphenylarsenic oxide, triethylphosphine, and tributyltin oxide) or elevated temperatures may cause polymerization. Reacts violently with warm water. Incompatible with strong acids, alcohols, glycols, amines, amides, ammonia, caprolactam, caustics, strong oxidizers. Attacks some plastics, rubber, or coatings.

METILMERCAPTANO (Italian) (74-93-1) May accumulate static electrical charges, and may cause ignition of its vapors. Forms explosive air–gas mixture (flash point 0°F/−18°C). Strong oxidizers may cause fire and explosions. Attacks some plastics, coatings, and rubber.

METIL METACRILATO (Italian) (80-62-6) Forms explosive mixture with air (flash point 50°F/10°C). Incompatible with caustics, nitrates, strong acids, aliphatic amines, alkanolamines, peroxides, and strong oxidizers. Heat and/or lack of appropriate inhibitor can cause polymerization. May accumulate static electrical charges, and may cause ignition of its vapors.

N-METIL-1-NAFTIL-CARBAMMATO (Italian) (63-25-2) Incompatible with strong oxidizers, strongly alkaline pesticides.

4-METILPENTAN-2-OLO (Italian) (108-11-2) Forms explosive mixture with air (flash point 106°F/41°C). Contact with alkali metals produces hydrogen gas. Incompatible with strong acids, caustics, aliphatic amines, isocyanates. Attacks some plastics, rubber, and coatings. May accumulate static electrical charges, and may cause ignition of its vapors.

METILTRIAZOTION (86-50-0) Contact with strong oxidizers may cause fire and explosions.

METMERCAPTURON (2032-65-7) Contact with strong oxidizers may cause fire and explosions.

METOKSYETYLOWY ALCOHOL (Polish) (109-86-4) Forms explosive mixture with air (flash point 103°F/39°C). Heat or oxidizers may cause formation of unstable peroxides. Attacks many metals. Strong oxidizers cause fire and explosions. Strong bases cause decomposition. Attacks some plastics, rubber, and coatings. May accumulate static electrical charges, and may cause ignition of its vapors.

METOLACHLOR (51218-45-0) Incompatible with strong acids, nitrates, oxidizers.

METOX (72-43-5) Contact with strong oxidizers may cause fire and explosions. Attacks some plastics, rubber, and coatings.

METRON (298-00-0) Mixtures with magnesium may be explosive.

METSO 20 (1344-09-8) Solution is a strong base; reacts with acids, organic anhydrides, alkylene oxides, epichlorohydrin, aldehydes, alcohols, glycols, phenols, cresols, caprolactam solution. Attacks chemically active metals.

METSO PENTABEAD 20 (1344-09-8) Solution is a strong base; reacts with acids, organic anhydrides, alkylene oxides, epichlorohydrin, aldehydes, alcohols, glycols, phenols, cresols, caprolactam solution. Attacks chemically active metals.

MET-SPAR (1305-62-0) Contact with maleic anhydride, phosphorus, nitroethane, nitromethane, nitroparaffins, nitropropane may cause explosion.

METYLAL (Polish) (109-87-5) May be able to form unstable and explosive peroxides. Forms explosive mixture with air (flash point −4°F/−18°C). Incompatible with strong oxidizers, acids. Attacks some plastics, rubber, and coatings.

METYLOAMINA (Polish) (74-89-5) Forms explosive gas mixture with air. Incompatible with acids, organic anhydrides, isocyanates, vinyl acetate, acrylates, substituted allyls, alkylene oxides, epichlorohydrin, ketones, aldehydes, alcohols, glycols, phenols, cresols, capro-lactam solution, strong oxidizers. Attacks aluminum, copper, lead, tin, zinc, and alloys, and some plastics, rubber, and coatings.

METYLOETYLOKETON (Polish) (78-93-3) Forms explosive mixture with air (flash point 16°F/−9°C). Extremely flammable. Incompatible with sulfuric acid, nitric acid, aliphatic amines, strong oxidizers, potassium *tert*-butoxide, 2-propanol, chlorosulfonic acid, oleum.

METYLOHYDRAZYNA (Polish) (60-34-4) Highly reactive reducing agent and a strong base. Forms explosive mixture with air (flash point 17°F/−8°C). Reacts violently with oxidizers, combustibles, acids. Incompatible with alcohols, glycols, isocyanates, phenols, cresols. Oxides of iron or copper, manganese, lead, copper, or their alloys can lead to fire and explosions. Attacks cork, some plastics, coatings, and rubber.

METYLOWY ALKOHOL (Polish) (67-56-1) May accumulate static electrical charges, and may cause ignition of its vapors. Forms explosive mixture with air (flash point 52°F/11°C). Incompatible with strong acids, strong oxidizers, caustics, aliphatic amines, isocyanates, chromic anhydride, lead perchlorate, perchloric acid, phosphorus trioxide. May react with metallic aluminum at high temperature. Attacks some plastics, rubber, and coatings.

METYLU CHLOREK (Polish) (74-87-3) Moisture causes decomposi-tion. Contact with aluminum causes formation of a product that may ignite spontaneously in air. Reacts with barium, lithium, sodium, magnesium, titanium. Attacks plastics, rubber, and coatings.

METYLU JODEK (Polish) (74-88-4) Contact with strong oxidizers may cause fire and explosions. Heat, light, and moisture contribute to instability.

MEVINFOS (Dutch) (7786-34-7) Contact with strong oxidizers may cause fire and explosions. Attacks some plastics, rubber, and coatings.

MEVINOX (7786-34-7) Contact with strong oxidizers may cause fire and explosions. Attacks some plastics, rubber, and coatings.

MEVINPHOS (7786-34-7) Contact with strong oxidizers may cause fire and explosions. Attacks some plastics, rubber, and coatings.

MEXACARBATE (315-18-4) Usually dissolved in a combustible liquid; avoid oxidizer contact.

MFB (462-06-6) Forms explosive mixture with air (flash point 5°F/−15°C). Incompatible with strong oxidizers, ammonium nitrate, chromic acid, halogens, hydrogen peroxide, nitric acid. Attacks some plastics, rubber, and coatings.

MH (123-33-1) Contact with strong oxidizers may cause fire and explosions.

MHA ACID (583-91-5) Incompatible with sulfuric acid, caustics, ammonia, amines, isocyanates, alkylene oxides, epichlorohydrin.

MHA-FA (583-91-5) Incompatible with sulfuric acid, caustics, ammo-nia, amines, isocyanates, alkylene oxides, epichlorohydrin.

MIBC (108-11-2) Forms explosive mixture with air (flash point 106°F/41°C). Contact with alkali metals produces hydrogen gas. Incompatible with strong acids, caustics, aliphatic amines, isocya-nates. Attacks some plastics, rubber, and coatings. May accumulate static electrical charges, and may cause ignition of its vapors.

MIBK (108-10-1) Forms explosive mixture with air (flash point 64°F/18°C). Incompatible with strong oxidizers, strong acids, aliphatic amines. Dissolves some plastics, resins, and rubber.

MIC (556-61-6) Forms explosive mixture with air (flash point 90°F/32°C). Incompatible with strong acids, caustics, ammonia, amines, amides, alcohols, glycols, caprolactam solution, strong oxidizers.

MIC (624-83-9) Forms explosive mixture with air (flash point 0°F/−18°C). Contact with iron, tin, copper (or salts of these elements) and with certain other catalysts (such as triphenylarsenic oxide, triethylphosphine, and tributyltin oxide) or elevated temperatures may cause polymerization. Reacts violently with warm water. Incompatible with strong acids, alcohols, glycols, amines, amides, ammonia, caprolactam, caustics, strong oxidizers. Attacks some plastics, rubber, or coatings.

3-MIC (108-11-2) Forms explosive mixture with air (flash point 106°F/41°C). Contact with alkali metals produces hydrogen gas. Incompatible with strong acids, caustics, aliphatic amines, isocyanates. Attacks some plastics, rubber, and coatings. May accumulate static electrical charges, and may cause ignition of its vapors.

MICRO DDT 75 (50-29-3) Incompatible with salts of iron or aluminum, and bases. Do not store in iron containers.

MICROFLOTOX (7704-34-9) Combustible solid. Liquid contact produces sulfur dioxide with air. Reacts violently with strong oxidizers. Forms explosive, shock-sensitive, or pyrophoric mixtures with ammonia, ammonium nitrate, bromates, calcium carbide, charcoal, chlorates, hydrocarbons, iodates, iron. Reacts violently with halogen compounds, sodium, tin, uranium, and other compounds. Attacks steel when moist. May accumulate static electrical charges, and may cause ignition of its vapors.

MIERENZUUR (Dutch) (64-18-6) Forms explosive mixture with air (flash point 156°F/(69°C). A strong reducing agent; reacts violently with oxidizers. Incompatible with sulfuric acid, bases, ammonia, aliphatic amines, alkanolamines, furfuryl alcohol, hydrogen peroxide, isocyanates, alkylene oxides, epichlorohydrin. Attacks aluminum, cast iron, steel, some plastics, rubber, and coatings.

MIGHTY 150 (91-20-3) Forms explosive mixture with air (flash point 174°F/79°C). Incompatible with strong oxidizers, chromium oxide (reacts violently). Attacks some plastics, rubber, and coatings. May accumulate static electrical charges, and may cause ignition of its vapors.

MIK (108-10-1) Forms explosive mixture with air (flash point 64°F/18°C). Incompatible with strong oxidizers, strong acids, aliphatic amines. Dissolves some plastics, resins, and rubber.

MIL-B-4394-B (74-97-5) Reacts with chemically active metals such as calcium, powdered aluminum, zinc, and magnesium. Liquid attacks some plastics, rubber, and coatings.

MILBOL 49 (58-89-9) Not combustible, but may be dissolved in a combustible solvent. If solvent comes in contact with oxidizers, fire and explosions may result.

MILK ACID (598-82-3) Incompatible with strong acids. Aqueous solution reacts with bases.

MILK WHITE (7446-14-2) Reacts violently with potassium. Incompatible with aluminum, magnesium.

MILLER'S FUMIGRAIN (107-13-1) Forms explosive mixture with air (flash point 32°F/0°C o.c.). Forms explosive peroxides; heat, light, caustics, silver nitrate, and peroxides can cause polymerization. Incompatible with strong acids, strong oxidizers, amines, 2-aminoethanol, bromine, chlorosulfonic acid, ethylene diamine, nitric acid, oleum, potassium hydroxide, sodium hydroxide, sulfuric acid. Attacks copper and copper alloys; attacks aluminum in high concentrations. May accumulate static electrical charges, and may cause ignition of its vapors.

MILOCEP (51218-45-0) Incompatible with strong acids, nitrates, oxidizers.

MILTON (7681-52-9) A powerful oxidizer and a strong base. Decomposes in sunlight, producing oxygen and increasing the risk of fire. Stability decreases with concentration, heat, light, decrease in pH, and contamination with metals. Incompatible with strong acids, reducing agents, combustible substances; all cause violent reaction, fire, and explosions. Contact with amines and ammonia salts produces explosive chloroamines. Corrodes many metals: steel, 12% and 17% chrome steel, cast iron, monel, aluminum, nickel, brass, bronze, iconel.

MINERAL CARBON activated (64365-11-3); **purified** (7440-44-0) Incompatible with strong oxidizers, strong acids, oxides, unsaturated oils.

MINERAL COLZA OIL (N/A) Oxidizers may cause fire and explosions. Incompatible with nitric acid. May accumulate static electrical charges, and may cause ignition of its vapors.

MINERAL NAPHTHA (71-43-2) Forms explosive mixture with air (Flash point 12°F/−11°C).) May accumulate static electrical charges, and may cause ignition of its vapors. Incompatible with strong oxidizers, nitric acid, oxygen, ozone, perchlorates. Attacks some forms of plastics, coatings, and rubber.

MINERAL OIL (8012-95-1) Oxidizers may cause fire and explosions. Incompatible with nitric acid. May accumulate static electrical charges, and may cause ignition of its vapors.

MINERAL PITCH (8052-42-4) Incompatible with nitric acid, fluorine, strong oxidizers.

MINERAL SEAL OIL (N/A) Oxidizers may cause fire and explosions. Incompatible with nitric acid. May accumulate static electrical charges, and may cause ignition of its vapors.

MINERAL SPIRITS (8052-41-3) Forms explosive mixture with air (flash point 102°F to 140°F/38.7°C to 60°C). Incompatible with nitric acid, strong oxidizers. Attacks some plastics, rubber, and coatings.

MINERAL SPIRITS (8032-32-4) Forms explosive mixture with air (flash point −40°F to −86°F/−40°C to −66°C). Incompatible with strong acids, strong oxidizers. Attacks some plastics, rubber, and coatings. May accumulate static electrical charges, and may cause ignition of its vapors.

MINERAL TURPENTINE (8032-32-4) Forms explosive mixture with air (flash point −40°F to −86°F/−40°C to −66°C). Incompatible with strong acids, strong oxidizers. Attacks some plastics, rubber, and coatings. May accumulate static electrical charges, and may cause ignition of its vapors.

MINERAL THINNER (8032-32-4) Forms explosive mixture with air (flash point −40°F to −86°F/−40°C to −66°C). Incompatible with strong acids, strong oxidizers. Attacks some plastics, rubber, and coatings. May accumulate static electrical charges, and may cause ignition of its vapors.

MIPAX (131-11-3) Incompatible with strong alkalies, strong acids, nitrates, oxidizers.

MIPK (563-80-4) Forms explosive mixture with air at 43°F/6°C. Reacts violently with strong oxidizers.

MIRACLE (94-75-7) Decomposes in sunlight. Incompatible with strong oxidizers; may cause fire and explosions.

MIRAMID (105-60-2) Contact with strong oxidizers may cause fire and explosions.

MIRBANE OIL (98-95-3) Forms explosive mixture with air (flash point 190°F/88°C). Incompatible with nitric acid, nitrogen tetroxide, caustics, ammonia, amines. Attacks some plastics, rubber, and coatings.

MIT (556-61-6) Forms explosive mixture with air (flash point 90°F/32°C). Incompatible with strong acids, caustics, ammonia, amines, amides, alcohols, glycols, caprolactam solution, strong oxidizers.

MITC (556-61-6) Forms explosive mixture with air (flash point 90°F/32°C). Incompatible with strong acids, caustics, ammonia, amines, amides, alcohols, glycols, caprolactam solution, strong oxidizers.

MITIS GREEN (12002-03-8) Contact with strong oxidizers may cause fire and explosions.

MIXED PRIMARY AMYL NITRATES (1002-16-0) Forms explosive mixture with air (flash point 120°F/49°C). An oxidizer. Reacts strongly with reducing agents, strong acids, esters, combustibles. Attacks some plastics, rubber, and coatings.

MIXTURE OF BENZENE, TOLUENE, XYLENES (8030-31-7) Oxidizers may cause fire and explosions. Incompatible with nitric acid. May accumulate static electrical charges, and may cause ignition of its vapors.

MLT (121-75-5) Incompatible with strong oxidizers, magnesium, alkaline pesticides. Attacks metals, some plastics, rubber, and coatings.

MME (80-62-6) Forms explosive mixture with air (flash point 50°F/10°C). Incompatible with caustics, nitrates, strong acids, aliphatic amines, alkanolamines, peroxides, and strong oxidizers. Heat and/or lack of appropriate inhibitor can cause polymerization. May accumulate static electrical charges, and may cause ignition of its vapors.

MMH (60-34-4) Highly reactive reducing agent and a strong base. Forms explosive mixture with air (flash point 17°F/−8°C). Reacts violently with oxidizers, combustibles, acids. Incompatible with alcohols, glycols, isocyanates, phenols, cresols. Oxides of iron or copper, manganese, lead, copper, or their alloys can lead to fire and explosions. Attacks cork, some plastics, coatings, and rubber.

2-MMT (12108-13-3) Incompatible with strong acids, strong oxidizers.

MNBK (591-78-6) Forms explosive mixture with air (flash point 77°F/25°C). Strong oxidizers may cause fire and explosions. Dissolves some plastics, resins, and rubber.

MNT (99-08-1) Incompatible with strong acids, caustics, ammonia, amines, reducing agents, oxidizers. Attacks some plastics, rubber, and coatings.

MODANE SOFT (119-36-8) Forms explosive mixture with air (flash point 205°F/96°C). Incompatible with strong acids, nitrates, oxidizers.

MOHR'S SALT (10045-89-3) Reacts violently with tetranitromethane, ammonium perchlorate, and mercury(II) dinitrate. Sulfates react with aluminum, magnesium.

MOLASSES ALCOHOL (64-17-5) Forms explosive mixture with air (flash point 65°F/18°C). May accumulate static electrical charges, and may cause ignition of its vapors. Reactions may be violent with oleum, sulfuric acid, nitric acid, bases, aliphatic amines, isocyanates, oxidizers.

MOLATOC (119-36-8) Forms explosive mixture with air (flash point 205°F/96°C). Incompatible with strong acids, nitrates, oxidizers.

MOLCER (119-36-8) Forms explosive mixture with air (flash point 205°F/96°C). Incompatible with strong acids, nitrates, oxidizers.

MOLECULAR BROMINE (7726-95-6) A powerful oxidizer. May cause fire and explosions in contact with organic or other readily oxidizable materials. Contact with ammonia, acetaldehyde, acetylene, acrylonitrile, or metals may cause violent reactions. Reacts violently with aluminum, titanium, mercury, or potassium. Also incompatible with alcohols, antimony, alkali hydroxides, arsenites, boron, calcium nitrite, cesium monoxide, carbonyls, dimethyl formamide, ethyl phosphine, fluorine, ferrous and mercurous salts, germanium, hypophosphites, iron carbide, isobutyronphenone, magnesium phosphide, methanol, nickel carbonyl, olefins, ozone, sodium, and many other substances. Attacks some coatings and some forms of plastic and rubber. Corrodes iron, steel, stainless steels, and copper.

MOLECULAR CHLORINE (7782-50-5) A powerful oxidizer. Reacts violently with combustible materials, reducing agents, and many other substances. Forms explosive mixtures with gasoline and petroleum products, turpentine, alcohols, acetylene, carbon disulfide, hydrogen, anhydrous ammonia, finely divided metals, organic compounds, phosphorus. Fire, explosion, and the formation of toxic fumes may result from contact with many substances including alkyphosphines, aluminum, antimony, arsenic compounds, arsine, bismuth, boron, brass, calcium compounds, carbon, diethyl zinc, fluorine, germanium, hydrocarbons, rubber. Attacks some plastics and coatings. In the presence of moisture, extremely corrosive to iron, steel, copper, bronze, zinc.

MOLE-DEATH (57-24-9) Contact with strong oxidizers may cause fire and explosions.

MOLOFAC (119-36-8) Forms explosive mixture with air (flash point 205°F/96°C). Incompatible with strong acids, nitrates, oxidizers.

MOLOL (8012-95-1) Oxidizers may cause fire and explosions. Incompatible with nitric acid. May accumulate static electrical charges, and may cause ignition of its vapors.

MOLYBDATE (7439-98-7) Finely divided particles react violently with strong oxidizers, heat, or ignition source.

MOLYBDENUM or **MOLYBDENUM, METALLIC** (7439-98-7) Finely divided particles react violently with strong oxidizers, heat, or ignition source.

MOLYBDENUM(VI) OXIDE (1313-27-5) Incompatible with alkali metals or molten magnesium.

MOLYBDENUM TRIOXIDE (1313-27-5) Incompatible with alkali metals or molten magnesium.

MOLYBDIC ANHYDRIDE (1313-27-5) Incompatible with alkali metals or molten magnesium.

MOLYBDIC TRIOXIDE (1313-27-5) Incompatible with alkali metals or molten magnesium.

MONDUR TDS (584-84-9) Incompatible with strong acids including non-oxidizing mineral and organic acids, caustics, ammonia, amines, amides, alcohols, glycols, caprolactam solution. Water contact causes violent foaming and spattering; produces carbon dioxide and an organic base. Attacks copper and its alloys, some plastics including polyethylene, and rubber.

MONOAMMONIUM ORTHOPHOSPHATE (7783-28-0) Incompatible with strong oxidizers, strong bases. Contact with air causes substance to give off corrosive anhydrous ammonia fumes.

MONOAMMONIUM PHOSPHATE (7783-28-0) Incompatible with strong oxidizers, strong bases. Contact with air causes substance to give off corrosive anhydrous ammonia fumes.

MONOAMMONIUM SALT OF SULFAMIC ACID (7773-06-0) Fires and explosions may result from contact with strong oxidizers, hot water, potassium, sodium, sodium nitrite, metal chlorates, hot acid solutions. Corrosive to mild steel.

MONOAMMONIUM SULFAMATE (7773-06-0) Fires and explosions may result from contact with strong oxidizers, hot water, potassium, sodium, sodium nitrite, metal chlorates, hot acid solutions. Corrosive to mild steel.

MONOBROMOACETONE (598-31-2) Forms explosive mixture with air (flash point 113°F/45°C). Contact with oxidizers may cause fire and explosions.

MONOBROMOBENZENE (108-86-1) Forms explosive mixture with air (flash point 124°F/51°C). Incompatible with strong oxidizers, alkaline earth metals (barium, calcium, magnesium, strontium, etc.), metallic salts, with risk of violent reactions. May accumulate static electrical charges, and may cause ignition of its vapors.

MONOBROMOETHANE (74-96-4) Forms explosive mixture with air (flash point less than −4°F/−20°C). Hydrolyzes in water, producing hydrogen bromide. Oxidizers may cause fire or explosions. Fire and explosions may be caused by contact with aluminum, magnesium, or zinc powders, and lithium, potassium, sodium. Attacks some plastic, rubber, and coatings.

MONOBROMOMETHANE (74-83-9) Incompatible with strong oxidizers, aluminum, dimethylsulfoxide, ethylene oxide, water. Attacks zinc, magnesium, alkali metals and their alloys, some plastics, rubber, and coatings. Forms aluminum alkyls in presence of aluminum; aluminum alkyls are spontaneously ignitable materials.

MONOBROMOTRIFLUORMETHANE (75-63-8) Reacts with chemically active metals, powdered aluminum, zinc, magnesium. May attack some plastics, rubber, and coatings.

555

MONOBUTYLAMINE (109-73-9) Forms explosive mixture with air (flash point 10°F/−12°C). May accumulate static electrical charges, and may cause ignition of its vapors. Incompatible with acids, organic anhydrides, isocyanates, vinyl acetate, acrylates, substituted allyls, alkylene oxides, epichlorohydrin, ketones, aldehydes, alcohols, glycols, phenols, cresols, caprolactam solution, strong oxidizers. In the presence of moisture, corrodes light metals and copper on contact.

MONOCALCIUM PHOSPHATE MONOHYDRATE (10103-46-5) Water contact produces acid; attacks metals.

MONOCHLOORBENZEEN (Dutch) (108-90-7) Forms explosive mixture with air (flash point 82°F/28°C). May accumulate static electrical charges, and may cause ignition of its vapors. Incompatible with oxidizers, dimethylsulfoxide, sodium powder, silver perchloride. Attacks some plastics, rubber, and coatings.

MONOCHLORBENZENE (108-90-7) Forms explosive mixture with air (flash point 82°F/28°C). May accumulate static electrical charges, and may cause ignition of its vapors. Incompatible with oxidizers, dimethylsulfoxide, sodium powder, silver perchloride. Attacks some plastics, rubber, and coatings.

MONOCHLORBENZOL (German) (108-90-7) Forms explosive mixture with air (flash point 82°F/28°C). May accumulate static electrical charges, and may cause ignition of its vapors. Incompatible with oxidizers, dimethylsulfoxide, sodium powder, silver perchloride. Attacks some plastics, rubber, and coatings.

MONOCHLORETHANE (75-00-3) Flammable gas. Reacts slowly with water; produces hydrogen chloride gas. Contact with moisture produces hydrochloric acid. May accumulate static electrical charges, and may cause ignition of its vapors. Forms explosive mixture with air (flash point −58°F/−55°C). Contact with aluminum, lithium, magnesium, sodium, potassium, zinc may cause fire and explosions. Attacks some plastics and rubber.

MONOCHLORETHANOIC ACID, ETHYL ESTER (105-39-5) Forms explosive mixture with air (flash point 100°F/38°C). Water contact produces toxic and corrosive fumes. Reacts violently with alkaline earth metals (e.g., barium, calcium, magnesium, strontium), alkaline metals, sodium cyanide. Attacks metals in the presence of moisture.

MONOCHLOROACETALDEHYDE (107-20-0) Incompatible with oxidizers, acids, water. May form a water-soluble polymer in storage.

MONOCHLOROACETIC ACID (79-11-8) Aqueous solution is a strong acid. Incompatible with strong oxidizers, bases. Attacks most common metals in the presence of moisture.

MONOCHLOROACETIC ACID, METHYL ESTER (96-34-4) Forms explosive mixture with air (flash point 135°F/57°C). Strong reaction with oxidizers.

MONOCHLOROACETYL CHLORIDE (79-04-9) Forms corrosive vapors with air. Strong reaction with water, producing hydrochloric acid. Incompatible with sulfuric acid, caustics, alkalis, alcohols, aliphatic amines, alkanolamines, ammonia, isocyanates, alkylene oxides, epichlorohydrin.

556

MONOCHLOROBENZENE (108-90-7) Forms explosive mixture with air (flash point 82°F/28°C). May accumulate static electrical charges, and may cause ignition of its vapors. Incompatible with oxidizers, dimethylsulfoxide, sodium powder, silver perchloride. Attacks some plastics, rubber, and coatings.

MONOCHLORODIFLUOROMETHANE (75-45-6) May accumulate static electrical charges, and may cause ignition of its vapors. Moisture and rust cause slow decomposition, producing toxic gases. Attacks some plastics, rubber, and coatings. Thermal decomposition occurs at high temperature with alkalies and alkaline earth metals.

MONOCHLORODIMETHYL ETHER (107-30-2) Forms explosive mixture with air (flash point 0°F/−17.8°C). May be able to form unstable and explosive peroxides. Produces hydrochloric acid on contact with moisture. Corrosive to metals in presence of moisture.

MONOCHLOROETHANOIC ACID (79-11-8) Aqueous solution is a strong acid. Incompatible with strong oxidizers, bases. Attacks most common metals in the presence of moisture.

2-MONOCHLOROETHANOL (107-07-3) Forms explosive mixture with air (flash point 140°F/60°C). Strong oxidizers may cause fire and explosions. Incompatible with strong caustics (with formation of ethylene gas), strong acids, aliphatic amines, isocyanates. Attacks some plastics, rubber, and coatings.

MONOCHLOROETHENE (75-01-4) Forms explosive gas mixture with air. Atmospheric oxygen and various contaminants or strong oxidizers causes formation of peroxides, which can initiate a violent polymerization action. Also able to polymerize when heated and exposed, long term, to light. Reacts violently with strong oxidizers or oxides of nitrogen. Contact with copper or other acetylide-forming metals produces explosive compounds. Attacks iron and steel in the presence of moisture.

MONOCHLOROETHYLENE (75-01-4) Forms explosive gas mixture with air. Atmospheric oxygen and various contaminants or strong oxidizers produce peroxides, which can initiate a violent polymerization action. Also able to polymerize when heated and exposed, long term, to light. Reacts violently with strong oxidizers or oxides of nitrogen. Contact with copper or other acetylide-forming metals produce explosive compounds. Attacks iron and steel in the presence of moisture.

MONOCHLOROMETHANE (74-87-3) Moisture causes decomposition. Contact with aluminum causes formation of a product that may ignite spontaneously in air. Reacts with barium, lithium, sodium, magnesium, titanium. Attacks plastics, rubber, and coatings.

MONOCHLOROMETHYL ETHER (107-30-2) Forms explosive mixture with air (flash point 0°F/−17.8°C). May be able to form unstable and explosive peroxides. Produces hydrochloric acid on contact with moisture. Corrosive to metals in presence of moisture.

MONO-CHLORO-MONO-BROMO-METHANE (74-97-5) Reacts with chemically active metals such as calcium, powdered aluminum, zinc, and magnesium. Liquid attacks some plastics, rubber, and coatings.

MONOCHLOROPENTAFLUOROETHANE (76-15-3) Thermal decomposition occurs at high temperatures with alkalies and alkaline earth metals.

o-**MONOCHLOROPHENOL** or *ortho*-**MONOCHLOROPHENOL** (95-57-8) Forms explosive mixture with air (flash point 147°F/64°C). Contact with strong oxidizers may cause fire and explosions. Attacks aluminum and copper.

b-MONOCHLOROPROPIONIC ACID or beta-MONOCHLORO-PROPIONIC ACID (107-94-8) Reacts with strong oxidizers, caustics, ammonia, amines, isocyanates, alkylene oxide, epichlorohydrin, strong oxidizers. Aqueous solution is acid; attacks metals. Only aluminum or stainless steel with protective lining or coating may contact liquid or vapor.

MONOCHLOROSULFURIC ACID (7790-94-5) Reacts violently with water, producing sulfuric and hydrochloric acids and dense fumes. Dangerously reactive; avoid contact with all other material. Strong oxidizer and strong acid; reacts violently with bases, reducing agents, combustibles, acids (especially sulfuric), alcohol, diphenyl ether, silver nitrate, water. Phosphorus contact may cause fire and explosions. Forms explosive material with ethyl alcohol. Incompatible with many materials. Attacks many metals.

MONOCHLOROTRIFLUOROMETHANE (75-72-9) Reacts with aluminum, magnesium, zinc, and their alloys.

MONOCHROMIUM OXIDE (1333-82-0) A strong oxidizer. Reacts with acetic acid, acetic anhydride, acetone, anthracene, chromous sulfide, diethyl ether, dimethyl formamide, ethanol, hydrogen sulfide, methanol, naphthalene, camphor, glycerol, potassium ferricyanide, pyridine, turpentine, combustibles, organics. Aqueous solution is strongly acidic. Attacks metals in presence of moisture.

MONOCHROMIUM TRIOXIDE (1333-82-0) A strong oxidizer. Reacts with acetic acid, acetic anhydride, acetone, anthracene, chromous sulfide, diethyl ether, dimethyl formamide, ethanol, hydrogen sulfide, methanol, naphthalene, camphor, glycerol, potassium ferricyanide, pyridine, turpentine, combustibles, organics. Aqueous solution is strongly acidic. Attacks metals in presence of moisture.

"MONOCITE" METHACRYLATE MONOMER (80-62-6) Forms explosive mixture with air (flash point 50°F/10°C). Incompatible with caustics, nitrates, strong acids, aliphatic amines, alkanolamines, peroxides, and strong oxidizers. Heat and/or lack of appropriate inhibitor can cause polymerization. May accumulate static electrical charges, and may cause ignition of its vapors.

MONOCYANOGEN (460-19-5) Reacts explosively with acids, water, liquid oxygen, oxidizers.

MONOETHANOLAMINE (141-43-5) Will oxidize in air; often shipped under a pad of inert gas. Forms explosive mixture with air (flash point 185°F/85°C). Contact with strong oxidizers may cause fire and explosions. A strong base. Incompatible with acids, organic anhydrides, isocyanates, alkylene oxides, epichlorohydrin, aldehydes, alcohols, glycols, phenols, cresols, caprolactam solution. Contact with strong acids may cause spattering. Attacks plastics, rubber, and aluminum, copper, tin, and their alloys.

MONOETHYLAMINE (75-04-7) Forms explosive mixture with air (flash point less than 0°F/−18°C). Incompatible with acids, organic anhydrides, isocyanates, vinyl acetate, acrylates, substituted allyls, alkylene oxides, epichlorohydrin, ketones, aldehydes, alcohols, glycols, phenols, cresols, caprolactam solution, strong oxidizers. Attacks aluminum, copper, lead, tin, zinc, and alloys, some plastics, rubber, and coatings.

MONOETHYLAMINE, ANHYDROUS (75-04-7) Forms explosive mixture with air (flash point less than 0°F/−18°C). Incompatible with acids, organic anhydrides, isocyanates, vinyl acetate, acrylates, substituted allyls, alkylene oxides, epichlorohydrin, ketones, aldehydes, alcohols, glycols, phenols, cresols, caprolactam solution, strong oxidizers. Attacks aluminum, copper, lead, tin, zinc, and alloys, and some plastics, rubber, and coatings.

MONOETHYLENE GLYCOL (107-21-1) Incompatible with strong acids, caustics, aliphatic amines, isocyanates, chlorosulfonic acid, oleum, strong oxidizers.

MONOETHYLENE GLYCOL ETHER (110-71-4) Forms explosive mixture with air (flash point 29°F/−2°C). May accumulate static electrical charges, and may cause ignition of its vapors. Incompatible with sulfuric acid, isocyanates, strong oxidizers.

MONOFLUOROBENZENE (462-06-6) Forms explosive mixture with air (flash point 5°F/−15°C). Incompatible with strong oxidizers, ammonium nitrate, chromic acid, halogens, hydrogen peroxide, nitric acid. Attacks some plastics, rubber, and coatings.

MONOFLUOROETHYLENE (75-02-5) Flammable gas. May be able to polymerize. Reacts violently with oxidizers. May accumulate static electrical charges, and may cause ignition of its vapors.

MONOFLUROTRICHLOROMETHANE (75-69-4) Reacts with barium, lithium, sodium, magnesium, titanium. Attacks some plastics, rubber, and coatings.

MONOGLYME (110-71-4) Forms explosive mixture with air (flash point 29°F/−2°C). May accumulate static electrical charges, and may cause ignition of its vapors. Incompatible with sulfuric acid, isocyanates, strong oxidizers.

MONOHYDROXY BENZENE (108-95-2) Forms explosive mixture with air (flash point 174°F/79°C). Incompatible with strong oxidizers, strong acids, caustics, aliphatic amines, amides, oxidizers, formaldehyde, butadiene, calcium hypochlorite. Liquid attacks some plastics, rubber, and coatings; hot liquid attacks aluminum, magnesium, lead, and zinc metals.

MONOHYDROXYMETHANE (67-56-1) May accumulate static electrical charges, and may cause ignition of its vapors. Forms explosive mixture with air (flash point 52°F/11°C). Incompatible with strong acids, strong oxidizers, caustics, aliphatic amines, isocyanates, chromic anhydride, lead perchlorate, perchloric acid, phosphorus trioxide. May react with metallic aluminum at high temperature. Attacks some plastics, rubber, and coatings.

MONOIODOMETHANE (74-88-4) Contact with strong oxidizers may cause fire and explosions. Heat, light, and moisture contribute to instability.

MONOISOBUTYLAMINE (78-81-9) Forms explosive mixture with air (flash point 15°F/−9°C). Incompatible with strong acids including mineral and organic acids, organic anhydrides, isocyanates, vinyl acetate, acrylates, substituted allyls, alkylene oxides, epichlorohydrin, ketones, aldehydes, alcohols, glycols, phenols, cresols, caprolactam solution, halogens, alkali metals, copper and its alloys.

MONOISOPROPANOLAMINE (78-96-6) Forms explosive mixture with air (flash point 171°F/77°C). Strong oxidizers may cause fire and explosions. Incompatible with acids, organic anhydrides, isocyanates, aldehydes.

MONOISOPROPYLAMINE (75-31-0) Forms explosive mixture with air (flash point −35°F/−37°C). A strong base. Reacts with alkali metals, alkaline earth, strong acids (with explosive spattering), strong oxidizers (with fire and explosions). Attacks some plastics, rubber, and coatings.

MONOMETHYLAMINE (74-89-5) Forms explosive gas mixture with air. Incompatible with acids, organic anhydrides, isocyanates, vinyl acetate, acrylates, substituted allyls, alkylene oxides, epichlorohydrin, ketones, aldehydes, alcohols, glycols, phenols, cresols, caprolactam solution, strong oxidizers. Attacks aluminum, copper, lead, tin, zinc and alloys, and some plastics, rubber, and coatings.

MONOMETHYLAMINE SOLUTION (40%) (74-89-5) Forms explosive mixture with air (flash point 10°F/−12°C). Incompatible with acids, organic anhydrides, isocyanates, vinyl acetate, acrylates, substituted allyls, alkylene oxides, epichlorohydrin, ketones, aldehydes, alcohols, glycols, phenols, cresols, caprolactam solution, strong oxidizers. Attacks aluminum, copper, lead, magnesium, tin, zinc, and alloys, and some plastics, rubber, and coatings.

MONOMETHYLAMINOETHANOL (109-83-1) Forms explosive mixture with air (flash point 165°F/74°C). Contact with strong oxidizers may cause fire and explosions. A strong base; reacts with acids. Attacks aluminum, copper, zinc, and alloys.

MONOMETHYL ANILINE (100-61-8) Forms explosive mixture with air (flash point 174°F/79°C). Incompatible with strong acids, strong oxidizers. Attacks some plastics, rubber, and coatings. May accumulate static electrical charges, and may cause ignition of its vapors.

N-MONOMETHYLANILINE (100-61-8) Forms explosive mixture with air (flash point 174°F/79°C). Incompatible with strong acids, strong oxidizers. Attacks some plastics, rubber, and coatings. May accumulate static electrical charges, and may cause ignition of its vapors.

MONOMETHYL ETHANOLAMINE (109-83-1) Forms explosive mixture with air (flash point 165°F/74°C). Contact with strong oxidizers may cause fire and explosions. A strong base; reacts with acids. Attacks aluminum, copper, zinc, and alloys.

MONOMETHYL HYDRAZINE (60-34-4) Highly reactive reducing agent and a strong base. Forms explosive mixture with air (flash point 17°F/−8°C). Reacts violently with oxidizers, combustibles, acids. Incompatible with alcohols, glycols, isocyanates, phenols, cresols. Oxides of iron or copper, manganese, lead, copper, or their alloys can lead to fire and explosions. Attacks cork, some plastics, coatings, and rubber.

MONONITROGEN MONOXIDE (10102-43-9) A strong oxidizer. Incompatible with all combustible materials, chlorinated hydrocarbons, ammonia, carbon disulfide, many metals, fluorine, and ozone. Attacks some plastics, rubber, and coatings.

MONO PE (115-77-5) Incompatible with organic acids, oxidizers. Forms explosive material with thiophosphoryl chloride + heat.

560

MONOPHENYLHYDRAZINE (100-63-0) A highly reactive reducing agent. Forms explosive mixture with air (flash point 190°F/88°C). Incompatible with oxides of iron or copper, and with manganese, lead, copper, or their alloys, and methyl iodide, oxidizers, organic compounds. Attacks some plastics, rubber, coatings, and cork.

MONOPLEX DOA (103-23-1) Incompatible with strong acids, nitrates, oxidizers.

MONOPROPYLAMINE (107-10-8) Forms explosive mixture with air. Incompatible with acids, organic anhydrides, halogenated hydrocarbons, isocyanates, nitroparaffins, vinyl acetate, acrylates, substituted allyls, alkylene oxides, epichlorohydrin, ketones, aldehydes, alcohols, glycols, mercury, phenols, cresols, triethynyl aluminum, caprolactam solution, strong oxidizers. Attacks aluminum, copper, lead, tin, zinc, and alloys.

MONO-N-PROPYLAMINE (107-10-8) Forms explosive mixture with air. Incompatible with acids, organic anhydrides, halogenated hydrocarbons, isocyanates, nitroparaffins, vinyl acetate, acrylates, substituted allyls, alkylene oxides, epichlorohydrin, ketones, aldehydes, alcohols, glycols, mercury, phenols, cresols, triethynyl aluminum, captrolactam solution, strong oxidizers. Attacks aluminum, copper, lead, tin, zinc and alloys.

MONOPROPYLENE GLYCOL (57-55-7) Incompatible with strong acids, caustics, aliphatic amines, isocyanates, strong oxidizers (with violent reaction).

MONOSAN (94-75-7) Decomposes in sunlight. Incompatible with strong oxidizers; may cause fire and explosions.

MONOSODIUM SALT OF SULFUROUS ACID (7631-90-5) Slowly oxidized to the sulfate on contact with air. Contact with oxidizers or acids produces sulfur dioxide gas.

MONOXIDE (630-08-0) Forms extremely explosive mixture with air. Reacts violently with strong oxidizers.

MONTAR (generic CAS for PCBs 1336-36-3) Incompatible with strong oxidizers, strong acids.

MONTREL (299-86-5) Decomposes in strongly alkaline (pH >7) and strongly acidic media. Unstable over long periods of time in water and at temperatures above 140°F/60°C.

MOON (56-81-5) Polymerizes at approximately 300°F/150°C. Incompatible with acetic anhydride, potassium permanganate, strong acids, caustics, aliphatic amines, isocyanates, oxidizers.

MOPARI (62-73-7) Attacks some plastics, rubber, and coatings.

MORBICID (50-00-0) May polymerize unless properly inhibited (usually with methanol). Forms explosive mixture with air (flash point 122°F/50°C). Incompatible with strong acids, amines, strong oxidizers, alkaline materials, nitrogen dioxide, performic acid. Reaction with hydrochloric acid produces bis-chloromethyl ether, a carcinogen.

MORPHOLINE (110-91-8) Forms explosive mixture with air (flash point 98°F/37°C). Incompatible with acids, organic anhydrides, isocyanates, vinyl acetate, acrylates, substituted allyls, alkylene oxides, epichlorohydrin, ketones, aldehydes, alcohols, glycols, phenols, cresols, caprolactam solution, strong oxidizers. Attacks aluminum, copper, lead, tin, zinc and alloys, and some plastics, rubber, and coatings.

MORTON WP-161-E (556-61-6) Forms explosive mixture with air (flash point 90°F/32°C). Incompatible with strong acids, caustics, ammonia, amines, amides, alcohols, glycols, caprolactam solution, strong oxidizers.

MORTOPAL (107-49-3) Decomposes above 300°F/150°C, producing flammable ethylene gas. Strong oxidizers may cause fire and explosions. Attacks some plastics, rubber, and coatings.

MOSCARDA (121-75-5) Incompatible with strong oxidizers, magnesium, alkaline pesticides. Attacks metals, some plastics, rubber, and coatings.

MOSS GREEN (12002-03-8) Contact with strong oxidizers may cause fire and explosions.

MOTH BALLS or MOTH FLAKES (91-20-3) Form explosive mixture with air (flash point 174°F/79°C). Incompatible with strong oxidizers, chromium oxide (react violently). Attack some plastics, rubber, and coatings. May accumulate static electrical charges, and may cause ignition of their vapors.

MOTOPAL (107-49-3) Decomposes above 300°F/150°C, producing flammable ethylene gas. Contact with strong oxidizers may cause fire and explosions. Attacks some plastics, rubber, and coatings.

MOTOR BENZOL (71-43-2) Forms explosive mixture with air (flash point 12°F/−11°C). May accumulate static electrical charges, and may cause ignition of its vapors. Incompatible with strong oxidizers, nitric acid, oxygen, ozone, perchlorates. Attacks some forms of plastics, coatings, and rubber.

MOTOR FUEL (8006-61-9) Forms explosive mixture with air (flash point 125°F/51°C). Contact with oxidizers may cause fire and explosions. Incompatible with nitric acid. May accumulate static electrical charges, and may cause ignition of its vapors.

MOTOR FUEL ANTI-KNOCK COMPOUND (75-74-1) Forms explosive mixture with air (flash point 100°F/38°C). Decomposes above 212°F/100°C. Contact with strong oxidizers and strong acids may cause fire and explosions. Attacks some plastics, rubber, and coatings.

MOTOR FUEL ANTI-KNOCK COMPOUND (78-00-2) Forms explosive mixture with air and decomposes above 200°F/93°C. Incompatible with strong oxidizers, concentrated acids; may cause fire and explosions. Attacks some plastics, rubber, and coatings.

MOTOR FUEL ANTI-KNOCK COMPOUNDS CONTAINING LEAD ALKYLS (N/A) Form explosive mixtures with air (flash point 89°F/22°C). Incompatible with strong oxidizers, rust, chemically active metals. Attack rubber, some plastics, and coatings.

MOTOR SPIRIT (8006-61-9) Forms explosive mixture with air (flash point 125°F/51°C). Contact with oxidizers may cause fire and explosions. Incompatible with nitric acid. May accumulate static electrical charges, and may cause ignition of its vapors.

MOTOX (8001-35-2) Reacts with oxidizers, with a risk of fire or explosions. Attacks metals in the presence of moisture.

MOUSE-TOX (57-24-9) Contact with strong oxidizers may cause fire and explosions.

MOXIE (72-43-5) Contact with strong oxidizers may cause fire and explosions. Attacks some plastics, rubber, and coatings.

MOXONE (94-75-7) Decomposes in sunlight. Incompatible with strong oxidizers; may cause fire and explosions.

MPK (107-87-9) May accumulate static electrical charges, and may cause ignition of its vapors. Forms explosive mixture with air (flash point 45°F/7°C). Reacts violently with strong oxidizers. Attacks some plastics, rubber, and coatings.

MPTD (676-97-1) Reacts violently with water, producing hydrochloric acid and hydrogen chloride fumes. Corrodes metals.

MROWCZAN ETYLU (Polish) (109-94-4) Forms explosive mixture with air (flash point −4°F/−20°C). May accumulate static electrical charges, and may cause ignition of its vapors. Incompatible with nitrates, strong oxidizers, strong alkalies, strong acids.

MSZYCOL (58-89-9) Not combustible, but may be dissolved in a combustible solvent. If solvent comes in contact with oxidizers, fire and explosions may result.

MTD (95-80-7) Incompatible with oxidizers, acids, organic anhydrides, isocyanates, aldehydes. Attacks aluminum, brass, bronze, copper, zinc.

MULHOUSE WHITE (7446-14-2) Reacts violently with potassium. Incompatible with aluminum, magnesium.

MULTRATHANE M (101-68-8) Incompatible with strong alkalies, acids, alcohols, ammonia, amines, amides, glycols, caprolactam. Unstable above 100°F/37.8°C. Attacks some plastics, rubber, and coatings.

MURFOS (56-38-2) Combustible liquid. Mixtures with endrin may be explosive. Contact with strong oxidizers may cause fire and explosions. Attacks some plastics, rubber, and coatings.

MURIACITE (7778-18-9) Contact with diazomethane, aluminum, phosphorus may cause explosion.

MURIATIC ACID (7647-01-0) Reacts with acetic anhydride, aliphatic amines, alkanolamines, alkylene oxides, aromatic amines, amides, 2-aminoethanol, ammonia, ammonium hydroxide, calcium phosphide, chlorosulfonic acid, ethylene diamine, ethyleneimine, epichlorohydrin, isocyanates, metal acetylides, oleum, organic anhydrides, perchloric acid, 3-propiolactone, uranium phosphide, sulfuric acid, sodium hydroxide and other bases, strong oxidizers, vinyl acetate, vinylidene fluoride. Attacks most metals (can cause formation of flammable hydrogen gas), and some plastics, rubber, and coatings.

MURIATIC ETHER (75-00-3) Flammable gas. Reacts slowly with water; prodces hydrogen chloride gas. Contact with moisture produces hydrochloric acid. May accumulate static electrical charges, and may cause ignition of its vapors. Forms explosive mixture with air (flash point −58°F/−55°C). Contact with aluminum, lithium, magnesium, sodium, potassium, zinc may cause fire and explosions. Attacks some plastics and rubber.

MUSCATOX (56-72-4) Contact with strong oxidizers may cause fire and explosions.

MUTHMANN'S LIQUID (79-27-6) Reacts with chemically active metals, caustics. Contact with hot iron, aluminum, or zinc in the presence of steam may produce toxic vapors. Softens or destroys most plastics and rubbers.

MUTOXIN (50-29-3) Incompatible with salts of iron or aluminum, and bases. Do not store in iron containers.

MYCROLYSIN (76-06-2) Can be self-reactive. Fast heating, shock, alkali metals, or alkaline earth may cause explosions. A strong oxidizer; reacts violently with reducing agents, aniline in presence of heat, alcoholic sodium hydroxide, combustible substances, sodium methoxide, propargyl bromide. Liquid attacks some plastics, rubber, and coatings.

MYOCON (55-63-0) Liquid portion of this chemical is an alcohol; reacts with strong acids, caustics, aliphatic amines, isocyanates. Dry portion of chemical is nitroglycerine, a shock- and friction-sensitive explosive. Acids, heat, or mechanical shock may result in explosions.

MYRCENE (123-35-3) Forms explosive mixture with air (flash point 103°F/40°C). Incompatible with strong oxidizers, sulfuric acid, nitric acid.

MYRISTIC ALCOHOL (75-50-3) Flammable gas (flash point, liquid 10°F/−12°C). Incompatible with acids, organic anhydrides, ethylene oxide, isocyanates, aldehydes, oxidizers, triethylaluminum.

MYRISTYL ALCOHOL (75-50-3) Flammable gas (flash point, liquid 10°F/−12°C). Incompatible with acids, organic anhydrides, ethylene oxide, isocyanates, aldehydes, oxidizers, triethylaluminum.

- N -

NABAM or NABAME (142-59-6) Boiling water causes formation of hydrogen sulfide and carbon disulfide vapors.

NAC (63-25-2) Incompatible with strong oxidizers, strongly alkaline pesticides.

NACAP (2492-26-4) Incompatible with oxidizers. Concentrated aqueous solutions are corrosive.

NACCANOL NR or NACCANOL SW (25155-30-0) Reacts with acids, including fumes.

NACCONATE 100 (584-84-9) Incompatible with strong acids including nonoxidizing mineral and organic acids, caustics, ammonia, amines, amides, alcohols, glycols, caprolactam solution. Water contact causes violent foaming and spattering; produces carbon dioxide and an organic base. Attacks copper and its alloys, some plastics including polyethylene and rubber.

NACCONATE 300 (101-68-8) Incompatible with strong alkalies, acids, alcohols, ammonia, amines, amides, glycols, caprolactam. Unstable above 100°F/37.8°C. Attacks some plastics, rubber, and coatings.

NACCONOL 988 A (27176-87-0) Forms explosive mixture with air (flash point 100°F/38°C). Strong oxidizers may cause fire and explosions. Do not store in carbon steel or aluminum.

NADONE (108-94-1) Forms explosive mixture with air (flash point 111°F/44°C). May accumulate static electrical charges, and may cause ignition of its vapors. Attacks red metals and lead. Incompatible with amines, sulfuric acid, nitric acid, oxidizers, strong acids, aliphatic amines. Dissolves plastics, resins, and rubber.

NAFTALEN (Polish) (91-20-3) Forms explosive mixture with air (flash point 174°F/79°C). Incompatible with strong oxidizers, chromium oxide (reacts violently). Attacks some plastics, rubber, and coatings. May accumulate static electrical charges, and may cause ignition of its vapors.

1-NAFTILAMINA (Spanish) (134-32-7) Oxidizes in air. Incompatible with nitrous acid, oxidizers, nitrates, organic anhydrides, isocyanates, aldehydes.

1-NAFTIL-TIOUREA (Italian) (86-88-4) Strong oxidizers may cause fires and explosion. Also reacts with silver nitrate.

alpha-NAFTYLAMIN (Czech) (134-32-7) Oxidizes in air. Incompatible with nitrous acid, oxidizers, nitrates, organic anhydrides, isocyanates, aldehydes.

1-NAFTYLAMINE (Dutch) (134-32-7) Oxidizes in air. Incompatible with nitrous acid, oxidizers, nitrates, organic anhydrides, isocyanates, aldehydes.

alfa-NAFTYLOAMINA (Polish) (134-32-7) Oxidizes in air. Incompatible with nitrous acid, oxidizers, nitrates, organic anhydrides, isocyanates, aldehydes.

alpha-NAFTYL-N-METHYLKARBAMAT (Czech) (63-25-2) Incompatible with strong oxidizers, strongly alkaline pesticides.

1-NAFTYLTHIOUREUM (Dutch) (86-88-4) Strong oxidizers may cause fires and explosion. Also reacts with silver nitrate.

565

NAH 80 (7646-69-7) Water contact is potentially explosive with release of hydrogen gas. Finely divided form or powders can self-ignite in moist air. Reacts with acetylene, oxygen, halogens, strong oxidizers, sulfur.

NAKO H (106-50-3) Incompatible with acids, organic anhydrides, isocyanates, aldehydes, strong oxidizers; may cause fire and explosions. Heat and light contribute to instability.

NAKO TGG (108-46-3) Reacts violently with strong oxidizers, nitric acid. Incompatible with acetanilide, albumin, alkalies, antipyrine, camphor, ferric salts, menthol, spirit nitrous ether. Absorbs moisture from air (hygroscopic). May accumulate static electrical charges, and may cause ignition of its vapors.

NALCO 7046 (2425-06-1) Incompatible with acids or acid vapor. Strongly alkaline conditions contribute to instability.

NALED (300-76-5) Unstable in the presence of iron. Reacts with acids, strong oxidizers in sunlight. Corrosive to metals. Attacks some plastics, rubber, and coatings.

NANCHOR (299-84-3) Temperatures above 300°F/150°C may cause explosive decomposition. Contact with strong oxidizers may cause fire and explosions. Attacks some plastics, rubber, and coatings.

NANKER (299-84-3) Temperatures above 300°F/150°C may cause explosive decomposition. Contact with strong oxidizers may cause fire and explosions. Attacks some plastics, rubber, and coatings.

NAPCHLOR-G (131-52-2) Contact with strong oxidizers may cause fire and explosions.

NAPHID (1338-24-5) Incompatible with sulfuric acid, caustics, ammonia, aliphatic amines, alkanolamines, isocyanates, alkylene oxides, epichlorohydrin, strong oxidizers. Generally corrosive to metals.

NAPHTHA (8030-30-6) Forms explosive mixture with air (flash point 100°F/38°C). Incompatible with nitric acid, strong oxidizers. Attacks some plastics, rubber, and coatings.

NAPHTHA (8030-31-7) Contact with oxidizers may cause fire and explosions. Incompatible with nitric acid. May accumulate static electrical charges, and may cause ignition of its vapors.

NAPHTHA, COAL TAR (8030-30-6) Forms explosive mixture with air (flash point 100°F/38°C). Incompatible with nitric acid, strong oxidizers. Attacks some plastics, rubber, and coatings.

NAPHTHA DISTILLATE (DOT) (8030-30-6) Forms explosive mixture with air (flash point 100°F/38°C). Incompatible with nitric acid, strong oxidizers. Attacks some plastics, rubber, and coatings.

1-NAPHTHALENAMINE, TECHNICAL GRADE (134-32-7) Oxidizes in air. Incompatible with nitrous acid, oxidizers, nitrates, organic anhydrides, isocyanates, aldehydes.

NAPHTHALENE (91-20-3) Forms explosive mixture with air (flash point 174°F/79°C). Incompatible with strong oxidizers, chromium oxide (reacts violently). Attacks some plastics, rubber, and coatings. May accumulate static electrical charges, and may cause ignition of its vapors.

NAPTHALENE, HEXACHLORO- (1335-87-1) Oxidizers may cause fire and explosions.

NAPHTHALENE, 1-METHYL- (90-12-0) Forms explosive mixture with air (flash point 180°F/82°C). Oxidizers may cause fire and explosions. Incompatible with nitric acid. May accumulate static electrical charges, and may cause ignition of its vapors.

NAPHTHALENE, OCTACHLORO- (2234-13-1) Strong oxidizers may cause fire and explosions.

NAPHTHALENE OIL (65996-93-2) Incompatible with oxidizers, strong acids, caustics, aliphatic amines, isocyanates.

NAPHTHALIDAM (134-32-7) Oxidizes in air. Incompatible with nitrous acid, oxidizers, nitrates, organic anhydrides, isocyanates, aldehydes.

NAPHTHALIDINE (134-32-7) Oxidizes in air. Incompatible with nitrous acid, oxidizers, nitrates, organic anhydrides, isocyanates, aldehydes.

NAPHTHANE (91-17-8) Forms explosive mixture with air (flash point 134°F/57°C). May accumulate static electrical charges, and may cause ignition of its vapors. Strong oxidizers may cause fire and explosions.

NAPHTHA PETROLEUM (DOT) (8030-30-6) Forms explosive mixture with air (flash point 100°F/38°C). Incompatible with nitric acid, strong oxidizers. Attacks some plastics, rubber, and coatings.

NAPHTHA SAFETY SOLVENT (8052-41-3) Forms explosive mixture with air (flash point 102°F to 140°F/38.7°C to 60°C). Incompatible with nitric acid, strong oxidizers. Attacks some plastics, rubber, and coatings.

NAPHTHA: SOLVENT (8002-05-9) Forms explosive mixture with air (flash point 100°F/38°C). Incompatible with nitric acid, strong oxidizers. Attacks some plastics, rubber, and coatings. May accumulate static electrical charges, and may cause ignition of its vapors.

NAPHTHA SOLVENT (DOT) (8030-30-6) Forms explosive mixture with air (flash point 100°F/38°C). Incompatible with nitric acid, strong oxidizers. Attacks some plastics, rubber, and coatings.

NAPHTHA: STODDARD SOLVENT (8052-41-3) Forms explosive mixture with air (flash point 102°F to 140°F/38.7°C to 60°C). Incompatible with nitric acid, strong oxidizers. Attacks some plastics, rubber, and coatings.

NAPHTHA: VM&P (8032-32-4) Forms explosive mixture with air (flash point −40°F to −86°F/−40°C to −66°C). Incompatible with strong acids, strong oxidizers. Attacks some plastics, rubber, and coatings. May accumulate static electrical charges, and may cause ignition of its vapors.

NAPHTHENIC ACID (1338-24-5) Incompatible with sulfuric acid, caustics, ammonia, aliphatic amines, alkanolamines, isocyanates, alkylene oxides, epichlorohydrin, strong oxidizers. Generally corrosive to metals.

NAPHTHENIC ACID, COPPER SALT (1338-02-9) Forms explosive mixture with air (flash point 100°F/38°C). Incompatible with strong oxidizers, strong acids.

NAPHTHENIC ACIDS (1338-24-5) Incompatible with sulfuric acid, caustics, ammonia, aliphatic amines, alkanolamines, isocyanates, alkylene oxides, epichlorohydrin, strong noxidizers. Generally corrosive to metals.

1-NAPHTHOL (63-25-2) Incompatible with strong oxidizers, strongly alkaline pesticides.

1-NAPHTHOL N-METHYLCARBAMATE (63-25-2) Incompatible with strong oxidizers, strongly alkaline pesticides.

alpha-NAPHTHOTHIOUREA (86-88-4) Strong oxidizers may cause fires and explosion. Also reacts with silver nitrate.

1-NAPHTHYLAMIN (German) (134-32-7) Oxidizes in air. Incompatible with nitrous acid, oxidizers, nitrates, organic anhydrides, isocyanates, aldehydes.

1-NAPHTHYLAMINE (134-32-7) Oxidizes in air. Incompatible with nitrous acid, oxidizers, nitrates, organic anhydrides, isocyanates, aldehydes.

a-NAPHTHYLAMINE or alpha-NAPHTHYLAMINE (134-32-7) Oxidizes in air. Incompatible with nitrous acid, oxidizers, nitrates, organic anhydrides, isocyanates, aldehydes.

1-NAPHTHYLMETHYLCARBAMATE (63-25-2) Incompatible with strong oxidizers, strongly alkaline pesticides.

1-NAPHTHYL N-METHYLCARBAMATE (63-25-2) Incompatible with strong oxidizers, strongly alkaline pesticides.

alpha-NAPHTHYL N-METHYLCARBAMATE (63-25-2) Incompatible with strong oxidizers, strongly alkaline pesticides.

alpha-NAPHTHYLTHIOCARBAMIDE (86-88-4) Strong oxidizers may cause fires and explosion. Also reacts with silver nitrate.

1-NAPHTHYL THIOUREA (86-88-4) Strong oxidizers may cause fires and explosion. Also reacts with silver nitrate.

1-(1-NAPHTHYL)-2-THIOUREA (86-88-4) Strong oxidizers may cause fires and explosion. Also reacts with silver nitrate.

N-(1-NAPHTHYL)-2-THIOUREA (86-88-4) Strong oxidizers may cause fires and explosion. Also reacts with silver nitrate.

alpha-NAPHTHYLTHIOUREA (86-88-4) Strong oxidizers may cause fires and explosion. Also reacts with silver nitrate.

1-NAPHTHYL-THIOUREE (French) (86-88-4) Strong oxidizers may cause fires and explosion. Also reacts with silver nitrate.

1-NAPHTHYL-THIOHARNSTOFF (German) (86-88-4) Strong oxidizers may cause fires and explosion. Also reacts with silver nitrate.

NAPHTOL AS-KG (106-49-0) Forms explosive mixture with air (flash point 188°F/87°C). Incompatible with strong oxidizers, strong acids. Attacks some plastics, rubber, and coatings.

NAPHTOX (86-88-4) Strong oxidizers may cause fires and explosion. Also reacts with silver nitrate.

1-NAPHTYL N-METHYL-CARBAMATE (63-25-2) Incompatible with strong oxidizers, strongly alkaline pesticides.

NAPTHALANE (do not confuse with naphthalene) (91-17-8) Forms explosive mixture with air (flash point 134°F/57°C). May accumulate static electrical charges, and may cause ignition of its vapors. Strong oxidizers may cause fire and explosions.

NAPTHALIN (91-20-3) Forms explosive mixture with air (flash point 174°F/79°C). Incompatible with strong oxidizers, chromium oxide (reacts violently). Attacks some plastics, rubber, and coatings. May accumulate static electrical charges, and may cause ignition of its vapors.

NAPTHALINE (91-20-3) Forms explosive mixture with air (flash point 174°F/79°C). Incompatible with strong oxidizers, chromium oxide (reacts violently). Attacks some plastics, rubber, and coatings. May accumulate static electrical charges, and may cause ignition of its vapors.

NARCOGEN (79-01-6) Contact with caustics produces a toxic and flammable gas. Reacts violently with chemically active metals. Contact with aluminum may produce a violent, self-accelerating polymerization reaction. Incompatible with acids, organic anhydrides, isocyanates, alkylene oxides, aldehydes, alcohols, glycols, phenols, cresols, caprolactam solution, epichlorohydrin, nitrogen tetroxide, metal powders, oxygen. May accumulate static electrical charges, and may cause ignition of its vapors.

NARCOTILE (75-00-3) Flammable gas. Reacts slowly reaction with water; produces hydrogen chloride gas. Contact with moisture produces hydrochloric acid. May accumulate static electrical charges, and may cause ignition of its vapors. Forms explosive mixture with air (flash point −58°F/−55°C). Contact with aluminum, lithium, magnesium, sodium, potassium, zinc may cause fire and explosions. Attacks some plastics and rubber.

NARCYLEN (74-86-2) A strong reducing agent that reacts violently with oxidizers. Forms explosive mixture with air (flash point 0°F/18°C). Forms shock sensitive mixture with copper and copper salts, mercury and mercury salts, silver and silver salts. Reacts with brass, bromine, cesium hydride, chlorine, cobalt, fluorine, iodine, mercuric nitrate, nitric acid, potassium, rubidium hydride, trifluoromethyl hypofluorite, sodium hydride.

NARKOSOID (79-01-6) Contact with caustics produces a toxic and flammable gas. Reacts violently with chemically active metals. Contact with aluminum may produce a violent, self-accelerating polymerization reaction. Incompatible with acids, organic anhydrides, isocyanates, alkylene oxides, aldehydes, alcohols, glycols, phenols, cresols, caprolactam solution, epichlorohydrin, nitrogen tetroxide, metal powders, oxygen. May accumulate static electrical charges, and may cause ignition of its vapors.

NATRIUM (German) (7440-23-5) A strong reducing agent. Water contact produces hydrogen gas, corrosive sodium hydroxide fumes, a caustic soda solution, and heat; fire often occurs. Forms unstable peroxides with moist air. May ignite spontaneously in moist air. Reacts violently with oxidizers, acids, air, carbon dioxide, carbon tetrachloride, chloroform, haloids, halogens, halogenated hydrocarbons, heavy metal oxides, methyl chloride, moisture, nitrous oxide, tetrachloroethane, trichloroethylene, water and many other substances. Separate from all other materials.

NATRIUMALUMINIUMFLUORID (German) (15096-52-3) Contact with strong acids may cause formation of toxic and corrosive hydrogen fluoride gas.

NATRIUMAZID (German) (26628-22-8) Incompatible with acids, some metals (i.e., lead, copper, silver, mercury). Forms explosion-sensitive compounds.

NATRIUMBICHROMAAT (Dutch) (10588-01-9) Aqueous solution is corrosive. A strong oxidizer; Reacts violently with reducing agents, acids, acetic anhydride, combustibles. Attacks copper, zinc, tin, brass, bronze.

NATRIUMCHLORAAT (Dutch) (7775-09-9) A powerful oxidizer; reacts violently with reducing agents and combustible matter. Explosions may be caused by contact with ammonia salts, carbon, oils, metal sulfides, nitrobenzene, powdered metals, sugar. Contact with strong acids produces carbon dioxide. Forms shock-sensitive mixtures with some organic materials. Solution (50%) decomposes at 300°F/149°C, liberating oxygen.

NATRIUMCHLORAT (German) (7775-09-9) A powerful oxidizer; reacts violently with reducing agents and combustible matter. Explosions may be caused by contact with ammonia salts, carbon, oils, metal sulfides, nitrobenzene, powdered metals, sugar. Contact with strong acids produces carbon dioxide. Forms shock-sensitive mixtures with some organic materials. Solution (50%) decomposes at 300°F/149°C, liberating oxygen.

NATRIUMDICHROMAAT (Dutch) (10588-01-9) Aqueous solution is corrosive. A strong oxidizer; Reacts violently with reducing agents, acids, acetic anhydride, combustibles. Attacks copper, zinc, tin, brass, bronze.

NATRIUMDICHROMAT (German) (10588-01-9) Aqueous solution is corrosive. A strong oxidizer; reacts violently with reducing agents, acids, acetic anhydride, combustibles. Attacks copper, zinc, tin, brass, bronze.

NATRIUMHEXAFLUOROALUMINATE (German) (15096-52-3) Contact with strong acids may cause formation of toxic and corrosive hydrogen fluoride gas.

NATRIUMHYDROXID (German) (1310-73-2) Contact with water, acids, flammable liquids, and organic halogens, especially trichloroethylene, may cause fires and explosions. Contact with metals such as aluminum, tin, and zinc causes corrosion and the formation of flammable hydrogen gas. Contact with nitromethane, and similar nitro compounds produces shock-sensitive salts. Water contact produces heat and corrosive fumes. Attacks some plastics, rubber, and coatings.

NATRIUMHYDROXYDE (Dutch) (1310-73-2) Contact with water, acids, flammable liquids, and organic halogens, especially trichloroethylene, may cause fires and explosions. Contact with metals such as aluminum, tin, and zinc causes corrosion and the formation of flammable hydrogen gas. Contact with nitromethane and similar nitro compounds produces shock-sensitive salts. Water contact produces heat and corrosive fumes. Attacks some plastics, rubber, and coatings.

NATRIUMMAZIDE (Dutch) (26628-22-8) Incompatible with acids, with some metals (i.e., lead, copper, silver, mercury). Forms explosion-sensitive compounds.

NATRIUMRHODANID (German) (540-72-7) Incompatible with acids, bases, ammonia, amines, amides, alcohols, glycols, caprolactam.

NATURAL ANHYDRITE (7778-18-9) Contact with diazomethane, aluminum, phosphorus may cause explosion.

NATURAL CALCIUM CARBONATE (1317-65-3) Incompatible with acids, alum, ammonium salts, fluorine.

NATURAL GAS or NATURAL GAS, REFRIGERATED LIQUID (74-82-8) Forms an explosive mixture with air. Reacts violently with strong oxidizers.

NATURAL GASOLINE (8006-61-9) Forms explosive mixture with air (flash point 125°F/51°C). Contact with oxidizers may cause fire and explosions. Incompatible with nitric acid. May accumulate static electrical charges, and may cause ignition of its vapors.

NATURAL IRON OXIDES (1309-37-1) Contact with hydrogen peroxide, ethylene oxide, calcium hypochlorite will cause explosion. Reacts violently with powdered aluminum, hydrazine, hydrogen trisulfide.

NATURAL LEAD SULFIDE (1314-87-0) Reacts violently with iodine monochloride or hydrogen peroxide.

NATURAL RED OXIDE (1309-37-1) Contact with hydrogen peroxide, ethylene oxide, calcium hypochlorite will cause explosion. Reacts violently with powdered aluminum, hydrazine, hydrogen trisulfide.

NAUGATUCK DO-14 (2312-35-8) Forms explosive mixture with air (flash point 82°F/28°C). Strong oxidizers may cause fire and explosions.

NAXOL (108-93-0) Forms explosive mixture with air (flash point 154°F/68°C). Incompatible with strong acids, caustics, aliphatic amines, isocyanates. Attacks some plastics, rubber, or coatings.

NBA (71-36-3) Forms explosive mixture with air (flash point 98°F/37°C). May react with aluminum above 120°F/49°C. Attacks some plastics, rubber, and coatings. Incompatible with strong acids, halogens, caustics, alkali metals, aliphatic amines, isocyanates.

NCI-C00044 (309-00-2) Incompatible with concentrated mineral acids, acid catalysts, acid oxidizing agents, phenols, reactive metals.

NCI-C00066 (86-50-0) Strong oxidizers may cause fire and explosions.

NCI-C00077 (133-06-2) Incompatible with tetraethyl pyrophosphate, parathion.

NCI-C00099 (57-74-9) Contact with strong oxidizers may cause fire and explosions. Attacks some plastics, rubber, and coatings.

NCI-C00113 (62-73-7) Attacks some plastics, rubber, and coatings.

NCI-C00124 (60-57-1) Incompatible with concentrated mineral acids, acid catalysts, acid oxidizing agents, phenols, reactive metals.

NCI-C00157 (72-20-8) Incompatible with parathion, strong acids (forms explosive vapors), strong oxidizers.

NCI-C00180 (76-44-8) Produces hydrogen chloride gas with iron and rust above 165°F/74°C.

NCI-C00204 (58-89-9) Not combustible, but may be dissolved in a combustible solvent. If solvent comes in contact with oxidizers fire and explosions may result.

NCI-C00215 (121-75-5) Incompatible with strong oxidizers, magnesium, alkaline pesticides. Attacks metals, some plastics, rubber, and coatings.

NCI-C00226 (56-38-2) Combustible liquid. Mixtures with endrin may be explosive. Strong oxidizers may cause fire and explosions. Attacks some plastics, rubber, and coatings.

NCI-C00259 (8001-35-2) Reacts with oxidizers, with a risk of fire or explosions. Attacks metals in the presence of moisture.

NCI-C00464 (50-29-3) Incompatible with salts of iron or aluminum, and bases. Do not store in iron containers.

NCI-C00475 (72-54-8) Contact with strong oxidizers may cause fire and explosions.

NCI-C00486 (115-32-2) Incompatible with strong acids, caustics, aliphatic amines, isocyanates.

NCI-C00497 (72-43-5) Contact with strong oxidizers may cause fire and explosions. Attacks some plastics, rubber, and coatings.

NCI-C00500 (96-12-8) Forms explosive mixture with air (flash point 170°F/77°C). Reacts with oxidizers and chemically active metals (i.e., aluminum, magnesium, and tin alloys). Attacks some rubber materials and coatings.

NCI-C00511 (107-06-2) May accumulate static electrical charges, and may cause ignition of its vapors. Contact with hot water may produce hydrochloric acid. Forms explosive mixture with air (flash point 55°F/13°C). Incompatible with strong oxidizers, strong caustics, chemically active metals. Corrosive to iron and other metals. Attacks some plastics, rubber, or coatings.

NCI-C00522 (106-93-4) Reacts with chemically active metals, liquid ammonia, strong oxidizers. Heat and light cause slow decomposition. Attacks some plastics and rubber.

NCI-C00554 (315-18-4) Usually dissolved in a combustible liquid; avoid oxidizer contact.

NCI-C00566 (115-29-7) Hydrolyzed by acids and alkalies. Corrosive to iron.

NCI-C00920 (107-21-1) Incompatible with strong acids, caustics, aliphatic amines, isocyanates, chlorosulfonic acid, oleum, strong oxidizers.

NCI-C01865 (121-14-2) Strong oxidizers or caustics may cause fire and explosions. Incompatible with nitric acid; forms an explosive material. Sodium oxide contact causes ignition. Attacks metals.

NCI-C02051 (95-79-4) Incompatible with acids, organic anhydrides, isocyanates, aldehydes, oxidizers.

NCI-C02084 (7632-00-0) A strong oxidizer; reacts violently with reducing agents and combustibles. Reacts with acids, ammonium salts + heat, butadiene, cyanides, lithium, phthalic acid, thiosulfate, sodium amide, and many other substances. Under certain conditions the chemical can also be a strong reducing agent.

NCI-C02119 (57-13-6) Heat forms anhydrous ammonia fumes. Contact with oxidizers, nitrates may cause fire and explosions. Contact with chlorinating agents including hypochlorite bleaches may cause formation of explosive nitrogen trichloride. Reacts with nitrosyl perchlorate.

NCI-C02200 (100-42-5) If inhibitor (often *tert*-butyl catechol) is not present in adequate concentrations, polymerization may occur and explode containers. Temperatures above 150°F/66°C speed up polymerization. Incompatible with strong oxidizers, acids, rust, catalysts for vinyl polymerization, such as peroxides, strong acids, and aluminum chloride. Corrodes copper, copper alloys, and dissolves rubber. Attacks some plastics, rubber, and coatings. May accumulate static electrical charges, and may cause ignition of its vapors.

NCI-C02302 (95-80-7) Incompatible with oxidizers, acids, organic anhydrides, isocyanates, aldehydes. Attacks aluminum, brass, bronze, copper, zinc.

NCI-C02551 (1306-19-0) May form an explosive mixture with magnesium + heat. May react with sulfur, selenium, zinc. Acid contact produces hydrogen gas.

NCI-C02686 (67-66-3) Decomposes in the presence of excess water or high temperatures, producing phosgene and hydrogen chloride. Can become explosive in the presence of strong alkalies and water. Can accumulate static electrical charges. In contact with water and high temperatures, it becomes corrosive; attacks iron and other metals. Incompatible with acetone, aluminum, strong oxidizers, potassium, sodium, chemically active metals, strong bases. Attacks plastics and rubber.

NCI-C02799 (50-00-0) May polymerize unless properly inhibited (usually with methanol). Forms explosive mixture with air (flash point 122°F/50°C). Incompatible with strong acids, amines, strong oxidizers, alkaline materials, nitrogen dioxide, performic acid. Reaction with hydrochloric acid produces bis-chloromethyl ether, a carcinogen.

NCI-C02937 (156-62-7) Contact with any form of moisture causes decomposition, liberating acetylene and ammonia. Contact with all solvents tested also causes decomposition.

NCI-C02971 (298-00-0) Mixtures with magnesium may be explosive.

NCI-C03134 (64-17-5) Forms explosive mixture with air (flash point 65°F/18°C). May accumulate static electrical charges, and may cause ignition of its vapors. Reactions may be violent with oleum, sulfuric acid, nitric acid, bases, aliphatic amines, isocyanates, oxidizers.

NCI-C03361 (92-87-5) Oxidizes on exposure to light and air. Reacts violently with strong oxidizers. Red fuming nitric acid may cause fire.

NCI-C03554 (79-34-5) Exposure to heat, light and air produces corrosive and toxic vapors. Reacts with strong caustics to produce explosive dichloroacetylene. Reacts violently with chemically active metals or sodium amide. In presence of steam, contact with hot iron, aluminum, or zinc may produce toxic vapors. Attacks some plastics, rubber, and coatings.

NCI-C03601 (85-44-9) Incompatible with strong acids, caustics, ammonia, amines, strong oxidizers. Attacks some plastics, rubber, and coatings.

NCI-C03689 (123-91-1) Forms explosive mixture with air (flash point 54°F/12°C). Moisture causes formation of unstable peroxides. May accumulate static electrical charges, and may cause ignition of its vapors. Incompatible with strong oxidizers, strong acids. Attacks many plastics.

NCI-C03736 (62-53-3) Forms explosive mixture with air (flash point 158°F/70°C). Unless inhibited (usually with methanol), readily able to polymerize. Fires and explosions may result from contact with halogens, strong acids, oxidizers, organic anhydrides, acetic anhydride, isocyanates, aldehydes, sodium peroxide. Reacts with alkali metals and alkali earth metals. Attacks some plastics, rubber, and coatings, and copper and copper alloys.

NCI-C03985 (542-75-6) Forms explosive mixture with air (flash point 95°F/35°C). Reacts violently with strong oxidizers. May accumulate static electrical charges, and may cause ignition of its vapors. Incompatible with strong acids, oxidizers, aluminum or magnesium compounds, aliphatic amines, alkanolamines, alkaline materials, or corrosives.

NCI-C04535 (75-34-3) Forms explosive mixture with air (flash point 2°F/−17°C). Incompatible with strong oxidizers, strong caustics. Attacks plastics and rubber.

NCI-C04546 (79-01-6) Contact with caustics produces a toxic and flammable gas. Reacts violently with chemically active metals. Contact with aluminum may produce a violent, self-accelerating polymerization reaction. Incompatible with acids, organic anhydrides, isocyanates, alkylene oxides, aldehydes, alcohols, glycols, phenols, cresols, caprolactam solution, epichlorohydrin, nitrogen tetroxide, metal powders, oxygen. May accumulate static electrical charges, and may cause ignition of its vapors.

NCI-C04579 (79-00-5) Incompatible with strong oxidizers, strong caustics, chemically active metals (especially powders), sodium amide; may cause fire and explosions. Attacks some plastics, rubber, and coatings.

NCI-C04591 (75-15-0) Highly reactive. Contact with many substances can cause fire and explosions. Forms explosive mixture with air (flash point −22°F/−30°C). Shock can cause explosive decomposition. Incompatible with alkali metals, aliphatic amines, alkanolamines, aluminum, azides, chlorine monoxide, combustible substances, ethylene diamine, ethyleneimine, lead azide, lithium azide, nitric oxide, nitrogen dioxide, potassium, potassium azide, reducing agents, rubidium azide, sodium azide, zinc.

NCI-C04615 (107-05-1) Forms explosive mixture with air (flash point −25°F/−32°C). Violent polymerization and explosion may be caused by heat, light, or contact with acid catalysts, ferric chloride, aluminum chloride, Lewis acids, or Ziegler catalyst (e.g., titanium tetrachloride + triethylaluminum monochloride). Incompatible with strong acids, amines, aluminum chloride, boron trifluoride, chlorosulfonic acid, ethylene diamine, ethyleneimine, ferric chloride, oleum, oxidizers, sodium hydroxide. Slow decomposition occurs with moisture; attacks some coatings, plastics, and rubber. Corrosive to steel. May accumulate static electrical charges, and may cause ignition of its vapors.

NCI-C04637 (75-69-4) Reacts with barium, lithium, sodium, magnesium, titanium. Attacks some plastics, rubber, and coatings.

NCI-C05970 (108-46-3) Reacts violently with strong oxidizers, nitric acid. Incompatible with acetanilide, albumin, alkalies, antipyrine, camphor, ferric salts, menthol, spirit nitrous ether. Absorbs moisture from air (hygroscopic). May accumulate static electrical charges, and may cause ignition of its vapors.

NCI-C06111 (100-51-6) Forms explosive mixture with air (flash point 213°F/101°C). Slowly oxidizes in air and oxygen. Incompatible with mineral acids, caustics, aliphatic amines, isocyanates. Reacts violently with strong oxidizers. Corrodes aluminum at high temperature. Attacks some nonfluorinated plastics; may not attack polypropylene.

NCI-C06155 (109-69-3) Forms explosive mixture with air (flash point 15°F/−9°C). May accumulate static electrical charges, and may cause ignition of its vapors. Water contact slowly produces hydrochloric acid. Incompatible with strong oxidizers, alkaline earth and alkali metals, finely divided metal. Attacks metals in presence of moisture. Attacks some plastics, rubber, and coatings. May accumulate static electrical charges, and may cause ignition of its vapors.

NCI-C06224 (75-00-3) Flammable gas. Reacts slowly with water; produces hydrogen chloride gas. Contact with moisture produces hydrochloric acid. May accumulate static electrical charges, and may cause ignition of its vapors. Forms explosive mixture with air (flash point −58°F/−55°C). Contact with aluminum, lithium, magnesium, sodium, potassium, zinc may cause fire and explosions. Attacks some plastics and rubber.

NCI-C06360 (100-44-7) Forms explosive mixture with air (flash point 153°F/67°C). Water contact produces hydrogen chloride fume. Strong oxidizers may cause fire and explosions. Violent polymerization may be caused by contact with copper, aluminum, iron, zinc, magnesium and tin. May accumulate static electrical charges, and may cause ignition of its vapors. Attacks some plastics and rubber.

NCI-C06462 (26628-22-8) Incompatible with acids, some metals (i.e, lead, copper, silver, mercury) Forms explosion-sensitive compounds.

NCI-C06508 (140-11-4) Forms explosive mixture with air (flash point 216°F/102°C). Reacts with strong acids, nitrates, oxidizers.

NCI-C07272 (108-88-3) Forms explosive mixture with air (flash point 40°F/4°C). Strong oxidizers may cause fire and explosions. Attacks some plastics, rubber, and coatings. May accumulate static electrical charges, and may cause ignition of its vapors.

NCI-C08662 (56-72-4) Contact with strong oxidizers may cause fire and explosions.

NCI-C08673 (333-41-5) Incompatible with water, copper-containing compounds, oxidizers, acids, or bases.

NCI-C50044 (108-60-1) Forms explosive mixture with air (flash point 170°F/77°C). Reacts violently with strong oxidizers. Incompatible with aluminum, copper, epoxy coatings.

NCI-C50055 (120-61-6) Incompatible with strong acids, nitrates, strong oxidizers.

NCI-C50077 (115-07-1) Forms explosive mixture with air (flash point −162°F/−108°C). Incompatible with strong oxidizers, strong acids.

NCI-C50088 (75-21-8) Forms explosive mixture with air (flash point 20°F/−6°C). Incompatible with alkali metal hydroxides, highly active catalysts (e.g., anhydrous chlorides of iron, tin, or aluminum and oxides of iron or aluminum). Avoid contact with copper. Protect container from physical damage, sun and heat. Attacks some plastics, rubber, or coatings.

NCI-C50099 (75-56-9) Forms explosive gas mixture with air (flash point −35°F/−37°C). Incompatible with anhydrous metal chlorides. strong acids, caustics, peroxides; cause polymerization. Reacts with ammonia, amines, acetylene-forming metals. Attacks some plastics, rubber, and coatings.

NCI-C50124 (108-95-2) Forms explosive mixture with air (flash point 174°F/79°C). Incompatible with strong oxidizers, strong acids, caustics, aliphatic amines, amides, oxidizers, formaldehyde, butadiene, calcium hypochlorite. Liquid attacks some plastics, rubber, and coatings; hot liquid attacks aluminum, magnesium, lead, and zinc metals.

NCI-C50135 (107-07-3) Forms explosive mixture with air (flash point 140°F/60°C). Incompatible with strong oxidizers (may cause fire and explosions), strong caustics (with formation of ethylene gas), strong acids, aliphatic amines, isocyanates. Attacks some plastics, rubber, and coatings.

NCI-C50168 (8001-29-4) Incompatible with strong acids, nitrates, oxidizers.

NCI-C50384 (140-88-5) Forms explosive mixture with air (flash point 48°F/9°C). Atmospheric moisture and strong alkalies may cause fire and explosions. Unless properly inhibited (note: Inert gas blanket not recommended), heat, light, or peroxides can cause polymerization. Incompatible with oxidizers (may be violent), strong acids, amines. May accumulate static electrical charges, and may cause ignition of its vapors.

NCI-C50533 (584-84-9) Incompatible with strong acids including nonoxidizing mineral and organic acids, caustics, ammonia, amines, amides, alcohols, glycols, caprolactam solution. Water contact causes violent foaming and spattering; produces carbon dioxide and an organic base. Attacks copper and its alloys, some plastics including polyethylene and rubber.

NCI-C50602 (106-99-0) Self-reactive. Forms explosive peroxides with air. Fires, explosions, or hazardous polymerization may result from contact with air, strong oxidizers, strong acids, ozone, nitrogen dioxide, copper and its alloys, phenol, chlorine dioxide, crotonaldehyde, or a free radical polymerization initiator such as hydroquinone. Add inhibitor (such as *tert*-butyl catechol) and monitor to ensure that effective levels are maintained at all times. May accumulate static electrical charges, and may cause ignition of its vapors.

NCI-C50646 (105-60-2) Contact with strong oxidizers may cause fire and explosions.

NCI-C50668 (101-68-8) Incompatible with strong alkalies, acids, alcohols, ammonia, amines, amides, glycols, caprolactam. Unstable above 100°F/37.8°C. Attacks some plastics, rubber, and coatings.

NCI-C50680 (80-62-6) Forms explosive mixture with air (flash point 50°F/10°C). Incompatible with caustics, nitrates, strong acids, aliphatic amines, alkanolamines, peroxides, and strong oxidizers. Heat and/or lack of appropriate inhibitor can cause polymerization. May accumulate static electrical charges, and may cause ignition of its vapors.

NCI-C50704 (75-78-5) Forms explosive gas mixture with air (flash point 15°F/−9°C). Water (vapor) or air form hydrochloric acid. Incompatible with acetone, amines, ammonia, alcohols, strong oxidizers, caustics. Attacks most metals.

NCI-C52733 (117-81-7) Incompatible with strong acids, strong alkalies, nitrates, oxidizers.

NCI-C52904 (91-20-3) Forms explosive mixture with air (flash point 174°F/79°C). Incompatible with strong oxidizers, chromium oxide (reacts violently). Attacks some plastics, rubber, and coatings. May accumulate static electrical charges, and may cause ignition of its vapors.

NCI-C53894 (76-01-7) Incompatible with water, producing dichloroacetic acid. May self-ignite. Reacts violently with alkali metals (i.e., lithium, sodium, potassium, rubidium, cesium, francium). Forms spontaneously explosive materials with alkalies, metals. Forms shock- and friction-sensitive material upon mixing with potassium.

NCI-C54262 (75-35-4) Forms explosive mixture with air (18°F/−28°C). Air or contaminants can cause formation of peroxides; may polymerize. Reacts violently with strong oxidizers, alkali metals. Incompatible with nitric acid.

NCI-C54375 (85-68-7) Incompatible with strong acids, nitrates, oxidizers.

NCI-C54386 (103-23-1) Incompatible with strong acids, nitrates, oxidizers.

NCI-C54773 (868-85-9) Incompatible with acids (nitric, sulfuric, nonoxidizing mineral, organic), nitrates, oxidizers.

NCI-C54820 (563-47-3) Forms explosive mixture with air (flash point 11°F/−12°C). Strong oxidizers may cause fire and explosions.

NCI-C54831 (52-68-6) Contact with strong oxidizers may cause fire and explosions.

NCI-C54853 (110-80-5) Forms explosive mixture with air (flash point 120°F/49°C). Strong oxidizers may cause fire and explosions. Attacks some plastics, rubber, and coatings. Able to form peroxides. Incompatible with strong acids, aluminum and its alloys.

NCI-C54886 (108-90-7) Forms explosive mixture with air (flash point 82°F/28°C). May accumulate static electrical charges, and may cause ignition of its vapors. Incompatible with oxidizers, dimethylsulfoxide, sodium powder, silver perchloride. Attacks some plastics, rubber, and coatings.

NCI-C54933 (87-86-5) Hot water causes decomposition, producing hydrochloric acid. Strong oxidizers may cause fire and explosions.

NCI-C54944 (95-50-1) Forms explosive mixture with air (flash point 151°F/66°C). Incompatible with strong oxidizers, hot aluminum or aluminum alloy. Attacks some plastics, rubber, and coatings.

NCI-C54955 (106-46-7) Forms explosive mixture with air (flash point 150°F/66°C). Incompatible with strong oxidizers, metal powders and alkali metals (i.e., lithium, sodium, potassium, rubidium, cesium, francium). Attacks some plastics, rubber, and coating.

NCI-C54988 (78-00-2) Forms explosive mixture with air and decomposes above 200°F/93°C. Incompatible with strong oxidizers, concentrated acids may cause fire and explosions. Attacks some plastics, rubber, and coatings.

NCI-C55130 (75-25-2) Reacts with chemically active metals, calcium, acetone, strong caustics. Attacks some plastics, rubber, and coatings.

NCI-C55141 (78-87-5) Forms explosive mixture with air (flash point 60°F/16°C). May accumulate static electrical charges, and may cause ignition of its vapors. Strong oxidizers may cause fire and explosions. Strong acids can cause decomposition and the formation of hydrogen chloride vapors. Corrodes aluminum. Attacks some plastics, rubber, and coatings.

NCI-C55174 (111-42-2) Incompatible with acids, organic anhydrides, isocyanates, vinyl acetate, acrylates, substituted allyls, alkylene oxides, epichlorohydrin, aldehydes, oxidizers. Corrosive to copper, copper alloys, zinc, and galvanized iron.

NCI-C55209 (144-62-7) Silver compounds may cause formation of explosive salt. Incompatible with caustics, furfuryl alcohol, oxidizers, mercury, silver, sodium chlorite, sodium hypochlorite.

NCI-C55276 (106-88-7) Forms explosive mixture with air (flash point −7°F/−22°C). Strong bases or metal chlorides cause polymerization. Reacts violently with oxidizers, acids. May accumulate static electrical charges, and may cause ignition of its vapors.

NCI-C55301 (110-86-1) Forms explosive mixture with air (flash point 68°F/20°C). Strong oxidizers may cause fire and explosions. Strong acids may cause violent spattering. Attacks some plastics, rubber, and coatings.

NCI-C55345 2,4-DICHLOROPHENOL (120-83-2) Reacts violently with strong oxidizers, acids. Incompatible with caustics. Quickly corrodes aluminum; slowly corrodes zinc, tin, brass, bronze, copper and its alloys. May accumulate static electrical charges, and may cause ignition of its vapors.

NCI-C55367 (75-65-0) Forms explosive mixture with air (flash point 52°F/11°C). May accumulate static electrical charges, and may cause ignition of its vapors. Incompatible with strong acids including mineral acids, strong oxidizers or caustics, aliphatic amines, isocyanates, alkali metals (i.e., lithium, sodium, potassium, rubidium, cesium, francium). Attacks many plastics and some coatings.

NCI-C55378 (87-86-5) Hot water causes decomposition, producing forming hydrochloric acid. Strong oxidizers may cause fire and explosions.

NCI-C55425 (111-30-8) Water contact produces a polymer solution. A strong reducing agent. Incompatible with strong acids, caustics, ammonia, amines, strong oxidizers.

NCI-C55447 (1338-23-4) Forms explosive mixture with air (flash point 125°F/52°C). Explosive decomposition occurs above 176°F/80°C. Pure substance is shock-sensitive. Strong oxidizer. Reacts violently with strong acids, strong bases, reducing agents, combustible substances, organic materials, oxides of heavy metals, salts, trace contaminants, amines. May accumulate static electrical charges, and may cause ignition of its vapors.

NCI-C55481 (74-96-4) Forms explosive mixture with air (flash point less than −4°F/−20°C)). Hydrolyzes in water, producing hydrogen bromide. Oxidizers may cause fire or explosions. Fire and explosions may result from contact with aluminum, magnesium, or zinc powders, and lithium, potassium, sodium. Attacks some plastic, rubber, and coatings.

NCI-C55549 (556-52-5) Forms explosive mixture with air (flash point 158°F/70°C). Reacts violently with strong oxidizers. Contact with caustics, acids, barium, lithium, sodium, magnesium, titanium can cause polymerization. Incompatible with nitrates. Attacks some plastics, rubber, and coatings.

NCI-C55607 (77-47-4) Incompatible with water, producing hydrochloric acid. Contact with sodium may be explosive. Corrodes iron and other metals in the presence of moisture.

NCI-C55658 (98-85-1) Forms explosive mixture with air (flash point 205°F/96°C). Incompatible with strong acids, caustics, aliphatic amines, isocyanates, oxidizers.

NCI-C55834 (123-31-9) Incompatible with strong oxidizers, caustics. May be oxidized to quinone at room temperatures in the presence of moisture. May explode on contact with oxygen.

NCI-C55845 (106-51-4) Forms explosive mixture with air (flash point 104°F/40°C). Incompatible with strong bases, reducing agents, strong oxidizers. Attacks some plastics, rubber, and coatings.

NCI-C55856 (120-80-9) Strong oxidizers may cause fire and explosion.

NCI-C55947 (509-14-8) Combustible organic matter wet with this chemical may be highly explosive, and susceptible to mild shock. Contact with hydrocarbons, alkalies, or metals produces explosive mixtures. Attacks some plastics, rubber, and coatings.

NCI-C55992 (100-02-7) Combustible solid. A strong oxidizer; reacts with reducing agents, combustibles, organic and other easily oxidizable materials. Incompatible with strong acids, caustics, aliphatic amines, amides.

NCI-C56031 (540-59-0) Forms explosive mixture with air (flash point 36°F/2.2°C). Incompatible with strong bases, oxidizers, difluoromethylene, dihypofluoride, nitrogen tetroxide (explosive). Attacks some plastics, rubber, and coatings.

NCI-C56133 (100-52-7) Forms explosive mixture with air (flash point 145°F/63°C). A strong reducing agent. Forms peroxides with air. Reacts violently with strong acids, aluminum, caustics, ammonia, amines, iron, strong oxidizers. Attacks some plastics, rubber, and coatings.

NCI-C56177 (98-01-1) Forms explosive mixture with air (flash point 140°F/60°C). Incompatible with strong acids, caustics, ammonia, aliphatic amines, alkanolamines, aromatic amines, oxidizers. Attacks many plastics and coatings.

NCI-C56188 (87-62-7) Forms explosive mixture with air (flash point 206°F/96.7°C). Contact with strong acids may cause fire and explosions. Contact with hypochlorite bleaches produces explosive chloroamines. Incompatible with strong acids, organic acids and anhydrides, isocyanates, aldehydes.

NCI-C56202 (110-00-9) Forms explosive mixture with air (flash point −58°F/−50°C). Reacts violently with acids, oxidizers. Unless stabilized with an inhibitor, air exposure produces unstable peroxides.

NCI-C56224 (98-00-0) Forms explosive mixture with air (flash point 170°F/77°C). Strong acids (including some organic acids) or acid catalysts may cause polymerization. Reacts strongly with oxidizers and acids. Incompatible with caustics, aliphatic amines, isocyanates. Attacks some plastics, coatings, and rubber.

NCI-C56279 (123-73-9) A strong reducing agent. Forms explosive mixture with air (flash point 55°F/12.8°C). Readily converted by oxygen to peroxides and acids; heat or contact with many other substances may cause polymerization. Incompatible with strong oxidizers, strong acids including nonoxidizing mineral acids, ammonia, aliphatic amines, aromatic amines, 1,3-butadiene, strong bases. Liquid attacks some plastics, rubber, and coatings.

NCI-C56291 (123-72-8) May accumulate static electrical charges, and may cause ignition of its vapors. Forms explosive mixture with air (flash point −8°F/−22°C). Incompatible with strong oxidizers, strong acids, caustics, ammonia, aliphatic amines, alkanolamines, aromatic amines.

NCI-C56326 (75-07-0) Slowly polymerizes to paraldehyde. Explodes when mixed with iodine. Contact with strong bases can cause explosive polymerization. A strong reducing agent; reacts violently with combustibles, strong acids, caustics, ammonia, aliphatic amines, alkanolamines, aromatic amines, organic substances, halogens, oxidizers. Forms explosive peroxides with air. May dissolve rubber. May accumulate static electrical charges, and may cause ignition of its vapors.

NCI-C56393 (100-41-4) May accumulate static electrical charges, and may cause ignition of its vapors. Forms explosive mixture with air (flash point 59°F/15°C). Incompatible with strong oxidizers, nitric acid.

NCI-C56406 (25013-15-4) Forms explosive mixture with air (flash point 17°F/53°C). Polymerization inhibitor (usually 10 to 50 ppm of *tert*-butyl catechol) must be present in adequate concentrations to avoid explosive polymerization. Oxidizers may cause fire and explosions. Also, the following should be avoided: peroxides, aluminum chloride, acids (non-oxidizing mineral, sulfuric, nitric), caustics, ammonia, aliphatic amines, alkanolamines.

NCI-C56417 (10043-35-3) Incompatible with bases, potassium. Acetic anhydride causes formation of heat-sensitive explosive.

NCI-C56439 (4016-08-6) Forms explosive mixture with air (flash point 92°F/33°C). Incompatible with strong oxidizers (fire), strong caustics (polymerization). Air and light cause formation of unstable and explosive peroxides. Attacks some rubbers and plastics.

NCI-C56655 (87-86-5) Hot water causes decomposition, producing hydrochloric acid. Strong oxidizers may cause fire and explosions.

NCI-C60048 (84-66-2) Reacts violently with strong acids, strong oxidizers, permanganates, water. Attacks some forms of plastic.

NCI-C60082 (98-95-3) Forms explosive mixture with air (flash point 190°F/88°C). Incompatible with nitric acid, nitrogen tetroxide, caustics, ammonia, amines. Attacks some plastics, rubber, and coatings.

NCI-C60173 (7487-94-7) Incompatible with light metals (aluminum, magnesium, beryllium, etc.), sodium, potassium.

NCI-C60219 (75-44-5) Incompatible with water, producing hydrochloric acid. Reacts violently with strong oxidizers, anhydrous ammonia, isopropyl alcohol and chemically active metals. Forms shock-sensitive material with potassium. Attacks most metals in moist conditions.

NCI-C60220 (96-18-4) Forms explosive mixture with air (flash point 164°F/74°C). Strong oxidizers, strong caustics, chemically active metals may cause fire and explosions. May decompose on contact with aluminum. Attacks some plastics, rubber, and coatings.

NCI-C60231 (79-11-8) Aqueous solution is a strong acid. Incompatible with strong oxidizers, bases. Attacks most common metals in the presence of moisture.

NCI-C60399 (7439-97-6) Acetylene, acetylene products, ammonia gases can form shock-sensitive solids that can initiate fires of combustibles. Reacts violently reactions with boron phosphodiiodide, chlorine, chlorine dioxide, methyl azide. Attacks copper and copper alloys.

NCI-C60402 (107-15-3) Forms explosive mixture with air (flash point 104°F/40°C). Incompatible with chlorinated organic compounds, silver perchlorate, 3-propiolactone, mesityl oxide, ethylene dichloride, acids, organic anhydrides, isocyanates, vinyl acetate, acrylates, substituted allyls, alkylene oxides, epichlorohydrin, ketones, aldehydes, alcohols, glycols, phenols, cresols, caprolactam solution, strong oxidizers. Attacks aluminum, copper, lead, tin, zinc, and alloys, and some plastics, rubber, and coatings.

NCI-C60537 (99-99-0) Incompatible with strong acids, caustics, ammonia, amines, reducing agents, oxidizers. Attacks some plastics, rubber, and coatings.

NCI-C60560 (109-99-9) Forms explosive mixture with air (flash point 6°F/−14°C). Unless inhibited, can form unstable and explosive peroxides. Incompatible with strong acids, strong oxidizers. Attacks some plastics. May accumulate static electric charges that can result in ignition of its vapors.

NCI-C60571 (110-54-3) Forms explosive mixture with air (flash point −7°F/−22°C). Strong oxidizers may cause fire and explosions. Attacks some plastics, rubber, and coatings. May accumulate static electrical charges, and may cause ignition of its vapors.

NCI-C60753 (97-02-9) Reacts violently with strong oxidizers. May detonate when heated under confinement.

NCI-C60822 (75-05-8) Forms explosive mixture with air (flash point 42°F/6°C). Incompatible with water (especially if acid or alkaline), acids, caustics, nitrating agents, indium, nitrogen tetroxide, *n*-fluoro compounds, sulfur trioxide, iron(III) salts of perchlorate, nitrogen-fluorine compounds. Reacts violently with oxidizers. May accumulate static electrical charges, and may cause ignition of its vapors.

NCI-C60866 (109-79-5) Forms explosive mixture with air (flash point 35°F/2°C). Incompatible with strong oxidizers and nitric acid. Attacks some plastics and rubber.

NCI-C60899 (105-58-8) Forms explosive mixture with air (flash point 77°F/25°C). Reacts violently with strong oxidizers.

NCI-C60913 (68-12-2) Forms explosive mixture with air (flash point 136°F/58°C). Contact with carbon tetrachloride and other halogenated compounds, particularly in presence of iron or strong oxidizers may cause fire and explosions. Reacts vigorously reaction with alkylalumi-nums. Incompatible with nonoxidizing mineral acids, strong acids, chlorinated hydrocarbons, isocyanates, nitrates, organic nitrates, phenols, cresols, ammonia, chromic anhydride, magnesium nitrate, methylene diisocyante, phosphorus trioxide, triethyaluminum. Attacks some plastics, rubber, and coatings.

NCI-C60935 (112-55-0) Contact with strong oxidizers may cause fire and explosions.

NCI-C60968 (78-84-2) Forms explosive gas mixture with air. Incompatible with strong acids, caustics, aliphatic amines, alkanolamines, aromatic amines, strong oxidizers.

NCI-C61029 (123-38-6) Forms explosive mixture with air (flash point −2°F/−30°C). Incompatible with strong acids, caustics, amines. Reacts violently with strong oxidizers. Can self-ignite if finely dispersed on porous or combustible material. Heat or ultraviolet can cause decomposition. May accumulate static electrical charges, and may cause ignition of its vapors.

NCI-C61187 (99-95-4) Contact with strong oxidizers may cause fire and explosions.

NCI-C61405 (124-09-4) Forms explosive mixture with air (flash point 176°F/80°C). A strong reducing agent. Reacts with ethylene dichloride, acids, organic anhydrides, isocyanates, vinyl acetate, acrylates, substituted allyls, alkylene oxides, epichlorohydrin, ketones, aldehydes, alcohols, glycols, phenols, cresols, caprolactam solution, strong oxidizers. Attacks aluminum, copper, lead, tin, zinc, and alloys.

NDMA (62-75-9) Ultraviolet light, strong oxidizers can cause reactions. Store in dark bottles.

NEANTINE (84-66-2) Reacts violently with strong acids, strong oxidizers, permanganates, water. Attacks some forms of plastic.

NEATSFOOT OIL (N/A) Oxidizers may cause fire and explosions. Incompatible with nitric acid. May accumulate static electrical charges, and may cause ignition of its vapors.

NECATORINA (56-23-5) Becomes corrosive when in contact with water. Corrodes metals. Reacts violently with many compounds. Decomposes on contact with chemically active metals such as sodium, potassium, and magnesium. Incompatible with allyl alcohol, fluorine gas, alkali metals, aluminum. Attacks some coatings, plastics, and rubber.

NECATORINE (56-23-5) Becomes corrosive when in contact with water. Corrodes metals. Reacts violently with many compounds. Decomposes on contact with chemically active metals such as sodium, potassium, and magnesium. Incompatible with allyl alcohol, fluorine gas, alkali metals, aluminum. Attacks some coatings, plastics, and rubber.

NECIDOL (333-41-5) Incompatible with water, copper-containing compounds, oxidizers, acids, or bases.

NEGUVON (52-68-6) Contact with strong oxidizers may cause fire and explosions.

NEKTAL (119-36-8) Forms explosive mixture with air (flash point 205°F/96°C). Incompatible with strong acids, nitrates, oxidizers.

NEMABROM (96-12-8) Forms explosive mixture with air (flash point 170°F/77°C). Reacts with oxidizers and chemically active metals (i.e., aluminum, magnesium, and tin alloys). Attacks some rubber materials and coatings.

NEMAFENE (8003-19-8) Incompatible with strong acids, oxidizers, aluminum or magnesium compounds, aliphatic amines, alkanolamines, alkaline materials, or corrosives.

NEMAGONE (96-12-8) Forms explosive mixture with air (flash point 170°F/77°C). Reacts with oxidizers and chemically active metals (i.e., aluminum, magnesium, and tin alloys). Attacks some rubber materials and coatings.

NEMAGON SOIL FUMAGANT (96-12-8) Forms explosive mixture with air (flash point 170°F/77°C). Reacts with oxidizers and chemically active metals (i.e., aluminum, magnesium, and tin alloys). Attacks some rubber materials and coatings.

NEMANAX (96-12-8) Forms explosive mixture with air (flash point 170°F/77°C). Reacts with oxidizers and chemically active metals (i.e., aluminum, magnesium and tin alloys). Attacks some rubber materials and coatings.

NEMAPAZ (96-12-8) Forms explosive mixture with air (flash point 170°F/77°C). Reacts with oxidizers and chemically active metals (i.e., aluminum, magnesium, and tin alloys). Attacks some rubber materials and coatings.

NEMASET (96-12-8) Forms explosive mixture with air (flash point 170°F/77°C). Reacts with oxidizers and chemically active metals (i.e., aluminum, magnesium, and tin alloys). Attacks some rubber materials and coatings.

NEMATOX (96-12-8) Forms explosive mixture with air (flash point 170°F/77°C). Reacts with oxidizers and chemically active metals (i.e., aluminum, magnesium, and tin alloys). Attacks some rubber materials and coatings.

NEMAZON (96-12-8) Forms explosive mixture with air (flash point 170°F/77°C). Reacts with oxidizers and chemically active metals (i.e., aluminum, magnesium, and tin alloys). Attacks some rubber materials and coatings.

NENDRIN (72-20-8) Incompatible with parathion, strong acids (forms explosive vapors), strong oxidizers.

NEOBAR (7727-43-7) Explosions may result from contact with aluminum in the presence of heat. Incompatible with potassium, phosphorus.

NEOCID (50-29-3) Incompatible with salts of iron or aluminum, and bases. Do not store in iron containers.

NEOCIDOL (333-41-5) Incompatible with water, copper-containing compounds, oxidizers, acids, or bases.

NEODECANOIC ACID (26896-20-8) Forms explosive mixture with air (flash point 201°F/94°C). Incompatible with sulfuric acid, caustics, ammonia, aliphatic amines, alkanolamines, isocyanates, isocyanates, alkylene oxides, epichlorohydrin, oxidizers.

NEODECANOIC ACID, VINYL ESTER (N/A) Forms explosive mixture with air (flash point 175°F/79°C). Polymerization inhibitor (usually monomethyl ether of hydroquinone) must be present in adequate concentrations to avoid explosive polymerization. The monomer is usually supplied in bulk or resin-lined drums, and may be stored in tin- or stainless steel-lined drums. Storage in plastic or other vessels is not recommended. Incompatible with acids, 2-aminoethanol, ammonia, aliphatic amines, alkanolamines, ethylene diamine, ethyleneimine, ozone.

NEO-FAT 8 (124-07-2) Forms explosive mixture with air (flash point 56°F/13°C). Reacts violently with strong oxidizers. May accumulate static electrical charges, and may cause ignition of its vapors.

NEO-FAT 10 (334-48-5) Incompatible with sulfuric acid, caustics, ammonia, aliphatic amines, alkanolamines, isocyanates, alkylene oxides, epichlorohydrin. Attacks most common metals.

NEOFAT 12 (143-07-7) Incompatible with sulfuric acid, bases, ammonia, amines, alkylene oxide, epichlorohydrin.

NEO-FAT 12-43 (143-07-7) Incompatible with sulfuric acid, bases, ammonia, amines,alkylene oxide,epichlorohydrin.

NEO-FAT 18-61 (57-11-4) Incompatible with strong oxidizers, sulfuric acid, caustics, ammonia, amines, isocyanates, alkylene oxides, epichlorohydrin. Attacks chemically active metals.

NEO-FAT 90-04 (112-80-1) Incompatible with strong oxidizers, perchloric acid, and aluminum powder; may cause explosions. Corrodes aluminum.

NEOHEXANE or NEOHEXANE (DOT) (75-83-2) Forms explosive mixture with air (flash point −54°F/−48°C). Reacts violently with strong oxidizers.

NEOL (126-30-7) Forms explosive mixture with air (flash point 265°F/129°C). Incompatible with strong acids, caustics, aliphatic amines, isocyanates, oxidizers.

NEOLID (8001-79-4) Incompatible with strong acids, oxidizers, nitrates.

NEOPENTANOIC ACID (75-98-9) Forms explosive mixture with air (flash point 147°F/64°C). Incompatible with sulfuric acid, caustics, ammonia, amines, isocyanates, alkylene oxides, epichlorohydrin, oxidizers.

NEOPENTYLENE GLYCOL (126-30-7) Forms explosive mixture with air (flash point 265°F/129°C). Incompatible with strong acids, caustics, aliphatic amines, isocyanates, oxidizers.

NEOPENTYL GLYCOL (126-30-7) Forms explosive mixture with air (flash point 265°F/129°C). Incompatible with strong acids, caustics, aliphatic amines, isocyanates, oxidizers.

NEOPRENE (126-99-8) Forms unstable peroxides that cause polymerization. Forms explosive mixture with air (flash point −4°F/−20°C). May accumulate static electrical charges, and may cause ignition of its vapors. Reacts violently with liquid or gaseous fluorine, alkali metals, metal powders, oxidizers. Attacks some plastics, rubber, and coatings.

NEO-SCABICIDOL (58-89-9) Not combustible, but may be dissolved in a combustible solvent. If solvent comes in contact with oxidizers, fire and explosions may result.

NEPHIS (106-93-4) Reacts with chemically active metals, liquid ammonia, strong oxidizers. Heat and light cause slow decomposition. Attacks some plastics and rubber.

NERKOL (62-73-7) Attacks some plastics, rubber, and coatings.

NERVANAID B ACID (60-00-4) Incompatible with sulfuric acid, bases, ammonia, aliphatic amines, alkanolamines, isocyanates, alkylene oxides, epichlorohydrin.

NESOL (138-86-3) Forms explosive mixture with air (flash point 115°F/46°C). Strong oxidizers may cause fire and explosions.

NETAGRONE (94-75-7) Decomposes in sunlight. Incompatible with strong oxidizers; may cause fire and explosions.

NETAGRONE 600 (94-75-7) Decomposes in sunlight. Incompatible with strong oxidizers; may cause fire and explosions.

NEURONILCA (50-78-2) Fires and explosions may result from contact with strong oxidizers. Alkali hydroxides or carbonates cause decomposition.

NEUTRAL AMMONIUM CHROMATE (7788-98-9) Contact with water produces an alkaline solution and evolves free ammonia.

NEUTRAL AMMONIUM FLUORIDE (12125-01-8) Produces hydrochloric acid with water. Incompatible with strong acids, caustics. Corrosive to glass, cement, and most metals.

NEUTRAL ANHYDROUS CALCIUM HYPOCHLORITE (7778-54-3) Decomposes in heat or sunlight. Incompatible with acids, moisture, reducing agents, combustible materials, all other chemicals, especially acetylene, aniline and all other amines, anthracene, carbon tetrachloride, iron oxide, manganese oxide, mercaptans, diethylene glycol monomethyl ether, nitromethane, organic matter, organic sulfides, phenol, 1-propanethiol, propyl mercaptan, sulfur, organic sulfur compounds.

NEUTRAL LEAD ACETATE (301-04-2) Contact with strong acids produces acetic acid. Incompatible with bases, ammonia, amines, cresols, isocyanates, alkylene oxides, epichlorohydrin, phenols, sulfites. Reacts violently with bromates, strong oxidizers.

NEUTRAL POTASSIUM CHROMATE (7789-00-6) A powerful oxidizer. Contact with combustibles, reducing agents, organic material, or finely divided metals may cause fire and explosions.

NEUTRAL SODIUM CHROMATE (7777-11-3) A strong oxidizer; reacts violently with reducing agents, combustibles, strong acids, organic materials.

NEUTRAL VERDIGRIS (142-71-2) Incompatible with strong acids, nitrates.

NEVAX (119-36-8) Forms explosive mixture with air (flash point 205°F/96°C). Incompatible with strong acids, nitrates, oxidizers.

NEX (1563-66-2) Incompatible with alkaline material, acids, strong oxidizers.

NEXEN FB (58-89-9) Not combustible, but may be dissolved in a combustible solvent. If solvent comes in contact with oxidizers, fire and explosions may result.

NEXIT (58-89-9) Not combustible, but may be dissolved in a combustible solvent. If solvent comes in contact with oxidizers, fire and explosions may result.

NEXIT-STARK (58-89-9) Not combustible, but may be dissolved in a combustible solvent. If solvent comes in contact with oxidizers, fire and explosions may result.

NEXOL-E (58-89-9) Not combustible, but may be dissolved in a combustible solvent. If solvent comes in contact with oxidizers, fire and explosions may result.

NG (55-63-0) Liquid portion of this chemical is an alcohol; reacts with strong acids, caustics, aliphatic amines, isocyanates. Dry portion of chemical is nitroglycerine, a shock- and friction-sensitive explosive. Acids, heat, or mechanical shock may cause explosions.

NI 270 (7440-02-0) Powder may self-ignite in air. Contact with strong acids produces flammable and explosive hydrogen gas. Sulfur contact evolves heat. Reacts violently with strong oxidizers, nitric acid, hydrazine, and many other chemicals.

NI 0901-S (7440-02-0) Powder may self-ignite in air. Contact with strong acids produces flammable and explosive hydrogen gas. Sulfur contact evolves heat. Reacts violently with strong oxidizers, nitric acid, hydrazine, and many other chemicals.

NI 4303T (7440-02-0) Powder may self-ignite in air. Contact with strong acids produces flammable and explosive hydrogen gas. Sulfur contact evolves heat. Reacts violently with strong oxidizers, nitric acid, hydrazine, and many other chemicals.

NIA 1240 (563-12-2) Incompatible with alkaline formulations. Mixtures with magnesium may be explosive.

NIA 5462 (115-29-7) Hydrolyzed by acids and alkalies. Corrosive to iron.

NIA 10242 (1563-66-2) Incompatible with alkaline material, acids, strong oxidizers.

NIAGARA 1240 (563-12-2) Incompatible with alkaline formulations. Mixtures with magnesium may be explosive.

NIAGARA 5462 (115-29-7) Hydrolyzed by acids and alkalis. Corrosive to iron.

NIAGARA 10242 (1563-66-2) Incompatible with alkaline material, acids, strong oxidizers.

NIALATE (563-12-2) Incompatible with alkaline formulations. Mixtures with magnesium may be explosive.

NIALK (79-01-6) Contact with caustics produces a toxic and flammable gas. Reacts violently with chemically active metals. Contact with aluminum may produce a violent, self-accelerating polymerization reaction. Incompatible with acids, organic anhydrides, isocyanates, alkylene oxides, aldehydes, alcohols, glycols, phenols, cresols, caprolactam solution, epichlorohydrin, nitrogen tetroxide, metal powders, oxygen. May accumulate static electrical charges, and may cause ignition of its vapors.

NIAX TDI (584-84-9) Incompatible with strong acids including nonoxidizing mineral and organic acids, caustics, ammonia, amines, amides, alcohols, glycols, caprolactam solution. Water contact causes violent foaming and spattering; produces carbon dioxide and an organic base. Attacks copper and its alloys, some plastics including polyethylene, and rubber.

NIAX TDI-P (584-84-9) Incompatible with strong acids including nonoxidizing mineral and organic acids, caustics, ammonia, amines, amides, alcohols, glycols, caprolactam solution. Water contact causes violent foaming and spattering; produces carbon dioxide and an organic base. Attacks copper and its alloys, some plastics including polyethylene, and rubber.

NICHEL (Italian) (7440-02-0) Powder may self-ignite in air. Contact with strong acids produces flammable and explosive hydrogen gas. Sulfur contact evolves heat. Reacts violently with strong oxidizers, nitric acid, hydrazine, and many other chemicals.

NICHEL TETRACARBONILE (Italian) (13463-39-3) Forms explosive mixture with air (flash point $-4°F/-20°C$). May explode when heated above $140°F/60°C$. In the presence of air, forms a deposit that becomes peroxidized; this tends to decompose and ignite. Contact with acids and oxidizers can cause fire and explosions. Vapor may promote the ignition of mixtures of combustible vapors (such as gasoline) and air. Attacks some plastics, rubber, and coatings. Store under inert gas blanket. May accumulate static electrical charges, and may cause ignition of its vapors.

NICKEL or NICKEL PARTICLES or NICKEL SPONGE (7440-02-0) Powder may self-ignite in air. Contact with strong acids produces flammable and explosive hydrogen gas. Sulfur contact evolves heat. Reacts violently with strong oxidizers, nitric acid, hydrazine, and many other chemicals.

NICKEL 270 (7440-02-0) Powder may self-ignite in air. Contact with strong acids produces flammable and explosive hydrogen gas. Sulfur contact evolves heat. Reacts violently with strong oxidizers, nitric acid, hydrazine and many other chemicals.

NICKEL ACETATE (373-02-4) Incompatible with strong acids, strong oxidizers, sulfur, selenium.

NICKEL(II) ACETATE (1:2) (373-02-4) Incompatible with strong acids, strong oxidizers, sulfur, selenium.

NICKEL ACETATE TETRAHYDRATE (373-02-4) Incompatible with strong acids, strong oxidizers, sulfur, selenium.

NICKEL AMMONIUM SULFATE (15699-18-0) Incompatible with strong acids, selenium.

NICKEL AMMONIUM SULFATE HEXAHYDRATE (15699-18-0) Incompatible with strong acids, selenium.

NICKEL BROMIDE (13462-88-9) Incompatible with strong acids, sulfur, selenium. Mixing with potassium produces impact-sensitive explosive.

NICKEL BROMIDE TRIHYDRATE (13462-88-9) Incompatible with strong acids, sulfur, selenium. Mixing with potassium produces impact-sensitive explosive.

NICKEL CARBONYL (13463-39-3) Forms explosive mixture with air (flash point −4°F/−20°C). May explode when heated above 140°F/60°C. In the presence of air, forms a deposit that becomes peroxidized; this tends to decompose and ignite. Contact with acids and oxidizers can cause fire and explosions. Vapor may promote the ignition of mixtures of combustible vapors (such as gasoline) and air. Attacks some plastics, rubber, and coatings. Store under inert gas blanket. May accumulate static electrical charges, and may cause ignition of its vapors.

NICKEL CARBONYLE (French) (13463-39-3) Forms explosive mixture with air (flash point −4°F/−20°C). May explode when heated above 140°F/60°C. In the presence of air, forms a deposit that becomes peroxidized; this tends to decompose and ignite. Contact with acids and oxidizers can cause fire and explosions. Vapor may promote the ignition of mixtures of combustible vapors (such as gasoline) and air. Attacks some plastics, rubber, and coatings. Store under inert gas blanket. May accumulate static electrical charges, and may cause ignition of its vapors.

NICKEL CATALYST, WET (7440-02-0) Powder may self-ignite in air. Contact with strong acids produces flammable and explosive hydrogen gas. Sulfur contact evolves heat. Reacts violently with strong oxidizers, nitric acid, hydrazine and many other chemicals.

NICKEL CHLORIDE (7718-54-9) Mixing with potassium produces an impact-sensitive explosive.

NICKEL(2+) CHLORIDE or NICKEL(II) CHLORIDE or NICKEL(II) CHLORIDE (1:2) (7718-54-9) Mixing with potassium produces an impact-sensitive explosive.

NICKEL CHLORIDE HEXAHYDRATE (7718-54-9) Mixing with potassium produces an impact-sensitive explosive.

NICKEL CYANIDE or NICKEL CYANIDE, SOLID (557-19-7) Incompatible with strong acids, sulfur, selenium.

NICKEL(2+) CYANIDE or NICKEL(II) CYANIDE (557-19-7) Incompatible with strong acids, sulfur, selenium.

NICKEL DIHYDROXIDE (12054-48-7) Combustible solid. Strong oxidizers may cause fire and explosions.

NICKEL HYDROXIDE (12054-48-7) Combustible solid. Strong oxidizers may cause fire and explosions.

NICKEL(2+) HYDROXIDE or NICKEL(II) HYDROXIDE (12054-48-7) Combustible solid. Strong oxidizers may cause fire and explosions.

NICKEL NITRATE HEXAHYDRATE (1347-00-7) Incompatible with strong acids, sulfur, wood, and other combustibles.

NICKEL(2+) NITRATE, HEXAHYDRATE or NICKEL(II) NITRATE, HEXAHYDRATE (1347-00-7) Incompatible with strong acids, sulfur, wood, and other combustibles.

NICKELOUS ACETATE (373-02-4) Incompatible with strong acids, strong oxidizers, sulfur, selenium.

NICKELOUS CHLORIDE (7718-54-9) Mixing with potassium produces an impact-sensitive explosive.

NICKELOUS HYDROXIDE (12054-48-7) Combustible solid. Strong oxidizers may cause fire and explosions.

NICKEL TETRACARBONYL (13463-39-3) Forms explosive mixture with air (flash point −4°F/−20°C). May explode when heated above 140°F/60°C. In the presence of air, forms a deposit which becomes peroxidized; this tends to decompose and ignite. Contact with acids and oxidizers can cause fire and explosions. Vapor may promote the ignition of mixtures of combustible vapors (such as gasoline) and air. Attacks some plastics, rubber, and coatings. Store under inert gas blanket. May accumulate static electrical charges, and may cause ignition of its vapors.

NICKEL TETRACARBONYLE (French) (13463-39-3) Forms explosive mixture with air (flash point −4°F/−20°C). May explode when heated above 140°F/60°C. In the presence of air, forms a deposit which becomes peroxidized; this tends to decompose and ignite. Contact with acids and oxidizers can cause fire and explosions. Vapor may promote the ignition of mixtures of combustible vapors (such as gasoline) and air. Attacks some plastics, rubber, and coatings. Store under inert gas blanket. May accumulate static electrical charges, and may cause ignition of its vapors.

NICOCHLORAN (58-89-9) Not combustible, but may be dissolved in a combustible solvent. If solvent comes in contact with oxidizers fire and explosions may result.

NICOCIDE (54-11-5) Incompatible with strong acids, strong oxidizers. Attacks some plastics, rubber, and coatings. May accumulate static electrical charges, and may cause ignition of its vapors.

NICODUST (54-11-5) Incompatible with strong acids, strong oxidizers. Attacks some plastics, rubber, and coatings. May accumulate static electrical charges, and may cause ignition of its vapors.

NICOFUME (54-11-5) Incompatible with strong acids, strong oxidizers. Attacks some plastics, rubber, and coatings. May accumulate static electrical charges, and may cause ignition of its vapors.

NICOTINE or NICOTINE, LIQUID or NICOTINE, SOLID (54-11-5) Incompatible with strong acids, strong oxidizers. Attacks some plastics, rubber, and coatings. May accumulate static electrical charges, and may cause ignition of its vapors.

NIESYMETRYCZNA DWUMETYLOHYDRAZYN (Polish) (57-14-7) Forms explosive mixture with air (flash point 5°F/−15°C). A strong reducing agent. Incompatible with strong acids, halogens, metallic mercury, strong oxidizers (with possible spontaneous ignition). Attacks some plastics, rubber, and coatings. May accumulate static electrical charges, and may cause ignition of its vapors.

NIFOS (107-49-3) Decomposes above 300°F/150°C, producing flammable ethylene gas. Strong oxidizers may cause fire and explosions. Attacks some plastics, rubber, and coatings.

NIFOS T (107-49-3) Decomposes above 300°F/150°C producing flammable ethylene gas. Strong oxidizers may cause fire and explosions. Attacks some plastics, rubber, and coatings.

NIFROST (107-49-3) Decomposes above 300°F/150°C producing flammable ethylene gas. Strong oxidizers may cause fire and explosions. Attacks some plastics, rubber, and coatings.

NIGLYCON (55-63-0) Liquid portion of this chemical is an alcohol; reacts with strong acids, caustics, aliphatic amines, isocyanates. Dry portion of chemical is nitroglycerine, a shock- and friction-sensitive explosive. Acids, heat, or mechanical shock may result in explosions.

NIKKEL TETRACARBONYL (Dutch) (13463-39-3) Forms explosive mixture with air (flash point −4°F/−20°C). May explode when heated above 140°F/60°C. In the presence of air, forms a deposit that becomes peroxidized; this tends to decompose and ignite. Contact with acids and oxidizers can cause fire and explosions. Attacks some plastics, rubber, and coatings. May accumulate static electrical charges, and may cause ignition of its vapors.

NIKKOL OTP 70 (119-36-8) Forms explosive mixture with air (flash point 205°F/96°C). Incompatible with strong acids, nitrates, oxidizers.

NINOL AA-62 EXTRA (143-07-7) Incompatible with sulfuric acid, bases, ammonia, amines, alkylene oxide, epichlorohydrin.

NIOBE OIL (93-58-3) Forms explosive mixture with air (flash point 181°F/83°C). Incompatible with strong acids, nitrates, oxidizers.

NIONG (55-63-0) Liquid portion of this chemical is an alcohol; reacts with strong acids, caustics, aliphatic amines, isocyanates. Dry portion of chemical is nitroglycerine, a shock- and friction-sensitive explosive. Acids, heat, or mechanical shock may cause explosions.

NIPAR S-20 (79-46-9) Forms explosive mixture with air (flash point 74°F/24°C). Incompatible with amines, strong acids, alkalies, strong oxidizers. Contact with some metal oxides may cause decomposition. Mixtures with hydrocarbons are extremely flammable. Attacks some plastics, rubber, and coatings.

NIPAR S-30 SOLVENT (79-46-9) Forms explosive mixture with air (flash point 74°F/24°C). Incompatible with amines, strong acids, alkalies, strong oxidizers. Contact with some metal oxides may cause decomposition. Mixtures with hydrocarbons are extremely flammable. Attacks some plastics, rubber, and coatings.

NIPHEN (100-02-7) Combustible solid. A strong oxidizer; reacts with reducing agents, combustibles, organic and other easily oxidizable materials. Incompatible with strong acids, caustics, aliphatic amines, amides.

NIPSAN (333-41-5) Incompatible with water, copper-containing compounds, oxidizers, acids, or bases.

NIRAN (57-74-9) Contact with strong oxidizers may cause fire and explosions. Attacks some plastics, rubber, and coatings.

NIRAN (56-38-2) Combustible liquid. Mixtures with endrin may be explosive. Strong oxidizers may cause fire and explosions. Attacks some plastics, rubber, and coatings.

NIRAN E-4 (56-38-2) Combustible liquid. Mixtures with endrin may be explosive. Strong oxidizers may cause fire and explosions. Attacks some plastics, rubber, and coatings.

NITAL (7697-37-2) A strong oxidizer that can react violently with reducing agents, combustible materials. Incompatible with many substances, including acrylates, aliphatic amines, alcohols, aldehydes, alkanolamines, alkylene oxides, anion exchange resins, aromatic amines, amides, bases, cresols, cyanides, cyclic ketones, epichlorohydrin, glycols, isocyanates, ketones, oleum, organic anhydrides, phenols, substituted allyls, sulfuric acid, strong oxidizers, terpenes. Attacks most metals, and some plastics, rubber, and coatings.

NITER (7757-79-1) A powerful oxidizer. Reacts violently with reducing agents and combustibles. Strong acid causes formation of toxic vapors. Forms explosive mixtures with many substances, including sodium acetate, metal powders, sodium hypophosphite, trichloroethylene, zinc, etc. Dangerously reactive, separate from all materials.

NITRAM (6484-52-2) A strong oxidizer. Contact with reducing agents, combustibles, organic materials, finely divided metals may produce explosive mixtures or cause fire and explosions.

NITRAM (13473-90-0) A strong oxidizer. Reacts violently with combustibles, organics, reducing agents. Aqueous solution is acidic. Attacks metals in the presence of moisture.

NITRAMINE (479-45-8) A powerful oxidizer and highly sensitive explosive. Protect containers from shock, friction, heat, oxidizers, and hydrazine.

NITRAN (298-00-0) Mixtures with magnesium may be explosive.

4-NITRANBINE (100-01-06) Strong oxidizers and moisture may cause spontaneous heating. Caustics and heat may cause formation of explosive mixtures. Attacks some plastics, rubber, and coatings. May accumulate static electrical charges, and may cause ignition of its vapors.

2-NITRANILINE (88-74-4) *ortho;* (99-09-2) *meta.* Incompatible with strong acids, combustibles, organics, moisture.

NITRAPYRIN (1929-82-4) Reacts with aluminum, magnesium, or their alloys.

NITRATE d'AMYLE (French) (1002-16-0) Forms explosive mixture with air (flash point 120°F/49°C). An oxidizer. Reacts strongly with reducing agents, strong acids, esters, combustibles. Attacks some plastics, rubber, and coatings.

NITRATE d'ARGENT (French) (7761-88-8) Contact with acetylene produces shock-sensitive material. Ammonia contact causes formation of compounds that are explosive when dry. Hydrogen peroxide causes violent decomposition to oxygen gas. Attacks some plastics, rubber, and coatings.

NITRATE de BARYUM (French) (10022-31-8) Incompatible with strong acids, aluminum–magnesium alloys, sulfur, combustibles, finely divided metals; may form shock-sensitive compounds.

NITRATE de SODIUM (French) (7631-99-4) Powerful oxidizer; reacts violently with reducing agents, combustible substances, strong acids, organic materials, powdered metals, bitumens.

NITRATE de ZINC (French) (7779-88-6) A strong oxidizer. Reacts violently with combustibles, organic compounds, reducing agents, carbon, metal sulfides, phosphorus, and sulfur.

NITRATE MERCUREUX (French) (10415-75-5) A powerful oxidizer. Incompatible with reducing agents, combustible materials. Forms shock-sensitive compounds with phosphorus. Aqueous solutions are corrosive.

NITRATE MERCURIQUE (French) (10045-94-0) A powerful oxidizer. Incompatible with reducing agents, combustibles; forms shock-sensitive compounds with acetylene, ethanol, isobutene, phosphine, sulfur. Forms heat-sensitive compound with potassium cyanide. Incompatible with strong acids, hydrocarbons. Aqueous solution corrodes metals.

NITRATINE (7631-99-4) Powerful oxidizer; reacts violently with reducing agents, combustible substances, strong acids, organic materials, powdered metals, bitumens.

NITRETAN (Polish) (79-24-3) Forms explosive mixture with air (flash point 82°F/28°C). Caustics, ammonia, strong acids, amines can cause sensitization, which can lead to explosions. Strong oxidizers may cause fire and explosions. Mixtures with hydrocarbons (or other combustible materials) are highly flammable. Attacks some plastics, rubber, and coatings.

NITREX NITROGEN SOLUTIONS (NON-PRESSURE) (6484-52-2) Incompatible with sulfuric acid, isocyanates.

NITRIC ACID (7697-37-2) A strong oxidizer that can react violently with reducing agents, combustible materials. Incompatible with many substances, including acrylates, aliphatic amines, alcohols, aldehydes, alkanolamines, alkylene oxides, anion exchange resins, aromatic amines, amides, bases, cresols, cyanides, cyclic ketones, epichlorohydrin, glycols, isocyanates, ketones, oleum, organic anhydrides, phenols, substituted allyls, sulfuric acid, strong oxidizers, terpenes. Attacks most metals, and some plastics, rubber, and coatings.

NITRIC ACID, ALUMINUM(3+) or NITRIC ACID, ALUMINUM(III) (13473-90-0) A strong oxidizer. Reacts violently with combustibles, organics, reducing agents. Aqueous solution is acidic. Attacks metals in the presence of moisture.

NITRIC ACID, ALUMINUM SALT (13473-90-0) A strong oxidizer. Reacts violently with combustibles, organics, reducing agents. Aqueous solution is acidic. Attacks metals in the presence of moisture.

NITRIC ACID, AMMONIUM SALT (6484-52-2) A strong oxidizer. Reducing agents, combustibles, organic materials, finely divided metals may form explosive mixtures or cause fire and explosions.

NITRIC ACID, BARIUM SALT (10022-31-8) Incompatible with strong acids, aluminum–magnesium alloys, sulfur, combustibles, finely divided metals; may form shock-sensitive compounds.

NITRIC ACID, BERYLLIUM SALT (778-75-5) Produces nitric acid with water. Incompatible with acids, caustics, chlorinated hydrocarbons, oxidizers, molten lithium. Corrodes metals in a moist environment.

NITRIC ACID, CADMIUM SALT, TETRAHYDRATE (10022-68-1) May react with strong oxidizers, selenium, sulfur or tellurium.

NITRIC ACID, CALCIUM SALT (10124-37-5) A strong oxidizer. Incompatible with combustible materials, reducing agents, organics and other oxidizable materials, chemically active metals, aluminum nitrate, ammonium nitrate.

NITRIC ACID, COBALT(2+) SALT (10141-05-6) A strong oxidizer; reacts with reducing agents, combustibles.

NITRIC ACID, COPPER(2+) SALT or NITRIC ACID, COPPER(II) SALT (3251-23-8) Aqueous solution is acidic. A strong oxidizer; Reacts violently with reducing agents, combustibles. Incompatible with bases, acetic anhydride, cyanides, esters. Attacks metals in the presence of moisture.

NITRIC ACID, IRON(3+) SALT or NITRIC ACID, Iron (III) SALT (10421-48-4) Solution is corrosive to metals. Contact with combustibles may cause fire.

NITRIC ACID, LEAD(2+) SALT or NITRIC ACID, LEAD(II) SALT (10099-74-8) A strong oxidizer; reacts violently with reducing agents and combustible materials. Incompatible with carbon, lead hypophosphite, ammonium thiocyanate, potassium acetate.

NITRIC ACID, MAGNESIUM SALT (10377-60-3) A powerful oxidizer. Reacts violently with dimethylformamide, reducing agents, combustibles, organic and easily oxidizable matter.

NITRIC ACID, MERCURY(I) SALT (10415-75-5) A powerful oxidizer. Incompatible with reducing agents, combustible materials. Produces shock-sensitive compounds with phosphorus. Aqueous solutions are corrosive.

NITRIC ACID, MERCURY(2+) SALT or NITRIC ACID, MERCURY(II) SALT (10045-94-0) A powerful oxidizer. Incompatible wih reducing agents, combustibles; produces shock-sensitive compounds with acetylene, ethanol, isobutene, phosphine, sulfur. Produces heat-sensitive compound with potassium cyanide. Incompatible with strong acids, hydrocarbons. Aqueous solution corrodes metals.

NITRIC ACID, NICKEL(2+) SALT, HEXAHYDRATE or NITRIC ACID, NITRIC ACID TRIESTER OF GLYCEROL (55-63-0) Liquid portion of this chemical is an alcohol; reacts with strong acids, caustics, aliphatic amines, isocyanates. Dry portion of chemical is nitroglycerine, a shock- and friction-sensitive explosive. Acids, heat, or mechanical shock may cause explosions.

NITRIC ACID, NICKEL(II) SALT, HEXAHYDRATE (1347-00-7) Incompatible with strong acids, sulfur, wood, and other combustibles.

NITRIC ACID, PENTYL ESTER (1002-16-0) Forms explosive mixture with air (flash point 120°F/49°C). An oxidizer. Reacts strongly with reducing agents, strong acids, esters, combustibles. Attacks some plastics, rubber, and coatings.

NITRIC ACID, POTASSIUM SALT (7757-79-1) A powerful oxidizer. Reacts violently with reducing agents and combustibles. Strong acid contact produces toxic vapors. Forms explosive mixtures with many substances including sodium acetate, metal powders, sodium hypophosphite, trichloroethylene, zinc, etc. Dangerously reactive; separate from all materials.

NITRIC ACID, PROPYL ESTER (627-13-4) Forms explosive mixture with air (flash point 68°F/20°C). Strong oxidizers or combustibles may cause fire and explosions. Attacks some plastics, rubber, and coatings.

NITRIC ACID, RED FUMING (7697-37-2) A strong oxidizer that can react violently with reducing agents, combustible materials. Incompatible with many substances, including acrylates, aliphatic amines, alcohols, aldehydes, alkanolamines, alkylene oxides, anion exchange resins, aromatic amines, amides, bases, cresols, cyanides, cyclic ketones, epichlorohydrin, glycols, isocyanates, ketones, oleum, organic anhydrides, phenols, substituted allyls, sulfuric acid, strong oxidizers, terpenes. Attacks most metals, and some plastics, rubber, and coatings.

NITRIC ACID, SILVER(1+) SALT or NITRIC ACID, SILVER(II) SALT (7761-88-8) Contact with acetylene produces shock-sensitive material. Ammonia contact produces compounds that are explosive when dry. Hydrogen peroxide causes violent decomposition to oxygen gas. Attacks some plastics, rubber, and coatings.

NITRIC ACID, SODIUM SALT (7631-99-4) Powerful oxidizer; reacts violently with reducing agents, combustible substances, strong acids, organic materials, powdered metals, bitumens.

NITRIC ACID, STRONTIUM SALT (10042-76-9) A strong oxidizer; reacts violently with reducing agents, combustibles, acids, organic and other easily oxidizable materials.

NITRIC ACID, THORIUM(4+) SALT or NITRIC ACID, THORIUM(IV) SALT (13823-29-5) Incompatible with water, producing weak nitric acid solution. A strong oxidizer; reacts violently with reducing agents, combustibles, organic materials. Attacks metals when wet.

NITRIC ACID, WHITE FUMING (7697-37-2) A strong oxidizer that can react violently with reducing agents, combustible materials. Incompatible with many substances including acrylates, aliphatic amines, alcohols, aldehydes, alkanolamines, alkylene oxides, anion exchange resins, aromatic amines, amides, bases, cresols, cyanides, cyclic ketones, epichlorohydrin, glycols, isocyanates, ketones, oleum, organic anhydrides, phenols, substituted allyls, sulfuric acid, strong oxidizers, terpenes. Attacks most metals, and some plastics, rubber, and coatings.

NITRIC ACID, ZINC SALT (7779-88-6) A strong oxidizer. Reacts violently with combustibles, organic compounds, reducing agents, carbon, metal sulfides, phosphorus, and sulfur.

NITRIC OXIDE (10102-43-9) A strong oxidizer. Avoid contact with all combustible materials, chlorinated hydrocarbons, ammonia, carbon disulfide, many metals, fluorine, and ozone. Attacks some plastics, rubber, and coatings.

NITRILE ACRILICO (Italian) (107-13-1) Forms explosive mixture with air (flash point 32°F/0°C o.c.). Forms explosive peroxides; heat, light, caustics, silver nitrate, and peroxides can cause polymerization. Incompatible with strong acids, strong oxidizers, amines, 2-aminoethanol, bromine, chlorosulfonic acid, ethylene diamine, nitric acid, oleum, potassium hydroxide, sodium hydroxide, sulfuric acid. Attacks copper and copper alloys; attacks aluminum in high concentrations. May accumulate static electrical charges, and may cause ignition of its vapors.

NITRILE ACRYLIQUE (French) (107-13-1) Forms explosive mixture with air (flash point 32°F/0°C o.c.). Forms explosive peroxides; heat, light, caustics, silver nitrate, and peroxides can cause polymerization. Incompatible with strong acids, strong oxidizers, amines, 2-aminoethanol, bromine, chlorosulfonic acid, ethylene diamine, nitric acid, oleum, potassium hydroxide, sodium hydroxide, sulfuric acid. Attacks copper and copper alloys; attacks aluminum in high concentrations. May accumulate static electrical charges, and may cause ignition of its vapors.

NITRILE ADIPICO (Italian) (111-69-3) May accumulate static electrical charges, and may cause ignition of its vapors. Decomposes at its flash point, producing poisonous cyanide gas. Forms explosive mixture with air (flash point 199°F/93°C). Incompatible with sulfuric acid, oxidizers.

NITRILOACETONITRILE (460-19-5) Explosive reaction occurs with acids, water, liquid oxygen, oxidizers.

2,2′,2′-NITRILO-TRIETHANOL (102-71-6) Aqueous solution is caustic. Incompatible with acids, organic anhydrides, isocyanates, vinyl acetate, acrylates, substituted allyls, alkylene oxides, epichlorohydrin, aldehydes, strong oxidizers. Corrodes copper and its alloys.

NITRINE-TDC (55-63-0) Liquid portion of this chemical is an alcohol; reacts with strong acids, caustics, aliphatic amines, isocyanates. Dry portion of chemical is nitroglycerine, a shock- and friction-sensitive explosive. Acids, heat, or mechanical shock may result in explosions.

4-NITROANILINE (100-01-06) Strong oxidizers and moisture may cause spontaneous heating. Caustics and heat may cause formation of explosive mixtures. Attacks some plastics, rubber, and coatings. May accumulate static electrical charges, and may cause ignition of its vapors.

o-**NITROANILINE or** *ortho*-**NITROANILINE** (88-74-4) Incompatible with strong acids, combustibles, organics, moisture.

p-**NITROANILINE or** *para*-**NITROANILINE** (100-01-06) Strong oxidizers and moisture may cause spontaneous heating. Caustics and heat may cause formation of explosive mixtures. Attacks some plastics, rubber, and coatings. May accumulate static electrical charges, and may cause ignition of its vapors.

NITROBARITE (10022-31-8) Incompatible with strong acids, aluminum–magnesium alloys, sulfur, combustibles, finely divided metals; may form shock-sensitive compounds.

NITROBENZEN (Polish) (98-95-3) Forms explosive mixture with air (flash point 190°F/88°C). Incompatible with nitric acid, nitrogen tetroxide, caustics, ammonia, amines. Attacks some plastics, rubber, and coatings.

NITROBENZENE (98-95-3) Forms explosive mixture with air (flash point 190°F/88°C). Incompatible with nitric acid, nitrogen tetroxide, caustics, ammonia, amines. Attacks some plastics, rubber, and coatings.

4-NITROBENZENAMINE (100-01-06) Strong oxidizers and moisture may cause spontaneous heating. Caustics and heat may cause formation of explosive mixtures. Attacks some plastics, rubber, and coatings. May accumulate static electrical charges, and may cause ignition of its vapors.

NITROBENZOL (98-95-3) Forms explosive mixture with air (flash point 190°F/88°C). Incompatible with nitric acid, nitrogen tetroxide, caustics, ammonia, amines. Attacks some plastics, rubber, and coatings.

NITROBENZOL, 1 (98-95-3) Forms explosive mixture with air (flash point 190°F/88°C). Incompatible with nitric acid, nitrogen tetroxide, caustics, ammonia, amines. Attacks some plastics, rubber, and coatings.

NITROCALCITE (10124-37-5) A strong oxidizer. Incompatible with combustible materials, reducing agents, organics and other oxidizable materials, chemically active metals, aluminum nitrate, ammonium nitrate.

NITROCARBOL (75-52-5) Forms explosive mixture with air (flash point 95°F/35°C). Caustics, ammonia, strong acids, amines can cause sensitization, which can cause explosions. Strong oxidizers may cause fire and explosions. Mixtures with hydrocarbons (and other combustible materials) are highly flammable. Contact with metallic oxides may cause decomposition and development of pressure. Attacks some plastics, rubber, and coatings.

NITROCELLULOSE (9004-70-0) If ether solution evaporates, dry nitrocellulose material is a shock-sensitive explosive and fire hazard. Forms explosive mixture with air (flash point −64°F/−53°C). Strong oxidizers may cause fire and explosions. May accumulate static electrical charges, and may cause ignition of its vapors. Attacks some plastics, rubber, and coatings.

NITROCELLULOSE GUM (9004-70-0) If ether solution evaporates, dry nitrocellulose material is a shock-sensitive explosive and fire hazard. Forms explosive mixture with air (flash point −64°F/−53°C). Strong oxidizers may cause fire and explosions. May accumulate static electrical charges, and may cause ignition of its vapors. Attacks some plastics, rubber, and coatings.

NITROCELLULOSE SOLUTION (9004-70-0) If ether solution evaporates, dry nitrocellulose material is a shock-sensitive explosive and fire hazard. Forms explosive mixture with air (flash point −64°F/−53°C). Strong oxidizers may cause fire and explosions. May accumulate static electrical charges, and may cause ignition of its vapors. Attacks some plastics, rubber, and coatings.

m-**NITROCHLOROBENZENE (DOT) or** *m*-**NITROCHLOROBENZENE, SOLID (DOT) or** *meta*-**NITROCHLOROBENZENE** (121-73-3) Particles form explosive mixture with air. A strong oxidizer; reacts violently with reducing agents, combustibles. Reacts with caustics, ammonia, amines. Attacks some plastics, rubber, and coatings.

o-**NITROCHLOROBENZENE (DOT) or** *o*-**NITROCHLOROBENZENE, LIQUID (DOT) or** *ortho*-**NITROCHLOROBENZENE** (88-73-3) Forms explosive mixture with air. A strong oxidizer; reacts violently with reducing agents, combustibles. Reacts with caustics, ammonia, amines. Attacks some plastics, rubber, and coatings.

p-**NITROCHLOROBENZENE (DOT) or** *para*-**NITROCHLOROBENZENE** (100-00-5) Forms explosive mixture with air. A strong oxidizer; reacts violently with reducing agents, combustibles. Reacts with caustics, ammonia, amines. Attacks some plastics, rubber, and coatings.

NITROCHLOROFORM (76-06-2) Can be self-reactive. Fast heating, shock, alkali metals or alkaline earth may cause explosions. A strong oxidizer; reacts violently with reducing agents, aniline in presence of heat, alcoholic sodium hydroxide, combustible substances, sodium methoxide, propargyl bromide. Liquid attacks some plastics, rubber, and coatings.

NITROETHANE (79-24-3) Forms explosive mixture with air (flash point 82°F/28°C). Caustics, ammonia, strong acids, amines can cause sensitization, which can lead to explosions. Strong oxidizers may cause fire and explosions. Mixtures with hydrocarbons (or other combustible materials) are highly flammable. Attacks some plastics, rubber, and coatings.

NITROFAN (534-52-1) Reacts with heat and strong oxidizers. Dust can form explosive mixtures with air.

4-NITROFENOL (Dutch) (100-02-7) Combustible solid. A strong oxidizer; reacts with reducing agents, combustibles, organic and other easily oxidizable materials. Incompatible with strong acids, caustics, aliphatic amines, amides.

NITROGEN CHLORIDE (10025-85-1) Shock, friction, light, or temperatures above 130°F/57°C can cause substance to explode. Contact with many substances can cause explosive decomposition, including organic substances, combustible materials, water, ammonia, reducing agents, phosphorus.

NITROGEN LIME (156-62-7) Contact with any form of moisture causes decomposition, liberating acetylene and ammonia. Contact with all solvents tested also causes decomposition.

NITROGEN MONOXIDE (10102-43-9) A strong oxidizer. Avoid contact with all combustible materials, chlorinated hydrocarbons, ammonia, carbon disulfide, many metals, fluorine, and ozone. Attacks some plastics, rubber, and coatings.

NITROGEN OXIDE (10024-97-2) May form explosive mixture with flammable gases. Nonflammable but supports combustion; as temperature increases, it becomes both a strong oxidizer and self-reactive. High temperatures may cause explosive decomposition. Incompatible with aluminum, anhydrous ammonia (may cause explosion), boron, carbon monoxide, hydrazine, hydrogen, hydrogen sulfide, lithium hydride, phosphine, sodium.

NITROGEN OXYCHLORIDE (696-92-6) Incompatible with water, producing acid solution and toxic red oxides of nitrogen. Corrodes aluminum and other metals in the presence of moisture.

NITROGEN TETROXIDE (10544-72-6); (10102-44-0 CAS for nitrogen dioxide). A powerful oxidizer. Water contact produces nitric acid and nitric oxide. Incompatible with nitric oxide; reacts with air, producing more nitrogen tetroxide. Explosive reaction occurs with alcohols, fuels and petroleum, nitrobenzene, organic matter, anhydrous ammonia, chlorinated hydrocarbons, toluene. Reacts vigorously with combustibles, reducing agents. Attacks metals in presence of moisture.

NITROGEN TRICHLORIDE (10025-85-1) Shock, friction, light, or temperatures above 130°F/57°C can cause substance to explode. Contact with many substances can cause explosive decomposition, including organic substances, combustible materials, water, ammonia, reducing agents, phosphorus.

NITROGLYCERIN (55-63-0) Liquid portion of this chemical is an alcohol; reacts with strong acids, caustics, aliphatic amines, isocyanates. Dry portion of chemical is nitroglycerine, a shock- and friction-sensitive explosive. Acids, heat, or mechanical shock may cause explosions.

NITROGLYCERIN AND ETHYLENE GLYCOL DINITRATE (N/A) Acids, heat, or mechanical shock may cause explosions.

NITROGLYCEROL (55-63-0) Liquid portion of this chemical is an alcohol; reacts with strong acids, caustics, aliphatic amines, isocyanates. Dry portion of chemical is nitroglycerine, a shock- and friction-sensitive explosive. Acids, heat, or mechanical shock may cause explosions.

NITROGLYN (55-63-0) Liquid portion of this chemical is an alcohol; reacts with strong acids, caustics, aliphatic amines, isocyanates. Dry portion of chemical is nitroglycerine, a shock- and friction-sensitive explosive. Acids, heat, or mechanical shock may cause explosions.

NITROISOPROPANE (79-46-9) Forms explosive mixture with air (flash point 74°F/24°C). Incompatible with amines, strong acids, alkalies, strong oxidizers. Contact with some metal oxides may cause decomposition. Mixtures with hydrocarbons are extremely flammable. Attacks some plastics, rubber, and coatings.

NITRO KLEENUP (51-28-5) Explosion may be caused by heat, friction, or shock. Contact with reducing agents, combustibles may cause fire and explosions. Produces explosive salts with ammonia or strong bases. May accumulate static electrical charges, and may cause ignition of its vapors.

NITROL (55-63-0) Liquid portion of this chemical is an alcohol; reacts with strong acids, caustics, aliphatic amines, isocyanates. Dry portion of chemical is nitroglycerine, a shock- and friction-sensitive explosive. Acids, heat, or mechanical shock may cause explosions.

NITROLIME (156-62-7) Contact with any form of moisture causes decomposition, liberating acetylene and ammonia. Contact with all solvents tested also causes decomposition.

NITRO, LIQUID (98-95-3) Forms explosive mixture with air (flash point 190°F/88°C). Incompatible with nitric acid, nitrogen tetroxide, caustics, ammonia, amines. Attacks some plastics, rubber, and coatings.

1,1′,1′-NITROLOTRI-2-PROPANOL (122-20-3) Incompatible with acids, organic anhydrides, isocyanates, vinyl acetate, acrylates, substituted allyls, alkylene oxides, epichlorohydrin, aldehydes.

NITROMAGNESITE (10377-60-3) A powerful oxidizer. Reacts violently with dimethylformamide, reducing agents, combustibles, organic and easily oxidizable matter.

NITROMETAN (Polish) (75-52-5) Forms explosive mixture with air (flash point 95°F/35°C). Caustics, ammonia, strong acids, amines can cause sensitization, which can cause explosions. Strong oxidizers may cause fire and explosions. Mixtures with hydrocarbons (and other combustible materials) are highly flammable. Contact with metallic oxides may cause decomposition and development of pressure. Attacks some plastics, rubber, and coatings.

NITROMETHANE (75-52-5) Forms explosive mixture with air (flash point 95°F/35°C). Caustics, ammonia, strong acids, amines can cause sensitization, which can cause explosions. Strong oxidizers may cause fire and explosions. Mixtures with hydrocarbons (and other combustible materials) are highly flammable. Contact with metallic oxides may cause decomposition and development of pressure. Attacks some plastics, rubber, and coatings.

1-NITROPENTANE (463-04-7) Forms explosive mixture with air (flash point 0°F/−18°C). Decomposes in light, air, or water; produces oxides of nitrogen. A strong oxidizer; reacts with reducing agents, combustibles, organics, and other strong oxidizers. Corrodes metal with moisture.

2-NITROPHENOL (88-75-5) A strong oxidizer; reacts violently reaction with reducing agents, combustibles, organic matter, chlorosulfonic acid, potassium hydroxide.

3-NITROPHENOL (554-84-7) A strong oxidizer; reacts with combustibles, reducing agents, organics, and other easily oxidizable materials.

4-NITROPHENOL (100-02-7) Combustible solid. A strong oxidizer; reacts with reducing agents, combustibles, organics and other easily oxidizable materials. Incompatible with strong acids, caustics, aliphatic amines, amides.

m-**NITROPHENOL or** *meta*-**NITROPHENOL** (554-84-7) A strong oxidizer; reacts with combustibles, reducing agents, organics, and other easily oxidizable materials.

o-**NITROPHENOL or** *ortho*-**NITROPHENOL** (88-75-5) A strong oxidizer; reacts violently with reducing agents, combustibles, organic matter, chlorosulfonic acid, potassium hydroxide.

p-**NITROPHENOL or** *para*-**NITROPHENOL** (100-02-7) Combustible solid. A strong oxidizer; reacts with reducing agents, combustibles, organics, and other easily oxidizable materials. Incompatible with strong acids, caustics, aliphatic amines, amides.

p-**NITROPHENYLAMINE or** *para*-**NITROPHENYLAMINE** (100-01-06) Strong oxidizers and moisture may cause spontaneous heating. Caustics and heat may cause formation of explosive mixtures. Attacks some plastics, rubber, and coatings. May accumulate static electrical charges, and may cause ignition of its vapors.

NITROPROPANE (108-03-2) Overheating in closed containers may cause violent explosion. Forms explosive mixture with air (flash point 96°F/36°C). Incompatible with aliphatic amines, strong acids, alkalies, strong oxidizers, isocyanates. Contact with some metal oxides may cause decomposition. Mixtures with hydrocarbons are extremely flammable. Attacks some plastics, rubber, and coatings.

1-NITROPROPANE (108-03-2) Overheating in closed containers may cause violent explosion. Forms explosive mixture with air (flash point 96°F/36°C). Incompatible with aliphatic amines, strong acids, alkalies, strong oxidizers, isocyanates. Contact with some metal oxides may cause decomposition. Mixtures with hydrocarbons are extremely flammable. Attacks some plastics, rubber, and coatings.

2-NITROPROPANE (79-46-9) Forms explosive mixture with air (flash point 74°F/24°C). Incompatible with amines, strong acids, alkalies, strong oxidizers. Contact with some metal oxides may cause decomposition. Mixtures with hydrocarbons are extremely flammable. Attacks some plastics, rubber, and coatings.

b-NITROPROPANE or **beta-NITROPROPANE** (79-46-9) Forms explosive mixture with air (flash point 74°F/24°C). Incompatible with amines, strong acids, alkalies, strong oxidizers. Contact with some metal oxides may cause decomposition. Mixtures with hydrocarbons are extremely flammable. Attacks some plastics, rubber, and coatings.

sec-**NITROPROPANE** or *secondary*-**NITROPROPANE** (79-46-9) Forms explosive mixture with air (flash point 74°F/24°C). Incompatible with amines, strong acids, alkalies, strong oxidizers. Contact with some metal oxides may cause decomposition. Mixtures with hydrocarbons are extremely flammable. Attacks some plastics, rubber, and coatings.

NITRO-SIL (7664-41-7) anhydrous (a flammable gas); (1336-21-6) solution in water. Reacts violently with strong oxidizers, acids. Shock-sensitive compounds may be formed with halogens, mercury oxide, silver oxide. Fire and explosions may be caused by trimethylammonium amide, 1-chloro-2,4-dinitrobenzene, *o*-chloronitrobenzene, platinum, selenium difluoride dioxide, boron halides, mercury, chlorine, iodine, bromine, hypochlorites, chlorine bleach, amides, organic anhydrides, isocyanates, vinyl acetate, alkylene oxides, epichlorohydrin, aldehydes. Attacks some coatings, plastics, and rubber. Attacks copper, brass, bronze, aluminum, steel, and their alloys.

NITROSODIMETHYLAMINE (62-75-9) Ultraviolet light, strong oxidizers can cause reactions. Store in dark bottles.

N-NITROSODIMETHYLAMINE (62-75-9) Ultraviolet light, strong oxidizers can cause reactions. Store in dark bottles.

N-NITROSO-N,N-DIMETHYLAMINE (62-75-9) Ultraviolet light, strong oxidizers can cause reactions. Store in dark bottles

NITROSTIGMIN (German) (56-38-2) Combustible liquid. Mixtures with endrin may be explosive. Strong oxidizers may cause fire and explosions. Attacks some plastics, rubber, and coatings.

NITROSTIGMINE (56-38-2) Combustible liquid. Mixtures with endrin may be explosive. Strong oxidizers may cause fire and explosions. Attacks some plastics, rubber, and coatings.

NITROSYL CHLORIDE (696-92-6) Incompatible with water, producing acid solution and toxic red oxides of nitrogen. Corrodes aluminum and other metals in the presence of moisture.

NITROSYL ETHOXIDE (109-95-5) Forms explosive mixture with air (flash point −31°F/−35°C). A powerful oxidizer. Heat above 190°F/88°C can cause explosive decomposition. Incompatible with nitric acid, reducing agents; can cause a violent reaction, fire, or explosions.

2-NITROTOLUENE (88-72-2) Incompatible with strong acids, caustics, ammonia, amines, reducing agents, oxidizers. Attacks some plastics, rubber, and coatings.

3-NITROTOLUENE (99-08-1) Incompatible with strong acids, caustics, ammonia, amines, reducing agents, oxidizers. Attacks some plastics, rubber, and coatings.

4-NITROTOLUENE (99-99-0) Incompatible with strong acids, caustics, ammonia, amines, reducing agents, oxidizers. Attacks some plastics, rubber, and coatings.

m-**NITROTOLUENE** or *meta*-**NITROTOLUENE** (99-08-1) Incompatible with strong acids, caustics, ammonia, amines, reducing agents, oxidizers. Attacks some plastics, rubber, and coatings.

o-**NITROTOLUENE** or *ortho*-**NITROTOLUENE** (88-72-2) Incompatible with strong acids, caustics, ammonia, amines, reducing agents, oxidizers. Attacks some plastics, rubber, and coatings.

p-**NITROTOLUENE or** *para*-**NITROTOLUENE** (99-99-0) Incompatible with strong acids, caustics, ammonia, amines, reducing agents, oxidizers. Attacks some plastics, rubber, and coatings.

NITROTOLUOL (88-72-2) Incompatible with strong acids, caustics, ammonia, amines, reducing agents, oxidizers. Attacks some plastics, rubber, and coatings.

3-NITROTOLUOL (99-08-1) Incompatible with strong acids, caustics, ammonia, amines, reducing agents, oxidizers. Attacks some plastics, rubber, and coatings.

4-NITROTOLUOL (99-99-0) Incompatible with strong acids, caustics, ammonia, amines, reducing agents, oxidizers. Attacks some plastics, rubber, and coatings.

p-**NITROTOLUOL or** *para*-**NITROTOLUOL** (99-99-0) Incompatible with strong acids, caustics, ammonia, amines, reducing agents, oxidizers. Attacks some plastics, rubber, and coatings.

NITROTRICHLOROMETHANE (76-06-2) Can be self-reactive. Fast heating, shock, alkali metals, or alkaline earth may cause explosions. A strong oxidizer; reacts violently with reducing agents, aniline in presence of heat, alcoholic sodium hydroxide, combustible substances, sodium methoxide, propargyl bromide. Liquid attacks some plastics, rubber, and coatings.

NITROUS ACID ETHYL ESTER (109-95-5) Forms explosive mixture with air (flash point −31°F/−35°C). A powerful oxidizer. Heat above 190°F/88°C can cause explosive decomposition. Incompatible with nitric acid, reducing agents; can cause a violent reaction, fire, or explosions.

NITROUS ACID-3-METHYLBUTYL ESTER (110-46-3) Forms explosive mixture with air (flash point 0°F/−18°C). Decomposes in light, air, or water; produces oxides of nitrogen. A strong oxidizer; reacts with reducing agents, combustibles, organics, and other strong oxidizers. Corrodes metals with moisture.

NITROUS ACID, PENTYL ESTER (463-04-7) Forms explosive mixture with air (flash point 0°F/−18°C). Decomposes in light, air, or water; produces oxides of nitrogen. A strong oxidizer; reacts with reducing agents, combustibles, organics, and other strong oxidizers. Corrodes metals with moisture.

NITROUS ACID, SODIUM SALT (7632-00-0) A strong oxidizer; reacts violently with reducing agents and combustibles. Reacts with acids, ammonium salts + heat, butadiene, cyanides, lithium, phthalic acid, thiosulfate, sodium amide, and many other substances. Under certain conditions the chemical can also be a strong reducing agent.

NITROUS ETHER (109-95-5) Forms explosive mixture with air (flash point −31°F/−35°C). A powerful oxidizer. Heat above 190°F/88°C can cause explosive decomposition. Incompatible with nitric acid, reducing agents; can cause a violent reaction, fire, or explosions.

NITROUS ETHYL ETHER (109-95-5) Forms explosive mixture with air (flash point −31°F/−35°C). A powerful oxidizer. Heat above 190°F/88°C can cause explosive decomposition. Incompatible with nitric acid, reducing agents; can cause a violent reaction, fire, or explosions.

NITROUS FUMES (7697-37-2) A strong oxidizer that can react violently with reducing agents, combustible materials. Incompatible with many substances, including acrylates, aliphatic amines, alcohols, aldehydes, alkanolamines, alkylene oxides, anion exchange resins, aromatic amines, amides, bases, cresols, cyanides, cyclic ketones, epichlorohydrin, glycols, isocyanates, ketones, oleum, organic anhydrides, phenols, substituted allyls, sulfuric acid, strong oxidizers, terpenes. Attacks most metals, and some plastics, rubber, and coatings.

NITROUS OXIDE (10024-97-2) May form explosive mixture with flammable gases. Nonflammable but supports combustion; as temperature increases, it becomes both a strong oxidizer and self-reactive. High temperatures may cause explosive decomposition. Incompatible with aluminum, anhydrous ammonia (may cause explosion), boron, carbon monoxide, hydrazine, hydrogen, hydrogen sulfide, lithium hydride, phosphine, sodium.

NITROUS(2+) OXIDE or NITROUS(II) OXIDE (10102-43-9) A strong oxidizer. Avoid contact with all combustible materials, chlorinated hydrocarbons, ammonia, carbon disulfide, many metals, fluorine, and ozone. Attacks some plastics, rubber, and coatings.

NITROX (298-00-0) Mixtures with magnesium may be explosive.

NITROXANTHIC ACID (88-89-1) Dry material is explosive. Impact-sensitive; protect from shock. Copper, lead, zinc, and other metals, or their salts can form salts that are initiators and much more sensitive to shock than this chemical. Shock-sensitive salts include ammonium salts and calcium salts; the calcium salt may be produced when picric acid comes into contact with plaster and concrete. May accumulate static electrical charges, and may cause explosion. Aqueous solution is a strong oxidizer and a strong acid; reacts violently with reducing agents, combustibles, organics and easily oxidized materials, caustics, aluminum, and other metal powders. Attacks metals.

NITRYL HYDROXIDE (7697-37-2) A strong oxidizer that can react violently with reducing agents, combustible materials. Incompatible with many substances, including acrylates, aliphatic amines, alcohols, aldehydes, alkanolamines, alkylene oxides, anion exchange resins, aromatic amines, amides, bases, cresols, cyanides, cyclic ketones, epichlorohydrin, glycols, isocyanates, ketones, oleum, organic anhydrides, phenols, substituted allyls, sulfuric acid, strong oxidizers, terpenes. Attacks most metals, and some plastics, rubber, and coatings.

NIUIF-100 (56-38-2) Combustible liquid. Mixtures with endrin may be explosive. Strong oxidizers may cause fire and explosions. Attacks some plastics, rubber, and coatings.

NMP (872-50-4) Forms explosive mixture with air (flash point 7°F/−14°C). A strong base. Incompatible with strong acids, organic anhydrides, isocyanates, aldehydes, light metals, flammable and porous materials. Attacks some plastics, rubber, and coatings.

NOBECUTAN (137-26-8) Combustible solid (flash point 192°F/89°C). Strong oxidizers may cause fire and explosions; contact with strong acid or oxidizable materials produces toxic gases.

NOBLEN (9003-07-0) Incompatible with strong acids, strong oxidizers, chlorine, potassium permanganate.

NO BUNT LIQUID (118-74-1) Reacts violently with strong oxidizers, formyldimethylamine.

NOFLAMOL (generic CAS for PCBs 1336-36-3) Incompatible with strong oxidizers, strong acids.

NOGOS (62-73-7) Attacks some plastics, rubber, and coatings.

NOKTAN (Polish) (111-65-9) Forms explosive mixture with air (flash point 56°F/13.3°C). Strong oxidizers may cause fire and explosions. Attacks some plastics, rubber, and coatings. May accumulate static electrical charges, and may cause ignition of its vapors.

NOMERSAN (137-26-8) Combustible solid (flash point 192°F/89°C). Strong oxidizers may cause fire and explosions; contact with strong acid or oxidizable materials produces toxic gases.

NONANE (111-84-2) Forms explosive mixture with air (flash point 88°F/31°C). Strong oxidizers may cause fire and explosions. May accumulate static electrical charges, and may cause ignition of its vapors.

N-NONANE (111-84-2) Forms explosive mixture with air (flash point 88°F/31°C). Strong oxidizers may cause fire and explosions. May accumulate static electrical charges, and may cause ignition of its vapors.

NONANE AND TRIMETHYLBENZENE MIXTURE (85:15) (8052-41-3) Forms explosive mixture with air (flash point 102°F to 140°F/38.7°C to 60°C). Incompatible with nitric acid, strong oxidizers. Attacks some plastics, rubber, and coatings.

1-NONANECARBOXYLIC ACID (334-48-5) Incompatible with sulfuric acid, caustics, ammonia, aliphatic amines, alkanolamines, isocyanates, alkylene oxides, epichlorohydrin. Attacks most common metals.

NONANOL (143-08-8) Forms explosive mixture with air (flash point 165°F/74°C). Incompatible with strong acids, caustics, aliphatic amines, isocyanates, strong oxidizers.

1-NONANOL (143-08-8) Forms explosive mixture with air (flash point 165°F/74°C). Incompatible with strong acids, caustics, aliphatic amines, isocyanates, strong oxidizers.

NONAN-1-OL (143-08-8) Forms explosive mixture with air (flash point 165°F/74°C). Incompatible with strong acids, caustics, aliphatic amines, isocyanates, strong oxidizers.

NONANOL ACETATE (143-13-5) Incompatible with strong acids, nitrates.

2-NONANONE (821-55-6) Forms explosive mixture with air (flash point 140°F/60°C). Incompatible with strong acids, aliphatic amines, oxidizers.

NONAN-2-ONE (821-55-6) Forms explosive mixture with air (flash point 140°F/60°C). Incompatible with strong acids, aliphatic amines, oxidizers.

NONENE (139-01-4) Forms explosive mixture with air (flash point 75°F/24°C). Incompatible with sulfuric acid, nitric acid, oxidizers.

1-NONENE (27214-95-8) Forms explosive mixture with air (flash point 78°F/26°C). Incompatible with sulfuric acid, nitric acid, oxidizers. Attacks some plastics, rubber, and coatings. May accumulate static electrical charges, and may cause ignition of its vapors.

NONENE (NONLINEAR) (139-01-4) Forms explosive mixture with air (flash point 75°F/24°C). Incompatible with sulfuric acid, nitric acid, oxidizers.

NONYL ACETATE (143-13-5) Incompatible with strong acids, nitrates.

N-NONYL ACETATE (143-13-5) Incompatible with strong acids, nitrates.

602

NONYL ALCOHOL (143-08-8) Forms explosive mixture with air (flash point 165°F/74°C). Incompatible with strong acids, caustics, aliphatic amines, isocyanates, strong oxidizers.

N-NONYL ALCOHOL (143-08-8) Forms explosive mixture with air (flash point 165°F/74°C). Incompatible with strong acids, caustics, aliphatic amines, isocyanates, strong oxidizers.

sec-**NONYL ALCOHOL** (108-82-7) Forms explosive mixture with air (flash point 165°F/74°C). Incompatible with strong acids, caustics, aliphatic amines, isocyanates, strong oxidizers.

NONYL PHENOL (25154-52-3) Incompatible with strong acids, caustics, aliphatic amines, amides, oxidizers.

NONYLCARBINOL (112-30-1) Forms explosive mixture with air (flash point 180°F/82°C). Incompatible with strong acids, caustics, aliphatic amines, isocyanates, strong oxidizers.

NONYLENE (27214-95-8) Forms explosive mixture with air (flash point 78°F/26°C). Incompatible with sulfuric acid, nitric acid, oxidizers. Attacks some plastics, rubber, and coatings. May accumulate static electrical charges, and may cause ignition of its vapors.

1-NONYLENE (27214-95-8) Forms explosive mixture with air (flash point 78°F/26°C). Incompatible with sulfuric acid, nitric acid, oxidizers. Attacks some plastics, rubber, and coatings. May accumulate static electrical charges, and may cause ignition of its vapors.

N-NONYLETHYLENE (N/A) Incompatible with strong oxidizers, strong acids.

NONYLHYDRIDE (111-84-2) Forms explosive mixture with air (flash point 88°F/31°C). Strong oxidizers may cause fire and explosions. May accumulate static electrical charges, and may cause ignition of its vapors.

NONYLPHENOL (25154-52-3) Incompatible with strong acids, caustics, aliphatic amines, amides, strong oxidizers. May accumulate static electrical charges, and may cause ignition of its vapors.

NO-PEST (62-73-7) Attacks some plastics, rubber, and coatings.

NO-PEST STRIP (62-73-7) Attacks some plastics, rubber, and coatings.

2-NORBORNENE, 5-ETHYLIDENE- (16219-75-3) Reacts violently with oxygen and strong oxidizers. Forms explosive mixture with air (flash point 100°F/38°C). May accumulate static electrical charges, and may cause ignition of its vapors. Inhibit peroxide formation with *tert*-butyl catechol.

NORCAMPHOR, SYNTHETIC CAMPHOR (76-22-2) Forms explosive mixture with air (flash point 155°F/66°C). Violent, possibly explosive, reaction occurs with strong oxidizers. May accumulate static electrical charges, and may cause ignition of its vapors.

NORKOOL (107-21-1) Incompatible with strong acids, caustics, aliphatic amines, isocyanates, chlorosulfonic acid, oleum, strong oxidizers.

NORMAL LEAD ACETATE (301-04-2) Contact with strong acids produces acetic acid. Incompatible with bases, ammonia, amines, cresols, isocyanates, alkylene oxides, epichlorohydrin, phenols, sulfites. Reacts violently with bromates, strong oxidizers.

NORMENTHANE (696-29-7) Forms explosive mixture with air (flash point 96°F/36°C). Incompatible with strong oxidizers, acids.

NORMERSAN (137-26-8) Combustible solid (flash point 192°F/89°C). Strong oxidizers may cause fire and explosions; contact with strong acid or oxidizable materials produces toxic gases.

NORVAL (119-36-8) Forms explosive mixture with air (flash point 205°F/96°C). Incompatible with strong acids, nitrates, oxidizers.

NORVALAMINE (109-73-9) Forms explosive mixture with air (flash point 10°F/−12°C). May accumulate static electrical charges, and may cause ignition of its vapors. Incompatible with acids, organic anhydrides, isocyanates, vinyl acetate, acrylates, substituted allyls, alkylene oxides, epichlorohydrin, ketones, aldehydes, alcohols, glycols, phenols, cresols, caprolactam solution, strong oxidizers. In the presence of moisture, corrodes light metals and copper on contact.

NORWAY SALTPETER (6484-52-2) A strong oxidizer. Reducing agents, combustibles, organic materials, finely divided metals may produce explosive mixtures or cause fire and explosions.

NORWAY SALTPETER (13473-90-0) A strong oxidizer. Reacts violently with combustibles, organics, reducing agents. Aqueous solution is acidic. Attacks metals in the presence of moisture.

NORWEGIAN SALTPETER (10124-37-5) A strong oxidizer. Incompatible with combustible materials, reducing agents, organics, and other oxidizable materials, chemically active metals, aluminum nitrate, ammonium nitrate.

NO SCALD (122-39-4) Incompatible with strong acids, aldehydes, organic anhydrides, isocyanates, oxidizers, hexachloromelamine, trichloromelamine.

NOURITHION (56-38-2) Combustible liquid. Mixtures with endrin may be explosive. Strong oxidizers may cause fire and explosions. Attacks some plastics, rubber, and coatings.

NOVIGAM (58-89-9) Not combustible, but may be dissolved in a combustible solvent. If solvent comes in contact with oxidizers, fire and explosions may result.

NOWERGAN (137-26-8) Combustible solid (flash point 192°F/89°C). Strong oxidizers may cause fire and explosions; contact with strong acid or oxidizable materials produces toxic gases.

1-NP (108-03-2) Overheating in closed containers may cause violent explosion. Forms explosive mixture with air (flash point 96°F/36°C). Incompatible with aliphatic amines, strong acids, alkalies, strong oxidizers, isocyanates. Contact with some metal oxides may cause decomposition. Mixtures with hydrocarbons are extremely flammable. Attacks some plastics, rubber, and coatings.

2-NP (79-46-9) Forms explosive mixture with air (flash point 74°F/24°C). Incompatible with amines, strong acids, alkalies, strong oxidizers. Contact with some metal oxides may cause decomposition. Mixtures with hydrocarbons are extremely flammable. Attacks some plastics, rubber, and coatings.

NSC 423 (94-75-7) Decomposes in sunlight. Incompatible with strong oxidizers; may cause fire and explosions.

NSC 763 (67-68-5) Forms explosive mixture with air (flash point 203°F/95°C). Reacts violently with oxidizers. Reacts with ethanoyl chloride, boron compounds, halides, metal alkoxides, oxidizers.

NSC 1532 (51-28-5) Explosion may be caused by heat, friction, or shock. Contact with reducing agents, combustibles may cause fire and explosions. Produces explosive salts with ammonia or strong bases. May accumulate static electrical charges, and may cause ignition of its vapors.

NSC 1771 (137-26-8) Combustible solid (flash point 192°F/89°C). Strong oxidizers may cause fire and explosions; contact with strong acid or oxidizable materials produces toxic gases.

NSC 2752 (110-17-8) Reacts with strong oxidizers. Incompatible with sulfuric acid, caustics, ammonia, amines, isocyanates, alkylene oxides, epichlorohydrin.

NSC 3072 (26628-22-8) Incompatible with acids, some metals (i.e, lead, copper, silver, mercury). Forms explosion-sensitive compounds.

NSC 3138 (127-19-5) Forms explosive mixture with air (flash point 158°F/70°C). Incompatible with non-oxidizing mineral acids, strong acids, ammonia, isocyanates, phenols, cresols, halogenated compounds, above 185°F/85°C. Attacks some plastics, rubber, and coatings.

NSC 5356 (68-12-2) Forms explosive mixture with air (flash point 136°F/58°C). Contact with carbon tetrachloride and other halogenated compounds, particularly in presence of iron or strong oxidizers may cause fire and explosions. Reacts vigorously with alkylaluminums. Incompatible with non-oxidizing mineral acids, strong acids, chlorinated hydrocarbons, isocyanates, nitrates, organic nitrates, phenols, cresols, ammonia, chromic anhydride, magnesium nitrate, methylene diisocyante, phosphorus trioxide, triethyaluminum. Attacks some plastics, rubber, and coatings.

NSC 6738 (62-73-7) Attacks some plastics, rubber, and coatings.

NSC 8819 (107-02-8) Forms explosive mixture with air (flash point −15°F/−26°C). Unstable and very reactive. In storage can produce heat- and shock-sensitive compounds. Unless inhibited (usually with hydroquinone), readily forms explosive peroxides. Able to polymerize. A strong reducing agent. Reacts violently with oxidizers, strong acids, caustics, amines, 2-aminoethanol, ammonia, ammonium hydroxide, ethylene diamine, ethyleneimine, hydroxides, metal salts, oxidizers, sulfur dioxide, thiourea. Attacks metals: cadmium and zinc. May accumulate static electrical charges, and may cause ignition of its vapors.

NSC 54739 (2238-07-5) Forms explosive mixture with air (flash point 147°F/64°C). Contact with strong oxidizers may cause fire and explosions. Ethers, as a class, tend to form peroxides upon contact with air and exposure to light. Attacks some forms of plastics, coatings, and rubber.

NSC 405124 (87-90-1) A powerful oxidizer. Forms explosive material with nitrogen compounds. Contact with organic materials or reducing agents may cause fire.

NTM (131-11-3) Incompatible with strong alkalies, strong acids, nitrates, oxidizers.

NTO (10544-72-6); (10102-44-0 CAS for nitrogen dioxide). A powerful oxidizer. Water contact produces nitric acid and nitric oxide. Incompatible with nitric oxide; reacts with air, producing more nitrogen tetroxide. Explosive reaction occurs with alcohols, fuels and petroleum, nitrobenzene, organic matter, anhydrous ammonia, chlorinated hydrocarbons, toluene. Reacts vigorously with combustibles, reducing agents. Attacks metals in presence of moisture.

NUCHAR 722 activated (64365-11-3); **purified** (7440-44-0). Incompatible with strong oxidizers, strong acids, oxides, unsaturated oils.

NUCIDOL (333-41-5) Incompatible with water, copper-containing compounds, oxidizers, acids, or bases.

NUJOL (8012-95-1) Oxidizers may cause fire and explosions. Incompatible with nitric acid. May accumulate static electrical charges, and may cause ignition of its vapors.

NULLAPON B ACID (60-00-4) Incompatible with sulfuric acid, bases, ammonia, aliphatic amines, alkanolamines, isocyanates, alkylene oxides, epichlorohydrin.

NULLAPON BF ACID (60-00-4) Incompatible with sulfuric acid, bases, ammonia, aliphatic amines, alkanolamines, isocyanates, alkylene oxides, epichlorohydrin.

NUOPLAZ (119-06-2) Incompatible with strong acids, nitrates, oxidizers.

NURELLE (99-95-4) Contact with strong oxidizers may cause fire and explosions.

NUVA (62-73-7) Attacks some plastics, rubber, and coatings.

NUX-VOMICA (57-24-9) Contact with strong oxidizers may cause fire and explosions.

NYLON X 1051 (105-60-2) Contact with strong oxidizers may cause fire and explosions.

- O -

OBELINE PICRATE (131-74-8) A self-reactive explosive. Heat above 250°F/121°C can cause explosions. A powerful oxidizer. Contact with metal, concrete, or plaster causes formation of salts that are more shock-sensitive than ammonium picrate. Water increases rate of reactivity with metals.

OBSTON (119-36-8) Forms explosive mixture with air (flash point 205°F/96°C). Incompatible with strong acids, nitrates, oxidizers.

OCBM (698-41-1) Incompatible with strong oxidizers, heat.

OCHRE (1309-37-1) Contact with hydrogen peroxide, ethylene oxide, calcium hypochlorite will cause explosion. Reacts violently with powdered aluminum, hydrazine, hydrogen trisulfide.

1,2,4,5,6,7,8,8-OCTACHLOOR-3A,4,7,7A-TETRAHYDRO-4,7-ENDO-METHANO-INDAAN (Dutch) (57-74-9) Attacks some plastics, rubber, and coatings. Contact with strong oxidizers may cause fire and explosions. Attacks some plastics, rubber, and coatings.

OCTACHLOR (57-74-9) Contact with strong oxidizers may cause fire and explosions. Attacks some plastics, rubber, and coatings.

OCTACHLOROCAMPHENE (8001-35-2) Reacts with oxidizers, with a risk of fire or explosions. Attacks metals in the presence of moisture.

OCTACHLORODIHYDRODICYCLOPENTADIENE (57-74-9) Contact with strong oxidizers may cause fire and explosions. Attacks some plastics, rubber, and coatings.

1,2,4,5,6,7,8,8-OCTACHLORO-2,3,3A,4,7,7A-HEXAHYDRO-4,7-METHANOINDENE (57-74-9) Contact with strong oxidizers may cause fire and explosions. Attacks some plastics, rubber, and coatings.

1,2,4,5,6,7,8,8-OCTACHLORO-2,3,3A,4,7,7A-HEXAHYDRO-4,7-METHANO-1H-INDENE (57-74-9) Contact with strong oxidizers may cause fire and explosions. Attacks some plastics, rubber, and coatings.

1,2,4,5,6,7,8,8-OCTACHLORO-3A,4,7,7A-HEXAHYDRO-4,7-METHYLENE INDANE (57-74-9) Contact with strong oxidizers may cause fire and explosions. Attacks some plastics, rubber, and coatings.

OCTACHLORO-4,7-METHANOHYDROINDANE (57-74-9) Contact with strong oxidizers may cause fire and explosions. Attacks some plastics, rubber, and coatings.

OCTACHLORO-4,7-METHANOTETRAHYDROINDANE (57-74-9) Contact with strong oxidizers may cause fire and explosions. Attacks some plastics, rubber, and coatings.

1,2,4,5,6,7,8,8-OCTACHLORO-4,7-METHANO-3A,4,7,7A-TETRAHYDROINDANE (57-74-9) Contact with strong oxidizers may cause fire and explosions. Attacks some plastics, rubber, and coatings.

OCTACHLORONAPHTHALENE (2234-13-1) Strong oxidizers may cause fire and explosions.

1,2,3,4,5,6,7,8-OCTACHLORONAPHTHALENE (2234-13-1) Strong oxidizers may cause fire and explosions.

1,2,4,5,6,7,8,8-OCTACHLORO-3A,4,7,7A-TETRAHYDRO-4,7-METHANOINDAN (57-74-9) Contact with strong oxidizers may cause fire and explosions. Attacks some plastics, rubber, and coatings.

1,2,4,5,6,7,8,8-OCTACHLORO-3A,4,7,7A-TETRAHYDRO-4,7-METHANOINDANE (57-74-9) Contact with strong oxidizers may cause fire and explosions. Attacks some plastics, rubber, and coatings.

OCTACHLORO-4,7,8,9-TETRAHYDRO-4,7-METHYLENEIN-DANE (57-74-9) Contact with strong oxidizers may cause fire and explosions. Attacks some plastics, rubber, and coatings.

1,2,4,5,6,7,8,8-OCTACHLOR-3A,4,7,7A-TETRAHYDRO-4,7-ENDO-METHANO-INDAN (German) (57-74-9) Contact with strong oxidizers may cause fire and explosions. Attacks some plastics, rubber, and coatings.

OCTADECANOIC ACID (57-11-4) Incompatible with strong oxidizers, sulfuric acid, caustics, ammonia, amines, isocyanates, alkylene oxides, epichlorohydrin. Attacks chemically active metals.

1-OCTADECANOL, CRUDE (112-92-5) Incompatible with strong acids, caustics, aliphatic amines, isocyanates.

CIS-9-OCTADECENOIC ACID (112-80-1) Incompatible with strong oxidizers, perchloric acid, and aluminum powder; may cause explosions. Corrodes aluminum.

N-OCTADECYLIC ACID (57-11-4) Incompatible with strong oxidizers, sulfuric acid, caustics, ammonia, amines, isocyanates, alkylene oxides, epichlorohydrin. Attacks chemically active metals.

1,6-OCTADIENE,7-METHYL-3-METHYLENE (123-35-3) Forms explosive mixture with air (flash point 103°F/40°C). Incompatible with strong oxidizers, sulfuric acid, nitric acid.

OCTA-KLOR (57-74-9) Contact with strong oxidizers may cause fire and explosions. Attacks some plastics, rubber, and coatings.

OCTALENE (309-00-2) Incompatible with concentrated mineral acids, acid catalysts, acid oxidizing agents, phenols, reactive metals.

OCTALOX (60-57-1) Incompatible with concentrated mineral acids, acid catalysts, acid oxidizing agents, phenols, reactive metals.

1-OCTANAL (124-13-0) Forms explosive mixture with air (flash point 125°F/52°C). Incompatible with strong oxidizers, caustics, ammonia, aliphatic amines, alkanolamines, aromatic amines, strong oxidizers.

OCTANALDEHYDE (124-13-0) Forms explosive mixture with air (flash point 125°F/52°C). Incompatible with strong oxidizers, caustics, ammonia, aliphatic amines, alkanolamines, aromatic amines, strong oxidizers.

OCTAN AMYLU (Polish) (628-63-7) Incompatible with strong alkalies, strong acids, nitrates, strong oxidizers. Attacks some plastics, coatings, and rubber.

OCTANE (111-65-9) Forms explosive mixture with air (flash point 56°F/13.3°C). Strong oxidizers may cause fire and explosions. Attacks some plastics, rubber, and coatings. May accumulate static electrical charges, and may cause ignition of its vapors.

n-**OCTANE or** *normal*-**OCTANE** (111-65-9) Forms explosive mixture with air (flash point 56°F/13.3°C). Strong oxidizers may cause fire and explosions. Attacks some plastics, rubber, and coatings. May accumulate static electrical charges, and may cause ignition of its vapors.

OCTAN N-BUTYLU (Polish) (123-86-4) Forms explosive mixture with air (flash point 72°F/22°C). Reacts with water on standing to produce acetic acid and *n*-butyl alcohol. Reacts violently with strong oxidizers and potassium-*tert*-butoxide. Incompatible with caustics, strong acids, nitrates. Dissolves rubber, many plastics, resins, and some coatings. May accumulate static electrical charges, and may cause ignition of its vapors.

OCTAN ETYLU (Polish) (141-78-6) Will hydrolyze on standing producing acetic acid and ethyl alcohol. This reaction is greatly accelerated by bases (alkalies). Forms explosive mixture with air (flash point 24°F/−4.4°C). Incompatible with strong acids, nitrates, oxidizers, chlorosulfonic acid, lithium aluminum hydride, oleum.

OCTAN MEDNATY (Czech) (142-71-2) Incompatible with strong acids, nitrates.

OCTAN METYLU (Polish) (79-20-9) Forms explosive mixture with air (flash point 14°F/−10°C). Incompatible with strong acids, nitrates, oxidizers, bases. Attacks some plastics. May accumulate static electrical charges, and may cause ignition of its vapors.

OCTANOIC ACID (124-07-2) Forms explosive mixture with air (flash point 56°F/13°C). Reacts violently with strong oxidizers. May accumulate static electrical charges, and may cause ignition of its vapors.

OCTANOL (111-87-5) Forms explosive mixture with air (flash point 178°F/81°C). Incompatible with strong acids, caustics, aliphatic amines, isocyanates, strong oxidizers.

1-OCTANOL (111-87-5) Forms explosive mixture with air (flash point 178°F/81°C). Incompatible with strong acids, caustics, aliphatic amines, isocyanates, strong oxidizers.

2-OCTANOL (123-96-6) Forms explosive mixture with air (flash point 190°F/88°C). Incompatible with strong acids, caustics, aliphatic amines, isocyanates, oxidizers.

n-**OCTANOL or** *normal*-**OCTANOL** (111-87-5) Forms explosive mixture with air (flash point 178°F/81°C). Incompatible with strong acids, caustics, aliphatic amines, isocyanates, strong oxidizers.

3-OCTANONE (541-85-5) Forms explosive mixture with air (flash point 110°F/43°C). Incompatible with sulfuric acid, nitric acid, aliphatic amines, oxidizers.

OCTAN PROPYLU (Polish) (109-60-4) Forms explosive mixture with air (flash point 58°F/14°C). Incompatible with strong acids, nitrates, strong acids, strong oxidizers. Attacks some plastics, rubber, and coatings. May accumulate static electrical charges, and may cause ignition of its vapors.

OCTAN WINYLU (Polish) (108-05-4) Forms explosive mixture with air (flash point 18°F/−8°C). Polymerizes readily if not inhibited; heat can initiate reaction. Reacts violently with oxidizers. Incompatible with nonoxidizing mineral acids, strong acids, ammonia, aliphatic amines, alkanolamines. Also reacts with 2-aminoethanol, chlorosulfonic acid, ethylene diamine, ethyeneimine, ozone, oleum, peroxides. May accumulate static electrical charges, and may cause ignition of its vapors.

1-OCTENE (111-66-0) Forms explosive mixture with air (flash point 70°F/21°C). Incompatible with strong acids, strong oxidizers.

OCTILIN (111-87-5) Forms explosive mixture with air (flash point 178°F/81°C). Incompatible with strong acids, caustics, aliphatic amines, isocyanates, strong oxidizers.

n-**OCTOIC ACID** or *normal*-**OCTOIC ACID** (124-07-2) Forms explosive mixture with air (flash point 56°F/13°C). Reacts violently with strong oxidizers. May accumulate static electrical charges, and may cause ignition of its vapors.

OCTOIL (117-84-0) Water contact causes foaming. Incompatible with strong acids, nitrates.

OCTOWY BEZWODNIK (Polish) (108-24-7) Forms explosive mixture with air (flash point 121°F/50°C). Water contact produces acetic acid and considerable heat. Reacts violently with potassium hydroxide, sodium hydroxide, alcohols amines. Incompatible with acids, ammonia, aniline, chromic anhydride, ethylene diamine, glycerol, hydrogen peroxide, nitrogen tetroxide, oxidizers, permanganates, sodium peroxide. Attacks some coatings and some plastics and rubber. Corrodes iron and steel. May accumulate static electrical charges, and may cause ignition of its vapors.

OCTOWY EHYD (Polish) (75-07-0) Slowly polymerizes to paraldehyde. Explodes when mixed with iodine. Contact with strong bases can cause explosive polymerization. A strong reducing agent; reacts violently with combustibles, strong acids, caustics, ammonia, aliphatic amines, alkanolamines, aromatic amines, organic substances, halogens, oxidizers. Forms explosive peroxides with air. May dissolve rubber. May accumulate static electrical charges, and may cause ignition of its vapors.

OCTOWY KWAS (Polish) (64-19-7) Vapor forms explosive mixture with air (flash point 105°F/40°C). Reacts violently with oxidizers and bases. Incompatible with strong acids, aliphatic amines, alkanolamines, isocyanates, alkylene oxides, epichlorohydrin, acetaldehyde, 2-aminoethanol, ammonia, ammonium nitrate, chlorosulfonic acid, chromic acid, ethylene diamine, ethyleneimine, perchloric acid, permanganates, phosphorus isocyanate, phosphorus trichloride, potassium *tert*-butoxide, xylene. Attacks cast iron and other metals, producing flammable hydrogen gas.

OCTYL ACETATE (103-09-3) Forms explosive mixture with air (flash point 160°F/71°C. Strong oxidizers may cause fire or explosions.

OCTYL ACRYLATE (103-11-7) Sunlight, heat, contaminants or peroxides can cause polymerization. Reacts violently with strong oxidizers, with risk of fire and explosions. Forms explosive mixture with air (flash point 180°F/82°C). Incompatible with strong acids, aliphatic amines, alkanolamines.

OCTYL ADIPATE (103-23-1) Incompatible with strong acids, nitrates, oxidizers.

OCTYL ALCOHOL (104-76-7) Forms explosive mixture with air (flash point 175°F/79°C. Incompatible with strong acids, caustics, aliphatic amines, isocyanates, oxidizers. May accumulate static electrical charges, and may cause ignition of its vapors.

n-**OCTYL ALCOHOL** or *normal*-**OCTYL ALCOHOL** (111-87-5) Forms explosive mixture with air (flash point 178°F/81°C). Incompatible with strong acids, caustics, aliphatic amines, isocyanates, strong oxidizers.

OCTYL ALDEHYDE (123-05-7) Forms explosive mixture with air (flash point 112°F/44°C). Reacts violently with oxidizers. May ignite spontaneously when spilled on clothing or other absorbent materials. Under certain conditions ignites spontaneously with air. Incompatible with strong acids, caustics, ammonia, amines. Forms unstable peroxides on contact with air.

n-**OCTYL ALDEHYDE** or *normal*-**OCTYL ALDEHYDE** (124-13-0) Forms explosive mixture with air (flash point 125°F/52°C). Incompatible with strong oxidizers, caustics, ammonia, aliphatic amines, alkanolamines, aromatic amines, strong oxidizers.

OCTYL ALDEHYDES (124-13-0) Form explosive mixture with air (flash point 125°F/52°C). Incompatible with strong oxidizers, caustics, ammonia, aliphatic amines, alkanolamines, aromatic amines, strong oxidizers.

OCTYL CARBINOL (143-08-8) Forms explosive mixture with air (flash point 165°F/74°C). Incompatible with strong acids, caustics, aliphatic amines, isocyanates, strong oxidizers.

alpha-OCTYLENE (111-66-0) Forms explosive mixture with air (flash point 70°F/21°C). Incompatible with strong acids, strong oxidizers.

OCTYL EPOXY TALLATE (N/A) Incompatible with strong acids, nitrates, oxidizers.

OCTYL NITRATES (N/A) Incompatible with nonoxidizing mineral acids, sulfuric acid, nitric acid, nitrates, oxidizers.

para-**OCTYL PHENOL** (1806-6-4) Incompatible with strong acids, caustics, aliphatic amines, amides, oxidizers. May accumulate static electrical charges, and may cause ignition of its vapors.

OCTYL PHTHALATE (117-81-7) Incompatible with strong acids, strong alkalies, nitrates, oxidizers.

ODB (95-50-1) Forms explosive mixture with air (flash point 151°F/66°C). Incompatible with strong oxidizers, hot aluminum, or aluminum alloy. Attacks some plastics, rubber, and coatings.

ODCB (95-50-1) Forms explosive mixture with air (flash point 151°F/66°C). Incompatible with strong oxidizers, hot aluminum, or aluminum alloy. Attacks some plastics, rubber, and coatings.

OIL MIST, MINERAL (8012-95-1) Oxidizers may cause fire and explosions. Incompatible with nitric acid. May accumulate static electrical charges, and may cause ignition of its vapors.

OIL OF BITTER ALMOND (100-52-7) Forms explosive mixture with air (flash point 145°F/63°C). A strong reducing agent. Forms peroxides with air. Reacts violently with strong acids, aluminum, caustics, ammonia, amines, iron, strong oxidizers. Attacks some plastics, rubber, and coatings.

OIL OF CASHEW NUTSHELL (8001-24-7) Incompatible with sulfuric acid, caustics, ammonia, amines, isocyanates, alkylene oxides, epichlorohydrin.

OIL OF MIRBANE (98-95-3) Forms explosive mixture with air (flash point 190°F/88°C). Incompatible with nitric acid, nitrogen tetroxide, caustics, ammonia, amines. Attacks some plastics, rubber, and coatings.

OIL OF MYRBANE (98-95-3) Forms explosive mixture with air (flash point 190°F/88°C). Incompatible with nitric acid, nitrogen tetroxide, caustics, ammonia, amines. Attacks some plastics, rubber, and coatings.

OIL OF NIOBE (93-58-3) Forms explosive mixture with air (flash point 181°F/83°C). Incompatible with strong acids, nitrates, oxidizers.

OIL OF PALMA CHRISTI (8001-79-4) Incompatible with strong acids, oxidizers, nitrates.

OIL OF PINE (8002-09-3) Forms explosive mixture with air (flash point 138°F/59°C). Incompatible with strong acids, strong oxidizers.

OIL OF TURPENTINE (8006-64-2) Forms explosive mixture with air (flash point 95°F/35°C). Incompatible with strong oxidizers, especially chlorine; may cause fire and explosions. Incompatible with strong acids, chromic anhydride, chromyl chloride, hexachloromelamine, stannic chloride. Attacks ordinary rubber.

OIL OF VITRIOL (7664-93-9) A strong oxidizer that can react violently with risk of fire and explosion with many substances, including reducing agents, organic and combustible substances, and bases. Incompatible with nonoxidizing mineral acids, organic acids, bases, acrylates, aldehydes, alcohols, alkylene oxides, ammonia, aliphatic amines, alkanolamines, aromatic amines, amides, chlorates, epichlorohydrin, fulminates, glycols, isocyanates, ketones, metals (powdered), organic anhydrides, perchlorates, picrates, substituted allyls, phenols, cresols, water, acetic anhydride, acetone cyanhydrin, acetonitrile, acrolein, acrylonitrile, allyl alcohol, allyl chloride, 2-aminoethanol, ammonium hydroxide, aniline, bromine pentafluoride, n-butyraldehyde, caprolactam solution, carbides, cesium acetylene carbide, chlorine trifluoride, chlorosulfonic acid, cuprous nitride, diisobutylene, ethylene cyanohydrin, ethylene diamine, ethylene glycol, ethyleneimine, hydrochloric acid, iodine heptafluoride, iron, isoprene, lithium silicide, mercuric nitride, mesityl oxide, nitric acid, p-nitrotoluene, perchloric acid, phosphorus, potassium *tert*-butoxide, potassium chlorate, potassium permanganate, propiolactone (beta-), propylene oxide, pyridine, rubidium acetylene, silver permanganate, sodium, sodium carbonate, sodium chlorate, sodium hydroxide, styrene monomer, vinyl acetate. Attacks most metals, and some plastics, rubber, and coatings.

OIL, PALM (8001-75-3) Incompatible with strong acids, oxidizers.

OILS: BABASSU (N/A) Incompatible with strong acids, oxidizers, nitrates.

OILS: CLARIFIED (64741-62-4) Oxidizers may cause fire and explosions. Incompatible with nitric acid. May accumulate static electrical charges, and may cause ignition of its vapors.

OILS: COAL (8008-20-6) Oxidizers may cause fire and explosions. Incompatible with nitric acid. May accumulate static electrical charges, and may cause ignition of its vapors.

OILS: CRUDE (8002-05-9) Oxidizers may cause fire and explosions. Incompatible with nitric acid. May accumulate static electrical charges, and may cause ignition of its vapors.

OILS: DIESEL (64741-44-2) Forms explosive mixture with air (flash point approximately 151°F/55°C). Incompatible with nitric acid. Strong oxidizers may cause fire and explosions. May accumulate static electrical charges, and may cause ignition of its vapors.

OILS, DIESEL FUEL: 1-D (68334-30-5) Oxidizers may cause fire and explosions. Incompatible with nitric acid. May accumulate static electrical charges, and may cause ignition of its vapors.

OILS, DIESEL FUEL: 2-D (68334-30-5) Oxidizers may cause fire and explosions. Incompatible with nitric acid. May accumulate static electrical charges, and may cause ignition of its vapors.

OILS, EDIBLE: BABASSU (N/A) Incompatible with strong acids, oxidizers, nitrates.

OILS, EDIBLE: CASTOR (8001-79-4) Incompatible with strong acids, oxidizers, nitrates.

OILS, EDIBLE: COCONUT (8001-31-8) Incompatible with strong acids, oxidizers, nitrates.

OILS, EDIBLE: CORN (N/A) Incompatible with strong acids, oxidizers, nitrates.

OILS, EDIBLE: COTTONSEED (8001-29-4) Incompatible with strong acids, nitrates, oxidizers.

OILS, EDIBLE: COTTONSEED PIGMENT GLANDS (8001-29-4) Incompatible with strong acids, nitrates, oxidizers.

OILS, EDIBLE: FISH (N/A) Incompatible with strong acids, oxidizers.

OILS, EDIBLE: LARD (N/A) Incompatible with strong acids, oxidizers.

OILS, EDIBLE: OLIVE (8001-25-0) Incompatible with strong acids, oxidizers.

OILS, EDIBLE: PALM (8001-75-3) Incompatible with strong acids, oxidizers.

OILS, EDIBLE: PEANUT (8002-03-7) Incompatible with strong acids, oxidizers.

OILS, EDIBLE: SAFFLOWER (8001-23-8) Incompatible with strong acids, oxidizers.

OILS, EDIBLE: SOYA BEAN (N/A) Incompatible with strong acids, oxidizers.

OILS, EDIBLE: TUCUM (N/A) Incompatible with strong acids, oxidizers.

OILS, EDIBLE: VEGETABLE (68956-68-3) Incompatible with strong acids, oxidizers, nitrates.

OILS, FUEL: 1-D (68334-30-5) Oxidizers may cause fire and explosions. Incompatible with nitric acid. May accumulate static electrical charges, and may cause ignition of its vapors.

OILS, FUEL: 2-D (68334-30-5) Oxidizers may cause fire and explosions. Incompatible with nitric acid. May accumulate static electrical charges, and may cause ignition of its vapors.

OILS, FUEL: NO. 1 (68476-31-3) Forms explosive mixture with air above 100°F/38°C. Oxidizers may cause fire and explosions. Incompatible with nitric acid. May accumulate static electrical charges, and may cause ignition of its vapors.

OILS, FUEL: NO. 2 (68476-30-2) Oxidizers may cause fire and explosions. Incompatible with nitric acid. May accumulate static electrical charges, and may cause ignition of its vapors.

OILS, FUEL: NO. 4 (68476-33-5) Forms explosive mixture with air (flash point 142°F/61°C). Oxidizers may cause fire and explosions. Incompatible with nitric acid. May accumulate static electrical charges, and may cause ignition of its vapors.

OILS, FUEL: NO. 5 (68476-33-5) Forms explosive mixture with air (flash point 142°F/61°C). Oxidizers may cause fire and explosions. Incompatible with nitric acid. May accumulate static electrical charges, and may cause ignition of its vapors.

OILS, FUEL: NO. 6 (68476-33-5) Forms explosive mixture with air (flash point 142°F/61°C). Oxidizers may cause fire and explosions. Incompatible with nitric acid. May accumulate static electrical charges, and may cause ignition of its vapors.

OILS, MISCELLANEOUS: ABSORPTION (N/A) Incompatible with nitric acid, oxidizers.

OILS, MISCELLANEOUS: CASHEW NUT SHELL (8001-24-7) Incompatible with sulfuric acid, caustics, ammonia, amines, isocyanates, alkylene oxides, epichlorohydrin.

OILS, MISCELLANEOUS: COAL TAR (8007-45-2) Forms explosive mixture with air (flash point > 60°F/15°C). Incompatible with strong acids, nitrates.

OILS, MISCELLANEOUS: LINSEED (8001-26-1) Oxidizers may cause fire and explosions. Incompatible with nitric acid. May accumulate static electrical charges, and may cause ignition of its vapors.

OILS, MISCELLANEOUS: LUBRICATING (N/A) Incompatible with nitric acid, strong oxidizers; poses risk of fire or explosions.

OILS, MISCELLANEOUS: MOTOR (N/A) Incompatible with nitric acid, strong oxidizers; poses risk of fire or explosions.

OILS, MISCELLANEOUS: MINERAL (8012-95-1) Oxidizers may cause fire and explosions. Incompatible with nitric acid. May accumulate static electrical charges, and may cause ignition of its vapors.

OILS, MISCELLANEOUS: MINERAL SEAL (N/A) Oxidizers may cause fire and explosions. Incompatible with nitric acid. May accumulate static electrical charges, and may cause ignition of its vapors.

OILS, MISCELLANEOUS: NEATSFOOT (N/A) Oxidizers may cause fire and explosions. Incompatible with nitric acid. May accumulate static electrical charges, and may cause ignition of its vapors.

OILS, MISCELLANEOUS: PINE (8002-09-3) Forms explosive mixture with air (flash point 138°F/59°C). Incompatible with strong acids, strong oxidizers.

OILS, MISCELLANEOUS: RESIN (N/A) Oxidizers may cause fire and explosions. Incompatible with nitric acid. May accumulate static electrical charges, and may cause ignition of its vapors.

OILS, MISCELLANEOUS: ROAD (8052-42-4) Incompatible with nitric acid, fluorine, strong oxidizers.

OILS, MISCELLANEOUS: SPINDLE (N/A) Oxidizers may cause fire and explosions. Incompatible with nitric acid. May accumulate static electrical charges, and may cause ignition of its vapors.

OILS, MISCELLANEOUS: SPRAY (N/A) Oxidizers may cause fire and explosions. Incompatible with nitric acid. May accumulate static electrical charges, and may cause ignition of its vapors.

OILS, MISCELLANEOUS: TALL (8002-26-4) Incompatible with strong acids, oxidizers, nitrates.

OILS, MISCELLANEOUS: TALL, FATTY ACID (N/A) Incompatible with strong acids, oxidizers, nitrates.

OILS, MISCELLANEOUS: TANNER'S (N/A) Oxidizers may cause fire and explosions. Incompatible with nitric acid. May accumulate static electrical charges, and may cause ignition of its vapors.

OILS, MISCELLANEOUS: TRANSFORMER (N/A) Oxidizers may cause fire and explosions. Incompatible with nitric acid. May accumulate static electrical charges, and may cause ignition of its vapors. May accumulate static electrical charges, and may cause ignition of its vapors.

OILS, MISCELLANEOUS: TURBINE (N/A) Incompatible with nitric acid, strong oxidizers; poses risk of fire or explosions.

OKO (62-73-7) Attacks some plastics, rubber, and coatings.

OKTANEN (Dutch) (111-65-9) Forms explosive mixture with air (flash point 56°F/13.3°C). Strong oxidizers may cause fire and explosions. Attacks some plastics, rubber, and coatings. May accumulate static electrical charges, and may cause ignition of its vapors.

OKTATERR (57-74-9) Contact with strong oxidizers may cause fire and explosions. Attacks some plastics, rubber, and coatings.

OLAMINE (141-43-5) Will oxidize in air; often shipped under a pad of inert gas. Forms explosive mixture with air (flash point 185°F/85°C). Strong oxidizers may cause fire and explosions. A strong base. Incompatible with acids, organic anhydrides, isocyanates, alkylene oxides, epichlorohydrin, aldehydes, alcohols, glycols, phenols, cresols, caprolactam solution. Contact with strong acids may cause spattering. Attacks plastics, rubber, and aluminum, copper, tin, and their alloys.

OLEFIANT GAS (74-85-1) A flammable gas. May accumulate static electrical charges, and may cause ignition of its vapors. Oxidizers may cause explosive polymerization and fire. Incompatible with acids, halogens, nitrogen oxides, hydrogen bromide, aluminum chloride, bromotrichloromethane, carbon tetrachloride, chlorine, chlorine dioxide, nitrogen dioxide.

OLEFIN C-13 (N/A) Incompatible with strong oxidizers, strong acids.

OLEIC ACID (112-80-1) Incompatible with strong oxidizers, perchloric acid, and aluminum powder; may cause explosions. Corrodes aluminum.

OLEIC ACID, AMMONIUM SALT (544-60-5) Incompatible with strong oxidizers, nitric acid.

OLEIC ACID, DISTILLED (112-80-1) Incompatible with strong oxidizers, perchloric acid and aluminum powder; may cause explosions. Corrodes aluminum.

OLEIC ACID, POTASSIUM SALT (143-48-0) Incompatible with nitric acid, sulfuric acid, nitrates.

OLEIC ACID, SODIUM SALT (143-19-1) Combustible solid. Strong oxidizers may cause fires or explosions.

OLEINIC ACID (112-80-1) Incompatible with strong oxidizers, perchloric acid and aluminum powder; may cause explosions. Corrodes aluminum.

OLEOFOS-20 (56-38-2) Combustible liquid. Mixtures with endrin may be explosive. Strong oxidizers may cause fire and explosions. Attacks some plastics, rubber, and coatings.

OLEOGESAPRIM (1912-24-9) Incompatible with strong acids.

OLEOPARATHION (56-38-2) Combustible liquid. Mixtures with endrin may be explosive. Strong oxidizers may cause fire and explosions. Attacks some plastics, rubber, and coatings.

OLEOVOFOTOX (298-00-0) Mixtures with magnesium may be explosive.

OLEUM (8014-95-7) A powerful oxidizer and strong acid. Reacts violently with reducing agents, organic materials. Reacts with air, producing corrosive fumes. Contact with cast iron may cause a violent reaction. Extremely hazardous on contact with many materials including chlorates, carbides, fulminates. Reacts vigorously reaction with metals (powders will ignite), releasing hydrogen. Attacks some plastics, rubber, and coatings on brief contact.

OLEUM ABIETIS (8002-09-3) Forms explosive mixture with air (flash point 138°F/59°C). Incompatible with strong acids, strong oxidizers.

OLEYL ALCOHOL (143-28-2) Incompatible with strong acids, caustics, aliphatic amines, isocyanates, oxidizers.

OLIVE OIL (8001-25-0) Incompatible with strong acids, oxidizers.

OMAIT (2312-35-8) Forms explosive mixture with air (flash point 82°F/28°C). Strong oxidizers may cause fire and explosions.

OMAL (99-95-4) Contact with strong oxidizers may cause fire and explosions.

OMITE (2312-35-8) Forms explosive mixture with air (flash point 82°F/28°C). Strong oxidizers may cause fire and explosions.

OMNITOX (58-89-9) Not combustible, but may be dissolved in a combustible solvent. If solvent comes in contact with oxidizers, fire and explosions may result.

OMS-14 (62-73-7) Attacks some plastics, rubber, and coatings.

OMS-29 (63-25-2) Incompatible with strong oxidizers, strongly alkaline pesticides.

OMS-93 (2032-65-7) Contact with strong oxidizers may cause fire and explosions.

OMS-570 (115-29-7) Hydrolyzed by acids and alkalis. Corrosive to iron.

OMS-971 (2921-88-) Incompatible with strong acids. Hydrolyzes from acid or alkaline solutions.

ONA (88-74-4) *ortho*; (99-09-2) *meta*. Incompatible with strong acids, combustibles, organics, moisture.

ONCB (100-00-5) Forms explosive mixture with air. A strong oxidizer; reacts violently with reducing agents, combustibles. Reacts with caustics, ammonia, amines. Attacks some plastics, rubber, and coatings.

ONP (88-75-5) A strong oxidizer; reacts violently with reducing agents, combustibles, organic matter, chlorosulfonic acid, potassium hydroxide.

ONT (88-72-2) Incompatible with strong acids, caustics, ammonia, amines, reducing agents, oxidizers. Attacks some plastics, rubber, and coatings.

ONTRACK 8E (51218-45-0) Incompatible with strong acids, nitrates, oxidizers.

OPLOSSINGEN (Dutch) (50-00-0) May polymerize unless properly inhibited (usually with methanol). Forms explosive mixture with air (flash point 122°F/50°C). Incompatible with strong acids, amines, strong oxidizers, alkaline materials, nitrogen dioxide, performic acid. Reaction with hydrochloric acid forms bis-chloromethyl ether, a carcinogen.

OPTAL (71-23-8) Forms explosive mixture with air (flash point 74°F/23°C). Attacks some plastics, rubber, and coatings. Incompatible with strong acids, caustics, aliphatic amines, isocyanates.

OP-THAL-ZIN (7733-02-0) Incompatible with strong bases.

ORANGE BASE CIBA 2 (88-74-4) *ortho*; (99-09-2) *meta*. Incompatible with strong acids, combustibles, organics, moisture.

ORATRAST (7727-43-7) Explosions may result from contact with aluminum in the presence of heat. Incompatible with potassium, phosphorus.

ORDINARY LACTIC ACID (598-82-3) Incompatible with strong acids. Aqueous solution reacts with bases.

ORGA-414 (61-82-5) Substance acts as a weak base to produce salts in contact with acids. Corrosive to iron, aluminum, copper, and copper alloys.

ORGAMIDE (105-60-2) Contact with strong oxidizers may cause fire and explosions.

ORPIMENT (1303-33-9) Water contact produces hydrogen sulfide. Incompatible with strong oxidizers, acids, halogens.

ORSIN (106-50-3) Incompatible with acids, organic anhydrides, isocyanates, aldehydes, strong oxidizers; may cause fire and explosions. Heat and light contribute to instability.

ORTHO 4355 (300-76-5) Unstable in the presence of iron. Reacts with acids, strong oxidizers in sunlight. Corrosive to metals. Attacks some plastics, rubber, and coatings.

ORTHO 5865 (2425-06-1) Incompatible with acids or acid vapor. Strongly alkaline conditions contribute to instability.

ORTHOARSENIC ACID (7778-39-4) Incompatible with sulfuric acid, caustics, ammonia, amines, isocyanates, alkylene oxides, oxidizers, epichlorohydrin, vinyl acetate, amides. Corrodes metals.

ORTHOBORIC ACID (10043-35-3) Incompatible with bases, potassium. Acetic anhydride forms heat-sensitive explosive.

ORTHO C-1 DEFOLIANT & WEED KILLER (7775-09-9) A powerful oxidizer; reacts violently with reducing agents and combustible matter. Explosions may be caused by contact with ammonia salts, carbon, oils, metal sulfides, nitrobenzene, powdered metals, sugar. Contact with strong acids form carbon dioxide. Forms shock-sensitive mixtures with some organic materials. Solution (50%) decomposes at 300°F/149°C, liberating oxygen.

ORTHOCIDE (133-06-2) Incompatible with tetraethyl pyrophosphate, parathion.

ORTHOCIDE 7.5 (133-06-2) Incompatible with tetraethyl pyrophosphate, parathion.

ORTHOCIDE 50 (133-06-2) Incompatible with tetraethyl pyrophosphate, parathion.

ORTHOCIDE 406 (133-06-2) Incompatible with tetraethyl pyrophosphate, parathion.

ORTHODIBROM (300-76-5) Unstable in the presence of iron. Reacts with acids, strong oxidizers in sunlight. Corrosive to metals. Attacks some plastics, rubber, and coatings.

ORTHODIBROMO (300-76-5) Unstable in the presence of iron. Reacts with acids, strong oxidizers in sunlight. Corrosive to metals. Attacks some plastics, rubber, and coatings.

ORTHODICHLOROBENZENE (95-50-1) Forms explosive mixture with air (flash point 151°F/66°C). Incompatible with strong oxidizers, hot aluminum or aluminum alloy. Attacks some plastics, rubber, and coatings.

ORTHODICHLOROBENZOL (95-50-1) Forms explosive mixture with air (flash point 151°F/66°C). Incompatible with strong oxidizers, hot aluminum or aluminum alloy. Attacks some plastics, rubber, and coatings.

ORTHO-KLOR (57-74-9) Contact with strong oxidizers may cause fire and explosions. Attacks some plastics, rubber, and coatings.

ORTHOPHOS (56-38-2) Combustible liquid. Mixtures with endrin may be explosive. Strong oxidizers may cause fire and explosions. Attacks some plastics, rubber, and coatings.

ORTHOPHOSPHORIC ACID (7664-38-2) Incompatible with aliphatic amines, alkanolamines, alkylene oxides, aromatic amines, amides, ammonia, ammonium hydroxide, bases, calcium oxide, epichlorohydrin, isocyanates, explosive with nitromethane, oleum, organic anhydrides, sulfuric acid, sodium tetrahydroborate, strong oxidizers, vinyl acetate, water. Contact with most metals form hydrogen gas. Attacks some plastics, rubber, and coatings, and glass and ceramics.

ORTHOSIL (1344-09-8) Solution is a strong base; reacts with acids, organic anhydrides, alkylene oxides, epichlorohydrin, aldehydes, alcohols, glycols, phenols, cresols, caprolactam solution. Attacks chemically active metals.

ORTHOTITANIC ACID, TETRABUTYL ESTER (5593-70-4) Forms explosive mixture with air (flash point 170°F/77°C). Reacts violently with oxidizers. Water contact produces butanol and titanium dioxide.

ORTIZON (124-43-6) Combustible solid. At 122°F/50°C reacts with dust, organics.

ORVINYLCARBINOL (107-18-6) May form unstable and explosive peroxides. Able to polymerize. May accumulate static electrical charges, and may cause ignition of its vapors. A strong reducing agent; reacts violently with oxidizers. Forms explosive mixture with air (flash point 70°F/21°C). Incompatible with strong acids, amines, isocyanates, carbon tetrachloride, chlorosulfonic acid, diallyl phosphide, oleum, sodium hydroxide, tri-*n*-bromomelamine, metal halides, caustic soda, sodium, and magnesium, aluminum, and their alloys. Attacks some coatings, some plastics, and rubber.

OS 1897 (96-12-8) Forms explosive mixture with air (flash point 170°F/77°C). Reacts with oxidizers and chemically active metals (i.e., aluminum, magnesium, and tin alloys). Attacks some rubber materials and coatings.

OS 2046 (7786-34-7) Contact with strong oxidizers may cause fire and explosions. Attacks some plastics, rubber, and coatings.

OSMIC ACID (20816-12-0) Contact with hydrochloric acid produces chlorine gas. Contact with organic and combustible materials may cause fires and explosions.

OSMIUM(VIII) OXIDE (20816-12-0) Contact with hydrochloric acid produces chlorine gas. Contact with organic and combustible materials may cause fires and explosions.

OSMIUM TETROXIDE (20816-12-0) Contact with hydrochloric acid produces chlorine gas. Contact with organic and combustible materials may cause fires and explosions.

OSMOSOL EXTRA (71-23-8) Forms explosive mixture with air (flash point 74°F/23°C). Attacks some plastics, rubber, and coatings. Incompatible with strong acids, caustics, aliphatic amines, isocyanates.

OTTANI (Italian) (111-65-9) Forms explosive mixture with air (flash point 56°F/13.3°C). Strong oxidizers may cause fire and explosions. Attacks some plastics, rubber, and coatings. May accumulate static electrical charges, and may cause ignition of its vapors.

1,2,4,5,6,7,8,8-OTTOCHLORO-3A,4,7,7A-TETRAIDRO-4,7-ENDO-METANO-INDANO (Italian) (57-74-9) Contact with strong oxidizers may cause fire and explosions. Attacks some plastics, rubber, and coatings.

OVADZIAK (58-89-9) Not combustible, but may be dissolved in a combustible solvent. If solvent comes in contact with oxidizers, fire and explosions may result.

OWADZIAK (58-89-9) Not combustible, but may be dissolved in a combustible solvent. If solvent comes in contact with oxidizers, fire and explosions may result.

OWANOL 33B (107-98-2) Forms explosive mixture with air (flash point 90°F/32°C). Incompatible with sulfuric acid, isocyanates, perchloric acid.

OXAALZUUR (Dutch) (144-62-7) Contact with silver compounds may produce explosive salt. Incompatible with caustics, furfuryl alcohol, oxidizers, mercury, silver, sodium chlorite, sodium hypochlorite.

OXACETIC ACID (298-12-4) Reacts with alkali metals; releases flammable hydrogen gas.

OXACYCLOPENTADIENE (110-00-9) Forms explosive mixture with air (flash point −58°F/−50°C). Reacts violently with acids, oxidizers. Unless stabilized with an inhibitor, air exposure causes formation of unstable peroxides.

OXACYCLOPENTANE (109-99-9) Forms explosive mixture with air (flash point 6°F/−14°C). Unless inhibited, can form unstable and explosive peroxides. Incompatible with strong acids, strong oxidizers. Attacks some plastics. May accumulate static electric charges that can result in ignition of its vapors.

OXACYCLOPROPANE (75-21-8) Forms explosive mixture with air (flash point 20°F/−6°C). Incompatible with alkali metal hydroxides, highly active catalysts (e.g., anhydrous chlorides of iron, tin, or aluminum and oxides of iron or aluminum). Avoid contact with copper. Protect container from physical damage, sun, and heat. Attacks some plastics, rubber, or coatings.

OXAL (107-22-2) Forms explosive mixture with air. A strong reducing agent; reacts violently with oxidizers. Heat or contact with water, strong bases causes polymerization. Incompatible with strong acids, caustics, ammonia, amines, chlorosulfonic acid, ethylene amine. Corrosive to metals.

OXALDEHYDE (107-22-2) Forms explosive mixture with air. A strong reducing agent; reacts violently with oxidizers. Heat or contact with water, strong bases causes polymerization. Incompatible with strong acids, caustics, ammonia, amines, chlorosulfonic acid, ethylene amine. Corrosive to metals.

OXALIC ACID (144-62-7) Contact with silver compounds may produce explosive salt. Incompatible with caustics, furfuryl alcohol, oxidizers, mercury, silver, sodium chlorite, sodium hypochlorite.

OXALIC ACID DIHYDRATE (144-62-7) Contact with silver compounds may produce explosive salt. Incompatible with caustics, furfuryl alcohol, oxidizers, mercury, silver, sodium chlorite, sodium hypochlorite.

OXALIC ACID DINITRILE (460-19-5) Explosive reaction occurs with acids, water, liquid oxygen, oxidizers.

OXALIC ALDEHYDE (107-22-2) Forms explosive mixture with air. A strong reducing agent; reacts violently with oxidizers. Heat or contact with water, strong bases causes polymerization. Incompatible with strong acids, caustics, ammonia, amines, chlorosulfonic acid, ethylene amine. Corrosive to metals.

OXALONITRILE (460-19-5) Explosive reaction occurs with acids, water, liquid oxygen, oxidizers.

OXALSAEURE (German) (144-62-7) Contact with silver compounds may produce explosive salt. Incompatible with caustics, furfuryl alcohol, oxidizers, mercury, silver, sodium chlorite, sodium hypochlorite.

OXALYL CYANIDE (460-19-5) Explosive reaction occurs with acids, water, liquid oxygen, oxidizers.

OXAMMONIUM (7803-49-8) Self-reactive; contaminants or temperatures above 150°F/65°C may cause explosive decomposition. Incompatible with strong acids, organic anhydrides, isocyanates, aldehydes, sodium, zinc. Aqueous solution is a base. Strong oxidizers may cause fire and explosions. Attacks some metals.

OXAMMONIUM SULFATE (10039-54-0) A strong oxidizer. Heat or friction may cause explosions. Violent, possibly explosive, reaction occurs with caustics, reducing agents, or combustibles. Incompatible with strong acids, organic anhydrides, isocyanates, aldehydes. Magnesium or aluminum contact may be explosive.

OXANE (75-21-8) Forms explosive mixture with air (flash point 20°F/−6°C). Incompatible with alkali metal hydroxides, highly active catalysts (e.g., anhydrous chlorides of iron, tin, or aluminum and oxides of iron or aluminum). Avoid contact with copper. Protect container from physical damage, sun and heat. Attacks some plastics, rubber, or coatings.

3-OXA-1,5-PENTANEDIOL (111-46-6) Combustible (flash point 255°F/124°C). Incompatible with sulfuric acid, isocyanates, strong oxidizers.

2H-1,4-OXAZINE, TETRAHYDRO- (110-91-8) Forms explosive mixture with air (flash point 98°F/37°C). Incompatible with acids, organic anhydrides, isocyanates, vinyl acetate, acrylates, substituted allyls, alkylene oxides, epichlorohydrin, ketones, aldehydes, alcohols, glycols, phenols, cresols, caprolactam solution, strong oxidizers. Attacks aluminum, copper, lead, tin, zinc, and alloys, and some plastics, rubber, and coatings.

2-OXETANONE (57-57-8) Polymerizes in storage. High temperatures cause ruptured containers. Decomposes at room temperature; hydrolyzes in water.

OXIDATE LE (93-58-3) Forms explosive mixture with air (flash point 181°F/83°C). Incompatible with strong acids, nitrates, oxidizers.

OXIDE (106-89-8) Forms explosive mixture with air (flash point 88°F/31°C). Heat, acids, alkalies and metallic halides can cause explosive polymerization. Aliphatic amines, alkaline earths, alkali metals, alkanolamines, powdered metals, strong oxidizers may cause fire and explosions. Decomposition produces highly toxic phosgene gas. Will pit steel in the presence of moisture. May accumulate static electrical charges, and may cause ignition of its vapors.

OXIDES OF NITROGEN (10544-72-6); (10102-44-0 CAS for nitrogen dioxide). Powerful oxidizer. Water contact produces nitric acid and nitric oxide. Incompatible with nitric oxide; reacts with air, producing more nitrogen tetroxide. Explosive reaction with alcohols, fuels and petroleum, nitrobenzene, organic matter, anhydrous ammonia, chlorinated hydrocarbons, toluene. Reacts vigorously with combustibles, reducing agents. Attacks metals in presence of moisture.

OXIDOETHANE (75-21-8) Forms explosive mixture with air (flash point 20°F/−6°C). Incompatible with alkali metal hydroxides, highly active catalysts (e.g., anhydrous chlorides of iron, tin, or aluminum and oxides of iron or aluminum). Avoid contact with copper. Protect container from physical damage, sun, and heat. Attacks some plastics, rubber, or coatings.

alpha, beta-OXIDOETHANE (75-21-8) Forms explosive mixture with air (flash point 20°F/−6°C). Incompatible with alkali metal hydroxides, highly active catalysts (e.g., anhydrous chlorides of iron, tin, or aluminum and oxides of iron or aluminum). Avoid contact with copper. Protect container from physical damage, sun, and heat. Attacks some plastics, rubber, or coatings.

OXIRAAN (Dutch) (75-21-8) Forms explosive mixture with air (flash point 20°F/−6°C). Incompatible with alkali metal hydroxides, highly active catalysts (e.g., anhydrous chlorides of iron, tin, or aluminum and oxides of iron or aluminum). Avoid contact with copper. Protect container from physical damage, sun, and heat. Attacks some plastics, rubber, or coatings.

OXIRAN (75-21-8) Forms explosive mixture with air (flash point 20°F/−6°C). Incompatible with alkali metal hydroxides, highly active catalysts (e.g., anhydrous chlorides of iron, tin, or aluminum and oxides of iron or aluminum). Avoid contact with copper. Protect container from physical damage, sun, and heat. Attacks some plastics, rubber, or coatings.

OXIRANE (75-21-8) Forms explosive mixture with air (flash point 20°F/−6°C). Incompatible with alkali metal hydroxides, highly active catalysts (e.g., anhydrous chlorides of iron, tin, or aluminum and oxides of iron or aluminum). Avoid contact with copper. Protect container from physical damage, sun, and heat. Attacks some plastics, rubber, or coatings.

OXIRANE, (CHLOROMETHYL)- (106-89-8) Forms explosive mixture with air (flash point 88°F/31°C). Heat, acids, alkalies and metallic halides can cause explosive polymerization. Aliphatic amines, alkaline earths, alkali metals, alkanolamines, powdered metals, strong oxidizers may cause fire and explosions. Decomposition produces highly

621

toxic phosgene gas. Will pit steel in the presence of moisture. May accumulate static electrical charges, and may cause ignition of its vapors.

OXIRANE, 2-(CHLOROMETHYL)- (106-89-8) Forms explosive mixture with air (flash point 88°F/31°C). Heat, acids, alkalies, and metallic halides can cause explosive polymerization. Aliphatic amines, alkaline earths, alkali metals, alkanolamines, powdered metals, strong oxidizers may cause fire and explosions. Decomposition produces highly toxic phosgene gas. Will pit steel in the presence of moisture. May accumulate static electrical charges, and may cause ignition of its vapors.

OXIRANEMETHANOL (556-52-5) Forms explosive mixture with air (flash point 158°F/70°C). Reacts violently with strong oxidizers. Contact with caustics, acids, barium, lithium, sodium, magnesium, titanium can cause polymerization. Incompatible with nitrates. Attacks some plastics, rubber, and coatings.

OXIRANEMETHOL (556-52-5) Forms explosive mixture with air (flash point 158°F/70°C). Reacts violently with strong oxidizers. Contact with caustics, acids, barium, lithium, sodium, magnesium, titanium can cause polymerization. Incompatible with nitrates. Attacks some plastics, rubber, and coatings.

OXIRANE, METHYL- (75-56-9) Forms explosive gas mixture with air (flash point −35°F/−37°C). Incompatible with anhydrous metal chlorides. Strong acids, caustics, peroxides cause polymerization. Reacts with ammonia, amines, acetylene-forming metals. Attacks some plastics, rubber, and coatings.

OXIRENE, DIHYDRO- (75-21-8) Forms explosive mixture with air (flash point 20°F/−6°C). Incompatible with alkali metal hydroxides, highly active catalysts (e.g., anhydrous chlorides of iron, tin, or aluminum and oxides of iron or aluminum). Avoid contact with copper. Protect container from physical damage, sun and heat. Attacks some plastics, rubber, or coatings.

OXITOL (110-80-5) Forms explosive mixture with air (flash point 120°F/49°C). Strong oxidizers may cause fire and explosions. Attacks some plastics, rubber, and coatings. Able to form peroxides. Incompatible with strong acids, aluminum and its alloys.

3-OXOBUTANOIC ACID ETHYL ESTER (141-97-9) Forms explosive mixture with air (flash point 135°F/57°C). Incompatible with strong acids, nitrates, oxidizers

3-OXOBUTANOIC ACID METHYL ESTER (105-45-3) Incompatible with oxidizers, strong acids, nitrates.

alpha-OXODIPHENYLMETHANE (119-61-9) Combustible solid. Strong oxidizers may cause fire and explosions. Attacks some plastics, rubber, and coatings.

alpha-OXODITANE (119-61-9) Combustible solid. Strong oxidizers may cause fire and explosions. Attacks some plastics, rubber, and coatings.

OXOETHANOIC ACID (298-12-4) Reacts with alkali metals; releases flammable hydrogen gas.

2-OXOHEXAMETHYLENIMINE (105-60-2) Contact with strong oxidizers may cause fire and explosions.

5-OXOHEXANENITRILE (10412-98-3) Incompatible with sulfuric acid.

622

OXOLANE (109-99-9) Forms explosive mixture with air (flash point 6°F/−14°C). Unless inhibited, can form unstable and explosive peroxides. Incompatible with strong acids, strong oxidizers. Attacks some plastics. May accumulate static electric charges that can result in ignition of its vapors.

OXOLE (110-00-9) Forms explosive mixture with air (flash point −58°F/−50°C). Reacts violently with acids, oxidizers. Unless stabilized with an inhibitor, air exposure causes formation of unstable peroxides.

OXOMETHANE (50-00-0) May polymerize unless properly inhibited (usually with methanol). Forms explosive mixture with air (flash point 122°F/50°C). Incompatible with strong acids, amines, strong oxidizers, alkaline materials, nitrogen dioxide, performic acid. Reaction with hydrochloric acid forms bis-chloromethyl ether, a carcinogen.

OXO OCTALDEHYDE (N/A) Forms explosive mixture with air (flash point 104°F/40°C). Incompatible with strong acids, caustics, ammonia, amines.

OXO OCTYL ALCOHOL (26952-21-6) Forms explosive mixture with air (flash point 180°F/82°C). Incompatible with strong acids, caustics, amines, isocyanates.

OXOTRIDECYL ALCOHOL (112-70-9) Incompatible with strong acids, caustics, aliphatic amines, isocyanates, strong oxidizers. Attacks aluminum. May accumulate static electrical charges, and may cause ignition of its vapors.

OXRALOX (60-57-1) Incompatible with concentrated mineral acids, acid catalysts, acid oxidizing agents, phenols, reactive metals.

OXY-5 (94-36-0) Confined storage of dry chemical may lead to decomposition and explosion. A strong oxidant; extremely reactive. Fires and explosion may result from heat or contamination, and from contact with strong acids, combustible materials, oxidizers, acids, bases, alcohols, reducing agents, metals, metal oxides, amines, accelerators, methyl methacrylate, organic matter, lithium aluminum carbide, dimethyl aniline, amines, metallic naphthenates. May attack some plastics, rubber, and coatings. Protect containers from shock and friction.

OXYBENZENE (108-95-2) Forms explosive mixture with air (flash point 174°F/79°C). Incompatible with strong oxidizers, strong acids, caustics, aliphatic amines, amides, oxidizers, formaldehyde, butadiene, calcium hypochlorite. Liquid attacks some plastics, rubber, and coatings; hot liquid attacks aluminum, magnesium, lead, and zinc metals.

1,1-OXYBISBENZENE (101-84-8) Incompatible with strong acids, strong oxidizers; may cause fire and explosions. Attacks some plastics, rubber, and coatings.

1,1'-OXYBIS(BUTANE) (142-96-1) Forms explosive mixture with air (flash point 77°F/25°C). May accumulate static electrical charges, and may cause ignition of its vapors. Incompatible with strong acids, oxidizers. Air contact or light may cause formation of unstable and explosive peroxides.

1,1-OXYBIS(2-CHLOROETHANE) (111-44-4) Water contact may produce hydrogen chloride fumes. Can form peroxides. Forms explosive mixture with air (flash point 131°F/55°C). Strong oxidizers may cause fire and explosions. Attacks some plastics, rubber, and coatings.

OXYBIS (CHLOROMETHANE) (542-88-1) Highly volatile. Heat may cause fire. Water contact produces hydrogen chloride and formaldehyde.

2,2′OXYBISETHANOL (111-46-6) Combustible (flash point 255°F/124°C). Incompatible with sulfuric acid, isocyanates, strong oxidizers.

2,2′-(OXYBIS(ETHYLENEOXY))DIETHANOL (112-60-7) Incompatible with sulfuric acid, isocyanates, perchloric acid. Attacks some plastics, rubber, and coatings.

OXYBIS METHYANE (115-10-6) Flammable gas. Forms explosive mixture with air. Forms unstable peroxides in storage. Reacts violently with strong oxidizers.

1,1′-[OXYBIS(METHYLENE)] BIS BENZENE (103-50-4) Forms unstable peroxides. Reacts violently with strong oxidizers. May accumulate static electrical charges, and may cause ignition of its vapors.

2,2′-OXYBISPROPANE (108-20-3) Forms explosive mixture with air (flash point −18°F/−28°C). May accumulate static electrical charges, and may cause ignition of its vapors. Air contact produces explosive peroxides that may detonate with heat or shock. Strong oxidizers may cause fire and explosions. Attacks some plastics, rubber, and coatings.

3,3′-OXYBIS(1-PROPENE) (557-40-4) Forms explosive mixture with air (flash point 20°F/−7°C). May accumulate static electrical charges, and may cause ignition of its vapors. Forms explosive peroxides with air. Incompatible with strong acids, oxidizers.

OXYCIL (7775-09-9) A powerful oxidizer; reacts violently with reducing agents and combustible matter. Explosions may be caused by contact with ammonia salts, carbon, oils, metal sulfides, nitrobenzene, powdered metals, sugar. Contact with strong acids produces carbon dioxide. Forms shock-sensitive mixtures with some organic materials. Solution (50%) decomposes at 300°F/149°C, liberating oxygen.

OXY DBCP (96-12-8) Forms explosive mixture with air (flash point 170°F/77°C). Reacts with oxidizers and chemically active metals (i.e., aluminum, magnesium, and tin alloys). Attacks some rubber materials and coatings.

OXYDE d'BARYUM (French) (1304-28-5) Incompatible with water, hydrogen sulfide, carbon dioxide, hydroxlamine, nitrogen tetroxide, sulfur trioxide, and triuranium.

OXYDE de CALCIUM (French) (1305-78-8) Reacts violently with water. Incompatible with ethanol, hydrogen fluoride, acids, halogens, metal halides, light metals, some oxides, boron trifluoride, chlorine trifluoride, liquid hydrofluoric acid, phosphorus pentoxide, boric oxide and calcium chloride mixtures, fluorine, chlorine trifluoride, carbon dioxide.

OXYDE de CARBONE (French) (630-08-0) Forms extremely explosive mixture with air. Reacts violently with strong oxidizers.

OXYDE d'ETHYLE (French) (60-29-7) Forms explosive mixture with air (flash point −49°F/−45°C). Incompatible with strong acids, strong oxidizers. Can form peroxides with exposure to air or light; may explode when container is unstoppered or otherwise opened. Attacks some plastics, rubber, and coatings. Being a nonconductor, chemical may accumulate static electric charges that may result in ignition of its vapor.

OXYDE de MESITYLE (French) (141-79-7) Forms explosive mixture with air (flash point 87°F/31°C). Forms peroxides. Incompatible with strong acids aliphatic amines, alkanolamines, 2-aminoethanol, ethylene diamine, chlorosulfonic acid, oleum, oxidizers. Dissolves some forms of plastics, resins, and rubber. Attacks copper.

OXYDE de PROPYLENE (French) (75-56-9) Forms explosive gas mixture with air (flash point −35°F/−37°C). Incompatible with anhydrous metal chlorides. Strong acids, caustics, peroxides cause polymerization. Reacts with ammonia, amines, acetylene-forming metals. Attacks some plastics, rubber, and coatings.

2,2′-OXYDIETHANOL (111-46-6) Combustible (flash point 255°F/124°C). Incompatible with sulfuric acid, isocyanates, strong oxidizers.

1,1′-OXYDI-2-PROPANOL (110-98-5) Incompatible with sulfuric acid, perchloric acid, isocyanates, strong oxidizers.

OXYDOL (7722-84-1) A powerful oxidizer; attacks many substances. Contact with most organic, readily oxidizable materials, reducing agents and combustibles causes fire and explosions. Contact with iron, copper, brass, bronze, chromium, zinc, lead, manganese, silver, and other catalytic metals (and their salts), especially in a basic (pH 7 or above) environment, causes rapid decomposition with evolution of oxygen gas. Attacks, and may ignite, some plastics, rubber, and coatings. Decomposes slowly at ordinary temperatures and builds up pressure in a closed container. The rate of decomposition doubles for each 50°F/10°C rise (1.5 times 10°C rise) in temperature and becomes self-sustaining at 285°F/141°C.

OXYGEN or OXYGEN, LIQUID (7782-44-7) Heat of water will vigorously vaporize liquid oxygen. A strong oxidizer. Reacts violently with reducing agents, combustibles, organic and easily oxidizable materials; contact may cause fire and explosions. The low temperature may cause brittleness on contact with some materials.

OXYGEN DIFLUORIDE (7783-41-7) Forms explosive mixture with water. A powerful oxidizer. Contact with all reducing agents, organic and combustible materials, ammonia, platinum and many other metals, metal oxides, and moist air may cause fire and explosions. Attacks some forms of plastics, rubber, and coatings. See 29 CFR 1910.101 for specific regulations on storage of compressed gas cylinders.

OXYGEN FLUORIDE (7783-41-7) Forms explosive mixture with water. A powerful oxidizer. Contact with all reducing agents, organic and combustible materials, ammonia, platinum and many other metals, metal oxides, and moist air may cause fire and explosions. Attacks some forms of plastics, rubber, and coatings. See 29 CFR 1910.101 for specific regulations on storage of compressed gas cylinders.

OXYLITE (94-36-0) Confined storage of dry chemical may lead to decomposition and explosion. A strong oxidant; extremely reactive. Fires and explosion may result from heat or contamination, and from contact with strong acids, combustible materials, oxidizers, acids, bases, alcohols, reducing agents, metals, metal oxides, amines, accelerators, methyl methacrylate, organic matter, lithium aluminum carbide, dimethyl aniline, amines, metallic naphthenates. May attack some plastics, rubber, and coatings. Protect containers from shock and friction.

OXYMETHYLENE (50-00-0) May polymerize unless properly inhibited (usually with methanol). Forms explosive mixture with air (flash point 122°F/50°C). Incompatible with strong acids, amines, strong oxidizers, alkaline materials, nitrogen dioxide, performic acid. Reaction with hydrochloric acid produces bis-chloromethyl ether, a carcinogen.

OXYMURIATE OF POTASH (3811-04-9) A powerful oxidizer. Reacts violently with reducing agents or combustibles, ammonia gas, ammonium salts, organic matter, hydrogen iodide, organic acids, sulfuric acid. Forms explosive mixtures with metallic powders, ammonium chloride, organic solids, including agricultural materials.

3-OXYPENTANE-1,5-DIOL (111-46-6) Combustible (flash point 255°F/124°C). Incompatible with sulfuric acid, isocyanates, strong oxidizers.

OXYPHENIC ACID (120-80-9) Strong oxidizers may cause fire and explosion.

OXYSULFATOVANADIUM (27774-13-6) Incompatible with strong oxidizers such as chlorine or fluorine.

OXYTOLUENES (1319-77-3) Incompatible with strong acids, oxidizers, alkalies, aliphatic amines, amides, chlorosulfonic acid, oleum. Liquid attacks some plastics, coatings, and rubber.

OXY WASH (94-36-0) Confined storage of dry chemical may lead to decomposition and explosion. A strong oxidant; extremely reactive. Fires and explosion may result from heat or contamination, and from contact with strong acids, combustible materials, oxidizers, acids, bases, alcohols, reducing agents, metals, metal oxides, amines, accelerators, methyl methacrylate, organic matter, lithium aluminum carbide, dimethyl aniline, amines, metallic naphthenates. May attack some plastics, rubber, and coatings. Protect containers from shock and friction.

OZON (Polish) (10028-15-6) A powerful oxidizer. Spontaneously decomposes to oxygen under ordinary conditions. Reacts with all reducing agents, combustibles, organic and inorganic oxidizable materials, and can form products that are highly explosive. Incompatible with alkenes, aniline, benzene, bromine, ether, ethylene, hydrogen bromide, nitric oxide, stibine. Contact with rubber may form shock- and friction-sensitive compounds.

OZONE (10028-15-6) A powerful oxidizer. Spontaneously decomposes to oxygen under ordinary conditions. Reacts with all reducing agents, combustibles, organic and inorganic oxidizable materials, and can form products that are highly explosive. Incompatible with alkenes, aniline, benzene, bromine, ether, ethylene, hydrogen bromide, nitric oxide, stibine. Contact with rubber may form shock- and friction-sensitive compounds.

- P -

PAA (79-21-0) A powerful oxidizer and extremely sensitive explosive. Forms explosive mixture with air (flash point 100°F/38°C). Reacts violently with many substances, including acetic anhydride, combustibles, strong bases, ether solvents, olefins, organic matter, magnesium, metal oxides, heavy metals, metal chloride solutions, reducing agents, nickel, phosphorus, sodium nitride, zinc.

PAC (56-38-2) Combustible liquid. Mixtures with endrin may be explosive. Strong oxidizers may cause fire and explosions. Attacks some plastics, rubber, and coatings.

PAINT DRIER (1338-02-9) Forms explosive mixture with air (flash point 100°F/38°C). Incompatible with strong oxidizers, strong acids.

PAINTERS NAPHTHA (8032-32-4) Forms explosive mixture with air (flash point −40°F to −86°F/−40°C to −66°C). Incompatible with strong acids, strong oxidizers. Attacks some plastics, rubber, and coatings. May accumulate static electrical charges, and may cause ignition of its vapors.

PALATINOL A (84-66-2) Reacts violently with strong acids, strong oxidizers, permanganates, water. Attacks some forms of plastic.

PALATINOL C (84-74-2) Incompatible with strong acids, nitrates, strong oxidizers, strong alkalies.

PALATINOL IC (84-69-5) Incompatible with strong acids, strong oxidizers, nitrates.

PALATINOL M (131-11-3) Incompatible with strong alkalies, strong acids, nitrates, oxidizers.

PALLADIUM BLACK (7440-06-4) Powdered form is highly reactive and may cause fire and explosions on contact with oxidizers, finely divided aluminum, strong acids, acetone, and other substances.

PALLADIUM SPONGE (7440-06-4) Powdered form is highly reactive and may cause fire and explosions on contact with oxidizers, finely divided aluminum, strong acids, acetone, and other substances.

PALM BUTTER (8001-75-3) Incompatible with strong acids, oxidizers.

PALM FRUIT OIL (8001-75-3) Incompatible with strong acids, oxidizers.

PALM OIL (8001-75-3) Incompatible with strong acids, oxidizers.

PAMOLYN (112-80-1) Incompatible with strong oxidizers, perchloric acid, and aluminum powder; may cause explosions. Corrodes aluminum.

PAN (85-44-9) Incompatible with strong acids, caustics, ammonia, amines, strong oxidizers. Attacks some plastics, rubber, and coatings.

PANAM (63-25-2) Incompatible with strong oxidizers, strongly alkaline pesticides.

PANORAM (60-57-1) Incompatible with concentrated mineral acids, acid catalysts, acid oxidizing agents, phenols, reactive metals.

PANORAM 75 (137-26-8) Combustible solid (flash point 192°F/89°C). Strong oxidizers may cause fire and explosions; contact with strong acid or oxidizable materials produces toxic gases.

PANORAM D-31 (60-57-1) Incompatible with concentrated mineral acids, acid catalysts, acid oxidizing agents, phenols, reactive metals.

PANOXYL (94-36-0) Confined storage of dry chemical may lead to decomposition and explosion. A strong oxidant; extremely reactive. Fires and explosion may result from heat or contamination, and from contact with strong acids, combustible materials, oxidizers, acids, bases, alcohols, reducing agents, metals, metal oxides, amines, accelerators, methyl methacrylate, organic matter, lithium aluminum carbide, dimethyl aniline, amines, metallic naphthenates. May attack some plastics, rubber, and coatings. Protect containers from shock and friction.

PANTHION (56-38-2) Combustible liquid. Mixtures with endrin may be explosive. Strong oxidizers may cause fire and explosions. Attacks some plastics, rubber, and coatings.

PAPER MAKER'S ALUM (10043-01-3) Incompatible with water, with formation of sulfuric acid; reacts strongly with bases and many other materials. Dry material is weakly corrosive to carbon steel; aqueous solution attacks metals, producing hydrogen gas.

PAPI (9016-87-9) Water contact produces scum and carbon dioxide gas. Incompatible with acids, caustics, ammonia, amines, amides, alcohols, glycols, captrolactam solution.

PARA (106-50-3) A strong reducing agent; reacts violently with oxidizers, combustibles, organic substances. Incompatible with strong acids, organic anhydrides, isocyanates, aldehydes. Heat and light contribute to instability.

PARAACETALDEHYDE (123-63-7) Forms explosive mixture with air (flash point 96°F/36°C). Reacts with strong acids, caustics, ammonia, amines, oxidizers. Contact with acids produces acetaldehyde.

PARACHLOROCIDUM (50-29-3) Incompatible with salts of iron or aluminum, and bases. Do not store in iron containers.

PARACHLOROPHENOL (106-48-9) Forms explosive mixture with air (flash point 147°F/64°C). Reacts with oxidizers, with a risk of fire or explosions. Attacks active metals, aluminum, copper, etc. May accumulate static electrical charges, and may cause ignition of its vapors.

PARACIDE (106-46-7) Forms explosive mixture with air (flash point 150°F/66°C). Incompatible with strong oxidizers, metal powders, and alkali metals (i.e., lithium, sodium, potassium, rubidium, cesium, francium). Attacks some plastics, rubber, and coating.

PARA CRYSTALS (106-46-7) Forms explosive mixture with air (flash point 150°F/66°C). Incompatible with strong oxidizers, metal powders, and alkali metals (i.e., lithium, sodium, potassium, rubidium, cesium, francium). Attacks some plastics, rubber, and coating.

PARA CYMENE (99-87-6) Forms explosive mixture with air (flash point 117°F/47°C). Incompatible with nitric acid, strong oxidizers. Attacks and softens rubber. May accumulate static electrical charges, and may cause ignition of its vapors.

PARA CYMOL (99-87-6) Forms explosive mixture with air (flash point 117°F/47°C). Incompatible with nitric acid, strong oxidizers. Attacks and softens rubber. May accumulate static electrical charges, and may cause ignition of its vapors.

PARADI (106-46-7) Forms explosive mixture with air (flash point 150°F/66°C). Incompatible with strong oxidizers, metal powders, and alkali metals (i.e., lithium, sodium, potassium, rubidium, cesium, francium). Attacks some plastics, rubber, and coating.

PARADICHLOROBENZENE (106-46-7) Forms explosive mixture with air (flash point 150°F/66°C). Incompatible with strong oxidizers, metal powders, and alkali metals (i.e., lithium, sodium, potassium, rubidium, cesium, francium). Attacks some plastics, rubber, and coating.

PARADICHLOROBENZOL (106-46-7) Forms explosive mixture with air (flash point 150°F/66°C). Incompatible with strong oxidizers, metal powders, and alkali metals (i.e., lithium, sodium, potassium, rubidium, cesium, francium). Attacks some plastics, rubber, and coating.

PARADOW (106-46-7) Forms explosive mixture with air (flash point 150°F/66°C). Incompatible with strong oxidizers, metal powders, and alkali metals (i.e., lithium, sodium, potassium, rubidium, cesium, francium). Attacks some plastics, rubber, and coating.

PARADUST (56-38-2) Combustible liquid. Mixtures with endrin may be explosive. Strong oxidizers may cause fire and explosions. Attacks some plastics, rubber, and coatings.

PARAFFIN (8002-74-2) Reacts with oxidizers, with a risk of fire or explosions.

PARAFFIN JELLY (8012-95-1) Oxidizers may cause fire and explosions. Incompatible with nitric acid. May accumulate static electrical charges, and may cause ignition of its vapors.

PARAFFIN OIL (8002-74-2) Reacts with oxidizers, with a risk of fire or explosions.

PARAFFIN SCALE or PARAFFIN SCALE FUME (8002-74-2) Reacts with oxidizers, with a risk of fire or explosions.

PARAFFIN WAX (8002-74-2) Reacts with oxidizers, with a risk of fire or explosions.

PARAFORM (50-00-0) May polymerize unless properly inhibited (usually with methanol). Forms explosive mixture with air (flash point 122°F/50°C). Incompatible with strong acids, amines, strong oxidizers, alkaline materials, nitrogen dioxide, performic acid. Reaction with hydrochloric acid produces bis-chloromethyl ether, a carcinogen.

PARAFORM 3 (30525-89-4) Forms explosive mixture with air (flash point 160°F/71°C). Aqueous solution produces formaldehyde. Reacts violently with strong oxidizers or liquid oxygen. May accumulate static electrical charges, and may cause ignition of its vapors.

PARAFORMALDEHYDE (30525-89-4) Forms explosive mixture with air (flash point 160°F/71°C). Aqueous solution produces formaldehyde. Reacts violently with strong oxidizers or liquid oxygen. May accumulate static electrical charges, and may cause ignition of its vapors.

PARA HYDROGEN (1333-74-0) Vapors form explosive or combustible mixture with air over a wide range of concentrations. Ignites easily with oxygen. Reacts violently with oxidizers, halogens, acetylene, bromine, chlorine, fluorine, nitrous oxide and other gases. Mild steel and most iron alloys become brittle at liquid hydrogen temperatures.

PARAL (123-63-7) Forms explosive mixture with air (flash point 96°F/36°C). Reacts with strong acids, caustics, ammonia, amines, oxidizers. Contact with acids produces acetaldehyde.

629

PARALDEHYD (German) (123-63-7) Forms explosive mixture with air (flash point 96°F/36°C). Reacts with strong acids, caustics, ammonia, amines, oxidizers. Contact with acids produces acetaldehyde.

PARALDEHYDE (123-63-7) Forms explosive mixture with air (flash point 96°F/36°C). Reacts with strong acids, caustics, ammonia, amines, oxidizers. Contact with acids produces acetaldehyde.

PARALDEIDE (Italian) (123-63-7) Forms explosive mixture with air (flash point 96°F/36°C). Reacts with strong acids, caustics, amines, oxidizers. Contact with acids produces acetaldehyde.

PARAMAR (56-38-2) Combustible liquid. Mixtures with endrin may be explosive. Strong oxidizers may cause fire and explosions. Attacks some plastics, rubber, and coatings.

PARAMAR 50 (56-38-2) Combustible liquid. Mixtures with endrin may be explosive. Strong oxidizers may cause fire and explosions. Attacks some plastics, rubber, and coatings.

PARAMINODIPHENYL (92-67-1) Contact with strong oxidizers may cause fire and explosions.

PARAMOTH (106-46-7) Forms explosive mixture with air (flash point 150°F/66°C). Incompatible with strong oxidizers, metal powders, and alkali metals (i.e., lithium, sodium, potassium, rubidium, cesium, francium). Attacks some plastics, rubber, and coatings.

PARANAPHTHALENE (120-12-7) Dust or fine powder forms an explosive mixture with air. Strong oxidizers may cause fire and explosions.

PARANITROFENOL (Dutch) (100-02-7) Combustible solid. A strong oxidizer; reacts with reducing agents, combustibles, organic and other easily oxidizable materials. Incompatible with strong acids, caustics, aliphatic amines, amides.

PARANITROFENOLO (Italian) (100-02-7) Combustible solid. A strong oxidizer; reacts with reducing agents, combustibles, organic and other easily oxidizable materials. Incompatible with strong acids, caustics, aliphatic amines, amides.

PARANITROPHENOL (French and German) (100-02-7) Combustible solid. A strong oxidizer; reacts with reducing agents, combustibles, organic and other easily oxidizable materials. Incompatible with strong acids, caustics, aliphatic amines, amides.

PARANUGGETS (106-46-7) Forms explosive mixture with air (flash point 150°F/66°C). Incompatible with strong oxidizers, metal powders, and alkali metals (i.e., lithium, sodium, potassium, rubidium, cesium, francium). Attacks some plastics, rubber, and coating.

PARAPEST M-50 (298-00-0) Mixtures with magnesium may be explosive.

PARAPHENYLEN-DIAMINE (106-50-3) A strong reducing agent; reacts violently with oxidizers, combustibles, organic substances. Incompatible with strong acids, organic anhydrides, isocyanates, aldehydes. Heat and light contribute to instability.

PARAPHOS (56-38-2) Combustible liquid. Mixtures with endrin may be explosive. Strong oxidizers may cause fire and explosions. Attacks some plastics, rubber, and coatings.

PARATAF (298-00-0) Mixtures with magnesium may be explosive.

PARATHENE (56-38-2) Combustible liquid. Mixtures with endrin may be explosive. Strong oxidizers may cause fire and explosions. Attacks some plastics, rubber, and coatings.

PARATHION (56-38-2) Combustible liquid. Mixtures with endrin may be explosive. Strong oxidizers may cause fire and explosions. Attacks some plastics, rubber, and coatings.

m-**PARATHION or** *meta*-**PARATHION** (298-00-0) Mixtures with magnesium may be explosive.

PARATHION ETHYL (56-38-2) Combustible liquid. Mixtures with endrin may be explosive. Strong oxidizers may cause fire and explosions. Attacks some plastics, rubber, and coatings.

PARATHION METILE (Italian) (298-00-0) Mixtures with magnesium may be explosive.

PARATHION METHYL (298-00-0) Mixtures with magnesium may be explosive.

PARATOX (298-00-0) Mixtures with magnesium may be explosive.

PARAWET (56-38-2) Combustible liquid. Mixtures with endrin may be explosive. Strong oxidizers may cause fire and explosions. Attacks some plastics, rubber, and coatings.

PARAZENE (106-46-7) Forms explosive mixture with air (flash point 150°F/66°C). Incompatible with strong oxidizers, metal powders, and alkali metals (i.e., lithium, sodium, potassium, rubidium, cesium, francium). Attacks some plastics, rubber, and coatings.

PARIDOL (298-00-0) Mixtures with magnesium may be explosive.

PARIS GREEN (12002-03-8) Contact with strong oxidizers may cause fire and explosions.

PAROL (8012-95-1) Oxidizers may cause fire and explosions. Incompatible with nitric acid. May accumulate static electrical charges, and may cause ignition of its vapors.

PARROT GREEN (12002-03-8) Contact with strong oxidizers may cause fire and explosions.

PARZATE (142-59-6) Boiling water contact produces hydrogen sulfide and carbon disulfide vapors.

PATENT ALUM (10043-01-3) Aqueous solution is a strong acid. Incompatible with caustics.

PATENT ALUMINUM (10043-01-3) Aqueous solution is a strong acid. Incompatible with caustics.

PATRON M (298-00-0) Mixtures with magnesium may be explosive.

PBI CROP SAVER (121-75-5) Incompatible with strong oxidizers, magnesium, alkaline pesticides. Attacks metals, some plastics, rubber, and coatings.

PBNA (90-30-2) Combustible. Incompatible with oxidizers, strong acids, organic anhydrides, isocyanates, aldehydes.

PBNA (135-88-6) Combustible. May accumulate static electrical charges, and may cause ignition of its vapors. Incompatible with oxidizers, strong acids, organic anhydrides, isocyanates, aldehydes.

PBX(AF) 108 (121-82-4) Contact with mercury fulminate amy cause detonation. Contact with combustibles or strong oxidizers may cause fire and explosions.

PCB (1336-36-3) Incompatible with strong oxidizers, strong acids.

PCC (8001-35-2) Reacts with oxidizers, with a risk of fire or explosions. Attacks metals in the presence of moisture.

PCHO (123-63-7) Forms explosive mixture with air (flash point 96°F/36°C). Reacts with strong acids, caustics, ammonia, amines, oxidizers. Contact with acids produces acetaldehyde.

PCL (77-47-4) Incompatible with water, producing hydrochloric acid. Contact with sodium may be explosive. Corrodes iron and other metals in the presence of moisture.

PCM (594-42-3) Incompatible with caustics or amines (with rapid decomposition), hot iron (with production of toxic carbon tetrachloride fumes), hot water (causes formation of hydrochloric acid), sulfur, and carbon dioxide. Attacks some plastics, rubber, and coatings.

PCP (87-86-5) Hot water causes decomposition, producing hydrochloric acid. Strong oxidizers may cause fire and explosions.

PDB (106-46-7) Forms explosive mixture with air (flash point 150°F/66°C). Incompatible with strong oxidizers, metal powders, and alkali metals (i.e., lithium, sodium, potassium, rubidium, cesium, francium). Attacks some plastics, rubber, and coating.

PDCB (106-46-7) Forms explosive mixture with air (flash point 150°F/66°C). Incompatible with strong oxidizers, metal powders, and alkali metals (i.e., lithium, sodium, potassium, rubidium, cesium, francium). Attacks some plastics, rubber, and coating.

PE (115-77-5) Incompatible with organic acids, oxidizers. Forms explosive material with thiophosphoryl chloride + heat.

PEANUT OIL (8002-03-7) Incompatible with strong acids, oxidizers.

PEARL ASH (3811-04-9) A powerful oxidizer. Reacts violently with reducing agents or combustibles, ammonia gas, ammonium salts, organic matter, hydrogen iodide, organic acids, sulfuric acid. Forms explosive mixtures with metallic powders, ammonium chloride, organic solids, including agricultural materials.

PEARL STEARIC (57-11-4) Incompatible with strong oxidizers, sulfuric acid, caustics, ammonia, amines, isocyanates, alkylene oxides, epichlorohydrin. Attacks chemically active metals.

PEAR OIL (123-92-2) Forms explosive mixture with air (flash point 77°F/25°C). Incompatible with strong alkalies, strong acids, nitrates, oxidizers. Reacts violently with reducing agents. Attacks asbestos; softens and dissolves many plastics, rubber, and coatings.

PEAR OIL (626-38-0) Incompatible with strong acids, nitrates, strong alkalies, strong oxidizers, heat. May soften or dissolve plastics.

PEAR OIL (628-63-7) Incompatible with strong alkalies, strong acids, nitrates, strong oxidizers. Attacks some plastics, coatings, and rubber.

PEARSALL (7446-70-0) Contact with air or water produces hydrochloric acid. Water, alcohol, alkenes cause polymerization. Incompatible with nitrobenzene, organic material, and bases. Attacks metals in presence of moisture.

PEB 1 (50-29-3) Incompatible with salts of iron or aluminum, and bases. Do not store in iron containers.

PEBBLE LIME (1305-78-8) Reacts violently with water. Incompatible with ethanol, hydrogen fluoride, acids, halogens, metal halides, light metals, some oxides, boron trifluoride, chlorine trifluoride, liquid hydrofluoric acid, phosphorus pentoxide, boric oxide and calcium chloride mixtures, fluorine, chlorine trifluoride, carbon dioxide.

PEDRACZAK (58-89-9) Not combustible, but may be dissolved in a combustible solvent. If solvent comes in contact with oxidizers, fire and explosions may result.

PELADOW (10043-52-4) Incompatible with water, bromine trifluoride, 2-furan, percarboxylic acid. Attacks metals.

PELAGOL D (106-50-3) A strong reducing agent; reacts violently with oxidizers, combustibles, organic substances. Incompatible with strong acids, organic anhydrides, isocyanates, aldehydes. Heat and light contribute to instability.

PELAGOL DR (106-50-3) A strong reducing agent; reacts violently with oxidizers, combustibles, organic substances. Incompatible with strong acids, organic anhydrides, isocyanates, aldehydes. Heat and light contribute to instability.

PELAGOL GREY C (120-80-9) Strong oxidizers may cause fire and explosion.

PELAGOL GREY D (106-50-3) A strong reducing agent; reacts violently with oxidizers, combustibles, organic substances. Incompatible with strong acids, organic anhydrides, isocyanates, aldehydes. Heat and light contribute to instability.

PELAGOL GREY J (95-80-7) Incompatible with oxidizers, acids, organic anhydrides, isocyanates, aldehydes. Attacks aluminum, brass, bronze, copper, zinc.

PELAGOL GREY RS (108-46-3) Reacts violently with strong oxidizers, nitric acid. Incompatible with acetanilide, albumin, alkalies, antipyrine, camphor, ferric salts, menthol, spirit nitrous ether. Absorbs moisture from air (hygroscopic). May accumulate static electrical charges, and may cause ignition of its vapors.

PELAGOL RS (108-46-3) Reacts violently with strong oxidizers, nitric acid. Incompatible with acetanilide, albumin, alkalies, antipyrine, camphor, ferric salts, menthol, spirit nitrous ether. Absorbs moisture from air (hygroscopic). May accumulate static electrical charges, and may cause ignition of its vapors.

PELARGIC ACID (112-05-0) Incompatible with sulfuric acid, caustics, ammonia, aliphatic amines, alkanolamines, isocyanates, alkylene oxides, epichlorohydrin.

PELARGONIC ACID (112-05-0) Incompatible with sulfuric acid, caustics, ammonia, aliphatic amines, alkanolamines, isocyanates, alkylene oxides, epichlorohydrin.

PELARGONIC ALCOHOL (143-08-8) Forms explosive mixture with air (flash point 165°F/74°C). Incompatible with strong acids, caustics, aliphatic amines, isocyanates, strong oxidizers.

PELTOL D (106-50-3) A strong reducing agent; reacts violently with oxidizers, combustibles, organic substances. Incompatible with strong acids, organic anhydrides, isocyanates, aldehydes. Heat and light contribute to instability.

PENATROL (1912-24-9) Incompatible with strong acids.

PENCHLOROL (87-86-5) Hot water causes decomposition, producing hydrochloric acid. Strong oxidizers may cause fire and explosions.

PENNAMINE (94-75-7) Decomposes in sunlight. Incompatible with strong oxidizers; may cause fire and explosions.

PENNAMINE D (94-75-7) Decomposes in sunlight. Incompatible with strong oxidizers; may cause fire and explosions.

PENNCAP-M (298-00-0) Mixtures with magnesium may be explosive.

PENNFLOAT M (112-55-0) Contact with strong oxidizers may cause fire and explosions.

PENNFLOAT S (112-55-0) Contact with strong oxidizers may cause fire and explosions.

PENTA (87-86-5) Hot water causes decomposition, producing hydrochloric acid. Strong oxidizers may cause fire and explosions.

1,4,7,10,13-PENTAAZATRIDECANE (112-57-2) Forms explosive mixture with air. Incompatible with acids, organic anhydrides, halogenated hydrocarbons, isocyanates, nitroparaffins, vinyl acetate, acrylates, substituted allyls, alkylene oxides, epichlorohydrin, ketones, aldehydes, alcohols, glycols, mercury, phenols, cresols, captrolactam solution, strong oxidizers. Attacks aluminum, copper, lead, tin, zinc, and alloys.

PENTABORANE (19624-22-7) Forms explosive mixture with air (flash point 95°F/35°C). May ignite spontaneously in moist air. Water contact produces boric acid and explosive hydrogen gas. Oxidizers causes formation of highly explosive mixtures. Halogens and halogenated compounds may cause explosions. Attacks some plastics, rubber, and coatings.

PENTABORANE(9) (19624-22-7) Forms explosive mixture with air (flash point 95°F/35°C). May ignite spontaneously in moist air. Water contact produces boric acid and explosive hydrogen gas. Oxidizers cause formation of highly explosive mixtures. Halogens and halogenated compounds may cause explosions. Attacks some plastics, rubber, and coatings.

PENTABORANE UNDECAHYDRIDE (19624-22-7) Forms explosive mixture with air (flash point 95°F/35°C). May ignite spontaneously in moist air. Water contact forms boric acid and explosive hydrogen gas. Oxidizers form highly explosive mixtures. Halogen and halogenated compounds may cause explosions. Attacks some plastics, rubber, and coatings.

PENTABORON NONAHYDRIDE (19624-22-7) Forms explosive mixture with air (flash point 95°F/35°C). May ignite spontaneously in moist air. Water contact produces boric acid and explosive hydrogen gas. Oxidizers cause formation of highly explosive mixtures. Halogens and halogenated compounds may cause explosions. Attacks some plastics, rubber, and coatings.

(9)-PENTABORON NONAHYDRIDE (19624-22-7) Forms explosive mixture with air (flash point 95°F/35°C). May ignite spontaneously in moist air. Water contact produces boric acid and explosive hydrogen gas. Oxidizers cause formation of highly explosive mixtures. Halogens and halogenated compounds may cause explosions. Attacks some plastics, rubber, and coatings.

PENT ACETATE (628-63-7) Incompatible with strong alkalies, strong acids, nitrates, strong oxidizers. Attacks some plastics, coatings, and rubber.

PENTACHLOORETHAAN (Dutch) (76-01-7) Incompatible with water, producing dichloroacetic acid. May self-ignite. Reacts violently with alkali metals (i.e., lithium, sodium, potassium, rubidium, cesium, francium). Forms spontaneously explosive materials with alkalies, metals. Forms shock- and friction-sensitive material on mixing with potassium.

PENTACHLOORFENOL (Dutch) (87-86-5) Hot water causes decomposition, producing hydrochloric acid. Strong oxidizers may cause fire and explosions.

PENTACHLORAETHAN (German) (76-01-7) Incompatible with water, producing dichloroacetic acid. May self-ignite. Reacts violently with alkali metals (i.e., lithium, sodium, potassium, rubidium, cesium, francium). Forms spontaneously explosive materials with alkalies, metals. Forms shock- and friction-sensitive material on mixing with potassium.

PENTACHLORETHANE (French) (76-01-7) Incompatible with water, producing dichloroacetic acid. May self-ignite. Reacts violently with alkali metals (i.e., lithium, sodium, potassium, rubidium, cesium, francium). Forms spontaneously explosive materials with alkalies, metals. Forms shock- and friction-sensitive material on mixing with potassium.

PENTACHLORIN (50-29-3) Incompatible with salts of iron or aluminum, and bases. Do not store in iron containers.

PENTACHLOROANTIMONY (7647-18-9) Water contact causes formation of hydrogen chloride gas. Reacts violently with ammonia, caustics. Attacks many metals. Contact with air produces heavier-than-air corrosive vapor.

PENTACHLOROETHANE (76-01-7) Incompatible with water, producing dichloroacetic acid. May self-ignite. Reacts violently with alkali metals (i.e., lithium, sodium, potassium, rubidium, cesium, francium). Forms spontaneously explosive materials with alkalies, metals. Forms shock- and friction-sensitive material on mixing with potassium.

PENTACHLOROFENOL (87-86-5) Hot water causes decomposition, producing hydrochloric acid. Strong oxidizers may cause fire and explosions.

PENTACHLORONAPHTHALENE (1321-64-8) Strong oxidizers may cause fire and explosions. Heat contributes to instability.

PENTACHLOROPHENATE (87-86-5) Hot water causes decomposition, producing hydrochloric acid. Strong oxidizers may cause fire and explosions.

PENTACHLOROPHENOL (87-86-5) Hot water causes decomposition, producing hydrochloric acid. Strong oxidizers may cause fire and explosions.

2,3,4,5,6-PENTACHLOROPHENOL (87-86-5) Hot water causes decomposition, producing hydrochloric acid. Strong oxidizers may cause fire and explosions.

PENTACHLOROPHENOL, DOWICIDE EC-7 (87-86-5) Hot water causes decomposition, producing hydrochloric acid. Strong oxidizers may cause fire and explosions.

PENTACHLOROPHENOL, DP-2 (87-86-5) Hot water causes decomposition, producing hydrochloric acid. Strong oxidizers may cause fire and explosions.

PENTACHLOROPHENOL, TECHNICAL (87-86-5) Hot water causes decomposition, producing hydrochloric acid. Strong oxidizers may cause fire and explosions.

PENTACHLOROPHENYL CHLORIDE (118-74-1) Reacts violently with strong oxidizers, formyldimethylamine.

PENTACHLORPHENOL (German) (87-86-5) Hot water causes decomposition, producing hydrochloric acid. Strong oxidizers may cause fire and explosions.

PENTACLOROETANO (Italian) (76-01-7) Incompatible with water, producing dichloroacetic acid. May self-ignite. Reacts violently with alkali metals (i.e., lithium, sodium, potassium, rubidium, cesium, francium). Forms spontaneously explosive materials with alkalies, metals. Form shock- and friction-sensitive material on mixing with potassium.

PENTACLOROFENOLO (Italian) (87-86-5) Hot water causes decomposition, producing hydrochloric acid. Strong oxidizers may cause fire and explosions.

PENTACON (87-86-5) Hot water causes decomposition forming hydrochloric acid. Strong oxidizers may cause fire and explosions.

O-PENTADECADIENYL SALICYLIC ACID (8001-24-7) Incompatible with sulfuric acid, caustics, ammonia, amines, isocyanates, alkylene oxides, epichlorohydrin.

PENTADECANOL (629-76-5) Incompatible with strong acids, caustics, aliphatic amines, isocyanates, oxidizers.

1-PENTADECANOL (629-76-5) Incompatible with strong acids, caustics, aliphatic amines, isocyanates, oxidizers.

PENTADECYL ALCOHOL (629-76-5) Incompatible with strong acids, caustics, aliphatic amines, isocyanates, oxidizers.

1,3-PENTADIENE (504-60-9) Forms explosive mixture with air (flash point −20°F/−29°C). Reacts violently with oxidizers. Attacks some plastics, rubber, and coatings. May accumulate static electrical charges, and may cause ignition of its vapors.

1,4-PENTADIENE (591-93-5) Forms explosive mixture with air (flash point 40°F/40°C). Strong oxidizers may cause fire and explosions.

PENTA-1,4-DIENE (591-93-5) Forms explosive mixture with air (flash point 40°F/40°C). Strong oxidizers may cause fire and explosions.

cis-**PENTADIENE-1,3** (504-60-9) Forms explosive mixture with air (flash point −20°F/−29°C). Reacts violently with oxidizers. Attacks some plastics, rubber, and coatings. May accumulate static electrical charges, and may cause ignition of its vapors.

trans-**PENTADIENE-1,3** (504-60-9) Forms explosive mixture with air (flash point −20°F/−29°C). Reacts violently with oxidizers. Attacks some plastics, rubber, and coatings. May accumulate static electrical charges, and may cause ignition of its vapors.

PENTAERYTHRITE (115-77-5) Incompatible with organic acids, oxidizers. Forms explosive material with thiophosphoryl chloride + heat.

PENTAERYTHRITOL (115-77-5) Organic acids, oxidizers. Forms explosive material with thiophosphoryl chloride + heat.

PENTAETHYLENE HEXAMINE (4067-16-7) Incompatible with acids, organic anhydrides, isocyanates, aldehydes, oxidizers.

PENTAETHYLENEHEXAMINE, TETRAETHYLENEPENTAMINE MIXTURE (CAS unknown) Forms explosive mixture with air. Incompatible with acids, organic anhydrides, chlorinated hydrocarbons, isocyanates, vinyl acetate, acrylates, substituted allyls, alkylene oxides, epichlorohydrin, ketones, aldehydes, alcohols, glycols, phenols, cresols, captrolactam solution, strong oxidizers. Attacks aluminum, copper, lead, tin, zinc, and alloys.

PENTAFLUOROANTIMONY (7783-70-2) Water contact produces hydrofluoric acid. Incompatible with reducing agents, organics, phosphates, siliceous materials. Severe corrosion of glass or metals (with formation of explosive hydrogen gas) occurs in presence of moisture.

PENTAFLUOROMONOCHLOROETHANE (76-15-3) Thermal decomposition occurs at high temperatures with alkalies and alkaline earth metals.

PENTA-KIL (87-86-5) Hot water causes decomposition, producing hydrochloric acid. Strong oxidizers may cause fire and explosions.

PENTALIN (76-01-7) Incompatible with water, producing dichloroacetic acid. May self-ignite. Reacts violently with alkali metals (i.e., lithium, sodium, potassium, rubidium, cesium, francium). Forms spontaneously explosive materials with alkalies, metals. Forms shock- and friction-sensitive material on mixing with potassium.

PENTAMETHYLENE (287-92-3) Forms explosive mixture with air (flash point less than 20°F/−7°C). May accumulate static electrical charges, and may cause ignition of its vapors. Contact with strong oxidizers may cause fire and explosions.

PENTAMETHYLENEIMINE (110-89-4) Forms explosive mixture with air (flash point 61°F/16°C). Aqueous solution is a strong base; reacts with acids, organic anhydrides, isocyantes, alkylene oxides, epichlorohydrin, aldehydes, alcohols, glycols, phenols, cresols, captrolactam solutions. A powerful reducing agent; reacts violently with oxidizers.

PENTAN (109-66-0) Forms explosive mixture with air (flash point −57°F/−49°C). Incompatible with strong oxidizers (may cause fire and explosions). Attacks some plastics, rubber, and coatings.

PENTANAL (110-62-3) Forms explosive mixture with air (flash point 54°F/12°C). Incompatible with strong acids, oxidizers, caustics, amines.

N-PENTANAL (110-62-3) Forms explosive mixture with air (flash point 54°F/12°C). Incompatible with strong acids, oxidizers, caustics, amines.

PENTAN-2,4-DIONE (123-54-6) Forms explosive mixture with air (flash point 95°F/35°C). Reacts violently with strong oxidizers. Incompatible with aliphatic amines, alkanolamines, organic acids, isocyanates.

PENTANE (109-66-0) Forms explosive mixture with air (flash point −57°F/−49°C). Incompatible with strong oxidizers (cause fire and explosions). Attacks some plastics, rubber, and coatings.

n-**PENTANE or** *normal*-**PENTANE** (109-66-0) Forms explosive mixture with air (flash point −57°F/−49°C). Incompatible with strong oxidizers (cause fire and explosions). Attacks some plastics, rubber, and coatings.

PENTANEDIAL (111-30-8) Water contact produces a polymer solution. A strong reducing agent. Incompatible with strong acids, caustics, ammonia, amines, strong oxidizers.

1,5-PENTANEDIAL (111-30-8) Water contact produces a polymer solution. A strong reducing agent. Incompatible with strong acids, caustics, ammonia, amines, strong oxidizers.

PENTANEDIOIC ACID (110-94-1) Reacts with oxidizers and strong bases.

2,4-PENTANEDIOL, 2-METHYL- (107-41-5) Forms explosive mixture with air (flash point 209°F/98°C). Incompatible with strong acids, caustics, aliphatic amines, isocyanates, strong oxidizers.

PENTANEDIONE (123-54-6) Forms explosive mixture with air (flash point 95°F/35°C). Reacts violently with strong oxidizers. Incompatible with aliphatic amines, alkanolamines, organic acids, isocyanates.

1,5-PENTANEDIONE (111-30-8) Water contact produces a polymer solution. A strong reducing agent. Incompatible with strong acids, caustics, ammonia, amines, strong oxidizers.

2,4-PENTANEDIONE (123-54-6) Forms explosive mixture with air (flash point 95°F/35°C). Reacts violently with strong oxidizers. Incompatible with aliphatic amines, alkanolamines, organic acids, isocyanates.

1-PENTANETHIOL (110-66-7) Forms explosive mixture with air (flash point 65°F/18°C). Incompatible with strong oxidizers, reducing agents, alkali metals, calcium hypochlorite.

PENTANOIC ACID (109-52-4) Forms explosive mixture with air (flash point 205°F/96°C). Incompatible with sulfuric acid, caustics, ammonia, amines, isocyanates, alkylene oxides, epichlorohydrin, strong oxidizers.

n-**PENTANOIC ACID** (109-52-4) Forms explosive mixture with air (flash point 205°F/96°C). Incompatible with sulfuric acid, caustics, ammonia, amines, isocyanates, alkylene oxides, epichlorohydrin, strong oxidizers.

PENTANOL (71-41-0) Forms explosive mixture with air (flash point 91°F/33°C). Incompatible with strong acids, caustics, aliphatic amines, isocyanates. Alkali metals and alkaline earth cause formation of hydrogen gas.

1-PENTANOL (71-41-0) Forms explosive mixture with air (flash point 91°F/33°C). Incompatible with strong acids, caustics, aliphatic amines, isocyanates. Alkali metals and alkaline earth cause formation of hydrogen gas.

PENTAN-1-OL (71-41-0) Forms explosive mixture with air (flash point 91°F/33°C). Incompatible with strong acids, caustics, aliphatic amines, isocyanates. Alkali metals and alkaline earth cause formation of hydrogen gas.

PENTANOL-1 (71-41-0) Forms explosive mixture with air (flash point 91°F/33°C). Incompatible with strong acids, caustics, aliphatic amines, isocyanates. Alkali metals and alkaline earth cause formation of hydrogen gas.

2-PENTANOL (6032-29-7) Forms explosive mixture with air (flash point 93°F/34°C). Reacts violently with strong oxidizers and alkali metals. May also react with alkali earth metals.

n-**PENTANOL or** *normal*-**PENTANOL** (71-41-0) Forms explosive mixture with air (flash point 91°F/33°C). Incompatible with strong acids, caustics, aliphatic amines, isocyanates. Alkali metals and alkaline earth cause formation of hydrogen gas.

tert-**PENTANOL or** *tertiary*-**PENTANOL** (75-85-4) Forms explosive mixture with air (flash point 77°F/25°C). May accumulate static electrical charges, and may cause ignition of its vapors. Incompatible with strong acids, caustics, aliphatic amines, isocyanates, strong oxidizers, alkali metals (i.e., lithium, sodium, potassium, rubidium, cesium, francium).

638

1-PENTANOL ACETATE (628-63-7) Incompatible with strong alkalies, strong acids, nitrates, strong oxidizers. Attacks some plastics, coatings, and rubber.

2-PENTANOL ACETATE (626-38-0) Incompatible with strong acids, nitrates, strong alkalies, strong oxidizers, heat. May soften or dissolve plastics.

2-PENTANOL, 4-METHYL- (108-11-2) Forms explosive mixture with air (flash point 106°F/41°C). Contact with alkali metals produces hydrogen gas. Incompatible with strong acids, caustics, aliphatic amines, isocyanates. Attacks some plastics, rubber, and coatings. May accumulate static electrical charges, and may cause ignition of its vapors.

PENTANONE-2 (107-87-9) May accumulate static electrical charges, and may cause ignition of its vapors. Forms explosive mixture with air (flash point 45°F/7°C). Reacts violently with strong oxidizers. Attacks some plastics, rubber, and coatings.

2-PENTANONE (107-87-9) May accumulate static electrical charges, and may cause ignition of its vapors. Forms explosive mixture with air (flash point 45°F/7°C). Reacts violently with strong oxidizers. Attacks some plastics, rubber, and coatings.

3-PENTANONE (96-22-0) Forms explosive mixture with air (flash point 55°F/13°C). Incompatible with strong acids, aliphatic amines, strong oxidizers. Attacks some plastics, rubber, and coatings. May accumulate static electrical charges, and may cause ignition of its vapors.

PENTANONE-3 (96-22-0) Forms explosive mixture with air (flash point 55°F/13°C). Incompatible with strong acids, aliphatic amines, strong oxidizers. Attacks some plastics, rubber, and coatings. May accumulate static electrical charges, and may cause ignition of its vapors.

3-PENTANONE DIMETHYL ACETONE (96-22-0) Forms explosive mixture with air (flash point 55°F/13°C). Incompatible with strong acids, aliphatic amines, strong oxidizers. Attacks some plastics, rubber, and coatings. May accumulate static electrical charges, and may cause ignition of its vapors.

2-PENTANONE, 4-HYDROXY-4-METHYL- (123-42-2) Incompatible with strong acids, strong alkalies (cause formation of flammable acetone vapors), aliphatic amines, isocyanates, oxidizers, alkali metals. Forms explosive mixture with air (flash point 136°F/58°C). Attacks some forms of plastics, resins, and rubber.

2-PENTANONE, 4-METHYL- (108-10-1) Forms explosive mixture with air (flash point 64°F/18°C). Incompatible with strong oxidizers, strong acids, aliphatic amines. Dissolves some plastics, resins, and rubber.

PENTAPHENE (80-46-6) Reacts with strong oxidizers. May accumulate static electrical charges, and may cause ignition of its vapors.

PENTASOL (87-86-5) Hot water causes decomposition, producing hydrochloric acid. Strong oxidizers may cause fire and explosions.

PENTASOL (71-41-0) Forms explosive mixture with air (flash point 91°F/33°C). Incompatible with strong acids, caustics, aliphatic amines, isocyanates. Alkali metals and alkaline earth cause formation of hydrogen gas.

639

PENTASULFURE de PHOSPHORE (French) (1314-80-3) Contact with water or alcohol yields heat and possible ignition. May self-ignite in air. Reacts violently with acids, oxidizers.

PENTECH (50-29-3) Incompatible with salts of iron or aluminum, and bases. Do not store in iron containers.

PENTEK (115-77-5) Incompatible with organic acids, oxidizers. Forms explosive material with thiophosphoryl chloride + heat.

1-PENTENE (109-67-1) Forms explosive mixture with air (flash point 0°F/−18°C). Incompatible with strong acids, oxidizers.

PENTIFORMIC ACID (142-62-1) Reacts with strong oxidizers. Incompatible with sulfuric acid, caustics, ammonia, amines, isocyanates, alkylene oxides, epichlorohydrin. Attacks common metals.

2-PENTYLACETATE (626-38-0) Incompatible with strong acids, nitrates, strong alkalies, strong oxidizers, heat. May soften or dissolve plastics.

2-PENTYL ACETATE (626-38-0) Incompatible with strong acids, nitrates, strong alkalies, strong oxidizers, heat. May soften or dissolve plastics.

tert-**PENTYL ACETATE** or *tertiary*-**PENTYL ACETATE** (625-16-1) Forms explosive mixture with air (flash point 77°F/25°C). May accumulate static electrical charges, and may cause ignition of its vapors. Incompatible with strong acids, nitrates.

PENTYL ACETATES (628-63-7) Incompatible with strong alkalies, strong acids, nitrates, strong oxidizers. Attacks some plastics, coatings, and rubber.

PENTYL ALCOHOL (71-41-0) Forms explosive mixture with air (flash point 91°F/33°C). Incompatible with strong acids, caustics, aliphatic amines, isocyanates. Alkali metals and alkaline earth cause formation of hydrogen gas.

sec-**PENTYL ALCOHOL** or *secondary*-**PENTYL ALCOHOL** (6032-29-7) Forms explosive mixture with air (flash point 93°F/34°C). Reacts violently with strong oxidizers and alkali metals. May also react with alkali earth metals.

2-PENTYLBROMIDE (107-81-3) Forms explosive mixture with air (flash point 90°F/32°C). Incompatible with strong oxidizers, strong acids.

PENTYL CARBINOL (111-27-3) Forms explosive mixture with air (flash point 145°F/63°C). Incompatible with strong acids, caustics, aliphatic amines, isocyanates, strong oxidizers.

3-PENTYLCARBINOL (97-95-0) Forms explosive mixture with air (flash point 70°F/21°C). Incompatible with strong acids, caustics, isocyanates, amines, isocyanates.

sec-**PENTYLCARBINOL** (97-95-0) Forms explosive mixture with air (flash point 70°F/21°C). Incompatible with strong acids, caustics, isocyanates, amines, isocyanates.

1-PENTYL CHLORIDE (543-59-9) Forms explosive mixture with air (flash point 34°F/1°C). Strong oxidizers may cause fire and explosions.

PENTYL ESTER OF ACETIC ACID (628-63-7) Incompatible with strong alkalies, strong acids, nitrates, strong oxidizers. Attacks some plastics, coatings, and rubber.

PENTYL ETHER (693-65-2) Forms explosive mixture with air (flash point 134°F/56°C). Reacts violently with strong oxidizers. Ethers can forms unstable and explosive peroxides.

PENTYLFORMIC ACID (142-62-1) Reacts with strong oxidizers. Incompatible with sulfuric acid, caustics, ammonia, amines, isocyanates, alkylene oxides, epichlorohydrin. Attacks common metals.

PENTYL MERCAPTAN (110-66-7) Forms explosive mixture with air (flash point 65°F/18°C). Incompatible with strong oxidizers, reducing agents, alkali metals, calcium hypochlorite.

PENTYL METHYL KETONE (110-43-0) Forms explosive mixture with air (flash point 102°F/39°C). Incompatible with strong acids, alkalies, aliphatic amines, oxidizers. Attacks some plastics and rubber.

PENTYL NITRITE (463-04-7) Forms explosive mixture with air (flash point 0°F/−18°C). Decomposes in light, air, or water; produces oxides of nitrogen. A strong oxidizer; reacts with reducing agents, combustibles, organics, and other strong oxidizers. Corrodes metals with moisture.

PENTYL PROPIONATE (624-54-4) Forms explosive mixture with air (flash point 106°F/41°C). Incompatible with sulfuric acid, nitric acid, nitrates, oxidizers.

n-**PENTYL PROPIONATE** (624-54-4) Forms explosive mixture with air (flash point 106°F/41°C). Incompatible with sulfuric acid, nitric acid, nitrates, oxidizers.

PENTYLSILICON TRICHLORIDE (107-72-2) Forms explosive mixture with air (flash point 145°F/62°C). Water contact produces hydrochloric acid. Incompatible with strong bases. Attacks metals in presence of moisture.

PENTYLTRICHLOROSILANE (107-72-2) Forms explosive mixture with air (flash point 145°F/62°C). Water contact produces hydrochloric acid. Incompatible with strong bases. Attacks metals in presence of moisture.

PENWAR (87-86-5) Hot water causes decomposition, producing hydrochloric acid. Strong oxidizers may cause fire and explosions.

PERACETIC ACID or PERACETIC ACID SOLUTION (79-21-0) A powerful oxidizer and extremely sensitive explosive. Forms explosive mixture with air (flash point 100°F/38°C). Reacts violently with many substances, including acetic anhydride, combustibles, strong bases, ether solvents, olefins, organic matter, magnesium, metal oxides, heavy metals, metal chloride solutions, reducing agents, nickel, phosphorus, sodium nitride, zinc.

PERATOX (87-86-5) Hot water causes decomposition, producing hydrochloric acid. Strong oxidizers may cause fire and explosions.

PERCARBAMIDE (124-43-6) Combustible solid. At 122°F/50°C reacts with dust, organics.

PERCARBAMITE (124-43-6) Combustible solid. At 122°F/50°C reacts with dust, organics.

PERCHLOR (127-18-4) Incompatible with strong oxidizers, finely divided metals, caustics. Stable up to 258°F/126°C; at this temperature product gives off poisonous fumes.

PERCHLORATE de MAGNESIUM (French) (10034-81-8) A powerful oxidizer. Reacts violently with reducing agents, organic matter, ethylene oxide, powdered metals, phosphorus, dimethylsulfoxide. Forms explosive material with ethyl alcohol. Incompatible with many materials. Attacks many metals.

PERCHLORATE SOLUTION (7601-90-3) Aqueous solution is a strong acid. Anhydrous material decomposes explosively at atmospheric pressure. Heat above 165°F/74°C may cause explosion. A strong oxidizer. Reacts violently with reducing agents, combustible materials, acetic acid, alcohols, glycols, hypophosphites, ketones, metal powders, strong acids, sulfoxides, with risk of fire and explosions. Contact with common materials (wood, fabric) can cause spontaneous ignition.

PERCHLORIC ACID (7601-90-3) Aqueous solution is a strong acid. Anhydrous material decomposes explosively at atmospheric pressure. Heat above 165°F/75°C may cause explosion. A strong oxidizer. Reacts violently with reducing agents, combustible materials, acetic acid, alcohols, glycols, hypophosphites, ketones, metal powders, strong acids, sulfoxides, with risk of fire and explosions. Contact with common materials (wood, fabric) can cause spontaneous ignition.

PERCHLORIC ACID, AMMONIUM SALT (7790-98-9) Explosion-sensitive to friction and heat. Contact with organic matter, sulfur, metal powders, ferrocene produces shock-sensitive materials.

PERCHLORIC ACID, BARIUM SALT or PERCHLORIC ACID, BARIUM SALT $3H_2O$ (13465-95-7) A strong oxidizer. Reacts violently with reducing agents and combustibles. Contact with strong acids produces explosive perchloric acid. Contact with ammonium compounds, metal powders (especially magnesium, aluminum, sulfur, calcium hydride, and strontium hydride), or sulfur produces friction-/impact-sensitive compounds. Explodes when heated or shocked.

PERCHLORIC ACID, MAGNESIUM SALT (10034-81-8) A powerful oxidizer. Reacts violently with reducing agents, organic matter, ethylene oxide, powdered metals, phosphorus, dimethylsulfoxide. Forms explosive material with ethyl alcohol. Incompatible with many materials. Attacks many metals.

PERCHLORIC ACID SOLUTION (7601-90-3) Aqueous solution is a strong acid. Anhydrous material decomposes explosively at atmospheric pressure. Heat above 165°F/75°C may cause explosion. A strong oxidizer. Reacts violently with reducing agents, combustible materials, acetic acid, alcohols, glycols, hypophosphites, ketones, metal powders, strong acids, sulfoxides, with risk of fire and explosions. Contact with common materials (wood, fabric) can cause spontaneous ignition.

PERCHLORIDE OF MERCURY (7487-94-7) Incompatible with light metals (e.g., aluminum, magnesium, beryllium), sodium, potassium.

PERCHLOROBENZENE (118-74-1) Reacts violently with strong oxidizers, formyldimethylamine.

PERCHLOROBUTADIENE (87-68-3) Forms explosive mixture with air (flash point 195°F/90°C). Reacts strongly with oxidizers, aluminum powder. Attacks aluminum. Attacks some plastics, rubber, and coatings.

PERCHLOROCYCLOPENTADIENE (77-47-4) Incompatible with water, producing hydrochloric acid. Contact with sodium may be explosive. Corrodes iron and other metals in the presence of moisture.

PERCHLOROETHANE (67-72-1) Incompatible with aluminum or zinc. Alkalies cause formation of spontaneously explosive chloroacetylene. Attacks some plastics, rubber, and coatings.

PERCHLOROETHYLENE (127-18-4) Incompatible with strong oxidizers, finely divided metals, caustics. Stable up to 258°F/126°C; at this temperature product gives off poisonous fumes.

PERCHLOROMETHANE (56-23-5) Becomes corrosive when in contact with water. Corrosive to metals. Reacts violently with many compounds. Decomposes on contact with chemically active metals such as sodium, potassium, and magnesium. Incompatible with allyl alcohol, fluorine gas, alkali metals, aluminum. Attacks some coatings, plastics, and rubber.

PERCHLOROMETHYL MERCAPTAN (594-42-3) Incompatible with caustics or amines (with rapid decomposition), hot iron (produces toxic carbon tetrachloride fumes), hot water (produces hydrochloric acid), sulfur, and carbon dioxide. Attacks some plastics, rubber, and coatings.

PERCHLORON (7778-54-3) Decomposes in heat or sunlight. Incompatible with acids, moisture, reducing agents, combustible materials, all other chemicals, especially acetylene, aniline and all other amines, anthracene, carbon tetrachloride, iron oxide, manganese oxide, mercaptans, diethylene glycol monomethyl ether, nitromethane, organic matter, organic sulfides, phenol, 1-propanethiol, propyl mercaptan, sulfur, organic sulfur compounds.

PERCHLORONAPHTALENE (2234-13-1) Strong oxidizers may cause fire and explosions.

PERCHLOROYL FLUORIDE (7616-94-6) A powerful oxidizer. Incompatible with strong bases, amines, finely divided metals. Contact with reducing agents, combustibles, organic materials and readily oxidizable materials may cause fire and explosions or yield explosive product. Attacks some plastics, rubber, and coatings.

PERCHLORURE d'ANTIMOINE (French) (7647-18-9) Water contact produces hydrogen chloride gas. Reacts violently with ammonia, caustics. Attacks many metals. Contact with air produces heavier-than-air corrosive vapor.

PERCHLORURE de FER (French) (7705-08-0) Incompatible with water. Solution is very acidic; reacts violently with bases. Forms shock- and friction-sensitive explosive material with potassium, sodium, and other active metals. Reacts strongly with allyl chloride. Attacks metals when wet.

PERCLENE (127-18-4) Incompatible with strong oxidizers, finely divided metals, caustics. Stable up to 258°F/126°C; at this temperature, product gives off poisonous fumes.

PERFLUROETHYLENE (116-14-3) A highly reactive flammable gas. Able to form unstable peroxides; if inhibitor is not present in adequate concentrations, explosive polymerization may occur. Reacts violently with air, oxygen, oxidizers, sulfur trioxide.

PERHYDROAZEPINE (111-49-9) Forms explosive mixture with air (flash point 99°F/37°C). Incompatible with acids, organic anhydrides, isocyanates, vinyl acetate, acrylates, substituted allyls, alkylene oxides, epichlorohydrin, ketones, aldehydes, alcohols, glycols, phenols, cresols, captrolactam solution, strong oxidizers. Attacks aluminum, copper, lead, tin, zinc, and alloys.

2-PERHYDROAZEPINONE (105-60-2) Contact with strong oxidizers may cause fire and explosions.

PERHYDROL (7722-84-1) A powerful oxidizer; attacks many substances. Contact with most organic, readily oxidizable materials, reducing agents, and combustibles causes fire and explosions. Contact with iron, copper, brass, bronze, chromium, zinc, lead, manganese, silver, and other catalytic metals (and their salts), especially in a basic (pH7 or above) environment, causes rapid decomposition with evolution of oxygen gas. Attacks, and may ignite, some plastics, rubber, and coatings. Decomposes slowly at ordinary temperatures and builds up pressure in a closed container. The rate of decomposition doubles for each 50°F/10°C rise (1.5 times 10°C rise) in temperature and becomes self-sustaining at 285°F/141°C.

PERHYDRONAPTHALENE (91-17-8) Forms explosive mixture with air (flash point 134°F/57°C). May accumulate static electrical charges, and may cause ignition of its vapors. Strong oxidizers may cause fire and explosions.

PERICLASE (1309-48-4) Reacts violently with chlorine trifluoride, bromine pentafluoride, phosphorus pentachloride. Incompatible with acids.

PERK (127-18-4) Incompatible with strong oxidizers, finely divided metals, caustics. Stable up to 258°F/126°C; at this temperature, product gives off poisonous fumes.

PERKLONE (127-18-4) Incompatible with strong oxidizers, finely divided metals, caustics. Stable up to 258°F/126°C; at this temperature, product gives off poisonous fumes.

PERL ALUM (10043-01-3) Aqueous solution is a strong acid. Incompatible with caustics.

PERM-A-CHLOR (79-01-6) Contact with produces form a toxic and flammable gas. Reacts violently with chemically active metals. Contact with aluminum may produce a violent, self-accelerating polymerization reaction. Incompatible with acids, organic anhydrides, isocyanates, alkylene oxides, aldehydes, alcohols, glycols, phenols, cresols, captrolactam solution, epichlorohydrin, nitrogen tetroxide, metal powders, oxygen. May accumulate static electrical charges, and may cause ignition of its vapors.

PERMACIDE (87-86-5) Hot water causes decomposition, producing hydrochloric acid. Strong oxidizers may cause fire and explosions.

PERMAGARD (87-86-5) Hot water causes decomposition, producing hydrochloric acid. Strong oxidizers may cause fire and explosions.

PERMA KLEER 50 ACID (60-00-4) Incompatible with sulfuric acid, bases, ammonia, aliphatic amines, alkanolamines, isocyanates, alkylene oxides, epichlorohydrin.

PERMANENT WHITE (7727-43-7) Explosions may result from contact with aluminum in the presence of heat. Incompatible with potassium, phosphorus.

PERMANGANATE de POTASSIUM (French) (7722-64-7) A strong oxidizer. Produces heat- and shock-sensitive compound with sulfuric acid. Reacts violently with combustibles or reducing agents. Reacts explosively with acetic acid, acetic anhydride, anhydrous ammonia, hydrogen peroxide, glycerol, hydroxylamine, organic matter, powdered sulfur, etc. Incompatible with nitric acid; producing toxic chlorine fumes. A dangerous fire and explosion hazard; isolate from all other materials.

PERMANGANATE OF POTASH (7722-64-7) A strong oxidizer. Produces heat- and shock-sensitive compound with sulfuric acid. Reacts violently with combustibles or reducing agents. Reacts explosively with acetic acid, acetic anhydride, anhydrous ammonia, hydrogen peroxide, glycerol, hydroxylamine, organic matter, powdered sulfur, etc. Incompatible with nitric acid, producing toxic chlorine fumes. A dangerous fire and explosion hazard; isolate from all other materials.

PERMANGANIC ACID, BARIUM SALT (7787-36-2) May be spontaneously combustible. A strong oxidizer; reacts violently with reducing agents, combustibles. Mixtures with combustibles may be ignited by friction or acids.

PERMASAN (87-86-5) Hot water causes decomposition, producing hydrochloric acid. Strong oxidizers may cause fire and explosions.

PERMATOX DP-2 (87-86-5) Hot water causes decomposition, producing hydrochloric acid. Strong oxidizers may cause fire and explosions.

PERMATOX PENTA (87-86-5) Hot water causes decomposition, producing hydrochloric acid. Strong oxidizers may cause fire and explosions.

PERMITE (87-86-5) Hot water causes decomposition, producing hydrochloric acid. Strong oxidizers may cause fire and explosions.

PERONE (7722-84-1) A powerful oxidizer; attacks many substances. Contact with most organic, readily oxidizable materials, reducing agents, and combustibles causes fire and explosions. Contact with iron, copper, brass, bronze, chromium, zinc, lead, manganese, silver, and other catalytic metals (and their salts), especially in a basic (pH7 or above) environment, causes rapid decomposition with evolution of oxygen gas. Attacks, and may ignite, some plastics, rubber, and coatings. Decomposes slowly at ordinary temperatures and builds up pressure in a closed container. The rate of decomposition doubles for each 50°F/10°C rise (1.5 times 10°C rise) in temperature and becomes self-sustaining at 285°F/141°C.

PEROXAN (7722-84-1) A powerful oxidizer; attacks many substances. Contact with most organic, readily oxidizable materials, reducing agents, and combustibles causes fire and explosions. Contact with iron, copper, brass, bronze, chromium, zinc, lead, manganese, silver, and other catalytic metals (and their salts), especially in a basic (pH7 or above) environment, causes rapid decomposition with evolution of oxygen gas. Attacks, and may ignite, some plastics, rubber, and coatings. Decomposes slowly at ordinary temperatures and builds up pressure in a closed container. The rate of decomposition doubles for each 50°F/10°C rise (1.5 times 10°C rise) in temperature and becomes self-sustaining at 285°F/141°C.

PEROXIDE (7722-84-1) A powerful oxidizer; attacks many substances. Contact with most organic, readily oxidizable materials, reducing agents, and combustibles causes fire and explosions. Contact with iron, copper, brass, bronze, chromium, zinc, lead, manganese, silver, and other catalytic metals (and their salts), especially in a basic (pH7 or above) environment, causes rapid decomposition with evolution of oxygen gas. Attacks, and may ignite, some plastics, rubber, and coatings. Decomposes slowly at ordinary temperatures and builds up pressure in a closed container. The rate of decomposition doubles for each 50°F/10°C rise (1.5 times 10°C rise) in temperature and becomes self-sustaining at 285°F/141°C.

PEROXIDE, DIBENZOYL (94-36-0) Confined storage of dry chemical may lead to decomposition and explosion. A strong oxidant; extremely reactive. Fires and explosion may result from heat or contamination, and from contact with strong acids, combustible materials, oxidizers, acids, bases, alcohols, reducing agents, metals, metal oxides, amines, accelerators, methyl methacrylate, organic matter, [carbon tetrachloride + ethylene], dimethylaniline, lithium aluminum carbide, dimethyl aniline, amines, metallic naphthenates. May attack some plastics, rubber, and coatings. Protect containers from shock and friction.

PEROXYACETIC ACID (79-21-0) A powerful oxidizer and extremely sensitive explosive. Forms explosive mixture with air (flash point 100°F/38°C). Reacts violently with many substances, including acetic anhydride, combustibles, strong bases, ether solvents, olefins, organic matter, magnesium, metal oxides, heavy metals, metal chloride solutions, reducing agents, nickel, phosphorus, sodium nitride, zinc.

PEROXYDE de BARYUM (French) (1304-29-6) Reacts with water, producing oxygen. A strong oxidizer. Reacts violently with combustibles, reducing agents, acids, finely divided metals. Contact with organic substances produces shock- and friction-sensitive compounds. Attacks metals in the presence of moisture.

PEROXYDICARBONIC ACID, BIS(1-METHYLETHYL) ESTER (105-64-6) Combustible solid. Contact with some metals may cause decomposition.

PEROXYDISULFANIC ACID, DIAMMONIUM SALT (7727-54-0) A strong oxidizer. Reacts violently with reducing agents, aluminum powder, iron. Mixing with sodium peroxide produces a friction-, heat-, and water-sensitive explosive.

PERSADOX (94-36-0) Confined storage of dry chemical may lead to decomposition and explosion. A strong oxidant; extremely reactive. Fires and explosion may result from heat or contamination, and from contact with strong acids, combustible materials, oxidizers, acids, bases, alcohols, reducing agents, metals, metal oxides, amines, accelerators, methyl methacrylate, organic matter, lithium aluminum carbide, dimethyl aniline, amines, metallic naphthenates. May attack some plastics, rubber, and coatings. Protect containers from shock and friction.

PERSEC (127-18-4) Incompatible with strong oxidizers, finely divided metals, caustics. Stable up to 258°F/126°C; at this temperature, product gives off poisonous fumes.

PERSIAN-INSECT POWDER (8003-34-7) Strong oxidizers may cause fire and explosions.

PERSIA-PERAZOL (106-46-7) Forms explosive mixture with air (flash point 150°F/66°C). Incompatible with strong oxidizers, metal powders, and alkali metals (i.e., lithium, sodium, potassium, rubidium, cesium, francium). Attacks some plastics, rubber, and coating.

PERSULFATE d'AMMONIUM (French) (7727-54-0) A strong oxidizer. Reacts violently with reducing agents, aluminum powder, iron. Mixing with sodium peroxide produces a friction-, heat-, and water-sensitive explosive.

PERTITE (88-89-1) Dry material is explosive. Impact-sensitive; protect from shock. Contact with copper, lead, zinc, and other metals or their salts can cause formation of other salts that are initiators and much more sensitive to shock than this chemical. Shock-sensitive salts include ammonium salts and calcium salts; the calcium salt may be produced when picric acid comes into contact with plaster and concrete. May accumulate static electrical charges, and may cause explosion. Aqueous solution is a strong oxidizer and a strong acid; reacts violently with reducing agents, combustibles, organics and easily oxidized materials, caustics, aluminum, and other metal powders. Attacks metals.

PESTMASTER (106-93-4) Reacts with chemically active metals, liquid ammonia, strong oxidizers. Heat and light cause slow decomposition. Attacks some plastics and rubber.

PESTMASTER EDB-85 (106-93-4) Reacts with chemically active metals, liquid ammonia, strong oxidizers. Heat and light cause slow decomposition. Attacks some plastics and rubber.

PESTOX PLUS (56-38-2) Combustible liquid. Mixtures with endrin may be explosive. Strong oxidizers may cause fire and explosions. Attacks some plastics, rubber, and coatings.

PETHION (56-38-2) Combustible liquid. Mixtures with endrin may be explosive. Strong oxidizers may cause fire and explosions. Attacks some plastics, rubber, and coatings.

PETROHOL (67-63-0) Forms explosive mixture with air (flash point 53°F/12°C). Incompatible with alkaline earth and alkali metals, crotonaldehyde, phosgene, strong acids, amines, ammonia, caustics, strong oxidizers. Attacks some plastics, rubber, and coatings. Reacts with metallic aluminum at high temperatures.

PETROL (8006-61-9) Forms explosive mixture with air (flash point 125°F/51°C). Oxidizers may cause fire and explosions. Incompatible with nitric acid. May accumulate static electrical charges, and may cause ignition of its vapors.

PETROLATUM (8012-95-1) Oxidizers may cause fire and explosions. Incompatible with nitric acid. May accumulate static electrical charges, and may cause ignition of its vapors.

PETROLATUM JELLY (8012-95-1) Oxidizers may cause fire and explosions. Incompatible with nitric acid. May accumulate static electrical charges, and may cause ignition of its vapors.

PETROLEUM (8002-05-9) Oxidizers may cause fire and explosions. Incompatible with nitric acid. May accumulate static electrical charges, and may cause ignition of its vapors.

PETROLEUM ASPHALT (8052-42-4) Incompatible with nitric acid, fluorine, strong oxidizers.

PETROLEUM BENZIN (8032-32-4) Forms explosive mixture with air (flash point −40°F to −86°F/−40°C to −66°C). Incompatible with strong acids, strong oxidizers. Attacks some plastics, rubber, and coatings. May accumulate static electrical charges, and may cause ignition of its vapors.

PETROLEUM CRUDE OIL (8002-05-9) Oxidizers may cause fire and explosions. Incompatible with nitric acid. May accumulate static electrical charges, and may cause ignition of its vapors.

PETROLEUM DISTILLATE (8002-05-9) Flash point −40°F to −86°F/−40°C to −66°C. Oxidizers may cause fire and explosions. Incompatible with nitric acid. May accumulate static electrical charges, and may cause ignition of its vapors.

PETROLEUM DISTILLATE (64741-43-1) Oxidizers may cause fire and explosions. Incompatible with nitric acid. May accumulate static electrical charges, and may cause ignition of its vapors.

PETROLEUM DISTILLATES (NAPHTHA) (8030-30-6) Forms explosive mixture with air (flash point 100°F/38°C). Incompatible with nitric acid, strong oxidizers. Attacks some plastics, rubber, and coatings.

PETROLEUM ETHER (8032-32-4) Forms explosive mixture with air (flash point −40°F to −86°F/−40°C to −66°C). Incompatible with strong acids, strong oxidizers. Attacks some plastics, rubber, and coatings. May accumulate static electrical charges, and may cause ignition of its vapors.

PETROLEUM ETHER (DOT) (8030-30-6) Forms explosive mixture with air (flash point 100°F/38°C). Incompatible with nitric acid, strong oxidizers. Attacks some plastics, rubber, and coatings.

PETROLEUM GAS, LIQUIFIED (68476-85-7) Forms explosive mixture with air. Strong oxidizers may cause fire and explosions. Attacks some plastics, rubber, and coatings.

PETROLEUM INSULATING OIL (N/A) Oxidizers may cause fire and explosions. Incompatible with nitric acid. May accumulate static electrical charges, and may cause ignition of its vapors.

PETROLEUM JELLY (8012-95-1) Oxidizers may cause fire and explosions. Incompatible with nitric acid. May accumulate static electrical charges, and may cause ignition of its vapors.

PETROLEUM NAPHTHA (DOT) (8030-30-6) Forms explosive mixture with air (flash point 100°F/38°C). Incompatible with nitric acid, strong oxidizers. Attacks some plastics, rubber, and coatings.

PETROLEUM NAPHTHA, ALIPHATIC (8002-05-9) Flash point −40°F to −86°F/−40°C to −66°C. Oxidizers may cause fire and explosions. Incompatible with nitric acid. May accumulate static electrical charges, and may cause ignition of its vapors.

PETROLEUM NAPHTHA, AROMATIC (8052-41-3) Forms explosive mixture with air (flash point 102°F to 140°F/38.7°C to 60°C). Incompatible with nitric acid, strong oxidizers. Attacks some plastics, rubber, and coatings.

PETROLEUM OIL (64741-44-2) Forms explosive mixture with air (flash point approximately 151°F/55°C). Incompatible with nitric acid. Strong oxidizers may cause fire and explosions. May accumulate static electrical charges, and may cause ignition of its vapors.

PETROLEUM PITCH (8052-42-4) Incompatible with nitric acid, fluorine, strong oxidizers.

PETROLEUM RESIDUE (8052-42-4) Incompatible with nitric acid, fluorine, strong oxidizers.

PETROLEUM ROOFING TAR (8052-42-4) Incompatible with nitric acid, fluorine, strong oxidizers.

PETROLEUM SOLVENT (8002-05-9) Forms explosive mixture with air (flash point 100°F/38°C). Incompatible with nitric acid, strong oxidizers. Attacks some plastics, rubber, and coatings. May accumulate static electrical charges, and may cause ignition of its vapors.

PETROLEUM SOLVENT (8032-32-4) Forms explosive mixture with air (flash point −40°F to −86°F/−40°C to −66°C). Incompatible with strong acids, strong oxidizers. Attacks some plastics, rubber, and coatings. May accumulate static electrical charges, and may cause ignition of its vapors.

PETROLEUM SOLVENT (8052-41-3) Forms explosive mixture with air (flash point 102°F to 140°F/38.7°C to 60°C). Incompatible with nitric acid, strong oxidizers. Attacks some plastics, rubber, and coatings.

PETROLEUM SPIRIT (DOT) (8030-30-6) Forms explosive mixture with air (flash point 100°F/38°C). Incompatible with nitric acid, strong oxidizers. Attacks some plastics, rubber, and coatings.

PETROLEUM SPIRITS (8032-32-4) Forms explosive mixture with air (flash point −40°F to −86°F/−40°C to −66°C). Incompatible with strong acids, strong oxidizers. Attacks some plastics, rubber, and coatings. May accumulate static electrical charges, and may cause ignition of its vapors.

PETROLEUM SPIRITS (8052-41-3) Forms explosive mixture with air (flash point 102°C to 140°F/38.7°F to 60°C). Incompatible with nitric acid, strong oxidizers. Attacks some plastics, rubber, and coatings.

PETROLEUM TAILINGS (8052-42-4) Incompatible with nitric acid, fluorine, strong oxidizers.

PETROLEUM THINNER (8052-41-3) Forms explosive mixture with air (flash point 102°F to 140°F/38.7°C to 60°C). Incompatible with nitric acid, strong oxidizers. Attacks some plastics, rubber, and coatings.

PETROLEUM WAX (8002-74-2) Reacts with oxidizers, with a risk of fire or explosions.

PETZINOL (79-01-6) Contact with caustics produces a toxic and flammable gas. Reacts violently with chemically active metals. Contact with aluminum may produce a violent, self-accelerating polymerization reaction. Incompatible with acids, organic anhydrides, isocyanates, alkylene oxides, aldehydes, alcohols, glycols, phenols, cresols, captro-lactam solution, epichlorohydrin, nitrogen tetroxide, metal powders, oxygen. May accumulate static electrical charges, and may cause ignition of its vapors.

PEZIFILM (137-26-8) Combustible solid (flash point 192°F/89°C). Strong oxidizers may cause fire and explosions; contact with strong acid or oxidizable materials produces toxic gases.

PFLANZOL (58-89-9) Not combustible, but may be dissolved in a combustible solvent. If solvent comes in contact with oxidizers, fire and explosions may result.

PG12 (57-55-7) Incompatible with strong acids, caustics, aliphatic amines, isocyanates, strong oxidizers (reacts violently).

PGE (122-60-1) (122-60-1) Forms explosive mixture with air (flash point more than 167°F/75°C). May form unstable peroxides if inhibitor is not maintained in adequate concentrations. Strong oxidizers may cause fire and explosions. Contact with amines, strong acids, and strong bases may cause polymerization with the liberation of heat and spattering. Exposure to light and air may result in the formation of explosive peroxides. Attacks some forms of plastics, coatings, and rubber.

PGME (107-98-2) Forms explosive mixture with air (flash point 90°F/32°C). Incompatible with sulfuric acid, isocyanates, perchloric acid.

PHELLANDRENE (138-86-3) Forms explosive mixture with air (flash point 115°F/46°C). Strong oxidizers may cause fire and explosions.

PHENACHLOR (99-95-4) Contact with strong oxidizers may cause fire and explosions.

PHENACIDE (8001-35-2) Reacts with oxidizers, with a risk of fire or explosions. Attacks metals in the presence of moisture.

PHENACYL CHLORIDE (532-27-4) Incompatible with water, or steam.

PHENADOR-X (92-52-4) Mist forms explosive mixture with air. Strong oxidizers may cause fire and explosions.

PHENATOX (8001-35-2) Reacts with oxidizers, with a risk of fire or explosions. Attacks metals in the presence of moisture.

PHENE (71-43-2) Forms explosive mixture with air (flash point 12°F/−11°C).) May accumulate static electrical charges, and may cause ignition of its vapors. Incompatible with strong oxidizers, nitric acid, oxygen, ozone, perchlorates. Attacks some forms of plastics, coatings, and rubber.

a-PHENETHYL ALCOHOL or alpha-PHENETHYL ALCOHOL (98-85-1) Forms explosive mixture with air (flash point 205°F/96°C). Incompatible with strong acids, caustics, aliphatic amines, isocyanates, oxidizers.

PHENETOLE (103-73-1) Forms explosive mixture with air (flash point 140°F/60°C). Able to form unstable and explosive peroxides. Reacts violently with strong acids, oxidizers.

PHENIC ACID (108-95-2) Forms explosive mixture with air (flash point 174°F/79°C). Incompatible with strong oxidizers, strong acids, caustics, aliphatic amines, amides, oxidizers, formaldehyde, butadiene, calcium hypochlorite. Liquid attacks some plastics, rubber, and coatings; hot liquid attacks aluminum, magnesium, lead and zinc metals.

PHENOCHLOR (generic CAS for PCBs 1336-36-3) Incompatible with strong oxidizers, strong acids.

PHENOL (108-95-2) Forms explosive mixture with air (flash point 174°F/79°C). Incompatible with strong oxidizers, strong acids, caustics, aliphatic amines, amides, oxidizers, formaldehyde, butadiene, calcium hypochlorite. Liquid attacks some plastics, rubber, and coatings; hot liquid attacks aluminum, magnesium, lead, and zinc metals.

PHENOL, 4-t-BUTYL-2-CHLORO-, ESTER WITH METHYL METHYLPHOSPHORAMIDATE (299-86-5) Decomposes in strongly alkaline (pH >7) and strongly acidic media. Unstable over long periods of time in water and at temperatures above 140°F/60°C).

PHENOL, 2-CHLORO- (95-57-8) Forms explosive mixture with air (flash point 147°F/64°C). Contact with strong oxidizers may cause fire and explosions. Attacks aluminum and copper.

PHENOL, *o*-CHLORO (95-57-8) Forms explosive mixture with air (flash point 147°F/64°C) Contact with strong oxidizers may cause fire and explosions. Attacks aluminum and copper.

PHENOL, DIMETHYL- (1300-71-6) Forms explosive mixture with air (flash point 186°F/86°C). Reacts violently with strong oxidizers.

PHENOLE (German) (108-95-2) Forms explosive mixture with air (flash point 174°F/79°C). Incompatible with strong oxidizers, strong acids, caustics, aliphatic amines, amides, oxidizers, formaldehyde, butadiene, calcium hypochlorite. Liquid attacks some plastics, rubber, and coatings; hot liquid attacks aluminum, magnesium, lead, and zinc metals.

PHENOL, *o*-ETHYL (90-00-6) Forms explosive mixture with air (flash point 173°F/78°C). Incompatible with strong acids, caustics, aliphatic amines, amides.

PHENOL, HEXAHYDRO- (108-93-0) Forms explosive mixture with air (flash point 154°F/68°C). Incompatible with strong acids, caustics, aliphatic amines, isocyanates. Attacks some plastics, rubber, or coatings.

PHENOL, *m*-HYDROXY- (108-46-3) Reacts violently with strong oxidizers, nitric acid. Incompatible with acetanilide, albumin, alkalies, antipyrine, camphor, ferric salts, menthol, spirit nitrous ether. Absorbs moisture from air (hygroscopic). May accumulate static electrical charges, and may cause ignition of its vapors.

PHENOL, METHYL- (9CI) (1319-77-3) Incompatible with strong acids, oxidizers, alkalies, aliphatic amines, amides, chlorosulfonic acid, oleum. Liquid attacks some plastics, coatings, and rubber.

PHENOL, 2-NITRO- (88-75-5) A strong oxidizer; reacts violently with reducing agents, combustibles, organic matter, chlorosulfonic acid, potassium hydroxide.

PHENOL, 4-NITRO (100-02-7) Combustible solid. A strong oxidizer; reacts with reducing agents, combustibles, organic and other easily oxidizable materials. Incompatible with strong acids, caustics, aliphatic amines, amides.

PHENOL, *o*-NITRO or PHENOL, *ortho*-NITRO (88-75-5) A strong oxidizer; reacts violently with reducing agents, combustibles, organic matter, chlorosulfonic acid, potassium hydroxide.

PHENOL, *p*-NITRO or PHENOL, *para*-NITRO (100-02-7) Combustible solid. A strong oxidizer; reacts with reducing agents, combustibles, organic and other easily oxidizable materials. Incompatible with strong acids, caustics, aliphatic amines, amides.

PHENOL, PENTACHLORO- (87-86-5) Hot water causes decomposition, producing hydrochloric acid. Strong oxidizers may cause fire and explosions.

PHENOL, PENTACHLORO-, SODIUM SALT (131-52-2) Contact with strong oxidizers may cause fire and explosions.

***o-p*-PHENOLSULFONIC ACID** (98-67-9) Incompatible with sulfuric acid, caustics, ammonia, aliphatic amines, alkanolamines, isocyanates, alkylene oxides, epichlorohydrin, steam, water.

PHENOL TRINITRATE (88-89-1) Dry material is explosive. Impact-sensitive; protect from shock. Contact wth copper, lead, zinc, and other metals or their salts can cause formation of other salts that are initiators and much more sensitive to shock than this chemical. Shock-sensitive salts include ammonium salts and calcium salts; the calcium salt may be produced when picric acid comes into contact with plaster and concrete. May accumulate static electrical charges, and may cause explosion. Aqueous solution is a strong oxidizer and a strong acid; reacts violently with reducing agents, combustibles, organics and easily oxidized materials, caustics, aluminum and other metal powders. Attacks metals.

PHENOL, 2,4,6-TRINITRO-, AMMONIUM SALT (9CI) (131-74-8) A self-reactive explosive. Heat above 250°F/121°C can cause explosions. A powerful oxidizer. Contact with metal, concrete, or plaster causes formation of salts that are more shock-sensitive than ammonium picrate. Water increases rate of reactivity with metals.

PHENOOXY ALCOHOL (122-99-6) Incompatible with sulfuric acid, isocyanates, strong oxidizers.

PHENOX (94-75-7) Decomposes in sunlight. Incompatible with strong oxidizers; may cause fire and explosions.

PHENOXY BENZENE (101-84-8) Incompatible with strong acids, strong oxidizers; may cause fire and explosions. Attacks some plastics, rubber, and coatings.

PHENOXYETHANE (103-73-1) Forms explosive mixture with air (flash point 140°F/60°C). Able to form unstable and explosive peroxides. Reacts violently with strong acids, oxidizers.

2-PHENOXYETHANOL (122-99-6) Incompatible with sulfuric acid, isocyanates, strong oxidizers.

(PHENOXYMETHYL)OXIRANE (122-60-1) Forms explosive mixture with air (flash point more than 167°F/75°C). May form unstable peroxides if inhibitor is not maintained in adequate concentrations. Strong oxidizers may cause fire and explosions. Contact with amines, strong acids, and strong bases may cause polymerization with the liberation of heat and spattering. Exposure to light and air may result in the formation of explosive peroxides. Attacks some forms of plastics, coatings, and rubber.

PHENOXY-PROPENE OXIDE (122-60-1) Forms explosive mixture with air (flash point more than 167°F/75°C). May form unstable peroxides if inhibitor is not maintained in adequate concentrations. Strong oxidizers may cause fire and explosions. Contact with amines, strong acids, and strong bases may cause polymerization with the liberation of heat and spattering. Exposure to light and air may result in the formation of explosive peroxides. Attacks some forms of plastics, coatings, and rubber.

PHENOXYTOL (122-99-6) Incompatible with sulfuric acid, isocyanates, strong oxidizers.

PHENYL ALLYL ETHER (1746-13-0) Forms explosive mixture with air (flash point 143°F/153°F/62°C). May form peroxides with air. Incompatible with strong acids, oxidizers.

PHENYLAMINE (62-53-3) Forms explosive mixture with air (flash point 158°F/70°C). Unless inhibited (usually with methanol), readily able to polymerize. Fires and explosions may result from contact with halogens, strong acids, oxidizers, organic anhydrides, acetic anhydride, isocyanates, aldehydes, sodium peroxide. Reacts with alkali metals and alkaline earth metals. Attacks some plastics, rubber, and coatings, and copper and copper alloys.

2-(PHENYLAMINO)ETHANOL (122-98-5) Incompatible with strong acids, caustics, aliphatic amines, isocyanates, oxidizers.

PHENYLANILINE or *n*-**PHENYLANILINE or** *normal*-**PHENYLANILINE** (122-39-4) Incompatible with strong acids, aldehydes, organic anhydrides, isocyanates, oxidizers, hexachloromelamine, trichloromelamine.

4-PHENYLANILINE (92-67-1) Contact with strong oxidizers may cause fire and explosions.

p-PHENYLANILINE (92-67-1) Contact with strong oxidizers may cause fire and explosions.

PHENYLARSENIC DICHLORIDE (696-28-6) Forms explosive mixture with air (flash point 60°F/−16°C). Water contact produces hydrochloric acid. Attacks metals in the presence of moisture.

PHENYLARSENOUS DICHLORIDE (696-28-6) Forms explosive mixture with air (flash point 60°F/−16°C). Water contact produces hydrochloric acid. Attacks metals in the presence of moisture.

PHENYL BENZENE (92-52-4) Mist forms explosive mixture with air. Strong oxidizers may cause fire and explosions.

n-PHENYLBENZENAMINE or *normal*-PHENYLBENZENAMINE (122-39-4) Incompatible with strong acids, aldehydes, organic anhydrides, isocyanates, oxidizers, hexachloromelamine, trichloromelamine.

PHENYL BROMIDE (108-86-1) Forms explosive mixture with air (flash point 124°F/51°C). Incompatible with strong oxidizers, alkaline earth metals (e.g., barium, calcium, magnesium, strontium), metallic salts, with risk of violent reactions. May accumulate static electrical charges, and may cause ignition of its vapors.

PHENYLCARBIMIDE (103-71-9) Forms explosive mixture with air (flash point 122°F/50°C). Incompatible with acids, caustics, ammonia, amines, amides, alcohols, glycols, captrolactam solution, strong oxidizers, water. Attacks metals.

PHENYLCARBINOL (100-51-6) Forms explosive mixture with air (flash point 213°F/101°C). Slowly oxidizes in air and oxygen. Incompatible with mineral acids, caustics, aliphatic amines, isocyanates. Reacts violently with strong oxidizers. Corrodes aluminum at high temperature. Attacks some nonfluorinated plastics; may not attack polypropylene.

PHENYL CARBONIMIDE (103-71-9) Forms explosive mixture with air (flash point 122°F/50°C). Incompatible with acids, caustics, ammonia, amines, amides, alcohols, glycols, captrolactam solution, strong oxidizers, water. Attacks metals.

PHENYL CELLOSOLVE (122-99-6) Incompatible with sulfuric acid, isocyanates, strong oxidizers.

PHENYL CHLORIDE (108-90-7) Forms explosive mixture with air (flash point 82°F/28°C). May accumulate static electrical charges, and may cause ignition of its vapors. Incompatible with oxidizers, dimethylsulfoxide, sodium powder, silver perchloride. Attacks some plastics, rubber, and coatings.

PHENYL CHLOROMETHYL KETONE (532-27-4) Incompatible with water, or steam.

PHENYL CYANIDE (100-47-0) Forms explosive mixture with air (flash point 167°F/75°C). Incompatible with strong acids, oxidizers. Attacks some plastics, rubber, and coatings.

L-PHENYLDECANE (104-72-2) Incompatible with nitric acid. Attacks some plastics, rubber, and coatings.

PHENYLDICHLOROARSINE (696-28-6) Forms explosive mixture with air (flash point 60°F/−16°C). Water contact produces hydrochloric acid. Attacks metals in the presence of moisture.

PHENYLDODECAN (123-01-3) Incompatible with nitric acid, strong oxidizers; poses risk of fire or explosions.

1-PHENYLDODECANE (123-01-3) Incompatible with nitric acid, strong oxidizers; poses risk of fire or explosions.

p-**PHENYLENEDIAMINE** (106-50-3) A strong reducing agent; reacts violently with oxidizers, combustibles, organic substances. Incompatible with strong acids, organic anhydrides, isocyanates, aldehydes. Heat and light contribute to instability.

o-**PHENYLENEDIOL or** *ortho*-**PHENYLENEDIOL** (120-80-9) Strong oxidizers may cause fire and explosion.

PHENYL 2,3-EPOXYPROPYL ETHER (122-60-1) Forms explosive mixture with air (flash point more than 167°F/75°C). May form unstable peroxides if inhibitor is not maintained in adequate concentrations. Strong oxidizers may cause fire and explosions. Contact with amines, strong acids, and strong bases may cause polymerization with the liberation of heat and spattering. Exposure to light and air may result in the formation of explosive peroxides. Attacks some forms of plastics, coatings, and rubber.

PHENYLETHANE (100-41-4) May accumulate static electrical charges, and may cause ignition of its vapors. Forms explosive mixture with air (flash point 59°F/15°C). Incompatible with strong oxidizers, nitric acid.

1-PHENYLETHANOL (98-85-1) Forms explosive mixture with air (flash point 205°F/96°C). Incompatible with strong acids, caustics, aliphatic amines, isocyanates, oxidizers.

PHENYLETHANOLAMINE or *n*-**PHENYLETHANOLAMINE** (122-98-5) Incompatible with strong acids, caustics, aliphatic amines, isocyanates, oxidizers.

1-PHENYLETHANONE (98-86-2) Incompatible with strong acids, aliphatic amines, oxidizers. Forms explosive mixture with air (flash point 170°F/77°C).

PHENYLETHENE (100-42-5) If inhibitor (often *tert*-butyl catechol) is not present in adequate concentrations, polymerization may occur and explode containers. Temperatures above 150°F/66°C speed up polymerization. Incompatible with strong oxidizers, acids, rust, catalysts for vinyl polymerization, such as peroxides, strong acids, and aluminum chloride. Corrodes copper, and copper alloys, and dissolves rubber. Attacks some plastics, rubber, and coatings. May accumulate static electrical charges, and may cause ignition of its vapors.

PHENYL ETHER (101-84-8) Incompatible with strong acids, strong oxidizers; may cause fire and explosions. Attacks some plastics, rubber, and coatings.

PHENYL ETHER–BIPHENYL MIXTURE (8004-13-5) Contact with strong oxidizers may cause fire and explosions. Attacks some plastics, rubber, and coatings.

1-PHENYL ETHYL ALCOHOL (98-85-1) Forms explosive mixture with air (flash point 205°F/96°C). Incompatible with strong acids, caustics, aliphatic amines, isocyanates, oxidizers.

a-PHENYL ETHYL ALCOHOL (98-85-1) Forms explosive mixture with air (flash point 205°F/96°C). Incompatible with strong acids, caustics, aliphatic amines, isocyanates, oxidizers.

PHENYLETHYLDICHLOROSILANE (1125-27-5) Forms explosive mixture with air (flash point 150°F/66°C). Incompatible with strong bases, water (produces hydrochloric acid). Corrodes most metals. Attacks some plastics, rubber, and coatings.

PHENYLETHYLENE (100-42-5) If inhibitor (often *tert*-butyl catechol) is not present in adequate concentrations, polymerization may occur and explode containers. Temperatures above 150°F/66°C speed up polymerization. Incompatible with strong oxidizers, acids, rust, catalysts for vinyl polymerization, such as peroxides, strong acids, and aluminum chloride. Corrodes copper and copper alloys, and dissolves rubber. Attacks some plastics, rubber, and coatings. May accumulate static electrical charges, and may cause ignition of its vapors.

PHENYL ETHYL ETHER (103-73-1) Forms explosive mixture with air (flash point 140°F/60°C). Able to form unstable and explosive peroxides. Reacts violently with strong acids, oxidizers.

PHENYL FLUORIDE (462-06-6) Forms explosive mixture with air (flash point 5°F/−15°C). Incompatible with strong oxidizers, ammonium nitrate, chromic acid, halogens, hydrogen peroxide, nitric acid. Attacks some plastics, rubber, and coatings.

PHENYL GLYCIDYL ETHER (122-60-1) (122-60-1) Forms explosive mixture with air (flash point more than 167°F/75°C). May form unstable peroxides if inhibitor is not maintained in adequate concentrations. Strong oxidizers may cause fire and explosions. Contact with amines, strong acids, and strong bases may cause polymerization with the liberation of heat and spattering. Exposure to light and air may result in the formation of explosive peroxides. Attacks some forms of plastics, coatings, and rubber.

PHENYL HYDRATE (108-95-2) Forms explosive mixture with air (flash point 174°F/79°C). Incompatible with strong oxidizers, strong acids, caustics, aliphatic amines, amides, oxidizers, formaldehyde, butadiene, calcium hypochlorite. Liquid attacks some plastics, rubber, and coatings; hot liquid attacks aluminum, magnesium, lead, and zinc metals.

PHENYLHYDRAZINE (100-63-0) A highly reactive reducing agent. Forms explosive mixture with air (flash point 190°F/88°C). Reacts with oxides of iron or copper, and with manganese, lead, copper, or their alloys, and methyl iodide, oxidizers, organic compounds. Attacks some plastics, rubber, and coatings, and cork.

PHENYLHYDRAZINE HYDROCHLORIDE (59-88-1) Combustible solid. Incompatible with strong oxidizers, alkali metals (i.e., lithium, sodium, potassium, rubidium, cesium, francium). Corrosive to metals, copper salts, nickel, chromates.

PHENYLHYDRAZINE MONOHYDROCHLORIDE (59-88-1) Combustible solid. Incompatible with strong oxidizers, alkali metals (i.e., lithium, sodium, potassium, rubidium, cesium, francium). Corrosive to metals, copper salts, nickel, chromates.

PHENYLHYDRAZINIUM CHLORIDE (59-88-1) Combustible solid. Incompatible with strong oxidizers, alkali metals (i.e., lithium, sodium, potassium, rubidium, cesium, francium). Corrosive to metals, copper salts, nickel, chromates.

PHENYL HYDRIDE (71-43-2) Forms explosive mixture with air (flash point 12°F/−11°C).) May accumulate static electrical charges, and may cause ignition of its vapors. Incompatible with strong oxidizers, nitric acid, oxygen, ozone, perchlorates. Attacks some forms of plastics, coatings, and rubber.

PHENYL HYDROXIDE (108-95-2) Forms explosive mixture with air (flash point 174°F/79°C). Incompatible with strong oxidizers, strong acids, caustics, aliphatic amines, amides, oxidizers, formaldehyde, butadiene, calcium hypochlorite. Liquid attacks some plastics, rubber, and coatings; hot liquid attacks aluminum, magnesium, lead, and zinc metals.

PHENYLIC ACID (108-95-2) Forms explosive mixture with air (flash point 174°F/79°C). Incompatible with strong oxidizers, strong acids, caustics, aliphatic amines, amides, oxidizers, formaldehyde, butadiene, calcium hypochlorite. Liquid attacks some plastics, rubber, and coatings; hot liquid attacks aluminum, magnesium, lead, and zinc metals.

PHENYL ISOCYANATE (103-71-9) Forms explosive mixture with air (flash point 122°F/50°C). Incompatible with acids, caustics, ammonia, amines, amides, alcohols, glycols, captrolactam solution, strong oxidizers, water. Attacks metals.

PHENYL KETONE (119-61-9) Combustible solid. Strong oxidizers may cause fire and explosions. Attacks some plastics, rubber, and coatings.

PHENYL MERCAPTAN (108-98-5) Forms explosive mixture with air (flash point 132°F/56°C). Incompatible with strong acids, caustics, alkali metals (i.e., lithium, sodium, potassium, rubidium, cesium, francium). Oxidizes on contact with air. Corrosive to carbon steel.

PHENYLMETHANAL (100-52-7) Forms explosive mixture with air (flash point 145°F/63°C). A strong reducing agent. Forms peroxides with air. Reacts violently with strong acids, aluminum, caustics, ammonia, amines, iron, strong oxidizers. Attacks some plastics, rubber, and coatings.

PHENYL METHANE (108-88-3) Forms explosive mixture with air (flash point 40°F/4°C). Strong oxidizers may cause fire and explosions. Attacks some plastics, rubber, and coatings. May accumulate static electrical charges, and may cause ignition of its vapors.

PHENYLMETHANOL (100-51-6) Forms explosive mixture with air (flash point 213°F/101°C). Slowly oxidizes in air and oxygen. Incompatible with mineral acids, caustics, aliphatic amines, isocyanates. Reacts violently with strong oxidizers. Corrodes aluminum at high temperature. Attacks some nonfluorinated plastics; may not attack polypropylene.

PHENYLMETHYL ACETATE (140-11-4) Forms explosive mixture with air (flash point 216°F/102°C). Reacts with strong acids, nitrates, oxidizers.

PHENYLMETHYL ALCOHOL (100-51-6) Forms explosive mixture with air (flash point 213°F/101°C). Slowly oxidizes in air and oxygen. Incompatible with mineral acids, caustics, aliphatic amines, isocyanates. Reacts violently with strong oxidizers. Corrodes aluminum at high temperature. Attacks some nonfluorinated plastics; may not attack polypropylene.

PHENYLMETHYL AMINE (100-46-9) Forms explosive mixture with air (flash point 168°F/76°C). Incompatible with strong acids, organic anhydrides, isocyanates, aldehydes. Slowly corrodes metals in a moist environment. Attacks some plastics, rubber, and coatings.

N-PHENYLMETHYLAMINE (100-61-8) Forms explosive mixture with air (flash point 174°F/79°C). Incompatible with strong acids, strong oxidizers. Attacks some plastics, rubber, and coatings. May accumulate static electrical charges, and may cause ignition of its vapors.

PHENYLMETHYL CARBINOL (98-85-1) Forms explosive mixture with air (flash point 205°F/96°C). Incompatible with strong acids, caustics, aliphatic amines, isocyanates, oxidizers.

PHENYL METHYL KETONE (98-86-2) Incompatible with strong acids, aliphatic amines, oxidizers. Forms explosive mixture with air (flash point 170°F/77°C).

PHENYL MONOGLYCOL ETHER (122-99-6) Incompatible with sulfuric acid, isocyanates, strong oxidizers.

N-PHENYL-1-NAPHTHYLAMINE (90-30-2) Combustible. Incompatible with oxidizers, strong acids, organic anhydrides, isocyanates, aldehydes.

N-PHENYL-2-NAPHTHYLAMINE (135-88-6) Combustible. May accumulate static electrical charges, and may cause ignition of its vapors. Incompatible with oxidizers, strong acids, organic anhydrides, isocyanates, aldehydes.

PHENYL-alpha-NAPHTHALAMINE (90-30-2) Combustible. Incompatible with oxidizers, strong acids, organic anhydrides, isocyanates, aldehydes.

4-PHENYLNITROBENZENE (92-93-3) Contact with strong oxidizers may cause fire and explosions.

p-**PHENYLNITROBENZENE** (92-93-3) Contact with strong oxidizers may cause fire and explosions.

PHENYL OXIDE (101-84-8) Incompatible with strong acids, strong oxidizers; may cause fire and explosions. Attacks some plastics, rubber, and coatings.

PHENYL PERCHLORYL (118-74-1) Reacts violently with strong oxidizers, formyldimethylamine.

PHENYL PHOSPHATE (115-86-6) Incompatible with strong oxidizers, strong acids, nitrates; may cause fire or explosions. Attacks some plastics, rubber, and coatings.

PHENYLPHOSPHINE DICHLORIDE (644-97-3) Water contact produces hydrochloric acid. Corrodes metals except 316 stainless steel, nickel, and Hastelloy.

PHENYLPHOSPHINE THIODICHLORIDE (14684-5-4) Water contact produces hydrochloric acid. Corrosive to metals.

PHENYLPHOSPHONOTHIOIC DICHLORIDE (14684-5-4) Water contact produces hydrochloric acid. Corrosive to metals.

PHENYLPHOSPHONOUS DICHLORIDE (644-97-3) Water contact produces hydrochloric acid. Corrodes metals except 316 stainless steel, nickel, and Hastelloy.

PHENYLPHOSPHORUS DICHLORIDE (644-97-3) Water contact forms hydrochloric acid. Corrodes metals except 316 stainless steel, nickel, and Hastelloy.

1-PHENYLPROPANE (103-65-1) Forms explosive mixture with air (flash point 86°F/30°C). Incompatible with nitric acid.

2-PHENYLPROPANE (98-82-8) Forms explosive mixture with air (flash point 99°F/37°C). Incompatible with strong acids, strong oxidizers. Air contact forms cumene hydroperoxide. May be able to form unstable peroxides. Attacks rubber. May accumulate static electrical charges, and may cause ignition of its vapors.

PHENYLPROPYLENE (98-83-9) Forms explosive mixture with air (flash point 129°F/54°C). Avoid heat and/or lack of appropriate inhibitor concentration, and/or contact with catalysts for vinyl or ionic polymerization such as aluminum or iron chloride. Incompatible with strong acids, oxidizers (including peroxides and halogens).

2-PHENYLPROPYLENE (98-83-9) Forms explosive mixture with air (flash point 129°F/54°C). Avoid heat and/or lack of appropriate inhibitor concentration, and/or contact with catalysts for vinyl or ionic polymerization such as aluminum or iron chloride. Incompatible with strong acids, oxidizers (including peroxides and halogens).

beta-PHENYLPROPYLENE (98-83-9) Forms explosive mixture with air (flash point 129°F/54°C). Avoid heat and/or lack of appropriate inhibitor concentration, and/or contact with catalysts for vinyl or ionic polymerization such as aluminum or iron chloride. Incompatible with strong acids, oxidizers (including peroxides and halogens).

1-PHENYLTETRADECANE (N/A) Incompatible with nitric acid. Attacks some plastics, rubber, and coatings.

PHENYLTHIOL (108-98-5) Forms explosive mixture with air (flash point 132°F/56°C). Incompatible with strong acids, caustics, alkali metals (i.e., lithium, sodium, potassium, rubidium, cesium, francium). Oxidizes on contact with air. Corrosive to carbon steel.

1-PHENYLTRIDECANE (123-02-4) Forms explosive mixture with air (flash point 230°F/110°C). Incompatible with nitric acid, strong oxidizers.

1-PHENYLUNDECANE (N/A) Combustible liquid. Incompatible with nitric acid, strong oxidizers. Attacks some plastics, rubber, and coatings.

PHENYL XYLYL ETHANE (6196-95-8) Incompatible with nitric acid, strong oxidizers; poses risk of fire or explosions.

1-PHENYL-1-XYLYL ETHANE (6196-95-8) Incompatible with nitric acid, strong oxidizers; poses risk of fire or explosions.

PHLOROL (90-00-6) Forms explosive mixture with air (flash point 173°F/78°C). Incompatible with strong acids, caustics, aliphatic amines, amides.

PHORONE (504-20-1) Forms explosive mixture with air (flash point 183°F/84°C). Incompatible with strong oxidizers, acids.

PHORSAEURELOESUNGEN (German) (7664-38-2) Incompatible with aliphatic amines, alkanolamines, alkylene oxides, aromatic amines, amides, ammonia, ammonium hydroxide, bases, calcium oxide, epichlorohydrin, isocyanates, nitromethane (explosive), oleum, organic anhydrides, sulfuric acid, sodium tetrahydroborate, strong oxidizers, vinyl acetate, water. Contact with most metals form hydrogen gas. Attacks some plastics, rubber, and coatings, and glass and ceramics.

PHORTOX (93-76-5) Sealed metal containers may burst in heat above 316°F/158°C. Incompatible with sulfuric acid, bases, ammonia, aliphatic amines, alkanolamines, isocyanates, alkylene oxides, epichlorohydrin.

PHOSDRIN (7786-34-7) Contact with strong oxidizers may cause fire and explosions. Attacks some plastics, rubber, and coatings.

PHOSDRIN 24 (7786-34-7) Contact with strong oxidizers may cause fire and explosions. Attacks some plastics, rubber, and coatings.

cis-PHOSDRIN (7786-34-7) Contact with strong oxidizers may cause fire and explosions. Attacks some plastics, rubber, and coatings.

PHOSFENE (7786-34-7) Contact with strong oxidizers may cause fire and explosions. Attacks some plastics, rubber, and coatings.

PHOSFLEX 179-C (78-30-8) Contact with magnesium or strong oxidizers may cause explosion. Attacks some plastics, rubber, and coatings.

PHOSGEN (German) (75-44-5) Incompatible with water, producing hydrochloric acid. Reacts violently with strong oxidizers, anhydrous ammonia, isopropyl alcohol and chemically active metals. Produces shock-sensitive material with potassium. Attacks most metals in moist conditions.

PHOSGENE (75-44-5) Incompatible with water, producing hydrochloric acid. Reacts violently with strong oxidizers, anhydrous ammonia, isopropyl alcohol and chemically active metals. Forms shock-sensitive material with potassium. Attacks most metals in moist conditions.

PHOSKIL (56-38-2) Combustible liquid. Mixtures with endrin may be explosive. Strong oxidizers may cause fire and explosions. Attacks some plastics, rubber, and coatings.

PHOSPHATE de O,O-DIMETHYLE et de O-(1,2-DIBROMO-2,2-DICHLORETHYLE) (French) (300-76-5) Unstable in the presence of iron. Reacts with acids, strong oxidizers in sunlight. Corrosive to metals. Attacks some plastics, rubber, and coatings.

PHOSPHATE de DIMETHYLE et de 2-DIMETHYLCARBAMOYL 1-METHYL VINYLE (French) (141-66-2) Corrosive to cast iron, mild steel, brass, and stainless steel 304.

PHOSPHENE (French) (7786-34-7) Contact with strong oxidizers may cause fire and explosions. Attacks some plastics, rubber, and coatings.

PHOSPHENYL CHLORIDE (644-97-3) Water contact produces hydrochloric acid. Corrodes metals except 316 stainless steel, nickel, and Hastelloy.

PHOSPHINE (7803-51-2) Flammable gas. Strong reducing agent; reacts violently with oxidizers with possibility of fire or explosions on contact. Incompatible with aliphatic amines, alkanolamines, aromatic amines, combustibles, organic materials. Ignites spontaneously in air at 212°F/100°C.

PHOSPHONIC ACID, DIMETHYL ESTER (868-85-9) Incompatible with acids (nitric, sulfuric, nonoxidizing mineral, organic), nitrates, oxidizers.

PHOSPHORAMIDIC ACID, METHYL-, 4-*tert*-BUTYL-2-CHLOROPHENYL (299-86-5) Decomposes in strongly alkaline (pH >7) and strongly acidic media. Unstable over long periods of time in water and at temperatures above 140°F/60°C.

PHOSPHORAMIDIC ACID, 4-*tert*-BUTYL-2-CHLOROPHENYL-PHOSPHORAMIDATE (299-86-5) Decomposes in strongly alkaline (pH >7) and strongly acidic media. Unstable over long periods of time in water and at temperatures above 140°F/60°C.

PHOSPHORAMIDIC ACID, METHYL-,2-CHLORO-4-(1,1-DIMETHYLETHYL)PHENYL METHYL ESTER (299-86-5) Decomposes in strongly alkaline (pH >7) and strongly acidic media. Unstable over long periods of time in water and at temperatures above 140°F/60°C.

PHOSPHORATED HYDROGEN (7803-51-2) Unusually reactive. Contact with air or other oxidizer may cause self-ignition and explosion. Violent reaction or forms explosive products with many substances, including acids, halogenated hydrocarbons, moisture. Heat may cause cylinders to explode.

PHOSPHORE (PENTACHLORURE de) (French) (10026-13-8) Moisture will cause development of pressure in a closed container. Water causes violent spattering and formation of toxic and corrosive hydrogen chloride gas and phosphoric acid mist. Reacts violently with magnesium oxide, chemically active metals, alkalies. Attacks some plastics, rubber, and coatings.

PHOSPHORE (TRICHLORURE de) (French) (7719-12-2) Contact with water or alcohol may cause fire and explosions, particularly in presence of combustible organic matter. Incompatible with ammonia, caustics, reducing agents, nitric acid, chemically active metals. In the presence of moisture, corrodes most metals. Attacks some plastics, rubber, and coatings.

PHOSPHORIC ACID (7664-38-2) Incompatible with aliphatic amines, alkanolamines, alkylene oxides, aromatic amines, amides, ammonia, ammonium hydroxide, bases, calcium oxide, epichlorohydrin, isocyanates, nitromethane (explosive), oleum, organic anhydrides, sulfuric acid, sodium tetrahydroborate, strong oxidizers, vinyl acetate, water. Contact with most metals produces hydrogen gas. Attacks some plastics, rubber, and coatings, and glass and ceramics.

m-**PHOSPHORIC ACID or** *meta*-**PHOSPHORIC ACID** (7664-38-2) Incompatible with aliphatic amines, alkanolamines, alkylene oxides, aromatic amines, amides, ammonia, ammonium hydroxide, bases, calcium oxide, epichlorohydrin, isocyanates, nitromethane (explosive), oleum, organic anhydrides, sulfuric acid, sodium tetrahydroborate, strong oxidizers, vinyl acetate, water. Contact with most metals produces hydrogen gas. Attacks some plastics, rubber, and coatings, and glass and ceramics.

PHOSPHORIC ACID, AQUEOUS (7664-38-2) Incompatible with aliphatic amines, alkanolamines, alkylene oxides, aromatic amines, amides, ammonia, ammonium hydroxide, bases, calcium oxide, epichlorohydrin, isocyanates, nitromethane (explosive), oleum, organic anhydrides, sulfuric acid, sodium tetrahydroborate, strong oxidizers, vinyl acetate, water. Contact with most metals produces hydrogen gas. Attacks some plastics, rubber, and coatings, and glass and ceramics.

PHOSPHORIC ACID, DIBUTYL ESTER (107-66-4) Incompatible with strong oxidizers. Attacks some plastics, rubber, and coatings.

PHOSPHORIC ACID, O,O-DIETHYL O-6-METHYL-2-(1-METHYLETHYL)-4-PYRIMIDINYL ESTER (333-41-5) Incompatible with water, copper-containing compounds, oxidizers, acids, or bases.

PHOSPHORIC ACID, DIMETHYL ESTER, WITH (E)-3-HYDROXY-N,N-DIMETHYLCROTONAMIDE (141-66-2) Corrosive to cast iron, mild steel, brass, and stainless steel 304.

PHOSPHORIC ACID, TETRAETHYL ESTER (107-49-3) Decomposes above 300°F/150°C, producing flammable ethylene gas. Strong oxidizers may cause fire and explosions. Attacks some plastics, rubber, and coatings.

PHOSPHORIC ACID, TRIBUTYL ESTER (126-73-8) Contact with water produces corrosive phosphoric acid. Incompatible with strong acids, strong oxidizers.

PHOSPHORIC ACID, TRIETHYL ESTER (78-40-0) Incompatible with strong acids, nitrates, oxidizers.

PHOSPHORIC ACID, TRIPHENYL ESTER (115-86-6) Incompatible with strong oxidizers, strong acids, nitrates; may cause fire or explosions. Attacks some plastics, rubber, and coatings.

PHOSPHORIC ACID, TRIS(2-METHYLPHENYL) ESTER (78-30-8) Contact with magnesium or strong oxidizers may cause explosion. Attacks some plastics, rubber, and coatings.

PHOSPHORIC ANHYDRIDE (1314-56-3) Aqueous solution is a strong acid; reacts violently with caustics, water. Attacks metals in the presence of moisture.

PHOSPHORIC CHLORIDE (10026-13-8) Moisture will cause development of pressure in a closed container. Water causes violent spattering and formation of toxic and corrosive hydrogen chloride gas and phosphoric acid mist. Reacts violently with magnesium oxide, chemically active metals, alkalies. Attacks some plastics, rubber, and coatings.

PHOSPHORIC SULFIDE (1314-80-3) Contact with water or alcohol yields heat and possible ignition. May self-ignite in air. Reacts violently with acids, oxidizers.

PHOSPHORODICHLORIDIC ACID, ETHYL ESTER (1498-51-7) Water contact produces hydrochloric acid. Attacks metals when wet.

PHOSPHORODIFLUORIDIC ACID (13779-41-4) Water contact forms hydrofluoric acid. Attacks glass and other siliceous materials. Corrodes metals.

PHOSPHORODIFLUORIDIC ACID (ANHYDROUS) (13779-41-4) Water contact produces hydrofluoric acid. Attacks glass and other siliceous materials. Corrodes metals.

PHOSPHORODITHIOIC ACID, O,O-DIETHYL ESTER, S,S-DIESTER WITH METHANEDITHIOL (563-12-2) Incompatible with alkaline formulations. Mixtures with magnesium may be explosive.

PHOSPHORODITHIONIC ACID, S-2-(ETHYLTHIO)ETHYL-O,O-DIETHYL ESTER (298-04-4) Forms explosive mixture with air (flash point 180°F/82°C). Incompatible with alkalis, strong oxidizers.

PHOSPHOROTHIOATE (56-72-4) Contact with strong oxidizers may cause fire and explosions.

PHOSPHOROTHIOIC ACID, O,O-DIETHYL ESTER,O-ESTER WITH 3-CHLORO-7-HYDROXY-4-METHYLCOUMARIN (56-72-4) Contact with strong oxidizers may cause fire and explosions.

PHOSPHOROTHIOIC ACID, O,O-DIETHYL O-2-(ETHYLTHI-O)ETHYL ESTER, MIXED WITH O,O-DIETHYL S-2-(ETHYLTHI-O)ETHYL PHOSPHOROTHIOATE (8065-48-3) Forms explosive mixture with air (flash point 113°F/45°C). Incompatible with water, strong oxidizers, caustics.

PHOSPHOROTHIOIC ACID, O,O-DIETHYL O-P-NITROPHE-NYL ESTER (56-38-2) Combustible liquid. Mixtures with endrin may be explosive. Strong oxidizers may cause fire and explosions. Attacks some plastics, rubber, and coatings.

PHOSPHOROTHIOIC ACID, O,O,-DIETHYLO-(P-NITROPHE-NYL)ESTER (56-38-2) Combustible liquid. Mixtures with endrin may be explosive. Strong oxidizers may cause fire and explosions. Attacks some plastics, rubber, and coatings.

PHOSPHOROTHIOIC ACID, O,O-DIMETHYL O-(4-NITROPHE-NYL)ESTER (298-00-0) Mixtures with magnesium may be explosive.

PHOSPHOROUS ACID, TRIETHYL ESTER (122-52-1) Incompatible with acids (nonoxidizing mineral, sulfuric, nitric, organic), nitrates, oxidizers, magnesium.

PHOSPHOROUS ACID, TRIMETHYL ESTER (121-45-9) Forms explosive mixture with air (flash point 82°F/28°C). Reacts violently with strong acids, nitrates, oxidizers, magnesium salts of perchloric acid.

PHOSPHOROUS CHLORIDE (7719-12-2) Contact with water or alcohol may cause fire and explosions, particularly in presence of combustible organic matter. Incompatible with ammonia, caustics, reducing agents, nitric acid, chemically active metals. In the presence of moisture, corrodes most metals. Attacks some plastics, rubber, and coatings.

PHOSPHOROUS OXIDE (1314-56-3) Aqueous solution is a strong acid; reacts violently with caustics, water. Attacks metals in the presence of moisture.

PHOSPHOROUS PENTOXIDE (1314-56-3) Aqueous solution is a strong acid; reacts violently with caustics, water. Attacks metals in the presence of moisture.

PHOSPHORPENTACHLORID (German) (10026-13-8) Moisture will cause development of pressure in a closed container. Water causes violent spattering and formation of toxic and corrosive hydrogen chloride gas and phosphoric acid mist. Reacts violently with magnesium oxide, chemically active metals, alkalies. Attacks some plastics, rubber, and coatings.

PHOSPHORTRICHLORID (German) (7719-12-2) Contact with water or alcohol may cause fire and explosions, particularly in presence of combustible organic matter. Incompatible with ammonia, caustics, reducing agents, nitric acid, chemically active metals. In the presence of moisture, corrodes most metals. Attacks some plastics, rubber, and coatings.

PHOSPHORUS, AMORPHOUS, RED (7723-14-0) Contact with all oxidizers may cause fire and explosions or produce shock-sensitive compounds. Reacts violently with caustics + heat, chlorosulfonic acid + heat, lead dioxide, performic acid, selenium oxychloride, chlorates, and other materials. Incompatible with many other substances, and forms phosphine gas on contact with moisture and oxygen; opened packages should be stored under inert gas blanket.

PHOSPHORUS BROMIDE (7789-60-8) Reacts violently with water, producinging hydrogen bromide. Incompatible with acids, caustics, alcohols, ammonia, oxidizers, alkali metals (i.e., lithium, sodium, potassium, rubidium, cesium, francium). Corrosive to most metals in the presence of moisture, except lead and nickel.

PHOSPHORUS CHLORIDE (10025-87-3) Water contact produces hydrochloric and phosphoric acids, phosphine, and may cause self-ignition. Air contact produces corrosive fumes. Reacts with acids, alcohols, alkali metals, caustics, combustible materials, carbon disulfide, dimethyl formamide, organic matter, strong bases, zinc powder. Rapid corrosion of steel and most metals, except lead, occurs in the presence of moisture.

PHOSPHORUS OXYCHLORIDE (10025-87-3) Water contact produces hydrochloric and phosphoric acids, phosphine, and may cause self-ignition. Air contact forms corrosive fumes. Reacts with acids, alcohols, alkali metals, caustics, combustible materials, carbon disulfide, dimethyl formamide, organic matter, strong bases, zinc powder. Rapid corrosion of steel and most metals, except lead, occurs in the presence of moisture.

PHOSPHORUS OXYTRICHLORIDE (10025-87-3) Water contact produces hydrochloric and phosphoric acids, phosphine, and may cause self-ignition. Air contact forms corrosive fumes. Reacts with acids, alcohols, alkali metals, caustics, combustible materials, carbon disulfide, dimethyl formamide, organic matter, strong bases, zinc powder. Rapid corrosion of steel and most metals, except lead, in the presence of moisture.

PHOSPHORUS PENTACHLORIDE (10026-13-8) Moisture will cause development of pressure in a closed container. Water causes violent spattering and formation of toxic and corrosive hydrogen chloride gas and phosphoric acid mist. Reacts violently with magnesium oxide, chemically active metals, alkalies. Attacks some plastics, rubber, and coatings.

PHOSPHORUS PENTASULFIDE (1314-80-3) Contact with water or alcohol yields heat and possible ignition. May self-ignite in air. Reacts violently with acids, oxidizers.

PHOSPHORUS PERCHLORIDE (10026-13-8) Moisture will cause development of pressure in a closed container. Water causes violent spattering and formation of toxic and corrosive hydrogen chloride gas and phosphoric acid mist. Reacts violently with magnesium oxide, chemically active metals, alkalies. Attacks some plastics, rubber, and coatings.

PHOSPHORUS PERSULFIDE (1314-80-3) Contact with water or alcohol yields heat and possible ignition. May self-ignite in air. Reacts violently with acids, oxidizers.

PHOSPHORUS, RED (7723-14-0) Contact with all oxidizers may cause fire and explosions or produce shock-sensitive compounds. Reacts violently with caustics + heat, chlorosulfonic acid + heat, lead dioxide, performic acid, selenium oxychloride, chlorates and other materials. Incompatible with many other substances, and produces phosphine gas on contact with moisture and oxygen; opened packages should be stored under inert gas blanket.

PHOSPHORUS SULFIDE (1314-80-3) Contact with water or alcohol yields heat and possible ignition. May self-ignite in air. Reacts violently with acids, oxidizers.

PHOSPHORUS TRIBROMIDE (7789-60-8) Reacts violently with water, producing hydrogen bromide. Incompatible with acids, caustics, alcohols, ammonia, oxidizers, alkali metals (i.e., lithium, sodium, potassium, rubidium, cesium, francium). Corrosive to most metals in the presence of moisture, except lead and nickel.

PHOSPHORUS TRICHLORIDE (7719-12-2) Contact with water or alcohol may cause fire and explosions, particularly in presence of combustible organic matter. Incompatible with ammonia, caustics, reducing agents, nitric acid, chemically active metals. In the presence of moisture, corrodes most metals. Attacks some plastics, rubber, and coatings.

PHOSPHORUS, WHITE or YELLOW (7723-14-0) A strong reducing agent and fire risk; store under water away from heat. Ignites spontaneously on contact with air (flash point at or above 86°F/30°C). Contact with all oxidizers (including elemental sulfur) will cause fire and explosions. Contact with strong caustics produces poisonous and flammable phosphine gas. Liquid attacks some plastics, rubber, and coatings.

PHOSPHORWASSERSTOFF (German) (7803-51-2) Unusually reactive. Contact with air or other oxidizer may cause self-ignition and explosion. Reacts violently or forms explosive products with many substances including acids, halogenated hydrocarbons, moisture. Heat may cause cylinders to explode.

PHOSPHORYL CHLORIDE (10025-87-3) Water contact produces hydrochloric and phosphoric acids, phosphine, and may cause self-ignition. Air contact produces corrosive fumes. Reacts with acids, alcohols, alkali metals, caustics, combustible materials, carbon disulfide, dimethyl formamide, organic matter, strong bases, zinc powder. Rapid corrosion of steel and most metals, except lead, occurs in the presence of moisture.

PHOSPHOSTIGMINE (56-38-2) Combustible liquid. Mixtures with endrin may be explosive. Strong oxidizers may cause fire and explosions. Attacks some plastics, rubber, and coatings.

PHOSPHOTOX E (563-12-2) Incompatible with alkaline formulations. Mixtures with magnesium may be explosive.

PHOSPHURES d'ALUMIUM (French) (20859-73-8) Contact with moisture (including atmospheric moisture) produces spontaneously combustible phosphine gas.

PHOSTOXIN (7803-51-2) Unusually reactive. Contact with air or other oxidizer may cause self-ignition and explosion. Reacts violently or forms explosive products with many substances, including acids, halogenated hydrocarbons, moisture. Heat may cause cylinders to explode.

PHOSTOXIN (20859-73-8) Contact with moisture (including atmospheric moisture) produces spontaneously combustible phosphine gas.

PHOSVIN (1314-84-7) Incompatible with oxidizers, acids, water.

PHOSVIT (62-73-7) Attacks some plastics, rubber, and coatings.

PHOTOPHOR (1305-99-3) A strong reducing agent. Forms phosphine in air. Water or acids can cause explosions. Incompatible with strong oxidizers, acids, chlorine, chlorine monoxide, oxygen, sulfur.

PHPH (92-52-4) Mist forms explosive mixture with air. Strong oxidizers may cause fire and explosions.

PHTHALANDIONE (85-44-9) Incompatible with strong acids, caustics, ammonia, amines, strong oxidizers. Attacks some plastics, rubber, and coatings.

1,3-PHTHALANDIONE (85-44-9) Incompatible with strong acids, caustics, ammonia, amines, strong oxidizers. Attacks some plastics, rubber, and coatings.

PHTHALIC ACID (88-99-3) Incompatible with strong acids, nitrates, oxidizers.

m-PHTHALIC ACID or *meta*-PHTHALIC ACID (88-99-3) Incompatible with strong acids, nitrates, oxidizers.

PHTHALIC ACID ANHYDRIDE (85-44-9) Incompatible with strong acids, caustics, ammonia, amines, strong oxidizers. Attacks some plastics, rubber, and coatings.

PHTHALIC ACID, BENZYL BUTYL ETHER (85-68-7) Incompatible with strong acids, nitrates, oxidizers.

PHTHALIC ACID, BIS(2-ETHYLHEXYL ESTER) (117-81-7) Incompatible with strong acids, strong alkalies, nitrates, oxidizers.

PHTHALIC ACID, BIS(2-ETHYLHEXYL ESTER) (117-84-0) Water contact causes foaming. Incompatible with strong acids, nitrates.

PHTHALIC ACID, BIS(2-METHOXYETHYL)ESTER (121-45-9) Forms explosive mixture with air (flash point 82°F/28°C). Reacts violently with strong acids, nitrates, oxidizers, magnesium salts of perchloric acid.

PHTHALIC ACID, BIS(8-METHYL-NONYL) ESTER (26761-40-0) Incompatible with strong acids, strong alkalies, nitrates, oxidizers.

PHTHALIC ACID, BIS(7-METHYLOCTYL) ESTER (88-99-3) Incompatible with strong acids, nitrates, oxidizers.

PHTHALIC ACID, DIAMYL ESTER (3648-21-3) Incompatible with strong acids, strong alkalies, nitrates, strong oxidizers. Attacks some plastics, rubber, and coatings.

PHTHALIC ACID, DIAMYL ESTER (131-18-0) Incompatible with strong acids, nitrates. Attacks some plastics, rubber, and coatings.

PHTHALIC ACID, DIBUTYL ESTERCELLUFLEX DBP (84-74-2) Incompatible with strong acids, nitrates, strong oxidizers, strong alkalies.

PHTHALIC ACID, DIBUTYL ESTER (84-74-2) Incompatible with strong acids, nitrates, strong oxidizers, strong alkalies.

PHTHALIC ACID, DIETHYL ESTER (84-66-2) Reacts violently with strong acids, strong oxidizers, permanganates, water. Attacks some forms of plastic.

PHTHALIC ACID, DIHEPTYL ESTER (3648-21-3) Incompatible with strong acids, strong alkalies, nitrates, strong oxidizers. Attacks some plastics, rubber, and coatings.

PHTHALIC ACID, DI-ISOBUTYL ESTER (84-69-5) Incompatible with strong acids, strong oxidizers, nitrates.

PHTHALIC ACID, DIISODECYL ESTER (26761-40-0) Incompatible with strong acids, strong alkalies, nitrates, oxidizers.

PHTHALIC ACID, DIMETHYL ESTER (131-11-3) Incompatible with strong alkalies, strong acids, nitrates, oxidizers.

PHTHALIC ACID, DINONYL ESTER (84-76-4) Incompatible with strong acids, nitrates, oxidizers.

PHTHALIC ACID, DIOCTYL ESTER (117-84-0) Water contact causes foaming. Incompatible with strong acids, nitrates.

PHTHALIC ACID, DIPENTYL ESTER (131-18-0) Incompatible with strong acids, nitrates. Attacks some plastics, rubber, and coatings.

PHTHALIC ACID, DITRIDECYL ESTER (119-06-2) Incompatible with strong acids, nitrates, oxidizers.

PHTHALIC ACID, DIUNDECYL ESTER (3648-20-2) Incompatible with strong acids, nitrates.

PHTHALIC ANHYDRIDE (85-44-9) Incompatible with strong acids, caustics, ammonia, amines, strong oxidizers. Attacks some plastics, rubber, and coatings.

PHTHALOL (84-66-2) Reacts violently with strong acids, strong oxidizers, permanganates, water. Attacks some forms of plastic.

PHTHALSAEUREANHYDRID (German) (85-44-9) Incompatible with strong acids, caustics, ammonia, amines, strong oxidizers. Attacks some plastics, rubber, and coatings.

PHTHALSAEUREDIMETHYLESTER (German) (131-11-3) Incompatible with strong alkalies, strong acids, nitrates, oxidizers.

PHYTAR (124-65-2) Corrodes common metals.

PHYTAR-560 (124-65-2) Corrodes common metals.

PIC-CHLOR (76-06-2) Can be self-reactive. Fast heating, shock, alkali metals, or alkaline earth may cause explosions. A strong oxidizer; reacts violently with reducing agents, aniline in presence of heat, alcoholic sodium hydroxide, combustible substances, sodium methoxide, propargyl bromide. Liquid attacks some plastics, rubber, and coatings.

PICKEL ALUM (10043-01-3) Aqueous solution is a strong acid. Incompatible with caustics.

PICFUME (76-06-2) Can be self-reactive. Fast heating, shock, alkali metals, or alkaline earth may cause explosions. A strong oxidizer; reacts violently with reducing agents, aniline in presence of heat, alcoholic sodium hydroxide, combustible substances, sodium methoxide, propargyl bromide. Liquid attacks some plastics, rubber, and coatings.

2-PICOLINE (109-06-8) Forms explosive mixture with air (flash point 79°F/26°C). Strong oxidizers may cause fire and explosions. Attacks copper/alloys.

3-PICOLINE (108-99-6) Forms explosive mixture with air (flash point 97°F/36°C). Incompatible with acids, organic anhydrides, isocyanates, aldehydes, oxidizers.

4-PICOLINE (108-89-4) Forms explosive mixture with air (flash point 134°F/57°C). Incompatible with acids, organic anhydrides, isocyanates, aldehydes, oxidizers.

a-PICOLINE or alpha-PICOLINE (109-06-8) Forms explosive mixture with air (flash point 79°F/26°C). Strong oxidizers may cause fire and explosions. Attacks copper and its alloys.

b-PICOLINE or beta-PICOLINE (108-99-6) Forms explosive mixture with air (flash point 97°F/36°C). Incompatible with acids, organic anhydrides, isocyanates, aldehydes, oxidizers.

gamma-PICOLINE (108-89-4) Forms explosive mixture with air (flash point 134°F/57°C). Incompatible with acids, organic anhydrides, isocyanates, aldehydes, oxidizers.

m-**PICOLINE or** *meta*-**PICOLINE** (108-99-6) Forms explosive mixture with air (flash point 97°F/36°C). Incompatible with acids, organic anhydrides, isocyanates, aldehydes, oxidizers.

PICRIC ACID (88-89-1) Dry material is explosive. Impact-sensitive; protect from shock. Reaction with copper, lead, zinc, and other metals or their salts can produce other salts that are initiators and much more sensitive to shock than this chemical. Shock-sensitive salts include ammonium salts and calcium salts; the calcium salt may be produced when picric acid comes into contact with plaster and concrete. May accumulate static electrical charges, and may cause explosion. Aqueous solution is a strong oxidizer and a strong acid; reacts violently with reducing agents, combustibles, organics and easily oxidized materials, caustics, aluminum, and other metal powders. Attacks metals.

PICRIC ACID, AMMONIUM SALT (131-74-8) A self-reactive explosive. Heat above 250°F/121°C can cause explosions. A powerful oxidizer. Contact with metal, concrete, or plaster produces salts that are more shock-sensitive than ammonium picrate. Water increases rate of reactivity with metals.

PICRIDE (76-06-2) Can be self-reactive. Fast heating, shock, alkali metals, or alkaline earth may cause explosions. A strong oxidizer; reacts violently with reducing agents, aniline in presence of heat, alcoholic sodium hydroxide, combustible substances, sodium methoxide, propargyl bromide. Liquid attacks some plastics, rubber, and coatings.

PICRONITRIC ACID (88-89-1) Dry material is explosive. Impact sensitive; protect from shock. Reaction with copper, lead, zinc, and other metals or their salts can produce other salts that are initiators and much more sensitive to shock than this chemical. Shock-sensitive salts include ammonium salts and calcium salts; the calcium salt may be produced when picric acid comes into contact with plaster and concrete. May accumulate static electrical charges, and may cause explosion. Aqueous solution is a strong oxidizer and a strong acid; reacts violently with reducing agents, combustibles, organics and easily oxidized materials, caustics, aluminum, and other metal powders. Attacks metals.

PICRYLNITROMETHYLAMINE (479-45-8) A powerful oxidizer and highly sensitive explosive. Protect containers from shock, friction, heat, oxidizers, and hydrazine.

PICTAROL (131-74-8) A self-reactive explosive. Heat above 250°F/121°C can cause explosions. A powerful oxidizer. Contact with metal, concrete, or plaster produces salts that are more shock-sensitive than ammonium picrate. Water increases rate of reactivity with metals.

PIECIOCHLOREK FOSFORU (Polish) (10026-13-8) Moisture will cause development of pressure in a closed container. Water causes violent spattering and formation of toxic and corrosive hydrogen chloride gas and phosphoric acid mist. Reacts violently with magnesium oxide, chemically active metals, alkalies. Attacks some plastics, rubber, and coatings.

PIELIK (94-75-7) Decomposes in sunlight. Incompatible with strong oxidizers; may cause fire and explosions.

PIKRINEZUUR (Dutch) (88-89-1) Dry material is explosive. Impact-sensitive; protect from shock. Reaction with copper, lead, zinc, and other metals or their salts can produce other salts that are initiators and much more sensitive to shock than this chemical. Shock-sensitive salts include ammonium salts and calcium salts; the calcium salt may form when picric acid comes into contact with plaster and concrete. May accumulate static electrical charges, and may cause explosion. Aqueous solution is a strong oxidizer and a strong acid; reacts violently with reducing agents, combustibles, organics and easily oxidized materials, caustics, aluminum, and other metal powders. Attacks metals.

PIKRINSAEURE (German) (88-89-1) Dry material is explosive. Impact-sensitive; protect from shock. Reaction with copper, lead, zinc, and other metals or their salts can produce other salts that are initiators and much more sensitive to shock than this chemical. Shock-sensitive salts include ammonium salts and calcium salts; the calcium salt may be produced when picric acid comes into contact with plaster and concrete. May accumulate static electrical charges, and may cause explosion. Aqueous solution is a strong oxidizer and a strong acid; reacts violently with reducing agents, combustibles, organics and easily oxidized materials, caustics, aluminum, and other metal powders. Attacks metals.

PIKRYNOWY KWAS (Polish) (88-89-1) Dry material is explosive. Impact-sensitive; protect from shock. Reaction with copper, lead, zinc, and other metals or their salts can produce other salts that are initiators and much more sensitive to shock than this chemical. Shock-sensitive salts include ammonium salts and calcium salts; the calcium salt may be produced when picric acid comes into contact with plaster and concrete. May accumulate static electrical charges, and may cause explosion. Aqueous solution is a strong oxidizer and a strong acid; reacts violently with reducing agents, combustibles, organics and easily oxidized materials, caustics, aluminum, and other metal powders. Attacks metals.

PILLARFURAN (1563-66-2) Incompatible with alkaline material, acids, strong oxidizers.

PILOT HD-90 (25155-30-0) Reacts with acids, including fumes.

PILOT SF-40 (25155-30-0) Reacts with acids, including fumes.

PIMELIC KETONE (108-94-1) Forms explosive mixture with air (flash point 111°F/44°C). May accumulate static electrical charges, and may cause ignition of its vapors. Attacks red metals and lead. Incompatible with amines, sulfuric acid, nitric acid, oxidizers, strong acids, aliphatic amines. Dissolves plastics, resins, and rubber.

PINAKON (107-41-5) Forms explosive mixture with air (flash point 209°F/98°C). Incompatible with strong acids, caustics, aliphatic amines, isocyanates, strong oxidizers.

PINENE (80-56-8) Incompatible with sulfuric acid, nitric acid.

2-PINENE (80-56-8) Incompatible with sulfuric acid, nitric acid.

a-PINENE or alpha-PINENE (80-56-8) Incompatible with sulfuric acid, nitric acid.

PINE OIL (8002-09-3) Forms explosive mixture with air (flash point 138°F/59°C). Incompatible with strong acids, strong oxidizers.

PINE OIL, STEAM DISTILLED (8002-09-3) Forms explosive mixture with air (flash point 138°F/59°C). Incompatible with strong acids, strong oxidizers.

PINON WOOD CHARCOAL activated (64365-11-3); purified (7440-44-0). Incompatible with strong oxidizers, strong acids, oxides, unsaturated oils.

PIPERAZIDINE (110-85-0) Forms explosive mixture with air (flash point 178°F/81°C). Incompatible with nitrogen compounds, carbon tetrachloride, strong oxidizers. Aqueous solutions react with acids. Attacks aluminum, copper, nickel, magnesium, and zinc.

PIPERAZIN (German) (110-85-0) Forms explosive mixture with air (flash point 178°F/81°C). Incompatible with nitrogen compounds, carbon tetrachloride, strong oxidizers. Aqueous solutions react with acids. Attacks aluminum, copper, nickel, magnesium, and zinc.

PIPERAZINE or PIPERAZINE, ANHYDROUS (110-85-0) Forms explosive mixture with air (flash point 178°F/81°C). Incompatible with nitrogen compounds, carbon tetrachloride, strong oxidizers. Aqueous solutions react with acids. Attacks aluminum, copper, nickel, magnesium, and zinc.

1-PIPERAZINE ETHANAMINE (140-31-8) Solution is a strong base. Forms explosive mixture with air (flash point 199°F/93°C). Incompatible with acids, organic anhydrides, isocyanates, vinyl acetate, acrylates, substituted allyls, alkylene oxides, epichlorohydrin, ketones, aldehydes, alcohols, glycols, phenols, cresols, captrolactam solution, strong oxidizers. Contact with copper alloys, zinc, or galvanized steel may cause violent reaction.

PIPERIDINE (110-89-4) Forms explosive mixture with air (flash point 61°F/16°C). Aqueous solution is a strong base; reacts with acids, organic anhydrides, isocyanates, alkylene oxides, epichlorohydrin, aldehydes, alcohols, glycols, phenols, cresols, captrolactam solution. A powerful reducing agent; reacts violently with oxidizers.

PIPERYLENE (504-60-9) Forms explosive mixture with air (flash point −20°F/−29°C). Reacts violently with oxidizers. Attacks some plastics, rubber, and coatings. May accumulate static electrical charges, and may cause ignition of its vapors.

PIROFOS (3689-24-5) Containers may burst in heat. Strong oxidizers may cause fire and explosions. Attacks some plastics, rubber, and coatings.

PIRYDYNA (Polish) (110-86-1) Forms explosive mixture with air (flash point 68°F/20°C). Strong oxidizers may cause fire and explosions. Strong acids may cause violent spattering. Attacks some plastics, rubber, and coatings.

PITTCHLOR (7778-54-3) Decomposes in heat or sunlight. Incompatible with acids, moisture, reducing agents, combustible materials, all other chemicals, especially acetylene, aniline, and all other amines, anthracene, carbon tetrachloride, iron oxide, manganese oxide, mercaptans, diethylene glycol monomethyl ether, nitromethane, organic matter, organic sulfides, phenol, 1-propanethiol, propyl mercaptan, sulfur, organic sulfur compounds.

PITTCIDE (7778-54-3) Decomposes in heat or sunlight. Incompatible with acids, moisture, reducing agents, combustible materials, all other chemicals, especially acetylene, aniline, and all other amines, anthracene, carbon tetrachloride, iron oxide, manganese oxide, mercaptans, diethylene glycol monomethyl ether, nitromethane, organic matter, organic sulfides, phenol, 1-propanethiol, propyl mercaptan, sulfur, organic sulfur compounds.

PIVALIC ACID (75-98-9) Forms explosive mixture with air (flash point 147°F/64°C). Incompatible with sulfuric acid, caustics, ammonia, amines, isocyanates, alkylene oxides, epichlorohydrin, oxidizers.

PLACIDOL E (84-66-2) Reacts violently with strong acids, strong oxidizers, permanganates, water. Attacks some forms of plastic.

PLANOTOX (94-75-7) Decomposes in sunlight. Incompatible with strong oxidizers; may cause fire and explosions.

PLANT DITHIO AEROSOL (3689-24-5) Containers may burst in heat. Strong oxidizers may cause fire and explosions. Attacks some plastics, rubber, and coatings.

PLANTFUME 103 SMOKE GENERATOR (3689-24-5) Containers may burst in heat. Strong oxidizers may cause fire and explosions. Attacks some plastics, rubber, and coatings.

PLANTGARD (94-75-7) Decomposes in sunlight. Incompatible with strong oxidizers; may cause fire and explosions.

PLANT SPRAY OIL (N/A) Oxidizers may cause fire and explosions. Incompatible with nitric acid. May accumulate static electrical charges, and may cause ignition of its vapors.

PLASKIN 8200 (105-60-2) Contact with strong oxidizers may cause fire and explosions.

PLASTICIZED DDP (26761-40-0) Incompatible with strong acids, strong alkalies, nitrates, oxidizers.

PLASTIC LATEX (9016-00-6) Incompatible with sulfuric acid, isocyanates, strong oxidizers.

PLASTOMOLL DOA (103-23-1) Incompatible with strong acids, nitrates, oxidizers.

PLATIN (German) (7440-06-4) Powdered form is highly reactive and may cause fire and explosions on contact with oxidizers, finely divided aluminum, strong acids, acetone, and other substances.

PLATINIC CHLORIDE (7440-06-4) Powdered form is highly reactive and may cause fire and explosions on contact with oxidizers, finely divided aluminum, strong acids, acetone, and other substances.

PLATINUM (7440-06-4) Powdered form is highly reactive and may cause fire and explosions on contact with oxidizers, finely divided aluminum, strong acids, acetone, and other substances.

PLEOPARAPHENE (56-38-2) Combustible liquid. Mixtures with endrin may be explosive. Strong oxidizers may cause fire and explosions. Attacks some plastics, rubber, and coatings.

PLOMB FLUORURE (French) (7783-46-2) Reacts violently with fluorine, calcium carbide.

PLUMBANE, TETRAETHYL- (78-00-2) Forms explosive mixture with air and decomposes above 200°F/93°C. Incompatible with strong oxidizers, concentrated acids; may cause fire and explosions. Attacks some plastics, rubber, and coatings.

PLUMBANE, TETRAMETHYL- (75-74-1) Forms explosive mixture with air (flash point 100°F/38°C). Decomposes above 212°F/100°C. Strong oxidizers and strong acids cause fire and explosions. Attacks some plastics, rubber, and coatings.

PLUMBOUS ARSENATE (7784-40-9) Contact with strong oxidizers may cause fire and explosions.

PLUMBOUS CHLORIDE (7758-95-4) Reacts violently with calcium when heated.

PLUMBOUS FLUORIDE (7783-46-2) Reacts violently with fluorine, calcium carbide.

PLUMBOUS OXIDE (1317-36-8) Reacts violently with aluminum powder, hydrogen peroxide, hydrogen trisulfide, lithium carbide, perchloric acid, strong oxidizers.

PLUMBOUS SULFIDE (1314-87-0) Reacts violently with iodine monochloride or hydrogen peroxide.

PLURACOL POLYOL (25322-69-4) Incompatible with sulfuric acid, isocyanates, perchloric acid, oxidizers.

PMM (594-42-3) Incompatible with caustics or amines (with rapid decomposition), hot iron causes formation of toxic carbon tetrachloride fumes), hot water produces hydrochloric acid), sulfur, and carbon dioxide. Attacks some plastics, rubber, and coatings.

PNA (100-01-06) Strong oxidizers and moisture may cause spontaneous heating. Caustics and heat may cause formation of explosive mixtures. Attacks some plastics, rubber, and coatings. May accumulate static electrical charges, and may cause ignition of its vapors.

PNB (92-93-3) Contact with strong oxidizers may cause fire and explosions.

PNP (100-02-7) Combustible solid. A strong oxidizer; reacts with reducing agents, combustibles, organic and other easily oxidizable materials. Incompatible with strong acids, caustics, aliphatic amines, amides.

PNT (99-99-0) Incompatible with strong acids, caustics, ammonia, amines, reducing agents, oxidizers. Attacks some plastics, rubber, and coatings.

POLYBUTENE (9003-29-6) Incompatible with sulfuric acid, nitric acid.

POLYCAT-8 (98-94-2) Forms explosive mixture with air. Incompatible with acids, organic anhydrides, chlorinated hydrocarbons, isocyanates, vinyl acetate, acrylates, substituted allyls, alkylene oxides, epichlorohydrin, ketones, aldehydes, alcohols, glycols, phenols, cresols, captrolactam solution, strong oxidizers. Attacks aluminum, copper, lead, tin, zinc, and alloys.

POLYCHLORCAMPHENE (8001-35-2) Reacts with oxidizers, with a risk of fire or explosions. Attacks metals in the presence of moisture.

POLYCHLORINATED BIPHENYLS (1336-36-3) Incompatible with strong oxidizers, strong acids.

POLYCHLORINATED CAMPHENE (8001-35-2) Reacts with oxidizers, with a risk of fire or explosions. Attacks metals in the presence of moisture.

POLYCHLOROBIPHENYL (1336-36-3) Incompatible with strong oxidizers, strong acids.

POLYCHLOROCAMPHENE (8001-35-2) Reacts with oxidizers, with a risk of fire or explosions. Attacks metals in the presence of moisture.

POLYCIZER 962-BPA (119-06-2) Incompatible with strong acids, nitrates, oxidizers.

POLYCIZER DBP (84-74-2) Incompatible with strong acids, nitrates, strong oxidizers, strong alkalies.

POLYDIMETHYL SILOXANE (9016-00-6) Incompatible with sulfuric acid, isocyanates, strong oxidizers.

POLYDIMETHYL SILYLENE (9016-00-6) Incompatible with sulfuric acid, isocyanates, strong oxidizers.

POLY(ETHYLENEIMINE) (26913-06-4) Forms explosive mixture with air. Incompatible with acids, organic anhydrides, isocyanates, vinyl acetate, acrylates, substituted allyls, alkylene oxides, epichlorohydrin, ketones, aldehydes, alcohols, glycols, phenols, cresols, captrolactam solution, strong oxidizers. Attacks aluminum, copper, lead, tin, zinc, and alloys.

POLYETHYLENEIMINE (26913-06-4) Forms explosive mixture with air. Incompatible with acids, organic anhydrides, isocyanates, vinyl acetate, acrylates, substituted allyls, alkylene oxides, epichlorohydrin, ketones, aldehydes, alcohols, glycols, phenols, cresols, captrolactam solution, strong oxidizers. Attacks aluminum, copper, lead, tin, zinc, and alloys.

POLYETHYLENE POLYAMINES (26913-06-4) Forms explosive mixture with air. Incompatible with acids, organic anhydrides, isocyanates, vinyl acetate, acrylates, substituted allyls, alkylene oxides, epichlorohydrin, ketones, aldehydes, alcohols, glycols, phenols, cresols, captrolactam solution, strong oxidizers. Attacks aluminum, copper, lead, tin, zinc, and alloys.

POLYFORMALDEHYDE (30525-89-4) Forms explosive mixture with air (flash point 160°F/71°C). Aqueous solution causes formation of formaldehyde. Reacts violently with strong oxidizers or liquid oxygen. May accumulate static electrical charges, and may cause ignition of its vapors.

POLYISOBUTYLENE PLASTICS (9003-29-6) Incompatible with sulfuric acid, nitric acid.

POLYISOBUTYLENE RESINS (9003-29-6) Incompatible with sulfuric acid, nitric acid.

POLYISOBUTYLENE WAXES (9003-29-6) Incompatible with sulfuric acid, nitric acid.

POLYMER GASOLINES (N/A) Forms explosive mixture with air (flash point 0°F to 73°F/−18°C to 3°C). Incompatible with acids, oxidizers, perchlorates.

POLYMETHYLENE POLYPHENYL ISOCYANATE (9016-87-9) Water contact produces scum and carbon dioxide gas. Incompatible with acids, caustics, ammonia, amines, amides, alcohols, glycols, captrolactam solution.

POLY(OXY(DIMETHYLSILYENE)) (9016-00-6) Incompatible with sulfuric acid, isocyanates.

POLY(OXYETHYL) DODECYL ETHER (9008-57-5) Incompatible with strong acids, caustics, aliphatic amines, isocyanates.

POLY(OXYETHYL) LAURYL ETHER (9008-57-5) Incompatible with strong acids, caustics, aliphatic amines, isocyanates.

POLY(OXYETHYL) MYRISTYL ETHER (N/A) Incompatible with strong acids, caustics, aliphatic amines, isocyanates.

POLY(OXYETHYL) PENTADECYL ETHER (N/A) Incompatible with sulfuric acid, nitric acid, caustics, aliphatic amines, isocyanates.

POLY(OXYETHYL) TETRADECYL ETHER (N/A) Incompatible with strong acids, caustics, aliphatic amines, isocyanates.

POLY(OXYETHYL) TRIDECYL ETHER (N/A) Incompatible with strong acids, caustics, aliphatic amines, isocyanates.

POLY PROPENE (9003-07-0) Incompatible with strong acids, strong oxidizers, chlorine, potassium permanganate.

POLY (PROPYLENE GLYCOL) METHYL ETHER (107-98-2) Forms explosive mixture with air (flash point 90°F/32°C). Incompatible with sulfuric acid, isocyanates, perchloric acid.

POLYOXYMETHYLENE (30525-89-4) Forms explosive mixture with air (flash point 160°F/71°C). Aqueous solution causes formation of formaldehyde. Reacts violently with strong oxidizers or liquid oxygen. May accumulate static electrical charges, and may cause ignition of its vapors.

POLYOXYMETHYLENE GLYCOL (30525-89-4) Forms explosive mixture with air (flash point 160°F/71°C). Aqueous solution causes formation of formaldehyde. Reacts violently with strong oxidizers or liquid oxygen. May accumulate static electrical charges, and may cause ignition of its vapors.

POLYOXYMETHYLENE GLYCOLS (50-00-0) May polymerize unless properly inhibited (usually with methanol). Forms explosive mixture with air (flash point 122°F/50°C). Incompatible with strong acids, amines, strong oxidizers, alkaline materials, nitrogen dioxide, performic acid. Reaction with hydrochloric acid produces bis-chloro-methyl ether, a carcinogen.

POLYOXPROPYLENE GLYCOL (25322-69-4) Incompatible with sulfuric acid, isocyanates, perchloric acid, oxidizers.

POLYOXYPROPYLENE GLYCOL METHYL ETHER (107-98-2) Forms explosive mixture with air (flash point 90°F/32°C). Incompatible with sulfuric acid, isocyanates, perchloric acid.

POLYOXYPROPYLENE GLYCOL (25322-69-4) Incompatible with sulfuric acid, isocyanates, perchloric acid, oxidizers.

POLYPHOSPHORIC ACID (7664-38-2) Incompatible with aliphatic amines, alkanolamines, alkylene oxides, aromatic amines, amides, ammonia, ammonium hydroxide, bases, calcium oxide, epichlorohy-drin, isocyanates, nitromethane (explosive), oleum, organic anhy-drides, sulfuric acid, sodium tetrahydroborate, strong oxidizers, vinyl acetate, water. Contact with most metals form hydrogen gas. Attacks some plastics, rubber, and coatings, and glass and ceramics.

POLYPROPYLENE (9003-07-0) Incompatible with strong acids, strong oxidizers, chlorine, potassium permanganate.

POLYPROPYLENE GLYCOL (25322-69-4) Incompatible with sulfuric acid, isocyanates, perchloric acid, oxidizers.

POLYPROPYLENE GLYCOL 750 (25322-69-4) Incompatible with sulfuric acid, isocyanates, perchloric acid, oxidizers.

POLYPROPYLENE GLYCOL METHYL ETHER (107-98-2) Forms explosive mixture with air (flash point 90°F/32°C). Incompatible with sulfuric acid, isocyanates, perchloric acid.

POLYPROPYLENE GLYCOLS P400 to P4000 (25322-69-4) Incom-patible with sulfuric acid, isocyanates, perchloric acid, oxidizers.

POLYRAM ULTRA (137-26-8) Combustible solid (flash point 192°F/89°C). Strong oxidizers may cause fire and explosions; contact with strong acid or oxidizable materials produces toxic gases.

POLY-SOLV (111-96-6) Forms explosive mixture on exposure to air, heat or light. Reacts violently with oxidizers, metal halides. Incompati-ble with sulfuric acid, isocyanates, perchloric acid. May accumulate static electrical charges, and may cause ignition of its vapors.

POLY-SOLV (111-90-0) Forms explosive mixture with air (flash point 201°F/94°C). Heat or contact with atmospheric air may cause formation of unstable peroxides. Incompatible with strong acids, isocyanates, strong oxidizers.

POLY-SOLV DB (112-34-5) Heat or contact with atmospheric air may cause formation of unstable peroxides. Forms explosive mixture with air (flash point 172°F/78°C). Incompatible with strong oxidizers, strong bases, sulfuric acid, isocyanates. Attacks light metals.

POLY-SOLV DM (111-77-3) Incompatible with sulfuric acid, isocyanates, perchloric acid.

POLY-SOLV E (110-80-5) Forms explosive mixture with air (flash point 120°F/49°C). Strong oxidizers may cause fire and explosions. Attacks some plastics, rubber, and coatings. Able to form peroxides. Incompatible with strong acids, aluminum and its alloys.

POLY-SOLV EB (111-76-2) Forms explosive mixture with air (flash point 143°F/62°C). Reacts violently with strong caustics and strong oxidizers. Attacks some coatings, plastics, and rubber. Attacks metallic aluminum at high temperatures.

POLY-SOLV EE (110-80-5) Forms explosive mixture with air (flash point 120°F/49°C). Strong oxidizers may cause fire and explosions. Attacks some plastics, rubber, and coatings. Able to form peroxides. Incompatible with strong acids, aluminum and its alloys.

POLY-SOLV EE ACETATE (111-15-9) Forms explosive mixture with air (flash point 117°F/47°C). Incompatible with strong acids, nitrates. Reacts violently with oxidizers. May form unstable peroxides. Softens many plastics.

POLY-SOLV EM (109-86-4) Forms explosive mixture with air (flash point 103°F/39°C). Heat or oxidizers may cause formation of unstable peroxides. Attacks many metals. Strong oxidizers cause fire and explosions. Strong bases cause decomposition. Attacks some plastics, rubber, and coatings. May accumulate static electrical charges, and may cause ignition of its vapors.

POLY-SOLVE MPM (107-98-2) Forms explosive mixture with air (flash point 90°F/32°C). Incompatible with sulfuric acid, isocyanates, perchloric acid.

POLY-SOLV TE (112-50-5) Incompatible with sulfuric acid, perchloric acid, isocyanates, strong oxidizers.

POMARSOL (137-26-8) Combustible solid (flash point 192°F/89°C). Strong oxidizers may cause fire and explosions; contact with strong acid or oxidizable materials produces toxic gases.

POMASOL (137-26-8) Combustible solid (flash point 192°F/89°C). Strong oxidizers may cause fire and explosions; contact with strong acid or oxidizable materials produces toxic gases.

PORAL (123-63-7) Forms explosive mixture with air (flash point 96°F/36°C). Reacts with strong acids, caustics, ammonia, amines, oxidizers. Contact with acids produces acetaldehyde.

PORTLAND STONE (1317-65-3) Incompatible with acids, alum, ammonium salts, fluorine.

POTASH CHLORATE (3811-04-9) A powerful oxidizer. Reacts violently with reducing agents or combustibles, ammonia gas, ammonium salts, organic matter, hydrogen iodide, organic acids, sulfuric acid. Forms explosive mixtures with metallic powders, ammonium chloride, organic solids, including agricultural materials.

POTASH NITRATE (7757-79-1) A powerful oxidizer. Reacts violently with reducing agents and combustibles. Strong acid causes formation of toxic vapors. Forms explosive mixtures with many substances, including sodium acetate, metal powders, sodium hypophosphite, trichloroethylene, zinc, etc. Dangerously reactive; separate from all materials.

POTASSA (1310-58-3) Aqueous solution is a strong base. Reacts violently with acids, organic anhydrides, isocyanates, alkylene oxides, epichlorohydrin, aldehydes, alcohols, glycols, phenols, cresols, captrolactam solution. Dissolves in water, yielding high heat, caustic fumes, and flammable hydrogen gas. Also reacts with halogenated hydrocarbons, maleic anhydride. Attacks metals such as aluminum, tin, lead, and zinc, producing flammable hydrogen gas.

POTASSE CAUSTIQUE (French) (1310-58-3) Aqueous solution is a strong base. Reacts violently with acids, organic anhydrides, isocyanates, alkylene oxides, epichlorohydrin, aldehydes, alcohols, glycols, phenols, cresols, captrolactam solution. Dissolves in water, yielding high heat, caustic fumes, and flammable hydrogen gas. Also reacts with halogenated hydrocarbons, maleic anhydride. Attacks metals such as aluminum, tin, lead, and zinc, producing flammable hydrogen gas.

POTASSIO (CHLORATODI) (Italian) (3811-04-9) A powerful oxidizer. Reacts violently with reducing agents or combustibles, ammonia gas, ammonium salts, organic matter, hydrogen iodide, organic acids, sulfuric acid. Forms explosive mixtures with metallic powders, ammonium chloride, organic solids, including agricultural materials.

POTASSIO (IDROSSIDO di) (Italian) (1310-58-3) Aqueous solution is a strong base. Reacts violently with acids, organic anhydrides, isocyanates, alkylene oxides, epichlorohydrin, aldehydes, alcohols, glycols, phenols, cresols, captrolactam solution. Dissolves in water, yielding high heat, caustic fumes, and flammable hydrogen gas. Also reacts with halogenated hydrocarbons, maleic anhydride. Attacks metals such as aluminum, tin, lead, and zinc, producing flammable hydrogen gas.

POTASSIO (PERMANGANATO di) (Italian) (7722-64-7) A strong oxidizer. Forms heat- and shock-sensitive compound with sulfuric acid. Reacts violently with combustibles or reducing agents. Reacts explosively with acetic acid, acetic anhydride, anhydrous ammonia, hydrogen peroxide, glycerol, hydroxylamine, organic matter, powdered sulfur, etc. Incompatible with nitric acid; produces toxic chlorine fumes. A dangerous fire and explosion hazard; isolate from all other materials.

POTASSIUM (7440-09-7) A dangerous explosion hazard. Air contact causes spontaneous ignition. Reacts violently with water. A powerful reducing agent. Reacts violently with oxidizers, organic materials, and many other substances. Forms unstable peroxides under normal conditions of temperature and storage.

POTASSIUM BICHROMATE (7778-50-9) Contact with powdered combustibles will cause ignition. Strong acids cause formation of toxic vapors.

POTASSIUM BROMATE (7758-01-2) A strong oxidizer; reacts vigorously with reducing agents, combustibles, organic materials. Incompatible with aluminum, powdered metals, phosphorus, sulfur, and many other substances.

POTASSIUM CHLORATE (3811-04-9) A powerful oxidizer. Reacts violently with reducing agents or combustibles, ammonia gas, ammonium salts, organic matter, hydrogen iodide, organic acids, sulfuric acid. Forms explosive mixtures with metallic powders, ammonium chloride, organic solids, including agricultural materials.

POTASSIUM (CHLORATE de) (French) (3811-04-9) A powerful oxidizer. Reacts violently with reducing agents or combustibles, ammonia gas, ammonium salts, organic matter, hydrogen iodide, organic acids, sulfuric acid. Forms explosive mixtures with metallic powders, ammonium chloride, organic solids, including agricultural materials.

POTASSIUM CHROMATE (7789-00-6) A powerful oxidizer. Contact with combustibles, reducing agents, organic material, or finely divided metals may cause fire and explosions.

POTASSIUM CHROMATE(VI) (7789-00-6) A powerful oxidizer. Contact with combustibles, reducing agents, organic material, or finely divided metals may cause fire and explosions.

POTASSIUM CYANIDE (151-50-8) Incompatible with acids, organic anhydrides, isocyanates, alkylene oxides, epichlorohydrin, aldehydes, alcohols, glycols, phenols, cresols, captrolactam, strong oxidizers, sodium chlorate. Attacks aluminum, copper, zinc in the presence of moisture.

POTASSIUM CYANIDE SOLUTION (151-50-8) Aqueous solution is a strong base. Incompatible with acids, organic anhydrides, isocyanates, alkylene oxides, epichlorohydrin, aldehydes, alcohols, glycols, phenols, cresols, captrolactam, strong oxidizers, sodium chlorate. Attacks aluminum, copper, zinc in the presence of moisture.

POTASSIUM DICHLORO ISOCYANURATE or POTASSIUM DICHLOROISOCYANURATE (2244-21-5) Incompatible with water, forming a bleach solution. A strong oxidizer; reacts violently with reducing agents, combustibles, organics, easily chlorinated materials.

POTASSIUM DICHLORO-*s*-TRIAZINETRIONE (2244-21-5) Incompatible with water, forming a bleach solution. A strong oxidizer; reacts violently with reducing agents, combustibles, organics, easily chlorinated materials.

POTASSIUM DICHROMATE (7778-50-9) A strong oxidizer; reacts violently with reducing agents and combustibles. Contact with powdered combustibles will cause ignition. Strong acids produce toxic vapors.

POTASSIUM DICHROMATE(6+) or POTASSIUM DICHROMATE(VI) (7778-50-9) A strong oxidizer; reacts violently with reducing agents and combustibles. Powdered combustibles will cause ignition. Strong acids produce toxic vapors.

POTASSIUM, ELEMENTAL (7440-09-7) A dangerous explosion hazard. Air contact causes spontaneous ignition. Reacts violently with water. A powerful reducing agent. Reacts violently with oxidizers, organic materials, and many other substances. Forms unstable peroxides under normal conditions of temperature and storage.

POTASSIUM HYDRATE (1310-58-3) Aqueous solution is a strong base. Reacts violently with acids, organic anhydrides, isocyanates, alkylene oxides, epichlorohydrin, aldehydes, alcohols, glycols, phenols, cresols, captrolactam solution. Dissolves in water, yieldinh high

676

heat, caustic fumes, and flammable hydrogen gas. Also reacts with halogenated hydrocarbons, maleic anhydride. Attacks metals such as aluminum, tin, lead, and zinc, producing flammable hydrogen gas.

POTASSIUM, HYDRIXYDE de (French) (1310-58-3) Aqueous solution is a strong base. Reacts violently with acids, organic anhydrides, isocyanates, alkylene oxides, epichlorohydrin, aldehydes, alcohols, glycols, phenols, cresols, captrolactam solution. Dissolves in water, yieldinh high heat, caustic fumes, and flammable hydrogen gas. Also reacts with halogenated hydrocarbons, maleic anhydride. Attacks metals such as aluminum, tin, lead, and zinc, producing flammable hydrogen gas.

POTASSIUM HYDROXIDE (1310-58-3) Aqueous solution is a strong base. Reacts violently with acids, organic anhydrides, isocyanates, alkylene oxides, epichlorohydrin, aldehydes, alcohols, glycols, phenols, cresols, captrolactam solution. Dissolves in water, yielding high heat, caustic fumes, and flammable hydrogen gas. Also reacts with halogenated hydrocarbons, maleic anhydride. Attacks metals such as aluminum, tin, lead, and zinc, producing flammable hydrogen gas.

POTASSIUM IODATE (7758-05-6) A strong oxidizer; reacts violently with reducing agents, combustibles, organic materials, phosphorus, sulfur.

POTASSIUM IODIDE (7681-11-0) Incompatible with water, producing a corrosive. Air causes decomposition to iodine. Reacts violently with strong oxidizers, bromotrifluorides, chlorotrifluorides, fluorine perchlorate, metallic salts. Attacks metals in a moist environment.

POTASSIUM METAL (7440-09-7) A dangerous explosion hazard. Air contact causes spontaneous ignition. Reacts violently with water. A powerful reducing agent. Reacts violently with oxidizers, organic materials, and many other substances. Forms unstable peroxides under normal conditions of temperature and storage.

POTASSIUM NITRATE (7757-79-1) A powerful oxidizer. Reacts violently with reducing agents and combustibles. Strong acid causes formation of vapors. Forms explosive mixtures with many substances, including sodium acetate, metal powders, sodium hypophosphite, trichloroethylene, and zinc. Dangerously reactive; separate from all materials.

POTASSIUM NITRIDE (29285-24-3) May self-ignite on contact with air. Forms a dangerous, possibly flammable, mixture with phosphorus or sulfur.

POTASSIUM NITRITE (7758-09-0) An oxidizer; reacts with ammonium sulfate, reducing agents, boron, combustibles, organic materials.

POTASSIUM *cis*-9-OCTADECENOIC ACID (143-48-0) Incompatible with nitric acid, sulfuric acids, nitrates.

POTASSIUM OLEATE (143-48-0) Incompatible with nitric acid, sulfuric acids, nitrates.

POTASSIUM OXYMURIATE (3811-04-9) A powerful oxidizer. Reacts violently with reducing agents or combustibles, ammonia gas, ammonium salts, organic matter, hydrogen iodide, organic acids, sulfuric acid. Forms explosive mixtures with metallic powders, ammonium chloride, organic solids, including agricultural materials.

POTASSIUM PERIODATE (7790-21-8) A strong oxidizer; reacts violently with reducing agents, combustibles substances, and organic matter.

POTASSIUM PERMANGANATE (7722-64-7) A strong oxidizer. Forms heat- and shock-sensitive compound with sulfuric acid. Reacts violently with combustibles or reducing agents. Reacts explosively with acetic acid, acetic anhydride, anhydrous ammonia, hydrogen peroxide, glycerol, hydroxylamine, organic matter, powdered sulfur, etc. Incompatible with nitric acid, producing toxic chlorine fumes. A dangerous fire and explosion hazard; isolate from all other materials.

POTASSIUM, PERMANGANATE de (French) (7722-64-7) A strong oxidizer. Forms heat- and shock-sensitive compound with sulfuric acid. Reacts violently with combustibles or reducing agents. Reacts explosively with acetic acid, acetic anhydride, anhydrous ammonia, hydrogen peroxide, glycerol, hydroxylamine, organic matter, powdered sulfur, etc. Incompatible with nitric acid, producing toxic chlorine fumes. A dangerous fire and explosion hazard; isolate from all other materials.

POTASSIUM PEROXIDE (17014-71-0) Reacts violently with water, with liberation of heat and oxygen, and forms a caustic solution. Self-ignites in air. A strong oxidizer; reacts violently with reducing agents, hydrocarbons. Can form explosive and self-igniting mixtures with wood, organic, and other combustible materials. Reacts violently with chemically active metals.

POTASSIUM PERSULFATE (7727-21-1) A strong oxidizer; reacts violently with reducing agents, combustible materials, and organic substances. Contact with potassium hydroxide and moisture may cause fire or explosions.

POTASSIUM SALT OF HYDROCYANIC ACID (151-50-8) Incompatible with acids, organic anhydrides, isocyanates, alkylene oxides, epichlorohydrin, aldehydes, alcohols, glycols, phenols, cresols, captrolactam, strong oxidizers, sodium chlorate. Attacks aluminum, copper, zinc in the presence of moisture.

POTASSIUM SULFATE or POTASSIUM SULFATE (2:1) (7778-80-5) Incompatible with aluminum, magnesium.

POTASSIUM SULFIDE (1312-73-8) Flammable solid. Reacts vigorously with oxidizers. May self-ignite or explode on contact with air, or if subjected to shock or rapid heating.

POTASSIUM SUPEROXIDE (17014-71-0) Reacts violently with water, with liberation of heat and oxygen, and forms a caustic solution. Self-ignites in air. A strong oxidizer; reacts violently with reducing agents, hydrocarbons. Can form explosive and self-igniting mixtures with wood, organic, and other combustible materials. Reacts violently with chemically active metals.

POTASSIUM TROCLOSENE (2244-21-5) Incompatible with water, forming a bleach solution. A strong oxidizer; reacts violently with reducing agents, combustibles, organics, easily chlorinated materials.

POTATO ALCOHOL (64-17-5) Forms explosive mixture with air (flash point 65°F/18°C). May accumulate static electrical charges, and may cause ignition of its vapors. Reactions may be violent with oleum, sulfuric acid, nitric acid, bases, aliphatic amines, isocyanates, oxidizers.

POTATO SPIRIT OIL (123-51-3) Forms explosive mixture with air (flash point 109°F/43°C). Strong oxidizers may cause fire and explosions. Attacks some plastics, rubber, and coatings.

POTENTIATED ACID GLUTARALDEHYDE (111-30-8) Water contact produces a polymer solution. A strong reducing agent. Incompatible with strong acids, caustics, ammonia, amines, strong oxidizers.

POWDERED ALUMINUM (7429-90-5) A strong reducing agent; reacts violently with many compounds, including oxidizers, acids, caustics, halogens, halogenated hydrocarbons, nitrates, sulfates, mercury and mercury compounds, methyl chloride, metal oxides, trichloroethylene, oxygen. May accumulate static electrical charges from friction or stirring, and may self-ignite.

PPD (106-50-3) A strong reducing agent; reacts violently with oxidizers, combustibles, organic substances. Incompatible with strong acids, organic anhydrides, isocyanates, aldehydes.

PPZEIDAN (50-29-3) Incompatible with salts of iron or aluminum, and bases. Do not store in iron containers.

PRECIPITATED BARIUM SULPHATE (7727-43-7) Explosions may result from contact with aluminum in the presence of heat. Incompatible with potassium, phosphorus.

PRECIPITATED SULFUR (7704-34-9) Combustible solid. Liquid causes formation of sulfur dioxide with air. Reacts violently with strong oxidizers. Forms explosive, shock-sensitive, or pyrophoric mixtures with ammonia, ammonium nitrate, bromates, calcium carbide, charcoal, chlorates, hydrocarbons, iodates, iron. Reacts violently with halogen compounds, sodium, tin, uranium, and other compounds. Attacks steel when moist. May accumulate static electrical charges, and may cause ignition of its vapors.

PREEGLONE (85-00-7) Concentrated solution attacks aluminum.

PRESERV-O-SOTE (65996-93-2) Incompatible with oxidizers, strong acids, caustics, aliphatic amines, isocyanates.

PRESPERSION, 75 UREA (57-13-6) Heat causes formation of anhydrous ammonia fumes. Contact with oxidizers, nitrates may cause fire and explosions. Contact with chlorinating agents including hypochlorite bleaches, may yield explosive nitrogen trichloride. Reacts with nitrosyl perchlorate.

PREVENTOL-1 (99-95-4) Contact with strong oxidizers may cause fire and explosions.

PRILTOX (87-86-5) Hot water causes decomposition, producing hydrochloric acid. Strong oxidizers may cause fire and explosions.

PRIMAGRAM (51218-45-0) Incompatible with strong acids, nitrates, oxidizers.

PRIMARY AMYL ACETATE (628-63-7) Incompatible with strong alkalies, strong acids, nitrates, strong oxidizers. Attacks some plastics, coatings, and rubber.

PRIMARY AMYL ALCOHOL (71-41-0) Forms explosive mixture with air (flash point 91°F/33°C). Incompatible with strong acids, caustics, aliphatic amines, isocyanates. Alkali metals and alkaline earth cause formation of hydrogen gas.

PRIMARY DECYL ALCOHOL (112-30-1) Forms explosive mixture with air (flash point 180°F/82°C). Incompatible with strong acids, caustics, aliphatic amines, isocyanates, strong oxidizers.

PRIMARY ISOAMYL ALCOHOL (123-51-3) Forms explosive mixture with air (flash point 109°F/43°C). Strong oxidizers may cause fire and explosions. Attacks some plastics, rubber, and coatings.

PRIMARY OCTYL ALCOHOL (111-87-5) Forms explosive mixture with air (flash point 178°F/81°C). Incompatible with strong acids, caustics, aliphatic amines, isocyanates, strong oxidizers.

PRIMATOL (1912-24-9) Incompatible with strong acids.

PRIMATOL A (1912-24-9) Incompatible with strong acids.

PRIMAZE (1912-24-9) Incompatible with strong acids.

PRIME STEAM LARD (N/A) Incompatible with strong acids, oxidizers.

PRIMEXTRA (51218-45-0) Incompatible with strong acids, nitrates, oxidizers.

PRINDINA (Italian) (110-86-1) Forms explosive mixture with air (flash point 68°F/20°C). Strong oxidizers may cause fire and explosions. Strong acids may cause violent spattering. Attacks some plastics, rubber, and coatings.

PRIODERM (121-75-5) Incompatible with strong oxidizers, magnesium, alkaline pesticides. Attacks metals, some plastics, rubber, and coatings.

PRIST (109-86-4) Forms explosive mixture with air (flash point 103°F/39°C). Heat or oxidizers may cause formation of unstable peroxides. Attacks many metals. Strong oxidizers cause fire and explosions. Strong bases cause decomposition. Attacks some plastics, rubber, and coatings. May accumulate static electrical charges, and may cause ignition of its vapors.

PRODOX 131 (88-69-7) Forms explosive mixture with air (flash point 220°F/104°C). Incompatible with strong oxidizers, acids.

PROFAX (9003-07-0) Incompatible with strong acids, strong oxidizers, chlorine, potassium permanganate.

PROFUME A (76-06-2) Can be self-reactive. Fast heating, shock, alkali metals or alkaline earth may cause explosions. A strong oxidizer; reacts violently with reducing agents, aniline in presence of heat, alcoholic sodium hydroxide, combustible substances, sodium methoxide, propargyl bromide. Liquid attacks some plastics, rubber, and coatings.

PROPADIENE (463-49-0) Flammable gas. Incompatible with oxidizers. May accumulate static electrical charges, and may cause ignition of its vapors.

PROPADIENE-ALLENE MIXTURE (59355-75-8) Forms explosive gas mixture with air. Incompatible with strong oxidizers, copper alloys containing more than 6% copper; may form explosive compounds. Attacks some plastics, rubber, and coatings. May accumulate static electrical charges, and may cause ignition of its vapors.

PROPADIENE-METHYLACETYLENE MIXTURE (59355-75-8) Forms explosive gas mixture with air. Incompatible with strong oxidizers, copper alloys containing more than 6% copper; may form explosive compounds. Attacks some plastics, rubber, and coatings. May accumulate static electrical charges, and may cause ignition of its vapors.

PROPALDEHYDE (123-38-6) Forms explosive mixture with air (flash point −2°F/−30°C). Incompatible with strong acids, caustics, amines. Reacts violently with strong oxidizers. Can self-ignite if finely dispersed on porous or combustible material. Heat or ultraviolet can cause decomposition. May accumulate static electrical charges, and may cause ignition of its vapors.

PROPANAL (123-38-6) Forms explosive mixture with air (flash point −2°F/−30°C). Incompatible with strong acids, caustics, amines. Reacts violently with strong oxidizers. Can self-ignite if finely dispersed on porous or combustible material. Heat or ultraviolet can cause decomposition. May accumulate static electrical charges, and may cause ignition of its vapors.

PROPANAMINE (107-10-8) Forms explosive mixture with air (flash point −35°F/−37°C).Incompatible with acids, organic anhydrides, halogenated hydrocarbons, isocyanates, nitroparaffins, vinyl acetate, acrylates, substituted allyls, alkylene oxides, epichlorohydrin, ketones, aldehydes, alcohols, glycols, mercury, phenols, cresols, triethyl aluminum, captrolactam solution, strong oxidizers. Attacks aluminum, copper, lead, tin, zinc, and alloys.

1-PROPANAMINE (107-10-8) Forms explosive mixture with air. Incompatible with acids, organic anhydrides, halogenated hydrocarbons, isocyanates, nitroparaffins, vinyl acetate, acrylates, substituted allyls, alkylene oxides, epichlorohydrin, ketones, aldehydes, alcohols, glycols, mercury, phenols, cresols, triethyl aluminum, captrolactam solution, strong oxidizers. Attacks aluminum, copper, lead, tin, zinc, and alloys.

2-PROPANAMINE (75-31-0) Forms explosive mixture with air (flash point −35°F/−37°C). A strong base. Reacts with alkali metals, alkaline earth, strong acids (with explosive spattering), strong oxidizers (with fire and explosions). Attacks some plastics, rubber, and coatings.

2-PROPANAMINE, N-(1-METHYLETHYL)- (108-18-9) Forms explosive mixture with air (flash point 30°F/−1°C). Incompatible with acids, organic anhydrides, isocyanates, vinyl acetate, acrylates, substituted allyls, alkylene oxides, epichlorohydrin, ketones, aldehydes, alcohols, glycols, mercury, phenols, cresols, captrolactam solution, strong oxidizers. Attacks aluminum, copper, lead, tin, zinc and alloys, and some plastics, rubber, and coatings.

1-PROPANAMINE, 2-METHYL-N-(2-METHYL PROPYL)- (110-96-3) Forms explosive mixture with air (flash point 85°F/29°C). Incompatible with acids, organic anhydrides, isocyanates, vinyl acetate, acrylates, substituted allyls, alkylene oxides, epichlorohydrin, ketones, aldehydes, alcohols, glycols, mercury, phenols, cresols, captrolactam solution, strong oxidizers. Attacks aluminum, copper, lead, tin, zinc, and alloys.

PROPANE or *n*-PROPANE or *normal*-PROPANE (74-98-6) Forms explosive gas mixture with air. Strong oxidizers may cause fire and explosions. Liquid attacks some plastics, rubber, and coatings. May accumulate static electrical charges, and may cause ignition of its vapors.

PROPANEACID (79-10-7) Forms explosive mixture with air (flash point 122°F/50°C). Forms explosive peroxides; light, heat and peroxides can cause polymerization. Incompatible with sulfuric acid, caustics, ammonia, amines, isocyanates, alkylene oxides, epichlorohydrin, oxidizers, toluenediamine, pyridine, methyl pyridine, *n*-methyl pyrrolidone, 2-methyl-6-ethyl aniline, aniline, ethylene diamine, ethyleneimine, 2-aminoethanol. Severely corrodes carbon steel and iron; attacks other metals. May accumulate static electrical charges, and may cause ignition of its vapors.

PROPANE-BUTANE-(PROPYLENE) (68476-85-7) Forms explosive mixture with air. Strong oxidizers may cause fire and explosions. Attacks some plastics, rubber, and coatings.

1-PROPANECARBOXYIC ACID (107-92-6) Forms explosive mixture with air (flash point 162°F/72°C). Incompatible with sulfuric acid, caustics, ammonia, aliphatic amines, isocyanates, strong oxidizers, alkylene oxides, epichlorohydrin.

PROPANE-2-CARBOXYLIC ACID (79-31-2) Forms explosive mixture with air (flash point 132°F/56°C). Incompatible with sulfuric acid, caustics, ammonia, amines, isocyanates, alkylene oxides, epichlorohydrin. Attacks aluminum and other metals.

PROPANE, CHLORO- (540-54-5) Forms explosive mixture with air (flash point 0°F/−18°C). Reacts violently with strong oxidizers.

PROPANE, 1-CHLORO-2,3-EPOXY (106-89-8) Forms explosive mixture with air (flash point 88°F/31°C). Heat, acids, alkalies, and metallic halides can cause explosive polymerization. Aliphatic amines, alkaline earths, alkali metals, alkanolamines, powdered metals, strong oxidizers may cause fire and explosions. Decomposition produces highly toxic phosgene gas. Will pit steel in the presence of moisture. May accumulate static electrical charges, and may cause ignition of its vapors.

PROPANE, 1-CHLORO-1-NITRO- (600-25-5) Contact with strong oxidizers may cause fire and explosions. Attacks some plastics, rubber, and coatings.

PROPANE CYANOHYDRIN (75-86-5) Forms explosive mixture with air (flash point 165°F/74°C). Reacts violently with strong oxidizers. Heat may cause decomposition. Incompatible with nonoxidizing mineral acids, sulfuric acid, nitric acid, organic acids, caustics, aliphatic amines, alkanolamines, aromatic amines, organic anhydrides, alkylene oxides, epichlorohydrin, captrolactam solution, ammonia, isocyanates, phenols, cresol.

1,3-PROPANE DICARBOXYLIC ACID (110-94-1) Reacts with oxidizers and strong bases.

PROPANE,1,1-DICHLORO- (78-99-9) Forms explosive mixture with air (flash point about 60°F/16°C). Strong oxidizers may cause fire and explosions. Corrodes aluminum. Attacks some plastics, rubber, and coatings.

PROPANE, 1,2-DICHLORO- (78-87-5) Forms explosive mixture with air (flash point 60°F/16°C). May accumulate static electrical charges, and may cause ignition of its vapors. Strong oxidizers may cause fire and explosions. Strong acids can cause decomposition and the formation of hydrogen chloride vapors. Corrodes aluminum. Attacks some plastics, rubber, and coatings.

1,2-PROPANEDIOL (57-55-7) Incompatible with strong acids, caustics, aliphatic amines, isocyanates, strong oxidizers (reacts violently).

PROPANE-1,2-DIOL (57-55-7) Incompatible with strong acids, caustics, aliphatic amines, isocyanates, strong oxidizers (reacts violently).

1,2-PROPANEDIOL-1-ACRYLATE (999-61-1) Incompatible with strong acids, nitrates.

1,3-PROPANEDIOL, 2,2-DIMETHYL (126-30-7) Forms explosive mixture with air (flash point 265°F/129°C). Incompatible with strong acids, caustics, aliphatic amines, isocyanates, oxidizers.

1,2-PROPANEDIOL 1-METHACRYLATE (N/A) Forms explosive mixture with air (flash point 250°F/121°C). Strong oxidizers may cause fire and explosions. May polymerize when hot or when exposed to ultraviolet light or free-radical catalysts.

PROPANEDINITRILE (109-77-3) Incompatible with sulfuric acid. Caustics or heat above 160°F/+171°C may cause polymerization or spontaneous combustion.

PROPANE, EPOXY- (75-56-9) Forms explosive gas mixture with air (flash point −35°F/−37°C). Incompatible with anhydrous metal chlorides. Strong acids, caustics, peroxides cause polymerization. Reacts with ammonia, amines, acetylene-forming metals. Attacks some plastics, rubber, and coatings.

PROPANE, 2-METHOXY-2-METHYL(9CI) (1634-04-4) Forms explosive mixture with air (flash point −14°F/−26°C). May be able to form unstable peroxides. Incompatible with strong acids. Reacts violently with strong oxidizers. May accumulate static electrical charges, and may cause ignition of its vapors.

PROPANE, 2-METHYL (75-28-5) Forms explosive gas mixture with air. Strong oxidizers may cause fire and explosions. May accumulate static electrical charges, and may cause ignition of its vapors.

PROPANENITRILE (107-12-0) Forms explosive mixture with air (flash point 36°F/2°C). Reacts violently with oxidizers. Water or acid contact produces hydrogen cyanide fumes.

PROPANENITRILE, 2-HYDROXY-2-METHYL- (75-86-5) Forms explosive mixture with air (flash point 165°F/74°C). Reacts violently with strong oxidizers. Heat may cause decomposition. Incompatible with nonoxidizing mineral acids, sulfuric acid, nitric acid, organic acids, caustics, aliphatic amines, alkanolamines, aromatic amines, organic anhydrides, alkylene oxides, epichlorohydrin, captrolactam solution, ammonia, isocyanates, phenols, cresol.

PROPANE, 1-NITRO- (108-03-2) Overheating in closed containers may cause violent explosion. Forms explosive mixture with air (flash point 96°F/36°C). Incompatible with aliphatic amines, strong acids, alkalies, strong oxidizers, isocyanates. Contact with some metal oxides may cause decomposition. Mixtures with hydrocarbons are extremely flammable. Attacks some plastics, rubber, and coatings.

PROPANE, 2-NITRO (79-46-9) Forms explosive mixture with air (flash point 74°F/24°C). Incompatible with amines, strong acids, alkalies, strong oxidizers. Contact with some metal oxides may cause decomposition. Mixtures with hydrocarbons are extremely flammable. Attacks some plastics, rubber, and coatings.

PROPANETHIOL (107-03-9) Incompatible with oxidizers, reducing agents, strong acids, caustics, alkali metal, calcium hypochlorite (reacts violently).

1-PROPANETHIOL (107-03-9) Incompatible with oxidizers, reducing agents, strong acids, caustics, alkali metal, calcium hypochlorite (reacts violently).

PROPANE-1-THIOL (107-03-9) Incompatible with oxidizers, reducing agents, strong acids, caustics, alkali metal, calcium hypochlorite (reacts violently).

2-PROPANETHIOL (75-33-2) Forms explosive mixture with air (flash point −30°F/−34°C). Reacts violently with strong oxidizers.

PROPANE-2-THIOL (75-33-2) Forms explosive mixture with air (flash point −30°F/−34°C). Reacts violently with strong oxidizers.

1,2,3-PROPANE TRICARBOXYLIC ACID, 2-HYDROXY-, AMMONIUM SALT 2- (3012-65-5) Combustible solid. Reacts with strong oxidizers.

PROPANE, 1,2,3-TRICHLORO (96-18-4) Forms explosive mixture with air (flash point 164°F/74°C). Strong oxidizers, strong caustics, chemically active metals may cause fire and explosions. May decompose on contact with aluminum. Attacks some plastics, rubber, and coatings.

1,2,3-PROPANETRIOL (56-81-5) Polymerizes at approximately 300°F/150°C. Incompatible with acetic anhydride, potassium permanganate, strong acids, caustics, aliphatic amines, isocyanates, oxidizers.

PROPANOIC ACID (79-09-4) Forms explosive mixture with air (flash point 126°F/52°C). Incompatible with sulfuric acid, strong bases, ammonia, aliphatic amines, alkanolamines, isocyanates, alkylene oxides, epichlorohydrin, oxidizers.

PROPANOIC ACID (75-98-9) Forms explosive mixture with air (flash point 147°F/64°C). Incompatible with sulfuric acid, caustics, ammonia, amines, isocyanates, alkylene oxides, epichlorohydrin, oxidizers.

tert-**PROPANOIC ACID** or *tertiary*-**PROPANOIC ACID** (75-98-9) Forms explosive mixture with air (flash point 147°F/64°C). Incompatible with sulfuric acid, caustics, ammonia, amines, isocyanates, alkylene oxides, epichlorohydrin, oxidizers.

PROPANOIC ACID, BUTYL ESTER (590-01-2) Forms explosive mixture with air (flash point 90°F/32°C). Incompatible with strong oxidizers, strong acids.

PROPANOIC ACID, BUTYL ESTER (9CI) (590-01-2) Forms explosive mixture with air (flash point 90°F/32°C). Incompatible with strong oxidizers, strong acids.

PROPANOIC ACID, 2-CHLORO- (598-78-7) Reacts with strong oxidizers, caustics, ammonia, amines, isocyanates, alkylene oxide, epichlorohydrin, strong oxidizers. Aqueous solution is acid; attacks metals. Only aluminum or stainless steel with protective lining or coating may contact liquid or vapor.

PROPANOIC ACID, 2,2-DICHLORO- (75-99-0) Corrosive to iron, aluminum, and copper.

PROPANOIC ACID, 2,2-DI-METHYL- (75-98-9) Forms explosive mixture with air (flash point 147°F/64°C). Incompatible with sulfuric acid, caustics, ammonia, amines, isocyanates, alkylene oxides, epichlorohydrin, oxidizers.

PROPANOIC ACID, ETHYL ESTER (105-37-3) Forms explosive mixture with air (flash point 54°F/12°C). May accumulate static electrical charges, and may cause ignition of its vapors. Incompatible with strong acids, nitrates, strong oxidizers.

PROPANOIC ACID, 2-HYDROXY-, BUTYL ESTER (9CI) (79-09-4) Forms explosive mixture with air (flash point 126°F/52°C). Incompatible with sulfuric acid, strong bases, ammonia, aliphatic amines, alkanolamines, isocyanates, alkylene oxides, epichlorohydrin, oxidizers.

PROPANOIC ACID, METHYL ESTER (554-12-1) Forms explosive mixture with air (28°F/−2°C). Strong oxidizers may cause fire and explosions. Incompatible with strong acids and nitrates. May accumulate static electrical charges, and may cause ignition of its vapors.

2-PROPANOIC ACID, 1-METHYL-, ETHYL ESTER (97-63-2) Forms explosive mixture with air (flash point 80°F/27°C). Incompatible with strong acids, amines, oxidizers. Corrodes some metals.

1-PROPANOL (71-23-8) Forms explosive mixture with air (flash point 74°F/23°C). Reacts violently with strong oxidizers. Attacks some plastics, rubber, and coatings. Incompatible with strong acids, caustics, aliphatic amines, isocyanates.

PROPANOL-1 (71-23-8) Forms explosive mixture with air (flash point 74°F/23°C). Reacts violently with strong oxidizers. Attacks some plastics, rubber, and coatings. Incompatible with strong acids, caustics, aliphatic amines, isocyanates.

2-PROPANOL (67-63-0) Forms explosive mixture with air (flash point 53°F/12°C). Incompatible with alkaline earth and alkali metals, crotonaldehyde, phosgene, strong acids, amines, ammonia, caustics, strong oxidizers. Attacks some plastics, rubber, and coatings. Reacts with metallic aluminum at high temperatures.

n-**PROPANOL** or *normal*-**PROPANOL** (71-23-8) Forms explosive mixture with air (flash point 74°F/23°C). Reacts violently with strong oxidizers. Attacks some plastics, rubber, and coatings. Incompatible with strong acids, caustics, aliphatic amines, isocyanates.

1,3-PROPANOLAMINE (156-87-6) Forms explosive mixture with air (flash point 175°F/79°C). Incompatible with acids, organic anhydrides, isocyanates, vinyl acetate, acrylates, substituted allyls, alkylene oxides, epichlorohydrin, aldehydes. Corrodes copper and its alloys.

2-PROPANOLAMINE (78-91-1) Forms explosive mixture with air (flash point 145°F/63°C). Incompatible with acids, organic anhydrides, isocyanates, vinyl acetate, acrylates, substituted allyls, alkylene oxides, epichlorohydrin, aldehydes.

3-PROPANOLAMINE (156-87-6) Forms explosive mixture with air (flash point 175°F/79°C). Incompatible with acids, organic anhydrides, isocyanates, vinyl acetate, acrylates, substituted allyls, alkylene oxides, epichlorohydrin, aldehydes. Corrodes copper and its alloys. (156-87-6) Forms explosive mixture with air (flash point 175°F/79°C). Incompatible with acids, organic anhydrides, isocyanates, vinyl acetate, acrylates, substituted allyls, alkylene oxides, epichlorohydrin, aldehydes. Corrodes copper and its alloys.

n-**PROPANOLAMINE** or *normal*-**PROPANOLAMINE** (156-87-6) Forms explosive mixture with air (flash point 175°F/79°C). Incompatible with acids, organic anhydrides, isocyanates, vinyl acetate, acrylates, substituted allyls, alkylene oxides, epichlorohydrin, aldehydes. Corrodes copper and its alloys.

1-PROPANOL, 2-AMINO- (78-91-1) Forms explosive mixture with air (flash point 145°F/63°C). Incompatible with acids, organic anhydrides, isocyanates, vinyl acetate, acrylates, substituted allyls, alkylene oxides, epichlorohydrin, aldehydes.

1-PROPANOL, 3-AMINO (156-87-6) Forms explosive mixture with air (flash point 175°F/79°C). Incompatible with acids, organic anhydrides, isocyanates, vinyl acetate, acrylates, substituted allyls, alkylene oxides, epichlorohydrin, aldehydes. Corrodes copper and its alloys.

1-PROPANOL, 2-AMINO-2-METHYL- (124-68-5) Forms explosive mixture with air (flash point 153°F/67°C). Incompatible with strong acids, mineral acids, organic acids, organic anhydrides, isocyanates, vinyl acetate, acrylates, substituted allyls, alkylene oxides, epichlorohydrin, aldehydes.

PROPANOLE (German) (71-23-8) Forms explosive mixture with air (flash point 74°F/23°C). Reacts violently with strong oxidizers. Attacks some plastics, rubber, and coatings. Incompatible with strong acids, caustics, aliphatic amines, isocyanates.

PROPANOLEN (Dutch) (71-23-8) Forms explosive mixture with air (flash point 74°F/23°C). Reacts violently with strong oxidizers. Attacks some plastics, rubber, and coatings. Incompatible with strong acids, caustics, aliphatic amines, isocyanates.

2-PROPANOL-1-ETHOXY (1569-02-4) Forms explosive mixture with air (flash point 109°F/43°C). Incompatible with sulfuric acid, isocyanates, perchlorates.

PROPANOLI (Italian) (71-23-8) Forms explosive mixture with air (flash point 74°F/23°C). Reacts violently with strong oxidizers. Attacks some plastics, rubber, and coatings. Incompatible with strong acids, caustics, aliphatic amines, isocyanates.

PROPANOLIDE (57-57-8) Forms explosive mixture with air (flash point 165°F/74°C). Incompatible with strong acids, 2-aminoethyl alcohol, caustics. Polymerizes in storage. High temperatures cause ruptured containers. Decomposes at room temperature; hydrolyzes in water.

PROPANOL, 3-(3-(3-METHOXY PROPOXY)PROPOXY)- (25498-49-1) Forms explosive mixture with air (flash point 250°F/121°C). Incompatible with sulfuric acid, perchloric acid, isocyanates, oxidizers.

2-PROPANOL, 2-METHYL- (75-65-0) Forms explosive mixture with air (flash point 52°F/11°C). May accumulate static electrical charges, and may cause ignition of its vapors. Incompatible with strong acids including mineral acids, strong oxidizers or caustics, aliphatic amines, isocyanates, alkali metals (i.e., lithium, sodium, potassium, rubidium, cesium, francium). Attacks many plastics and some coatings.

2-PROPANOL, 1,1',1'-NITRILOTRI- (122-20-3) Incompatible with acids, organic anhydrides, isocyanates, vinyl acetate, acrylates, substituted allyls, alkylene oxides, epichlorohydrin, aldehydes.

PROPANOL, OXYBIS-, METHYL ETHER (34590-94-8) Ethers, as a class, can form dangerous peroxides on standing. Strong oxidizers may cause fire and explosions. Attacks some plastics, rubber, and coatings.

PROPANONE (67-64-1) Forms explosive mixture with air (flash point −4°F/−20°C). Reacts violently with chloroform. Incompatible with strong acids, aliphatic amines, chloroform, chromic anhydride, chromyl chloride, hexachloromelamine, hydrogen peroxide, nitrosyl chloride, nitrosyl perchlorate, nitryl perchlorate, permonosulfuric acid, potassium *tert*-butoxide. Unstable and explosive peroxides are formed with strong oxidizers. May accumulate static electrical charges, and may cause ignition of its vapors. Dissolves most rubber, resins, and plastics.

2-PROPANONE (67-64-1) Forms explosive mixture with air (flash point −4°F/−20°C). Reacts violently with chloroform. Incompatible with strong acids, aliphatic amines, chloroform, chromic anhydride, chromyl chloride, hexachloromelamine, hydrogen peroxide, nitrosyl chloride, nitrosyl perchlorate, nitryl perchlorate, permonosulfuric acid, potassium *tert*-butoxide, strong oxidizers (forms explosive peroxide); sodium hypobromite, trichloromelamine. Dissolves most rubber, resins, and plastics.

PROPARGIL (2312-35-8) Forms explosive mixture with air (flash point 82°F/28°C). Strong oxidizers may cause fire and explosions.

PROPARGITE (2312-35-8) Forms explosive mixture with air (flash point 82°F/28°C). Strong oxidizers may cause fire and explosions.

PROPARGYL ALCOHOL (107-19-7) Incompatible with strong acids, aliphatic amines, phosphorus pentoxide, caustics, isocyanates, strong oxidizers.

PROPASOL SOLVENT M (107-98-2) Forms explosive mixture with air (flash point 90°F/32°C). Incompatible with sulfuric acid, isocyanates, perchloric acid.

PROPELLANT 11 (75-69-4) Reacts with barium, lithium, sodium, magnesium, titanium. Attacks some plastics, rubber, and coatings.

PROPELLANT 12 (75-71-8) Reacts violently with liquid aluminum. Incompatible with chemically active metals. Attacks some plastics, rubber, and coatings.

PROPELLANT 14 (76-14-2) Reacts with barium, lithium, sodium, magnesium, titanium. Attacks some plastics, rubber, and coatings.

PROPELLENT 22 (75-45-6) May accumulate static electrical charges, and may cause ignition of its vapors. Moisture and rust cause slow decomposition, with formation of toxic gases. Attacks some plastics, rubber, and coatings. Thermal decomposition occurs at high temperature with alkalies and alkaline earth metals.

PROPENAL or PROPENAL (Czech) (107-02-8) Forms explosive mixture with air (flash point −15°F/−26°C). Unstable and very reactive. In storage can form heat- and shock-sensitive compounds. Unless inhibited (usually with hydroquinone), readily forms explosive peroxides. Able to polymerize. A strong reducing agent. Reacts violently with oxidizers, strong acids, caustics, amines, 2-aminoethanol, ammonia, ammonium hydroxide, ethylene diamine, ethyleneimine, hydroxides, metal salts, oxidizers, sulfur dioxide, thiourea. Attacks metals: cadmium and zinc. May accumulate static electrical charges, and may cause ignition of its vapors.

2-PROPENAL (107-02-8) Forms explosive mixture with air (flash point −15°F/−26°C). Unstable and very reactive. In storage can form heat- and shock-sensitive compounds. Unless inhibited (usually with hydroquinone), readily forms explosive peroxides. Able to polymerize. A strong reducing agent. Reacts violently with oxidizers, strong acids, caustics, amines, 2-aminoethanol, ammonia, ammonium hydroxide, ethylene diamine, ethyleneimine, hydroxides, metal salts, oxidizers, sulfur dioxide, thiourea. Attacks metals: cadmium and zinc. May accumulate static electrical charges, and may cause ignition of its vapors.

PROP-2-EN-1-AL (107-02-8) Forms explosive mixture with air (flash point −15°F/−26°C). Unstable and very reactive. In storage can form heat and shock-sensitive compounds. Unless inhibited (usually with hydroquinone), readily forms explosive peroxides. Able to polymerize. A strong reducing agent. Reacts violently with oxidizers, strong acids, caustics, amines, 2-aminoethanol, ammonia, ammonium hydroxide, ethylene diamine, ethyleneimine, hydroxides, metal salts, oxidizers, sulfur dioxide, thiourea. Attacks metals: cadmium and zinc. May accumulate static electrical charges, and may cause ignition of its vapors.

PROPENAMIDE (79-06-1) Unless inhibited, ultraviolet light and heat (above 184°F/85°C) can cause polymerization. Incompatible with nonoxidizing mineral acids, strong acids, ammonia, oleum, oxidizers, isocyanates.

2-PROPENAMIDE (79-06-1) Unless inhibited, ultraviolet light and heat (above 184°F/85°C) can cause polymerization. Incompatible with nonoxidizing mineral acids, strong acids, ammonia, oleum, oxidizers, isocyanates.

PROPENE or 1-PROPENE (9CI) (115-07-1) Forms explosive mixture with air (flash point −162°F/−108°C). Incompatible with strong oxidizers, strong acids.

1-PROPENE, 2-METHYL TRIMER (7756-94-7) Incompatible with sulfuric or nitric acids.

PROPENENITRILE (107-13-1) Forms explosive mixture with air (flash point 32°F/0°C o.c.). Forms explosive peroxides; heat, light, caustics, silver nitrate and peroxides can cause polymerization. Incompatible with strong acids, strong oxidizers, amines, 2-aminoethanol, bromine, chlorosulfonic acid, ethylene diamine, nitric acid, oleum, potassium hydroxide, sodium hydroxide, sulfuric acid. Attacks copper and copper alloys; attacks aluminum in high concentrations. May accumulate static electrical charges, and may cause ignition of its vapors.

2-PROPENENITRILE (107-13-1) Forms explosive mixture with air (flash point 32°F/0°C o.c.). Forms explosive peroxides; heat, light, caustics, silver nitrate, and peroxides can cause polymerization. Incompatible with strong acids, strong oxidizers, amines, 2-aminoethanol, bromine, chlorosulfonic acid, ethylene diamine, nitric acid, oleum, potassium hydroxide, sodium hydroxide, sulfuric acid. Attacks copper and copper alloys; attacks aluminum in high concentrations. May accumulate static electrical charges, and may cause ignition of its vapors.

2-PROPENENITRILE, 2-METHYL- (126-98-7) Forms explosive mixture with air (flash point 34°F/1°C). Incompatible with aliphatic amines, alkanolamines. Reacts violently with oxidizers. Incompatible with strong acids, strong bases, or light exposure; may cause polymerization.

PROPENEOXIDE or PROPENE OXIDE (75-56-9) Forms explosive gas mixture with air (flash point −35°F/−37°C). Incompatible with anhydrous metal chlorides. Strong acids, caustics, peroxides cause polymerization. Reacts with ammonia, amines, acetylene-forming metals. Attacks some plastics, rubber, and coatings.

PROPENE POLYMER (9003-07-0) Incompatible with strong acids, strong oxidizers, chlorine, potassium permanganate.

PROPENE TETRAMER (6842-15-5) Forms explosive mixture with air (flash point 144°F/62°C). Incompatible with strong acids, oxidizers.

PROPENE, TRIMER (139-01-4) Forms explosive mixture with air (flash point 75°F/24°C). Incompatible with sulfuric acid, nitric acid, oxidizers. May accumulate static electrical charges, and may cause ignition of its vapors.

2-PROPENIC ACID, 2-METHYL-, BUTYL ESTER (97-88-1) Unless inhibitor is maintained at the proper level, oxidizers, heat, ultraviolet light, or moisture may cause polymerization. May accumulate static electrical charges, and may cause ignition of its vapors.

2-PROPENOIC ACID (79-10-7) Forms explosive mixture with air (flash point 122°F/50°C). Forms explosive peroxides; light, heat, and peroxides can cause polymerization. Incompatible with sulfuric acid, caustics, ammonia, amines, isocyanates, alkylene oxides, epichlorohydrin, oxidizers, toluenediamine, pyridine, methyl pyridine, n-methyl pyrrolidone, 2-methyl-6-ethyl aniline, aniline, ethylene diamine, ethyleneimine, 2-aminoethanol. Severely corrodes carbon steel and iron; attacks other metals. May accumulate static electrical charges, and may cause ignition of its vapors.

2-PROPENOIC ACID, BUTYL ESTER (141-32-2) Forms explosive mixture with air (flash point 120°F/49°C). Heat, sparks, open flame, light, or peroxides may cause explosive polymerization. Incompatible with strong acids, amines, halogens, hydrogen compounds, oxidizers, sunlight, or other catalysts.

2-PROPENOIC ACID, DECYL ESTER (2156-96-6) Incompatible with strong acids, aliphatic amines, alkanolamines, strong oxidizers, polymerization initiators. Attacks copper and its alloys, zinc, galvanized steel, alloys having more than 10% zinc by weight. Swells some rubbers and softens some paints and coatings.

2-PROPENOIC ACID, ETHYL ESTER (140-88-5) Forms explosive mixture with air (flash point 48°F/9°C). Atmospheric moisture and strong alkalies may cause fire and explosions. Unless properly inhibited (note: inert gas blanket not recommended), heat, light or peroxides can cause polymerization. Incompatible with oxidizers (reaction may be violent), strong acids, amines. May accumulate static electrical charges, and may cause ignition of its vapors.

PROPENOIC ACID, METHYL ESTER (96-33-3) Heat, light and/or lack of appropriate inhibitor concentration can cause polymerization. Forms explosive mixture with air (flash point 27°F/−3°C). Incompatible with strong acids, oxidizers, aliphatic amines, alkanolamines.

2-PROPENOIC ACID, METHYL ESTER (96-33-3) Heat, light and/or lack of appropriate inhibitor concentration can cause polymerization. Forms explosive mixture with air (flash point 27°F/−3°C). Incompatible with strong acids, oxidizers, aliphatic amines, alkanolamines.

2-PROPENOIC ACID, 2-METHYL-, METHYL ESTER (80-62-6) Forms explosive mixture with air (flash point 50°F/10°C). Incompatible with caustics, nitrates, strong acids, aliphatic amines, alkanolamines, peroxides, and strong oxidizers. Heat and/or lack of appropriate inhibitor can cause polymerization. May accumulate static electrical charges, and may cause ignition of its vapors.

PROPENOL (107-18-6) May form unstable and explosive peroxides. Able to polymerize. May accumulate static electrical charges, and may cause ignition of its vapors. A strong reducing agent; reacts violently with oxidizers. Forms explosive mixture with air (flash point 70°F/21°C). Incompatible with strong acids, amines, isocyanates, carbon tetrachloride, chlorosulfonic acid, diallyl phosphide, oleum, sodium hydroxide, tri-N-bromomelamine, metal halides, caustic soda, sodium, and magnesium, aluminum, and their alloys. Attacks some coatings, some plastics, and rubber.

PROPEN-1-OL-3 (107-18-6) May form unstable and explosive peroxides. Able to polymerize. May accumulate static electrical charges, and may cause ignition of its vapors. A strong reducing agent; reacts violently with oxidizers. Forms explosive mixture with air (flash point 70°F/21°C). Incompatible with strong acids, amines, isocyanates, carbon tetrachloride, chlorosulfonic acid, diallyl phosphide, oleum, sodium hydroxide, tri-*n*-bromomelamine, metal halides, caustic soda, sodium, and magnesium, aluminum, and their alloys. Attacks some coatings, some plastics, some rubber.

1-PROPEN-3-OL (75-56-9) Forms explosive gas mixture with air (flash point −35°F/−37°C). Incompatible with anhydrous metal chlorides. Strong acids, caustics, peroxides cause polymerization. Reacts with ammonia, amines, acetylene-forming metals. Attacks some plastics, rubber, and coatings.

1-PROPEN-3-OL (107-18-6) May form unstable and explosive peroxides. Able to polymerize. May accumulate static electrical charges, and may cause ignition of its vapors. A strong reducing agent; reacts violently with oxidizers. Forms explosive mixture with air (flash point 70°F/21°C). Incompatible with strong acids, amines, isocyanates, carbon tetrachloride, chlorosulfonic acid, diallyl phosphide, oleum, sodium hydroxide, tri-*n*-bromomelamine, metal halides, caustic soda, sodium, and magnesium, aluminum, and their alloys. Attacks some coatings, some plastics, and rubber.

2-PROPENOL (107-18-6) May form unstable and explosive peroxides. Able to polymerize. May accumulate static electrical charges, and may cause ignition of its vapors. A strong reducing agent; reacts violently with oxidizers. Forms explosive mixture with air (flash point 70°F/21°C). Incompatible with strong acids, amines, isocyanates, carbon tetrachloride, chlorosulfonic acid, diallyl phosphide, oleum, sodium hydroxide, tri-*n*-bromomelamine, metal halides, caustic soda, sodium, and magnesium, aluminum, and their alloys. Attacks some coatings, some plastics, and rubber.

2-PROPEN-1-OL (107-18-6) May form unstable and explosive peroxides. Able to polymerize. May accumulate static electrical charges, and may cause ignition of its vapors. A strong reducing agent; reacts violently with oxidizers. Forms explosive mixture with air (flash point 70°F/21°C). Incompatible with strong acids, amines, isocyanates, carbon tetrachloride, chlorosulfonic acid, diallyl phosphide, oleum, sodium hydroxide, tri-*n*-bromomelamine, metal halides, caustic soda, sodium, and magnesium, aluminum, and their alloys. Attacks some coatings, some plastics, and rubber.

2-PROPEN-1-ONE (107-02-8) Forms explosive mixture with air (flash point −15°F/−26°C). Unstable and very reactive. In storage can form heat- and shock-sensitive compounds. Unless inhibited (usually with hydroquinone), readily forms explosive peroxides. Able to polymerize. A strong reducing agent. Reacts violently with oxidizers, strong acids, caustics, amines, 2-aminoethanol, ammonia, ammonium hydroxide, ethylene diamine, ethyleneimine, hydroxides, metal salts, oxidizers, sulfur dioxide, thiourea. Attacks metals: cadmium and zinc. May accumulate static electrical charges, and may cause ignition of its vapors.

PROPENYL ALCOHOL (107-18-6) May form unstable and explosive peroxides. Able to polymerize. May accumulate static electrical charges, and may cause ignition of its vapors. A strong reducing agent; reacts violently with oxidizers. Forms explosive mixture with air (flash point 70°F/21°C). Incompatible with strong acids, amines, isocyanates, carbon tetrachloride, chlorosulfonic acid, diallyl phosphide, oleum, sodium hydroxide, tri-*n*-bromomelamine, metal halides, caustic soda, sodium, and magnesium, aluminum, and their alloys. Attacks some coatings, some plastics, and rubber.

2-PROPENYL ALCOHOL (107-18-6) May form unstable and explosive peroxides. Able to polymerize. May accumulate static electrical charges, and may cause ignition of its vapors. A strong reducing agent; reacts violently with oxidizers. Forms explosive mixture with air (flash point 70°F/21°C). Incompatible with strong acids, amines, isocyanates, carbon tetrachloride, chlorosulfonic acid, diallyl phosphide, oleum, sodium hydroxide, tri-*n*-bromomelamine, metal halides, caustic soda, sodium, and magnesium, aluminum, and their alloys. Attacks some coatings, some plastics, and rubber.

2-PROPENYLAMINE (107-11-9) Forms explosive mixture with air (flash point −20°F/−29°C). Able to form unstable peroxides and polymerize. May accumulate static electrical charges, and may cause ignition of its vapors. A strong base. Incompatible with acids, strong oxidizers, hypochlorites, nitrosyl perchlorate. Corrodes metal.

2-PROPENYL CHLORIDE (107-05-1) Forms explosive mixture with air (flash point −25°F/−32°C). Violent polymerization and explosion may occur from heat, light, or contact with acid catalysts, ferric chloride, aluminum chloride, Lewis acids, or Ziegler catalyst (e.g., titanium tetrachloride + triethylaluminum monochloride). Incompatible with strong acids, amines, aluminum chloride, boron trifluoride, chlorosulfonic acid, ethylene diamine, ethyleneimine, ferric chloride, oleum, oxidizers, sodium hydroxide. Slow decomposition occurs with moisture; attacks some coatings, plastics, and rubber. Corrosive to steel. May accumulate static electrical charges, and may cause ignition of its vapors.

PROPENYL ETHER (557-40-4) Forms explosive mixture with air (flash point 20°F/−7°C). May accumulate static electrical charges, and may cause ignition of its vapors. Forms explosive peroxides with air. Incompatible with strong acids, oxidizers.

(2-PROPENYLOXY)BENZENE (1746-13-0) Forms explosive mixture with air (flash point 143°F/153°F/62°C). May form peroxides with air. Incompatible with strong acids, oxidizers.

(2-PROPENYLOXY)METHYL))OXIRAINE (106-92-3) Forms explosive mixture with air (flash point 135°F/57°C). Explosive peroxides may be formed in light or air. Unless inhibited, may polymerize on contact with acids or bases. Strong acids, amines, oxidizers may cause fire and explosions. Attacks some forms of plastics, coatings, and rubber.

b-PROPIOLACTONE or beta-PROPIOLACTONE (57-57-8) Forms explosive mixture with air (flash point 165°F/74°C). Incompatible with strong acids, 2-aminoethyl alcohol, caustics. Polymerizes in storage. High temperatures cause ruptured containers. Decomposes at room temperature; hydrolyzes in water.

PROPIOLIC ALCOHOL (107-19-7) Incompatible with strong acids, aliphatic amines, phosphorus pentoxide, caustics, isocyanates, strong oxidizers.

PROPIONALDEHYDE (123-38-6) Forms explosive mixture with air (flash point −2°F/−30°C). Incompatible with strong acids, caustics, amines. Reacts violently with strong oxidizers. Can self-ignite if finely dispersed on porous or combustible material. Heat or ultraviolet can cause decomposition. May accumulate static electrical charges, and may cause ignition of its vapors.

PROPIONE (96-22-0) Forms explosive mixture with air (flash point 55°F/13°C). Incompatible with strong acids, aliphatic amines, strong oxidizers. Attacks some plastics, rubber, and coatings. May accumulate static electrical charges, and may cause ignition of its vapors.

PROPIONIC ACID (79-09-4) Forms explosive mixture with air (flash point 126°F/52°C). Incompatible with sulfuric acid, strong bases, ammonia, aliphatic amines, alkanolamines, isocyanates, alkylene oxides, epichlorohydrin, oxidizers.

PROPIONIC ACID ANHYDRIDE (123-62-6) (123-62-6) Forms explosive mixture with air (flash point 145°F/63°C). Water causes formation of propionic acid. Incompatible with strong ammonia, amines, acids, oxidizers. Contact with caustics may cause fire and explosions.

PROPIONIC ACID BUTYL ESTER (590-01-2) Forms explosive mixture with air (flash point 90°F/32°C). Incompatible with strong oxidizers, strong acids.

PROPIONIC ACID, 3-CHLORO- (107-94-8) Reacts with strong oxidizers, caustics, ammonia, amines, isocyanates, alkylene oxide, epichlorohydrin, strong oxidizers. Aqueous solution is acid; attacks metals. Only aluminum or stainless steel with protective lining or coating may contact liquid or vapor.

PROPIONIC ACID, 3-ETHOXYETHYL ESTER (763-69-9) Forms explosive mixture with air (flash point 138°F/59°C). Incompatible with strong acids, nitrates, oxidizers.

PROPIONIC ACID GRAIN PRESERVER (79-09-4) Forms explosive mixture with air (flash point 126°F/52°C). Incompatible with sulfuric acid, strong bases, ammonia, aliphatic amines, alkanolamines, isocyanates, alkylene oxides, epichlorohydrin, oxidizers.

PROPIONIC ACID, 2-METHYLENE- (79-41-4) Forms explosive mixture with air (flash point 152°F/67°C). A reducing agent; reacts with oxidizers. Forms unstable peroxides; can polymerize violently, especially if stored above 120°F/49°C. Incompatible with strong acids, caustics, ammonia, amines, isocyanates, alkylene oxides, epichlorohydrin.

PROPIONIC ALDEHYDE (123-38-6) Forms explosive mixture with air (flash point −2°F/−30°C). Incompatible with strong acids, caustics, amines. Reacts violently with strong oxidizers. Can self-ignite if finely dispersed on porous or combustible material. Heat or ultraviolet can cause decomposition. May accumulate static electrical charges, and may cause ignition of its vapors.

PROPIONIC ANHYDRIDE (123-62-6) (123-62-6) Forms explosive mixture with air (flash point 145°F/63°C). Water causes formation of propionic acid. Incompatible with strong ammonia, amines, acids, oxidizers. Contact with caustics may cause fire and explosions.

PROPIONIC ETHER (105-37-3) Forms explosive mixture with air (flash point 54°F/12°C). May accumulate static electrical charges, and may cause ignition of its vapors. Incompatible with strong acids, nitrates, strong oxidizers.

PROPIONIC NITRILE (107-12-0) Forms explosive mixture with air (flash point 36°F/2°C). Reacts violently with oxidizers. Water or acid contact produces hydrogen cyanide fumes.

PROPIONITRILE (107-12-0) Forms explosive mixture with air (flash point 36°F/2°C). Reacts violently with oxidizers. Water or acid contact produces hydrogen cyanide fumes.

3-PROPIONOLACTONE (57-57-8) Forms explosive mixture with air (flash point 165°F/74°C). Incompatible with strong acids, 2-aminoethyl alcohol, caustics. Polymerizes in storage. High temperatures cause ruptured containers. Decomposes at room temperature; hydrolyzes in water.

b-PROPIONOLACTONE or beta-PROPIONOLACTONE (57-57-8) Forms explosive mixture with air (flash point 165°F/74°C). Incompatible with strong acids, 2-aminoethyl alcohol, caustics. Polymerizes in storage. High temperatures cause ruptured containers. Decomposes at room temperature; hydrolyzes in water.

PROPIONYL OXIDE (123-62-6) (123-62-6) Forms explosive mixture with air (flash point 145°F/63°C). Water causes formation of propionic acid. Incompatible with strong ammonia, amines, acids, oxidizers. Contact with caustics may cause fire and explosions.

2-PROPOXYETHANOL (2807-30-9) Incompatible with sulfuric acid, isocyanates, strong oxidizers.

N-PROPOXYPROPANOL (N/A) Forms explosive mixture with air 128°F/53°C). Incompatible with sulfuric acid, isocyanates, perchloric acid, oxidizers.

PROPROP (75-99-0) Corrosive to iron, aluminum, and copper.

PROPYL ACETATE (109-60-4) Forms explosive mixture with air (flash point 58°F/14°C). Incompatible with strong acids, nitrates, strong oxidizers. Attacks some plastics, rubber, and coatings. May accumulate static electrical charges, and may cause ignition of its vapors.

1-PROPYL ACETATE (109-60-4) Forms explosive mixture with air (flash point 58°F/14°C). Incompatible with strong acids, nitrates, strong oxidizers. Attacks some plastics, rubber, and coatings. May accumulate static electrical charges, and may cause ignition of its vapors.

2-PROPYL ACETATE (108-21-4) Forms explosive mixture with air (flash point 40°F/4.4°C). Incompatible with strong alkalies, strong acids, nitrates, strong oxidizers. Dissolves rubber, and many plastic materials. Contact with steel may cause slow decomposition. May accumulate static electrical charges, and may cause ignition of its vapors.

n-**PROPYL ACETATE or** *normal*-**PROPYL ACETATE** (109-60-4) Forms explosive mixture with air (flash point 58°F/14°C). Incompatible with strong acids, nitrates, strong oxidizers. Attacks some plastics, rubber, and coatings. May accumulate static electrical charges, and may cause ignition of its vapors.

sec-**PROPYL ACETATE or** *secondary*-**PROPYL ACETATE** (108-21-4) Forms explosive mixture with air (flash point 40°F/4.4°C). Incompatible with strong alkalies, strong acids, nitrates, strong oxidizers. Dissolves rubber, and many plastic materials. Contact with steel may cause slow decomposition. May accumulate static electrical charges, and may cause ignition of its vapors.

PROPYLACETIC ACID (109-52-4) Forms explosive mixture with air (flash point 205°F/96°C). Incompatible with sulfuric acid, caustics, ammonia, amines, isocyanates, alkylene oxides, epichlorohydrin, strong oxidizers.

PROPYLACETONE (591-78-6) Forms explosive mixture with air (flash point 77°F/25°C). Strong oxidizers may cause fire and explosions. Dissolves some plastics, resins, and rubber.

PROPYL ALCOHOL or PROPYL ALCOHOL, NORMAL (71-23-8) Forms explosive mixture with air (flash point 74°F/23°C). Reacts violently with strong oxidizers. Attacks some plastics, rubber, and coatings. Incompatible with strong acids, caustics, aliphatic amines, isocyanates.

1-PROPYL ALCOHOL (71-23-8) Forms explosive mixture with air (flash point 74°F/23°C). Reacts violently with strong oxidizers. Attacks some plastics, rubber, and coatings. Incompatible with strong acids, caustics, aliphatic amines, isocyanates.

n-**PROPYL ALCOHOL or** *normal*-**PROPYL ALCOHOL** (71-23-8) Forms explosive mixture with air (flash point 74°F/23°C). Reacts violently with strong oxidizers. Attacks some plastics, rubber, and coatings. Incompatible with strong acids, caustics, aliphatic amines, isocyanates.

sec-**PROPYL ALCOHOL or** *secondary*-**PROPYL ALCOHOL** (67-63-0) Forms explosive mixture with air (flash point 53°F/12°C). Incompatible with alkaline earth and alkali metals, crotonaldehyde, phosgene, strong acids, amines, ammonia, caustics, strong oxidizers. Attacks some plastics, rubber, and coatings. Reacts with metallic aluminum at high temperatures.

PROPYL ALDEHYDE (123-38-6) Forms explosive mixture with air (flash point −2°F/−30°C). Incompatible with strong acids, caustics, amines. Reacts violently with strong oxidizers. Can self-ignite if finely dispersed on porous or combustible material. Heat or ultraviolet can cause decomposition. May accumulate static electrical charges, and may cause ignition of its vapors.

N-PROPYL ALKOHOL (German) (71-23-8) Forms explosive mixture with air (flash point 74°F/23°C). Reacts violently with strong oxidizers. Attacks some plastics, rubber, and coatings. Incompatible with strong acids, caustics, aliphatic amines, isocyanates.

PROPYLAMINE (107-10-8) Forms explosive mixture with air. Incompatible with acids, organic anhydrides, halogenated hydrocarbons, isocyanates, nitroparaffins, vinyl acetate, acrylates, substituted allyls, alkylene oxides, epichlorohydrin, ketones, aldehydes, alcohols, glycols, mercury, phenols, cresols, triethynyl aluminum, captrolactam solution, strong oxidizers. Attacks aluminum, copper, lead, tin, zinc, and alloys.

2-PROPYLAMINE (107-10-8) Forms explosive mixture with air. Incompatible with acids, organic anhydrides, halogenated hydrocarbons, isocyanates, nitroparaffins, vinyl acetate, acrylates, substituted allyls, alkylene oxides, epichlorohydrin, ketones, aldehydes, alcohols, glycols, mercury, phenols, cresols, triethynyl aluminum, captrolactam solution, strong oxidizers. Attacks aluminum, copper, lead, tin, zinc, and alloys.

n-**PROPYLAMINE or** *normal*-**PROPYLAMINE** (107-10-8) Forms explosive mixture with air. Incompatible with acids, organic anhydrides, halogenated hydrocarbons, isocyanates, nitroparaffins, vinyl acetate, acrylates, substituted allyls, alkylene oxides, epichlorohydrin, ketones, aldehydes, alcohols, glycols, mercury, phenols, cresols, triethynyl aluminum, captrolactam solution, strong oxidizers. Attacks aluminum, copper, lead, tin, zinc, and alloys.

sec-**PROPYLAMINE** or *secondary*-**PROPYLAMINE** (75-31-0) Forms explosive mixture with air (flash point −35°F/−37°C). A strong base. Reacts with alkali metals, alkaline earth, strong acids (with explosive spattering), strong oxidizers (with fire and explosions). Attacks some plastics, rubber, and coatings.

PROPYL BENZENE (103-65-1) Forms explosive mixture with air (flash point 86°F/30°C). Incompatible with nitric acid.

N-PROPYLBENZENE (103-65-1) Forms explosive mixture with air (flash point 86°F/30°C). Incompatible with nitric acid.

PROPYLBROMIDE (106-94-5) Forms explosive mixture with air (flash point 78°F/26°C). Strong oxidizers may cause fire and explosions.

N-PROPYLBROMIDE (106-94-5) Forms explosive mixture with air (flash point 78°F/26°C). Strong oxidizers may cause fire and explosions.

PROPYL CARBINOL (71-36-3) Forms explosive mixture with air (flash point 98°F/37°C). May react with aluminum above 120°F/49°C. Attacks some plastics, rubber, and coatings. Incompatible with strong acids, halogens, caustics, alkali metals, aliphatic amines, isocyanates.

N-PROPYL CARBINOL (71-36-3) Forms explosive mixture with air (flash point 98°F/37°C). May react with aluminum above 120°F/49°C. Attacks some plastics, rubber, and coatings. Incompatible with strong acids, halogens, caustics, alkali metals, aliphatic amines, isocyanates.

n-**PROPYLCARBINYL CHLORIDE** (109-69-3) Forms explosive mixture with air (flash point 15°F/−9°C). May accumulate static electrical charges, and may cause ignition of its vapors. Water contact slowly causes formation of hydrochloric acid. Incompatible with strong oxidizers, alkaline earth and alkali metals, finely divided metal. Attacks metals in presence of moisture. Attacks some plastics, rubber, and coatings. May accumulate static electrical charges, and may cause ignition of its vapors.

PROPYL CELLOSOLVE (2807-30-9) Incompatible with sulfuric acid, isocyanates, strong oxidizers.

n-**PROPYL CHLORIDE** or *normal*-**PROPYL CHLORIDE** (540-54-5) Forms explosive mixture with air (flash point 0°F/−18°C). Reacts violently with strong oxidizers.

PROPYL CYANIDE (109-74-0) Forms explosive mixture with air (flash point 79°F/26°C). Reacts violently with strong acids (forms hydrogen cyanide gas), strong oxidizers (with possible fire and explosions). May accumulate static electrical charges, and may cause ignition of its vapors.

PROPYLENE (115-07-1) Forms explosive mixture with air (flash point −162°F/−108°C). Incompatible with strong oxidizers, strong acids.

1-PROPYLENE (115-07-1) Forms explosive mixture with air (flash point −162°F/−108°C). Incompatible with strong oxidizers, strong acids.

PROPYLENE ALDEHYDE (123-73-9) A strong reducing agent. Forms explosive mixture with air (flash point 55°F/12.8°C). Readily converted by oxygen to peroxides and acids; heat or contact with many other substances may cause polymerization. Incompatible with strong oxidizers, strong acids including nonoxidizing mineral acids, ammonia, aliphatic amines, aromatic amines, 1,3-butadiene, strong bases. Liquid attacks some plastics, rubber, and coatings.

PROPYLENE CHLORIDE (78-87-5) Forms explosive mixture with air (flash point 60°F/16°C). May accumulate static electrical charges, and may cause ignition of its vapors. Strong oxidizers may cause fire and explosions. Strong acids can cause decomposition and the formation of hydrogen chloride vapors. Corrodes aluminum. Attacks some plastics, rubber, and coatings.

PROPYLENE DICHLORIDE (78-87-5) Forms explosive mixture with air (flash point 60°F/16°C). May accumulate static electrical charges, and may cause ignition of its vapors. Strong oxidizers may cause fire and explosions. Strong acids can cause decomposition and the formation of hydrogen chloride vapors. Corrodes aluminum. Attacks some plastics, rubber, and coatings.

alpha, beta-PROPYLENE DICHLORIDE (78-87-5) Forms explosive mixture with air (flash point 60°F/16°C). May accumulate static electrical charges, and may cause ignition of its vapors. Strong oxidizers may cause fire and explosions. Strong acids can cause decomposition and the formation of hydrogen chloride vapors. Corrodes aluminum. Attacks some plastics, rubber, and coatings.

PROPYLENE GLYCOL (57-55-7) Incompatible with strong acids, caustics, aliphatic amines, isocyanates, strong oxidizers (reacts violently).

1,2-PROPYLENE GLYCOL (57-55-7) Incompatible with strong acids, caustics, aliphatic amines, isocyanates, strong oxidizers (reacts violently).

PROPYLENE GLYCOL ETHYL ETHER (1569-02-4) Forms explosive mixture with air (flash point 109°F/43°C). Incompatible with sulfuric acid, isocyanates, perchlorates.

PROPYLENE GLYCOL METHYL ETHER (107-98-2) Forms explosive mixture with air (flash point 90°F/32°C). Incompatible with sulfuric acid, isocyanates, perchloric acid.

PROPYLENE GLYCOL METHYL ETHER ACETATE (108-65-6) Forms explosive mixture with air (flash point 108°F/42°C). Incompatible with strong acids, nitrates, oxidizers.

PROPYLENE GLYCOL MONOACRYLATE (999-61-1) Incompatible with strong acids, nitrates.

PROPYLENE GLYCOL MONOMETHACRYLATE (N/A) Forms explosive mixture with air (flash point 250°F/121°C). Strong oxidizers may cause fire and explosions. May polymerize when hot or when exposed to ultraviolet light or free-radical catalysts.

PROPYLENE GLYCOL MONOMETHYL ETHER (107-98-2) Forms explosive mixture with air (flash point 90°F/32°C). Incompatible with sulfuric acid, isocyanates, perchloric acid.

a-PROPYLENE GLYCOL MONOMETHYL ETHER or alpha-PROPYLENE GLYCOL MONOMETHYL ETHER (107-98-2) Forms explosive mixture with air (flash point 90°F/32°C). Incompatible with sulfuric acid, isocyanates, perchloric acid.

PROPYLENEIMINE (75-55-8) Forms explosive mixture with air (flash point 25°F/−4°C). Contact with acids or high heat can cause violent polymerization. Strong oxidizers may cause fire and explosions. Attacks some plastics, rubber, and coatings.

1,2-PROPYLENIMINE (75-55-8) Forms explosive mixture with air (flash point 25°F/−4°C). Contact with acids or high heat can cause violent polymerization. Strong oxidizers may cause fire and explosions. Attacks some plastics, rubber, and coatings.

PROPYLENE OXIDE (75-56-9) Forms explosive gas mixture with air (flash point −35°F/−37°C). Incompatible with anhydrous metal chlorides. Strong acids, caustics, peroxides cause polymerization. Reacts with ammonia, amines, acetylene-forming metals. Attacks some plastics, rubber, and coatings.

1,2-PROPYLENE OXIDE (75-56-9) Forms explosive gas mixture with air (flash point −35°F/−37°C). Incompatible with anhydrous metal chlorides. Strong acids, caustics, peroxides cause polymerization. Reacts with ammonia, amines, acetylene-forming metals. Attacks some plastics, rubber, and coatings.

PROPYLENE TETRAMER (6842-15-5) Forms explosive mixture with air (flash point 144°F/62°C). Incompatible with strong acids, oxidizers.

PROPYLENE TRIMER (139-01-4) Forms explosive mixture with air (flash point 75°F/24°C). Incompatible with sulfuric acid, nitric acid, oxidizers. May accumulate static electrical charges, and may cause ignition of its vapors.

PROPYLENGLYKOL–MONOMETHYLAETHER (German) (107-98-2) Forms explosive mixture with air (flash point 90°F/32°C). Incompatible with sulfuric acid, isocyanates, perchloric acid.

n-**PROPYL ESTER OF ACETIC ACID or** *normal*-**PROPYL ESTER OF ACETIC ACID** Forms explosive mixture with air (flash point 58°F/14°C). Incompatible with strong acids, nitrates, strong acids, strong oxidizers. Attacks some plastics, rubber, and coatings. May accumulate static electrical charges, and may cause ignition of its vapors.

PROPYL ESTER OF NITRIC ACID (627-13-4) Forms explosive mixture with air (flash point 68°F/20°C). Strong oxidizers or combustibles may cause fire and explosions. Attacks some plastics, rubber, and coatings.

PROPYL ETHER (111-43-3) Forms explosive mixture with air (flash point 70°F/21°C). Forms explosive peroxides, especially when anhydrous. Incompatible with strong acids, strong oxidizers.

n-**PROPYL ETHER or** *normal*-**PROPYL ETHER** (111-43-3) Forms explosive mixture with air (flash point 70°F/21°C). Forms explosive peroxides, especially when anhydrous. Incompatible with strong acids, strong oxidizers.

PROPYLETHYLENE (109-67-1) Forms explosive mixture with air (flash point 0°F/−18°C). Incompatible with strong acids, oxidizers.

PROPYLFORMIC ACID (107-92-6) Forms explosive mixture with air (flash point 162°F/72°C). Incompatible with sulfuric acid, caustics, ammonia, aliphatic amines, isocyanates, strong oxidizers, alkylene oxides, epichlorohydrin.

PROPYL HYDRATE (74-98-6) Forms explosive gas mixture with air. Strong oxidizers may cause fire and explosions. Liquid attacks some plastics, rubber, and coatings. May accumulate static electrical charges, and may cause ignition of its vapors.

PROPYL HYDRIDE (74-98-6) Forms explosive gas mixture with air. Strong oxidizers may cause fire and explosions. Liquid attacks some plastics, rubber, and coatings. May accumulate static electrical charges, and may cause ignition of its vapors.

PROPYLIC ALCOHOL (71-23-8) Forms explosive mixture with air (flash point 74°F/23°C). Reacts violently with strong oxidizers. Attacks some plastics, rubber, and coatings. Incompatible with strong acids, caustics, aliphatic amines, isocyanates.

PROPYLIC ALDEHYDE (123-38-6) Forms explosive mixture with air (flash point −2°F/−30°C). Incompatible with strong acids, caustics, amines. Reacts violently with strong oxidizers. Can self-ignite if finely dispersed on porous or combustible material. Heat or ultraviolet can cause decomposition. May accumulate static electrical charges, and may cause ignition of its vapors.

PROPYLIDENE CHLORIDE (78-99-9) Forms explosive mixture with air (flash point about 60°F/16°C). Strong oxidizers may cause fire and explosions. Corrodes aluminum. Attacks some plastics, rubber, and coatings.

PROPYLIDENE DICHLORIDE (78-99-9) Forms explosive mixture with air (flash point about 60°F). Strong oxidizers may cause fire and explosions. Corrodes aluminum. Attacks some plastics, rubber, and coatings.

PROPYL KETONE (123-19-3) Oxidizers may cause violent reaction.

PROPYL MERCAPTAN (107-03-9) Incompatible with oxidizers, reducing agents, strong acids, caustics, alkali metal, calcium hypochlorite (reacts violently).

n-**PROPYL MERCAPTAN or** *normal*-**PROPYL MERCAPTAN** (107-03-9) Incompatible with oxidizers, reducing agents, strong acids, caustics, alkali metal, calcium hypochlorite (reacts violently).

PROPYLMETHANOL (71-36-3) Forms explosive mixture with air (flash point 98°F/37°C). May react with aluminum above 120°F/49°C. Attacks some plastics, rubber, and coatings. Incompatible with strong acids, halogens, caustics, alkali metals, aliphatic amines, isocyanates.

PROPYL NITRATE (627-13-4) Forms explosive mixture with air (flash point 68°F/20°C). Strong oxidizers or combustibles may cause fire and explosions. Attacks some plastics, rubber, and coatings.

n-**PROPYL NITRATE or** *normal*-**PROPYL NITRATE** (627-13-4) Forms explosive mixture with air (flash point 68°F/20°C). Strong oxidizers or combustibles may cause fire and explosions. Attacks some plastics, rubber, and coatings.

PROPYLNITRILE (107-12-0) Forms explosive mixture with air (flash point 36°F/2°C). Reacts violently with oxidizers. Water or acid contact produces hydrogen cyanide fumes.

PROPYLOWY ALKOHOL (Polish) (71-23-8) Forms explosive mixture with air (flash point 74°F/23°C). Reacts violently with strong oxidizers. Attacks some plastics, rubber, and coatings. Incompatible with strong acids, caustics, aliphatic amines, isocyanates.

N-PROPYL-1-PROPANAMINE (142-84-7) Forms explosive mixture with air (flash point 63°F/17°C). Incompatible with acids, organic anhydrides, isocyanates, vinyl acetate, acrylates, substituted allyls, alkylene oxides, epichlorohydrin, ketones, aldehydes, alcohols, glycols, mercury, phenols, cresols, captrolactam solution, strong oxidizers. Attacks aluminum, copper, lead, tin, zinc, and alloys.

1-PROPYNE-3-OL (107-19-7) Incompatible with strong acids, aliphatic amines, phosphorus pentoxide, caustics, isocyanates, strong oxidizers.

PROPYNE (74-99-7) Forms explosive gas mixture with air. Can form explosive peroxide. Strong oxidizers may cause fire and explosions. Forms shock-sensitive compounds with copper, magnesium, silver, and their alloys. Copper or copper alloys containing more than 67% copper should not be used in handling equipment. Attacks some plastics, rubber, and coatings. May accumulate static electrical charges, and may cause ignition of its vapors.

1-PROPYNE (74-99-7) Forms explosive gas mixture with air. Can form explosive peroxide. Strong oxidizers may cause fire and explosions. Forms shock-sensitive compounds with copper, magnesium, silver, and their alloys. Copper or copper alloys containing more than 67% copper should not be used in handling equipment. Attacks some plastics, rubber, and coatings. May accumulate static electrical charges, and may cause ignition of its vapors.

PROPYNE–ALLENE MIXTURE (59355-75-8) Forms explosive gas mixture with air. Incompatible with strong oxidizers, copper alloys containing more than 6% copper; may form explosive compounds. Attacks some plastics, rubber, and coatings. May accumulate static electrical charges, and may cause ignition of its vapors.

PROPYNE MIXED WITH PROPADIENE (59355-75-8) Forms explosive gas mixture with air. Incompatible with strong oxidizers, copper alloys containing more than 6% copper; may form explosive compounds. Attacks some plastics, rubber, and coatings. May accumulate static electrical charges, and may cause ignition of its vapors.

PROPYNE–PROPADIENE MIXTURE (59355-75-8) Forms explosive gas mixture with air. Incompatible with strong oxidizers, copper alloys containing more than 6% copper; may form explosive compounds. Attacks some plastics, rubber, and coatings. May accumulate static electrical charges, and may cause ignition of its vapors.

2-PROPYN-1-OL (107-19-7) Incompatible with strong acids, aliphatic amines, phosphorus pentoxide, caustics, isocyanates, strong oxidizers.

3-PROPYNOL (107-19-7) Incompatible with strong acids, aliphatic amines, phosphorus pentoxide, caustics, isocyanates, strong oxidizers.

2-PROPYNYL ALCOHOL (107-19-7) Incompatible with strong acids, aliphatic amines, phosphorus pentoxide, caustics, isocyanates, strong oxidizers.

PROTOCHLORURE d'IODE (French) (7790-99-0) Heat may cause explosions. Reacts violently with water, organic matter, aluminum foil, cadmium sulfide, lead sulfide, phosphorus, phosphorus trichloride, potassium, rubber, silver sulfide, sodium, zinc sulfide, and other metals.

PROXEL EF (2425-06-1) Incompatible with acids or acid vapor. Strongly alkaline conditions contribute to instability.

PROXOL (52-68-6) Contact with strong oxidizers may cause fire and explosions.

PROZOIN (79-09-4) Forms explosive mixture with air (flash point 126°F/52°C). Incompatible with sulfuric acid, strong bases, ammonia, aliphatic amines, alkanolamines, isocyanates, alkylene oxides, epichlorohydrin, oxidizers.

PRUSSIAN BROWN (1309-37-1) Contact with hydrogen peroxide, ethylene oxide, calcium hypochlorite will cause explosion. Reacts violently with powdered aluminum, hydrazine, hydrogen trisulfide.

PRUSSIC ACID (74-90-8) Unless stabilized and maintained, samples stored more than 90 days are hazardous. Samples containing more than 2–5% water are less stable than dry material. Can be self-reactive, forming an explosive mixture with air (flash point 0°F/−18°C). Heat or contact with amines or strong bases can cause polymerization. Incompatible with acetaldehyde. Oxidizers may cause fire and explosions. Attacks some plastics, rubber, and coatings.

PRUSSITE (460-19-5) Explosive reaction occurs with acids, water, liquid oxygen, oxidizers.

PS (76-06-2) Can be self-reactive. Fast heating, shock, alkali metals or alkaline earth may cause explosions. A strong oxidizer; reacts violently with reducing agents, aniline in presence of heat, alcoholic sodium hydroxide, combustible substances, sodium methoxide, propargyl bromide. Liquid attacks some plastics, rubber, and coatings.

PSEUDOACETIC ACID (79-09-4) Forms explosive mixture with air (flash point 126°F/52°C). Incompatible with sulfuric acid, strong bases, ammonia, aliphatic amines, alkanolamines, isocyanates, alkylene oxides, epichlorohydrin, oxidizers.

PSEUDOCUMENE (95-63-6) Forms explosive mixture with air (flash point 112°F/44°C). Oxidizers may cause fire and explosions. Incompatible with nitric acid. May accumulate static electrical charges, and may cause ignition of its vapors.

PSEUDOCUMOL (95-63-6) Forms explosive mixture with air (flash point 112°F/44°C). Oxidizers may cause fire and explosions. Incompatible with nitric acid. May accumulate static electrical charges, and may cause ignition of its vapors.

PSEUDOHEXYL ALCOHOL (97-95-0) Forms explosive mixture with air (flash point 70°F/21°C). Incompatible with strong acids, caustics, isocyanates, amines.

PSEUDOTHIOUREA (62-56-6) Aqueous solution is a base; reacts strongly with acids. Incompatible with nitric acid, hydrogen peroxide, acrolein. Incompatible with metals.

PSEUDOUREA (57-13-6) Heat causes formation of anhydrous ammonia fumes. Contact with oxidizers, nitrates may produce fire and explosions. Contact with chlorinating agents, including hypochlorite bleaches may form explosive nitrogen trichloride. Reacts with nitrosyl perchlorate.

PURALIN (137-26-8) Combustible solid (flash point 192°F/89°C). Strong oxidizers may cause fire and explosions; contact with strong acid or oxidizable materials form toxic gases.

PURATRONIC CHROMIUM TRIOXIDE (1333-82-0) A strong oxidizer. Reacts with acetic acid, acetic anhydride, acetone, anthracene, chromous sulfide, diethyl ether, dimethyl formamide, ethanol, hydrogen sulfide, methanol, naphthalene, camphor, glycerol, potassium ferricyanide, pyridine, turpentine, combustibles, organics. Aqueous solution is strongly acidic. Attacks metals in presence of moisture.

PURE GRAIN ALCOHOL (64-17-5) Forms explosive mixture with air (flash point 65°F/18°C). May accumulate static electrical charges, and may cause ignition of its vapors. Reactions may be violent with oleum, sulfuric acid, nitric acid, bases, aliphatic amines, isocyanates, oxidizers.

PURPLE SALT (7722-64-7) A strong oxidizer. Forms heat- and shock-sensitive compound with sulfuric acid. Reacts violently with combustibles or reducing agents. Reacts explosively with acetic acid, acetic anhydride, anhydrous ammonia, hydrogen peroxide, glycerol, hydroxylamine, organic matter, powdered sulfur, etc. Incompatible with nitric acid, producing toxic chlorine fumes. A dangerous fire and explosion hazard; isolate from all other materials.

PX 104 (84-74-2) Incompatible with strong acids, nitrates, strong oxidizers, strong alkalies.

PX-238 (103-23-1) Incompatible with strong acids, nitrates, oxidizers.

PYRALENE (generic CAS for PCBs 1336-36-3) Incompatible with strong oxidizers, strong acids.

PYRANOL (generic CAS for PCBs 1336-36-3) Incompatible with strong oxidizers, strong acids.

PYRANOL 1478 (87-61-6) Flash point 210°F/99°C. Strong oxidizers may cause fire and explosions.

PYRAZINE HEXAHYDRIDE (110-85-0) Forms explosive mixture with air (flash point 178°F/81°C). Incompatible with nitrogen compounds, carbon tetrachloride, strong oxidizers. Aqueous solutions react with acids. Attacks aluminum, copper, nickel, magnesium, and zinc.

PYRAZINE, HEXAHYDRO- (110-85-0) Forms explosive mixture with air (flash point 178°F/81°C). Incompatible with nitrogen compounds, carbon tetrachloride, strong oxidizers. Aqueous solutions react with acids. Attacks aluminum, copper, nickel, magnesium, and zinc.

PYRENE (129-00-0) Strong oxidizers may cause fire and explosions.

PYRETHRINS (8003-34-7) Strong oxidizers may cause fire and explosions.

PYRETHRUM or PYRETHRUM I or PYRETHRUM II or PYRETHRUM FLOWERS (8003-34-7) Strong oxidizers may cause fire and explosions.

PYRIDIN (German) (110-86-1) Forms explosive mixture with air (flash point 68°F/20°C). Strong oxidizers may cause fire and explosions. Strong acids may cause violent spattering. Attacks some plastics, rubber, and coatings.

2-PYRIDINAMINE (504-29-0) Reacts with strong oxidizers, strong acids, sodium nitrate.

4-PYRIDINAMINE (504-24-2) Forms a strong base with water. Incompatible with strong oxidizers, acids.

alpha-PYRIDINAMINE (504-29-0) Reacts with strong oxidizers, strong acids, sodium nitrate.

PYRIDINE (110-86-1) Forms explosive mixture with air (flash point 68°F/20°C). Strong oxidizers may cause fire and explosions. Strong acids may cause violent spattering. Attacks some plastics, rubber, and coatings.

PYRIDINE, 2-AMINO (504-29-0) Reacts with strong oxidizers, strong acids, sodium nitrate.

PYRIDINE, 3-METHYL (108-99-6) Forms explosive mixture with air (flash point 97°F/36°C). Incompatible with acids, organic anhydrides, isocyanates, aldehydes, oxidizers.

PYRIDINE, 3-(1-METHYL-2-PYRROLIDINYL)- (54-11-5) Incompatible with strong acids, strong oxidizers. Attacks some plastics, rubber, and coatings. May accumulate static electrical charges, and may cause ignition of its vapors.

PYRIDINE, (S)-3-(1-METHYL-2-PYRROLIDINYL)- AND SALTS (54-11-5) Incompatible with strong acids, strong oxidizers. Attacks some plastics, rubber, and coatings. May accumulate static electrical charges, and may cause ignition of its vapors.

2-PYRIDYLAMINE (504-29-0) Reacts with strong oxidizers, strong acids, sodium nitrate.

4-PYRIDYLAMINE (504-24-2) Forms a strong base with water. Incompatible with strong oxidizers, acids.

alpha-PYRIDYLAMINE (504-29-0) Reacts with strong oxidizers, strong acids, sodium nitrate.

PYRINEX (2921-88-) Incompatible with strong acids. Hydrolyzes from acid or alkaline solutions.

PYROACETIC ACID (67-64-1) Forms explosive mixture with air (flash point −4°F/−20°C). Reacts violently with chloroform. Incompatible with strong acids, aliphatic amines, chloroform, chromic anhydride, chromyl chloride, hexachloromelamine, hydrogen peroxide, nitrosyl chloride, nitrosyl perchlorate, nitryl perchlorate, permonosulfuric acid, potassium *tert*-butoxide. Unstable and explosive peroxides are formed with strong oxidizers. May accumulate static electrical charges, and may cause ignition of its vapors. Dissolves most rubber, resins, and plastics.

PYROACETIC ETHER (67-64-1) Forms explosive mixture with air (flash point −4°F/−20°C). Reacts violently with chloroform. Incompatible with strong acids, aliphatic amines, chloroform, chromic anhydride, chromyl chloride, hexachloromelamine, hydrogen peroxide, nitrosyl chloride, nitrosyl perchlorate, nitryl perchlorate, permonosulfuric acid, potassium *tert*-butoxide. Unstable and explosive peroxides are formed with strong oxidizers. May accumulate static electrical charges, and may cause ignition of its vapors. Dissolves most rubber, resins, and plastics.

PYROBENZOL (71-43-2) Forms explosive mixture with air (flash point 12°F/−11°C. May accumulate static electrical charges, and may cause ignition of its vapors. Incompatible with strong oxidizers, nitric acid, oxygen, ozone, perchlorates. Attacks some forms of plastics, coatings, and rubber.

PYROCATECHIN (120-80-9) Strong oxidizers may cause fire and explosion.

PYROCATECHINE (120-80-9) Strong oxidizers may cause fire and explosion.

PYROCATECHINIC ACID (120-80-9) Strong oxidizers may cause fire and explosion.

PYROCATECHOL (120-80-9) Strong oxidizers may cause fire and explosion.

PYROCATECHOL MONOMETHYL ETHER (90-05-1) Forms explosive mixture with air (flash point 82°F/180°C). Reacts violently with strong oxidizers.

PYROCATECHUIC ACID (120-80-9) Strong oxidizers may cause fire and explosion.

PYROCELLULOSE (9004-34-6) Incompatible with water, bromine pentafluoride, hydrogen peroxide, sodium hypochlorite, sodium nitrate, fluorine, or strong oxidizers.

PYROFAX (68476-85-7) Forms explosive mixture with air. Strong oxidizers may cause fire and explosions. Attacks some plastics, rubber, and coatings.

PYROGALLIC ACID (87-66-1) Incompatible with strong oxidizers, caustics, ammonia, amines, isocyanates, alkylene oxides, epichlorohydrin.

PYROGALLOL (87-66-1) Incompatible with strong oxidizers, caustics, ammonia, amines, isocyanates, alkylene oxides, epichlorohydrin.

PYROGENTISIC ACID (123-31-9) Incompatible with strong oxidizers, caustics. May be oxidized to quinone at room temperatures in the presence of moisture. May explode on contact with oxygen.

M-PYROL (872-50-4) Forms explosive mixture with air (flash point 7°F/−14°C). A strong base. Incompatible with strong acids, organic anhydrides, isocyanates, aldehydes, light metals, flammable and porous materials. Attacks some plastics, rubber, and coatings.

PYROLIGNEOUS SPIRIT (67-56-1) May accumulate static electrical charges, and may cause ignition of its vapors. Forms explosive mixture with air (flash point 52°F/11°C). Incompatible with strong acids, strong oxidizers, caustics, aliphatic amines, isocyanates, chromic anhydride, lead perchlorate, perchloric acid, phosphorus trioxide. May react with metallic aluminum at high temperature. Attacks some plastics, rubber, and coatings.

PYROMUCIC ALDEHYDE (98-01-1) Forms explosive mixture with air (flash point 140°F/60°C). Incompatible with strong acids, caustics, ammonia, aliphatic amines, alkanolamines, aromatic amines, oxidizers. Attacks many plastics and coatings.

PYROPHOSPHATE de TETRAETHYLE (French) (107-49-3) Decomposes above 300°F/150°C forming flammable ethylene gas. Strong oxidizers may cause fire and explosions. Attacks some plastics, rubber, and coatings.

PYROPHOSPHORODITHIOIC ACID, TETRAETHYL ESTER (3689-24-5) Containers may burst in heat. Strong oxidizers may cause fire and explosions. Attacks some plastics, rubber, and coatings.

PYROPHOSPHORODITHIOIC ACID, O,O,O,O-TETRAETHYL ESTER (3689-24-5) Containers may burst in heat. Strong oxidizers may cause fire and explosions. Attacks some plastics, rubber, and coatings.

PYROSULPHURIC ACID (8014-95-7) A powerful oxidizer and strong acid. Reacts violently with reducing agents, organic materials. Reacts with air, producing corrosive fumes. Contact with cast iron may cause a violent reaction. Extremely hazardous on contact with many materials including chlorates, carbides, fulminates. Reacts vigorously with metals (powders will ignite), releasing hydrogen. Attacks some plastics, rubber, and coatings on brief contact.

PYROXYLIC SPIRIT (67-56-1) May accumulate static electrical charges, and may cause ignition of its vapors. Forms explosive mixture with air (flash point 52°F/11°C). Incompatible with strong acids, strong oxidizers, caustics, aliphatic amines, isocyanates, chromic anhydride, lead perchlorate, perchloric acid, phosphorus trioxide. May react with metallic aluminum at high temperature. Attacks some plastics, rubber, and coatings.

PYROXYLIN SOLUTION (9004-70-0) If ether solution evaporates, dry nitrocellulose material is a shock-sensitive explosive and fire hazard. Forms explosive mixture with air (flash point −64°F/−53°C). Strong oxidizers may cause fire and explosions. May accumulate static electrical charges, and may cause ignition of its vapors. Attacks some plastics, rubber, and coatings.

PYRROLYLENE (106-99-0) Self-reactive. Forms explosive peroxides with air. Fires, explosions, or hazardous polymerization may result from contact with air, strong oxidizers, strong acids, ozone, nitrogen dioxide, copper and its alloys, phenol, chlorine dioxide, crotonaldehyde, or a free radical polymerization initiator such as hydroquinone. Add inhibitor (such as *tert*-butyl catechol) and monitor to ensure that effective levels are maintained at all times. May accumulate static electrical charges, and may cause ignition of its vapors.

- Q -

QUAKERAL (98-01-1) Forms explosive mixture with air (flash point 140°F/60°C). Incompatible with strong acids, caustics, ammonia, aliphatic amines, alkanolamines, aromatic amines, oxidizers. Attacks many plastics and coatings.

QUARTZ (14808-60-7) Powerful oxidizers (fluorine, chlorine trifluoride, manganese trioxide, oxygen difluoride, etc.) may cause fire and explosions. This chemical is attacked by hydrogen fluoride (hydrofluoric acid).

QUAZO PURO (Italian) (14808-60-7) Powerful oxidizers (fluorine, chlorine trifluoride, manganese trioxide, oxygen difluoride, etc.) may cause fire and explosions. This chemical is attacked by hydrogen fluoride (hydrofluoric acid).

QUECKSILBER (German) (7439-97-6) Acetylene, acetylene products, ammonia gases can form shock-sensitive solids that can initiate fires of combustibles. Reacts violently with boron phosphodiiodide, chlorine, chlorine dioxide, methyl azide. Attacks copper and copper alloys.

QUECKSILBER CHLORID (German) (7487-94-7) Incompatible with light metals (aluminum, magnesium, beryllium, etc.), sodium, potassium.

QUELLADA (58-89-9) Not combustible, but may be dissolved in a combustible solvent. If solvent comes in contact with oxidizers, fire and explosions may result.

QUICKLIME (1305-78-8) Reacts violently with water. Incompatible with ethanol, hydrogen fluoride, acids, halogens, metal halides, light metals, some oxides, boron trifluoride, chlorine trifluoride, liquid hydrofluoric acid, phosphorus pentoxide, boric oxide and calcium chloride mixtures, fluorine, chlorine trifluoride, carbon dioxide.

QUICKPHOS (20859-73-8) Contact with moisture (including atmospheric moisture) causes formation of spontaneously combustible phosphine gas.

QUICKSET EXTRA (1338-23-4) Forms explosive mixture with air (flash point 125°F/52°C). Explosive decomposition occurs above 176°F/80°C. Pure substance is shock-sensitive. Strong oxidizer. Reacts violently with strong acids, strong bases, reducing agents, combustible substances, organic materials, oxides of heavy metals, salts, trace contaminants, amines. May accumulate static electrical charges, and may cause ignition of its vapors.

QUICK SILVER (7439-97-6) Acetylene, acetylene products, ammonia gases can form shock-sensitive solids that can initiate fires of combustibles. Reacts violently with boron phosphodiiodide, chlorine, chlorine dioxide, methyl azide. Attacks copper and copper alloys.

QUINOL (123-31-9) Incompatible with strong oxidizers, caustics. May be oxidized to quinone at room temperatures in the presence of moisture. May explode on contact with oxygen.

b-QUINOL (123-31-9) Incompatible with strong oxidizers, caustics. May be oxidized to quinone at room temperatures in the presence of moisture. May explode on contact with oxygen.

QUINOLINE (91-22-5) Contact with strong oxidizers may cause fire and explosions. Attacks some plastics, rubber, and coatings.

QUINOLOR (94-36-0) Confined storage of dry chemical may lead to decomposition and explosion. A strong oxidant; extremely reactive. Fires and explosion may result from heat or contamination, and from contact with strong acids, combustible materials, oxidizers, acids, bases, alcohols, reducing agents, metals, metal oxides, amines, accelerators, methyl methacrylate, organic matter, lithium aluminum carbide, dimethyl aniline, amines, metallic naphthenates. May attack some plastics, rubber and coatings. Protect containers from shock and friction.

QUINONE (106-51-4) Forms explosive mixture with air (flash point 104°F/40°C). Incompatible with strong bases, reducing agents, strong oxidizers. Attacks some plastics, rubber, and coatings.

p-**QUINONE or** *para*-**QUINONE** (106-51-4) Forms explosive mixture with air (flash point 104°F/40°C). Incompatible with strong bases, reducing agents, strong oxidizers. Attacks some plastics, rubber, and coatings.

QUINTOX (60-57-1) Incompatible with concentrated mineral acids, acid catalysts, acid oxidizing agents, phenols, reactive metals.

- R -

R-10 (56-23-5) Becomes corrosive when in contact with water. Corrosive to metals. Reacts violently with many compounds. Decomposes on contact with chemically active metals such as sodium, potassium and magnesium. Incompatible with allyl alcohol, fluorine gas, alkali metals, aluminum. Attacks some coatings, plastics, and rubber.

R-11 (75-69-4) Reacts with barium, lithium, sodium, magnesium, titanium. Attacks some plastics, rubber, and coatings.

R-12 (75-71-8) Reacts violently with liquid aluminum. Incompatible with chemically active metals. Attacks some plastics, rubber, and coatings.

R-13 (75-72-9) Reacts with aluminum, magnesium, zinc, and their alloys.

R-13B1 (75-63-8) Reacts with chemically active metals, powdered aluminum, zinc, magnesium. May attack some plastics, rubber, and coatings.

R-14 (76-14-2) Reacts with barium, lithium, sodium, magnesium, titanium. Attacks some plastics, rubber, and coatings.

R-20 (67-66-3) Decomposes in the presence of excess water or at high temperatures, producing phosgene and hydrogen chloride. Can become explosive in the presence of strong alkalies and water. Can accumulate static electrical charges. In contact with water and at high temperatures, it becomes corrosive; attacks iron and other metals. Incompatible with acetone, aluminum, strong oxidizers, potassium, sodium, chemically active metals, strong bases. Attacks plastics and rubber.

R-21 (75-43-4) Reacts with water and chemically active metals. Attacks some plastics, rubber, and coatings.

R-22 (75-45-6) May accumulate static electrical charges, and may cause ignition of its vapors. Moisture and rust cause slow decomposition, with formation of toxic gases. Attacks some plastics, rubber, and coatings. Thermal decomposition occurs at high temperature with alkalies and alkaline earth metals.

R-112 (76-12-0) Reacts with chemically active metals, powdered aluminum, zinc, and magnesium. Attacks some forms of plastics, rubber, and coatings.

R-113 (76-13-1) Reacts with barium, lithium, sodium, magnesium, titanium. Contact with alloys containing more than 2% magnesium may cause decomposition (with hydrogen chloride, hydrogen fluoride, and carbon monoxide released). Attacks some plastics, rubber, and coatings.

R-152A (75-37-6) Flammable gas. May accumulate static electrical charges, and may cause ignition of its vapors. Reacts violently with strong oxidizers. Attacks metals in presence of moisture.

R-717 (7664-41-7) **anhydrous (a flammable gas);** (1336-21-6) **solution in water.** Reacts violently with strong oxidizers, acids. Shock-sensitive compounds may be formed with halogens, mercury oxide, silver oxide. Fire and explosions may be caused by trimethylammonium amide, 1-chloro-2,4-dinitrobenzene, *o*-chloronitrobenzene, platinum, selenium difluoride dioxide, boron halides, mercury, chlorine, iodine, bromine, hypochlorites, chlorine bleach, amides, organic anhydrides, isocyanates, vinyl acetate, alkylene oxides, epichlorohydrin, aldehydes. Attacks some coatings, plastics, and rubber, and copper, brass, bronze, aluminum, steel, and their alloys.

R-1113 (79-38-9) Flammable gas. Forms explosive mixture with air (flash point −18°F/−28°C). Reacts violently with oxidizers. Contact with ethylene may cause explosive polymerization.

R-1582 (86-50-0) Strong oxidizers may cause fire and explosions.

R-40B1 (74-83-9) Incompatible with strong oxidizers, aluminum, dimethylsulfoxide, ethylene oxide, water. Attacks zinc, magnesium, alkali metals, and their alloys, and some plastics, rubber, and coatings. Forms aluminum alkyls in presence of aluminum; aluminum aikyls are spontaneously ignitable materials.

RACEMIC-LACTIC ACID (598-82-3) Incompatible with strong acids. Aqueous solution reacts with bases.

RAD-E-CATE 16 (124-65-2) Corrodes common metals.

RADAPON (75-99-0) Corrosive to iron, aluminum, and copper.

RADAZIN or RADIZINE (1912-24-9) Incompatible with strong acids.

RAFEX (534-52-1) Incompatible with heat and strong oxidizers. Dust can cause formation of explosive mixtures with air.

RAMOR (7440-28-0) Reacts violently with fluorine.

RANGE OIL (8008-20-6) Forms explosive mixture with air (flash point 100°F/38°C). Oxidizers may cause fire and explosions. Incompatible with nitric acid. Causes steel to rust. May accumulate static electrical charges, and may cause ignition of its vapors.

RANNEY NICKEL (7440-02-0) Flammable solid; may self-ignite on contact with air. A strong reducing agent; reacts violently with strong oxidizers, acids, hydrogen, sulfur, and many other substances.

RAPESEED OIL (N/A) Incompatible with strong acids, oxidizers.

RAPISOL (119-36-8) Forms explosive mixture with air (flash point 205°F/96°C). Incompatible with strong acids, nitrates, oxidizers.

RASIKAL (7775-09-9) A powerful oxidizer; reacts violently with reducing agents and combustible matter. Explosions may be caused by contact with ammonia salts, carbon, oils, metal sulfides, nitrobenzene, powdered metals, sugar. Contact with strong acids produces carbon dioxide. Forms shock-sensitive mixtures with some organic materials. Solution (50%) decomposes at 300°F/149°C, liberating oxygen.

RATBANE-1080 (62-74-8) Incompatible with strong acids.

RAVYON (63-25-2) Incompatible with strong oxidizers, strongly alkaline pesticides.

RAW LINSEED OIL (8001-26-1) Oxidizers may cause fire and explosions. Incompatible with nitric acid. May accumulate static electrical charges, and may cause ignition of its vapors.

RB (56-38-2) Combustible liquid. Mixtures with endrin may be explosive. Strong oxidizers may cause fire and explosions. Attacks some plastics, rubber, and coatings.

RC PLASTICIZER DBP (84-74-2) Incompatible with strong acids, nitrates, strong oxidizers, strong alkalies.

RCRA NO. D026 (N/A) Incompatible with acids, amides, organic anhydrides, isocyanates, alkylene oxides, epichlorohydrin, aldehydes, alcohols, glycols, phenols, cresols, caprolactam, ammonium nitrate, bromates, calcium carbide, charcoal, chlorates, hydrocarbons, iodates, iron. Reacts violently with halogen compounds, sodium, tin, uranium, and other compounds. Attacks steel when moist.

RCRA NO. P003 (107-02-8) Forms explosive mixture with air (flash point −15°F/−26°C). Unstable and very reactive. In storage can form heat- and shock-sensitive compounds. Unless inhibited (usually with hydroquinone), readily forms explosive peroxides. Able to polymerize. A strong reducing agent. Reacts violently with oxidizers, strong acids, caustics, amines, 2-aminoethanol, ammonia, ammonium hydroxide, ethylene diamine, ethyleneimine, hydroxides, metal salts, oxidizers, sulfur dioxide, thiourea. Attacks metals: cadmium and zinc. May accumulate static electrical charges, and may cause ignition of its vapors.

RCRA NO. P004 (309-00-2) Incompatible with concentrated mineral acids, acid catalysts, acid oxidizing agents, phenols, reactive metals.

RCRA NO. P005 (107-18-6) May form unstable and explosive peroxides. Able to polymerize. May accumulate static electrical charges, and may cause ignition of its vapors. A strong reducing agent; reacts violently with oxidizers. Forms explosive mixture with air (flash point 70°F/21°C). Incompatible with strong acids, amines, isocyanates, carbon tetrachloride, chlorosulfonic acid, diallyl phosphide, oleum, sodium hydroxide, tri-*n*-bromomelamine, metal halides, caustic soda, sodium, and magnesium, aluminum, and their alloys. Attacks some coatings, some plastics, and rubber.

RCRA NO. P006 (20859-73-8) Contact with moisture (including atmospheric moisture) produces spontaneously combustible phosphine gas.

RCRA NO. P008 (504-24-2) Forms a strong base with water. Incompatible with strong oxidizers, acids.

RCRA NO. P009 (131-74-8) A self-reactive explosive. Heat above 250°F/121°C can cause explosions. A powerful oxidizer. Contact with metal, concrete or plaster produces salts that are more shock-sensitive than ammonium picrate. Water increases rate of reactivity with metals.

RCRA NO. P010 (7778-39-4) Incompatible with sulfuric acid, caustics, ammonia, amines, isocyanates, alkylene oxides, oxidizers, epichlorohydrin, vinyl acetate, amides. Corrodes metals.

RCRA NO. P011 (1303-28-2) Incompatible with acids, aluminum, halogens, rubidium carbide, zinc. Aqueous solutions are acidic.

RCRA NO. P012 (1327-53-3) Incompatible with acids, fluorine, fluorides, sodium chlorate.

RCRA NO. P013 (542-62-1) Incompatible with nitric acid, nitrates, nitrites, chlorates, magnesium, strong oxidizers. Corrosive to metals in the presence of moisture.

RCRA NO. P014 (108-98-5) Forms explosive mixture with air (flash point 132°F/56°C). Incompatible with strong acids, caustics, alkali metals (i.e., lithium, sodium, potassium, rubidium, cesium, francium). Oxidizes on contact with air. Corrosive to carbon steel.

RCRA NO. P015 (7440-41-7) Contact with acids produces flammable hydrogen gas. Incompatible with alkalies, chlorinated hydrocarbons, oxidizable agents, carbon tetrachloride, trichloroethylene, lithium, phosphorus.

709

RCRA NO. P017 (598-31-2) Forms explosive mixture with air (flash point 113°F/45°C). Contact with oxidizers may cause fire and explosions.

RCRA NO. P018 (357-57-3) Contact with strong oxidizers may cause fire and explosions. May accumulate static electrical charges, and may cause ignition of its vapors.

RCRA NO. P021 (592-01-8) Incompatible with water, strong acids, fluorine, magnesium, nitrates, nitrites. Attacks aluminum, copper, zinc.

RCRA NO. P022 (75-15-0) Highly reactive. Contact with many substances can cause fire and explosions. Forms explosive mixture with air (flash point −22°F/−30°C). Shock can cause explosive decomposition. Incompatible with alkali metals, aliphatic amines, alkanolamines, aluminum, azides, chlorine monoxide, combustible substances, ethylene diamine, ethyleneimine, lead azide, lithium azide, nitric oxide, nitrogen dioxide, potassium, potassium azide, reducing agents, rubidium azide, sodium azide, zinc.

RCRA NO. P023 (107-20-0) Incompatible with oxidizers, acids, water. May form a water-soluble polymer in storage.

RCRA NO. P028 (100-44-7) Forms explosive mixture with air (flash point 153°F/67°C). Water contact causes formation of hydrogen chloride fume. Strong oxidizers may cause fire and explosions. Violent polymerization may be caused by contact with copper, aluminum, iron, zinc, magnesium, and tin. May accumulate static electrical charges, and may cause ignition of its vapors. Attacks some plastics and rubber.

RCRA NO. P029 (544-92-3) May produce hydrogen cyanide with water and contact. Incompatible with chlorine, fluorine, peroxides.

RCRA NO. P030 (372-09-8) Incompatible with oxidizers, strong acids, organic acids, caustics, reducing agents.

RCRA NO. P031 (460-19-5) Explosive reaction occurs with acids, water, liquid oxygen, oxidizers.

RCRA NO. P033 (506-77-4) Violent polymerization can be caused by chlorine. Water, steam, alcohols, acids, acid salts, amines, strong alkalies, olefins, strong oxidizers may cause fire and explosion. In crude form chemical trimerizes violently if catalyzed by traces of hydrogen chloride or ammonium chloride. Corrodes brass, copper, bronze.

RCRA NO. P034 (131-89-5) Reacts with oxidizers, with a risk of fire or explosions.

RCRA NO. P036 (696-28-6) Forms explosive mixture with air (flash point 60°F/−16°C). Water contact produces hydrochloric acid. Attacks metals in the presence of moisture.

RCRA NO. P039 (298-04-4) Forms explosive mixture with air (flash point 180°F/82°C). Incompatible with alkalies, strong oxidizers.

RCRA NO. P047 (534-52-1) Incompatible with heat and strong oxidizers. Dust can cause formation of explosive mixtures with air.

RCRA NO. P048 (51-28-5) Explosion is caused by heat, friction, or shock. Contact with reducing agents, combustibles may cause fire and explosions. Forms explosive salts with ammonia or strong bases. May accumulate static electrical charges, and may cause ignition of its vapors.

RCRA NO. P050 (115-29-7) Hydrolyzed by acids and alkalies. Corrosive to iron.

RCRA NO. P051 (72-20-8) Incompatible with parathion, strong acids (with formation of explosive vapors), strong oxidizers.

RCRA NO. P054 (151-56-4) Forms explosive mixture with air (flash point 12°F/−11°C). May accumulate static electrical charges, and may cause ignition of its vapors. Contact with acid, aluminum, carbon dioxide or silver may cause explosive polymerization. Attacks rubber, coatings, plastics, and chemically active metals. Self-reactive with heat or atmospheric carbon dioxide.

RCRA NO. P056 (7782-41-4) A powerful oxidizer and a dangerously reactive gas. Reacts violently with reducing agents and combustible materials. Contact with water can produce corrosive hydrogen fluoride, hydrochloric acid, oxygen, and oxygen difluoride. Reacts vigorously reaction with most oxidizable materials at room temperature, frequently with ignition. Corrodes most metals (except the cylinders in which it is shipped), especially in the presence of moisture. Contact with nitric acid produces explosive gas. Attacks some plastics, rubber, and coatings. Explosive reaction with ammonia, graphite, halocarbons, hydrocarbons, hydrogen, and many other substances. Reacts with nearly every known element.

RCRA NO. P058 (62-74-8) Incompatible with strong acids.

RCRA NO. P059 (76-44-8) Forms hydrogen chloride gas with iron and rust above 165°F/74°C.

RCRA NO. P063 (74-90-8) Unless stabilized and maintained, samples stored more than 90 days are hazardous. Samples containing more than 2–5% water are less stable than dry material. Can be self-reactive, forming an explosive mixture with air (flash point 0°F/−18°C). Heat or contact with amines or strong bases can cause polymerization. Incompatible with acetaldehyde. Oxidizers may cause fire and explosions. Attacks some plastics, rubber, and coatings.

RCRA NO. P064 (624-83-9) Forms explosive mixture with air (flash point 0°F/−18°C). Contact with iron, tin, copper (or salts of these elements) and with certain other catalysts (such as triphenylarsenic oxide, triethylphosphine, and tributyltin oxide) or elevated temperatures may cause polymerization. Reacts violently with warm water. Incompatible with strong acids, alcohols, glycols, amines, amides, ammonia, caprolactam, caustics, strong oxidizers. Attacks some plastics, rubber, or coatings.

RCRA NO. P067 (75-55-8) Forms explosive mixture with air (flash point 25°F/−4°C). Contact with acids or high heat can cause violent polymerization. Strong oxidizers may cause fire and explosions. Attacks some plastics, rubber, and coatings.

RCRA NO. P068 (60-34-4) Highly reactive reducing agent and a strong base. Forms explosive mixture with air (flash point 17°F/−8°C). Reacts violently with oxidizers, combustibles, acids. Incompatible with alcohols, glycols, isocyanates, phenols, cresols. Oxides of iron or copper, manganese, lead, copper, or their alloys can lead to fire and explosions. Attacks cork, some plastics, coatings, and rubber.

RCRA NO. P069 (75-86-5) Forms explosive mixture with air (flash point 165°F/74°C). Reacts violently with strong oxidizers. Heat may cause decomposition. Incompatible with nonoxidizing mineral acids, sulfuric acid, nitric acid, organic acids, caustics, aliphatic amines, alkanolamines, aromatic amines, organic anhydrides, allylene oxides, epichlorohydrin, caprolactam solution, ammonia, isocyanates, phenols, cresol.

RCRA NO. P071 (298-00-0) Mixtures with magnesium may be explosive.

RCRA NO. P074 (557-19-7) Incompatible with strong acids, sulfur, selenium.

RCRA NO. P075 (54-11-5) Incompatible with strong acids, strong oxidizers. Attacks some plastics, rubber, and coatings. May accumulate static electrical charges, and may cause ignition of its vapors.

RCRA NO. P076 (10102-43-9) A strong oxidizer. Avoid contact with all combustible materials, chlorinated hydrocarbons, ammonia, carbon disulfide, many metals, fluorine, and ozone. Attacks some plastics, rubber, and coatings.

RCRA NO. P077 (100-01-06) Strong oxidizers and moisture may cause spontaneous heating. Caustics and heat may cause formation of explosive mixtures. Attacks some plastics, rubber, and coatings. May accumulate static electrical charges, and may cause ignition of its vapors.

RCRA NO. P089 (56-38-2) Combustible liquid. Mixtures with endrin may be explosive. Strong oxidizers may cause fire and explosions. Attacks some plastics, rubber, and coatings.

RCRA NO. P095 (75-44-5) Incompatible with water, producing hydrochloric acid. Reacts violently with strong oxidizers, anhydrous ammonia, isopropyl alcohol and chemically active metals. Forms shock-sensitive material with potassium. Attacks most metals in moist conditions.

RCRA NO. P098 (151-50-8) Incompatible with acids, organic anhydrides, isocyanates, alkylene oxides, epichlorohydrin, aldehydes, alcohols, glycols, phenols, cresols, caprolactam, strong oxidizers, sodium chlorate. Attacks aluminum, copper, zinc in the presence of moisture.

RCRA NO. P101 (107-12-0) Forms explosive mixture with air (flash point 36°F/2°C). Reacts violently with oxidizers. Water or acid contact produces hydrogen cyanide fumes.

RCRA NO. P102 (107-19-7) Incompatible with strong acids, aliphatic amines, phosphorus pentoxide, caustics, isocyanates, strong oxidizers.

RCRA NO. P105 (26628-22-8) Incompatible with acids, some metals (i.e, lead, copper, silver, mercury). Forms explosion-sensitive compounds.

RCRA NO. P106 (143-33-9) Incompatible with acids, organic anhydrides, isocyanates, alkylene oxides, epichlorohydrin, aldehydes, alcohols, glycols, phenols, cresols, caprolactam solution, strong oxidizers, nitrates, nitrites. Attacks aluminum, copper, zinc.

RCRA NO. P108 (57-24-9) Contact with strong oxidizers may cause fire and explosions.

RCRA NO. P109 (3689-24-5) Containers may burst in heat. Strong oxidizers may cause fire and explosions. Attacks some plastics, rubber, and coatings.

RCRA NO. P110 (78-00-2) Forms explosive mixture with air and decomposes above 200°F/93°C. Incompatible with strong oxidizers, concentrated acids; may cause fire and explosions. Attacks some plastics, rubber, and coatings.

RCRA NO. P111 (107-49-3) Decomposes above 300°F/150°C yielding flammable ethylene gas. Strong oxidizers may cause fire and explosions. Attacks some plastics, rubber, and coatings.

RCRA NO. P112 (509-14-8) Combustible organic matter wet with this chemical may be highly explosive, and susceptible to mild shock. Contact with hydrocarbons, alkalies, or metals produces explosive mixtures. Attacks some plastics, rubber, and coatings.

RCRA NO. P118 (594-42-3) Incompatible with caustics or amines (with rapid decomposition), hot iron (produces toxic carbon tetrachloride fumes), hot water (produces hydrochloric acid), sulfur, and carbon dioxide. Attacks some plastics, rubber, and coatings.

RCRA NO. P120 (1314-62-1) Aqueous solution is acidic. Reacts with lithium at high temperature.

RCRA NO. P121 (557-21-1) Contact with acids and acid salts liberates hydrogen cyanide gas.

RCRA NO. P122 (1314-84-7) Incompatible with oxidizers, acids, water.

RCRA NO. P123 (8001-35-2) Reacts with oxidizers, with a risk of fire or explosions. Attacks metals in the presence of moisture.

RCRA NO. U001 (75-07-0) Slowly polymerizes to paraldehyde. Explodes when mixed with iodine. Contact with strong bases can cause explosive polymerization. A strong reducing agent; reacts violently with combustibles, strong acids, caustics, ammonia, aliphatic amines, alkanolamines, aromatic amines, organic substances, halogens, oxidizers. Forms explosive peroxides with air. May dissolve rubber. May accumulate static electrical charges, and may cause ignition of its vapors.

RCRA NO. U002 (67-64-1) Forms explosive mixture with air (flash point −4°F/−20°C). Reacts violently with chloroform. Incompatible with strong acids, aliphatic amines, chloroform, chromic anhydride, chromyl chloride, hexachloromelamine, hydrogen peroxide, nitrosyl chloride, nitrosyl perchlorate, nitryl perchlorate, permonosulfuric acid, potassium *tert*–butoxide. Forms unstable and explosive peroxides with strong oxidizers. May accumulate static electrical charges, and may cause ignition of its vapors. Dissolves most rubber, resins, and plastics.

RCRA NO. U003 (75-05-8) Forms explosive mixture with air (flash point 42°F/6°C). Incompatible with water (especially if acid or alkaline), acids, caustics, nitrating agents, indium, nitrogen tetroxide, N-fluoro compounds, sulfur trioxide, iron(III) salts of perchlorate, indium, nitrogen-fluorine compounds. Reacts violently with oxidizers. May accumulate static electrical charges, and may cause ignition of its vapors.

RCRA NO. U004 (98-86-2) Incompatible with strong acids, aliphatic amines, oxidizers. Forms explosive mixture with air (flash point 170°F/77°C).

RCRA NO. U006 (75-36-5) Forms explosive mixture with air (flash point 40°F/4°C). May accumulate static electrical charges, and may cause ignition of its vapors. Reacts violently with water, oxidizers, strong bases, alcohols (especially ethanol), dimethyl sulfoxide, phosphorus trichloride; produces corrosive hydrochloric acid with air.

RCRA NO. U007 (79-06-1) Unless inhibited, ultraviolet light and heat (above 184°F/85°C) can cause polymerization. Incompatible with nonoxidizing mineral acids, strong acids, ammonia, oleum, oxidizers, isocyanates.

RCRA NO. U008 (79-10-7) Forms explosive mixture with air (flash point 122°F/50°C). Forms explosive peroxides; light, heat and peroxides can cause polymerization. Incompatible with sulfuric acid, caustics, ammonia, amines, isocyanates, alkylene oxides, epichlorohydrin, oxidizers, toluenediamine, pyridine, methyl pyridine, N-methyl pyrrolidone, 2-methyl-6-ethyl aniline, aniline, ethylene diamine, ethyleneimine, 2-aminoethanol. Severely corrodes carbon steel and iron; attacks other metals. May accumulate static electrical charges, and may cause ignition of its vapors.

RCRA NO. U009 (107-13-1) Forms explosive mixture with air (flash point 32°F/0°C o.c.). Forms explosive peroxides; heat, light, caustics, silver nitrate and peroxides can cause polymerization. Incompatible with strong acids, strong oxidizers, amines, 2-aminoethanol, bromine, chlorosulfonic acid, ethylene diamine, nitric acid, oleum, potassium hydroxide, sodium hydroxide, sulfuric acid. Attacks copper and copper alloys; attacks aluminum in high concentrations. May accumulate static electrical charges, and may cause ignition of its vapors.

RCRA NO. U012 (62-53-3) Forms explosive mixture with air (flash point 158°F/70°C). Unless inhibited (usually with methanol), readily able to polymerize. Fires and explosions may result from contact with halogens, strong acids, oxidizers, organic anhydrides, acetic anhydride, isocyanates, aldehydes, sodium peroxide. Reacts with alkali metals and alkaline earth metals. Attacks some plastics, rubber, and coatings, and copper and copper alloys.

RCRA NO. U013 (77-78-1) Forms explosive mixture with air (flash point 182°F/83°C). Incompatible with strong oxidizers, strong acids, strong alkalies, strong ammonia solutions. Attacks some plastics, rubber, and coatings.

RCRA NO. U017 (117-84-0) Water contact causes foaming. Incompatible with strong acids, nitrates.

RCRA NO. U017 (98-87-3) Forms explosive mixture with air (flash point 153°F/67°C). Reacts with acids, bases, strong oxidizers. Produces acid fumes with air. Attacks plastics and coatings.

RCRA NO. U020 (98-09-9) Incompatible with ammonia, aliphatic amines, water solutions. Water contact produces hydrochloric and chlorosulfonic acids; aqueous solutions react violently with bases. Attacks metals in presence of moisture.

RCRA NO. U021 (92-87-5) Oxidizes on exposure to light and air. Reacts violently with strong oxidizers. Red fuming nitric acid may cause fire.

RCRA NO. U021 (108-46-3) Reacts violently with strong oxidizers, nitric acid. Incompatible with acetanilide, albumin, alkalies, antipyrine, camphor, ferric salts, menthol, spirit nitrous ether. Absorbs moisture from air (hygroscopic). May accumulate static electrical charges, and may cause ignition of its vapors.

RCRA NO. U025 (111-44-4) Water contact may produce hydrogen chloride fumes. Can form peroxides. Forms explosive mixture with air (flash point 131°F/55°C). Strong oxidizers may cause fire and explosions. Attacks some plastics, rubber, and coatings.

RCRA NO. U027 (108-60-1) Forms explosive mixture with air (flash point 170°F/77°C). Reacts violently with strong oxidizers. Incompatible with aluminum, copper, epoxy coatings.

RCRA NO. U028 (117-81-7) Incompatible with strong acids, strong alkalies, nitrates, oxidizers.

RCRA NO. U029 (74-83-9) Incompatible with strong oxidizers, aluminum, dimethylsulfoxide, ethylene oxide, water. Attacks zinc, magnesium, alkali metals, and their alloys, and some plastics, rubber, and coatings. Forms aluminum alkyls in presence of aluminum; aluminum alkyls are spontaneously ignitible materials.

RCRA NO. U031 (71-36-3) Forms explosive mixture with air (flash point 98°F/37°C). May react with aluminum above 120°F/49°C. Attacks some plastics, rubber, and coatings. Incompatible with strong acids, halogens, caustics, alkali metals, aliphatic amines, isocyanates.

RCRA NO. U032 (13765-19-0) A strong oxidizer. Incompatible with boron (reacts violently), combustible material.

RCRA NO. U033 (353-50-4) Moisture causes formation of hydrogen fluoride gas and carbon dioxide.

RCRA NO. U034 (75-87-6) Contact with acids or exposure to light may cause polymerization. Reacts with water, producing corrosive and toxic chloral hydrate. Contact with oxidizers may cause fire and explosions.

RCRA NO. U036 (57-74-9) Contact with strong oxidizers may cause fire and explosions. Attacks some plastics, rubber, and coatings.

RCRA NO. U037 (60-57-1) Incompatible with concentrated mineral acids, acid catalysts, acid oxidizing agents, phenols, reactive metals.

RCRA NO. U037 (108-90-7) Forms explosive mixture with air (flash point 82°F/28°C). May accumulate static electrical charges, and may cause ignition of its vapors. Incompatible with oxidizers, dimethylsulfoxide, sodium powder, silver perchloride. Attacks some plastics, rubber, and coatings.

RCRA NO. U041 (106-89-8) Forms explosive mixture with air (flash point 88°F/31°C). Heat, acids, alkalies, and metallic halides can cause explosive polymerization. Aliphatic amines, alkaline earths, alkali metals, alkanolamines, powdered metals, strong oxidizers may cause fire and explosions. Decomposition produces highly toxic phosgene gas. Will pit steel in the presence of moisture. May accumulate static electrical charges, and may cause ignition of its vapors.

RCRA NO. U043 (75-01-4) Forms explosive gas mixture with air. Atmospheric oxygen and various contaminants or strong oxidizers cause formation of peroxides, which can initiate a violent polymerization action. Also able to polymerize when heated and exposed, long term, to light. Reacts violently with strong oxidizers or oxides of nitrogen. Contact with copper or other acetylide-forming metals produces explosive compounds. Attacks iron and steel in the presence of moisture.

RCRA NO. U044 (67-66-3) Decomposes in the presence of excess water or at high temperatures, producing phosgene and hydrogen chloride. Can become explosive in the presence of strong alkalies and water. Can accumulate static electrical charges. In contact with water and at high temperatures, it becomes corrosive; attacks iron and other metals. Incompatible with acetone, aluminum, strong oxidizers, potassium, sodium, chemically active metals, strong bases. Attacks plastics and rubber.

RCRA NO. U045 (74-87-3) Moisture causes decomposition. Contact with aluminum yields a product that may ignite spontaneously in air. Reacts with barium, lithium, sodium, magnesium, titanium. Attacks plastics, rubber, and coatings.

RCRA NO. U046 (107-30-2) Forms explosive mixture with air (flash point 0°F/−17.8°C). May be able to form unstable and explosive peroxides. Produces hydrochloric acid on contact with moisture. Corrosive to metals in presence of moisture.

RCRA NO. U048 (95-57-8) Forms explosive mixture with air (flash point 147°F/64°C). Contact with strong oxidizers may cause fire and explosions. Attacks aluminum and copper.

RCRA NO. U051 (65996-93-2) Incompatible with oxidizers, strong acids, caustics, aliphatic amines, isocyanates.

RCRA NO. U052 (108-39-4) Incompatible with strong acids, oxidizers, alkalies, aliphatic amines, amides, chlorosulfonic acid, oleum. Liquid attacks some plastics, coatings, and rubber.

RCRA NO. U052 (106-44-5) Incompatible with strong acids, oxidizers, alkalies, aliphatic amines, amides, chlorosulfonic acid, oleum. Liquid attacks some plastics, coatings, and rubber.

RCRA NO. U052 (95-48-7) Forms explosive mixture with air (flash point 178°F/81°C). Incompatible with strong acids, oxidizers, alkalies, aliphatic amines, amides, chlorosulfonic acid, oleum. Liquid attacks some plastics and rubber.

RCRA NO. U053 (4170-30-3) A strong reducing agent. Forms explosive mixture with air (flash point 55°F/12.8°C). Readily converted by oxygen to peroxides and acids; heat or contact with many other substances may cause polymerization. Incompatible with strong oxidizers, strong acids including nonoxidizing mineral acids, ammonia, aliphatic amines, aromatic amines, 1,3-butadiene, strong bases. Liquid attacks some plastics, rubber, and coatings.

RCRA NO. U053 (123-73-9) A strong reducing agent. Forms explosive mixture with air (flash point 55°F/12.8°C). Readily converted by oxygen to peroxides and acids; heat or contact with many other substances may cause polymerization. Incompatible with strong oxidizers, Strong acids including nonoxidizing mineral acids, ammonia, aliphatic amines, aromatic amines, 1,3-butadiene, strong bases. Liquid attacks some plastics, rubber, and coatings.

RCRA NO. U055 (98-82-8) Forms explosive mixture with air (flash point 99°F/37°C). Incompatible with strong acids, strong oxidizers. Air contact produces cumene hydroperoxide. May be able to form unstable peroxides. Attacks rubber. May accumulate static electrical charges, and may cause ignition of its vapors.

RCRA NO. U056 (110-82-7) Forms explosive mixture with air (flash point −4°F/−20°C). Incompatible with strong oxidizers, nitrogen dioxide.

RCRA NO. U057 (108-94-1) Forms explosive mixture with air (flash point 111°F/44°C). May accumulate static electrical charges, and may cause ignition of its vapors. Attacks red metals and lead. Incompatible with amines, sulfuric acid, nitric acid, oxidizers, strong acids, aliphatic amines. Dissolves plastics, resins, and rubber.

RCRA NO. U061 (50-29-3) Incompatible with salts of iron or aluminum, and bases. Do not store in iron containers.

716

RCRA NO. U067 (106-93-4) Reacts with chemically active metals, liquid ammonia, strong oxidizers. Heat and light cause slow decomposition. Attacks some plastics and rubber.

RCRA NO. U068 (74-95-3) Mixing with potassium produces a shock-sensitive explosive.

RCRA NO. U069 (84-74-2) Incompatible with strong acids, nitrates, strong oxidizers, strong alkalies.

RCRA NO. U070 (95-50-1) Forms explosive mixture with air (flash point 151°F/66°C). Incompatible with strong oxidizers, hot aluminum, or aluminum alloy. Attacks some plastics, rubber, and coatings.

RCRA NO. U071 (541-73-1) Forms explosive mixture with air (flash point 151°F/66°C). Incompatible with strong oxidizers, alkali metals (i.e., lithium, sodium, potassium, rubidium, cesium, francium).

RCRA NO. U072 (106-46-7) Forms explosive mixture with air (flash point 150°F/66°C). Incompatible with strong oxidizers, metal powders, and alkali metals (i.e., lithium, sodium, potassium, rubidium, cesium, francium). Attacks some plastics, rubber, and coating.

RCRA NO. U074 (764-41-0) Forms explosive mixture with air (flash point 126°F/52°C). Water contact produces hydrochloric acid. Incompatible with strong oxidizers, bases. Corrodes metals in the presence of moisture.

RCRA NO. U075 (75-71-8) Reacts violently with liquid aluminum. Incompatible with chemically active metals. Attacks some plastics, rubber, and coatings.

RCRA NO. U076 (75-34-3) Forms explosive mixture with air (flash point 2°F/−17°C). Incompatible with strong oxidizers, strong caustics. Attacks plastics and rubber.

RCRA NO. U077 (107-06-2) May accumulate static electrical charges, and may cause ignition of its vapors. Contact with hot water may produce hydrochloric acid. Forms explosive mixture with air (flash point 55°F/13°C). Incompatible with strong oxidizers, strong caustics, chemically active metals. Corrosive to iron and other metals. Attacks some plastics, rubber, or coatings.

RCRA NO. U078 (75-35-4) Forms explosive mixture with air (18°F/−28°C). Air or contaminants can cause formation of peroxides; may polymerize. Reacts violently with strong oxidizers, alkali metals. Incompatible with nitric acid.

RCRA NO. U080 (75-09-2) Incompatible with strong oxidizers, strong caustics, chemically active metals. Attacks some plastics, rubber, and coatings.May accumulate static electrical charges, and may cause ignition of its vapors.

RCRA NO. U081 2,4-DICHLOROPHENOL (120-83-2) Reacts violently with strong oxidizers, acids. Incompatible with caustics. Quickly corrodes aluminum; slowly corrodes zinc, tin, brass, bronze, copper and its alloys. May accumulate static electrical charges, and may cause ignition of its vapors.

RCRA NO. U083 (78-87-5) Forms explosive mixture with air (flash point 60°F/16°C). May accumulate static electrical charges, and may cause ignition of its vapors. Strong oxidizers may cause fire and explosions. Strong acids can cause decomposition and the formation of hydrogen chloride vapors. Corrodes aluminum. Attacks some plastics, rubber, and coatings.

RCRA NO. U084 (542-75-6) Forms explosive mixture with air (flash point 95°F/35°C). Reacts violently with strong oxidizers. May accumulate static electrical charges, and may cause ignition of its vapors. Incompatible with strong acids, oxidizers, aluminum or magnesium compounds, aliphatic amines, alkanolamines, alkaline materials, or corrosives.

RCRA NO. U088 (84-66-2) Reacts violently with strong acids, strong oxidizers, permanganates, water. Attacks some forms of plastic.

RCRA NO. U092 (124-40-3) A gas. Incompatible with acids, organic anhydrides, isocyanates, vinyl acetate, acrylates, substituted allyls, alkylene oxides, epichlorohydrin, ketones, aldehydes, alcohols, glycols, mercury, phenols, cresols, caprolactam solution, strong oxidizers. Attacks aluminum, copper, lead, tin, zinc, and alloys, and some plastics, rubber, and coatings.

RCRA NO. U097 (79-44-7) Rapidly hydrolyzed in water. Incompatible with strong acids, oxidizers.

RCRA NO. U098 (57-14-7) Forms explosive mixture with air (flash point 5°F/−15°C). A strong reducing agent. Incompatible with strong acids, halogens, metallic mercury, strong oxidizers (with possible spontaneous ignition). Attacks some plastics, rubber, and coatings. May accumulate static electrical charges, and may cause ignition of its vapors.

RCRA NO. U099 (540-73-8) Forms explosive mixture with air (flash point 73°F/23°C). A strong reducing agent. Reacts violently with strong oxidizers, strong acids. Attacks some plastics, rubber, and coatings. May accumulate static electrical charges, and may cause ignition of its vapors.

RCRA NO. U102 (131-11-3) Incompatible with strong alkalies, strong acids, nitrates, oxidizers.

RCRA NO. U105 (121-14-2) Strong oxidizers or caustics may cause fire and explosions. Incompatible with nitric acid, producing an explosive material. Sodium oxide contact causes ignition. Attacks metals.

RCRA NO. U106 (606-20-2) Strong oxidizers or caustics may cause fire and explosions. Contact with nitric acid produces an explosive material. Sodium oxide contact causes ignition. Attacks some metals.

RCRA NO. U108 (123-91-1) Forms explosive mixture with air (flash point 54°F/12°C). Moisture causes formation of unstable peroxides. May accumulate static electrical charges, and may cause ignition of its vapors. Incompatible with strong oxidizers, strong acids. Attacks many plastics.

RCRA NO. U109 (71-43-2) Forms explosive mixture with air (flash point 12°F/−11°C).) May accumulate static electrical charges, and may cause ignition of its vapors. Incompatible with strong oxidizers, nitric acid, oxygen, ozone, perchlorates. Attacks some forms of plastics, coatings, and rubber.

RCRA NO. U110 (142-84-7) Forms explosive mixture with air (flash point 63°F/17°C). Incompatible with acids, organic anhydrides, isocyanates, vinyl acetate, acrylates, substituted allyls, alkylene oxides, epichlorohydrin, ketones, aldehydes, alcohols, glycols, mercury, phenols, cresols, caprolactam solution, strong oxidizers. Attacks aluminum, copper, lead, tin, zinc, and alloys.

RCRA NO. U112 (141-78-6) Will hydrolyze on standing, to yield acetic acid and ethyl alcohol. This reaction is greatly accelerated by bases (alkalies). Forms explosive mixture with air (flash point 24°F/−4.4°C). Incompatible with strong acids, nitrates, oxidizers, chlorosulfonic acid, lithium aluminum hydride, oleum.

RCRA NO. U113 (140-88-5) Forms explosive mixture with air (flash point 48°F/9°C). Atmospheric moisture and strong alkalies may cause fire and explosions. Unless properly inhibited (note: inert gas blanket not recommended), heat, light, or peroxides can cause polymerization. Incompatible with oxidizers (may react violently), strong acids, amines. May accumulate static electrical charges, and may cause ignition of its vapors.

RCRA NO. U115 (75-21-8) Forms explosive mixture with air (flash point 20°F/−6°C). Incompatible with alkali metal hydroxides, highly active catalysts (e.g., anhydrous chlorides of iron, tin, or aluminum and oxides of iron or aluminum). Avoid contact with copper. Protect container from physical damage, sun, and heat. Attacks some plastics, rubber, or coatings.

RCRA NO. U117 (60-29-7) Forms explosive mixture with air (flash point −49°F/−45°C). Incompatible with strong acids, strong oxidizers. Can form peroxides with air or light; may explode when container is unstoppered or otherwise opened. Attacks some plastics, rubber, and coatings. Being a nonconductor, chemical may accumulate static electric charges that may result in ignition of its vapor.

RCRA NO. U118 (97-63-2) Forms explosive mixture with air (flash point 80°F/27°C). Incompatible with strong acids, amines, oxidizers. Corrodes some metals.

RCRA NO. U121 (75-69-4) Reacts with barium, lithium, sodium, magnesium, titanium. Attacks some plastics, rubber, and coatings.

RCRA NO. U122 (50-00-0) May polymerize unless properly inhibited (usually with methanol). Forms explosive mixture with air (flash point 122°F/50°C). Incompatible with strong acids, amines, strong oxidizers, alkaline materials, nitrogen dioxide, performic acid. Reaction with hydrochloric acid produces bis-chloromethyl ether, a carcinogen.

RCRA NO. U123 (64-18-6) Forms explosive mixture with air (flash point 156°F/69°C). A strong reducing agent; reacts violently with oxidizers. Incompatible with sulfuric acid, bases, ammonia, aliphatic amines, alkanolamines, furfuryl alcohol, hydrogen peroxide, isocyanates, alkylene oxides, epichlorohydrin. Attacks aluminum, cast iron and steel, some plastics, rubber, and coatings.

RCRA NO. U124 (110-00-9) Forms explosive mixture with air (flash point −58°F/−50°C). Reacts violently with acids, oxidizers. Unless stabilized with an inhibitor, air exposure causes formation of unstable peroxides.

RCRA NO. U125 (98-01-1) Forms explosive mixture with air (flash point 140°F/60°C). Incompatible with strong acids, caustics, ammonia, aliphatic amines, alkanolamines, aromatic amines, oxidizers. Attacks many plastics and coatings.

RCRA NO. U127 (118-74-1) Reacts violently with strong oxidizers, formyldimethylamine.

RCRA NO. U128 (87-68-3) Forms explosive mixture with air (flash point 195°F/90°C). Reacts strongly with oxidizers, aluminum powder. Attacks aluminum. Attacks some plastics, rubber, and coatings.

RCRA NO. U130 (77-47-4) Incompatible with water, producing hydrochloric acid. Contact with sodium may be explosive. Corrodes iron and other metals in the presence of moisture.

RCRA NO. U131 (67-72-1) Incompatible with aluminum or zinc. Alkalies cause formation of spontaneously explosive chloroacetylene. Attacks some plastics, rubber, and coatings.

RCRA NO. U133 (302-01-2) Forms explosive mixture with air (flash point 100°F/38°C). A highly reactive reducing agent and a strong base. Incompatible with oxides of iron or copper, and with manganese, lead, copper or their alloys; can cause fire and explosions. Reacts violently with acids and halogens. Attacks cork, glass, some plastics, rubber, and coatings.

RCRA NO. U134 (7664-39-3) A strong acid. Reacts—possibly violently—with acetic anhydride, aliphatic amines, alcohols, alkanolamines, alkylene oxides, aromatic amines, amides, 2-aminoethanol, ammonia, ammonium hydroxide, arsenic trioxide, bismuthic acid, calcium oxide, ethylene diamine, ethyleneimine, epichlorohydrin, isocyanates, metal acetylides, nitrogen trifluoride, oleum, organic anhydrides, oxygen difluoride, phosphorus pentoxide, sulfuric acid, sodium hydroxide and other bases, strong oxidizers, vinyl acetate, vinylidene fluoride. Attacks glass, concrete, ceramics, metals, some plastics, rubber, and coatings.

RCRA NO. U135 (7783-06-4) A highly flammable and reactive gas. Incompatible with acetaldehyde, barium pentafluoride, chlorine monoxide, chlorine trifluoride, chromic anhydride, copper, lead dioxide, nitric acid, nitrogen iodide, nitrogen trichloride, nitrogen trifluoride, oxygen difluoride, oxidizers, phenyl diazonium chloride, sodium, sodium peroxide. Reacts with alkali metals and alkali earth metals. Attacks metals.

RCRA NO. U138 (74-88-4) Contact with strong oxidizers may cause fire and explosions. Heat, light, and moisture contribute to instability.

RCRA NO. U140 (78-83-1) Forms explosive mixture with air (flash point 82°F/28°C). Incompatible with strong acids, strong oxidizers, caustics, aliphatic amines, isocyanates, alkali metals, and alkaline earth. Attacks some plastics, rubber, and coatings. May react with aluminum at high temperatures.

RCRA NO. U144 (301-04-2) Contact with strong acids produces acetic acid. Incompatible with bases, ammonia, amines, cresols, isocyanates, alkylene oxides, epichlorohydrin, phenols, sulfites. Reacts violently with bromates, strong oxidizers.

RCRA NO. U147 (108-31-7) Aqueous solution is a strong acid. Contact with strong oxidizers may cause fire and explosions. Contact with amines or alkali metals cause polymerization if temperature is greater than 150°F/66°C. Attacks some plastics, rubber, and coatings.

RCRA NO. U148 (123-33-1) Contact with strong oxidizers may cause fire and explosions.

RCRA NO. U149 (109-77-3) Incompatible with sulfuric acid. Caustics or heat above 160°F/71°C may cause polymerization or spontaneous combustion.

RCRA NO. U151 (7439-97-6) Acetylene, acetylene products, ammonia gases can form shock-sensitive solids that can initiate fires of combustibles. Reacts violently with boron phosphodiiodide, chlorine, chlorine dioxide, methyl azide. Attacks copper and copper alloys.

RCRA NO. U152 (126-98-7) Forms explosive mixture with air (flash point 34°F/1°C). Incompatible with aliphatic amines, alkanolamines. Reacts violently with oxidizers. Incompatible with strong acids, strong bases, or light exposure; may cause polymerization.

RCRA NO. U153 (74-93-1) May accumulate static electrical charges, and may cause ignition of its vapors. Forms explosive air–gas mixture (flash point 0°F/−18°C). Strong oxidizers may cause fire and explosions. Attacks some plastics, coatings, and rubber.

RCRA NO. U154 (67-56-1) May accumulate static electrical charges, and may cause ignition of its vapors. Forms explosive mixture with air (flash point 52°F/11°C). Incompatible with strong acids, strong oxidizers, caustics, aliphatic amines, isocyanates, chromic anhydride, lead perchlorate, perchloric acid, phosphorus trioxide. May react with metallic aluminum at high temperature. Attacks some plastics, rubber, and coatings.

RCRA NO. U156 (79-22-1) Forms explosive mixture with air (flash point 54°F/12°C). Water contact produces hydrochloric acid. Corrodes metals in the presence of moisture. Attacks some plastics, rubber, and coatings.

RCRA NO. U159 (78-93-3) Forms explosive mixture with air (flash point 16°F/−9°C). Extremely flammable. Incompatible with sulfuric acid, nitric acid, aliphatic amines, strong oxidizers, potassium *tert*-butoxide, 2-propanol, chlorosulfonic acid, oleum.

RCRA NO. U160 (1338-23-4) Forms explosive mixture with air (flash point 125°F/52°C). Explosive decomposition occurs above 176°F/80°C. Pure substance is shock-sensitive. Strong oxidizer. Reacts violently with strong acids, strong bases, reducing agents, combustible substances, organic materials, oxides of heavy metals, salts, trace contaminants, amines. May accumulate static electrical charges, and may cause ignition of its vapors.

RCRA NO. U161 (108-10-1) Forms explosive mixture with air (flash point 64°F/18°C). Incompatible with strong oxidizers, strong acids, aliphatic amines. Dissolves some plastics, resins, and rubber.

RCRA NO. U162 (80-62-6) Forms explosive mixture with air (flash point 50°F/10°C). Incompatible with caustics, nitrates, strong acids, aliphatic amines, alkanolamines, peroxides, and strong oxidizers. Heat and/or lack of appropriate inhibitor can cause polymerization. May accumulate static electrical charges, and may cause ignition of its vapors.

RCRA NO. U165 (91-20-3) Forms explosive mixture with air (flash point 174°F/79°C). Incompatible with strong oxidizers, chromium oxide (reacts violently). Attacks some plastics, rubber, and coatings. May accumulate static electrical charges, and may cause ignition of its vapors.

RCRA No. U167 (134-32-7) Oxidizes in air. Incompatible with nitrous acid, oxidizers, nitrates, organic anhydrides, isocyanates, aldehydes.

RCRA NO. U169 (98-95-3) Forms explosive mixture with air (flash point 190°F/88°C). Incompatible with nitric acid, nitrogen tetroxide, caustics, ammonia, amines. Attacks some plastics, rubber, and coatings.

RCRA NO. U170 (100-02-7) Combustible solid. A strong oxidizer; reacts with reducing agents, combustibles, organic and other easily oxidizable materials. Incompatible with strong acids, caustics, aliphatic amines, amides.

RCRA NO. U171 (79-46-9) Forms explosive mixture with air (flash point 74°F/24°C). Incompatible with amines, strong acids, alkalies, strong oxidizers. Contact with some metal oxides may cause decomposition. Mixtures with hydrocarbons are extremely flammable. Attacks some plastics, rubber, and coatings.

RCRA NO. U182 (123-63-7) Forms explosive mixture with air (flash point 96°F/36°C). Reacts with strong acids, caustics, ammonia, amines, oxidizers. Contact with acids produces acetaldehyde.

RCRA NO. U188 (108-95-2) Forms explosive mixture with air (flash point 174°F/79°C). Incompatible with strong oxidizers, strong acids, caustics, aliphatic amines, amides, oxidizers, formaldehyde, butadiene, calcium hypochlorite. Liquid attacks some plastics, rubber, and coatings; hot liquid attacks aluminum, magnesium, lead, and zinc metals.

RCRA NO. U190 (85-44-9) Incompatible with strong acids, caustics, ammonia, amines, strong oxidizers. Attacks some plastics, rubber, and coatings.

RCRA NO. U191 (109-06-8) Forms explosive mixture with air (flash point 79°F/26°C). Strong oxidizers may cause fire and explosions. Attacks copper/alloys.

RCRA NO. U194 (107-10-8) Forms explosive mixture with air. Incompatible with acids, organic anhydrides, halogenated hydrocarbons, isocyanates, nitroparaffins, vinyl acetate, acrylates, substituted allyls, alkylene oxides, epichlorohydrin, ketones, aldehydes, alcohols, glycols, mercury, phenols, cresols, triethyl aluminum, caprolactam solution, strong oxidizers. Attacks aluminum, copper, lead, tin, zinc, and alloys.

RCRA NO. U196 (110-86-1) Forms explosive mixture with air (flash point 68°F/20°C). Strong oxidizers may cause fire and explosions. Strong acids may cause violent spattering. Attacks some plastics, rubber, and coatings.

RCRA NO. U197 (106-51-4) Forms explosive mixture with air (flash point 104°F/40°C). Incompatible with strong bases, reducing agents, strong oxidizers. Attacks some plastics, rubber, and coatings.

RCRA NO. U209 (79-34-5) Exposure to heat, light, or air produces corrosive and toxic vapors. Reacts with strong caustics to form explosive dichloroacetylene. Reacts violently with chemically active metals or sodium amide. In presence of steam, contact with hot iron, aluminum, or zinc may produce toxic vapors. Attacks some plastics, rubber, and coatings.

RCRA NO. U210 (127-18-4) Incompatible with strong oxidizers, finely divided metals, caustics. Stable up to 258°F/126°C; at this temperature product gives off poisonous fumes.

RCRA NO. U213 (109-99-9) Forms explosive mixture with air (flash point 6°F/−14°C). Unless inhibited, can form unstable and explosive peroxides. Incompatible with strong acids, strong oxidizers. Attacks some plastics. May accumulate static electric charges that can result in ignition of its vapors.

RCRA NO. U214 (563-68-8) Incompatible with strong acids, nitrates.

RCRA NO. U215 (6533-73-9) Incompatible with strong oxidizers, strong acids.

RCRA NO. U219 (62-56-6) Aqueous solution is a base; reacts strongly with acids. Incompatible with nitric acid, hydrogen peroxide, acrolein. Incompatible with metals.

RCRA NO. U220 (108-88-3) Forms explosive mixture with air (flash point 40°F/4°C). Strong oxidizers may cause fire and explosions. Attacks some plastics, rubber, and coatings. May accumulate static electrical charges, and may cause ignition of its vapors.

RCRA NO. U221 (95-80-7) Incompatible with oxidizers, acids, organic anhydrides, isocyanates, aldehydes. Attacks aluminum, brass, bronze, copper, zinc.

RCRA NO. U223 (584-84-9) Incompatible with strong acids, including non-oxidizing mineral and organic acids, caustics, ammonia, amines, amides, alcohols, glycols, caprolactam solution. Water contact causes violent foaming and spattering; produces carbon dioxide and an organic base. Attacks copper and its alloys, some plastics including polyethylene, and rubber.

RCRA NO. U225 (75-25-2) Reacts with chemically active metals, calcium, acetone, strong caustics. Attacks some plastics, rubber, and coatings.

RCRA NO. U226 (71-55-6) Reacts with acetone, strong caustics, alkaline earth and alkali metals, dinitrogen tetroxide, sodium hydroxide, amides, strong oxidizers, sodium, potassium. Reacts violently with copper, bronze, and other metal powders. The uninhibited grade is corrosive to aluminum. Attacks some plastics, rubber, and coatings.

RCRA NO. U227 (79-00-5) Incompatible with strong oxidizers, strong caustics, chemically active metals (especially powders), sodium amide; may cause fire and explosions. Attacks some plastics, rubber, and coatings.

RCRA NO. U228 (79-01-6) Contact with caustics produces a toxic and flammable gas. Reacts violently with chemically active metals. Contact with aluminum may produce a violent, self-accelerating polymerization reaction. Incompatible with acids, organic anhydrides, isocyanates, alkylene oxides, aldehydes, alcohols, glycols, phenols, cresols, caprolactam solution, epichlorohydrin, nitrogen tetroxide, metal powders, oxygen. May accumulate static electrical charges, and may cause ignition of its vapors.

RCRA NO. U230 (99-95-4) Contact with strong oxidizers may cause fire and explosions.

RCRA NO. U232 (93-76-5) Sealed metal containers may burst in heat above 316°F/158°C.

RCRA NO. U233 (93-72-1) A powerful oxidizer; reacts violently with reducing agents, combustibles.

RCRA NO. U239 (95-47-6) Forms explosive mixture with air (flash point 90°F/32°C). Strong oxidizers may cause fire and explosions. Attacks some plastics, rubber, and coatings. May accumulate static electrical charges, and may cause ignition of its vapors.

RCRA NO. U239 (106-42-3) Forms explosive mixture with air (flash point 81°F/27.2°C). Strong oxidizers may cause fire and explosions. Attacks some plastics, rubber, and coatings. May accumulate static electrical charges, and may cause ignition of its vapors.

RCRA NO. U239 (108-38-3) Forms explosive mixture with air (flash point 84°F/28.9°). Strong oxidizers may cause fire and explosions. Attacks some plastics, rubber, and coatings. May accumulate static electrical charges, and may cause ignition of its vapors.

RCRA NO. U240 (94-75-7) Decomposes in sunlight. Incompatible with strong oxidizers; may cause fire and explosions.

RCRA NO. U242 (87-86-5) Hot water causes decomposition, producing hydrochloric acid. Strong oxidizers may cause fire and explosions.

RCRA NO. U244 (137-26-8) Combustible solid (flash point 192°F/89°C). Strong oxidizers may cause fire and explosions; contact with strong acid or oxidizable materials produces toxic gases.

RCRA NO. U246 (506-68-3) May be unstable unless dry and pure. Reacts violently with acids, ammonia, amines. Water contact produces hydrogen cyanide and hydrogen bromide.

RCRA NO. U274 (72-43-5) Contact with strong oxidizers may cause fire and explosions. Attacks some plastics, rubber, and coatings.

RCRA NO. U328 (95-53-4) Forms explosive mixture with air (flash point 185°F/85°C). Incompatible with strong acids, mineral acids, organic anhydrides, isocyanates, aldehydes, oxidizers. Attacks some plastics, rubber, and coatings.

RCRA NO. U353 (106-49-0) Forms explosive mixture with air (flash point 188°F/87°C). Incompatible with strong oxidizers, strong acids. Attacks some plastics, rubber, and coatings.

RCRA NO. U359 (110-80-5) Forms explosive mixture with air (flash point 120°F/49°C). Strong oxidizers may cause fire and explosions. Attacks some plastics, rubber, and coatings. Able to form peroxides. Incompatible with strong acids, aluminum and its alloys.

RE-4355 (300-76-5) Unstable in the presence of iron. Reacts with acids, strong oxidizers in sunlight. Corrosive to metals. Attacks some plastics, rubber, and coatings.

REALGAR (7440-38-2) Contact with hydrogen gas produces arsine. Incompatible with strong acids, strong oxidizers, peroxides, bromine pentafluoride, bromine trifluoride, cesium acetylene carbide, chromium trioxide, nitrogen trichloride, silver nitrate.

RED CHROMATE OF POTASH (7778-50-9) Strong oxidizer; reacts violently with reducing agents and combustibles. Powdered combustibles will ignite. Strong acids cause formation of toxic vapors.

REDDON (93-76-5) Sealed metal containers may burst in heat above 316°F/158°C. Incompatible with sulfuric acid, bases, ammonia, aliphatic amines, alkanolamines, isocyanates, alkylene oxides, epichlorohydrin.

RED FUMING NITRIC ACID (7697-37-2) A strong oxidizer that can react violently with reducing agents, combustible materials. Incompatible with many substances, including acrylates, aliphatic amines, alcohols, aldehydes, alkanolamines, alkylene oxides, anion exchange resins, aromatic amines, amides, bases, cresols, cyanides, cyclic ketones, epichlorohydrin, glycols, isocyanates, ketones, oleum, organic anhydrides, phenols, substituted allyls, sulfuric acid, strong oxidizers, terpenes. Attacks most metals, and some plastics, rubber, and coatings.

RED MERCURIC IODIDE (7774-29-0) Reacts violently with strong oxidizers, chlorine, fluorine.

RED OIL (112-80-1) Incompatible with strong oxidizers, perchloric acid, and aluminum powder; may cause explosions. Corrodes aluminum.

RED OIL, DISTILLED (112-80-1) Incompatible with strong oxidizers, perchloric acid, and aluminum powder; may cause explosions. Corrodes aluminum.

RED OXIDE OF MERCURY (21908-53-2) A strong oxidizer; reacts violently with reducing agents and combustible materials. Incompatible with alcohols, chlorine, phosphorus, sulfur, alkali metals (i.e., lithium, sodium, potassium, rubidium, cesium, francium).

RED OXIDE OF NITROGEN (7723-14-0) Contact with all oxidizers may cause fire and explosions or produce shock-sensitive compounds. Reacts violently with caustics + heat, chlorosulfonic acid + heat, lead dioxide, performic acid, selenium oxychloride, chlorates and other materials. Incompatible with many other substances, and produces phosphine gas on contact with moisture and oxygen; opened packages should be stored under inert gas blanket.

RED PHOSPHORUS (7723-14-0) Contact with all oxidizers may cause fire and explosions or produce shock-sensitive compounds. Reacts violently with caustics + heat, chlorosulfonic acid + heat, lead dioxide, performic acid, selenium oxychloride, chlorates, and other materials. Incompatible with many other substances, and produces phosphine gas on contact with moisture and oxygen; opened packages should be stored under inert gas blanket.

RED TR BASE (95-79-4) Incompatible with acids, organic anhydrides, isocyanates, aldehydes, oxidizers.

REFINED SOLVENT NAPHTHA (8032-32-4) Forms explosive mixture with air (flash point −40°F to −86°F/−40°C to −66°C). Incompatible with strong acids, strong oxidizers. Attacks some plastics, rubber, and coatings. May accumulate static electrical charges, and may cause ignition of its vapors.

REFORMATES, GASOLINE BLENDING STOCKS (N/A) Forms explosive mixture with air (flash point 0°F to 73°F/−18°C to 23°C). Incompatible with strong oxidizers, nitric acid.

REFRIGERANT 11 (75-69-4) Reacts with barium, lithium, sodium, magnesium, titanium. Attacks some plastics, rubber, and coatings.

REFRIGERANT 12 (75-71-8) Reacts violently with liquid aluminum. Incompatible with chemically active metals. Attacks some plastics, rubber, and coatings.

REFRIGERANT 13B1 (75-63-8) Reacts with chemically active metals, powdered aluminum, zinc, magnesium. May attack some plastics, rubber, and coatings.

REFRIGERANT 21 (75-43-4) Reacts with water and chemically active metals. Attacks some plastics, rubber, and coatings.

REFRIGERANT 112 (76-12-0) Reacts with chemically active metals, powdered aluminum, zinc, and magnesium. Attacks some forms of plastics, rubber, and coatings.

REFRIGERANT 113 (76-13-1) Reacts with barium, lithium, sodium, magnesium, titanium. Contact with alloys containing more than 2% magnesium may cause decomposition (with hydrogen chloride, hydrogen fluoride, and carbon monoxide released). Attacks some plastics, rubber, and coatings.

REFRIGERANT 114 (76-14-2) Reacts with barium, lithium, sodium, magnesium, titanium. Attacks some plastics, rubber, and coatings.

REFRIGERANT 152a (75-37-6) Flammable gas. May accumulate static electrical charges, and may cause ignition of its vapors. Reacts violently with strong oxidizers. Attacks metals in presence of moisture.

REGALON (85-00-7) Concentrated solution attacks aluminum.

REGLON or REGLONE (85-00-7) Concentrated solution attacks aluminum.

REGLOX (85-00-7) Concentrated solution attacks aluminum.

REGULOX (123-33-1) Contact with strong oxidizers may cause fire and explosions.

REGUTOL (119-36-8) Forms explosive mixture with air (flash point 205°F/96°C). Incompatible with strong acids, nitrates, oxidizers.

REOFOS 95 (25155-23-1) Incompatible with sulfuric acid, nitric acid, nitrates, strong oxidizers.

REOMOL DOA (103-23-1) Incompatible with strong acids, nitrates, oxidizers.

REQUTOL (119-36-8) Forms explosive mixture with air (flash point 205°F/96°C). Incompatible with strong acids, nitrates, oxidizers.

RESIDUAL ASPHALT (8052-42-4) Incompatible with nitric acid, fluorine, strong oxidizers.

RESIDUAL FUEL OIL NO. 4 (68476-33-5) Forms explosive mixture with air (flash point 142°F/61°C). Oxidizers may cause fire and explosions. Incompatible with nitric acid. May accumulate static electrical charges, and may cause ignition of its vapors.

RESIDUAL FUEL OIL NO. 5 (68476-33-5) Forms explosive mixture with air (flash point 142°F/61°C). Oxidizers may cause fire and explosions. Incompatible with nitric acid. May accumulate static electrical charges, and may cause ignition of its vapors.

RESIDUAL FUEL OIL NO. 6 (68476-33-5) Forms explosive mixture with air (flash point 142°F/61°C). Oxidizers may cause fire and explosions. Incompatible with nitric acid. May accumulate static electrical charges, and may cause ignition of its vapors.

RESIDUAL OIL (8052-42-4) Incompatible with nitric acid, fluorine, strong oxidizers.

RESIN OIL (N/A) Oxidizers may cause fire and explosions. Incompatible with nitric acid. May accumulate static electrical charges, and may cause ignition of its vapors.

RESITOX (56-72-4) Contact with strong oxidizers may cause fire and explosions.

RESORCIN or RESORCINE (108-46-3) Reacts violently with strong oxidizers, nitric acid. Incompatible with acetanilide, albumin, alkalies, antipyrine, camphor, ferric salts, menthol, spirit nitrous ether. Absorbs moisture from air (hygroscopic). May accumulate static electrical charges, and may cause ignition of its vapors.

RESORCINE BROWN J (1300-73-8) Forms explosive mixture with air (flash point 206°F/96.7°C). Contact with strong acids may cause fire and explosions. Contact with hypochlorite bleaches produces explosive chloroamines. Incompatible with strong acids, organic acids and anhydrides, isocyanates, aldehydes.

RESORCINE BROWN R (1300-73-8) Forms explosive mixture with air (flash point 206°F/96.7°C). Contact with strong acids may cause fire and explosions. Contact with hypochlorite bleaches produces explosive chloramines. Incompatible with strong acids, organic acids and anhydrides, isocyanates, aldehydes.

RESORCINOL (108-46-3) Reacts violently with strong oxidizers, nitric acid. Incompatible with acetanilide, albumin, alkalies, antipyrine, camphor, ferric salts, menthol, spirit nitrous ether. Absorbs moisture from air (hygroscopic). May accumulate static electrical charges, and may cause ignition of its vapors.

RETARDER AK (85-44-9) Incompatible with strong acids, caustics, ammonia, amines, strong oxidizers. Attacks some plastics, rubber, and coatings.

RETARDER ESEN (85-44-9) Incompatible with strong acids, caustics, ammonia, amines, strong oxidizers. Attacks some plastics, rubber, and coatings.

RETARDER PD (85-44-9) Incompatible with strong acids, caustics, ammonia, amines, strong oxidizers. Attacks some plastics, rubber, and coatings.

RETARDER W (69-72-7) Reacts with strong oxidizers, ethyl nitrite, iodine, iron salts, lead acetate.

RETINOL (N/A) Oxidizers may cause fire and explosions. Incompatible with nitric acid. May accumulate static electrical charges, and may cause ignition of its vapors.

REVAC (119-36-8) Forms explosive mixture with air (flash point 205°F/96°C). Incompatible with strong acids, nitrates, oxidizers.

REVENGE (75-99-0) Corrosive to iron, aluminum, and copper.

REXENE (9003-07-0) Incompatible with strong acids, strong oxidizers, chlorine, potassium permanganate.

REZIFILM (137-26-8) Combustible solid (flash point 192°F/89°C). Strong oxidizers may cause fire and explosions; contact with strong acid or oxidizable materials produces toxic gases.

RFNA (7697-37-2) A strong oxidizer that can react violently with reducing agents, combustible materials. Incompatible with many substances, including acrylates, aliphatic amines, alcohols, aldehydes, alkanolamines, alkylene oxides, anion exchange resins, aromatic amines, amides, bases, cresols, cyanides, cyclic ketones, epichlorohydrin, glycols, isocyanates, ketones, oleum, organic anhydrides, phenols, substituted allyls, sulfuric acid, strong oxidizers, terpenes. Attacks most metals, and some plastics, rubber, and coatings.

RHODANATE (540-72-7) Incompatible with acids, bases, ammonia, amines, amides, alcohols, glycols, caprolactam.

RHODIA (94-75-7) Decomposes in sunlight. Incompatible with strong oxidizers; may cause fire and explosions.

RHODIACHLOR (76-44-8) Forms hydrogen chloride gas with iron and rust above 165°F/74°C.

RHODIASOL (56-38-2) Combustible liquid. Mixtures with endrin may be explosive. Strong oxidizers may cause fire and explosions. Attacks some plastics, rubber, and coatings.

RHODIATOX (56-38-2) Combustible liquid. Mixtures with endrin may be explosive. Strong oxidizers may cause fire and explosions. Attacks some plastics, rubber, and coatings.

RHODIATROX (56-38-2) Combustible liquid. Mixtures with endrin may be explosive. Strong oxidizers may cause fire and explosions. Attacks some plastics, rubber, and coatings.

RHOPLEX AC-33 (97-63-2) Forms explosive mixture with air (flash point 80°F/27°C). Incompatible with strong acids, amines, oxidizers. Corrodes some metals.

RHOTHANE (72-54-8) Contact with strong oxidizers may cause fire and explosions.

RICE BRAN OIL (N/A) Incompatible with strong acids, oxidizers.

RICHONATE 1850 (25155-30-0) Reacts with acids, including fumes.

RICINUS OIL (8001-79-4) Incompatible with strong acids, oxidizers, nitrates.

RIMSO-50 (67-68-5) Forms explosive mixture with air (flash point 203°F/95°C). Reacts violently with oxidizers. Reacts with ethanoyl chloride, boron compounds, halides, metal alkoxides, oxidizers.

ROACH SALT (7681-49-4) Incompatible with water, producing a corrosive. Reacts with acids.

ROAD ASPHALT (8052-42-4) Incompatible with nitric acid, fluorine, strong oxidizers.

ROAD BINDER (8052-42-4) Incompatible with nitric acid, fluorine, strong oxidizers.

ROAD OIL (8052-42-4) Incompatible with nitric acid, fluorine, strong oxidizers.

ROAD TAR (8052-42-4) Incompatible with nitric acid, fluorine, strong oxidizers.

ROCK CANDY (57-50-1) Reacts with potassium hydroxide, strong acids, strong oxidizers.

ROCK OIL (8002-05-9) Oxidizers may cause fire and explosions. Incompatible with nitric acid. May accumulate static electrical charges, and may cause ignition of its vapors.

RODOCID (563-12-2) Incompatible with alkaline formulations. Mixtures with magnesium may be explosive.

ROMAN VITRIOL (7758-98-7) Aqueous solution is an acid. Incompatible with strong bases, hydroxylamine, magnesium.

ROSE ETHER (122-99-6) Incompatible with sulfuric acid, isocyanates, strong oxidizers.

ROSIN OIL (N/A) Oxidizers may cause fire and explosions. Incompatible with nitric acid. May accumulate static electrical charges, and may cause ignition of its vapors.

ROSINOL (N/A) Oxidizers may cause fire and explosions. Incompatible with nitric acid. May accumulate static electrical charges, and may cause ignition of its vapors.

ROTOX (74-83-9) Incompatible with strong oxidizers, aluminum, dimethylsulfoxide, ethylene oxide, water. Attacks zinc, magnesium, alkali metals and their alloys, some plastics, rubber, and coatings. Forms aluminum alkyls in presence of aluminum; aluminum alkyls are spontaneously ignitable materials.

ROYAL MH 30 (123-33-1) Contact with strong oxidizers may cause fire and explosions.

ROYALTAC (112-30-1) Forms explosive mixture with air (flash point 180°F/82°C). Incompatible with strong acids, caustics, aliphatic amines, isocyanates, strong oxidizers.

ROYAL TMTD (137-26-8) Combustible solid (flash point 192°F/89°C). Strong oxidizers may cause fire and explosions; contact with strong acid or oxidizable materials produces toxic gases.

RP 8167 (563-12-2) Incompatible with alkaline formulations. Mixtures with magnesium may be explosive.

RTEC (Polish) (7439-97-6) Acetylene, acetylene products, ammonia gases can form shock-sensitive solids that can initiate fires of combustibles. Reacts violently with boron phosphodiiodide, chlorine, chlorine dioxide, methyl azide. Attacks copper and copper alloys.

RUBBING ALCOHOL (67-63-0) Forms explosive mixture with air (flash point 53°F/12°C). Incompatible with alkaline earth and alkali metals, crotonaldehyde, phosgene, strong acids, strong oxidizers. Attacks some plastics, rubber, and coatings. Reacts with metallic aluminum at high temperatures.

RUBINATE (101-68-8) Incompatible with strong alkalies, acids, alcohols, ammonia, amines, amides, glycols, caprolactam. Unstable above 100°F/37.8°C. Attacks some plastics, rubber, and coatings.

RUCOFLEX PLASTICIZER DOA (103-23-1) Incompatible with strong acids, nitrates, oxidizers.

RUELENE (299-86-5) Decomposes in strongly alkaline (pH >7) and strongly acidic media. Unstable over long periods of time in water and at temperatures above 140°F/60°C.

RUELENE DRENCH (299-86-5) Decomposes in strongly alkaline (pH >7) and strongly acidic media. Unstable over long periods of time in water and at temperatures above 140°F/60°C.

RUKSEAM (50-29-3) Incompatible with salts of iron or aluminum, and bases. Do not store in iron containers.

RUMETAN (1314-84-7) Incompatible with oxidizers, acids, water.

- S -

S95 (75-99-0) Corrosive to iron, aluminum, and copper.

S 276 (298-04-4) Forms explosive mixture with air (flash point 180°F/82°C). Incompatible with alkalis, strong oxidizers.

SA (69-72-7) Reacts with strong oxidizers, ethyl nitrite, iodine, iron salts, lead acetate.

SACCHAROSE (57-50-1) Reacts with potassium hydroxide, strong acids, strong oxidizers.

SACCHARUM (57-50-1) Reacts with potassium hydroxide, strong acids, strong oxidizers.

SADOFOS (121-75-5) Incompatible with strong oxidizers, magnesium, alkaline pesticides. Attacks metals, some plastics, rubber, and coatings.

SADOPHOS (121-75-5) Incompatible with strong oxidizers, magnesium, alkaline pesticides. Attacks metals, some plastics, rubber, and coatings.

SADOPLON (137-26-8) Combustible solid (flash point 192°F/89°C). Strong oxidizers may cause fire and explosions; contact with strong acid or oxidizable materials produces toxic gases.

SAEURE FLUORIDE (German) (7782-41-4) A powerful oxidizer and a dangerously reactive gas. Reacts violently with reducing agents and combustible materials. Contact with water can produce corrosive hydrogen fluoride, hydrochloric acid, oxygen and oxygen difluoride. Reacts vigorously with most oxidizable materials at room temperature, frequently with ignition. Corrodes most metals (except the cylinders in which it is shipped), especially in the presence of moisture. Contact with nitric acid produces explosive gas. Attacks some plastics, rubber, and coatings. Reacts explosively with ammonia, graphite, halocarbons, hydrocarbons, hydrogen, and many other substances. Reacts with nearly every known element.

SAFFLOWER OIL or SAFFLOWER SEED OIL (8001-23-8) Incompatible with strong acids, oxidizers.

SAH (90-02-8) Forms explosive mixture with air (flash point 172°F/78°C). Reacts violently with strong oxidizers.

SALACETIN (50-78-2) Fires and explosions may result from contact with strong oxidizers. Alkali hydroxides or carbonates cause decomposition.

SAL AMMONIAC (12125-02-9) Contact with boron trifluoride, boron pentafluoride, iodine heptafluoride, or potassium chlorate causes violent reactions. Incompatible with alkalies, acids, salts of lead or silver. At fire temperatures fume corrodes metals.

SALCETOGEN (50-78-2) Fires and explosions may result from contact with strong oxidizers. Alkali hydroxides or carbonates cause decomposition.

SALETIN (50-78-2) Fires and explosions may result from contact with strong oxidizers. Alkali hydroxides or carbonates cause decomposition.

SALICYLAL (90-02-8) Forms explosive mixture with air (flash point 172°F/78°C). Reacts violently with strong oxidizers.

SALICYLALDEHYDE (90-02-8) Forms explosive mixture with air (flash point 172°F/78°C). Reacts violently with strong oxidizers.

SALICYLIC ACID (69-72-7) Reacts with strong oxidizers, ethyl nitrite, iodine, iron salts, lead acetate.

SALICYLIC ACID, ACETATE (50-78-2) Fires and explosions may result from contact with strong oxidizers. Alkali hydroxides or carbonates cause decomposition.

SALICYLIC ALDEHYDE (90-02-8) Forms explosive mixture with air (flash point 172°F/78°C). Reacts violently with strong oxidizers.

SALMIAC (12125-02-9) Contact with boron trifluoride, boron pentafluoride, iodine heptafluoride, or potassium chlorate causes violent reactions. Incompatible with alkalies, acids, salts of lead or silver. At fire temperatures fume corrodes metals.

SALPETERSAEURE (German) (7697-37-2) A strong oxidizer that can react violently with reducing agents, combustible materials. Incompatible with many substances, including acrylates, aliphatic amines, alcohols, aldehydes, alkanolamines, alkylene oxides, anion exchange resins, aromatic amines, amides, bases, cresols, cyanides, cyclic ketones, epichlorohydrin, glycols, isocyanates, ketones, oleum, organic anhydrides, phenols, substituted allyls, sulfuric acid, strong oxidizers, terpenes. Attacks most metals, and some plastics, rubber, and coatings.

SALPETERZUUROPLOSSINGEN (Dutch) (7697-37-2) A strong oxidizer that can react violently with reducing agents, combustible materials. Incompatible with many substances, including acrylates, aliphatic amines, alcohols, aldehydes, alkanolamines, alkylene oxides, anion exchange resins, aromatic amines, amides, bases, cresols, cyanides, cyclic ketones, epichlorohydrin, glycols, isocyanates, ketones, oleum, organic anhydrides, phenols, substituted allyls, sulfuric acid, strong oxidizers, terpenes. Attacks most metals, and some plastics, rubber, and coatings.

SALT OF SATURN (301-04-2) Contact with strong acids produces acetic acid. Incompatible with bases, ammonia, amines, cresols, isocyanates, alkylene oxides, epichlorohydrin, phenols, sulfites. Reacts violently with bromates, strong oxidizers.

SALT OF TARTER (3811-04-9) A powerful oxidizer. Reacts violently with reducing agents or combustibles, ammonia gas, ammonium salts, organic matter, hydrogen iodide, organic acids, sulfuric acid. Forms explosive mixtures with metallic powders, ammonium chloride, organic solids, including agricultural materials.

SALTPETER (7757-79-1) A powerful oxidizer. Reacts violently with reducing agents and combustibles. Strong acid causes formation of toxic vapors. Forms explosive mixtures with many substances including sodium acetate, metal powders, sodium hypophosphite, trichloroethylene, zinc, etc. Dangerously reactive; separate from all materials.

SALVO (94-75-7) Decomposes in sunlight. Incompatible with strong oxidizers; may cause fire and explosions.

SAND ACID (16961-83-4) Incompatible with aliphatic amines, alkanolamines, alkylene oxides, aromatic amines, amides, ammonia, ammonium hydroxide, bases, calcium oxide, epichlorohydrin, isocyanates, oleum, organic anhydrides, sulfuric acid, strong oxidizers, vinyl acetate, water.

SANG GAMMA (58-89-9) Not combustible, but may be dissolved in a combustible solvent. If solvent comes in contact with oxidizers, fire and explosions may result.

SANMORIN OT 70 (119-36-8) Forms explosive mixture with air (flash point 205°F/96°C). Incompatible with strong acids, nitrates, oxidizers.

SANOCIDE (118-74-1) Reacts violently with strong oxidizers, formyl-dimethylamine.

SANSPOR (2425-06-1) Incompatible with acids or acid vapor. Strongly alkaline conditions contribute to instability.

SANTAR (21908-53-2) A strong oxidizer; reacts violently with reducing agents and combustible materials. Incompatible with alcohols, chlorine, phosphorus, sulfur, alkali metals (i.e., lithium, sodium, potassium, rubidium, cesium, francium).

SANTICIZER 711 (3648-20-2) Incompatible with strong acids, nitrates.

SANTOBANE (50-29-3) Incompatible with salts of iron or aluminum, and bases. Do not store in iron containers.

SANTOBRITE (87-86-5) Hot water causes decomposition producing hydrochloric acid. Strong oxidizers may cause fire and explosions.

SANTOCHLOR (106-46-7) Forms explosive mixture with air (flash point 150°F/66°C). Incompatible with strong oxidizers, metal powders, and alkali metals (i.e., lithium, sodium, potassium, rubidium, cesium, francium). Attacks some plastics, rubber, and coatings.

SANTOMERSE-3 (25155-30-0) Reacts with acids, including fumes.

SANTOPHEN (87-86-5) Hot water causes decomposition, producing hydrochloric acid. Strong oxidizers may cause fire and explosions.

SANTOPHEN 20 (87-86-5) Hot water causes decomposition producing hydrochloric acid. Strong oxidizers may cause fire and explosions.

SANTOTHERM (generic CAS for PCBs 1336-36-3) Incompatible with strong oxidizers, strong acids.

SANTOTHERM FR (generic CAS for PCBs 1336-36-3) Incompatible with strong oxidizers, strong acids.

SAPONATED CRESOL SOLUTION (1319-77-3) Incompatible with strong acids, oxidizers, alkalies, aliphatic amines, amides, chlorosulfonic acid, oleum. Liquid attacks some plastics, coatings, and rubber.

SAROLEX (333-41-5) Incompatible with water, copper-containing compounds, oxidizers, acids, or bases.

SATICIZER 160 (85-68-7) Incompatible with strong acids, nitrates, oxidizers.

SAX (69-72-7) Reacts with strong oxidizers, ethyl nitrite, iodine, iron salts, lead acetate.

SBA or S.B.A. (78-92-2) Forms explosive mixture with air (flash point 75°F/4°C). Attacks some plastics, rubber, and coatings. Forms an explosive peroxide in air. Ignites with chromium trioxide. Incompatible with strong oxidizers, strong acids, aliphatic amines, isocyanates, organic peroxides.

SBO (119-36-8) Forms explosive mixture with air (flash point 205°F/96°C). Incompatible with strong acids, nitrates, oxidizers.

SCALDIP (122-39-4) Incompatible with strong acids, aldehydes, organic anhydrides, isocyanates, oxidizers, hexachloromelamine, trichloromelamine.

SCHWEFELDDIOXYD (German) (7446-09-5) Produces corrosive hydrochloric acid fumes with air. Contact with copper, bronze, or alkali metals may cause fire and explosions. Reacts violently with alcohols, caustics, amines, water. Decomposes above 140°F; produces toxic and corrosive oxides of sulfur. Attacks some plastics, rubber, and coatings.

SCHWEFELKOHLENSTOFF (German) (75-15-0) Highly reactive. Contact with many substances can cause fire and explosions. Forms explosive mixture with air (flash point −22°F/−30°C). Shock can cause explosive decomposition. Incompatible with alkali metals, aliphatic amines, alkanolamines, aluminum, azides, chlorine monoxide, combustible substances, ethylene diamine, ethyleneimine, lead azide, lithium azide, nitric oxide, nitrogen dioxide, potassium, potassium azide, reducing agents, rubidium azide, sodium azide, zinc.

SCHWEFELWASSERSTOFF (German) (7783-06-4) A highly flammable and reactive gas. Incompatible with acetaldehyde, barium pentafluoride, chlorine monoxide, chlorine trifluoride, chromic anhydride, copper, lead dioxide, nitric acid, nitrogen iodide, nitrogen trichloride, nitrogen trifluoride, oxygen difluoride, oxidizers, phenyl diazonium chloride, sodium, sodium peroxide. Reacts with alkali metals and alkali earth metals. Attacks metals.

SCHWEINFURTH GREEN (12002-03-8) Contact with strong oxidizers may cause fire and explosions.

SCINTILLAR (106-42-3) Forms explosive mixture with air (flash point 81°F/27.2°C). Strong oxidizers may cause fire and explosions. Attacks some plastics, rubber, and coatings. May accumulate static electrical charges, and may cause ignition of its vapors.

SCONATEX (75-35-4) Forms explosive mixture with air (18°F/−28°C). Air or contaminants can causes formation of peroxides; may polymerize. Reacts violently with strong oxidizers, alkali metals. Incompatible with nitric acid.

SCYAN (540-72-7) Incompatible with acids, bases, ammonia, amines, amides, alcohols, glycols, caprolactam.

SD-1750 (62-73-7) Attacks some plastics, rubber, and coatings.

SD 3562 (141-66-2) Corrosive to cast iron, mild steel, brass, and stainless steel 304.

SD 5532 (57-74-9) Contact with strong oxidizers may cause fire and explosions. Attacks some plastics, rubber, and coatings.

SD ALCOHOL 23-HYDROGEN (64-17-5) Forms explosive mixture with air (flash point 65°F/18°C). May accumulate static electrical charges, and may cause ignition of its vapors. Reactions may be violent with oleum, sulfuric acid, nitric acid, bases, aliphatic amines, isocyanates, oxidizers.

SDIC (2893-78-9) Contact with water causes the release of chlorine gas. A powerful oxidizer; reacts violently with reducing agents, organic matter, easily chlorinated or oxidized materials. Incompatible with ammonium salts, amines. Explosive reaction with sodium hypochlorite and water.

SDMH (540-73-8) Forms explosive mixture with air (flash point 73°F/23°C). A strong reducing agent. Reacts violently with strong oxidizers, strong acids. Attacks some plastics, rubber, and coatings. May accumulate static electrical charges, and may cause ignition of its vapors.

733

SEAL-COATING MATERIAL (8052-42-4) Incompatible with nitric acid, fluorine, strong oxidizers.

SEAWATER MAGNESIA (1309-48-4) Reacts violently with chlorine trifluoride, bromine pentafluoride, phosphorus pentachloride. Incompatible with acids.

SEDRIN (309-00-2) Incompatible with concentrated mineral acids, acid catalysts, acid oxidizing agents, phenols, reactive metals.

SEEDRIN (309-00-2) Incompatible with concentrated mineral acids, acid catalysts, acid oxidizing agents, phenols, reactive metals.

SELENIC ANHYDRIDE (13768-86-0) Water contact causes formation of selenic acid. Attacks most metals.

SELENINYL CHLORIDE (7791-23-3) Reacts violently with antimony, cesium, francium, lithium, metal oxides, potassium, sodium, white phosphorus, reducing agents.

SELENIOUS ANHYDRIDE (7446-08-4) An oxidizer; reacts with reducing agents, combustible materials, acids. Attacks most metals.

SELENIUM CHLORIDE OXIDE (7791-23-3) Reacts violently with antimony, cesium, francium, lithium, metal oxides, potassium, sodium, white phosphorus, reducing agents.

SELENIUM DIOXIDE (7446-08-4) An oxidizer; reacts with reducing agents, combustible materials, acids. Attacks most metals.

SELENIUM(+4) DIOXIDE(1:2) or SELENIUM(IV) DIOXIDE(1:2) (7446-08-4) An oxidizer; reacts with reducing agents, combustible materials, acids. Attacks most metals.

SELENIUM HYDRIDE (7783-07-5) Flammable gas; forms explosive mixture with air. A strong reducing agent; reacts violently with oxidizers, combustible materials. Incompatible with acids, water, halogenated hydrocarbons; may cause fire and explosions.

SELENIUM OXIDE (7446-08-4) An oxidizer; reacts with reducing agents, combustible materials, acids. Attacks most metals.

SELENIUM(6+) OXIDE or SELENIUM(VI) OXIDE (13768-86-0) Water contact causes formation of selenic acid. Attacks most metals.

SELENIUM OXYCHLORIDE (7791-23-3) Reacts violently with antimony, cesium, francium, lithium, metal oxides, potassium, sodium, white phosphorus, reducing agents.

SELENIUM TRIOXIDE (13768-86-0) Water contact causes formation of selenic acid. Attacks most metals.

SELEPHOS (56-38-2) Combustible liquid. Mixtures with endrin may be explosive. Strong oxidizers may cause fire and explosions. Attacks some plastics, rubber, and coatings.

SENECA OIL (8002-05-9) Oxidizers may cause fire and explosions. Incompatible with nitric acid. May accumulate static electrical charges, and may cause ignition of its vapors.

SENTRY (7778-54-3) Decomposes in heat or sunlight. Incompatible with acids, moisture, reducing agents, combustible materials, all other chemicals, especially acetylene, aniline and all other amines, anthracene, carbon tetrachloride, iron oxide, manganese oxide, mercaptans, diethylene glycol monomethyl ether, nitromethane, organic matter, organic sulfides, phenol, 1-propanethiol, propyl mercaptan, sulfur, organic sulfur compounds.

SENTRY GRAIN PRESERVER (79-09-4) Forms explosive mixture with air (flash point 126°F/52°C). Incompatible with sulfuric acid, strong bases, ammonia, aliphatic amines, alkanolamines, isocyanates, alkylene oxides, epichlorohydrin, oxidizers.

SEPTENE (63-25-2) Incompatible with strong oxidizers, strongly alkaline pesticides.

SEQ-100 (60-00-4) Incompatible with sulfuric acid, bases, ammonia, aliphatic amines, alkanolamines, isocyanates, alkylene oxides, epichlorohydrin.

SEQUESTRENE AA (60-00-4) Incompatible with sulfuric acid, bases, ammonia, aliphatic amines, alkanolamines, isocyanates, alkylene oxides, epichlorohydrin.

SEQUESTRIC ACID (60-00-4) Incompatible with sulfuric acid, bases, ammonia, aliphatic amines, alkanolamines, isocyanates, alkylene oxides, epichlorohydrin.

SEQUESTROL (60-00-4) Incompatible with sulfuric acid, bases, ammonia, aliphatic amines, alkanolamines, isocyanates, alkylene oxides, epichlorohydrin.

SESONE (136-78-7) Contact with strong oxidizers may cause fire and explosions.

SEVIMOL (63-25-2) Incompatible with strong oxidizers, strongly alkaline pesticides.

SEVIN (63-25-2) Incompatible with strong oxidizers, strongly alkaline pesticides.

SEWER GAS (7783-06-4) A highly flammable and reactive gas. Incompatible with acetaldehyde, barium pentafluoride, chlorine monoxide, chlorine trifluoride, chromic anhydride, copper, lead dioxide, nitric acid, nitrogen iodide, nitrogen trichloride, nitrogen trifluoride, oxygen difluoride, oxidizers, phenyl diazonium chloride, sodium, sodium peroxide. Reacts with alkali metals and alkali earth metals. Attacks metals.

SEXTONE (108-94-1) Forms explosive mixture with air (flash point 111°F/44°C). May accumulate static electrical charges, and may cause ignition of its vapors. Attacks red metals and lead. Incompatible with amines, sulfuric acid, nitric acid, oxidizers, strong acids, aliphatic amines. Dissolves plastics, resins, and rubber.

SEXTONE B (108-87-2) Forms explosive mixture with air (flash point 25°F/−3.9°C). Strong oxidizers may cause fire and explosions. Attacks some plastics, rubber, and coatings.

SF 60 (121-75-5) Incompatible with strong oxidizers, magnesium, alkaline pesticides. Attacks metals, some plastics, rubber, and coatings.

SFA (62-74-8) Incompatible with strong acids.

SHED-A-LEAF (7775-09-9) A powerful oxidizer; reacts violently with reducing agents and combustible matter. Explosions may be caused by contact with ammonia salts, carbon, oils, metal sulfides, nitrobenzene, powdered metals, sugar. Contact with strong acids produces carbon dioxide. Forms shock-sensitive mixtures with some organic materials. Solution (50%) decomposes at 300°F/149°C, liberating oxygen.

SHED-A-LEAF 'L' (7775-09-9) A powerful oxidizer; reacts violently with reducing agents and combustible matter. Explosions may be caused by contact with ammonia salts, carbon, oils, metal sulfides, nitrobenzene, powdered metals, sugar. Contact with strong acids produces carbon dioxide. Forms shock-sensitive mixtures with some organic materials. Solution (50%) decomposes at 300°F/149°C liberating oxygen.

SHELL 5520 (9003-07-0) Incompatible with strong acids, strong oxidizers, chlorine, potassium permanganate.

SHELL CHARCOAL activated (64365-11-3); **purified** (7440-44-0) Incompatible with strong oxidizers, strong acids, oxides, unsaturated oils.

SHELL MIBK (108-10-1) Forms explosive mixture with air (flash point 64°F/18°C). Incompatible with strong oxidizers, strong acids, aliphatic amines. Dissolves some plastics, resins, and rubber.

SHELL SD-3562 (141-66-2) Corrosive to cast iron, mild steel, brass, and stainless steel 304.

SHELL SD-5532 (57-74-9) Contact with strong oxidizers may cause fire and explosions. Attacks some plastics, rubber, and coatings.

SHELLSOL 140 (111-84-2) Forms explosive mixture with air (flash point 88°F/31°C). Strong oxidizers may cause fire and explosions. May accumulate static electrical charges, and may cause ignition of its vapors.

SHELL UNKRAUTTED A (107-18-6) May form unstable and explosive peroxides. Able to polymerize. May accumulate static electrical charges, and may cause ignition of its vapors. A strong reducing agent; reacts violently with oxidizers. Forms explosive mixture with air (flash point 70°F/21°C). Incompatible with strong acids, amines, isocyanates, carbon tetrachloride, chlorosulfonic acid, diallyl phosphide, oleum, sodium hydroxide, tri-n-bromomelamine, metal halides, caustic soda, sodium, and magnesium, aluminum, and their alloys. Attacks some coatings, some plastics, and rubber.

SHINNIPPON FAST RED GG BASE (100-01-06) Strong oxidizers and moisture may cause spontaneous heating. Caustics and heat may cause formation of explosive mixtures. Attacks some plastics, rubber, and coatings. May accumulate static electrical charges, and may cause ignition of its vapors.

SIARKI DWUTLENEK (Polish) (7446-09-5) Produces corrosive hydrochloric acid fumes with air. Contact with copper, bronze, or alkali metals may cause fire and explosions. Reacts violently with alcohols, caustics, amines, water. Decomposes above 140°F/160°C; forms toxic and corrosive oxides of sulfur. Attacks some plastics, rubber, and coatings.

SIARKOWODOR (Polish) (7783-06-4) A highly flammable and reactive gas. Incompatible with acetaldehyde, barium pentafluoride, chlorine monoxide, chlorine trifluoride, chromic anhydride, copper, lead dioxide, nitric acid, nitrogen iodide, nitrogen trichloride, nitrogen trifluoride, oxygen difluoride, oxidizers, phenyl diazonium chloride, sodium, sodium peroxide. Reacts with alkali metals and alkali earth metals. Attacks metals.

SICOL 160 (85-68-7) Incompatible with strong acids, nitrates, oxidizers.

SICOL 250 (103-23-1) Incompatible with strong acids, nitrates, oxidizers.

736

SIENNA (1309-37-1) Contact with hydrogen peroxide, ethylene oxide, calcium hypochlorite will cause explosion. Reacts violently with powdered aluminum, hydrazine, hydrogen trisulfide.

SIGNAL OIL (N/A) Oxidizers may cause fire and explosions. Incompatible with nitric acid. May accumulate static electrical charges, and may cause ignition of its vapors.

SILANE, CHLOROTRIMETHYL- (75-77-4) Forms explosive mixture with air (flash point −18°F/−28°C). Reacts violently with water. Vigorous reaction with aluminum. Moisture and air contact produce hydrochloric acid.

SILANE, DICHLORODIMETHYL- (75-78-5) Forms explosive gas mixture with air (flash point 15°F/−9°C). Water (vapor) or air contact produces hydrochloric acid. Incompatible with acetone, amines, ammonia, alcohols, strong oxidizers, caustics. Attacks most metals.

SILANE, DICHLORODIPHENYL- (80-10-4) Water contact produces hydrochloric acid. Strong oxidizers may cause fire and explosions. Corrodes metals in the presence of moisture.

SILANE, DICHLOROETHYL- (1789-58-8) Forms explosive mixture with air (flash point 30°F/1°C). Water contact produces hydrochloric acid. Incompatible with strong bases. Attacks most common metals.

SILANE, DICHLOROETHYLPHENYL- (1125-27-5) Forms explosive mixture with air (flash point 150°F/66°C). Incompatible with strong bases, water (producing hydrochloric acid). Corrodes most metals. Attacks some plastics, rubber, and coatings.

SILANE, DICHLOROMETHYL- (75-54-7) Forms explosive mixture with air (flash point 15°F/−9°C). Water and air contact causes formation of corrosive fumes. Incompatible with strong bases and oxidizers. Corrodes most metals. Attacks some plastics, rubber, and coatings.

SILANE, DODECYLTRICHLORO- (4484-7-4) Forms explosive mixture with air (flash point 150°F/66°C). Water and air contact causes formation of hydrochloric acid fumes. Incompatible with strong bases and oxidizers. Corrodes metals.

SILANE, METHYLTRICHLORO- (75-79-6) Forms explosive gas mixture with air (flash point 15°F/−9°C). Water or air contact produces hydrochloric acid. Incompatible with strong oxidizers (with fire and explosions), bases. Corrodes most metals. Attacks some plastics, rubber, and coatings.

SILANE, PENTYLTRICHLORO- (107-72-2) Forms explosive mixture with air (flash point 145°F/62°C). Water contact produces hydrochloric acid. Incompatible with strong bases. Attacks metal in presence of moisture.

SILANE, TRICHLOROALLYL- (107-37-9) Forms explosive mixture with air (flash point 95°F/35°C). Reacts violently with water, producing hydrochloric acid. Attacks metals in presence of moisture.

SILANE, TRICHLORODODECYL- (4484-7-4) Forms explosive mixture with air (flash point 150°F/66°C). Water and air contact produces hydrochloric acid fumes. Incompatible with strong bases and oxidizers. Corrodes metals.

SILANE, TRICHLOROETHYL- (115-1-9) Forms explosive mixture with air (flash point 72°F/22°C). Water contact produces hydrochloric acid. Strong oxidizers may cause fire and explosions. Corrodes most common metals.

SILANE, TRICHLOROMETHYL- (75-79-6) Forms explosive gas mixture with air (flash point 15°F/−9°C). Water or air contact produces hydrochloric acid. Incompatible with strong oxidizers (with fire and explosions), bases. Corrodes most metals. Attacks some plastics, rubber, and coatings.

SILANE, TRICHLOROPENTYL- (107-72-2) Forms explosive mixture with air (flash point 145°F/62°C). Water contact produces hydrochloric acid. Incompatible with strong bases. Attacks metal in presence of moisture.

SILANE, TRICHLORO-2-PROPENYL- (107-37-9) Forms explosive mixture with air (flash point 95°F/35°C). Reacts violently with water producing hydrochloric acid. Attacks metal in presence of moisture.

SILANE, TRIMETHYLCHLORO- (75-77-4) Forms explosive mixture with air (flash point −18°F/−28°C). Reacts violently with water. Reacts vigorously reaction with aluminum. Moisture and air contact produces hydrochloric acid.

SILANE, VINYL TRICHLORO 1-150 (75-94-5) Incompatible with water, (violent reaction). Corrodes common metals and produces hydrogen.

SILBERNITRAT (German) (7761-88-8) Contact with acetylene produces shock-sensitive material. Ammonia contact causes formation of compounds that are explosive when dry. Hydrogen peroxide causes violent decomposition to oxygen gas. Attacks some plastics, rubber, and coatings.

SILIBOND (78-10-4) Forms explosive mixture with air (flash point 99°F/37°C). Incompatible with strong oxidizers, strong acids, water (produces volatile ethyl alcohol). Attacks some plastics and rubber.

SILICANE, CHLOROTRIMETHYL- (75-77-4) Forms explosive mixture with air (flash point −18°F/−28°C). Reacts violently with water. Reacts vigorously with aluminum. Moisture and air contact produces hydrochloric acid.

SILICANE, TRICHLOROETHYL- (115-1-9) Forms explosive mixture with air (flash point 72°F/22°C). Water contact produces hydrochloric acid. Strong oxidizers may cause fire and explosions. Corrodes most common metals.

SILICATE d'ETHYLE (French) (78-10-4) Forms explosive mixture with air (flash point 99°F/37°C). Incompatible with strong oxidizers, strong acids, water (produces volatile ethyl alcohol). Attacks some plastics and rubber.

SILICIC ACID TETRAETHYL ESTER (78-10-4) Forms explosive mixture with air (flash point 99°F/37°C). Incompatible with strong oxidizers, strong acids, water (produces volatile ethyl alcohol). Attacks some plastics and rubber.

SILICOCHLOROFORM (10025-78-5) Forms explosive mixture with air (flash point 7°F/-14°C). Water contact produces hydrogen chloride fumes. Incompatible with strong oxidizers, acids. Corrodes metals in the presence of moisture.

SILICOFLUORIC ACID (16961-83-4) Incompatible with aliphatic amines, alkanolamines, alkylene oxides, aromatic amines, amides, ammonia, ammonium hydroxide, bases, calcium oxide, epichlorohydrin, isocyanates, oleum, organic anhydrides, sulfuric acid, strong oxidizers, vinyl acetate, water.

738

SILICON CHLORIDE (10026-04-7) Water contact produces hydrochloric acid. Reacts violently with potassium or sodium. Attacks metals in the presence of moisture.

SILICON OXIDE (10097-28-6) May self-ignite in air. A strong reducing agent; reacts violently with strong oxidizers, combustible materials, acids.

SILICON(2+) OXIDE or SILICON(II) OXIDE (10097-28-6) May self-ignite in air. A strong reducing agent; reacts violently with strong oxidizers, combustible materials, acids.

SILICON TETRACHLORIDE (10026-04-7) Water contact produces hydrochloric acid. Reacts violently with potassium or sodium. Attacks metals in the presence of moisture.

SILIWAX (119-36-8) Forms explosive mixture with air (flash point 205°F/96°C). Incompatible with strong acids, nitrates, oxidizers.

SILVANOL (58-89-9) Not combustible, but may be dissolved in a combustible solvent. If solvent comes in contact with oxidizers, fire and explosions may result.

SILVER NITRATE (7761-88-8) Contact with acetylene produces shock-sensitive material. Ammonia contact forms compounds that are explosive when dry. Hydrogen peroxide causes violent decomposition to oxygen gas. Attacks some plastics, rubber, and coatings.

SILVER ACETATE (563-63-3) An oxidizer; reacts with reducing agents, combustibles, acids, organic materials.

SILVER DIFLUORIDE (7783-95-1) Contact with acetylene produces shock-sensitive material. Ammonia contact produces compounds that are explosive when dry. Hydrogen peroxide causes violent decomposition to oxygen gas. Soluble silver compounds attack some forms of plastics, rubber, and coatings.

SILVER FLUORIDE (7783-95-1) Contact with acetylene produces shock-sensitive material. Ammonia contact produces compounds that are explosive when dry. Hydrogen peroxide causes violent decomposition to oxygen gas. Soluble silver compounds attack some forms of plastics, rubber, and coatings.

SILVER(2+) FLUORIDE or SILVER(II) FLUORIDE (7783-95-1) Contact with acetylene produces shock-sensitive material. Ammonia contact produces compounds that are explosive when dry. Hydrogen peroxide causes violent decomposition to oxygen gas. Soluble silver compounds attack some forms of plastics, rubber, and coatings.

SILVER NITRATE (7761-88-8) Contact with acetylene produces shock-sensitive material. Ammonia contact produces compounds that are explosive when dry. Hydrogen peroxide causes violent decomposition to oxygen gas. Attacks some plastics, rubber, and coatings.

SILVER(1+) NITRATE or SILVER(I) NITRATE (7761-88-8) Contact with acetylene produces shock-sensitive material. Ammonia contact produces compounds that are explosive when dry. Hydrogen peroxide causes violent decomposition to oxygen gas. Attacks some plastics, rubber, and coatings.

SILVER(1+) NITRATE(1:1) or SILVER(I) NITRATE(1:1) (7761-88-8) Contact with acetylene produces shock-sensitive material. Ammonia contact produces compounds that are explosive when dry. Hydrogen peroxide causes violent decomposition to oxygen gas. Attacks some plastics, rubber, and coatings.

SILVER OXIDE (20667-12-3) A strong oxidizer; reacts with reducing agents, acids, amines, combustible materials. Reacts violently with ammonia.

SILVER(1+) OXIDE or SILVER(I) OXIDE (20667-12-3) A strong oxidizer; reacts with reducing agents, acids, amines, combustible materials. Reacts violently with ammonia.

SILVEX or SILVEX HERBICIDE (93-72-1) A powerful oxidizer; reacts violently with reducing agents, combustibles.

SILVEX, 2,4,5-TP (93-72-1) A powerful oxidizer; reacts violently with reducing agents, combustibles.

SILVIRHAP (93-72-1) A powerful oxidizer; reacts violently with reducing agents, combustibles.

SILVISAR (124-65-2) Corrodes common metals.

SILVISAR 510 (75-60-5) Aqueous solution reacts with chemically active metals. Incompatible with sulfuric acid, caustics, ammonia, amines, isocyanates, alkylene oxides, epichlorohydrin.

SIMPLA (2893-78-9) Contact with water causes the release of chlorine gas. A powerful oxidizer; reacts violently with reducing agents, organic matter, easily chlorinated or oxidized materials. Incompatible with ammonium salts, amines. Reacts explosively with sodium hypochlorite and water.

SINAFID M-48 (298-00-0) Mixtures with magnesium may be explosive.

SINITUHO (87-86-5) Hot water causes decomposition, producing hydrochloric acid. Strong oxidizers may cause fire and explosions.

SINOX (534-52-1) Incompatible with heat and strong oxidizers. Dust contact can cause formation of explosive mixtures with air.

SIONIT (50-70-4) Incompatible with strong acids, bases, aliphatic amines, isocyanates, strong oxidizers.

SIONON (50-70-4) Incompatible with strong acids, bases, aliphatic amines, isocyanates, strong oxidizers.

SIPOL L8 (111-87-5) Forms explosive mixture with air (flash point 178°F/81°C). Incompatible with strong acids, caustics, aliphatic amines, isocyanates, strong oxidizers.

SIPOL L10 (112-30-1) Forms explosive mixture with air (flash point 180°F/82°C). Incompatible with strong acids, caustics, aliphatic amines, isocyanates, strong oxidizers.

SIPOL L12 (112-53-8) Incompatible with strong acids, caustics, aliphatic amines, isocyanates, oxidizers.

SIPTOX 1 (121-75-5) Incompatible with strong oxidizers, magnesium, alkaline pesticides. Attacks metals, some plastics, rubber, and coatings.

SIRLENE (57-55-7) Incompatible with strong acids, caustics, aliphatic amines, isocyanates, strong oxidizers (violent reaction).

SIRNIK FOSFORECNY (Czech) (1314-80-3) Incompatible with water, alcohols and strong oxidizers; may cause fire and explosions.

SKDN (63394-00-3) Forms explosive mixture with air (flash point 104°F/40°C). Oxidizers may cause fire and explosions. Incompatible with nitric acid. May accumulate static electrical charges, and may cause ignition of its vapors.

SKELLY-SOLVE-B (110-54-3) Forms explosive mixture with air (flash point −7°F/−22°C). Strong oxidizers may cause fire and explosions. Attacks some plastics, rubber, and coatings. May accumulate static electrical charges, and may cause ignition of its vapors.

SKELLY-SOLVE C (142-82-5) Forms explosive mixture with air (flash point 25°F/−4°C). Strong oxidizers may cause fire and explosions. Attacks some plastics, rubber, and coatings. May accumulate static electric charges that can ignite its vapors.

SLAKED LIME (1305-62-0) Contact with maleic anhydride, phosphorus, nitroethane, nitromethane, nitroparaffins or nitropropane, phosphorus may cause explosion.

SLIMICIDE (107-02-8) Forms explosive mixture with air (flash point −15°F/−26°C). Unstable and very reactive. In storage can form heat and shock-sensitive compounds. Unless inhibited (usually with hydroquinone), readily forms explosive peroxides. Able to polymerize. A strong reducing agent. Reacts violently with oxidizers, strong acids, caustics, amines, 2-aminoethanol, ammonia, ammonium hydroxide, ethylene diamine, ethyleneimine, hydroxides, metal salts, oxidizers, sulfur dioxide, thiourea. Attacks metals: cadmium and zinc. May accumulate static electrical charges, and may cause ignition of its vapors.

SLO-GRO (123-33-1) Contact with strong oxidizers may cause fire and explosions.

SLOW CURING ASPHALT (8052-42-4) Incompatible with nitric acid, fluorine, strong oxidizers.

SLOW-FE (7720-78-7) Aqueous solution is acidic. Contact with bases form iron.

SMEESANA (86-88-4) Strong oxidizers may cause fires and explosion. Also reacts with silver nitrate.

SMUT-GO (118-74-1) Reacts violently with strong oxidizers, formyldimethylamine.

SNIECIOTOX (118-74-1) Reacts violently with strong oxidizers, formyldimethylamine.

SNOMELT (10043-52-4) Incompatible with water, bromine trifluoride, 2-furan, percarboxylic acid. Attacks metals.

SNP (56-38-2) Combustible liquid. Mixtures with endrin may be explosive. Strong oxidizers may cause fire and explosions. Attacks some plastics, rubber, and coatings.

SOBITOL (119-36-8) Forms explosive mixture with air (flash point 205°F/96°C). Incompatible with strong acids, nitrates, oxidizers.

SOBUTANOIC ACID (79-31-2) Forms explosive mixture with air (flash point 132°F/56°C). Incompatible with sulfuric acid, caustics, ammonia, amines, isocyanates, alkylene oxides, epichlorohydrin. Attacks aluminum and other metals.

SODA ASH (497-19-8) Aqueous solution is a strong base. Incompatible with acids, aluminum, organic anhydrides, isocyanates, alkylene oxides, epichlorohydrin, aldehydes, alcohols, glycols, phenols, cresols, caprolactam solution. Attacks metals.

SODA CHLORATE (7775-09-9) A powerful oxidizer; reacts violently with reducing agents and combustible matter. Explosions may be caused by contact with ammonia salts, carbon, oils, metal sulfides, nitrobenzene, powdered metals, sugar. Contact with strong acids produces carbon dioxide. Forms shock-sensitive mixtures with some organic materials. Solution (50%) decomposes at 300°F/149°C, liberating oxygen.

SODA CHLORIC ACID, SODIUM SALT (7775-09-9) A powerful oxidizer; reacts violently with reducing agents and combustible matter. Explosions may be caused by contact with ammonia salts, carbon, oils, metal sulfides, nitrobenzene, powdered metals, sugar. Contact with strong acids produces carbon dioxide. Forms shock-sensitive mixtures with some organic materials. Solution (50%) decomposes at 300°F/149°C, liberating oxygen.

SODA LYE (1310-73-2) Contact with water, acids, flammable liquids, and organic halogens, especially trichloroethylene, may cause fires and explosions. Contact with metals such as aluminum, tin, and zinc causes corrosion and the formation of flammable hydrogen gas. Contact with nitromethane, and similar nitro compounds produces shock-sensitive salts. Water contact produces heat and corrosive fumes. Attacks some plastics, rubber, and coatings.

SODAMIDE (7782-92-5) Reacts violently with water, and may burst into flames; produces heat and caustic soda solution. May self-ignite in moist air, or produce corrosive hydrogen chloride and ammonia fumes. Friction or heat may cause substance to ignite or explode. Reacts violently with oxidizers, acids, sodium nitrite, and other compounds. May accumulate static electrical charges, and may cause ignition of its vapors.

SODAMIDE (26628-22-8) Contact with moisture may cause explosion. May self-ignite in moist air. Able to form unstable peroxides. A strong base; incompatible with acids, organic anhydrides, halogenated hydrocarbons, isocyanates, alkylene oxides, epichlorohydrin, aldehydes, alcohols, glycols, phenols, cresols, caprolactam solution, oxidizers.

SODA NITER (7631-99-4) Powerful oxidizer; reacts violently with reducing agents, combustible substances, strong acids, organic materials, powdered metals, bitumens.

SODIO (CLORATO DI) (ITALIAN) (7775-09-9) A powerful oxidizer; reacts violently with reducing agents and combustible matter. Explosions may be caused by contact with ammonia salts, carbon, oils, metal sulfides, nitrobenzene, powdered metals, sugar. Contact with strong acids produces carbon dioxide. Forms shock-sensitive mixtures with some organic materials. Solution (50%) decomposes at 300°F/149°C, liberating oxygen.

SODIO (DICROMATO di) (Italian) (10588-01-9) Aqueous solution is corrosive. A strong oxidizer; reacts violently with reducing agents, acids, acetic anhydride, combustibles. Attacks copper, zinc, tin, brass, bronze.

SODIO (IDROSSIDO di) (Italian) (1310-73-2) Contact with water, acids, flammable liquids, and organic halogens, especially trichloroethylene, may cause fires and explosions. Contact with metals such as aluminum, tin, and zinc causes corrosion and the formation of flammable hydrogen gas. Contact with nitromethane, and similar nitro compounds, produces shock-sensitive salts. Water contact produces heat and corrosive fumes. Attacks some plastics, rubber, and coatings.

SODIUM (7440-23-5) A strong reducing agent. Water contact produces hydrogen gas, corrosive sodium hydroxide fumes, a caustic soda solution, and heat; fire often occurs. Forms unstable peroxides with moist air. May ignite spontaneously in moist air. Reacts violently with oxidizers, acids, air, carbon dioxide, carbon tetrachloride, chloroform, haloids, halogens, halogenated hydrocarbons, heavy metal oxides, methyl chloride, moisture, nitrous oxide, tetrachloroethane, trichloroethylene, water, and many other substances. Separate from all other materials.

SODIUM ACID BISULFITE or SODIUM ACID BISULFITE SOLUTION (7631-90-5) Slowly oxidized to the sulfate on contact with air. Oxidizers or acids causes formation of sulfur dioxide gas. Attacks metals.

SODIUM ACID FLUORIDE (1333-83-1) Aqueous solution is highly acidic; corrodes glass, concrete, and metals containing silica, such as cast iron. Incompatible with sulfuric acid, caustics, ammonia, amines, organic anhydrides, isocyanates, vinyl acetate, alkylene oxides, epichlorohydrin. Attacks natural rubber and organic materials.

SODIUM ACID SULFITE (7631-90-5) Slowly oxidized to the sulfate on contact with air. Oxidizers or acids cause formation of sulfur dioxide gas. Attacks metals.

SODIUM ALUMINATE or SODIUM ALUMINATE SOLUTION (45% OR LESS) (1302-42-7) Incompatible with acids, organic anhydrides, isocyanates, alkylene oxides, epichlorohydrin, aldehydes, alcohols, glycols, caprolactam, chlorocarbons. Reacts with copper, tin, aluminum, and zinc.

SODIUM AMIDE (7782-92-5) Reacts violently with water, and may burst into flames; produces heat and caustic soda solution. May self-ignite in moist air, or produce corrosive hydrogen chloride and ammonia fumes. Friction or heat may cause substance to ignite or explode. Reacts violently with oxidizers, acids, sodium nitrite, and other compounds. May accumulate static electrical charges, and may cause ignition of its vapors.

SODIUM AZIDE (26628-22-8) Incompatible with acids, some metals (i.e, lead, copper, silver, mercury). Produces explosion-sensitive compounds.

SODIUM, AZOTURE de (French) (26628-22-8) Incompatible with acids, some metals (i.e, lead, copper, silver, mercury). Produces explosion-sensitive compounds.

SODIUM AZOTURO (Italian) (26628-22-8) Incompatible with acids, some metals (i.e, lead, copper, silver, mercury). Produces explosion-sensitive compounds.

SODIUM-2-BENZOTHIAZOLETHIOATE (2492-26-4) Incompatible with oxidizers. Concentrated aqueous solutions are corrosive.

SODIUM BIBORATE (1303-96-4) Reacts with acids, metallic salts.

SODIUM BICHROMATE (10588-01-9) Aqueous solution is corrosive. A strong oxidizer; reacts violently with reducing agents, acids, acetic anhydride, combustibles. Attacks copper, zinc, tin, brass, bronze.

SODIUM BIFLUORIDE (1333-83-1) Aqueous solution is highly acidic; corrodes glass, concrete, and metals containing silica, such as cast iron. Incompatible with sulfuric acid, caustics, ammonia, amines, organic anhydrides, isocyanates, vinyl acetate, alkylene oxides, epichlorohydrin. Attacks natural rubber and organic materials.

SODIUM BISULFIDE (16721-80-5) **Flammable solid** (flash point, 45% solution: 73°F/23°C). Moisture may cause self-ignition or produce heat sufficient to ignite combustible materials. Incompatible with acids, amides, organic anhydrides, isocyanates, alkylene oxides, epichlorohydrin, aldehydes, alcohols, glycols, phenols, cresols, caprolactam solution, oxidizers. Corrodes steel above 150°F/66°C, and aluminum.

SODIUM BISULFITE or SODIUM BISULFITE, SOLID or SODIUM BISULFITE, SOLUTION (7631-90-5) Slowly oxidized to the sulfate on contact with air. Oxidizers or acids cause formation of sulfur dioxide gas. Attacks metals.

SODIUM BORATE (1303-96-4) Reacts with acids, metallic salts.

SODIUM BOROHYDRIDE (16940-66-2) Produces hydrogen gas on contact with water or moist air. Reacts violently with oxidizers, dimethyl formamide, aldehydes, acids (forms diborane gas), ketones, metal powders. Polymerization may be caused by contact with bases, acrylonitrile. Corrodes glass slowly.

SODIUM BROMATE (7789-38-0) A strong oxidizer; reacts violently with strong reducing agents, combustible materials, organic substances, aluminum powder, copper powder, grease or oily substances, phosphorus, metal sulfides.

SODIUM CACODYLATE (124-65-2) Corrodes common metals.

SODIUM CARBONATE (2:1) (497-19-8) Aqueous solution is a strong base. Incompatible with acids, aluminum, organic anhydrides, isocyanates, alkylene oxides, epichlorohydrin, aldehydes, alcohols, glycols, phenols, cresols, caprolactam solution. Attacks metals.

SODIUM CHLORATE (7775-09-9) A powerful oxidizer; reacts violently with reducing agents and combustible matter. Explosions may be caused by contact with ammonia salts, carbon, oils, metal sulfides, nitrobenzene, powdered metals, sugar. Contact with strong acids produces carbon dioxide. Forms shock-sensitive mixtures with some organic materials. Solution (50%) decomposes at 300°F/149°C liberating oxygen.

SODIUM (CHLORATE de) (French) (7775-09-9) A powerful oxidizer; reacts violently with reducing agents and combustible matter. Explosions may be caused by contact with ammonia salts, carbon, oils, metal sulfides, nitrobenzene, powdered metals, sugar. Contact with strong acids produces carbon dioxide. Forms shock-sensitive mixtures with some organic materials. Solution (50%) decomposes at 300°F/149°C, liberating oxygen.

SODIUM CHLORITE (7758-19-2) A strong oxidizer; reacts violently with reducing agents, combustible materials, organic substances. Friction, shock, heat or contact with acids, oils, and grease and many other materials may cause explosion.

SODIUM CHROMATE (10588-01-9) Aqueous solution is corrosive. A strong oxidizer; reacts violently with reducing agents, acids, acetic anhydride, combustibles. Attacks copper, zinc, tin, brass, bronze.

SODIUM CHROMATE (7777-11-3) Strong oxidizer; reacts violently with reducing agents, combustibles, strong acids, organic materials.

SODIUM CHROMATE(VI) (7777-11-3) Strong oxidizer; reacts violently with reducing agents, combustibles, strong acids, organic materials.

SODIUM CYANIDE (143-33-9) Incompatible with acids, organic anhydrides, isocyanates, alkylene oxides, epichlorohydrin, aldehydes, alcohols, glycols, phenols, cresols, caprolactam solution, strong oxidizers, nitrates, nitrites. Attacks aluminum, copper, zinc.

SODIUM DICHLORISOCYANURATE (2893-78-9) Contact with water causes the release of chlorine gas. A powerful oxidizer; reacts violently with reducing agents, organic matter, easily chlorinated or oxidized materials. Incompatible with ammonium salts, amines. Explosive reaction with sodium hypochlorite and water.

SODIUM DICHLOROISOCYANURATE (2893-78-9) Contact with water causes the release of chlorine gas. A powerful oxidizer; reacts violently with reducing agents, organic matter, easily chlorinated or oxidized materials. Incompatible with ammonium salts, amines. Reacts explosively with sodium hypochlorite and water.

SODIUM 2-(2,4-DICHLOROPHENOXY)ETHYL SULFATE (136-78-7) Contact with strong oxidizers may cause fire and explosions.

SODIUM-1,3-DICHLORO-1,3,5-TRIAZINE-2,4-DIONE-6-OXIDE (2893-78-9) Contact with water causes the release of chlorine gas. A powerful oxidizer; reacts violently with reducing agents, organic matter, easily chlorinated or oxidized materials. Incompatible with ammonium salts, amines. Explosive reaction with sodium hypochlorite and water.

SODIUM DICHLORO-*s*-TRIAZINETRIONE DIHYDRATE (2893-78-9) Contact with water causes the release of chlorine gas. A powerful oxidizer; reacts violently with reducing agents, organic matter, easily chlorinated or oxidized materials. Incompatible with ammonium salts, amines. Explosive reaction with sodium hypochlorite and water.

SODIUM DICHLORO-*s*-TRIAZINETRIONE (2893-78-9) Contact with water causes the release of chlorine gas. A powerful oxidizer; reacts violently with reducing agents, organic matter, easily chlorinated or oxidized materials. Incompatible with ammonium salts, amines. Explosive reaction with sodium hypochlorite and water.

1-SODIUM-3,5-DICHLORO-1,3,5-TRIAZINE-2,4,6-TRIONE (2893-78-9) Contact with water causes the release of chlorine gas. A powerful oxidizer; reacts violently with reducing agents, organic matter, easily chlorinated or oxidized materials. Incompatible with ammonium salts, amines. Reacts explosively with sodium hypochlorite and water.

1-SODIUM-3,5-DICHLORO-*s*-TRIAZINE-2,4,6-TRIONE (2893-78-9) Contact with water causes the release of chlorine gas. A powerful oxidizer; reacts violently with reducing agents, organic matter, easily chlorinated or oxidized materials. Incompatible with ammonium salts, amines. Reacts explosively with sodium hypochlorite and water.

SODIUM DICHROMATE (10588-01-9) Aqueous solution is corrosive. A strong oxidizer; reacts violently with reducing agents, acids, acetic anhydride, combustibles. Attacks copper, zinc, tin, brass, bronze.

SODIUM (DICHROMATE de) (French) (10588-01-9) Aqueous solution is corrosive. A strong oxidizer; reacts violently with reducing agents, acids, acetic anhydride, combustibles. Attacks copper, zinc, tin, brass, bronze.

SODIUM DICHROMATE(VI) (10588-01-9) Aqueous solution is corrosive. A strong oxidizer; reacts violently with reducing agents, acids, acetic anhydride, combustibles. Attacks copper, zinc, tin, brass, bronze.

SODIUM DIFLUORIDE (1333-83-1) Aqueous solution is highly acidic; corrodes glass, concrete, and metals containing silica, such as cast iron. Incompatible with sulfuric acid, caustics, ammonia, amines, organic anhydrides, isocyanates, vinyl acetate, alkylene oxides, epichlorohydrin. Attacks natural rubber and organic materials.

SODIUM DIMETHYLARSENATE (124-65-2) Corrodes common metals.

SODIUM DIOXIDE (1313-60-6) Reacts violently with water, producing a strong caustic solution. Aqueous solution is incompatible with acids, organic anhydrides, isocyanates, alkylene oxides, epichlorohydrin, aldehydes, alcohols, glycols, phenols, cresols, caprolactam solution. A powerful oxidizer; reacts violently with reducing agents, combustible materials, organic substances, acids, metal powders.

SODIUM DODECYLBENZENE-SULFONATE (25155-30-0) Reacts with acids, including fumes.

SODIUM FLUOACETIC ACID (62-74-8) Incompatible with strong acids.

SODIUM FLUORIDE (7681-49-4) Aqueous solution is highly corrosive. Reacts with acids.

SODIUM FLUOROACETATE (62-74-8) Incompatible with strong acids.

SODIUM HYDRATE (1310-73-2) Contact with water, acids, flammable liquids, and organic halogens, especially trichloroethylene, may cause fires and explosions. Contact with metals such as aluminum, tin, and zinc causes corrosion and the formation of flammable hydrogen gas. Contact with nitromethane, and similar nitro compounds, produces shock-sensitive salts. Water contact produces heat and corrosive fumes. Attacks some plastics, rubber, and coatings.

SODIUM HYDRATE SOLUTION (1310-73-2) Contact with water, acids, flammable liquids, and organic halogens, especially trichloroethylene, may cause fires and explosions. Contact with metals such as aluminum, tin, and zinc causes corrosion and the formation of flammable hydrogen gas. Contact with nitromethane, and similar nitro compounds, produces shock-sensitive salts. Water contact produces heat and corrosive fumes. Attacks some plastics, rubber, and coatings.

SODIUM HYDRIDE (7646-69-7) Water contact is potentially explosive with release of hydrogen gas. Finely divided form or powders can self-ignite in moist air. Reacts with acetylene, oxygen, halogens, strong oxidizers, sulfur.

SODIUM HYDROGEN DIFLUORIDE (1333-83-1) Aqueous solution is highly acidic; corrodes glass, concrete, and metals containing silica, such as cast iron. Incompatible with sulfuric acid, caustics, ammonia, amines, organic anhydrides, isocyanates, vinyl acetate, alkylene oxides, epichlorohydrin. Attacks natural rubber and organic materials.

SODIUM HYDROGEN FLUORIDE (1333-83-1) Aqueous solution is highly acidic; corrodes glass, concrete, and metals containing silica, such as cast iron. Incompatible with sulfuric acid, caustics, ammonia, amines, organic anhydrides, isocyanates, vinyl acetate, alkylene oxides, epichlorohydrin. Attacks natural rubber and organic materials.

SODIUM HYDROGEN SULFIDE (16721-80-5) **Flammable solid** (flash point, 45% solution: 73°F/23°C). Moisture may cause self-ignition or produce heat sufficient to ignite combustible materials. Incompatible with acids, amides, organic anhydrides, isocyanates, alkylene oxides, epichlorohydrin, aldehydes, alcohols, glycols, phenols, cresols, caprolactam solution, oxidizers. Corrodes steel above 150°F/66°C, and aluminum.

SODIUM HYDROGEN SULFITE or SODIUM HYDROGEN SULFITE, SOLID or SODIUM HYDROGEN SULFITE SOLUTION (7631-90-5) Slowly oxidized to the sulfate on contact with air. Oxidizers or acids cause formation of sulfur dioxide gas. Attacks metals.

SODIUM HYDROSULFIDE or SODIUM HYDROSULFIDE, SOLUTION or SODIUM HYDROSULFIDE, SOLID (16721-80-5) **Flammable solid** (flash point, 45% solution: 73°F/23°C). Moisture may cause self-ignition or produce heat sufficient to ignite combustible materials. Incompatible with acids, amides, organic anhydrides, isocyanates, alkylene oxides, epichlorohydrin, aldehydes, alcohols, glycols, phenols, cresols, caprolactam solution, oxidizers. Corrodes steel above 150°F/66°C, and aluminum.

SODIUM HYDROXIDE CAUSTIC SODA SOLUTION (1310-73-2) Contact with water, acids, flammable liquids, and organic halogens, especially trichloroethylene, may cause fires and explosions. Contact with metals such as aluminum, tin, and zinc causes corrosion and the formation of flammable hydrogen gas. Contact with nitromethane, and similar nitro compounds, produces shock-sensitive salts. Water contact produces heat and corrosive fumes. Attacks some plastics, rubber, and coatings.

SODIUM HYDROXIDE or SODIUM HYDROXIDE, BEAD or SODIUM HYDROXIDE, DRY or SODIUM HYDROXIDE, FLAKE OR SODIUM HYDROXIDE, GRANULAR or SODIUM HYDROXIDE, LIQUID or SODIUM HYDROXIDE, SOLID or SODIUM HYDROXIDE, SOLUTION (1310-73-2) Contact with water, acids, flammable liquids, and organic halogens, especially trichloroethylene, may cause fires and explosions. Contact with metals such as aluminum, tin, and zinc causes corrosion and the formation of flammable hydrogen gas. Contact with nitromethane, and similar nitro compounds, produces shock-sensitive salts. Water contact produces heat and corrosive fumes. Attacks some plastics, rubber, and coatings.

SODIUM (HYDROXYDE de) (French) (1310-73-2) Contact with water, acids, flammable liquids, and organic halogens, especially trichloroethylene, may cause fires and explosions. Contact with metals such as aluminum, tin, and zinc causes corrosion and the formation of flammable hydrogen gas. Contact with nitromethane, and similar nitro compounds, produces shock-sensitive salts. Water contact produces heat and corrosive fumes. Attacks some plastics, rubber, and coatings.

SODIUM HYPOCHLORITE or **SODIUM HYPOCHLORITE SO-LUTION** (7681-52-9) A strong oxidizer and a strong base. Decomposes in sunlight, producing oxygen and increasing the risk of fire. Stability decreases with concentration, heat, light, decrease in pH, and contamination with metals. Incompatible with strong acids, reducing agents, combustible substances; all cause violent reaction, fire, and explosions. Contact with amines and ammonia salts produces explosive chloroamines. Corrodes many metals: steel, 12% and 17% chrome steel, cast iron, monel, aluminum, nickel, brass, bronze, iconel.

SODIUM HYPOSULFITE (10102-17-7) Incompatible with acids, halogenated compounds, oxidizers, metal nitrates, nitrites.

SODIUM ISOTHIOCYANATE (540-72-7) Incompatible with acids, bases, ammonia, amines, amides, alcohols, glycols, caprolactam.

SODIUM MERCAPTAN (16721-80-5) Flammable solid (flash point, 45% solution: 73°F/23°C). Moisture may cause self-ignition or produce heat sufficient to ignite combustible materials. Incompatible with acids, amides, organic anhydrides, isocyanates, alkylene oxides, epichlorohydrin, aldehydes, alcohols, glycols, phenols, cresols, caprolactam solution, oxidizers. Corrodes steel above 150°F/66°C, and aluminum.

SODIUM-2-MERCAPTOBENZOTHIAZOL (2492-26-4) Incompatible with oxidizers. Concentrated aqueous solutions are corrosive.

SODIUM META-ALUMINATE SOLUTION (1302-42-7) Incompatible with acids, organic anhydrides, isocyanates, alkylene oxides, epichlorohydrin, aldehydes, alcohols, glycols, caprolactam, chlorocarbons. Reacts with copper, tin, aluminum and zinc.

SODIUM METABISULFITE (7631-90-5) Slowly oxidized to the sulfate on contact with air. Oxidizers or acids cause formation of sulfur dioxide gas. Attacks metals.

SODIUMMETABORATE PEROXIHYDRATE (1113-47-9) A strong oxidizer; reacts violently with reducing agents, combustible materials, organic substances. Incompatible with acids.

SODIUM METAL (7440-23-5) A strong reducing agent. Water contact produces hydrogen gas, corrosive sodium hydroxide fumes, a caustic soda solution, and heat; fire often occurs. Forms unstable peroxides with moist air. May ignite spontaneously in moist air. Reacts violently with oxidizers, acids, air, carbon dioxide, carbon tetrachloride, chloroform, haloids, halogens, halogenated hydrocarbons, heavy metal oxides, methyl chloride, moisture, nitrous oxide, tetrachloroethane, trichloroethylene, water, and many other substances. Separate from all other materials.

SODIUM, METAL LIQUID ALLOY (7440-23-5) A strong reducing agent. Water contact produces hydrogen gas, corrosive sodium hydroxide fumes, a caustic soda solution and heat; fire often occurs. Forms unstable peroxides with moist air. May ignite spontaneously in moist air. Reacts violently with oxidizers, acids, air, carbon dioxide, carbon tetrachloride, chloroform, haloids, halogens, halogenated hydrocarbons, heavy metal oxides, methyl chloride, moisture, nitrous oxide, tetrachloroethane, trichloroethylene, water, and many other substances. Separate from all other materials.

SODIUM METASILICATE, ANHYDROUS (1344-09-8) Solution is a strong base; reacts with acids, organic anhydrides, alkylene oxides, epichlorohydrin, aldehydes, alcohols, glycols, phenols, cresols, caprolactam solution. Attacks chemically active metals.

SODIUM METHOXIDE (124-41-4) Flammable solid; may ignite spontaneously in moist air. A strong reducing agent; reacts violently with oxidizers, water (ignites), chloroform, chlorine oxyfluoride, aluminum, magnesium, beryllium. Attacks some plastics, rubber, and coatings.

SODIUM METHYLATE (124-41-4) Flammable solid; may ignite spontaneously in moist air. A strong reducing agent; reacts violently with oxidizers, water (ignites), chloroform, chlorine oxyfluoride, aluminum, magnesium, beryllium. Attacks some plastics, rubber, and coatings.

SODIUM MONOFLUOROACETATE (62-74-8) Incompatible with strong acids.

SODIUM MONOHYDRIDE (7646-69-7) Water contact is potentially explosive with release of hydrogen gas. Finely divided form or powders can self-ignite in moist air. Reacts with acetylene, oxygen, halogens, strong oxidizers, sulfur.

SODIUM MONOSULFIDE (1313-82-2) Once moist, can self-ignite upon drying in air. Incompatible with acids, carbon, oxidizers, diazonium salts. Attacks metals.

SODIUM NITRATE (7631-99-4) A powerful oxidizer; reacts violently with reducing agents, combustible substances, strong acids, organic materials, powdered metals, bitumens.

SODIUM NITRATE (1:1) (7631-99-4) A powerful oxidizer; reacts violently with reducing agents, which cause combustion or spontaneous combustion.

SODIUM NITRITE (7632-00-0) A strong oxidizer; reacts violently with reducing agents and combustibles. Reacts with acids, ammonium salts + heat, butadiene, cyanides, lithium, phthalic acid, thiosulfate, sodium amide, and many other substances. Under certain conditions the chemical can also be a strong reducing agent.

SODIUM OF CACODYLIC ACID (124-65-2) Corrodes common metals.

SODIUM OLEATE (143-19-1) Combustible solid. Strong oxidizers may cause fires or explosions.

SODIUM OXYCHLORIDE (7681-52-9) A strong oxidizer and a strong base. Decomposes in sunlight, producing oxygen and increasing the risk of fire. Stability decreases with concentration, heat, light, decrease in pH, and contamination with metals. Incompatible with strong acids, reducing agents, combustible substances all cause violent reaction, fire, and explosions. Contact with amines and ammonia salts produces explosive chloroamines. Corrodes many metals: steel, 12% and 17% chrome steel, cast iron, monel, aluminum, nickel, brass, bronze, iconel.

SODIUM PCP (131-52-2) Contact with strong oxidizers may cause fire and explosions.

SODIUM PENTACHLOROPHENATE (131-52-2) Contact with strong oxidizers may cause fire and explosions.

SODIUM PENTALCHLOROPHENOL (131-52-2) Contact with strong oxidizers may cause fire and explosions.

SODIUM PERBORATE (1113-47-9) A strong oxidizer; reacts violently with reducing agents, combustible materials, organic substances. Incompatible with acids.

SODIUM PEROXIDE (1313-60-6) Reacts violently with water forming a strong caustic solution. Aqueous solution incompatible with acids, organic anhydrides, isocyanates, alkylene oxides, epichlorohydrin, aldehydes, alcohols, glycols, phenols, cresols, caprolactam solution. A powerful oxidizer; reacts violently with reducing agents, combustible materials, organic substances, acids, metal powders.

SODIUM PEROXYDISULFATE (7775-27-2) A strong oxidizer; reacts violently with reducing agents, combustible materials, organic substances.

SODIUM PERSULFATE (7775-27-2) A strong oxidizer; reacts violently with reducing agents, combustible materials, organic substances.

SODIUM PYROBORATE (1303-96-4) Reacts with acids, metallic salts.

SODIUM PYROSULFITE (7631-90-5) Slowly oxidized to the sulfate on contact with air. Oxidizers or acids cause formation of sulfur dioxide gas. Attacks metals.

SODIUM RHODANIDE (540-72-7) Incompatible with acids, bases, ammonia, amines, amides, alcohols, glycols, caprolactam.

SODIUM SALT OF HYDRAZOIC ACID (26628-22-8) Incompatible with acids, some metals (i.e, lead, copper, silver, mercury). Forms explosion-sensitive compounds.

SODIUM SILICATE (1344-09-8) Solution is a strong base; reacts with acids, organic anhydrides, alkylene oxides, epichlorohydrin, aldehydes, alcohols, glycols, phenols, cresols, caprolactam solution. Attacks chemically active metals.

SODIUM SULFHYDRATE (16721-80-5) **Flammable solid** (flash point, 45% solution: 73°F/23°C). Moisture may cause self-ignition or produce heat sufficient to ignite combustible materials. Incompatible with acids, amides, organic anhydrides, isocyanates, alkylene oxides, epichlorohydrin, aldehydes, alcohols, glycols, phenols, cresols, caprolactam solution, oxidizers. Corrodes steel above 150°F/66°C, and aluminum.

SODIUM SULFHYDRATE (7631-90-5) Slowly oxidized to the sulfate on contact with air. Oxidizers or acids cause formation of sulfur dioxide gas. Attacks metals.

SODIUM SULFIDE (1313-82-2) Once moist, can self-ignite upon drying in air. Incompatible with acids, carbon, oxidizers, diazonium salts. Attacks metals.

SODIUM SULFIDE, ANHYDROUS (1313-82-2) Once moist, can self-ignite upon drying in air. Incompatible with acids, carbon, oxidizers, diazonium salts. Attacks metals.

SODIUM SULFITE or SODIUM SULFITE(2:1) or SODIUM SULFITE, ANHYDROUS (7757-83-7) A reducing agent. Incompatible with strong oxidizers, combustibles, acids, organic materials.

SODIUM SULFOCYANATE (540-72-7) Incompatible with acids, bases, ammonia, amines, amides, alcohols, glycols, caprolactam.

SODIUM SULFURET (1313-82-2) Once moist, can self-ignite upon drying in air. Incompatible with acids, carbon, oxidizers, diazonium salts. Attacks metals.

SODIUM SULPHIDE (1313-82-2) Once moist, can self-ignite upon drying in air. Incompatible with acids, carbon, oxidizers, diazonium salts. Attacks metals.

SODIUM TETRABORATE (1303-96-4) Reacts with acids, metallic salts.

SODIUM TETRABORATE, ANHYDROUS (1303-96-4) Reacts with acids, metallic salts.

SODIUM TETRABORATE DECAHYDRATE (1303-96-4) Reacts with acids, metallic salts.

SODIUM TETRAHYDROBORATE(1-) (16940-66-2) Produces hydrogen gas on contact with water or moist air. Reacts violently with oxidizers, dimethyl formamide, aldehydes, acids (produces diborane gas), ketones, metal powders. Polymerization may be caused by contact with bases, acrylonitrile. Corrodes glass slowly.

SODIUM THIOCYANATE (540-72-7) Incompatible with acids, bases, ammonia, amines, amides, alcohols, glycols, caprolactam.

SODIUM THIOSULFATE (10102-17-7) Incompatible with acids, halogenated compounds, oxidizers, metal nitrates, nitrites.

SOFRIL (7704-34-9) Combustible solid. Liquid causes formation of sulfur dioxide with air. Reacts violently with strong oxidizers. Forms explosive, shock-sensitive or pyrophoric mixtures with ammonia, ammonium nitrate, bromates, calcium carbide, charcoal, chlorates, hydrocarbons, iodates, iron. Reacts violently with halogen compounds, sodium, tin, uranium, and other compounds. Attacks steel when moist. May accumulate static electrical charges, and may cause ignition of its vapors.

SOFTIL (119-36-8) Forms explosive mixture with air (flash point 205°F/96°C). Incompatible with strong acids, nitrates, oxidizers.

SOHNHOFEN STONE (1317-65-3) Incompatible with acids, alum, ammonium salts, fluorine.

SOILBROM-40 (106-93-4) Reacts with chemically active metals, liquid ammonia, strong oxidizers. Heat and light cause slow decomposition. Attacks some plastics and rubber.

SOILBROM-85 (106-93-4) Reacts with chemically active metals, liquid ammonia, strong oxidizers. Heat and light cause slow decomposition. Attacks some plastics and rubber.

SOILBROM-90EC (106-93-4) Reacts with chemically active metals, liquid ammonia, strong oxidizers. Heat and light cause slow decomposition. Attacks some plastics and rubber.

SOILBROME-85 (106-93-4) Reacts with chemically active metals, liquid ammonia, strong oxidizers. Heat and light cause slow decomposition. Attacks some plastics and rubber.

SOILFUME (106-93-4) Reacts with chemically active metals, liquid ammonia, strong oxidizers. Heat and light cause slow decomposition. Attacks some plastics and rubber.

SOK (63-25-2) Incompatible with strong oxidizers, strongly alkaline pesticides.

SOLACTOL (97-64-3) Fire or explosion hazard (flash point 115°F/46°C). Strong oxidizers may cause fire and explosions.

SOLAR-40 (25155-30-0) Reacts with acids, including fumes.

SOLAR NITROGEN SOLUTION (15978-77-5) Incompatible with oxidizers, combustibles, organics, acids, amides, organic anhydrides, isocyanates, vinyl acetate, alkylene oxides, epichlorohydrin, aldehydes, sulfur. Reacts violently with copper and copper alloys. Attacks lead, zinc.

751

SOLAR WINTER BAN (57-55-7) Incompatible with strong acids, caustics, aliphatic amines, isocyanates, strong oxidizers (violent reaction).

SOLBAR (7727-43-7) Explosions may result from contact with aluminum in the presence of heat. Incompatible with potassium, phosphorus.

SOLFO BLACK 2B SUPRA (51-28-5) Explosion may be caused by heat, friction or shock. Contact with reducing agents, combustibles may cause fire and explosions. Produces explosive salts with ammonia or strong bases. May accumulate static electrical charges, and may cause ignition of its vapors.

SOLFO BLACK B (51-28-5) Explosion may be caused by heat, friction, or shock. Contact with reducing agents, combustibles may cause fire and explosions. Produces explosive salts with ammonia or strong bases. May accumulate static electrical charges, and may cause ignition of its vapors.

SOLFO BLACK BB (51-28-5) Explosion may be caused by heat, friction or shock. Contact with reducing agents, combustibles may cause fire and explosions. Produces explosive salts with ammonia or strong bases. May accumulate static electrical charges, and may cause ignition of its vapors.

SOLFO BLACK G (51-28-5) Explosion may be caused by heat, friction or shock. Contact with reducing agents, combustibles may cause fire and explosions. Produces explosive salts with ammonia or strong bases. May accumulate static electrical charges, and may cause ignition of its vapors.

SOLFO BLACK SB (51-28-5) Explosion may be caused by heat, friction or shock. Contact with reducing agents, combustibles may cause fire and explosions. Produces explosive salts with ammonia or strong bases. May accumulate static electrical charges, and may cause ignition of its vapors.

SOLFURO di CARBONIO (Italian) (75-15-0) Highly reactive. Contact with many substances can cause fire and explosions. Forms explosive mixture with air (flash point $-22°F/-30°C$). Shock can cause explosive decomposition. Incompatible with alkali metals, aliphatic amines, alkanolamines, aluminum, azides, chlorine monoxide, combustible substances, ethylene diamine, ethyleneimine, lead azide, lithium azide, nitric oxide, nitrogen dioxide, potassium, potassium azide, reducing agents, rubidium azide, sodium azide, zinc.

SOLOZONE (1313-60-6) Reacts violently with water, forming a strong caustic solution. Aqueous solution is incompatible with acids, organic anhydrides, isocyanides, alkylene oxides, epichlorohydrin, aldehydes, alcohols, glycols, phenols, cresols, caprolactam solution. A powerful oxidizer; reacts violently with reducing agents, combustible materials, organic substances, acids, metal powders.

SOLPYRON (50-78-2) Fires and explosions may result from contact with strong oxidizers. Alkali hydroxides or carbonates cause decomposition.

SOLUBLE GLASS (1344-09-8) Solution is a strong base; reacts with acids, organic anhydrides, alkylene oxides, epichlorohydrin, aldehydes, alcohols, glycols, phenols, cresols, caprolactam solution. Attacks chemically active metals.

SOLUSOL-75% (119-36-8) Forms explosive mixture with air (flash point 205°F/96°C). Incompatible with strong acids, nitrates, oxidizers.

SOLUSOL-100% (119-36-8) Forms explosive mixture with air (flash point 205°F/96°C). Incompatible with strong acids, nitrates, oxidizers.

SOLVANOL (84-66-2) Reacts violently with strong acids, strong oxidizers, permanganates, water. Attacks some forms of plastic.

SOLVANOM (131-11-3) Incompatible with strong alkalies, strong acids, nitrates, oxidizers.

SOLVARONE (131-11-3) Incompatible with strong acids, nitrates, oxidizers.

SOLVENT ETHER (60-29-7) Forms explosive mixture with air (flash point −49°F/−45°C). Incompatible with strong acids, strong oxidizers. Can form peroxides with exposure to air or light; may explode when container is unstoppered or otherwise opened. Attacks some plastics, rubber, and coatings. Being a nonconductor, chemical may accumulate static electric charges that may result in ignition of vapor.

SOLVENT NAPHTHA (8032-32-4) Forms explosive mixture with air (flash point −40°F to −86°F/−40°C to −66°C). Incompatible with strong acids, strong oxidizers. Attacks some plastics, rubber, and coatings. May accumulate static electrical charges, and may cause ignition of its vapors.

SOLVIREX (298-04-4) Forms explosive mixture with air (flash point 180°F/82°C). Incompatible with alkalis, strong oxidizers.

SOLVOSOL (111-90-0) Forms explosive mixture with air (flash point 201°F/94°C). Heat or contact with atmospheric air may cause formation of unstable peroxides. Incompatible with strong acids, isocyanates, strong oxidizers.

SOMI-PRONT (67-68-5) Forms explosive mixture with air (flash point 203°F/95°C). Reacts violently with oxidizers. Reacts with ethanoyl chloride, boron compounds, halides, metal alkoxides, oxidizers

SONACIDE (111-30-8) Water contact causes formation of a polymer solution. A strong reducing agent. Incompatible with strong acids, caustics, ammonia, amines, strong oxidizers.

SOPRABEL (7784-40-9) Contact with strong oxidizers may cause fire and explosions.

SOPRATHION (563-12-2) Incompatible with alkaline formulations. Mixtures with magnesium may be explosive.

SOPRATHION (56-38-2) Combustible liquid. Mixtures with endrin may be explosive. Strong oxidizers may cause fire and explosions. Attacks some plastics, rubber, and coatings.

SORBITE (50-70-4) Incompatible with strong acids, bases, aliphatic amines, isocyanates, strong oxidizers.

SORBITOL (50-70-4) Incompatible with strong acids, bases, aliphatic amines, isocyanates, strong oxidizers.

SORBO (50-70-4) Incompatible with strong acids, bases, aliphatic amines, isocyanates, strong oxidizers.

SORBOL (50-70-4) Incompatible with strong acids, bases, aliphatic amines, isocyanates, strong oxidizers.

SOVOL (generic CAS for PCBs 1336-36-3) Incompatible with strong oxidizers, strong acids.

SOY BEAN OIL or SOY OIL or SOYABEAN OIL or SOYBEAN OIL (EPOXIDIZED) (N/A) Incompatible with strong acids, oxidizers.

SPECIAL TERMITE FLUID (95-50-1) Forms explosive mixture with air (flash point 151°F/66°C). Incompatible with strong oxidizers, hot aluminum, or aluminum alloy. Attacks some plastics, rubber, and coatings.

SPECTRACIDE (333-41-5) Incompatible with water, copper-containing compounds, oxidizers, acids, or bases.

SPECULAR IRON (1309-37-1) Contact with hydrogen peroxide, ethylene oxide, calcium hypochlorite will cause explosion. Reacts violently with powdered aluminum, hydrazine, hydrogen trisulfide.

SPENCER 401 (105-60-2) Contact with strong oxidizers may cause fire and explosions.

SPERLOX-S (7704-34-9) Combustible solid. Liquid causes formation of sulfur dioxide with air. Reacts violently with strong oxidizers. Forms explosive, shock-sensitive, or pyrophoric mixtures with ammonia, ammonium nitrate, bromates, calcium carbide, charcoal, chlorates, hydrocarbons, iodates, iron. Reacts violently with halogen compounds, sodium, tin, uranium and other compounds. Attacks steel when moist. May accumulate static electrical charges, and may cause ignition of its vapors.

SPERSUL (7704-34-9) Combustible solid. Liquid causes formation of sulfur dioxide with air. Reacts violently with strong oxidizers. Forms explosive, shock-sensitive, or pyrophoric mixtures with ammonia, ammonium nitrate, bromates, calcium carbide, charcoal, chlorates, hydrocarbons, iodates, iron. Reacts violently with halogen compounds, sodium, tin, uranium and other compounds. Attacks steel when moist. May accumulate static electrical charges, and may cause ignition of its vapors.

SPERSUL THIOVIT (7704-34-9) Combustible solid. Liquid causes formation of sulfur dioxide with air. Reacts violently with strong oxidizers. Forms explosive, shock-sensitive or pyrophoric mixtures with ammonia, ammonium nitrate, bromates, calcium carbide, charcoal, chlorates, hydrocarbons, iodates, iron. Reacts violently with halogen compounds, sodium, tin, uranium, and other compounds. Attacks steel when moist. May accumulate static electrical charges, and may cause ignition of its vapors.

SPINDLE OIL (N/A) Oxidizers may cause fire and explosions. Incompatible with nitric acid. May accumulate static electrical charges, and may cause ignition of its vapors.

SPIRIT (64-17-5) Forms explosive mixture with air (flash point 65°F/18°C). May accumulate static electrical charges, and may cause ignition of its vapors. Reactions may be violent with oleum, sulfuric acid, nitric acid, bases, aliphatic amines, isocyanates, oxidizers.

SPIRIT OF ETHER NITRITE (109-95-5) Forms explosive mixture with air (flash point −31°F/−35°C). A powerful oxidizer. Heat above 190°F/88°C can cause explosive decomposition. Incompatible with nitric acid, reducing agents; can cause a violent reaction, fire, or explosions.

SPIRIT OF HARTSHORN (7664-41-7) anhydrous (a flammable gas); (1336-21-6) solution in water. Reacts violently with strong oxidizers, acids. Shock-sensitive compounds may be produced with halogens, mercury oxide, silver oxide. Fire and explosions may be caused by trimethylammonium amide, 1-chloro-2,4-dinitrobenzene, o-chloronitrobenzene, platinum, selenium difluoride dioxide, boron halides, mercury, chlorine, iodine, bromine, hypochlorites, chlorine bleach, amides, organic anhydrides, isocyanates, vinyl acetate, alkylene oxides, epichlorohydrin, aldehydes. Attacks some coatings, plastics, and rubber. Attacks copper, brass, bronze, aluminum, steel, and their alloys.

SPIRITS OF SALT (7647-01-0) Reacts with acetic anhydride, aliphatic amines, alkanolamines, alkylene oxides, aromatic amines, amides, 2-aminoethanol, ammonia, ammonium hydroxide, calcium phosphide, chlorosulfonic acid, ethylene diamine, ethyleneimine, epichlorohydrin, isocyanates, metal acetylides, oleum, organic anhydrides, perchloric acid, 3-propiolactone, uranium phosphide, sulfuric acid, sodium hydroxide and other bases, strong oxidizers, vinyl acetate, vinylidene fluoride. Attacks most metals (can produce flammable hydrogen gas), and some plastics, rubber, and coatings.

SPIRITS OF TURPENTINE (8006-64-2) Forms explosive mixture with air (flash point 95°F/35°C). Incompatible with strong oxidizers, especially chlorine; may cause fire and explosions. Incompatible with strong acids, chromic anhydride, chromyl chloride, hexachloromelamine, stannic chloride. Attacks ordinary rubber.

SPIRITS OF WINE (64-17-5) Forms explosive mixture with air (flash point 65°F/18°C). May accumulate static electrical charges, and may cause ignition of its vapors. Reactions may be violent with oleum, sulfuric acid, nitric acid, bases, aliphatic amines, isocyanates, oxidizers.

SPONTOX (93-76-5) Sealed metal containers may burst in heat above 316°F/158°C. Incompatible with sulfuric acid, bases, ammonia, aliphatic amines, alkanolamines, isocyanates, alkylene oxides, epichlorohydrin.

SPOTRETE (137-26-8) Combustible solid (flash point 192°F/89°C). Strong oxidizers may cause fire and explosions; contact with strong acid or oxidizable materials produces toxic gases.

SPOTTING NAPHTHA (8052-41-3) Forms explosive mixture with air (flash point 102°F to 140°F/38.7°C to 60°C). Incompatible with nitric acid, strong oxidizers. Attacks some plastics, rubber, and coatings.

SPRAY OIL (N/A) Oxidizers may cause fire and explosions. Incompatible with nitric acid. May accumulate static electrical charges, and may cause ignition of its vapors.

SPRAYSET MEKP (1338-23-4) Forms explosive mixture with air (flash point 125°F/52°C). Explosive decomposition occurs above 176°F/80°C. Pure substance is shock-sensitive. Strong oxidizer. Reacts violently with strong acids, strong bases, reducing agents, combustible substances, organic materials, oxides of heavy metals, salts, trace contaminants, amines. May accumulate static electrical charges, and may cause ignition of its vapors.

SPRING-BAK (142-59-6) Boiling water causes formation of hydrogen sulfide and carbon disulfide vapors.

SPRITZ-HORMIN/2,4-D or SPRITZ-HORMIT/2,4-D (94-75-7) Decomposes in sunlight. Incompatible with strong oxidizers; may cause fire and explosions.

SPRITZ-RAPIDIN (58-89-9) Not combustible, but may be dissolved in a combustible solvent. If solvent comes in contact with oxidizers, fire and explosions may result.

SPRUEHPFLANZOL (German) (58-89-9) Not combustible, but may be dissolved in a combustible solvent. If solvent comes in contact with oxidizers, fire and explosions may result.

SQ 1489 (137-26-8) Combustible solid (flash point 192°F/89°C). Strong oxidizers may cause fire and explosions; contact with strong acid or oxidizable materials produces toxic gases.

SQ 9453 (67-68-5) Forms explosive mixture with air (flash point 203°F/95°C). Reacts violently with oxidizers. Reacts with ethanoyl chloride, boron compounds, halides, metal alkoxides, oxidizers.

SR406 (133-06-2) Incompatible with tetraethyl pyrophosphate, parathion.

STABLE PENTABORANE (19624-22-7) Forms explosive mixture with air (flash point 95°F/35°C). May ignite spontaneously in moist air. Water contact produces boric acid and explosive hydrogen gas. Oxidizers cause formation of highly explosive mixtures. Halogen and halogenated compounds may cause explosions. Attacks some plastics, rubber, and coatings.

STA-FAST (93-72-1) A powerful oxidizer; reacts violently with reducing agents, combustibles.

STAFLEX DBP (84-74-2) Incompatible with strong acids, nitrates, strong oxidizers, strong alkalies.

STAFLEX DTDP (119-06-2) Incompatible with strong acids, nitrates, oxidizers.

STANNOUS FLUORIDE (7783-47-3) Acid contact causes formation of hydrogen fluoride fumes.

STATHION (56-38-2) Combustible liquid. Mixtures with endrin may be explosive. Strong oxidizers may cause fire and explosions. Attacks some plastics, rubber, and coatings.

STAUFFER CAPTAN (133-06-2) Incompatible with tetraethyl pyrophosphate, parathion.

STCC (56-23-5) Becomes corrosive when in contact with water. Corrosive to metals. Reacts violently with many compounds. Decomposes on contact with chemically active metals such as sodium, potassium and magnesium. Incompatible with allyl alcohol, fluorine gas, alkali metals, aluminum. Attacks some coatings, plastics, and rubber.

STEAM DISTILLED TURPENTINE (8006-64-2) Forms explosive mixture with air (flash point 95°F/35°C). Incompatible with strong oxidizers, especially chlorine; may cause fire and explosions. Incompatible with strong acids, chromic anhydride, chromyl chloride, hexachloromelamine, stannic chloride. Attacks ordinary rubber.

STEAM TURBINE LUBE OIL or STEAM TURBINE OIL (N/A) Incompatible with nitric acid, strong oxidizers; poses risk of fire or explosions.

STEAREX (57-11-4) Incompatible with strong oxidizers, sulfuric acid, caustics, ammonia, amines, isocyanates, alkylene oxides, epichlorohydrin. Attacks chemically active metals.

STEARIC ACID (57-11-4) Incompatible with strong oxidizers, sulfuric acid, caustics, ammonia, amines, isocyanates, alkylene oxides, epichlorohydrin. Attacks chemically active metals.

STEARIC ACID, AMMONIUM SALT (1002-89-7) Combustible solid (flash point 140°F/60°C). Strong oxidizers may cause fire and explosions.

STEARIC ACID, ZINC SALT (557-05-1) Combustible solid. May accumulate static electrical charges, and may cause ignition of its vapors.

STEAROPHANIC ACID (57-11-4) Incompatible with strong oxidizers, sulfuric acid, caustics, ammonia, amines, isocyanates, alkylene oxides, epichlorohydrin. Attacks chemically active metals.

STEARYL ALCOHOL, CRUDE (112-92-5) Incompatible with strong acids, caustics, aliphatic amines, isocyanates.

STEARYLDIMETHYLBENZYL-AMMONIUM CHLORIDE (959-55-7) Incompatible with strong oxidizers, strong acids.

STEINBUHL YELLOW (13765-19-0) A strong oxidizer. Incompatible with boron (violent reaction), combustible material.

STIBINE, TRIBROMO- (7789-61-9) Contact with air, alcohol or water produces corrosive hydrogen bromide.

STIBINE, TRICHLORO- (10025-91-9) Water contact produces corrosive solution. Air contact produces corrosive vapor.

STIBIUM (7440-36-0) Reacts violently with strong oxidizers, strong acids. Forms explosive mixtures with chloric and perchloric acid. Oxidizers and acids, especially halogenated acids, can produce toxic stibine gas (antimony hydride).

STICKDIOXYD (German) (10024-97-2) May form explosive mixture with flammable gases. Nonflammable but supports combustion; as temperature increases, it becomes both a strong oxidizer and self-reactive. High temperatures may cause explosive decomposition. Incompatible with aluminum, anhydrous ammonia (may cause explosion), boron, carbon monoxide, hydrazine, hydrogen, hydrogen sulfide, lithium hydride, phosphine, sodium.

STINK DAMP (7783-06-4) A highly flammable and reactive gas. Incompatible with acetaldehyde, barium pentafluoride, chlorine monoxide, chlorine trifluoride, chromic anhydride, copper, lead dioxide, nitric acid, nitrogen iodide, nitrogen trichloride, nitrogen trifluoride, oxygen difluoride, oxidizers, phenyl diazonium chloride, sodium, sodium peroxide. Reacts with alkali metals and alkali earth metals. Attacks metals.

STIROLO (Italian) (100-42-5) If inhibitor (often *tert*-butyl catechol) is not present in adequate concentrations, polymerization may occur and explode containers. Temperatures above 150°F/66°C speed up polymerization. Incompatible with strong oxidizers, acids, rust, catalysts for vinyl polymerization, such as peroxides, strong acids, and aluminum chloride. Corrodes copper, copper alloys, and dissolves rubber. Attacks some plastics, rubber, and coatings. May accumulate static electrical charges, and may cause ignition of its vapors.

STONE RED (1309-37-1) Contact with hydrogen peroxide, ethylene oxide, calcium hypochlorite will cause explosion. Reacts violently with powdered aluminum, hydrazine, hydrogen trisulfide.

STOP-GRO (123-33-1) Contact with strong oxidizers may cause fire and explosions.

STRAIGHT RUN GASOLINE (68606-11-1); (64741-43-1). Oxidizers may cause fire and explosions. Incompatible with nitric acid. May accumulate static electrical charges, and may cause ignition of its vapors.

STRAIGHT RUN KEROSENE (8008-20-6) Forms explosive mixture in air above 100°F/38°C. Oxidizers may cause fire and explosions. Incompatible with nitric acid. May accumulate static electrical charges, and may cause ignition of its vapors.

STRATHION (56-38-2) Combustible liquid. Mixtures with endrin may be explosive. Strong oxidizers may cause fire and explosions. Attacks some plastics, rubber, and coatings.

STRAZINE (1912-24-9) Incompatible with strong acids.

STREUNEX (58-89-9) Not combustible, but may be dissolved in a combustible solvent. If solvent comes in contact with oxidizers fire and explosions may result.

STROBANE-T (8001-35-2) Reacts with oxidizers, with a risk of fire or explosions. Attacks metals in the presence of moisture.

STROBANE-T-90 (8001-35-2) Reacts with oxidizers, with a risk of fire or explosions. Attacks metals in the presence of moisture.

STRONTIUM(2+) NITRATE(1:2) or STRONTIUM(II) NITRATE(1:2) (10042-76-9) A strong oxidizer; reacts violently with reducing agents, combustibles, acids, organic and other easily oxidizable materials.

STRONTIUM CHROMATE or STRONTIUM CHROMATE(1:1) or STRONTIUM CHROMATE A (7789-06-2) Incompatible with water, acids, bases.

STRONTIUM CHROMATE(6+) or STRONTIUM CHROMATE(VI) (7789-06-2) Incompatible with water, acids, bases.

STRONTIUM CHROMATE 12170 (7789-06-2) Incompatible with water, acids, bases.

STRONTIUM CHROMATE X-2396 (7789-06-2) Incompatible with water, acids, bases.

STRONTIUM NITRATE (10042-76-9) A strong oxidizer; reacts violently with reducing agents, combustibles, acids, organic and other easily oxidizable materials.

STRONTIUM YELLOW (7789-06-2) Incompatible with water, acids, bases.

STRYCHNINE (57-24-9) Contact with strong oxidizers may cause fire and explosions.

STRYCHNOS (57-24-9) Contact with strong oxidizers may cause fire and explosions.

STUNTMAN (123-33-1) Contact with strong oxidizers may cause fire and explosions.

STYRALLYL ALCOHOL or STYRALYL ALCOHOL (98-85-1) Forms explosive mixture with air (flash point 205°F/96°C). Incompatible with strong acids, caustics, aliphatic amines, isocyanates, oxidizers.

STYREEN (Dutch) (100-42-5) If inhibitor (often *tert*-butyl catechol) is not present in adequate concentrations, polymerization may occur and explode containers. Temperatures above 150°F/66°C speed up polymerization. Incompatible with strong oxidizers, acids, rust, catalysts for vinyl polymerization, such as peroxides, strong acids, and aluminum chloride. Corrodes copper, copper and alloys, and dissolves rubber. Attacks some plastics, rubber, and coatings. May accumulate static electrical charges, and may cause ignition of its vapors.

STYREN (Czech) (100-42-5) If inhibitor (often *tert*-butyl catechol) is not present in adequate concentrations, polymerization may occur and explode containers. Temperatures above 150°F/66°C speed up polymerization. Incompatible with strong oxidizers, acids, rust, catalysts for vinyl polymerization, such as peroxides, strong acids, and aluminum chloride. Corrodes copper and copper alloys, and dissolves rubber. Attacks some plastics, rubber, and coatings. May accumulate static electrical charges, and may cause ignition of its vapors.

STYRENE (100-42-5) If inhibitor (often *tert*-butyl catechol) is not present in adequate concentrations, polymerization may occur and explode containers. Temperatures above 150°F/66°C speed up polymerization. Incompatible with strong oxidizers, acids, rust, catalysts for vinyl polymerization, such as peroxides, strong acids, and aluminum chloride. Corrodes copper and copper alloys, and dissolves rubber. Attacks some plastics, rubber, and coatings. May accumulate static electrical charges, and may cause ignition of its vapors.

STYRENE EPOXIDE (96-09-3) Forms explosive mixture with air (flash point 165°F/74°C). Reacts with acids, caustics, oxidizers, possibly causing explosive polymerization.

STYRENE MONOMER or STYRENE MONOMER, INHIBITED (100-42-5) If inhibitor (often *tert*-butyl catechol) is not present in adequate concentrations, polymerization may occur and explode containers. Temperatures above 150°F/66°C speed up polymerization. Incompatible with strong oxidizers, acids, rust, catalysts for vinyl polymerization, such as peroxides, strong acids, and aluminum chloride. Corrodes copper and copper alloys, and dissolves rubber. Attacks some plastics, rubber, and coatings. May accumulate static electrical charges, and may cause ignition of its vapors.

STYRENE OXIDE or STYRENE-7,8-OXIDE (96-09-3) Forms explosive mixture with air (flash point 165°F/74°C). Reacts with acids, caustics, oxidizers, possibly causing explosive polymerization.

STYROL (German) (100-42-5) If inhibitor (often *tert*-butyl catechol) is not present in adequate concentrations, polymerization may occur and explode containers. Temperatures above 150°F/66°C speed up polymerization. Incompatible with strong oxidizers, acids, rust, catalysts for vinyl polymerization, such as peroxides, strong acids, and aluminum chloride. Corrodes copper and copper alloys, and dissolves rubber. Attacks some plastics, rubber, and coatings. May accumulate static electrical charges, and may cause ignition of its vapors.

STYROLE (100-42-5) If inhibitor (often *tert*-butyl catechol) is not present in adequate concentrations, polymerization may occur and explode containers. Temperatures above 150°F/66°C speed up polymerization. Incompatible with strong oxidizers, acids, rust, catalysts for vinyl polymerization, such as peroxides, strong acids, and aluminum chloride. Corrodes copper and copper alloys, and dissolves rubber. Attacks some plastics, rubber, and coatings. May accumulate static electrical charges, and may cause ignition of its vapors.

STYROLENE (100-42-5) If inhibitor (often *tert*-butyl catechol) is not present in adequate concentrations, polymerization may occur and explode containers. Temperatures above 150°F/66°C speed up polymerization. Incompatible with strong oxidizers, acids, rust, catalysts for vinyl polymerization, such as peroxides, strong acids, and aluminum chloride. Corrodes copper and copper alloys, and dissolves rubber. Attacks some plastics, rubber, and coatings. May accumulate static electrical charges, and may cause ignition of its vapors.

STYRON (100-42-5) If inhibitor (often *tert*-butyl catechol) is not present in adequate concentrations, polymerization may occur and explode containers. Temperatures above 150°F/66°C speed up polymerization. Incompatible with strong oxidizers, acids, rust, catalysts for vinyl polymerization, such as peroxides, strong acids, and aluminum chloride. Corrodes copper and copper alloys, and dissolves rubber. Attacks some plastics, rubber, and coatings. May accumulate static electrical charges, and may cause ignition of its vapors.

STYROPOL (100-42-5) If inhibitor (often *tert*-butyl catechol) is not present in adequate concentrations, polymerization may occur and explode containers. Temperatures above 150°F/66°C speed up polymerization. Incompatible with strong oxidizers, acids, rust, catalysts for vinyl polymerization, such as peroxides, strong acids, and aluminum chloride. Corrodes copper and copper alloys, and dissolves rubber. Attacks some plastics, rubber, and coatings. May accumulate static electrical charges, and may cause ignition of its vapors.

STYROPOR (100-42-5) If inhibitor (often *tert*-butyl catechol) is not present in adequate concentrations, polymerization may occur and explode containers. Temperatures above 150°F/66°C speed up polymerization. Incompatible with strong oxidizers, acids, rust, catalysts for vinyl polymerization, such as peroxides, strong acids, and aluminum chloride. Corrodes copper and copper alloys, and dissolves rubber. Attacks some plastics, rubber, and coatings. May accumulate static electrical charges, and may cause ignition of its vapors.

STYRYL OXIDE (96-09-3) Forms explosive mixture with air (flash point 165°F/74°C). Reacts with acids, caustics, oxidizers, possibly causing explosive polymerization.

SUBERANE (291-64-5) Forms explosive mixture with air (flash point 43°F/6°C). Strong oxidizers may cause fire and explosions.

SUBLIMAT (Czech) (7487-94-7) Incompatible with light metals (aluminum, magnesium, beryllium, etc.), sodium, potassium.

SUBLIMED SULFUR (7704-34-9) Combustible solid. Liquid causes formation of sulfur dioxide with air. Reacts violently with strong oxidizers. Forms explosive, shock-sensitive or pyrophoric mixtures with ammonia, ammonium nitrate, bromates, calcium carbide, charcoal, chlorates, hydrocarbons, iodates, iron. Reacts violently with halogen compounds, sodium, tin, uranium, and other compounds. Attacks steel when moist. May accumulate static electrical charges, and may cause ignition of its vapors.

SUCROSE (57-50-1) Reacts with potassium hydroxide, strong acids, strong oxidizers.

SUGAR (57-50-1) Reacts with potassium hydroxide, strong acids, strong oxidizers.

SUGAR OF LEAD (301-04-2) Contact with strong acids produces acetic acid. Incompatible with bases, ammonia, amines, cresols, isocyanates, alkylene oxides, epichlorohydrin, phenols, sulfites. Reacts violently with bromates, strong oxidizers.

SULEMA (Russian) (7487-94-7) Incompatible with light metals (aluminum, magnesium, beryllium, etc.), sodium, potassium.

SULFAMATE (7773-06-0) Fires and explosions may result from contact with strong oxidizers, hot water, potassium, sodium, sodium nitrite, metal chlorates, hot acid solutions. Corrosive to mild steel.

SULFAMIC ACID (5329-14-6) Incompatible with nitric acid, bases, ammonia, amines, isocyanates, alkylene oxides, epichlorohydrin, oxidizers, water.

SULFAMIC ACID, COBALT SALT (16107-41-3) An oxidizer; reacts with reducing agents and combustibles.

760

SULFAMIC ACID, MONOAMMONIUM SALT (7773-06-0) Fires and explosions may result from contact with strong oxidizers, hot water, potassium, sodium, sodium nitrite, metal chlorates, hot acid solutions. Corrosive to mild steel.

SULFAMINSAEURE (German) (7773-06-0) Fires and explosions may result from contact with strong oxidizers, hot water, potassium, sodium, sodium nitrite, metal chlorates, hot acid solutions. Corrosive to mild steel.

SULFAN (7446-11-9) Can produce sulfuric acid fumes in moist air or steam. Reacts violently with water, producing sulfuric acid. A strong oxidizer; reacts violently with reducing agents, combustible materials, organic substances. Incompatible with strong acids, caustics, ammonia, amines, organic anhydrides, isocyanates, vinyl acetate, metal oxides, acrylates, substituted allyls, alkylene oxides, ketones, aldehydes, alcohols, glycols, phenols, cresols, caprolactam solution, phosphorus and many other substances. Should be stabilized with an inhibitor to prevent a dangerous and potentially explosive change in the vapor pressure of this material. Attacks metals in the presence of moisture.

SULFAPOL (25155-30-0) Reacts with acids, including fumes.

SULFATE de CUIVRE (French) (7758-98-7) Aqueous solution is an acid. Incompatible with strong bases, hydroxylamine, magnesium.

SULFATE de METHYLE (French) (77-78-1) Forms explosive mixture with air (flash point 182°F/83°C). Incompatible with strong oxidizers, strong acids, strong alkalies, strong ammonia solutions. Attacks some plastics, rubber, and coatings.

SULFATE de PLOMB (French) (7446-14-2) Reacts violently with potassium. Incompatible with aluminum, magnesium.

SULFATE de ZINC (French) (7733-02-0) Incompatible with strong bases.

SULFATE DIMETHYLIQUE (French) (77-78-1) Forms explosive mixture with air (flash point 182°F/83°C). Incompatible with strong oxidizers, strong acids, strong alkalies, strong ammonia solutions. Attacks some plastics, rubber, and coatings.

SULFATED NEATSFOOT OIL (N/A) Oxidizers may cause fire and explosions. Incompatible with nitric acid. May accumulate static electrical charges, and may cause ignition of its vapors.

SULFATE MERCURIQUE (French) (7783-35-9) Water contact produces sulfuric acid. Light may cause decomposition. Incompatible with aluminum or magnesium. Reacts violently with hydrogen chloride above 250°F/121°C. Attacks metals when wet.

SULFATE OF COPPER (7758-98-7) Aqueous solution is an acid. Incompatible with strong bases, hydroxylamine, magnesium.

SULFATEP (3689-24-5) Containers may burst in heat. Strong oxidizers may cause fire and explosions. Attacks some plastics, rubber, and coatings.

SULFATE TURPENTINE (8006-64-2) Forms explosive mixture with air (flash point 95°F/35°C). Incompatible with strong oxidizers, especially chlorine; may cause fire and explosions. Incompatible with strong acids, chromic anhydride, chromyl chloride, hexachloromelamine, stannic chloride. Attacks ordinary rubber.

761

SULFATE WOOD TURPENTINE (8006-64-2) Forms explosive mixture with air (flash point 95°F/35°C). Incompatible with strong oxidizers, especially chlorine; may cause fire and explosions. Incompatible with strong acids, chromic anhydride, chromyl chloride, hexachloromelamine, stannic chloride. Attacks ordinary rubber.

SULFERROUS (7720-78-7) Aqueous solution is acidic. Contact with bases produces iron.

SULFICYLBIS(METHANE) (67-68-5) Forms explosive mixture with air (flash point 203°F/95°C). Reacts violently with oxidizers. Reacts with ethanoyl chloride, boron compounds, halides, metal alkoxides, oxidizers.

SULFIDAL (7704-34-9) Combustible solid. Liquid causes formation of sulfur dioxide with air. Reacts violently with strong oxidizers. Forms explosive, shock-sensitive, or pyrophoric mixtures with ammonia, ammonium nitrate, bromates, calcium carbide, charcoal, chlorates, hydrocarbons, iodates, iron. Reacts violently with halogen compounds, sodium, tin, uranium, and other compounds. Attacks steel when moist. May accumulate static electrical charges, and may cause ignition of its vapors.

SULFIMEL DOS (119-36-8) Forms explosive mixture with air (flash point 205°F/96°C). Incompatible with strong acids, nitrates, oxidizers.

SULFITE CELLULOSE (9004-34-6) Incompatible with water, bromine pentafluoride, hydrogen peroxide, sodium hypochlorite, sodium nitrate, fluorine, or strong oxidizers.

SULFOCARBOLIC ACID (98-67-9) Incompatible with sulfuric acid, caustics, ammonia, aliphatic amines, alkanolamines, isocyanates, alkylene oxides, epichlorohydrin, steam, water.

SULFONIC ACID, MONOCHLORIDE (7790-94-5) Reacts violently with water producing sulfuric and hydrochloric acids and dense fumes. Dangerously reactive; avoid contact with all other material. Strong oxidizer and strong acid; reacts violently with bases, reducing agents, combustibles, acids (especially sulfuric), alcohol, diphenyl ether, silver nitrate, water. Phosphorus contact may cause fire and explosions. Forms explosive material with ethyl alcohol. Incompatible with many materials. Attacks many metals.

SULFONIMIDE (2425-06-1) Incompatible with acids or acid vapor. Strongly alkaline conditions contribute to instability.

SULFONYL CHLORIDE (7791-25-5) Water contact produces sulfuric acid and hydrogen chloride vapors. Reacts violently with caustics, alkali metals, lead dioxide, phosphorus. Attacks metals in the presence of moisture, producing hydrogen gas.

SULFORON (7704-34-9) Combustible solid. Liquid causes formation of sulfur dioxide with air. Reacts violently with strong oxidizers. Forms explosive, shock-sensitive, or pyrophoric mixtures with ammonia, ammonium nitrate, bromates, calcium carbide, charcoal, chlorates, hydrocarbons, iodates, iron. Reacts violently with halogen compounds, sodium, tin, uranium, and other compounds. Attacks steel when moist. May accumulate static electrical charges, and may cause ignition of its vapors.

SULFOTEP or SULFOTEPP (3689-24-5) Containers may burst in heat. Strong oxidizers may cause fire and explosions. Attacks some plastics, rubber, and coatings.

SULFOXYL (94-36-0) Confined storage of dry chemical may lead to decomposition and explosion. A strong oxidant; extremely reactive. Fires and explosion may result from heat or contamination, and from contact with strong acids, combustible materials, oxidizers, acids, bases, alcohols, reducing agents, metals, metal oxides, amines, accelerators, methyl methacrylate, organic matter, lithium aluminum carbide, dimethyl aniline, amines, metallic naphthenates. May attack some plastics, rubber, and coatings. Protect containers from shock and friction.

SULFRAMIN 40 (25155-30-0) Reacts with acids, including fumes.

SULFRAMIN 85 (25155-30-0) Reacts with acids, including fumes.

SULFTECH (7757-83-7) A reducing agent. Incompatible with strong oxidizers, combustibles, acids, organic materials.

SULFUR or SULFUR, MOLTEN or SULFUR, SOLID (7704-34-9) Combustible solid. Liquid causes formation of sulfur dioxide with air. Reacts violently with strong oxidizers. Forms explosive, shock-sensitive, or pyrophoric mixtures with ammonia, ammonium nitrate, bromates, calcium carbide, charcoal, chlorates, hydrocarbons, iodates, iron. Reacts violently with halogen compounds, sodium, tin, uranium and other compounds. Attacks steel when moist. May accumulate static electrical charges, and may cause ignition of its vapors.

SULFUR CHLORIDE (10025-67-9) Water contact produces hydrochloric acid, sulfur dioxide, sulfur, sulfur thiosulfite, hydrogen sulfide; solution is a strong acid. Incompatible with strong oxidizers, strong bases, oxides of phosphorus, organics; may cause fires. Attacks some plastics, rubber, and coatings.

SULFUR CHLORIDE or SULFUR CHLORIDE (MONO) (DOT) (10545-99-0) Contact with air produces hydrochloric acid fumes. Incompatible with water, acetone, aluminum, ammonia, amines, caustics, oxidizers, metal powders, potassium, sodium.

SULFUR DICHLORIDE (10545-99-0) Contact with air produces hydrochloric acid fumes. Incompatible with water, acetone, aluminum, ammonia, amines, caustics, oxidizers, metal powders, potassium, sodium.

SULFUR DIOXIDE or SULFUR DIOXIDE, LIQUIFIED (7446-09-5) Forms corrosive hydrochloric acid fumes with air. Contact with copper, bronze, or alkali metals may cause fire and explosions. Reacts violently with alcohols, caustics, amines, water. Decomposes above 140°F/60°C; forms toxic and corrosive oxides of sulfur. Attacks some plastics, rubber, and coatings.

SULFURETTED HYDROGEN (7783-06-4) A highly flammable and reactive gas. Incompatible with acetaldehyde, barium pentafluoride, chlorine monoxide, chlorine trifluoride, chromic anhydride, copper, lead dioxide, nitric acid, nitrogen iodide, nitrogen trichloride, nitrogen trifluoride, oxygen difluoride, oxidizers, phenyl diazonium chloride, sodium, sodium peroxide. Reacts with alkali metals and alkaline earth metals. Attacks metals.

SULFURE de METHYLE (French) (75-18-3) Forms explosive mixture with air (flash point −36°F/−38°C). Reacts violently with strong oxidizers. May accumulate static electrical charges, and may cause ignition of its vapors.

SULFUR HYDRIDE (7783-06-4) A highly flammable and reactive gas. Acetaldehyde, barium pentafluoride, chlorine monoxide, chlorine trifluoride, chromic anhydride, copper, lead dioxide, nitric acid, nitrogen iodide, nitrogen trichloride, nitrogen trifluoride, oxygen difluoride, oxidizers, phenyl diazonium chloride, sodium, sodium peroxide. Reacts with alkali metals and alkaline earth metals. Attacks metals.

SULFUR HYDROXIDE (7783-06-4) A highly flammable and reactive gas. Incompatible with acetaldehyde, barium pentafluoride, chlorine monoxide, chlorine trifluoride, chromic anhydride, copper, lead dioxide, nitric acid, nitrogen iodide, nitrogen trichloride, nitrogen trifluoride, oxygen difluoride, oxidizers, phenyl diazonium chloride, sodium, sodium peroxide. Reacts with alkali metals and alkali earth metals. Attacks metals.

SULFURIC ACID (7664-93-9) A strong oxidizer that can react violently, with risk of fire and explosion with many substances including reducing agents, organic and combustible substances and bases. Incompatible with nonoxidizing mineral acids, organic acids, bases, acrylates, aldehydes, alcohols, alkylene oxides, ammonia, aliphatic amines, alkanolamines, aromatic amines, amides, chlorates, epichlorohydrin, fulminates, glycols, isocyanates, ketones, metals (powdered), organic anhydrides, perchlorates, picrates, substituted allyls, phenols, cresols, water, acetic anhydride, acetone cyanhydrin, acetonitrile, acrolein, acrylonitrile, allyl alcohol, allyl chloride, 2-aminoethanol, ammonium hydroxide, aniline, bromine pentafluoride, *n*-butyraldehyde, caprolactam solution, carbides, cesium acetylene carbide, chlorine trifluoride, chlorosulfonic acid, cuprous nitride, diisobutylene, ethylene cyanohydrin, ethylene diamine, ethylene glycol, ethyleneimine, hydrochloric acid, iodine heptafluoride, iron, isoprene, lithium silicide, mercuric nitride, mesityl oxide, nitric acid, *p*-nitrotoluene, perchloric acid, phosphorus, potassium *tert*-butoxide, potassium chlorate, potassium permanganate, propiolactone (beta-), propylene oxide, pyridine, rubidium acetylene, silver permanganate, sodium, sodium carbonate, sodium chlorate, sodium hydroxide, styrene monomer, vinyl acetate, Attacks most metals, and some plastics, rubber, and coatings.

SULFURIC ACID, ALUMINUM SALT (3:2) (10043-01-3) Incompatible with water, producing sulfuric acid; react strongly with bases and many other materials. Dry material is weakly corrosive to carbon steel; aqueous solution attacks metals, producing hydrogen gas.

SULFURIC ACID, AMMONIUM IRON(2+) SALT (2:2:1) or SULFURIC ACID, AMMONIUM IRON(II) SALT (2:2:1) (10045-89-3) Reacts violently with tetranitromethane, ammonium perchlorate and mercury(II) dinitrate. Sulfates react with aluminum, magnesium.

SULFURIC ACID, AMMONIUM NICKEL(2+) SALT (2:2:1) or SULFURIC ACID, AMMONIUM NICKEL(II) SALT (2:2:1) (15699-18-0) Incompatible with strong acids, selenium.

SULFURIC ACID, BARIUM SALT (1:1) (7727-43-7) Explosions may result from contact with aluminum in the presence of heat. Incompatible with potassium, phosphorus.

SULFURIC ACID, CALCIUM SALT (7778-18-9) Contact with diazomethane, aluminum, phosphorus may cause explosion.

SULFURIC ACID, COBALT(2+) SALT(1:1) or SULFURIC ACID, COBALT(II) SALT(1:1) (10124-43-3) Incompatible with oxidizers, acetylene.

SULFURIC ACID, COPPER(2+) SALT (1:1) (7758-98-7) Aqueous solution is an acid. Incompatible with strong bases, hydroxylamine, magnesium.

SULFURIC ACID, DIAMMONIUM SALT (7783-20-2) Aqueous solution is a strong acid; reacts with bases, producing ammonia. Attacks metals. Hot material reacts with nitrates, nitrites, chlorates.

SULFURIC ACID, DIETHYL ESTER (64-67-5) Vigorous reaction with strong oxidizers or water. Incompatible with strong acids, strong alkalies, nitrates, water.

SULFURIC ACID, DIMETHYL ESTER (77-78-1) Forms explosive mixture with air (flash point 182°F/83°C). Incompatible with strong oxidizers, strong acids, strong alkalies, strong ammonia solutions. Attacks some plastics, rubber, and coatings.

SULFURIC ACID, DIPOTASSIUM SALT (7778-80-5) Incompatible with aluminum, magnesium.

SULFURIC ACID, FUMING (8014-95-7) A powerful oxidizer and strong acid. Reacts violently with reducing agents, organic materials. Reacts with air, with formation of corrosive fumes. Contact with cast iron may cause a violent reaction. Extremely hazardous on contact with many materials including chlorates, carbides, fulminates. Vigorous reaction with metals (powders will ignite) releasing hydrogen. Attacks some plastics, rubber, and coatings on brief contact.

SULFURIC ACID, IRON SALT (1:1) (7720-78-7) Aqueous solution is acidic. Contact with bases produces iron.

SULFURIC ACID, IRON(2+) SALT (1:1) or SULFURIC ACID, IRON(II) SAL (1:1) (7720-78-7) Aqueous solution is acidic. Contact with bases produces iron.

SULFURIC ACID, IRON(3+) SALT (3:2) or SULFURIC ACID, IRON(III) SALT (3:2) (10028-22-5) Corrosive to copper and its alloys, mild and galvanized steel.

SULFURIC ACID, LEAD(2+) SALT (1:1) or SULFURIC ACID, LEAD(II) SALT (1:1) (7446-14-2) Reacts violently with potassium. Incompatible with aluminum, magnesium.

SULFURIC ACID, MERCURY(2+) SALT (1:1) or SULFURIC ACID, MERCURY(II) SALT (1:1) (7783-35-9) Water contact produces sulfuric acid. Light may cause decomposition. Violent reaction may occur with air, oxygen, oxidizers, sulfur trioxide.

SULFURIC ACID MIXED WITH SULFUR TRIOXIDE (8014-95-7) A powerful oxidizer and strong acid. Reacts violently with reducing agents, organic materials. Reacts with air, producing corrosive fumes. Contact with cast iron may cause a violent reaction. Extremely hazardous on contact with many materials including chlorates, carbides, fulminates. Vigorous reaction with metals (powders will ignite) releasing hydrogen. Attacks some plastics, rubber, and coatings on brief contact.

SULFURIC ACID, ZINC SALT (1:1) (7733-02-0) Reacts with strong bases.

SULFURIC CHLOROHYDRIN (7790-94-5) Reacts violently with water producing sulfuric and hydrochloric acids and dense fumes. Dangerously reactive; avoid contact with all other material. Strong oxidizer and strong acid; reacts violently with bases, reducing agents, combustibles, acids (especially sulfuric), alcohol, diphenyl ether, silver nitrate, water. phosphorus contact may cause fire and explosions. Forms explosive material with ethyl alcohol. Incompatible with many materials. Attacks many metals.

SULFURIC ETHER (60-29-7) Forms explosive mixture with air (flash point −49°F/−45°C). Incompatible with strong acids, strong oxidizers. Can form peroxides with exposure to air or light; may explode when container is unstoppered or otherwise opened. Attacks some plastics, rubber, and coatings. Being a nonconductor, chemical may accumulate static electric charges that may result in ignition of vapor.

SULFURIC OXIDE (7446-11-9) Can form sulfuric acid fumes in moist air or steam. Reacts violently with water, with formation of sulfuric acid. A strong oxidizer; reacts violently with reducing agents, combustible materials, organic substances. Incompatible with strong acids, caustics, ammonia, amines, organic anhydrides, isocyanates, vinyl acetate, metal oxides, acrylates, substituted allyls, alkylene oxides, ketones, aldehydes, alcohols, glycols, phenols, cresols, caprolactam solution, phosphorus and many other substances. Should be stabilized with an inhibitor to prevent a dangerous and potentially explosive change in the vapor pressure of this material. Attacks metals in the presence of moisture.

SULFURIC OXYCHLORIDE (7791-25-5) Water contact produces sulfuric acid and hydrogen chloride vapors. Reacts violently with caustics, alkali metals, lead dioxide, phosphorus. Attacks metals in the presence of moisture producing hydrogen gas.

SULFUR MONOCHLORIDE (10025-67-9) Water contact produces hydrochloric acid, sulfur dioxide, sulfur, sulfur thiosulfite, hydrogen sulfide; solution is a strong acid. Incompatible with strong oxidizers, strong bases, oxides of phosphorus, organics; may cause fires. Attacks some plastics, rubber, and coatings.

SULFUROUS ACID ANHYDRIDE (7446-09-5) Forms corrosive sulfuric acid fumes with air. Contact with copper, bronze, or alkali metals may cause fire and explosions. Reacts violently with alcohols, caustics, amines, water. Decomposes above 140°F/60°C; forms toxic and corrosive oxides of sulfur. Attacks some plastics, rubber, and coatings.

SULFUROUS ACID, MONOSODIUM SALT (7631-90-5) Slowly oxidized to the sulfate on contact with air. Oxidizers or acids cause formation of sulfur dioxide gas. Attacks metals.

SULFUROUS ACID, SODIUM SALT(1:2) (7757-83-7) A reducing agent. Incompatible with strong oxidizers, combustibles, acids, organic materials.

SULFUROUS ANHYDRIDE (7446-09-5) Forms corrosive surfuric acid fumes with air. Contact with copper, bronze, or alkali metals may cause fire and explosions. Reacts violently with alcohols, caustics, amines, water. Decomposes above 140°F/60°C; produces toxic and corrosive oxides of sulfur. Attacks some plastics, rubber, and coatings.

SULFUROUS OXIDE (7446-09-5) Forms corrosive sulfuric acid fumes with air. Contact with copper, bronze, or alkali metals may cause fire and explosions. Reacts violently with alcohols, caustics, amines, water. Decomposes above 140°F/60°C; produces toxic and corrosive oxides of sulfur. Attacks some plastics, rubber, and coatings.

SULFUR OXIDE (7446-09-5) Produces corrosive sulfuric acid fumes with air. Contact with copper, bronze, or alkali metals may cause fire and explosions. Reacts violently with alcohols, caustics, amines, water. Decomposes above 140°F/60°C; forms toxic and corrosive oxides of sulfur. Attacks some plastics, rubber, and coatings.

SULFUR PHOSPHIDE (1314-80-3) Incompatible with water, alcohols and strong oxidizers; may cause fire and explosions.

SULFUR SUBCHLORIDE (10025-67-9) Water contact produces hydrochloric acid, sulfur dioxide, sulfur, sulfur thiosulfite, hydrogen sulfide; solution is a strong acid. Incompatible with strong oxidizers, strong bases, oxides of phosphorus, organics; may cause fires. Attacks some plastics, rubber, and coatings.

SULFUR TRIOXIDE or SULFUR TRIOXIDE (STABILIZED) (7446-11-9) Can produce sulfuric acid fumes in moist air or steam. Reacts violently with water, with formation of sulfuric acid. A strong oxidizer; reacts violently with reducing agents, combustible materials, organic substances. Incompatible with strong acids, caustics, ammonia, amines, organic anhydrides, isocyanates, vinyl acetate, metal oxides, acrylates, substituted allyls, alkylene oxides, ketones, aldehydes, alcohols, glycols, phenols, cresols, caprolactam solution, phosphorus and many other substances. Should be stabilized with an inhibitor to prevent a dangerous and potentially explosive change in the vapor pressure of this material. Attacks metals in the presence of moisture.

SULFURYL CHLORIDE (7791-25-5) Water contact produces sulfuric acid and hydrogen chloride vapors. Reacts violently with caustics, alkali metals, lead dioxide, phosphorus. Attacks metals in the presence of moisture producing hydrogen gas.

SULKOL (7704-34-9) Combustible solid. Liquid causes formation of sulfur dioxide with air. Reacts violently with strong oxidizers. Forms explosive, shock-sensitive or pyrophoric mixtures with ammonia, ammonium nitrate, bromates, calcium carbide, charcoal, chlorates, hydrocarbons, iodates, iron. Reacts violently with halogen compounds, sodium, tin, uranium and other compounds. Attacks steel when moist. May accumulate static electrical charges, and may cause ignition of its vapors.

SULOUREA (62-56-6) Aqueous solution is a base; strong reaction with acids. Incompatible with nitric acid, hydrogen peroxide, acrolein, metals.

SULPHAMIC ACID (DOT) (5329-14-6) Incompatible with nitric acid, bases, ammonia, amines, isocyanates, alkylene oxides, epichlorohydrin, oxidizers, water.

SULPHEIMIDE (2425-06-1) Reacts with acids or acid vapor. Strongly alkaline conditions contribute to instability.

SULPHOCARBONIC ANHYDRIDE (75-15-0) Highly reactive. Contact with many substances can cause fire and explosions. Forms explosive mixture with air (flash point −22°F/−30°C). Shock can cause explosive decomposition. Incompatible with alkali metals, aliphatic amines, alkanolamines, aluminum, azides, chlorine monoxide, combustible substances, ethylene diamine, ethyleneimine, lead azide, lithium azide, nitric oxide, nitrogen dioxide, potassium, potassium azide, reducing agents, rubidium azide, sodium azide, zinc.

SULPHOS (56-38-2) Combustible liquid. Mixtures with endrin may be explosive. Strong oxidizers may cause fire and explosions. Attacks some plastics, rubber, and coatings.

SULPHUR (7704-34-9) Combustible solid. Liquid causes formation of sulfur dioxide with air. Reacts violently with strong oxidizers. Forms explosive, shock-sensitive or pyrophoric mixtures with ammonia, ammonium nitrate, bromates, calcium carbide, charcoal, chlorates, hydrocarbons, iodates, iron. Reacts violently with halogen compounds, sodium, tin, uranium, and other compounds. Attacks steel when moist. May accumulate static electrical charges, and may cause ignition of its vapors.

SULPHURETTED HYDROGEN (7783-06-4) A highly flammable and reactive gas. Incompatible with acetaldehyde, barium pentafluoride, chlorine monoxide, chlorine trifluoride, chromic anhydride, copper, lead dioxide, nitric acid, nitrogen iodide, nitrogen trichloride, nitrogen trifluoride, oxygen difluoride, oxidizers, phenyl diazonium chloride, sodium, sodium peroxide. Reacts with alkali metals and alkaline earth metals. Attacks metals.

SULPHURIC ACID, CADMIUM SALT(1:1) (10124-36-4) May react with strong oxidizers, sulfur, selenium, tellurium, zinc.

SULSOL (7704-34-9) Combustible solid. Liquid causes formation of sulfur dioxide with air. Reacts violently with strong oxidizers. Forms explosive, shock-sensitive, or pyrophoric mixtures with ammonia, ammonium nitrate, bromates, calcium carbide, charcoal, chlorates, hydrocarbons, iodates, iron. Reacts violently with halogen compounds, sodium, tin, uranium and other compounds. Attacks steel when moist. May accumulate static electrical charges, and may cause ignition of its vapors.

SUMITOX (121-75-5) Incompatible with strong oxidizers, magnesium, alkaline pesticides. Attacks metals, some plastics, rubber, and coatings.

SUNATIPIC ACID-B or SUNATIPIC ACID-C (1338-24-5) Incompatible with sulfuric acid, caustics, ammonia, aliphatic amines, alkanolamines, isocyanates, alkylene oxides, epichlorohydrin, strong oxidizers. Generally corrosive to metals.

SUNFLOWER SEED OIL (N/A) Incompatible with strong acids, oxidizers.

SUNTOL (56-72-4) Contact with strong oxidizers may cause fire and explosions.

SUPERCEL 3000 (57-13-6) Heat causes formation of anhydrous ammonia fumes. Contact with oxidizers, nitrates may cause fire and explosions. Contact with chlorinating agents, including hypochlorite bleaches, may form explosive nitrogen trichloride. Reacts with nitrosyl perchlorate.

SUPER COSAN (7704-34-9) Combustible solid. Liquid causes formation of sulfur dioxide with air. Reacts violently with strong oxidizers. Forms explosive, shock-sensitive, or pyrophoric mixtures with ammonia, ammonium nitrate, bromates, calcium carbide, charcoal, chlorates, hydrocarbons, iodates, iron. Reacts violently with halogen compounds, sodium, tin, uranium and other compounds. Attacks steel when moist. May accumulate static electrical charges, and may cause ignition of its vapors.

SUPER-DE-SPROUT (123-33-1) Contact with strong oxidizers may cause fire and explosions.

SUPER D WEEDONE (93-76-5) Sealed metal containers may burst in heat above 316°F/158°C. Incompatible with sulfuric acid, bases, ammonia, aliphatic amines, alkanolamines, isocyanates, alkylene oxides, epichlorohydrin.

SUPER D WEEDONE (94-75-7) Decomposes in sunlight. Incompatible with strong oxidizers; may cause fire and explosions.

SUPERFLAKE, ANHYDROUS (10043-52-4) Incompatible with water, bromine trifluoride, 2-furan, percarboxylic acid. Attacks metals.

SUPERLYSOFORM (50-00-0) May polymerize unless properly inhibited (usually with methanol). Forms explosive mixture with air (flash point 122°F/50°C). Incompatible with strong acids, amines, strong oxidizers, alkaline materials, nitrogen dioxide, performic acid. Reaction with hydrochloric acid produces bis-chloromethyl ether, a carcinogen.

SUPERORMONE CONCENTRE (94-75-7) Decomposes in sunlight. Incompatible with strong oxidizers; may cause fire and explosions.

SUPEROXOL (7722-84-1) A powerful oxidizer; attacks many substances. Contact with most organic, readily oxidizable materials, reducing agents, and combustibles causes fire and explosions. Contact with iron, copper, brass, bronze, chromium, zinc, lead, manganese, silver and other catalytic metals (and their salts), especially in a basic (pH 7 or above) environment, cause rapid decomposition with evolution of oxygen gas. Attacks, and may ignite, some plastics, rubber, and coatings. Decomposes slowly at ordinary temperatures and builds up pressure in a closed container. The rate of decomposition doubles for each 50°F/10°C rise (1.5 times 10°C rise) in temperature and becomes self-sustaining at 285°F/141°C.

SUP'R FLO (330-54-1) Hydrolyzes in fairly strong acids.

SUPER RODIATOX (56-38-2) Combustible liquid. Mixtures with endrin may be explosive. Strong oxidizers may cause fire and explosions. Attacks some plastics, rubber, and coatings.

SUPRA (1309-37-1) Contact with hydrogen peroxide, ethylene oxide, calcium hypochlorite will cause explosion. Reacts violently with powdered aluminum, hydrazine, hydrogen trisulfide.

SUPRAMIKE (7727-43-7) Explosions may result from contact with aluminum in the presence of heat. Incompatible with potassium, phosphorus.

SURCHLOR (7681-52-9) A strong oxidizer and a strong base. Decomposes in sunlight, producing oxygen and increasing the risk of fire. Stability decreases with concentration, heat, light, decrease in Ph, and contamination with metals. Incompatible with strong acids, reducing agents, combustible substances all cause violent reaction, fire, and explosions. Contact with amines and ammonia salts produces explosive chloroamines. Corrodes many metals: steel, 12% and 17% chrome steel, cast iron, monel, aluminum, nickel, brass, bronze, iconel.

SWEET BIRCH OIL (119-36-0) Incompatible with strong acids, nitrates, strong oxidizers.

SWEET OIL (8001-25-0) Incompatible with strong acids, oxidizers.

SWEET SPIRITS OF NITRE (109-95-5) Forms explosive mixture with air (flash point −31°F/−35°C). A powerful oxidizer. Heat above 190°F/88°C can cause explosive decomposition. Incompatible with nitric acid, reducing agents; can cause a violent reaction, fire or explosions.

SYMCLOSENE (87-90-1) A powerful oxidizer. Forms explosive material with nitrogen compounds. Contact with organic materials or reducing agents may cause fire.

SYMETRYCZNA DWUMETYLOHYDRAZYNA (Polish) (540-73-8) Forms explosive mixture with air (flash point 73°F/23°C). A strong reducing agent. Reacts violently with strong oxidizers, strong acids. Attacks some plastics, rubber, and coatings. May accumulate static electrical charges, and may cause ignition of its vapors.

SYNKLOR (57-74-9) Contact with strong oxidizers may cause fire and explosions. Attacks some plastics, rubber, and coatings.

SYNTEXAN (67-68-5) Forms explosive mixture with air (flash point 203°F/95°C). Reacts violently with oxidizers. Reacts with ethanoyl chloride, boron compounds, halides, metal alkoxides, oxidizers.

SYNTHETIC 3956 (8001-35-2) Reacts with oxidizers, with a risk of fire or explosions. Attacks metals in the presence of moisture.

SYNTHETIC GLYCERIN (56-81-5) Polymerizes at approximately 300°F/150°C. Incompatible with acetic anhydride, potassium permanganate, strong acids, caustics, aliphatic amines, isocyanates, oxidizers.

SYNTHETIC IRON OXIDE (1309-37-1) Contact with hydrogen peroxide, ethylene oxide, calcium hypochlorite will cause explosion. Reacts violently with powdered aluminum, hydrazine, hydrogen trisulfide.

SYNTHETIC RUBBER LATEX (9016-00-6) Incompatible with sulfuric acid, isocyanates, strong oxidizers.

SYSTEMOX (8065-48-3) Forms explosive mixture with air (flash point 113°F/45°C). Incompatible with water, strong oxidizers, caustics.

SYSTOX (8065-48-3) Forms explosive mixture with air (flash point 113°F/45°C). Incompatible with water, strong oxidizers, caustics.

- T -

2,4,5-T (93-76-5) Sealed metal containers may burst in heat above 316°F/158°C. Incompatible with sulfuric acid, bases, ammonia, aliphatic amines, alkanolamines, isocyanates, alkylene oxides, epichlorohydrin.

b-T or beta-T (79-00-5) Incompatible with strong oxidizers, strong caustics, chemically active metals (especially powders), or sodium amide; may cause fire and explosions. Attacks some plastics, rubber, and coatings.

T4 (121-82-4) Contact with mercury fulminate may cause detonation. Contact with combustibles or strong oxidizers may cause fire and explosions.

T-47 (56-38-2) Combustible liquid. Mixtures with endrin may be explosive. Strong oxidizers may cause fire and explosions. Attacks some plastics, rubber, and coatings.

TAA (102-70-5) Forms explosive mixture with air (flash point 140°F/60°C). A strong reducing agent and base; reacts violently with strong oxidizers, acids. Incompatible with many substances including organic anhydrides, isocyanates, alkylene oxides, epichlorohydrin, aldehydes, alcohols, glycols, phenols, cresols, caprolactam solution. Attacks aluminum, copper, copper alloys, tin, zinc.

TAC 121 (7705-07-9) Reacts with water, producing hydrogen chloride fumes. A strong reducing agent; reacts violently with oxidizers. Attacks metals in the presence of moisture.

TAC 131 (7705-07-9) Reacts with water, producing hydrogen chloride fumes. A strong reducing agent; reacts violently with oxidizers. Attacks metals in the presence of moisture.

2,4,5-T, ACID (93-76-5) Sealed metal containers may burst in heat above 316°F/158°C. Incompatible with sulfuric acid, bases, ammonia, aliphatic amines, alkanolamines, isocyanates, alkylene oxides, epichlorohydrin.

TALL OIL, FATTY ACID (8002-26-4) Incompatible with strong acids, oxidizers, nitrates.

TALLEOL (8002-26-4) Incompatible with strong acids, oxidizers, nitrates.

TALLOL (8002-26-4) Incompatible with strong acids, oxidizers, nitrates.

TALLOW (N/A) Incompatible with strong acids, oxidizers.

TALLOW BENZYL DIMETHYLAMMONIUM CHLORIDE (959-55-7) Incompatible with strong oxidizers, strong acids.

TALLOW FATTY ALCOHOL (112-92-5) Incompatible with strong acids, caustics, aliphatic amines, isocyanates.

TALLOW NITRILE (N/A) Incompatible with strong acids, oxidizers.

TALLOW OIL (N/A) Incompatible with strong acids, oxidizers.

TANGANTANGAN OIL (8001-79-4) Incompatible with strong acids, oxidizers, nitrates.

TANNERS OIL (N/A) Oxidizers may cause fire and explosions. Incompatible with nitric acid. May accumulate static electrical charges, and may cause ignition of its vapors.

771

TANOL SECONDAIRE (French) (78-92-2) Forms explosive mixture with air (flash point 75°F/24°C). Attacks some plastics, rubber, and coatings. Forms an explosive peroxide in air. Ignites with chromium trioxide. Incompatible with strong oxidizers, strong acids, aliphatic amines, isocyanates, organic peroxides.

TAP 9 VP (62-73-7) Attacks some plastics, rubber, and coatings.

TAP 85 (58-89-9) Not combustible, but may be dissolved in a combustible solvent. If solvent comes in contact with oxidizers, fire and explosions may result.

TAR (8007-45-2) Forms explosive mixture with air (flash point 60°F/15°C). Incompatible with strong acids, nitrates.

TAR ACIDS (1319-77-3) Incompatible with sulfuric acid, nitric acid, bases, aliphatic amines, amides, oxidizers, chlorosulfonic acid, oleum.

TARAPACAITE (7789-00-6) A powerful oxidizer. Contact with combustibles, reducing agents, organic material or finely divided metals may cause fire and explosions.

TAR CAMPHOR (91-20-3) Forms explosive mixture with air (flash point 174°F/79°C). Incompatible with strong oxidizers, chromium oxide (violent). Attacks some plastics, rubber, and coatings. May accumulate static electrical charges, and may cause ignition of its vapors.

TAR, COAL (8007-45-2) Forms explosive mixture with air (flash point > 60°F/15°C). Incompatible with strong acids, nitrates.

TAR, LIQUID (8007-45-2) Forms explosive mixture with air (flash point > 60°F/15°C). Incompatible with strong acids, nitrates.

TARLON XB (105-60-2) Contact with strong oxidizers may cause fire and explosions.

TAR OIL (65996-93-2) Incompatible with oxidizers, strong acids, caustics, aliphatic amines, isocyanates.

TARS LIQUID (8052-42-4) Incompatible with nitric acid, fluorine, strong oxidizers.

TARTARIC ACID (87-69-4) Aqueous solution reacts with bases, ammonia, amines, isocyanates, alkylene oxides, epichlorohydrin, strong oxidizers.

TARTARIC ACID, COPPER SALT (815-82-7) Acetylene and nitromethane may causes formation of explosive compounds.

TASK (62-73-7) Attacks some plastics, rubber, and coatings.

TASK TABS (62-73-7) Attacks some plastics, rubber, and coatings.

TAT CHLOR 4 (57-74-9) Contact with strong oxidizers may cause fire and explosions. Attacks some plastics, rubber, and coatings.

TBA (75-64-9) Forms explosive mixture with air (flash point 48°F/−9°C). May accumulate static electrical charges, and may cause ignition of its vapors. Incompatible with acids, organic anhydrides, isocyanates, vinyl acetate, acrylates, substituted allyls, alkylene oxides, epichlorohydrin, ketones, aldehydes, alcohols, glycols, phenols, cresols, caprolactam solution, strong oxidizers.

TBA (75-65-0) Forms explosive mixture with air (flash point 52°F/11°C). May accumulate static electrical charges, and may cause ignition of its vapors. Incompatible with strong acids, including mineral acids, strong oxidizers or caustics, aliphatic amines, isocyanates, alkali metals (i.e., lithium, sodium, potassium, rubidium, cesium, francium). Attacks many plastics and some coatings.

TBA (102-82-9) Forms explosive mixture with air (flash point 187°F/83°F). A strong base. Incompatible with strong oxidizers, acids, organic anhydrides, isocyanates, aldehydes. Attacks aluminum, copper, copper alloys, tin, zinc.

TBE (79-27-6) Reacts with chemically active metals, caustics. Hot iron, aluminum or zinc in the presence of steam may produce toxic vapors. Softens or destroys most plastics and rubbers.

TBHP-70 (75-91-2) Forms explosive mixture with air (flash point 100°F/38°C). A powerful oxidizer; reacts violently with reducing agents, organic materials, ethylene dichloride.

TBP (126-73-8) Contact with water produces corrosive phosphoric acid. Incompatible with strong acids, strong oxidizers.

T5 BRUSH KIL (93-76-5) Sealed metal containers may burst in heat above 316°F/158°C. Incompatible with sulfuric acid, bases, ammonia, aliphatic amines, alkanolamines, isocyanates, alkylene oxides, epichlorohydrin.

TBT (98-51-1) Forms explosive mixture with air (flash point 155°F/67°C). Reacts with strong oxidizers. May accumulate static electrical charges, and may cause ignition of its vapors.

p-**TBT** (98-51-1) Forms explosive mixture with air (flash point 155°F/67°C). Reacts with strong oxidizers. May accumulate static electrical charges, and may cause ignition of its vapors.

TCE (79-01-6) Contact with caustics produces a toxic and flammable gas. Reacts violently with chemically active metals. Contact with aluminum may produce a violent, self-accelerating polymerization reaction. Incompatible with acids, organic anhydrides, isocyanates, alkylene oxides, aldehydes, alcohols, glycols, phenols, cresols, caprolactam solution, epichlorohydrin, nitrogen tetroxide, metal powders, oxygen.

TCE (71-55-6) Reacts with acetone, strong caustics, alkaline earth and alkali metals, dinitrogen tetroxide, sodium hydroxide, amides, strong oxidizers, sodium, potassium. Reacts violently with copper, bronze and other metal powders. The uninhibited grade is corrosive to aluminum. Attacks some plastics, rubber, and coatings.

TCE (79-34-5) Exposure to heat, light and air causes formation of corrosive and toxic vapors. Reacts with strong caustics to produce explosive dichloroacetylene. Reacts violently with chemically active metals or sodium amide. In presence of steam, contact with hot iron, aluminum or zinc may produce toxic vapors. Attacks some plastics, rubber, and coatings.

1,1,1-TCE (71-55-6) Reacts with acetone, strong caustics, alkaline earth and alkali metals, dinitrogen tetroxide, sodium hydroxide, amides, strong oxidizers, sodium, potassium. Reacts violently with copper, bronze and other metal powders. The uninhibited grade is corrosive to aluminum. Attacks some plastics, rubber, and coatings.

a-TCE or alpha-TCE (71-55-6) Reacts with acetone, strong caustics, alkaline earth and alkali metals, dinitrogen tetroxide, sodium hydroxide, amides, strong oxidizers, sodium, potassium. Reacts violently with copper, bronze and other metal powders. The uninhibited grade is corrosive to aluminum. Attacks some plastics, rubber, and coatings.

TCM (67-66-3) Decomposes in the presence of excess water or at high temperatures, producing phosgene and hydrogen chloride. Can become explosive in the presence of strong alkalies and water. Can accumulate static electrical charges. In contact with water and at high

773

temperatures, it becomes corrosive; attacks iron and other metals. Incompatible with acetone, aluminum, strong oxidizers, potassium, sodium, chemically active metals, strong bases. Attacks plastics and rubber.

TCP (78-30-8) Contact with magnesium or strong oxidizers may cause fire and explosions. Attacks some plastics, rubber, and coatings.

TDA (95-80-7) Incompatible with oxidizers, acids, organic anhydrides, isocyanates, aldehydes. Attacks aluminum, brass, bronze, copper, zinc.

TDE (72-54-8) Contact with strong oxidizers may cause fire and explosions.

P,P'-TDE (72-54-8) Contact with strong oxidizers may cause fire and explosions.

TDI (584-84-9) Incompatible with strong acids including nonoxidizing mineral and organic acids, caustics, ammonia, amines, amides, alcohols, glycols, caprolactam solution. Water contact causes violent foaming and spattering; produces carbon dioxide and an organic base. Attacks copper and its alloys, some plastics including polyethylene, and rubber.

TDI-80 (584-84-9) Incompatible with strong acids including nonoxidizing mineral and organic acids, caustics, ammonia, amines, amides, alcohols, glycols, caprolactam solution. Water contact causes violent foaming and spattering; produces carbon dioxide and an organic base. Attacks copper and its alloys, some plastics including polyethylene and rubber.

2,4-TDI (584-84-9) Incompatible with strong acids, including nonoxidizing mineral and organic acids, caustics, ammonia, amines, amides, alcohols, glycols, caprolactam solution. Water contact causes violent foaming and spattering; produces carbon dioxide and an organic base. Attacks copper and its alloys, some plastics including polyethylene, and rubber.

TEA (97-93-8) A dangerous explosion hazard. Ignites spontaneously, at all temperatures, in air. A powerful reducing agent; reacts violently with oxidizers, water (with formation of ethane gas), alcohols, amines, carbon dioxide, carbon tetrachloride, halogenated hydrocarbons, oxides of nitrogen or sulfur, phenols, and many other substances. May accumulate static electrical charges, and may cause ignition of its vapors.

TEA (102-71-6) Aqueous solution is caustic. Incompatible with acids, organic anhydrides, isocyanates, vinyl acetate, acrylates, substituted allyls, alkylene oxides, epichlorohydrin, aldehydes, strong oxidizers. Corrodes copper and its alloys.

TEA (107-10-8) Forms explosive mixture with air. Incompatible with acids, organic anhydrides, halogenated hydrocarbons, isocyanates, nitroparaffins, vinyl acetate, acrylates, substituted allyls, alkylene oxides, epichlorohydrin, ketones, aldehydes, alcohols, glycols, mercury, phenols, cresols, triethyl aluminum, caprolactam solution, strong oxidizers. Attacks aluminum, copper, lead, tin, zinc, and alloys.

TEA (121-44-8) Forms explosive mixture with air (flash point 20°F/−6.7°C). Incompatible with strong acids, mineral acids, organic acids (may cause violent spattering), and with organic anhydrides, halogenated hydrocarbons, isocyanates, nitroparaffins, nitrogen tetroxide, vinyl acetate, acrylates, substituted allyls, alkylene oxides, epichlorohydrin, ketones, aldehydes, alcohols, glycols, phenols, cresols, caprolactam solution, strong oxidizers. Attacks aluminum, copper, lead, tin, zinc, and alloys, and some plastics, rubber, and coatings.

TEABERRY OIL (119-36-0) Incompatible with strong acids, nitrates, strong oxidizers.

TEAR GAS (532-27-4) Incompatible with water or steam.

TECHNETIUM TC 99M SULFUR COLLOID (7704-34-9) Combustible solid. Liquid causes formation of sulfur dioxide with air. Reacts violently with strong oxidizers. Forms explosive, shock-sensitive, or pyrophoric mixtures with ammonia, ammonium nitrate, bromates, calcium carbide, charcoal, chlorates, hydrocarbons, iodates, iron. Reacts violently with halogen compounds, sodium, tin, uranium, and other compounds. Attacks steel when moist. May accumulate static electrical charges, and may cause ignition of its vapors.

90 TECHNICAL GLYCERIN (56-81-5) Polymerizes at approximately 300°F/150°C. Incompatible with acetic anhydride, potassium permanganate, strong acids, caustics, aliphatic amines, isocyanates, oxidizers.

TECQUINOL (123-31-9) Incompatible with strong oxidizers, caustics. May be oxidized to quinone at room temperatures in the presence of moisture. May explode on contact with oxygen.

TECZA (112-24-3) Forms explosive mixture with air (flash point 275°F/135°C). Incompatible with acids, organic anhydrides, chlorinated hydrocarbons, isocyanates, vinyl acetate, acrylates, substituted allyls, alkylene oxides, epichlorohydrin, ketones, aldehydes, alcohols, glycols, phenols, cresols, caprolactam solution, strong oxidizers. Attacks aluminum, copper, lead, tin, zinc, and alloys.

TEDP (3689-24-5) Containers may burst in heat. Strong oxidizers may cause fire and explosions. Attacks some plastics, rubber, and coatings.

TEDTP (3689-24-5) Containers may burst in heat. Strong oxidizers may cause fire and explosions. Attacks some plastics, rubber, and coatings.

TEFLON MONOMER (116-14-3) A highly reactive flammable gas. Able to form unstable peroxides; if inhibitor is not present in adequate concentrations, explosive polymerization may occur. Reacts violently with air, oxygen, oxidizers, sulfur trioxide.

TEG (112-27-6) Incompatible with sulfuric acid, isocyanates, perchloric acid, strong oxidizers.

TEG (112-60-7) Incompatible with sulfuric acid, isocyanates, perchloric acid. Attacks some plastics, rubber, and coatings.

TEKRESOL (1319-77-3) Incompatible with sulfuric acid, nitric acid, bases, aliphatic amines, amides, oxidizers, chlorosulfonic acid, oleum.

TEKWAISA (298-00-0) Mixtures with magnesium may be explosive.

TEL (78-00-2) Forms explosive mixture with air and decomposes above 200°F/93°C. Incompatible with strong oxidizers, concentrated acids may cause fire and explosions. Attacks some plastics, rubber, and coatings.

TELONE (542-75-6) Forms explosive mixture with air (flash point 95°F/35°C). Reacts violently with strong oxidizers. May accumulate static electrical charges, and may cause ignition of its vapors.

Incompatible with strong acids, oxidizers, aluminum or magnesium compounds, aliphatic amines, alkanolamines, alkaline materials, or corrosives.

TELONE (8003-19-8) Incompatible with strong acids, oxidizers, aluminum or magnesium compounds, aliphatic amines, alkanolamines, alkaline materials, or corrosives.

TELONE II SOIL FUMIGANT (542-75-6) Forms explosive mixture with air (flash point 95°F/35°C). Reacts violently with strong oxidizers. May accumulate static electrical charges, and may cause ignition of its vapors. Incompatible with strong acids, oxidizers, aluminum or magnesium compounds, aliphatic amines, alkanolamines, alkaline materials, or corrosives.

TELVAR DIURON WEED KILLER (330-54-1) Hydrolyzes in fairly strong acids.

TEN (121-44-8) Forms explosive mixture with air (flash point 20°F/−6.7°C). Incompatible with strong acids, mineral acids, organic acids (may cause violent spattering), and with organic anhydrides, halogenated hydrocarbons, isocyanates, nitroparaffins, nitrogen tetroxide, vinyl acetate, acrylates, substituted allyls, alkylene oxides, epichlorohydrin, ketones, aldehydes, alcohols, glycols, phenols, cresols, caprolactam solution, strong oxidizers. Attacks aluminum, copper, lead, tin, zinc, and alloys, and some plastics, rubber, and coatings.

TENAC (62-73-7) Attacks some plastics, rubber, and coatings.

TENDUST (54-11-5) Incompatible with strong acids, strong oxidizers. Attacks some plastics, rubber, and coatings. May accumulate static electrical charges, and may cause ignition of its vapors.

TENITE 423 (9003-07-0) Incompatible with strong acids, strong oxidizers, chlorine, potassium permanganate.

TENOX HQ (123-31-9) Incompatible with strong oxidizers, caustics. May be oxidized to quinone at room temperatures in the presence of moisture. May explode on contact with oxygen.

TENOX P GRAIN PRESERVATIVE (79-09-4) Forms explosive mixture with air (flash point 126°F/52°C). Incompatible with sulfuric acid, strong bases, ammonia, aliphatic amines, alkanolamines, isocyanates, alkylene oxides, epichlorohydrin, oxidizers.

TENTACHLORURE d'ANTIMOINE (French) (7647-18-9) Water contact produces hydrogen chloride gas. Reacts violently with ammonia, caustics. Attacks many metals. Contact with air forms heavier-than-air corrosive vapor.

TEOS (78-10-4) Forms explosive mixture with air (flash point 99°F/37°C). Incompatible with strong oxidizers, strong acids, water (produces volatile ethyl alcohol). Attacks some plastics and rubber.

TEP (78-40-0) Incompatible with strong acids, nitrates, oxidizers.

TEP (107-49-3) Decomposes above 300°F/150°C producing flammable ethylene gas. Strong oxidizers may cause fire and explosions. Attacks some plastics, rubber, and coatings.

TEPP (107-49-3) Decomposes above 300°F/150°C producing flammable ethylene gas. Strong oxidizers may cause fire and explosions. Attacks some plastics, rubber, and coatings.

TERABOL (74-83-9) Incompatible with strong oxidizers, aluminum, dimethylsulfoxide, ethylene oxide, water. Attacks zinc, magnesium, alkali metals and their alloys, and some plastics, rubber, and coatings.

TERCYL (63-25-2) Incompatible with strong oxidizers, strongly alkaline pesticides.

TEREBENTHINE (French) (8006-64-2) Forms explosive mixture with air (flash point 95°F/35°C). Incompatible with strong oxidizers, especially chlorine; may cause fire and explosions. Incompatible with strong acids, chromic anhydride, chromyl chloride, hexachloromelamine, stannic chloride. Attacks ordinary rubber.

TEREPHTHALIC ACID, DIMETHYL ESTER (120-61-6) Incompatible with strong acids, nitrates, strong oxidizers.

TEREPHTHALIC ACID, METHYL ESTER (120-61-6) Incompatible with strong acids, nitrates, strong oxidizers.

TERETON (79-20-9) Forms explosive mixture with air (flash point 14°F/−10°C). Incompatible with strong acids, nitrates, oxidizers, bases. Attacks some plastics. May accumulate static electrical charges, and may cause ignition of its vapors.

TERGITOL NONIONIC 3-A-6 (N/A) Incompatible with strong acids, caustics, aliphatic amines, isocyanates.

TERGITOL NONIONIC 45-S-10 (N/A) Incompatible with strong acids, caustics, aliphatic amines, isocyanates.

TERGITOL NONIONIC TMN (9008-57-5) Incompatible with strong acids, caustics, aliphatic amines, isocyanates.

TERMITKIL (95-50-1) Forms explosive mixture with air (flash point 151°F/66°C). Incompatible with strong oxidizers, hot aluminum or aluminum alloy. Attacks some plastics, rubber, and coatings.

TERM-I-TROL (87-86-5) Hot water causes decomposition, producing hydrochloric acid. Strong oxidizers may cause fire and explosions.

TERPENTIN OEL (German) (8006-64-2) Forms explosive mixture with air (flash point 95°F/35°C). Incompatible with strong oxidizers, especially chlorine; may cause fire and explosions. Incompatible with strong acids, chromic anhydride, chromyl chloride, hexachloromelamine, stannic chloride. Attacks ordinary rubber.

TERPINENE (138-86-3) Forms explosive mixture with air (flash point 115°F/46°C). Strong oxidizers may cause fire and explosions.

d-1,8-TERPODIENE or delta-1,8-TERPODIENE (138-86-3) Forms explosive mixture with air (flash point 115°F/46°C). Strong oxidizers may cause fire and explosions.

TERR-O-GAS 100 (74-83-9) Incompatible with strong oxidizers, aluminum, dimethylsulfoxide, ethylene oxide, water. Attacks zinc, magnesium, alkali metals and their alloys, some plastics, rubber, and coatings.

TERSAN (137-26-8) Combustible solid (flash point 192°F/89°C). Strong oxidizers may cause fire and explosions; contact with strong acid or oxidizable materials produces toxic gases.

TERSAN 1991 (17804-35-2) Heat, water, strong acids and strong alkalies can cause decomposition and formation of toxic oxides of nitrogen.

TERTAL D (106-50-3) A strong reducing agent; reacts violently with oxidizers, combustibles, organic substances. Incompatible with strong acids, organic anhydrides, isocyanates, aldehydes. Heat and light contribute to instability.

TERTROSULPHUR BLACK PB (51-28-5) Explosion is caused by heat, friction or shock. Contact with reducing agents, combustibles may cause fire and explosions. Forms explosive salts with ammonia or strong bases. May accumulate static electrical charges, and may cause ignition of its vapors.

TERTROSULSULPHUR PBR (51-28-5) Explosion is caused by heat, friction or shock. Contact with reducing agents, combustibles may cause fire and explosions. Forms explosive salts with ammonia or strong bases. May accumulate static electrical charges, and may cause ignition of its vapors.

TESCOL (64-17-5) Forms explosive mixture with air (flash point 65°F/18°C). May accumulate static electrical charges, and may cause ignition of its vapors. Reactions may be violent with oleum, sulfuric acid, nitric acid, bases, aliphatic amines, isocyanates, oxidizers.

TESCOL (107-21-1) Incompatible with strong acids, caustics, aliphatic amines, isocyanates, chlorosulfonic acid, oleum, strong oxidizers.

TESULOID (7704-34-9) Combustible solid. Liquid causes formation of sulfur dioxide with air. Reacts violently with strong oxidizers. Forms explosive, shock-sensitive, or pyrophoric mixtures with ammonia, ammonium nitrate, bromates, calcium carbide, charcoal, chlorates, hydrocarbons, iodates, iron. Reacts violently with halogen compounds, sodium, tin, uranium, and other compounds. Attacks steel when moist. May accumulate static electrical charges, and may cause ignition of its vapors.

TETA (112-24-3) Forms explosive mixture with air (flash point 275°F/135°C). Incompatible with acids, organic anhydrides, chlorinated hydrocarbons, isocyanates, vinyl acetate, acrylates, substituted allyls, alkylene oxides, epichlorohydrin, ketones, aldehydes, alcohols, glycols, phenols, cresols, caprolactam solution, strong oxidizers. Attacks aluminum, copper, lead, tin, zinc, and alloys.

TETAN (509-14-8) Combustible organic matter wet with this chemical may be highly explosive, and susceptible to mild shock. Contact with hydrocarbons, alkalies, or metals produces explosive mixtures. Attacks some plastics, rubber, and coatings.

O,O,O',O'-TETRAAETHYL-BIS(DITHIOPHOSPHAT) (German) (563-12-2) Incompatible with alkaline formulations. Mixtures with magnesium may be explosive.

O,O,O,O-TETRAAETHYL-DIPHOSPHAT, TETRACAP (German) (127-18-4) Incompatible with strong oxidizers, finely divided metals, caustics. Stable up to 258°F/126°C; at this temperature product gives off poisonous fumes.

TETRAAMINE COPPER SULFATE (10380-29-7) Water contact causes formation of ammonia. Incompatible with chlorine, fluorine, peroxides, hydroxylamine, magnesium.

1,4,7,10-TETRAAZADECANE (112-24-3) Forms explosive mixture with air (flash point 275°F/135°C). Incompatible with acids, organic anhydrides, chlorinated hydrocarbons, isocyanates, vinyl acetate, acrylates, substituted allyls, alkylene oxides, epichlorohydrin, ketones, aldehydes, alcohols, glycols, phenols, cresols, caprolactam solution, strong oxidizers. Attacks aluminum, copper, lead, tin, zinc, and alloys.

3,6,9,12-TETRAAZATETRADECANE-1,14-DIAMINE (4067-16-7) Incompatible with acids, organic anhydrides, isocyanates, aldehydes, oxidizers.

1,1,2,2-TETRABROMAETHAN (German) (79-27-6) Reacts with chemically active metals, caustics. Hot iron, aluminum, or zinc in the presence of steam may produce toxic vapors. Softens or destroys most plastics and rubbers.

TETRABROMOACETYLENE (79-27-6) Reacts with chemically active metals, caustics. Hot iron, aluminum, or zinc in the presence of steam may produce toxic vapors. Softens or destroys most plastics and rubbers.

1,1,2,2-TETRABROMOETANO (Italian) (79-27-6) Reacts with chemically active metals, caustics. Hot iron, aluminum, or zinc in the presence of steam may produce toxic vapors. Softens or destroys most plastics and rubbers.

TETRABROMOETHANE (79-27-6) Reacts with chemically active metals, caustics. Hot iron, aluminum, or zinc in the presence of steam may produce toxic vapors. Softens or destroys most plastics and rubbers.

s-TETRABROMOETHANE or *sym*-TETRABROMOETHANE or *symmetrical*-TETRABROMOETHANE (79-27-6) Reacts with chemically active metals, caustics. Hot iron, aluminum, or zinc in the presence of steam may produce toxic vapors. Softens or destroys most plastics and rubbers.

TETRABROMOMETHANE (558-13-4) Lithium or hexylcyclohexyldilead may cause an explosion.

1,1,2,2-TETRABROOMETHAAN (Dutch) (79-27-6) Reacts with chemically active metals, caustics. Hot iron, aluminum, or zinc in the presence of steam may produce toxic vapors. Softens or destroys most plastics and rubbers.

TETRABUTYLSTANNINE (1461-25-2) Strong oxidizers may cause fire and explosions. Attacks some forms of plastics, rubber, and coatings.

TETRABUTYLTIN or *n*-TETRABUTYL TIN (1461-25-2) Strong oxidizers may cause fire and explosions. Attacks some forms of plastics, rubber, and coatings.

TETRACARBONYL NICKEL (13463-39-3) Forms explosive mixture with air (flash point −4°F/−20°C). May explode when heated above 140°F/60°C. In the presence of air, forms a deposit that becomes peroxidized; this tends to decompose and ignite. Contact with acids and oxidizers can cause fire and explosions. Vapor may promote the ignition of mixtures of combustible vapors (such as gasoline) and air. Attacks some plastics, rubber, and coatings. Store under inert gas blanket. May accumulate static electrical charges, and may cause ignition of its vapors.

TETRACHLORO-1,2-DIFLUOROETHANE (76-12-0) Reacts with chemically active metals, powdered aluminum, zinc, and magnesium. Attacks some forms of plastics, rubber, and coatings.

1,1,2,2-TETRACHLOORETHAAN (Dutch) (127-18-4) Incompatible with strong oxidizers, finely divided metals, caustics. Stable up to 258°F/126°C; at this temperature product gives off poisonous fumes.

TETRACHLOORKOOLSTOF (Dutch) (56-23-5) Becomes corrosive when in contact with water. Corrosive to metals. Reacts violently with many compounds. Decomposes on contact with chemically active metals such as sodium, potassium, and magnesium. Incompatible with allyl alcohol, fluorine gas, alkali metals, aluminum. Attacks some coatings, plastics, and rubber.

TETRACHLOORMETAN (Dutch) (56-23-5) Becomes corrosive when in contact with water. Corrosive to metals. Reacts violently with many compounds. Decomposes on contact with chemically active metals such as sodium, potassium and magnesium. Incompatible with allyl alcohol, fluorine gas, alkali metals, aluminum. Attacks some coatings, plastics, and rubber.

1,1,2,2-TETRACHLORAETHAN (German) (127-18-4) Incompatible with strong oxidizers, finely divided metals, caustics. Stable up to 258°F/126°C; at this temperature product gives off poisonous fumes.

N-(1,1,2,2-TETRACHLORAETHYLTHIO)-CYCLOHEX-4-EN-1,4-DIACARBOXIMID (German) (2425-06-1) Incompatible with acids or acid vapor. Strongly alkaline conditions contribute to instability.

TETRACHLORETHANE (127-18-4) Incompatible with strong oxidizers, finely divided metals, caustics. Stable up to 258°F/126°C; at this temperature product gives off poisonous fumes.

1,1,2,2-TETRACHLORETHANE (French) (127-18-4) (127-18-4) Incompatible with strong oxidizers, finely divided metals, caustics. Stable up to 258°F/126°C; at this temperature product gives off poisonous fumes.

TETRACHLORKOHLENSTOFF, TETRA (German) (56-23-5) Becomes corrosive when in contact with water. Corrosive to metals. Reacts violently with many compounds. Decomposes on contact with chemically active metals such as sodium, potassium and magnesium. Incompatible with allyl alcohol, fluorine gas, alkali metals, aluminum. Attacks some coatings, plastics, and rubber.

TETRACHLORMETHAN (German) (56-23-5) Becomes corrosive when in contact with water. Corrosive to metals. Reacts violently with many compounds. Decomposes on contact with chemically active metals such as sodium, potassium and magnesium. Allyl alcohol, fluorine gas, alkali metals, aluminum. Attacks some coatings, plastics, and rubber.

1,1,2,2-TETRACHLORO- (127-18-4) Incompatible with strong oxidizers, finely divided metals, caustics. Stable up to 258°F/126°C; at this temperature product gives off poisonous fumes.

1,1,1,2-TETRACHLORO-2,2-DIFLUOROETHANE (76-11-9) Reacts with chemically active metals, oxidizers. Attacks some forms of plastics, rubber, and coatings.

1,1,2,2-TETRACHLORO-1,2-DIFLUOROETHANE (76-12-0) Reacts with chemically active metals, powdered aluminum, zinc, and magnesium.

sym-**TETRACHLORO-1,2-DIFLUOROETHANE** (76-12-0) Reacts with chemically active metals, powdered aluminum, zinc, and magnesium. Attacks some forms of plastics, rubber, and coatings.

N-1,1,2,2-TETRACHLOROETHYLMERCAPTO-4-CYCLOHEXENE-1,2-CARBOXIMIDE (2425-06-1) Incompatible with acids or acid vapor. Strong alkaline conditions contribute to instability.

TETRACHLOROCARBON (56-23-5) Becomes corrosive when in contact with water. Corrosive to metals. Reacts violently with many compounds. Decomposes on contact with chemically active metals such as sodium, potassium and magnesium. Incompatible with allyl alcohol, fluorine gas, alkali metals, aluminum. Attacks some coatings, plastics, and rubber.

TETRACHLORODIPHENYLETHANE (72-54-8) Contact with strong oxidizers may cause fire and explosions.

TETRACHLOROETHANE (79-34-5) Exposure to heat, light, and air causes formation of corrosive and toxic vapors. Reacts with strong caustics to produce explosive dichloroacetylene. Reacts violently with chemically active metals or sodium amide. In presence of steam, contact with hot iron, aluminum, or zinc may causes formation of toxic vapors. Attacks some plastics, rubber, and coatings.

sym-**TETRACHLOROETHANE** or *symmetrical*-**TETRACHLORO-ETHANE** (79-34-5) Exposure to heat, light and air causes formation of corrosive and toxic vapors. Reacts with strong caustics to produce explosive dichloroacetylene. Reacts violently with chemically active metals or sodium amide. In presence of steam, contact with hot iron, aluminum, or zinc may causes formation of toxic vapors. Attacks some plastics, rubber, and coatings.

TETRACHLOROETHYLENE (127-18-4) Incompatible with strong oxidizers, finely divided metals, caustics. Stable up to 258°F/126°C; at this temperature product gives off poisonous fumes.

N-(1,1,2,2-TETRACHLOROETHYL)SULFENYL)-*cis*-4-CYCLO-HEXENE-1,2-DIC ARBOXIMIDE (2425-06-1) Incompatible with acids or acid vapor. Strongly alkaline conditions contribute to instability.

N-(1,1,2,2-TETRACHLOROETHYLTHIO)-4-CYCLOHEXENE-1,2-DICARBOXIMIDE (2425-06-1) Acids or acid vapor. Strong alkaline conditions contribute to instability.

TETRACHLOROMETHANE (56-23-5) Becomes corrosive when in contact with water. Corrosive to metals. Reacts violently with many compounds. Decomposes on contact with chemically active metals such as sodium, potassium and magnesium. Incompatible with allyl alcohol, fluorine gas, alkali metals, aluminum. Attacks some coatings, plastics, and rubber.

TETRACHLOROSILANE (10026-04-7) Water contact produces hydrochloric acid. Reacts violently with potassium or sodium. Attacks metals in the presence of moisture.

TETRACHLORO ZIRCONIUM (10026-11-6) Decomposes in the presence of moist air. Reacts with water or moisture to produce hydrochloric acid fume. Alkali metals may cause explosions. Attacks some plastics, rubber, and coatings.

TETRACHLORURE d'ACETYLENE (French) (79-34-5) Exposure to heat, light and air causes formation of corrosive and toxic vapors. Reacts with strong caustics to form explosive dichloroacetylene. Reacts violently with chemically active metals or sodium amide. In presence of steam, contact with hot iron, aluminum, or zinc may cause formation of toxic vapors. Attacks some plastics, rubber, and coatings.

TETRACHLORURE de CARBONE (French) (56-23-5) Becomes corrosive when in contact with water. Corrosive to metals. Reacts violently with many compounds. Decomposes on contact with chemically active metals such as sodium, potassium, and magnesium. Incompatible with allyl alcohol, fluorine gas, alkali metals, aluminum. Attacks some coatings, plastics, and rubber.

1,1,2,2-TETRACLOROETANO (Italian) (79-34-5) Exposure to heat, light and air causes formation of corrosive and toxic vapors. Reacts with strong caustics to form explosive dichloroacetylene. Reacts violently with chemically active metals or sodium amide. In presence of steam, contact with hot iron, aluminum, or zinc may causes formation of toxic vapors. Attacks some plastics, rubber, and coatings.

TETRACLOROMETANO (Italian) (56-23-5) Becomes corrosive when in contact with water. Corrosive to metals. Reacts violently with many compounds. Decomposes on contact with chemically active metals such as sodium, potassium and magnesium. Incompatible with allyl alcohol, fluorine gas, alkali metals, aluminum. Attacks some coatings, plastics, and rubber.

TETRACLORURO di CARBONIO (Italian) (56-23-5) Becomes corrosive when in contact with water. Corrosive to metals. Reacts violently with many compounds. Decomposes on contact with chemically active metals such as sodium, potassium and magnesium. Incompatible with allyl alcohol, fluorine gas, alkali metals, aluminum. Attacks some coatings, plastics, and rubber.

TETRADECANOL or TETRADECANOL, MIXED ISOMERS (75-50-3) Flammable gas (flash point,liquid 10°F/−12°C). Incompatible with acids, organic anhydrides, ethylene oxide, isocyanates, aldehydes, oxidizers, triethylaluminum.

1-TETRADECANOL (75-50-3) Flammable gas (flash point,liquid 10°F/−12°C). Incompatible with acids, organic anhydrides, ethylene oxide, isocyanates, aldehydes, oxidizers, triethylaluminum.

N-TETRADECYL ALCOHOL (75-50-3) Flammable gas (flash point,liquid 10°F/−12°C). Incompatible with acids, organic anhydrides, ethylene oxide, isocyanates, aldehydes, oxidizers, triethylaluminum.

TETRADECYLBENZENE (N/A) Incompatible with nitric acid. Attacks some plastics, rubber, and coatings.

TETRAETHOXYSILANE (78-10-4) Forms explosive mixture with air (flash point 99°F/37°C). Incompatible with strong oxidizers, strong acids, water (produces volatile ethyl alcohol). Attacks some plastics and rubber.

O,O,O,O-TETRAETHYL-DIFOSFAAT (Dutch) (107-49-3) Decomposes above 300°F/150°C producing flammable ethylene gas. Strong oxidizers may cause fire and explosions. Attacks some plastics, rubber, and coatings.

O,O,O,O-TETRAETHYL-DITHIO-DIFOSFAAT (Dutch) (3689-24-5) Containers may burst in heat. Strong oxidizers may cause fire and explosions. Attacks some plastics, rubber, and coatings.

TETRAETHYL DITHIOPYROPHOSPHATE (3689-24-5) Containers may burst in heat. Strong oxidizers may cause fire and explosions. Attacks some plastics, rubber, and coatings.

O,O,O,O-TETRAETHYLDITHIOPYROPHOSPHATE (3689-24-5) Containers may burst in heat. Strong oxidizers may cause fire and explosions. Attacks some plastics, rubber, and coatings.

TETRAETHYLDITHIOPYROPHOSPHATE, LIQUID (3689-24-5) Containers may burst in heat. Strong oxidizers may cause fire and explosions. Attacks some plastics, rubber, and coatings.

TETRAETHYLENE GLYCOL: (112-60-7) Incompatible with sulfuric acid, isocyanates, perchloric acid. Attacks some plastics, rubber, and coatings.

TETRAETHYLENE PENTAMINE (112-57-2) Forms explosive mixture with air. Incompatible with acids, organic anhydrides, halogenated hydrocarbons, isocyanates, nitroparaffins, vinyl acetate, acrylates, substituted allyls, alkylene oxides, epichlorohydrin, ketones, aldehydes, alcohols, glycols, mercury, phenols, cresols, caprolactam solution, strong oxidizers. Attacks aluminum, copper, lead, tin, zinc, and alloys.

TETRAETHYLE PLUMB (French) (78-00-2) Forms explosive mixture with air and decomposes above 200°F/93°C. Incompatible with strong oxidizers, concentrated acids; may cause fire and explosions. Attacks some plastics, rubber, and coatings.

TETRAETHYL LEAD (78-00-2) Forms explosive mixture with air and decomposes above 200°F/93°C. Incompatible with strong oxidizers, concentrated acids may cause fire and explosions. Attacks some plastics, rubber, and coatings.

O,O,O,O-TETRAETHYL S,S'-METHYLENEBIS(DITHIO-PHOSPHATE) (563-12-2) Incompatible with alkaline formulations. Mixtures with magnesium may be explosive.

O,O,O',O'-TETRAETHYL S,S'-METHYLENEBISPHOSPHORDI-THIOATE (563-12-2) Incompatible with alkaline formulations. Mixtures with magnesium may be explosive.

TETRAETHYL S,S'-METHYLENE BIS(PHOSPHOROTHIO-LOTHIONATE) (563-12-2) Incompatible with alkaline formulations. Mixtures with magnesium may be explosive.

O,O,O',O'-TETRAETHYL S,S'-METHYLENE DI(PHOS-PHORODITHIOATE) (563-12-2) Incompatible with alkaline formulations. Mixtures with magnesium may be explosive.

TETRAETHYL ORTHOSILICATE (78-10-4) Forms explosive mixture with air (flash point 99°F/37°C). Incompatible with strong oxidizers, strong acids, water (produces volatile ethyl alcohol). Attacks some plastics and rubber.

TETRAETHYLPLUMBANE (78-00-2) Forms explosive mixture with air and decomposes above 200°F/93°C. Incompatible with strong oxidizers, concentrated acids may cause fire and explosions. Attacks some plastics, rubber, and coatings.

TETRAETHYL PYROFOSFAAT (Belgian) (107-49-3) Decomposes above 300°F/150°C producing flammable ethylene gas. Strong oxidizers may cause fire and explosions. Attacks some plastics, rubber, and coatings.

TETRAETHYL PYROPHOSPHATE (107-49-3) Decomposes above 300°F/150°C, producing flammable ethylene gas. Strong oxidizers may cause fire and explosions. Attacks some plastics, rubber, and coatings.

TETRAETHYLPYROPHOSPHATE (107-49-3) Decomposes above 300°F/150°C), producing flammable ethylene gas. Strong oxidizers may cause fire and explosions. Attacks some plastics, rubber, and coatings.

TETRAETHYL PYROPHOSPHATE, LIQUID (107-49-3) Decomposes above 300°F/150°C, producing flammable ethylene gas. Strong oxidizers may cause fire and explosions. Attacks some plastics, rubber, and coatings.

TETRAETHYL SILICATE (78-10-4) Forms explosive mixture with air (flash point 99°F/37°C). Incompatible with strong oxidizers, strong acids, water (produces volatile ethyl alcohol). Attacks some plastics and rubber.

O,O,O,O-TETRAETIL-DITIO-PIROFOSFATO (Italian) (3689-24-5) Containers may burst in heat. Strong oxidizers may cause fire and explosions. Attacks some plastics, rubber, and coatings.

O,O,O,O-TETRAETIL-PIROFOSFATO (Italian) (107-49-3) Decomposes above 300°F/150°C producing flammable ethylene gas. Strong oxidizers may cause fire and explosions. Attacks some plastics, rubber, and coatings.

TETRAFINOL (56-23-5) Becomes corrosive when in contact with water. Corrosive to metals. Reacts violently with many compounds. Decomposes on contact with chemically active metals such as sodium, potassium and magnesium. Incompatible with allyl alcohol, fluorine gas, alkali metals, aluminum. Attacks some coatings, plastics, and rubber.

TETRAFLUROBORATE(1-) LEAD(2+) (13814-96-5) Aqueous solution is acidic; corrodes many metals.

TETRAFLUOROETHENE (116-14-3) A highly reactive flammable gas. Able to form unstable peroxides; if inhibitor is not present in adequate concentrations, explosive polymerization may occur. Reacts violently with air, oxygen, oxidizers, sulfur trioxide.

TETRAFLUOROETHYLENE or TETRAFLUOROETHYLENE, INHIBITED (116-14-3) A highly reactive flammable gas. Able to form unstable peroxides; if inhibitor is not present in adequate concentrations, explosive polymerization may occur. Reacts violently with air, oxygen, oxidizers, sulfur trioxide.

1,1,2,2-TETRAFLUOROETHYLENE (116-14-3) A highly reactive flammable gas. Able to form unstable peroxides; if inhibitor is not present in adequate concentrations, explosive polymerization may occur. Reacts violently with air, oxygen, oxidizers, sulfur trioxide.

TETRAFORM (56-23-5) Becomes corrosive when in contact with water. Corrosive to metals. Reacts violently with many compounds. Decomposes on contact with chemically active metals such as sodium, potassium, and magnesium. Incompatible with allyl alcohol, fluorine gas, alkali metals, aluminum. Attacks some coatings, plastics and rubber.

TETRAHYDROBENZENE (110-83-8) Forms explosive mixture with air (flash point less than 20°F/−7°C. May polymerize from buildup of unstable peroxides. May accumulate static electrical charges, and may cause ignition of its vapors.

1,2,3,4-TETRAHYDROBENZENE (110-83-8) Forms explosive mixture with air (flash point less than 20°F/−7°C. May polymerize from buildup of unstable peroxides. May accumulate static electrical charges, and may cause ignition of its vapors.

3A,4,7,7A-TETRAHYDRODIMETHYL-4,7-METHANOINDENE (26472-00-4) Forms explosive mixture with air (flash point 80°F/27°C). Incompatible with strong acids, strong oxidizers.

TETRAHYDRO-1,4-DIOXIN (123-91-1) Forms explosive mixture with air (flash point 54°F/12°C). Moisture causes formation of unstable peroxides. May accumulate static electrical charges, and may cause ignition of its vapors. Incompatible with strong oxidizers, strong acids. Attacks many plastics.

TETRAHYDRO-p-DIOXIN (123-91-1) Forms explosive mixture with air (flash point 54°F/12°C). Moisture causes formation of unstable peroxides. May accumulate static electrical charges, and may cause ignition of its vapors. Incompatible with strong oxidizers, strong acids. Attacks many plastics.

TETRAHYDROFURAAN (Dutch) (109-99-9) Forms explosive mixture with air (flash point 6°F/−14°C). Unless inhibited, can form unstable and explosive peroxides. Incompatible with strong acids, strong oxidizers. Attacks some plastics. May accumulate static electric charges that can result in ignition of its vapors.

TETRAHYDROFURAN (109-99-9) Forms explosive mixture with air (flash point 6°F/−14°C). Unless inhibited, can form unstable and explosive peroxides. Incompatible with strong acids, strong oxidizers. Attacks some plastics. May accumulate static electric charges that can result in ignition of its vapors.

TETRAHYDRO-2-FURANMETHANOL (97-99-4) Forms explosive mixture with air (flash point 162°F/72°C). Incompatible with strong acids, caustics, aliphatic amines, isocyanates, oxidizers. Unless inhibited, can form unstable and explosive peroxides.

TETRAHYDROFURANNE (French) (109-99-9) Forms explosive mixture with air (flash point 6°F/−14°C). Unless inhibited, can form unstable and explosive peroxides. Incompatible with strong acids, strong oxidizers. Attacks some plastics. May accumulate static electric charges that can result in ignition of its vapors.

TETRAHYDROFURFURYL ALCOHOL (97-99-4) Forms explosive mixture with air (flash point 162°F/72°C). Incompatible with strong acids, caustics, aliphatic amines, isocyanates, oxidizers. Unless inhibited, can form unstable and explosive peroxides.

TETRAHYDROFURFURYL CARBINOL (97-99-4) Forms explosive mixture with air (flash point 162°F/72°C). Incompatible with strong acids, caustics, aliphatic amines, isocyanates, oxidizers. Unless inhibited, can form unstable and explosive peroxides.

TETRAHYDRO-1,4-ISOXAZINE (110-91-8) Forms explosive mixture with air (flash point 98°F/37°C). Incompatible with acids, organic anhydrides, isocyanates, vinyl acetate, acrylates, substituted allyls, alkylene oxides, epichlorohydrin, ketones, aldehydes, alcohols, glycols, phenols, cresols, caprolactam solution, strong oxidizers. Attacks aluminum, copper, lead, tin, zinc, and alloys; some plastics, rubber, and coatings.

TETRAHYDRONAPHTHALENE (119-64-2) Forms explosive mixture with air (flash point 160°F/71°F). Strong oxidizers may cause fire and explosions. Attacks some plastics, rubber, and coatings. May accumulate static electrical charges, and may cause ignition of its vapors.

1,2,3,4-TETRAHYDRONAPHTHALENE (119-64-2) Forms explosive mixture with air (flash point 160°F/71°F). Strong oxidizers may cause fire and explosions. Attacks some plastics, rubber, and coatings. May accumulate static electrical charges, and may cause ignition of its vapors.

TETRAHYDRO-1,4-OXAZINE (110-91-8) Forms explosive mixture with air (flash point 98°F/37°C). Incompatible with acids, organic anhydrides, isocyanates, vinyl acetate, acrylates, substituted allyls, alkylene oxides, epichlorohydrin, ketones, aldehydes, alcohols, glycols, phenols, cresols, caprolactam solution, strong oxidizers. Attacks aluminum, copper, lead, tin, zinc, and alloys; some plastics, rubber, and coatings.

TETRAHYDRO-2H-1,4-OXAZINE (110-91-8) Forms explosive mixture with air (flash point 98°F/37°C). Incompatible with acids, organic anhydrides, isocyanates, vinyl acetate, acrylates, substituted allyls, alkylene oxides, epichlorohydrin, ketones, aldehydes, alcohols, glycols, phenols, cresols, caprolactam solution, strong oxidizers. Attacks aluminum, copper, lead, tin, zinc, and alloys; some plastics, rubber, and coatings.

TETRAHYDRO-P-OXAZINE (110-91-8) Forms explosive mixture with air (flash point 98°F/37°C). Incompatible with acids, organic anhydrides, isocyanates, vinyl acetate, acrylates, substituted allyls, alkylene oxides, epichlorohydrin, ketones, aldehydes, alcohols, glycols, phenols, cresols, caprolactam solution, strong oxidizers. Attacks aluminum, copper, lead, tin, zinc, and alloys; some plastics, rubber, and coatings.

TETRAHYDROXYMETHYLMETHANE (115-77-5) Incompatible with organic acids, oxidizers. Forms explosive material with thiophosphoryl chloride + heat.

TETRAIDROFURANO (Italian) (109-99-9) Forms explosive mixture with air (flash point 6°F/−14°C). Unless inhibited, can form unstable and explosive peroxides. Incompatible with strong acids, strong oxidizers. Attacks some plastics. May accumulate static electric charges that can result in ignition of its vapors.

TETRALENO (127-18-4) Incompatible with strong oxidizers, finely divided metals, caustics. Stable up to 258°F/126°C; at this temperature product gives off poisonous fumes.

TETRALEX (127-18-4) Incompatible with strong oxidizers, finely divided metals, caustics. Stable up to 258°F/126°C; at this temperature product gives off poisonous fumes.

TETRALIN (119-64-2) Forms explosive mixture with air (flash point 160°F/71°F). Strong oxidizers may cause fire and explosions. Attacks some plastics, rubber, and coatings. May accumulate static electrical charges, and may cause ignition of its vapors.

TETRALINE (119-64-2) Forms explosive mixture with air (flash point 160°F/71°F). Strong oxidizers may cause fire and explosions. Attacks some plastics, rubber, and coatings. May accumulate static electrical charges, and may cause ignition of its vapors.

1,2,3,5-TETRAMETHYLBENZENE (527-53-7) Forms explosive mixture with air (flash point 160°F/71°C). Strong oxidizers may cause fire and explosions.

TETRAMETHYLDIURANE SULPHITE (137-26-8) Combustible solid (flash point 192°F/89°C). Strong oxidizers may cause fire and explosions; contact with strong acid or oxidizable materials produces toxic gases.

TETRAMETHYLENE CYANIDE (111-69-3) May accumulate static electrical charges, and may cause ignition of its vapors. Decomposes at its flash point, producing poisonous cyanide gas. Forms explosive mixture with air (flash point 199°F/93°C). Incompatible with sulfuric acid, oxidizers.

TETRAMETHYLENE DICHLORIDE, DCB (764-41-0) Forms explosive mixture with air (flash point 126°F/52°C). Water causes formation of hydrochloric acid. Incompatible with strong oxidizers, bases. Corrodes metals in the presence of moisture.

TETRAMETHYLENE GLYCOL (110-63-4) Incompatible with strong acids, caustics, aliphatic amines, isocyanates, oxidizers.

1,4-TETRAMETHYLENE GLYCOL (110-63-4) Incompatible with strong acids, caustics, aliphatic amines, isocyanates, oxidizers.

TETRAMETHYLENE OXIDE (109-99-9) Forms explosive mixture with air (flash point 6°F/−14°C). Unless inhibited, can form unstable and explosive peroxides. Incompatible with strong acids, strong oxidizers. Attacks some plastics. May accumulate static electric charges that can result in ignition of its vapors.

TETRAMETHYLENETHIURAM DISULPHIDE (137-26-8) Combustible solid (flash point 192°F/89°C). Strong oxidizers may cause fire and explosions; contact with strong acid or oxidizable materials produces toxic gases.

TETRAMETHYL LEAD (75-74-1) Forms explosive mixture with air (flash point 100°F/38°C). Decomposes above 212°F/100°C. Strong oxidizers and strong acids cause fire and explosions. Attacks some plastics, rubber, and coatings.

TETRAMETHYL PLUMBANE (75-74-1) Forms explosive mixture with air (flash point 100°F/38°C). Decomposes above 212°F/100°C. Strong oxidizers and strong acids cause fire and explosions. Attacks some plastics, rubber, and coatings.

TETRAMETHYLTHIOCARBAMOYLDISULPHIDE (137-26-8) Combustible solid (flash point 192°F/89°C). Strong oxidizers may cause fire and explosions; contact with strong acid or oxidizable materials form toxic gases.

TETRAMETHYLTHIOPEROXYDICARBONIC DIAMIDE (137-26-8) Combustible solid (flash point 192°F/89°C). Strong oxidizers may cause fire and explosions; contact with strong acid or oxidizable materials form toxic gases.

TETRAMETHYL THIORAM DISULFIDE (Dutch) (137-26-8) Combustible solid (flash point 192°F/89°C). Strong oxidizers may cause fire and explosions; contact with strong acid or oxidizable materials produces toxic gases.

TETRAMETHYL-THIRAM DISULFID (German) (137-26-8) Combustible solid (flash point 192°F/89°C). Strong oxidizers may cause fire and explosions; contact with strong acid or oxidizable materials produces toxic gases.

TETRAMETHYLTHIURAM BISULFIDE (137-26-8) Combustible solid (flash point 192°F/89°C). Strong oxidizers may cause fire and explosions; contact with strong acid or oxidizable materials form toxic gases.

TETRAMETHYLTHIURAM DISULFIDE (137-26-8) Combustible solid (flash point 192°F/89°C). Strong oxidizers may cause fire and explosions; contact with strong acid or oxidizable materials form toxic gases.

TETRAMETHYL THIURAM DISULFIDE (137-26-8) Combustible solid (flash point 192°F/89°C). Strong oxidizers may cause fire and explosions; contact with strong acid or oxidizable materials produces toxic gases.

N,N,N',N'-TETRAMETHYL THIURAM DISULFIDE (137-26-8) Combustible solid (flash point 192°F/89°C). Strong oxidizers may cause fire and explosions; contact with strong acid or oxidizable materials form toxic gases.

TETRAMETHYL THIURANE DISULFIDE (137-26-8) Combustible solid (flash point 192°F/89°C). Strong oxidizers may cause fire and explosions; contact with strong acid or oxidizable materials form toxic gases.

TETRAMP (119-64-2) Forms explosive mixture with air (flash point 160°F/71°F). Strong oxidizers may cause fire and explosions. Attacks some plastics, rubber, and coatings. May accumulate static electrical charges, and may cause ignition of its vapors.

TETRANAP (119-64-2) Forms explosive mixture with air (flash point 160°F/71°F). Strong oxidizers may cause fire and explosions. Attacks some plastics, rubber, and coatings. May accumulate static electrical charges, and may cause ignition of its vapors.

TETRANITROMETHANE (509-14-8) Combustible organic matter wet with this chemical may be highly explosive, and susceptible to mild shock. Contact with hydrocarbons, alkalies, or metals produces explosive mixtures. Attacks some plastics, rubber, and coatings.

N,2,4,5-TETRANITRO-N-METHYLANILINE (479-45-8) A powerful oxidizer and highly sensitive explosive. Protect containers from shock, friction, heat, oxidizers, and hydrazine.

TETRA OLIVE N2G (120-12-7) Dust or fine powder forms an explosive mixture with air. Strong oxidizers may cause fire and explosions.

TETRAOXYMETHYLENE (50-00-0) May polymerize unless properly inhibited (usually with methanol). Forms explosive mixture with air (flash point 122°F/50°C). Incompatible with strong acids, amines, strong oxidizers, alkaline materials, nitrogen dioxide, performic acid. Reaction with hydrochloric acid produces bis-chloromethyl ether, a carcinogen.

TETRAPROPYLENE (6842-15-5) Forms explosive mixture with air (flash point 144°F/62°C). Incompatible with strong acids, oxidizers.

TETRASODIUM SALT OF EDTA SOLUTION (N/A) Incompatible with strong acids, nitrates, oxidizers.

TETRASOL (56-23-5) Becomes corrosive when in contact with water. Corrosive to metals. Reacts violently with many compounds. Decomposes on contact with chemically active metals such as sodium, potassium and magnesium. Incompatible with allyl alcohol, fluorine gas, alkali metals, aluminum. Attacks some coatings, plastics, and rubber.

TETRASTIGMINE (107-49-3) Decomposes above 300°F/150°C, producing flammable ethylene gas. Strong oxidizers may cause fire and explosions. Attacks some plastics, rubber, and coatings.

TETRAVOS (62-73-7) Attacks some plastics, rubber, and coatings.

TETRINE ACID (60-00-4) Incompatible with sulfuric acid, bases, ammonia, aliphatic amines, alkanolamines, isocyanates, alkylene oxides, epichlorohydrin.

TETROLE (110-00-9) Forms explosive mixture with air (flash point −58°F/−50°C). Reacts violently with acids, oxidizers. Unless stabilized with an inhibitor, air exposure forms unstable peroxides.

TETRON (107-49-3) Decomposes above 300°F/150°C) forming flammable ethylene gas. Strong oxidizers may cause fire and explosions. Attacks some plastics, rubber, and coatings.

TETRON-100 (107-49-3) Decomposes above 300°F/150°C, producing flammable ethylene gas. Strong oxidizers may cause fire and explosions. Attacks some plastics, rubber, and coatings.

TETROPIL (127-18-4) Incompatible with strong oxidizers, finely divided metals, caustics. Stable up to 258°F/126°C; at this temperature product gives off poisonous fumes.

TETROSIN LY (92-52-4) Contact with strong oxidizers may cause fire and explosions.

TETRYL (479-45-8) A powerful oxidizer and highly sensitive explosive. Protect containers from shock, friction, heat, oxidizers, and hydrazine.

2,4,6-TETRYL (479-45-8) A powerful oxidizer and highly sensitive explosive. Protect containers from shock, friction, heat, oxidizers, and hydrazine.

TEXANOL (25265-77-4) Incompatible with strong acids, nitrates, oxidizers.

TEX-WET (119-36-8) Forms explosive mixture with air (flash point 205°F/96°C). Incompatible with strong acids, nitrates, oxidizers.

TEXWET 1001 (119-36-8) Forms explosive mixture with air (flash point 205°F/96°C). Incompatible with strong acids, nitrates, oxidizers.

TFE (116-14-3) A highly reactive flammable gas. Able to form unstable peroxides; if inhibitor is not present in adequate concentrations, explosive polymerization may occur. Reacts violently with air, oxygen, oxidizers, sulfur trioxide.

THALLIUM (7440-28-0) Reacts violently with fluorine.

THALLIUM ACETATE (563-68-8) Incompatible with strong acids, nitrates.

THALLIUM(1+) ACETATE or THALLIUM(I) ACETATE (563-68-8) Incompatible with strong acids, nitrates.

THALLIUM CARBONATE (6533-73-9) Incompatible with strong oxidizers, strong acids.

THALLIUM(1+) CARBONATE(2:1) or THALLIUM(I) CARBONATE(2:1) (6533-73-9) Incompatible with strong oxidizers, strong acids.

THALLIUM MONOACETATE (563-68-8) Incompatible with strong acids, nitrate.

THALLOUS ACETATE (563-68-8) Incompatible with strong acids, nitrates

THALLOUS CARBONATE (6533-73-9) Incompatible with strong oxidizers, strong acids.

THANOL PPG (25322-69-4) Incompatible with sulfuric acid, isocyanates, perchloric acid, oxidizers.

THERMACURE (1338-23-4) Forms explosive mixture with air (flash point 125°F/52°C). Explosive decomposition occurs above 176°F/80°C. Pure substance is shock-sensitive. Strong oxidizer. Reacts violently with strong acids, strong bases, reducing agents, combustible substances, organic materials, oxides of heavy metals, salts, trace contaminants, amines. May accumulate static electrical charges, and may cause ignition of its vapors.

THERMAL BLACK (1333-86-4) Dust can form an explosive mixture in air. Oxidizers may cause fire and explosions.

THERMALOX (1304-56-9) Incompatible with acids, bases, chlorinated hydrocarbons, oxidizers, molten lithium.

THERMINOL FR-1 (generic CAS for PCBs 1336-36-3) Incompatible with strong oxidizers, strong acids.

THF (109-99-9) Forms explosive mixture with air (flash point 6°F/−14°C). Unless inhibited, can form unstable and explosive peroxides. Incompatible with strong acids, strong oxidizers. Attacks some plastics. May accumulate static electric charges that can result in ignition of its vapors.

THFA (97-99-4) Forms explosive mixture with air (flash point 162°F/72°C). Incompatible with strong acids, caustics, aliphatic amines, isocyanates, oxidizers. Unless inhibited, can form unstable and explosive peroxides.

THIFOR (115-29-7) Hydrolyzed by acids and alkalies. Corrosive to iron.

THILLATE (137-26-8) Combustible solid (flash point 192°F/89°C). Strong oxidizers may cause fire and explosions; contact with strong acid or oxidizable materials produces toxic gases.

THIMER (137-26-8) Combustible solid (flash point 192°F/89°C). Strong oxidizers may cause fire and explosions; contact with strong acid or oxidizable materials produces toxic gases.

THIMUL (115-29-7) Hydrolyzed by acids and alkalis. Corrosive to iron.

THIOBUTYL ALCOHOL (109-79-5) Forms explosive mixture with air (flash point 35°F/2°C). Incompatible with strong oxidizers and nitric acid. Attacks some plastics and rubber.

THIOCARBAMIDE (62-56-6) Aqueous solution is a base; reacts strongly with acids. Incompatible with nitric acid, hydrogen peroxide, acrolein, metals.

THIOCARBONYL CHLORIDE (463-71-8) Water contact produces hydrochloric acid, carbon disulfide and carbon dioxide. Corrodes most metals.

THIOCARBONYL DICHLORIDE (463-71-8) Water contact produces hydrochloric acid, carbon disulfide and carbon dioxide. Corrodes most metals.

THIOCARBONYL TETRACHLORIDE (594-42-3) Reacts with caustics or amines (with rapid decomposition), hot iron (produces toxic carbon tetrachloride fumes), hot water (produces hydrochloric acid), sulfur, and carbon dioxide. Attacks some plastics, rubber, and coatings.

THIODIETHYLENE GLYCOL (111-48-8) Incompatible with strong acids, bases, aliphatic amines, isocyanates.

THIODAN (115-29-7) Hydrolyzed by acids and alkalies. Corrosive to iron.

THIODEMETON (298-04-4) Forms explosive mixture with air (flash point 180°F/82°C). Incompatible with alkalies, strong oxidizers.

THIODEMETRON (298-04-4) Forms explosive mixture with air (flash point 180°F/82°C). Incompatible with alkalies, strong oxidizers.

THIODIGLYCOL (111-48-8) Incompatible with strong acids, bases, aliphatic amines, isocyanates.

THIOETHANOL (75-08-1) Forms explosive mixture with air (flash point less than 0°F/−18°C). May accumulate static electrical charges, and may cause ignition of its vapors. Contact with strong oxidizers may cause fire and explosions. Attacks some forms of plastics, coatings, and rubber.

2-THIOETHANOL (60-24-2) Forms explosive mixture with air (flash point 165°F/74°C). Incompatible with oxidizers, strong acids, caustics, aliphatic amines, isocyanates.

THIOETHYL ALCOHOL (75-08-1) Forms explosive mixture with air (flash point less than 0°F/−18°C). May accumulate static electrical charges, and may cause ignition of its vapors. Contact with strong oxidizers may cause fire and explosions. Attacks some forms of plastics, coatings, and rubber.

THIOETHYLENE GLYCOL (60-24-2) Forms explosive mixture with air (flash point 165°F/74°C). Incompatible with oxidizers, strong acids, caustics, aliphatic amines, isocyanates.

THIOFACO T-35 (102-71-6) Aqueous solution is caustic. Incompatible with acids, organic anhydrides, isocyanates, vinyl acetate, acrylates, substituted allyls, alkylene oxides, epichlorohydrin, aldehydes, strong oxidizers. Corrodes copper and its alloys.

THIOFOSGEN (463-71-8) Water contact produces hydrochloric acid, carbon disulfide and carbon dioxide. Corrodes most metals.

THIOFURAN (110-02-1) Forms explosive mixture with air (flash point 30°F/−1°C). reacts violently with strong oxidizers, nitric acid. May accumulate static electrical charges, and may cause ignition of its vapors.

THIOFURFURAN (110-02-1) Forms explosive mixture with air (flash point 30°F/−1°C). reacts violently with strong oxidizers, nitric acid. May accumulate static electrical charges, and may cause ignition of its vapors.

THIOGLYCOL (60-24-2) Forms explosive mixture with air (flash point 165°F/74°C). Incompatible with oxidizers, strong acids, caustics, aliphatic amines, isocyanates.

THIOLE (110-02-1) Forms explosive mixture with air (flash point 30°F/−1°C). reacts violently with strong oxidizers, nitric acid. May accumulate static electrical charges, and may cause ignition of its vapors.

THIOLITE (7778-18-9) Contact with diazomethane, aluminum, phosphorus may cause explosion.

THIOLUX (7704-34-9) Combustible solid. Liquid causes formation of sulfur dioxide with air. Reacts violently with strong oxidizers. Forms explosive, shock-sensitive, or pyrophoric mixtures with ammonia, ammonium nitrate, bromates, calcium carbide, charcoal, chlorates, hydrocarbons, iodates, iron. Reacts violently with halogen compounds, sodium, tin, uranium and other compounds. Attacks steel when moist. May accumulate static electrical charges, and may cause ignition of its vapors.

791

THIOMETHANOL (74-93-1) May accumulate static electrical charges, and may cause ignition of its vapors. Forms explosive air-gas mixture (flash point 0°F/−18°C). Strong oxidizers may cause fire and explosions. Attacks some plastics, coatings, and rubber.

THIOMETHYL ALCOHOL (74-93-1) May accumulate static electrical charges, and may cause ignition of its vapors. Forms explosive air-gas mixture (flash point 0°F/−18°C). Strong oxidizers may cause fire and explosions. Attacks some plastics, coatings, and rubber.

THIONEX (115-29-7) Hydrolyzed by acids and alkalies. Corrosive to iron.

THIONYL CHLORIDE (7719-09-7) Highly reactive. Reacts with water, and on exposure to air and light, producing sulfur dioxide and strong acids. Reacts with alcohols, ammonia, many organic compounds, combustible materials with risk of fire and explosions. Attacks some plastics, rubber, and coatings.

THIONYLDICHLORIDE (7719-09-7) Highly reactive. Reacts with water, and on exposure to air and light, producing sulfur dioxide and strong acids. Reacts with alcohols, ammonia, many organic compounds, combustible materials with risk of fire and explosions. Attacks some plastics, rubber, and coatings.

THIOPHENE (110-02-1) Forms explosive mixture with air (flash point 30°F/−1°C). reacts violently with strong oxidizers, nitric acid. May accumulate static electrical charges, and may cause ignition of its vapors.

THIOPHENIT (298-00-0) Mixtures with magnesium may be explosive.

THIOPHENOL (108-98-5) Forms explosive mixture with air (flash point 132°F/56°C). Incompatible with strong acids, caustics, alkali metals (i.e., lithium, sodium, potassium, rubidium, cesium, francium). Oxidizes on contact with air. Corrosive to carbon steel.

THIOPHOS (56-38-2) Combustible liquid. Mixtures with endrin may be explosive. Strong oxidizers may cause fire and explosions. Attacks some plastics, rubber, and coatings.

THIOPHOS 3422 (56-38-2) Combustible liquid. Mixtures with endrin may be explosive. Strong oxidizers may cause fire and explosions. Attacks some plastics, rubber, and coatings.

THIOPHOSGENE (463-71-8) Water contact produces hydrochloric acid, carbon disulfide and carbon dioxide. Corrodes most metals.

THIOPHOSPHATE de O,O-DIETHYLE et de O-(3-CHLORO-4-METHYL-7-COUMARINYLE) (French) (56-72-4) Contact with strong oxidizers may cause fire and explosions.

THIOPHOSPHORIC ANHYDRIDE (1314-80-3) Contact with water or alcohol yields heat and possible ignition. May self-ignite in air. Reacts violently with acids, oxidizers.

2-THIOPROPANE (75-18-3) Forms explosive mixture with air (flash point −36°F/−38°C). Reacts violently with strong oxidizers. May accumulate static electrical charges, and may cause ignition of its vapors.

THIOSAN (137-26-8) Combustible solid (flash point 192°F/89°C). Strong oxidizers may cause fire and explosions; contact with strong acid or oxidizable materials produces toxic gases.

THIOSULFURIC ACID, LEAD SALT (13478-50-7) Aqueous solution is acidic; corrodes many metals.

THIOSULFUROUS DICHLORIDE (10025-67-9) Water contact produces hydrochloric acid, sulfur dioxide, sulfur, sulfur thiosulfite, hydrogen sulfide; solution is a strong acid. Incompatible with strong oxidizers, strong bases, oxides of phosphorus, organics; may cause fires. Attacks some plastics, rubber, and coatings.

THIOTEPP (3689-24-5) Containers may burst in heat. Strong oxidizers may cause fire and explosions. Attacks some plastics, rubber, and coatings.

THIOTOX (137-26-8) Combustible solid (flash point 192°F/89°C). Strong oxidizers may cause fire and explosions; contact with strong acid or oxidizable materials produces toxic gases.

THIOUREA (62-56-6) Aqueous solution is a base; reacts strongly with acids. Incompatible with nitric acid, hydrogen peroxide, acrolein, metals.

2-THIOUREA (62-56-6) Aqueous solution is a base; reacts strongly with acids. Incompatible with nitric acid, hydrogen peroxide, acrolein, metals.

THIOUREA, 1-NAPHTHALENYL- (86-88-4) Strong oxidizers may cause fires and explosion. Also reacts with silver nitrate.

THIOVIT (7704-34-9) Combustible solid. Liquid causes formation of sulfur dioxide with air. Reacts violently with strong oxidizers. Forms explosive, shock-sensitive, or pyrophoric mixtures with ammonia, ammonium nitrate, bromates, calcium carbide, charcoal, chlorates, hydrocarbons, iodates, iron. Reacts violently with halogen compounds, sodium, tin, uranium, and other compounds. Attacks steel when moist. May accumulate static electrical charges, and may cause ignition of its vapors.

THIRAM (137-26-8) Combustible solid (flash point 192°F/89°C). Strong oxidizers may cause fire and explosions; contact with strong acid or oxidizable materials produces toxic gases.

THIRAMAD (137-26-8) Combustible solid (flash point 192°F/89°C). Strong oxidizers may cause fire and explosions; contact with strong acid or oxidizable materials produces toxic gases.

THIRAME (French) (137-26-8) Combustible solid (flash point 192°F/89°C). Strong oxidizers may cause fire and explosions; contact with strong acid or oxidizable materials produces toxic gases.

THISULFAN (115-29-7) Hydrolyzed by acids and alkalies. Corrosive to iron.

THITETROLE (110-02-1) Forms explosive mixture with air (flash point 30°F/−1°C). reacts violently with strong oxidizers, nitric acid. May accumulate static electrical charges, and may cause ignition of its vapors.

THIURAM (137-26-8) Combustible solid (flash point 192°F/89°C). Strong oxidizers may cause fire and explosions; contact with strong acid or oxidizable materials produces toxic gases.

THIURAMYL (137-26-8) Combustible solid (flash point 192°F/89°C). Strong oxidizers may cause fire and explosions; contact with strong acid or oxidizable materials produces toxic gases.

THORIUM NITRATE (13823-29-5) Incompatible with water, with formation of weak nitric acid solution. A strong oxidizer; reacts violently with reducing agents, combustibles, organic materials. Attacks metals when wet.

THORIUM(4+) NITRATE or THORIUM(IV) NITRATE (13823-29-5) Incompatible with water, with formation weak nitric acid solution. A strong oxidizer; reacts violently with reducing agents, combustibles, organic materials. Attacks metals when wet.

THORIUM NITRATE TETRAHYDRATE (13823-29-5) Incompatible with water, with formation of weak nitric acid solution. A strong oxidizer; reacts violently with reducing agents, combustibles, organic materials. Attacks metals when wet.

THORIUM TETRANITRATE (13823-29-5) Incompatible with water, with formation of weak nitric acid solution. A strong oxidizer; reacts violently with reducing agents, combustibles, organic materials. Attacks metals when wet.

THREE ELEPHANT (10043-35-3) Incompatible with bases, potassium. Acetic anhydride causes formation of heat-sensitive explosive.

THRETHYLENE (79-01-6) Contact with caustics produces a toxic and flammable gas. Reacts violently with chemically active metals. Contact with aluminum may produce a violent, self-accelerating polymerization reaction. Incompatible with acids, organic anhydrides, isocyanates, alkylene oxides, aldehydes, alcohols, glycols, phenols, cresols, caprolactam solution, epichlorohydrin, nitrogen tetroxide, metal powders, oxygen. May accumulate static electrical charges, and may cause ignition of its vapors.

THU (62-56-6) Aqueous solution is a base; reacts strongly with acids. Incompatible with nitric acid, hydrogen peroxide, acrolein, metals.

THYLATE (137-26-8) Combustible solid (flash point 192°F/89°C). Strong oxidizers may cause fire and explosions; contact with strong acid or oxidizable materials produces toxic gases.

THYLFAR M-50 (298-00-0) Mixtures with magnesium may be explosive.

TIBA (100-99-2) Ignites spontaneously in air. Water contact produces hydrocarbons. Reacts with oxidizers, carbon dioxide, alcohols, ammonia, halogenated hydrocarbons. Attacks silicone and urethane rubbers.

TIBAL (100-99-2) Ignites spontaneously in air. Water contact forms hydrocarbons. Reacts with oxidizers, carbon dioxide, alcohols, ammonia, halogenated hydrocarbons. Attacks silicone and urethane rubbers.

TIN or TIN FLAKE or TIN POWDER (7440-31-5) Incompatible with strong acids, strong bases, strong oxidizers, sulfur.

TIN BIFLUORIDE (7783-47-3) Acid contact produces hydrogen fluoride fumes.

TIN(II) CHLORIDE(1:2) (7772-99-8) A strong reducing agent; reacts violently with oxidizers, nitrates, potassium, sodium, with risk of fire and explosions.

TIN(IV) CHLORIDE (7646-78-8) Water contact produces hydrogen chloride. Contact with alcohols, amines, potassium, sodium, turpentine may cause fire and explosions. Attacks some plastics, rubber, and coatings.

TIN DICHLORIDE (7772-99-8) A strong reducing agent; reacts violently with oxidizers, nitrates, potassium, sodium, with risk of fire and explosions.

TIN DIFLUORIDE (7783-47-3) Acid contact produces hydrogen fluoride fumes.

TINNING GLUX (7646-85-7) Aqueous solution is strongly acidic; reacts with strong bases, and attacks metals in the presence of moisture.

TINPROTOCHLORIDE (7772-99-8) A strong reducing agent; reacts violently with oxidizers, nitrates, potassium, sodium, with risk of fire and explosions.

TIN TETRACHLORIDE (7646-78-8) Water contact produces hydrogen chloride. Contact with alcohols, amines, potassium, sodium, turpentine may cause fire and explosions. Attacks some plastics, rubber, and coatings.

TIOFOS (56-38-2) Combustible liquid. Mixtures with endrin may be explosive. Strong oxidizers may cause fire and explosions. Attacks some plastics, rubber, and coatings.

TIOVEL (115-29-7) Hydrolyzed by acids and alkalies. Corrosive to iron.

TIPPON (93-76-5) Sealed metal containers may burst in heat above 316°F/158°C. Incompatible with sulfuric acid, bases, ammonia, aliphatic amines, alkanolamines, isocyanates, alkylene oxides, epichlorohydrin.

TITANIUM BUTOXIDE (5593-70-4) Forms explosive mixture with air (flash point 170°F/77°C). Reacts violently with oxidizers. Water contact produces butanol and titanium dioxide.

TITANIUM CHLORIDE (7550-45-0) Contact with moisture in the air produces hydrogen chloride fumes and dense white smoke. Reacts violently with water, producing hydrochloric acid; similar reaction occurs with alcohols. Attacks metals in the presence of moisture.

TITANIUM(III) CHLORIDE (7705-07-9) Reacts with water producing hydrogen chloride fumes. A strong reducing agent; reacts violently with oxidizers. Attacks metals in the presence of moisture.

TITANIUM(4+) CHLORIDE or TITANIUM(IV) CHLORIDE (7550-45-0) Contact with moisture in the air produces hydrogen chloride fumes and dense white smoke. Reacts violently with water producing hydrochloric acid; similar reaction occurs with alcohols. Attacks metals in the presence of moisture.

TITANIUM TETRABUTOXIDE (5593-70-4) Forms explosive mixture with air (flash point 170°F/77°C). Reacts violently with oxidizers. Water contact produces butanol and titanium dioxide.

TITANIUM TETRACHLORIDE (7550-45-0) Contact with moisture in the air produces hydrogen chloride fumes and dense white smoke. Reacts violently with water, producing hydrochloric acid; similar reaction with alcohols. Attacks metals in the presence of moisture.

TITANIUM TRICHLORIDE (7705-07-9) Reacts with water, producing hydrogen chloride fumes. A strong reducing agent; reacts violently with oxidizers. Attacks metals in the presence of moisture.

TITANOUS CHLORIDE (7705-07-9) Reacts with water, producing hydrogen chloride fumes. A strong reducing agent; reacts violently with oxidizers. Attacks metals in the presence of moisture.

TITRIPLEX (60-00-4) Incompatible with sulfuric acid, bases, ammonia, aliphatic amines, alkanolamines, isocyanates, alkylene oxides, epichlorohydrin.

TIURAM (Polish) (137-26-8) Combustible solid (flash point 192°F/89°C). Strong oxidizers may cause fire and explosions; contact with strong acid or oxidizable materials produces toxic gases.

TIURAMYL (137-26-8) Combustible solid (flash point 192°F/89°C). Strong oxidizers may cause fire and explosions; contact with strong acid or oxidizable materials produces toxic gases.

TK (121-75-5) Incompatible with strong oxidizers, magnesium, alkaline pesticides. Attacks metals, some plastics, rubber, and coatings.

TL 4N (111-46-6) Combustible (flash point 255°F/124°C). Incompatible with sulfuric acid, isocyanates, strong oxidizers.

TL 69 (696-28-6) Forms explosive mixture with air (flash point 60°F/−16°C). Water contact produces hydrochloric acid. Attacks metals in the presence of moisture.

TL 314 (107-13-1) Forms explosive mixture with air (flash point 32°F/0°C o.c.). Forms explosive peroxides; heat, light, caustics, silver nitrate, and peroxides can cause polymerization. Incompatible with strong acids, strong oxidizers, amines, 2-aminoethanol, bromine, chlorosulfonic acid, ethylene diamine, nitric acid, oleum, potassium hydroxide, sodium hydroxide, sulfuric acid. Attacks copper and copper alloys, and aluminum in high concentrations. May accumulate static electrical charges, and may cause ignition of its vapors.

TL 337 (151-56-4) Forms explosive mixture with air (flash point 12°F/−11°C). May accumulate static electrical charges, and may cause ignition of its vapors. Contact with acid, aluminum, carbon dioxide or silver may cause explosive polymerization. Attacks rubber, coatings, plastics, and chemically active metals. Self-reactive with heat or atmospheric carbon dioxide.

TL 389 (79-44-7) Rapidly hydrolyzed in water. Incompatible with strong acids, oxidizers.

TL 423 (541-43-3) Forms explosive mixture with air (flash point 61°F/16°C). Strong oxidizers may cause fire and explosions. Attacks metals in the presence of moisture. Water and air contact produce hydrochloric acid.

TL 822 (506-68-3) May be unstable unless dry and pure. Reacts violently with acids, ammonia, amines. Water contact produces hydrogen cyanide and hydrogen bromide.

TL 869 (62-74-8) Incompatible with strong acids.

TL 898 (7487-94-7) Incompatible with light metals (aluminum, magnesium, beryllium, etc.), sodium, potassium.

TL 1163 (75-77-4) Forms explosive mixture with air (flash point −18°F/−28°C). Reacts violently with water. Vigorous reaction with aluminum. Moisture and air contact produces hydrochloric acid.

TL 1450 (624-83-9) Forms explosive mixture with air (flash point 0°F/−18°C). Contact with iron, tin, copper (or salts of these elements) and with certain other catalysts (such as triphenylarsenic oxide, triethylphosphine, and tributyltin oxide) or elevated temperatures may cause polymerization. Reacts violently with warm water. Incompatible with strong acids, alcohols, glycols, amines, amides, ammonia, caprolactam, caustics, strong oxidizers. Attacks some plastics, rubber, or coatings.

TLA (540-88-5) Incompatible with strong acids, oxidizers, alkalies, nitrates. Softens and dissolves plastics.

TM-4049 (121-75-5) Incompatible with strong oxidizers, magnesium, alkaline pesticides. Attacks metals, some plastics, rubber, and coatings.

796

TMA (75-50-3) Forms explosive mixture with air (flash point 10°F/−12°C). Incompatible with strong oxidizers, acids. Attacks chemically active metals: aluminum, copper, zinc, and their alloys.

TMA (75-24-1) A dangerous explosion hazard. Ignites spontaneously in air. A powerful reducing agent; reacts violently with oxidizers, water, alcohols, amines, carbon dioxide, carbon tetrachloride, halogenated hydrocarbons, oxides of nitrogen or sulfur, phenols, and many other substances. May accumulate static electrical charges, and may cause ignition of its vapors.

TML (75-74-1) Forms explosive mixture with air (flash point 100°F/38°C). Decomposes above 212°F/100°C. Strong oxidizers and strong acids cause fire and explosions. Attacks some plastics, rubber, and coatings.

TMP (121-45-9) Forms explosive mixture with air (flash point 82°F/28°C). Reacts violently with strong acids, nitrates, oxidizers, magnesium salts of perchloric acid.

TMTD (137-26-8) Combustible solid (flash point 192°F/89°C). Strong oxidizers may cause fire and explosions; contact with strong acid or oxidizable materials produces toxic gases.

TNCS 53 (7758-98-7) Aqueous solution is an acid. Incompatible with strong bases, hydroxylamine, magnesium.

TNM (509-14-8) Combustible organic matter wet with this chemical may be highly explosive, and susceptible to mild shock. Contact with hydrocarbons, alkalies or metals produces explosive mixtures. Attacks some plastics, rubber, and coatings.

TNPA (102-69-2) Incompatible with strong acids, strong oxidizers.

TNT (118-96-7) Explosive; protect from shock and heat. Slow decomposition occurs above 356°F/356°F/180°C. Exposure to light may increase impact sensitivity. Rapid heating may cause detonation. Strong oxidizers may cause fire. Contact with ammonia or with strong alkalies may increase sensitivity to shock. Reacts vigorously with oxidizable materials.

TOCP (78-30-8) Contact with magnesium may cause explosion. Attacks some plastics, rubber, and coatings.

TOFK (78-30-8) Contact with magnesium may cause explosion. Attacks some plastics, rubber, and coatings.

2,4-TOLAMINE (95-80-7) Incompatible with oxidizers, acids, organic anhydrides, isocyanates, aldehydes. Attacks aluminum, brass, bronze, copper, zinc.

o-**TOLIDIN or** *ortho*-**TOLIDIN** (95-53-4) Forms explosive mixture with air (flash point 185°F/85°C). Incompatible with strong acids, mineral acids, organic anhydrides, isocyanates, aldehydes, oxidizers. Attacks some plastics, rubber, and coatings.

2-TOLIDIN (German) (95-53-4) Forms explosive mixture with air (flash point 185°F/85°C). Incompatible with strong acids, mineral acids, organic anhydrides, isocyanates, aldehydes, oxidizers. Attacks some plastics, rubber, and coatings.

2-TOLIDINA (Italian) (95-53-4) Forms explosive mixture with air (flash point 185°F/85°C). Incompatible with strong acids, mineral acids, organic anhydrides, isocyanates, aldehydes, oxidizers. Attacks some plastics, rubber, and coatings.

TOLIDINE (95-53-4) Forms explosive mixture with air (flash point 185°F/85°C). Incompatible with strong acids, mineral acids, organic anhydrides, isocyanates, aldehydes, oxidizers.

2-TOLIDINE (95-53-4) Forms explosive mixture with air (flash point 185°F/85°C). Incompatible with strong acids, mineral acids, organic anhydrides, isocyanates, aldehydes, oxidizers.

3,3′-TOLIDINE (95-53-4) Forms explosive mixture with air (flash point 185°F/85°C). Incompatible with strong acids, mineral acids, organic anhydrides, isocyanates, aldehydes, oxidizers.

O,O′-TOLIDINE (95-53-4) Forms explosive mixture with air (flash point 185°F/85°C). Incompatible with strong acids, mineral acids, organic anhydrides, isocyanates, aldehydes, oxidizers. Attacks some plastics, rubber, and coatings.

TOLL (298-00-0) Mixtures with magnesium may be explosive.

TOLUEEN-DISOCYANAAT (Dutch) (584-84-9) Incompatible with strong acids including nonoxidizing mineral and organic acids, caustics, ammonia, amines, amides, alcohols, glycols, caprolactam solution. Water contact causes violent foaming and spattering; produces carbon dioxide and an organic base. Attacks copper and its alloys, some plastics including polyethylene and rubber.

TOLUEEN (Dutch) (108-88-3) Forms explosive mixture with air (flash point 40°F/4°C). Strong oxidizers may cause fire and explosions. Attacks some plastics, rubber, and coatings. May accumulate static electrical charges, and may cause ignition of its vapors.

TOLUEN-DISOCIANATO (Italian) (584-84-9) Incompatible with strong acids including nonoxidizing mineral and organic acids, caustics, ammonia, amines, amides, alcohols, glycols, caprolactam solution. Water contact causes violent foaming and spattering; produces carbon dioxide and an organic base. Attacks copper and its alloys, some plastics including polyethylene and rubber.

TOLUEN (Czech) (108-88-3) Forms explosive mixture with air (flash point 40°F/4°C). Strong oxidizers may cause fire and explosions. Attacks some plastics, rubber, and coatings. May accumulate static electrical charges, and may cause ignition of its vapors.

TOLUENE (108-88-3) Forms explosive mixture with air (flash point 40°F/4°C). Strong oxidizers may cause fire and explosions. Attacks some plastics, rubber, and coatings. May accumulate static electrical charges, and may cause ignition of its vapors.

2,4-TOLUENEDIAMINE (95-80-7) Incompatible with oxidizers, acids, organic anhydrides, isocyanates, aldehydes. Attacks aluminum, brass, bronze, copper, zinc.

TOLUENE, *p-tert*-BUTYL or TOLUENE, *para-tert*-BUTYL (98-51-1) Forms explosive mixture with air (flash point 155°F/67°C). Reacts with strong oxidizers. May accumulate static electrical charges, and may cause ignition of its vapors.

TOLUENE, *o*-CHLORO- (95-49-8) Forms explosive mixture with air (flash point 96°F/36°C). Strong oxidizers may cause fire and explosions. Produces hydrochloric acid with water. Corrodes metal in the presence of moisture. Attacks some plastics, rubber, and coatings. May accumulate static electrical charges, and may cause ignition of its vapors.

2,4-TOLUENE-2,4-DIAMINE (95-80-7) Incompatible with oxidizers, acids, organic anhydrides, isocyanates, aldehydes. Attacks aluminum, brass, bronze, copper, zinc.

m-**TOLUENEDIAMINE** (95-80-7) Incompatible with oxidizers, acids, organic anhydrides, isocyanates, aldehydes. Attacks aluminum, brass, bronze, copper, zinc.

TOLUENE,alpha,alpha-DICHLORO- (98-87-3) Forms explosive mixture with air (flash point 153°F/67°C). Reacts with acids, bases, strong oxidizers. Produces acid fumes with air. Attacks plastics and coatings.

TOLUENE DI-ISOCYANATE (584-84-9) Incompatible with strong acids, including nonoxidizing mineral and organic acids, caustics, ammonia, amines, amides, alcohols, glycols, caprolactam solution. Water contact causes violent foaming and spattering; produces carbon dioxide and an organic base. Attacks copper and its alloys, some plastics including polyethylene and rubber.

TOLUENE DIISOCYANATE (584-84-9) Incompatible with strong acids, including nonoxidizing mineral and organic acids, caustics, ammonia, amines, amides, alcohols, glycols, caprolactam solution. Water contact causes violent foaming and spattering; produces carbon dioxide and an organic base. Attacks copper and its alloys, some plastics including polyethylene, and rubber.

TOLUENE-2,4-DIISOCYANATE (584-84-9) Incompatible with strong acids including nonoxidizing mineral and organic acids, caustics, ammonia, amines, amides, alcohols, glycols, caprolactam solution. Water contact causes violent foaming and spattering; produces carbon dioxide and an organic base. Attacks copper and its alloys, some plastics including polyethylene and rubber.

2,4-TOLUENEDIISOCYANATE (584-84-9) Incompatible with strong acids, including nonoxidizing mineral and organic acids, caustics, ammonia, amines, amides, alcohols, glycols, caprolactam solution. Water contact causes violent foaming and spattering; produces carbon dioxide and an organic base. Attacks copper and its alloys, some plastics including polyethylene, and rubber.

TOLUENE, 2,6-DINITRO- (606-20-2) Strong oxidizers or caustics may cause fire and explosions. Contact with nitric acid produces an explosive material. Sodium oxide contact causes ignition. Attacks some metals.

TOLUENE, 2,4-DINITRO- (121-14-2) Strong oxidizers or caustics may cause fire and explosions. Incompatible with nitric acid; forms an explosive material. Sodium oxide contact causes ignition. Attacks metals.

TOLUENE HEXAHYDRIDE (108-87-2) Forms explosive mixture with air (flash point 25°F/−3.9°C). Strong oxidizers may cause fire and explosions. Attacks some plastics, rubber, and coatings.

3,4-TOLUENE, 3,4-DINITRO- (610-39-9) Strong oxidizers or caustics may cause fire and explosions. Contact with nitric acid produces an explosive material. Sodium oxide contact causes ignition. Attacks some metals.

TOLUENE, *o*-NITRO or TOLUENE, *ortho*-NITRO (88-72-2) Incompatible with strong acids, caustics, ammonia, amines, reducing agents, oxidizers. Attacks some plastics, rubber, and coatings.

TOLUENE, *p*-NITRO- or TOLUENE, *para*-NITRO- (99-99-0) Incompatible with strong acids, caustics, ammonia, amines, reducing agents, oxidizers. Attacks some plastics, rubber, and coatings.

4-TOLUENESULFONIC ACID (104-15-4) Solution is a strong acid. Incompatible with sulfuric acid, caustics, ammonia, amines, amides, organic anhydrides, isocyanates, vinyl acetate, alkylene oxides, epichlorohydrin. Attacks metals in the presence of moisture and produces hydrogen gas.

***p*-TOLUENESULFONIC ACID** or ***para*-TOLUENESULFONIC ACID** (104-15-4) Solution is a strong acid. Incompatible with sulfuric acid, caustics, ammonia, amines, amides, organic anhydrides, isocyanates, vinyl acetate, alkylene oxides, epichlorohydrin. Attacks metals in the presence of moisture and produces hydrogen gas.

***m*-TOLUIDIN (Czech)** (108-44-1) Forms explosive mixture with air (flash point 188°F/87°C). Incompatible with strong oxidizers, strong acids. Attacks some plastics, rubber, and coatings.

3-TOLUIDINE (108-44-1) Forms explosive mixture with air (flash point 188°F/87°C). Incompatible with strong oxidizers, strong acids. Attacks some plastics, rubber, and coatings.

4-TOLUIDINE (106-49-0) Forms explosive mixture with air (flash point 188°F/87°C). Incompatible with strong oxidizers, strong acids. Attacks some plastics, rubber, and coatings.

***m*-TOLUIDINE or *meta*-TOLUIDINE** (108-44-1) Forms explosive mixture with air (flash point 188°F/87°C). Incompatible with strong oxidizers, strong acids. Attacks some plastics, rubber, and coatings.

***o*-TOLUIDINE or *ortho*-TOLUIDINE** (95-53-4) Forms explosive mixture with air (flash point 185°F/85°C). Incompatible with strong acids, mineral acids, organic anhydrides, isocyanates, aldehydes, oxidizers. Attacks some plastics, rubber, and coatings.

***p*-TOLUIDINE or *para*-TOLUIDINE** (106-49-0) Forms explosive mixture with air (flash point 188°F/87°C). Incompatible with strong oxidizers, strong acids. Attacks some plastics, rubber, and coatings.

TOLUILENODWUIZOCYJANIAN (Polish) (584-84-9) Incompatible with strong acids including nonoxidizing mineral and organic acids, caustics, ammonia, amines, amides, alcohols, glycols, caprolactam solution. Water contact causes violent foaming and spattering; produces carbon dioxide and an organic base. Attacks copper and its alloys, some plastics including polyethylene, and rubber.

TOLUOL (108-88-3) Forms explosive mixture with air (flash point 40°F/4°C). Strong oxidizers may cause fire and explosions. Attacks some plastics, rubber, and coatings. May accumulate static electrical charges, and may cause ignition of its vapors.

***o*-TOLUOL or *ortho*-TOLUOL** (95-48-7) Forms explosive mixture with air (flash point 178°F/81°C). Incompatible with strong acids, oxidizers, alkalies, aliphatic amines, amides, chlorosulfonic acid, oleum. Liquid attacks some plastics and rubber.

***p*-TOLUOL or *para*-TOLUOL** (106-44-5) Incompatible with strong acids, oxidizers, alkalies, aliphatic amines, amides, chlorosulfonic acid, oleum. Liquid attacks some plastics, coatings, and rubber.

TOLUOLO (108-88-3) Forms explosive mixture with air (flash point 40°F/4°C). Strong oxidizers may cause fire and explosions. Attacks some plastics, rubber, and coatings. May accumulate static electrical charges, and may cause ignition of its vapors.

TOLU-SOL (108-88-3) Forms explosive mixture with air (flash point 40°F/4°C). Strong oxidizers may cause fire and explosions. Attacks some plastics, rubber, and coatings. May accumulate static electrical charges, and may cause ignition of its vapors.

m-**TOLUYLENEDIAMINE** or *meta*-**TOLUYLENEDIAMINE** (95-80-7) Incompatible with oxidizers, acids, organic anhydrides, isocyanates, aldehydes. Attacks aluminum, brass, bronze, copper, zinc.

TOLUYLENE-2,4-DIISOCYANATE (584-84-9) Incompatible with strong acids including nonoxidizing mineral and organic acids, caustics, ammonia, amines, amides, alcohols, glycols, caprolactam solution. Water contact causes violent foaming and spattering; produces carbon dioxide and an organic base. Attacks copper and its alloys, some plastics including polyethylene, and rubber.

TOLYENE 2,4-DISOCYANATE (584-84-9) Incompatible with strong acids, including nonoxidizing mineral and organic acids, caustics, ammonia, amines, amides, alcohols, glycols, caprolactam solution. Water contact causes violent foaming and spattering; forms carbon dioxide and an organic base. Attacks copper and its alloys, some plastics including polyethylene, and rubber.

TOLYLAMINE (106-49-0) Forms explosive mixture with air (flash point 188°F/87°C). Incompatible with strong oxidizers, strong acids. Attacks some plastics, rubber, and coatings.

m-**TOLYLAMINE** or *meta*-**TOLYLAMINE** (108-44-1) Forms explosive mixture with air (flash point 188°F/87°C). Incompatible with strong oxidizers, strong acids. Attacks some plastics, rubber, and coatings.

p-**TOLYLAMINE** or *para*-**TOLYLAMINE** (106-49-0) Forms explosive mixture with air (flash point 188°F/87°C). Incompatible with strong oxidizers, strong acids. Attacks some plastics, rubber, and coatings.

TOLYL CHLORIDE (100-44-7) Forms explosive mixture with air (flash point 153°F/67°C). Water contact produces hydrogen chloride fume. Strong oxidizers may cause fire and explosions. Violent polymerization may be caused by contact with copper, aluminum, iron, zinc, magnesium and tin. May accumulate static electrical charges, and may cause ignition of its vapors. Attacks some plastics and rubber.

m-**TOLYL CHLORIDE** or *meta*-**TOLYL CHLORIDE** (108-41-8) Forms explosive mixture with air (flash point 123°F/51°C). Incompatible with strong oxidizers, rare earth metals.

o-**TOLYL CHLORIDE** or *ortho*-**TOLYL CHLORIDE** (95-49-8) Forms explosive mixture with air (flash point 96°F/36°C). Strong oxidizers may cause fire and explosions. Produces hydrochloric acid with water. Corrodes metal in the presence of moisture. Attacks some plastics, rubber, and coatings. May accumulate static electrical charges, and may cause ignition of its vapors.

o-**TOLYLCHLORIDE** or *ortho*-**TOLYLCHLORIDE** (95-49-8) Forms explosive mixture with air (flash point 96°F/36°C). Strong oxidizers may cause fire and explosions. Forms hydrochloric acid with water. Corrodes metal in the presence of moisture. Attacks some plastics, rubber, and coatings. May accumulate static electrical charges, and may cause ignition of its vapors.

p-**TOLYL CHLORIDE** or *para*-**TOLYL CHLORIDE** (106-43-4) Reacts with oxidizers, with a risk of fire or explosions.

TOLYLENE-2,4-DIISOCYANATE (584-84-9) Incompatible with strong acids, including non-oxidizing mineral and organic acids, caustics, ammonia, amines, amides, alcohols, glycols, caprolactam solution. Water contact causes violent foaming and spattering; forms carbon dioxide and an organic base. Attacks copper and its alloys, some plastics including polyethylene and rubber.

2,4-TOLYLENE DIISOCYANATE (584-84-9) Incompatible with strong acids, including nonoxidizing mineral and organic acids, caustics, ammonia, amines, amides, alcohols, glycols, caprolactam solution. Water contact causes violent foaming and spattering; produces carbon dioxide and an organic base. Attacks copper and its alloys, some plastics including polyethylene, and rubber.

4-*m*-TOLYLENEDIAMINE (95-80-7) Incompatible with oxidizers, acids, organic anhydrides, isocyanates, aldehydes. Attacks aluminum, brass, bronze, copper, zinc.

TOLYL EPOXYPROPYL ETHER (26447-14-3) Forms explosive mixture with air (flash point 200°F/93°C). Incompatible with strong oxidizers, alkalies, sodium azide, acetylene, strong acids, aliphatic amines, amides. Attacks some plastics, rubber, and coatings.

***m*-TOLYL FLUORIDE or *meta*-TOLYL FLUORIDE** (352-70-5) Forms explosive mixture with air (flash point 49°F/9°C). Reacts violently with strong oxidizers.

***o*-TOLYL FLUORIDE or *ortho*-TOLYL FLUORIDE** (95-52-3) Forms explosive mixture with air (flash point 55°F/13°C). Reacts violently with strong oxidizers.

***p*-TOLYL FLUORIDE or *para*-TOLYL FLUORIDE** (352-32-9) Forms explosive mixture with air (flash point 105°F/49°C). Reacts violently with oxidizers.

TOLYL GLYCIDYL ETHER (26447-14-3) Forms explosive mixture with air (flash point 200°F/93°C). Incompatible with strong oxidizers, alkalies, sodium azide, acetylene, strong acids, aliphatic amines, amides. Attacks some plastics, rubber, and coatings.

***o*-TOLYL PHOSPHATE or *ortho*-TOLYL PHOSPHATE** (78-30-8) Contact with magnesium may cause explosion. Attacks some plastics, rubber, and coatings.

***o*-TOLYL PHOSPHATE PHOSPHORIC ACID or *ortho*-TOLYL PHOSPHATE PHOSPHORIC ACID** (78-30-8) Contact with magnesium may cause explosion. Attacks some plastics, rubber, and coatings.

***o*-TOLYLPHOSPHATE PHOSPHORIC ACID** (78-30-8) Contact with magnesium or strong oxidizers may cause fire and explosions. Attacks some plastics, rubber, and coatings.

TOPANEL (123-73-9) A strong reducing agent. Forms explosive mixture with air (flash point 55°F/12.8°C). Readily converted by oxygen to peroxides and acids; heat or contact with many other substances may cause polymerization. Incompatible with strong oxidizers, strong acids including nonoxidizing mineral acids, ammonia, aliphatic amines, aromatic amines, 1,3-butadiene, strong bases. Liquid attacks some plastics, rubber, and coatings.

TOPEX (94-36-0) Confined storage of dry chemical may lead to decomposition and explosion. A strong oxidant; extremely reactive. Fires and explosion may result from heat or contamination, and from contact with strong acids, combustible materials, oxidizers, acids, bases, alcohols, reducing agents, metals, metal oxides, amines, accelerators, methyl methacrylate, organic matter, lithium aluminum carbide, dimethyl aniline, amines, metallic naphthenates. May attack some plastics, rubber and coatings. Protect containers from shock and friction.

TOPICHLOR 20 (57-74-9) Contact with strong oxidizers may cause fire and explosions. Attacks some plastics, rubber, and coatings.

TOPICLOR (57-74-9) Contact with strong oxidizers may cause fire and explosions. Attacks some plastics, rubber, and coatings.

TOPICLOR 20 (57-74-9) Contact with strong oxidizers may cause fire and explosions. Attacks some plastics, rubber, and coatings.

TOPSYM (67-68-5) Forms explosive mixture with air (flash point 203°F/95°C). Reacts violently with oxidizers. Reacts with ethanoyl chloride, boron compounds, halides, metal alkoxides, oxidizers

TORMONA (93-76-5) Sealed metal containers may burst in heat above 316°F/158°C. Incompatible with sulfuric acid, bases, ammonia, aliphatic amines, alkanolamines, isocyanates, alkylene oxides, epichlorohydrin.

TOTP (78-30-8) Contact with magnesium may cause explosion. Attacks some plastics, rubber, and coatings.

TOX 47 (56-38-2) Combustible liquid. Mixtures with endrin may be explosive. Strong oxidizers may cause fire and explosions. Attacks some plastics, rubber, and coatings.

TOXADUST (8001-35-2) Reacts with oxidizers, with a risk of fire or explosions. Attacks metals in the presence of moisture.

TOXAFEEN (Dutch) (8001-35-2) Reacts with oxidizers, with a risk of fire or explosions. Attacks metals in the presence of moisture.

TOXAKIL (8001-35-2) Reacts with oxidizers, with a risk of fire or explosions. Attacks metals in the presence of moisture.

TOXAPHEN (German) (8001-35-2) Reacts with oxidizers, with a risk of fire or explosions. Attacks metals in the presence of moisture.

TOXAPHENE (8001-35-2) Reacts with oxidizers, with a risk of fire or explosions. Attacks metals in the presence of moisture.

TOXASPRAY (8001-35-2) Reacts with oxidizers, with a risk of fire or explosions. Attacks metals in the presence of moisture.

TOXICHLOR (57-74-9) Contact with strong oxidizers may cause fire and explosions. Attacks some plastics, rubber, and coatings.

TOXILIC ACID (110-16-7) May accumulate static electrical charges, and may cause ignition of its vapors. Aqueous solution is an acid. Incompatible with strong oxidizers, bases. Corrodes metal if wet.

TOXILIC ANHYDRIDE (108-31-7) Aqueous solution is a strong acid. Contact with strong oxidizers may cause fire and explosions. Contact with amines or alkali metals cause polymerization if temperature is greater than 150°F/66°C. Attacks some plastics, rubber, and coatings.

TOXOL (3) (56-38-2) Combustible liquid. Mixtures with endrin may be explosive. Strong oxidizers may cause fire and explosions. Attacks some plastics, rubber, and coatings.

TOXON 63 (8001-35-2) Reacts with oxidizers, with a risk of fire or explosions. Attacks metals in the presence of moisture.

TOXYPHEN (8001-35-2) Reacts with oxidizers, with a risk of fire or explosions. Attacks metals in the presence of moisture.

2,4,5-TP (93-72-1) A powerful oxidizer; reacts violently with reducing agents, combustibles.

2,4,5-TP ACID ESTERS (N/A) Incompatible with sulfuric acid, nitric acid, nitrates, oxidizers.

TPP (115-86-6) Incompatible with strong oxidizers, strong acids, nitrates; may cause fire or explosions. Attacks some plastics, rubber, and coatings.

TRABUTYL TITANATE (5593-70-4) Forms explosive mixture with air (flash point 170°F/77°C). Reacts violently with oxidizers. Water contact produces butanol and titanium dioxide.

TRANSAMINE (93-76-5) Sealed metal containers may burst in heat above 316°F/158°C. Incompatible with sulfuric acid, bases, ammonia, aliphatic amines, alkanolamines, isocyanates, alkylene oxides, epichlorohydrin.

TRANSAMINE (94-75-7) Decomposes in sunlight. Incompatible with strong oxidizers may cause fire and explosions.

TRANS-2-BUTENOIC ACID (374-65-0) Forms explosive mixture with air (flash point 190°F/88°C). A powerful reducing agent; reacts violently with oxidizers, combustibles. Incompatible with bases. Moisture or strong sunlight may cause explosive polymerization. May accumulate static electrical charges, and may cause ignition of its vapors.

TRANSFORMER OIL (N/A) Oxidizers may cause fire and explosions. Incompatible with nitric acid. May accumulate static electrical charges, and may cause ignition of its vapors.

TRANSMISSION OIL (N/A) Incompatible with nitric acid, strong oxidizers; risk of fire or explosions.

TRAPEX (556-61-6) Forms explosive mixture with air (flash point 90°F/32°C). Incompatible with strong acids, caustics, ammonia, amines, amides, alcohols, glycols, caprolactam solution, strong oxidizers.

TRAPEXIDE (556-61-6) Forms explosive mixture with air (flash point 90°F/32°C). Incompatible with strong acids, caustics, ammonia, amines, amides, alcohols, glycols, caprolactam solution, strong oxidizers.

TRAVAD (7727-43-7) Explosions may result from contact with aluminum in the presence of heat. Incompatible with potassium, phosphorus.

TRAVEX (7775-09-9) A powerful oxidizer; reacts violently with reducing agents and combustible matter. Explosions may be caused by contact with ammonia salts, carbon, oils, metal sulfides, nitrobenzene, powdered metals, sugar. Contact with strong acids produces carbon dioxide. Forms shock-sensitive mixtures with some organic materials. Solution (50%) decomposes at 300°F/149°C, liberating oxygen.

TRI (79-01-6) Contact with caustics produces a toxic and flammable gas. Reacts violently with chemically active metals. Contact with aluminum may produce a violent, self-accelerating polymerization reaction. Incompatible with acids, organic anhydrides, isocyanates, alkylene oxides, aldehydes, alcohols, glycols, phenols, cresols, caprolactam solution, epichlorohydrin, nitrogen tetroxide, metal powders, oxygen. May accumulate static electrical charges, and may cause ignition of its vapors.

TRI-6 (58-89-9) Not combustible, but may be dissolved in a combustible solvent. If solvent comes in contact with oxidizers, fire and explosions may result.

TRIACETALDEHYDE (French) (123-63-7) Forms explosive mixture with air (flash point 96°F/36°C). Reacts with strong acids, caustics, ammonia, amines, oxidizers. Contact with acids produces acetaldehyde.

TRIAD (79-01-6) Contact with caustics produces a toxic and flammable gas. Reacts violently with chemically active metals. Contact with aluminum may produce a violent, self-accelerating polymerization reaction. Incompatible with acids, organic anhydrides, isocyanates, alkylene oxides, aldehydes, alcohols, glycols, phenols, cresols, caprolactam solution, epichlorohydrin, nitrogen tetroxide, metal powders, oxygen. May accumulate static electrical charges, and may cause ignition of its vapors.

TRIAETHYLAMIN (German) (121-44-8) Forms explosive mixture with air (flash point 20°F/−6.7°C). Incompatible with strong acids, mineral acids, and organic acids (may cause violent spattering), and with organic anhydrides, halogenated hydrocarbons, isocyanates, nitroparaffins, nitrogen tetroxide, vinyl acetate, acrylates, substituted allyls, alkylene oxides, epichlorohydrin, ketones, aldehydes, alcohols, glycols, phenols, cresols, caprolactam solution, strong oxidizers. Attacks aluminum, copper, lead, tin, zinc, and alloys, and some plastics, rubber, and coatings.

TRIALLYLAMINE (102-70-5) Forms explosive mixture with air (flash point 140°F/60°C). A strong reducing agent and base; reacts violently with strong oxidizers, acids. Incompatible with many substances including organic anhydrides, isocyanates, alkylene oxides, epichlorohydrin, aldehydes, alcohols, glycols, phenols, cresols, caprolactam solution. Attacks aluminum, copper, copper alloys, tin, zinc.

TRIANGLE (7758-98-7) Aqueous solution is an acid. Incompatible with strong bases, hydroxylamine, magnesium.

TRIARYLPHOSPHATE (N/A) Incompatible with strong acids, nitrates, oxidizers.

TRIASOL (79-01-6) Contact with caustics produces a toxic and flammable gas. Reacts violently with chemically active metals. Contact with aluminum may produce a violent, self-accelerating polymerization reaction. Incompatible with acids, organic anhydrides, isocyanates, alkylene oxides, aldehydes, alcohols, glycols, phenols, cresols, caprolactam solution, epichlorohydrin, nitrogen tetroxide, metal powders, oxygen. May accumulate static electrical charges, and may cause ignition of its vapors.

TRIATOMIC OXYGEN (10028-15-6) A powerful oxidizer. Spontaneously decomposes to oxygen under ordinary conditions. Reacts with all reducing agents, combustibles, organic and inorganic oxidizable materials, and can form products that are highly explosive. Incompatible with alkenes, aniline, benzene, bromine, ether, ethylene, hydrogen bromide, nitric oxide, stibine. Contact with rubber may form shock- and friction-sensitive compounds.

1,4,7-TRIAZAHEPTANE (111-40-0) Forms explosive mixture with air (flash point 208°F/98°C). Incompatible with acids, organic anhydrides, isocyanates, vinyl acetate, acrylates, substituted allyls, alkylene oxides, epichlorohydrin, ketones, aldehydes, alcohols, glycols, mercury, phenols, cresols, caprolactam solution, strong oxidizers. Attacks aluminum, copper, lead, tin, zinc and alloys.

TRIAZINE A1294 (1912-24-9) Incompatible with strong acids.

1,3,5-TRIAZINE-2,4-DIAMINE, 6-CHLORO-N-ETHYL-N'-(1-METHYLETHYL)-(9CI) (1912-24-9) Incompatible with strong acids.

s-**TRIAZINE, EXAHYDRO-1,3,5-TRINITRO-** (121-82-4) Contact with mercury fulminate may cause detonation. Contact with combustibles or strong oxidizers may cause fire and explosions.

1,3,5-TRIAZINE-2,4,6-(1H,3H,5H)-TRIONE, 1,3-DICHLORO-, POTASSIUM SALT (2244-21-5) Incompatible with water, forms a bleach solution. A strong oxidizer; reacts violently with reducing agents, combustibles, organics, easily chlorinated materials.

S-TRIAZINE-2,4,6(1H,3H,5H)-TRIONE, DICHLORO-, POTASSIUM DERIVATIVE (2244-21-5) Incompatible with water; forms a bleach solution. A strong oxidizer; reacts violently with reducing agents, combustibles, organics, easily chlorinated materials.

1,3,5-TRIAZINE-2,4,6(1H,3H,5H)-TRIONE, 1,3,5-TRICHLORO- (87-90-1) A powerful oxidizer. Forms explosive material with nitrogen compounds. Contact with organic materials or reducing agents may cause fire.

TRIAZOLAMINE (61-82-5) Substance acts as a weak base to produce salts in contact with acids. Corrosive to iron, aluminum, copper and copper alloys.

TRIBROMMETHAAN (Dutch) (75-25-2) Reacts with chemically active metals, calcium, acetone, strong caustics. Attacks some plastics, rubber, and coatings.

TRIBROMMETHAN (German) (75-25-2) Reacts with chemically active metals, calcium, acetone, strong caustics. Attacks some plastics, rubber, and coatings.

TRIBROMOMETAN (Italian) (75-25-2) Reacts with chemically active metals, calcium, acetone, strong caustics. Attacks some plastics, rubber, and coatings.

TRIBROMOMETHANE (75-25-2) Reacts with barium, lithium, sodium, magnesium, titanium. Attacks some plastics, rubber, and coatings.

TRIBROMO-PHOSPHINE (7789-60-8) reacts violently with water forming hydrogen bromide. Incompatible with acids, caustics, alcohols, ammonia, oxidizers, alkali metals (i.e., lithium, sodium, potassium, rubidium, cesium, francium). Corrosive to most metals in the presence of moisture, except lead and nickel.

TRIBUTILFOSFATO (Italian) (126-73-8) Forms explosive mixture with air. Incompatible with strong acids, oxidizers, water.

TRIBUTON (93-76-5) Sealed metal containers may burst in heat above 316°F/158°C. Incompatible with sulfuric acid, bases, ammonia, aliphatic amines, alkanolamines, isocyanates, alkylene oxides, epichlorohydrin.

TRIBUTON (94-75-7) Decomposes in sunlight. Incompatible with strong oxidizers; may cause fire and explosions.

TRIBUTYLAMINE (102-82-9) Forms explosive mixture with air (flash point 187°F/83°F). A strong base. Incompatible with strong oxidizers, acids, organic anhydrides, isocyanates, aldehydes. Attacks aluminum, copper, copper alloys, tin, zinc.

TRI-N-BUTYLAMINE (102-82-9) Forms explosive mixture with air (flash point 187°F/83°F). A strong base. Incompatible with strong oxidizers, acids, organic anhydrides, isocyanates, aldehydes. Attacks aluminum, copper, copper alloys, tin, zinc.

TRIBUTYLE (PHOSPHATE de) (French) (126-73-8) Forms explosive mixture with air. Incompatible with strong acids, oxidizers, water.

TRIBUTYLFOSFAAT (Dutch) (126-73-8) Forms explosive mixture with air. Incompatible with strong acids, oxidizers, water.

TRIBUTYLPHOSPHAT (German) (126-73-8) Forms explosive mixture with air. Incompatible with strong acids, oxidizers, water.

TRIBUTYL PHOSPHATE (126-73-8) Forms explosive mixture with air. Incompatible with strong acids, oxidizers, water.

TRI-N-BUTYL PHOSPHATE (126-73-8) Forms explosive mixture with air. Incompatible with strong acids, oxidizers, water.

TRICALCIUM DIPHOSPHIDE (1305-99-3) A strong reducing agent. Produces phosphine in air. Water or acids can cause explosions. Incompatible with strong oxidizers, acids, chlorine, chlorine monoxide, oxygen, sulfur.

TRICARNAM (63-25-2) Incompatible with strong oxidizers, strongly alkaline pesticides.

1,1,1-TRICHLOOR-2,2-BIS(4-CHLOORFENYL)-ETHAAN (Dutch) (50-29-3) Incompatible with salts of iron or aluminum, and bases. Do not store in iron containers.

TRICHLOORETHEEN (Dutch) (79-01-6) Contact with caustics produces a toxic and flammable gas. Reacts violently with chemically active metals. Contact with aluminum may produce a violent, self-accelerating polymerization reaction. Incompatible with acids, organic anhydrides, isocyanates, alkylene oxides, aldehydes, alcohols, glycols, phenols, cresols, caprolactam solution, epichlorohydrin, nitrogen tetroxide, metal powders, oxygen. May accumulate static electrical charges, and may cause ignition of its vapors.

(2,4,5-TRICHLOOR-FENOXY)-AZIJNZUUR (Dutch) (93-76-5) Sealed metal containers may burst in heat above 316°F/158°C. Incompatible with sulfuric acid, bases, ammonia, aliphatic amines, alkanolamines, isocyanates, alkylene oxides, epichlorohydrin.

TRICHLOORMETHAAN (Dutch) (67-66-3) Decomposes in the presence of excess water or at high temperatures, producing phosgene and hydrogen chloride. Can become explosive in the presence of strong alkalies and water. Can accumulate static electrical charges. In contact with water and at high temperatures, becomes corrosive; attacks iron and other metals. Incompatible with acetone, aluminum, strong oxidizers, potassium, sodium, chemically active metals, strong bases. Attacks plastics and rubber.

TRICHLOR (76-06-2) Can be self-reactive. Fast heating, shock, alkali metals or alkaline earth may cause explosions. A strong oxidizer; reacts violently with reducing agents, aniline in presence of heat, alcoholic sodium hydroxide, combustible substances, sodium methoxide, propargyl bromide. Liquid attacks some plastics, rubber, and coatings.

TRICHLOR (79-01-6) Contact with caustics produces a toxic and flammable gas. Reacts violently with chemically active metals. Contact with aluminum may produce a violent, self-accelerating polymerization reaction. Incompatible with acids, organic anhydrides, isocyanates, alkylene oxides, aldehydes, alcohols, glycols, phenols, cresols, caprolactam solution, epichlorohydrin, nitrogen tetroxide, metal powders, oxygen. May accumulate static electrical charges, and may cause ignition of its vapors.

TRICHLORAETHEN (German) (79-01-6) Contact with caustics produces a toxic and flammable gas. Reacts violently with chemically active metals. Contact with aluminum may produce a violent, self-accelerating polymerization reaction. Incompatible with acids, organic anhydrides, isocyanates, alkylene oxides, aldehydes, alcohols, glycols, phenols, cresols, caprolactam solution, epichlorohydrin, nitrogen tetroxide, metal powders, oxygen. May accumulate static electrical charges, and may cause ignition of its vapors.

TRICHLORAMINE (10025-85-1) Shock, friction, light or temperatures above 130°F/57°C can cause substance to explode. Contact with many substances can cause explosive decomposition, including organic substances, combustible materials, water, ammonia, reducing agents, phosphorus.

1,1,1-TRICHLOR-2,2-BIS(4-CHLOR-PHENYL)-AETHAN (German) (50-29-3) Incompatible with salts of iron or aluminum, and bases. Do not store in iron containers.

TRICHLORESSIGSAEURE (German) (76-03-9) A strong acid. Incompatible with sulfuric acid, bases, ammonia, amines, isocyanates, alkylene oxides, epichlorohydrin, oxidizers.

1,1,2-TRICHLORETHANE (79-00-5) Incompatible with strong oxidizers, strong caustics, chemically active metals (especially powders), or sodium amide may cause fire and explosions. Attacks some plastics, rubber, and coatings.

TRICHLORETHANOIC ACID (76-03-9) A strong acid. Incompatible with sulfuric acid, bases, ammonia, amines, isocyanates, alkylene oxides, epichlorohydrin, oxidizers.

TRICHLORETHENE (French or Italian) (79-01-6) Contact with caustics produces a toxic and flammable gas. Reacts violently with chemically active metals. Contact with aluminum may produce a violent, self-accelerating polymerization reaction. Incompatible with acids, organic anhydrides, isocyanates, alkylene oxides, aldehydes, alcohols, glycols, phenols, cresols, caprolactam solution, epichlorohydrin, nitrogen tetroxide, metal powders, oxygen. May accumulate static electrical charges, and may cause ignition of its vapors.

TRICHLORFON (52-68-6) Contact with strong oxidizers may cause fire and explosions.

TRICHLORINATED ISOCYANURIC ACID (87-90-1) A powerful oxidizer. Forms explosive material with nitrogen compounds. Contact with organic materials or reducing agents may cause fire.

TRICHLORINE NITRIDE (10025-85-1) Shock, friction, light or temperatures above 130°F/57°C can cause substance to explode. Contact with many substances can cause explosive decomposition, including organic substances, combustible materials, water, ammonia, reducing agents, phosphorus.

TRICHLORMETHAN (Czech) (67-66-3) Decomposes in the presence of excess water or at high temperatures, producing phosgene and hydrogen chloride. Can become explosive in the presence of strong alkalies and water. Can accumulate static electrical charges. In contact with water and at high temperatures, becomes corrosive;

attacks iron and other metals. Incompatible with acetone, aluminum, strong oxidizers, potassium, sodium, chemically active metals, strong bases. Attacks plastics and rubber.

TRICHLORMETHYL SULFUR CHLORIDE (594-42-3) Incompatible with caustics or amines (with rapid decomposition), hot iron (produces toxic carbon tetrachloride fumes), hot water (produces hydrochloric acid), sulfur and carbon dioxide. Attacks some plastics, rubber, and coatings.

TRICHLOROACETALDEHYDE (75-87-6) Contact with acids or exposure to light may cause polymerization. Reacts with water, producing corrosive and toxic chloral hydrate. Contact with oxidizers may cause fire and explosions.

TRICHLOROACETALDEHYDE MONOHYDRATE (302-17-0) Reacts with strong oxidizers, strong bases. Attacks metals in the presence of moisture.

TRICHLOROACETIC ACID SOLUTION (76-03-9) A strong acid. Incompatible with sulfuric acid, bases, ammonia, amines, isocyanates, alkylene oxides, epichlorohydrin, oxidizers.

TRICHLOROALLYLSILANE (107-37-9) Forms explosive mixture with air (flash point 95°F/35°C). Reacts violently with water, producing hydrochloric acid. Attacks metals in presence of moisture.

TRICHLOROALUMINUM or TRICHLOROALUMINUM SOLUTION (7446-70-0) Contact with air or water produces hydrochloric acid. Water, alcohol, alkenes cause polymerization. Incompatible with nitrobenzene, organic material, and bases. Attacks metal in presence of moisture.

TRICHLOROAMYLSILANE (107-72-2) Forms explosive mixture with air (flash point 145°F/62°C). Water contact produces hydrochloric acid. Incompatible with strong bases. Attacks metals in presence of moisture.

TRICHLOROARSINE (7784-34-1) Water contact produces corrosive mixture of acids. Light exposure produces toxic gas. Reacts violently with anhydrous ammonia, strong acids, strong oxidizers, and halogens. Corrodes metals in the presence of moisture.

1,2,3-TRICHLOROBENZENE (87-61-6) (Flash point 210°F/99°C.) Strong oxidizers may cause fire and explosions.

1,2,4-TRICHLOROBENZENE (120-82-1) Contact with strong oxidizers may cause fire and explosions. Attacks most rubbers.

unsym-**TRICHLOROBENZENE** or *unsymmetrical*-**TRICHLORO-BENZENE** (120-82-1) Contact with strong oxidizers may cause fire and explosions. Attacks most rubbers.

v-TRICHLOROBENZENE or vic-TRICHLOROBENZENE (87-61-6) (Flash point 210°F/99°C.) Strong oxidizers may cause fire and explosions.

1,2,4-TRICHLOROBENZENE, LIQUID (120-82-1) Contact with strong oxidizers may cause fire and explosions. Attacks most rubbers.

1,2,4-TRICHLOROBENZOL (120-82-1) Contact with strong oxidizers may cause fire and explosions. Attacks most rubbers.

TRICHLOROBIS(4-CHLOROPHENYL)ETHANE (50-29-3) Incompatible with salts of iron or aluminum, and bases. Do not store in iron containers.

1,1,1-TRICHLORO-2,2-BIS(p-CHLOROPHENYL)ETHANE (50-29-3) Incompatible with salts of iron or aluminum, and bases. Do not store in iron containers.

1,1,1-TRICHLORO-2,2-BIS(p-METHOXYPHENYL)ETHANE (72-43-5) Contact with strong oxidizers may cause fire and explosions. Attacks some plastics, rubber, and coatings.

1,1,1-TRICHLORO-2,2-BIS(4-CLORO-FENIL)-ETANO (Italian) (50-29-3) Incompatible with salts of iron or aluminum, and bases. Do not store in iron containers.

TRICHLOROCYANURIC ACID (87-90-1) A powerful oxidizer. Forms explosive material with nitrogen compounds. Contact with organic materials or reducing agents may cause fire.

1,1,1-TRICHLORO-2,2-DI(4-CHLOROPHENYL)-ETHANE (50-29-3) Incompatible with salts of iron or aluminum, and bases. Do not store in iron containers.

TRICHLORODODECYLSILANE (4484-7-4) Forms explosive mixture with air (flash point 150°F/66°C). Water and air vapor cause formation of hydrochloric acid fumes. Incompatible with strong bases and oxidizers. Corrodes metals.

TRICHLOROETHANE or TRICHLOROETHANE, STABILIZED (71-55-6) Reacts with acetone, strong caustics, alkaline earth, alkali metals, dinitrogen tetroxide, sodium hydroxide, amides, strong oxidizers, sodium, potassium. Reacts violently with copper, bronze, and other metal powders. The uninhibited grade is corrosive to aluminum. Attacks some plastics, rubber, and coatings.

1,1,1-TRICHLOROETHANE (71-55-6) Reacts with acetone, strong caustics, alkaline earth and alkali metals, dinitrogen tetroxide, sodium hydroxide, amides, strong oxidizers, sodium, potassium. Reacts violently with copper, bronze, and other metal powders. The uninhibited grade is corrosive to aluminum. Attacks some plastics, rubber, and coatings.

1,1,2-TRICHLOROETHANE (79-00-5) Incompatible with strong oxidizers, strong caustics, chemically active metals (especially powders), or sodium amide; may cause fire and explosions. Attacks some plastics, rubber, and coatings.

1,2,2-TRICHLOROETHANE (79-00-5) Incompatible with strong oxidizers, strong caustics, chemically active metals (especially powders), or sodium amide; may cause fire and explosions. Attacks some plastics, rubber, and coatings.

a-TRICHLOROETHANE or alpha-TRICHLOROETHANE (71-55-6) Reacts with acetone, strong caustics, alkaline earth and alkali metals, dinitrogen tetroxide, sodium hydroxide, amides, strong oxidizers, sodium, potassium. Reacts violently with copper, bronze, and other metal powders. The uninhibited grade is corrosive to aluminum. Attacks some plastics, rubber, and coatings.

b-TRICHLOROETHANE or beta-TRICHLOROETHANE (79-00-5) Incompatible with strong oxidizers, strong caustics, chemically active metals (especially powders), or sodium amide; may cause fire and explosions. Attacks some plastics, rubber, and coatings.

TRICHLOROETHANOIC ACID SOLUTION (76-03-9) A strong acid. Incompatible with sulfuric acid, bases, ammonia, amines, isocyanates, alkylene oxides, epichlorohydrin, oxidizers.

TRICHLOROETHENE (79-01-6) Contact with caustics produces a toxic and flammable gas. Reacts violently with chemically active metals. Contact with aluminum may produce a violent, self-accelerating polymerization reaction. Incompatible with acids, organic anhydrides, isocyanates, alkylene oxides, aldehydes, alcohols, glycols, phenols, cresols, caprolactam solution, epichlorohydrin, nitrogen tetroxide, metal powders, oxygen. May accumulate static electrical charges, and may cause ignition of its vapors.

TRICHLOROETHYLENE (79-01-6) Contact with caustics produces a toxic and flammable gas. Reacts violently with chemically active metals. Contact with aluminum may produce a violent, self-accelerating polymerization reaction. Incompatible with acids, organic anhydrides, isocyanates, alkylene oxides, aldehydes, alcohols, glycols, phenols, cresols, caprolactam solution, epichlorohydrin, nitrogen tetroxide, metal powders, oxygen. May accumulate static electrical charges, and may cause ignition of its vapors.

1,1,2-TRICHLOROETHYLENE (79-01-6) Contact with caustics produces a toxic and flammable gas. Reacts violently with chemically active metals. Contact with aluminum may produce a violent, self-accelerating polymerization reaction. Incompatible with acids, organic anhydrides, isocyanates, alkylene oxides, aldehydes, alcohols, glycols, phenols, cresols, caprolactam solution, epichlorohydrin, nitrogen tetroxide, metal powders, oxygen. May accumulate static electrical charges, and may cause ignition of its vapors.

TRICHLOROETHYLENE TRI (French) (79-01-6) Contact with caustics produces a toxic and flammable gas. Reacts violently with chemically active metals. Contact with aluminum may produce a violent, self-accelerating polymerization reaction. Incompatible with acids, organic anhydrides, isocyanates, alkylene oxides, aldehydes, alcohols, glycols, phenols, cresols, caprolactam solution, epichlorohydrin, nitrogen tetroxide, metal powders, oxygen. May accumulate static electrical charges, and may cause ignition of its vapors.

TRICHLOROETHYLIDENE GLYCOL (302-17-0) Reacts with strong oxidizers, strong bases. Attacks metals in the presence of moisture.

TRICHLOROETHYLSILANE (115-1-9) Forms explosive mixture with air (flash point 72°F/22°C). Water contact produces hydrochloric acid. Strong oxidizers may cause fire and explosions. Corrodes most common metals.

TRICHLOROETHYLSILICANE (115-1-9) Forms explosive mixture with air (flash point 72°F/22°C). Water contact produces hydrochloric acid. Strong oxidizers may cause fire and explosions. Corrodes most common metals.

TRICHLOROFLUOROMETHANE (75-69-4) Reacts with barium, lithium, sodium, magnesium, titanium. Attacks some plastics, rubber, and coatings.

TRICHLOROFORM (67-66-3) Decomposes in the presence of excess water or at high temperatures, producing phosgene and hydrogen chloride. Can become explosive in the presence of strong alkalies and water. Can accumulate static electrical charges. In contact with water and at high temperatures, it becomes corrosive; attacks iron and other metals. Incompatible with acetone, aluminum, strong oxidizers, potassium, sodium, chemically active metals, strong bases. Attacks plastics and rubber.

TRICHLOROHYDRIN (96-18-4) Forms explosive mixture with air (flash point 164°F/74°C). Incompatible with strong oxidizers, strong caustics, chemically active metals; may cause fire and explosions. May decompose on contact with aluminum. Attacks some plastics, rubber, and coatings.

(2,2,2-TRICHLORO-1-HYDROXYETHYL)DIMETHYLPHOS-PHONATE (52-68-6) Contact with strong oxidizers may cause fire and explosions.

TRICHLOROIMINOISOCYANURIC ACID (87-90-1) A powerful oxidizer. Forms explosive material with nitrogen compounds. Contact with organic materials or reducing agents may cause fire.

TRICHLOROISOCYANURIC ACID (87-90-1) A powerful oxidizer. Forms explosive material with nitrogen compounds. Contact with organic materials or reducing agents may cause fire.

1,3,5-TRICHLOROISOCYANURIC ACID (87-90-1) A powerful oxidizer. Forms explosive material with nitrogen compounds. Contact with organic materials or reducing agents may cause fire.

TRICHLOROMETAPHOS (299-84-3) Temperatures above 300°F/150°C may cause explosive decomposition. Contact with strong oxidizers may cause fire and explosions. Attacks some plastics, rubber, and coatings.

TRICHLOROMETHANE (67-66-3) Decomposes in the presence of excess water or at high temperatures, producing phosgene and hydrogen chloride. Can become explosive in the presence of strong alkalies and water. Can accumulate static electrical charges. In contact with water and high temperatures, it becomes corrosive; attacks iron and other metals. Incompatible with acetone, aluminum, strong oxidizers, potassium, sodium, chemically active metals, strong bases. Attacks plastics and rubber.

TRICHLOROMETHANESULFENYL CHLORIDE (594-42-3) Incompatible with caustics or amines (with rapid decomposition), hot iron (produces toxic carbon tetrachloride fumes), hot water (produces hydrochloric acid), sulfur and carbon dioxide. Attacks some plastics, rubber, and coatings.

TRICHLOROMETHANE SULFURYL CHLORIDE (594-42-3) Incompatible with caustics or amines (rapid decomposition), hot iron (produces toxic carbon tetrachloride fumes), hot water (produces hydrochloric acid), sulfur, and carbon dioxide. Attacks some plastics, rubber, and coatings.

N-TRICHLOROMETHYLMERCAPTO-4-CYCLOHEXENE-1,2-DI-CARBOXIMIDE (133-06-2) Incompatible with tetraethyl pyrophosphate, parathion.

N-(TRICHLOROMETHYLMERCAPTO)-delta(SUP4)-TETRAHY-DROPHTHALIMIDE (133-06-2) Incompatible with tetraethyl pyrophosphate, parathion.

TRICHLOROMETHYLSILANE (75-79-6) Forms explosive gas mixture with air (flash point 15°F/−9°C). Water or air contact causes formation of hydrochloric acid. Strong oxidizers may cause fire and explosions. Incompatible with bases. Corrodes most metals.

TRICHLOROMETHYL SULFOCHLORIDE (594-42-3) Caustics or amines (rapid decomposition), hot iron (produces toxic carbon tetrachloride fumes), hot water (produces hydrochloric acid), sulfur, and carbon dioxide. Attacks some plastics, rubber, and coatings.

TRICHLOROMETHYL SULFUR CHLORIDE (594-42-3) Incompatible with caustics or amines (with rapid decomposition), hot iron (produces toxic carbon tetrachloride fumes), hot water (produces hydrochloric acid), sulfur, and carbon dioxide. Attacks some plastics, rubber, and coatings.

N-TRICHLOROMETHYLTHIOCYCLOHEX-4-ENE-1,2-DICAR-BOXIMIDE (133-06-2) Incompatible with tetraethyl pyrophosphate, parathion.

N-(TRICHLOROMETHYLTHIO)-4-CYCLOHEXENE-1,2-DICAR-BOXIMIDE (133-06-2) Incompatible with tetraethyl pyrophosphate, parathion.

N-TRICHLOROMETHYLTHIO-*cis*-delta(SUP4)-CYCLOHEXENE-1,2-DICARBOXIMIDE (133-06-2) Incompatible with tetraethyl pyrophosphate, parathion.

N-(TRICHLOROMETHYLTHIO)TETRAHYDROPHTHALIMIDE (133-06-2) Incompatible with tetraethyl pyrophosphate, parathion.

N-TRICHLOROMETHYLTHIO-3A,4,7,7A-TETRAHYDRO-PHTHALIMIDE (133-06-2) Incompatible with tetraethyl pyrophosphate, parathion.

TRICHLOROMONOFLUOROMETHANE (75-69-4) Reacts with barium, lithium, sodium, magnesium, titanium. Attacks some plastics, rubber, and coatings.

TRICHLOROMONOSILANE (10025-78-5) Forms explosive mixture with air (flash point 7°C/14°C). Water contact produces hydrogen chloride fumes. Incompatible with strong oxidizers, acids. Corrodes metals in the presence of moisture.

TRICHLORONITROMETHANE (76-06-2) Can be self reactive. Fast heating, shock, alkali metals, or alkaline earth may cause explosions. A strong oxidizer; reacts violently with reducing agents, aniline in presence of heat, alcoholic sodium hydroxide, combustible substances, sodium methoxide, propargyl bromide. Liquid attacks some plastics, rubber, and coatings.

TRICHLOROOXOVANADIUM (7727-18-6) Incompatible with water, producing hydrochloric acid solution. Air contact causes formation of corrosive hydrogen chloride fumes. Reacts violently with bases, sodium, potassium.

TRICHLOROPENTYLSILANE (107-72-2) Forms explosive mixture with air (flash point 145°F/62°C). Water contact produces hydrochloric acid. Incompatible with strong bases. Attacks metal in presence of moisture.

TRICHLOROPHENOL (99-95-4) Contact with strong oxidizers may cause fire and explosions.

2,4,5-TRICHLOROPHENOL (99-95-4) Contact with strong oxidizers may cause fire and explosions.

2,4,5-TRICHLOROPHENOL, O-ESTER WITH O,O-DIMETHYL-PHOSPHOROTHIOATE (299-84-3) Temperatures above 300°F/150°C may cause explosive decomposition. Contact with strong oxidizers may cause fire and explosions. Attacks some plastics, rubber, and coatings.

2,4,5-TRICHLOROPHENOXY-ACETIC ACID (93-76-5) Sealed metal containers may burst in heat above 316°F/158°C. Incompatible with sulfuric acid, bases, ammonia, aliphatic amines, alkanolamines, isocyanates, alkylene oxides, epichlorohydrin.

2,4,5-TRICHLOROPHENOXY-ACETIC ACID, ESTERS (8002-26-4) Water contact causes foaming. Strong oxidizers may cause fire and explosions. Attack some plastics, rubber, and coatings.

2,4,5-TRICHLOROPHENOXYACETIC ACID, SODIUM SALT (93-79-8) Incompatible with strong acids, nitrates, strong oxidizers.

2-(2,4,5-TRICHLOROPHENYOXY) PROPANOIC ACID (93-72-1) A powerful oxidizer; reacts violently with reducing agents, combustibles.

2-(2,4,5-TRICHLOROPHENOXY) PROPANOIC ACID, ISOOCTYL ESTER (N/A) Incompatible with sulfuric acid, nitric acid, nitrates, oxidizers.

TRICHLOROPHON (52-68-6) Contact with strong oxidizers may cause fire and explosions.

TRICHLOROPHOSPHINE (7719-12-2) Contact with water or alcohol may cause fire and explosions particularly in presence of combustible organic matter. Incompatible with ammonia, caustics, reducing agents, nitric acid, chemically active metals. In the presence of moisture, corrodes most metals. Attacks some plastics, rubber, and coatings.

1,2,3-TRICHLOROPROPANE (96-18-4) Forms explosive mixture with air (flash point 164°F/74°C). Incompatible with strong oxidizers (cause a violent reaction), strong bases, chemically active metals.

3,5,6-TRICHLORO-2-PYRIDOL-O-ESTER WITH O,O-DIETYL-PHOSPHOROTHIOATE (2921-88-) Incompatible with strong acids. Hydrolyzes from acid or alkaline solutions.

TRICHLORORAN (79-01-6) Contact with caustics produces a toxic and flammable gas. Reacts violently with chemically active metals. Contact with aluminum may produce a violent, self-accelerating polymerization reaction. Incompatible with acids, organic anhydrides, isocyanates, alkylene oxides, aldehydes, alcohols, glycols, phenols, cresols, caprolactam solution, epichlorohydrin, nitrogen tetroxide, metal powders, oxygen. May accumulate static electrical charges, and may cause ignition of its vapors.

TRICHLOROSILANE (10025-78-5) Forms explosive mixture with air (flash point 7°C/14°C). Water contact produces hydrogen chloride fumes. Incompatible with strong oxidizers, acids. Corrodes metals in the presence of moisture.

TRICHLOROSTIBINE (10025-91-9) Water contact produces corrosive solution. Air contact causes formation of corrosive vapor.

TRICHLORO TITANIUM (7705-07-9) Reacts with water, producing hydrogen chloride fumes. A strong reducing agent; reacts violently with oxidizers. Attacks metals in the presence of moisture.

1,1,2-TRICHLORO-1,2,2-TRIFLUOROETHANE (76-13-1) Reacts with barium, lithium, sodium, magnesium, titanium. Contact with alloys containing more than 2% magnesium may cause decomposition (with hydrogen chloride, hydrogen fluoride, and carbon monoxide released). Attacks some plastics, rubber, and coatings.

TRICHLORO-*s*-TRIAZINETRIONE (87-90-1) A powerful oxidizer. Forms explosive material with nitrogen compounds. Contact with organic materials or reducing agents may cause fire.

TRICHLORO-*s*-TRIAZINE-2,4,6(1H,3H,5H)-TRIONE (87-90-1) A powerful oxidizer. Forms explosive material with nitrogen compounds. Contact with organic materials or reducing agents may cause fire.

1,3,5-TRICHLORO-1,2,5-TRIAZINE-2,4,6(1H,3H,5H)-TRIONE (87-90-1) A powerful oxidizer. Forms explosive material with nitrogen compounds. Contact with organic materials or reducing agents may cause fire.

1,3,5-TRICHLORO-*s*-TRIAZINE-2,4,6(1H,3H,5H)-TRIONE (87-90-1) A powerful oxidizer. Forms explosive material with nitrogen compounds. Contact with organic materials or reducing agents may cause fire.

TRICHLORO-*s*-TRIAZINETRIONE (87-90-1) A powerful oxidizer. Forms explosive material with nitrogen compounds. Contact with organic materials or reducing agents may cause fire.

TRICHLORO-*s*-TRIAZINETRIONE, DRY (87-90-1) A powerful oxidizer. Forms explosive material with nitrogen compounds. Contact with organic materials or reducing agents may cause fire.

1,1,2-TRICHLORO-1,2,2-TRIFLUOROETHANE (76-13-1) Reacts with barium, lithium, sodium, magnesium, titanium. Contact with alloys containing more than 2% magnesium may cause decomposition (with hydrogen chloride, hydrogen fluoride, and carbon monoxide released). Attacks some plastics, rubber, and coatings.

1,3,5-TRICHLORO-2,4,6-TRIOXOHEXAHYDRO-S-TRIAZINE (87-90-1) A powerful oxidizer. Forms explosive material with nitrogen compounds. Contact with organic materials or reducing agents may cause fire.

TRICHLORO(VINYL)SILANE (75-94-5) Incompatible with water (violent reaction). Corrodes common metals and produces hydrogen.

TRICHLOROVINYL SILICANE (75-94-5) Incompatible with water, (violent reaction). Corrodes common metals and produces hydrogen.

(2,4,5-TRICHLOR-PHENOXY)-ESSIGSAEURE (German) (93-76-5) Sealed metal containers may burst in heat above 316°F/158°C. Incompatible with sulfuric acid, bases, ammonia, aliphatic amines, alkanolamines, isocyanates, alkylene oxides, epichlorohydrin.

TRICHLORURE d'ANTIMOINE (French) (10025-91-9) Water contact causes formation of corrosive solution. Air contact produces corrosive vapor.

TRI-CLENE (79-01-6) Contact with caustics produces a toxic and flammable gas. Reacts violently with chemically active metals. Contact with aluminum may produce a violent, self-accelerating polymerization reaction. Incompatible with acids, organic anhydrides, isocyanates, alkylene oxides, aldehydes, alcohols, glycols, phenols, cresols, caprolactam solution, epichlorohydrin, nitrogen tetroxide, metal powders, oxygen. May accumulate static electrical charges, and may cause ignition of its vapors.

TRICLOORAZIJNZUUR (Dutch) (76-03-9) A strong acid. Incompatible with sulfuric acid, bases, ammonia, amines, isocyanates, alkylene oxides, epichlorohydrin, oxidizers.

TRICLOROETILENE (Italian) (79-01-6) Contact with caustics produces a toxic and flammable gas. Reacts violently with chemically active metals. Contact with aluminum may produce a violent, self-accelerating polymerization reaction. Incompatible with acids, organic anhydrides, isocyanates, alkylene oxides, aldehydes, alcohols, glycols, phenols, cresols, caprolactam solution, epichlorohydrin, nitrogen tetroxide, metal powders, oxygen. May accumulate static electrical charges, and may cause ignition of its vapors.

TRICLOROMETANO (Italian) (67-66-3) Decomposes in the presence of excess water or at high temperatures, with formation of phosgene and hydrogen chloride. Can become explosive in the presence of strong alkalies and water. Can accumulate static electrical charges. In contact with water and high temperatures, becomes corrosive; attacks iron and other metals. Incompatible with acetone, aluminum, strong oxidizers, potassium, sodium, chemically active metals, strong bases. Attacks plastics and rubber.

TRICON BW (60-00-4) Incompatible with sulfuric acid, bases, ammonia, aliphatic amines, alkanolamines, isocyanates, alkylene oxides, epichlorohydrin.

TRICRESOL (1319-77-3) Incompatible with sulfuric acid, nitric acid, bases, aliphatic amines, amides, oxidizers, chlorosulfonic acid, oleum

TRI-*o*-CRESYL ESTER (78-30-8) Contact with magnesium or strong oxidizers may cause fire and explosions. Attacks some plastics, rubber, and coatings.

TRICRESYL PHOSPHATE (78-30-8) Contact with magnesium may cause explosion. Attacks some plastics, rubber, and coatings.

TRICYCLO[5.2.1.0 2,6-DECA-3,8-DIENE (77-73-6) Forms explosive mixture with air (flash point 90°F/32°C). Forms peroxides with air. May polymerize unless inhibited and maintained under inert atmosphere. reacts violently with strong oxidizers. May accumulate static electrical charges, and may cause ignition of its vapors.

TRIDECANE (629-50-5) Incompatible with strong acids, nitrates, oxidizers.

n-**TRIDECANE or** *normal*-**TRIDECANE** (629-50-5) Incompatible with strong acids, nitrates, oxidizers.

TRIDECANOL (112-70-9) Incompatible with strong acids, caustics, aliphatic amines, isocyanates, strong oxidizers. Attacks aluminum. May accumulate static electrical charges, and may cause ignition of its vapors.

1-TRIDECANOL (112-70-9) Incompatible with strong acids, caustics, aliphatic amines, isocyanates, strong oxidizers. Attacks aluminum. May accumulate static electrical charges, and may cause ignition of its vapors.

1-TRIDECANOL, PHTHALATE (119-06-2) Incompatible with strong acids, nitrates, oxidizers.

1-TRIDECENE (N/A) Forms explosive mixture above 175°F/79°C. Incompatible with strong acids.

TRIDECYL ALCOHOL (112-70-9) Incompatible with strong acids, caustics, aliphatic amines, isocyanates, strong oxidizers. Attacks aluminum. May accumulate static electrical charges, and may cause ignition of its vapors.

TRIDECYLBENZENE (123-02-4) Forms explosive mixture with air (flash point 230°F/110°C). Incompatible with nitric acid, strong oxidizers.

TRIDIMETHYLPHENYL PHOSPHATE (25155-23-1) Incompatible with sulfuric acid, nitric acid, nitrates, strong oxidizers.

TRIEN (112-24-3) Forms explosive mixture with air (flash point 275°F/135°C). Incompatible with acids, organic anhydrides, chlorinated hydrocarbons, isocyanates, vinyl acetate, acrylates, substituted allyls, alkylene oxides, epichlorohydrin, ketones, aldehydes, alcohols, glycols, phenols, cresols, caprolactam solution, strong oxidizers. Attacks aluminum, copper, lead, tin, zinc, and alloys.

TRIENTINE (112-24-3) Forms explosive mixture with air (flash point 275°F/135°C). Incompatible with acids, organic anhydrides, chlorinated hydrocarbons, isocyanates, vinyl acetate, acrylates, substituted allyls, alkylene oxides, epichlorohydrin, ketones, aldehydes, alcohols, glycols, phenols, cresols, caprolactam solution, strong oxidizers. Attacks aluminum, copper, lead, tin, zinc, and alloys.

TRIESTE FLOWERS (8003-34-7) Strong oxidizers may cause fire and explosions.

TRI-ETHANE (71-55-6) Incompatible with strong caustics, strong oxidizers, chemically active metals, sodium, potassium. Attacks some plastics, rubber, and coatings.

TRIETHANOLAMINE (102-71-6) Aqueous solution is caustic. Incompatible with acids, organic anhydrides, isocyanates, vinyl acetate, acrylates, substituted allyls, alkylene oxides, epichlorohydrin, aldehydes, strong oxidizers. Corrodes copper and its alloys.

TRIETHYL ALUMINUM (97-93-8) A dangerous explosion hazard. Ignites spontaneously, at all temperatures, in air. A powerful reducing agent; reacts violently with oxidizers, water (producing ethane gas), alcohols, amines, carbon dioxide, carbon tetrachloride, halogenated hydrocarbons, oxides of nitrogen or sulfur, phenols, and many other substances. May accumulate static electrical charges, and may cause ignition of its vapors.

TRIETHYLAMINE (121-44-8) Forms explosive mixture with air (flash point 20°F/−6.7°C). Incompatible with strong acids, mineral acids, and organic acids may cause violent spattering), and with organic anhydrides, halogenated hydrocarbons, isocyanates, nitroparaffins, nitrogen tetroxide, vinyl acetate, acrylates, substituted allyls, alkylene oxides, epichlorohydrin, ketones, aldehydes, alcohols, glycols, phenols, cresols, caprolactam solution, strong oxidizers. Attacks aluminum, copper, lead, tin, zinc, and alloys, and some plastics, rubber, and coatings.

TRIETHYLBENZENE (102-25-0) Forms explosive mixture with air (flash point 181°F/83°C). Incompatible with nitric acid, strong oxidizers. May accumulate static electrical charges, and may cause ignition of its vapors.

1,3,5-TRIETHYLBENZENE (102-25-0) Forms explosive mixture with air (flash point 181°F/83°C). Incompatible with nitric acid, strong oxidizers. May accumulate static electrical charges, and may cause ignition of its vapors.

sym-**TRIETHYLBENZENE** (102-25-0) Forms explosive mixture with air (flash point 181°F/83°C). Incompatible with nitric acid, strong oxidizers. May accumulate static electrical charges, and may cause ignition of its vapors.

TRIETHYLBORANE (97-94-9) Forms explosive mixture with air (flash point less than 194°F/90°C). May self-ignite on contact with air. A strong reducing agent; reacts violently with strong oxidizers, halogens. Incompatible with alcohols, glycols.

TRIETHYLBORINE (97-94-9) Forms explosive mixture with air (flash point less than 194°F/90°C). May self-ignite on contact with air. A strong reducing agent; reacts violently with strong oxidizers, halogens. Incompatible with alcohols, glycols.

817

TRIETHYLENE GLYCOL (112-27-6) Incompatible with sulfuric acid, isocyanates, perchloric acid, strong oxidizers.

TRIETHYLENE GLYCOL BUTYL ETHER MIXTURE (N/A) Incompatible with sulfuric acid, isocyanates.

TRIETHYLENE GLYCOL DI-(2-ETHYLBUTYRATE) (95-08-9) Incompatible with strong acids, strong oxidizers, nitrates.

TRIETHYLENE GLYCOL ETHYL ETHER (112-50-5) Incompatible with sulfuric acid, isocyanates, perchloric acid, oxidizers.

TRIETHYLENE GLYCOL METHYL ETHER (112-35-6) Forms explosive mixture with air (flash point 230°F/110°C). Incompatible with sulfuric acid, isocyanates, perchloric acid, oxidizers.

TRIETHYLENE GLYCOL MONOETHYLETHER (112-50-5) Incompatible with sulfuric acid, isocyanates, perchloric acid, oxidizers.

TRIETHYLENE GLYCOL MONOMETHYL ETHER (112-35-6) Forms explosive mixture with air (flash point 230°F/110°C). Incompatible with sulfuric acid, isocyanates, perchloric acid, oxidizers.

TRIETHYLENE TETRAMINE (107-15-3) Forms explosive mixture with air (flash point 104°F/40°C). Incompatible with chlorinated organic compounds, silver perchlorate, 3-propiolactone, mesityl oxide, ethylene dichloride, acids, organic anhydrides, isocyanates, vinyl acetate, acrylates, substituted allyls, alkylene oxides, epichlorohydrin, ketones, aldehydes, alcohols, glycols, phenols, cresols, caprolactam solution, strong oxidizers. Attacks aluminum, copper, lead, tin, zinc, and alloys, and some plastics, rubber, and coatings.

TRIETHYLOLAMINE (102-71-6) Aqueous solution is caustic. Incompatible with acids, organic anhydrides, isocyanates, vinyl acetate, acrylates, substituted allyls, alkylene oxides, epichlorohydrin, aldehydes, strong oxidizers. Corrodes copper and its alloys.

TRIETHYL PHOSPHATE (78-40-0) Incompatible with strong acids, nitrates, oxidizers.

TRIETHYL PHOSPHITE (122-52-1) Incompatible with acids (nonoxidizing mineral, sulfuric, nitric, organic), nitrates, oxidizers, magnesium.

TRIETILAMINA (Italian) (121-44-8) Forms explosive mixture with air (flash point 20°F/−6.7°C). Incompatible with strong acids, mineral acids, and organic acids (may cause violent spattering), and with organic anhydrides, halogenated hydrocarbons, isocyanates, nitroparaffins, nitrogen tetroxide, vinyl acetate, acrylates, substituted allyls, alkylene oxides, epichlorohydrin, ketones, aldehydes, alcohols, glycols, phenols, cresols, caprolactam solution, strong oxidizers. Attacks aluminum, copper, lead, tin, zinc, and alloys, and some plastics, rubber, and coatings.

TRIFLUOROACETIC ACID or TRIFLUOROACETIC ACID (DOT) (76-05-1) A strong acid; reacts violently with strong bases. Incompatible with sulfuric acid (causes decomposition), ammonia, amines, isocyanates, alkylene oxides, epichlorohydrin. Attacks metals.

TRIFLUOROANTIMONY, STIBINE, TRIFLUORO- (7783-56-4) Contact with strong oxidizers may cause fire and explosions. Explosive mixture may be formed with perchloric acid.

TRIFLUOROBORANE (7637-07-2) Decomposes in heat, water and moist air. Fire and explosions may result from contact with alkali metals, alkaline earth metals (except magnesium), alkyl nitrate, calcium oxide. Incandesces when heated with alkali metals or alkaline earth metals (except magnesium). Corrodes most metals in the presence of moisture.

TRIFLUOROBORANE (7787-71-5) A powerful oxidizer. Water contact produces hydrogen fluoride gas. Reacts violently with reducing agents, organic materials, acids, halogens, salts, alkalies, metal oxides, and many other materials. Attacks some plastics, rubber, and coatings.

TRIFLUOROBORON (7637-07-2) Decomposes in heat, water, and moist air. Fire and explosions may result from contact with alkali metals, alkaline earth metals (except magnesium), alkyl nitrate, calcium oxide. Incandesces when heated with alkali metals or alkaline earth metals (except magnesium). Corrodes most metals in the presence of moisture.

TRIFLUOROBROMOMETHANE (75-63-8) Reacts with chemically active metals, powdered aluminum, zinc, magnesium. May attack some plastics, rubber, and coatings.

TRIFLUOROCHLOROETHYLENE (79-38-9) Flammable gas. Forms explosive mixture with air (flash point −18°F/−28°C). Reacts violently with oxidizers. Contact with ethylene may cause explosive polymerization.

1,1,2-TRIFLUORO-2-CHLOROETHYLENE (79-38-9) Flammable gas. Forms explosive mixture with air (flash point −18°F/−28°C). Reacts violently with oxidizers. Contact with ethylene may cause explosive polymerization.

TRIFLUOROCHLOROMETHANE (75-72-9) Reacts with aluminum, magnesium, zinc, and their alloys.

TRIFLUOROETHANOIC ACID (76-05-1) A strong acid; reacts violently with strong bases. Incompatible with sulfuric acid (causes decomposition), ammonia, amines, isocyanates, alkylene oxides, epichlorohydrin. Attacks metals.

TRIFLUOROMETHYL CHLORIDE (75-72-9) Reacts with aluminum, magnesium, zinc, and their alloys.

TRIFLUOROMONOBROMOMETHANE (75-63-8) Reacts with chemically active metals, powdered aluminum, zinc, magnesium. May attack some plastics, rubber, and coatings.

TRIFLUOROMONOCHLOROETHYLENE (79-38-9) Flammable gas. Forms explosive mixture with air (flash point −18°F/−28°C). Reacts violently with oxidizers. Contact with ethylene may cause explosive polymerization.

TRIFLUOROVINYL CHLORIDE (79-38-9) Flammable gas. Forms explosive mixture with air (flash point −18°F/−28°C). Reacts violently with oxidizers. Contact with ethylene may cause explosive polymerization.

TRIFLUORURE de CHLORE (French) (7790-91-2) Will cause most combustible materials to ignite spontaneously. Dangerously reactive with many materials and substances, including water, acids, aluminum oxide, ammonia, arsenic trioxide, asbestos, benzene, chromic anhydride, chromic oxide, copper, most elements, fuels, organic matter, silicon-containing compounds, sand, glass and glass wool, graphite, oxides. Attacks all forms of plastics, rubber, coatings and resins except for highly fluorinated polymers such as 'Teflon' and 'Kel-F.'

TRIFORMOL (30525-89-4) Forms explosive mixture with air (flash point 160°F/71°C). Aqueous solution causes formation of formaldehyde. Reacts violently with strong oxidizers or liquid oxygen. May accumulate static electrical charges, and may cause ignition of its vapors.

TRIFRINA (534-52-1) Incompatible with heat and strong oxidizers. Dust can form explosive mixtures with air.

TRIGEN (112-27-6) Incompatible with sulfuric acid, isocyanates, perchloric acid, strong oxidizers.

TRIGLYCOL (112-27-6) Incompatible with sulfuric acid, isocyanates, perchloric acid, strong oxidizers.

TRIGLYCOL DICAPROATE (95-08-9) Incompatible with strong acids, strong oxidizers, nitrates.

TRIGLYCOL DIHEXOATE (95-08-9) Incompatible with strong acids, strong oxidizers, nitrates.

TRIGLYCOL METHYL ETHER (112-35-6) Forms explosive mixture with air (flash point 230°F/110°C). Incompatible with sulfuric acid, isocyanates, perchloric acid, oxidizers.

TRIGLYCOL MONOETHYL ETHER (112-50-5) Incompatible with sulfuric acid, isocyanates, perchloric acid, strong oxidizers.

TRI(2-HYDROXYETHYL)AMINE (102-71-6) Aqueous solution is caustic. Incompatible with acids, organic anhydrides, isocyanates, vinyl acetate, acrylates, substituted allyls, alkylene oxides, epichlorohydrin, aldehydes, strong oxidizers. Corrodes copper and its alloys.

1,2,3-TRIHYDROXYBENZENE (87-66-1) Incompatible with strong oxidizers, caustics, ammonia, amines, isocyanates, alkylene oxides, epichlorohydrin.

3,4,5-TRIHYDROXYBENZOIC ACID (149-91-7) Incompatible with sulfuric acid, caustics, ammonia, aliphatic and alkanol amines, isocyanates, alkylene oxides, epichlorohydrin, strong oxidizers.

TRIHYDROXYPROPANE (56-81-5) Polymerizes at approximately 300°F/150°C. Incompatible with acetic anhydride, potassium permanganate, strong acids, caustics, aliphatic amines, isocyanates, oxidizers.

1,2,3-TRIHYDROXYPROPANE (56-81-5) Polymerizes at approximately 300°F/150°C. Incompatible with acetic anhydride, potassium permanganate, strong acids, caustics, aliphatic amines, isocyanates, oxidizers.

TRI(HYDROXYTRIETHYL)AMINE (102-71-6) Aqueous solution is caustic. Incompatible with acids, organic anhydrides, isocyanates, vinyl acetate, acrylates, substituted allyls, alkylene oxides, epichlorohydrin, aldehydes, strong oxidizers. Corrodes copper and its alloys.

TRIISOBUTENE (7756-94-7) Incompatible with sulfuric and or nitric acid.

TRIISOBUTYLALANE (100-99-2) Ignites spontaneously in air. Water contact causes formation of hydrocarbons. Reacts with oxidizers, carbon dioxide, alcohols, ammonia, halogenated hydrocarbons. Attacks silicone and urethane rubbers.

TRIISOBUTYLALUMINUM (100-99-2) Ignites spontaneously in air. Water contact causes formation of hydrocarbons. Reacts with oxidizers, carbon dioxide, alcohols, ammonia, halogenated hydrocarbons. Attacks silicone and urethane rubbers.

TRIISOBUTYLENE (7756-94-7) Incompatible with sulfuric or nitric acid.

TRIISOOCTYL TRIMELLITATE (N/A) Incompatible with strong acids, nitrates, oxidizers.

TRIISOPROPANOLAMINE or TRI-ISO-PROPANOLAMINE (122-20-3) Incompatible with acids, organic anhydrides, isocyanates, vinyl acetate, acrylates, substituted allyls, alkylene oxides, epichlorohydrin, aldehydes.

TRIISOPROPANOLAMINE SALT OF 2,4-DICHLOROPHENOXY ACETIC ACID SOLUTION (N/A) Incompatible with sulfuric acid, isocyanates.

o-**TRIKRESYLPHOSPHAT (German)** (78-30-8) Contact with magnesium or strong oxidizers may cause fire and explosions. Attacks some plastics, rubber, and coatings.

TRILENE (79-01-6) Contact with caustics produces a toxic and flammable gas. Reacts violently with chemically active metals. Contact with aluminum may produce a violent, self-accelerating polymerization reaction. Incompatible with acids, organic anhydrides, isocyanates, alkylene oxides, aldehydes, alcohols, glycols, phenols, cresols, caprolactam solution, epichlorohydrin, nitrogen tetroxide, metal powders, oxygen. May accumulate static electrical charges, and may cause ignition of its vapors.

TRILON B (60-00-4) Incompatible with sulfuric acid, bases, ammonia, aliphatic amines, alkanolamines, isocyanates, alkylene oxides, epichlorohydrin.

TRILON BW (60-00-4) Incompatible with sulfuric acid, bases, ammonia, aliphatic amines, alkanolamines, isocyanates, alkylene oxides, epichlorohydrin.

TRIMAR (79-01-6) Contact with caustics produces a toxic and flammable gas. Reacts violently with chemically active metals. Contact with aluminum may produce a violent, self-accelerating polymerization reaction. Incompatible with acids, organic anhydrides, isocyanates, alkylene oxides, aldehydes, alcohols, glycols, phenols, cresols, caprolactam solution, epichlorohydrin, nitrogen tetroxide, metal powders, oxygen. May accumulate static electrical charges, and may cause ignition of its vapors.

TRIMETHOXYPHOSPHINE (121-45-9) Forms explosive mixture with air (flash point 82°F/28°C). Reacts violently with strong acids, nitrates, oxidizers, magnesium salts of perchloric acid.

1,7,7-TRIMETHYL- (76-22-2) Forms explosive mixture with air (flash point 155°F/66°C). Violent, possibly explosive, reaction with occurs strong oxidizers. May accumulate static electrical charges, and may cause ignition of its vapors.

TRIMETHYLACETIC ACID (75-98-9) Forms explosive mixture with air (flash point 147°F/64°C). Incompatible with sulfuric acid, caustics, ammonia, amines, isocyanates, alkylene oxides, epichlorohydrin, oxidizers.

TRIMETHYLALANE (75-24-1) A dangerous explosion hazard. Ignites spontaneously in air. A powerful reducing agent; reacts violently with oxidizers, water, alcohols, amines, carbon dioxide, carbon tetrachloride, halogenated hydrocarbons, oxides of nitrogen or sulfur, phenols, and many other substances. May accumulate static electrical charges, and may cause ignition of its vapors.

TRIMETHYLALUMINUM (75-24-1) A dangerous explosion hazard. Ignites spontaneously in air. A powerful reducing agent; reacts violently with oxidizers, water, alcohols, amines, carbon dioxide, carbon tetrachloride, halogenated hydrocarbons, oxides of nitrogen or sulfur, phenols, and many other substances. May accumulate static electrical charges, and may cause ignition of its vapors.

TRIMETHYLAMINE (75-50-3) Forms explosive mixture with air (flash point 10°F/−12°C). Incompatible with strong oxidizers, acids. Attacks chemically active metals: aluminum, copper, zinc, and their alloys.

TRIMETHYLAMINOMETHANE (75-64-9) Forms explosive mixture with air (flash point 48°F/−9°C). May accumulate static electrical charges, and may cause ignition of its vapors. Incompatible with acids, organic anhydrides, isocyanates, vinyl acetate, acrylates, substituted allyls, alkylene oxides, epichlorohydrin, ketones, aldehydes, alcohols, glycols, phenols, cresols, caprolactam solution, strong oxidizers.

1,2,4-TRIMETHYLBENZENE (95-63-6) Forms explosive mixture with air (flash point 112°F/44°C). Oxidizers may cause fire and explosions. Incompatible with nitric acid. May accumulate static electrical charges, and may cause ignition of its vapors.

1,3,5-TRIMETHYLBENZENE (108-67-8) Forms explosive mixture with air (flash point 117°F/47°C). Reacts violently with strong oxidizers. May accumulate static electrical charges, and may cause ignition of its vapors.

asym-**TRIMETHYL BENZENE or** *asymmetrical*-**TRIMETHYL BENZENE** (95-63-6) Forms explosive mixture with air (flash point 112°F/44°C). Oxidizers may cause fire and explosions. Incompatible with nitric acid. May accumulate static electrical charges, and may cause ignition of its vapors.

TRIMETHYLBENZOL (108-67-8) Forms explosive mixture with air (flash point 117°F/47°C). Reacts violently with strong oxidizers. May accumulate static electrical charges, and may cause ignition of its vapors.

TRIMETHYLBENZYLAMMONIUM CHLORIDE (56-93-9) Incompatible with strong acids, oxidizers.

1,7,7-TRIMETHYLBICYCLO(2.2.1)-2-HEPTANONE (76-22-2) Forms explosive mixture with air (flash point 155°F/66°C). Violent, possibly explosive, reaction occurs with strong oxidizers. May accumulate static electrical charges, and may cause ignition of its vapors.

1,7,7-TRIMETHYLBICYCLO(2,2,1)HEPTANONE-2 (76-22-2) Forms explosive mixture with air (flash point 155°F/66°C). Violent, possibly explosive, reaction with strong oxidizers. May accumulate static electrical charges, and may cause ignition of its vapors.

2,6,6-TRIMETHYLBICYCLO[3.1.1]HEPT-2-ENE (80-56-8) Incompatible with sulfuric acid, nitric acid.

2,6,6-TRIMETHYLBICYCLO[3.1.1]-2-HEPT-2-ENE (80-56-8) Incompatible with sulfuric acid, nitric acid.

TRIMETHYL CARBINOL (75-65-0) Forms explosive mixture with air (flash point 52°F/11°C). May accumulate static electrical charges, and may cause ignition of its vapors. Incompatible with strong acids including mineral acids, strong oxidizers or caustics aliphatic amines, isocyanates, alkali metals (i.e., lithium, sodium, potassium, rubidium, cesium, francium). Attacks many plastics and some coatings.

TRIMETHYLCARBINYLAMINE (75-64-9) Forms explosive mixture with air (flash point 48°F/−9°C). May accumulate static electrical charges, and may cause ignition of its vapors. Incompatible with acids, organic anhydrides, isocyanates, vinyl acetate, acrylates, substituted allyls, alkylene oxides, epichlorohydrin, ketones, aldehydes, alcohols, glycols, phenols, cresols, caprolactam solution, strong oxidizers.

TRIMETHYLCHLOROSILANE (75-77-4) Forms explosive mixture with air (flash point −18°F/−28°C). Reacts violently with water. Vigorous reaction with aluminum. Moisture and air contact forms hydrochloric acid.

3,5,5-TRIMETHYL-2-CYCLOHEXANE-1-ONE (78-59-1) Forms explosive mixture with air (flash point 184°F/84°C). Incompatible with strong oxidizers, aliphatic amines. Strong oxidizers may cause fire and explosions. Attacks some plastics, resins, and rubber.

1,1,3-TRIMETHYL-3-CYCLOHEXEN-5-ONE (78-59-1) Forms explosive mixture with air (flash point 184°F/84°C). Incompatible with strong oxidizers, aliphatic amines. Strong oxidizers may cause fire and explosions. Attacks some plastics, resins, and rubber.

3,5,5-TRIMETHYL-2-CYCLOPHENONE (78-59-1) Forms explosive mixture with air (flash point 184°F/84°C). Incompatible with strong oxidizers, aliphatic amines. Strong oxidizers may cause fire and explosions. Attacks some plastics, resins, and rubber.

TRIMETHYLEENTRINITRAMINE (Dutch) (121-82-4) Contact with mercury fulminate may cause detonation. Contact with combustibles or strong oxidizers may cause fire and explosions.

TRIMETHYLENE (75-19-4) Flammable gas. May accumulate static electrical charges, and may cause ignition of its vapors. Strong oxidizers may cause fire and explosions.

TRIMETHYLENE CHLORIDE (142-28-9) Forms explosive mixture with air (flash point 70°F/21°C). Incompatible with strong oxidizers, light metals. Attacks some plastics, rubber, and coatings. May accumulate static electrical charges, and may cause ignition of its vapors.

TRIMETHYLENE DICHLORIDE (142-28-9) Forms explosive mixture with air (flash point 70°F/21°C). Incompatible with strong oxidizers, light metals. Attacks some plastics, rubber, and coatings. May accumulate static electrical charges, and may cause ignition of its vapors.

alpha,alpha,alpha'-TRIMETHYLENE GLYCOL (107-41-5) Forms explosive mixture with air (flash point 209°F/98°C). Incompatible with strong acids, caustics, aliphatic amines, isocyanates, strong oxidizers.

TRIMETHYLENETRINITRAMINE (121-82-4) Contact with mercury fulminate may cause detonation. Contact with combustibles or strong oxidizers may cause fire and explosions.

sym-**TRIMETHYLENETRINITRAMINE** (121-82-4) Contact with mercury fulminate amy cause detonation. Contact with combustibles or strong oxidizers may cause fire and explosions.

TRIMETHYL GLYCOL (57-55-7) Incompatible with strong acids, caustics, aliphatic amines, isocyanates, strong oxidizers (violent reaction).

TRIMETHYLHEPTANALS (N/A) Forms explosive mixture with air (flash point 185°F/85°C). Incompatible with strong acids, caustics, ammonia, amines.

TRIMETHYL HEXAMETHYLENE DIAMINE (25513-64-8) Incompatible with nitric, sulfuric, mineral and organic acids, and with organic anhydrides, isocyanates, vinyl acetate, acrylates, substituted allyls, alkylene oxides, ketones, aldehydes, alcohols, glycols, phenols, cresols, caprolactam solution.

TRIMETHYLMETHANE (75-28-5) Forms explosive gas mixture with air. Strong oxidizers may cause fire and explosions. May accumulate static electrical charges, and may cause ignition of its vapors.

TRIMETHYLMETHANOL (75-65-0) Forms explosive mixture with air (flash point 52°F/11°C). May accumulate static electrical charges, and may cause ignition of its vapors. Incompatible with strong acids including mineral acids, strong oxidizers or caustics, aliphatic amines, isocyanates, alkali metals (i.e., lithium, sodium, potassium, rubidium, cesium, francium). Attacks many plastics and some coatings.

1,7,7-TRIMETHYLNORCAMPHOR (76-22-2) Forms explosive mixture with air (flash point 155°F/66°C). Violent, possibly explosive, reaction occurs with strong oxidizers. May accumulate static electrical charges, and may cause ignition of its vapors.

2,2,4-TRIMETHYLPENTANE (540-84-1) Forms explosive mixture with air (flash point 40°F/5°C). Reacts with oxidizers. May accumulate static electrical charges, and may cause ignition of its vapors.

2,2,4-TRIMETHYL-1,3-PENTANEDIOL-1,3-DIISOBUTYRATE (N/A) Incompatible with strong acids, nitrates, oxidizers.

2,2,4-TRIMETHYL-1,3-PENTANEDIOL-1-ISOBUTYRATE (25265-77-4) Incompatible with strong acids, nitrates, oxidizers.

2,2,4-TRIMETHYL-1,3-PENTANEDIOLMONOISOBUTYRATE (25265-77-4) Incompatible with strong acids, nitrates, oxidizers.

2,4-TRIMETHYL-1-PENTENE (11071-47-9) Forms explosive mixture with air (flash point 23°F/−5°C). May be able to form unstable and explosive peroxides. Incompatible with strong acids, strong oxidizers. May accumulate static electrical charges, and may cause ignition of its vapors.

2,4,4-TRIMETHYLPENTENE-1 (25167-70-8) Forms explosive mixture with air (flash point 23°F/−5°C). May form unstable and explosive peroxides. Incompatible with strong acids, strong oxidizers. May accumulate static electrical charges, and may cause ignition of its vapors.

2,4,4-TRIMETHYLPENTENE-2 (25167-70-8) Forms explosive mixture with air (flash point 23°F/−5°C). May form unstable and explosive peroxides. Incompatible with strong acids, strong oxidizers. May accumulate static electrical charges, and may cause ignition of its vapors.

TRIMETHYLPHOSPHITE (121-45-9) Forms explosive mixture with air (flash point 82°F/28°C). Reacts violently with strong acids, nitrates, oxidizers, magnesium salts of perchloric acid.

TRIMETHYLSILYL CHLORIDE (75-77-4) Forms explosive mixture with air (flash point −18°F/−28°C). Reacts violently with water. Vigorous reaction with aluminum. Moisture and air contact causes formation of hydrochloric acid.

2,4,6-TRIMETHYL-1,3,5-TRIOXAAN (Dutch) (123-63-7) Forms explosive mixture with air (flash point 96°F/36°C). Reacts with strong acids, caustics, ammonia, amines, oxidizers. Contact with acids produces acetaldehyde.

2,4,6-TRIMETHYL-1,3,5-TRIOXANE (123-63-7) Forms explosive mixture with air (flash point 96°F/36°C). Reacts with strong acids, caustics, ammonia, amines, oxidizers. Contact with acids produces acetaldehyde.

2,4,6-TRIMETHYLTRIOXYMETHYLENE (123-63-7) Forms explosive mixture with air (flash point 96°F/36°C). Reacts with strong acids, caustics, ammonia, amines, oxidizers. Contact with acids produces acetaldehyde.

s-**TRIMETHYLTRIOXYMETHYLENE** (123-63-7) Forms explosive mixture with air (flash point 96°F/36°C). Reacts with strong acids, caustics, ammonia, amines, oxidizers. Contact with acids produces acetaldehyde.

2,4,6-TRIMETIL-1,3,5-TRIOSSANO (Italian) (123-63-7) Forms explosive mixture with air (flash point 96°F/36°C). Reacts with strong acids, caustics, ammonia, amines, oxidizers. Contact with acids form acetaldehyde.

TRIMETYLENE GLYCOL (504-63-2) Incompatible with strong acids, strong bases, aliphatic amines, isocyanates, oxidizers.

TRINEX (52-68-6) Contact with strong oxidizers may cause fire and explosions.

TRINITROCYCLOTRIMETHYLENE TRIAMINE (121-82-4) Contact with mercury fulminate may cause detonation. Contact with combustibles or strong oxidizers may cause fire and explosions.

TRINITROGLYCERIN (55-63-0) Liquid portion of this chemical is an alcohol; reacts with strong acids, caustics, aliphatic amines, isocyanates. Dry portion of chemical is nitroglycerine, a shock- and friction-sensitive explosive. Acids, heat, or mechanical shock may cause explosions.

TRINITROPHENYLMETHYLNITRAMINE (479-45-8) A powerful oxidizer and highly sensitive explosive. Protect containers from shock, friction, heat, oxidizers, and hydrazine.

2,4,6-TRINITROPHENYLMETHYLNITRAMINE (479-45-8) A powerful oxidizer and highly sensitive explosive. Protect containers from shock, friction, heat, oxidizers, and hydrazine.

2,4,6-TRINITROPHENYL-N-METHYLNITRAMINE (479-45-8) A powerful oxidizer and highly sensitive explosive. Protect containers from shock, friction, heat, oxidizers, and hydrazine.

TRINITROTOLUOL (118-96-7) Explosive; protect from shock and heat. Slow decomposition occurs above 356°F/180°C. Exposure to light may increase impact sensitivity. Rapid heating may cause detonation. Strong oxidizers may cause fire. Contact with ammonia or with strong alkalies may increase sensitivity to shock. Reacts vigorously with oxidizable materials.

TRINITROTOLUENE or *sym*-TRINITROTOLUENE (118-96-7) Explosive; protect from shock and heat. Slow decomposition occurs above 356°F/180°C. Exposure to light may increase impact sensitivity. Rapid heating may cause detonation. Strong oxidizers may cause fire. Contact with ammonia or with strong alkalies may increase sensitivity to shock. Reacts vigorously with oxidizable materials.

2,4,6-TRINITROTOLUENE (118-96-7) Explosive; protect from shock and heat. Slow decomposition occurs above 356°F/180°C). Exposure to light may increase impact sensitivity. Rapid heating may cause detonation. Strong oxidizers may cause fire. Contact with ammonia or with strong alkalies may increase sensitivity to shock. Reacts vigorously with oxidizable materials.

1,3,5-TRINITRO-1,3,5-TRIAZACYCLOHEXANE (121-82-4) Contact with mercury fulminate may cause detonation. Contact with combustibles or strong oxidizers may cause fire and explosions.

TRINOXOL (93-76-5) Sealed metal containers may burst in heat above 316°F/158°C. Incompatible with sulfuric acid, bases, ammonia, aliphatic amines, alkanolamines, isocyanates, alkylene oxides, epichlorohydrin.

TRIORTHOCRESYL PHOSPHATE (78-30-8) Contact with magnesium may cause explosion. Attacks some plastics, rubber, and coatings.

TRIOXANE (50-00-0) May polymerize unless properly inhibited (usually with methanol). Forms explosive mixture with air (flash point 122°F/50°C). Incompatible with strong acids, amines, strong oxidizers, alkaline materials, nitrogen dioxide, performic acid. Reaction with hydrochloric acid produces bis-chloromethyl ether, a carcinogen.

TRIOXANE (110-88-3) Forms explosive mixture with air (flash point 113°F/35°C). Able to form unstable and explosive peroxides. Heat, shock, friction can cause material to explode. Incompatible with oxidizers, acids with the risk of fires and explsion. May accumulate static electrical charges, and may cause ignition of its vapors.

1,3,5-TRIOXANE (110-88-3) Forms explosive mixture with air (flash point 113°F/35°C). Able to form unstable and explosive peroxides. Heat, shock, friction can cause material to explode. Incompatible with oxidizers, acids with the risk of fires and explsion. May accumulate static electrical charges, and may cause ignition of its vapors.

s-**TRIOXANE** (110-88-3) Forms explosive mixture with air (flash point 113°F/35°C). Able to form unstable and explosive peroxides. Heat, shock, friction can cause material to explode. Incompatible with oxidizers, acids with the risk of fires and explosion. May accumulate static electrical charges, and may cause ignition of its vapors.

5,8,11-TRIOXAPENTADECANE (112-73-2) Incompatible with sulfuric acid, isocyanates, strong oxidizers. May accumulate static electrical charges, and may cause ignition of its vapors.

3,6,9-TRIOXAUNDECAN-1-OL (112-50-5) Incompatible with sulfuric acid, isocyanates, perchloric acid, oxidizers.

3,6,9-TRIOXAUNDECAN-1,11-DIOL (112-60-7) Incompatible with sulfuric acid, isocyanates, perchloric acid. Attacks some plastics, rubber, and coatings.

3,6,9-TRIOXAUNDECANOL,11-DIOL (112-60-7) Incompatible with sulfuric acid, isocyanates, perchloric acid. Attacks some plastics, rubber, and coatings.

TRIOXON (93-76-5) Sealed metal containers may burst in heat above 316°F/158°C. Incompatible with sulfuric acid, bases, ammonia, aliphatic amines, alkanolamines, isocyanates, alkylene oxides, epichlorohydrin.

TRIOXYMETHYLENE (30525-89-4) Forms explosive mixture with air (flash point 160°F/71°C). Aqueous solution forms formaldehyde. Reacts violently with strong oxidizers or liquid oxygen. May accumulate static electrical charges, and may cause ignition of its vapors.

TRIOZONE (8002-26-4) Water contact causes foaming. Strong oxidizers may cause fire and explosions. Attacks some plastics, rubber, and coatings.

TRIOZONE (93-76-5) Sealed metal containers may burst in heat above 316°F/158°C. Incompatible with sulfuric acid, bases, ammonia, aliphatic amines, alkanolamines, isocyanates, alkylene oxides, epichlorohydrin.

TRIPHENYL ESTER OF PHOSPHORIC ACID ATE (115-86-6) Strong oxidizers may cause fire or explosions. Attacks some plastics, rubber, and coatings.

TRIPHENYL PHOSPHATE (115-86-6) Incompatible with strong oxidizers, strong acids, nitrates may cause fire or explosions. Attacks some plastics, rubber, and coatings.

TRI-PLUS (79-01-6) Contact with caustics produces a toxic and flammable gas. Reacts violently with chemically active metals. Contact with aluminum may produce a violent, self-accelerating polymerization reaction. Incompatible with acids, organic anhydrides, isocyanates, alkylene oxides, aldehydes, alcohols, glycols, phenols, cresols, caprolactam solution, epichlorohydrin, nitrogen tetroxide, metal powders, oxygen. May accumulate static electrical charges, and may cause ignition of its vapors.

TRI-2-PROPENYL-2-PROPEN-1-AMINE (102-70-5) Forms explosive mixture with air (flash point 140°F/60°C). A strong reducing agent and base; reacts violently with strong oxidizers, acids. Incompatible with many substances including organic anhydrides, isocyanates, alkylene oxides, epichlorohydrin, aldehydes, alcohols, glycols, phenols, cresols, caprolactam solution. Attacks aluminum, copper, copper alloys, tin, zinc.

TRIPROPYLAMINE (102-69-2) Incompatible with strong acids, strong oxidizers.

TRI-N-PROPYLAMINE (102-69-2) Incompatible with strong acids, strong oxidizers.

TRIPROPYLENE (139-01-4) Forms explosive mixture with air (flash point 75°F/24°C). Incompatible with sulfuric acid, nitric acid, oxidizers.

TRIPROPYLENE GLYCOL (24800-44-0) Incompatible with sulfuric acid, isocyanates, perchloric acid, strong oxidizers. Attacks some plastics, rubber, and coatings.

TRIPROPYLENE GLYCOL METHYL ETHER (25498-49-1) Forms explosive mixture with air (flash point 250°F/121°C). Incompatible with sulfuric acid, perchloric acid, isocyanates, oxidizers.

TRIS-N-BUTYLAMINE (102-82-9) Forms explosive mixture with air (flash point 187°F/83°F). A strong base. Incompatible with strong oxidizers, acids, organic anhydrides, isocyanates, aldehydes. Attacks aluminum, copper, copper alloys, tin, zinc.

TRIS(HYDROXYETHYL)AMINE (102-71-6) Aqueous solution is caustic. Incompatible with acids, organic anhydrides, isocyanates, vinyl acetate, acrylates, substituted allyls, alkylene oxides, epichlorohydrin, aldehydes, strong oxidizers. Corrodes copper and its alloys.

TRIS(2-HYDROXYPROPYL) AMINE (122-20-3) Incompatible with acids, organic anhydrides, isocyanates, vinyl acetate, acrylates, substituted allyls, alkylene oxides, epichlorohydrin, aldehydes.

TRIS(o-METHYLPHENYL)PHOSPHATE (78-30-8) Contact with magnesium or strong oxidizers may cause fire and explosions. Attacks some plastics, rubber, and coatings.

TRISODIUM PHOSPHATE (7601-54-9) Aqueous solution is caustic; reacts with acids. Attacks aluminum, copper, zinc, and related alloys in the presence of moisture.

TRITHENE (79-38-9) Flammable gas. Forms explosive mixture with air (flash point −18°F/−28°C). Reacts violently with oxidizers. Contact with ethylene may cause explosive polymerization.

o-**TRITOLYL PHOSPHATE** (78-30-8) Contact with magnesium or strong oxidizers may cause fire and explosions. Attacks some plastics, rubber, and coatings.

TRITON GR-5 (119-36-8) Forms explosive mixture with air (flash point 205°F/96°C). Incompatible with strong acids, nitrates, oxidizers.

TRIXYLENYL PHOSPHATE (25155-23-1) Incompatible with sulfuric acid, nitric acid, nitrates, strong oxidizers.

TRIXYLYL PHOSPHATE (25155-23-1) Incompatible with sulfuric acid, nitric acid, nitrates, strong oxidizers.

TROCLOSENE POTASSIUM (2244-21-5) Incompatible with water, producing a bleach solution. A strong oxidizer; reacts violently with reducing agents, combustibles, organics, easily chlorinated materials.

TROJCHLOREK FOSFORU (Polish) (7719-12-2) Contact with water or alcohol may cause fire and explosions particularly in presence of combustible organic matter. Incompatible with ammonia, caustics, reducing agents, nitric acid, chemically active metals. In the presence of moisture, corrodes most metals. Attacks some plastics, rubber, and coatings.

TROJCHLOROBENZEN (Polish) (120-82-1) Contact with strong oxidizers may cause fire and explosions. Attacks most rubbers.

TROJCHLOROETAN(1,1,2) (Polish) (79-00-5) Incompatible with strong oxidizers, strong caustics, chemically active metals (especially powders), or sodium amide; may cause fire and explosions. Attacks some plastics, rubber, and coatings.

TROJKREZYLU FOSFORAN (Polish) (78-30-8) Contact with magnesium or strong oxidizers may cause fire and explosions. Attacks some plastics, rubber, and coatings.

TROLAMINE (102-71-6) Aqueous solution is caustic. Incompatible with acids, organic anhydrides, isocyanates, vinyl acetate, acrylates, substituted allyls, alkylene oxides, epichlorohydrin, aldehydes, strong oxidizers. Corrodes copper and its alloys.

TROLEN (299-84-3) Temperatures above 300°F/150°C may cause explosive decomposition. Contact with strong oxidizers may cause fire and explosions. Attacks some plastics, rubber, and coatings.

TROLENE (299-84-3) Temperatures above 300°F/150°C may cause explosive decomposition. Contact with strong oxidizers may cause fire and explosions. Attacks some plastics, rubber, and coatings.

TROMETE (7601-54-9) Aqueous solution is caustic; reacts with acids. Attacks aluminum, copper, zinc, and related alloys in the presence of moisture.

TRONA (10294-33-4) Incompatible with water, steam, alcohols, potassium, sodium. Attacks metals, rubber.

TRONAMANG (7439-96-5) Water contact produces hydrogen gas. Oxidizers, nitric acid, nitrogen, finely divided aluminum and other metals, sulfur dioxide may cause fire and explosions.

TROVIDUER (75-01-4) Forms explosive gas mixture with air. Atmospheric oxygen and various contaminants or strong oxidizers cause formation of peroxides, which can initiate a violent polymerization action. Also able to polymerize when heated and exposed, long term, to light. Reacts violently with strong oxidizers or oxides of nitrogen. Contact with copper or other acetylide-forming metals produces explosive compounds. Attacks iron and steel in the presence of moisture.

TRUE AMMONIUM SULFIDE (12124-99-1) Forms explosive mixture with air (flash point 72°F/22°C). Incompatible with strong oxidizers, strong acids. Contact with strong bases produces ammonia. Corrodes aluminum, brass, copper, zinc, and related alloys; attacks some plastics, rubber and coatings. May accumulate static electrical charges, and may cause ignition of its vapors.

TRUFLEX DOA (103-23-1) Incompatible with strong acids, nitrates, oxidizers.

2,4,5-T, SODIUM SALT (93-79-8) Incompatible with strong acids, nitrates, strong oxidizers.

TSP (7601-54-9) Aqueous solution is caustic; reacts with acids. Attacks aluminum, copper, zinc, and related alloys in the presence of moisture.

T-STUFF (7722-84-1) A powerful oxidizer; attacks many substances. Contact with most organic, readily oxidizable materials, reducing agents, and combustibles causes fire and explosions. Contact with iron, copper, brass, bronze, chromium, zinc, lead, manganese, silver, and other catalytic metals (and their salts), especially in a basic (pH7 or above) environment, cause rapid decomposition with evolution of oxygen gas. Attacks, and may ignite, some plastics, rubber, and coatings. Decomposes slowly at ordinary temperatures and builds up pressure in a closed container. The rate of decomposition doubles for each 50°F/10°C rise (1.5 times 10°C rise) in temperature and becomes self-sustaining at 285°F/141°C.

TTD (137-26-8) Combustible solid (flash point 192°F/89°C). Strong oxidizers may cause fire and explosions; contact with strong acid or oxidizable materials produces toxic gases.

TTE (76-13-1) Reacts with barium, lithium, sodium, magnesium, titanium. Contact with alloys containing more than 2% magnesium may cause decomposition (with hydrogen chloride, hydrogen fluoride, and carbon monoxide released). Attacks some plastics, rubber, and coatings.

TUADS (137-26-8) Combustible solid (flash point 192°F/89°C). Strong oxidizers may cause fire and explosions; contact with strong acid or oxidizable materials produces toxic gases.

TUBERCUPROSE (544-19-4) Contact with strong oxidizers may cause fire and explosions.

TUCUM OIL (N/A) Incompatible with strong acids, oxidizers.

TUEX (137-26-8) Combustible solid (flash point 192°F/89°C). Strong oxidizers may cause fire and explosions; contact with strong acid or oxidizable materials produces toxic gases.

TUFFLITE (9003-07-0) Incompatible with strong acids, strong oxidizers, chlorine, potassium permanganate.

TUGON (52-68-6) Contact with strong oxidizers may cause fire and explosions.

TULISAN (137-26-8) Combustible solid (flash point 192°F/89°C). Strong oxidizers may cause fire and explosions; contact with strong acid or oxidizable materials produces toxic gases.

TULUYLEN DISOCYANAT (German) (584-84-9) Incompatible with strong acids including nonoxidizing mineral and organic acids, caustics, ammonia, amines, amides, alcohols, glycols, caprolactam solution. Water contact causes violent foaming and spattering; produces carbon dioxide and an organic base. Attacks copper and its alloys, some plastics including polyethylene, and rubber.

TULUYLENE-2,4-DISOCYANAT (German) (584-84-9) Incompatible with strong acids, including nonoxidizing mineral and organic acids, caustics, ammonia, amines, amides, alcohols, glycols, caprolactam solution. Water contact causes violent foaming and spattering; produces carbon dioxide and an organic base. Attacks copper and its alloys, some plastics including polyethylene, and rubber.

TUMBLEAF (7775-09-9) A powerful oxidizer; reacts violently with reducing agents and combustible matter. Explosions may be caused by contact with ammonia salts, carbon, oils, metal sulfides, nitrobenzene, powdered metals, sugar. Contact with strong acids produces carbon dioxide. Forms shock-sensitive mixtures with some organic materials. Solution (50%) decomposes at 300°F/149°C liberating oxygen.

TURBINE OIL (N/A) Incompatible with nitric acid, strong oxidizers; poses risk of fire or explosions.

TURKEY-RED OIL (SULFATED CASTOR OIL) (8001-79-4) Incompatible with strong acids, oxidizers, nitrates.

TURPENTINE or TURPENTINE STEAM DISTILLED (8006-64-2) Forms explosive mixture with air (flash point 95°F/35°C). Incompatible with strong oxidizers, especially chlorine; may cause fire and explosions. Incompatible with strong acids, chromic anhydride, chromyl chloride, hexachloromelamine, stannic chloride. Attacks ordinary rubber.

TURPENTINE SUBSTITUTE (8032-32-4) Forms explosive mixture with air (flash point −40°F to −86°F/−40°C to −66°C). Incompatible with strong acids, strong oxidizers. Attacks some plastics, rubber, and coatings. May accumulate static electrical charges, and may cause ignition of its vapors.

TURPS (8006-64-2) Forms explosive mixture with air (flash point 95°F/35°C). Incompatible with strong oxidizers, especially chlorine; may cause fire and explosions. Incompatible with strong acids, chromic anhydride, chromyl chloride, hexachloromelamine, stannic chloride. Attacks ordinary rubber.

TUTANE (13952-84-6) Forms explosive mixture with air. Incompatible with acids, organic anhydrides, isocyanates, vinyl acetate, acrylates, substituted allyls, alkylene oxides, epichlorohydrin, ketones, aldehydes, alcohols, glycols, phenols, cresols, caprolactam solution, strong oxidizers.

TYRANTON (123-42-2) Incompatible with strong acids, strong alkalies (cause formation of flammable acetone vapors), aliphatic amines, isocyanates, oxidizers, alkali metals. Forms explosive mixture with air (flash point 136°F/58°C).

- U -

U-46 (93-76-5) Sealed metal containers may burst in heat above 316°F/158°C. Incompatible with sulfuric acid, bases, ammonia, aliphatic amines, alkanolamines, isocyanates, alkylene oxides, epichlorohydrin.

U-46 (94-75-7) Decomposes in sunlight. Incompatible with strong oxidizers may cause fire and explosions.

U-46DP (94-75-7) Decomposes in sunlight. Incompatible with strong oxidizers may cause fire and explosions.

U-46KW (8002-26-4) Water contact causes foaming. Strong oxidizers may cause fire and explosions. Attacks many plastics, rubber, and coatings.

U-1149 (110-17-8) Reacts with strong oxidizers. Incompatible with sulfuric acid, caustics, ammonia, amines, isocyanates, alkylene oxides, epichlorohydrin.

U-3886 (26628-22-8) Incompatible with acids, some metals (i.e, lead, copper, silver, mercury). Forms explosion-sensitive compounds.

U-4224 (68-12-2) Forms explosive mixture with air (flash point 136°F/58°C). Contact with carbon tetrachloride and other halogenated compounds, particularly in presence of iron or strong oxidizers, may cause fire and explosions. Vigorous reaction with alkylaluminums. Incompatible with nonoxidizing mineral acids, strong acids, chlorinated hydrocarbons, isocyanates, nitrates, organic nitrates, phenols, cresols, ammonia, chromic anhydride, magnesium nitrate, methylene diisocyanate, phosphorous trioxide, triethyaluminum. Attacks some plastics, rubber, and coatings.

U-5043 (94-75-7) Decomposes in sunlight. Incompatible with strong oxidizers; may cause fire and explosions.

U-5227 (86-88-4) Strong oxidizers may cause fires and explosion. Also reacts with silver nitrate.

U-5954 (127-19-5) Forms explosive mixture with air (flash point 158°F/70°C). Incompatible with nonoxidizing mineral acids, strong acids, ammonia, isocyanates, phenols, cresols, halogenated compounds above 185°F/85°C. Attacks some plastics, rubber, and coatings.

UAN-NITROGEN SOLUTION (15978-77-5) Incompatible with oxidizers, combustibles, organics, acids, amides, organic anhydrides, isocyanates, vinyl acetate, alkylene oxides, epichlorohydrin, aldehydes, sulfur. Reacts violently with copper and copper alloys. Attacks lead, zinc.

UAN SOLUTION (15978-77-5) Incompatible with oxidizers, combustibles, organics, acids, amides, organic anhydrides, isocyanates, vinyl acetate, alkylene oxides, epichlorohydrin, aldehydes, sulfur. Reacts violently with copper and copper alloys. Attacks lead, zinc.

UC-7744 (63-25-2) Incompatible with strong oxidizers, strongly alkaline pesticides.

UCANE ALKYLATE 12 (123-01-3) Incompatible with nitric acid, strong oxidizers; poses risk of fire or explosions.

UCAR-17 (107-21-1) Incompatible with strong acids, caustics, aliphatic amines, isocyanates, chlorosulfonic acid, oleum, strong oxidizers.

UCAR BISPHENOL HP (80-05-7) Combustible. Fine dust can form an explosive mixture with air. Contact with strong oxidizers may cause fire and explosions.

UCAR BUTYLPHENOL 4-T (98-54-4) Incompatible with strong acids, caustics, aliphatic amines, amides, oxidizers.

UCAR SOLVENT 2LM (34590-94-8) Ethers, as a class, can form dangerous peroxides on standing. Strong oxidizers may cause fire and explosions. Attacks some plastics, rubber, and coatings.

UCON-11 (75-69-4) Reacts with barium, lithium, sodium, magnesium, titanium. Attacks some plastics, rubber, and coatings.

UCON-12 (75-71-8) Reacts violently with liquid aluminum. Incompatible with chemically active metals. Attacks some plastics, rubber, and coatings.

UCON-113 (76-13-1) Reacts with barium, lithium, sodium, magnesium, titanium. Contact with alloys containing more than 2% magnesium may cause decomposition (with hydrogen chloride, hydrogen fluoride, and carbon monoxide released). Incompatible with acetanilide, albumin, alkalies, antipyrine, camphor, ferric salts, menthol, spirit nitrous ether. Absorbs moisture from air (hygroscopic).

UCONN-13 (75-72-9) Reacts with aluminum, magnesium, zinc, and their alloys.

UCONN-22 (75-45-6) Thermal decomposition occurs at high temperature with alkalies and alkaline earth metals.

UCON REFRIGERANT-11 (75-69-4) Reacts with barium, lithium, sodium, magnesium, titanium. Attacks some plastics, rubber, and coatings.

UDMH (57-14-7) Forms explosive mixture with air (flash point $5°F/-15°C$). A strong reducing agent. Incompatible with strong acids, halogens, metallic mercury, strong oxidizers (with possible spontaneous ignition). Attacks many plastics, rubber, and coatings. May accumulate static electrical charges, and may cause ignition of its vapors.

UF OXYLIGNIN (68514-06-7) Water contact generates heat. Incompatible with acids, organic anhydrides, isocyanates, alkylene oxides, epichlorohydrin, aldehydes, alcohols, glycols, phenols, cresols, caprolactam solution, oxidizers. Contact with acids produces carbon dioxide and alkaline mists. Incompatible with aluminum, zinc, tin; reactions produce heat and hydrogen gas.

ULSTRONG (9003-07-0) Incompatible with strong acids, strong oxidizers, chlorine, potassium permanganate.

ULTRAMID (105-60-2) Contact with strong oxidizers may cause fire and explosions.

ULTRA WET K (25155-30-0) Reacts with acids, including fumes.

UMBETHION (56-72-4) Contact with strong oxidizers may cause fire and explosions.

UNDECANOIC ACID (112-37-8) Incompatible with sulfuric acid, caustics, ammonia, amines, isocyanates, epichlorohydrin. Attacks many common metals.

n-**UNDECANOIC ACID** (112-37-8) Incompatible with sulfuric acid, caustics, ammonia, amines, isocyanates, epichlorohydrin. Attacks many common metals.

UNDECANOL (103-08-2) Incompatible with strong acids, caustics, aliphatic amines, isocyanates, acetaldehyde, ethylene oxide, nitrogen tetroxide, tri-isobutyl aluminum.

1-UNDECANOL (103-08-2) Incompatible with strong acids, caustics, aliphatic amines, isocyanates, acetaldehyde, ethylene oxide, nitrogen tetroxide, tri-isobutyl aluminum.

n-**UNDECOIC ACID** (112-37-8) Incompatible with sulfuric acid, caustics, ammonia, amines, isocyanates, epichlorohydrin. Attacks many common metals.

UNDECYL ALCOHOL (103-08-2) Incompatible with strong acids, caustics, aliphatic amines, isocyanates, acetaldehyde, ethylene oxide, nitrogen tetroxide, tri-isobutyl aluminum.

n-**UNDECYLBENZENE** (N/A) Combustible liquid. Incompatible with nitric acid, strong oxidizers. Attacks many plastics, rubber, and coatings.

UNDECYLETHYLENE (N/A) Forms explosive mixture above 175°F/79°C. Incompatible with strong acids.

n-**UNDECYLIC ACID** (112-37-8) Incompatible with sulfuric acid, caustics, ammonia, amines, isocyanates, epichlorohydrin. Attacks many common metals.

UNIBARYT (7727-43-7) Explosions may result from contact with aluminum in the presence of heat. Incompatible with potassium, phosphorus.

UNIDRON (330-54-1) Hydrolyzes in fairly strong acids.

UNIFUME (106-93-4) Reacts with chemically active metals, liquid ammonia, strong oxidizers. Heat and light cause slow decomposition. Attacks some plastics, and rubber.

UNION CARBIDE 7,744 (63-25-2) Incompatible with strong oxidizers, strongly alkaline pesticides.

UNION CARBIDE A-150 (75-94-5) Incompatible with water (reacts violently). Corrodes common metals and produces hydrogen.

UNIPINE (8002-09-3) Forms explosive mixture with air (flash point 138°F/59°C). Incompatible with strong acids, strong oxidizers.

UNIPON (75-99-0) Corrosive to iron, aluminum, and copper.

UNIROYAL D-014 (2312-35-8) Forms explosive mixture with air (flash point 82°F/28°C). Strong oxidizers may cause fire and explosions.

UNITED CHEMICAL DEFOLIANT NO. 1 (7775-09-9) A powerful oxidizer; reacts violently with reducing agents and combustible matter. Explosions may be caused by contact with ammonia salts, carbon, oils, metal sulfides, nitrobenzene, powdered metals, sugar. Contact with strong acids produces carbon dioxide. Forms shock-sensitive mixtures with some organic materials. Solution (50%) decomposes at 300°F/149°C liberating oxygen.

UNIVERM (56-23-5) Becomes corrosive when in contact with water. Corrosive to metals. Reacts violently with many compounds. Decomposes on contact with chemically active metals such as sodium, potassium and magnesium. Incompatible with allyl alcohol, fluorine gas, alkali metals, aluminum. Attacks some coatings, plastics, and rubber.

UNSLAKED LIME (1305-78-8) Reacts violently with water. Incompatible with ethanol, hydrogen fluoride, acids, halogens, metal halides, light metals, some oxides, boron trifluoride, chlorine trifluoride, liquid hydrofluoric acid, phosphorus pentoxide, boric oxide and calcium chloride mixtures, fluorine, chlorine trifluoride, carbon dioxide.

URANIUM or URANIUM, METALLIC or URANIUM POWDER (7440-61-1) Turnings and fines stored outdoors in closed containers under water or water-soluble oil will partially convert to uranium hydride and eventually ignite during hot weather. Contact with air in the presence of moisture may cause self-ignition. Strong oxidizers, strong acids, carbon dioxide, carbon tetrachloride, or nitric acid may cause fire and explosions. Fine particles are a fire and explosion hazard.

URANIUM HEXAFLUORIDE (10049-14-6) Water contact produces hydrofluoric acid. Attacks some plastics, rubber, and coatings; aqueous solution attacks glass, ceramics and silica-containing substances such as cast iron.

URANIUM HYDRIDE or URANIUM(III) HYDRIDE (13598-56-6) Water contact with powder leads to spontaneous ignition. Reacts violently with halocarbon or halogenated hydrocarbons. Strong oxidizers may cause fire and explosions.

URANIUM NITRATE (10102-06-4) Reacts with combustible materials (with fire and explosions). Water contact produces hydrochloric acid; solution attacks most metals. Reacts with esters. Attacks many plastics, rubber, and coatings.

URAN, RUSTICA (15978-77-5) Incompatible with oxidizers, combustibles, organics, acids, amides, organic anhydrides, isocyanates, vinyl acetate, alkylene oxides, epichlorohydrin, aldehydes, sulfur. Reacts violently with copper and copper alloys. Attacks lead, zinc.

URANYL NITRATE (10102-06-4) Reacts with combustible materials (with fire and explosions). Water contact produces hydrochloric acid; solution attacks most metals. Reacts with esters. Attacks many plastics, rubber, and coatings.

UREA (57-13-6) Heat causes formation of anhydrous ammonia fumes. Contact with oxidizers, nitrates may cause fire and explosions. Contact with chlorinating agents including hypochlorite bleaches, may produce explosive nitrogen trichloride. Reacts with nitrosyl perchlorate.

UREA-AMMONIUM NITRATE SOLUTION (W/AQUA AMMONIA) (15978-77-5) Incompatible with oxidizers, combustibles, organics, acids, amides, organic anhydrides, isocyanates, vinyl acetate, alkylene oxides, epichlorohydrin, aldehydes, sulfur. Reacts violently with copper and copper alloys. Attacks lead, zinc.

UREA DIOXIDE (124-43-6) Combustible solid. At 122°F/50°C reacts with dust, organics.

UREA HYDROGEN PEROXIDE (124-43-6) Combustible solid. At 122°F/50°C reacts with dust, organics.

UREA, HYDROGEN PEROXIDE SALT (124-43-6) Combustible solid. At 122°F/50°C reacts with dust, organics.

UREA, 1-(1-NAPHTHYL)-2-THIO- (86-88-4) Strong oxidizers may cause fires and explosion. Also reacts with silver nitrate.

UREA PEROXIDE (124-43-6) Combustible solid. At 122°F/50°C reacts with dust, organics.

UREAPHIL (57-13-6) Heat causes formation of anhydrous ammonia fumes. Contact with oxidizers, nitrates may cause fire and explosions. Contact with chlorinating agents, including hypochlorite bleaches may produce explosive nitrogen trichloride. Reacts with nitrosyl perchlorate.

UREA, THIO- (62-56-6) Aqueous solution is a base; reacts strongly with acids. Incompatible with nitric acid, hydrogen peroxide, acrolein, metals.

UREOPHIL (57-13-6) Heat causes formation of anhydrous ammonia fumes. Contact with oxidizers, nitrates may cause fire and explosions. Contact with chlorinating agents including hypochlorite bleaches, may produce explosive nitrogen trichloride. Reacts with nitrosyl perchlorate.

UREVERT (57-13-6) Heat causes formation of anhydrous ammonia fumes. Contact with oxidizers, nitrates may cause fire and explosions. Contact with chlorinating agents, including hypochlorite bleaches, may produce explosive nitrogen trichloride. Reacts with nitrosyl perchlorate.

UROTROPIN (100-97-0) Incompatible with acids, organic anhydrides, isocyanates, vinyl acetate, acrylates, substituted allyls, alkylene oxides, epichlorohydrin, ketones, aldehydes, alcohols, glycols, phenols, cresols, caprolactam solution, sodium peroxide, strong oxidizers. Attacks aluminum, copper, lead, tin, zinc, and alloys, and some plastics, rubber, and coatings.

UROX D (330-54-1) Hydrolyzes in fairly strong acids.

U.S. RUBBER D-014 (2312-35-8) Forms explosive mixture with air (flash point 82°F/28°C). Strong oxidizers may cause fire and explosions.

URSOL D (106-50-3) A strong reducing agent; reacts violently with oxidizers, combustibles, organic substances. Incompatible with strong acids, organic anhydrides, isocyanates, aldehydes. Heat and light contribute to instability.

USAF A-4600 (109-77-3) Incompatible with sulfuric acid. Caustics or heat above 160°F/71°C may cause polymerization or spontaneous combustion.

USAF B-30 (137-26-8) Combustible solid (flash point 192°F/89°C). Strong oxidizers may cause fire and explosions; contact with strong acid or oxidizable materials produces toxic gases.

USAF CB-22 (91-59-8) Forms explosive mixture with air (flash point 300°F/149°C). Strong oxidizers, strong acids may cause fire and explosions. Oxidizes in the presence of air and light.

USAF CY-2 (156-62-7) Contact with any form of moisture causes decomposition, liberating acetylene and ammonia. Contact with all solvents tested also causes decomposition.

USAF DO-45 (105-57-7) Forms explosive mixture with air (flash point −5°F/−21°C). Reacts violently with oxidizers. Forms unstable and explosive peroxides with heat and light. May accumulate static electrical charges, and may cause ignition of its vapors.

USAF DO-461 (140-31-8) Solution is a strong base. Forms explosive mixture with air (flash point 199°F/93°C). Incompatible with acids, organic anhydrides, isocyanates, vinyl acetate, acrylates, substituted allyls, alkylene oxides, epichlorohydrin, ketones, aldehydes, alcohols, glycols, phenols, cresols, caprolactam solution, strong oxidizers. Contact with copper alloys, zinc, or galvanized steel may cause violent reaction.

USAF EK-218 (91-22-5) Contact with strong oxidizers may cause fire and explosions. Attacks some plastics, rubber, and coatings.

USAF EK-356 (123-31-9) Incompatible with strong oxidizers, caustics. May be oxidized to quinone at room temperatures in the presence of moisture. May explode on contact with oxygen.

USAF EK-394 (106-50-3) A strong reducing agent; reacts violently with oxidizers, combustibles, organic substances. Incompatible with strong acids, organic anhydrides, isocyanates, aldehydes. Heat and light contribute to instability.

USAF EK-488 (75-05-8) Forms explosive mixture with air (flash point 42°F/6°C). Incompatible with water (especially if acid or alkaline), acids, caustics, nitrating agents, indium, dinitrogen tetroxide, n-fluoro compounds, sulfur trioxide, iron(III) salts of perchlorate, indium, nitrogen-fluorine compounds. Reacts violently with oxidizers. May accumulate static electrical charges, and may cause ignition of its vapors.

USAF EK-496 (98-86-2) Incompatible with strong acids, aliphatic amines, oxidizers. Forms explosive mixture with air (flash point 170°F/77°C).

USAF EK-497 (62-56-6) Aqueous solution is a base; reacts strongly with acids. Incompatible with nitric acid, hydrogen peroxide, acrolein, metals.

USAF EK-1850 (110-02-1) Forms explosive mixture with air (flash point 30°F/−1°C). Reacts violently with strong oxidizers, nitric acid. May accumulate static electrical charges, and may cause ignition of its vapors.

USAF EK-2089 (137-26-8) Combustible solid (flash point 192°F/89°C). Strong oxidizers may cause fire and explosions; contact with strong acid or oxidizable materials produces toxic gases.

USAF EK-P-583 (110-17-8) Reacts with strong oxidizers. Incompatible with sulfuric acid, caustics, ammonia, amines, isocyanates, alkylene oxides, epichlorohydrin.

USAF EK-P-5976 (86-88-4) Strong oxidizers may cause fires and explosion. Also reacts with silver nitrate.

USAF EK-T-434 (540-72-7) Incompatible with acids, bases, ammonia, amines, amides, alcohols, glycols, caprolactam.

USAF P-5 (137-26-8) Combustible solid (flash point 192°F/89°C). Strong oxidizers may cause fire and explosions; contact with strong acid or oxidizable materials produces toxic gases.

USAF P-7 (330-54-1) Hydrolyzes in fairly strong acids.

USAF P-220 (106-51-4) Forms explosive mixture with air (flash point 104°F/40°C). Incompatible with strong bases, reducing agents, strong oxidizers. Attacks some plastics, rubber, and coatings.

USAF RH-7 (109-78-4) Forms explosive mixture with air (flash point 140°F/60°C). Avoid basic contamination to prevent polymerization. A powerful reducing agent. Hot water causes formation of cyanide gas. Reacts violently with strong oxidizers or sodium hydroxide. Reacts with acids, acid salts, chlorates, nitrates. Attacks mild steel, copper, and copper alloys.

USAF RH-8 (75-86-5) Forms explosive mixture with air (flash point 165°F/74°C). Reacts violently with strong oxidizers. Heat may cause decomposition. In compatible with non-oxidizing mineral acids, sulfuric acid, nitric acid, organic acids, caustics, aliphatic amines, alkanolamines, aromatic amines, organic anhydrides, allylene oxides, epichlorohydrin, caprolactam solution, ammonia, isocyanates, phenols, cresol.

USAF ST-40 (126-98-7) Forms explosive mixture with air (flash point 34°F/1°C). Incompatible with aliphatic amines, alkanolamines. Reacts violently with oxidizers. Incompatible with strong acids, strong bases or light exposure; may cause polymerization.

USAF XR-22 (61-82-5) Substance acts as a weak base to form salts on contact with acids. Corrosive to iron, aluminum, copper, and copper alloys.

USAF XR-42 (330-54-1) Hydrolyzes in fairly strong acids.

- V -

VAC (108-05-4) Forms explosive mixture with air (flash point 18°F/−8°C). Polymerizes readily if not inhibited; heat can initiate reaction. Reacts violently with oxidizers. Incompatible with nonoxidizing mineral acids, strong acids, ammonia, aliphatic amines, alkanolamines. Also reacts with 2-aminoethanol, chlorosulfonic acid, ethylene diamine, ethyeneimine, ozone, oleum, peroxides. May accumulate static electrical charges, and may cause ignition of its vapors.

VAL-DROP (7775-09-9) A powerful oxidizer; reacts violently with reducing agents and combustible matter. Explosions may be caused by contact with ammonia salts, carbon, oils, metal sulfides, nitrobenzene, powdered metals, sugar. Contact with strong acids produces carbon dioxide. Forms shock-sensitive mixtures with some organic materials. Solution (50%) decomposes at 300°F/149°C, liberating oxygen.

VALERAL (110-62-3) Forms explosive mixture with air (flash point 54°F/12°C). Incompatible with strong acids, oxidizers, caustics, amines.

VALERALDEHYDE (110-62-3) Forms explosive mixture with air (flash point 54°F/12°C). Incompatible with strong acids, oxidizers, caustics, amines.

N-VALERALDEHYDE (110-62-3) Forms explosive mixture with air (flash point 54°F/12°C). Incompatible with strong acids, oxidizers, caustics, amines.

VALERIANIC ALDEHYDE (110-62-3) Forms explosive mixture with air (flash point 54°F/12°C). Incompatible with strong acids, oxidizers, caustics, amines.

VALERIC ACID (109-52-4) Forms explosive mixture with air (flash point 205°F/96°C). Incompatible with sulfuric acid, caustics, ammonia, amines, isocyanates, alkylene oxides, epichlorohydrin, strong oxidizers.

VALERIC ACID ALDEHYDE (110-62-3) Forms explosive mixture with air (flash point 54°F/12°C). Incompatible with strong acids, oxidizers, caustics, amines.

VALERIC ALDEHYDE (110-62-3) Forms explosive mixture with air (flash point 54°F/12°C). Incompatible with strong acids, oxidizers, caustics, amines.

N-VALERIC ALDEHYDE (110-62-3) Forms explosive mixture with air (flash point 54°F/12°C). Incompatible with strong acids, oxidizers, caustics, amines.

VALERONE (108-83-8) Forms explosive mixture with air (flash point 140°F/60°C). Incompatible with strong acids, aliphatic amines, strong oxidizers. Attacks plastics, coatings, and rubber.

VALINE ALDEHYDE (78-84-2) Forms explosive mixture with air (flash point −40°F/−40°C). Incompatible with strong acids, caustics, ammonia, amines.

VAM (108-05-4) Forms explosive mixture with air (flash point 18°F/−8°C). Polymerizes readily if not inhibited; heat can initiate reaction. Reacts violently with oxidizers. Incompatible with nonoxidizing mineral acids, strong acids, ammonia, aliphatic amines, alkanolamines, oxidizers. Also reacts with 2-aminoethanol, chlorosulfonic acid, ethylene diamine, ethyeneimine, ethyleneimine, ozone, oleum, peroxides. May accumulate static electrical charges, and may cause ignition of its vapors.

VANADIC ANHYDRIDE (1314-62-1) Aqueous solution is acidic. Reacts with lithium at high temperature.

VANADIUM OXIDE (1314-62-1) Aqueous solution is acidic. Reacts with lithium at high temperature.

VANADIUM(5+) OXIDE or VANADIUM(V) OXIDE or VANADIUM(V) OXIDE FUME (1314-62-1) Aqueous solution is acidic. Reacts with lithium at high temperature.

VANADIUM OXYSULFATE (27774-13-6) Incompatible with strong oxidizers such as chlorine or fluorine.

VANADIUM OXYTRICHLORIDE (7727-18-6) Incompatible with water, producing hydrochloric acid solution. Air contact causes formation of corrosive hydrogen chloride fumes. Reacts violently with bases, sodium, potassium.

VANADIUM PENTOXIDE or VANADIUM PENTOXIDE FUME or VANADIUM PENTOXIDE DUST (1314-62-1) Aqueous solution is acidic. Reacts with lithium at high temperature.

VANADIUM SESQUIOXIDE (1314-43-7) May self-ignite in air. Incompatible with lithium, halogenated compounds.

VANADIUM TRICHLORIDE OXIDE (7727-18-6) Incompatible with water, producing hydrochloric acid solution. Air contact causes formation of corrosive hydrogen chloride fumes. Reacts violently with bases, sodium, potassium.

VANADIUM TRIOXIDE (1314-43-7) May self-ignite in air. Incompatible with lithium, halogenated compounds.

VANADYL CHLORIDE (7727-18-6) Incompatible with water, producing hydrochloric acid solution. Air contact causes formation corrosive hydrogen chloride fumes. Reacts violently with bases, sodium, potassium.

VANADYL SULFATE (27774-13-6) Incompatible with strong oxidizers such as chlorine or fluorine.

VANADYL SULFATE DIHYDRATE (27774-13-6) Incompatible with strong oxidizers such as chlorine or fluorine.

VANADYL TRICHLORIDE (7727-18-6) Incompatible with water, producing hydrochloric acid solution. Air contact causes formation of corrosive hydrogen chloride fumes. Reacts violently with bases, sodium, potassium.

VANCIDA TM-95 (137-26-8) Combustible solid (flash point 192°F/89°C). Strong oxidizers may cause fire and explosions; contact with strong acid or oxidizable materials produces toxic gases.

VANCIDE (133-06-2) Incompatible with tetraethyl pyrophosphate, parathion.

VANCIDE 89 (133-06-2) Incompatible with tetraethyl pyrophosphate, parathion.

VANCIDE TM (137-26-8) Combustible solid (flash point 192°F/89°C). Strong oxidizers may cause fire and explosions; contact with strong acid or oxidizable materials produces toxic gases.

VANGARD K (133-06-2) Incompatible with tetraethyl pyrophosphate, parathion.

VANILLAN BLACK LIQUOR (68514-06-7) Water contact generates heat. Incompatible with acids, organic anhydrides, isocyanates, alkylene oxides, epichlorohydrin, aldehydes, alcohols, glycols, phenols, cresols, caprolactam solution, oxidizers. Contact with acids produces carbon dioxide and alkaline mists. Incompatible with not compatible with aluminum, zinc, tin; reactions produce heat and hydrogen gas.

VANOXIDE (94-36-0) Confined storage of dry chemical may lead to decomposition and explosion. A strong oxidant; extremely reactive. Fires and explosion may result from heat or contamination, and from contact with strong acids, combustible materials, oxidizers, acids, bases, alcohols, reducing agents, metals, metal oxides, amines, accelerators, methyl methacrylate, organic matter, lithium aluminum carbide, dimethyl aniline, metallic naphthenates. May attack some plastics, rubber, and coatings. Protect containers from shock and friction.

VANSIDE P-75 (133-06-2) Incompatible with tetraethyl pyrophosphate, parathion.

VAPONA (62-73-7) Attacks some plastics, rubber, and coatings.

VAPONITE (62-73-7) Attacks some plastics, rubber, and coatings.

VAPOPHOS (56-38-2) Combustible liquid. Mixtures with endrin may be explosive. Strong oxidizers may cause fire and explosions. Attacks many plastics, rubber, and coatings.

VAPOROLE (110-46-3) Forms explosive mixture with air (flash point 0°F/−18°C). Decomposes in light, air, or water; produces oxides of nitrogen. A strong oxidizer; reacts with reducing agents, combustibles, organics, and other strong oxidizers. Corrodes metal with moisture.

VAPOTONE (107-49-3) Decomposes above 300°F/150°C, producing flammable ethylene gas. Strong oxidizers may cause fire and explosions. Attacks many plastics, rubber, and coatings.

VARIOFORM I (13473-90-0) A strong oxidizer. Reacts violently with combustibles, organics, reducing agents. Aqueous solution is acidic. Attacks metals in the presence of moisture.

VARIOFORM II (57-13-6) Heat causes formation of anhydrous ammonia fumes. Contact with oxidizers, nitrates may cause fire and explosions. Contact with chlorinating agents, including hypochlorite bleaches, may produce explosive nitrogen trichloride. Reacts with nitrosyl perchlorate.

VARNISH MAKERS AND PAINTERS NAPHTHA (8032-32-4) Forms explosive mixture with air (flash point −40°F to −86°F/−40°C to −66°C). Incompatible with strong acids, strong oxidizers. Attacks some plastics, rubber, and coatings. May accumulate static electrical charges, and may cause ignition of its vapors.

VARNOLINE (8052-41-3) Forms explosive mixture with air (flash point 102°F to 140°F/38.7°C to 60°C). Incompatible with nitric acid, strong oxidizers. Attacks many plastics, rubber, and coatings.

VARSOL (8032-32-4) Forms explosive mixture with air (flash point −40°F to −86°F/−40°C to −66°C). Incompatible with strong acids, strong oxidizers. Attacks some plastics, rubber, and coatings. May accumulate static electrical charges, and may cause ignition of its vapors.

VASELINE (8012-95-1) Oxidizers may cause fire and explosions. Incompatible with nitric acid. May accumulate static electrical charges, and may cause ignition of its vapors.

VATERITE (1317-65-3) Incompatible with acids, alum, ammonium salts, fluorine.

VATSOL OT (119-36-8) Forms explosive mixture with air (flash point 205°F/96°C). Incompatible with strong acids, nitrates, oxidizers.

VC (75-01-4) Forms explosive gas mixture with air. Atmospheric oxygen and various contaminants or strong oxidizers cause formation of peroxides, which can initiate a violent polymerization action. Also able to polymerize when heated and exposed, long term, to light. Reacts violently with strong oxidizers or oxides of nitrogen. Contact with copper or other acetylide-forming metals produces explosive compounds. Attacks iron and steel in the presence of moisture.

VCL (75-01-4) Forms explosive gas mixture with air. Atmospheric oxygen and various contaminants or strong oxidizers cause formation of peroxides, which can initiate a violent polymerization action. Also able to polymerize when heated and exposed, long term, to light. Reacts violently with strong oxidizers or oxides of nitrogen. Contact with copper or other acetylide-forming metals produces explosive compounds. Attacks iron and steel in the presence of moisture.

VCM (75-01-4) Forms explosive gas mixture with air. Atmospheric oxygen and various contaminants or strong oxidizers cause formation of peroxides, which can initiate a violent polymerization action. Also able to polymerize when heated and exposed, long term, to light. Reacts violently with strong oxidizers or oxides of nitrogen. Contact with copper or other acetylide-forming metals produces explosive compounds. Attacks iron and steel in the presence of moisture.

VCN (107-13-1) Forms explosive mixture with air (flash point 32°F/0°C o.c.). Forms explosive peroxides; heat, light, caustics, silver nitrate, and peroxides can cause polymerization. Incompatible with strong acids, strong oxidizers, amines, 2-aminoethanol, bromine, chlorosulfonic acid, ethylene diamine, nitric acid, oleum, potassium hydroxide, sodium hydroxide, sulfuric acid. Attacks copper and copper alloys; attacks aluminum in high concentrations. May accumulate static electrical charges, and may cause ignition of its vapors.

VDC (75-35-4) Forms explosive mixture with air (flash point 18°F/−28°C). Air or contaminants can cause formation of peroxides; may polymerize. Reacts violently with strong oxidizers, alkali metals. Incompatible with nitric acid.

VECTAL (1912-24-9) Incompatible with strong acids.

VECTAL SC (1912-24-9) Incompatible with strong acids.

VEGETABLE CARBON activated (64365-11-3); **purified** (7440-44-0) Incompatible with strong oxidizers, strong acids, oxides, unsaturated oils.

VEGETABLE MIST (68956-68-3) Incompatible with strong acids, oxidizers, nitrates.

VEGETABLE OIL MIST (68956-68-3) Incompatible with strong acids, oxidizers, nitrates.

VEGFRU (Indian) (121-75-5) Incompatible with strong oxidizers, magnesium, alkaline pesticides. Attacks metals, some plastics, rubber, and coatings.

VEGFRUFOSMITE (563-12-2) Incompatible with alkaline formulations. Mixtures with magnesium may be explosive.

VEGFRU MALATOX (121-75-5) Incompatible with strong oxidizers, magnesium, alkaline pesticides. Attacks metals, some plastics, rubber, and coatings.

VELMOL (119-36-8) Forms explosive mixture with air (flash point 205°F/96°C). Incompatible with strong acids, nitrates, oxidizers.

VENETIAN RED (1309-37-1) Contact with hydrogen peroxide, ethylene oxide, calcium hypochlorite will cause explosion. Reacts violently with powdered aluminum, hydrazine, hydrogen trisulfide.

VENTOX (107-13-1) Forms explosive mixture with air (flash point 32°F/0°C o.c.). Forms explosive peroxides; heat, light, caustics, silver nitrate and peroxides can cause polymerization. Incompatible with strong acids, strong oxidizers, amines, 2-aminoethanol, bromine, chlorosulfonic acid, ethylene diamine, nitric acid, oleum, potassium hydroxide, sodium hydroxide, sulfuric acid. Attacks copper and copper alloys; attacks aluminum in high concentrations. May accumulate static electrical charges, and may cause ignition of its vapors.

VEON (93-76-5) Sealed metal containers may burst in heat above 316°F/158°C. Incompatible with sulfuric acid, bases, ammonia, aliphatic amines, alkanolamines, isocyanates, alkylene oxides, epichlorohydrin.

VEON 245 (93-76-5) Sealed metal containers may burst in heat above 316°F/158°C. Incompatible with sulfuric acid, bases, ammonia, aliphatic amines, alkanolamines, isocyanates, alkylene oxides, epichlorohydrin.

VERAZINC (7733-02-0) Incompatible with strong bases.

VERDICAN (62-73-7) Attacks some plastics, rubber, and coatings.

VERDIPOR (62-73-7) Attacks some plastics, rubber, and coatings.

VERGEMASTER (94-75-7) Decomposes in sunlight. Incompatible with strong oxidizers; may cause fire and explosions.

VERMICIDE BAYER 2349 (52-68-6) Contact with strong oxidizers may cause fire and explosions.

VERMOESTRICID (56-23-5) Becomes corrosive when in contact with water. Corrosive to metals. Reacts violently with many compounds. Decomposes on contact with chemically active metals such as sodium, potassium and magnesium. Incompatible with allyl alcohol, fluorine gas, alkali metals, aluminum. Attacks some coatings, plastics, and rubber.

VERSENE (60-00-4) Incompatible with sulfuric acid, bases, ammonia, aliphatic amines, alkanolamines, isocyanates, alkylene oxides, epichlorohydrin.

VERSENE ACID (60-00-4) Incompatible with sulfuric acid, bases, ammonia, aliphatic amines, alkanolamines, isocyanates, alkylene oxides, epichlorohydrin.

VERSNELLER NL 63/10 (121-69-7) Forms explosive mixture with air (flash point 145°F/63°C). Contact with strong oxidizers may cause fire and explosions; contact with strong acids can cause violent spattering. Contact with benzoyl peroxide, diisopropyl perdicarbonate, and other material may cause explosions. Attacks some plastics, rubber, and coatings.

VERTAC 90% (8001-35-2) Reacts with oxidizers, with a risk of fire or explosions. Attacks metals in the presence of moisture.

VERTAC TOXAPHENE 90 (8001-35-2) Reacts with oxidizers, with a risk of fire or explosions. Attacks metals in the presence of moisture.

VERTON (94-75-7) Decomposes in sunlight. Incompatible with strong oxidizers may cause fire and explosions.

VERTON D (94-75-7) Decomposes in sunlight. Incompatible with strong oxidizers; may cause fire and explosions.

VERTON 2D (94-75-7) Decomposes in sunlight. Incompatible with strong oxidizers may cause fire and explosions.

VERTON 2T (93-76-5) Sealed metal containers may burst in heat above 316°F/158°C. Incompatible with sulfuric acid, bases, ammonia, aliphatic amines, alkanolamines, isocyanates, alkylene oxides, epichlorohydrin.

VERTRON 2D (94-75-7) Decomposes in sunlight. Incompatible with strong oxidizers; may cause fire and explosions.

VESTINOL OA (103-23-1) Incompatible with strong acids, nitrates, oxidizers.

VESTROL (79-01-6) Contact with caustics produces a toxic and flammable gas. Reacts violently with chemically active metals. Contact with aluminum may produce a violent, self-accelerating polymerization reaction. Incompatible with acids, organic anhydrides, isocyanates, alkylene oxides, aldehydes, alcohols, glycols, phenols, cresols, caprolactam solution, epichlorohydrin, nitrogen tetroxide, metal powders, oxygen. May accumulate static electrical charges, and may cause ignition of its vapors.

VETIOL (121-75-5) Incompatible with strong oxidizers, magnesium, alkaline pesticides. Attacks metals, some plastics, rubber, and coatings.

VI-CAD (10108-64-2) Incompatible with strong oxidizers, elemental sulfur, selenium, tellurium.

VICKNITE (7757-79-1) A powerful oxidizer. Reacts violently with reducing agents and combustibles. Strong acid causes formation of toxic vapors. Forms explosive mixtures with many substances including sodium acetate, metal powders, sodium hypophosphite, trichloroethylene, and zinc. Dangerously reactive, separate from all materials.

VIDDEN D (542-75-6) Forms explosive mixture with air (flash point 95°F/35°C). Reacts violently with strong oxidizers. May accumulate static electrical charges, and may cause ignition of its vapors. Incompatible with strong acids, oxidizers, aluminum or magnesium compounds, aliphatic amines, alkanolamines, alkaline materials, or corrosives.

VIDDEN D (8003-19-8) Incompatible with strong acids, oxidizers, aluminum or magnesium compounds, aliphatic amines, alkanolamines, alkaline materials, or corrosives.

VIDLON (105-60-2) Contact with strong oxidizers may cause fire and explosions.

VIDON 638 (94-75-7) Decomposes in sunlight. Incompatible with strong oxidizers; may cause fire and explosions.

VIENNA GREEN (12002-03-8) Contact with strong oxidizers may cause fire and explosions.

VIKANE or VIKANE FUMIGANT (2699-79-8) Reacts with moisture. A compressed gas. High temperatures may cause cylinders to burst. See 29 CFR 1910.101 for specific storage regulations.

844

VILLIAUMITE (7681-49-4) Aqueous solution is highly corrosive. Reacts with acids.

VILLIAUMITE (15096-52-3) Contact with strong acids may produce toxic and corrosive hydrogen fluoride gas.

VINAMAR (109-92-2) Incompatible with acids (nonoxidizing, sulfuric, nitric), ammonia, aliphatic amines, alkanolamines.

VINEGAR ACID (64-19-7) Vapor forms explosive mixture with air (flash point 105°F/40°C). Reacts violently with oxidizers and bases. Incompatible with strong acids, aliphatic amines, alkanolamines, isocyanates, alkylene oxides, epichlorohydrin, acetaldehyde, 2-aminoethanol, ammonia, ammonium nitrate, chlorosulfonic acid, chromic acid, ethylene diamine, ethyleneimine, perchloric acid, permanganates, phosphorus isocyanate, phosphorus trichloride, potassium *tert*-butoxide, xylene. Attacks cast iron and other metals, producing flammable hydrogen gas.

VINEGAR NAPHTHA (141-78-6) Will hydrolyze on standing with formation of acetic acid and ethyl alcohol. This reaction is greatly accelerated by bases (alkalies). Forms explosive mixture with air (flash point 24°F/−4.4°C). Incompatible with strong acids, nitrates, oxidizers, chlorosulfonic acid, lithium aluminum hydride, oleum

VINICIZER (117-84-0) Water contact causes foaming. Incompatible with strong acids, nitrates.

VINILE (ACETATO di) (Italian) (108-05-4) Forms explosive mixture with air (flash point 18°F/−8°C). Polymerizes readily if not inhibited; heat can initiate reaction. Reacts violently with oxidizers. Incompatible with nonoxidizing mineral acids, strong acids, ammonia, aliphatic amines, alkanolamines. Also reacts with, 2-aminoethanol, chlorosulfonic acid, ethylene diamine, ethyeneimine, ethyleneimine, ozone, oleum, peroxides.

VINYLACETAAT (Dutch) (108-05-4) Forms explosive mixture with air (flash point 18°F/−8°C). Polymerizes readily if not inhibited; heat can initiate reaction. Reacts violently with oxidizers. Incompatible with nonoxidizing mineral acids, strong acids, ammonia, aliphatic amines, alkanolamines. Also reacts with, 2-aminoethanol, chlorosulfonic acid, ethylene diamine, ethyeneimine, ethyleneimine, ozone, oleum, peroxides. May accumulate static electrical charges, and may cause ignition of its vapors.

VINYLACETAT (German) (108-05-4) Forms explosive mixture with air (flash point 18°F/−8°C). Polymerizes readily if not inhibited; heat can initiate reaction. Reacts violently with oxidizers. Incompatible with nonoxidizing mineral acids, strong acids, ammonia, aliphatic amines, alkanolamines. Also reacts with, 2-aminoethanol, chlorosulfonic acid, ethylene diamine, ethyeneimine, ethyleneimine, ozone, oleum, peroxides. May accumulate static electrical charges, and may cause ignition of its vapors.

VINYL ACETATE or VINYL ACETATE H.Q. or VINYL ACETATE MONOMER (108-05-4) Forms explosive mixture with air (flash point 18°F/−8°C). Polymerizes readily if not inhibited; heat can initiate reaction. Reacts violently with oxidizers. Incompatible with nonoxidizing mineral acids, strong acids, ammonia, aliphatic amines, alkanolamines, oxidizers. Also reacts with, 2-aminoethanol, chlorosulfonic acid, ethylene diamine, ethyleneimine, ozone, oleum, peroxides. May accumulate static electrical charges, and may cause ignition of its vapors.

VINYL AMIDE (79-06-1) Unless inhibited, ultraviolet light and heat (above 184°F/85°C) can cause polymerization. Incompatible with nonoxidizing mineral acids, strong acids, ammonia, oleum, oxidizers, isocyanates.

VINYL A MONOMER (108-05-4) Forms explosive mixture with air (flash point 18°F/−8°C). Polymerizes readily if not inhibited; heat can initiate reaction. Reacts violently with oxidizers. Incompatible with nonoxidizing mineral acids, strong acids, ammonia, aliphatic amines, alkanolamines, oxidizers. Also reacts with, 2-aminoethanol, chlorosulfonic acid, ethylene diamine, ethyleneimine, ozone, oleum, peroxides. May accumulate static electrical charges, and may cause ignition of its vapors.

VINYLBENZEN (Czech) (100-42-5) If inhibitor (often *tert*-butyl catechol) is not present in adequate concentrations, polymerization may occur and explode containers. Temperatures above 150°F/66°C speed up polymerization. Incompatible with strong oxidizers, acids, rust, catalysts for vinyl polymerization, such as peroxides, strong acids, and aluminum chloride. Corrodes copper and copper alloys, and dissolves rubber. Attacks many plastics, rubber, and coatings. May accumulate static electrical charges, and may cause ignition of its vapors.

VINYL BENZENE (100-42-5) If inhibitor (often *tert*-butyl catechol) is not present in adequate concentrations, polymerization may occur and explode containers. Temperatures above 150°F/66°C speed up polymerization. Incompatible with strong oxidizers, acids, rust, catalysts for vinyl polymerization, such as peroxides, strong acids, and aluminum chloride. Corrodes copper and copper alloys, and dissolves rubber. Attacks many plastics, rubber, and coatings. May accumulate static electrical charges, and may cause ignition of its vapors.

VINYLBENZOL (100-42-5) If inhibitor (often *tert*-butyl catechol) is not present in adequate concentrations, polymerization may occur and explode containers. Temperatures above 150°F/66°C speed up polymerization. Incompatible with strong oxidizers, acids, rust, catalysts for vinyl polymerization, such as peroxides, strong acids, and aluminum chloride. Corrodes copper, copper alloys, and dissolves rubber. Attacks many plastics, rubber, and coatings. May accumulate static electrical charges, and may cause ignition of its vapors.

VINYL BROMIDE (593-60-2) Flammable gas. A strong reducing agent; reacts violently with oxidizers. May accumulate static electrical charges, and may cause ignition of its vapors.

VINYL CARBINOL (107-18-6) May form unstable and explosive peroxides. Able to polymerize. May accumulate static electrical charges, and may cause ignition of its vapors. A strong reducing agent; reacts violently with oxidizers. Forms explosive mixture with air (flash point 70°F/21°C). Incompatible with strong acids, amines, isocyanates, carbon tetrachloride, chlorosulfonic acid, diallyl phosphide, oleum, sodium hydroxide, tri-*n*-bromomelamine, metal halides, caustic soda, sodium, and magnesium, aluminum, and their alloys. Attacks some coatings, some plastics, and rubber.

VINYL CARBINOL,2-PROPENOL (107-18-6) May form unstable and explosive peroxides. Able to polymerize. May accumulate static electrical charges, and may cause ignition of its vapors. A strong reducing agent; reacts violently with oxidizers. Forms explosive mixture with air (flash point 70°F/21°C). Incompatible with strong acids, amines, isocyanates, carbon tetrachloride, chlorosulfonic acid, diallyl phosphide, oleum, sodium hydroxide, tri-*n*-bromomelamine, metal halides, caustic soda, sodium, and magnesium, aluminum, and their alloys. Attacks some coatings, some plastics, and rubber.

VINYL CHLORIDE or VINYL CHLORIDE MONOMER (75-01-4) Forms explosive gas mixture with air. Atmospheric oxygen and various contaminants or strong oxidizers cause formation peroxides, which can initiate a violent polymerization action. Also able to polymerize when heated and exposed, long term, to light. Reacts violently with strong oxidizers or oxides of nitrogen. Contact with copper or other acetylide-forming metals produces explosive compounds. Attacks iron and steel in the presence of moisture.

VINYL C MONOMER (75-01-4) Forms explosive gas mixture with air. Atmospheric oxygen and various contaminants or strong oxidizers form peroxides, which can initiate a violent polymerization action. Also able to polymerize when heated and exposed, long term, to light. Reacts violently with strong oxidizers or oxides of nitrogen. Contact with copper or other acetylide-forming metals produces explosive compounds. Attacks iron and steel in the presence of moisture.

VINYL CYANIDE (107-13-1) Forms explosive mixture with air (flash point 32°F/0°C o.c.). Forms explosive peroxides; heat, light, caustics, silver nitrate, and peroxides can cause polymerization. Incompatible with strong acids, strong oxidizers, amines, 2-aminoethanol, bromine, chlorosulfonic acid, ethylene diamine, nitric acid, oleum, potassium hydroxide, sodium hydroxide, sulfuric acid. Attacks copper and copper alloys; attacks aluminum in high concentrations. May accumulate static electrical charges, and may cause ignition of its vapors.

VINYL CYANIDE, PROPENENITRILE (107-13-1) Forms explosive mixture with air (flash point 32°F/0°C o.c.). Forms explosive peroxides; heat, light, caustics, silver nitrate and peroxides can cause polymerization. Incompatible with strong acids, strong oxidizers, amines, 2-aminoethanol, bromine, chlorosulfonic acid, ethylene diamine, nitric acid, oleum, potassium hydroxide, sodium hydroxide, sulfuric acid. Attacks copper and copper alloys; attacks aluminum in high concentrations. May accumulate static electrical charges, and may cause ignition of its vapors.

VINYL CYCLOHEXENE (100-40-3) Forms explosive mixture with air (flash point 68°F/20°C). Hydrolyzes in water. Incompatible with alcohols, amines, oxidizers.

VINYL-1-CYCLOHEXENE (100-40-3) Forms explosive mixture with air (flash point 68°F/20°C). Hydrolyzes in water. Incompatible with alcohols, amines, oxidizers.

4-VINYL CYCLOHEXENE (100-40-3) Forms explosive mixture with air (flash point 68°F/20°C). Hydrolyzes in water. Incompatible with alcohols, amines, oxidizers.

4-VINYL-1-CYCLOHEXENE (100-40-3) Forms explosive mixture with air (flash point 68°F/20°C). Hydrolyzes in water. Incompatible with alcohols, amines, oxidizers.

VINYLE (ACETATE de) (French) (108-05-4) Forms explosive mixture with air (flash point 18°F/−8°C). Polymerizes readily if not inhibited; heat can initiate reaction. Reacts violently with oxidizers. Incompatible with nonoxidizing mineral acids, strong acids, ammonia, aliphatic amines, alkanolamines. Also reacts with 2-aminoethanol, chlorosulfonic acid, ethylene diamine, ethyleneimine, ozone, oleum, peroxides. May accumulate static electrical charges, and may cause ignition of its vapors.

VINYLETHYLENE (106-99-0) Self-reactive. Forms explosive peroxides with air. Fires, explosions, or hazardous polymerization may result from contact with air, strong oxidizers, strong acids, ozone, nitrogen dioxide, copper and its alloys, phenol, chlorine dioxide, crotonaldehyde, or a free radical polymerization initiator such as hydroquinone. Add inhibitor (such as *tert*-butyl catechol) and monitor to ensure that effective levels are maintained at all times. May accumulate static electrical charges, and may cause ignition of its vapors.

VINYL ETHYL ETHER (109-92-2) Incompatible with acids (nonoxidizing, sulfuric, nitric), ammonia, aliphatic amines, alkanolamines.

VINYL FLUORIDE or VINYL FLUORIDE MONOMER (75-02-5) Flammable gas. May be able to polymerize. Reacts violently with oxidizers. May accumulate static electrical charges, and may cause ignition of its vapors.

VINYL FORMIC ACID (79-10-7) Forms explosive mixture with air (flash point 122°F/50°C). Forms explosive peroxides; light, heat and peroxides can cause polymerization. Incompatible with sulfuric acid, caustics, ammonia, amines, isocyanates, alkylene oxides, epichlorohydrin, oxidizers, toluenediamine, pyridine, methyl pyridine, *n*-methyl pyrrolidone, 2-methyl-6-ethyl aniline, aniline, ethylene diamine, ethyleneimine, 2-aminoethanol. Severely corrodes carbon steel and iron; attacks other metals.

VINYLFOS (62-73-7) Attacks some plastics, rubber, and coatings.

VINYLIDENE CHLORIDE or VINYLIDENE CHLORIDE, INHIBITED (75-35-4) Forms explosive mixture with air (flash point 18°F/−28°C). Air or contaminants can cause formation of peroxides; may polymerize. Reacts violently with strong oxidizers, alkali metals. Incompatible with nitric acid.

VINYLIDENE CHLORIDE(II) or VINYLIDENE CHLORIDE(2+) (75-35-4) Forms explosive mixture with air (flash point 18°F/−28°C). Air or contaminants can cause formation of peroxides; may polymerize. Reacts violently with strong oxidizers, alkali metals. Incompatible with nitric acid.

VINYLIDINE CHLORIDE MONOMER (75-35-4) Forms explosive mixture with air (flash point 18°F/−28°C). Air or contaminants can cause formation of peroxides; may polymerize. Reacts violently with strong oxidizers, alkali metals. Incompatible with nitric acid.

VINYLIDENE DICHLORIDE (75-35-4) Forms explosive mixture with air (flash point 18°F/−28°C). Air or contaminants can cause formation of peroxides; may polymerize. Reacts violently with strong oxidizers, alkali metals. Incompatible with nitric acid.

VINYL METHYL ETHER or VINYL METHYL ETHER, INHIBITED (107-25-5) Forms explosive mixture with air (flash point −69°F/−56°C). Water contact slowly causes formation of acetaldehyde and methyl alcohol. Forms unstable peroxides; able to polymerize. Acids, oxidizers may cause fire and explosions. May accumulate static electrical charges, and may cause ignition of its vapors.

VINYL METHYL KETONE (78-94-4) Forms explosive mixture with air (flash point 20°F/−7°C). May accumulate static electrical charges, and may cause ignition of its vapors. Heat can cause polymerization. Reacts violently with strong oxidizers.

VINYL NEODECANOATE (N/A) Forms explosive mixture with air (flash point 175°F/79°C). Polymerization inhibitor (usually monomethyl ether of hydroquinone) must be present in adequate concentrations to avoid explosive polymerization. The monomer is usually supplied in bulk or resin-lined drums, and may be stored in tin- or stainless steel-lined drums. Storage in plastic or other vessels is not recommended. Incompatible with acids, 2-aminoethanol, ammonia, aliphatic amines, alkanolamines, ethylene diamine, ethyleneimine, ozone.

VINYLOPHOS (62-73-7) Attacks some plastics, rubber, and coatings.

VINYLSILICON TRICHLORIDE (75-94-5) Incompatible with water, (violent reaction). Corrodes common metals and produces hydrogen.

VINYL TOLUENE or VINYL TOLUENE, INHIBITED or VINYL TOLUENE, MIXED ISOMERS (25013-15-4) Forms explosive mixture with air (flash point 17°F/53°C). Polymerization inhibitor (usually 10 to 50 ppm of *tert*-butyl catechol) must be present in adequate concentrations to avoid explosive polymerization. Oxidizers may cause fire and explosions. Also the following should be avoided: peroxides, aluminum chloride, acids (nonoxidizing mineral, sulfuric, nitric), caustics, ammonia, aliphatic amines, alkanolamines.

m-**VINYL TOLUENE or *meta*-VINYL TOLUENE** (100-80-1) Forms explosive mixture with air (flash point 17°F/53°C). Polymerization inhibitor (usually 10 to 50 ppm of *tert*-butyl catechol) must be present in adequate concentrations to avoid explosive polymerization. Oxidizers may cause fire and explosions. Also the following should be avoided: peroxides, aluminum chloride, acids (nonoxidizing mineral, sulfuric, nitric), caustics, ammonia, aliphatic amines, alkanolamines.

o-**VINYL TOLUENE or *ortho*-VINYL TOLUENE** (611-15-4) Forms explosive mixture with air (flash point 17°F/53°C). Polymerization inhibitor (usually 10 to 50 ppm of *tert*-butyl catechol) must be present in adequate concentrations to avoid explosive polymerization. Oxidizers may cause fire and explosions. Also the following should be avoided: peroxides, aluminum chloride, acids (non-oxidizing mineral, sulfuric, nitric), caustics, ammonia, aliphatic amines, alkanolamines.

p-**VINYL TOLUENE or *para*-VINYL TOLUENE** (622-97-9) Forms explosive mixture with air (flash point 17°F/53°C). Polymerization inhibitor (usually 10 to 50 ppm of *tert*-butyl catechol) must be present in adequate concentrations to avoid explosive polymerization. Oxidizers may cause fire and explosions. Also the following should be avoided: peroxides, aluminum chloride, acids (nonoxidizing mineral, sulfuric, nitric), caustics, ammonia, aliphatic amines, alkanolamines.

VINYL TRICHLORIDE (79-00-5) Incompatible with strong oxidizers, strong caustics, chemically active metals (especially powders), or sodium amide; may cause fire and explosions. Attacks many plastics, rubber, and coatings.

VINYL TRICHLOROSILANE (75-94-5) Incompatible with water (reacts violently). Corrodes common metals and produces hydrogen.

VINYLTRICHLOROSILANE, INHIBITED (75-94-5) Incompatible with water, (react violently). Corrodes common metals and produces hydrogen.

VISCOL 350P (9003-07-0) Incompatible with strong acids, strong oxidizers, chlorine, potassium permanganate.

VISKO (94-75-7) Decomposes in sunlight. Incompatible with strong oxidizers; may cause fire and explosions.

VISKO-RHAP (94-75-7) Decomposes in sunlight. Incompatible with strong oxidizers; may cause fire and explosions.

VISKO-RHAP LOW DRIFT HERBICIDES (94-75-7) Decomposes in sunlight and above 356°F/180°C. Incompatible with strong oxidizers.

VISKO RHAP LOW VOLATILE ESTER (93-76-5) Sealed metal containers may burst in heat above 316°F/158°C. Incompatible with sulfuric acid, bases, ammonia, aliphatic amines, alkanolamines, isocyanates, alkylene oxides, epichlorohydrin.

VISKO-RHAP LOW VOLATILE 4L (94-75-7) Decomposes in sunlight. Incompatible with strong oxidizers; may cause fire and explosions.

VITON (58-89-9) Corrodes metals in the presence of moisture.

VITRAN (79-01-6) Contact with caustics produces a toxic and flammable gas. Reacts violently with chemically active metals. Contact with aluminum may produce a violent, self-accelerating polymerization reaction. Incompatible with acids, organic anhydrides, isocyanates, alkylene oxides, aldehydes, alcohols, glycols, phenols, cresols, caprolactam solution, epichlorohydrin, nitrogen tetroxide, metal powders, oxygen. May accumulate static electrical charges, and may cause ignition of its vapors.

VITREX (56-38-2) Combustible liquid. Mixtures with endrin may be explosive. Strong oxidizers may cause fire and explosions. Attacks many plastics, rubber, and coatings.

VITRIOL BROWN ACID (7664-93-9) A strong oxidizer that can react violently with risk of fire and explosion, with many substances, including reducing agents, organic and combustible substances and bases. Incompatible with nonoxidizing mineral acids, organic acids, bases, acrylates, aldehydes, alcohols, alkylene oxides, ammonia, aliphatic amines, alkanolamines, aromatic amines, amides, chlorates, epichlorohydrin, fulminates, glycols, isocyanates, ketones, metals (powdered), organic anhydrides, perchlorates, picrates, substituted allyls, phenols, cresols, water, acetic anhydride, acetone cyanhydrin, acetonitrile, acrolein, acrylonitrile, allyl alcohol, allyl chloride, 2-aminoethanol, ammonium hydroxide, aniline, bromine pentafluoride, *n*-butyraldehyde, caprolactam solution, carbides, cesium acetylene carbide, chlorine trifluoride, chlorosulfonic acid, cuprous nitride, diisobutylene, ethylene cyanohydrin, ethylene diamine, ethylene glycol, ethyleneimine, hydrochloric acid, iodine heptafluoride, iron, isoprene, lithium silicide, mercuric nitride, mesityl oxide, nitric acid, *p*-nitrotoluene, perchloric acid, phosphorus, potassium *tert*-butoxide, potassium chlorate, potassium permanganate, propiolactone (beta-), propylene oxide, pyridine, rubidium acetylene, silver permanganate, sodium, sodium carbonate, sodium chlorate, sodium hydroxide, styrene monomer, vinyl acetate. Attacks most metals, and some plastics, rubber, and coatings.

VITROL RED (1309-37-1) Contact with hydrogen peroxide, ethylene oxide, calcium hypochlorite will cause explosion. Reacts violently with powdered aluminum, hydrazine, hydrogen trisulfide.

VIZENE (299-84-3) Temperatures above 300°F/150°C may cause explosive decomposition. Contact with strong oxidizers may cause fire and explosions. Attacks some plastics, rubber, and coatings.

VM&P NAPHTHA (8032-32-4) Forms explosive mixture with air (flash point −40°F to −86°F/−40°C to −66°C). Incompatible with strong acids, strong oxidizers. Attacks some plastics, rubber, and coatings. May accumulate static electrical charges, and may cause ignition of its vapors.

VOFATOX (298-00-0) Mixtures with magnesium may be explosive.

VOGEL'S IRON RED (1309-37-1) Contact with hydrogen peroxide, ethylene oxide, calcium hypochlorite will cause explosion. Reacts violently with powdered aluminum, hydrazine, hydrogen trisulfide.

VOLFARTOL (52-68-6) Contact with strong oxidizers may cause fire and explosions.

VONDALDHYDE (123-33-1) Contact with strong oxidizers may cause fire and explosions.

VONDCAPTAN (133-06-2) Incompatible with tetraethyl pyrophosphate, parathion.

VONDRAX (123-33-1) Contact with strong oxidizers may cause fire and explosions.

VONDURON (330-54-1) Hydrolyzes in fairly strong acids.

VORLEX (556-61-6) Forms explosive mixture with air (flash point 90°F/32°C). Incompatible with strong acids, caustics, ammonia, amines, amides, alcohols, glycols, caprolactam solution, strong oxidizers.

VOROX (61-82-5) Substance acts as a weak base to produce salts on contact with acids. Corrosive to iron, aluminum, copper, and copper alloys.

VORTEX (556-61-6) Forms explosive mixture with air (flash point 90°F/32°C). Incompatible with strong acids, caustics, ammonia, amines, amides, alcohols, glycols, caprolactam solution, strong oxidizers.

VOTEXIT (52-68-6) Contact with strong oxidizers may cause fire and explosions.

VUAGT-1-4 (137-26-8) Combustible solid (flash point 192°F/89°C). Strong oxidizers may cause fire and explosions; contact with strong acid or oxidizable materials produces toxic gases.

VULCAFOR TMTD (137-26-8) Combustible solid (flash point 192°F/89°C). Strong oxidizers may cause fire and explosions; contact with strong acid or oxidizable materials produces toxic gases.

VULKACIT HX (5459-93-8) Forms explosive mixture with air (flash point 86°F/30°C). Incompatible with acids, organic anhydrides, isocyanates, vinyl acetate, acrylates, substituted allyls, alkylene oxides, epichlorohydrin, ketones, aldehydes, alcohols, glycols, mercury, phenols, cresols, caprolactam solution, strong oxidizers. Attacks aluminum, copper, lead, tin, zinc, and alloys.

VULKACIT MTIC (137-26-8) Combustible solid (flash point 192°F/89°C). Strong oxidizers may cause fire and explosions; contact with strong acid or oxidizable materials produces toxic gases.

VULKACIT THIURAD (137-26-8) Combustible solid (flash point 192°F/89°C). Strong oxidizers may cause fire and explosions; contact with strong acid or oxidizable materials produces toxic gases.

VULKACIT THIURAM/C (137-26-8) Combustible solid (flash point 192°F/89°C). Strong oxidizers may cause fire and explosions; contact with strong acid or oxidizable materials produces toxic gases.

VULKANOX 4020 (106-50-3) A strong reducing agent; reacts violently with oxidizers, combustibles, organic substances. Incompatible with strong acids, organic anhydrides, isocyanates, aldehydes. Heat and light contribute to its instability.

VULKAUT THIRAM (137-26-8) Combustible solid (flash point 192°F/89°C). Strong oxidizers may cause fire and explosions; contact with strong acid or oxidizable materials produces toxic gases.

VV 10 VINYL MONOMER (N/A) Forms explosive mixture with air (flash point 175°F/79°C). Polymerization inhibitor (usually monomethyl ether of hydroquinone) must be present in adequate concentrations to avoid explosive polymerization. The monomer is usually supplied in bulk or resin-lined drums, and may be stored in tin- or stainless steel-lined drums. Storage in plastic or other vessels is not recommended. Incompatible with acids, 2-aminoethanol, ammonia, aliphatic amines, alkanolamines, ethylene diamine, ethyleneimine, ozone.

VYAC (108-05-4) Forms explosive mixture with air (flash point 18°F/−8°C). Polymerizes readily if not inhibited; heat can initiate reaction. Reacts violently with oxidizers. Incompatible with nonoxidizing mineral acids, strong acids, ammonia, aliphatic amines, alkanolamines. Also reacts with 2-aminoethanol, chlorosulfonic acid, ethylene diamine, ethyeneimine, ethyleneimine, ozone, oleum, peroxides. May accumulate static electrical charges, and may cause ignition of its vapors.

- W -

W 101 (9003-07-0) Incompatible with strong acids, strong oxidizers, chlorine, potassium permanganate.

WAPNIOWY TLENEK (Polish) (1305-78-8) Reacts violently with water. Incompatible with ethanol, hydrogen fluoride, acids, halogens, metal halides, light metals, some oxides, boron trifluoride, chlorine trifluoride, liquid hydrofluoric acid, phosphorus pentoxide, boric oxide and calcium chloride mixtures, fluorine, chlorine trifluoride, carbon dioxide.

WARFARIN (81-81-2) Strong oxidizers may cause fire and explosions.

WARF COMPOUND (81-81-2) Strong oxidizers may cause fire and explosions.

WARKEELATE ACID (60-00-4) Incompatible with sulfuric acid, bases, ammonia, aliphatic amines, alkanolamines, isocyanates, alkylene oxides, epichlorohydrin

WASH OIL (65996-93-2) Incompatible with oxidizers, strong acids, caustics, aliphatic amines, isocyanates.

WASSERSTOFFPEROXIDE (German) (7722-84-1) A powerful oxidizer; attacks many substances. Contact with most organic, readily oxidizable materials, reducing agents, and combustibles causes fire and explosions. Contact with iron, copper, brass, bronze, chromium, zinc, lead, manganese, silver, and other catalytic metals (and their salts), especially in a basic (pH 7 or above) environment, causes rapid decomposition with evolution of oxygen gas. Attacks, and may ignite, some plastics, rubber, and coatings. Decomposes slowly at ordinary temperatures and builds up pressure in a closed container. The rate of decomposition doubles for each 50°F/10°C rise (1.5 times 10°C rise) in temperature and becomes self-sustaining at 285°F/141°C.

WATER GLASS (1344-09-8) Solution is a strong base; reacts with acids, organic anhydrides, alkylene oxides, epichlorohydrin, aldehydes, alcohols, glycols, phenols, cresols, caprolactam solution. Attacks chemically active metals.

WATERSTOFPEROXYDE (Dutch) (7722-84-1) A powerful oxidizer; attacks many substances. Contact with most organic, readily oxidizable materials, reducing agents and combustibles cause fire and explosions. Contact with iron, copper, brass, bronze, chromium, zinc, lead, manganese, silver and other catalytic metals (and their salts), especially in a basic (pH 7 or above) environment, cause rapid decomposition with evolution of oxygen gas. Attacks, and may ignite, some plastics, rubber, and coatings. Decomposes slowly at ordinary temperatures and builds up pressure in a closed container. The rate of decomposition doubles for each 50°F/10°C rise (1.5 times 10°C rise) in temperature and becomes self-sustaining at 285°F/141°C.

WAX: CARNAUBA (N/A) Incompatible with sulfuric acid, nitric acid, oxidizers, nitrates.

WAXOL (119-36-8) Forms explosive mixture with air (flash point 205°F/96°C). Incompatible with strong acids, nitrates, oxidizers.

WAX: PARAFFIN (8002-74-2) Reacts with oxidizers, with a risk of fire or explosions.

WEC-50 (52-68-6) Contact with strong oxidizers may cause fire and explosions.

WECOLINE OO (112-80-1) Incompatible with strong oxidizers, perchloric acid, and aluminum powder; may cause explosions. Corrodes aluminum.

WECOLINE 1295 (143-07-7) Incompatible with sulfuric acid, bases, ammonia, amines, alkylene oxide, epichlorohydrin.

WEED-AG-BAR (94-75-7) Decomposes in sunlight and above 356°F/180°C. Incompatible with strong oxidizers.

WEEDAR (93-76-5) Sealed metal containers may burst in heat above 316°F/158°C. Incompatible with sulfuric acid, bases, ammonia, aliphatic amines, alkanolamines, isocyanates, alkylene oxides, epichlorohydrin.

WEEDAR or WEEDAR-64 (94-75-7) Decomposes in sunlight. Incompatible with strong oxidizers; may cause fire and explosions.

WEEDAR ADS (61-82-5) Substance acts as a weak base to produce salts on contact with acids. Corrosive to iron, aluminum, copper, and copper alloys.

WEEDAZIN (61-82-5) Substance acts as a weak base to form salts on contact with acids. Corrosive to iron, aluminum, copper, and copper alloys.

WEEDAZOL (61-82-5) Substance acts as a weak base to produce salts on contact with acids. Corrosive to iron, aluminum, copper, and copper alloys.

WEEDBEADS (131-52-2) Contact with strong oxidizers may cause fire and explosions.

WEED-B-GON (94-75-7) Decomposes in sunlight. Incompatible with strong oxidizers; may cause fire and explosions.

WEED DRENCH (107-18-6) May form unstable and explosive peroxides. Able to polymerize. May accumulate static electrical charges, and may cause ignition of its vapors. A strong reducing agent; reacts violently with oxidizers. Forms explosive mixture with air (flash point 70°F/21°C). Incompatible with strong acids, amines, isocyanates, carbon tetrachloride, chlorosulfonic acid, diallyl phosphide, oleum, sodium hydroxide, tri-*n*-bromomelamine, metal halides, caustic soda, sodium, and magnesium, aluminum, and their alloys. Attacks some coatings, some plastics, and rubber.

WEEDEX A (1912-24-9) Incompatible with strong acids.

WEEDEX GRANULAT (61-82-5) Substance acts as a weak base to produce salts on contact with acids. Corrosive to iron, aluminum, copper, and copper alloys.

WEEDEZ WONDER BAR (94-75-7) Decomposes in sunlight. Incompatible with strong oxidizers; may cause fire and explosions.

WEEDOCLOR (61-82-5) Substance acts as a weak base to produce salts on contact with acids. Corrosive to iron, aluminum, copper, and copper alloys.

WEEDONE (87-86-5) Hot water causes decomposition with formation of hydrochloric acid. Strong oxidizers may cause fire and explosions.

WEEDONE (93-76-5) Sealed metal containers may burst in heat above 316°F/158°C. Incompatible with sulfuric acid, bases, ammonia, aliphatic amines, alkanolamines, isocyanates, alkylene oxides, epichlorohydrin.

WEEDONE (94-75-7) Decomposes in sunlight. Incompatible with strong oxidizers; may cause fire and explosions.

WEEDONE 128 (94-11-1) Forms explosive mixture with air (flash point 175°F/79°C). Incompatible with strong oxidizers, strong acids, nitrates. Attacks many plastics, rubber, and coatings.

WEEDONE LV4 (94-75-7) Decomposes in sunlight. Incompatible with strong oxidizers; may cause fire and explosions.

WEED-RHAP (94-75-7) Decomposes in sunlight. Incompatible with strong oxidizers; may cause fire and explosions.

WEED TOX (94-75-7) Decomposes in sunlight. Incompatible with strong oxidizers; may cause fire and explosions.

WEEDTRINE-D (85-00-7) Concentrated solution attacks aluminum.

WEEDTROL (94-75-7) Decomposes in sunlight. Incompatible with strong oxidizers; may cause fire and explosions.

WEEVILTOX (75-15-0) Highly reactive. Contact with many substances can cause fire and explosions. Forms explosive mixture with air (flash point −22°F/−30°C). Shock can cause explosive decomposition. Incompatible with alkali metals, aliphatic amines, alkanolamines, aluminum, azides, chlorine monoxide, combustible substances, ethylene diamine, ethyleneimine, lead azide, lithium azide, nitric oxide, nitrogen dioxide, potassium, potassium azide, reducing agents, rubidium azide, sodium azide, zinc.

WEGLA DWUSIARCZEK (Polish) (75-15-0) Highly reactive. Contact with many substances can cause fire and explosions. Forms explosive mixture with air (flash point −22°F/−30°C). Shock can cause explosive decomposition. Incompatible with alkali metals, aliphatic amines, alkanolamines, aluminum, azides, chlorine monoxide, combustible substances, ethylene diamine, ethyleneimine, lead azide, lithium azide, nitric oxide, nitrogen dioxide, potassium, potassium azide, reducing agents, rubidium azide, sodium azide, zinc.

WELDING GAS (74-86-2) A strong reducing agent that reacts violently with oxidizers. Forms explosive mixture with air (flash point 0°F/18°C). Forms shock sensitive mixture with copper and copper salts, mercury and mercury salts, silver and silver salts. Reacts with brass, bromine, cesium hydride, chlorine, cobalt, cuprous acetylise, fluorine, iodine, mercuric nitrate, nitric acid, potassium, rubidium hydride, trifluoromethyl hypofluorite, sodium hydride.

WESTRONG (79-34-5) Exposure to heat, light and air causes formation of corrosive and toxic vapors. Reacts with strong caustics to produce explosive dichloroacetylene. Reacts violently with chemically active metals or sodium amide. In presence of steam, contact with hot iron, aluminum or zinc may produce toxic vapors. Attacks many plastics, rubber, and coatings.

WESTROSOL (79-01-6) Contact with caustics produces a toxic and flammable gas. Reacts violently with chemically active metals. Contact with aluminum may produce a violent, self-accelerating polymerization reaction. Incompatible with acids, organic anhydrides, isocyanates, alkylene oxides, aldehydes, alcohols, glycols, phenols, cresols, caprolactam solution, epichlorohydrin, nitrogen tetroxide, metal powders, oxygen. May accumulate static electrical charges, and may cause ignition of its vapors.

WETAID SR (119-36-8) Forms explosive mixture with air (flash point 205°F/96°C). Incompatible with strong acids, nitrates, oxidizers.

WEX 1242 (9003-07-0) Incompatible with strong acids, strong oxidizers, chlorine, potassium permanganate.

WFNA (7697-37-2) A strong oxidizer that can react violently with reducing agents, combustible materials. Incompatible with many substances, including acrylates, aliphatic amines, alcohols, aldehydes, alkanolamines, alkylene oxides, anion exchange resins, aromatic amines, amides, bases, cresols, cyanides, cyclic ketones, epichlorohydrin, glycols, isocyanates, ketones, oleum, organic anhydrides, phenols, substituted allyls, sulfuric acid, strong oxidizers, terpenes. Attacks most metals, plastics, rubber, and coatings.

WHITE ARSENIC (1327-53-3) Incompatible with acids, fluorine, fluorides, sodium chlorate.

WHITE CAMPHOR OIL (76-22-2) Forms explosive mixture with air (flash point 155°F/66°C). Violent, possibly explosive, reaction occurs with strong oxidizers. May accumulate static electrical charges, and may cause ignition of its vapors.

WHITE CAUSTIC (1310-73-2) Contact with water, acids, flammable liquids, and organic halogens, especially trichloroethylene, may cause fires and explosions. Contact with metals such as aluminum, tin, and zinc causes corrosion and the formation of flammable hydrogen gas. Contact with nitromethane and similar nitro compounds produces shock-sensitive salts. Water contact produces heat and corrosive fumes. Attacks many plastics, rubber, and coatings.

WHITE COPPERAS (7733-02-0) Incompatible with strong bases.

WHITE FUMING NITRIC ACID (7697-37-2) A strong oxidizer that can react violently with reducing agents, combustible materials. Incompatible with many substances including acrylates, aliphatic amines, alcohols, aldehydes, alkanolamines, alkylene oxides, anion exchange resins, aromatic amines, amides, bases, cresols, cyanides, cyclic ketones, epichlorohydrin, glycols, isocyanates, ketones, oleum, organic anhydrides, phenols, substituted allyls, sulfuric acid, strong oxidizers, terpenes. Attacks most metals, plastics, rubber, and coatings.

WHITE LEAD (7446-14-2) Reacts with aluminum, magnesium, potassium (violently).

WHITE MERCURY PRECIPITATE (10124-48-8) Reacts violently with halogens and metal salts of amines.

WHITE MINERAL OIL (8012-95-1) Oxidizers may cause fire and explosions. Incompatible with nitric acid. May accumulate static electrical charges, and may cause ignition of its vapors.

WHITE OIL (8012-95-1) Oxidizers may cause fire and explosions. Incompatible with nitric acid. May accumulate static electrical charges, and may cause ignition of its vapors.

WHITE OIL OF CAMPHOR (76-22-2) Forms explosive mixture with air (flash point 155°F/66°C). Violent, possibly explosive, reaction occurs with strong oxidizers. May accumulate static electrical charges, and may cause ignition of its vapors.

WHITE PHOSPHORIC ACID (7664-38-2) Incompatible with aliphatic amines, alkanolamines, alkylene oxides, aromatic amines, amides, ammonia, ammonium hydroxide, bases, calcium oxide, epichlorohydrin, isocyanates, nitromethane, (explosive) oleum, organic anhydrides, sulfuric acid, sodium tetrahydroborate, strong oxidizers, vinyl acetate, water. Contact with most metals produces hydrogen gas. Attacks many plastics, rubber, and coatings, and glass and ceramics.

WHITE PHOSPHORUS (7723-14-0) A strong reducing agent and fire risk; store under water away from heat. Ignites spontaneously on contact with air (flash point at or above 86°F/30°C). Contact with all oxidizers (including elemental sulfur) will cause fire and explosions. Contact with strong caustics produces poisonous and flammable phosphine gas. Liquid attacks some plastics, rubber, and coatings.

WHITE SPIRIT (8052-41-3) Forms explosive mixture with air (flash point 102°F to 140°F/38.7°C to 60°C). Incompatible with nitric acid, strong oxidizers. Attacks many plastics, rubber, and coatings.

WHITE SPIRIT (low [15-20%] aromatic) (63394-00-3) Forms explosive mixture with air (flash point 104°F/40°C). Oxidizers may cause fire and explosions. Incompatible with nitric acid. May accumulate static electrical charges, and may cause ignition of its vapors.

WHITE SPIRITS (8032-32-4) Forms explosive mixture with air (flash point varies by manufacturer: regular flash 50°F/10°C; high flash 85°F/29°C). Incompatible with nitric acid, strong oxidizers.

WHITE TAR (91-20-3) Forms explosive mixture with air (flash point 174°F/79°C). Incompatible with strong oxidizers, chromium oxide (violent). Attacks some plastics, rubber, and coatings. May accumulate static electrical charges, and may cause ignition of its vapors.

WHITE VITRIOL (7733-02-0) Incompatible with strong bases.

WICKENOL 158 (103-23-1) Incompatible with strong acids, nitrates, oxidizers.

WIJS' CHLORIDE (7790-99-0) Heat may cause explosions. Reacts violently with water, organic matter, aluminum foil, cadmium sulfide, lead sulfide, phosphorus, phosphorus trichloride, potassium, rubber, silver sulfide, sodium, zinc sulfide, and other metals.

WINTERGREEN OIL (119-36-0) Incompatible with strong acids, nitrates, strong oxidizers.

WINYLU CHLORED (Polish) (75-01-4) Forms explosive gas mixture with air. Atmospheric oxygen and various contaminants or strong oxidizers cause formation of peroxides, which can initiate a violent polymerization. Also able to polymerize when heated and exposed, long term, to light. Reacts violently with strong oxidizers or oxides of nitrogen. Contact with copper or other acetylide-forming metals produces explosive compounds. Attacks iron and steel in the presence of moisture.

WITCIZER-300 (84-74-2) Incompatible with strong acids, nitrates, strong oxidizers, strong alkalies.

WITCIZER-312 (117-81-7) Incompatible with strong acids, strong alkalies, nitrates, oxidizers.

WITOMOL-320 (103-23-1) Incompatible with strong acids, nitrates, oxidizers.

WITTOX-C (1338-02-9) Forms explosive mixture with air (flash point 100°F/38°C). Incompatible with strong oxidizers, strong acids.

WN 12 (556-61-6) Forms explosive mixture with air (flash point 90°F/32°C). Incompatible with strong acids, caustics, ammonia, amines, amides, alcohols, glycols, caprolactam solution, strong oxidizers.

WOCHEM-320 (112-80-1) Incompatible with strong oxidizers, perchloric acid, and aluminum powder; may cause explosions. Corrodes aluminum.

WOFATOX (298-00-0) Mixtures with magnesium may be explosive.

WONUK (1912-24-9) Incompatible with strong acids.

WOOD ALCOHOL (67-56-1) May accumulate static electrical charges, and may cause ignition of its vapors. Forms explosive mixture with air (flash point 52°F/11°C). Incompatible with strong acids, strong oxidizers, caustics, aliphatic amines, isocyanates, chromic anhydride, lead perchlorate, perchloric acid, phosphorus trioxide. May react with metallic aluminum at high temperature. Attacks some plastics, rubber, and coatings.

WOOD CHARCOAL activated (64365-11-3); **purified** (7440-44-0). Incompatible with strong oxidizers, strong acids, oxides, unsaturated oils.

WOOD ETHER (115-10-6) Flammable gas. Forms explosive mixture with air. Forms unstable peroxides in storage. Reacts violently with strong oxidizers.

WOOD NAPHTHA (67-56-1) May accumulate static electrical charges, and may cause ignition of its vapors. Forms explosive mixture with air (flash point 52°F/11°C). Incompatible with strong acids, strong oxidizers, caustics, aliphatic amines, isocyanates, chromic anhydride, lead perchlorate, perchloric acid, phosphorus trioxide. May react with metallic aluminum at high temperature. Attacks some plastics, rubber, and coatings.

WOOD SPIRIT (67-56-1) May accumulate static electrical charges, and may cause ignition of its vapors. Forms explosive mixture with air (flash point 52°F/11°C). Incompatible with strong acids, strong oxidizers, caustics, aliphatic amines, isocyanates, chromic anhydride, lead perchlorate, perchloric acid, phosphorus trioxide. May react with metallic aluminum at high temperature. Attacks some plastics, rubber, and coatings.

WOOD TURPENTINE (8006-64-2) Forms explosive mixture with air (flash point 95°F/35°C). Incompatible with strong oxidizers, especially chlorine; may cause fire and explosions. Incompatible with strong acids, chromic anhydride, chromyl chloride, hexachloromelamine, stannic chloride. Attacks ordinary rubber.

WOTEXIT (52-68-6) Contact with strong oxidizers may cause fire and explosions.

WP (7723-14-0) Pyrophoric solid; must be kept under water to prevent spontaneous combustion, but may ignite in moist air. A strong reducing agent; may explode on contact with oxidizers, including elemental sulfur and strong caustics. In addition to air, reacts with halogens, halides, alkali hydroxides (produces phosphine gas).

WT-27 (119-36-8) Forms explosive mixture with air (flash point 205°F/96°C). Incompatible with strong acids, nitrates, oxidizers.

- X -

XENENE (92-52-4) Fine particles may explode in air. Incompatible with strong oxidizers, nitric acid.

XERAC (94-36-0) Confined storage of dry chemical may lead to decomposition and explosion. A strong oxidant; extremely reactive. Fires and explosion may result from heat or contamination, and from contact with strong acids, combustible materials, oxidizers, acids, bases, alcohols, reducing agents, metals, metal oxides, amines, accelerators, methyl methacrylate, organic matter, lithium aluminum carbide, dimethyl aniline, amines, metallic naphthenates. May attack some plastics, rubber, and coatings. Protect containers from shock and friction.

XILENOLI (Italian) (1300-71-6) Forms explosive mixture with air (flash point 186°F/86°C). Reacts violently with strong oxidizers.

XILIDINE (Italian) (1300-73-8) Forms explosive mixture with air (flash point 206°F/96.7°C). Contact with strong acids may cause fire and explosions. Contact with hypochlorite bleaches produces explosive chloroamines. Incompatible with strong acids, organic acids and anhydrides, isocyanates, aldehydes. Attacks some plastics, rubber, and coatings.

XL ALL INSECTICIDE (54-11-5) Incompatible with strong acids, strong oxidizers. Attacks some plastics, rubber, and coatings. May accumulate static electrical charges, and may cause ignition of its vapors.

XYLENE or XYLENE, MIXED ISOMERS (1330-20-7) Forms explosive mixture with air (flash point approximately 84°F/29°C). Strong oxidizers may cause fire and explosions. Attacks many plastics, rubber, and coatings. May accumulate static electrical charges, and may cause ignition of its vapors.

1,2-XYLENE (95-47-6) Forms explosive mixture with air (flash point 90°F/32°C). Strong oxidizers may cause fire and explosions. Attacks many plastics, rubber, and coatings. May accumulate static electrical charges, and may cause ignition of its vapors.

1,3-XYLENE (108-38-3) Forms explosive mixture with air (flash point 84°F/28.9°C). Strong oxidizers may cause fire and explosions. Attacks many plastics, rubber, and coatings. May accumulate static electrical charges, and may cause ignition of its vapors.

1,4-XYLENE (106-42-3) Forms explosive mixture with air (flash point 81°F/27.2°C). Strong oxidizers may cause fire and explosions. Attacks many plastics, rubber, and coatings. May accumulate static electrical charges, and may cause ignition of its vapors.

m-**XYLENE or** *meta*-**XYLENE** (108-38-3) Forms explosive mixture with air (flash point 84°F/28.9°C). Strong oxidizers may cause fire and explosions. Attacks many plastics, rubber, and coatings. May accumulate static electrical charges, and may cause ignition of its vapors.

o-**XYLENE or** *ortho*-**XYLENE** (95-47-6) Forms explosive mixture with air (flash point 90°F/32°C). Strong oxidizers may cause fire and explosions. Attacks many plastics, rubber, and coatings. May accumulate static electrical charges, and may cause ignition of its vapors.

p-XYLENE or *para*-XYLENE (106-42-3) Forms explosive mixture with air (flash point 81°F/27.2°C). Strong oxidizers may cause fire and explosions. Attacks many plastics, rubber, and coatings. May accumulate static electrical charges, and may cause ignition of its vapors.

2,6-XYLENOL (1300-71-6) Forms explosive mixture with air (flash point 186°F/86°C). Reacts violently with strong oxidizers.

XYLENOLEN (Dutch) (1300-71-6) Forms explosive mixture with air (flash point 186°F/86°C). Reacts violently with strong oxidizers.

3,5-XYLENOL, 4-(METHYLTHIO)-, METHYLCARBAMATE (2032-65-7) Contact with strong oxidizers may cause fire and explosions.

XYLENOL, PHOSPHATE (3:1) (25155-23-1) Incompatible with sulfuric acid, nitric acid, nitrates, strong oxidizers.

XYLIDINE, MIXED ISOMERS (1300-73-8) Forms explosive mixture with air (flash point 206°F/96.7°C). Contact with strong acids may cause fire and explosions. Contact with hypochlorite bleaches produces explosive chloroamines. Incompatible with strong acids, organic acids and anhydrides, isocyanates, aldehydes. Attacks some plastics, rubber, and coatings.

2,4-XYLIDINE (95-68-1) Forms explosive mixture with air (flash point 206°F/96.7°C). Contact with strong acids may cause fire and explosions. Contact with hypochlorite bleaches produces explosive chloroamines. Incompatible with anhydrides, isocyanates, aldehydes. Attacks some plastics, rubber, and coatings.

2,5-XYLIDINE (95-78-3) Forms explosive mixture with air (flash point 206°F/96.7°C). Contact with strong acids may cause fire and explosions. Contact with hypochlorite bleaches produces explosive chloroamines. Incompatible with strong acids, organic acids and anhydrides, isocyanates, aldehydes. Attacks some plastics, rubber, and coatings.

2,6-XYLIDINE (87-62-7) Forms explosive mixture with air (flash point 206°F/96.7°C). Contact with strong acids may cause fire and explosions. Contact with hypochlorite bleaches produces explosive chloroamines. Incompatible with strong acids, organic acids and anhydrides, isocyanates, aldehydes.

m-4-XYLIDINE (95-68-1) Forms explosive mixture with air (flash point 206°F/96.7°C). Contact with strong acids may cause fire and explosions. Contact with hypochlorite bleaches produces explosive chloroamines. Incompatible with anhydrides, isocyanates, aldehydes. Attacks some plastics, rubber, and coatings.

m-XYLIDINE or *meta*-XYLIDINE (95-68-1) Forms explosive mixture with air (flash point 206°F/96.7°C). Contact with strong acids may cause fire and explosions. Contact with hypochlorite bleaches produces explosive chloroamines. Incompatible with anhydrides, isocyanates, aldehydes. Attacks some plastics, rubber, and coatings.

o-XYLIDINE or *ortho*-XYLIDINE (87-62-7) Forms explosive mixture with air (flash point 206°F/96.7°C). Contact with strong acids may cause fire and explosions. Contact with hypochlorite bleaches produces explosive chloroamines. Incompatible with strong acids, organic acids and anhydrides, isocyanates, aldehydes.

p-XYLIDINE or *para*-XYLIDINE (95-78-3) Forms explosive mixture with air (flash point 206°F/96.7°C). Contact with strong acids may cause fire and explosions. Contact with hypochlorite bleaches produces explosive chloroamines. Incompatible with strong acids, organic acids and anhydrides, isocyanates, aldehydes. Attacks some plastics, rubber, and coatings.

XYLIDINEN (Dutch) (1300-73-8) Forms explosive mixture with air (flash point 206°F/96.7°C). Contact with strong acids may cause fire and explosions. Contact with hypochlorite bleaches produces explosive chloroamines. Incompatible with strong acids, organic acids and anhydrides, isocyanates, aldehydes. Attacks some plastics, rubber, and coatings.

XYLODIDIN (9004-70-0) If ether solution evaporates, dry nitrocellulose material is a shock-sensitive explosive and fire hazard. Forms explosive mixture with air (flash point −64°F/−53°C). Strong oxidizers may cause fire and explosions. May accumulate static electrical charges, and may cause ignition of its vapors. Attacks many plastics, rubber, and coatings.

m-**XYLOL or** *meta*-**XYLOL** (108-38-3) Forms explosive mixture with air (flash point 84°F/28.9°C). Strong oxidizers may cause fire and explosions. Attacks many plastics, rubber, and coatings. May accumulate static electrical charges, and may cause ignition of its vapors.

o-**XYLOL or** *ortho*-**XYLOL** (95-47-6) Forms explosive mixture with air (flash point 90°F/32°C). Strong oxidizers may cause fire and explosions. Attacks many plastics, rubber, and coatings. May accumulate static electrical charges, and may cause ignition of its vapors.

p-**XYLOL or** *para*-**XYLOL** (106-42-3) Forms explosive mixture with air (flash point 81°F/27.2°C). Strong oxidizers may cause fire and explosions. Attacks many plastics, rubber, and coatings. May accumulate static electrical charges, and may cause ignition of its vapors.

2,6-XYLYLAMINE (87-62-7) Forms explosive mixture with air (flash point 206°F/96.7°C). Contact with strong acids may cause fire and explosions. Contact with hypochlorite bleaches produces explosive chloroamines. Incompatible with strong acids, organic acids and anhydrides, isocyanates, aldehydes.

XYLYL PHOSPHATE (25155-23-1) Incompatible with sulfuric acid, nitric acid, nitrates, strong oxidizers.

- Y -

YALTOX (1563-66-2) Incompatible with alkaline material, acids, strong oxidizers.

YARMOR (8002-09-3) Forms explosive mixture with air (flash point 138°F/59°C). Incompatible with strong acids, strong oxidizers.

YARMOR PINE OIL (8002-09-3) Forms explosive mixture with air (flash point 138°F/59°C). Incompatible with strong acids, strong oxidizers.

YASO-KNOCK (62-74-8) Incompatible with strong acids.

YELLOW ARSENIC SULFIDE (1303-33-9) Water contact produces hydrogen sulfide. Incompatible with strong oxidizers, acids, halogens.

YELLOW FERRIC OXIDE (1309-37-1) Contact with hydrogen peroxide, ethylene oxide, calcium hypochlorite will cause explosion. Reacts violently with powdered aluminum, hydrazine, hydrogen trisulfide.

YELLOW LEAD OCHER (1317-36-8) Reacts violently with aluminum powder, hydrogen peroxide, hydrogen trisulfide, lithium carbide, perchloric acid, strong oxidizers.

YELLOW OXIDE OF IRON (1309-37-1) Contact with hydrogen peroxide, ethylene oxide, calcium hypochlorite will cause explosion. Reacts violently with powdered aluminum, hydrazine, hydrogen trisulfide.

YELLOW OXIDE OF MERCURY (21908-53-2) A strong oxidizer; reacts violently with reducing agents and combustible materials. Incompatible with alcohols, chlorine, phosphorus, sulfur, alkali metals (i.e., lithium, sodium, potassium, rubidium, cesium, francium).

YELLOW PETROLATUM (8012-95-1) Oxidizers may cause fire and explosions. Incompatible with nitric acid. May accumulate static electrical charges, and may cause ignition of its vapors.

YELLOW PHOSPHORUS (7723-14-0) Pyrophoric solid; must be kept under water to prevent spontaneous combustion, but may ignite in moist air. A strong reducing agent; may explode on contact with oxidizers including elemental sulfur and strong caustics. In addition to air, reacts with halogens, halides, alkali hydroxides (produces phosphine gas).

- Z -

ZACLON DISCOIDS (74-90-8) Unless stabilized and maintained, samples stored more than 90 days are hazardous. Samples containing more than 2-5% water are less stable than dry material. Can be self-reactive, forming an explosive mixture with air (flash point 0°F/−18°C). Heat or contact with bases can cause polymerization. Incompatible with acetaldehyde. Oxidizers may cause fire and explosions. Attacks some plastics, rubber, and coatings.

ZEAZINE (1912-24-9) Incompatible with strong acids.

ZECTANE (315-18-4) Usually dissolved in a combustible liquid; avoid oxidizer contact.

ZEIDANE (50-29-3) Incompatible with salts of iron or aluminum, and bases. Do not store in iron containers.

ZELAZA THLENK (Polish) (1309-37-1) Contact with hydrogen peroxide, ethylene oxide, calcium hypochlorite will cause explosion. Reacts violently with powdered aluminum, hydrazine, hydrogen trisulfide.

ZERDANE (50-29-3) Incompatible with salts of iron or aluminum, and bases. Do not store in iron containers.

ZESET T (108-05-4) Forms explosive mixture with air (flash point 18°F/−8°C). Polymerizes readily if not inhibited; heat can initiate reaction. Reacts violently with oxidizers. Incompatible with nonoxidizing mineral acids, strong acids, ammonia, aliphatic amines, alkanolamines. Also reacts with 2-aminoethanol, chlorosulfonic acid, ethylene diamine, ethyleneimine, ozone, oleum, peroxides. May accumulate static electrical charges, and may cause ignition of its vapors.

ZETAR (8007-45-2) Forms explosive mixture with air (flash point 60°F/15°C). Incompatible with strong acids, nitrates.

ZEXTRAN (315-18-4) Usually dissolved in a combustible liquid; avoid oxidizer contact.

ZINC or ZINC POWDER or ZINC, PYROPHORIC (7440-66-6) May self-ignite in air. A strong reducing agent; reacts violently with oxidizers. Incompatible with acids, caustics, water, sulfur. May accumulate static electrical charges, and may cause ignition of its vapors.

ZINC ACETATE (557-34-6) Incompatible with strong acids, strong bases, nitrates.

ZINC ACETATE DIHYDRATE (557-34-6) Incompatible with strong acids, strong bases, nitrates.

ZINC BICHROMATE (14018-95-2) An oxidizer; reacts with reducing agents, combustibles, organic materials.

ZINC BROMIDE (7699-45-8) Incompatible with potassium, sodium.

ZINC BROMIDE, CALCIUM BROMIDE SOLUTION (N/A) Incompatible with sulfuric acid, isocyanates.

ZINC BUTTER (7646-85-7) Aqueous solution is strongly acidic; reacts with strong bases, and attacks metals in the presence of moisture.

ZINC CHLORIDE (7646-85-7) Aqueous solution is strongly acidic; reacts with strong bases, and attacks metals in the presence of moisture.

ZINC (CHLORURE de) (French) (7646-85-7) Aqueous solution is strongly acidic; reacts with strong bases, and attacks metals in the presence of moisture.

ZINC CYANIDE (557-21-1) Contact with acids and acid salts liberates hydrogen cyanide gas.

ZINC DIACETATE (557-34-6) Incompatible with strong acids, strong bases, nitrates.

ZINC DIALKYLDITHIOPHOSPHATE (N/A) Incompatible with sulfuric acid, nitric acid, nitrates. Attacks many plastics, rubber, and coatings.

ZINC DIBROMIDE (7699-45-8) Incompatible with potassium, sodium.

ZINC-O,O-DI-N-BUTYLPHOSPHORODITHIOATE (N/A) Incompatible with sulfuric acid, nitric acid, nitrates. Attacks many plastics, rubber, and coatings.

ZINC DICHLORIDE (7646-85-7) Aqueous solution is strongly acidic; reacts with strong bases, and attacks metals in the presence of moisture.

ZINC DICHROMATE (14018-95-2) An oxidizer; reacts with reducing agents, combustibles, organic materials.

ZINC DICYANIDE (557-21-1) Contact with acids and acid salts liberates hydrogen cyanide gas.

ZINC DIETHYL- (557-20-0) Ignites spontaneously on contact with air or oxidizers. Explosive decomposition occurs at 245°F/120°C. Reacts violently with hydrazine, sulfur dioxide, some alcohols, ozone, possible fire and explosions. Water causes formation of ethane gas.

ZINC DIFLUORIDE (7783-49-5) Incompatible with potassium.

ZINC DIHEXYLDITHIOPHOSPHATE (N/A) Incompatible with sulfuric acid, nitric acid, nitrates. Attacks many plastics, rubber, and coatings.

ZINC DIHEXYLPHOSPHORODITHIOATE (N/A) Incompatible with sulfuric acid, nitric acid, nitrates. Attacks many plastics, rubber, and coatings.

ZINC DIMETHYL (544-97-8) Ignites spontaneously with air. Water contact produces methane gas.

ZINC DISTEARATE (557-05-1) Combustible solid. May accumulate static electrical charges, and may cause ignition of its vapors.

ZINC DITHIONITE (7779-86-4) Water contact produces oxides of sulfur. Incompatible with strong oxidizers, acids.

ZINC ETHIDE (557-20-0) Ignites spontaneously on contact with air or oxidizers. Explosive decomposition occurs at 245°F/120°C. Reacts violently with hydrazine, sulfur dioxide, some alcohols, ozone, with possible fire and explosions. Water causes formation of ethane gas.

ZINC ETHYL (557-20-0) Ignites spontaneously on contact with air or oxidizers. Explosive decomposition occurs at 245°F/120°C. Reacts violently with hydrazine, sulfur dioxide, some alcohols, ozone; possible fire and explosions. Water causes formation of ethane gas.

ZINC ETHYLENEDIAMINETETRAACETATE (12519-36-7) Incompatible with sulfuric acid, isocyanates.

ZINC FLUORIDE (7783-49-5) Incompatible with potassium.

ZINC FLUORURE (French) (7783-49-5) Incompatible with potassium.

ZINC HYDROSULFITE (7779-86-4) Water contact produces oxides of sulfur. Incompatible with strong oxidizers, acids.

ZINC METHYL (544-97-8) Ignites spontaneously with air. Water contact produces methane gas.

ZINC MURIATE SOLUTION (7646-85-7) Aqueous solution is strongly acidic; reacts with strong bases, and attacks metals in the presence of moisture.

ZINC NITRATE (7779-88-6) A strong oxidizer. Reacts violently with reducing agents, combustibles, organic materials, carbon.

ZINC NITRATE HEXAHYDRATE (7779-88-6) A strong oxidizer. Reacts violently with reducing agents, combustibles, organic materials, carbon.

ZINCO (CLORURO di) (Italian) (7646-85-7) Aqueous solution is strongly acidic; reacts with strong bases, and attacks metals in the presence of moisture.

ZINC OCTADECANOATE (557-05-1) Combustible solid. May accumulate static electrical charges, and may cause ignition of its vapors.

ZINC PHOSPHIDE (1314-84-7) Incompatible with oxidizers, acids, water.

ZINC STEARATE (557-05-1) Combustible solid. May accumulate static electrical charges, and may cause ignition of its vapors.

ZINC SULFATE (7733-02-0) Incompatible with strong bases.

ZINC SULFATE HEPTAHYDRATE (7733-02-0) Incompatible with strong bases.

ZINC TOX (1314-84-7) Incompatible with oxidizers, acids, water.

ZINC VITRIOL (7733-02-0) Incompatible with strong bases.

ZINFOS-METHYL (Dutch) (86-50-0) Strong oxidizers may cause fire and explosions.

ZINKCHLORID (German) (7646-85-7) Aqueous solution is strongly acidic; reacts with strong bases, and attacks metals in the presence of moisture.

ZINKCHLORIDE (Dutch) (7646-85-7) Aqueous solution is strongly acidic; reacts with strong bases, and attacks metals in the presence of moisture.

ZINKOSITE (7733-02-0) Incompatible with strong bases.

ZINN (German) (7440-31-5) Incompatible with strong acids, strong bases, strong oxidizers, sulfur.

ZIRCAT (7440-67-7) Finely divided material is spontaneously flammable in air. Explosions and fire may result from contact with strong oxidizers, alkali metals, carbon tetrachloride, cupric oxide, lead, lead oxide, potassium chlorate, potassium nitrate, sodium borate.

ZIRCONIUM (7440-67-7) Finely divided material is spontaneously flammable in air. Explosions and fire may result from contact with strong oxidizers, alkali metals, carbon tetrachloride, cupric oxide, lead, lead oxide, phosphorus, potassium chlorate, potassium nitrate, sodium borate.

ZIRCONIUM CARBIDE (N/A) Finely divided material is spontaneously flammable in air.

ZIRCONIUM CHLORIDE (10026-11-6) Decomposes in the presence of moist air. Reacts with water or moisture to produce hydrochloric acid fume. Alkali metals may cause explosions. Attacks many plastics, rubber, and coatings.

ZIRCONIUM(4+) CHLORIDE(1:4) or ZIRCONIUM(IV) CHLO-RIDE(1:4) (10026-11-6) Decomposes in the presence of moist air. Reacts with water or moisture to produce hydrochloric acid fume. Alkali metals may cause explosions. Attacks many plastics, rubber, and coatings.

ZIRCONIUM DIHYDRIDE (7740-99-6) A powerful reducing agent; reacts violently with oxidizers. Spontaneous ignition occurs when heated in air.

ZIRCONIUM HYDRIDE (7740-99-6) Flammable solid; dangerous when wet. A powerful reducing agent. Spontaneous ignition occurs when heated in air.

ZIRCONIUM NITRATE (13746-89-9) Flammable solid; dangerous when wet. A powerful oxidizer. Incompatible with reducing agents, esters, many other substances. Attacks metals.

ZIRCONIUM NITRATE PENTAHYDRATE (13746-89-9) A powerful oxidizer. Incompatible with reducing agents, esters, many other substances. Attacks metals.

ZIRCONIUM TETRACHLORIDE or ZIRCONIUM TETRACHLO-RIDE, SOLID (DOT) (10026-11-6) Decomposes in the presence of moist air. Reacts with water or moisture to produce hydrochloric acid fume. Alkali metals may cause explosions. Attacks many plastics, rubber, and coatings.

ZITHIOL (121-75-5) Incompatible with strong oxidizers, magnesium, alkaline pesticides. Attacks metals, some plastics, rubber, and coatings.

ZOBA BLACK D (106-50-3) A strong reducing agent; reacts violently with oxidizers, combustibles, organic substances. Incompatible with strong acids, organic anhydrides, isocyanates, aldehydes. Heat and light contribute to its instability.

ZOBA GKE (95-80-7) Incompatible with oxidizers, acids, organic anhydrides, isocyanates, aldehydes. Attacks aluminum, brass, bronze, copper, zinc.

ZOGEN DEVELOPER-H Incompatible with oxidizers, acids, organic anhydrides, isocyanates, aldehydes. Attacks aluminum, brass, bronze, copper, zinc.

ZOOCOUMARING (Russian) (81-81-2) Strong oxidizers may cause fire and explosions.

ZOTOX (7778-39-4) Incompatible with sulfuric acid, caustics, ammonia, amines, isocyanates, alkylene oxides, oxidizers, epichlorohydrin, vinyl acetate, amides. Corrodes metals.

Z-P (1314-84-7) Incompatible with oxidizers, acids, water.

ZWAVELWATERSTOF (Dutch) (7783-06-4) A highly flammable and reactive gas. Incompatible with acetaldehyde, barium pentafluoride, chlorine monoxide, chlorine trifluoride, chromic anhydride, copper, lead dioxide, nitric acid, nitrogen iodide, nitrogen trichloride, nitrogen trifluoride, oxygen difluoride, oxidizers, phenyl diazonium chloride, sodium, sodium peroxide. Reacts with alkali metals and alkaline earth metals. Attacks metals, some plastics, rubber, and coatings.

ZYTOX (74-83-9) Incompatible with strong oxidizers, aluminum, dimethylsulfoxide, ethylene oxide, water. Attacks zinc, magnesium, alkali metals and their alloys, some plastics, rubber, and coatings.